Instructor's Presentation CD-ROM (0-471-37671-X)

For lecture presentation purposes, this cross-platform CD-ROM includes 200 pieces of art from the new edition. The images can be edited and customized to meet the individual needs of instructors.

Atlases

Atlas of the Human Skeleton

This brief, four-color photographic atlas of the components of the human skeleton features photographs by Mark Nielsen, University of Utah. It provides students with an effective supplement to the illustrations in the text, as well as a guide for learning the skeletal system in the laboratory. Originally prepared by Gerard J. Tortora to accompany *Principles of Anatomy and Physiology, Eighth Edition,* this atlas should prove an especially useful supplement for students and instructors using this new edition of *Principles of Human Anatomy.*

For Students

Learning Guide (0-471-36760-5)

Written by Roberto Amitrano of Bergen Community College, this helpful student study guide features a chapter synopsis, topic outline, and objectives; an etymological exercise on scientific terminology; and study questions cross-referenced to the pages in the text. Answers to selected questions are included to allow students to check their learning progress. Each chapter concludes with a Self Quiz to provide a final self-check of the student's mastery of the material. With an emphasis on active learning, the Guide encourages students to learn concepts through a wide variety of activities and exercises.

Principles of Human Anatomy

Eighth Edition

Gerard J. Tortora

Bergen Community College

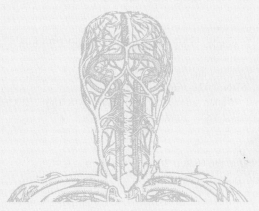

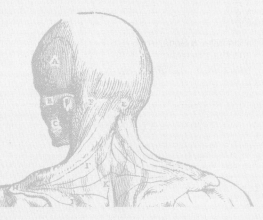

JOHN WILEY & SONS, INC.

New York • Chichester • Weinheim • Brisbane • Singapore • Toronto

Credits appear following the Glossary.

About the cover: The image of the heart and great arteries reproduced within each of the cover's color panels is taken from Plate 64, number 4, of *De Humani Corporis Fabrica* (1543) by Andreas Vesalius. A milestone in the history of medicine, Vesalius' work marked the transition of the study of human anatomy from the realm of folklore to a modern observational science.

Library of Congress Cataloging-in-Publication Data

Tortora, Gerard J.
    Principles of human anatomy / Gerard J. Tortora. — 8th ed.
      p. cm.
    Includes index.
    ISBN 0-471-36686-2
    1. Human anatomy.   2. Human physiology.   I. Title.
    QM23.2.T87   1998
    611—dc21   98-7560
                                        CIP

Printed in the United States of America
ISBN: 0-471-36686-2
10 9 8 7 6 5 4 3 2 1

John Wiley & Sons, Inc.
605 Third Avenue
New York, NY 10158

About the Author

Gerard J. Tortora

Jerry Tortora is Professor of Biology and former Biology Coordinator at Bergen Community College in Paramus, New Jersey, where he teaches human anatomy and physiology as well as microbiology. He received his bachelor's degree in biology from Fairleigh Dickinson University and his master's degree in science education from Montclair State College. He is a member of many professional organizations, such as the Human Anatomy and Physiology Society (HAPS), the American Society of Microbiology (ASM), American Association for the Advancement of Science (AAAS), National Education Association (NEA), and the Metropolitan Association of College and University Biologists (MACUB).

Above all, Jerry is devoted to his students and their aspirations. In recognition of this commitment, Jerry was the recipient of MACUB's 1992 President's Memorial Award. In 1996, he was named Distinguished Faculty Scholar at Bergen CC. Also in 1996, he received a National Institute for Staff and Organizational Development (NISOD) excellence award from the University of Texas and was selected to represent Bergen Community College in a campaign to increase awareness of the contributions of community colleges to higher education.

Jerry is the author of several best-selling science textbooks and laboratory manuals, a calling that often requires an additional 40 hours per week beyond his teaching responsibilities. Nevertheless, he still makes time for four or five weekly aerobics workouts that include biking and running. On many an afternoon in the biology department, the sound of Jerry's voice can be heard through the office door, singing solo passages from Verdi.

To my wife, Melanie, and to my children, Christopher, Anthony, and Andrew, who make it all worthwhile.

Preface

The study of human anatomy is an inquiry into the fascinating complexity of one's own body. From a practical standpoint, however, studying anatomy also can be the gateway to a rewarding career in a host of health-related sciences, dance and athletic training, and advanced scientific studies. True to the heritage of its predecessors, I wrote the eighth edition of *Principles of Human Anatomy* to meet the exacting requirements of introductory anatomy courses, yet I also wrote it in a manner intended to advance the personal and professional aspirations of all its readers.

CORE VALUES

My students have taught me to value simplicity, directness, and the power of clear illustrations. The noteworthy success of this book over the years, and its transformation into a standard-bearer in education, is a testament to the fact that many of my colleagues value these qualities too. Chapter after chapter, I have striven to build these core teaching values into *Principles of Human Anatomy* by providing:

- clear, accurate, straightforward, and complete anatomy discussions
- expertly executed and generously sized art
- thoughtful pedagogy
- outstanding student study support.

Although this book offers a host of useful features that will appeal to a variety of learning styles, experience has shown me that features alone cannot help students learn unless every anatomy discussion itself is easy to comprehend. By any measure, I trust that readers will find this edition of *Principles of Human Anatomy* to be easier than ever to read.

CLASSROOM-TESTED METHODOLOGY

Human anatomy is a complex body of knowledge to present in an introductory course, and mastering the subject can require a great deal of memory work on a student's part. Hence, I've woven four interrelated strands of information throughout *Principles of Human Anatomy* to help readers develop a working knowledge of the subject, to enable them to speak about anatomy with confidence, and to reveal the practical application of anatomical knowledge to their career choices or in their everyday lives.

Developing Unifying Concepts

From the outset, students need to understand how individual structures relate to the composition of the entire body. For this reason, the book begins with concepts such as regional names, directional terms, and the planes and sections that enable anatomists to isolate individual parts or to discuss their relationship to other body structures. Furthermore, I have found that most students benefit from a quick brush-up on the basics of cellular biology. Accordingly, *Principles of Human Anatomy* establishes early on all of the fundamental principles needed to study anatomy and redevelops them as necessary throughout the book.

Connecting Structure to Function

I am convinced from years of student feedback that readers learn anatomy more readily when they understand the relationship between structure and function. Indeed, the lack of such information in undergraduate textbooks was one reason that led me to write *Principles of Human Anatomy* in the first place. The challenge, of course, is to build connections to functions without overwhelming readers with unnecessary physiological detail. In this edition, I carefully trimmed and revised the functional anatomy coverage to ensure that it fulfills its pedagogic goal but that it does not eclipse the primary purpose of learning anatomy.

Building a Professional Vocabulary

Students—even the best—generally find it difficult at first to read and pronounce anatomical terms. Moreover, as an educator in the largest metropolitan region of the United States, I am sympathetic to the needs of the growing ranks of college students who speak English as a second language. For these reasons, I have striven since the book's inception to ensure that it has a strong and helpful vocabulary component. Anatomical terminology in *Principles of Human Anatomy* is prominently highlighted in boldface type, and most highlighted terms in the text, tables, and exhibits are accompanied by easy-to-understand pronunciation guides. As a further mnemonic aid, the most important highlighted terms include word roots.

Revealing Clinical and Practical Relevance

Principles of anatomy and functional anatomy sometimes are better understood by considering variations from normal. Second, most students are naturally curious about how the information in this book or in their anatomy course pertains to a potential career choice, or how it relates to them or family members. For these reasons, a variety of clinical perspectives—many proposed by students—are introduced as "Clinical Applications" within each chapter, directly following the discussions to which they relate. In addition, an *Applications to Health* companion is included with the text at no additional cost. The "Applications to Health" are concise discussions of major diseases, disorders, and medical conditions that illustrate departures from normal anatomy and function.

ORGANIZATION

Like most undergraduate anatomy courses, *Principles of Human Anatomy* follows a systems approach to the topic. The first two chapters introduce the nomenclature and conventions used to study human anatomy and review the fundamentals of cell biology. In Chapters 3 and 4, I present tissues and that most obvious of all bodily tissues, the skin. Because skeletal and muscle anatomy are studied intensively in most anatomy courses, these topics are covered meticulously in Chapters 5 through 10, chapters whose art and presentational style are now regarded as hallmarks of the book itself. A transitional topic in many anatomy courses, the surface anatomy coverage in Chapter 11 was completely rewritten for this edition.

Subsequent divisions of coverage include the anatomy of the cardiovascular and nervous systems. The anatomy of the blood, heart, blood vessels, and lymphatic system are presented in Chapters 12 through 15. The second most extensive coverage focus in the book, the anatomy of the nervous system, is presented in Chapters 16 through 21. In a departure from the previous edition, I now divide sensory and motor anatomy and the special senses into separate chapters to make this coverage eas-

ier for students to handle. Endocrine, respiratory, digestive, and urinary system anatomy follows in Chapters 22 through 25, and the book concludes with human reproductive anatomy and developmental anatomy in Chapters 26 and 27.

Special Themes

Developmental Anatomy   I often tell my students that they can better appreciate the "logic" of human anatomy by becoming aware of how various structures developed in the first place. As in previous editions, illustrated discussions of developmental anatomy are found near the conclusion of most body system chapters. Placing this coverage at the end of chapters enables students to master the anatomical terminology they need in order to learn about embryonic and fetal structures.

Aging   Students need to be reminded from time to time that anatomy is not static. As the body ages, its anatomy and related functions subtly change. Moreover, aging is a professionally relevant topic for the majority of this book's readers, who will go on to careers in health-related fields in which the median age of the client population is steadily advancing. For these reasons, many of the body system chapters explore age-related changes in anatomy and function, coverage that can be found after a distinctively designed heading at the end of most chapters.

Exercise   Echoing the previous theme, physical exercise can produce favorable changes in some anatomical structures, most notably those associated with the musculoskeletal and cardiovascular systems. This information has special relevance to readers embarking on careers in physical education, sports training, and dance. Hence, key chapters in *Principles of Human Anatomy* include brief discussions of exercise, which are signaled by a distinctive "running shoe" icon.

IMPROVED COVERAGE

Every chapter in the eighth edition of *Principles of Human Anatomy* incorporates a host of improvements to both the text and the art suggested by reviewers, educators, or students. Here are just some of the more noteworthy changes:

Chapter 3: Tissues   The much-imitated tissue tables in this chapter are redesigned to promote student interpretation of the histology figures. Now, every photomicrograph is accompanied by an analogous illustration. Moreover, I have added an inset next to each photomicrograph that indicates a primary location where each tissue type occurs in the body.

Chapter 8: Joints   New discussions in each exhibit strongly correlate each joint to the movements it permits and with the muscles that produce those movements. Equally important, the exhibits themselves are completely redesigned to make the coverage easier to study and to better correlate each discussion with its associated illustrations.

Chapter 10: The Muscular System   The exhibits that are a hallmark of this chapter now feature a complete narrative tour of each muscle group. Clinical Applications are easier to spot, and a new feature, "Relating Muscles to Movements," encourages readers to group related muscles according to their common actions. Repetitive innervation information now is handled as a discussion, a change that makes each exhibit's table simpler and easier to study.

Chapter 11: Surface Anatomy   Completely rewritten, this chapter now offers a helpful narrative tour of all the major surface landmarks of each region of the body. Most of the photographs are new to this edition.

Chapter 14: Blood Vessels   In this chapter's exhibits, I provide a more consistent level of detail when discussing the arterial and venous circulation, especially in the limbs. New cadaver photos better illustrate the relevant blood vessel anatomy, and extended flow diagrams visually reinforce learning of relevant anatomy and the blood vessel pathways.

HALLMARK FEATURES OF THE EIGHTH EDITION

The eighth edition of *Principles of Human Anatomy* builds on the legacy of thoughtful and clearly designed features that distinguished its predecessors. Some regular features are totally redesigned for this edition. For example, the tables are much easier to read and use. Some features are completely new. Here are the most noteworthy changes and additions:

More Helpful Exhibits   Students of anatomy need extra help in learning the numerous structures that constitute certain body systems—most notably skeletal muscles, articulations, blood vessels, lymph nodes, and nerves. As in previous editions, the chapters that present these topics are organized around **Exhibits,** each of which consists of an extended overview, a table summarizing the relevant anatomy, and an associated suite of illustrations. I trust that you will agree that the newly revised versions of the exhibits, along with their spacious new design, make them ideal study vehicles for learning anatomically complex body systems.

Effective Chapter Openers   Featuring a Renaissance anatomy illustration from Andreas Vesalius, each chapter begins with **Contents at a Glance,** an outline that now includes the page numbers of principal topics. In addition, readers will benefit from the carefully focused **Student Objectives.**

Updated Clinical Focus   A perennial favorite among anatomy students, the numerous **Clinical Applications** in every chapter explore the clinical, professional, or personal relevance of a particular anatomical structure or its related function. Some applications are new to this edition, and all have been reviewed for accuracy and relevance by a consulting panel of nurses.

New or Revised Study Aids   At the end of each chapter, readers will benefit from the popular **Study Outline** that is cross-referenced to the chapter discussions. The **Review Questions** test how well readers understand the anatomical concepts set forth in the student objectives. As a further feedback opportunity, each chapter includes a newly written **Self Quiz,** which offers questions written in a variety of styles calculated to appeal to readers' different testing preferences. A wholly new feature, the **Critical Thinking Questions** with hints, asks readers to apply anatomical concepts to real-life problems and challenges. The inimitable style of these questions ought to make students smile on occasion as well as think! Answers to the Self Quiz and Critical Thinking questions are found in Appendix II.

Study Tools for Mastering Vocabulary   Every chapter highlights **key terms in boldface** type. **Pronunciation guides** are offered at the point that major (or especially hard-to-pronounce) structures are introduced in the text, tables, or exhibits. **Word roots** that cite the Greek or Latinate derivations of anatomical terms are offered as a further mnemonic aid. More pronunciation guides and word roots have been added throughout the eighth edition. As a further service, *Principles of Human Anatomy* defines **Key Medical Terms** at the end of chapters and provides a comprehensive **Glossary** at the back of the book. Moreover, the basic building blocks of medical terminology are listed in **Combining Forms, Word Roots, Prefixes, and Suffixes** found inside the back cover.

ENHANCEMENTS TO THE ILLUSTRATION PROGRAM

Teaching human anatomy is both a visual and descriptive enterprise. Countless students have now benefited from the work of Leonard Dank, whose magnificent bone and muscle art is the touchstone for the detailed and generously proportioned art that characterizes *Principles of Human Anatomy*. Continuing our fruitful collaboration, we have added additional bone and muscle art to some figures, and subtly revised others. Beyond this, over 25% of the art is new to this edition, and nearly every figure has been revised in some way. Here are the highlights:

New Histology-Based Art   As part of a plan of continuous improvement, many of the anatomical illustrations based on histological preparations have been replaced in this edition. See, for example, the beautifully rendered illustration of the layers of the gastrointestinal tract on page 724.

New Cadaver Photos Most of the cadaver photos have been replaced with larger and clearer examples. Moreover, several additional cadaver photos have been added, such as those of the heart on page 404.

Helpful Orientation Insets Anatomy students sometimes need help figuring out the plane of view of the illustrations—descriptions alone don't always help. Hence, in the eighth edition, every major anatomy illustration now is accompanied by a thumbnail-size orientation diagram that depicts and explains the perspective of the view represented in the figure.

New Key Concept Statements This new art-related feature is a summation of an idea that is stated in the text and then amplified in a figure. Each statement is positioned above the figure and is signaled by a "key" icon.

Revised Figure Questions This well-received feature requires readers to synthesize verbal and visual information, think critically, or draw conclusions. Each figure question appears with its illustration and is highlighted in this edition with a blue "Q" icon. Answers appear at the end of each chapter.

New Overviews of Functional Anatomy This new art-related feature succinctly lists the functions of the anatomical structure or system depicted in the adjacent figure. (See page 109.) This juxtaposition of text and art further reinforces the connection between structure and function.

A.D.A.M.® Interactive Anatomy As an additional service to students who have access to A.D.A.M.® Interactive Anatomy software, a distinctive new "ADAM" icon appears alongside some figures and Critical Thinking questions. Chapter-by-chapter discussions that explain the supporting AIA images, as well as clear directions about how to access them, are located in the *Applications to Health* companion that is included with each copy of the text.

COMPLETE TEACHING AND LEARNING PACKAGE

Continuing the tradition of providing a complete teaching and learning package, the eighth edition of *Principles of Human Anatomy* is available with a host of carefully planned supplementary materials that will help instructors and students to maximize the benefits of the text. A complete description of the supplements listed below can be found on the inside front cover. Please contact your John Wiley representative for additional information about any of these resources, most of which are available on a complimentary basis to qualified adopters of the text.

For Instructors:

- **Professor's Resource Manual** by Izak J. Paul (0-471-36759-1).

- **Test Bank** by Janice Meeking (0-471-37469-5).
- **Test Gen-EQ Computerized Test Bank** (Mac: 0-471-37464-4/ Win: 0-471-37465-2).
- **Full-Color Acetate Transparencies** (0-471-37463-6).
- **Instructor's Presentation CD-ROM** (0-471-37459-8).

Atlases:

- **Atlas of the Human Skeleton** by Gerard J. Tortora (0-471-37474-1).

For Students:

- **Learning Guide** by Robert J. Amitrano (0-471-36760-5).
- Tortora web site at **http://www.wiley.com/college/tortora**

* * * * *

Like each of my children, this book has a life of its own. The structure, content, and production values of *Principles of Human Anatomy* are shaped as much by its relationship with educators and readers as by the vision that initiated the book seven editions ago. Today, you, my readers, are the "heart" of Tortora. I invite each of you to continue the tradition of sending your suggestions to me so that I can include them in the ninth edition.

Gerard J. Tortora
Bergen Community College
Dept. of Sciences and Health
Paramus, NJ 07652

Acknowledgments

The production of the eighth edition of *Principles of Human Anatomy,* my first edition with Benjamin/Cummings, has been the occasion of many pleasant surprises for me. I have enjoyed the opportunity of collaborating with a fresh team of very enthusiastic, dedicated, and talented publishing professionals. Accordingly, I would like to recognize and thank the members of the Tortora team—a true team in every sense of the word.

Let me begin with two writers who provided invaluable assistance to me. Janice Meeking of Mt. Royal College, Ontario, revised and updated the end-of-chapter Self Quizzes, providing many new, challenging, and diverse test questions. Joan Barber of Delaware Technical and Community College provided a feature new to this edition called Critical Thinking Questions. These questions not only challenge students to think critically, but they encourage them to do so through their spirited and often humorous tone. In addition, my thanks to Wojciech Pawlina of the University of Florida for developing the new AIA correlation guide, which cross-references the text and figures to analogous images in A.D.A.M.® Interactive Anatomy.

The assistance from editorial and production has been equally valuable. Mark Wales, Senior Developmental Editor, guided me to write a book in a fresh light. He taught me how to focus on pedagogy that directed and reinforced student learning. His insights, careful attention to detail, and years of experience have transformed a very good textbook into a truly outstanding one. Mark's imprint pervades my book. Kay Ueno, Senior Project Editor and Tortora A & P Book Team Manager, has worked very closely with Mark to guide the project to completion. Her directed comments, editorial wisdom, and continuous support and encouragement have been timely and very much appreciated. Nicki Richesin, Editorial Assistant, facilitated the flow of manuscript and communications between me and other team members, which has proved very useful to the timely completion of this book.

Wendy Earl, Managing Editor of Wendy Earl Productions, assumed the responsibility for producing my *Principles of Human Anatomy.* Her experience, creativity, and professionalism are her trademarks and have meant so much to the book. Sharon Montooth, Production Editor, expertly coordinated the evolution of manuscript and galleys and then pages with efficiency and accuracy. Her responsibilities were facilitated by David Novak, Production Assistant, who helped to prepare the working draft of the manuscript.

Claudia Durrell, Art Coordinator, has worked with me for many years, and her attention to design and knowledge of my art preferences have made my job so much easier and have enhanced the graphic elements of this book in ways I could never imagine. Mira Schachne, another long-time associate, is the Photo Coordinator. If the photo exists, Mira will find it—or pose for it herself! Her tenacity and sense of humor have been her signature for all the years I have known her. Heena Pitchaikani, Associate Art Coordinator, has ably assisted Claudia with the development of the art program. Andrew Ogus, the Text and Cover Designer, has provided the pleasing and effective page design that augments all the changes to this edition. Andrew also designed the striking and memorable cover based on an image from Vesalius.

Alan Titche, Copyeditor, helped me in numerous ways to make the manuscript consistent in presentation and style and helped me to polish my transitions. Martha Ghent, Proofreader, alerted me to discrepancies, typographical errors, omissions, and all the other "bugs" that can creep into a manuscript. Katharine Pitcoff, our Indexer, assembled the comprehensive new index that should ably serve the needs of students and instructors alike.

The illustrations and photographs have always been a signature feature of all of my textbooks. Our customary standard of excellence has been continued and heightened by the beautiful and pedagogically oriented medical illustrations provided by Leonard Dank, Sharon Ellis, Lauren Keswick, Jedin Jackson, Lynn O'Reilley, Biaggio John Melloni, Hilda Muinos, Nadine Sokol, Kevin Sommerville, Beth Willert, David Schneidmen Design, and Page Two Associates. Mark Nielsen of the University of Utah provided the beautiful new cadaver photos used throughout the book.

I am always extremely grateful to my colleagues who have reviewed the manuscript and offered numerous suggestions for improvement. The reviewers who have provided their time and expertise to maintain this book's accuracy are noted in the Reviewers list that follows.

As always, my wife, Melanie, and my children, other family members, colleagues, and friends have supported my writing activities in more ways than I could ever mention here. Their understanding and encouragement will always be appreciated and will never be taken for granted.

Jerry Tortora

REVIEWERS

Eighth Edition Reviewers

Kathleen Andersen, University of Iowa College of Medicine
Alfonse R. Burdi, University of Michigan Medical School
David T. Deutsch, Kent State University
Leslie Hendon, University of Alabama, Birmingham
Bert H. Jacobson, Oklahoma State University
Laura Mudd, Barry University
Virginia L. Naples, Northern Illinois University
Christine Peters, West Valley College
Clarence Thompson, University of North Dakota School
  of Medicine

Clinical Applications Consultants

Kathleen Zellner, Bellin College of Nursing
Christine Vandenhouten, Bellin College of Nursing

Seventh Edition Reviewers

Andy Anderson, Utah State University
Jane Aloi, Saddleback College
Leann Blem, Virginia Commonwealth University
Alphonse Burdi, University of Michigan
Neil Cumberlidge, Northern Michigan University
Charles Ferguson, University of Colorado, Denver
Charles Leavell, Fullerton College
Sherwin Mizell, Indiana University
Jerome Montvilo, Bryant College
John Moore, Parkland College
Izak Paul, Mt. Royal College
DeLoris M. Wenzel, University of Georgia

Illustration Review, Seventh Edition

Michael C. Kennedy, Alleghney University

Successful Learning Strategies: Seven Steps to Success!

A Visual Guide to Principles of Human Anatomy

A variety of special learning features are built into your book to help you study anatomy successfully and with confidence. Each feature has been developed and refined based on feedback from students and instructors using previous editions. Here is a student-tested, step-by-step method for mastering the study of human anatomy.

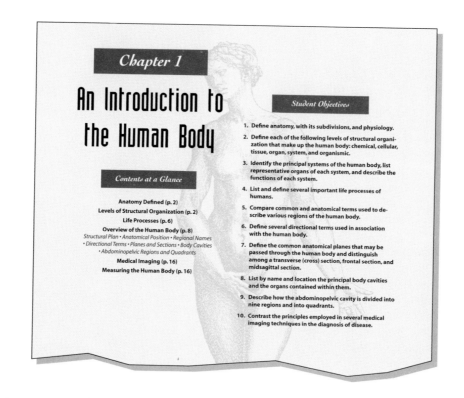

Chapter 1

An Introduction to the Human Body

Contents at a Glance

Anatomy Defined (p. 2)
Levels of Structural Organization (p. 2)
Life Processes (p. 6)
Overview of the Human Body (p. 8)
Structural Plan • Anatomical Position • Regional Names • Directional Terms • Planes and Sections • Body Cavities • Abdominopelvic Regions and Quadrants
Medical Imaging (p. 16)
Measuring the Human Body (p. 16)

Student Objectives

1. Define anatomy, with its subdivisions, and physiology.

2. Define each of the following levels of structural organization that make up the human body: chemical, cellular, tissue, organ, system, and organismic.

3. Identify the principal systems of the human body, list representative organs of each system, and describe the functions of each system.

4. List and define several important life processes of humans.

5. Compare common and anatomical terms used to describe various regions of the human body.

6. Define several directional terms used in association with the human body.

7. Define the common anatomical planes that may be passed through the human body and distinguish among a transverse (cross) section, frontal section, and midsagittal section.

8. List by name and location the principal body cavities and the organs contained within them.

9. Describe how the abdominopelvic cavity is divided into nine regions and into quadrants.

10. Contrast the principles employed in several medical imaging techniques in the diagnosis of disease.

 ACQUAINT yourself with the mission of each chapter. Study the **contents at a glance**, which lists the chapter's major topics and the pages on which each discussion begins. The **student objectives** reveal the competencies you will gain after studying the chapter.

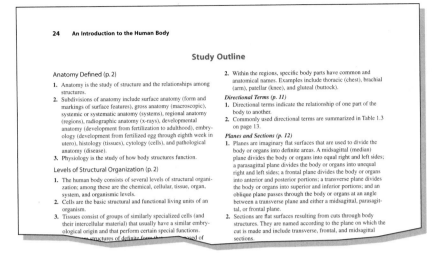

24   An Introduction to the Human Body

Study Outline

Anatomy Defined (p. 2)

1. Anatomy is the study of structure and the relationships among structures.
2. Subdivisions of anatomy include surface anatomy (form and markings of surface features), gross anatomy (macroscopic), systemic or systematic anatomy (systems), regional anatomy (regions), radiographic anatomy (x-rays), developmental anatomy (development from fertilization to adulthood), embryology (development from fertilized egg through eighth week in utero), histology (tissues), cytology (cells), and pathological anatomy (disease).
3. Physiology is the study of how body structures function.

Levels of Structural Organization (p. 2)

1. The human body consists of several levels of structural organization; among these are the chemical, cellular, tissue, organ, system, and organismic levels.
2. Cells are the basic structural and functional living units of an organism.
3. Tissues consist of groups of similarly specialized cells (and their intercellular material) that usually have a similar embryological origin and that perform certain special functions.

2. Within the regions, specific body parts have common and anatomical names. Examples include thoracic (chest), brachial (arm), patellar (knee), and gluteal (buttock).

Directional Terms (p. 11)

1. Directional terms indicate the relationship of one part of the body to another.
2. Commonly used directional terms are summarized in Table 1.3 on page 13.

Planes and Sections (p. 12)

1. Planes are imaginary flat surfaces that are used to divide the body or organs into definite areas. A midsagittal (median) plane divides the body or organs into equal right and left sides; a parasagittal plane divides the body or organs into unequal right and left sides; a frontal plane divides the body or organs into anterior and posterior portions; a transverse plane divides the body or organs into superior and inferior portions; and an oblique plane passes through the body or organs at an angle between a transverse plane and either a midsagittal, parasagittal, or frontal plane.
2. Sections are flat surfaces resulting from cuts through body structures. They are named according to the plane on which the cut is made and include transverse, frontal, and midsagittal sections.

2   **ANTICIPATE** important ideas by reading the **study outline** at the end of the chapter. There's a lot to learn and remember when studying human anatomy! Use this balanced synopsis of the chapter to increase your confidence and decrease your studying time by "pre-learning" the topic.

3   **SCAN** all the figures before reading a chapter. You need to look at anatomy carefully in order to learn it. Start by reading the **legend**, which explains what the image is about.

Next, read the **key concept statement,** which reveals a basic idea portrayed in the figure. Be sure to study *all* the parts of a figure and read the labels.

An **orientation diagram** located within some figures shows the viewing angle of the featured image or where the structure is found in the body.

Next, consider the **figure question,** which asks you to synthesize information, think critically, or draw conclusions about the figure.

Beneath some figures, you'll see an **A.D.A.M.®** "walking man" icon. This alerts you to additional views of the same structure that are available in the A.D.A.M.® Interactive Anatomy software. Check the cross-reference guide in your *Applications to Health* companion that came with your copy of the book for more details about using this feature.

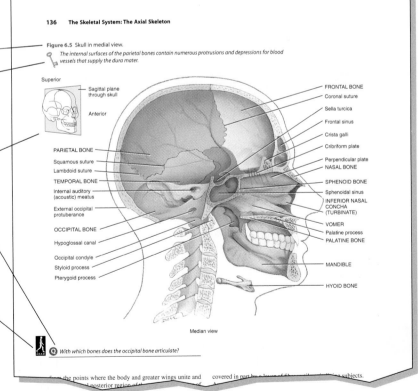

136   The Skeletal System: The Axial Skeleton

Figure 6.5 Skull in medial view.

The internal surfaces of the parietal bones contain numerous protrusions and depressions for blood vessels that supply the dura mater.

Superior

Sagittal plane through skull

Anterior

FRONTAL BONE
Coronal suture
Sella turcica
Frontal sinus
Crista galli
Cribriform plate
Perpendicular plate
NASAL BONE
SPHENOID BONE
Sphenoidal sinus
INFERIOR NASAL CONCHA (TURBINATE)
VOMER
Palatine process
PALATINE BONE
MANDIBLE
HYOID BONE

PARIETAL BONE
Squamous suture
Lambdoid suture
TEMPORAL BONE
Internal auditory (acoustic) meatus
External occipital protuberance
OCCIPITAL BONE
Hypoglossal canal
Occipital condyle
Styloid process
Pterygoid process

Median view

 With which bones does the occipital bone articulate?

xi

READ the chapter. Notice the following aids that are built in to help you learn. A **boldfaced term** is your signal that a word and the ideas associated with it are important to learn. **Pronunciation guides** appear directly after the most important—or hard-to-pronounce—terms.

Some terms are followed by **word roots,** which can help you remember a word by telling you about its original meaning or how it was put together. If you need more help while you read, be sure to check the **glossary** at the back of the book.

As you read, you will discover short **clinical applications** from time to time. Study these features carefully, because they will acquaint you with the clinical or professional significance of the preceding anatomy discussion.

Maxillae

The paired **maxillae** (mak-SIL-ē; *macerae* = to chew; singular is *maxilla*) unite to form the upper jawbone (Figure 6.10) and articulate with every bone of the face except the mandible, or lower jawbone. They form part of the floors of the orbits, part of the lateral walls and floor of the nasal cavity, and most of the hard palate. The hard palate is a bony partition, formed by the maxillae and palatine bones, that forms the roof of the mouth.

Each maxilla contains a *maxillary sinus* that empties into the nasal cavity (see Figure 6.11). The *alveolar* (al-VĒ-ō-lar; *alveolus* = hollow) *process* is an arch that contains the *alveoli* (sockets) for the maxillary (upper) teeth. The *palatine process* is a horizontal projection of the maxilla that forms the anterior three-quarters of the hard palate. The maxillary bones unite, and the fusion is normally completed before birth.

 *Clinical Application*

Cleft Palate and Cleft Lip

Usually the palatine processes of the maxillary bones unite during weeks 10 to 12 of embryonic development. Failure to do so can result in a condition called **cleft palate.** The condition may also involve incomplete fusion of the horizontal plates of the

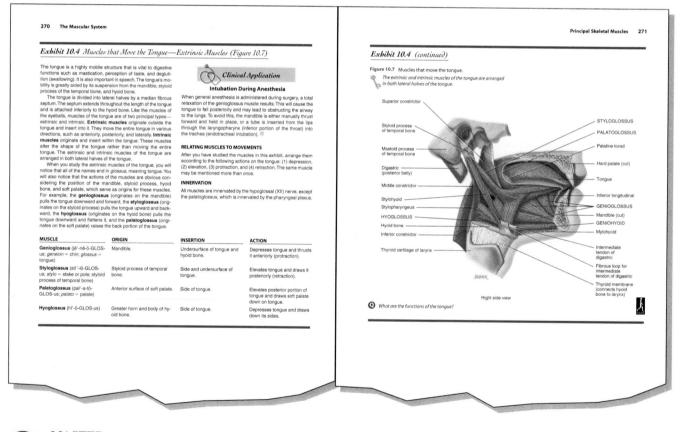

Exhibit 10.4 Muscles that Move the Tongue—Extrinsic Muscles (Figure 10.7)

The tongue is a highly mobile structure that is vital to digestive functions such as mastication, perception of taste, and deglutition (swallowing). It is also important in speech. The tongue's mobility is greatly aided by its suspension from the mandible, styloid process of the temporal bone, and hyoid bone.

The tongue is divided into lateral halves by a median fibrous septum. The septum extends throughout the length of the tongue and is attached inferiorly to the hyoid bone. Like the muscles of the eyeballs, muscles of the tongue are of two principal types—extrinsic and intrinsic. **Extrinsic muscles** originate outside the tongue and insert into it. They move the entire tongue in various directions, such as anteriorly, posteriorly, and laterally. **Intrinsic muscles** originate and insert within the tongue. These muscles alter the shape of the tongue rather than moving the entire tongue. The extrinsic and intrinsic muscles of the tongue are arranged in both lateral halves of the tongue.

When you study the extrinsic muscles of the tongue, you will notice that all of the names end in *glossus,* meaning tongue. You will also notice that the actions of the muscles are obvious considering the position of the mandible, styloid process, hyoid bone, and soft palate, which serve as origins for these muscles. For example, the **genioglossus** (originates on the mandible) pulls the tongue downward and forward, the **styloglossus** (originates on the styloid process) pulls the tongue upward and backward, the **hyoglossus** (originates on the hyoid bone) pulls the tongue downward and flattens it, and the **palatoglossus** (originates on the soft palate) raises the back portion of the tongue.

 *Clinical Application*

Intubation During Anesthesia

When general anesthesia is administered during surgery, a total relaxation of the genioglossus muscle results. This will cause the tongue to fall posteriorly and may lead to obstructing the airway to the lungs. To avoid this, the mandible is either manually thrust forward and held in place, or a tube is inserted from the lips through the laryngopharynx (inferior portion of the throat) into the trachea (endotracheal intubation).

RELATING MUSCLES TO MOVEMENTS

After you have studied the muscles in this exhibit, arrange them according to the following actions on the tongue: (1) depression, (2) elevation, (3) protraction, and (4) retraction. The same muscle may be mentioned more than once.

INNERVATION

All muscles are innervated by the hypoglossal (XII) nerve, except the palatoglossus, which is innervated by the pharyngeal plexus.

MUSCLE	ORIGIN	INSERTION	ACTION
Genioglossus (jē'-nē-ō-GLOS-us; *geneion* = chin; *glossus* = tongue)	Mandible.	Undersurface of tongue and hyoid bone.	Depresses tongue and thrusts it anteriorly (protraction).
Styloglossus (stī'-lō-GLOS-us; *stylo* = stake or pole; styloid process of temporal bone)	Styloid process of temporal bone.	Side and undersurface of tongue.	Elevates tongue and draws it posteriorly (retraction).
Palatoglossus (pal'-a-tō-GLOS-us; *palato* = palate)	Anterior surface of soft palate.	Side of tongue.	Elevates posterior portion of tongue and draws soft palate down on tongue.
Hyoglossus (hī'-ō-GLOS-us)	Greater horn and body of hyoid bone.	Side of tongue.	Depresses tongue and draws down its sides.

Exhibit 10.4 (continued)

Figure 10.7 Muscles that move the tongue.

The extrinsic and intrinsic muscles of the tongue are arranged in both lateral halves of the tongue.

Superior constrictor
Styloid process of temporal bone
Mastoid process of temporal bone
Digastric (posterior belly)
Middle constrictor
Stylohyoid
Stylopharyngeus
HYOGLOSSUS
Hyoid bone
Inferior constrictor
Thyroid cartilage of larynx

STYLOGLOSSUS
PALATOGLOSSUS
Palatine tonsil
Hard palate (cut)
Tongue
Inferior longitudinal
GENIOGLOSSUS
Mandible (cut)
GENIOHYOID
Mylohyoid
Intermediate tendon of digastric
Fibrous loop for intermediate tendon of digastric
Thyroid membrane (connects hyoid bone to larynx)

Right side view

What are the functions of the tongue?

MASTER the Exhibits. Some body systems are comprised of hundreds of structures, which can make your learning tasks more challenging. To help you, chapters that teach complex anatomy have special **exhibits.** Each exhibit is a self-contained learning exercise that consists of a narrative tour, a table summarizing all the structures you need to learn, and an associated group of illustrations. Spend as much time as you need mastering one exhibit before moving to the next.

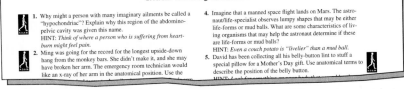

Review Questions

1. Define anatomy. List and define the various subdivisions of anatomy. Define physiology. (p. 2)
2. Give several examples of how structure and function are related. (p. 2)
3. Define each of the following terms: cell, tissue, organ, system, and organism. (p. 4)
4. Using Table 1.2 on page 4 as a guide, outline the functions of each system of the body; then list several organs that com-

9. What is a directional term? Why are these terms important? Use each of the directional terms listed in Table 1.3 on page 13 in a complete sentence.
10. Define the various planes that may be passed through the body. Explain how each plane divides the body. Describe the meaning of transverse, frontal, and midsagittal sections. (p. 12)
11. Define a body cavity. List the body cavities discussed and tell which major organs are located in each. What is the medi-

Self Quiz

Complete the following:

1. The four basic types of tissues in the body are ___, ___, ___ and ___.
2. Three serous membranes located in the ventral body cavity are the pericardium, peritoneum, and ___.
3. From superior to inferior, the three abdominopelvic regions on the right side are right ___, right ___, and right ___.
4. A ___ plane divides the body into anterior and posterior parts.

10. The diaphragm divides the abdominopelvic cavity into abdominal and pelvic portions.
11. The brain and the spinal cord are located in the dorsal body cavity.
12. The breakdown of food by the digestive system involves chemical reactions that are catabolic.
13. Match the following common and anatomical terms:
   (a) armpit   (1) thoracic
   (b) arm   (2) inguinal

Critical Thinking Questions

1. Why might a person with many imaginary ailments be called a "hypochondriac"? Explain why this region of the abdominopelvic cavity was given this name.
   HINT: *Think of where a person who is suffering from heartburn might feel pain.*
2. Ming was going for the record for the longest upside-down hang from the monkey bars. She didn't make it, and she may have broken her arm. The emergency room technician would like an x-ray of her arm in the anatomical position. Use the

4. Imagine that a manned space flight lands on Mars. The astronaut/life-specialist observes lumpy shapes that may be either life-forms or mud balls. What are some characteristics of living organisms that may help the astronaut determine if these are life-forms or mud balls?
   HINT: *Even a couch potato is "livelier" than a mud ball.*
5. David has been collecting all his belly-button lint to stuff a special pillow for a Mother's Day gift. Use anatomical terms to describe the position of the belly button.
   HINT: *Look for something on your body that you would use*

6 VALIDATE your understanding. Each chapter has a three-part review structure. Start by answering the **review** questions, which test your mastery of the student objectives. Or try taking the **self quiz**, which poses questions in a variety of formats like those you might encounter on a test. Finally, be sure you can answer at least three of the **critical thinking** questions, which ask you to apply anatomical concepts to real-life problems. The "walking man" icon refers you to the cross-reference guide in your *Applications to Health* companion for additional help if you're using A.D.A.M.® Interactive Anatomy Software. Answers to the self quiz and critical thinking questions are in Appendix II at the back of the book.

7 REINFORCE the connection between abnormalities in anatomy and function and the theoretical material in the text by referring to the discussions in your *Applications to Health* companion. This booklet, following the chapter organization of your text, focuses on major diseases, disorders, and medical conditions that illustrate departures from the normal anatomy and function discussed throughout your text.

Need additional study help?

Tortora Web Site
The Benjamin/Cummings web site address is listed at the end of every chapter next to our distinctive web site icon. Visit the Tortora book site for web links, study resources, practice tests, contests, and articles.

Learning Guide
This helpful study companion has chapter summaries, topic outlines and objectives, terminology exercises, and study questions cross-referenced to the book. Each chapter also includes a self quiz section. Ask for the *Learning Guide to Principles of Human Anatomy* (ISBN: 0471-36760-5) at your bookstore!

Contents in Brief

Contents

An Introduction to the Human Body

Student Objectives

1. Define anatomy, with its subdivisions, and physiology.

2. Define each of the following levels of structural organization that make up the human body: chemical, cellular, tissue, organ, system, and organismic.

3. Identify the principal systems of the human body, list representative organs of each system, and describe the functions of each system.

4. List and define several important life processes of humans.

5. Compare common and anatomical terms used to describe various regions of the human body.

6. Define several directional terms used in association with the human body.

7. Define the common anatomical planes that may be passed through the human body and distinguish among a transverse (cross) section, frontal section, and midsagittal section.

8. List by name and location the principal body cavities and the organs contained within them.

9. Describe how the abdominopelvic cavity is divided into nine regions and into quadrants.

10. Contrast the principles employed in several medical imaging techniques in the diagnosis of disease.

You are about to begin a study of the human body in order to learn how it is organized and how it functions. This study involves many branches of science, each of which contributes to an understanding of how your body normally works. In order to understand what happens to the body when it is injured, diseased, or placed under stress, you must first have a basic understanding of how the body is organized and how its different parts normally work. Much of what you study in this chapter will help you visualize the body as anatomists do, and you will learn a basic anatomical vocabulary that will help you talk about the body in a way that is understood by professionals in various fields.

ANATOMY DEFINED

Anatomy (a-NAT-ō-mē; *ana* = upward; *tome* = to cut) is the study of *structure* and the relationships among structures. Several subdivisions of anatomy are described in Table 1.1.

Whereas anatomy deals with structures of the body, **physiology** (fiz′-ē-OL-ō-jē) deals with *functions* of body parts—that is, how they work. Because function cannot be completely separated from structure, you will learn about the human body by studying first its anatomy and then the necessary physiology. You will see how each structure of the body is designed to carry out a particular function and how the structure of a part often determines the functions it performs. For example, the hairs lining the nose filter air that you inhale. The bones of the skull are tightly joined to protect the brain. The bones of the fingers, by contrast, are more loosely joined to permit various movements. The external ear is shaped in such a way as to collect sound waves, which facilitates hearing. The lungs are filled with millions of air sacs that are so thin that they permit both the movement of oxygen into the blood for use by body cells and the movement of carbon dioxide out of the blood to be exhaled.

LEVELS OF STRUCTURAL ORGANIZATION

The human body consists of several levels of structural organization that are associated with one another (Figure 1.1).

❶ The *chemical level* includes all substances needed to maintain life. Chemicals are made up of atoms, the smallest units of matter, and certain of these, such as carbon (C), hydrogen (H), oxygen (O), nitrogen (N), calcium (Ca), potassium (K), and sodium (Na), are essential for maintaining life. Atoms combine to form molecules, two or more atoms joined together. Familiar examples of molecules are proteins, carbohydrates, lipids, and vitamins.

Table 1.1 *Subdivisions of Anatomy*

SUBDIVISION	DESCRIPTION
Surface anatomy	Study of the form (morphology) and markings of the surface of the body.
Gross (macroscopic) anatomy	Study of structures that can be examined without the use of a microscope.
Systemic (systematic) anatomy	Study of specific systems of the body such as the nervous system or respiratory system.
Regional anatomy	Study of a specific region of the body such as the head or chest.
Radiographic (rā′-dē-ō-GRAF-ik; *radio* = ray; *graph* = to write) **anatomy**	Study of the structure of the body that includes the use of x-rays.
Developmental anatomy	Study of development from the fertilized egg to adult form.
Embryology (em′-brē-OL-ō-jē; *logos* = study of)	Study of development from the fertilized egg through the eighth week in utero.
Histology (hiss′-TOL-ō-jē; *histo* = tissue)	Microscopic study of the structure of tissues.
Cytology (sī-TOL-ō-jē; *cyto* = cell)	Chemical and microscopic study of the structure of cells.
Pathological (path′-ō-LOJ-i-kal; *patho* = disease) **anatomy**	Study of structural changes (from gross to microscopic) associated with disease.

Figure 1.1 Levels of structural organization that compose the human body.

🔑 *The levels of structural organization are chemical, cellular, tissue, organ, system, and organismic levels.*

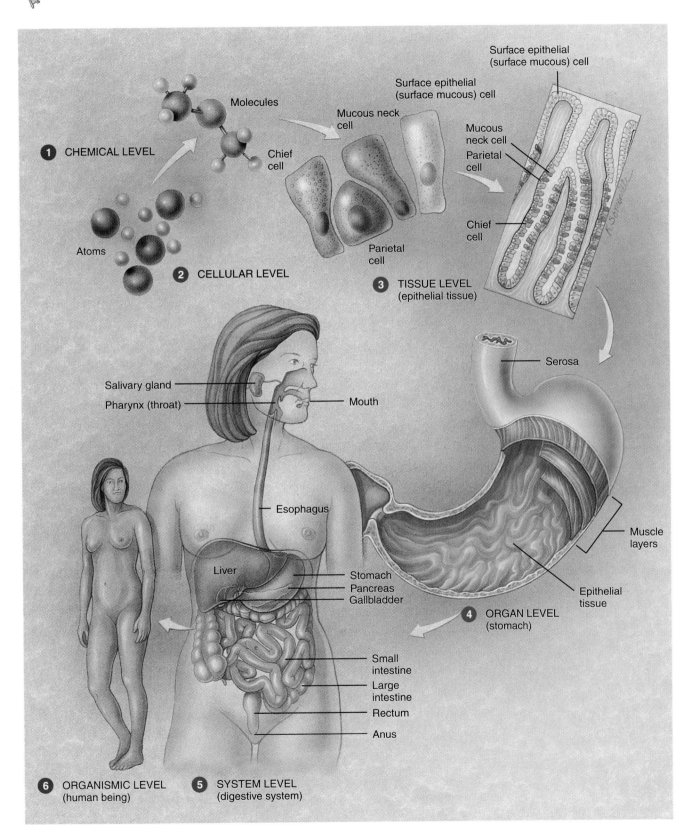

1 CHEMICAL LEVEL

Molecules

Atoms

2 CELLULAR LEVEL

Mucous neck cell

Chief cell

Surface epithelial (surface mucous) cell

Parietal cell

3 TISSUE LEVEL (epithelial tissue)

Surface epithelial (surface mucous) cell

Mucous neck cell

Parietal cell

Chief cell

Serosa

Muscle layers

Epithelial tissue

4 ORGAN LEVEL (stomach)

Salivary gland

Pharynx (throat)

Mouth

Esophagus

Liver

Stomach

Pancreas

Gallbladder

Small intestine

Large intestine

Rectum

Anus

6 ORGANISMIC LEVEL (human being)

5 SYSTEM LEVEL (digestive system)

Q *Which level of structural organization is composed of two or more different types of tissues and has a usually recognizable shape?*

❷ Molecules, in turn, combine to form the next higher level of organization: the *cellular level.* **Cells** are the basic structural and functional living units of an organism. Among the many kinds of cells in your body are muscle cells, nerve cells, and blood cells. As you will see in Chapter 2, cells contain specialized structures called *organelles,* such as the nucleus and mitochondria, that perform specific functions. Figure 1.1 shows four different types of cells from the lining of the stomach. Each has a different structure, and each performs a different function.

❸ The next higher level of structural organization is the *tissue level.* **Tissues** are groups of similar cells that, together with their intercellular material (substance between cells),

usually have a similar origin in an embryo and perform specialized functions. The four basic types of tissues in the body are *epithelial tissue, muscle tissue, connective tissue,* and *nervous tissue.* When the cells shown in Figure 1.1 are joined together, they form an epithelial tissue that lines the stomach. Each type of cell in the tissue has a specific function in digestion (Chapter 24).

❹ When different kinds of tissues are joined together they form the next higher level of organization: the *organ level.* **Organs** are structures that are composed of two or more different types of tissues, have specific functions, and usually have recognizable shapes. Examples of organs are the heart, liver, lungs, brain, and stomach. Figure 1.1 shows how sev-

Table 1.2 Principal Systems of the Human Body: Representative Organs and Functions

Integumentary system

Components: The skin and structures derived from it, such as hair, nails, and sweat and oil glands.

Functions: Helps regulate body temperature, protects the body, eliminates some wastes, helps produce vitamin D, and monitors certain stimuli such as changes in temperature and pressure.

Skeletal system

Components: All the bones of the body, their associated cartilages, and the joints of the body.

Functions: Supports and protects the body, assists in body movements, houses cells that give rise to blood cells, and stores minerals and lipids.

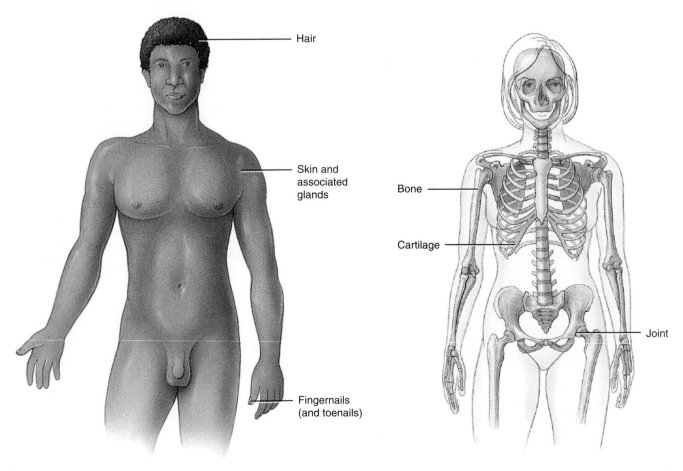

Hair

Skin and associated glands

Fingernails (and toenails)

Bone

Cartilage

Joint

eral tissues make up one organ, the stomach. (1) The outer covering is the *serosa,* a layer of epithelial tissue and connective tissue that protects the stomach and reduces friction when the stomach moves and rubs against other organs. (2) The *muscle tissue layers* under the serosa contract to churn and mix food and push it on to the next digestive organ (the small intestine). (3) The *epithelial tissue layer* lining the stomach, as noted earlier, participates in digestion.

❺ The fifth highest level of structural organization in the body is the *system level.* A **system** consists of related organs that have a common function. The digestive system, which functions in the breakdown and absorption of food, is composed of these organs: mouth, saliva-producing glands called salivary glands, pharynx (throat), esophagus, stomach, small intestine, large intestine, rectum, liver, gallbladder, and pancreas. An organ can be part of more than one system. The pancreas, for example, is part of both the digestive system and the hormone-producing endocrine system.

❻ The highest level of structural organization is the *organismic level.* All the body systems functioning with one another constitute the total **organism**—one living individual.

In the chapters that follow, you will examine the anatomy and physiology of the major body systems. Table 1.2 describes these systems in terms of their representative organs and their general functions, and presents them in the order in which they are discussed in later chapters.

Table 1.2 (continued)

Muscular system

Components: Specifically refers to skeletal muscle tissue, which is muscle usually attached to bones. Other muscle tissues are smooth and cardiac.

Functions: Powers movements of the body, such as walking or throwing a ball, stabilizes body positions (posture), and generates heat.

Cardiovascular system

Components: Blood, heart, and blood vessels.

Functions: Distributes oxygen and nutrients to cells, carries carbon dioxide and wastes away from cells, helps maintain the acid–base balance of the body, protects against disease, prevents hemorrhage by forming blood clots, and helps regulate body temperature.

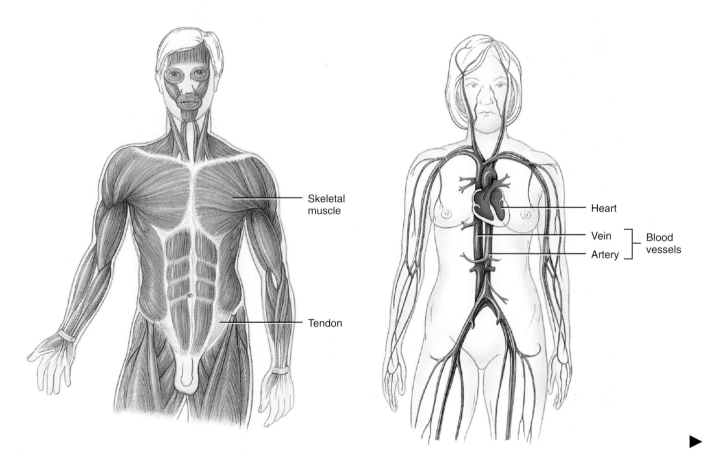

Skeletal muscle

Tendon

Heart

Vein

Artery

Blood vessels

▶

LIFE PROCESSES

All living organisms carry on certain processes that set them apart from nonliving things. Following are several of the more important life processes of humans:

1. **Metabolism** (*metabole* = change) is the sum of all the chemical reactions that occur in the body. One phase of metabolism, called **catabolism** (*cata* = downward), provides the energy needed to sustain life by breaking down substances such as food molecules. The other phase of metabolism, called **anabolism** (*ana* = upward), uses the energy from catabolism to make various substances that form body structures and enable them to function.

Table 1.2 *Principal Systems of the Human Body: Representative Organs and Functions (continued)*

Lymphatic system

Components: Lymph, lymphatic vessels, and structures or organs containing lymphatic tissue (large numbers of white blood cells called lymphocytes), such as the spleen, thymus gland, lymph nodes, and tonsils.

Functions: Returns proteins and plasma (liquid portion of blood) to the cardiovascular system, transports triglycerides (fats) from the gastrointestinal tract to the cardiovascular system, serves as a site of maturation and proliferation of certain white blood cells, and helps protect against disease through the production of proteins called antibodies, as well as other responses.

Nervous system

Components: Brain, spinal cord, nerves, and special sense organs, such as the eyes and ears.

Functions: Regulates body activities through action potentials (nerve impulses) by detecting changes in the internal and external environments, interpreting the changes, and responding to the changes by inducing muscular contractions or glandular secretions.

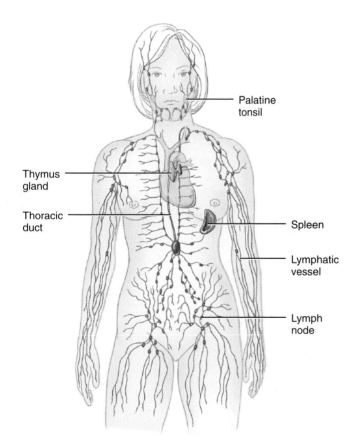

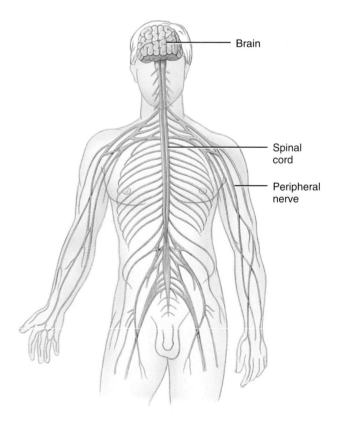

2. **Responsiveness** is the ability to detect and respond to changes outside or inside the body. For example, neurons (nerve cells) respond by generating electrical signals, known as nerve impulses. Muscle cells respond by contracting—usually becoming shorter—to move body parts.

3. **Movement** includes motion of the whole body, individual organs, single cells, or even structures inside cells. For example, the coordinated contraction of several leg muscles moves your whole body from place to place when you walk or run. After you eat a meal that contains triglycerides (fats), your gallbladder contracts and releases bile to help in their digestion. When a body tissue is damaged or infected, certain white blood cells move from the blood into the tissue to help clean up and repair the area. And inside individual cells, various cell parts move from one position to another.

4. **Growth** refers to an increase in the size of existing cells, the number of cells, or the amount of substance surrounding cells.

5. **Differentiation** is the process whereby unspecialized cells become specialized cells. Specialized cells differ in structure and function from the cells from which they

---

Table 1.2 *(continued)*

Endocrine system

Components: All hormone-producing cells and glands such as the pituitary gland, thyroid gland, and pancreas.

Functions: Regulates body activities through hormones, chemicals transported in the blood to various target organs of the body.

Respiratory system

Components: Lungs and a series of associated passageways leading into and out of them.

Functions: Supplies oxygen, eliminates carbon dioxide, helps regulate the acid–base balance of the body, and helps produce vocal sounds.

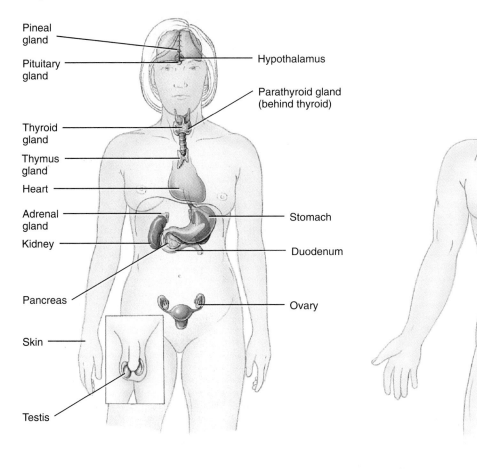

▶

originated. For example, following the union of a sperm and ovum, the fertilized egg undergoes tremendous differentiation and progresses through various stages to develop into a unique individual who is similar to, yet quite different from, either of the parents.

6. **Reproduction** refers either to the formation of new cells for growth, repair, or replacement or to the production of a new individual.

OVERVIEW OF THE HUMAN BODY
Structural Plan

The human body has certain general **anatomical characteristics** that will help you understand its overall structural plan. For example, humans have a *backbone (vertebral column)*, a characteristic that places them in a large group of organisms called *vertebrates*. Another characteristic is the body's *tube-within-a-tube* construction. The outer tube is

---

Table 1.2 *Principal Systems of the Human Body: Representative Organs and Functions (continued)*

Digestive system

Components: A long tube called the gastrointestinal tract and associated organs that include the salivary glands, liver, gallbladder, and pancreas.

Functions: Performs the physical and chemical breakdown of food and absorption of nutrients for use by cells and helps eliminate solid and other wastes.

Urinary system

Components: Kidneys, ureters, urinary bladder, and urethra that together produce, store, and eliminate urine.

Functions: Regulates the volume and chemical composition of blood, eliminates wastes, regulates fluid and electrolyte balance, helps maintain the acid–base and calcium balance of the body, and secretes a hormone that regulates red blood cell production.

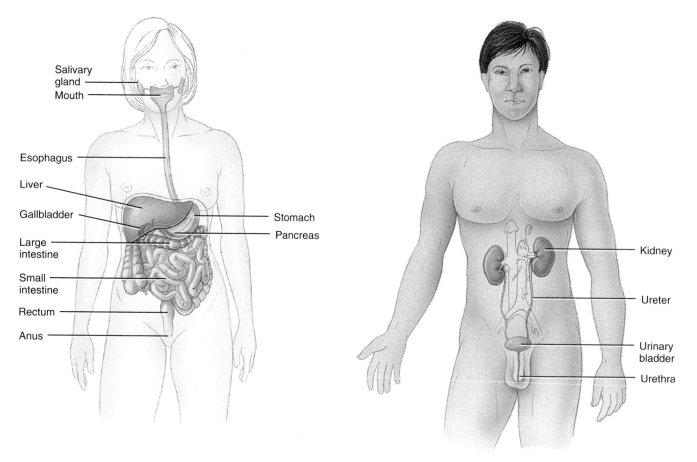

formed by the body wall; the inner tube is part of the gastrointestinal tract. Moreover, humans are for the most part *bilaterally symmetrical;* that is, essentially the left and right sides of the body are mirror images.

Anatomical Position

In anatomy, descriptions of any region or part of the human body assume that the body is in a specific position called the **anatomical position.** In the anatomical position (Figure 1.2), the subject stands erect (upright position) facing the observer with the head level and eyes facing forward, the feet are flat on the floor and directed forward, and the arms are placed at the sides with the palms turned forward. Once the body is in the anatomical position, it is easier to visualize and understand how it is organized into various regions. Having one standard anatomical position allows directional terms to be clear; any body part or region can be described relative to any other part.

Table 1.2 *(continued)*

Reproductive systems

Components: Organs (testes and ovaries) that produce reproductive cells or gametes (sperm and ova), and other organs such as the uterine (Fallopian) tubes and uterus in females and the epididymis, ductus (vas) deferens, and penis in males that transport and store reproductive cells.

Functions: Produces gametes, which can unite to form a new organism, and hormones that help regulate metabolism.

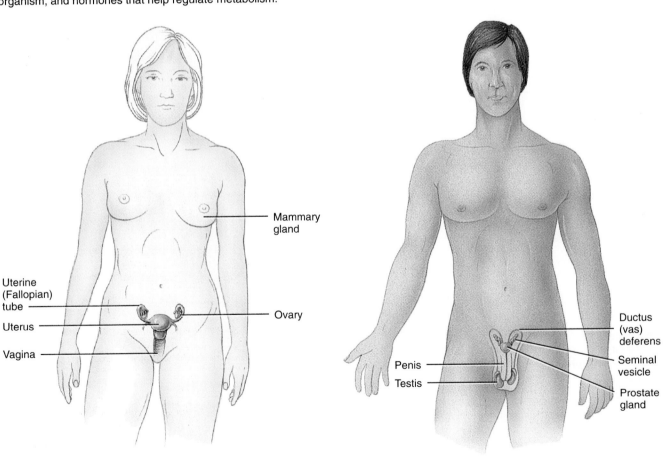

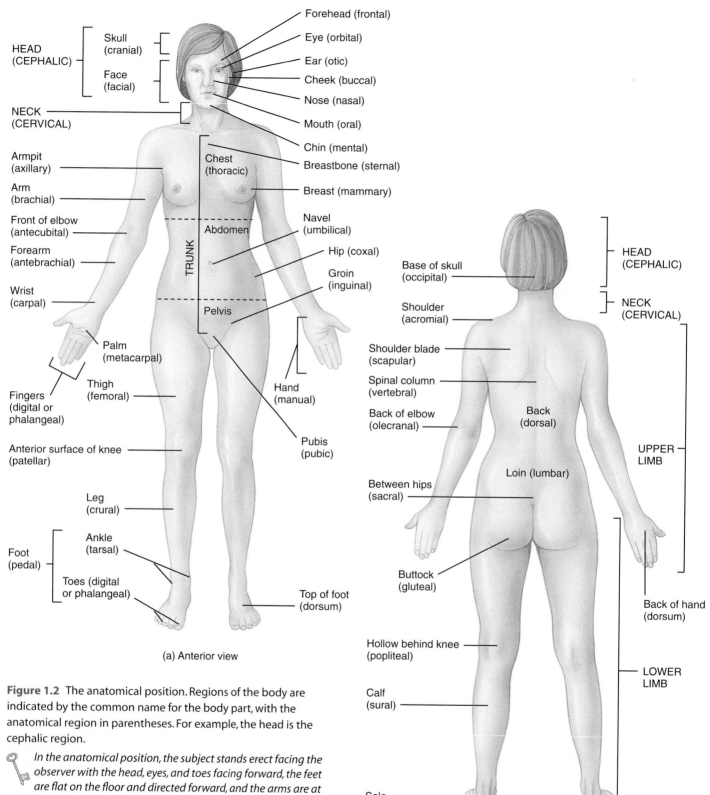

(a) Anterior view

(b) Posterior view

Figure 1.2 The anatomical position. Regions of the body are indicated by the common name for the body part, with the anatomical region in parentheses. For example, the head is the cephalic region.

In the anatomical position, the subject stands erect facing the observer with the head, eyes, and toes facing forward, the feet are flat on the floor and directed forward, and the arms are at the sides with the palms turned forward.

Q *What is the usefulness of defining one standard anatomical position?*

Figure 1.3 Directional terms. Study Table 1.3 with this figure to understand the directional terms: superior, inferior, anterior, posterior, medial, lateral, intermediate, ipsilateral, contralateral, proximal, distal, superficial, and deep.

Directional terms precisely locate various parts of the body in relation to one another.

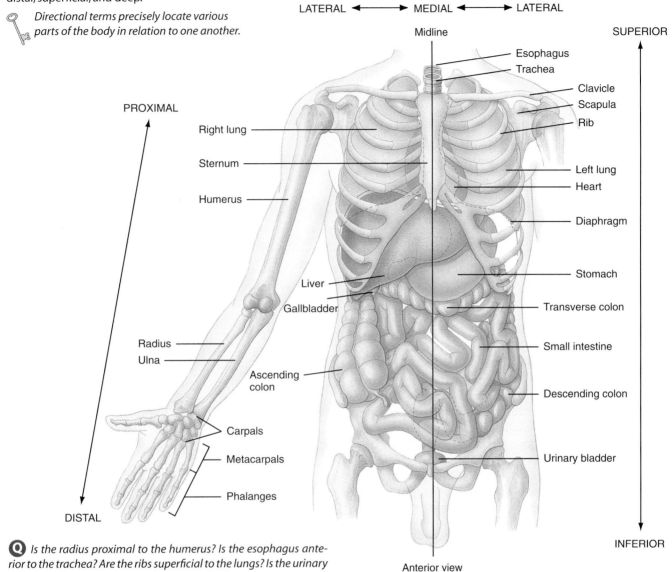

Q *Is the radius proximal to the humerus? Is the esophagus anterior to the trachea? Are the ribs superficial to the lungs? Is the urinary bladder medial to the ascending colon? Is the sternum lateral to the descending colon?*

Regional Names

The human body is divided into several major regions that can be identified externally. The principal regions are the head, neck, trunk, upper limbs (extremities), and lower limbs (extremities). The *head* consists of the skull and face. Whereas the skull (cranium) encloses and protects the brain, the *face* is the anterior portion of the head that includes the eyes, nose, mouth, forehead, cheeks, and chin. The *neck* supports the head and attaches it to the trunk. The *trunk* consists of the chest, abdomen, and pelvis. Each *upper limb (extremity)* is attached to the trunk and consists of the shoulder, armpit, arm (portion of the limb from the shoulder to the elbow), forearm (portion of the limb from the elbow to the wrist), wrist, and hand. Each *lower limb (extremity)* is also attached to the trunk and consists of the buttock, thigh (portion of the limb from the hip to the knee), knee, leg (portion of the limb from the knee to the ankle), ankle, and foot. The *groin* is the point of attachment between the trunk and lower limb that is marked by a depression (crease).

The common names and anatomical terms of the principal body parts within the principal regions are given in Figure 1.2. These terms designate particular areas of the body, such as the arm (brachial region) or the nose (nasal region).

Directional Terms

To locate various body structures in relation to one another, anatomists use certain **directional terms** (Figure 1.3). Such terms are precise and avoid the use of unnecessary words. There are many pairs of directional terms with opposite meanings—for example, superior and inferior. Important directional terms are defined in Table 1.3, and the parts of the body referred to in the examples are labeled in Figure 1.3. If you study the table and the figure together, the directional relations among various body parts will be made clear.

Planes and Sections

You will also study parts of the body relative to **planes** (imaginary flat surfaces) that pass through them (Figure 1.4). A **sagittal** (SAJ-i-tal; *sagittalis* = arrow) **plane** is a vertical plane that divides the body or an organ into right and left sides. More specifically, if such a plane passes through the midline of the body or organ and divides it into *equal* right and left sides, it is called a **midsagittal (median) plane.** If the sagittal plane does not pass through the midline but instead divides the body or an organ into *unequal* right and left sides, it is called a **parasagittal** (*para* = near) **plane.** A **frontal** or **coronal** (kō-RŌ-nal; *corona* = crown) **plane** divides the body or an organ into anterior (front) and posterior (back) portions. A **transverse (cross-sectional or horizontal) plane** divides the body or an organ into superior (top) and inferior (bottom) portions. Sagittal, frontal, and transverse planes are all at right angles to one another. An **oblique plane,** in contrast, passes through the body or an organ at an angle between the transverse plane and either a sagittal or frontal plane.

When you study a body region, you will often view it in *section,* meaning that you look at only one flat surface of the three-dimensional structure. It is important to know the plane of the section so that you can understand the anatomical relationship of one part to another. Figure 1.5 indicates how three different sections—a *transverse (cross) section,* a *frontal section,* and a *midsagittal section*—give different views of the brain to enable you to learn the anatomy of the brain from several distinct perspectives.

Body Cavities

Body cavities are spaces within the body that contain internal organs. The cavities help to protect, separate, and support the organs. The various body cavities may be separated from each other by structures such as muscles, bones, or ligaments. Figure 1.6 shows the two principal ones, the dorsal and ventral body cavities. The **dorsal body cavity** is located near the dorsal (back) surface of the body. It is further subdivided into a **cranial cavity,** which is formed by the cranial (skull) bones and contains the brain, and a **vertebral (spinal) canal,** which is formed by the vertebral column (backbone) and contains the spinal cord and the beginnings (roots) of spinal nerves.

Figure 1.4 Planes of the human body.

Frontal, transverse, sagittal, and oblique planes divide the body in specific ways.

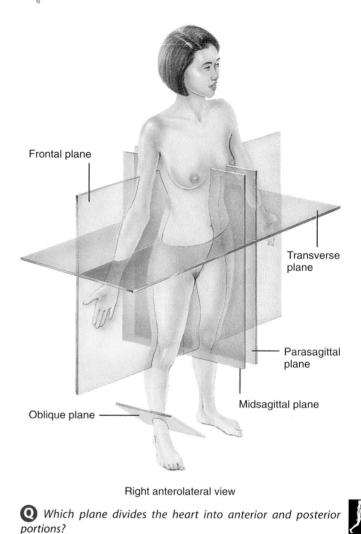

Frontal plane

Transverse plane

Parasagittal plane

Midsagittal plane

Oblique plane

Right anterolateral view

Q *Which plane divides the heart into anterior and posterior portions?*

The other principal body cavity is the **ventral body cavity.** This cavity is located on the ventral (front) aspect of the body. A thin, slippery membrane called a **serous membrane** lines the wall of the ventral body cavity and covers the organs within it. The organs inside are called **viscera** (VIS-er-a). The ventral body cavity also has two principal subdivisions: a superior portion, called the **thoracic** (thor-AS-ik) **cavity** (or chest cavity), and an inferior portion, called the **abdominopelvic** (ab-dom'-i-nō-PEL-vik) **cavity.** The structure that divides the ventral body cavity into the thoracic and abdominopelvic cavities is the diaphragm (DĪ-a-fram; *diaphragma* = partition or wall), an important muscle for breathing.

Table 1.3 Directional Terms[a]

TERM	DEFINITION	EXAMPLE
Superior (soo′-PEER-ē-or) (cephalic or cranial)	Toward the head or the upper part of a structure.	The heart is superior to the liver.
Inferior (in′-FEER-ē-or) (caudal)	Away from the head or toward the lower part of a structure.	The stomach is inferior to the lungs.
Anterior[b] (an-TEER-ē-or) (ventral)	Nearer to or at the front of the body. In the **prone position,** the body lies anterior side down. In the **supine position,** the body lies anterior side up.	The sternum (breastbone) is anterior to the heart.
Posterior[b] (pos-TEER-ē-or) (dorsal)	Nearer to or at the back of the body.	The esophagus is posterior to the trachea.
Medial (MĒ-dē-al) (mesial)	Nearer to the midline (median) of the body or a structure. The **midline** is an imaginary vertical line that divides the body into equal left and right sides.	The ulna is on the medial side of the forearm.
Lateral (LAT-er-al)	Farther from the midline of the body or a structure.	The lungs are lateral to the heart.
Intermediate (in′-ter-MĒ-dē-at)	Between two structures.	The ring finger is intermediate between the little and middle fingers.
Ipsilateral (ip-si-LAT-er-al)	On the same side of the midline of the body.	The gallbladder and ascending colon of the large intestine are ipsilateral.
Contralateral (CON-tra-lat-er-al)	On the opposite side of the midline of the body.	The ascending and descending colons of the large intestine are contralateral.
Proximal (PROK-si-mal)	Nearer to the attachment of a limb, the trunk, or a structure; nearer to the point of origin.	The humerus is proximal to the radius.
Distal (DIS-tal)	Farther from the attachment of a limb, the trunk, or a structure; farther from the point of origin.	The phalanges (bones of the fingers) are distal to the carpals (wrist bones).
Superficial (soo′-per-FISH-al)	Toward or on the surface of the body.	The muscles of the thoracic wall are superficial to the organs in the thoracic cavity. (See Figure 1.7.)
Deep (DĒP)	Away from the surface of the body.	The ribs are deep to the skin of the chest. (See Figure 1.7.)
Parietal (pa-RĪ-e-tal)	Pertaining to or forming the outer wall of a body cavity.	The parietal pleura forms the outer layer of the pleural sacs that surround the lungs. (See Figure 23.7a.)
Visceral (VIS-er-al)	Pertaining to the covering of an organ (viscus) within the ventral body cavity.	The visceral pleura forms the inner layer of the pleural sacs and covers the external surface of the lungs. (See Figure 23.7a.)

[a]Study this table with Figures 1.3 and 1.7 to visualize the examples given.
[b]Although anterior and ventral mean the same in humans, in four-legged animals ventral refers to the belly side (inferior surface). Similarly, posterior and dorsal mean the same in humans, but in four-legged animals dorsal refers to the animal's back surface (posterior surface).

Figure 1.5 Planes and sections through different parts of the brain. The planes are shown in the diagrams on the left and the resulting sections are shown in the photographs on the right.

Planes divide the body in various ways to produce sections.

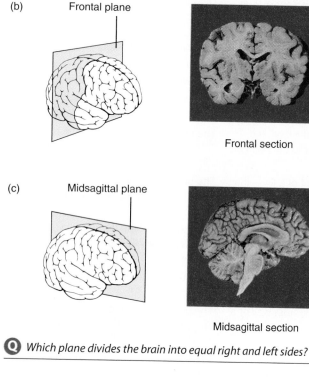

(a)

Superior (top)

Anterior (front)

Transverse (cross-sectional or horizontal) plane

Transverse (cross) section

(b) Frontal plane

Frontal section

(c) Midsagittal plane

Midsagittal section

Q *Which plane divides the brain into equal right and left sides?*

The thoracic cavity has three compartments: two pleural cavities and one pericardial cavity. Each of the two **pleural** (*pleur* = rib, side) **cavities** surrounds a lung (Figure 1.7). Each pleural cavity is a small, fluid-filled space between the part of the serous membrane that covers the lung and the part that lines the wall of the thoracic cavity. As you will see in Chapter 3, a serous membrane is a double-layered mem-

brane that (1) *lines* a body cavity that does not open directly to the exterior and (2) *covers* the organs within that cavity. The serous membrane associated with the lungs is called the **pleura.** Between the lungs is the **pericardial** (per´-i-KAR-dē-al; *peri* = around; *cardi* = heart) **cavity.** It is a fluid-filled space between the part of the serous membrane that covers the heart and the part that lines the thoracic cavity. The serous membrane associated with the heart is called the **pericardium.**

The **mediastinum** (mē´-dē-as-TĪ-num; *media* = middle; *stare* = stand in) in the thoracic cavity is a broad, median partition—actually a mass of tissues—between the coverings of lungs (pleurae), extending from the sternum (breastbone) to the vertebral column (backbone), as shown in Figures 1.7 and 1.8. The mediastinum includes all the contents of the thoracic cavity except the lungs themselves. Among the structures in the mediastinum are the heart, esophagus, trachea, thymus gland, and many large blood vessels such as the aorta, and lymphatic vessels.

The abdominopelvic cavity, as the name suggests, is divided into two portions, although no wall separates them. The superior portion of the abdominopelvic cavity, the **abdominal** (*abdere* means to hide) **cavity** (because it hides the viscera), contains the stomach, spleen, liver, gallbladder, pancreas, small intestine, and most of the large intestine. The serous membrane that lines the abdominal cavity and covers the organs within it is called the **peritoneum.** The inferior portion of the abdominopelvic cavity, the **pelvic cavity,** contains the urinary bladder, portions of the large intestine, and the internal organs of reproduction. The pelvic cavity is the region between two imaginary planes, shown by dashed lines in Figure 1.6a.

Abdominopelvic Regions and Quadrants

To describe the location of organs easily, the abdominopelvic cavity is divided into nine **abdominopelvic regions** (Figure 1.9a on page 18). Note which organs and parts of organs are in the different regions by carefully examining Figure 1.9b–d and Table 1.4 on page 19. Figure 1.9b–d illustrates successively deeper views of the abdominopelvic contents in their respective regions. Although some parts of the body in these illustrations are most likely to be unfamiliar to you at this point, they will be discussed in detail in later chapters.

The abdominopelvic cavity may also be divided more simply into **quadrants** (*quad* = four). These are shown in Figure 1.10 on page 20. In this method, a horizontal line and a vertical line are passed through the umbilicus (um-BIL-i-kus), or navel. These two lines divide the abdomen into a **right upper quadrant (RUQ), left upper quadrant (LUQ), right lower quadrant (RLQ),** and **left lower quadrant (LLQ).** Whereas the nine-region division is more

Figure 1.6 Body cavities.

🔑 *The two principal cavities are the dorsal and ventral body cavities.*

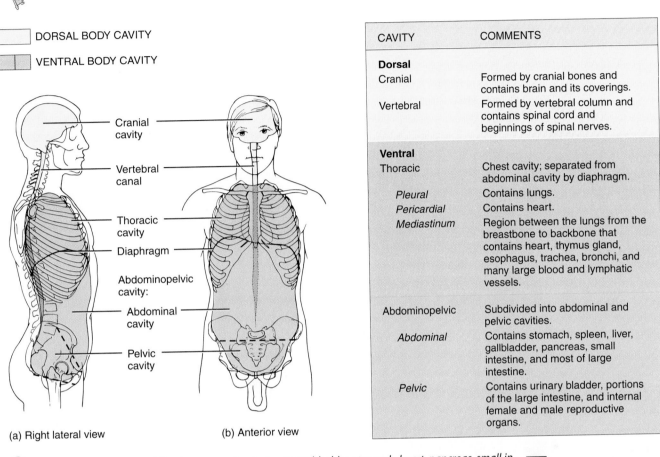

☐ DORSAL BODY CAVITY

▨ VENTRAL BODY CAVITY

(a) Right lateral view

(b) Anterior view

CAVITY	COMMENTS
Dorsal	
Cranial	Formed by cranial bones and contains brain and its coverings.
Vertebral	Formed by vertebral column and contains spinal cord and beginnings of spinal nerves.
Ventral	
Thoracic	Chest cavity; separated from abdominal cavity by diaphragm.
Pleural	Contains lungs.
Pericardial	Contains heart.
Mediastinum	Region between the lungs from the breastbone to backbone that contains heart, thymus gland, esophagus, trachea, bronchi, and many large blood and lymphatic vessels.
Abdominopelvic	Subdivided into abdominal and pelvic cavities.
Abdominal	Contains stomach, spleen, liver, gallbladder, pancreas, small intestine, and most of large intestine.
Pelvic	Contains urinary bladder, portions of the large intestine, and internal female and male reproductive organs.

Q *In which cavities are the following organs located: urinary bladder, stomach, heart, pancreas, small intestine, lungs, internal female reproductive organs, thymus gland, spleen, rectum, liver? Use the following symbols for your response: T = thoracic cavity, A = abdominal cavity, or P = pelvic cavity.*

widely used for anatomical studies, clinicians find that the quadrant division is better suited for locating the site of an abdominopelvic pain, tumor, or other abnormality. Compare Figure 1.10 with Figure 1.9b–d to note which organs are located in the various quadrants and at what depth.

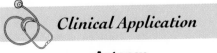 *Clinical Application*

Autopsy

To determine the cause of death accurately, it is sometimes necessary to perform an **autopsy** (AW-top-sē; *auto* = self; *opsis* = to see with one's own eyes). An autopsy can uncover the existence of diseases not detected during life. It also may support the accuracy of diagnostic tests, establish the beneficial and adverse ef-

fects of drugs, reveal the impact of environmental influences on the body, and educate health-care students. Moreover, an autopsy can reveal conditions (such as congenital heart defects) that may affect offspring or siblings. An autopsy may be needed in a criminal investigation, but it may also help resolve disputes between beneficiaries and insurance companies.

A typical autopsy consists of three principal phases. The first phase is examination of the exterior of the body for the presence of wounds, scars, tumors, or other abnormalities. The second phase includes the dissection and gross examination of the major body organs, looking for pathological changes or evidence of violent destruction. The third phase of autopsy consists of microscopic examination of tissues to ascertain any pathology. Depending on the circumstances, techniques may also be used to detect and/or recover microbes or toxic chemicals and to determine the presence of foreign substances in the body.

Figure 1.7 Location of pleural and pericardial cavities within the thoracic cavity. The borders of the mediastinum are indicated by a dashed line. The arrow in the inset indicates the direction from which the thoracic cavity is viewed (superiorly). This aid is used throughout the book.

Whereas the pleural cavities surround the lungs, the pericardial cavity surrounds the heart.

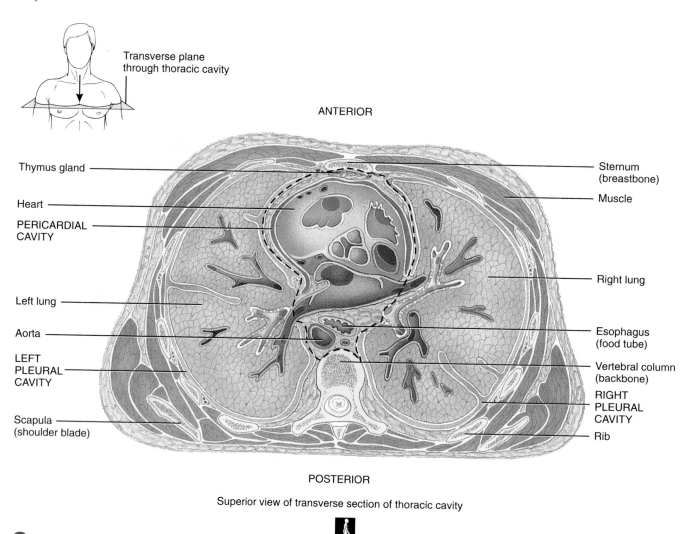

Transverse plane through thoracic cavity

ANTERIOR

Thymus gland

Heart

PERICARDIAL CAVITY

Left lung

Aorta

LEFT PLEURAL CAVITY

Scapula (shoulder blade)

Sternum (breastbone)

Muscle

Right lung

Esophagus (food tube)

Vertebral column (backbone)

RIGHT PLEURAL CAVITY

Rib

POSTERIOR

Superior view of transverse section of thoracic cavity

Q *What is a serous membrane?*

MEDICAL IMAGING

Various kinds of **medical imaging** techniques allow visualization of structures inside our bodies. The images provide clues to both abnormal anatomy and deviations from normal physiology. Imaging technology is increasingly helpful for precise diagnosis of a wide range of disorders. The grandfather of all medical imaging techniques is conventional radiography, in therapeutic use since the late 1940s. The newer techniques not only contribute to diagnosis of disease, but they also are advancing our understanding of normal physi-

ology. Table 1.5 on page 20 describes some commonly used medical imaging techniques.

MEASURING THE HUMAN BODY

An important aspect of describing the body and understanding how it works is *measurement*—what the dimensions of an organ are, how much it weighs, how long it takes for a physiological event to occur. Such measurements also have

Figure 1.8 Mediastinum. The mediastinal borders are indicated by a dashed line. Some of the structures shown and labeled may be unfamiliar to you now. However, they are discussed in detail in later chapters.

🔑 *The mediastinum is between the coverings (pleurae) of the lungs, extending from the sternum to the vertebral column.*

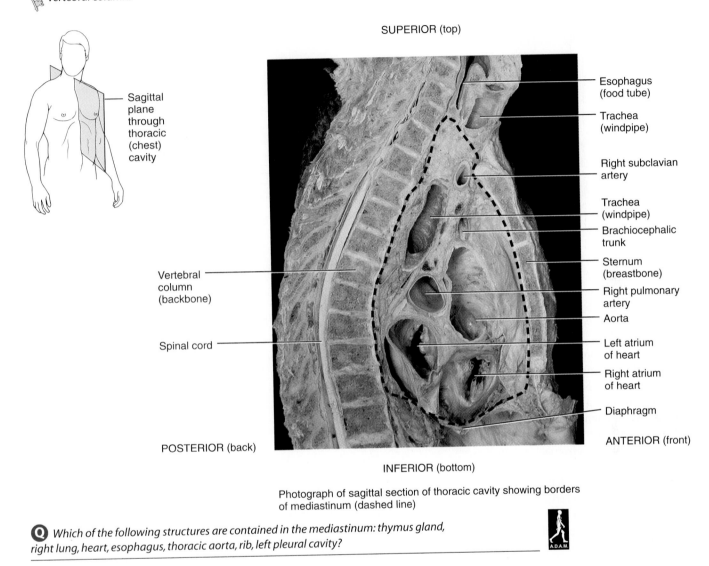

SUPERIOR (top)

Sagittal plane through thoracic (chest) cavity

Esophagus (food tube)

Trachea (windpipe)

Right subclavian artery

Trachea (windpipe)

Brachiocephalic trunk

Sternum (breastbone)

Right pulmonary artery

Aorta

Left atrium of heart

Right atrium of heart

Diaphragm

Vertebral column (backbone)

Spinal cord

POSTERIOR (back)

ANTERIOR (front)

INFERIOR (bottom)

Photograph of sagittal section of thoracic cavity showing borders of mediastinum (dashed line)

Q *Which of the following structures are contained in the mediastinum: thymus gland, right lung, heart, esophagus, thoracic aorta, rib, left pleural cavity?*

clinical importance, for example, in determining how much of a given medication should be administered. As you will see, measurements involving time, weight, temperature, size, and volume are a routine part of your studies in a medical science program.

Whenever you come across a measurement in this text, the measurement will be given in metric units. The metric system is standardly used in sciences. To help you compare the metric unit to a familiar unit, the approximate U.S. equivalent will also be given in parentheses directly after the metric unit. For example, you might be told that the length of a particular part of the body is 2.54 cm (1 in.).

As a first step in helping you understand the correlation between the metric system and the U.S. system of measurement, three exhibits have been prepared and are located in Appendix I: (1) metric units of length and some U.S. equivalents, (2) metric units of mass and some U.S. equivalents, and (3) metric units of volume and some U.S. equivalents.

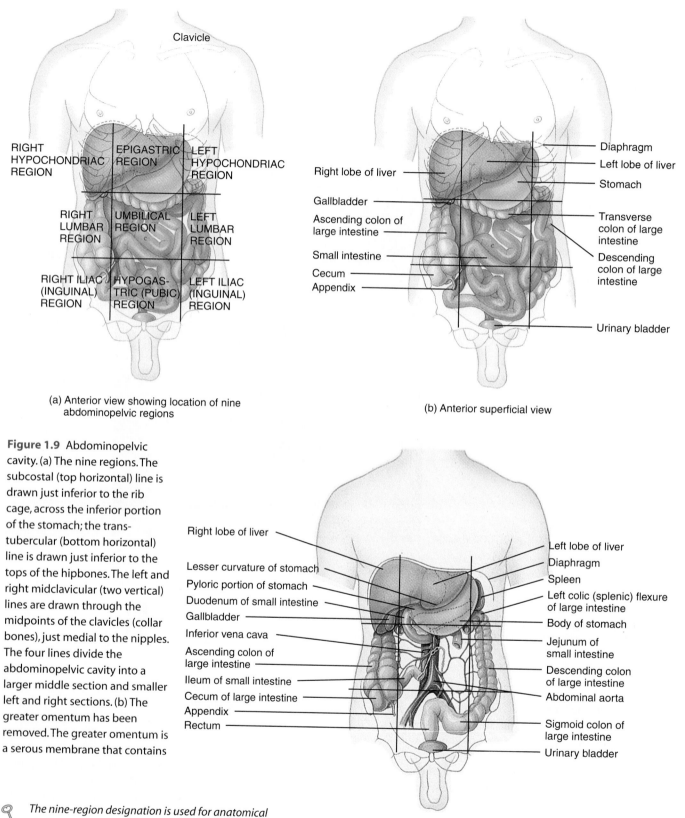

Clavicle

RIGHT HYPOCHONDRIAC REGION

EPIGASTRIC REGION

LEFT HYPOCHONDRIAC REGION

RIGHT LUMBAR REGION

UMBILICAL REGION

LEFT LUMBAR REGION

RIGHT ILIAC (INGUINAL) REGION

HYPOGAS-TRIC (PUBIC) REGION

LEFT ILIAC (INGUINAL) REGION

(a) Anterior view showing location of nine abdominopelvic regions

Right lobe of liver

Gallbladder

Ascending colon of large intestine

Small intestine

Cecum

Appendix

Diaphragm

Left lobe of liver

Stomach

Transverse colon of large intestine

Descending colon of large intestine

Urinary bladder

(b) Anterior superficial view

Figure 1.9 Abdominopelvic cavity. (a) The nine regions. The subcostal (top horizontal) line is drawn just inferior to the rib cage, across the inferior portion of the stomach; the trans-tubercular (bottom horizontal) line is drawn just inferior to the tops of the hipbones. The left and right midclavicular (two vertical) lines are drawn through the midpoints of the clavicles (collar bones), just medial to the nipples. The four lines divide the abdominopelvic cavity into a larger middle section and smaller left and right sections. (b) The greater omentum has been removed. The greater omentum is a serous membrane that contains

Right lobe of liver

Lesser curvature of stomach

Pyloric portion of stomach

Duodenum of small intestine

Gallbladder

Inferior vena cava

Ascending colon of large intestine

Ileum of small intestine

Cecum of large intestine

Appendix

Rectum

Left lobe of liver

Diaphragm

Spleen

Left colic (splenic) flexure of large intestine

Body of stomach

Jejunum of small intestine

Descending colon of large intestine

Abdominal aorta

Sigmoid colon of large intestine

Urinary bladder

(c) Anterior deep view

The nine-region designation is used for anatomical studies.

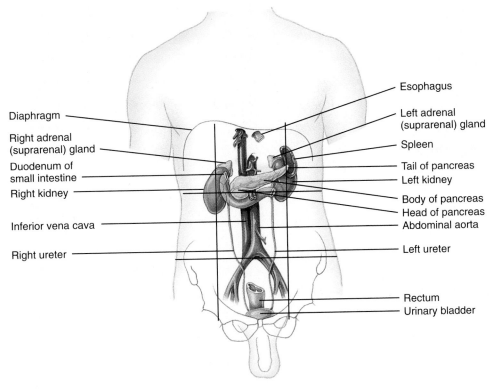

Esophagus

Left adrenal (suprarenal) gland

Spleen

Tail of pancreas

Left kidney

Body of pancreas

Head of pancreas

Abdominal aorta

Left ureter

Rectum

Urinary bladder

Diaphragm

Right adrenal (suprarenal) gland

Duodenum of small intestine

Right kidney

Inferior vena cava

Right ureter

(d) Anterior deeper view

Figure 1.9 (continued) fatty tissue which covers some of the abdominal organs. (c) Some anterior organs have been removed, exposing the more posterior structures. (d) All anterior organs removed, fully revealing posterior structures. The internal reproductive organs in the pelvic cavity are shown in Figs. 26.1 and 26.11.

Q *In which abdominopelvic region is each of the following found: left lobe of the liver, transverse colon, urinary bladder, spleen?*

Table 1.4 *Representative Structures Found in the Abdominopelvic Regions*

REGION	REPRESENTATIVE STRUCTURES
Right hypochondriac (hī-pō-KON-drē-ak; *hypo* = under; *chondro* = cartilage)	Right lobe of liver, gallbladder, and upper superior third of right kidney.
Epigastric (ep-i-GAS-trik; *epi* = above; *gaster* = stomach)	Left lobe and medial part of right lobe of liver, pyloric portion and lesser curvature of stomach, superior and descending portions of duodenum, body and superior part of head of pancreas, and right and left adrenal (suprarenal) glands.
Left hypochondriac	Body and fundus of stomach, spleen, left colic (splenic) flexure, superior two-thirds of left kidney, and tail of pancreas.
Right lumbar (*lumbus* = loin)	Superior part of cecum, ascending colon, right colic (hepatic) flexure, inferior lateral portion of right kidney, and part of small intestine.
Umbilical (um-BIL-i-kul; *umbilikus* = navel)	Middle portion of transverse colon, part of small intestine, and bifurcations (branching) of abdominal aorta and inferior vena cava.
Left lumbar	Descending colon, inferior third of left kidney, and part of small intestine.
Right iliac (inguinal) (IL-ē-ak; iliac refers to superior part of hipbone)	Lower end of cecum, appendix, and part of small intestine.
Hypogastric (pubic)	Urinary bladder when full, small intestine, and part of sigmoid colon.
Left iliac (inguinal)	Junction of descending and sigmoid parts of colon and part of small intestine.

Figure 1.10 Quadrants of the abdominopelvic cavity. The two lines intersect at right angles at the umbilicus (navel).

🔑 *The quadrant designation is used to locate the site of pain, a tumor, or some other abnormality.*

Q *In which quadrant would the pain from appendicitis (inflammation of the appendix) be felt?*

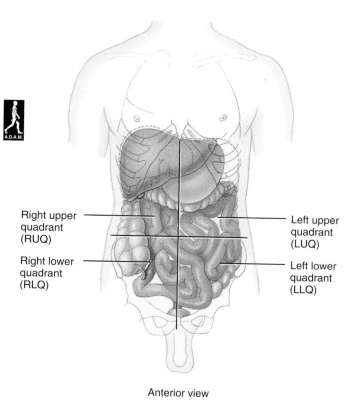

Right upper quadrant (RUQ)

Left upper quadrant (LUQ)

Right lower quadrant (RLQ)

Left lower quadrant (LLQ)

Anterior view

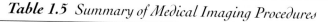

Table 1.5 *Summary of Medical Imaging Procedures*

PROCEDURE	DESCRIPTION
Conventional radiography	A single barrage of x-rays passes through the body and produces a two-dimensional image of the interior of the body called a *radiograph* (RĀ-dē-ō-graf) or *x-ray*. *Comment:* Overlap of structures can make diagnosis difficult, and subtle differences in tissue density cannot always be discerned.

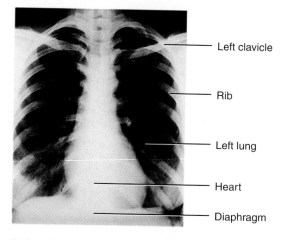

Left clavicle

Rib

Left lung

Heart

Diaphragm

Radiograph of thorax (chest) (anterior view)

Table 1.5 *(continued)*

PROCEDURE

Computed tomography (CT) scanning
[formerly called computerized axial tomography (CAT) scanning]

Note: In keeping with radiographic convention, as in CT scans, the section is viewed from the inferior aspect. Thus the left side of the body appears on the right side of the photograph.

DESCRIPTION

An x-ray beam moves in an arc around the body, producing an image of a transverse section called a *CT scan* on a video monitor hooked to a computer. By stacking images one on top of the other, the computer can construct three-dimensional images that can be rotated, for example, to plan plastic surgery.

Comment: Quick, painless CT scans have replaced exploratory surgery in many cases. Detailed images can reveal tumors, aneurysms (bulges in blood vessels or the heart), kidney stones, gallstones, infections, tissue damage, and deformities.

ANTERIOR

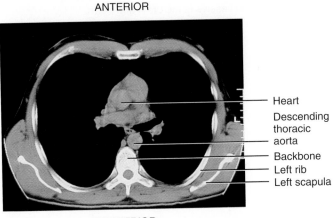

Heart
Descending thoracic aorta
Backbone
Left rib
Left scapula

POSTERIOR

CT scan of a transverse section of the thorax (chest) (inferior view)

Dynamic spatial reconstruction (DSR)

A highly sophisticated x-ray machine produces moving, three-dimensional, life-size images from any view.

Comment: By using a computer, an image can be rotated, tipped, "sliced open," enlarged, replayed, and viewed in slow motion or at high speeds; good procedure for imaging the heart, lung, and blood vessels; measuring movements and volumes; and assessing tissue damage.

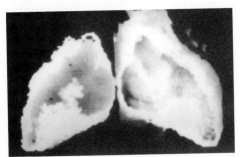

DSR image of "opened" heart

▶

Table 1.5 Summary of Medical Imaging Procedures (continued)

PROCEDURE	DESCRIPTION
Digital subtraction angiography (DSA)	A computer compares a radiograph of a region of the body before and after a contrast dye has been injected into a blood vessel. Tissues around the blood vessel in the first image can be subtracted (erased) from the second image, leaving an unobstructed view of the blood vessel. *Comment:* DSA is used primarily to study blood vessels in the brain and heart.

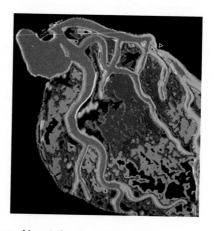

DSA image of heart showing narrowed blood vessel (triangle)

Positron emission tomography (PET)	A radioactive substance that emits positively charged particles called positrons is injected into the body. The collision of positrons with negatively charged electrons in body tissues releases gamma rays, which are similar to x-rays. A computer receives signals from gamma cameras positioned around the patient and constructs an image called a PET scan that can be displayed in color on a video monitor. The PET scan shows where the radioactive substance is being used in the body. *Comment:* PET scans provide information on function as well as structure and are very useful in detecting metabolic changes in healthy or diseased organs such as the brain and heart.

ANTERIOR

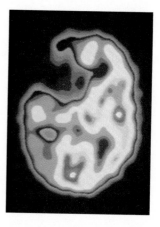

POSTERIOR

PET image showing blood flow through the brain (The darkened area at the upper left of the photo indicates where a stroke has occurred)

Table 1.5 (continued)

PROCEDURE	DESCRIPTION
Magnetic resonance imaging (MRI) [formerly called nuclear magnetic resonance (NMR) imaging]	Protons (small positive particles within atoms, such as hydrogen) in tissues respond to a pulse of radio waves while they are being magnetized. Their response is measured and the colored image that results is a two- or three-dimensional blueprint of cellular chemistry.

Comment: MRI is used to detect tumors and artery-clogging fatty plaques, assess mental disorders, reveal brain changes, measure blood flow, and study metabolism. Although MRI is noninvasive and employs radiation thought to be harmless (radio waves), it isn't recommended for pregnant women; because of the magnetism needed, MRI can't be used in persons with artificial pacemakers or metal joints.

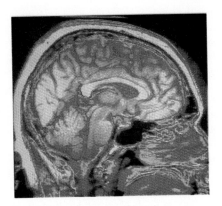

MRI of brain (sagittal view)

Ultrasound (US)

High-frequency sound waves produced by a hand-held device reflect from body tissues and are detected by the same instrument, which transmits the signals to a video monitor. The image, called a *sonogram* (SO-nō-gram) or *ultrasonogram,* may be still or it may be moving.

Comment: US is used to image the fetus during amniocentesis (see Fig. 27.10a) and to study abdominal and pelvic organs, blood flow, the heart (echocardiography), and the developing fetus (fetal ultrasound). The procedure uses sound waves, which have no obvious harmful effects, is noninvasive, painless, and uses no dyes. Obesity and scars interfere with US.

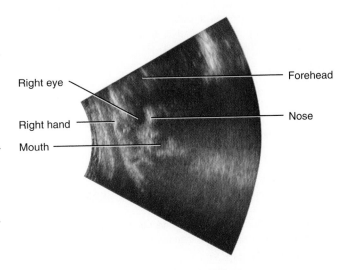

Sonogram (Courtesy of Anthony Gerard Tortora 12/4/94)

Study Outline

Anatomy Defined (p. 2)

1. Anatomy is the study of structure and the relationships among structures.
2. Subdivisions of anatomy include surface anatomy (form and markings of surface features), gross anatomy (macroscopic), systemic or systematic anatomy (systems), regional anatomy (regions), radiographic anatomy (x-rays), developmental anatomy (development from fertilization to adulthood), embryology (development from fertilized egg through eighth week in utero), histology (tissues), cytology (cells), and pathological anatomy (disease).
3. Physiology is the study of how body structures function.

Levels of Structural Organization (p. 2)

1. The human body consists of several levels of structural organization; among these are the chemical, cellular, tissue, organ, system, and organismic levels.
2. Cells are the basic structural and functional living units of an organism.
3. Tissues consist of groups of similarly specialized cells (and their intercellular material) that usually have a similar embryological origin and that perform certain special functions.
4. Organs are structures of definite form that are composed of two or more different tissues and have specific functions.
5. Systems consist of related organs that have a common function.
6. The human organism is a collection of structurally and functionally integrated systems.
7. The eleven systems of the human body are the integumentary, skeletal, muscular, cardiovascular, lymphatic, nervous, endocrine, respiratory, digestive, urinary, and reproductive systems (see Table 1.2 on page 4).

Life Processes (p. 6)

1. All living forms have certain processes that distinguish them from nonliving things.
2. Among the life processes in humans are metabolism, responsiveness, movement, growth, differentiation, and reproduction.

Overview of the Human Body (p. 8)

Structural Plan (p. 8)

1. The human body has certain general characteristics that are part of its overall organizational plan.
2. Among the characteristics are a backbone, a tube-within-a-tube organization, and bilateral symmetry.

Anatomical Position (p. 9)

1. Descriptions of any region of the body assume the body is in the anatomical position.
2. When in the anatomical position, the subject stands erect facing the observer with the head, eyes, and toes facing forward, the feet are flat on the floor and directed forward, and the arms are placed at the sides with the palms turned forward.

Regional Names (p. 11)

1. Regional names are terms given to specific regions of the body for reference. They include the head, neck, trunk, upper limbs, and lower limbs.

2. Within the regions, specific body parts have common and anatomical names. Examples include thoracic (chest), brachial (arm), patellar (knee), and gluteal (buttock).

Directional Terms (p. 11)

1. Directional terms indicate the relationship of one part of the body to another.
2. Commonly used directional terms are summarized in Table 1.3 on page 13.

Planes and Sections (p. 12)

1. Planes are imaginary flat surfaces that are used to divide the body or organs into definite areas. A midsagittal (median) plane divides the body or organs into equal right and left sides; a parasagittal plane divides the body or organs into unequal right and left sides; a frontal plane divides the body or organs into anterior and posterior portions; a transverse plane divides the body or organs into superior and inferior portions; and an oblique plane passes through the body or organs at an angle between a transverse plane and either a midsagittal, parasagittal, or frontal plane.
2. Sections are flat surfaces resulting from cuts through body structures. They are named according to the plane on which the cut is made and include transverse, frontal, and midsagittal sections.

Body Cavities (p. 12)

1. Spaces in the body that contain internal organs are called body cavities.
2. The dorsal and ventral cavities are the two principal body cavities. The dorsal cavity contains the brain and spinal cord.
3. The dorsal cavity is subdivided into the cranial cavity, which contains the brain, and the vertebral (spinal) canal, which contains the spinal cord and beginnings of spinal nerves.
4. The ventral body cavity is subdivided by the diaphragm into a superior thoracic cavity and an inferior abdominopelvic cavity.
5. The thoracic cavity contains two pleural cavities and a mediastinum, which includes the pericardial cavity.
6. The mediastinum is a broad, median partition—actually a mass of tissues—between the coverings of the lungs that extends from the sternum to the vertebral column; it contains all contents of the thoracic cavity except the lungs.
7. The abdominopelvic cavity is divided into a superior abdominal and an inferior pelvic cavity.
8. Viscera of the abdominal cavity include the stomach, spleen, pancreas, liver, gallbladder, small intestine, and most of the large intestine.
9. Viscera of the pelvic cavity include the urinary bladder, portions of the large intestine, and internal female and male reproductive structures.

Abdominopelvic Regions and Quadrants (p. 14)

1. To describe the location of organs easily, the abdominopelvic cavity may be divided into nine regions by drawing four imaginary lines (left midclavicular, right midclavicular, subcostal, and transtubercular).
2. The names of the nine abdominopelvic regions are right hypochondriac, epigastric, left hypochondriac, right lumbar,

umbilical, left lumbar, right iliac (inguinal), hypogastric (pubic), and left iliac (inguinal).

3. To locate the site of an abdominopelvic abnormality in clinical studies, the abdominopelvic cavity may be divided into quadrants by passing imaginary horizontal and vertical lines through the umbilicus.

4. The names of the abdominopelvic quadrants are right upper quadrant (RUQ), left upper quadrant (LUQ), right lower quadrant (RLQ), and left lower quadrant (LLQ).

Medical Imaging (p. 16)

1. Medical imaging techniques allow physicians to visualize structures inside the body to diagnose abnormal anatomy and deviations from normal physiology.

2. Table 1.5 on page 20 summarizes several imaging techniques.

Measuring the Human Body (p. 16)

1. Measurements involving time, weight, temperature, size, and volume are used in clinical situations.

2. Measurements in this book are given in metric units followed by U.S. equivalents in parentheses.

Review Questions

1. Define anatomy. List and define the various subdivisions of anatomy. Define physiology. (p. 2)

2. Give several examples of how structure and function are related. (p. 2)

3. Define each of the following terms: cell, tissue, organ, system, and organism. (p. 4)

4. Using Table 1.2 on page 4 as a guide, outline the functions of each system of the body; then list several organs that compose each system.

5. List and define the life processes of humans. (p. 6)

6. What is a vertebrate? Why is the body considered to be a tube-within-a-tube? What does bilateral symmetry mean? (p. 8)

7. Define the anatomical position. Why is the anatomical position used? (p. 9)

8. Review Figure 1.2 on page 10. See whether you can locate each region on your own body and name each by its common and anatomical term.

9. What is a directional term? Why are these terms important? Use each of the directional terms listed in Table 1.3 on page 13 in a complete sentence.

10. Define the various planes that may be passed through the body. Explain how each plane divides the body. Describe the meaning of transverse, frontal, and midsagittal sections. (p. 12)

11. Define a body cavity. List the body cavities discussed and tell which major organs are located in each. What is the mediastinum? (p. 12)

12. Describe how the abdominopelvic area is subdivided into nine regions. Name and locate each region and list the organs, or parts of organs, in each. (p. 14)

13. Describe how the abdominopelvic cavity is divided into quadrants and name each quadrant. (p. 14)

14. Why is an autopsy performed? Describe the three phases. (p. 15)

Self Quiz

Complete the following:

1. The four basic types of tissues in the body are ——, ——, ——, and ——.

2. Three serous membranes located in the ventral body cavity are the pericardium, peritoneum, and ——.

3. From superior to inferior, the three abdominopelvic regions on the right side are right ——, right ——, and right ——.

4. A —— plane divides the body into anterior and posterior parts.

5. Body structures known as —— are composed of two or more different tissues, have specific functions, and usually have recognizable shapes.

6. Because the stomach and the spleen are both located on the left side of the abdomen, they could be described as ——-lateral.

Are the following statements true or false?

7. The joints of the body are part of the muscular system.

8. The thoracic cavity is the same as the pleural cavity.

9. You would *not* look in the pelvic cavity to find the thymus gland.

10. The diaphragm divides the abdominopelvic cavity into abdominal and pelvic portions.

11. The brain and the spinal cord are located in the dorsal body cavity.

12. The breakdown of food by the digestive system involves chemical reactions that are catabolic.

13. Match the following common and anatomical terms:

   (a) armpit   (1) thoracic
   (b) arm   (2) inguinal
   (c) head   (3) axillary
   (d) chest   (4) tarsal
   (e) groin   (5) cephalic
   (f) leg   (6) brachial
   (g) wrist   (7) carpal
   (h) ankle   (8) crural

Choose the one best answer to the following questions:

14. You would look in the mediastinum to find the
   a. heart
   b. liver
   c. pleura
   d. diaphragm
   e. lungs

15. Which is most inferiorly located?
    a. abdominal cavity
    b. pleural cavity
    c. mediastinum
    d. diaphragm
    e. pelvic cavity
16. The spleen, tonsils, and thymus gland are all organs in which system?
    a. nervous
    b. lymphatic
    c. cardiovascular
    d. digestive
    e. endocrine
17. Which of the following five statements are true of a parasagittal plane?
    (1) It must pass through the midline of the body.
    (2) It is a longitudinal (vertical) plane.
    (3) It divides the body into unequal left and right parts.
    (4) It divides the body into anterior and posterior parts.
    (5) It is also known as a coronal plane.
    a. 1 and 2
    b. 2 and 3
    c. 3 and 5
    d. 2 and 5
    e. 2, 4, and 5
18. Any part of the body that is away from the midline is said to be
    a. medial
    b. distal
    c. lateral
    d. superior
    e. posterior
19. The system responsible for providing mechanical support, protecting internal organs, assisting in movement, storing of minerals and lipids, and housing of tissue that produces blood cells is the
    a. cardiovascular system
    b. skeletal system
    c. muscular system
    d. digestive system
    e. nervous system
20. In the anatomical position
    a. the palm is posterior
    b. the dorsum of the hand is anterior
    c. the body may be either in a vertical or horizontal position
    d. both a and b
    e. none of the above

Critical Thinking Questions

1. Why might a person with many imaginary ailments be called a "hypochondriac"? Explain why this region of the abdominopelvic cavity was given this name.
   HINT: *Think of where a person who is suffering from heartburn might feel pain.*

2. Ming was going for the record for the longest upside-down hang from the monkey bars. She didn't make it, and she may have broken her arm. The emergency room technician would like an x-ray of her arm in the anatomical position. Use the proper anatomical terms to describe the position of Ming's arm in the x-ray.
   HINT: *A tennis player hitting a backhand does not have her arm in anatomical position.*

3. Your friend is rushed to the hospital suffering from acute abdominal pain. A decision is made to remove the appendix. Using anatomical terminology, describe the location of the appendix within the abdominopelvic cavity.
   HINT: *The pain of appendicitis usually localizes in an area near the hip.*

4. Imagine that a manned space flight lands on Mars. The astronaut/life-specialist observes lumpy shapes that may be either life-forms or mud balls. What are some characteristics of living organisms that may help the astronaut determine if these are life-forms or mud balls?
   HINT: *Even a couch potato is "livelier" than a mud ball.*

5. David has been collecting all his belly-button lint to stuff a special pillow for a Mother's Day gift. Use anatomical terms to describe the position of the belly button.
   HINT: *Look for something on your body that resembles the "navel" on a navel orange.*

6. Janelle volunteered for a research project on the effects of sleep deprivation on brain function. She spent all night studying anatomy and will have brain imaging done this morning. But now she can't remember the name of the procedure. What type of imaging method studies brain function?
   HINT: *The size of your cranium won't give you any clues concerning how the brain is working inside it.*

Answers to Figure Questions

1.1 Organ level.
1.2 It allows directional terms to be clear, and any body part can be described in relation to any other body part.
1.3 No, No, Yes, Yes, No.
1.4 Frontal plane.
1.5 Midsagittal plane.
1.6 P, A, T, A, A, T, P, T, A, P, A.
1.7 A double-layered membrane that lines a body cavity that does not open to the exterior and covers the organs within the cavity.
1.8 Thymus gland, heart, esophagus, thoracic aorta.
1.9 Epigastric, umbilical, hypogastric, left hypochondriac.
1.10 Right lower quadrant (RLQ).

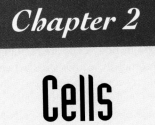

Chapter 2

Cells

Student Objectives

1. Define a cell and list its principal parts.

2. Explain the structure and functions of the plasma membrane.

3. Describe the various processes that move materials across plasma membranes.

4. Define cytoplasm and describe the chemical composition and functions of the cytosol.

5. Describe the structure and functions of the following cellular structures: nucleus, ribosomes, endoplasmic reticulum, Golgi complex, lysosomes, peroxisomes, mitochondria, cytoskeleton, flagella and cilia, and centrosome.

6. Define a cell inclusion and give several examples.

7. Explain the importance of programmed cell death.

8. Explain the relation of aging to cells and cell death.

A **cell** is the basic, living, structural and functional unit of the body. Cells are composed of characteristic parts, the work of which is coordinated in a way that enables each type of cell in the body to fulfill a unique biochemical or structural role. **Cytology** (sī-TOL-ō-jē; *cyt* = cell; *logos* = study of) is the branch of science concerned with the study of cells. Your understanding of human anatomy must be founded on a knowledge of the cellular level of organization. Hence, in this chapter you will study the structure, function, and reproduction of typical human cells.

CELLULAR DIVERSITY

The body of an average adult human is composed of nearly 100 trillion cells, but all of these cells can be classified into about 200 different cell types. The sizes and shapes of various body cells are related to the functions they perform. In order to see the smallest cells of the body, high-powered microscopes are necessary; the largest cell, a single ovum (the female reproductive cell), is just about visible to the unaided eye. The sizes of cells are measured in units called *micrometers*. One micrometer (μm) is equal to 1/10,000 of a centimeter (1/25,000 of an inch). Whereas a red blood cell has a diameter of 8 μm, an ovum has a diameter of about 140 μm. The shapes of cells vary from round, to oval, to flat, to cube-shaped, to column-shaped, to elongated, to star-shaped, to cylindrical, to disc-shaped.

Following are a few examples of how the structure of a cell is related to its function (Figure 2.1). Sperm cells have long whiplike tails (flagella) that they use for locomotion. Disc-shaped red blood cells provide a large surface area to transport and exchange oxygen and carbon dioxide. Nerve cells have long extensions that permit them to transmit nerve impulses over great distances. Muscle cells have complex internal protein structures that permit them to contract. Some epithelial cells contain small projections called microvilli, structures that greatly increase the cell's surface area. Mi-

Figure 2.1 Diverse shapes and sizes of human cells. The cells are not drawn to scale.

🔑 *The nearly 100 trillion cells of an average adult can be classified into about 200 different cell types.*

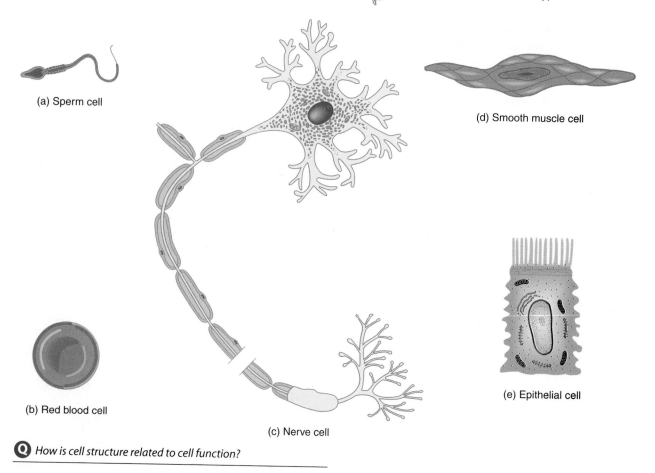

(a) Sperm cell

(d) Smooth muscle cell

(b) Red blood cell

(e) Epithelial cell

(c) Nerve cell

Q *How is cell structure related to cell function?*

crovilli are common in the epithelial cells lining the small intestine. They enhance the absorption of digested food from the intestine into the blood or lymph.

GENERALIZED CELL

The *generalized cell* illustrated in Figure 2.2 is a composite of many different cells in the body. Even though most cells have many of the features shown in this diagram, few cells have all the illustrated features. For ease of study, we can divide a cell into four principal parts:

1. **Plasma (cell) membrane.** Outer, limiting membrane separating the cell's internal components from the extracellular materials. Extracellular materials, which are substances external to the cell surface, will be examined together with tissues in Chapter 3.

2. **Cytoplasm** (SĪ-tō-plazm′; *cyto* = cell; *plassein* = to form). A general term for all cellular contents located between the plasma membrane and the nucleus. It includes three portions: cytosol, organelles (except the nucleus), and inclusions (explained shortly). The **cytosol** (SĪ-tō-sol) is the thick, semifluid portion of cytoplasm. It is also

Figure 2.2 Generalized cell based on electron microscope studies.

The cell is the basic, living, structural and functional unit of the body.

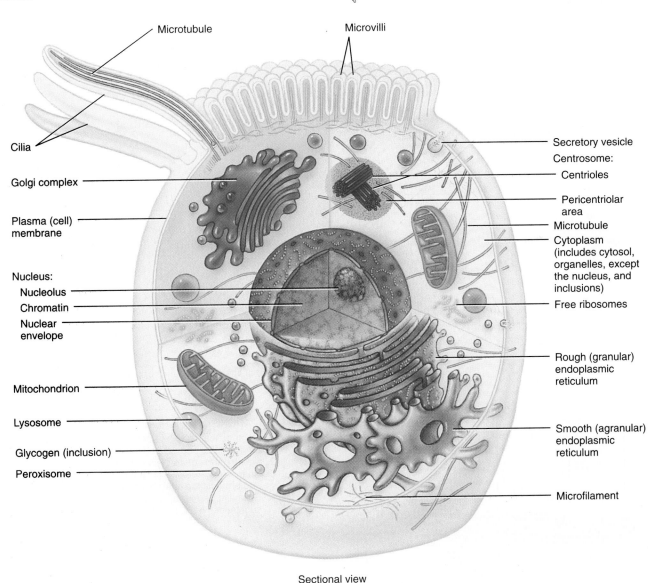

Sectional view

Q *List the four principal parts of a cell.*

referred to as **intracellular fluid** and contains mostly water, many dissolved proteins and enzymes, nutrients, ions, and other small molecules, all of which participate in various phases of cellular metabolism. Organelles and inclusions, described next, are suspended in the cytosol.

3. **Organelles.** Highly organized structures with characteristic shapes that are highly specialized for specific cellular activities.

4. **Inclusions.** Temporary structures that contain secretions and storage products of the cell.

PLASMA (CELL) MEMBRANE

The thin barrier that separates the internal components of a cell from the extracellular materials and external environment is the **plasma membrane** (also called the *cell membrane*). The plasma membrane is the gatekeeper that regulates passage of substances into and out of the cell (Figure 2.3).

The **fluid mosaic model** of membrane structure describes the molecular arrangement of the plasma membrane and other membranes in living organisms. A mosaic is a pattern of many small pieces fitted together. According to this

Figure 2.3 Structure of the plasma membrane.

The plasma membrane consists mostly of phospholipids and proteins.

Q *What types of lipids are present in plasma membranes of animal cells?*

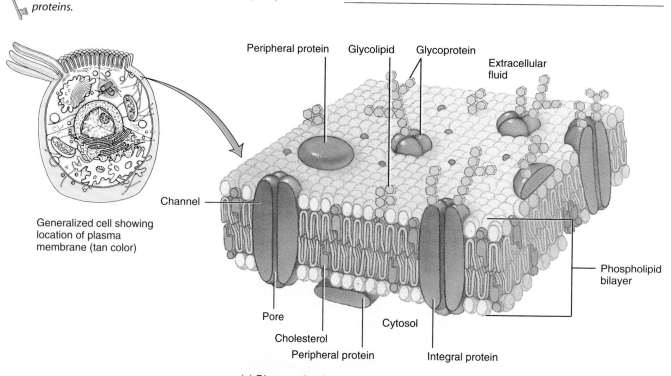

Generalized cell showing location of plasma membrane (tan color)

(a) Diagram showing arrangement of lipids and proteins of plasma membrane

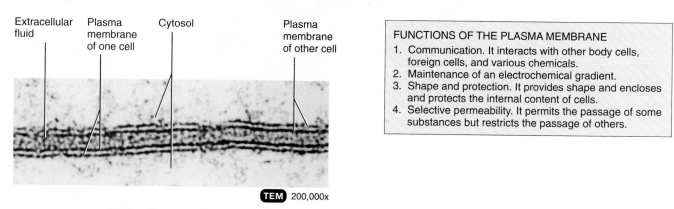

(b) Two plasma membranes

TEM 200,000x

FUNCTIONS OF THE PLASMA MEMBRANE
1. Communication. It interacts with other body cells, foreign cells, and various chemicals.
2. Maintenance of an electrochemical gradient.
3. Shape and protection. It provides shape and encloses and protects the internal content of cells.
4. Selective permeability. It permits the passage of some substances but restricts the passage of others.

model, the membrane is a mosaic of proteins floating like icebergs in a sea of lipids.

Chemistry and Anatomy

Plasma membranes of typical animal cells are about a 50:50 mix by weight of proteins and lipids. About 75% of the lipids are **phospholipids,** lipids that contain phosphorus. Present in smaller amounts are cholesterol (a steroid) and glycolipids, which are lipids with one or more sugar groups attached. The phospholipids line up in two parallel layers, forming a **phospholipid bilayer** (Figure 2.3). The bilayer forms the basic framework of plasma membranes. **Glycolipids,** about 5% of membrane lipids, appear only in the layer that faces the extracellular fluid. They are important for adhesion among cells and tissues, may mediate cell-to-cell recognition and communication, and contribute to regulation of cellular growth and development. The remaining 20% of membrane lipids are **cholesterol** molecules, which are located among the phospholipids in both sides of the bilayer in animal cells. Plant cell membranes lack cholesterol. Cholesterol strengthens the membrane but decreases its flexibility.

Membrane proteins are of two types: integral and peripheral (Figure 2.3). **Integral proteins** extend across the phospholipid bilayer. Most (maybe all) integral proteins are *glycoproteins,* combinations of sugar and protein. The sugar portion of a glycoprotein faces the extracellular fluid. **Peripheral proteins** do not extend across the phospholipid bilayer. They are loosely attached to the inner and outer surfaces of the membrane and are easily separated from it. Some integral proteins (or glycoproteins) are *channels* that have a *pore* (hole) through which certain substances flow into or out of the cell. Others act as *transporters* (carriers) to move a substance from one side of the membrane to the other. Integral proteins also serve as recognition sites called *receptors.* These are molecules that can identify and attach to a specific molecule, such as a hormone, a neurotransmitter, or a nutrient, that is important for some cellular function. Some integral and peripheral proteins are *enzymes.* Other peripheral proteins serve as *cytoskeleton anchors,* forming an attachment between the plasma membrane and filaments of the cytoskeleton. Membrane glycoproteins and glycolipids often are *cell identity markers.* They may enable a cell to recognize other cells of its own kind during tissue formation or to recognize and respond to potentially dangerous foreign cells.

Functions

Based on the discussion of the chemistry and structure of the plasma membrane, we can now describe several of its important functions:

1. **Communication.** This includes interactions with other body cells, foreign cells, and chemicals such as hor-

mones, neurotransmitters, enzymes, nutrients, and antibodies.

2. **Shape and protection.** The plasma membrane provides a flexible boundary that gives shape to a cell and protects the cellular contents.

3. **Electrochemical gradient.** The plasma membrane maintains a chemical and electrical gradient (difference) between the inside and outside of the cell. The gradient is due to the arrangement of ions (electrically charged particles) inside and outside the cell and is very important for the proper functioning of most cells, especially nerve and muscle cells.

4. **Selective permeability.** This function permits the passage of certain substances and restricts the passage of others.

Movement of Materials Across Plasma Membranes

Before discussing how materials move into and out of cells, we will first describe the locations of the various fluids through which the substances move. Fluid outside body cells is called **extracellular** (*extra* = outside) **fluid (ECF)** and is found in two principal places. The fluid filling the microscopic spaces between the cells of tissues is called **interstitial** (in´-ter-STISH-al) **fluid** (*inter* = between) or *intercellular fluid.* The extracellular fluid in blood vessels is termed **plasma** (Figure 2.4); in lymphatic vessels it is called **lymph.** Fluid within cells is called **intracellular** (*intra* = within, inside) **fluid (ICF)** or **cytosol.** Among the substances in extracellular fluid are gases, nutrients, and ions—all needed for the maintenance of life.

Extracellular fluid circulates through the blood and lymphatic vessels and from there moves into the spaces between the tissue cells. Thus, it is in constant motion throughout the body. Essentially, all body cells are surrounded by the same fluid environment. The movement of substances across a plasma membrane, or across membranes within cells, is essential to the life of the cell. Certain substances—oxygen, for example—must move into the cell to support life, whereas waste materials or harmful substances must be moved out. Plasma membranes regulate the movements of such materials.

The processes involved in the movement of substances across plasma membranes are classified as either passive or active. In *passive processes,* substances move across plasma membranes without the use of energy from the breakdown of ATP (adenosine triphosphate) by the cell. ATP is a molecule that stores energy for various cellular uses. The movement of substances in passive processes involves the *kinetic energy* (the energy of motion) of individual molecules or ions. The substances move on their own down a concentration gradient—that is, from an area where their concentration is

Figure 2.4 Body fluids. Intracellular fluid (ICF) is the fluid within cells. Extracellular fluid (ECF) is found outside cells, in blood vessels as plasma, and between tissue cells as interstitial fluid.

🗝️ *Plasma membranes regulate fluid movements from one compartment to another.*

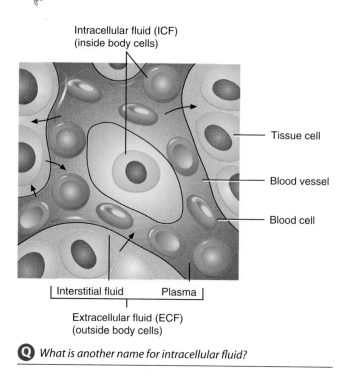

Intracellular fluid (ICF)
(inside body cells)

Tissue cell

Blood vessel

Blood cell

Interstitial fluid   Plasma

Extracellular fluid (ECF)
(outside body cells)

Q *What is another name for intracellular fluid?*

high to an area where their concentration is low. The substances may also be forced across the plasma membrane by pressure from an area where the pressure is high to an area where it is low. In *active processes,* the cell uses some of its own energy (from the breakdown of ATP) in moving the substance across the membrane since the substance typically moves against a concentration gradient—that is, from an area where its concentration is low to an area where its concentration is high.

Passive Processes

The passive processes considered here are simple diffusion, facilitated diffusion, osmosis, and filtration.

Simple Diffusion Simple diffusion (*diffus* = spreading) is a *net* (greater) movement of molecules or ions from a region of their higher concentration to a region of their lower concentration—that is, when the molecules move from an area where there are more of them to an area where there are fewer of them. The movement continues until the molecules are evenly distributed (equilibrium). At this point they move in both directions at an *equal* rate; there is no net movement.

A good example of diffusion in the body is the movement of oxygen from the blood into the cells and the movement of carbon dioxide from the cells back into the blood. This movement ensures that cells receive adequate amounts of oxygen and eliminate carbon dioxide as part of their normal metabolism.

Facilitated Diffusion Facilitated diffusion is accomplished with the assistance of integral proteins in the membrane that function as transporters (carriers). In this process, some chemical substances that are large molecules and insoluble in lipids can still pass through the plasma membrane. Among these are various sugars, especially glucose. In facilitated diffusion, the transporter attaches to glucose, making it more soluble. Then, by altering its shape the transporter moves the glucose across the plasma membrane.

Osmosis Osmosis (oz-MŌ-sis) is the net movement of water molecules through a selectively permeable membrane from an area of high water concentration to an area of low water concentration. The water molecules pass through pores in integral proteins in the membrane, and the net movement continues until equilibrium is reached. Water moves between various compartments of the body by osmosis.

Filtration Filtration involves the movements of solvents such as water and dissolved substances such as sugar across a selectively permeable membrane by gravity or mechanical pressure, usually hydrostatic (water) pressure. In the body, the driving pressure usually is blood pressure, generated by the pumping action of the heart. Most small- to medium-sized molecules such as nutrients, gases, ions, hormones, and vitamins can be forced through cell membranes, but large proteins cannot. Filtration is a very important process in which water and nutrients in the blood are pushed into interstitial fluid for use by body cells. It is also the primary process in the initial stage of urine formation.

Active Processes

The active processes considered here are active transport, endocytosis (phagocytosis, pinocytosis, and receptor-mediated endocytosis), and exocytosis.

Active Transport The process by which substances are transported across plasma membranes typically from an area of their low concentration to an area of their high concentration is called **active transport.** In active transport the substance being moved enters a pore in an integral membrane protein and makes contact with a specific site in the channel. Then the ATP splits, and the energy from the breakdown of ATP causes a change in the shape of the integral membrane protein that expels the substance on the opposite side of the membrane.

Active transport is vitally important in maintaining the concentrations of some ions inside body cells and other ions outside body cells. For example, before a nerve cell can conduct a nerve impulse, the concentration of potassium ions (K^+) must be considerably higher inside the nerve cell than outside. At the same time, the nerve cells must have a higher concentration of sodium ions (Na^+) outside than inside.

Endocytosis Large molecules and particles pass through plasma membranes by a process called **endocytosis** (*endo* = into; *cyto* = cell; *osis* = process). In this process, a segment of the plasma membrane surrounds the substance, encloses it, and then brings it into the cell. Endocytosis and exocytosis (described shortly) are important because molecules and particles of material that would normally be restricted from crossing the plasma membrane because of their large size can be brought in or removed from the cell. There are three basic kinds of endocytosis: phagocytosis, pinocytosis, and receptor-mediated endocytosis.

In **phagocytosis** (fag'-ō-sī-TŌ-sis) or "cell eating," projections of the plasma membrane and cytoplasm, called **pseudopods** (SOO-dō-pods; *pseudo* = false; *pous* = foot), surround large solid particles outside the cell and then engulf them (Figure 2.5). Then, the pseudopods fuse, forming a membrane sac around the particle called a **phagocytic vesicle (phagosome),** which enters the cytoplasm. The solid material inside the vesicle is digested by enzymes provided by lysosomes (described shortly). Phagocytic white blood cells and cells in other tissues engulf and destroy bacteria and other foreign substances (Figure 2.5b, c), which constitutes a vital defense mechanism to protect us from disease.

In **pinocytosis** (pi'-nō-sī-TŌ-sis) or "cell drinking," the engulfed material is a tiny droplet of extracellular fluid rather than a solid. Moreover, no pseudopods are formed. Instead, the membrane folds inward, forming a **pinocytic vesicle** that allows the liquid to flow inward. The membrane then surrounds the liquid. The pinocytic vesicle next detaches from the rest of the intact membrane. While only certain types of cells are capable of phagocytosis, most cells are capable of pinocytosis.

Although similar to pinocytosis, **receptor-mediated endocytosis** is a highly selective process in which cells can take up *specific* molecules or particles. Some of these substances bind to specific protein receptors in the plasma membrane and serve many functions. For example, cholesterol, iron, and vitamins are needed for chemical reactions that sustain life. Others are either hormones that deliver messages to cells so cells can respond in specific ways, or waste products that certain cells have the ability to break down. Although receptor-mediated endocytosis exists to import *needed* materials, some viruses, like the AIDS virus, can also enter body cells in this way.

Receptor-mediated endocytosis occurs as follows (Figure 2.6):

❶ On the extracellular side of the plasma membrane the ligand (the substance to be ingested) binds to a specific receptor, which is an integral membrane protein.

❷ This interaction causes the membrane to fold inward, forming an **endocytic vesicle** that contains the ligand bound to its receptor.

❸ After a vesicle moves inward from the plasma membrane, it merges with other endocytic vesicles to form a larger structure called an **endosome.**

❹ Within the endosome, receptors separate from ingested substances.

❺ The portion of the endosome that contains the receptors pinches off and moves back to the plasma membrane. There the receptors are reinserted into the plasma membrane by exocytosis (described shortly).

❻ The remaining portion of the endosome, containing the ingested material, merges with a lysosome. The ingested materials are then degraded by digestive enzymes from the lysosome or released for use by the cell.

Exocytosis The export of substances from a cell occurs by a process called **exocytosis** (*exo* = out of). During exocytosis membrane-enclosed structures called **secretory vesicles** that form inside the cell fuse with the plasma membrane and release their contents into the extracellular fluid. Exocytosis occurs in all cells but is especially important in nerve cells, which release their neurotransmitter substances by this process, and in secretory cells such as cells that secrete digestive enzymes.

The various passive and active processes by which substances move across plasma membranes are summarized in Table 2.1 on page 36.

CYTOPLASM

As noted earlier, the term **cytoplasm** (SĪ-tō-plazm') refers to all cellular contents between the plasma membrane and nucleus. It includes cytosol, organelles (except the nucleus), and inclusions. The **cytosol** (SĪ-tō-sol) is the thick semifluid portion of cytoplasm and is also referred to as intracellular fluid. At this point we will concentrate the discussion on just the cytosol. Detailed descriptions of the organelles and inclusions will follow shortly.

Physically, cytosol is a thick, semitransparent, elastic fluid. Chemically, it is 75% to 90% water plus solid components. Proteins, carbohydrates, lipids, and inorganic substances comprise most of the solids. Inorganic substances and smaller organic substances, such as simple sugars and amino acids, are soluble in water and are present as solutes. Larger organic compounds, such as proteins and the polysaccharide glycogen, are found as *colloids*—particles that

Figure 2.5 Phagocytosis.

Phagocytic cells destroy microbes and other foreign substances, an activity that helps protect against disease.

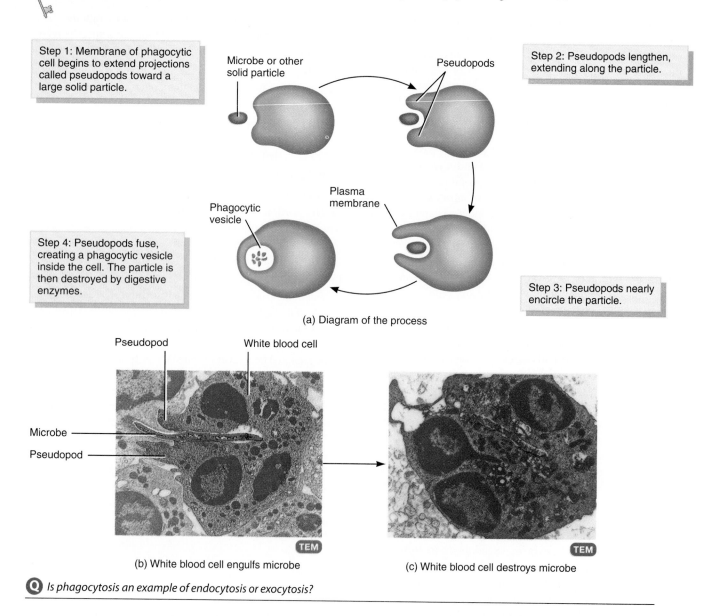

Step 1: Membrane of phagocytic cell begins to extend projections called pseudopods toward a large solid particle.

Microbe or other solid particle

Pseudopods

Step 2: Pseudopods lengthen, extending along the particle.

Phagocytic vesicle

Plasma membrane

Step 4: Pseudopods fuse, creating a phagocytic vesicle inside the cell. The particle is then destroyed by digestive enzymes.

Step 3: Pseudopods nearly encircle the particle.

(a) Diagram of the process

Pseudopod

White blood cell

Microbe

Pseudopod

TEM

TEM

(b) White blood cell engulfs microbe

(c) White blood cell destroys microbe

Q *Is phagocytosis an example of endocytosis or exocytosis?*

remain suspended in the surrounding medium but are not dissolved. The colloids bear electrical charges that repel each other and thus remain suspended and separated.

Functionally, cytosol is the medium in which many metabolic reactions occur. The cytosol receives raw materials from the external environment by way of extracellular fluid and obtains usable energy from them via decomposition reactions.

ORGANELLES

Despite the numerous chemical activities occurring simultaneously in the cell, there is little interference of one reaction with another. This is because the cell has a system of compartments provided by structures called **organelles** ("little organs"). These are specialized structures usually surrounded by one or two membranes. They have characteristic shapes and assume specific roles in growth, maintenance, repair, and control. The number and types of organelles vary among different cells, depending on their functions.

A series of illustrations accompanies each cell structure that you will study. A diagram of a generalized cell shows the location of the structure within the cell. An electron micrograph shows the actual appearance of the structure. An *electron micrograph (EM)* is a photograph taken with an electron microscope. Some electron microscopes can magnify objects up to 1 million times. In comparison, the light microscope that you probably use in your laboratory magnifies objects up to 1000 times their size. A diagram of the

Figure 2.6 Receptor-mediated endocytosis.

Normally, receptor-mediated endocytosis imports materials that are needed by cells.

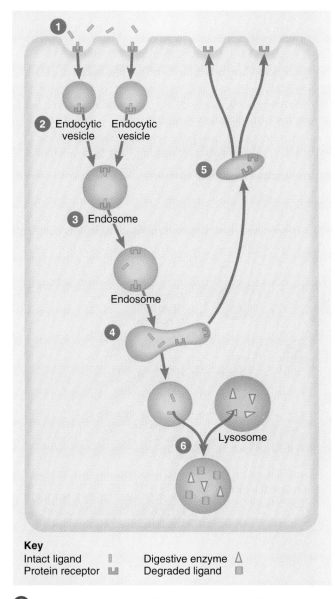

Key

Intact ligand		Digestive enzyme	△
Protein receptor		Degraded ligand	

Q *How does receptor-mediated endocytosis differ from pinocytosis?*

electron micrograph clarifies some of the small details by exaggerating their outlines. Finally, an enlarged diagram of the structure shows its details.

Nucleus

The **nucleus** (NOO-klē-us; *nucleus* = kernel) is usually a spherical or oval organelle and is the largest structure in the cell (Figure 2.7). It contains the hereditary units of the cell, called **genes,** which are arranged in single file along structures called **chromosomes** (*chromo* = color; *soma* = body). The genes determine cellular structure and direct many cellular activities. Most body cells contain a nucleus, although some, such as mature red blood cells, do not. Skeletal muscle cells and a few other cells contain several nuclei.

A double membrane called the **nuclear envelope (membrane)** separates the nucleus from the cytoplasm. Both layers of the nuclear envelope are phospholipid bilayers similar to the plasma membrane. The surface of the nuclear envelope is studded with ribosomes (described next) and is continuous at certain points with rough (granular) endoplasmic reticulum (also described shortly). **Nuclear pores** in the envelope allow most ions and water-soluble molecules to shuttle between the nucleus and the cytoplasm. Nuclear pores are about ten times larger in diameter than pores in the plasma membrane and thus also permit passage of large molecules such as RNA and various proteins.

One or more spherical bodies called **nucleoli** (noo-KLĒ-ō-lī′; singular, **nucleolus**) are present inside the nucleus. They consist of protein and RNA that are not covered by a membrane. Nucleoli are the sites of assembly of ribosomes (described shortly), which contain a type of RNA called ribosomal RNA. Ribosomes play a key role in protein synthesis. Nucleoli disperse and disappear during cell division and reorganize once new cells are formed.

A chromosome is a very long DNA molecule that is coiled and packed into an amazingly compact structure together with several proteins. In a cell that is not dividing, the 46 chromosomes are a tangled mass of stringy threads, which is called **chromatin** and has a fuzzy appearance under the light microscope. Electron micrographs reveal that chromatin has a "beads-on-a-string" structure. Each "bead" is a **nucleosome** and consists of double-stranded DNA wrapped twice around a core of eight proteins called **histones** (Figure 2.8). Histones help organize the coiling and folding of DNA. The "string" between "beads" is **linker DNA,** which holds adjacent nucleosomes together. Another histone promotes coiling of nucleosomes into a larger diameter **chromatin fiber,** which then folds into large **loops.** In cells that are not dividing, this is the extent of DNA packing. However, before cell division the DNA replicates (duplicates) and the loops condense even more into **chromatids.** During this stage of cell division, a pair of chromatids constitutes a chromosome that is easily seen as a rod-shaped structure under the light microscope (see Figure 2.17 on page 48).

Ribosomes

Ribosomes (RĪ-bō-sōms) are tiny granules that contain *ribosomal RNA (rRNA)* and many ribosomal proteins. The rRNA is synthesized by DNA in the nucleolus. Ribosomes were so named because of their high content of *ribo*nucleic acid. This particular type of RNA was subsequently named rRNA

Table 2.1 *Processes by Which Substances Move Across Plasma Membranes*

PROCESS	DESCRIPTION
Passive processes	Substances move down a concentration gradient from an area of higher to lower concentration or pressure; cell does not expend energy by the breakdown of ATP.
Simple diffusion	Net movement of molecules or ions due to their kinetic energy from an area of higher to lower concentration until an equilibrium is reached.
Facilitated diffusion	Movement of larger molecules across a selectively permeable membrane with the assistance of proteins in the membrane that serve as transporters (carriers).
Osmosis	Net movement of water molecules due to kinetic energy across a selectively permeable membrane from an area of higher to lower concentration of water until an equilibrium is reached.
Filtration	Movement of solvents (such as water) and solutes (such as glucose) across a selectively permeable membrane as a result of gravity or hydrostatic (water) pressure from an area of higher to lower pressure.

Figure 2.7 Nucleus. In (a) the nucleolus has been "lifted" out of the nucleus so that its shape can be seen.

The nucleus contains most of the genes, which are located on chromosomes.

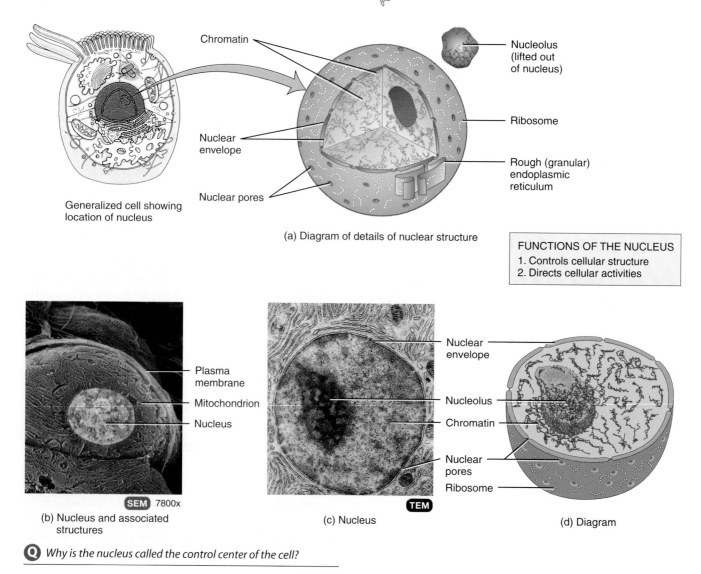

Chromatin

Nucleolus (lifted out of nucleus)

Ribosome

Nuclear envelope

Rough (granular) endoplasmic reticulum

Nuclear pores

Generalized cell showing location of nucleus

(a) Diagram of details of nuclear structure

FUNCTIONS OF THE NUCLEUS
1. Controls cellular structure
2. Directs cellular activities

Plasma membrane
Mitochondrion
Nucleus

SEM 7800x

(b) Nucleus and associated structures

(c) Nucleus

TEM

Nuclear envelope
Nucleolus
Chromatin
Nuclear pores
Ribosome

(d) Diagram

Q *Why is the nucleus called the control center of the cell?*

Table 2.1 *(continued)*

Active processes	Substances move against a concentration gradient from an area of lower to higher concentration; cell must expend energy released by the breakdown of ATP.
Active transport	Movement of substances, usually ions, across a selectively permeable membrane from a region of lower to higher concentration by an interaction with integral proteins in the membrane.
Endocytosis	Movement of large molecules and particles through plasma membranes in which the membrane surrounds the substance, encloses it, and brings it into the cell. Examples include phagocytosis ("cell eating"), pinocytosis ("cell drinking"), and receptor-mediated endocytosis.
Exocytosis	Export of substances from the cell by reverse endocytosis.

Figure 2.8 Packing of DNA into chromosomes.

A chromosome is a highly coiled and folded DNA molecule that is combined with protein molecules.

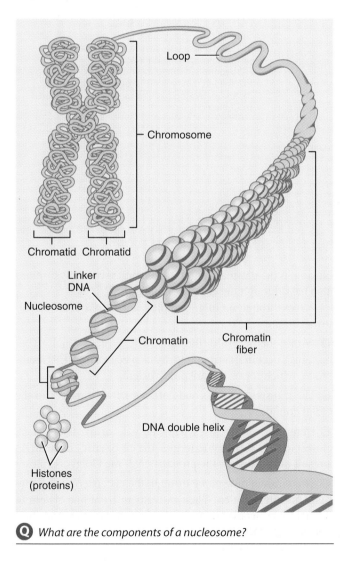

Q *What are the components of a nucleosome?*

to distinguish it from the other types of RNA in the cell. Structurally, a ribosome consists of two subunits, one about half the size of the other (Figure 2.9). Functionally, ribosomes are the sites of protein synthesis.

Some ribosomes, called *free ribosomes,* float in the cytosol; they have no attachments to other organelles (see Figure 2.9b). The free ribosomes occur singly or in clusters, and they are concerned primarily with synthesizing proteins for use inside the cell. Other ribosomes attach to a cellular structure called the endoplasmic reticulum (discussed next). These ribosomes are involved in the synthesis of proteins destined for insertion in the plasma membrane or for export from the cell.

Endoplasmic Reticulum

The **endoplasmic reticulum** (en'-dō-PLAS-mik rē-TIK-yoo-lum; *endo* = within; *plasmic* = cytoplasm; *reticulum* = network), or **ER,** is a network in the cytoplasm of membrane-enclosed channels of varying shapes called **cisterns** (SIS-terns; *cistern* = reservoir or cavity) or **cisternae** (Figure 2.10). The ER is continuous with the nuclear envelope. Based on its association with ribosomes, the ER is divided into two types. **Rough** *(granular)* **ER** is studded with ribosomes; **smooth** *(agranular)* **ER** has no ribosomes.

Ribosomes associated with rough ER synthesize proteins. The rough ER also serves as a temporary storage area for newly synthesized molecules and may add sugar groups to certain proteins, thus forming glycoproteins. Together, the rough ER and the Golgi complex (another organelle, described next) synthesize and package molecules that will be secreted from the cell.

Smooth ER is the site of fatty acid, phospholipid, and steroid synthesis. Also, in certain cells, enzymes within the smooth ER can inactivate or detoxify a variety of chemicals, including alcohol, pesticides, and carcinogens (cancer-causing agents). In muscle cells the sarcoplasmic reticulum, which is analogous to the smooth ER, releases the calcium ions that trigger muscle contraction.

Figure 2.9 Ribosomes.

Ribosomes are so named because of their high content of ribonucleic acid.

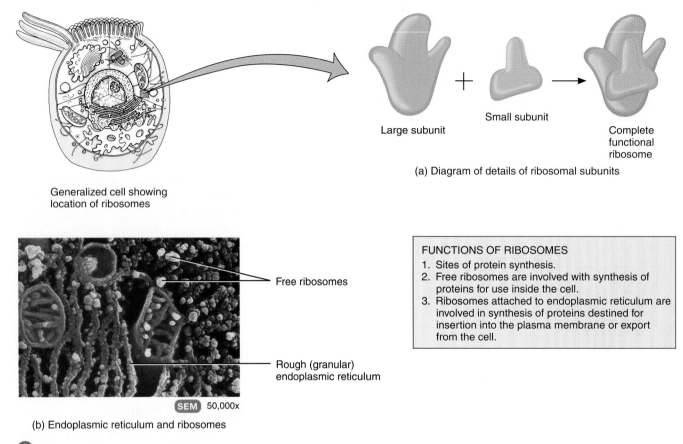

Generalized cell showing location of ribosomes

Large subunit   Small subunit   Complete functional ribosome

(a) Diagram of details of ribosomal subunits

Free ribosomes

Rough (granular) endoplasmic reticulum

SEM 50,000x

(b) Endoplasmic reticulum and ribosomes

Q *Where are ribosomes assembled?*

> **FUNCTIONS OF RIBOSOMES**
> 1. Sites of protein synthesis.
> 2. Free ribosomes are involved with synthesis of proteins for use inside the cell.
> 3. Ribosomes attached to endoplasmic reticulum are involved in synthesis of proteins destined for insertion into the plasma membrane or export from the cell.

Golgi Complex

The **Golgi** (GOL-jē) **complex (apparatus)** is an organelle that is generally located near the nucleus. In cells with high secretory activity, the Golgi complex is extensive. It consists of flattened sacs called **cisterns (cisternae),** stacked upon each other like a pile of plates with expanded bulges at their edges (Figure 2.11). Associated with the cisterns are small **Golgi vesicles,** which cluster along the expanded edges of the cisterns.

The Golgi complex processes, sorts, packages, and delivers proteins and lipids to the plasma membrane and forms lysosomes and secretory vesicles. All proteins destined for export from the cell follow a similar route: ribosomes (site of protein synthesis) → rough ER cistern → transport vesicles → Golgi complex → secretory vesicles → release to exterior of the cell by exocytosis. Proteins and lipids destined for in-

clusion in the plasma membrane or for use inside lysosomes also pass through the Golgi complex.

A trip through a Golgi complex normally occurs as follows (Figure 2.12):

❶ Within the cistern of the rough ER, proteins (and glycoproteins) become surrounded by a piece of the ER membrane, which forms a **transport vesicle.**

❷ The transport vesicle buds off the ER and moves to the Golgi complex.

❸ Here, the vesicle fuses with the side of the Golgi stack closest to the ER, which is termed the **cis** or **entry cistern.** As a result of this fusion, the proteins enter the Golgi complex.

❹ Once inside, they move through the **cis** cistern into a **medial (middle) cistern.** Their movement from one cistern to the next is probably by way of Golgi vesicles that bud from the edges of each cistern in the stack.

Figure 2.10 Endoplasmic reticulum.

The endoplasmic reticulum is a system of membrane-enclosed channels called cisterns that run throughout the cytoplasm and connect to the nucleus.

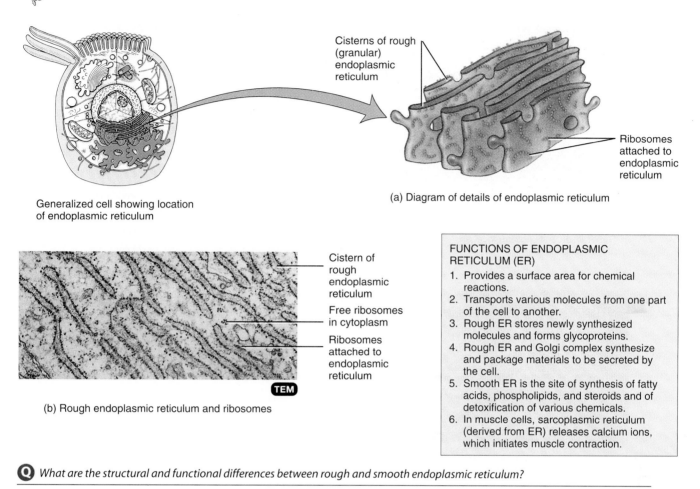

Generalized cell showing location of endoplasmic reticulum

Cisterns of rough (granular) endoplasmic reticulum

Ribosomes attached to endoplasmic reticulum

(a) Diagram of details of endoplasmic reticulum

Cistern of rough endoplasmic reticulum

Free ribosomes in cytoplasm

Ribosomes attached to endoplasmic reticulum

TEM

(b) Rough endoplasmic reticulum and ribosomes

FUNCTIONS OF ENDOPLASMIC RETICULUM (ER)

1. Provides a surface area for chemical reactions.
2. Transports various molecules from one part of the cell to another.
3. Rough ER stores newly synthesized molecules and forms glycoproteins.
4. Rough ER and Golgi complex synthesize and package materials to be secreted by the cell.
5. Smooth ER is the site of synthesis of fatty acids, phospholipids, and steroids and of detoxification of various chemicals.
6. In muscle cells, sarcoplasmic reticulum (derived from ER) releases calcium ions, which initiates muscle contraction.

Q *What are the structural and functional differences between rough and smooth endoplasmic reticulum?*

❺ As the proteins pass through the Golgi cisterns, they are modified in various ways depending on their function and destination. After moving through one or more medial cisterns, the proteins enter the **trans** or **exit cistern.**

❻ The **trans** cistern packages the modified proteins into vesicles.

❼ Some vesicles become **secretory vesicles,** which undergo exocytosis to discharge their contents into the extracellular fluid. Certain cells of the pancreas secrete digestive enzymes in this way. The vesicle membrane prevents the enzymes within from digesting the contents of the pancreatic cell itself.

❽ Other vesicles that leave the Golgi complex are loaded with digestive enzymes intended for use within the cell. They become organelles called lysosomes (described next).

Lysosomes

Lysosomes (LĪ-sō-sōms; *lysis* = dissolution; *soma* = body) are membrane-enclosed sacs (Figure 2.13 on page 42) that form in the Golgi complex. Inside are as many as 40 kinds of powerful digestive enzymes capable of breaking down a wide variety of molecules.

Lysosomal enzymes digest bacteria and other substances that enter the cell in phagocytic vesicles during phagocytosis, in pinocytic vesicles during pinocytosis, or in endosomes during receptor-mediated endocytosis. Eventually, the products of digestion are small enough to pass out of the lysosome into the cytosol. There they are recycled to synthesize various molecules needed by the cell.

Lysosomes also use their enzymes to recycle the cell's own molecules. A lysosome can engulf another organelle, digest it, and return the digested components to the cytosol

Figure 2.11 Golgi complex.

All proteins destined for export from the cell pass into secretory vesicles.

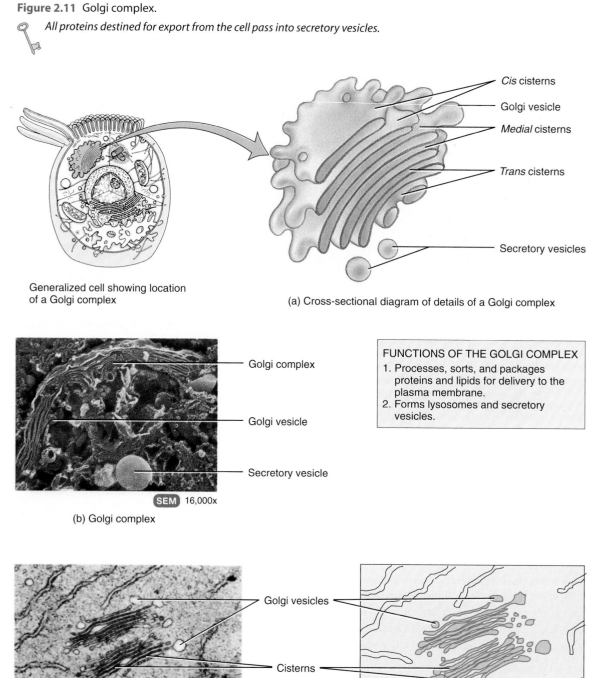

Generalized cell showing location
of a Golgi complex

Cis cisterns
Golgi vesicle
Medial cisterns
Trans cisterns
Secretory vesicles

(a) Cross-sectional diagram of details of a Golgi complex

Golgi complex

Golgi vesicle

Secretory vesicle

SEM 16,000x

(b) Golgi complex

FUNCTIONS OF THE GOLGI COMPLEX
1. Processes, sorts, and packages proteins and lipids for delivery to the plasma membrane.
2. Forms lysosomes and secretory vesicles.

Golgi vesicles

Cisterns

TEM 62,000x

(c) Golgi complex

(d) Diagram of electron micrograph

Q *What are the origin and destination of proteins that are processed by the Golgi complex?*

Figure 2.12 Packaging of synthesized proteins into secretory vesicles and lysosomes by the Golgi complex.

All proteins exported from the cell are processed in the Golgi complex.

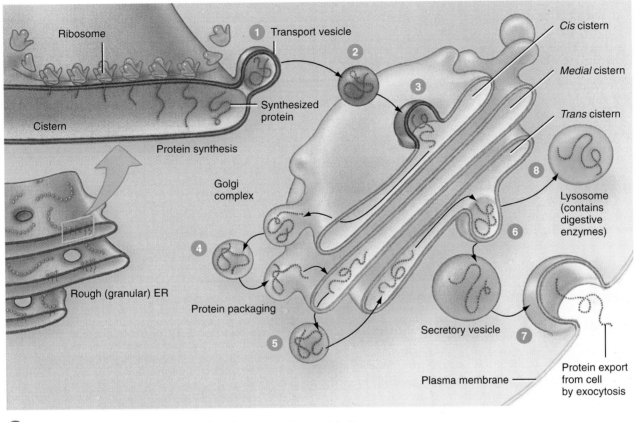

Q *What is the advantage of packaging digestive enzymes into vesicles?*

for reuse. In this way, old organelles are continually replaced. The process by which worn-out organelles are digested is called **autophagy** (aw-TOF-a-jē; *auto* = self; *phagio* = to eat). A human liver cell recycles about half its contents every week.

Lysosomal enzymes may also destroy their host cell, a process known as **autolysis** (aw-TOL-i-sis; *lysis* = dissolution). Autolysis occurs after death and in some pathological conditions.

Lysosomes also function in extracellular digestion. Lysosomal enzymes released at sites of injury help digest cellular debris, which prepares the injured area for effective repair.

Clinical Application

Tay–Sachs Disease

Some disorders are caused by faulty lysosomes. For example, **Tay–Sachs disease,** which often affects children of Jewish descent, is an inherited absence of a single lysosomal enzyme. This enzyme normally breaks down a membrane glycolipid (called ganglioside GM_2) that is especially prevalent in nerve cells. As GM_2 accumulates, the nerve cells function less efficiently. Children with the disease have seizures and muscle rigidity and become blind, demented, and uncoordinated and die, usually before the age of 5. ■

Peroxisomes

Another group of organelles similar in structure to lysosomes, but smaller, are called **peroxisomes** (pe-ROKS-i-sōms; *perox* = peroxide; *soma* = body). (See Figure 2.2 on page 29) They are so named because they usually contain one or more enzymes that use molecular oxygen to oxidize (remove hydrogen atoms from) various organic substances. Such reactions produce hydrogen peroxide (H_2O_2). One of the enzymes in peroxisomes, called *catalase*, uses the H_2O_2 generated by other enzymes to oxidize a variety of other substances, including phenol, formic acid, formaldehyde, and alcohol, toxic substances that may enter the bloodstream. This type of oxidation is especially important in liver and kidney cells whose peroxisomes detoxify the potentially harmful substances.

Figure 2.13 Lysosomes.

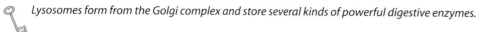

Lysosomes form from the Golgi complex and store several kinds of powerful digestive enzymes.

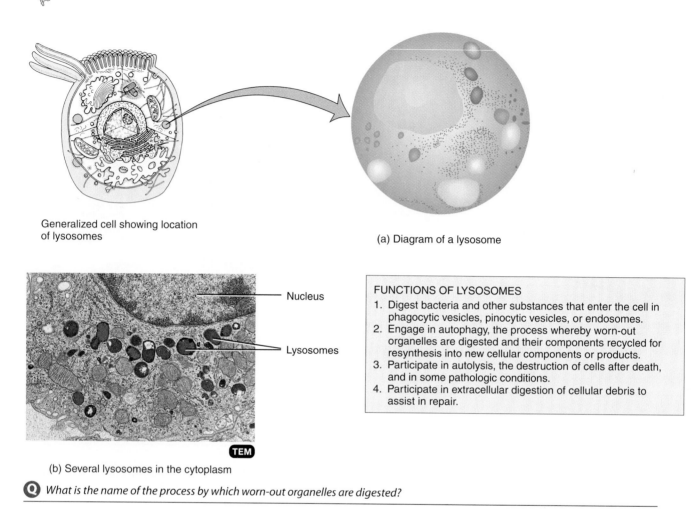

Generalized cell showing location
of lysosomes

(a) Diagram of a lysosome

Nucleus

Lysosomes

FUNCTIONS OF LYSOSOMES

1. Digest bacteria and other substances that enter the cell in phagocytic vesicles, pinocytic vesicles, or endosomes.
2. Engage in autophagy, the process whereby worn-out organelles are digested and their components recycled for resynthesis into new cellular components or products.
3. Participate in autolysis, the destruction of cells after death, and in some pathologic conditions.
4. Participate in extracellular digestion of cellular debris to assist in repair.

(b) Several lysosomes in the cytoplasm

Q *What is the name of the process by which worn-out organelles are digested?*

Mitochondria

Because of their function in generating ATP, an energy-rich molecule, **mitochondria** (mī-tō-KON-drē-a; *mitos* = thread; *chondros* = granule) are called the "powerhouses" of the cell. A mitochondrion (singular) consists of two membranes, each of which is similar in structure to the plasma membrane (Figure 2.14). The outer mitochondrial membrane is smooth, but the inner membrane is arranged in a series of folds called **cristae** (KRIS-tē). The singular is **crista** (KRIS-ta; *crista* = ridge). The central cavity of a mitochondrion, enclosed by the inner membrane and cristae, is called the **matrix.**

The cristae provide an enormous surface area for a group of chemical reactions known as **cellular respiration.** Enzymes that catalyze these reactions are located on the cristae. Cellular respiration occurs only if oxygen (O_2) is present. It results in the catabolism of nutrient molecules, such as glucose, to generate ATP. Active cells, such as muscle, liver, and kidney tubule cells, have a large number of mitochondria and use ATP at a high rate.

Mitochondria self-replicate; that is, they divide to increase in number. The replication process is controlled by DNA that is part of the mitochondrial structure and usually occurs in response to increased cellular need for ATP and at the time of cell division.

Figure 2.14 Mitochondria.

Chemical reactions within mitochondria generate ATP.

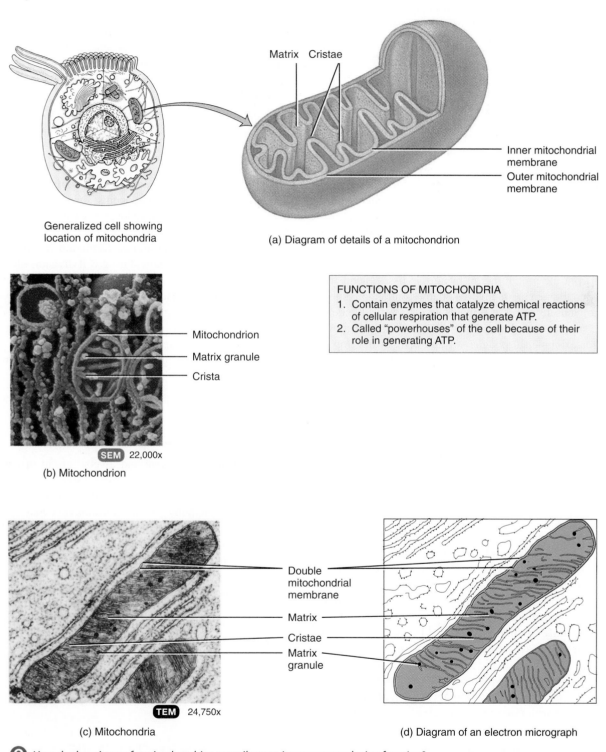

Generalized cell showing
location of mitochondria

Matrix   Cristae

Inner mitochondrial
membrane

Outer mitochondrial
membrane

(a) Diagram of details of a mitochondrion

Mitochondrion

Matrix granule

Crista

SEM 22,000x

(b) Mitochondrion

FUNCTIONS OF MITOCHONDRIA
1. Contain enzymes that catalyze chemical reactions
   of cellular respiration that generate ATP.
2. Called "powerhouses" of the cell because of their
   role in generating ATP.

TEM 24,750x

(c) Mitochondria

Double
mitochondrial
membrane

Matrix

Cristae

Matrix
granule

(d) Diagram of an electron micrograph

Q *How do the cristae of a mitochondrion contribute to its energy-producing function?*

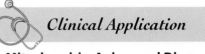

The Cytoskeleton

Cellular shape and the capability to carry out a variety of coordinated cellular movements depend on a complex network of filamentous proteins in the cytoplasm called the **cytoskeleton.** The cytoskeleton is responsible for movement of whole cells, such as phagocytes, and for movement of organelles and chemicals within the cell. Three main types of protein filaments—microfilaments, microtubules, and intermediate filaments—comprise the cytoskeleton (see Figure 2.2).

Microfilaments are rodlike structures of various lengths formed from the protein *actin.* In muscle tissue, actin filaments (thin filaments) are involved in the contraction (shortening) of muscle cells. In nonmuscle cells actin filaments provide support and shape. They also assist in cell movement (for example, in phagocytes and cells of developing embryos) and movements within cells (secretion, phagocytosis, pinocytosis).

Microtubules are larger than microfilaments. They are relatively straight, slender, cylindrical structures that consists of a protein called *tubulin.* Microtubules develop from the centrosome (described shortly). Together with microfilaments, microtubules help support and shape cells. Microtubules also function like a conveyor belt to move various substances and organelles through the cytosol, and they assist in the movement of pseudopods that are characteristic of phagocytes. As you will see shortly, microtubules form the structure of flagella and cilia (cellular appendages involved in motility), centrioles, and the mitotic spindle (a structure that is involved in cell division). Because of their role in cell division, they are targets of certain drugs (certain cancer chemotherapy agents and antifungal agents) designed to inhibit cell reproduction.

Intermediate filaments, so-named because their size is in between that of microfilaments and microtubules, are exceptionally strong and tough. Several proteins form intermediate filaments. They provide structural reinforcement inside cells, hold organelles (such as the nucleus) in place, and associate closely with microtubules to give shape to the cell.

Flagella and Cilia

Some body cells have projections for moving the entire cell or for moving substances along the surface of the cell. If the projections are few and long in proportion to the size of the cell (occurring singly or in pairs), they are called **flagella** (fla-JEL-a; *flagellum* = whip). The only example of a flagellum in the human body is the tail of a sperm cell, used for locomotion (see Figure 26.5). If the projections are numerous and short, resembling hairs, they are called **cilia** (SIL-ē-a; *cilio* = eyelashes). Ciliated cells of the respiratory tract remove mucus that has trapped foreign particles from the airways (see Figure 23.6b). Their movement is paralyzed by the nicotine in cigarette smoke. Both flagella and cilia contain nine pairs of microtubules that form a ring around two single microtubules in the center. They also contain cytosol and are covered by the plasma membrane.

Centrosome

Near the nucleus is a structure called a **centrosome** (Figure 2.15). It consists of two components: the pericentriolar area and centrioles. The **pericentriolar area** is a dense area of cytosol composed of small protein fibers. This area is the organizing center for microtubule formation in nondividing cells and the mitotic spindle during cell division (described shortly). Within the pericentriolar area is a pair of cylindrical structures called **centrioles.** Each centriole is composed of nine clusters of three microtubules (triplets) arranged in a circular pattern. Centrioles lack the two central single microtubules found in flagella and cilia. The long axis of one centriole is at a right angle to the long axis of the other. Centrioles assume a role in the formation or regeneration of flagella and cilia.

CELL INCLUSIONS

Cell inclusions are a large and diverse group of chemical substances produced by cells. Although some have recognizable shapes, they are not bounded by a membrane. These products are principally organic molecules and may appear or disappear at various times in the life of a cell. Examples include melanin, glycogen, and triglycerides. **Melanin** is a pigment stored in certain cells of the skin, hair, and eyes. It protects the body by screening out harmful ultraviolet rays from the sun. **Glycogen** is a polysaccharide that is stored in the liver, skeletal muscle, and the inner lining of the uterus and vagina. When the body requires quick energy, liver cells can break down the glycogen into glucose and release it into the blood. **Triglycerides (fats),** which are stored in adipocytes (fat cells), may be broken down to synthesize ATP.

The major parts of a cell and their functions are summarized in Table 2.2.

NORMAL CELL DIVISION

Most of the cell activities mentioned thus far maintain the life of the cell on a day-to-day basis. However, somatic cells ultimately become damaged, diseased, or worn out. New

Figure 2.15 Centrosome.

The pericentriolar area of a centrosome organizes the mitotic spindle during cell division while the centrioles help form or regenerate flagella and cilia.

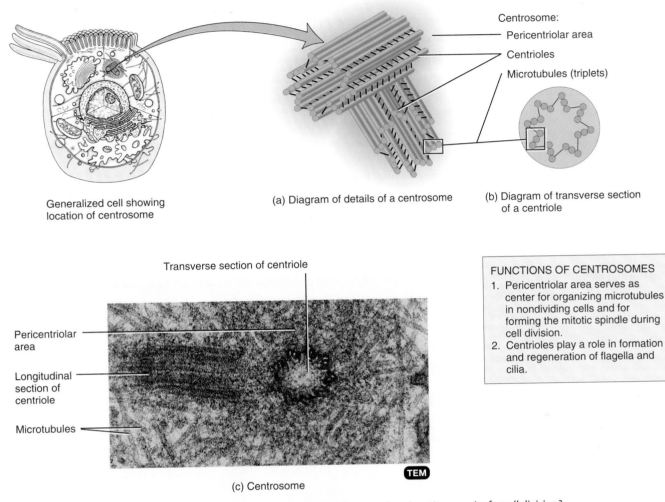

Generalized cell showing
location of centrosome

(a) Diagram of details of a centrosome

(b) Diagram of transverse section
of a centriole

Centrosome:
Pericentriolar area
Centrioles
Microtubules (triplets)

Transverse section of centriole

Pericentriolar
area

Longitudinal
section of
centriole

Microtubules

TEM

(c) Centrosome

FUNCTIONS OF CENTROSOMES
1. Pericentriolar area serves as center for organizing microtubules in nondividing cells and for forming the mitotic spindle during cell division.
2. Centrioles play a role in formation and regeneration of flagella and cilia.

Q *If you observed that a cell did not have a centrosome, what could you predict about its capacity for cell division?*

cells must be produced by cell division as replacements. In a 24-hour period, the average adult loses billions of cells from different parts of the body. Cells that have a short life span, such as cells of the outer layer of skin, are completely replaced every few days. In addition, cell division produces new somatic cells for tissue growth and new germ cells, which are specialized to form sperm and egg cells. At this point we examine normal cell division; later in the chapter we discuss abnormal cell division, as occurs in cancer.

Cell division is the process by which cells reproduce themselves. It consists of a nuclear division and a cytoplasmic division. Because *nuclear* division can be either one of two types, two kinds of cell division are recognized.

Kinds of Cell Division

In the first kind of division, **somatic cell division,** a single starting cell called a *parent cell* divides to produce two identical cells called *daughter cells*. This process consists of a nuclear division called **mitosis** and a cytoplasmic division called **cytokinesis.** The process ensures that each daughter cell has the same *number* and *kind* of chromosomes as the original parent cell. This kind of cell division replaces dead or injured cells and adds new ones for tissue growth.

The second type of cell division, **reproductive cell division,** is the mechanism by which sperm and egg cells are produced. These are the cells needed to form a new

Table 2.2 Cell Parts and Their Functions

PART	FUNCTIONS
Plasma Membrane	Protein-studded phospholipid bilayer that surrounds the cell; protects cellular contents; makes contact with other cells; provides receptors for hormones, enzymes, and antibodies; mediates the entrance and exit of materials.
Cytoplasm	Thick, semitransparent, elastic intracellular fluid containing water, ions, and many types of organic molecules such as enzymes; medium in which many of the cell's chemical reactions occur. Cytoplasm includes cytosol, all organelles (except the nucleus), and inclusions.
Organelles	Organized compartments within the cytoplasm that have specific structures and functions.
Nucleus	Surrounded by the nuclear envelope; contains genes and controls cellular activities.
Ribosomes	Sites of protein synthesis; may be either free in the cytosol or attached to endoplasmic reticulum.
Endoplasmic reticulum (ER)	Membrane-bounded network of channels that provides surface area for many types of chemical reactions; ribosomes attached to rough ER synthesize proteins that will be secreted; smooth ER (no attached ribosomes) synthesizes lipids and detoxifies certain molecules.
Golgi complex	Stack of disc-shaped membrane-enclosed cisterns; packages proteins and lipids into secretory vesicles for export or insertion into the plasma membrane; forms lysosomes.
Lysosomes	Contain digestive enzymes that break down molecules and microbes.
Peroxisomes	Contain enzymes that use molecular oxygen to oxidize various organic substances.
Mitochondria	Sites of production of most ATP during cellular respiration.
Cytoskeleton	Includes three types of protein filaments (microfilaments, microtubules, and intermediate filaments) that give the cell shape and allow coordinated movements of organelles; in some cells the cytoskeleton is responsible for movement of the cell itself.
Flagella and cilia	Allow movement of entire cell (flagella) or movement of substances along surface of cell (cilia).
Centrosome	Consists of pericentriolar area and centriole pair. Pericentriolar area helps organize microtubules in nondividing cell and forms mitotic spindle during cell division; centrioles play a role in the formation and regeneration of flagella and cilia.
Inclusions	Chemical substances produced by cells that are not surrounded by a membrane—for example, melanin, glycogen, and triglycerides (fats and oils).

organism. The process consists of a special nuclear division called **meiosis** (reduction division) followed by cytokinesis. We will discuss somatic cell division here and reproductive cell division in Chapter 26.

The Cell Cycle in Somatic Cells

Human cells, except for eggs and sperm, contain 23 pairs of chromosomes. Two chromosomes that belong to a pair, one contributed by the mother and one contributed by the father, are called **homologous chromosomes.** They have similar genes, which are usually arranged in the same order. When a cell reproduces, it must replicate (duplicate) all its chromosomes so that its hereditary traits may be passed on to the next generation of cells.

The **cell cycle** is a series of activities through which a cell passes from the time it is formed through its division into daughter cells. It is the growth and division of a single cell into two daughter cells. The cell cycle consists of two major activities: interphase and cell division (Figure 2.16).

Interphase

When a cell is between divisions, it is said to be in **interphase.** During this stage the replication of DNA and centrosomes occurs and the RNA and protein needed to produce structures required for doubling all cellular components are manufactured.

Interphase consists of three distinct phases: G_1, S, and G_2 (Figure 2.16). During the G_1- (for gap or growth) *phase*, cells are engaged in growth, metabolism, and the production of substances required for division. The period of interphase during which chromosomes replicate is called the *S*- (for synthesis) *phase.* After the S-phase, there is another growth phase called the G_2-*phase.* Because the G-phases are periods lacking events related to chromosomal replication, they are thought of as gaps or interruptions in DNA synthesis. Cells that are destined never to divide again are permanently arrested in the G_1-phase, a state termed the G_0-phase. Most nerve cells are in this state. Once a cell enters the S-phase, it is committed to go through cell division.

A microscopic view of a cell during interphase shows a clearly defined nuclear envelope, nucleoli, and chromatin (Figure 2.17a). The absence of visible chromosomes is another physical characteristic of interphase. Once a cell completes its replication of DNA, centrosomes, and centrioles and its production of RNA and proteins during interphase, mitosis begins.

Nuclear Division: Mitosis

The events that take place during mitosis and cytokinesis are plainly visible under a microscope because chromatin condenses into chromosomes. The process called **mitosis** is the

Figure 2.16 The cell cycle.

In a complete cell cycle, one parent cell divides into two daughter cells.

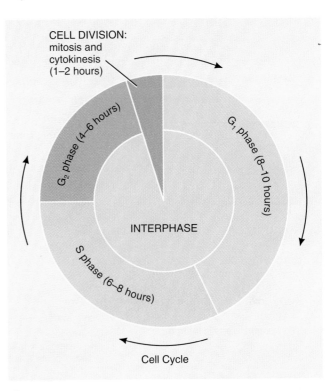

Cell Cycle

Q *During which phase of interphase do the chromosomes replicate?*

distribution of the two sets of chromosomes into two separate and equal nuclei. It results in the *exact* duplication of genetic information. For convenience, biologists divide the process into four stages: prophase, metaphase, anaphase, and telophase. Mitosis is a continuous process, however, with one stage merging imperceptibly into the next.

Prophase The first stage of mitosis is called **prophase** (*pro* = before) (Figure 2.17b). During early prophase, the chromatin fibers condense and shorten into visible chromosomes. Condensation may prevent entangling of the long DNA strands as they move during mitosis or meiosis. Because DNA replication took place during interphase, each prophase chromosome consists of a pair of identical double-stranded chromatids. Each pair of chromatids is held together by a small spherical body called a **centromere** that is required for the proper segregation of chromosomes. Attached to the outside of each centromere is a protein complex known as the **kinetochore** (ki-NET-ō-kor), whose function will be described shortly.

Figure 2.17 Cell division: mitosis and cytokinesis, as seen in photomicrographs and diagrammatic representations of the various stages of cell division in whitefish eggs. Read the sequence starting at (a) and move clockwise until you complete the cycle.

In somatic cell division, a single diploid parent cell divides to produce two identical diploid daughter cells.

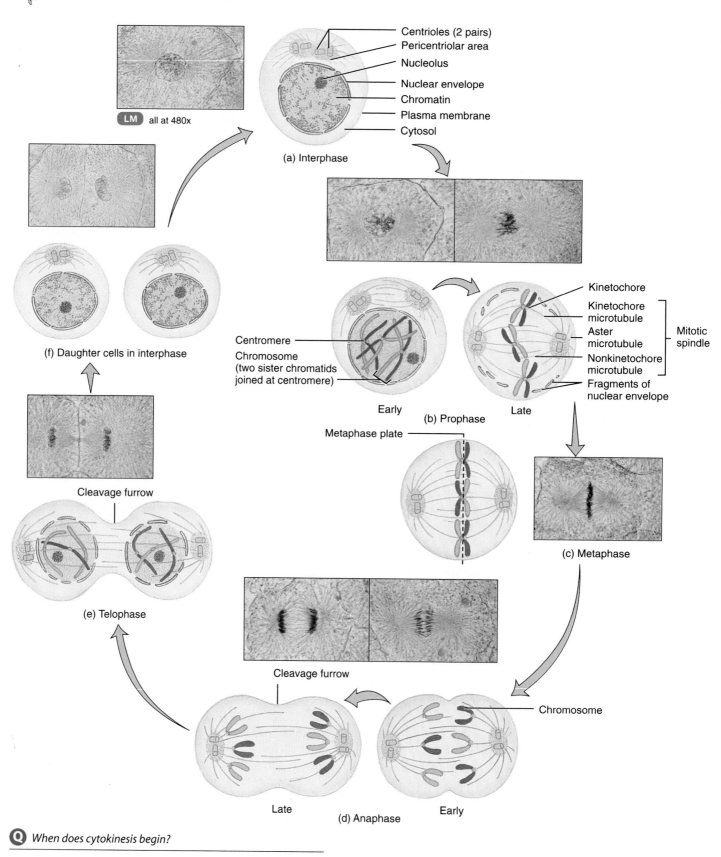

LM all at 480x

Centrioles (2 pairs)
Pericentriolar area
Nucleolus
Nuclear envelope
Chromatin
Plasma membrane
Cytosol

(a) Interphase

Kinetochore
Kinetochore microtubule
Aster microtubule
Nonkinetochore microtubule
Fragments of nuclear envelope

Mitotic spindle

Centromere
Chromosome (two sister chromatids joined at centromere)

Early   Late

(b) Prophase

Metaphase plate

(c) Metaphase

Chromosome

(f) Daughter cells in interphase

Cleavage furrow

(e) Telophase

Cleavage furrow

Late   Early

(d) Anaphase

Q *When does cytokinesis begin?*

Later in prophase, the nucleoli disappear, during which time synthesis of RNA is temporarily halted, and the nuclear envelope breaks down and is absorbed into the cytosol. In addition, each centrosome and its centrioles move to opposite poles (ends) of the cell. As they do so, the centrosomes start to form the **mitotic spindle,** a football-shaped assembly of microtubules that are responsible for the movement of chromosomes. The lengthening of microtubules between centrosomes pushes the centrosomes to the poles of the cell so that the spindle extends from pole to pole. As the mitotic spindle continues to develop, three types of microtubules form: (1) *nonkinetochore microtubules* grow from centrosomes, extend inward, but do not bind to kinetochores; (2) *kinetochore microtubules* grow from centrosomes, extend inward, and attach to kinetochores; and (3) *aster microtubules* grow out of centrosomes but radiate outward from the mitotic spindle. The spindle is an attachment site for chromosomes, and it also distributes chromosomes to opposite poles of the cell.

Metaphase During **metaphase** (*meta* = after), the second stage of mitosis, the centromeres of the chromatid pairs line up at the exact center of the mitotic spindle. This midpoint region is called the *metaphase plate* or *equatorial plane region* (Figure 2.17c).

Anaphase The third stage of mitosis, **anaphase** (*ana* = upward), is characterized by the splitting and separation of centromeres and the movement of the two sister chromatids of each pair toward opposite poles of the cell (Figure 2.17d). Once separated, the sister chromatids are referred to as daughter **chromosomes.** The movement of chromosomes is due to shortening of kinetochore microtubules and elongation of the nonkinetochore microtubules, processes that increase the distance between separated chromosomes. As the chromosomes are pulled by the microtubules during anaphase, they appear V-shaped as the centromeres lead the way and seem to drag the trailing parts of the chromosomes toward opposite poles of the cell.

Telophase The final stage of mitosis, **telophase** (*telos* = far or end), begins as soon as chromosomal movement stops. Telophase is essentially the opposite of prophase. During telophase, the identical sets of chromosomes at opposite poles of the cell uncoil and revert to the threadlike chromatin form. A new nuclear envelope forms around each chromatin mass, new nucleoli reappear in the daughter nuclei, and eventually the mitotic spindle breaks up (Figure 2.17e).

Cytoplasmic Division: Cytokinesis

Division of a parent cell's cytoplasm and organelles is called **cytokinesis** (sī-tō-ki-NĒ-sis; *cyto* = cell; *kinesis* = motion). It begins in late anaphase or early telophase with formation of a *cleavage furrow,* a slight indentation of the plasma membrane that extends around the center of the cell (Figure 2.17d and e). The furrow gradually deepens until opposite surfaces of the cell make contact and the cell is split in two. When cytokinesis is complete, interphase begins. The result of cytokinesis is two separate daughter cells (Figure 2.17f), each with separate portions of cytoplasm and organelles and its own set of identical chromosomes.

If we consider the cell cycle in its entirety, the sequence of events is G_1-phase → S-phase → G_2-phase → mitosis → cytokinesis (see Figure 2.16). A summary of the events that occur during interphase and cell division is presented in Table 2.3.

Length of the Cell Cycle

The time required for one cell cycle varies with the kind of cell, its location, and other factors such as temperature. Furthermore, the different stages of mitosis are not equal in duration. The G_1-phase is the most variable, ranging from almost nonexistent in rapidly dividing cells to days, weeks, years, or a lifetime. Mammalian cells studied in laboratory cultures often have the following time intervals during a cell cycle (see Figure 2.16).

G_1-phase	8–10 hours
S-phase	6–8 hours
G_2-phase	4–6 hours
Mitosis and cytokinesis	1–2 hours

Within the mitosis and cytokinesis time interval, prophase is longest and anaphase is shortest. As you can see, mitosis and cytokinesis represent only a small part of the life cycle of a cell. Together, the various phases of the cell cycle require 18–24 hours in many cultured mammalian cells.

APOPTOSIS: PROGRAMMED CELL DEATH

Throughout the lifetime of an organism, certain cells undergo an orderly, genetically programmed death called **apoptosis** (ap-ō-TŌ-sis; *apoptosis* = a falling off, like dead leaves from a tree). This process of suicide on demand occurs as a necessary event during embryological and fetal development. In addition, it occurs postnatally (after birth) to regulate the number of cells in a tissue and eliminate potentially dangerous cells, such as cancer cells.

In a cell undergoing apoptosis, a triggering agent from either outside a cell or within a cell induces the activation of enzymes by "cell-suicide" genes. These enzymes damage cells in several ways, including disruption of the cytoskeleton and nucleus. As a result of the damage, the cell shrinks and pulls away from neighboring cells. Then the nucleus and the cell itself break up into fragments. Phagocytes in the vicinity ingest the cell fragments.

Table 2.3 *Events of the Somatic Cell Cycle*

Interphase:	Cell is between divisions; chromosomes are not visible under a light microscope.

PHASE	ACTIVITY
G_1-phase	Cell engages in growth, metabolism, and production of substances required for division; no chromosomal replication.
S-phase	Chromosomes replicate (synthesis of new DNA and its associated proteins).
G_2-phase	Same as for G_1-phase.

Cell Division:	Single parent cell produces two chromosomally identical daughter cells; chromosomes are visible under light microscope.

PHASE	ACTIVITY
Mitosis	Nuclear division; distribution of two sets of chromosomes into separate nuclei.
Prophase	Chromatin fibers shorten and coil into visible chromosomes (paired chromatids). Nucleoli and nuclear envelope disappear. Each centrosome moves to opposite poles of cell. Pericentriolar area forms mitotic spindle.
Metaphase	Centromeres of chromatid pairs line up at metaphase plate of cell.
Anaphase	Centromeres divide; identical sets of chromosomes move to opposite poles of cell.
Telophase	Nuclear envelope reappears and encloses chromosomes. Chromosomes resume chromatin fiber form. Nucleoli reappear. Mitotic spindle breaks up.
Cytokinesis	Cytoplasmic division: cleavage furrow forms around center of cell, progresses inward, and divides cytoplasm into separate and usually equal portions.

Medical researchers speculate that abnormalities of apoptosis lead to the formation of too many or two few cells in the body, perhaps contributing to the development of diseases such as cancer, AIDS, and Alzheimer's disease. New insights into the process of apoptosis may lead to more effective therapies for such diseases.

Cells and Aging

Aging is a normal process accompanied by a progressive alteration of the body's homeostatic adaptive responses. It produces observable changes in structure and function and increases vulnerability to environmental stress and disease. The specialized branch of medicine that deals with the medical problems and care of elderly persons is called **geriatrics** (jer′-ē-AT-riks; *geras* = old age; *iatrike* = surgery, medicine).

The obvious characteristics of aging are well known: graying and loss of hair, loss of teeth, wrinkling of skin, de-

creased muscle mass, and increased fat deposits. The physiological signs of aging are gradual deterioration in both function and capacity to respond to environmental stresses. Metabolism slows, as does the ability to maintain a constant internal environment (homeostasis) in response to changes in temperature, diet, and oxygen supply. These signs of aging are related to a net decrease in the number of cells in the body and to dysfunctions in cells that remain.

Glucose, the most abundant sugar in the body, may play a role in the aging process. Glucose is haphazardly added to proteins inside and outside cells, forming irreversible cross-links between adjacent protein molecules. With advancing age, more cross-links form, and this may contribute to the stiffening and loss of elasticity that occur in aging tissues.

Although many millions of new cells normally are produced each minute, several kinds of cells in the body—heart cells, skeletal muscle cells, nerve cells—are not replaced. Experiments have shown that many other cell types have only a limited capability to divide. Cells grown outside the body divide only a certain number of times and then stop. These observations suggest that cessation of mitosis is a normal, genetically programmed event. According to this view, "aging" genes are part of the genetic blueprint at birth, and they turn on at preprogrammed times, slowing down or halting processes vital to life.

Another theory of aging is the *free radical theory.* Free radicals are electrically charged molecules that have an unpaired electron—for example, *superoxide.* Such molecules are unstable and highly reactive. A free radical produces oxidative damage in a nearby lipid, protein, or nucleic acid by "stealing" an electron to accompany its unpaired electron. Some effects are wrinkled skin, stiff joints, and hardened arteries. Normal cellular metabolism—for example, cellular respiration in mitochondria—produces some free radicals. Others are present in air pollution, radiation, and certain foods we eat. Naturally occurring enzymes, such as *superoxide dismutase, peroxidase,* and *catalase* in peroxisomes and in the cytosol, normally dispose of free radicals. Dietary substances, such as vitamin E, vitamin C, beta-carotene, and selenium, are antioxidants that inhibit free radical formation. Besides accelerating the aging process, free radicals contribute to reperfusion damage after a stroke or heart attack and are probably one factor in the development of cancer.

Two discoveries bolster the free radical theory of aging. Strains of fruit flies bred for longevity produce larger-than-normal amounts of superoxide dismutase, which functions to neutralize free radicals. Also, injection of genes that lead to production of superoxide dismutase into fruit fly embryos prolongs their average lifetime.

Whereas some theories of aging explain the process at the cellular level, others concentrate on regulatory mechanisms operating within the entire organism. For example, the immune system, which manufactures antibodies against foreign invaders, may start to attack the body's own cells. This autoimmune response might be caused by changes in cell identity markers at the surface of cells, causing antibodies to attach and mark the cell for destruction. As surface changes in cells increase, the autoimmune response intensifies, producing the well-known signs of aging.

The effects of aging on the various body systems are discussed in their respective chapters.

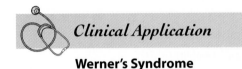

Clinical Application

Werner's Syndrome

Werner's syndrome is a rare, inherited disease that causes a rapid acceleration of aging, usually while the person is only in his or her twenties. It is characterized by wrinkling of the skin, graying of the hair and baldness, cataracts, muscular atrophy, and a tendency to develop diabetes mellitus, cancer, and cardiovascular disease. Most afflicted individuals die before age 50. Recently, the gene that causes Werner's syndrome has been identified. Researchers hope to use the information to gain insight into the mechanism of aging. ■

Key Medical Terms
Associated with Cells

Note to the Student

Each chapter in this text that discusses a major system of the body is followed by a glossary of key medical terms. Both normal and pathological conditions of the system are included in these glossaries. You should familiarize yourself with the terms because they will play an essential role in your medical vocabulary.

Some of these disorders, as well as disorders discussed in the text, are referred to as local or systemic. A *local disease* is one that affects one part or a limited area of the body. A *systemic disease* affects either the entire body or several parts.

The science that deals with why, when, and where diseases occur and how they are transmitted in a human community is known as **epidemiology** (ep′-i-dē-mē-OL-ō-jē; *epidemios* = prevalent; *logos* =study of). The science that deals with the effects and uses of drugs in the treatment of disease is called **pharmacology** (far′-ma-KOL-ō-jē; *pharmakon* = medicine; *logos* = study of).

Anaplasia (an′-a-PLĀ-zē′a; *an* = not; *plas* = to grow) The loss of tissue differentiation and function that is characteristic of most malignancies.

Dysplasia (dis-PLĀ-zē-a; *dys* = abnormal; *plas* = to grow) Alteration in the size, shape, and organization of cells due to chronic irritation or inflammation; may progress to neoplasia (tumor formation, usually malignant) or revert to normal if the stress is removed.

Hyperplasia (hī-per-PLĀ-zē-a; *hyper* = over) Increase in the number of cells due to an increase in the frequency of cell division.

Hypertrophy (hī-PER-trō-fē) Increase in the size of cells without cell division.

Metaplasia (met′-a-PLĀ-zē-a; *meta* = change) The transformation of one type of cell into another.

Metastasis (me-TAS-ta-sis; *stasis* = standing still) The transfer of disease from one part of the body to another that is not directly connected with it.

Neoplasm (NĒ-ō-plazm; *neo* = new) Any abnormal formation or growth, usually a malignant tumor.

Progeny (PROJ-e-nē; *progignere* = to bring forth) Offspring or descendants.

Study Outline

Cellular Diversity (p. 28)

1. An average human adult has about 100 trillion cells.
2. The many sizes and shapes of cells are related to their functions.

Generalized Cell (p. 29)

1. A cell is the basic, living, structural and functional unit of the body.
2. A generalized cell is a composite that represents various cells of the body.
3. Cytology is the science concerned with the study of cells.
4. The principal parts of a cell are the plasma (cell) membrane, cytoplasm, organelles, and inclusions.

Plasma (Cell) Membrane (p. 30)

Chemistry and Anatomy (p. 31)

1. The plasma (cell) membrane surrounds the cell and separates it from other cells and the external environment.
2. It is composed primarily of phospholipids and proteins. The proteins are integral and peripheral.
3. According to the fluid mosaic model, the membrane is a mosaic of proteins floating like icebergs in a sea of lipids.

Functions (p. 31)

1. The plasma membrane functions in cellular communication, shape and protection, establishment of an electrochemical gradient, and selective permeability.
2. The membrane's selectively permeable nature restricts the passage of certain substances.

Movement of Materials Across Plasma Membranes (p. 31)

1. Passive processes depend on the concentration of substances and their kinetic energy.
2. Simple diffusion is the net movement of molecules or ions from an area of higher concentration to an area of lower concentration until an equilibrium is reached.
3. In facilitated diffusion, certain molecules, such as glucose, move through the membrane with the help of a transporter.
4. Osmosis is the movement of water through a selectively permeable membrane from an area of higher water concentration to an area of lower water concentration.
5. Filtration is the movement of water and dissolved substances across a membrane due to gravity or hydrostatic pressure.
6. Active processes depend on the use of ATP by the cell.
7. Active transport is the movement of a substance across a cell membrane from lower to higher concentration using energy derived from ATP.
8. Endocytosis (phagocytosis, pinocytosis, receptor-mediated endocytosis) is the movement of substances through plasma membranes in which the membrane surrounds the substance, encloses it, and brings it into the cell.
9. Phagocytosis is the ingestion of solid particles by pseudopods. It is an important process used by some white blood cells to destroy bacteria that enter the body.
10. Pinocytosis is the ingestion of a liquid by the plasma membrane. In this process the liquid becomes surrounded by a pinocytic vesicle.
11. Receptor-mediated endocytosis is the selective uptake of large molecules by cells.
12. Exocytosis is the export of substances from a cell.

Cytoplasm (p. 33)

1. Cytoplasm is all cellular components inside the cell between the plasma membrane and nucleus. It includes the cytosol, organelles (except the nucleus), and inclusions.
2. Cytosol is composed mostly of water, many dissolved proteins and enzymes, nutrients, ions, and other small molecules.
3. Functionally, cytosol is the medium in which some chemical reactions occur.

Organelles (p. 34)

1. Organelles are specialized structures suspended in the cytosol that have characteristic appearances and functions.
2. They play specific roles in cellular growth, maintenance, repair, and control.

Nucleus (p. 35)

1. Usually the largest organelle, the nucleus controls cellular activities and contains the genetic information.
2. Most body cells have a single nucleus; some (red blood cells) have none, whereas others (skeletal muscle cells) have several.
3. The parts of the nucleus include the double nuclear envelope, nucleoli, and chromosomes.
4. Chromosomes consist of DNA (genetic material) and proteins.

Ribosomes (p. 35)

1. Ribosomes are tiny granules consisting of ribosomal RNA and protein.
2. Some ribosomes (free) float in the cytosol and have no attachments to other organelles; others attach to endoplasmic reticulum.
3. Functionally, ribosomes are the sites of protein synthesis.

Endoplasmic Reticulum (p. 37)

1. The ER is a network of membrane-enclosed channels continuous with the nuclear membrane.
2. Rough (granular) ER has ribosomes attached to it. Smooth (granular) ER does not contain ribosomes.
3. The ER transports substances, stores newly synthesized molecules, synthesizes and packages molecules, detoxifies chemicals, and releases calcium ions involved in muscle contraction.

Golgi Complex (p. 38)

1. The Golgi complex consists of several stacked, flattened membranous sacs (cisterns).
2. The principal function of the Golgi complex is to process, sort, and deliver proteins and lipids to the plasma membrane, lysosomes, and secretory vesicles.

Lysosomes (p. 39)

1. Lysosomes are spherical structures that contain digestive enzymes. They are formed by the Golgi complex.
2. They are found in large numbers in white blood cells, which carry on phagocytosis.

3. Lysosomes function in intracellular digestion, digestion of worn-out organelles (autophagy), digestion of cellular contents (autolysis), and extracellular digestion.

Peroxisomes (p. 41)
1. Peroxisomes are similar to lysosomes but smaller.
2. They contain enzymes (for example, catalase) that use molecular oxygen to oxidize various organic substances.

Mitochondria (p. 42)
1. Mitochondria consist of a smooth outer membrane and a folded inner membrane surrounding the interior matrix. The inner folds are called cristae.
2. The mitochondria are called "powerhouses" of the cell because ATP is produced in them.

The Cytoskeleton (p. 44)
1. Together, microfilaments, microtubules, and intermediate filaments form the cytoskeleton.
2. The cytoskeleton provides organization for chemical reactions and assists in transporting chemicals and organelles through the cytosol.

Flagella and Cilia (p. 44)
1. These cellular projections have the same basic structure and are used in movement.
2. If projections are few (typically occurring singly or in pairs) and long, they are called flagella. If they are numerous and short, they are called cilia.
3. The flagellum of a sperm cell moves the entire cell. The cilia on cells of the respiratory tract move foreign matter trapped in mucus along the cell surfaces toward the throat for elimination.

Centrosome (p. 44)
1. A centrosome consists of a pericentriolar area and centrioles. The dense area of cytosol containing protein fibers is the pericentriolar area. It serves as a center for organizing microtubules in interphase cells and the mitotic spindle during cell division.
2. Centrioles are paired cylinders arranged at right angles to one another that assume a role in the formation or regeneration of flagella and cilia.

Cell Inclusions (p. 44)
1. Cell inclusions are chemical substances produced by cells. They are usually organic and may have recognizable shapes.

2. Examples of cell inclusions are melanin, glycogen, and triglycerides (neutral fats).

Normal Cell Division (p. 44)
1. Cell division is the process by which cells reproduce themselves. It consists of nuclear division (mitosis and meiosis) and cytoplasmic division (cytokinesis).

Kinds of Cell Division (p. 45)
1. Cell division that results in an increase in the number of body cells is called somatic cell division and involves a nuclear division called mitosis plus cytokinesis.
2. Cell division that results in the production of sperm and eggs is called reproductive cell division and consists of a nuclear division called meiosis plus cytokinesis.

The Cell Cycle in Somatic Cells (p. 47)
1. Prior to mitosis and cytokinesis, the DNA molecules, or chromosomes, replicate themselves so the same chromosomal complement can be passed on to future generations of cells.
2. A cell between divisions carrying on every life process except division is said to be in interphase.
3. Mitosis is the distribution of two sets of chromosomes into separate and equal nuclei following their replication.
4. Mitosis consists of prophase, metaphase, anaphase, and telophase.
5. Cytokinesis usually begins in late anaphase and ends in telophase.
6. A cleavage furrow forms at the cell's metaphase plate and progresses inward, cutting through the cell to form two separate portions of cytoplasm.

Apoptosis: Programmed Cell Death (p. 49)
1. Apoptosis is an orderly, genetically programmed death of cells.
2. Abnormalities of apoptosis may contribute to diseases such as cancer, AIDS, and Alzheimer's disease.

Cells and Aging (p. 50)
1. Aging is a normal process accompanied by progressive alteration of the body's homeostatic adaptive responses.
2. Many theories of aging have been proposed, including genetically programmed cessation of cell division, excessive immune responses, and buildup of free radicals.

Review Questions

1. Define a cell. How is cell structure related to cell function? What are the four principal portions of a cell? (p. 28)
2. Discuss the chemistry and structure of the plasma membrane. (p. 31)
3. Describe the various functions of the plasma membrane. (p. 31)
4. What are the major differences between passive processes and active processes in moving substances across plasma membranes? (p. 31)

5. Define and give an example of each of the following: simple diffusion, facilitated diffusion, osmosis, filtration, active transport, phagocytosis, pinocytosis, and receptor-mediated endocytosis. (p. 32)
6. What is cytoplasm? Discuss the chemical composition and physical nature of the cytosol. What is its function? (p. 33)
7. What is an organelle? (p. 34) By means of a labeled diagram, indicate the parts of a generalized animal cell.

8. Describe the structure and functions of the nucleus of a cell. (p. 35)
9. Discuss the distribution of ribosomes. What is their function? (p. 37)
10. Distinguish between rough (granular) and smooth (agranular) endoplasmic reticulum (ER). What are the functions of ER? (p. 37)
11. Describe the structure and functions of the Golgi complex. (p. 38)
12. List and describe the various functions of lysosomes. (p. 39)
13. What is the importance of peroxisomes? (p. 41)
14. Why are mitochondria referred to as "powerhouses" of the cell? (p. 42)
15. Contrast the structure and functions of microfilaments, microtubules, and intermediate filaments. (p. 44)

16. How are flagella and cilia distinguished on the basis of structure and function? (p. 44)
17. Describe the structure and function of centrosomes. (p. 44)
18. Define a cell inclusion. Provide examples and indicate their functions. (p. 44)
19. Distinguish between the two types of cell division. Why is each important? (p. 45)
20. Define interphase. (p. 47)
21. Describe the principal events of each stage of mitosis. (p. 47)
22. What is apoptosis? Why is it important? (p. 49)
23. What is aging? List some of the characteristics of aging. (p. 50)
24. Refer to the glossary of key medical terms associated with cells. Be sure that you can define each term. (p. 51)

Self Quiz

1. Match the following:
   ___ (1) centrosome
   ___ (2) cilia
   ___ (3) endoplasmic reticulum
   ___ (4) flagella
   ___ (5) lysosomes
   ___ (6) Golgi complex
   ___ (7) microfilaments
   ___ (8) mitochondria
   ___ (9) nucleus
   ___ (10) microtubules
   ___ (11) ribosomes
   ___ (12) peroxisomes

   (a) directs cellular activities by means of genes located here
   (b) sites of protein synthesis; may occur attached to ER or scattered freely in cytoplasm
   (c) system of membranous channels providing pathways for transport within cell and surface areas for chemical reactions; may be rough or smooth
   (d) stacks of cisterns; involved in packaging and secretion of proteins and lipids
   (e) may release enzymes that lead to autolysis of the cell
   (f) similar to lysosomes, but smaller; contain the enzyme catalase
   (g) cristae-containing structures, called "powerhouses of the cell" because ATP production occurs here
   (h) part of cytoskeleton; involved with cell movement and contraction
   (i) part of cytoskeleton; provide support and give shape to cell; form flagella, cilia, centrioles, and mitotic spindle
   (j) helps organize mitotic spindle used in cell division
   (k) long, hairlike structures that help move entire cell, as in a sperm cell
   (l) short, hairlike structures that move particles over cell surface

Complete the following:

2. The process of cell eating is called ___.
3. The plasma membrane is composed of two main chemical components: a bilayer of ___ and protein.
4. The microscopic study of the structure of cells is called ___.
5. A membrane that permits the passage of only certain materials is described as ___.

Choose the one best answer to the following questions:

6. A red blood cell has a diameter of 8 micrometers. How many cells would fit on a line 8 centimeters long?
   a. 100
   b. 1000
   c. 10,000
   d. 100,000
   e. 1,000,000
7. Which of the following is *not* part of the cytoplasm?
   a. nutrients
   b. enzymes
   c. mitochondria
   d. nucleus
   e. inclusions

8. Respiratory gases move between the lungs and blood as a result of
   a. simple diffusion
   b. osmosis
   c. active transport
   d. phagocytosis
   e. pinocytosis
9. Which process does *not* belong with the others?
   a. phagocytosis
   b. osmosis
   c. simple diffusion
   d. facilitated diffusion
   e. filtration
10. The fluid located inside body cells is
   a. lymph
   b. intracellular fluid
   c. interstitial fluid
   d. intercellular fluid
   e. both c and d

11. As a result of somatic cell division, each daughter cell has
    a. half as many chromosomes as its parent cell
    b. twice as many chromosomes as its parent cell
    c. exactly the same number of chromosomes as its parent cell
    d. one-quarter as many chromosomes as its parent cell
    e. none of the above
12. The process by which worn-out cell organelles are digested is called
    a. autophagy
    b. hemolysis
    c. endocytosis
    d. autolysis
    e. exocytosis
13. Which of the following is a cell inclusion?
    a. nucleus
    b. peroxisome
    c. microtubule
    d. centrosome
    e. glycogen
14. In the life cycle of a cell, chromosomes are replicated during
    a. prophase
    b. G_1 phase
    c. G_2 phase
    d. S phase
    e. anaphase
15. Cytokinesis is the equal division of the cytoplasm of a cell. A similar equal division of the nucleus is a process called
    a. apoptosis
    b. mitosis
    c. meiosis
    d. metastasis
    e. pinocytosis

Are the following statements true or false?

16. Free radicals are naturally occurring enzymes that prevent oxidation of cellular components.
17. A noncancerous growth is called a neoplasm.
18. Sarcoma is a general term for any cancer arising from muscle cells or connective tissues.
19. Kinetochore microtubules attach to kinetochores during prophase.
20. Match the following:
    (a) chromosomes move toward opposite poles of the cell
    (b) nuclear envelope disappears; chromatin thickens into distinct chromosomes; centrosomes move to opposite ends of cell and mitotic spindle forms
    (c) the series of events is essentially the reverse of prophase; cytokinesis occurs
    (d) chromatids line up on the metaphase plate
    (e) the cell is not involved in cell division; it is in a phase that follows telophase

    (1) anaphase
    (2) interphase
    (3) metaphase
    (4) prophase
    (5) telophase

Critical Thinking Questions

1. Mucin is a protein present in saliva and other secretions. When mixed with water, mucin becomes the slippery substance known as mucus. Trace the route taken by mucin through the cell from its synthesis to its secretion, listing all the organelles and processes involved.
   HINT: *All proteins are produced initially by the same organelle.*
2. Tadpoles have gills and a fishlike tail, but adult frogs have no tail at all. What explanation can you give for the disappearance of the tail?
   HINT: *Human fetuses have froglike webbed hands in utero, but not at birth.*
3. Nicotine and other compounds present in cigarette smoke can inhibit the action of the cilia within the respiratory system. What problems could result from this and other effects of smoking?
   HINT: *Smokers often have a bout of "smoker's cough" first thing in the morning.*
4. If the cell cycle in an actively growing region of the body lasted 24 hours, how many total daughter cells would be produced from a single parent cell at the end of one week?
   HINT: *You'll have twice as many daughters on Monday as there were parent cell(s) on Sunday.*
5. A child was brought to the emergency room after eating rat poison containing arsenic. Arsenic kills rats by blocking the function of the mitochondria. What effect would the poison have on the child's body functions?
   HINT: *The child may appear lethargic.*
6. Imagine that a new chemotherapeutic agent has been discovered that disrupts microtubules in cancerous cells but leaves normal cells unaffected. What effect would this agent have on cancer cells? (See *Applications to Health*, p.1.)
   HINT: *Think of microtubules as the cell's Lego building set.*

Answers to Figure Questions

2.1 The structure of each cell part determines its function.

2.2 Plasma membrane, cytoplasm, organelles, inclusions.

2.3 Phospholipids, cholesterol, and glycolipids.

2.4 Cytosol.

2.5 Endocytosis, because it is a process that brings material into the cell.

2.6 Pinocytosis is a less specific process than receptor-mediated endocytosis.

2.7 It controls cellular activity by regulating which proteins will be synthesized.

2.8 Double-stranded DNA wrapped twice around a core of eight histones (proteins).

2.9 In the nucleoli inside the nucleus.

2.10 Rough ER has attached ribosomes, whereas smooth ER does not. Rough ER synthesizes proteins that will be exported from the cell, whereas smooth ER is associated with lipid synthesis and other metabolic reactions.

2.11 The proteins originate in the ER and become enclosed within a secretory vesicle or lysosome.

2.12 The membrane of the vesicle holds the digestive enzymes inside so they cannot inappropriately break down molecules within the cytosol.

2.13 Autophagy.

2.14 They increase surface area for chemical reactions.

2.15 It probably will not be able to undergo cell division.

2.16 S phase.

2.17 Usually in late anaphase or early telophase.

Chapter 3

Tissues

Student Objectives

1. Define tissues, compare their origins, and classify the tissues of the body into four major types.

2. Discuss the distinguishing characteristics of epithelial tissue.

3. List the structure, location, and function for the basic types of epithelial tissue.

4. Define a gland and distinguish between exocrine and endocrine glands.

5. Identify the distinguishing characteristics of connective tissue.

6. Discuss the cells, ground substance, and fibers that constitute connective tissue.

7. List the structure, function, and location of the basic types of connective tissue.

8. Define a membrane and describe the location and function of mucous, serous, and synovial membranes.

9. Contrast the three types of muscle tissue with regard to structure and location.

10. Describe the structural features and functions of nervous tissue.

Cells are highly organized, but typically they do not function as isolated units in your body. They work together in groups called tissues. A **tissue** (*texere* = to weave) is a group of similar cells that usually have a common embryonic origin and function together to carry out specialized activities. The structure and properties of a specific tissue are influenced by such things as the nature of the extracellular material that surrounds the tissue's cells and connections between the cells that compose the tissue. Tissues may be quite hard (e.g., bone), semisolid (e.g., fat), and even liquid (e.g., blood). In addition, tissues vary tremendously with respect to the kinds of cells present, how the cells are arranged, and the types of fibers present, if any. The science that deals with the study of tissues is **histology** (hiss'-TOL-ō-jē; *histio* = tissue; *logos* = study of).

TYPES OF TISSUES AND THEIR ORIGINS

Body tissues can be classified into four principal types according to their function and structure:

1. **Epithelial tissue,** which covers body surfaces; lines hollow organs, body cavities, and ducts (tubes); and forms glands.

2. **Connective tissue,** which protects and supports the body and its organs; binds organs together; stores energy reserves as fat; and provides immunity.

3. **Muscle tissue,** which is responsible for movement and generation of physical force.

4. **Nervous tissue,** which initiates and transmits action potentials (nerve impulses) that help coordinate body activities.

About 8 days after a sperm fertilizes an egg, the mass of cells that results from several cell divisions embeds in the lining of the uterus and begins to form three **primary germ layers: ectoderm, endoderm,** and **mesoderm.** These are the embryonic tissues from which all tissues and organs of the body develop. Epithelial tissues develop from all three germ layers. Connective tissues and most muscle tissues derive from mesoderm. Nervous tissue develops from ectoderm. (See Table 27.1 for a more detailed list of structures derived from the primary germ layers.)

Epithelial tissue and connective tissue, except for bone and blood, are discussed in detail in this chapter. The general features of bone tissue and blood will be introduced here, but their detailed discussion occurs later in Chapters 5 and 12,

respectively. Similarly, the structure and function of muscle tissue and nervous tissue are examined in detail in Chapters 9 and 16, respectively.

Normally, most cells within a tissue remain in place, anchored to other cells, basement membranes (described shortly), and connective tissues. A few cells, such as phagocytes, routinely move through the body, searching for invaders or cellular debris. Before birth, certain cells migrate extensively as part of the growth and development process.

Clinical Application

Medical Studies of Tissues

A **pathologist** (pa-THOL-ō-gist; *pathos* = disease) is a physician who specializes in laboratory studies of cells and tissues to help other physicians reach accurate diagnoses. One of the principal functions of a pathologist is to examine tissues for any changes that might indicate disease. Pathologists also conduct autopsies (see page 15) and supervise laboratory personnel who examine blood and other body fluids.

A **biopsy** (BĪ-op-sē; *bio* = life; *opsis* = vision) is the removal of a sample of living tissue for microscopic examination. It is used to help diagnose many disorders, especially cancer, and to discover the cause of unexplained infections and inflammations. Both normal and suspected diseased tissues are removed for purposes of comparison. Once the tissue sample is removed, it may be preserved, stained to highlight special properties, and cut into thin sections for microscopic observation. Sometimes a biopsy is conducted while a patient is anesthetized during surgery to help a physician determine the most appropriate treatment. For example, if a breast biopsy indicates malignant cells, the physician can proceed with the appropriate surgical procedure while the patient is anesthetized rather than have her return on another day. ■

CELL JUNCTIONS

As you will see, epithelial cells, and certain other body cells, are tightly joined to form a close functional unit. An important reason that tissue cells adhere to one another is that they contain specialized points of contact between their adjacent plasma membranes called **cell junctions.** In addition to providing cell-to-cell attachment and cell-to-extracellular material attachment, cell junctions also prevent the movement of molecules between some cells and provide channels for molecular communication between others.

For purposes of discussion, we will classify cell junctions into three principal types: (1) tight junctions, (2) plaque-bearing junctions, and (3) gap junctions (Figure 3.1).

Figure 3.1 Cell junctions.

Most epithelial cells and some muscle and nerve cells contain cell junctions.

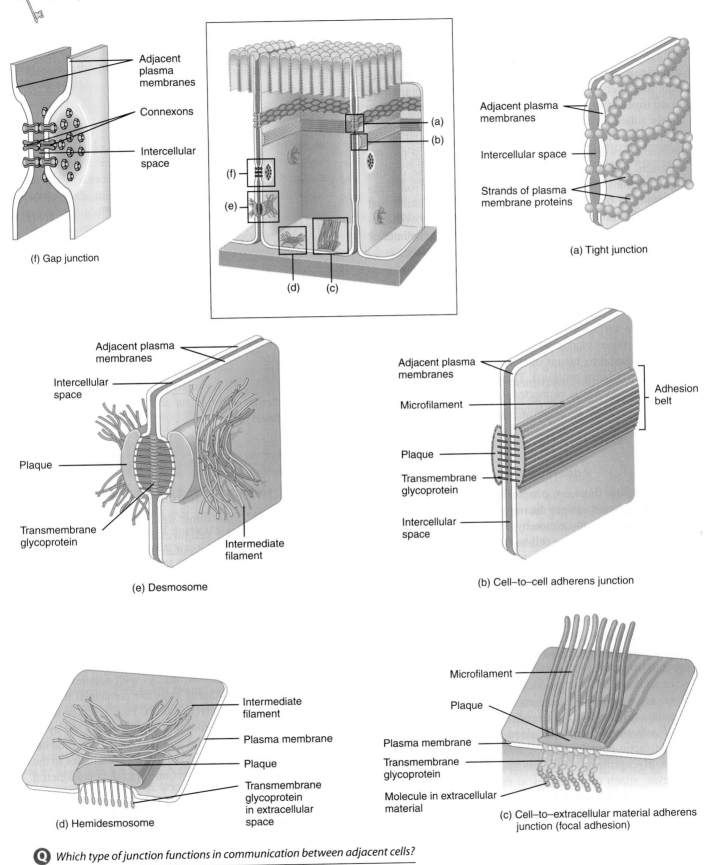

Adjacent plasma membranes

Connexons

Intercellular space

(f) Gap junction

(a)
(b)
(f)
(e)
(d)
(c)

Adjacent plasma membranes

Intercellular space

Strands of plasma membrane proteins

(a) Tight junction

Adjacent plasma membranes

Intercellular space

Plaque

Transmembrane glycoprotein

Intermediate filament

(e) Desmosome

Adjacent plasma membranes

Microfilament

Adhesion belt

Plaque

Transmembrane glycoprotein

Intercellular space

(b) Cell–to–cell adherens junction

Intermediate filament

Plasma membrane

Plaque

Transmembrane glycoprotein in extracellular space

(d) Hemidesmosome

Microfilament

Plaque

Plasma membrane

Transmembrane glycoprotein

Molecule in extracellular material

(c) Cell–to–extracellular material adherens junction (focal adhesion)

Q *Which type of junction functions in communication between adjacent cells?*

the gastrointestinal, urinary, and respiratory tracts. *Stable cells* cease regeneration at about the time of puberty but may resume regeneration under special conditions, such as injury. Epithelial cells of the liver, pancreas, kidneys, and thyroid glands are examples. There are also cells of the body called *permanent cells* that cease to regenerate after birth and remain that way throughout life. An example is a nerve cell. For this reason nerve cell damage or death in the brain or spinal cord results in permanent loss of function.

Covering and Lining Epithelium

Arrangement of Layers

The arrangement of covering and lining epithelium reflects its location and function. If the function of the epithelium is absorption or filtration and it is in an area that has minimal wear and tear, the cells of the tissue form a single layer. Such an arrangement is called *simple epithelium*. If the epithelium is in an area with a high degree of wear and tear, then the cells are stacked in several layers. This type of tissue is called *stratified epithelium*. A third, less common arrangement of epithelium is *pseudostratified columnar epithelium*. Like simple epithelium, pseudostratified columnar epithelium has only one layer of cells. However, some of the cells do not reach the surface, which gives the tissue a multilayered, or stratified, appearance.

Cell Shapes

Besides classifying covering and lining epithelium according to the number of its layers, we may also categorize it by cell shape. The cells may be flat, cubelike, columnar, or a combination of shapes. *Squamous* (SKWĀ-mus; *squama* = scale) cells are flat and attach to each other like floor tiles. *Cuboidal* cells are usually cube-shaped. *Columnar* cells are tall and cylindrical. *Transitional* cells readily change shape and are found where there is a great degree of distention (stretching) in the body. Transitional cells in the basal (deepest) layer of an epithelial tissue may range in shape from cuboidal to columnar. In the intermediate layer, they may be cuboidal or polyhedral (having many sides). In the superficial layer, they may range in shape from cuboidal to squamous, depending on how much they are stretched during certain body functions.

Classification

Considering layers and cell shapes in combination, we may classify covering and lining epithelium as follows:

I. Simple
   A. Squamous
   B. Cuboidal
   C. Columnar

II. Stratified
   A. Squamous*
   B. Cuboidal*
   C. Columnar*
   D. Transitional

III. Pseudostratified Columnar

Each of these epithelial tissues is described in the following sections and illustrated in Table 3.1. Each figure in the table consists of a photomicrograph, a corresponding diagram, and an inset that identifies a principal location of the tissue in the body. Along with the illustrations are descriptions, locations, and functions of the tissues.

Simple Epithelium

Simple Squamous Epithelium This consists of a single layer of flat cells (Table 3.1A). Its surface resembles a tiled floor when viewed from above. The nucleus of each cell is oval or spherical. Because simple squamous epithelium has only one layer of cells, it is highly adapted for the movement of molecules by processes such as diffusion, osmosis, and filtration. Simple squamous epithelium is found in body areas that are subject to little wear and tear.

Simple squamous epithelium that lines the heart, blood vessels, and lymphatic vessels is known as **endothelium** (*endo* = within; *thelium* = covering). Simple squamous epithelium that forms the epithelial layer of serous membranes is called **mesothelium** (*meso* = middle). Endothelium and mesothelium are derived from the embryonic mesoderm.

Simple Cuboidal Epithelium The cuboidal shape of the cells (Table 3.1B) is obvious only when the tissue is viewed from the side. The nuclei are usually round. Simple cuboidal epithelium performs the functions of secretion and absorption. **Secretion** is the production and release by cells of a fluid that may contain a variety of substances such as mucus, perspiration, or enzymes. **Absorption** is the intake of fluids or other substances by cells.

Simple Columnar Epithelium When viewed from the side, the cells appear rectangular with oval nuclei. Simple columnar epithelium exists in two forms: nonciliated simple columnar epithelium and ciliated simple columnar epithelium. *Nonciliated simple columnar epithilium* (Table 3.1C) contains microvilli and goblet cells. **Microvilli** (*micro* = small; *villus* = tuft of hair) (see Figure 2.2) are microscopic fingerlike cytoplasmic projections that serve to increase the

Text continues on page 67

---

*This classification is based on the shape of the outermost surface cells.

Table 3.1 *Epithelial Tissues*

COVERING AND LINING EPITHELIUM

A. Simple squamous epithelium

Description: Single layer of flat cells; centrally located nucleus.

Location: Lines heart, blood vessels, lymphatic vessels, air sacs of lungs, glomerular (Bowman's) capsule of kidneys, and inner surface of the tympanic membrane (eardrum); forms epithelial layer of serous membranes.

Function: Filtration, diffusion, osmosis, and secretion in serous membranes.

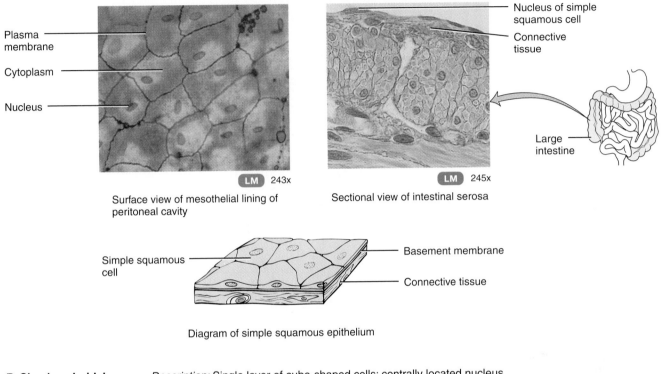

Plasma membrane

Cytoplasm

Nucleus

LM 243x

Surface view of mesothelial lining of peritoneal cavity

Nucleus of simple squamous cell

Connective tissue

LM 245x

Sectional view of intestinal serosa

Large intestine

Simple squamous cell

Basement membrane

Connective tissue

Diagram of simple squamous epithelium

B. Simple cuboidal epithelium

Description: Single layer of cube-shaped cells; centrally located nucleus.

Location: Covers surface of ovary, lines anterior surface of capsule of the lens of the eye, forms the pigmented epithelium at the back of the eye, lines kidney tubules and smaller ducts of many glands, and makes up the secreting portion of some glands such as the thyroid gland.

Function: Secretion and absorption.

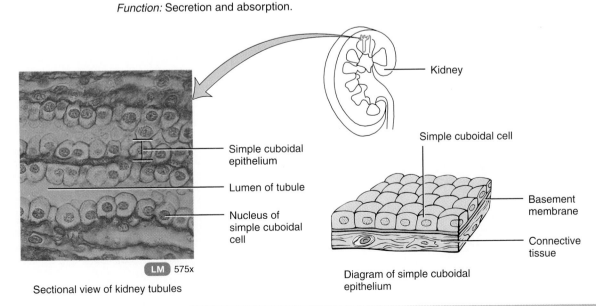

Kidney

Simple cuboidal epithelium

Lumen of tubule

Nucleus of simple cuboidal cell

LM 575x

Sectional view of kidney tubules

Simple cuboidal cell

Basement membrane

Connective tissue

Diagram of simple cuboidal epithelium

Table 3.1 *Epithelial Tissues (continued)*

COVERING AND LINING EPITHELIUM

C. Nonciliated simple columnar epithelium

Description: Single layer of nonciliated rectangular cells; nucleus at base of cell; contains goblet cells and microvilli in some locations.

Location: Lines the gastrointestinal tract from the stomach to the anus, ducts of many glands, and gallbladder.

Function: Secretion and absorption.

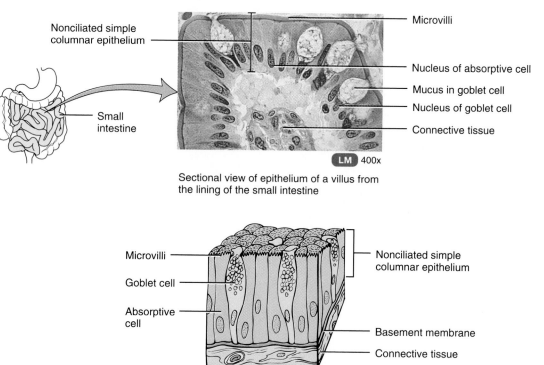

Sectional view of epithelium of a villus from the lining of the small intestine

Diagram of nonciliated simple columnar epithelium

D. Ciliated simple columnar epithelium

Description: Single layer of ciliated rectangular cells; nucleus at base of cell; contains goblet cells in some locations.

Location: Lines a few portions of upper respiratory tract, uterine (Fallopian) tubes, uterus, some paranasal sinuses, and central canal of spinal cord.

Function: Moves mucus and other substances by ciliary action.

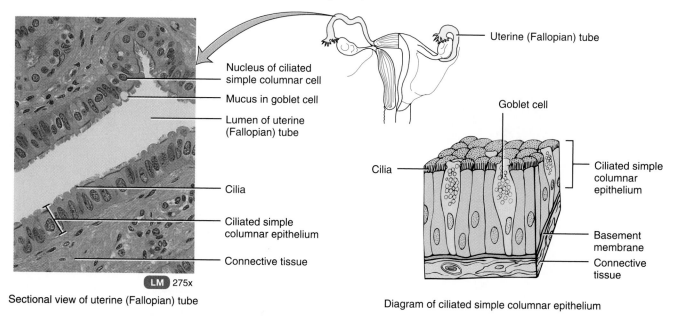

Sectional view of uterine (Fallopian) tube

Diagram of ciliated simple columnar epithelium

Table 3.1 (continued)

COVERING AND LINING EPITHELIUM

E. Stratified squamous epithelium

Description: Several layers of cells; cuboidal to columnar shape in deep layers; squamous cells in superficial layers; basal cells replace surface cells as they are lost.

Location: Keratinized variety forms superficial layer of skin; nonkeratinized variety lines wet surfaces such as lining of the mouth, esophagus, part of epiglottis, and vagina and covers the tongue.

Function: Protection.

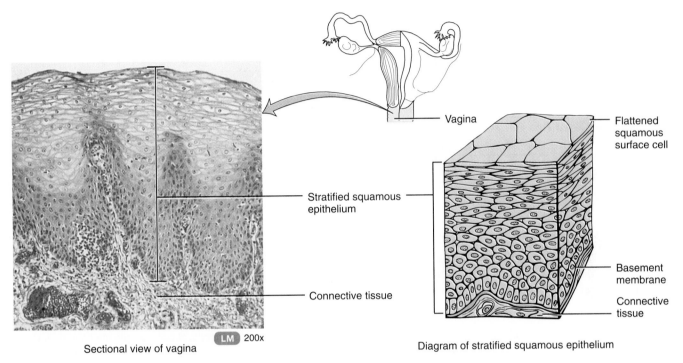

Vagina

Flattened squamous surface cell

Stratified squamous epithelium

Basement membrane

Connective tissue

Connective tissue

LM 200x

Sectional view of vagina

Diagram of stratified squamous epithelium

F. Stratified cuboidal epithelium

Description: Two or more layers of cells in which the superficial cells are cube-shaped.

Location: Ducts of adult sweat glands and part of male urethra.

Function: Protection.

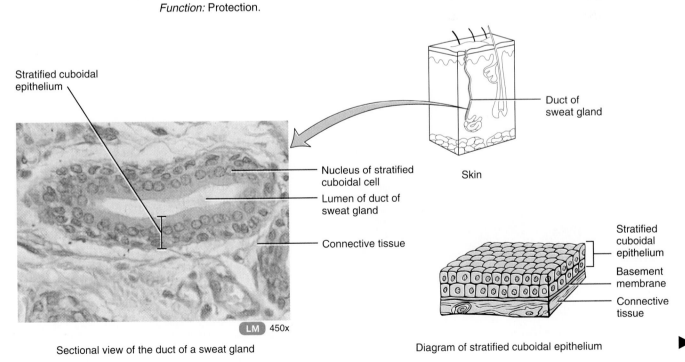

Stratified cuboidal epithelium

Duct of sweat gland

Nucleus of stratified cuboidal cell

Lumen of duct of sweat gland

Skin

Connective tissue

Stratified cuboidal epithelium

Basement membrane

Connective tissue

LM 450x

Sectional view of the duct of a sweat gland

Diagram of stratified cuboidal epithelium

Table 3.1 Epithelial Tissues *(continued)*

COVERING AND LINING EPITHELIUM

G. Stratified columnar epithelium

Description: Several layers of polyhedral cells; columnar cells are only in the superficial layer.

Location: Lines part of urethra, large excretory ducts of some glands, small areas in anal mucous membrane, and portion of conjunctiva of eye.

Function: Protection and secretion.

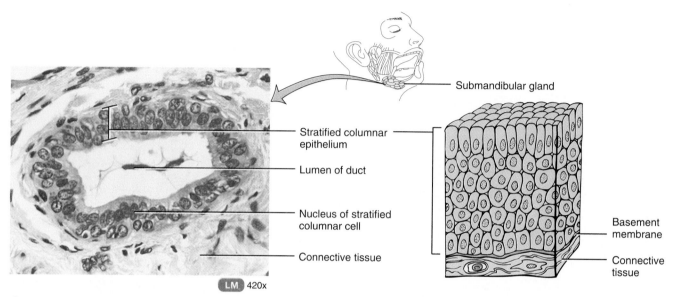

Submandibular gland

Stratified columnar epithelium

Lumen of duct

Nucleus of stratified columnar cell

Connective tissue

Basement membrane

Connective tissue

LM 420x

Sectional view of the duct of a submandibular salivary gland

Diagram of stratified columnar epithelium

H. Transitional epithelium

Description: Appearance is variable (transitional); shape of superficial cells ranges from squamous (when stretched) to cuboidal (when relaxed).

Location: Lines urinary bladder and portions of ureters and urethra.

Function: Permits distention.

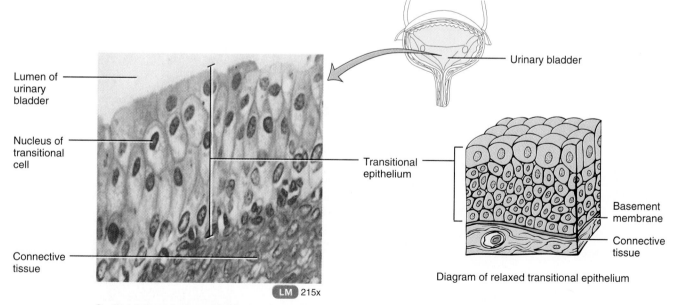

Urinary bladder

Lumen of urinary bladder

Nucleus of transitional cell

Connective tissue

Transitional epithelium

Basement membrane

Connective tissue

LM 215x

Sectional view of urinary bladder in relaxed state

Diagram of relaxed transitional epithelium

Table 3.1 *(continued)*

COVERING AND LINING EPITHELIUM

I. Pseudostratified columnar epithelium

Description: Not a true stratified tissue; nuclei of cells are at different levels; all cells are attached to basement membrane, but not all reach the free surface.

Location: Pseudostratified ciliated columnar epithelium lines most of upper respiratory tract; pseudostratified nonciliated columnar epithelium lines larger ducts of many glands, epididymis, and part of male urethra.

Function: Secretion and movement of mucus by ciliary action.

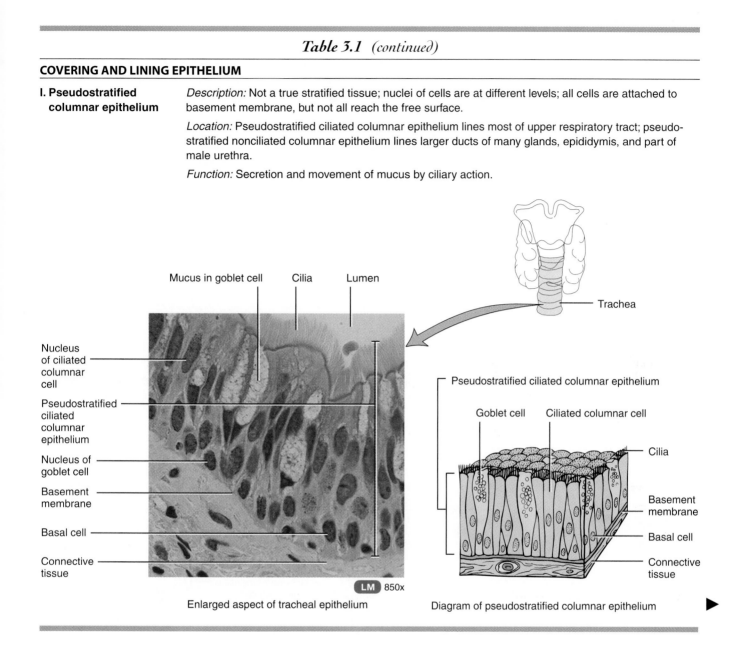

Enlarged aspect of tracheal epithelium

LM 850x

Diagram of pseudostratified columnar epithelium

surface area of the plasma membrane. This has the effect of increasing the rate at which digested nutrients and fluids can be absorbed into the cell. **Goblet cells** are modified columnar cells that secrete mucus (a slightly sticky fluid) that accumulates in the upper portion of the cell, causing that area to bulge out. The whole cell then resembles a goblet or wine glass. The secreted mucus serves as a lubricant for the linings of the digestive, respiratory, reproductive, and most of the urinary tracts. Mucus also helps to trap dust that enters the body, especially in the respiratory tract. In the gastrointestinal tract, mucus prevents destruction of the stomach lining by digestive enzymes.

Ciliated simple columnar epithelium (Table 3.1D) contains cells with hairlike processes called **cilia** (*cilio* = eyelashes). In a few parts of the upper respiratory tract, ciliated columnar cells are interspersed with goblet cells. Mucus secreted by the goblet cells forms a film over the respiratory surface that traps inhaled foreign particles. The cilia wave in unison and move the mucus, and any foreign particles, toward the throat, where it can be swallowed or spit out. Cilia also help to move an ovum down the uterine (Fallopian) tubes into the uterus for implantation.

Table 3.1 *Epithelial Tissues (continued)*

GLANDULAR EPITHELIUM

J. Exocrine glands

Description: Secretory products released into ducts.

Location: Sweat, oil, ear wax, and mammary glands of the skin; digestive glands such as salivary glands, which secrete into mouth cavity, and pancreas, which secretes into the small intestine.

Function: Produce mucus, perspiration, oil, ear wax, milk, or digestive enzymes.

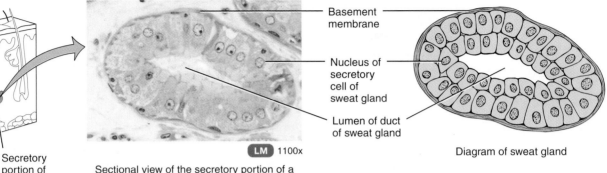

Skin   Secretory portion of sweat gland

Sectional view of the secretory portion of a sweat gland

Basement membrane

Nucleus of secretory cell of sweat gland

Lumen of duct of sweat gland

LM 1100x

Diagram of sweat gland

K. Endocrine glands

Description: Secretory products (hormones) diffuse into blood after passing through extracellular fluid.

Location: Examples include pituitary gland at base of brain, pineal gland in brain, thyroid and parathyroid glands near larynx (voice box), adrenal glands superior to kidneys, pancreas near stomach, ovaries in pelvic cavity, testes in scrotum, and thymus gland in thoracic cavity.

Function: Produce hormones that regulate various body activities.

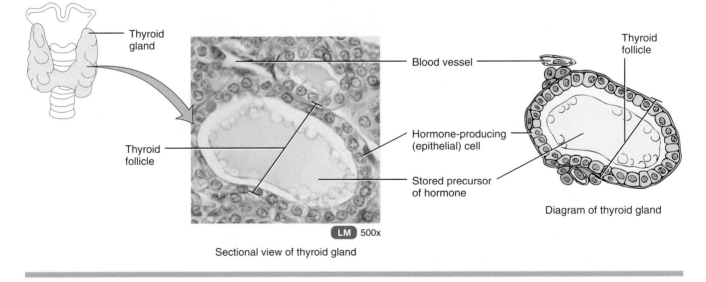

Thyroid gland

Thyroid follicle

Sectional view of thyroid gland

LM 500x

Blood vessel

Hormone-producing (epithelial) cell

Stored precursor of hormone

Thyroid follicle

Diagram of thyroid gland

Stratified Epithelium

In contrast to simple epithelium, stratified epithelium has at least two layers of cells. Thus it is more durable and can protect underlying tissues from the external environment and from wear and tear. Some cells of stratified epithelia also produce secretions. The name of the specific kind of stratified epithelium depends on the shape of the *surface* cells.

Stratified Squamous Epithelium In the superficial layers of this type of epithelium, the cells are flat, whereas in the deep layers, cells vary in shape from cuboidal to columnar (Table 3.1E). The basal (deepest) cells continually replicate by cell division. As new cells grow, the cells of the basal layer continually shift upward toward the surface. As they move farther from the deep layer and their blood supply (in the underlying connective tissue), they become dehydrated, shrunken, and harder. At the surface, the cells lose their cell junctions and are rubbed off. Old cells are sloughed off and replaced as new cells continually emerge.

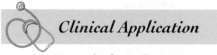

Clinical Application

Papanicolaou Smears

A **Papanicolaou** (pa-pa-NI-kō-lō) or **Pap smear** involves collecting and microscopically examining cells that have been sloughed off into secretions of the cervix and vagina. The test is performed mainly to detect early changes in the cells of the female reproductive system that may indicate cancer or a precancerous condition. Annual Pap smears are recommended for women over age 18 (earlier if sexually active) as part of a routine pelvic exam. ■

Stratified squamous epithelium exists in two forms: keratinized and nonkeratinized. In *keratinized stratified squamous epithelium,* a tough layer of keratin is deposited in the surface cells. **Keratin** (*kerato* = horny) is a protein that is resistant to friction and helps repel bacteria. *Nonkeratinized stratified squamous epithelium* does not contain keratin and remains moist.

Stratified Cuboidal Epithelium This fairly rare type of epithelium sometimes consists of more than two layers of cells (Table 3.1F). Its function is mainly protective.

Stratified Columnar Epithelium Like stratified cuboidal epithelium, this tissue also is uncommon in the body. Usually the basal layer or layers consist of shortened, irregularly polyhedral cells; only the superficial cells are columnar in form (Table 3.1G). It functions in protection and secretion.

Transitional Epithelium This kind of epithelium is variable in appearance, depending on whether it is relaxed or distended (stretched). In the relaxed state (Table 3.1H), it looks similar to stratified cuboidal epithelium except that the superficial cells tend to be large and rounded. This allows the tissue to be distended without the outer cells breaking apart from one another. When stretched, they are drawn out into squamous-shaped cells, giving the appearance of stratified squamous epithelium. Because of this capability, transitional epithelium lines hollow structures that are subjected to expansion from within, such as the urinary bladder. Its function is to help prevent rupture of these organs.

Pseudostratified Columnar Epithelium

The third category of covering and lining epithelium is called pseudostratified columnar epithelium (Table 3.1I). The nuclei of the cells are at various depths. Even though all the cells are attached to the basement membrane in a single layer, some cells do not reach the free surface. These features give the false impression of a multilayered tissue when viewed from the side, and thus the name *pseudo*stratified epithelium (*pseudo* = false). In *pseudostratified ciliated columnar epithelium,* the cells that reach the surface either secrete mucus or bear cilia that sweep away mucus and trapped foreign particles for eventual elimination from the body. *Pseudostratified nonciliated columnar epithelium* contains no cilia or goblet cells.

Glandular Epithelium

The function of glandular epithelium is secretion, which is accomplished by glandular cells that often lie in clusters deep to the covering and lining epithelium. A **gland** may consist of one cell or a group of highly specialized epithelial cells that secrete substances into ducts, onto a surface, or into the blood. The production of such substances always requires active work by the cells and results in an expenditure of energy. All glands of the body are classified as exocrine or endocrine.

Exocrine (*exo* = outside; *krin* = to secrete) **glands** (Table 3.1J) secrete their products into ducts (tubes) that empty at the surface of covering and lining epithelium or directly onto a free surface. The product of an exocrine gland may be released at the skin surface or into the lumen (cavity) of a hollow organ. The secretions of exocrine glands include mucus, perspiration, oil, wax, and digestive enzymes. Examples of exocrine glands are sweat glands, which produce perspiration to cool the skin, and salivary glands, which secrete mucus and digestive enzymes.

The secretions of **endocrine** (*endo* = within) **glands** (Table 3.1K) enter the extracellular fluid and then diffuse directly into the bloodstream without flowing through a duct. These secretions, called hormones, regulate many metabolic and physiological activities to maintain homeostasis. The pituitary, thyroid, and adrenal glands are examples of endocrine glands. Endocrine glands will be described in detail in Chapter 22.

Structural Classification of Exocrine Glands

Exocrine glands are classified into two structural types: multicellular and unicellular. Most glands are **multicellular glands,** which have many cells that form a distinctive microscopic structure or macroscopic organ. Examples are sweat, oil, and salivary glands.

Unicellular glands are single-celled. An important unicellular exocrine gland is the goblet cell (see page 64, nonciliated simple columnar epithelium). Although goblet cells do not contain ducts, they are often classified as unicellular mucus-secreting exocrine glands. Goblet cells are present in the epithelial lining of the digestive, respiratory, urinary, and

Figure 3.3 Functional classification of multicellular exocrine glands.

Functional classification is based on whether a secretion is a product of a cell or consists of an entire or partial glandular cell.

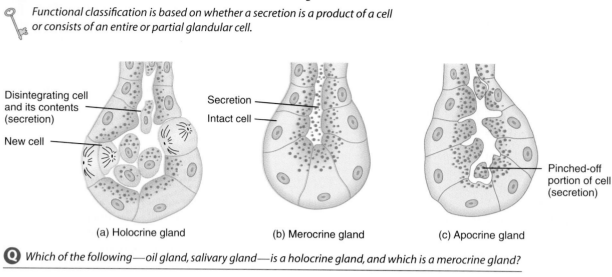

Disintegrating cell and its contents (secretion)

New cell

Secretion

Intact cell

Pinched-off portion of cell (secretion)

(a) Holocrine gland

(b) Merocrine gland

(c) Apocrine gland

Q *Which of the following—oil gland, salivary gland—is a holocrine gland, and which is a merocrine gland?*

reproductive systems. They produce mucus, which lubricates the apical surface of these tissues.

Functional Classification of Exocrine Glands

The functional classification of exocrine glands is based on whether a secretion is a product of a cell or consists of entire or partial glandular cells themselves. **Holocrine** (HŌ-lō-krin; *holos* = entire; *krin* = to secrete) **glands** accumulate a secretory product in their cytosol. The cell then dies and is discharged with its contents as the glandular secretion (Figure 3.3a). The discharged cell is replaced by a new cell. One example of a holocrine gland is a sebaceous (oil) gland of the skin. **Merocrine** (MER-ō-krin; *mero* = a part) **glands** simply form the secretory product and discharge it from the cell by exocytosis (Figure 3.3b). Most exocrine glands of the body are merocrine. Examples of merocrine glands are the salivary glands and pancreas. **Apocrine** (AP-ō-krin; *apo* = from) **glands** accumulate their secretory product at the apical (free) surface of the secreting cell. That portion of the cell pinches off from the rest of the cell to form the secretion (Figure 3.3c). The remaining part of the cell repairs itself and repeats the process. Based on electron micrograph studies, there is some question as to whether humans in fact have apocrine glands. What were once thought to be apocrine glands—for example, mammary glands—are probably merocrine glands.

CONNECTIVE TISSUE

The most abundant and most widely distributed tissue in the body is **connective tissue.** It binds together, supports, and strengthens other body tissues; protects and insulates internal organs; and compartmentalizes structures such as skeletal muscles. Blood, a fluid connective tissue, is the major transport system within the body, whereas adipose (fat) tissue is the major site of stored energy reserves.

General Features

Following are the general features of connective tissue:

1. Connective tissue consists of three basic elements: cells, ground substance, and fibers. The ground substance is the component of a connective tissue that occupies the space between the cells and fibers. Together, the ground substance and fibers, which are outside the cells, form the **matrix** (*matrix* = womb). Unlike epithelial cells, connective tissue cells rarely touch one another; they are separated by a considerable amount of matrix.

2. In contrast to epithelia, connective tissues do not usually occur on free surfaces, such as the covering or lining of internal organs, the lining of a body cavity, or the external surface of the body. However, joint cavities are lined by a type of connective tissue called areolar connective tissue.

3. Except for cartilage, connective tissue, like epithelium, has a nerve supply.

4. Unlike epithelium, connective tissue usually is highly vascular (has a rich blood supply). Exceptions include cartilage, which is avascular (has no blood supply), and tendons, which have a scanty blood supply.

5. The matrix of a connective tissue, which may be fluid, semifluid, gelatinous, fibrous, or calcified, is usually secreted by the connective tissue cells and adjacent cells and determines the tissue's qualities. In blood, the matrix,

which is not secreted by blood cells, is liquid. In cartilage, it is firm but pliable. In bone, it is considerably harder and not pliable.

Connective Tissue Cells

The cells in connective tissue derive from mesodermal embryonic cells called *mesenchymal cells.* Each major type of connective tissue contains an immature class of cell whose name ends in *-blast,* which means to bud, sprout, or produce.

These immature cells are called *fibroblasts* in loose and dense connective tissue, *chondroblasts* in cartilage, and *osteoblasts* in bone. These cells retain the capacity for mitosis and secrete the matrix that is characteristic of the tissue. In cartilage and bone, once the matrix is produced, the immature cells differentiate. The mature cells have names that end in *-cyte,* which means cell (chondrocytes and osteocytes). Mature cells have reduced capacity for cell division and matrix formation and are mostly involved in maintaining the matrix.

Following is a description of some of the cells contained in various types of connective tissue. The specific tissues to which they belong will be described shortly.

1. **Fibroblasts** (FĪ-brō-blasts; *fibro* = fibers) are large, flat, spindle-shaped cells with branching processes; they secrete the matrix.

2. **Macrophages** (MAK-rō-fā-jez; *macro* = large; *phagein* = to eat), or *histiocytes,* develop from *monocytes,* a type of white blood cell (see Figure 12.2). Macrophages have an irregular shape with short branching projections and are capable of engulfing bacteria and cellular debris by phagocytosis. Thus they provide a vital defense for the body. Whereas wandering macrophages leave the blood and migrate to infected tissues, fixed macrophages remain in certain tissues and organs of the body.

3. **Plasma cells** are small and either round or irregular in shape. They develop from a type of white blood cell called a *B lymphocyte (B cell).* Plasma cells secrete antibodies, proteins that attack or neutralize foreign substances in the body. Thus, plasma cells are an important part of the body's immune system. Although they are found in many places in the body, most plasma cells reside in connective tissues, especially in the gastrointestinal tract and the mammary glands.

4. **Mast cells** are abundant alongside blood vessels. They produce *histamine,* a chemical that dilates small blood vessels as part of the body's reaction to injury or infection. Mast cells also contain *heparin.* When produced by other cells of the body, heparin functions as an anticoagulant that prevents blood from clotting in blood vessels. However, heparin obtained from mast cells is a poor anticoagulant. The heparin in mast cells is in the form of heparin proteoglycan, which may serve to bind certain intracellular constituents of mast cells.

Connective Tissue Matrix

Each type of connective tissue has unique properties, due to accumulation of specific matrix materials between the cells. Matrix contains protein *fibers* embedded in a fluid, gel, or solid *ground substance.* The ground substance is the component of a connective tissue between the cells and fibers. It is colorless, transparent, and amorphous, meaning it has no specific shape. Connective tissue cells usually produce the ground substance and deposit it in the space between the cells.

Ground Substance

Ground substance contains a diversity of large molecules, many of which are complex combinations of polysaccharides (complex sugars) and proteins. For example, **hyaluronic** (hī-a-loo-RON-ik) **acid** is a viscous, slippery substance that binds cells together, lubricates joints, and helps maintain the shape of the eyeballs. It also appears to play a role in helping phagocytes migrate through connective tissue during development and wound repair. **Chondroitin** (kon-DROY-tin) **sulfate** is a jellylike substance that provides support and adhesiveness in cartilage, bone, skin, and blood vessels. The skin, tendons, blood vessels, and heart valves contain **dermatan sulfate,** whereas bone, cartilage, and the cornea of the eye contain **keratan sulfate.** Another class of molecules in matrix are **adhesion proteins,** which interact with receptors on plasma membranes to anchor cells in positions and to provide traction for the movement of cells.

The ground substance supports cells, binds them together, and provides a medium through which substances are exchanged between the blood and cells. Until recently, it was thought to function mainly as an inert scaffolding to support tissues. Now it is clear that the ground substance is quite active in tissue development, migration, proliferation, altering shape, and even metabolic functions.

Fibers

Fibers in the matrix provide strength and support for tissues. Three types of fibers are embedded in the matrix between the cells of connective tissue: collagen, elastic, and reticular fibers.

Collagen (*kolla* = glue) **fibers,** of which there are at least five different types, are very tough and resistant to a pulling force, yet allow some flexibility in the tissue because they are not taut. These fibers often occur in bundles made up of many tiny fibrils lying parallel to one another. The bundle arrangement affords great strength. Chemically, collagen fibers consist of the protein *collagen.* This is the most abundant protein in your body, representing about 25% of the total protein. Collagen fibers are found in most types of connective tissues, especially bone, cartilage, tendons, and ligaments.

Elastic fibers are smaller than collagen fibers and freely branch and rejoin one another. They consist of a protein called *elastin.* Like collagen fibers, elastic fibers provide strength. In addition, they can be stretched up to 150% of their relaxed length without breaking. Elastic fibers are plentiful in the skin, blood vessels, and lungs.

Reticular (*rete* = net) **fibers,** consisting of the protein collagen and a coating of glycoprotein, provide support in the walls of blood vessels and form a network around fat cells, nerve fibers, and skeletal and smooth muscle cells. Produced by fibroblasts, they are much thinner than collagen fibers and form branching networks. Like collagen fibers, reticular fibers provide support and strength and also form the **stroma** (*stroma* = bed covering) or framework of many soft organs, such as the spleen and lymph nodes. These fibers also help form the basement membrane.

Classification

Classification of connective tissues is difficult because of the diversity of cells, ground substance, and fibers and the differences in their relative proportions. Thus the separation of connective tissues into categories is not always clear-cut. We will classify them in the following way:

I. Embryonic connective tissue
   A. Mesenchyme
   B. Mucous connective tissue
II. Mature connective tissue
   A. Loose connective tissue
      1. Areolar connective tissue
      2. Adipose tissue
      3. Reticular connective tissue
   B. Dense connective tissue
      1. Dense regular connective tissue
      2. Dense irregular connective tissue
      3. Elastic connective tissue
   C. Cartilage
      1. Hyaline cartilage
      2. Fibrocartilage
      3. Elastic cartilage
   D. Bone (osseous) tissue
   E. Blood (vascular tissue)

Each of the connective tissues described in the following sections is illustrated in either Table 3.2 or Table 3.3. As with epithelium, each figure in the connective tissue tables consists of a photomicrograph, a corresponding diagram, and an inset that identifies a principal location of the tissue in the body. Along with the illustrations are description, locations, and functions of the tissues.

Embryonic Connective Tissue

Connective tissue that is present primarily in the embryo or fetus is called **embryonic connective tissue.** The term *embryo* refers to a developing human from fertilization through the first two months of pregnancy; *fetus* refers to a developing human from the third month of pregnancy to birth.

One example of embryonic connective tissue found almost exclusively in the embryo is **mesenchyme** (MEZ-en-kīm), the tissue from which all other connective tissues eventually arise (Table 3.2A). It is composed of irregularly shaped mesenchymal cells, a semifluid ground substance, and delicate reticular fibers. Mesenchymal cells in mature connective tissue differentiate into fibroblasts that assist in wound healing

Another kind of embryonic connective tissue is **mucous connective tissue** (*Wharton's jelly*), found primarily in the fetus. It is a form of mesenchyme that contains widely scattered fibroblasts, a more viscous, jellylike ground substance, and collagen fibers (Table 3.2B).

Mature Connective Tissue

Mature connective tissue exists in the newborn, has cells produced from mesenchyme, and does not change after birth. It is subdivided into several kinds: (1) loose connective tissue, (2) dense connective tissue, (3) cartilage, (4) bone, and (5) blood.

Loose Connective Tissue

In **loose connective tissue** the fibers are loosely woven, and there are many cells. The types are areolar connective tissue, adipose tissue, and reticular connective tissue.

Areolar Connective Tissue   Areolar (a-RĒ-ō-lar; *areola* = a small space) connective tissue is one of the most widely distributed connective tissues in the body. It contains several kinds of cells (Table 3.3A), including fibroblasts, macrophages, plasma cells, mast cells, adipocytes, and a few white blood cells. All three types of fibers—collagen, elastic, and reticular—are present and arranged randomly. The fluid, semifluid, or gelatinous ground substance contains hyaluronic acid, chondroitin sulfate, dermatan sulfate, and keratan sulfate.

The matrix materials normally aid the passage of nutrients from the blood vessels of the connective tissue into adjacent cells and tissues. The thick consistency of hyaluronic acid, however, may impede the movement of some drugs. If

Table 3.2 Embryonic Connective Tissue

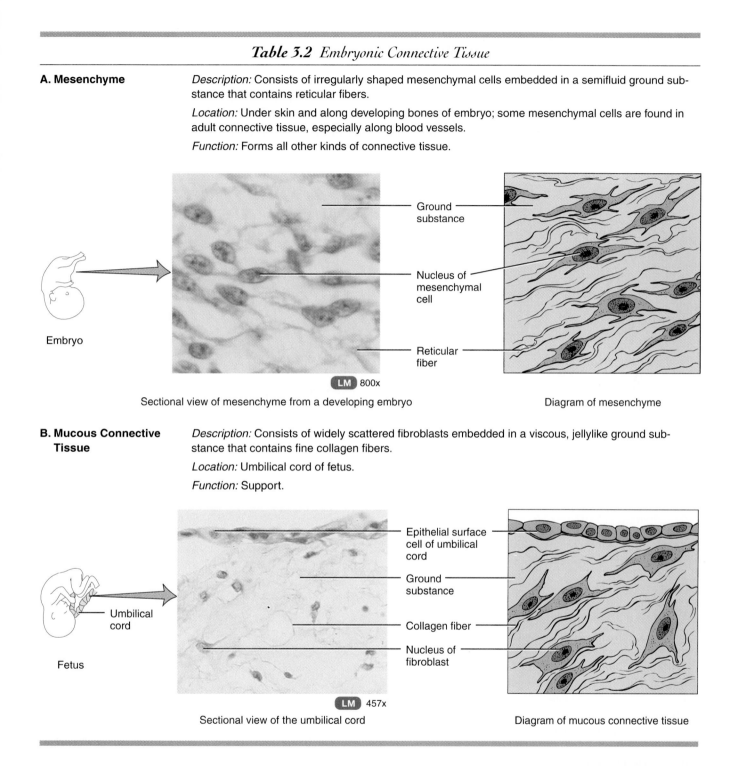

A. Mesenchyme

Description: Consists of irregularly shaped mesenchymal cells embedded in a semifluid ground substance that contains reticular fibers.

Location: Under skin and along developing bones of embryo; some mesenchymal cells are found in adult connective tissue, especially along blood vessels.

Function: Forms all other kinds of connective tissue.

Embryo

Ground substance

Nucleus of mesenchymal cell

Reticular fiber

LM 800x

Sectional view of mesenchyme from a developing embryo

Diagram of mesenchyme

B. Mucous Connective Tissue

Description: Consists of widely scattered fibroblasts embedded in a viscous, jellylike ground substance that contains fine collagen fibers.

Location: Umbilical cord of fetus.

Function: Support.

Fetus

Umbilical cord

Epithelial surface cell of umbilical cord

Ground substance

Collagen fiber

Nucleus of fibroblast

LM 457x

Sectional view of the umbilical cord

Diagram of mucous connective tissue

an enzyme called *hyaluronidase* is injected into the tissue, the ground substance changes to a watery consistency. This feature is of clinical importance because the reduced viscosity hastens the absorption and diffusion of injected drugs and fluids through the tissue and thus can lessen tension and pain. White blood cells, sperm, and some bacteria produce hyaluronidase. Combined with adipose tissue, areolar connective tissue forms the *subcutaneous* (sub'-kyoo-TĀ-nē-us; *sub* = under; *cut* = skin) *layer*—the layer of tissue that attaches the skin to underlying tissues and organs.

Table 3.3 *Mature Connective Tissue*

LOOSE CONNECTIVE TISSUE

A. Areolar connective tissue

Description: Consists of fibers (collagen, elastic, and reticular) and several kinds of cells (fibroblasts, macrophages, plasma cells, adipocytes, and mast cells) embedded in a semifluid ground substance.

Location: Subcutaneous layer of skin, papillary (superficial) region of dermis of skin, mucous membranes, blood vessels, nerves, and around body organs.

Function: Strength, elasticity, and support.

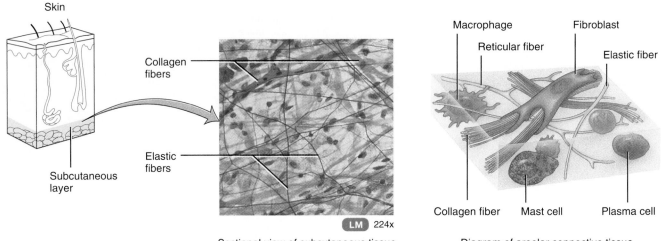

Sectional view of subcutaneous tissue

Diagram of areolar connective tissue

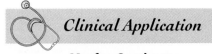

Clinical Application

Marfan Syndrome

Marfan (MAR-fan) **syndrome** is an inherited disorder caused by a defective gene. The result is abnormal elastic fibers. Tissues rich in elastic fibers are malformed or weakened. Structures affected most seriously are the covering layer of bones (periosteum), the ligament that suspends the lens of the eye, and the walls of the large arteries. People with Marfan syndrome tend to be tall and have disproportionately long arms, legs, fingers, and toes. A common sign is blurred vision caused by displacement of the lens of the eye. The most life-threatening complication of Marfan syndrome is weakening of the aorta (main artery that emerges from the heart), which can suddenly burst. ▪

Adipose Tissue Adipose tissue is a loose connective tissue in which the cells, called **adipocytes** (*adeps* = fat) are specialized for storage of triglycerides (fats). Adipocytes are derived from fibroblasts. Because of the accumulation of a single large triglyceride droplet, the cytoplasm and nucleus are pushed to the edge (periphery) of the cell (Table 3.3B). Adipose tissue is found wherever areolar connective tissue is located. Adipose tissue is a good insulator and can therefore reduce heat loss through the skin. It is a major energy reserve and generally supports and protects various organs.

Clinical Application

Suction Lipectomy (Liposuction)

A surgical procedure, called **suction lipectomy** (*lipo* = fat; *ectomy* = removal of) or **liposuction,** involves suctioning out small amounts of fat from certain areas of the body. The technique can be used in a number of areas of the body in which fat accumulates, such as the thighs, buttocks, arms, breasts, and abdomen. However, suction lipectomy is not a treatment for obesity, because it will not result in a permanent reduction in fat in the area. Rather, it is a body-contouring procedure. Moreover, certain complications may develop, including fat emboli (clots), infection, fluid depletion, and injury to internal structures. ▪

Table 3.3 (continued)

LOOSE CONNECTIVE TISSUE

B. Adipose tissue

Description: Consists of adipocytes, cells that are specialized to store triglycerides (fats) in a large central area in their cytoplasm; nuclei are peripherally located.

Location: Subcutaneous layer of skin, around heart and kidneys, yellow marrow of long bones, and padding around joints and behind eyeball in eye socket.

Function: Reduces heat loss through skin, serves as an energy reserve, supports, and protects.

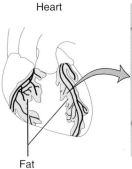
Heart

Fat

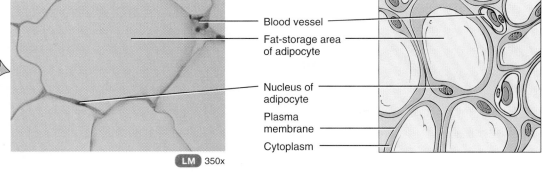

LM 350x

Sectional view of adipocytes of white fat

Blood vessel
Fat-storage area of adipocyte
Nucleus of adipocyte
Plasma membrane
Cytoplasm

Diagram of adipose tissue

C. Reticular connective tissue

Description: Consists of a network of interlacing reticular fibers and reticular cells.

Location: Stroma (framework) of liver, spleen, lymph nodes; portion of bone marrow that gives rise to blood cells; reticular lamina of the basement membrane; and around blood vessels and muscle.

Function: Forms stroma of organs; binds together smooth muscle tissue cells.

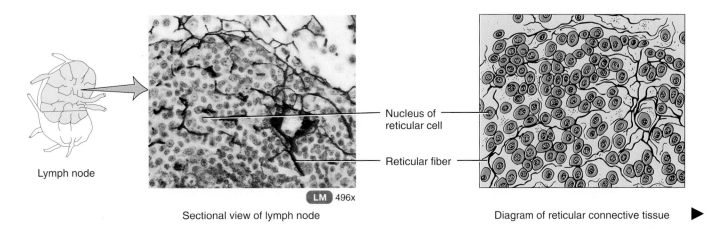

Lymph node

LM 496x

Sectional view of lymph node

Nucleus of reticular cell

Reticular fiber

Diagram of reticular connective tissue

▶

Reticular Connective Tissue Reticular connective tissue consists of fine interlacing reticular fibers and reticular cells (Table 3.3C). Reticular connective tissue forms the stroma (framework) of certain organs and helps to bind together cells of smooth muscle.

Dense Connective Tissue

Dense connective tissue contains more numerous, thicker, and denser fibers but considerably fewer cells than loose connective tissue. The types are dense regular connective tissue, dense irregular connective tissue, and elastic connective tissue.

Table 3.3 *Mature Connective Tissue (continued)*

DENSE CONNECTIVE TISSUE

D. Dense regular connective tissue

Description: Matrix looks shiny white; consists of predominantly collagen fibers arranged in bundles; fibroblasts present in rows between bundles.

Location: Forms tendons (attach muscle to bone), most ligaments (attach bone to bone), and aponeuroses (sheetlike tendons that attach muscle to muscle or muscle to bone).

Function: Provides strong attachment between various structures.

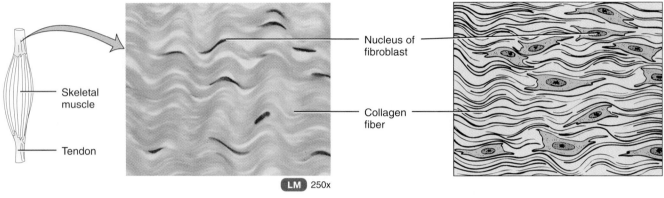

Skeletal muscle

Tendon

Nucleus of fibroblast

Collagen fiber

LM 250x

Sectional view of a tendon

Diagram of dense regular connective tissue

E. Dense irregular connective tissue

Description: Consists of predominantly collagen fibers, randomly arranged, and a few fibroblasts.

Location: Fasciae (tissue beneath skin and around muscles and other organs), reticular (deeper) region of dermis of skin, periosteum of bone, joint capsules, membrane capsules around various organs (kidneys, liver, testes, lymph nodes), pericardium of the heart, and heart valves.

Function: Provides strength.

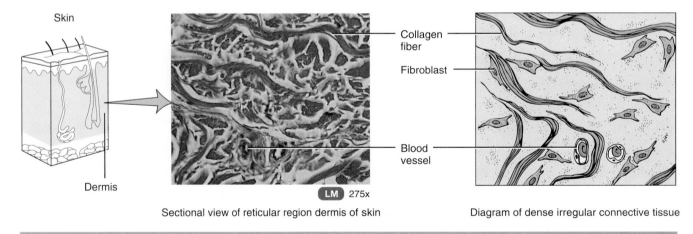

Skin

Dermis

Collagen fiber

Fibroblast

Blood vessel

LM 275x

Sectional view of reticular region dermis of skin

Diagram of dense irregular connective tissue

Dense Regular Connective Tissue In this tissue, bundles of collagen fibers have a *regular* (orderly), parallel arrangement that confers great strength (Table 3.3D). The tissue structure withstands pulling in one direction. Fibroblasts, which produce the fibers and ground substance, appear in rows between the fibers. The tissue is silvery white, tough, yet somewhat pliable.

Dense Irregular Connective Tissue This tissue contains collagen fibers that are *irregularly* arranged (Table 3.3E) and is found in parts of the body where tensions are exerted in various directions. The tissue usually occurs in sheets. Heart valves and the perichondrium, a membrane around cartilage, although classified as dense irregular connective tissue, have a fairly orderly arrangement of collagen fibers.

Table 3.3 (continued)

DENSE CONNECTIVE TISSUE

F. Elastic connective tissue

Description: Consists of predominantly freely branching elastic fibers; fibroblasts present in spaces between fibers.

Location: Lung tissue, walls of elastic arteries, trachea, bronchial tubes, true vocal cords, suspensory ligament of penis, and ligamenta flava of vertebrae (ligaments between vertebrae).

Function: Allows stretching of various organs.

Aorta

Heart

LM 335x

Sectional view of aorta (largest artery in the body)

Nucleus of fibroblast

Elastic fiber

Diagram of elastic connective tissue

CARTILAGE

G. Hyaline cartilage

Description: Consists of a bluish white, shiny ground substance with fine collagen fibers; contains numerous chondrocytes; most abundant type of cartilage.

Location: Ends of long bones, anterior ends of ribs, nose, parts of larynx, trachea, bronchi, bronchial tubes, and embryonic skeleton.

Function: Provides smooth surfaces for movement at joints, flexibility, and support.

Trachea

LM 512x

Sectional view of hyaline cartilage from trachea

Nucleus of chondrocyte

Lacuna containing chondrocyte

Ground substance

Diagram of hyaline cartilage

Table 3.3 *Mature Connective Tissue (continued)*

CARTILAGE

H. Fibrocartilage

Description: Consists of chondrocytes scattered among bundles of collagen fibers within the matrix.

Location: Pubic symphysis (point where hipbones join anteriorly), intervertebral discs (discs between vertebrae), menisci (cartilage pads) of knee, and portions of tendons that insert into cartilage.

Function: Support and fusion.

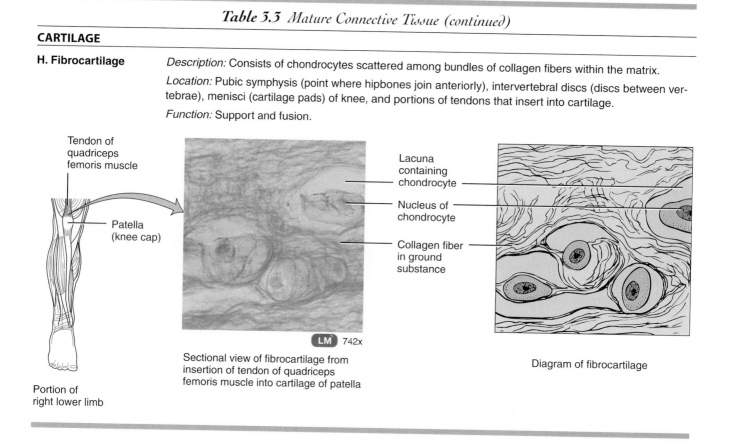

Tendon of quadriceps femoris muscle

Patella (knee cap)

Portion of right lower limb

Lacuna containing chondrocyte

Nucleus of chondrocyte

Collagen fiber in ground substance

LM 742x

Sectional view of fibrocartilage from insertion of tendon of quadriceps femoris muscle into cartilage of patella

Diagram of fibrocartilage

Elastic Connective Tissue Elastic connective tissue has a predominance of freely branching elastic fibers (Table 3.3F). These fibers give the unstained tissue a yellowish color. Fibroblasts are present in the spaces between the fibers. Elastic connective tissue can be stretched and will snap back into shape (elasticity). Elastic connective tissue provides stretch and strength, allowing structures to perform their functions efficiently.

Cartilage

Cartilage is capable of enduring considerably more stress than the connective tissues just discussed. Cartilage consists of a dense network of collagen fibers and elastic fibers firmly embedded in chondroitin sulfate, a rubbery component of the ground substance. Whereas the strength of cartilage is due to its collagen fibers, its resilience (ability to assume its original shape after deformation) is due to chondroitin sulfate.

The cells of mature cartilage, called **chondrocytes** (KON-drō-sīts; *chondros* = cartilage), occur singly or in groups within spaces called **lacunae** (la-KOO-nē; *lacuna* = little lake) in the matrix. The surface of most cartilage is surrounded by a membrane of dense irregular connective tissue

called the **perichondrium** (per'-i-KON-drē-um; *peri* = around). Unlike other connective tissues, cartilage has no blood vessels or nerves, except for those in the perichondrium. Three kinds of cartilage are recognized: hyaline cartilage, fibrocartilage, and elastic cartilage.

Hyaline Cartilage This cartilage (also called gristle) contains a resilient gel as its ground substance (Table 3.3G) and appears in the body as a bluish-white, shiny substance. The fine collagen fibers, although present, are not visible with ordinary staining techniques, and the prominent chondrocytes are found in lacunae. Hyaline cartilage has a perichondrium, except for articular cartilage and epiphyseal plates. Hyaline cartilage is the most abundant cartilage in the body. It affords flexibility and support and, at joints, reduces friction and absorbs shock. Hyaline cartilage is the weakest of the three types of cartilage.

Fibrocartilage Chondrocytes are scattered among clearly visible bundles of collagen fibers within the matrix of this type of cartilage (Table 3.3H). Fibrocartilage does not have a perichondrium. This tissue combines strength and rigidity and is the strongest of the three types of cartilage.

Table 3.3 (continued)

CARTILAGE

I. Elastic cartilage

Description: Consists of chondrocytes located in a threadlike network of elastic fibers within the matrix.

Location: Lid on top of larynx (epiglottis), external ear (auricle), and auditory (Eustachian) tubes.

Function: Gives support and maintains shape.

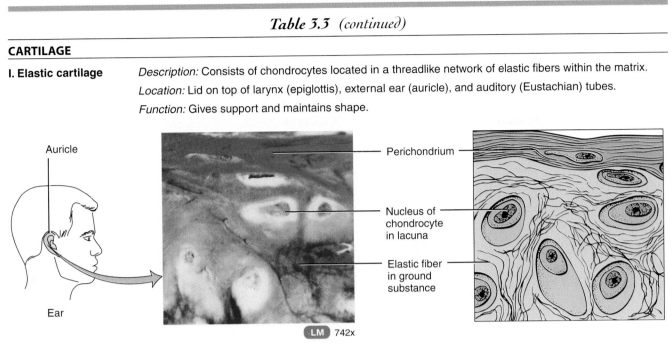

Auricle

Ear

Perichondrium

Nucleus of
chondrocyte
in lacuna

Elastic fiber
in ground
substance

LM 742x

Sectional view of elastic cartilage from auricle of ear

Diagram of elastic cartilage

BONE (OSSEOUS) TISSUE

Description: Compact bone tissue consists of osteons (Haversian systems) that contain lamellae, lacunae, osteocytes, canaliculi, and central (Haversian) canals. See also Figure 5.3a. Spongy bone tissue consists of thin columns called trabeculae; spaces between trabeculae are filled with red bone marrow. See Figure 5.3b and c.

Location: Both compact and spongy bone tissue comprise the various parts of bones of the body.

Function: Support, protection, storage, houses blood-forming tissue, and serves as levers that act together with muscle tissue to provide movement.

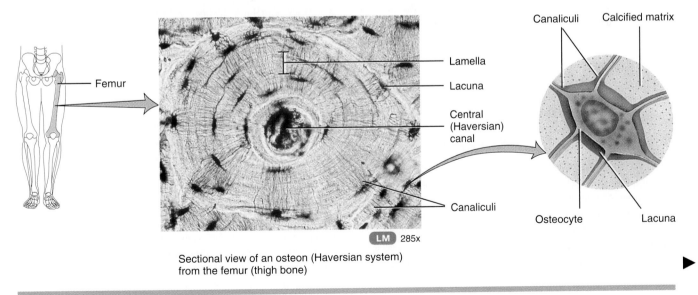

Femur

Canaliculi   Calcified matrix

Lamella

Lacuna

Central
(Haversian)
canal

Canaliculi

Osteocyte   Lacuna

LM 285x

Sectional view of an osteon (Haversian system)
from the femur (thigh bone)

▶

Table 3.3 Mature Connective Tissue (continued)

BLOOD (VASCULAR) TISSUE

Description: Consists of plasma and formed elements. The formed elements are erythrocytes (red blood cells), leukocytes (white blood cells), and platelets. See also Figure 12.2.

Location: Within blood vessels (arteries, arterioles, capillaries, venules, and veins) and the heart.

Function: Erythrocytes transport oxygen and carbon dioxide; leukocytes carry on phagocytosis and are involved in allergic reactions and immunity; platelets are essential for the clotting of blood.

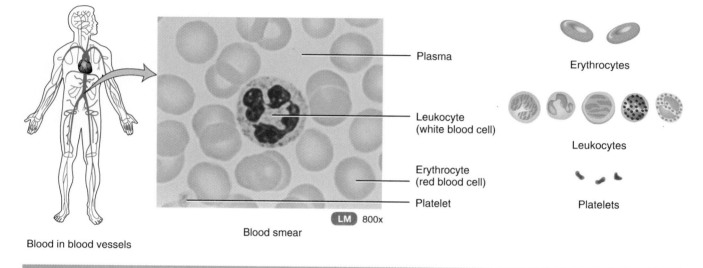

Blood in blood vessels   Blood smear   **LM** 800x

Elastic Cartilage In this tissue, chondrocytes are located in a threadlike network of elastic fibers within the matrix (Table 3.3I). A perichondrium is present. Elastic cartilage provides strength and elasticity and maintains the shape of certain organs.

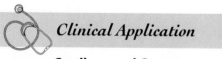

Clinical Application

Cartilage and Cancer

When cartilage grows, it does not allow the growth of blood vessels into it. Researchers found that this is due to the production by cartilage of chemicals called *antiangiogenesis factors (AAFs)*. Because cancers are the result of uncontrolled cell division, which is supported by blood vessel growth, AAFs may have great potential for treating cancer. Some AAFs have been isolated and are now in clinical trials for treatment of various types of cancers. They are also being tested for the treatment of stomach ulcers.

Bone (Osseous) Tissue

Together, cartilage, joints, and bones comprise the skeletal system. Bones are organs composed of several different connective tissues. These include bone or **osseous** (OS-ē-us) **tissue,** the *periosteum* (covering around a bone), red and yellow bone marrow, and the endosteum (lining of a space in a bone that stores yellow marrow).

Bone tissue is classified as either compact (dense) or spongy (cancellous), depending on how the matrix and cells are organized. The basic unit of compact bone is called an **osteon (Haversian system)** (Table 3.3). Each osteon is composed of **lamellae** (*lamella* = little plate or layer), concentric rings of matrix that consist of mineral salts (mostly tricalcium phosphate and calcium carbonate) that give bone its hardness, and collagen fibers that give bone its strength; **lacunae,** small spaces between lamellae that contain mature bone cells called **osteocytes; canaliculi** (*canaliculi* = small channels or canals), minute canals that project from lacunae and contain projections from osteocytes that provide numerous routes so that nutrients can reach osteocytes and wastes can be removed from them; and a **central (Haversian) canal** that contains blood vessels and nerves. Spongy bone has no osteons; instead, it consists of columns of bone called **trabeculae** (*trabecula* = little beam), which contain lamellae, osteocytes, lacunae, and canaliculi. Spaces between trabeculae are filled with red bone marrow. The details of the histology of bone are presented on page 111.

The skeletal system has several functions. It supports soft tissues, protects delicate structures, and works with skeletal muscles to generate movement. Bone stores calcium and phosphorus; houses red bone marrow, which produces several kinds of blood cells, and houses yellow bone marrow, which contains triglycerides as an energy source.

Blood (Vascular) Tissue

Blood (*vascular*) **tissue** is a connective tissue with a liquid matrix called plasma (Table 3.3). Suspended in the plasma are formed elements—cells and cell fragments. **Plasma** is a straw-colored liquid that consists mostly of water with a wide variety of dissolved substances (nutrients, wastes, enzymes, hormones, respiratory gases, and ions). The formed elements are erythrocytes (red blood cells), leukocytes (white blood cells), and platelets. (See also Figure 12.2.) **Erythrocytes** function in transporting oxygen to body cells and removing carbon dioxide from them. **Leukocytes** are involved in phagocytosis, immunity, and allergic reactions. **Platelets** function in blood clotting.

The details of blood are considered in Chapter 12.

MEMBRANES

The combination of an epithelial layer and an underlying connective tissue layer constitutes an **epithelial membrane.** The principal epithelial membranes of the body are mucous membranes, serous membranes, and the cutaneous membrane, or skin. The cutaneous membrane is an organ of the integumentary system and is discussed in Chapter 4. Another kind of membrane, a **synovial membrane,** contains only connective tissue and no epithelium.

Mucous Membranes

A **mucous membrane,** or **mucosa,** lines a body cavity that opens directly to the exterior. Mucous membranes line the entire digestive, respiratory, and reproductive systems and much of the urinary system. They consist of a lining layer of epithelium and an underlying layer of connective tissue.

The epithelial layer of a mucous membrane is an important aspect of the body's defense mechanisms. It is a barrier that microbes and other pathogens have difficulty penetrating. Usually, tight junctions connect the cells, so materials cannot leak in between. Certain cells of the epithelial layer of a mucous membrane secrete mucus, which prevents the cavities from drying out. It also traps particles in the respiratory passageways and lubricates food as it moves through

the gastrointestinal tract. In addition, the epithelial layer secretes some of the enzymes needed for digestion and is the site of food absorption in the gastrointestinal tract. As you will see in later discussions the epithelium of a mucous membrane varies greatly in different parts of the body. For example, the epithelium of the mucous membrane of the small intestine is simple columnar (see Chapter 24), whereas that of the large airways to the lungs is pseudostratified ciliated columnar.

The connective tissue layer of a mucous membrane is called the **lamina propria** (LAM-i-na; PRŌ-prē-a). The lamina propria is so named because it belongs to the mucous membrane (*proprius* = one's own). It binds the epithelium to the underlying structures and allows some flexibility of the membrane. It also holds the blood vessels in place and protects underlying muscles from abrasion or puncture. Oxygen and nutrients diffuse from the lamina propria to the epithelium covering it, whereas carbon dioxide and wastes diffuse in the opposite direction.

Serous Membranes

A **serous** (*serous* = watery) **membrane,** or **serosa,** lines a body cavity that does not open directly to the exterior, and it covers the organs that lie within the cavity. Serous membranes consist of areolar connective tissue covered by mesothelium (simple squamous epithelium) and are composed of two layers. The portion of a serous membrane attached to the cavity wall is called the *parietal* (pa-RĪ-e-tal; *paries* = wall) *layer.* The portion that covers and attaches to the organs inside these cavities is the *visceral* (*viscus* = body organ) *layer.* The mesothelium of a serous membrane secretes **serous fluid,** a watery lubricating fluid that allows organs to glide easily against one another or against the walls of cavities.

The serous membrane lining the thoracic cavity and covering the lungs is called the **pleura.** The serous membrane lining the heart cavity and covering the heart is the **pericardium** (*cardio* = heart). (See Figure 13.3a.) The serous membrane lining the abdominal cavity and covering the abdominal organs is called the **peritoneum.**

Synovial Membranes

Synovial (sin-Ō-vē-al) **membranes** line the cavities of freely movable joints (see Figure 8.1a). *Syn* means together or junction, here referring to a place where bones come together. Like serous membranes, they line structures that do not open to the exterior. Unlike mucous, serous, and cutaneous membranes, they do not contain epithelium and are therefore not epithelial membranes. They are composed of areolar connective tissue with elastic fibers and varying amounts of fat. Synovial membranes secrete **synovial fluid,**

Table 3.4 Muscle Tissue

Skeletal muscle tissue   *Description:* Long, cylindrical, striated fibers with many peripherally located nuclei; voluntary control.

Location: Usually attached to bones by tendons.

Function: Motion, posture, heat production (thermogenesis).

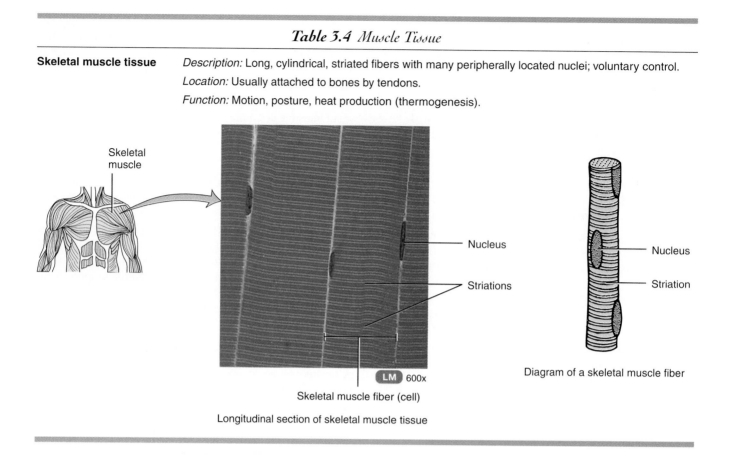

Skeletal muscle

Nucleus

Striations

Skeletal muscle fiber (cell)

LM 600x

Longitudinal section of skeletal muscle tissue

Nucleus

Striation

Diagram of a skeletal muscle fiber

---

which lubricates the cartilage at the ends of bones during their movements and nourishes the cartilage covering the bones at joints. These are articular synovial membranes. Other synovial membranes line cushioning sacs, called bursae, and tendon sheaths in hands and feet that ease the movement of muscle tendons.

MUSCLE TISSUE

Muscle tissue consists of fibers (cells) that are beautifully constructed to generate force for contraction. As a result of this characteristic, muscle tissue provides motion, maintains posture, and generates heat. Based on location and certain structural and functional characteristics, muscle tissue is classified into three types: skeletal, cardiac, and smooth (Table 3.4).

Skeletal muscle tissue is named for its location—usually attached to bones by tendons. It is also *striated;* that is, the fibers (cells) contain alternating light and dark bands (striations) that are perpendicular to the long axes of the fibers. The striations are visible under a microscope. Skeletal muscle is also *voluntary* because it can be made to con-

tract or relax by conscious control. A single skeletal muscle fiber is roughly cylindrical with many peripherally located nuclei. Within a muscle, the individual fibers are parallel to each other.

Cardiac muscle tissue forms the bulk of the wall of the heart. Like skeletal muscle, it is striated. However, unlike skeletal muscle tissue, it is *involuntary;* its contraction is usually not under conscious control. Cardiac muscle fibers are branched cylinders. The fibers usually have only one nucleus that is centrally located; sometimes there are two nuclei. Cardiac muscle fibers attach to each other end to end by transverse thickenings of the plasma membrane called **intercalated** (*intercalare* = to insert between) **discs,** which contain both desmosomes and gap junctions. Intercalated discs are unique to cardiac muscle. The desmosomes strengthen the tissue, and the gap junctions provide a route for quick conduction of muscle impulses throughout the heart.

Smooth muscle tissue is located in the walls of hollow internal structures such as blood vessels, airways to the lungs, the stomach, intestines, gallbladder, and urinary bladder. Its contraction helps constrict blood vessels, break down food, move food and fluids through the body, and eliminate wastes. Smooth muscle fibers are usually *involuntary,* and

Table 3.4 *(continued)*

Cardiac muscle tissue   *Description:* Branched striated fibers with one or two centrally located nuclei; contains intercalated discs; mainly involuntary control.

Location: Heart wall.

Function: Pumps blood to all parts of the body.

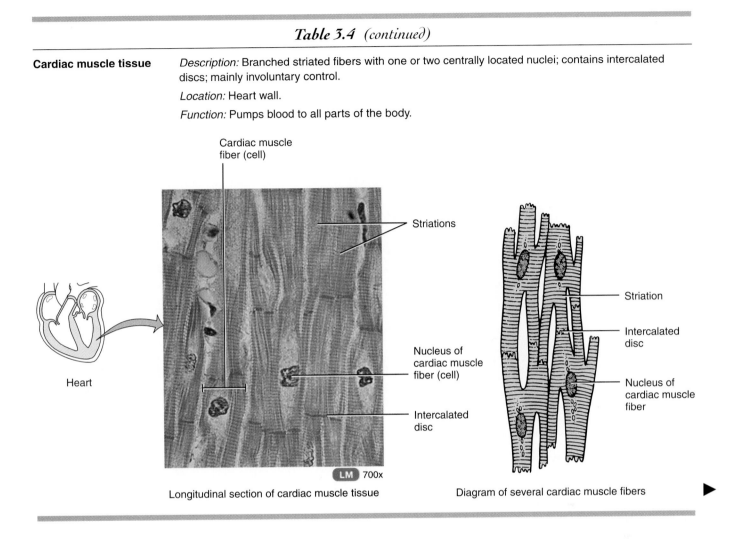

Longitudinal section of cardiac muscle tissue   Diagram of several cardiac muscle fibers

they are *nonstriated (smooth).* A smooth muscle fiber is small and spindle-shaped (thickest in the middle with each end tapering to a point) and contains a single, centrally located nucleus. Many individual fibers in some smooth muscle tissue—for example, in the wall of the intestines—are connected by gap junctions and thus contract as a single unit. In other locations—for example, the iris of the eye—smooth muscle fibers contract individually, like skeletal muscle fibers, because gap junctions are absent.

Chapter 9 provides a more detailed discussion of muscle tissue.

NERVOUS TISSUE

Despite the awesome complexity of the nervous system, it consists of only two principal kinds of cells: neurons and neuroglia. **Neurons** (*neuro* = nerve) or nerve cells are highly specialized cells that are the structural and functional units of the nervous system. They are sensitive to various stimuli, they convert stimuli into nerve impulses, and they conduct these impulses to other neurons, muscle fibers, or glands. Most neurons consist of three basic portions: a cell body and two kinds of processes called dendrites and axons (Table 3.5). The **cell body** contains the nucleus and other organelles. **Dendrites** (*dendro* = tree) are usually highly branched, tapered processes of the cell body that usually conduct nerve impulses toward the cell body. **Axons** (*axon* = axis) are typically single, long processes of the cell body that usually conduct nerve impulses away from the cell body.

Neuroglia do not generate or conduct nerve impulses. They do, however, have many important functions (see Table 16.1 on page 504). They are of clinical interest because they are often the sites of tumors of the nervous system.

The detailed structure and function of neurons and neuroglia are considered in Chapter 16.

Table 3.4 *Muscle Tissue (continued)*

Smooth muscle tissue

Description: Spindle-shaped, nonstriated fibers with one centrally located nucleus; usually involuntary control.

Location: Walls of hollow internal structures such as blood vessels, airways to the lungs, stomach, intestines, gallbladder, urinary bladder, and uterus.

Function: Motion (constriction of blood vessels and airways, propulsion of foods through gastrointestinal tract, contraction of urinary bladder and gallbladder).

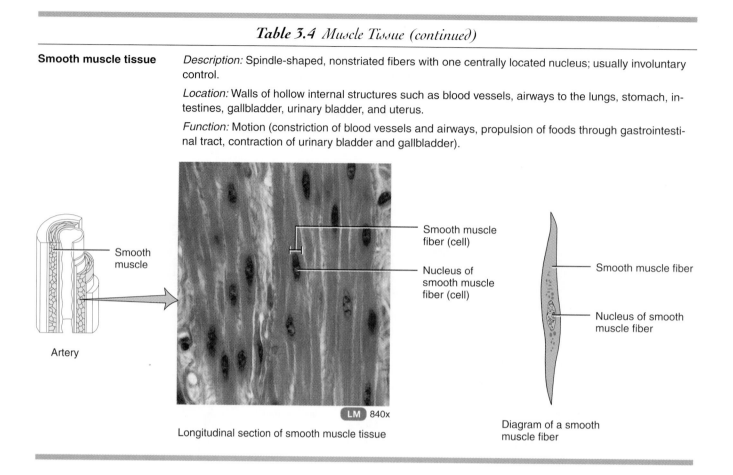

Artery

Smooth muscle

Smooth muscle fiber (cell)

Nucleus of smooth muscle fiber (cell)

LM 840x

Longitudinal section of smooth muscle tissue

Smooth muscle fiber

Nucleus of smooth muscle fiber

Diagram of a smooth muscle fiber

Table 3.5 *Nervous Tissue*

Description: Consists of neurons (nerve cells) and neuroglia. Neurons consist of a cell body and processes extending from the cell body called dendrites or axons. Neuroglia do not generate or conduct nerve impulses but have other important functions.

Location: Nervous system.

Function: Exhibits sensitivity to various types of stimuli, converts stimuli into nerve impulses, and conducts nerve impulses to other neurons, muscle fibers, or glands.

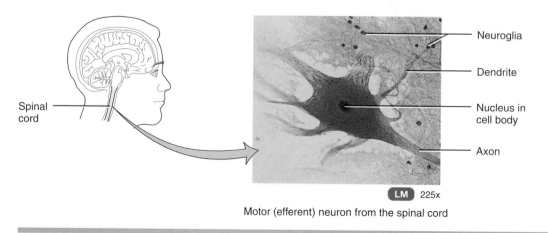

Spinal cord

Neuroglia

Dendrite

Nucleus in cell body

Axon

LM 225x

Motor (efferent) neuron from the spinal cord

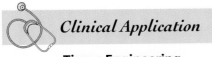

Clinical Application

Tissue Engineering

In recent years, a new technology called **tissue engineering** has emerged. Using a wide range of materials, scientists are learning to grow new tissues in the laboratory to replace damaged tissues in the body that cannot replace themselves. Tissue engineers have already developed laboratory-grown versions of skin and cartilage. In the procedure, scaffolding beds of biodegradable synthetic materials or collagen are used as substrates that permit body cells to be cultured and then implanted. Cells such as skin cells or cartilage cells are added to the scaffolding. As the cells divide and assemble, the scaffolding degrades and the new, permanent tissue is implanted in the patient. Other structures being developed by tissue engineers include bones, tendons, heart valves, bone marrow, and intestines. Work is also underway to develop insulin-producing cells (pancreas) for diabetics, dopamine-producing cells (brain) for Parkinson's disease patients, and even entire livers and kidneys. ▪

Key Medical Terms
Associated with Tissues

Atrophy (AT-rō-fe; *a* = without; *troph* = nourish) Wasting away of a tissue, organ, or even the entire body; a decrease in the size of a tissue or an organ; may be due to a disease, diminished blood supply, disuse of a tissue or an organ, or part of the normal aging process.

Necrosis (ne-KRŌ-sis; *necros* = death) Tissue death that may be caused by injury, physical agents (heat, chemicals, radiant energy), or diminished blood supply.

Polymyositis (pol-ē-mī-ō-SĪ-tis; *poly* = many; *myo* = muscle; *itis* = inflammation) An autoimmune disease of unknown cause characterized by inflammation of connective tissue and muscle tissue; signs and symptoms include skeletal muscle weakness and pain, joint pain, rash, fever, fatigue, and weight loss; the disorder is more common in females than males by a ratio of 2:1 and typically occurs between the ages of 40–60.

Sjögrens syndrome An autoimmune disorder that causes inflammation and dysfunction of exocrine glands, especially the lacrimal (tear) glands and salivary glands. It is characterized by dry eyes and mouth, and salivary gland enlargement; systemic effects include arthritis, difficulty in swallowing, pancreatitis, pleuritis, and migraine; the disorder affects more females than males by a ratio of 9:1; individuals with the condition have a greatly increased risk of developing malignant lymphoma.

Tissue rejection An immune response of the body directed at foreign proteins in a transplanted tissue or organ; immunosuppressive drugs, such as cyclosporine, have largely overcome tissue rejection in heart, kidney, and liver transplants.

Tissue transplantation The replacement of a diseased or injured tissue or organ; the most successful transplants involve use of a person's own tissues or those from an identical twin; less successful transplants involve tissues from an unrelated person or an animal.

Study Outline

Types of Tissues and Their Origins (p. 58)

1. A tissue is a group of similar cells that usually have a similar embryological origin and are specialized for a particular function.
2. Depending on their function and structure, the various tissues of the body are classified into four principal types: epithelial, connective, muscle, and nervous.

Cell Junctions (p. 58)

1. Cell junctions are points of contact between plasma membranes of adjacent cells or between cells and extracellular material.
2. Types of cell junctions include (a) tight junctions, which prevent the passage of molecules between cells; (b) plaque-bearing junctions (cell-to-cell adherens junctions, cell-to-extracellular material junctions or focal adhesions, desmosomes, and hemidesmosomes, which attach cells to each other or to extracellular materials); and (c) gap junctions, which permit electrical or chemical signals to pass between cells.

Epithelial Tissue (p. 61)

1. The subtypes of epithelium include covering and lining epithelium and glandular epithelium.
2. Some general characteristics of epithelium are as follows. It consists mostly of cells with little extracellular material, is arranged in sheets, is attached to a basement membrane, is avascular, has a nerve supply, has a high capacity for renewal, and is derived from all three primary germ layers.

Covering and Lining Epithelium (p. 62)

1. Epithelial layers are arranged as simple (one layer), stratified (several layers), and pseudostratified (one layer that appears as several); cell shapes include squamous (flat), cuboidal (cube-shaped), columnar (rectangular), and transitional (variable).
2. Simple squamous epithelium consists of a single layer of flat cells. It is adapted for diffusion and filtration and is found in the lungs and kidneys. Endothelium lines the heart and blood vessels. Mesothelium lines the thoracic and abdominopelvic cavities and covers the organs within them.

3. Simple cuboidal epithelium consists of a single layer of cube-shaped cells. It is adapted for secretion and absorption. It is found covering the ovaries, in the kidneys and eyes, and lining some glandular ducts.

4. Nonciliated simple columnar epithelium consists of a single layer of nonciliated rectangular cells. It lines most of the gastrointestinal tract. Specialized cells containing microvilli perform absorption. Goblet cells secrete mucus. Ciliated simple columnar epithelium consists of a single layer of ciliated rectangular cells. It is found in a few portions of the upper respiratory tract where it moves foreign particles trapped in mucus out of the respiratory tract.

5. Stratified squamous epithelium consists of several layers of cells in which the superficial layer is flat. It is protective. The nonkeratinized variety lines the mouth, esophagus, and vagina; the keratinized variety forms the outer layer of skin.

6. Stratified cuboidal epithelium consists of several layers of cells in which the superficial layer is cube-shaped. It is found in adult sweat glands and a portion of the male urethra.

7. Stratified columnar epithelium consists of several layers of cells in which the superficial layer is rectangular. It protects and secretes. It is found in a portion of the male urethra and large excretory ducts of some glands.

8. Transitional epithelium consists of several layers of cells whose appearance is variable. It lines the urinary bladder and is capable of stretching.

9. Pseudostratified columnar epithelium has only one layer but gives the appearance of many. It lines most of the upper respiratory tract (ciliated) and larger excretory ducts, part of the male urethra, and the epididymis (nonciliated).

Glandular Epithelium (p. 69)

1. A gland is a single cell or a mass of epithelial cells adapted for secretion.

2. Exocrine glands (sweat, oil, and digestive glands) secrete into ducts or directly onto a free surface.

3. Structural classification of glands includes unicellular and multicellular glands.

4. Functional classification of glands includes holocrine and merocrine glands.

5. Endocrine glands secrete hormones into the blood.

Connective Tissue (p. 70)

1. Connective tissue is the most abundant body tissue.

2. Some general characteristics of connective tissue are as follows. It consists of cells, ground substance, and fibers; has abundant matrix with relatively few cells; does not usually occur on free surfaces; has a nerve supply (except for cartilage); and is highly vascular (except for cartilage and tendons).

Connective Tissue Cells (p. 71)

1. Cells in connective tissue are derived from mesenchyme.

2. Types include fibroblasts (secrete matrix), macrophages (phagocytes), plasma cells (secrete antibodies), mast cells (produce histamine), adipocytes, and white blood cells.

Connective Tissue Matrix (p. 71)

1. The ground substance and fibers comprise the matrix.

2. Substances found in the ground substance include hyaluronic acid, chondroitin sulfate, dermatan sulfate, and keratan sulfate.

3. The ground substance supports, binds, provides a medium for the exchange of materials, and is active in influencing cell functions.

4. The fibers provide strength and support and are of three types.

5. Collagen fibers (composed of collagen) are found in large amounts in bone, tendons, and ligaments; elastic fibers (composed of elastin) are found in skin, blood vessels, and lungs; and reticular fibers (composed of collagen and glycoprotein) are found around fat cells, nerve fibers, and skeletal and smooth muscle cells.

Embryonic Connective Tissue (p. 72)

1. Mesenchyme forms all other connective tissues.

2. Mucous connective tissue is found in the umbilical cord of the fetus, where it gives support.

Mature Connective Tissue (p. 72)

1. Mature connective tissue is connective tissue that differentiates from mesenchyme and exists in the newborn and does not change after birth. It is subdivided into several kinds: loose connective tissue, dense connective tissue, cartilage, bone, and blood.

2. Loose connective tissue includes areolar connective tissue, adipose tissue, and reticular connective tissue.

3. Areolar connective tissue consists of the three types of fibers, several cells, and a semifluid ground substance. It is found in the subcutaneous layer and mucous membranes and around blood vessels, nerves, and body organs.

4. Adipose tissue consists of adipocytes that store triglycerides. It is found in the subcutaneous layer, around organs, and in the yellow bone marrow of long bones.

5. Reticular connective tissue consists of reticular fibers and reticular cells and is found in the liver, spleen, and lymph nodes.

6. Dense connective tissue includes dense regular connective tissue, dense irregular connective tissue, and elastic connective tissue.

7. Dense regular connective tissue consists of bundles of collagen fibers and fibroblasts. It forms tendons, most ligaments, and aponeuroses.

8. Dense irregular connective tissue consists of randomly arranged collagen fibers and few fibroblasts. It is found in fasciae, the dermis of skin, and membrane capsules.

9. Elastic connective tissue consists of elastic fibers and fibroblasts. It is found in the lungs, walls of arteries, and bronchial tubes.

10. Cartilage has a rubbery matrix (chondroitin sulfate) containing collagen and elastic fibers and chondrocytes.

11. Hyaline cartilage is found in the embryonic skeleton, at the ends of bones, in the nose, and in respiratory structures. It is flexible, allows movement, and provides support.

12. Fibrocartilage is found in the pubic symphysis, intervertebral discs, menisci (cartilage pads) of the knee joint, and portions of tendons that insert into cartilage.

13. Elastic cartilage maintains the shape of organs such as the epiglottis of the larynx, auditory (Eustachian) tubes, and external ear.

14. Bone (osseous) tissue consists of mineral salts and collagen fibers, which contribute to the hardness of bone, and cells called osteocytes. It supports, protects, helps provide movement, stores minerals, and houses blood-forming tissue.

15. Blood (vascular) tissue consists of plasma and formed elements (erythrocytes, leukocytes, and platelets). Functionally, its cells transport oxygen and carbon dioxide, carry on phagocytosis, participate in allergic reactions, provide immunity, and bring about blood clotting.

Membranes (p. 81)

1. An epithelial membrane consists of an epithelial layer overlying a connective tissue layer. Examples are mucous, serous, and cutaneous membranes.
2. Mucous membranes line cavities that open to the exterior, such as the gastrointestinal tract.
3. Serous membranes (pleura, pericardium, peritoneum) line closed cavities and cover the organs in the cavities. These membranes consist of parietal and visceral portions.
4. The cutaneous membrane is the skin.
5. Synovial membranes line joint cavities, bursae, and tendon sheaths and do not contain epithelium.

Muscle Tissue (p. 82)

1. Muscle tissue is modified for contraction and thus provides motion, maintenance of posture, and heat production.
2. Skeletal muscle tissue is attached to bones, is striated, and is voluntary.
3. Cardiac muscle tissue forms most of the heart wall, is striated, and is usually involuntary.
4. Smooth muscle tissue is found in the walls of hollow internal structures (blood vessels and viscera), is nonstriated, and is usually involuntary.

Nervous Tissue (p. 83)

1. The nervous system is composed of neurons (nerve cells) and neuroglia (protective and supporting cells).
2. Most neurons consist of a cell body and two types of processes called dendrites and axons.
3. Neurons are sensitive to stimuli, convert stimuli into nerve impulses, and conduct nerve impulses.

Review Questions

1. Define a tissue. What are the four basic types of human tissue? (p. 58)
2. Describe the three basic types of cell junctions and the functions of each. (p. 58)
3. Distinguish covering and lining epithelium from glandular epithelium. (p. 61)
4. What characteristics are common to all epithelial tissue? Describe the structure of the basement membrane. (p. 61)
5. Describe the various layering arrangements and cell shapes of epithelium. (p. 62)
6. How is epithelium classified? List the various types. (p. 62)
7. For each of the following kinds of epithelium, briefly describe the microscopic appearance, location in the body, and functions: simple squamous, simple cuboidal, simple columnar (nonciliated and ciliated), stratified squamous (keratinized and nonkeratinized), stratified cuboidal, stratified columnar, transitional, and pseudostratified columnar (ciliated and nonciliated). (pp. 63–67)
8. Define the following terms: endothelium, mesothelium, secretion, absorption, goblet cell, and keratin. (p. 62)
9. What is a gland? Distinguish between endocrine and exocrine glands. (p. 69)
10. Describe the classification of exocrine glands according to structure and function and give at least one example of each. (p. 69)
11. Enumerate the ways in which connective tissue differs from epithelium. (p. 70)
12. Describe the cells, ground substance, and fibers that comprise connective tissue. (p. 71)
13. How are connective tissues classified? List the various types. (p. 72)
14. How are embryonic connective tissue and mature connective tissue distinguished? (p. 72)
15. Describe the following connective tissues with regard to microscopic appearance, location in the body, and function: areolar connective tissue, adipose tissue, reticular connective tissue, dense regular connective tissue, dense irregular connective tissue, elastic connective tissue, hyaline cartilage, fibrocartilage, elastic cartilage, bone tissue, and blood. (pp. 74–80)
16. Define the following terms: matrix, ground substance, hyaluronic acid, chondroitin sulfate, dermatan sulfate, keratan sulfate, collagen fiber, elastic fiber, reticular fiber, fibroblast, macrophage, plasma cell, mast cell, adipocyte, chondrocyte, lacuna, osteocyte, lamella, canaliculus, and osteon (Haversian system). (p. 71)
17. Define the following kinds of membranes: mucous, serous, cutaneous, and synovial. Where is each located in the body? What are their functions? (p. 81)
18. How is muscle tissue classified? What are its functions? (p. 82)
19. Distinguish between neurons and neuroglia. Describe the structure and function of neurons. (p. 83)
20. Following are some descriptions of various tissues of the body. For each description, name the tissue described.
    a. An epithelium that permits distention (stretching).
    b. A single layer of flat cells concerned with filtration and absorption.
    c. Forms all other kinds of connective tissue.
    d. Specialized for triglyceride (fat) storage.
    e. An epithelium with waterproofing qualities.
    f. Forms the framework of many organs.
    g. Produces perspiration, wax, oil, or digestive enzymes.
    h. Cartilage that shapes the external ear.
    i. Contains goblet cells and lines the intestine.
    j. Most widely distributed connective tissue.
    k. Forms tendons, ligaments, and aponeuroses.
    l. Specialized for the secretion of hormones.
    m. Provides support in the umbilical cord.

n. Lines kidney tubules and is specialized for absorption and secretion.

o. Permits extensibility of lung tissue.

p. Stores red bone marrow, protects, supports.

q. Nonstriated, usually involuntary muscle tissue.

r. Consists of erythrocytes, leukocytes, and platelets.

s. Composed of a cell body, dendrites, and axon.

Self Quiz

Choose the one best answer to the following questions:

1. The microscopic examination of tissues is referred to as
   a. embryology
   b. histology
   c. cytology
   d. radiography
   e. none of the above

2. An abdominal incision will pass through the skin, then subcutaneous tissue, then muscle, to reach which membrane lining the inside wall of the abdomen?
   a. visceral pericardium
   b. parietal pericardium
   c. visceral peritoneum
   d. parietal peritoneum
   e. parietal pleura

3. A group of cells that has a similar embryological origin and operates together to perform a specialized activity is called a/an
   a. organ
   b. tissue
   c. system
   d. organelle
   e. organism

4. The basement membrane is classified as
   a. an epithelial membrane
   b. a plasma membrane
   c. a mucous membrane
   d. a synovial membrane
   e. none of the above

5. Which tissue is characterized by the presence of cell bodies, dendrites, and axons?
   a. muscle
   b. nervous
   c. vascular
   d. epithelial
   e. connective

6. Which of the following structures forms a secretory product and discharges it by exocytosis?
   a. holocrine gland
   b. connexon
   c. trabecula
   d. basal lamina
   e. merocrine gland

7. Mucous membranes
   a. line cavities of the body that are not open to the outside
   b. secrete a thin watery serous fluid
   c. cover the outside of such organs as the kidneys and stomach
   d. are found lining the respiratory and urinary passages
   e. are found on the outside surface of the body

8. Which of the following best describes connective tissue?
   a. usually contains a large amount of matrix
   b. is always arranged in a single layer of cells
   c. is primarily concerned with secretion
   d. cells are always very closely packed together
   e. is avascular

9. Which of the following is involuntary and striated?
   a. skeletal muscle tissue
   b. cardiac muscle tissue
   c. smooth muscle tissue
   d. both a and b
   e. none of the above

10. Which of the following statements best describes covering and lining epithelium?
    a. it is always arranged in a single layer of cells
    b. it contains large amounts of intercellular substance
    c. it has an abundant blood supply
    d. its free surface is exposed to the exterior of the body or to the interior of a hollow structure
    e. its cells are widely scattered

Are the following statements true or false?

11. Tight junctions are common between epithelial cells lining the stomach or urinary bladder, where they prevent fluid from passing between the lining cells.

12. Dense regular connective tissue contains a parallel arrangement of bundles of collagen fibers and is located in tendons and ligaments.

13. Focal adhesions are plaque-bearing adherens junctions that utilize transmembrane proteins to attach adjacent cells together.

14. Match the following:
    (a) modified columnar epithelial cell that secretes mucus; a unicellular gland
    (b) large phagocytic cell derived from monocyte; engulfs bacteria and cleans up debris; wandering or fixed
    (c) derived from fibroblast; specialized for triglyceride storage
    (d) a blood cell; functions in oxygen transport
    (e) large, branched cell; secretes the matrix of connective tissues

    ___ (1) fibroblast
    ___ (2) mast cell
    ___ (3) plasma cell
    ___ (4) osteocyte
    ___ (5) adipocyte
    ___ (6) chondrocyte
    ___ (7) goblet cell
    ___ (8) macrophage
    ___ (9) erythrocyte

(f) located along walls of blood vessels; secretes histamine and heparin
(g) cartilage cell; resides in space called a lacuna
(h) derived from B lymphocyte; produces antibodies as part of an immune response
(i) a mature bone cell; resides in a space called a lacuna

Complete the following:

15. Bone tissue is known as _____ tissue. Compact bone tissue consists of concentric rings called _____.
16. A gland that secretes its product into a duct is referred to as a/an _____ gland.
17. Blood, or _____ tissue, consists of a fluid called _____ containing three types of formed elements.
18. The embryonic tissue from which all connective tissues arise is called _____.

19. The connective tissue layer of a mucous membrane is called the _____.
20. Match the following:
(a) lines inner surface of the stomach and intestine
(b) lines urinary bladder, permitting distention
(c) forms fasciae and dermis of skin
(d) all cells attached to the basement membrane, but not all cells reach the surface of tissue
(e) lines mouth; present on outer surface of skin
(f) lines air sacs of lungs; suited for diffusion of gases

___ (1) dense irregular connective tissue
___ (2) simple squamous epithelium
___ (3) nonciliated simple columnar epithelium
___ (4) stratified squamous epithelium
___ (5) transitional epithelium
___ (6) pseudostratified columnar epithelium

Critical Thinking Questions

1. You've gone out to eat at your favorite fast food joint: General Bellisimo's Fried Chicken. A health-food zealot waves a chicken drumstick at you and declares, "This is all fat!" Using your knowledge of tissues, defend the contents of the chicken leg.
HINT: *Before it was a drumstick, it was a chicken's leg.*
2. Two college students return home at the end of their freshman year. The two students had weighed exactly the same when they played on their high school soccer team; now one is 20 pounds heavier than the other. Discuss the possible reasons for their weight difference.
HINT: *You are what you eat.*
3. Besides giving children something to do with their fingers, the mucous membrane of the nasal cavity has several important functions. Name some of these functions.
HINT: *The nose is also part of the respiratory system.*
4. Terra was reading her *Principles of Human Anatomy* at the All-night All-you-can-eat Salad Bar. She was having trouble visu-

alizing tissues until she took a scoop of some gelatin salad containing shredded carrots and grapes. As the salad quivered on her plate, she suddenly understood connective tissue. Relate the composition of the salad to the structure of connective tissue.
HINT: *The fruit is suspended in the gelatin.*
5. The neighborhood kids are walking around with common pins and sewing needles stuck through their fingertips. There is no visible bleeding. What type of tissue have they pierced? How do you know?
HINT: *If you pierced all the way through your earlobe it would bleed.*
6. Viola is fastidious about personal hygiene. She's decided that body secretions on the skin surface are unsanitary and unattractive. She wants to have all her exocrine glands removed. Is this a good idea?
HINT: *Consider the function and distribution of these glands.*

Answers to Figure Questions

3.1 Gap junctions.
3.2 Provides physical support for epithelium, provides for cell attachment, serves as a filter in the kidneys, and guides cell migration during development and tissue repair.
3.3 Oil gland is a holocrine gland, and salivary gland is a merocrine gland.

Chapter 4

The Integumentary System

Student Objectives

1. Define the integumentary system.

2. List the various layers of the epidermis and describe their structure and functions.

3. Describe the composition and functions of the dermis.

4. Explain the basis for skin color.

5. Describe the blood supply of the skin.

6. Compare the structure, distribution, and functions of hair, skin, glands, and nails.

7. Describe the development of the epidermis, its derivatives, and the dermis.

8. Describe the effects of aging on the integumentary system.

Of all the body's organs, none is more easily inspected or more exposed to infection, disease, and injury than the skin. Because of its visibility, skin reflects our emotions and some aspects of normal physiology, as evidenced by frowning, blushing, and sweating. Changes in skin color may indicate disorders in the body; for example, a bluish skin color is one sign of heart failure. Abnormal skin eruptions or rashes such as chicken pox, cold sores, or measles may reveal systemic infections or diseases of internal organs. Other disorders may involve just the skin itself, such as warts, age spots, or pimples. The skin's location makes it vulnerable to damage from trauma, sunlight, microbes, and pollutants in the environment.

Skin and its accessory organs—the hair and nails, along with various glands, muscles, and nerve endings—comprise the **integumentary** (in-teg-yoo-MEN-tar-ē; *integumentum* = covering) **system.** The organs that form the integumentary system guard the body's physical and biochemical integrity, help maintain a constant body temperature, and provide sensory information about the surrounding environment. This chapter discusses the body's most visible organ system, and the one that often defines our very self-image.

SKIN

The **skin** is an organ because it consists of different tissues that are joined to perform specific activities. It is one of the largest organs of the body in surface area and weight. In adults, the skin covers an area of about 2 square meters (22 square feet) and weighs 4.5 to 5 kg (10 to 11 lb). It ranges in thickness from 0.5 to 4.0 mm (0.02 to 0.16 in.), depending on location. The skin is not just a simple, thin coat that keeps the body together and provides protection. It performs several essential functions that will be described shortly. **Dermatology** (der'-ma-TOL-ō-jē; *dermato* = skin; *logos* = study of) is the medical specialty that deals with diagnosing and treating skin disorders.

Anatomy

Structurally, the skin consists of two principal parts (Figure 4.1). The superficial, thinner portion, which is composed of *epithelial tissue,* is called the **epidermis.** The epidermis is attached to the deeper, thicker, *connective tissue* part called the **dermis.**

Deep to the dermis and not part of the skin is a **subcutaneous (subQ) layer.** This layer, also called the **superficial fascia** or **hypodermis,** consists of areolar and adipose tissues. Fibers from the dermis extend into the subcutaneous layer and anchor the skin to it. The subcutaneous layer, in turn, attaches to underlying tissues and organs.

Epidermis

The **epidermis** (*epi* = above) is composed of keratinized stratified squamous epithelium and contains four principal types of cells (Figure 4.2b and c). About 90% of the epidermal cells are **keratinocytes** (ker-a-TIN-ō-sīts; *kerato* = horny). They produce the protein keratin that helps protect the skin and underlying tissues from light, heat, microbes, and many chemicals.

Melanocytes (MEL-a-nō-sīts; *melan* = black), which produce the pigment melanin, constitute about 8% of the epidermal cells. Their long, slender projections extend between and transfer melanin granules to keratinocytes. **Melanin** is a brown-black pigment that contributes to skin color and absorbs ultraviolet (UV) light. Once inside keratinocytes, the melanin granules cluster to form a protective veil over the nucleus, on the side toward the skin surface. In this way they shield the genetic material from damaging UV light.

The third type of cell in the epidermis is known as a **Langerhans** (LANG-er-hans) **cell.** These cells arise from red bone marrow and migrate to the epidermis, primarily the stratum spinosum. They interact with white blood cells called helper T cells in immune responses and are easily damaged by UV radiation.

A fourth type of cell found in the epidermis is called a **Merkel cell.** These cells are located in the deepest layer (stratum basale) of the epidermis of hairless skin, where they are attached to keratinocytes by anchoring junctions. Merkel cells make contact with the flattened portion of the ending of a sensory neuron (nerve cell), called a **tactile (Merkel) disc,** and are thought to function in the sensation of touch.

Four or five distinct layers of cells form the epidermis. In most regions of the body the epidermis is about 0.1 mm thick and has four layers. Where exposure to friction is greatest, such as in the palms and soles, the epidermis is thicker (1 to 2 mm) and has five layers. Constant exposure of thin or thick skin to friction or pressure stimulates formation of a **callus,** an abnormal thickening of the epidermis.

The names of the five layers (strata), from the deepest to the most superficial, are:

1. **Stratum basale** (*basale* = base) This layer is composed of a single row of cuboidal or columnar-shaped keratinocytes that are capable of continuous cell division. The nuclei of these cells are large, and the cytoplasm contains numerous ribosomes, a small Golgi complex, few mitochondria, and some granular endoplasmic reticulum. The cells of the stratum basale also contain cytoskeletal intermediate filaments composed of the protein keratin. These filaments attach to desmosomes, which bind stratum basale cells to each other and to the cells of

Text continues on page 94

Figure 4.1 Sectional view showing structure of the skin and underlying subcutaneous tissue. The stratum lucidum shown is not apparent on hairy skin but is included so that you can see the relationship of all epidermal strata.

The skin consists of a superficial, thinner epidermis (epithelial tissue) and a deep, thicker dermis (connective tissue).

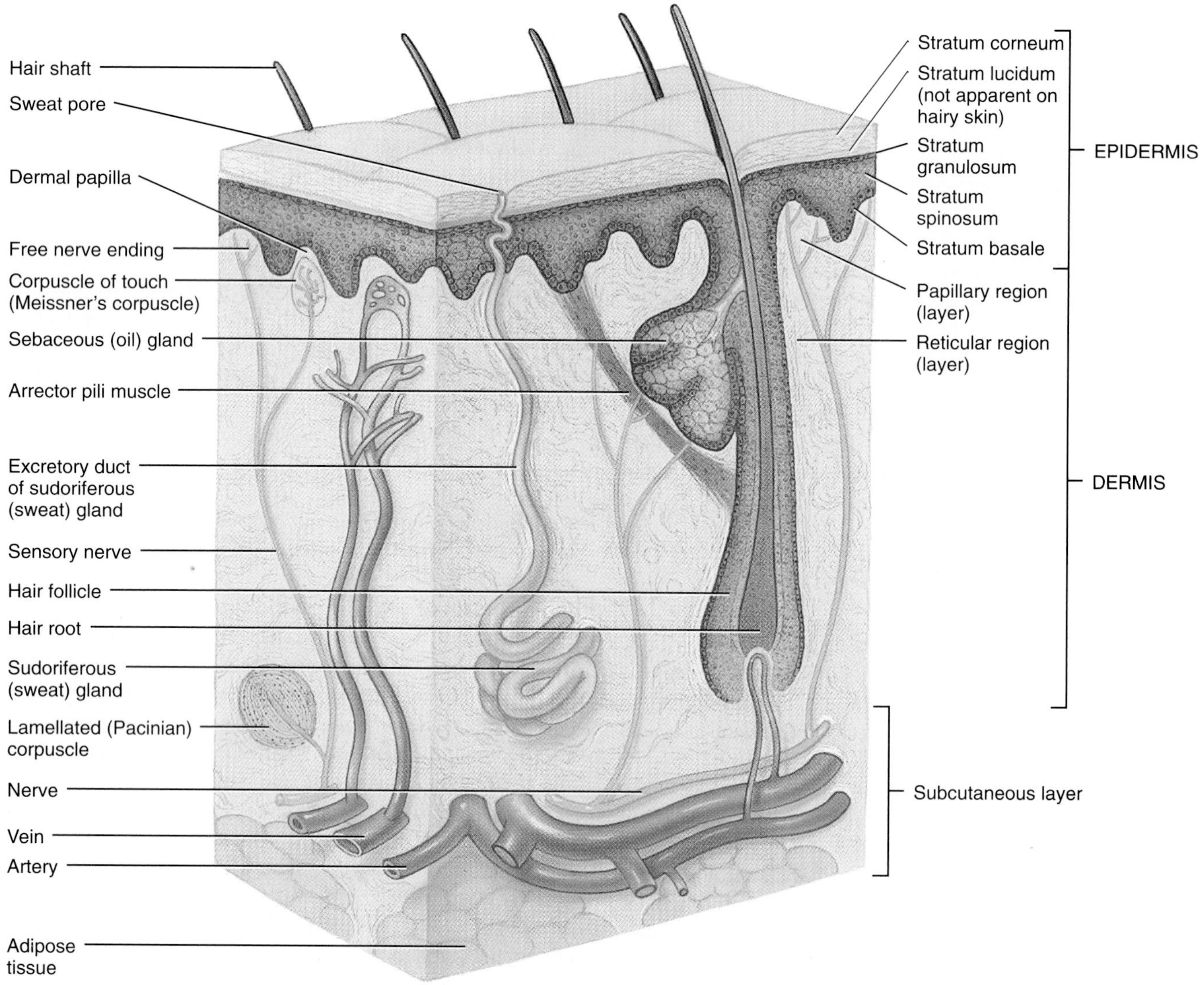

Hair shaft

Sweat pore

Dermal papilla

Free nerve ending

Corpuscle of touch (Meissner's corpuscle)

Sebaceous (oil) gland

Arrector pili muscle

Excretory duct of sudoriferous (sweat) gland

Sensory nerve

Hair follicle

Hair root

Sudoriferous (sweat) gland

Lamellated (Pacinian) corpuscle

Nerve

Vein

Artery

Adipose tissue

Stratum corneum

Stratum lucidum (not apparent on hairy skin)

Stratum granulosum

Stratum spinosum

Stratum basale

Papillary region (layer)

Reticular region (layer)

EPIDERMIS

DERMIS

Subcutaneous layer

Sectional view illustrating the complexity of the skin

OVERVIEW OF FUNCTIONS OF THE SKIN
1. Regulating body temperature.
2. Protecting from physical abrasion, bacteria, dehydration, and ultraviolet radiation.
3. Detecting stimuli related to temperature, touch, pressure, and pain.
4. Removing water, salts, and several organic compounds.
5. Assisting in immune responses (Langerhans cells).
6. Blood reservoir.
7. Beginning the synthesis of vitamin D.

Q *Which stratum of the epidermis is continually shed? Which is capable of prolific cell division?*

Figure 4.2 Cell layers and types in the epidermis of thick skin. Thin skin is similar, but the stratum lucidum is less apparent and the stratum corneum is thinner.

 The epidermis is keratinized stratified squamous epithelium and contains four different types of cells.

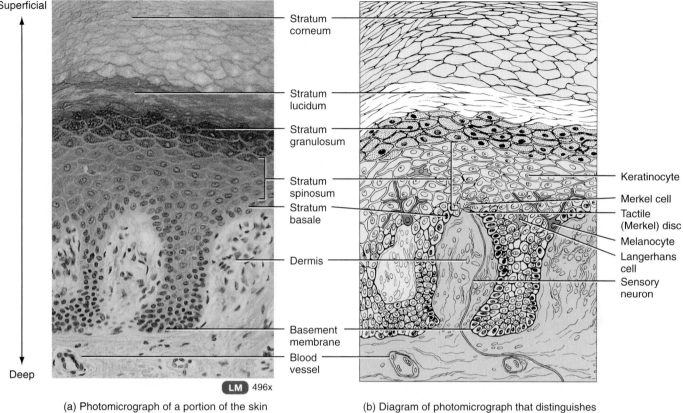

Superficial

Stratum corneum

Stratum lucidum

Stratum granulosum

Stratum spinosum

Stratum basale

Dermis

Basement membrane

Blood vessel

Deep

LM 496x

Keratinocyte

Merkel cell

Tactile (Merkel) disc

Melanocyte

Langerhans cell

Sensory neuron

(a) Photomicrograph of a portion of the skin

(b) Diagram of photomicrograph that distinguishes four principal types of cells

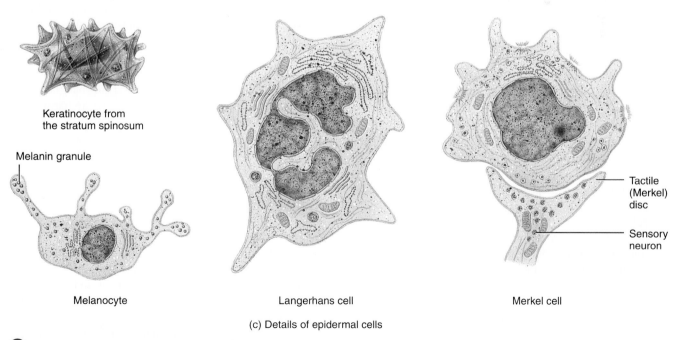

Keratinocyte from the stratum spinosum

Melanin granule

Melanocyte

Langerhans cell

Merkel cell

Tactile (Merkel) disc

Sensory neuron

(c) Details of epidermal cells

Q *Which epidermal cell functions in immunity?*

the next stratum, and to hemidesmosomes, which bind the keratinocytes to the basement membrane. The intermediate filaments also protect deeper layers from injury and microbial invasion. Melanocytes, Langerhans cells, and Merkel cells are scattered among the keratinocytes of the basal layer. The stratum basale is sometimes referred to as the **stratum germinativum** (jer′-mi-na-TĒ-vum; *germ* = sprout) to indicate its role in germinating new cells. The stratum basale also contains tactile (Merkel) discs that are sensitive to touch.

2. **Stratum spinosum** (*spinosum* = thornlike or prickly) This layer of the epidermis contains 8–10 rows of polyhedral (many-sided) keratinocytes that fit closely together. Cells in the more superficial portions of this layer become somewhat flattened. Cells of the stratum spinosum have the same organelles as cells of the stratum basale, and it is believed that some cells in this layer may retain their ability to undergo division. When cells of this tissue layer are prepared for microscopic examination, they shrink apart so that they appear to be covered with thornlike spines, although this isn't the case with living tissue. At each spinelike projection, bundles of intermediate filaments of the cytoskeleton insert into desmosomes, which tightly join the cells to one another. Long projections of the melanocytes extend among the keratinocytes, which take in melanin by phagocytosis of these melanocyte projections. Langerhans cells also appear in this stratum.

3. **Stratum granulosum** (*granulum* = little grain) This layer of the epidermis consists of three to five rows of flattened keratinocytes. Their nuclei and other organelles begin to degenerate, and intermediate filaments become more apparent. A distinctive feature of cells in this layer is the presence of darkly staining nonmembranous granules of a substance called **keratohyalin** (ker′-a-tō-HĪ-a-lin), which helps organize intermediate filaments into even coarser bundles in the superficial rows. Also present in the keratinocytes are membrane-enclosed **lamellar granules.** They produce a lipid-rich secretion that fills the spaces between cells of the stratum granulosum and forms a barrier between cells of this stratum and more superficial cells of the epidermis. The lipid-rich secretion functions as a waterproof sealant, preventing water loss and the entrance of foreign materials. As the nuclei of the cells break down, the cells can no longer carry on vital metabolic reactions and they die. This stratum marks the transition between the deeper, metabolically active strata and the dead cells of the more superficial strata.

4. **Stratum lucidum** (*lucidus* = clear) Typically, this layer is more apparent in the thick skin of the palms and soles. It consists of three to five rows of clear, flat, dead keratinocytes that contain densely packed intermediate filaments and thickened plasma membranes.

5. **Stratum corneum** (*corneum* = horny) This layer consists of 25 to 30 rows of dead, flat keratinocytes. The interior of the cells contains mostly densely packed intermediate filaments embedded in a thick matrix derived from keratohyalin. The cells also contain lipids from lamellar granules that help to waterproof this layer. These cells are continuously shed and replaced by cells from the deeper layers. The stratum corneum serves as an effective waterproofing barrier and also protects against light, heat, bacteria, and many chemicals.

In the process of **keratinization,** cells newly formed in the basal layers undergo a developmental process as they are pushed to the surface. As the cells relocate, they accumulate more and more keratin. At the same time the cytoplasm, nucleus, and other organelles disappear, and the cells die. Eventually, the keratinized cells slough off and are replaced by underlying cells that, in turn, become keratinized. The whole process by which a cell forms in the basal layer, rises to the surface, becomes keratinized, and sloughs off takes two to four weeks depending on age and the area of the body.

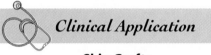

Clinical Application

Skin Grafts

Sometimes, the germinal portion of epidermis is destroyed, and new skin cannot regenerate. Such wounds require **skin grafts.** The most successful type of skin graft involves the transplantation of a segment of skin from a donor site to a recipient site of the same individual (*autograft*) or an identical twin (*isograft*).

If skin loss is so extensive that conventional grafting is impossible, a self-donation procedure called **autologous** (aw-TOL-ō-gus) **skin transplantation** may be employed. In this procedure, used most often on severely burned patients, small amounts of an individual's epidermis are removed and grown in the laboratory to produce thin sheets of skin. Then, they are transplanted back to the patient where they adhere to burn wounds and generate a permanent skin to cover burned areas. The process normally takes three to four weeks. ▨

Dermis

The second principal part of the skin, the **dermis** (*derm* = skin), is composed of connective tissue containing collagen and elastic fibers (see Figure 4.1). The few cells in the dermis include fibroblasts, macrophages, and adipocytes. The dermis is very thick in the palms and soles and very thin in the eyelids, penis, and scrotum. It also tends to be thicker on the posterior than on the anterior aspects of the body and thicker on the lateral than on the medial aspects of the limbs. Blood vessels, nerves, glands, and hair follicles are embedded in the dermis.

Papillary Region The superficial portion of the dermis, about one-fifth of the thickness of the total layer, is named the **papillary region (layer).** It consists of areolar connective tissue containing fine elastic fibers. Its surface area is greatly increased by small, fingerlike projections called **dermal papillae** (pa-PIL-ē; *papilla* = nipple). These nipple-shaped structures indent the epidermis, and many contain loops of capillaries. Some dermal papillae also contain tactile receptors called **corpuscles of touch (Meissner's corpuscles),** nerve endings that are sensitive to touch. Dermal papillae cause ridges in the overlying epidermis. It is secretions produced on the tops of these ridges that leave fingerprints on objects that are handled.

Reticular Region The deeper portion of the dermis is called the **reticular** (*rete* = net) **region (layer).** It consists of dense, irregular connective tissue that contains interlacing bundles of collagen and some coarse elastic fibers. Within the reticular region, bundles of collagen fibers interlace in a netlike manner. The spaces between the fibers are occupied by a small quantity of adipose tissue, hair follicles, nerves, oil glands, and the ducts of sweat glands. Varying thicknesses of the reticular region contribute to the differences in the thickness of skin.

The combination of collagen and elastic fibers in the reticular region provides the skin with strength, extensibility (ability to stretch), and elasticity (ability to return to original shape after stretching). The ability of the skin to stretch can readily be seen in pregnancy, obesity, and edema. Small tears that occur in the dermis during extreme stretching are initially red and remain visible afterward as silvery white streaks called *striae* (STRĪ-ē; *stria* = streak), or stretch marks.

The reticular region is attached to underlying organs, such as bone and muscle, by the **subcutaneous layer,** also called the **hypodermis** or **superficial fascia.** In addition to areolar connective tissue and adipose tissue, the subcutaneous layer also contains nerve endings called **lamellated** or **Pacinian** (pa-SIN-ē-an) **corpuscles** that are sensitive to pressure (see Figure 19.1). Nerve endings sensitive to cold are found in and just deep to the dermis, whereas those sensitive to heat are located in the intermediate and superficial dermis.

Clinical Application

Lines of Cleavage and Surgery

In certain regions of the body, collagen fibers tend to orient more in one direction than another. **Lines of cleavage (tension lines)** in the skin indicate the predominant direction of underlying collagen fibers (Figure 4.3). The lines are especially evident on the

Figure 4.3 Lines of cleavage.

Lines of cleavage in the skin indicate the predominant direction of underlying collagen fibers in the reticular region.

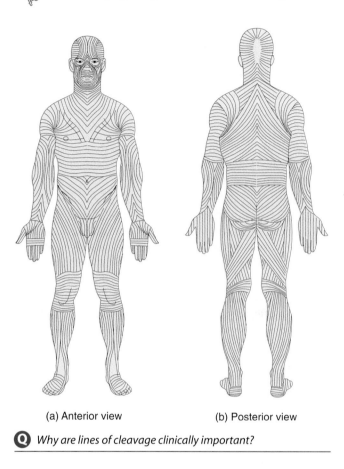

(a) Anterior view　　　　(b) Posterior view

Q *Why are lines of cleavage clinically important?*

palmar surfaces of the fingers, where they are arranged parallel to the long axis of the digits. An incision running parallel to the collagen fibers will heal with only a fine scar. An incision made across the rows of fibers disrupts the collagen, and the wound tends to gape open and heal in a broad, thick scar.

Functions

Among the numerous functions of the integumentary system (mainly the skin) are the following:

1. **Regulation of body temperature.** In response to high environmental temperature or strenuous exercise, the evaporation of sweat from the skin surface helps lower an elevated body temperature to normal. In response to low environmental temperature, production of sweat is decreased, which helps conserve heat. Sweat is produced by sudoriferous (sweat) glands, whose structure will be described shortly. Changes in the flow of blood to the skin also help regulate body temperature.

2. **Protection.** The skin covers the body and provides a physical barrier that protects underlying tissues from physical abrasion, bacterial invasion, dehydration, and ultraviolet (UV) radiation. Hair and nails also have protective functions, as described shortly.

3. **Sensation.** The skin contains abundant nerve endings and receptors that detect stimuli related to temperature, touch, pressure, and pain.

4. **Excretion.** Besides removing heat and some water from the body, sweat also is the vehicle for excretion of a small amount of salts and several organic compounds.

5. **Immunity.** Certain cells of the epidermis (Langerhans cells) are important components of the immune system, which fends off foreign invaders.

6. **Blood reservoir.** The dermis houses extensive networks of blood vessels that carry 8% to 10% of the total blood flow in a resting adult. In moderate exercise, skin blood flow may increase, which helps dissipate heat from the body. During strenuous exercise, however, skin blood vessels constrict (narrow) somewhat, and more blood is able to circulate to contracting muscles.

7. **Synthesis of vitamin D.** What is commonly called vitamin D is actually a group of closely related compounds. Synthesis of vitamin D begins with activation of a precursor molecule in the skin by ultraviolet (UV) rays in sunlight. Enzymes in the liver and kidneys then modify the molecule, finally producing vitamin D. This vitamin aids in the absorption of calcium in foods from the digestive system into the blood.

Skin Color

Three pigments—melanin, carotene, and hemoglobin—give skin a wide variety of colors. Melanin is located mostly in the epidermis. The amount of **melanin** varies the skin color from pale yellow to black. Melanocytes are most plentiful in the mucous membranes, penis, nipples of the breasts and the area just around the nipples (areola), face, and limbs. The *number* of melanocytes is about the same in all races. Differences in skin color are due mainly to the *amount of pigment* the melanocytes produce and disperse to keratinocytes. In some people, melanin tends to accumulate in patches called **freckles.** As one grows older, **liver (age) spots** may develop. These are flat skin patches that look like freckles and range in color from light brown to black. Like freckles, liver spots are accumulations of melanin, and neither tends to become cancerous.

Exposure of the skin to UV radiation leads to increased melanin production. Both the amount and the darkness of melanin increase, which tans the skin and further protects the body against UV radiation. Thus, within limits, melanin serves a protective function. However, as you will see later, excessive skin exposure to UV radiation can lead to skin cancer. A tan is lost when the increased melanin is lost in keratinocytes as they are shed.

An inherited inability of an individual of any race to produce melanin results in **albinism** (AL-bin-izm; *albus* = white). Most **albinos** (al-BĪ-nōs), individuals affected by albinism, have melanocytes, but the cells are unable to synthesize melanin. Melanin is thus absent in hair, eyes, and skin. In another condition, called **vitiligo** (vit-i-LĪ-gō), the partial or complete loss of melanocytes from patches of skin produces irregular white spots.

Carotene (KAR-o-tēn; *karaton* = carrot), a yellow-orange pigment, gives carrots and egg yolks their color. It is the precursor of vitamin A, which is used to synthesize pigments needed for vision. Carotene is found in the stratum corneum and fatty areas of the dermis and subcutaneous layer.

When little melanin or carotene are present, the epidermis is translucent. The epidermis itself has no blood vessels, a characteristic of all epithelia. Thus, the skin of Caucasians appears pink to red, depending on the amount and quality of the blood moving through capillaries in the dermis. The red color is due to **hemoglobin,** the pigment in red blood cells that carries oxygen in blood.

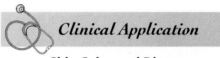

Clinical Application

Skin Color and Disease

The color of skin and mucous membranes can provide clues for diagnosing certain problems. When blood is not picking up an adequate amount of oxygen in the lungs, such as in a baby who has stopped breathing, mucous membranes, nail beds, and light-colored skin appear bluish or **cyanotic** (sī-an-OT-ic; *kyanos* = blue). This is because hemoglobin that is depleted of oxygen looks deep, purplish blue. **Jaundice** (JON-dis; *jaune* = yellow), a yellowed appearance of the whites of the eyes and of light-colored skin, usually indicates liver disease. It is due to a buildup of bilirubin, a yellow pigment found in bile. In **Addison's disease** there is increased skin pigmentation due to excessive secretion of hormones that stimulate melanocytes. **Erythema** (er-e-THĒ-ma; *erythros* = red), redness of the skin, is caused by engorgement of capillaries in the dermis with blood. In people who have light-colored skin, exercise and embarrassment may cause noticeable erythema, especially of the face. Exercise and embarrassment produce the same, though often unobservable, reaction in dark-colored skin. Erythema also occurs with skin injury, infection, inflammation, or allergic reactions.

Epidermal Ridges

The superficial surface of the skin of the palms, fingers, soles, and toes is marked by a series of ridges and grooves. They appear either as straight lines or as a pattern of loops and whorls, as on the tips of the digits. **Epidermal ridges** develop during the third and fourth fetal months as the epidermis conforms to the contours of the underlying dermal papillae. The function of the ridges is to increase the grip of the hand or foot by increasing friction and acting like tiny suction cups. Because the ducts of sweat glands open on the tops of the epidermal ridges as sweat pores, the sweat and ridges form fingerprints (or footprints) when a smooth object is touched. The ridge pattern, which is genetically determined, is unique for each individual. Normally, it does not change throughout life, except to enlarge, and thus can serve as identification through fingerprints or footprints.

EPIDERMAL DERIVATIVES

Organs that develop from the embryonic epidermis—hair, glands, nails—have a host of important functions. Hair and nails protect the body. The sweat glands help regulate body temperature. Tooth enamel is also an epidermal derivative.

Hair

Hairs, or *pili* (PI-lē), are growths of the epidermis variously distributed over the body. One function of hairs is protection. Although the protection is limited, hair on the head guards the scalp from injury and the sun's rays. It also decreases heat loss. Eyebrows and eyelashes protect the eyes from foreign particles. Hair in the nostrils protects against inhaling insects and foreign particles. Hair serves a similar protective function in the external ear canal. Touch receptors associated with hair follicles (hair root plexuses) are activated whenever a hair is even slightly moved. Thus, hairs function in sensing light touch. Hairs also aid in visual identification and stimulation, and in pheromone dispersal. **Pheromones** (FER-ō-mōns) are chemical substances secreted by organisms for communication with members of their own species. They are important in attracting the opposite sex and in sex recognition in many species. The surfaces of hairs are coated with sebum, an oily substance that contains pheromones.

Normal hair loss in an adult scalp is about 70 to 100 hairs per day. Both the rate of growth and the replacement cycle may be altered by illness, diet, high fever, surgery, blood loss, or severe emotional stress. Rapid weight-loss diets that severely restrict calories or protein increase hair loss. An increase in the rate of shedding can also occur for three to four months after childbirth and with certain drugs and radiation therapy for cancer.

Anatomy

A hair is composed of columns of dead, keratinized cells welded together. The **shaft** is the superficial portion of the hair that projects from the surface of the skin (Figure 4.4a).

The shaft of straight hair is round in cross section, that of wavy hair is oval, and that of woolly hair is elliptical or kidney-shaped. The **root** is the portion of the hair deep to the surface that penetrates into the dermis and sometimes into the subcutaneous layer (see Figure 4.1). The shaft and root both consist of three concentric layers (Figure 4.4a). The inner *medulla* is composed of two or three rows of polyhedral cells containing pigment granules and air spaces. The middle *cortex* forms the major part of the shaft and consists of elongated cells that contain pigment granules in dark hair but mostly air in white hair. The *cuticle of the hair,* the outermost layer, consists of a single layer of thin, flat cells that are the most heavily keratinized. Cuticle cells are arranged like shingles on the side of a house, with their free edges pointing toward the end of the hair (Figure 4.4b).

Surrounding the root of the hair is the **hair follicle,** which is made up of an external root sheath and an internal root sheath. The *external root sheath* is a downward continuation of the epidermis. Near the surface, it contains all the epidermal layers. At the base of the hair follicle, the external root sheath contains only the stratum basale. The *internal root sheath* forms a cellular tubular sheath between the external root sheath and the hair.

At the base of each hair follicle is an onion-shaped structure, the *bulb.* This structure houses a nipple-shaped indentation, the *papilla of the hair,* which contains areolar connective tissue. The papilla of the hair also contains many blood vessels and provides nourishment for the growing hair. The bulb also contains a region of cells called the *matrix,* which is the germinal layer. The cells of the matrix derive from the stratum basale. They are responsible for the growth of existing hairs and produce new hairs by cell division when older hairs are shed. This replacement occurs within the same follicle. Matrix cells also give rise to the cells of the internal root sheath.

Clinical Application

Hair Removal

A substance that removes superfluous hair is called a **depilatory.** It dissolves the protein in the hair shaft, turning it into a gelatinous mass that can be wiped away. Because the hair root is not affected, regrowth of the hair occurs. In **electrolysis,** the hair matrix is destroyed by an electric current so that the hair cannot regrow. ■

Sebaceous (oil) glands and a bundle of smooth muscle cells are also associated with hairs. Details of the sebaceous glands will be discussed shortly. The smooth muscle is called **arrector** (*arrector* = to raise) **pili** (plural is *arrectores pilorum*); it extends from the superficial dermis of the skin to the side of the hair follicle. In its normal position, hair

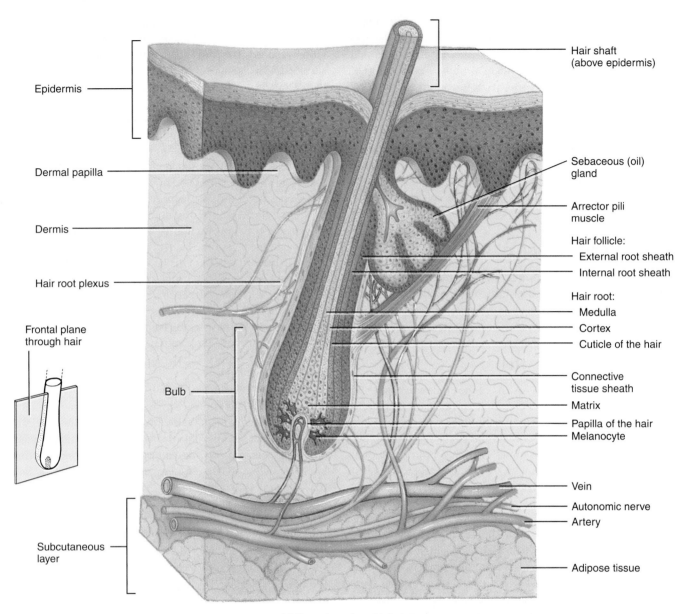

Epidermis

Dermal papilla

Dermis

Hair root plexus

Frontal plane
through hair

Bulb

Subcutaneous
layer

Hair shaft
(above epidermis)

Sebaceous (oil)
gland

Arrector pili
muscle

Hair follicle:
External root sheath
Internal root sheath

Hair root:
Medulla
Cortex
Cuticle of the hair

Connective
tissue sheath
Matrix
Papilla of the hair
Melanocyte

Vein
Autonomic nerve
Artery

Adipose tissue

(a) Frontal section of a hair

Figure 4.4 Hair. Shown here is the relationship of a hair to the epidermis, sebaceous glands, and arrector pili muscle. See also Figure 4.1.

Hairs are growths of epidermis composed of dead, keratinized cells.

Q *Why does it hurt when you pluck a hair out but not when you have a haircut?*

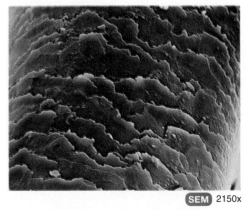

SEM 2150x

(b) Surface of a hair shaft showing the shinglelike cuticular scales emerges at an angle to the surface of the skin. Under physiologic or emotional stress, such as cold or fright, autonomic nerve endings along the arrector pili muscles stimulate them to contract, which pulls the hair shafts into a vertical position. This causes "goose bumps" or "gooseflesh" because the skin around the shaft forms slight elevations.

Around each hair follicle are nerve endings, called **hair root plexuses,** that are sensitive to touch. They respond if a hair shaft is moved.

Growth

Each hair follicle goes through a *growth cycle,* which consists of a *growth stage* and a *resting stage.* During the growth stage, a hair is formed by cells of the matrix that differentiate, become keratinized, and die. As new cells are added at the base of the hair root, the hair grows longer. In time, the growth of the hair stops and the resting stage begins. After the resting stage, a new growth cycle begins in which a new hair replaces the old hair and the old hair is pushed out of the hair follicle. In general, scalp hair grows for about three years and rests for about one to two years. At any given time, most hair is in the growth stage.

Color

The color of hair is due primarily to melanin. It is synthesized by melanocytes scattered in the matrix of the bulb and passes into cells of the cortex and medulla (Figure 4.4a). Dark-colored hair contains mostly true melanin. Blond and red hair contain variants of melanin in which there is iron and more sulfur. Gray hair occurs with a progressive decline in tyrosinase, an enzyme necessary for synthesis of melanin. White hair results from accumulation of air bubbles in the medullary shaft.

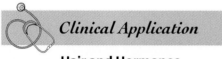

Clinical Application

Hair and Hormones

At puberty, when their testes begin secreting significant quantities of androgens (male sex hormones), males develop the typical male pattern of hair growth, including a beard and a hairy chest. In females, both the ovaries and the adrenal glands produce small quantities of androgens. Surprisingly, androgens also must be present for the most common form of baldness, **male-pattern baldness,** to occur. In genetically predisposed males, androgens somehow inhibit hair growth.

Occasionally, a tumor of one of these glands oversecretes androgens and causes **hirsutism** (hur-SOO-tiz-um; *hirsutus* = shaggy). This is a condition of excessive hairiness of the upper lip, chin, chest, inner thighs, and abdomen in females or prepubertal males. ▦

Glands

Several kinds of glands are associated with the skin: sebaceous (oil) glands, sudoriferous (sweat) glands, ceruminous glands, and mammary glands. The anatomy and physiology of the mammary glands, which are modified sudoriferous glands, will be discussed in Chapter 26 as components of the female reproductive system.

Sebaceous (Oil) Glands

Sebaceous (se-BĀ-shus; *sebo* = grease) or **oil glands,** with few exceptions, are connected to hair follicles (see Figures 4.1 and 4.4a). The secreting portions of the glands lie in the dermis and open into the necks of hair follicles or directly onto a skin surface (lips, glans penis, labia minora, and tarsal glands of the eyelids). Absent in the palms and soles, sebaceous glands vary in size and shape in other regions of the body. For example, they are small in most areas of the trunk and limbs, but large in the skin of the breasts, face, neck, and upper chest.

Sebaceous glands secrete an oily substance called **sebum** (SĒ-bum), which is a mixture of fats, cholesterol, proteins, inorganic salts, and pheromones. Sebum coats the surface of hairs and helps keep them from drying and becoming brittle. Sebum also prevents excessive evaporation of water from the skin, keeps the skin soft and pliable, and inhibits the growth of certain bacteria. When sebaceous glands of the face become enlarged because of accumulated sebum, *blackheads* develop. The color of blackheads is due to melanin and oxidized oil, not dirt. Because sebum is nutritive to certain bacteria despite its antibacterial properties, inflammatory reactions such as pimples or boils may occur.

Sudoriferous (Sweat) Glands

Three to four million **sudoriferous** (soo'-dor-IF-er-us; *sudor* = sweat; *ferre* = to bear) or **sweat glands** empty their secretions onto the skin surface. They are divided into two principal types, eccrine and merocrine, based on their structure, location, and type of secretion.

Eccrine (*ekkrinein* = to secrete) **sweat glands** are much more common than apocrine sweat glands. They are distributed throughout the skin except for the margins of the lips, nail beds of the fingers and toes, glans penis, glans clitoris, labia minora, and eardrums. Eccrine sweat glands are most numerous in the skin of the palms and the soles; their density can be as high as 450 per square centimeter (3000 per square inch) in the palms. The secretory portion of eccrine sweat glands is located in the subcutaneous layer, and the excretory duct projects into the dermis and epidermis. It ends as a pore at the surface of the epidermis (see Figure 4.1).

Merocrine (*meros* = a part) **sweat glands,** once believed to be apocrine glands, are found mainly in the skin of the axilla (armpit), pubic region, and areolae (pigmented areas around the nipples) of the breasts. The secretory portion of these sweat glands is located in the dermis or subcutaneous layer, and the excretory duct opens into hair follicles. Merocrine sweat glands begin to function at puberty and produce a more viscous secretion than eccrine sweat glands. They also produce pheromones. They are stimulated during emotional stresses and sexual excitement, and the secretions are commonly known as a "cold sweat."

Sweat (perspiration) is the fluid produced by sweat glands, mainly eccrine sweat glands because they are so much more numerous. Sweat is a mixture of water, salt, urea, uric acid, amino acids, ammonia, glucose, and lactic acid. Its principal function is to help regulate body temperature by providing a cooling mechanism. As sweat evaporates, large quantities of heat energy leave the body surface. It also plays a small role in eliminating wastes.

Ceruminous Glands

In the ear, modified sweat glands called **ceruminous** (se-ROO-mi-nus; *cera* = wax) **glands** produce a waxy secretion. The secretory portions of the ceruminous glands lie in the subcutaneous layer, deep to sebaceous glands. Their excretory ducts open either directly onto the surface of the external auditory canal (ear canal) or into ducts of sebaceous glands. The combined secretion of the ceruminous and sebaceous glands is called **cerumen.** Cerumen, together with hairs in the external auditory canal, provides a sticky barrier that prevents the entrance of foreign bodies.

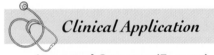

Clinical Application

Impacted Cerumen (Earwax)

Some people produce an abnormal amount of cerumen, or earwax, in the external auditory canal. It then becomes impacted and prevents sound waves from reaching the eardrum. The treatment for **impacted cerumen** is usually periodic ear irrigation or removal of the wax with a blunt instrument by trained medical personnel. The use of Q-tips is not recommended for this purpose since they may push the cerumen further into the external auditory canal. ■

Nails

Nails are plates of tightly packed, hard, keratinized cells of the epidermis. The cells form a clear, solid covering over the dorsal surfaces of the terminal portions of the fingers and toes. Each nail consists of a nail body, a free edge, and a nail root (Figure 4.5). The **nail body** is the portion of the nail that is visible, the **free edge** is the part that may extend past the distal end of the digit, and the **nail root** is the portion that is buried in a fold of skin. Most of the nail body is pink because of blood flowing through underlying capillaries. The free

edge is white because there are no underlying capillaries. The whitish semilunar area of the proximal end of the body is called the **lunula** (LOO-nyoo-la; *lunula* = little moon). It appears whitish because the vascular tissue underneath does not show through owing to the thickened stratum basale in the area. The **eponychium** (ep′-ō-NIK-ē-um; *onyx* = nail) or *cuticle* is a narrow band of epidermis that extends from the margin of the nail wall (lateral border), adhering to it. It occupies the proximal border of the nail and consists of stratum corneum.

The epithelium deep to the nail root is known as the **nail matrix.** Its function is to bring about the growth of nails. Growth occurs by the transformation of superficial cells of the matrix into nail cells. In the process, the outer, harder layer is pushed forward over the stratum basale. The average growth in the length of fingernails is about 1 mm (0.04 in.) per week. The growth rate is somewhat slower in toenails. The longer the digit, the faster the nail grows. Adding supplements such as gelatin to an otherwise healthy diet has no effect on making nails grow faster or stronger.

Functionally, nails help us grasp and manipulate small objects in various ways, provide protection against trauma to the ends of the digits, and help us scratch various parts of the body.

BLOOD SUPPLY OF THE INTEGUMENTARY SYSTEM

Although the epidermis is avascular, the dermis is well supplied with blood (see Figure 4.1). The arteries supplying the dermis are generally derived from branches of arteries supplying skeletal muscles in a particular region. Some arteries supply the skin directly. One plexus (network) of arteries, the **cutaneous plexus,** is located at the junction of the dermis and subcutaneous layer and sends branches that supply the sebaceous (oil) and sudoriferous (sweat) glands, the deep portions of hair follicles, and adipose tissue. The **papillary plexus,** formed at the level of the papillary region, sends branches that supply the capillary loops in the dermal papillae, sebaceous (oil) glands, and superficial portions of hair follicles. The arterial plexuses are accompanied by venous plexuses that drain blood from the dermis into larger subcutaneous veins.

Developmental Anatomy of the Integumentary System

Throughout this book, we will discuss the developmental anatomy of systems at the end of each discussion of the system. Because the principal features of embryonic development are not treated in detail until Chapter 27, it is necessary

Figure 4.5 Structure of nails. Shown is a fingernail.

Nail cells arise by transformation of superficial cells of the nail matrix into nail cells.

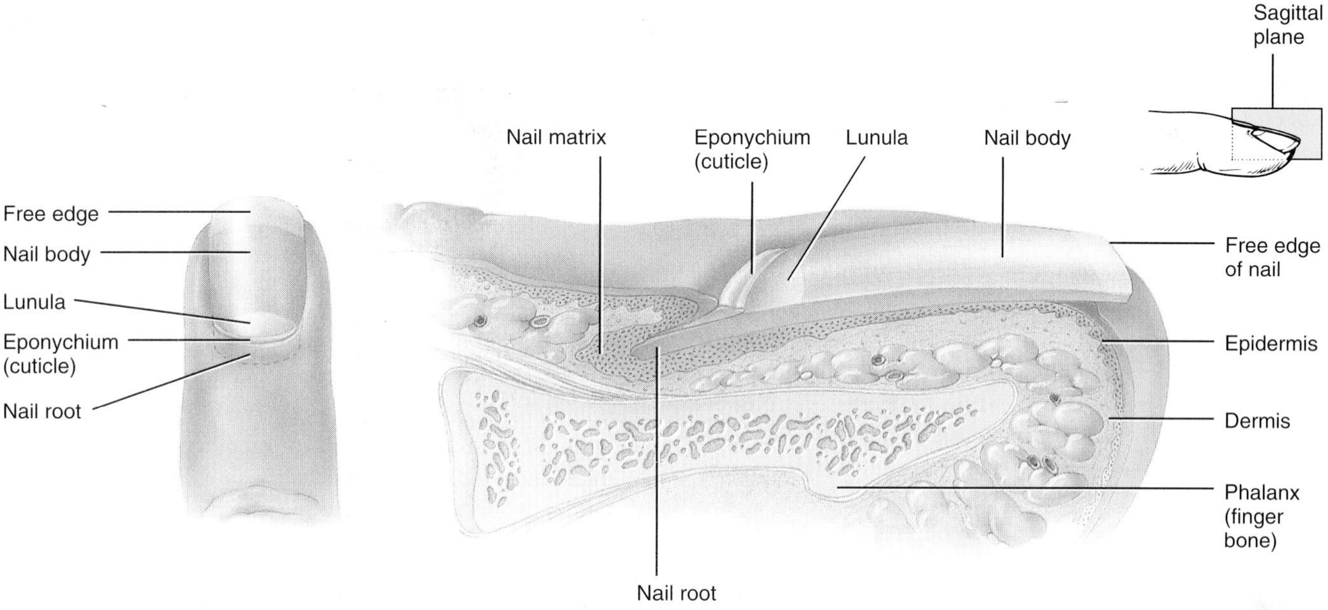

(a) Viewed from above

(b) Sagittal section showing internal detail

Q *Why are nails so hard?*

to explain and review a few terms at this point so that you can follow the development of organ systems.

As part of the early development of a fertilized egg, a portion of the developing embryo differentiates into three layers of tissue called **primary germ layers.** On the basis of position, the primary germ layers are referred to as **ectoderm** (*ecto* = outside), **mesoderm** (*meso* = middle), and **endoderm** (*endo* = within). They are the embryonic tissues from which all tissues and organs of the body will eventually develop (see Exhibit 27.1).

The *epidermis* is derived from the **ectoderm.** At the beginning of the eighth week the ectoderm consists of simple cuboidal epithelium. These cells become flattened and are known as the **periderm.** By the fourth month all layers of the epidermis are formed and each layer assumes its characteristic structure.

The *dermis* is derived from **mesodermal cells** in a zone beneath the ectoderm. There they undergo a process that changes them into the connective tissues that begin to form the dermis at about 11 weeks.

Nails are developed at about 10 weeks. Initially, they consist of a thick layer of epithelium called the **primary nail field.** The nail itself is keratinized epithelium and grows distally from its base. It is not until the ninth month that the nails actually reach the tips of the digits.

Hair follicles develop between the ninth and twelfth weeks as ingrowths of the stratum basale of the epidermis into the deeper dermis. The ingrowths soon differentiate into the bulb, papilla of the hair, beginnings of the epithelial portions of sebaceous glands, and other structures associated with hair follicles. By the fifth or sixth month, the follicles produce **lanugo** (delicate fetal hair), first on the head and then on other parts of the body. The lanugo is usually shed prior to birth.

The epithelial (secretory) portions of *sebaceous (oil) glands* develop from the sides of the hair follicles at about 16 weeks and remain connected to the follicles.

The epithelial portions of *sudoriferous (sweat) glands* are also derived from ingrowths of the stratum basale of the epidermis into the dermis. They appear at about 20 weeks on the palms and soles and a little later in other regions. The connective tissue and blood vessels associated with the glands develop from **mesoderm.** ■

Aging and the Integumentary System

At about the sixth month of fetal development, secretions from sebaceous (oil) glands mix with sloughed off epidermal cells and hairs to form a fatty substance called **vernix caseosa** (VER-niks kā-sē-Ō-sa; *vernix* = varnish; *caseus* = cheese). This substance covers and protects the skin of the fetus from the constant exposure to amniotic fluid in which it is bathed. In addition, the vernix caseosa facilitates the birth of the fetus because of its slippery nature.

For most infants and children, relatively few problems are encountered with the skin as it ages. With the arrival of adolescence, some individual develop acne.

The pronounced effects of aging do not become noticeable until people reach their late forties. Around that time, collagen fibers in the dermis begin to decrease in number, stiffen, break apart, and form into a shapeless, matted tangle. Elastic fibers lose some of their elasticity, thicken into clumps, and fray. Fibroblasts, which produce both collagen and elastic fibers, decrease in number. As a result, the skin forms crevices and furrows known as wrinkles.

With further aging, Langerhans cells dwindle in number and macrophages become less efficient phagocytes, thus decreasing the skin's immune responsiveness. Moreover, decreased size of sebaceous (oil) glands leads to dry and broken skin that is more susceptible to infection. Production of sweat diminishes, which probably contributes to the increased incidence of heat stroke in the elderly. The hair and nails grow more slowly. There is a decrease in the number of functioning melanocytes, resulting in gray hair and atypical skin pigmentation. An increase in the size of some melanocytes produces pigmented blotching (liver spots). Blood vessels in the dermis become thicker walled and less permeable, and subcutaneous fat is lost. For this reason, aged skin is thinner than young skin, especially the dermis, and the migration of cells from the basal layer to the epidermal surface slows considerably.

With the onset of old age, skin heals poorly and becomes more susceptible to pathological conditions such as skin cancer, itching, and pressure sores. The ultraviolet (UV) rays in sunlight can damage the DNA in epidermal cells and the extracellular matrix materials, such as collagen and elastic fibers, in the dermis. Over the long term, chronic UV exposure accelerates the aging of skin (photoaging) and is an important factor in development of nearly all skin cancers (see Applications to Health, p. 3). Protecting your skin from UV rays is a wise approach.

Key Medical Terms
Associated with the Integumentary System

Abrasion (a-BRĀ-shun; *ab* = away; *rasion* = scraped) A portion of the skin that has been scraped away.

Athlete's foot A superficial fungus infection of the skin of the foot.

Chicken pox Highly contagious disease that begins in the respiratory system, caused by the varicella-zoster virus and characterized by vesicular eruptions that fill with pus, rupture, and form a scab before healing. Also called *varicella* (var′-i-SEL-a). Shingles is a disease caused by reactivation of latent chicken pox viruses.

Cold sore A lesion, usually in oral mucous membrane, caused by type 1 herpes simplex virus (HSV), transmitted by oral or respiratory routes. Triggering factors include UV radiation, hormonal changes, and emotional stress. Also called a *fever blister.*

Contact dermatitis (der-ma-TĪ-tis; *derma* = skin; *itis* = inflammation of) Inflammation of the skin characterized by redness, itching, and swelling and caused by contact with chemicals (allergens) that bring about an allergic reaction, such as poison ivy.

Contusion (kon-TOO-shun; *contundere* = to bruise) Condition in which tissue deep to the skin is damaged, but the epidermis is not broken.

Corn A painful conical thickening of the stratum corneum of the epidermis found principally over toe joints and between the toes and often caused by pressure. It may be hard or soft, depending on the location. Hard corns are usually found over toe joints, and soft corns are usually found between the fourth and fifth toes.

Cyst (SIST; *cyst* = sac containing fluid) A sac with a distinct connective tissue wall, containing a fluid or other material.

German measles Highly contagious disease that begins in the respiratory system, caused by the rubella virus and characterized by a rash of small red spots on the skin. Also called **rubella** (roo-BEL-a).

Hemangioma (hē-man′-jē-Ō-ma; *hemo* = blood; *angio* = blood vessel; *oma* = tumor) Localized tumor of the skin and subcutaneous layer that results from an abnormal increase in blood vessels; one type is a *portwine stain,* a flat, pink, red, or purple lesion present at birth, usually at the nape of the neck.

Hives (HĪVZ) Condition of the skin marked by reddened elevated patches that are often itchy. Most commonly caused by infections, physical trauma, medications, emotional stress, food additives, and certain foods. Also called *urticaria* (yoor-ti-KAR-ē-a).

Impetigo (im′-pe-TĪ-gō) Superficial skin infection caused by staphylococci or streptococci; most common in children.

Intradermal (in′-tra-DER-mal; *intra* = within) Within the skin. Also called *intracutaneous.*

Keloid (KĒ-loid; *kelis* = tumor) An elevated, irregular darkened area of excess scar tissue caused by collagen formation during healing. It extends beyond the original injury and is tender and frequently painful. It occurs in the dermis and underlying subcutaneous tissue, usually after trauma, surgery, a burn, or severe acne; more common in people of African descent.

Laceration (las′-er-Ā-shun; *lacerare* = to tear) Wound or irregular tear of the skin.

Measles Highly contagious disease caused by the measles virus that begins in the respiratory system and is characterized by a papular rash on the skin. Also called *rubeola* (roo-bē-Ō-la).

Nevus (NE-vus) A round, pigmented, flat, or raised skin area that may be present at birth or develop later. Varies in color from yellow-brown to black. Also called a *mole* or *birthmark.*

Pruritus (proo-RĪ-tus; *pruire* = to itch) Itching, one of the most common dermatological disorders. It may be caused by skin disorders (infections), systemic disorders (cancer, kidney failure), or psychogenic factors (emotional stress).

Topical (TOP-i-kal) Pertaining to a definite area; local. Also in reference to a medication, applied to the surface rather than ingested or injected.

Wart (WORT) Mass produced by uncontrolled growth of epithelial skin cells; caused by a virus (papilloma virus). Most warts are noncancerous.

Study Outline

Skin (p. 91)

1. The skin and the organs derived from it (hair, glands, and nails) constitute the integumentary system.
2. The skin is one of the largest organs of the body. The principal parts of the skin are the superficial epidermis and deeper dermis. The dermis overlies the subcutaneous layer.
3. The epidermis consists of keratinocytes, melanocytes, Langerhans cells, and Merkel cells.
4. The epidermal layers, from deepest to most superficial, are the stratum basale, spinosum, granulosum, lucidum, and corneum. The stratum basale undergoes continuous cell division and produces all other layers.
5. The dermis consists of a papillary region and a reticular region. The papillary region is composed of areolar connective tissue containing fine elastic fibers, dermal papillae, and corpuscles of touch (Meissner's corpuscles). The reticular region is composed of dense irregular connective tissue containing randomly scattered collagen fibers and some elastic fibers, adipose tissue, hair follicles, nerves, sebaceous (oil) glands, and ducts of sudoriferous (sweat) glands.
6. Following are the functions of the skin: regulation of body temperature, protection, sensation, excretion, immunity, blood reservoir, and synthesis of vitamin D.
7. The color of skin is due to melanin, carotene, and hemoglobin.
8. Epidermal ridges increase friction for better grasping ability and provide the basis for fingerprints and footprints.

Epidermal Derivatives (p. 97)

1. Epidermal derivatives are structures developed from the embryonic epidermis.

2. Among the epidermal derivatives are hair, skin glands (sebaceous, sudoriferous, and ceruminous), and nails.

Hair (p. 97)

1. Hairs are epidermal growths that function in protection.
2. Hair consists of a shaft superficial to the surface, a root that penetrates the dermis and subcutaneous layer, and a hair follicle.
3. Associated with hairs are sebaceous (oil) glands, arrectores pilorum muscles, and hair root plexuses.
4. New hairs develop from cell division of the matrix in the bulb; hair replacement and growth occur in a cyclic pattern.

Glands (p. 99)

1. Sebaceous (oil) glands are usually connected to hair follicles; they are absent in the palms and soles. Sebaceous glands produce sebum, which moistens hairs and waterproofs the skin. Enlarged sebaceous glands may produce blackheads, pimples, and boils.
2. Sudoriferous (sweat) glands are divided into eccrine and merocrine (formerly thought to be apocrine). Eccrine sweat glands have an extensive distribution; their ducts terminate at pores at the surface of the epidermis. Apocrine sweat glands are limited in distribution to the skin of the axilla, pubis, and areolae; their ducts open into hair follicles. Sudoriferous glands produce perspiration, which carries small amounts of wastes to the surface and assists in maintaining body temperature.
3. Ceruminous glands are modified sudoriferous glands that secrete cerumen. They are found in the external auditory canal (ear canal).

Nails (p. 100)

1. Nails are hard, keratinized epidermal cells over the dorsal surfaces of the terminal portions of the fingers and toes.
2. The principal parts of a nail are the body, free edge, root, lunula, eponychium, and matrix. Cell division of the matrix cells produces new nails.

Blood Supply of the Integumentary System (p. 100)

1. The epidermis is avascular.
2. The dermis is supplied by the cutaneous and papillary plexuses.

Developmental Anatomy of the Integumentary System (p. 100)

1. The epidermis, hair, nails, and skin glands are ectodermal derivatives.
2. The dermis is derived from mesodermal cells.

Aging and the Integumentary System (p. 102)

1. Most effects of aging begin to occur when people reach their late forties.
2. Among the effects of aging are wrinkling, loss of subcutaneous fat, atrophy of sebaceous glands, and decrease in the number of melanocytes and Langerhans cells.

Review Questions

1. What is the integumentary system? (p. 91)
2. Compare the structure of epidermis and dermis. What is the subcutaneous layer? (p. 91)
3. Describe the various cells that constitute the epidermis. List and describe the epidermal layers from the deepest to superficial. What is the importance of each layer? (p. 91)
4. Compare the structure of the papillary and reticular regions of the dermis. (p. 95)
5. List the seven principal functions of the skin. (p. 95)
6. Explain the factors that produce skin color. What is an albino? (p. 96)
7. How are epidermal ridges formed? Why are they important? (p. 96)
8. List the receptors in the epidermis, dermis, and subcutaneous (subQ) layers and indicate the location and role of each. (p. 94)
9. Describe the structure of a hair. How are hairs moistened? What produces "goose bumps" or "gooseflesh"? (p. 97)
10. Contrast the locations and functions of sebaceous (oil) glands, sudoriferous (sweat) glands, and ceruminous glands. (p. 99)
11. Distinguish between eccrine and merocrine sweat glands. (p. 99)
12. Describe the principal parts of a nail. (p. 100)
13. Describe the blood supply of the skin. (p. 100)
14. Describe the origins of the epidermis, its derivatives, and the dermis. (p. 101)
15. Describe the effects of aging on the integumentary system. (p. 102)
16. Refer to the glossary of key medical terms associated with the integumentary system. Be sure that you can define each term. (p. 103)

Self Quiz

Complete the following:

1. In the lining of the external auditory meatus are located ___ that secretes earwax, also called ___.
2. The deeper portion of the skin, called the ___, is made of ___ tissue.
3. The epidermis is derived from the ___-derm. All its layers are formed by the ___ month. The dermis arises from ___-derm.
4. Sweat glands are also known as ___ glands.

Are the following statements true or false?

5. A typical sunburn is an example of a first-degree burn.
6. The internal root sheath is a downward continuation of the epidermis.
7. The dermis consists of two regions; the papillary region is most superficial, and the reticular region is deeper.
8. Racial differences in skin color are mainly due to the number of melanocytes in the epidermis.

9. Match the following terms with their definitions:
   ___ (1) melanocytes
   ___ (2) keratinocytes
   ___ (3) Merkel cells
   ___ (4) Langerhans cells
   ___ (5) Meissner's corpuscles (corpuscles of touch)
   ___ (6) lamellated (Pacinian) corpuscles

   (a) assist helper T cells in immune responses
   (b) produce pigment that shields cell nuclei from UV light
   (c) located in subcutaneous tissue; sensitive to pressure
   (d) most abundant epidermal cell
   (e) located in stratum basale of hairless skin; probably play a role in tactile sensation.
   (f) tactile receptors located in dermal papillae

Choose the one best answer to the following questions:

10. Which of the following statements about the function of skin is *not* true?
   a. it helps regulate body temperature
   b. it protects against bacterial invasion and dehydration
   c. it absorbs water and salts
   d. it participates in the synthesis of vitamin D
   e. it detects stimuli related to temperature and pain

11. The layer of the skin from which new epidermal cells are derived is the
   a. stratum corneum
   b. stratum basale
   c. stratum lucidum
   d. stratum granulosum
   e. stratum spinosum

12. The outermost layer of a hair is the
   a. lunula
   b. cortex
   c. eponychium
   d. medulla
   e. none of the above

13. The region of the dermis that is in direct contact with the epidermis is the
   a. papillary region
   b. stratum corneum
   c. stratum basale
   d. hypodermis
   e. reticular region

14. Match the following:
   ___ (1) stratum spinosum
   ___ (2) stratum corneum
   ___ (3) stratum basale
   ___ (4) stratum lucidum
   ___ (5) stratum granulosum

   (a) deepest layer, consisting of a single layer of cuboidal- to columnar-shaped keratinocytes
   (b) eight to ten rows of polyhedral keratinocytes that exhibit spinelike projections
   (c) layer consisting of three to five rows of flattened, degenerating keratinocytes
   (d) layer of flat, dead keratinocytes that are normally apparent only in thick skin of palms and soles
   (e) most superficial layer of skin, consisting of 25 to 30 rows of flat, dead keratinocytes

Critical Thinking Questions

1. Your 65-year-old aunt has spent every sunny day at the beach for as long as you can remember. Her skin looks a lot like a comfortable lounge chair—brown and wrinkled. Recently her dermatologist removed a suspicious growth from the skin of her face. What would you suspect is the problem and its likely cause?
HINT: *Tomatoes should be sun-dried, but not people.*

2. Fair-skinned Lars has a new job as a lifeguard. He wants a gorgeous tan NOW, so he does not wear any sun protection during his first 8-hour shift. Predict the result of Lars's action, and explain what he should have done.
HINT: *Lars would make a great poster boy for the Boiled Maine Lobster Fan Club.*

3. A new mother complains of excessive hair loss. What would you tell her about the cause and possible duration of this condition?
HINT: *The baby's still too young for its mom to be pulling her hair out; consider the pregnancy.*

4. Your nephew has been learning about cells in science class. He asks, "If all cells have a semipermeable membrane, and all tissues are made out of cells, then how come I don't swell up and pop when I take a bath?"
HINT: *The semipermeable membrane functions in living cells.*

5. Fifteen-year-old Suzi has a bad case of "blackheads." According to her Aunt Freida, Suzi's skin problems are from too much late-night TV, chicken nuggets, and cheddar popcorn. Explain the real cause of blackheads to Aunt Freida.
HINT: *The oil in the junk food does not go directly to the face.*

6. An accident while slicing a bagel lands you in the emergency room. After examining the injury, the nurse state that you've cut through to the subQ layer. Discuss the structures and functions of the skin layers that you've exposed.
HINT: *A little deeper and you'd have hit muscle.*

Answers to Figure Questions

4.1 Stratum corneum; stratum basale.

4.2 Langerhans cell.

4.3 Surgical incisions parallel to lines of cleavage leave only fine scars.

4.4 Plucking a hair stimulates hair root plexuses in the dermis, some of which are sensitive to pain. Because the cells of a hair shaft are already dead and because there are no nerves in the hair shaft, cutting hair is not painful.

4.5 They are composed of tightly packed, hard, keratinized epidermal cells.

Chapter 5

Bone Tissue

Student Objectives

1. Discuss the components and functions of the skeletal system.

2. List and describe the gross features of a long bone.

3. Describe the histological features of compact and spongy bone tissue.

4. Describe the process of bone formation.

5. Describe the conditions necessary for normal bone growth and replacement.

6. Describe the blood and nerve supply of bone tissue.

7. Describe the development of the skeletal system.

8. Explain the effects of exercise and aging on bone tissue.

Despite its rather simple and lifeless appearance, bone is actually a very complex and dynamic living tissue. Moreover, bone tissue is involved in a continuous process of building new bone and dissolving old bone.

In addition to bone tissue, several other tissues are associated with bones. These include cartilage, dense connective tissues, epithelium, various blood-forming tissues, adipose tissue, and nervous tissue. Because a bone is made up of several different tissues working together, each individual bone is an organ. The framework of bones and their cartilages together constitute the **skeletal system.** The study of bone structure and the treatment of bone disorders is called **osteology** (os-tē-OL-ō-jē; *osteo* = bone; *logos* = study of).

This chapter will survey the various components of bones so you can understand how bones form, how they age, and how exercise affects the density and strength of bones. It will also describe several disorders related to bone tissue.

FUNCTIONS OF BONE

Bone (osseous) tissue and the skeletal system perform several basic functions:

1. **Support.** Bone serves as a framework for the body by supporting soft tissues and providing points of attachment for many skeletal muscles.

2. **Protection.** Bones protect many internal organs from injury. For example, cranial bones protect the brain, vertebrae protect the spinal cord, and the rib cage protects the heart and lungs.

3. **Assistance in movement.** Skeletal muscles attach to bones. When these muscles contract, they pull on bones and together they produce movement. Movements of muscles, bones, and joints are discussed in detail in later chapters.

4. **Mineral storage and release.** Bone tissue stores several minerals, especially calcium and phosphorus, which are important in muscle contraction and nerve activity, among other functions. On demand, bone releases minerals into the blood to maintain critical mineral balances and to distribute them to other parts of the body.

5. **Blood cell production.** Within certain parts of bones, a connective tissue called red bone marrow produces blood cells, a process called **hemopoiesis** (hēm-ō-poy-Ē-sis). **Red bone marrow,** which is one type of bone marrow, consists of blood cells in immature stages, adipose cells, and macrophages. Red bone marrow produces red blood cells, white blood cells, and platelets.

6. **Storage of energy.** In the newborn, all bone marrow is red and is involved in hemopoiesis. However, with increasing age, blood cell production decreases and most of the bone marrow changes from red to yellow. **Yellow bone marrow** consists primarily of adipose cells and a few scattered blood cells. Lipids stored in the adipose cells are an important chemical energy reserve. Under certain conditions, such as severe bleeding or hypoxia (cellular oxygen deficiency), yellow bone marrow can convert back to red bone marrow.

ANATOMY: STRUCTURE OF BONE

The structure of bone may be analyzed by first considering the parts of a long bone such as the humerus, the arm bone (Figure 5.1). A long bone is one that has greater length than width. A typical long bone consists of the following parts:

1. **Diaphysis** (dī-AF-i-sis; *dia* = through; *physis* = growth). The shaft or body, the long, main, cylindrical portion of the bone.

2. **Epiphyses** (e-PIF-i-sēz; *epi* = above). The extremities or ends of the bone (singular is **epiphysis**).

3. **Metaphyses** (me-TAF-i-sēz; *meta* = after). The regions in a mature bone where the diaphysis joins the epiphysis. In a growing bone, they are the regions that contain the epiphyseal plates. The epiphyseal plate is a layer of hyaline cartilage that allows the diaphysis of the bone to grow in length. This is described later in the chapter.

4. **Articular cartilage.** A thin layer of hyaline cartilage covering the epiphysis where the bone forms an articulation (joint) with another bone. The cartilage reduces friction and absorbs shock at freely movable joints.

5. **Periosteum** (per'-ē-OS-tē-um). The periosteum (*peri* = around; *osteo* = bone) is a tough, white, fibrous membrane around the surface of the bone not covered by articular cartilage. It consists of two layers (see Figure 5.3a). The outer *fibrous layer* is composed of dense, irregular connective tissue that contains blood vessels, lymphatic vessels, and nerves that pass into the bone. The inner osteogenic (os'-tē-ō-JEN-ik) layer contains elastic fibers, blood vessels, and various types of bone cells. The periosteum helps bone to grow in diameter or thickness, but not in length. It also protects the bone, assists in fracture repair, helps nourish the bone, and serves as a point of attachment for ligaments and tendons.

6. **Medullary** (MED-yoo-lar'-ē; *medulla* = central part of an organ) or *marrow* **cavity.** This is the space within the

Figure 5.1 Parts of a long bone.

A long bone is covered by periosteum (diaphysis) and articular cartilage (epiphysis).

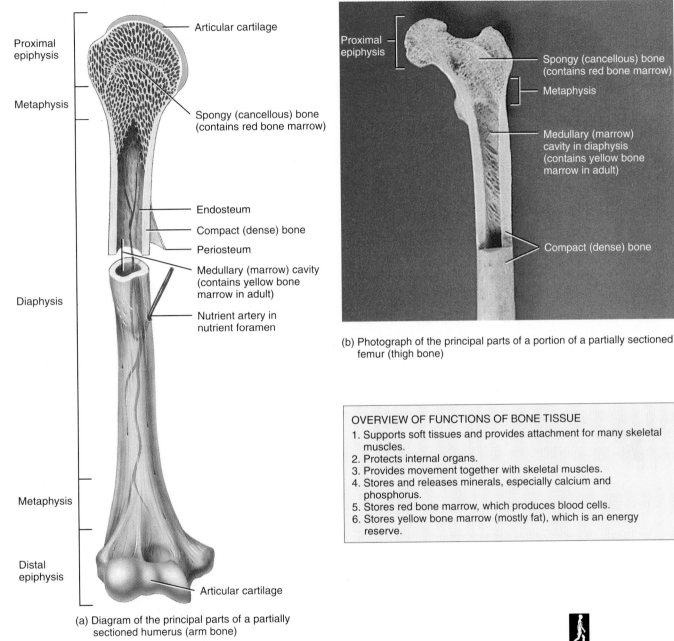

Proximal epiphysis

Metaphysis

Diaphysis

Metaphysis

Distal epiphysis

Articular cartilage

Spongy (cancellous) bone (contains red bone marrow)

Endosteum

Compact (dense) bone

Periosteum

Medullary (marrow) cavity (contains yellow bone marrow in adult)

Nutrient artery in nutrient foramen

Articular cartilage

(a) Diagram of the principal parts of a partially sectioned humerus (arm bone)

Proximal epiphysis

Spongy (cancellous) bone (contains red bone marrow)

Metaphysis

Medullary (marrow) cavity in diaphysis (contains yellow bone marrow in adult)

Compact (dense) bone

(b) Photograph of the principal parts of a portion of a partially sectioned femur (thigh bone)

OVERVIEW OF FUNCTIONS OF BONE TISSUE
1. Supports soft tissues and provides attachment for many skeletal muscles.
2. Protects internal organs.
3. Provides movement together with skeletal muscles.
4. Stores and releases minerals, especially calcium and phosphorus.
5. Stores red bone marrow, which produces blood cells.
6. Stores yellow bone marrow (mostly fat), which is an energy reserve.

Q *Which part of the bone reduces friction at joints? Produces blood cells? Is the shaft? Lines the medullary cavity?*

diaphysis that contains the fatty yellow bone marrow in adults.

7. **Endosteum** (end-OS-tē-um; *endo* = within). Lining the medullary cavity is the endosteum, a membrane that contains osteoprogenitor cells and osteoclasts (described shortly).

HISTOLOGY OF BONE TISSUE

Like other connective tissues, **bone** or **osseous** (OS-ē-us) **tissue** contains an abundant matrix (intercellular substance) surrounding widely separated cells. The matrix of bone tissue is about 25% water, 25% protein fibers, and 50% mineral salts. There are four basic types of cells in bone tissues:

Figure 5.2 Types of cells in bone tissue.

Osteoprogenitor cells undergo mitosis and develop into osteoblasts, which secrete the matrix materials.

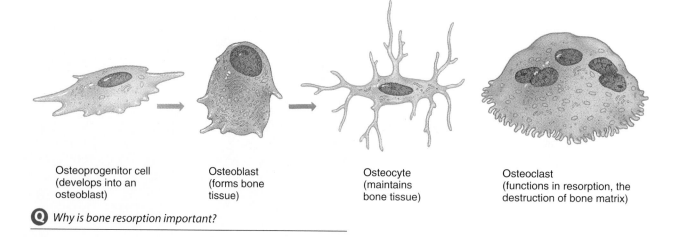

Osteoprogenitor cell
(develops into an
osteoblast)

Osteoblast
(forms bone
tissue)

Osteocyte
(maintains
bone tissue)

Osteoclast
(functions in resorption, the
destruction of bone matrix)

Q *Why is bone resorption important?*

osteoprogenitor (osteogenic) cells, osteoblasts, osteocytes, and osteoclasts (Figure 5.2).

1. **Osteoprogenitor** (os'-tē-ō-prō-JEN-i-tor; *osteo* = bone; *pro* = precursor; *gen* = to produce) **cells** are unspecialized cells derived from mesenchyme, the tissue from which all connective tissues are derived. They can undergo mitosis and develop into osteoblasts. Osteoprogenitor cells are found in the inner portion of the periosteum, in the endosteum, and in canals (perforating and central) in bone that contain blood vessels.

2. **Osteoblasts** (OS-tē-ō-blasts'; *blast* = germ or bud) are the cells that form bone, but they have lost the ability to divide by mitosis. They secrete collagen and other organic components needed to build bone tissue.

3. **Osteocytes** (OS-tē-ō-sīts'; *cyte* = cell) are mature bone cells that are derived from osteoblasts; they are the principal cells of bone tissue. Like osteoblasts, osteocytes have no mitotic potential. Osteoblasts are found on the surface of bone, but as they surround themselves with matrix materials they become trapped in their secretions and become osteocytes. Osteocytes no longer secrete matrix materials. Whereas osteoblasts initially form bone tissue, osteocytes maintain daily cellular activities of bone tissue, such as the exchange of nutrients and wastes with the blood.

4. **Osteoclasts** (OS-tē-ō-clasts'; *clast* = to break) are believed to develop from circulating monocytes (one type of white blood cell). They settle on the surfaces of bone

and function in bone resorption (destruction of matrix), which is important in the development, growth, maintenance, and repair of bone.

Unlike other connective tissues, the matrix of bone contains abundant mineral salts, primarily a crystallized form of tricalcium phosphate called *hydroxyapatite* and some calcium carbonate. In addition, there are small amounts of magnesium hydroxide, fluoride, and sulfate. As these salts are deposited in the framework formed by the collagen fibers of the matrix, they crystallize and the matrix hardens. This process is called **calcification** or *mineralization.*

Although the *hardness* of bone depends on the crystallized inorganic mineral salts, without the organic collagen fibers it would be very brittle. Like reinforcing metal rods in concrete, the collagen fibers and other organic molecules provide bone with *tensile strength,* which is resistance to being stretched or torn apart. Collagen fibers make bone less brittle than other calcium-based organic structures, such as egg shells and oyster shells. If the inorganic minerals are removed, for example, by soaking a bone in weak acid such as vinegar, the result is a rubbery, flexible structure.

Bone is not completely solid but has many small spaces (sometimes microscopic only) between its hard components. The spaces provide channels for blood vessels that supply bone cells with nutrients. The spaces also make bones lighter. Depending on the size and distribution of the spaces, the regions of a bone may be categorized as compact or spongy.

Figure 5.3 Histology of bone. A photomicrograph of compact bone tissue is provided in Table 3.3 on page 79.

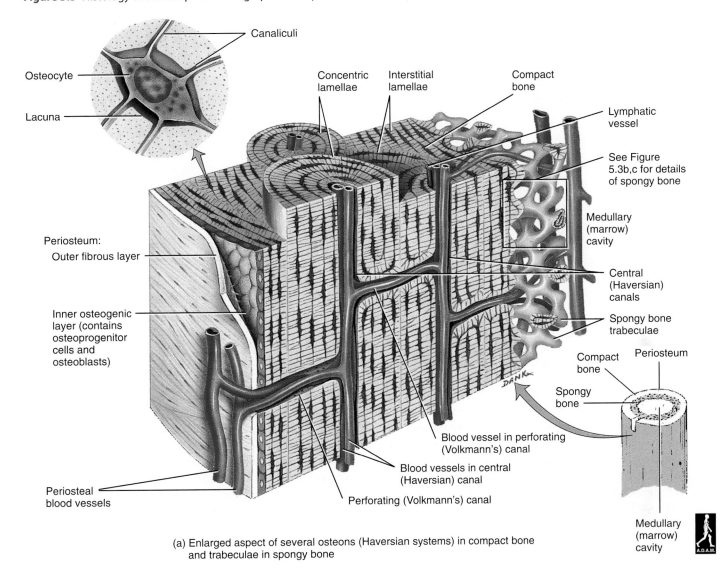

(a) Enlarged aspect of several osteons (Haversian systems) in compact bone and trabeculae in spongy bone

Compact Bone Tissue

Compact (dense) bone tissue contains few spaces. It forms the external layer of all bones of the body and the bulk of the diaphyses of long bones. Compact bone tissue provides protection and support and helps the long bones resist the stress of weight placed on them.

Compare the differences between spongy and compact bone tissues by looking at the diagram of a highly magnified section in Figure 5.3. As you will see shortly, compact bone is composed of concentric rings of bone tissue called *osteons,* whereas the lattice-work of spongy bone is composed of columns of bone tissue called *trabeculae.*

Blood vessels, lymphatic vessels, and nerves from the periosteum penetrate the compact bone through **perforating (Volkmann's) canals.** The blood vessels and nerves of these canals connect with those of the medullary cavity, periosteum, and **central (Haversian) canals.** The central canals run longitudinally through the bone. Around the canals are **concentric lamellae** (la-MEL-ē)—rings of hard, calcified matrix. Between the lamellae are small spaces called **lacunae** (la-KOO-nē; *lacuna* = little lake), which contain osteocytes. **Osteocytes,** as noted earlier, are mature bone cells that no longer secrete matrix materials.

Radiating in all directions from the lacunae are minute canals called **canaliculi** (kan′-a-LIK-yoo-lī; *canaliculi* =

Figure 5.3 (continued)

Osteocytes lie in the lacunae arranged in concentric circles around a central canal in compact bone and in lacunae arranged irregularly in trabeculae of spongy bone.

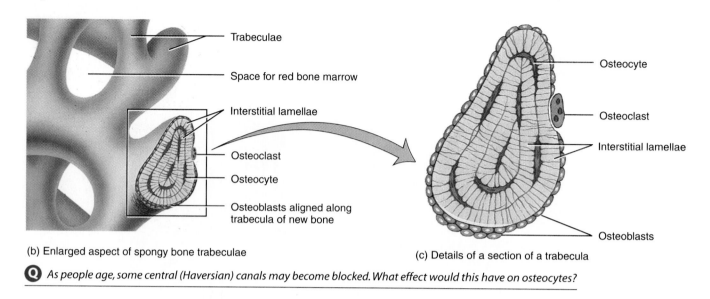

(b) Enlarged aspect of spongy bone trabeculae

(c) Details of a section of a trabecula

Q As people age, some central (Haversian) canals may become blocked. What effect would this have on osteocytes?

small channels or canals), which are filled with extracellular fluid. Inside the canaliculi are slender fingerlike processes of osteocytes (see inset at top of Figure 5.3a). The canaliculi connect lacunae with one another and, eventually, with the central canals. Thus there is an intricate, miniature canal system throughout the bone. This branching network of canaliculi provides many routes for nutrients and oxygen to reach the osteocytes and for wastes to diffuse away. Osteocytes from neighboring lacunae form gap junctions with each other, facilitating easy movement of materials from cell to cell.

Each central canal, with its surrounding lamellae, lacunae, osteocytes, and canaliculi, forms an **osteon (Haversian system).** Compact bone tissue is the only connective tissue that has a basic structural unit—the osteon—associated with it. The areas between osteons contain **interstitial lamellae.** These also have lacunae with osteocytes and canaliculi. Interstitial lamellae are fragments of older osteons that have been partially destroyed during bone rebuilding or growth.

Spongy Bone Tissue

In contrast to compact bone, **spongy (cancellous) bone tissue** does not contain true osteons (Figure 5.3b and c). It consists of lamellae arranged in an irregular latticework of thin columns of bone called **trabeculae** (tra-BEK-yoo-lē; *trabecula* = little beam). The macroscopic spaces between the trabeculae of some bones are filled with red bone marrow,

which produces blood cells. Within the trabeculae are osteocytes that lie in lacunae. Radiating from the lacunae are canaliculi. Blood vessels from the periosteum penetrate through to the spongy bone. Osteocytes in the trabeculae receive nourishment directly from the blood circulating through the marrow cavities. Osteons are not necessary in spongy bone because osteocytes are not deeply buried (as in compact bone) and have access to nutrients directly from the blood.

Spongy bone makes up most of the bone tissue of short, flat, and irregularly shaped bones and most of the epiphyses of long bones. Spongy bone tissue in the hipbones, ribs, breastbone (sternum), backbones (vertebrae), and ends of some long bones is the only site of red bone marrow storage and thus hemopoiesis in adults.

Most people think of all bone as a very hard, rigid material. Yet the bones of an infant are quite soft and become rigid only after growth stops during late adolescence. Even then, bone is constantly broken down and rebuilt. It is a dynamic, living tissue.

Clinical Application

Bone Scan

A **bone scan** is a very important diagnostic test that takes advantage of the fact that bone is a living tissue. The test is used to determine whether cancer, such as breast or prostate cancer, has metastasized to bone; to evaluate an unexplained bone pain or possible fracture; to monitor response to therapy of a cancer that has spread to bone; to monitor the progress of bone grafts and degenerative bone disorders; and to identify bone infections. A bone scan not only detects pathologies from three to six months sooner than standard x-ray procedures, but also exposes the patient to less radiation.

In the procedure, a small amount of a radioactive tracer compound, which is readily absorbed by bone, is injected intravenously. A scanning camera measures the radiation emitted from the bones. The information is translated into a photograph or diagram that can be read like an x-ray. Normal bone tissue is identified by a consistent gray color throughout because of its uniform uptake of the radioactive tracer. Darker or lighter areas indicate abnormal findings. For example, darker areas are called "hot spots," areas of increased metabolism that absorb more of the radioactive tracer. These areas may indicate bone cancer or metastasis, abnormal healing of fractures, or abnormal bone growth. Lighter areas are called "cold spots," areas of decreased metabolism that absorb less of the radioactive tracer. These areas may indicate abnormalities such as degenerative bone disease, decalcification of bone, fractures, osteomyelitis (bone infection), Paget's disease, and rheumatoid arthritis. ■

BONE FORMATION: OSSIFICATION

The process by which bone forms is called **ossification** (os′-i-fi-KĀ-shun; *facere* = to make). The "skeleton" of a human embryo is composed of fibrous connective tissue membranes formed by embryonic connective tissue (mesenchyme) or hyaline cartilage that are loosely shaped like bones. They provide the supporting structures for ossification. Ossification begins around the sixth or seventh week of embryonic life and continues throughout adulthood.

Two patterns of bone formation occur. The first pattern is called **intramembranous** (in′-tra-MEM-bra-nus; *intra* = within; *membranous* = membrane) **ossification.** This refers to the formation of bone directly on or within fibrous connective tissue membranes. The second pattern, **endochondral** (en′-dō-KON-dral; *endo* = within; *chondro* cartilage) **ossification,** refers to the formation of bone within a cartilage model. These two patterns of ossification do *not* lead to differences in the gross structure of mature bones. They are simply different methods of bone development. Both mechanisms involve the replacement of a preexisting connective tissue with bone. The fontanels ("soft spots") that help the fetal skull pass through the birth canal are replaced by bone through intramembranous ossification.

The first stage in the development of bone is the appearance of osteoprogenitor cells that undergo mitosis to produce osteoblasts, which in turn produce bony matrix by either intramembranous or endochondral ossification.

Intramembranous Ossification

Of the two patterns of bone formation, **intramembranous ossification** is easier to understand. The flat bones of the skull and the mandible (lower jawbone) are formed in this way. The process occurs as follows (Figure 5.4):

❶ At the site where bone will develop, cells in mesenchyme come together and develop into osteoprogenitor cells and then into osteoblasts. This site is called a **center of ossification.** Osteoblasts secrete the organic matrix of bone until they are completely surrounded by it.

❷ Then secretion of matrix stops and the cells, now called osteocytes, are in lacunae and extend narrow cytoplasmic processes into canaliculi. Within a few days, calcium and other mineral salts are deposited and the matrix hardens (calcification).

❸ As the bone matrix forms, it develops into *trabeculae* that fuse with one another to form spongy bone. The spaces between trabeculae fill with red bone marrow. On the outside of the bone, the mesenchyme condenses.

❹ The mesenchyme develops into the periosteum. Eventually, most surface layers of the spongy bone are replaced by compact bone, but spongy bone remains in the center of the bone. Much of this newly formed bone will be remodeled (destroyed and reformed) so that the bone may reach its final adult size and shape.

Endochondral Ossification

The replacement of cartilage by bone is called **endochondral ossification.** Most bones of the body are formed this way, but the process is best observed in a long bone. It occurs as follows (Figure 5.5):

❶ **Development of the cartilage model.** At the site where the bone will develop, cells in mesenchyme come together in the shape of the future bone. The cells then develop into cartilage-producing cells that change the model into hyaline cartilage. In addition, a membrane called the **perichondrium** (per-i-KON-drē-um) develops around the cartilage.

❷ **Growth of the cartilage model.** The cartilage model grows in length and thickness. As the cartilage model continues to grow, chondrocytes (cartilage cells) in its midregion bring about chemical changes that trigger calcification. Once the cartilage becomes calcified, other cartilage cells die because nutrients no longer diffuse quickly enough through the matrix. The lacunae of the cells that have died are now empty, and the thin partitions between them break down, forming small cavities. In the meantime, a nutrient artery penetrates the bone through a hole (nutrient foramen) in the midregion of the model. This stimulates osteoprogenitor cells in the perichondrium to develop into osteoblasts. The cells produce a thin layer of compact bone under the

Figure 5.4 Intramembranous ossification.

🔑 *Intramembranous ossification involves the formation of bone directly on or within loose fibrous connective tissue membranes.*

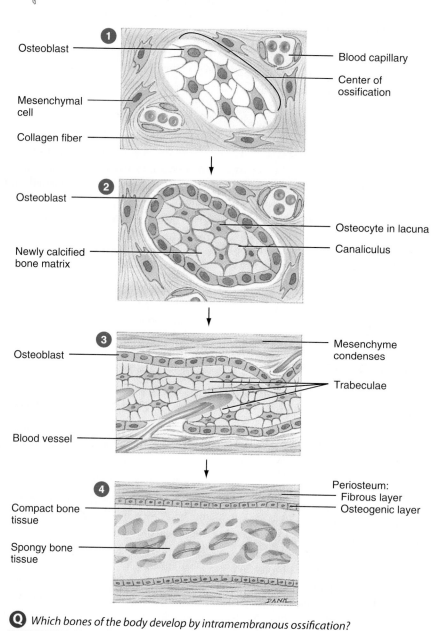

Ⓠ *Which bones of the body develop by intramembranous ossification?*

perichondrium. Once the perichondrium starts to form bone, it is known as the **periosteum.**

❸ **Development of the primary ossification center.** Near the middle of the model, capillaries of the periosteum grow into the disintegrating calcified cartilage. The capillaries stimulate growth of a *primary ossification center,* a region where bone tissue will replace most of the cartilage. Osteoblasts then begin to deposit bone matrix over the rem-

nants of calcified cartilage, forming spongy bone trabeculae. As the ossification center enlarges toward the ends of the bone, osteoclasts break down the newly formed spongy bone trabeculae. This activity leaves a cavity, the medullary (marrow) cavity, in the core of the model. The cavity then fills with red bone marrow.

❹ **Development of the diaphysis and epiphysis.** The diaphysis (shaft), which was once a solid mass of hyaline

Figure 5.5 Endochondral ossification of the tibia (shinbone).

Endochondral ossification involves the replacement of cartilage by bone.

Q *Which structure signals that bone growth in length has stopped?*

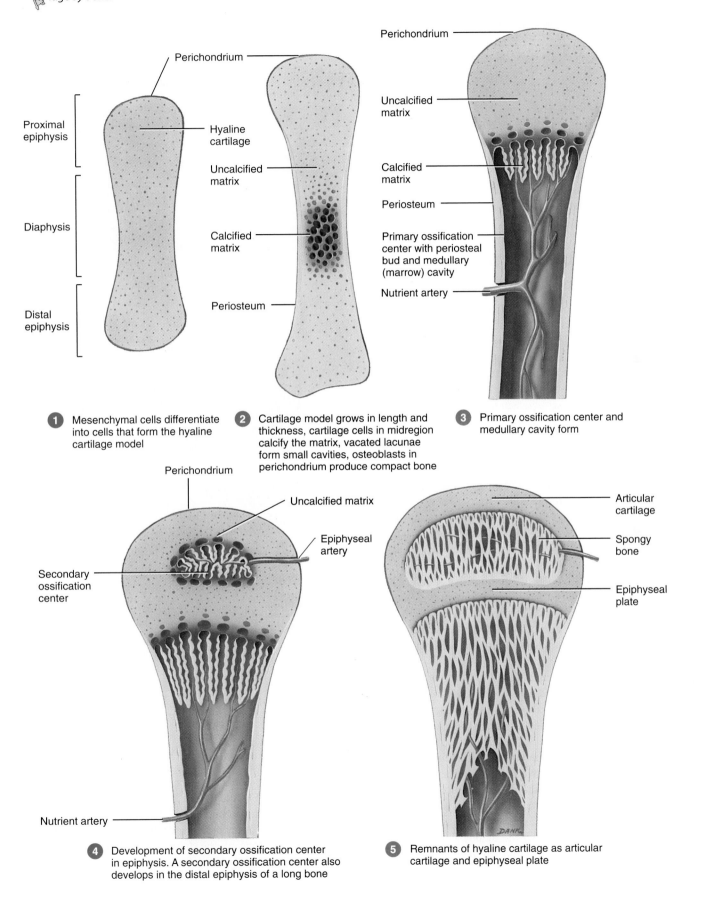

Proximal epiphysis

Diaphysis

Distal epiphysis

Perichondrium

Hyaline cartilage

Uncalcified matrix

Calcified matrix

Periosteum

1 Mesenchymal cells differentiate into cells that form the hyaline cartilage model

2 Cartilage model grows in length and thickness, cartilage cells in midregion calcify the matrix, vacated lacunae form small cavities, osteoblasts in perichondrium produce compact bone

Perichondrium

Uncalcified matrix

Calcified matrix

Periosteum

Primary ossification center with periosteal bud and medullary (marrow) cavity

Nutrient artery

3 Primary ossification center and medullary cavity form

Perichondrium

Uncalcified matrix

Epiphyseal artery

Secondary ossification center

Nutrient artery

4 Development of secondary ossification center in epiphysis. A secondary ossification center also develops in the distal epiphysis of a long bone

Articular cartilage

Spongy bone

Epiphyseal plate

5 Remnants of hyaline cartilage as articular cartilage and epiphyseal plate cartilage, is replaced by compact bone, the core of which contains a red bone marrow-filled medullary cavity. When blood vessels (epiphyseal arteries) enter the epiphyses, secondary ossification centers develop, usually around the time of birth.

❺ In the secondary ossification centers, bone formation is similar to that in the primary ossification centers. One difference, however, is that spongy bone remains in the interior of the epiphyses (no medullary cavities are formed in the epiphyses). Also, hyaline cartilage remains covering the epiphyses as the articular cartilage and between the diaphysis and epiphysis as the epiphyseal plate, which is responsible for the lengthwise growth of long bones.

BONE GROWTH
Growth in Length

Growth in length of bones is usually completed by about age 25, although bones may continue to thicken. To understand how a bone grows in length, you need to know some of the details of the structure of the epiphyseal plate.

The **epiphyseal** (ep′-i-FIZ-ē-al; *epiphyein* = to grow upon) **plate** is a layer of hyaline cartilage in a growing bone located in the metaphysis between the epiphysis and diaphysis. It is leftover cartilage from endochondral ossification. The epiphyseal plate consists of four zones (Figure 5.6). The *zone of resting cartilage* is near the epiphysis and consists of small, scattered chondrocytes. The cells do not function in bone growth (thus the term "resting"); they anchor the epiphyseal plate to the bone of the epiphysis.

The *zone of proliferating cartilage* consists of slightly larger chondrocytes arranged like stacks of coins. Chondrocytes divide to replace those that die at the diaphyseal surface of the epiphyseal plate.

The *zone of hypertrophic* (hī-per-TRŌF-ik) or *maturing cartilage* consists of even larger chondrocytes that are also arranged in columns. The lengthwise expansion of the epiphyseal plate is the result of cell divisions in the zone of proliferating cartilage and maturation of the cells in the zone of hypertrophic cartilage.

The *zone of calcified cartilage* is only a few cells thick and consists mostly of dead cells because the matrix around them has calcified. The calcified matrix is taken up by osteoclasts, and the area is invaded by osteoblasts and capillaries from the bone in the diaphysis. These cells lay down bone on the calcified cartilage that persists. As a result, the diaphyseal border of the epiphyseal plate is firmly cemented to the bone of the diaphysis. The activity of the epiphyseal plate is the only mechanism by which the diaphysis can increase in length.

The epiphyseal plate allows the diaphysis of the bone to increase in length until early adulthood. It also shapes the ar-

Figure 5.6 Epiphyseal plate.

The epiphyseal plate allows the diaphysis of a bone to increase in length.

Radiograph of a portion of the tibia and fibula of a 10-year-old child

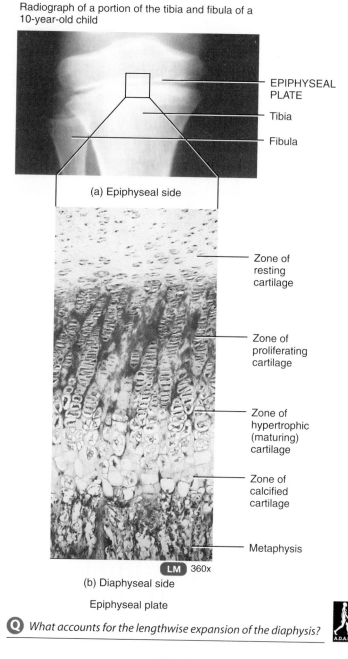

EPIPHYSEAL PLATE

Tibia

Fibula

(a) Epiphyseal side

Zone of resting cartilage

Zone of proliferating cartilage

Zone of hypertrophic (maturing) cartilage

Zone of calcified cartilage

Metaphysis

LM 360x

(b) Diaphyseal side

Epiphyseal plate

Q *What accounts for the lengthwise expansion of the diaphysis?*

ticular surfaces. The rate of growth is controlled by hormones such as human growth hormone (hGH) produced by the pituitary gland, and sex hormones (estrogens and testosterone) produced by the ovaries and testes. As a child grows, cartilage cells are produced by mitosis on the epiphyseal side of the plate. They are then destroyed, and the cartilage is replaced by bone on the diaphyseal side of the plate. In this way, the thickness of the epiphyseal plate remains almost constant, but the bone on the diaphyseal side increases in length. This pattern of growth that results in an increase in length is called **interstitial** (in′-ter-STISH-al) **growth,** which means growth from within.

Eventually, the epiphyseal cartilage cells stop dividing, and bone replaces the cartilage. The newly formed bony structure is called the **epiphyseal line,** a remnant of the once active epiphyseal plate. With the appearance of the epiphyseal line, bone stops growing in length. In general, lengthwise growth in bones in females is completed before that in males.

Growth in Diameter

The pattern of growth that results in an increase in diameter or thickness is called **appositional** (a-pō-ZISH-i-nal) **growth,** which means growth from outside. It occurs as follows: First, bone lining the medullary cavity is destroyed by osteoclasts in the endosteum so that the cavity increases in diameter. At the same time, osteoblasts from the periosteum add new bone tissue to the outer surface. Initially, diaphyseal and epiphyseal ossification produce only spongy bone. Later, the outer region of spongy bone is reorganized into compact bone.

BLOOD AND NERVE SUPPLY

Bone is richly supplied with blood, and blood vessels are especially abundant in portions of bone containing red bone marrow. Blood vessels pass into bones from the periosteum. Here we will consider the blood supply to a long bone only.

Because the artery to the diaphysis of a long bone is usually the largest, it is referred to as the **nutrient artery.** The artery first enters the bone during the early development and passes through an opening in the diaphysis, the **nutrient foramen** (see Figure 5.1a). On entering the medullary cavity, the nutrient artery divides into a proximal and a distal branch, each supplying most of the bone marrow, the inner portion of compact bone of the diaphysis, and the metaphysis. As the nutrient artery passes through compact bone on its way to the medullary cavity, it sends branches into the central (Haversian) canals. Branches of the **epiphyseal arteries** enter many vascular foramina in the epiphysis and supply the bone marrow, metaphysis, and bony tissue of the epiphysis. **Periosteal arteries,** accompanied by nerves, enter the diaphysis at numerous points through perforating (Volkmann's) canals (see Figure 5.3a). These blood vessels run in the central canals and supply the outer part of the compact bone of the diaphysis. Veins accompany the several types of arteries; the principal veins leave the bone by numerous openings at the epiphyses of the bone.

Although the nerve supply to bones is not extensive, some nerves accompany blood vessels, and some occur in the periosteum. The nerves of the periosteum are primarily concerned with pain, as might be associated with a fracture or tumor.

BONE REPLACEMENT

Bone, like skin, forms before birth but continually renews itself thereafter. **Remodeling** is the ongoing replacement of old bone tissue by new bone tissue. Bone constantly remodels and redistributes its matrix along lines of mechanical stress. Compact bone is formed from spongy bone. However, even after bones have reached their adult shapes and sizes, old bone is continually destroyed, and new bone tissue is formed in its place. Remodeling also removes worn and injured bone, replacing it with new bone tissue. This process allows bone to serve as the body's reservoir for calcium. Many other tissues need calcium to perform their functions. Several hormones continually regulate exchanges of calcium between blood and bones.

Remodeling takes place at different rates in various body regions. The distal portion of the thighbone (femur) is replaced about every four months. By contrast, bone in certain areas of the shaft will not be completely replaced during an individual's life. Osteoclasts are responsible for bone resorption (destruction of matrix). A delicate balance exists between the actions of osteoclasts in removing minerals and collagen and of bone-making osteoblasts in depositing minerals and collagen. Should too much new tissue be formed, the bones become abnormally thick and heavy. If too much mineral is deposited in the bone, the surplus may form thick bumps, or spurs, on the bone that interfere with movement at joints. A loss of too much calcium or tissue weakens the bones, and they may break, as occurs in osteoporosis, or become too flexible, as in rickets and osteomalacia. Abnormal acceleration of the remodeling process results in a condition called Paget's disease (see Applications to Health, p. 4).

Normal bone growth in the young and bone replacement in the adult depend on the presence of (1) several minerals: calcium, phosphorus, magnesium, boron, and manganese; (2) several vitamins: A, B_{12}, C, and D; and (3) several hormones: human growth hormone, sex hormones (estrogens and testosterone), insulin, thyroid hormones, calcitonin, and parathyroid hormone.

FRACTURE AND REPAIR OF BONE

A **fracture** is any break in a bone. Although bone has a generous blood supply, healing sometimes takes months. Sufficient calcium and phosphorous to strengthen and harden new bone is deposited only gradually. Bone cells generally

Figure 5.7 Steps involved in repair of a bone fracture.

Bone heals more rapidly than cartilage because its blood supply is more plentiful.

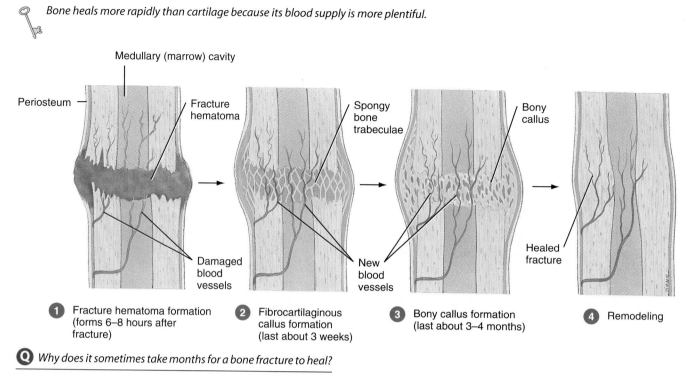

1 Fracture hematoma formation (forms 6–8 hours after fracture)

2 Fibrocartilaginous callus formation (last about 3 weeks)

3 Bony callus formation (last about 3–4 months)

4 Remodeling

Q *Why does it sometimes take months for a bone fracture to heal?*

grow and reproduce slowly. Moreover, in a fractured bone the blood supply is decreased, which helps to explain the difficulty in the healing of an infected bone. Recall that cartilage injuries heal even more slowly, however, because capillaries are present only in the perichondrium of cartilage. The following steps occur in the repair of a bone fracture (Figure 5.7):

1 As a result of the fracture, blood vessels crossing the fracture line are broken. These include vessels in the periosteum, osteons (Haversian systems), and medullary cavity. As blood pours from the torn ends of the vessels, it forms a clot in and about the site of the fracture. This clot, called a **fracture hematoma** (hē′-ma-TŌ-ma), usually occurs 6–8 hours after the injury. Because the circulation of blood ceases when the fracture hematoma forms, bone cells and periosteal cells at the fracture site die. The hematoma serves as a focus for the cellular invasion that follows. Swelling and inflammation occur after formation of the fracture hematoma, and there is a considerable amount of cell death and debris. Blood capillaries grow into the blood clot, and phagocytes (neutrophils and macrophages) and osteoclasts begin to remove the traumatized tissue in and around the fracture hematoma. This may take up to several weeks.

2 The infiltration of blood capillaries into the fracture hematoma helps organize it into an actively growing connective tissue called a **procallus.** Next, fibroblasts from the periosteum and osteoprogenitor cells from the periosteum,

endosteum, and marrow invade the procallus. The fibroblasts produce collagen fibers, which help connect the broken ends of the bones. Osteoprogenitor cells develop into chondroblasts in areas farther away from healthy bone tissue where the environment is avascular. Here chondroblasts begin to produce fibrocartilage, and the procallus is transformed into a **fibrocartilaginous (soft) callus.** A callus is actually a mass of repair tissue that bridges the broken ends of the bones. The stage of the fibrocartilaginous callus lasts about three weeks.

3 In areas closer to healthy bone tissue where the environment is more vascular, osteoprogenitor cells develop into osteoblasts, which begin to produce spongy bone trabeculae. The trabeculae join living and dead portions of the original bone fragments. In time, the fibrocartilage is converted to spongy bone and the callus is then referred to as a **bony (hard) callus.** The stage of the bony callus lasts about three to four months.

4 The final phase of fracture repair is **remodeling** of the callus. Dead portions of the original fragments are gradually resorbed by osteoclasts. Compact bone replaces spongy bone around the periphery of the fracture. Sometimes, the healing is so complete that the fracture line is undetectable, even in a radiograph (x-ray). However, a thickened area on the surface of the bone usually remains as evidence of a healed fracture site.

Although fractures may be classified in several different ways, the following list gives common categories.

1. **Partial:** the break across the bone is incomplete.

2. **Complete:** the break across the bone is complete, so that the bone is broken into two or more pieces.

3. **Closed (simple):** the bone does not break through the skin.

4. **Open (compound):** the broken ends of the bone protrude through the skin (Figure 5.8a).

5. **Comminuted** (KOM-i-nyoo-ted): the bone has splintered at the site of impact, and smaller fragments of bone lie between the two main fragments (Figure 5.8b).

6. **Greenstick:** a partial fracture in which one side of the bone is broken and the other side bends; occurs only in children (Figure 5.8c).

7. **Spiral:** the bone usually is twisted apart.

8. **Transverse:** a fracture at right angles to the long axis of the bone.

9. **Impacted:** one fragment is firmly driven into the other (Figure 5.8d).

10. **Displaced:** the anatomical alignment of the bone fragments is not preserved.

11. **Nondisplaced:** the anatomical alignment of the bone fragments is preserved.

12. **Stress:** microscopic fractures resulting from inability to withstand repeated stressful impact. Usually, they result from repeated, strenuous activities such as running, basketball, jumping, or aerobic dancing. About 25% of all stress fractures involve the shinbone (tibia).

13. **Pathologic:** weakening of a bone caused by disease processes such as osteogenic sarcoma (bone cancer), osteomyelitis (inflammation of a bone), osteoporosis (decreased bone mass), or osteomalacia.

14. **Pott's:** a fracture of the distal end of the lateral leg bone (fibula), with serious injury of the distal tibial articulation (Figure 5.8e).

15. **Colles'** (KOL-ez): a fracture of the distal end of the lateral forearm bone (radius) in which the distal fragment is displaced posteriorly (Figure 5.8f).

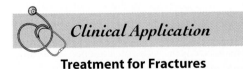

Clinical Application

Treatment for Fractures

Treatments for fractures depend on the person's age, type of fracture, the bone(s) involved, and the treatment options available. In order for bones to unite, the fractured ends must be brought into alignment, a process called **reduction,** and they must be held relatively immobile for the time required to heal the specific type of fracture. In **closed reduction,** the fractured

Figure 5.8 Types of fractures.

 A fracture is any break in a bone.

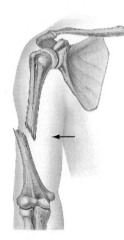
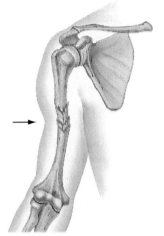

(a) Open fracture   (b) Comminuted fracture

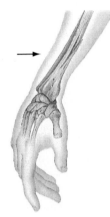
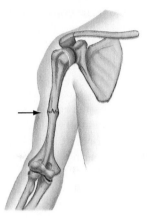

(c) Greenstick fracture   (d) Impacted fracture

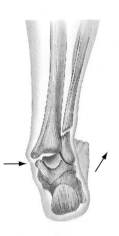
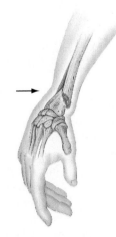

(e) Pott's fracture   (f) Colles' fracture

Q *What is a bony callus?*

ends of a bone are brought into alignment by manual manipulation and the skin remains intact. Traction may be applied to hasten closed reduction. In **open reduction,** the fractured ends of a bone are brought into alignment by a surgical procedure in which internal fixation devices such as screws, plates, pins, rods, and wires are used. Following reduction, the fractured bone is held immobile with a cast, sling, splint, elastic bandage, external fixation device, or a combination of these. ▪

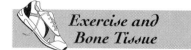

Exercise and Bone Tissue

Within limits, bone has the ability to alter its strength in response to mechanical stress. When placed under such stress, bone tissue becomes stronger with time, through increased deposition of mineral salts and production of collagen fibers. Another effect of stress is to increase the production of calcitonin, a hormone that inhibits bone resorption. Without mechanical stress, bone does not remodel normally because resorption outstrips bone formation. Removal of mechanical stress weakens bone through demineralization (loss of bone minerals) and collagen reduction.

The main mechanical stresses on bone are those that result from the pull of skeletal muscles and the pull of gravity. If a person is bedridden or has a fractured bone in a cast, the strength of the unstressed bones diminishes. Astronauts subjected to the weightlessness of space also lose bone mass. In both cases, the bone loss can be dramatic, as much as 1% per week. Bones of athletes, which are repetitively and highly stressed, become notably thicker than those of nonathletes. Weight-bearing activities, such as walking or moderate weight lifting, help build and retain bone mass. In recent years it has become increasingly clear that adolescent and young-adult females should engage in regular weight-bearing exercise prior to the sealing of the epiphyseal plates. This helps to build total bone mass prior to its inevitable reduction through adulthood.

Developmental Anatomy of the Skeletal System

Both intramembranous and endochondral ossification begin when **mesenchymal (mesodermal) cells** migrate into the area where bone formation will occur. In some skeletal structures, mesenchymal cells develop into **chondroblasts** that form *cartilage*. In other skeletal structures, mesenchymal cells develop into **osteoblasts** that form *bone tissue* by intramembranous or endochondral ossification.

Discussion of the development of the skeletal system provides us an excellent opportunity to trace the development of the limbs. The limbs make their appearance about the fifth

week as small elevations at the sides of the trunk called **limb buds** (Figure 5.9a). They consist of masses of general **mesoderm** covered by **ectoderm.** At this point, a mesenchymal skeleton exists in the limbs; some of the masses of mesoderm surrounding the developing bones will become the skeletal muscles of the limbs.

By the sixth week, the limb buds develop a constriction around the middle portion. The constriction produces distal segments of the upper buds called **hand plates** and distal segments of the lower buds called **foot plates** (Figure 5.9b). These plates represent the beginnings of the hands and feet, respectively. At this stage of limb development, a cartilaginous skeleton is present. By the seventh week (Figure 5.9c), the arm, forearm, and hand are evident in the upper limb bud, and the thigh, leg, and foot appear in the lower limb bud. Endochondral ossification has begun. By the eighth week (Figure 5.9d), as the shoulder, elbow, and wrist areas become apparent, the upper limb bud is appropriately called the upper limb and the lower limb bud is referred to as the lower limb.

The **notochord** is a flexible rod of mesoderm that defines the midline of the embryo and also gives it some rigidity. With respect to the development of the skeletal system, the notochord lies in a position where the future vertebral column will develop (see Figure 9.11b). As the vertebrae form, the notochord becomes surrounded by the developing vertebral bodies. The notochord eventually disappears except for remnants that persist as the *nucleus pulposus* of the intervertebral discs (see Figure 6.24).

Aging and Bone Tissue

There are two principal effects of aging on bone tissue. The first is the loss of calcium and other minerals from bone matrix (demineralization). This loss usually begins after age 30 in females, accelerates greatly around age 40 to 45 as levels of estrogens decrease, and continues until as much as 30% of the calcium in bones is lost by age 70. In males, calcium loss typically does not begin until after age 60. The loss of calcium from bones is one of the problems in a condition called osteoporosis (see Applications to Health, p. 4).

The second principal effect of aging on the skeletal system results from a decrease in the rate of protein synthesis. This results in decreased ability to produce the organic portion of bone matrix, mainly collagen, which normally gives bone its tensile strength. As a consequence, inorganic minerals gradually constitute a greater proportion of the bone matrix. The loss of tensile strength causes the bones to become very brittle and susceptible to fracture. In some elderly people collagen synthesis slows, in part, due to diminished production of human growth hormone.

Figure 5.9 External features of a developing human embryo at various stages of development. Many of the labeled structures are discussed in later chapters; they are indicated here to help you orient to the figure.

🔑 *Whereas chondroblasts derived from mesenchyme develop into cartilage, osteoblasts derived from mesenchyme develop into bone.*

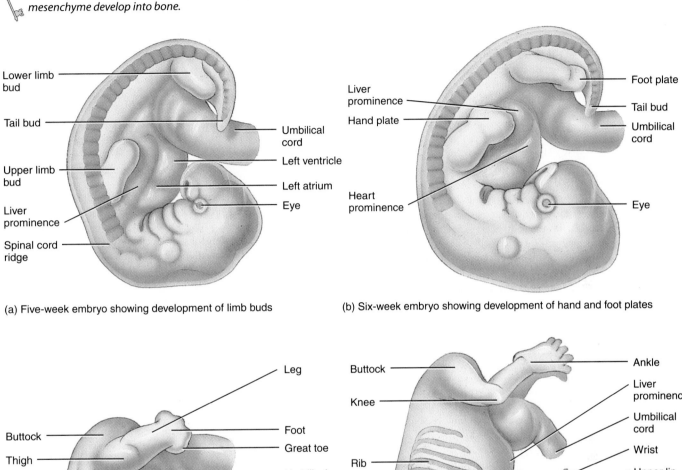

(a) Five-week embryo showing development of limb buds

(b) Six-week embryo showing development of hand and foot plates

(c) Seven-week embryo showing development of arm, forearm, and hand in upper limb bud and thigh, leg, and foot in lower limb bud

(d) Eight-week embryo in which limb buds have developed into upper and lower limbs

❓ *The skeletal system develops from which basic embryonic tissue?*

Key Medical Terms
Associated with Bone Tissue

Osteoarthritis (os′-tē-ō-ar-THRĪ-tis; *osteo* = bone; *arthro* = joint) The degeneration of articular cartilage, allowing the bony ends to touch and, from the friction of bone against bone, worsening the condition; usually associated with the elderly.

Osteogenic (os′-tē-Ō-JEN-ik) **sarcoma** (sar-KŌ-ma; *sarcoma* = connective tissue tumor) Bone cancer that primarily affects osteoblasts and occurs most often in the bones of teenagers during their growth spurt; most common sites are the metaphyses of the thighbone (femur), shinbone (tibia), and arm bone (humerus); metastases occur most often in lungs; treatment consists of multidrug chemotherapy and removal of the malignant growth or amputation of the limb.

Osteopenia (os′-tē-ō-PĒ-nē-a; *penia* = poverty) Reduced bone mass due to a decrease in the rate of bone synthesis to a level insufficient to compensate for normal bone breakdown; any decrease in bone mass below normal. An example is osteoporosis.

Study Outline

Functions of Bone (p. 108)

1. Bone (osseous) tissue and the skeletal system function in support, protection, movement, mineral storage and release, blood cell production, and storage of energy (lipids in adipose cells are an important chemical reserve).

Anatomy: Structure of Bone (p. 108)

1. Parts of a typical long bone are the diaphysis (shaft), epiphyses (ends), metaphyses, articular cartilage, periosteum, medullary (marrow) cavity, and endosteum.

Histology of Bone Tissue (p. 109)

1. Bone tissue consists of widely separated cells surrounded by large amounts of matrix.
2. The four principal types of cells are osteoprogenitor cells, osteoblasts, osteocytes, and osteoclasts.
3. The matrix of bone contains abundant mineral salts (mostly hydroxyapatite) and collagen fibers.

Compact Bone Tissue (p. 111)
1. Compact (dense) bone tissue consists of osteons (Haversian systems) with little space between them.
2. Compact bone lies over spongy (cancellous) bone and composes most of the bone tissue of the diaphysis. Functionally, compact bone tissue protects and supports the body and resists mechanical stress.

Spongy Bone Tissue (p. 112)
1. Spongy (cancellous) bone tissue does not contain osteons. It consists of trabeculae surrounding many red bone marrow–filled spaces.
2. It forms most of the structure of short, flat, and irregular bones, and the epiphyses of long bones.
3. Functionally, spongy bone tissue stores red bone marrow and provides some support.

Bone Formation: Ossification (p. 113)

1. Bone forms by a process called ossification, which begins when mesenchymal cells become transformed into osteoprogenitor cells. These undergo cell division and give rise to cells that differentiate into osteoblasts and osteoclasts.
2. The process begins during the sixth or seventh week of embryonic life and continues throughout adulthood. The two types of ossification, intramembranous and endochondral, involve the replacement of a preexisting connective tissue with bone.
3. Intramembranous ossification occurs within loose fibrous connective tissue membranes of the embryo and the adult.
4. Endochondral ossification occurs within a hyaline cartilage model. The primary ossification center of a long bone is in the diaphysis. Cartilage degenerates, leaving cavities that merge to form the medullary cavity. Osteoblasts lay down bone. Next, ossification occurs in the epiphyses, where bone replaces cartilage, except for the epiphyseal plate.

Bone Growth (p. 116)

1. The anatomical zones of the epiphyseal plate are the zones of resting cartilage, proliferating cartilage, hypertrophic (maturing) cartilage, and calcified cartilage.
2. Because of the activity of the epiphyseal plate, the diaphysis of a bone increases in length by interstitial growth.
3. Bone grows in diameter by appositional growth as a result of the addition of new bone tissue by periosteal osteoblasts around the outer surface of the bone.

Blood and Nerve Supply (p. 117)

1. Long bones are supplied by nutrient, epiphyseal, and periosteal arteries; veins accompany the arteries.
2. The nerve supply to bones is not extensive; some nerves accompany blood vessels, whereas others occur in the periosteum.

Bone Replacement (p. 117)

1. Remodeling is the replacement of old bone tissue by new bone tissue.
2. Old bone tissue is constantly destroyed by osteoclasts, whereas new bone is constructed by osteoblasts.
3. Remodeling requires minerals (calcium, phosphorus, magnesium, boron, and manganese), vitamins (D, C, A, and B_{12}), and hormones (human growth hormone, sex hormones, insulin, thyroid hormones, parathyroid hormone, and calcitonin).

Fracture and Repair of Bone (p. 117)

1. A fracture is any break in a bone.
2. Fracture repair involves formation of a fracture hematoma, fibrocartilaginous callus, bone callus, and remodeling.

3. The types of fractures include partial, complete, closed (simple), open (compound), comminuted, greenstick, spiral, transverse, impacted, Pott's, Colles', displaced, nondisplaced, pathologic, and stress.

Exercise and Bone Tissue (p. 120)

1. Bone can alter its strength in response to mechanical stress by increasing depositon of mineral salts and production of collagen fibers.
2. Removal of mechanical stress weakens bone through demineralization and collagen reduction.

Developmental Anatomy of the Skeletal System (p. 120)

1. Bone forms from mesoderm by intramembranous or endochondral ossification.
2. The limbs develop from limb buds, which consist of mesoderm and ectoderm.

Aging and Bone Tissue (p. 120)

1. The principal effect of aging is a loss of calcium from bones, which may result in osteoporosis.
2. Another effect is a decreased production of matrix (mostly collagen), which makes bones more susceptible to fracture.

Review Questions

1. Define the skeletal system. What are its six principal functions? (p. 108)
2. Diagram the parts of a long bone and list the functions of each part. (p. 108)
3. Why is bone tissue considered a connective tissue? Describe the cells present and the composition of the matrix. (p. 109)
4. Distinguish between spongy and compact bone tissue in terms of microscopic appearance, location, and function. (p. 111)
5. Diagram the microscopic appearance of compact bone tissue and indicate the functions of the various components. (p. 111)
6. What is meant by ossification? Describe the two major types. (p. 113)
7. Describe the histology of the various zones of the epiphyseal plate. How does the plate grow? What is the significance of the epiphyseal line? (p. 116)
8. Contrast bone growth in length with bone growth in diameter. (p. 116)
9. Describe the blood and nerve supplies of a long bone. (p. 117)
10. What is remodeling? Why is it important? (p. 117)
11. What is a fracture? Outline the three basic steps of fracture repair. Distinguish among several principal kinds of fractures. (p. 117)
12. Explain the effects of exercise on bone tissue. (p. 120)
13. Describe the development of the skeletal system. (p. 120)
14. Explain the effects of aging on bone tissue. (p. 120)
15. Refer to the glossary of key medical terms associated with bone tissue. Be sure that you can define each term. (p. 122)

Self Quiz

Complete the following:

1. Most bones of the body are formed by the process of ___ ossification.
2. The tensile strength of bone depends on the presence of ___ fibers in the matrix.
3. Typical of all connective tissues, bone consists mainly of ___.
4. Bones and cartilage are formed from ___-derm that later differentiates into the embryonic connective tissue called ___.
5. A ___ fracture is a partial fracture that only occurs in children.
6. Match the following.
   (a) thin layer of hyaline cartilage at end of long bone
   (b) region of mature bone where diaphysis joins epiphysis
   (c) covering over bone to which ligaments and tendons attach
   (d) inner layer of covering over bone; contains osteoprogenitor cells and osteoblasts
   (e) layer of osteoprogenitor cells and osteoblasts lining the medullary cavity

   (1) articular cartilage
   (2) endosteum
   (3) fibrous periosteum
   (4) metaphysis
   (5) osteogenic periosteum

Choose the best answers to these questions:

7. Which of the following statements is true?
   a. The outer connective tissue covering of a bone is the perichondrium.
   b. Spongy bone tissue contains osteons.
   c. The medullary cavity of a long bone contains red marrow.
   d. The spaces in spongy bone are called the medullary cavities.
   e. None of these statements is true.
8. The cells responsible for the resorption (destruction) of bone tissue are
   a. osteocytes
   b. osteoblasts
   c. osteoclasts
   d. chondrocytes
   e. chondroblasts
9. Which of the following is *not* considered a function of bone tissue and the skeletal system?
   a. attachment of muscles, tendons, and ligaments
   b. secretion of calcitonin that affects calcium levels in the blood
   c. formation of blood cells and fat storage
   d. support and protection of soft organs and tissue
   e. storage area for calcium and phosphorus
10. Consider the following four sets of vitamins and minerals.
    (1) vitamins B and K
    (2) calcium and phosphorus

Without bones, you would be unable to perform movements such as walking or grasping. The slightest jar to your head or chest could damage the brain or heart. It would even be impossible to chew food. Because the skeletal system forms the framework of the body, a familiarity with the names, shapes, and positions of individual bones will help you understand some of the other organ systems. For example, the radial artery, the site where pulse is usually taken, is named for its proximity to the radius, the lateral bone of the forearm. The frontal lobe of the brain lies deep to the frontal (forehead) bone. The tibialis anterior muscle is located near the anterior surface of the tibia (shinbone). The ulnar nerve is named for its proximity to the ulna, the medial bone of the forearm.

Movements such as throwing a ball, biking, and walking require an interaction between bones and muscles. To understand how muscles produce different movements, you need to learn where on bones the muscles attach and the types of joints acted on by the contracting muscles. The bones, muscles, and joints together form an integrated system called the **musculoskeletal system.** Many bones also serve as anatomical and surgical landmarks. For example, parts of certain bones serve to locate structures within the skull and to outline the lungs and heart and abdominal and pelvic organs. Blood vessels and nerves often run parallel to bones. These structures can be located more easily if the bone is identified first.

DIVISIONS OF THE SKELETAL SYSTEM

The adult human skeleton consists of 206 named bones grouped in two principal divisions: the **axial skeleton** and the **appendicular skeleton.** Refer to Figure 6.1 to see how the two divisions join to form the complete skeleton. The bones of the axial skeleton are shown in blue. The longitudinal *axis,* or center, of the human body is a straight line that runs through the body's center of gravity. This imaginary line extends through the head and down to the space between the feet. The axial skeleton consists of the bones that lie around the axis: skull bones, auditory ossicles (ear bones), hyoid bone, ribs, breastbone, and bones of the backbone. Although the auditory ossicles are not considered part of the axial or appendicular skeleton, but rather as a separate group of bones, they are placed with the axial skeleton for convenience. The middle portion of each ear contains three auditory ossicles held together by a series of ligaments. The auditory ossicles vibrate in response to sound waves that strike the eardrum and have a key role in the mechanism of hearing. This is described in detail in Chapter 20.

The appendicular skeleton contains the bones of the **upper** and **lower limbs (extremities),** plus the bones called **girdles** that connect the limbs to the axial skeleton. Table 6.1 on page 129 presents the standard grouping of the 80 bones of the axial skeleton and the 126 bones of the appendicular skeleton.

We will study bones by examining the various regions of the body. First, we will look at the skull and see how its bones relate to each other. We will then move on to the vertebral column (backbone) and the chest. In Chapter 7 we will examine the bones of the pectoral (shoulder) girdle and upper limbs, then the pelvic (hip) girdle and lower limbs. This regional approach will allow you to see how the many bones of the body relate to one another.

TYPES OF BONES

Almost all the bones of the body may be classified into five principal types on the basis of shape: long, short, flat, irregular, and sesamoid (Figure 6.2). **Long bones** have greater length than width and consist of a shaft and a variable number of extremities (ends). They are slightly curved for strength. A curved bone absorbs the stress of the body's weight at several different points so that the stress is evenly distributed. If such bones were straight, the weight of the body would be unevenly distributed and the bone would easily fracture. Long bones consist mostly of *compact bone tissue,* which is dense and has few spaces, but also contain considerable amounts of *spongy bone tissue,* which has larger spaces. Long bones include those in the thigh (femur), leg (tibia and fibula), toes (phalanges), arm (humerus), forearm (ulna and radius), and fingers (phalanges).

Short bones are somewhat cube-shaped and nearly equal in length and width. They are spongy bone except at the surface, where there is a thin layer of compact bone. Examples of short bones are the wrist or carpal bones (except for the pisiform, which is a sesamoid bone) and the ankle or tarsal bones (except for the calcaneus, which is an irregular bone).

Flat bones are generally thin and composed of two nearly parallel plates of compact bone enclosing a layer of spongy bone. The layers of compact bone are called external and internal *tables,* whereas the spongy bone is referred to as *diploe* (DIP-lō-ē). Flat bones afford considerable protection and provide extensive areas for muscle attachment. Flat bones include the cranial bones (which protect the brain), the breastbone (sternum) and ribs (which protect organs in the thorax), and the shoulder blades (scapulae).

Irregular bones have complex shapes and cannot be grouped into any of the three categories just described. They also vary in the amount of spongy and compact bone present. Such bones include the backbones (vertebrae) and certain facial bones.

Figure 6.1 Divisions of the skeletal system. The axial skeleton is indicated in blue. Note the position of the hyoid bone in Figure 6.4a.

The adult human skeleton consists of 206 bones grouped into axial and appendicular divisions.

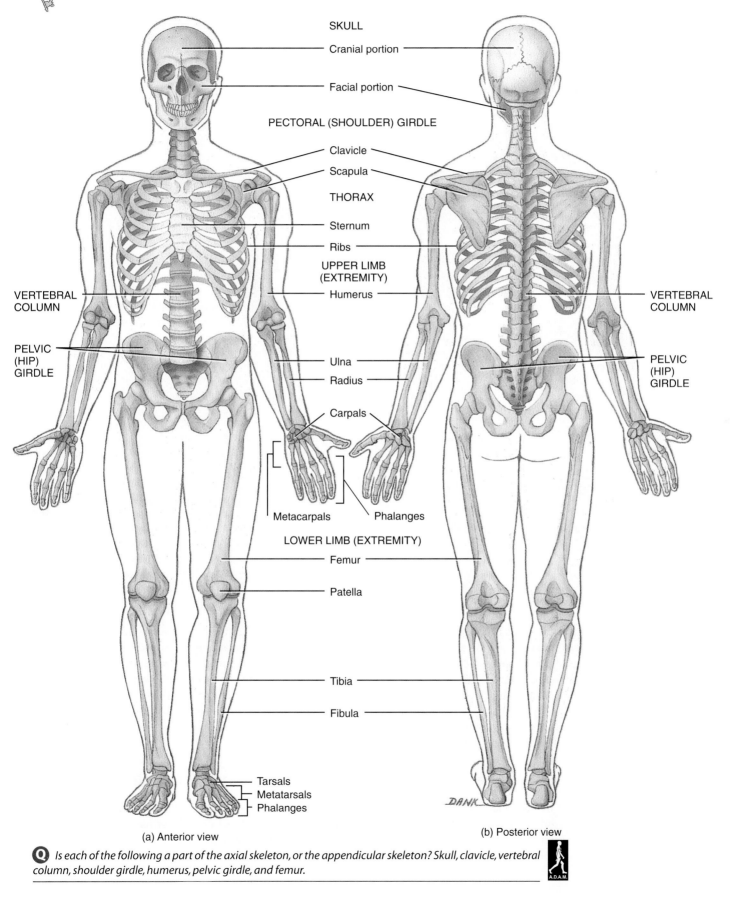

(a) Anterior view

(b) Posterior view

Q *Is each of the following a part of the axial skeleton, or the appendicular skeleton? Skull, clavicle, vertebral column, shoulder girdle, humerus, pelvic girdle, and femur.*

Figure 6.2 Types of bones based on shape. The bones are not drawn to scale.

The shapes of bones largely determine their functions.

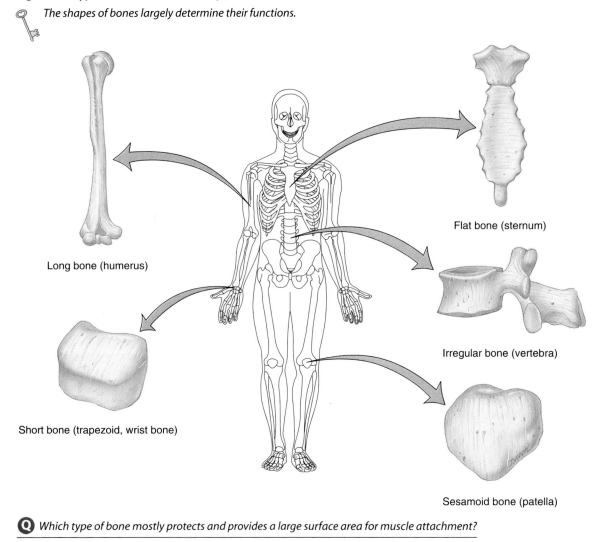

Long bone (humerus)

Flat bone (sternum)

Short bone (trapezoid, wrist bone)

Irregular bone (vertebra)

Sesamoid bone (patella)

Q *Which type of bone mostly protects and provides a large surface area for muscle attachment?*

Sesamoid (meaning shaped like a sesame seed) **bones** develop in certain tendons where there is considerable friction, tension, and physical stress. They are not always completely ossified and measure only a few millimeters in diameter; exceptions are the two patellae (kneecaps), which are large sesamoid bones. Sesamoid bones vary in number from person to person; exceptions again are the patellae, which are normally present in all individuals in the quadriceps femoris tendon. Functionally, sesamoid bones protect tendons from excessive wear and tear, and they often change the direction of pull of a tendon, which improves the mechanical advantage at a joint.

In the upper limbs, sesamoid bones normally occur only in the joints of the palmar surface of the hands. The two constant sesamoid bones are in the tendons of the adductor pollicis and flexor pollicis brevis muscles at the metacarpophalangeal joint of the thumb (see Figure 7.7). In the lower limbs, aside from the patellae, two constant sesamoid bones occur on the plantar surface of the foot in the tendons of the flexor hallucis brevis muscle at the metacarpophalangeal joint of the great (big) toe (see Figure 7.15).

An additional type of bone is not included in this classification by shape, but instead is classified by location. **Sutural** (SOO-chur-al; *sutura* = seam) or **Wormian** (named after a Danish anatomist, O. Worm, 1588–1654) **bones** are small bones located within the sutures (joints) of certain cranial bones. Their number varies greatly from person to person.

BONE SURFACE MARKINGS

The surfaces of bones have various structural features adapted to specific functions. These features are called **surface markings.** Long bones that bear a lot of weight have

Table 6.1 *Divisions of the Adult Skeletal System*

REGIONS OF THE SKELETON	NUMBER OF BONES
Axial Skeleton	
Skull	
Cranium	8
Face	14
Hyoid	1
Auditory ossicles	6
Vertebral column	26
Thorax	
Sternum	1
Ribs	24
Subtotal = 80	
Appendicular Skeleton	
Pectoral (shoulder) girdles	
Clavicle	2
Scapula	2
Upper limbs (extremities)	
Humerus	2
Ulna	2
Radius	2
Carpals	16
Metacarpals	10
Phalanges	28
Pelvic (hip) girdle	
Hip, pelvic, or coxal bone	2
Lower limbs (extremities)	
Femur	2
Fibula	2
Tibia	2
Patella	2
Tarsals	14
Metatarsals	10
Phalanges	28
Subtotal = 126	
Total = 206	

large, rounded ends that can form sturdy joints and provide adequate surface area for the attachment of ligaments and muscles. Other bones have depressions that receive the rounded ends. Rough areas serve as points of attachment for muscles, tendons, and ligaments. Grooves on the surfaces of bones provide for the passage of blood vessels. Openings occur where blood vessels and nerves pass through the bone. Table 6.2 describes the different bone surface markings and provides examples of each.

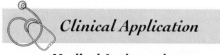

Clinical Application

Medical Anthropology

Skeletal remains may persist for thousands of years after a person has died. They make it possible to trace patterns of disease and nutrition, evaluate the effects of certain social and economic changes, and deduce patterns of reproduction and mortality. Skeletal remains also may reveal an individual's sex, age, height, and race.

Many disorders can leave permanent effects on skeletal material. Three of the many common causes of bone pathologies are malnutrition, tumors, and infections. Each may cause specific changes in bone that permit diagnosis from skeletal remains. ▪

SKULL
Divisions

The **skull,** which contains 22 bones, rests on the superior end of the vertebral column. It includes two sets of bones: cranial bones and facial bones (Table 6.3). The **cranial** (*cranium* = brain case) **bones** form the cranial cavity and enclose and protect the brain. The eight cranial bones are the frontal bone, parietal bones (2), temporal bones (2), occipital bone, sphenoid bone, and ethmoid bone. There are 14 **facial bones** that form the face: nasal bones (2), maxillae or maxillas (2), zygomatic bones (2), mandible, lacrimal bones (2), palatine bones (2), inferior nasal conchae (2), and vomer. Be sure you can locate all the skull bones in the various views of the skull in Figures 6.3 through 6.8.

General Features

Besides forming the large cranial cavity, the skull also forms several smaller cavities. These include the nasal cavity and orbits (eye sockets), which open to the exterior. Certain skull bones also contain cavities that are lined with mucous membranes and are called paranasal sinuses. The sinuses open into the nasal cavity. Also within the skull are small cavities that house the structures that are involved in hearing and equilibrium.

Table 6.2 Bone Surface Markings

Depressions and Openings: Sites allowing the passage of soft tissues (nerves, blood vessels, ligaments, tendons)

MARKING	DESCRIPTION	EXAMPLE
Fissure (FISH-ur)	Narrow slit between adjacent parts of bones through which blood vessels or nerves pass.	Superior orbital fissure of the sphenoid bone (Figure 6.13).
Foramen (fō-RĀ-men)	Opening (*foramen* = hole) through which blood vessels, nerves, or ligaments pass.	Optic foramen of the sphenoid bone (Figure 6.13).
Fossa (FOS-a)	Shallow depression (*fossa* = basin).	Coronoid fossa of the humerus (Figure 7.3a).
Sulcus (SUL-kus)	Furrow (*sulcus* = groove) along bone surface that accommodates a blood vessel, nerve, or tendon.	Intertubercular sulcus of the humerus (Figure 7.3a).
Meatus (mē-Ā-tus)	Tubelike opening (*meatus* = passageway).	External auditory (acoustic) meatus of the temporal bone (Figure 6.4a).

Processes: Projections or outgrowths on bone that form joints or attachment points for connective tissue.

MARKING	DESCRIPTION	EXAMPLE
Processes that form joints:		
Condyle (KON-dīl)	Large, round protuberance (*condylus* = knuckle) at the end of a bone.	Lateral condyle of the femur (Figure 7.12b).
Facet	Smooth flat surface.	Superior articular facet of a vertebra (Figure 6.17c).
Head	Rounded articular projection supported on the neck (constricted portion) of a bone.	Head of the femur (Figure 7.12a).
Ramus (RĀ-mus)	Part of the bone that forms an angle with the main body of the bone.	Ramus of the mandible (Figure 6.12a).
Processes that form attachment points for connective tissue:		
Crest	Prominent ridge or elongated projection.	Iliac crest of the hipbone (Figure 7.9a).
Epicondyle	Projection above (*epi* = above) a condyle.	Medial epicondyle of the femur (Figure 7.12b).
Line	Long, narrow ridge or border (less prominent than a crest).	Linea aspera of the femur (Figure 7.12b).
Spinous process	Sharp, slender projection.	Spinous process of a vertebra (Figure 6.17a).
Trochanter (trō-KAN-ter)	Very large projection on the femur.	Greater trochanter of the femur (Figure 7.12b).
Tubercle (TOO-ber-kul)	Small, rounded projection (*tube* = knob).	Greater tubercle of the humerus (Figure 7.3a).
Tuberosity	Large, rounded, usually roughened projection.	Ischial tuberosity of the hipbone (Figure 7.9a).

Other than the auditory ossicles, the mandible is the only movable bone of the skull. Most of the skull bones are held together by immovable joints called sutures. These are especially noticeable on the outer surfaces of the bones.

The skull has numerous surface markings such as foramina and fissures through which blood vessels and nerves pass. As various skull bones are described, you will learn the names of important bone surface markings.

In addition to protecting the brain, the cranial bones also have other functions. Their inner surfaces attach to membranes (meninges) that stabilize the positions of the brain, blood vessels, and nerves. Their outer surfaces provide large areas of attachment for muscles that move various parts of the head.

Besides forming the framework of the face, the facial bones protect and provide support for the entrances to the digestive and respiratory systems. The bones also provide attachment for some muscles that are involved in producing various facial expressions.

Together, the cranial and facial bones protect and support the delicate special sense organs for vision, taste, smell, hearing, and equilibrium (balance).

Table 6.3 *Summary of Bones of the Adult Skull*

CRANIAL BONES	FACIAL BONES
Frontal (1)	Nasal (2)
Parietal (2)	Maxillae (2)
Temporal (2)	Zygomatic (2)
Occipital (1)	Mandible (1)
Sphenoid (1)	Lacrimal (2)
Ethmoid (1)	Palatine (2)
	Inferior nasal conchae (2)
	Vomer (1)

The numbers in parentheses indicate how many of each bone are present.

Sutures

A **suture** (SOO-chur; *sutura* = seam) is an immovable joint that is found only between skull bones. Sutures hold all skull bones together. There are several types of sutures that can be distinguished according to how the margins of the bones unite. In some sutures, the margins of the bones are fairly smooth. In other sutures, the margins overlap. In still other sutures, the margins interlock in a jigsaw fashion. This latter arrangement provides sutures added strength and decreases their chance of fracturing.

The names of many sutures reflect the bones that they unite. For example, the frontozygomatic suture is between the frontal bone and zygomatic bone. Similarly, the spheno-parietal suture is between the sphenoid bone and parietal bone. In other cases, however, the names of sutures are not so obvious.

Of the many sutures that are found in the skull, we will identify only four prominent ones:

1. **Coronal** (kō-RŌ-nal; *corona* = crown) **suture.** This suture unites the frontal bone and two parietal bones. See Figure 6.4.

2. **Sagittal** (SAJ-i-tal; *sagitta* = arrow) **suture.** This suture unites the two parietal bones. The sagittal suture is so named because in the infant, before the bones of the skull are firmly united, the suture and the fontanels (soft spots) associated with it somewhat resemble an arrow. See Figure 6.6.

3. **Lambdoid** (LAM-doyd) **suture.** This suture unites the parietal bones and occipital bones. The suture is so named because of its resemblance to the Greek letter lambda (Λ), as can be seen in Figure 6.6.

4. **Squamous** (SKWĀ-mus; *squama* = flat) **suture.** This suture unites the parietal bones and temporal bones. It is so named because the part of the temporal bone that unites with the parietal bone to form the suture is a thin, flat region called the temporal squama. See Figure 6.4.

There may be sutural bones in the sagittal and lambdoid sutures.

Cranial Bones

Frontal Bone

The **frontal bone** forms the forehead (the anterior part of the cranium), the roofs of the *orbits* (eye sockets), and most of the anterior part of the cranial floor. Soon after birth, the left and right sides of the frontal bone are united by the *frontal suture* (see Figure 6.14a), which usually disappears by age 6. If it persists throughout life, it is referred to as the *metopic suture*.

If you examine the anterior view of the skull in Figure 6.3, you will note the *frontal squama,* or *vertical plate.* This scalelike plate, which corresponds to the forehead, gradually slopes inferiorly from the coronal suture, then turns abruptly inferiorly. Superior to the orbits, the frontal bone thickens, forming the *supraorbital* (*supra* = above) *margin.* From this margin the frontal bone extends posteriorly to form the roof of the orbit and part of the floor of the cranial cavity. Within the supraorbital margin, slightly medial to its midpoint, is a hole called the *supraorbital* (*supra* = above; *orbital* = orbit) *foramen* (plural: *foramina*). As the various foramina associated with cranial bones are discussed, refer to Table 6.4 on page 142 to note which structures pass through them. The *frontal sinuses* lie deep to the frontal squama. Among other functions, paranasal sinuses act as sound chambers that give the voice resonance.

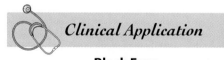

Clinical Application

Black Eyes

Just above the supraorbital margin is a sharp ridge. A blow to the ridge often lacerates the skin over it, resulting in bleeding. Bruising of the skin over the ridge causes tissue fluid and blood to accumulate in the surrounding connective tissue and gravitate into the upper eyelid. The resulting swelling and discoloration are called **black eye (lid contusion).** It may also be caused by fractures of the bones in the area. ■

Parietal Bones

The two **parietal** (pa-RĪ-e-tal; *paries* = wall) **bones** form the greater portion of the sides and roof of the cranial cavity

Figure 6.3 Skull in anterior view.

The skull consists of two sets of bones: cranial and facial.

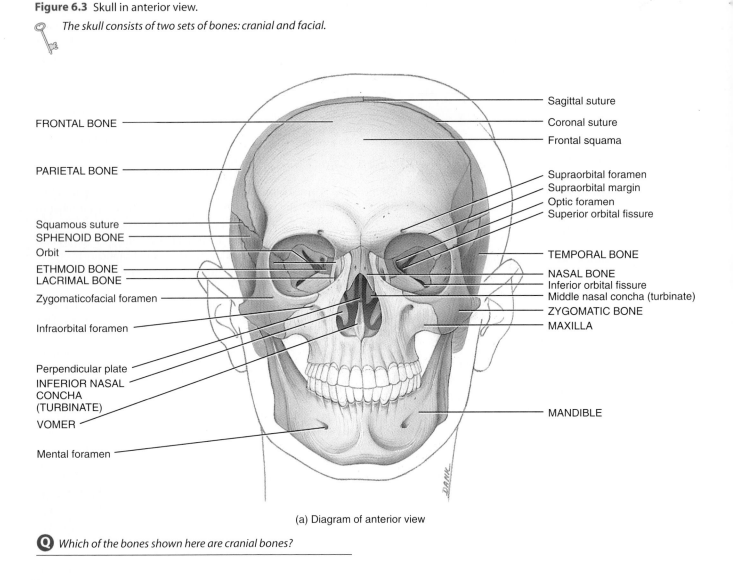

FRONTAL BONE

PARIETAL BONE

Squamous suture
SPHENOID BONE

Orbit
ETHMOID BONE
LACRIMAL BONE

Zygomaticofacial foramen

Infraorbital foramen

Perpendicular plate
INFERIOR NASAL
CONCHA
(TURBINATE)

VOMER

Mental foramen

Sagittal suture

Coronal suture

Frontal squama

Supraorbital foramen
Supraorbital margin
Optic foramen
Superior orbital fissure

TEMPORAL BONE

NASAL BONE
Inferior orbital fissure
Middle nasal concha (turbinate)
ZYGOMATIC BONE
MAXILLA

MANDIBLE

(a) Diagram of anterior view

Q *Which of the bones shown here are cranial bones?*

(see Figures 6.4 and 6.5). The internal surfaces of the parietal bones contain many protrusions and depressions that accommodate the blood vessels supplying the superficial meninx (covering) of the brain called the dura mater.

Temporal Bones

The two **temporal** (*tempora* = temples) **bones** form the inferior lateral aspects of the cranium and part of the cranial floor.

In the lateral view of the skull (Figure 6.4) it is possible to observe additional aspects of the skull. The *temporal squama* is a thin, flat portion of the temporal bone that forms the anterior and superior part of the temple. Projecting from the inferior portion of the temporal squama is the *zygomatic process,* which articulates with the temporal process of the zygomatic (cheek) bone. The zygomatic process of the temporal bone and the temporal process of the zygomatic bone form the *zygomatic arch* (see Figure 6.4).

At the floor of the cranial cavity (see Figure 6.8a) is the *petrous* (*petrous* = rocky) *portion* of the temporal bone. This portion is triangular and located at the base of the skull between the sphenoid and occipital bones. The petrous portion houses the internal and middle ear, structures involved in hearing and equilibrium (balance). It also contains the *carotid* (relating to the carotid artery in the neck) *foramen* (see Figure 6.7). Posterior to the carotid foramen and anterior to the occipital bone is the *jugular* (*jugulum* = throat just above the clavicle) *foramen.*

On the inferior surface of the zygomatic process of the temporal bone is a socket called the *mandibular (glenoid) fossa.* Anterior to the mandibular fossa is a rounded elevation, the *articular tubercle* (see Figures 6.4 and 6.7). The

Figure 6.3 (continued)

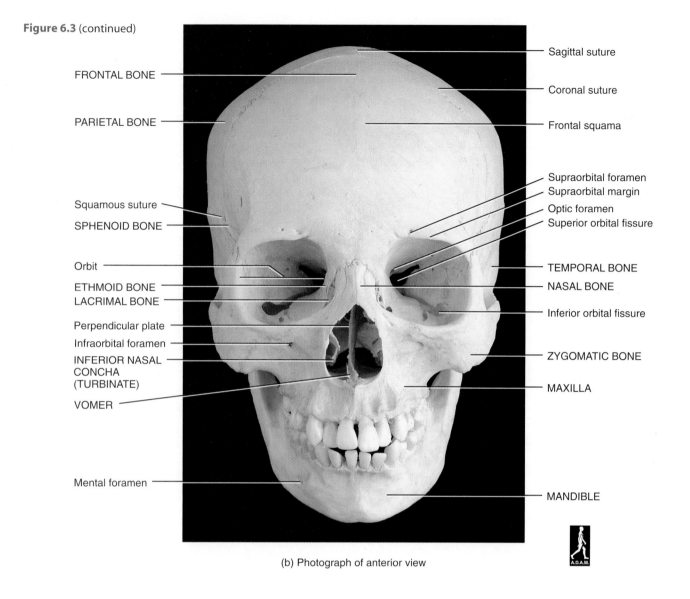

FRONTAL BONE

PARIETAL BONE

Squamous suture
SPHENOID BONE

Orbit
ETHMOID BONE
LACRIMAL BONE

Perpendicular plate
Infraorbital foramen
INFERIOR NASAL
CONCHA
(TURBINATE)

VOMER

Mental foramen

Sagittal suture

Coronal suture

Frontal squama

Supraorbital foramen
Supraorbital margin
Optic foramen
Superior orbital fissure

TEMPORAL BONE
NASAL BONE

Inferior orbital fissure

ZYGOMATIC BONE

MAXILLA

MANDIBLE

(b) Photograph of anterior view

mandibular fossa and articular tubercle articulate (form a joint) with the condylar process of the mandible (lower jawbone) to form the *temporomandibular joint (TMJ)*.

In the lateral view of the skull (see Figure 6.4), you will see the *mastoid (mastoid* = breast-shaped) *portion* of the temporal bone. It is located posterior and inferior to the *external auditory (acoustic) meatus,* or ear canal, which directs sound waves into the ear. In the adult, this portion of the bone contains several *mastoid air "cells."* These are air spaces, separated from the brain only by thin bony partitions. If **mastoiditis** (inflammation of these bony cells) occurs, the infection may spread from the throat to the middle ear to the brain or its outer covering.

The *mastoid process* is a rounded projection of the temporal bone posterior to the external auditory (acoustic) mea-

tus. It serves as a point of attachment for several neck muscles. Near the posterior border of the mastoid process is the *mastoid foramen* (see Figure 6.7). The external auditory (acoustic) meatus is the canal in the temporal bone that leads to the middle ear. The *internal auditory (acoustic) meatus* is superior to the jugular foramen (see Figure 6.5). The meatus transmits the facial (VII) and vestibulocochlear (VIII) cranial nerves. The *styloid (styloid* = stake or pole) *process* (see Figure 6.4a) projects inferiorly from the undersurface of the temporal bone and serves as a point of attachment for muscles and ligaments of the tongue and neck. Between the styloid process and the mastoid process is the *stylomastoid foramen* (see Figure 6.7).

Figure 6.4 Skull in lateral view. Although the hyoid bone is not part of the skull, it is included in the illustration for reference.

The zygomatic arch is formed by the zygomatic process of the temporal bone and the temporal process of the zygomatic bone.

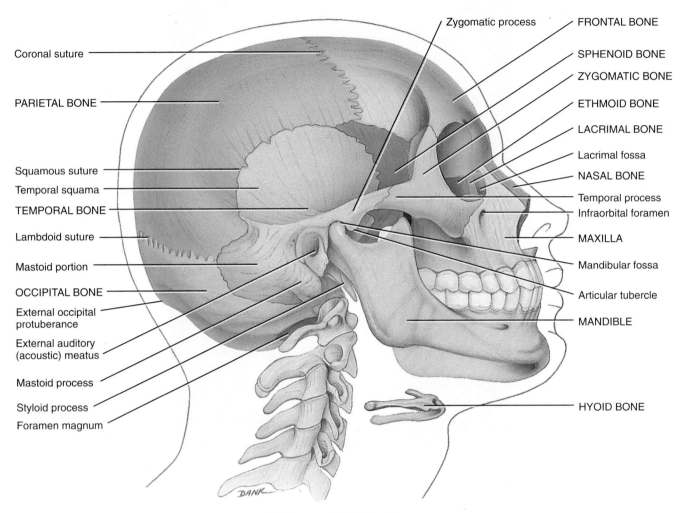

(a) Diagram of right lateral view

Occipital Bone

When the skull is examined in posterior view (Figure 6.6), the **occipital** (ok-SIP-i-tal; *occipital* = back of head) **bone** is easily seen. It forms the posterior part and most of the base of the cranium (Figures 6.4 and 6.6).

The *foramen magnum,* which means a large hole (*magnum* = large), is in the inferior part of the bone. Within this foramen, the medulla oblongata (inferior part of the brain) connects with the spinal cord. The vertebral and spinal arteries pass through this foramen.

The *occipital condyles* are oval processes with convex surfaces, one on either side of the foramen magnum, that articulate with depressions on the first cervical vertebra to form the *atlanto-occipital joint.* Superior to each occipital condyle on the inferior surface of the skull is the *hypoglossal* (*hypo* = under; *glossus* = tongue) *canal* (see Figure 6.5).

The *external occipital protuberance* is a prominent projection on the posterior surface of the bone just superior to the foramen magnum. You may be able to feel this structure as a definite bump on the back of your head, just above your neck. The protuberance is also shown in Figure 6.4. A large fibrous, elastic ligament, the *ligamentum nuchae,* extends from the external occipital protuberance to the seventh cervical vertebra. It helps to support the head, especially in animals. Extending laterally from the protuberance are two curved lines, the *superior nuchal lines,* and below these two *inferior nuchal lines,* which are areas of muscle attachment.

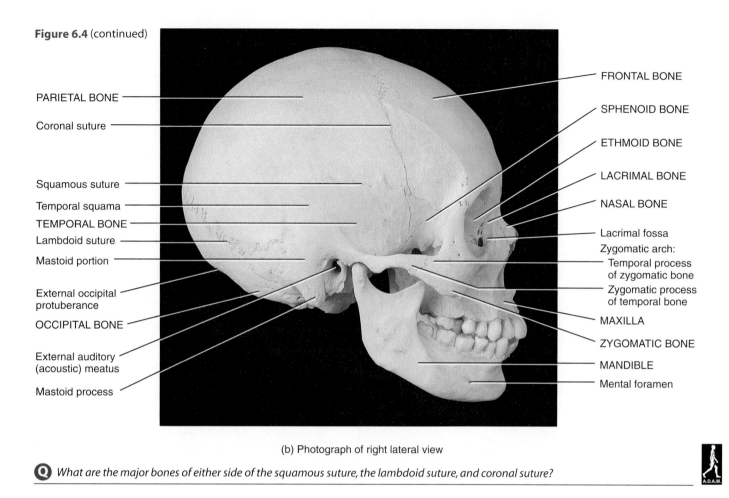

Figure 6.4 (continued)

PARIETAL BONE

Coronal suture

Squamous suture

Temporal squama

TEMPORAL BONE

Lambdoid suture

Mastoid portion

External occipital protuberance

OCCIPITAL BONE

External auditory (acoustic) meatus

Mastoid process

FRONTAL BONE

SPHENOID BONE

ETHMOID BONE

LACRIMAL BONE

NASAL BONE

Lacrimal fossa

Zygomatic arch:
Temporal process of zygomatic bone
Zygomatic process of temporal bone

MAXILLA

ZYGOMATIC BONE

MANDIBLE

Mental foramen

(b) Photograph of right lateral view

Q *What are the major bones of either side of the squamous suture, the lambdoid suture, and coronal suture?*

In the inferior view of the skull in Figure 6.7 on page 138, it is possible to examine the parts of the occipital bone, as well as surrounding structures.

Sphenoid Bone

The **sphenoid** (SFĒ-noyd; *spheno* = wedge-shaped) **bone** lies at the middle part of the base of the skull (see Figure 6.7 on page 138 and Figure 6.8 on page 140). This bone is called the keystone of the cranial floor because it articulates with all the other cranial bones, holding them together. Viewing the floor of the cranium from above (Figure 6.8a), note the sphenoid articulations: anteriorly with the frontal bone, laterally with the temporal bones, and posteriorly with the occipital bone. It lies posterior and slightly superior to the nasal cavity and forms part of the floor, sidewalls, and rear wall of the orbit (see Figure 6.13).

The shape of the sphenoid resembles a bat with outstretched wings (Figure 6.8c). The *body* of the sphenoid is the cubelike central portion between the ethmoid and occipital bones. It contains the *sphenoidal sinuses,* which drain into the nasal cavity (see Figure 6.11). On the superior surface of the body of the sphenoid is a depression called the *sella turcica* (SEL-a TUR-si-ka; = Turkish saddle), which cradles the pituitary gland.

The *greater wings* of the sphenoid project laterally from the body, forming the anterolateral floor of the cranium. The greater wings also form part of the lateral wall of the skull just anterior to the temporal bone. The *lesser wings,* which are smaller than the greater wings, form a ridge of bone anterior and superior to the greater wings. They form part of the floor of the cranium and the posterior part of the orbit.

Between the body and lesser wing just anterior to the sella turcica is the *optic* (*optikus* = eye) *foramen.* Lateral to the body between the greater and lesser wings is a somewhat triangular slit called the *superior orbital fissure.* This fissure may also be seen in the anterior view of the orbit in Figure 6.13.

You can see the *pterygoid* (TER-i-goyd; *pterygoid* = winglike) *processes* (Figure 6.8c) by looking on the inferior part of the sphenoid bone. These structures project inferiorly

Figure 6.5 Skull in medial view.

The internal surfaces of the parietal bones contain numerous protrusions and depressions for blood vessels that supply the dura mater.

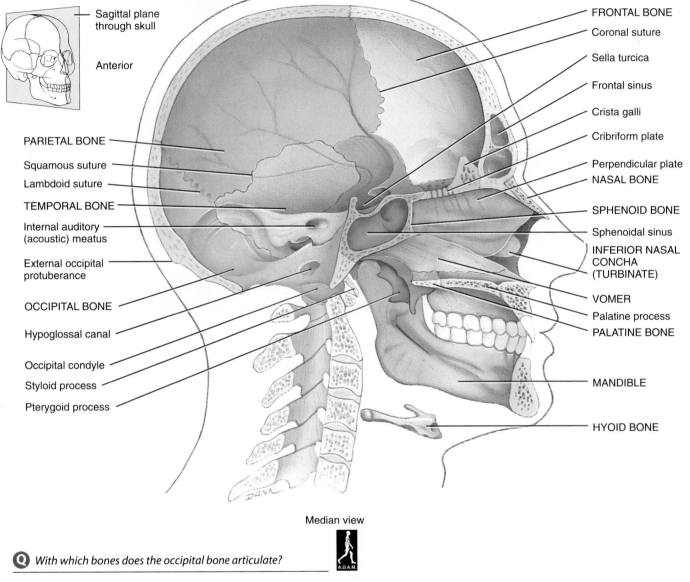

Median view

Q *With which bones does the occipital bone articulate?*

from the points where the body and greater wings unite and form the lateral posterior region of the nasal cavity. Some of the muscles that move the mandible attach to the pterygoid processes.

At the base of the lateral pterygoid process in the greater wing is the *foramen ovale* (= oval). Another foramen, the *foramen spinosum* (= resembling a spine), lies at the posterior angle of the sphenoid. The *foramen lacerum* (= lacerated) is bounded anteriorly by the sphenoid bone and medially by the sphenoid and occipital bones. The foramen is

covered in part by a layer of fibrocartilage in living subjects. Another foramen associated with the sphenoid bone is the *foramen rotundum* (= round opening) located at the junction of the anterior and medial parts of the sphenoid bone.

Ethmoid Bone

The **ethmoid bone** is a light, spongelike bone located on the midline in the anterior part of the cranial floor medial to the orbits. It is anterior to the sphenoid and posterior to the nasal bones (Figure 6.9 on page 141). The ethmoid bone forms

Figure 6.6 Skull in posterior view. The sutures are exaggerated for emphasis.

Within the foramen magnum of the occipital bone, the medulla oblongata connects with the spinal cord.

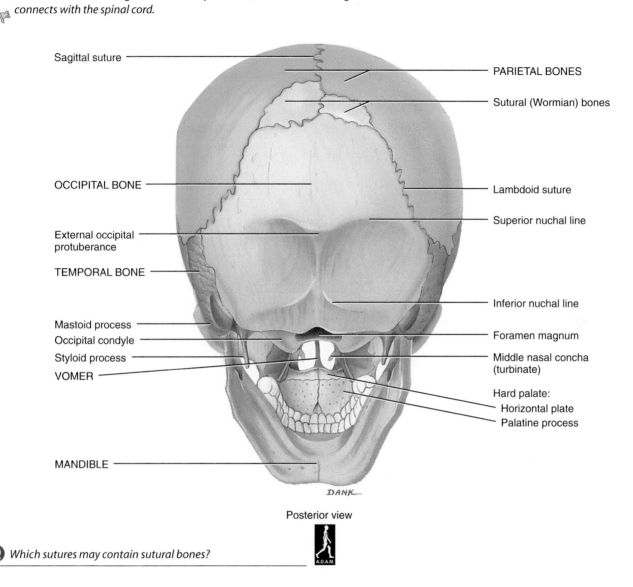

Posterior view

Q *Which sutures may contain sutural bones?*

---

(1) part of the anterior portion of the cranial floor, (2) the medial wall of the orbits, (3) the superior portions of the nasal septum, a partition that divides the nasal cavity into right and left sides, and (4) most of the sidewalls of the nasal roof. The ethmoid is a major supporting structure of the nasal cavity.

Its *lateral masses (labyrinths)* compose most of the wall between the nasal cavity and the orbits. They contain several air spaces, or "cells," ranging in number from 3 to 18. It is from these "cells" that the bone derives its name (*ethmos* = sieve). The ethmoidal cells together form the *ethmoidal sinuses.* The sinuses are shown in Figure 6.11. The *perpendicular plate* forms the superior portion of the nasal septum (see Figure 6.9). The *cribriform (horizontal) plate* lies in the an-

terior floor of the cranium and forms the roof of the nasal cavity. *Cribriform* means perforated like a sieve. The cribriform plate contains the *olfactory (olfacere* = to smell) *foramina.* Projecting superiorly from the cribriform plate is a triangular process called the *crista galli* (= cock's comb). This structure serves as a point of attachment for the membranes (meninges) that cover the brain.

The lateral masses contain two thin, scroll-shaped projections on either side of the nasal septum. These are called the *superior nasal concha* (KONG-ka; *concha* = shell) or *turbinate* and the *middle nasal concha (turbinate).* A third pair of **conchae** (plural), the inferior nasal conchae, will be discussed on page 147. The conchae cause turbulence in

Figure 6.7 Skull in inferior view. The arrow in the inset indicates the direction from which the skull is viewed (inferior).

A suture is an immovable joint between skull bones that holds them together.

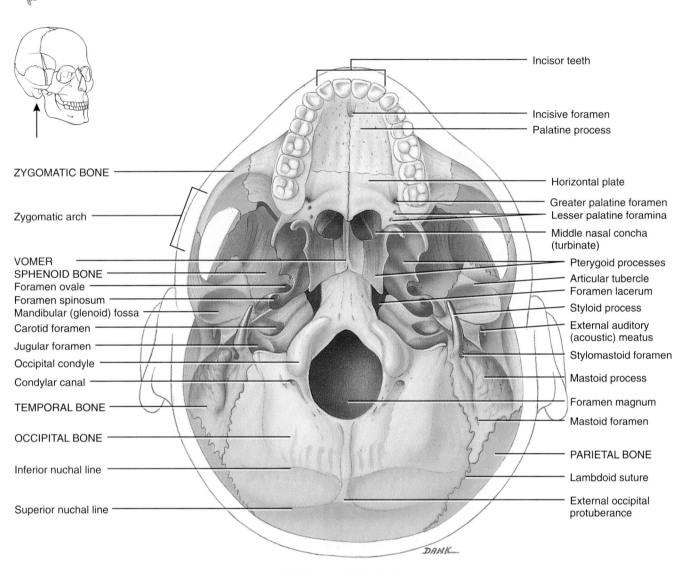

(a) Diagram of inferior view

inhaled air, which results in many inhaled particles striking and becoming trapped in the mucus found lining the nasal passageways. This turbulence thus cleanses inhaled air before it passes into the rest of the respiratory tract.

Cranial Fossae

The floor of the cranium contains three distinct levels, from anterior to posterior, called **cranial fossae** (see Figure 6.8b). The fossae contain depressions for the various brain convolutions, grooves for cranial blood vessels, and numerous

foramina. The highest level, the *anterior cranial fossa,* is formed largely by the portion of the frontal bone that constitutes the roof of the orbits and nasal cavity, the crista galli and cribriform plate of the ethmoid bone, and the lesser wings and part of the body of the sphenoid bone. This fossa houses the frontal lobes of the cerebral hemispheres of the brain. The rough surface of the frontal bone can lead to tearing of the frontal lobes of the cerebral hemispheres during head trauma. The next fossa is inferior to the level of the anterior cranial fossa, is posterior, and is called the *middle cranial fossa.* It is shaped like a butterfly and has a small median

Figure 6.7 (continued)

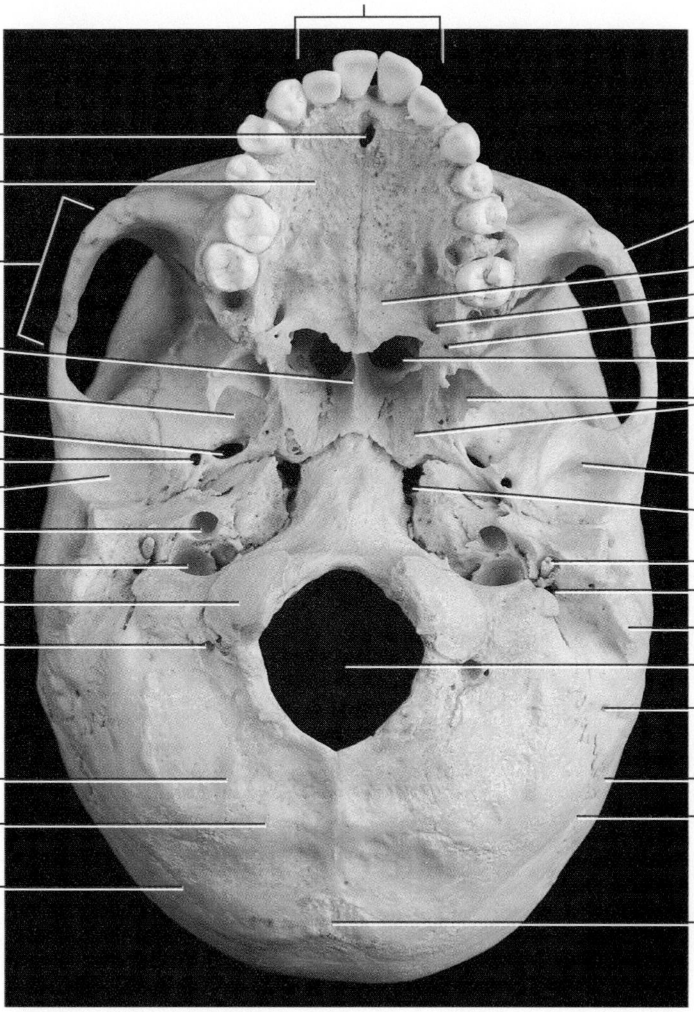

Incisor teeth

MAXILLA:
  Incisive foramen

  Palatine process

Zygomatic arch

VOMER

SPHENOID BONE
Foramen ovale

Foramen spinosum

Mandibular (glenoid) fossa

Carotid foramen

Jugular foramen

Occipital condyle

Condylar canal

OCCIPITAL BONE

Inferior nuchal line

Superior nuchal line

ZYGOMATIC BONE

PALATINE BONE:
  Horizontal plate
  Greater palatine foramen
  Lesser palatine foramen

Middle nasal concha
(turbinate)

Pterygoid processes

Articular tubercle
Foramen lacerum

Styloid process
Stylomastoid foramen

Mastoid process

Foramen magnum

Mastoid foramen

TEMPORAL BONE

Lambdoid suture

External occipital
protuberance

(b) Photograph of inferior view

Q *Which bone contains the carotid foramen, stylomastoid foramen, and mastoid foramen? Which of these foramina is the largest opening?*

portion and two expanded lateral portions. The median portion is formed by part of the body of the sphenoid bone, and the lateral portions are formed by the greater wings of the sphenoid bone, temporal squama, and parietal bone. The middle cranial fossa cradles the temporal lobes of the cerebral hemispheres. The last fossa, at the most inferior level, is the *posterior cranial fossa,* the largest of the fossae. It is formed largely by the occipital bone and the petrous and mastoid portions of the temporal bone. It is a very deep fossa that accommodates the cerebellum, pons, and medulla oblongata of the brain.

Facial Bones

The shape of the face changes dramatically during the first two years after birth. The brain and cranial bones expand, the teeth form and erupt, and the paranasal sinuses increase in size. Growth of the face ceases at about 16 years of age.

Nasal Bones

The paired **nasal bones** meet at the midline (see Figure 6.3) and form part of the bridge of the nose. The major portion of the nose consists of cartilage.

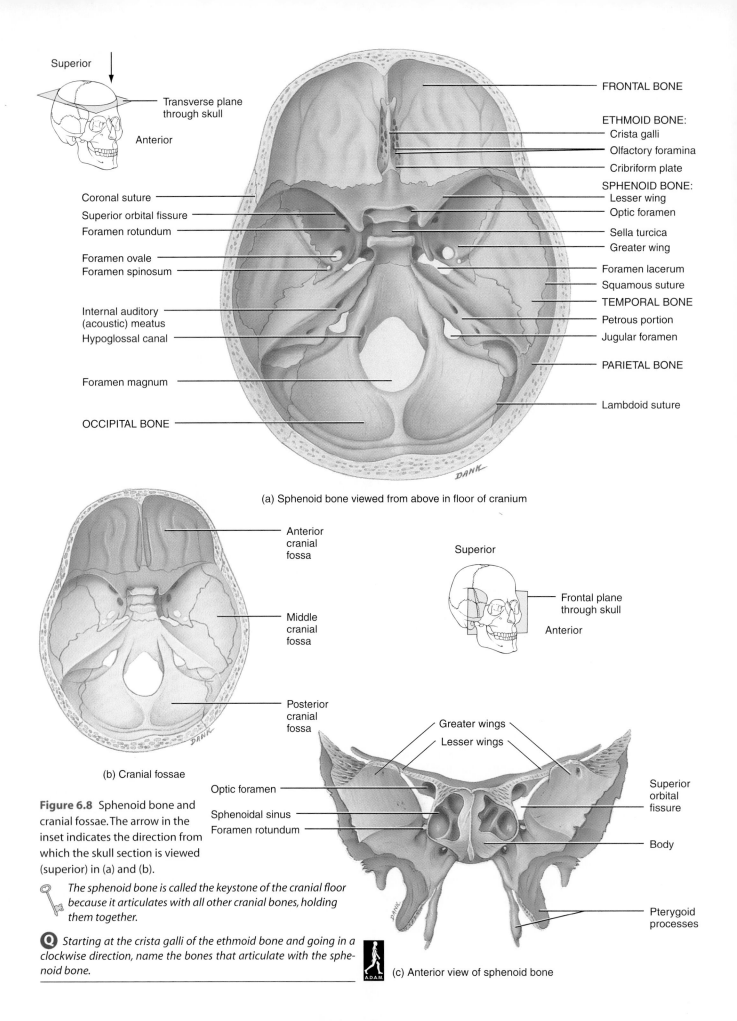

Superior

Transverse plane
through skull

Anterior

FRONTAL BONE

ETHMOID BONE:
Crista galli
Olfactory foramina
Cribriform plate

SPHENOID BONE:
Lesser wing
Optic foramen
Sella turcica
Greater wing
Foramen lacerum
Squamous suture
TEMPORAL BONE
Petrous portion
Jugular foramen
PARIETAL BONE
Lambdoid suture

Coronal suture
Superior orbital fissure
Foramen rotundum
Foramen ovale
Foramen spinosum
Internal auditory
(acoustic) meatus
Hypoglossal canal
Foramen magnum
OCCIPITAL BONE

(a) Sphenoid bone viewed from above in floor of cranium

Anterior
cranial
fossa

Middle
cranial
fossa

Posterior
cranial
fossa

(b) Cranial fossae

Superior

Frontal plane
through skull

Anterior

Greater wings
Lesser wings

Optic foramen
Sphenoidal sinus
Foramen rotundum

Superior
orbital
fissure

Body

Pterygoid
processes

Figure 6.8 Sphenoid bone and cranial fossae. The arrow in the inset indicates the direction from which the skull section is viewed (superior) in (a) and (b).

The sphenoid bone is called the keystone of the cranial floor because it articulates with all other cranial bones, holding them together.

Q *Starting at the crista galli of the ethmoid bone and going in a clockwise direction, name the bones that articulate with the sphenoid bone.*

(c) Anterior view of sphenoid bone

Figure 6.9 Ethmoid bone.

The ethmoid bone is the major supporting structure of the nasal cavity.

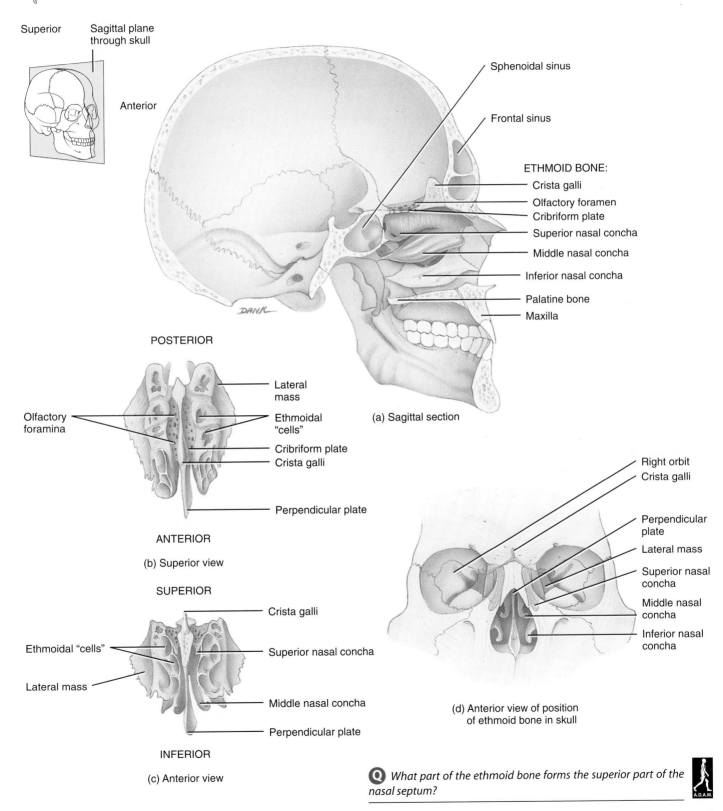

Superior   Sagittal plane through skull

Anterior

Sphenoidal sinus

Frontal sinus

ETHMOID BONE:
Crista galli
Olfactory foramen
Cribriform plate
Superior nasal concha
Middle nasal concha
Inferior nasal concha
Palatine bone
Maxilla

(a) Sagittal section

POSTERIOR

Lateral mass
Ethmoidal "cells"
Olfactory foramina
Cribriform plate
Crista galli
Perpendicular plate

ANTERIOR

(b) Superior view

SUPERIOR

Crista galli
Ethmoidal "cells"
Superior nasal concha
Lateral mass
Middle nasal concha
Perpendicular plate

INFERIOR

(c) Anterior view

Right orbit
Crista galli
Perpendicular plate
Lateral mass
Superior nasal concha
Middle nasal concha
Inferior nasal concha

(d) Anterior view of position of ethmoid bone in skull

Q *What part of the ethmoid bone forms the superior part of the nasal septum?*

Table 6.4 *Summary of the Foramina of the Skull*

FORAMEN	LOCATION	STRUCTURES PASSING THROUGH
Carotid (relating to carotid artery in neck)	Petrous portion of temporal bone (Figure 6.7a).	Internal carotid artery and sympathetic nerves for eyes.
Greater palatine (*palatum* = palate)	Posterior angle of hard palate of palatine bones (Figure 6.7a).	Greater palatine nerve and blood vessels.
Hypoglossal (*hypo* = under; *glossus* = tongue)	Superior to base of occipital condyles (Figure 6.8a).	Cranial nerve XII (hypoglossal) and branch of ascending pharyngeal artery.
Incisive (*incisive* = pertaining to incisor teeth)	Posterior to incisor teeth in maxilla (Figure 6.7a).	Branches of greater palatine blood vessels and nasopalatine nerve.
Infraorbital (*infra* = below)	Inferior to orbit in maxilla (Figure 6.13).	Infraorbital nerve and blood vessels and a branch of the maxillary division of cranial nerve V (trigeminal).
Jugular (*jugular* = pertaining to jugular vein)	Posterior to carotid canal between petrous portion of temporal bone and occipital bone (Figure 6.8a).	Internal jugular vein, cranial nerves IX (glossopharyngeal), X (vagus), and XI (accessory).
Lacerum (*lacerum* = lacerated)	Bounded anteriorly by sphenoid bone, posteriorly by petrous portion of temporal bone, and medially by the sphenoid bone and occipital bone (Figure 6.8a).	Branch of ascending pharyngeal artery in palatine bones.
Lesser palatine (*palatum* = palate)	Posterior to greater palatine foramen in palatine bones (Figure 6.7a).	Lesser palatine nerves and artery.
Magnum (*magnum* = large)	Occipital bone (Figure 6.7a).	Medulla oblongata and its membranes (meninges), cranial nerve XI (accessory), and vertebral and spinal arteries.
Mandibular (*mandere* = to chew)	Medial surface of ramus of mandible (Figure 6.12a).	Inferior alveolar nerve and blood vessels.
Mastoid (*mastoid* = breast-shaped)	Posterior border of mastoid process of temporal bone (Figure 6.7a).	Emissary vein to transverse sinus and branch of occipital artery to dura mater.
Mental (*mentum* = chin)	Inferior to second premolar tooth in mandible (Figure 6.12a).	Mental nerve and vessels.
Olfactory (*olfacere* = to smell)	Cribriform plate of ethmoid bone (Figure 6.8a).	Cranial nerve I (olfactory).
Optic (*optikas* = eye)	Between superior and inferior portions of small wing of sphenoid bone (Figure 6.13).	Cranial nerve II (optic) and ophthalmic artery.
Ovale (*ovale* = oval)	Greater wing of sphenoid bone (Figure 6.8a).	Mandibular branch of cranial nerve V (trigeminal).
Rotundum (*rotundum* = round opening)	Junction of anterior and medial parts of sphenoid bone (Figure 6.8a).	Maxillary branch of cranial nerve V (trigeminal).
Spinosum (*spinosum* = resembling a spine)	Posterior angle of sphenoid bone (Figure 6.8a).	Middle meningeal blood vessels.
Stylomastoid (*stylo* = stake or pole)	Between styloid and mastoid processes of temporal bone (Figure 6.7a).	Cranial nerve VII (facial) and stylomastoid artery.
Supraorbital (*supra* = above)	Supraorbital margin of orbit in frontal bone (Figure 6.13).	Supraorbital nerve and artery.
Zygomaticofacial (*zygoma* = cheekbone; *facial* = face)	Zygomatic bone (Figure 6.3).	Zygomaticofacial nerve and blood vessels.

Figure 6.10 Maxillae in relation to surrounding bones.

The maxillae articulate with every bone of the face except the mandible.

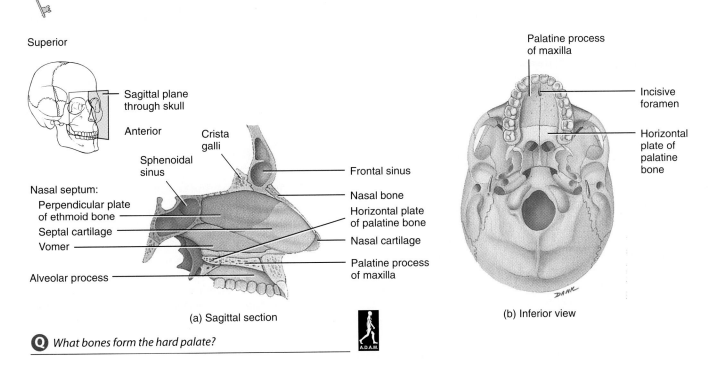

Superior

Sagittal plane through skull

Anterior

Crista galli

Sphenoidal sinus

Nasal septum:
Perpendicular plate of ethmoid bone
Septal cartilage
Vomer

Alveolar process

Frontal sinus

Nasal bone

Horizontal plate of palatine bone

Nasal cartilage

Palatine process of maxilla

Palatine process of maxilla

Incisive foramen

Horizontal plate of palatine bone

(a) Sagittal section

(b) Inferior view

Q *What bones form the hard palate?*

Maxillae

The paired **maxillae** (mak-SIL-ē; *macerae* = to chew; singular is *maxilla*) unite to form the upper jawbone (Figure 6.10) and articulate with every bone of the face except the mandible, or lower jawbone. They form part of the floors of the orbits, part of the lateral walls and floor of the nasal cavity, and most of the hard palate. The hard palate is a bony partition, formed by the maxillae and palatine bones, that forms the roof of the mouth.

Each maxilla contains a *maxillary sinus* that empties into the nasal cavity (see Figure 6.11). The *alveolar* (al-VĒ-ō-lar; *alveolus* = hollow) *process* is an arch that contains the *alveoli* (sockets) for the maxillary (upper) teeth. The *palatine process* is a horizontal projection of the maxilla that forms the anterior three-quarters of the hard palate. The maxillary bones unite, and the fusion is normally completed before birth.

Clinical Application

Cleft Palate and Cleft Lip

Usually the palatine processes of the maxillary bones unite during weeks 10 to 12 of embryonic development. Failure to do so can result in a condition called **cleft palate.** The condition may also involve incomplete fusion of the horizontal plates of the

palatine bones (see Figure 6.10). Another form of this condition, called **cleft lip,** involves a split in the upper lip. Cleft lip and cleft palate often occur together. Depending on the extent and position of the cleft, speech and swallowing may be affected. In addition, children with cleft palate have many ear infections that can lead to hearing loss. Facial and oral surgeons recommend closure of cleft lip during the first few weeks of life, and surgical results are excellent. Repair of cleft palate is done between 12 and 18 months of age, ideally before the child begins to talk. Orthodontic therapy may be needed to align the teeth, and speech therapy is needed since the palate is important in pronouncing many consonants. Again, results are usually excellent. ■

The *infraorbital* (*infra* = below; *orbital* = orbit) *foramen*, which can be seen in the anterior view of the skull in Figure 6.3, is an opening in the maxilla inferior to the orbit. Another prominent foramen in the maxilla is the *incisive* (= incisor teeth) *foramen* just posterior to the incisor teeth (see Figure 6.10b). A final structure associated with the maxilla and sphenoid bone is the *inferior orbital fissure*. It is located between the greater wing of the sphenoid and the maxilla (see Figure 6.13).

Paranasal Sinuses

Although they are not cranial or facial bones, this is an appropriate point to discuss the **paranasal** (*para* = beside) **sinuses.** These are paired cavities in certain cranial and facial bones near the nasal cavity and are clearly seen in a sagittal section of the skull (Figure 6.11). The paranasal sinuses are lined with mucous membranes that are continuous with the lining of the nasal cavity. Skull bones containing paranasal sinuses are the frontal, sphenoid, ethmoid, and maxillae. (The paranasal sinuses were described in the discussion of each of these bones.) Besides producing mucus, the paranasal sinuses serve as resonating chambers for sound as we speak or sing.

Sinusitis

Secretions produced by the mucous membranes of the paranasal sinuses drain into the nasal cavity. An inflammation of the membranes due to an allergic reaction or infection is called **sinusitis**. If the membranes swell enough to block drainage into the nasal cavity, fluid pressure builds up in the paranasal sinuses, and a sinus headache results. Chronic sinusitis may also be caused by nasal polyps, which can be removed surgically. ▨

Zygomatic Bones

The two **zygomatic** (*zygoma* = cheekbone) **bones,** commonly called cheekbones, form the prominences of the cheeks and part of the lateral wall and floor of each orbit (see Figure 6.13). They articulate with the frontal, maxilla, sphenoid, and temporal bones.

The *temporal process* of the zygomatic bone projects posteriorly and articulates with the zygomatic process of the temporal bone to form the *zygomatic arch* (see Figure 6.7). A foramen associated with the zygomatic bone is the *zygomaticofacial foramen* near the center of the bone (see Figure 6.3a).

Mandible

The **mandible** (*mandere* = to chew), or lower jawbone, is the largest, strongest facial bone (Figure 6.12). It is the only movable skull bone (other than the auditory ossicles).

In the lateral view, you can see that the mandible consists of a curved, horizontal portion, the *body,* and two perpendicular portions, the *rami* (RĀ-mī; singular is *ramus,* which means branch). The *angle* of the mandible is the area where each ramus meets the body. Each ramus has a posterior *condylar* (KON-di-lar) *process* that articulates with the mandibular fossa and articular tubercle of the temporal bone to form the *temporomandibular joint (TMJ).* It also has an anterior *coronoid* (KOR-ō-noyd) *process* to which the temporalis muscle attaches. The depression between the coronoid and condylar processes is called the *mandibular notch.* The *alveolar process* is an arch containing the *alveoli* (sockets) for the mandibular (lower) teeth.

The *mental* (*mental* = chin) *foramen* is approximately inferior to the second premolar tooth. It is near this foramen that dentists reach the mental nerve when injecting anesthetics. Another foramen associated with the mandible is the *mandibular foramen* on the medial surface of the ramus, another site often used by dentists to inject anesthetics. The mandibular foramen is the beginning of the *mandibular canal,* which runs anteriorly in the ramus deep to the roots of the teeth.

Temporomandibular Joint Syndrome

One problem associated with the temporomandibular joint (TMJ) is **TMJ syndrome.** It is characterized by dull pain around the ear, tenderness of the jaw muscles, a clicking or popping noise when opening or closing the mouth, limited or abnormal opening of the mouth, headache, tooth sensitivity, and abnormal wearing of the teeth. TMJ syndrome can be caused by improperly aligned teeth, grinding or clenching the teeth, trauma to the head and neck, or arthritis. Treatment may involve applying moist heat or ice, eating a soft diet, taking aspirin, muscle retraining, adjusting or reshaping the teeth, orthodontic treatment, or surgery. ▨

Lacrimal Bones

The paired **lacrimal** (LAK-ri-mal; *lacrima* = tear) **bones** are thin and roughly resemble a fingernail in size and shape. They are the smallest bones of the face. These bones are posterior and lateral to the nasal bones and form a part of the medial wall of each orbit. The lacrimal bones each contain a *lacrimal fossa,* a vertical groove formed with the maxilla, that houses the lacrimal sac, a structure that gathers tears and passes them into the naval cavity (see Figure 6.13). The lacrimal bones can be seen in the anterior and lateral views of the skull in Figures 6.3 and 6.4.

Palatine Bones

The two **palatine** (PAL-a-tīn; *palatum* = palate) **bones** are L-shaped. They form the posterior portion of the hard palate, part of the floor and lateral wall of the nasal cavity, and a small portion of the floors of the orbits. The posterior portion of the hard palate, which separates the nasal cavity from the oral cavity, is formed by the *horizontal plates* of the palatine bones. These can be seen in Figure 6.7.

Two foramina associated with the palatine bones are the greater and lesser palatine foramina (see Figure 6.7). The *greater palatine foramen* is at the posterior angle of the hard palate. The *lesser palatine foramina,* usually two or more on each side, are posterior to the greater palatine foramina.

Figure 6.11 Paranasal sinuses. (a and b) Location of the sinuses. (c) Openings of the paranasal sinuses into the nasal cavity. Portions of the conchae have been sectioned.

🔑 *Paranasal sinuses are mucous-membrane–lined spaces in certain skull bones that are connected to the nasal cavity.*

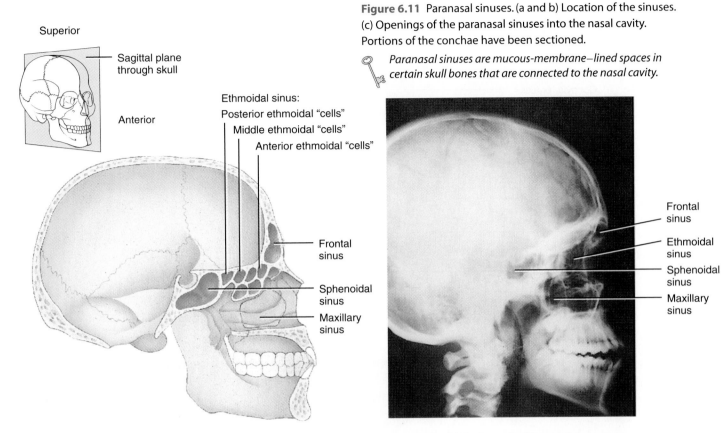

(a) Sagittal section

(b) Radiograph of lateral view

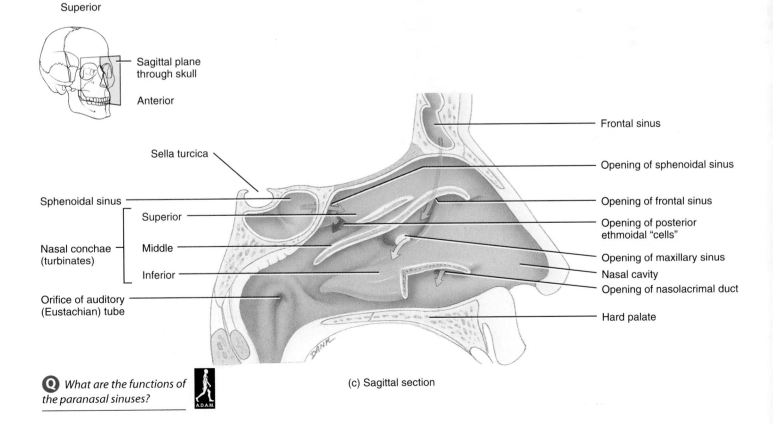

(c) Sagittal section

Q *What are the functions of the paranasal sinuses?*

Figure 6.12 Mandible.

The mandible is the largest and strongest facial bone.

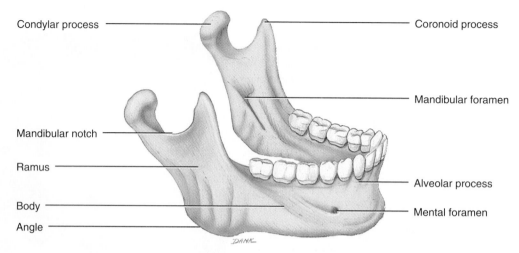

Condylar process

Coronoid process

Mandibular foramen

Mandibular notch

Ramus

Body

Angle

Alveolar process

Mental foramen

DANK

(a) Diagram of right lateral view

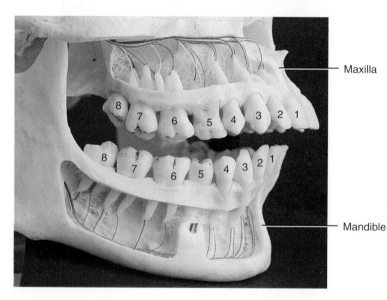

Maxilla

Mandible

1 - Central incisor
2 - Lateral incisor
3 - Cuspid (canine)
4 - First premolar (bicuspid)

5 - Second premolar
6 - First molar
7 - Second molar
8 - Third molar (wisdom tooth)

(b) Photograph of mandible (and maxilla) in right lateral view showing blood and nerve supply to permanent teeth

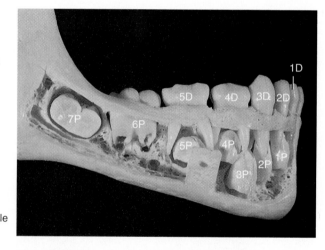

Deciduous teeth:
1D - Central incisor
2D - Lateral incisor
3D - Cuspid (canine)
4D - First molar (bicuspid)
5D - Second molar

Permanent teeth:
1P - Central incisor
2P - Lateral incisor
3P - Cuspid (canine)
4P - First premolar (bicuspid)
5P - Second premolar
6P - First molar
7P - Second molar

(c) Photograph of mandible of a six year old child in right lateral view showing erupted deciduous teeth and unerupted permanent teeth (labels provided by Roger P. Santise, D.D.S.)

Q *What is the distinctive functional feature of the mandible among all the skull bones?*

Figure 6.13 Details of the orbit (eye socket).

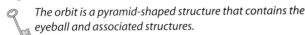

The orbit is a pyramid-shaped structure that contains the eyeball and associated structures.

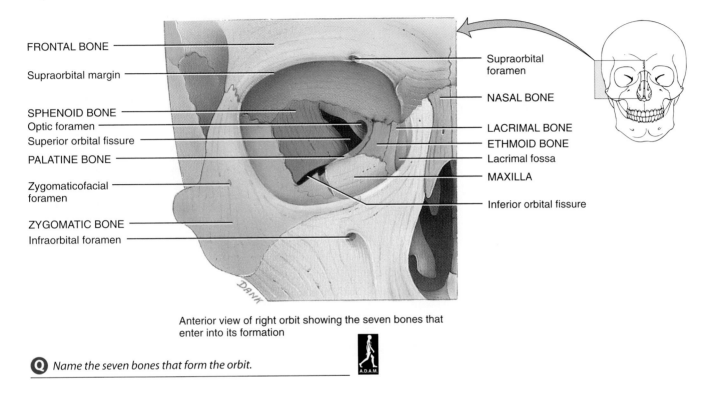

Anterior view of right orbit showing the seven bones that enter into its formation

Q *Name the seven bones that form the orbit.*

Inferior Nasal Conchae

Refer to the views of the skull in Figures 6.3 and 6.9a. The two **inferior nasal conchae** (KONG-kē) or **turbinates** are scroll-like bones that form a part of the lateral wall of the nasal cavity and project into the nasal cavity inferior to the superior and middle nasal conchae of the ethmoid bone. They serve the same function as the superior and middle nasal conchae of the ethmoid bone; that is, they promote turbulent circulation and filtration of air before it passes into the lungs. The inferior nasal conchae are separate bones and not part of the ethmoid.

Vomer

The **vomer** (= plowshare) is a roughly triangular bone that forms the inferior and posterior part of the nasal septum. It is clearly seen in the anterior view of the skull in Figure 6.3 and the inferior view in Figure 6.7.

The anterior border of the vomer articulates with the septal cartilage of the nasal septum that divides the external nose into right and left sides. Its superior border articulates with the perpendicular plate of the ethmoid bone. The structures that form the *nasal septum* (*septum* = partition) are the perpendicular plate of the ethmoid, septal cartilage, vomer, and parts of the palatine bones and maxillae (Figure 6.10a).

Clinical Application

Deviated Nasal Septum

A **deviated nasal septum (DNS)** is one that is deflected laterally from the midline of the nose. The deviation usually occurs at the junction of bone with the septal cartilage. A DNS may occur as a result of a developmental abnormality or trauma. If the deviation is severe, it may entirely block the nasal passageway. Even a partial blockage may lead to infection. If inflammation occurs, it may cause nasal congestion, blockage of the paranasal sinus openings, chronic sinusitis, headache, and nosebleeds. This condition can be corrected surgically. ▨

A summary of bones of the skull is presented in Table 6.3 on page 131.

Orbits

Each **orbit** (eye socket) is a pyramid-shaped structure that contains the eyeball and associated structures. It is formed by seven bones of the skull (Figure 6.13) and has four regions that converge posteriorly to form an *apex* (posterior end):

Figure 6.14 Fontanels of the skull at birth.

Fontanels are membrane-filled spaces between cranial bones that are present at birth.

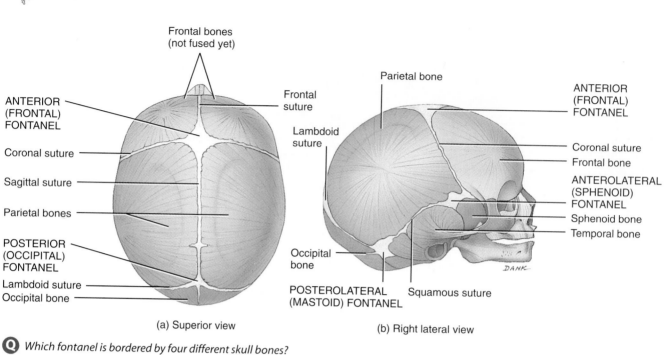

Frontal bones (not fused yet)

ANTERIOR (FRONTAL) FONTANEL

Coronal suture

Sagittal suture

Parietal bones

POSTERIOR (OCCIPITAL) FONTANEL

Lambdoid suture

Occipital bone

Frontal suture

Lambdoid suture

Occipital bone

POSTEROLATERAL (MASTOID) FONTANEL

(a) Superior view

Parietal bone

ANTERIOR (FRONTAL) FONTANEL

Coronal suture

Frontal bone

ANTEROLATERAL (SPHENOID) FONTANEL

Sphenoid bone

Temporal bone

DANK

Squamous suture

(b) Right lateral view

Q *Which fontanel is bordered by four different skull bones?*

1. The *roof* of the orbit consists of parts of the frontal and sphenoid bones.
2. Portions of the zygomatic and sphenoid bones form the *lateral wall.*
3. The *floor* of the orbit is formed by parts of the maxilla, zygomatic, and palatine bones.
4. Portions of the maxilla, lacrimal, ethmoid, and sphenoid bones form the *medial wall.*

Associated with the orbit are several openings:

1. Optic foramen at the junction of the roof and medial wall.
2. Superior orbital fissure at the superior lateral angle of the apex.
3. Inferior orbital fissure at the junction of the lateral wall and floor.
4. Supraorbital foramen on the medial side of the supra-orbital margin of the frontal bone.
5. Lacrimal fossa in the lacrimal bone.

Fontanels

The skeleton of a newly formed embryo consists of cartilage or fibrous connective tissue membrane structures shaped like bones. Gradually, bone replaces the cartilage or fibrous connective tissue membranes in a process called ossifica-tion. At birth, membrane-filled spaces called **fontanels** (fon′-ta-NELZ; = little fountains) are found between cranial bones (Figure 6.14). These "soft spots" are areas of fibrous connective tissue membranes that will eventually be re-placed with bone by intramembranous ossification and be-come sutures. Functionally, the fontanels enable the fetal skull to modify its size and shape as it passes through the birth canal and permit rapid growth of the brain during in-fancy. In addition, fontanels help a physician to gauge the degree of brain development by their state of closure and serve as landmarks (anterior fontanel) for withdrawal of blood for analysis from the superior sagittal sinus (a large vein around the brain). Although an infant may have many fontanels at birth, the form and location of several are fairly constant (Table 6.5).

Foramina

Many **foramina** associated with the skull were mentioned along with the descriptions of the cranial and facial bones they penetrate. As preparation for studying other systems of the body, especially the nervous and cardiovascular systems, these foramina, plus some additional ones, and the structures passing through them are listed in Table 6.4 on page 142. For your convenience and for future reference, the foramina are listed alphabetically.

Table 6.5 Fontanels (See also Figure 6.14)

FONTANEL		LOCATION	DESCRIPTION
Anterior (frontal)	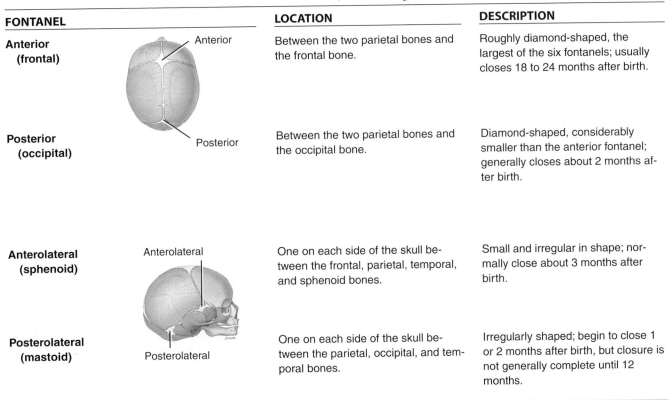 Anterior	Between the two parietal bones and the frontal bone.	Roughly diamond-shaped, the largest of the six fontanels; usually closes 18 to 24 months after birth.
Posterior (occipital)	Posterior	Between the two parietal bones and the occipital bone.	Diamond-shaped, considerably smaller than the anterior fontanel; generally closes about 2 months after birth.
Anterolateral (sphenoid)	Anterolateral	One on each side of the skull between the frontal, parietal, temporal, and sphenoid bones.	Small and irregular in shape; normally close about 3 months after birth.
Posterolateral (mastoid)	Posterolateral	One on each side of the skull between the parietal, occipital, and temporal bones.	Irregularly shaped; begin to close 1 or 2 months after birth, but closure is not generally complete until 12 months.

HYOID BONE

The single **hyoid bone** (*hyoedes* = U-shaped) is a unique component of the axial skeleton because it does not articulate with any other bone (see Figure 6.4a). Rather, it is suspended from the styloid processes of the temporal bones by ligaments and muscles. The hyoid is located in the neck between the mandible and larynx. It supports the tongue and provides attachment for some of its muscles and for muscles of the neck and pharynx. The hyoid consists of a horizontal *body* and paired projections called the *lesser horns* or *cornua* and the *greater horns* or *cornua* (Figure 6.15). (*Cornu*, which is singular, means horn.) Muscles and ligaments attach to these paired projections.

The hyoid bone, as well as cartilages of the larynx and trachea, are often fractured during strangulation. As a result, they are carefully examined in an autopsy when strangulation is suspected.

Figure 6.15 Hyoid bone.

The hyoid bone provides attachment for muscles of the tongue, neck, and pharynx.

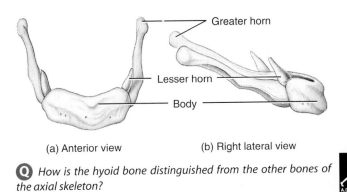

Greater horn
Lesser horn
Body

(a) Anterior view   (b) Right lateral view

Q *How is the hyoid bone distinguished from the other bones of the axial skeleton?*

VERTEBRAL COLUMN
Divisions

The **vertebral column** (*spine* or *backbone*), together with the sternum and ribs, forms the skeleton of the *trunk* of the body. Whereas the vertebral column consists of bone and connective tissue, the spinal cord consists of nervous tissue. The vertebral column makes up about two-fifths of the total height of the body and is composed of a series of bones called **vertebrae** (VER-te-brē; singular is *vertebra*). The length of the column is about 71 cm (28 in.) in an average adult male and about 61 cm (24 in.) in an average adult female. In effect, the vertebral column is a strong, flexible rod that bends anteriorly, posteriorly, and laterally and rotates. It encloses and protects the spinal cord, supports the head, and serves as a point of attachment for the ribs and the muscles of the back. Between vertebrae are openings called *intervertebral foramina*. The nerves that connect the spinal cord to various parts of the body pass through these openings.

The adult vertebral column typically contains 26 vertebrae (Figure 6.16a and b). These are distributed as follows: 7 *cervical vertebrae* (*cervix* = neck) in the neck region; 12 *thoracic vertebrae* (*thorax* = chest) posterior to the thoracic cavity; 5 *lumbar vertebrae* (*lumbus* = loin) supporting the lower back; 5 *sacral vertebrae* fused into one bone called the *sacrum* (SĀ-krum; *sacrum* = holy or sacred bone); and usually 4 *coccygeal* (kok-SIJ-ē-al) *vertebrae* fused into one or two bones called the *coccyx* (KOK-six; *kokkyx* = resembling the bill of a cuckoo). Before fusion of the sacral and coccygeal vertebrae, the total number of vertebrae is 33. Whereas the cervical, thoracic, and lumbar vertebrae are movable, the sacrum and coccyx are immovable.

Between adjacent vertebrae from the second cervical vertebra (axis) to the sacrum are **intervertebral discs.** Each disc has an outer fibrous ring consisting of fibrocartilage called the *annulus* (*annulus* = ring) *fibrosus* and an inner soft, pulpy, highly elastic structure called the *nucleus pulposus* (*pulposus* = pulp) (see Figures 6.16d and 6.24). The discs form strong joints, permit various movements of the vertebral column, and absorb vertical shock. Under compression, they flatten, broaden, and bulge from their intervertebral spaces. Superior to the sacrum, the intervertebral discs constitute about one-fourth the length of the vertebral column.

Normal Curves

When viewed from the side, the vertebral column shows four **normal curves** (Figure 6.16b). The *cervical* and *lumbar curves* are anteriorly convex (bulging out), whereas the *thoracic* and *sacral curves* are anteriorly concave (cupping in). The curves of the vertebral column are important because they increase its strength, help maintain balance in the upright position, absorb shocks from walking, and help protect the column from fracture.

In the fetus, there is only a single anteriorly concave curve (Figure 6.16c). At approximately the third month after birth, when an infant begins to hold its head erect, the cervical curve develops. Later, when the child sits up, stands, and walks, the lumbar curve develops. The thoracic and sacral curves are called *primary curves* because they form first during fetal development. The cervical and lumbar curves are referred to as *secondary curves* because they begin to form later, several months after birth. All curves are fully developed by age 10.

Typical Vertebra

Vertebrae in different regions of the spinal column vary in size, shape, and detail, but they are similar enough that we can discuss the parts and functions of a typical vertebra (Figure 6.17).

1. The *body* is the thick, disc-shaped anterior portion that is the weight-bearing part of a vertebra. Its superior and inferior surfaces are roughened for the attachment of cartilaginous intervertebral discs. The anterior and lateral surfaces contain nutrient foramina for blood vessels.

2. The *vertebral (neural) arch* extends posteriorly from the body of the vertebra. With the body of the vertebra, it surrounds the spinal cord. It is formed by two short, thick processes, the *pedicles* (PED-i-kuls; *pedicles* = little feet), which project posteriorly from the body to unite with the laminae. The *laminae* (LAM-i-nē; *lamina* = thin layer) are the flat parts that join to form the posterior portion of the vertebral arch. The space that lies between the vertebral arch and body contains the spinal cord, fat, areolar connective tissue, and blood vessels. This space is known as the *vertebral foramen*. The vertebral foramina of all vertebrae together form the *vertebral (spinal) canal*. The pedicles exhibit superior and inferior notches called *vertebral notches*. When they are stacked on top of one another, there is an opening between adjoining vertebrae on each side of the column. Each opening, called an *intervertebral foramen,* permits the passage of a single spinal nerve.

3. Seven *processes* arise from the vertebral arch. At the point where a lamina and pedicle join, a *transverse process* extends laterally on each side. A single *spinous process (spine)* projects from the junction of the laminae. These three processes serve as points of attachment for muscles. The remaining four processes form joints with other vertebrae above or below. The two *superior articular processes* of a vertebra articulate with the two inferior articular processes of the vertebra immediately superior to them. The two *inferior articular processes* of a vertebra articulate with the two superior articular processes of the vertebra immediately inferior to them. The articulating

Figure 6.16 Vertebral column. The numbers in parentheses in (a) indicate the number of vertebrae in each region. In (d), the relative size of the disc has been enlarged for emphasis. A "window" has been cut in the annulus fibrosus so that the nucleus pulposus can be seen.

The adult vertebral column typically contains 26 vertebrae.

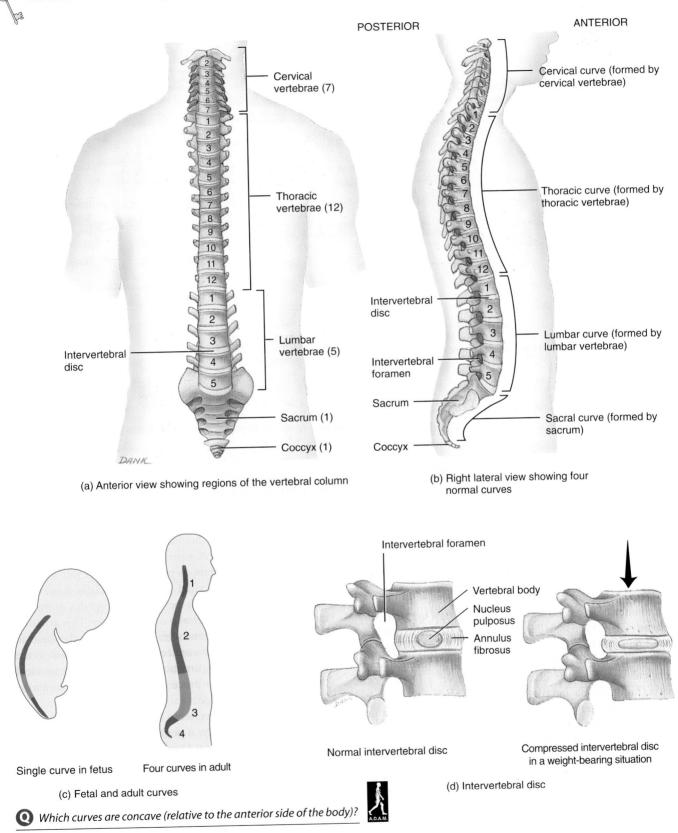

POSTERIOR

ANTERIOR

Cervical vertebrae (7)

Thoracic vertebrae (12)

Intervertebral disc

Lumbar vertebrae (5)

Sacrum (1)

Coccyx (1)

DANK

(a) Anterior view showing regions of the vertebral column

Cervical curve (formed by cervical vertebrae)

Thoracic curve (formed by thoracic vertebrae)

Intervertebral disc

Lumbar curve (formed by lumbar vertebrae)

Intervertebral foramen

Lumbar curve (formed by lumbar vertebrae)

Sacrum

Sacral curve (formed by sacrum)

Coccyx

(b) Right lateral view showing four normal curves

Single curve in fetus   Four curves in adult

(c) Fetal and adult curves

Intervertebral foramen

Vertebral body

Nucleus pulposus

Annulus fibrosus

Normal intervertebral disc

Compressed intervertebral disc in a weight-bearing situation

(d) Intervertebral disc

Q *Which curves are concave (relative to the anterior side of the body)?*

Figure 6.18 Cervical vertebrae.

The cervical vertebrae are found in the neck region.

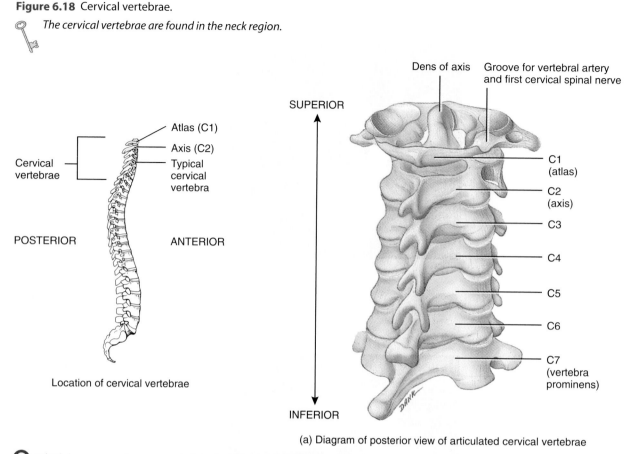

(a) Diagram of posterior view of articulated cervical vertebrae

Q *Which bones permit movement of the head to signify no?*

thoracic vertebrae also have facets or demifacets for articulation with the heads of the ribs. The articulations between the thoracic vertebrae and ribs are called *vertebrocostal joints.* As you can see in Figure 6.19, T1 has a superior facet and an inferior demifacet on each side. T2–T8 have a superior and inferior demifacet on each side. T9 has a superior demifacet on each side, and T10–T12 have a superior facet on each side. Movements of the thoracic region are limited by thin intervertebral (*inter* = between) discs and by the attachment of the ribs to the sternum.

Lumbar Region

The **lumbar vertebrae** (L1–L5) are the largest and strongest in the column (Figure 6.20) because the amount of body weight supported by the vertebrae increases toward the inferior end of the backbone. Their various projections are short and thick. The superior articular processes are directed medially instead of superiorly. The inferior articular processes are directed laterally instead of inferiorly. The spinous processes are quadrilateral in shape, thick, and broad and project nearly straight posteriorly. The spinous

processes are well-adapted for the attachment of the large back muscles.

A summary of the major structural differences among cervical, thoracic, and lumbar vertebrae is presented in Table 6.6 on page 158.

Sacrum

The **sacrum** is a triangular bone formed by the union of five sacral vertebrae. These are indicated in Figure 6.21 on page 159 as S1–S5. The sacral vertebrae begin to fuse between 16 and 18 years of age, a process usually completed by age 30. The sacrum serves as a strong foundation for the pelvic girdle. It is positioned at the posterior portion of the pelvic cavity medial to the two hipbones. The female's sacrum is shorter, wider, and more curved between S2 and S3 than the male's sacrum.

The concave anterior side of the sacrum faces the pelvic cavity. It is smooth and contains four *transverse lines (ridges)* that mark the joining of the sacral vertebral bodies. At the ends of these lines are four parts of *anterior (pelvic)*

Figure 6.18 (continued)

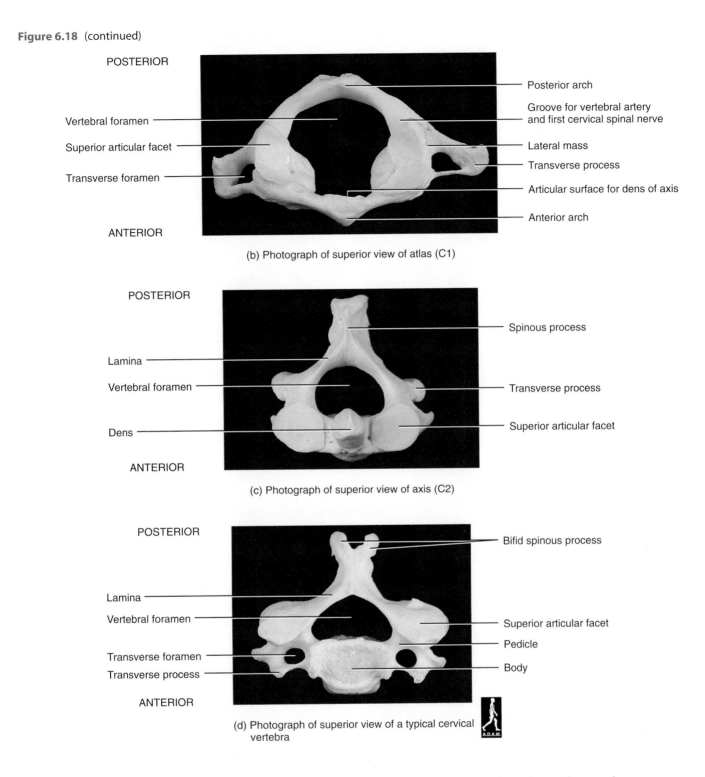

(b) Photograph of superior view of atlas (C1)

(c) Photograph of superior view of axis (C2)

(d) Photograph of superior view of a typical cervical vertebra

sacral foramina. The lateral portion of the superior surface contains a smooth surface called the *sacral ala* (wing), which is formed by the fused transverse processes of the first sacral vertebra (S1).

The convex, posterior surface of the sacrum contains a *median sacral crest,* the fused spinous processes of the up-

per sacral vertebrae; a *lateral sacral crest,* the transverse processes of the sacral vertebrae; and four pairs of *posterior (dorsal) sacral foramina.* These foramina communicate with the anterior sacral foramina through which nerves and blood vessels pass. The *sacral canal* is a continuation of the vertebral canal. The laminae of the fifth sacral vertebra, and

Figure 6.19 Thoracic vertebrae. Photographs of thoracic vertebrae are shown in Figure 6.17c and d.

The thoracic vertebrae are found in the chest region and articulate with the ribs.

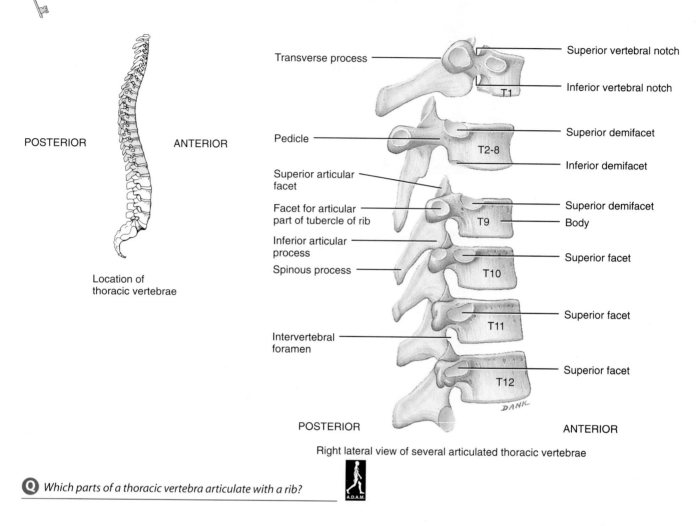

POSTERIOR   ANTERIOR

Location of
thoracic vertebrae

Transverse process

Pedicle

Superior articular
facet

Facet for articular
part of tubercle of rib

Inferior articular
process

Spinous process

Intervertebral
foramen

Superior vertebral notch

Inferior vertebral notch

Superior demifacet

Inferior demifacet

Superior demifacet

Body

Superior facet

Superior facet

Superior facet

T1

T2-8

T9

T10

T11

T12

POSTERIOR   ANTERIOR

Right lateral view of several articulated thoracic vertebrae

Q *Which parts of a thoracic vertebra articulate with a rib?*

sometimes the fourth, fail to meet. This leaves an inferior entrance to the vertebral canal called the *sacral hiatus* (hī-Ā-tus; *hiatus* = opening). On either side of the sacral hiatus are the *sacral cornua,* the inferior articular processes of the fifth sacral vertebra, connected by ligaments to the coccyx.

The narrow inferior portion of the sacrum is known as the *apex.* The broad superior portion of the sacrum is called the *base.* The anteriorly projecting border of the base is called the *sacral promontory* (PROM-on-tō′-rē). It is one of the points used for measurements of the pelvis. On both lateral surfaces, the sacrum has a large *auricular surface* that articulates with the ilium of each hipbone to form the *sacroiliac joint.* Posterior to the auricular surface is a roughened surface, the *sacral tuberosity,* that contains depressions for the attachment of ligaments. The sacral tuberosity is another surface of the sacrum that unites with the hipbone to form the sacroiliac joint. The *superior articular processes* of the

sacrum articulate with the fifth lumbar vertebra, and the base of the sacrum articulates with the body of the fifth lumbar vertebra, to form the *lumbosacral joint.*

Coccyx

The **coccyx** (KOK-siks) is also triangular in shape and is formed by the fusion of usually four coccygeal vertebrae. These are indicated in Figure 6.21 as Co1–Co4. The coccygeal vertebrae fuse between 20 and 30 years of age. The dorsal surface of the body of the coccyx contains two long *coccygeal cornua* that are connected by ligaments to the sacral cornua. The coccygeal cornua are the pedicles and superior articular processes of the first coccygeal vertebra. On the lateral surfaces of the coccyx are a series of *transverse processes,* the first pair being the largest. The coccyx articulates superiorly with the apex of the sacrum. In females, the coccyx points inferiorly; in males, it points anteriorly.

Figure 6.20 Lumbar vertebrae.

Lumbar vertebrae are found in the lower back.

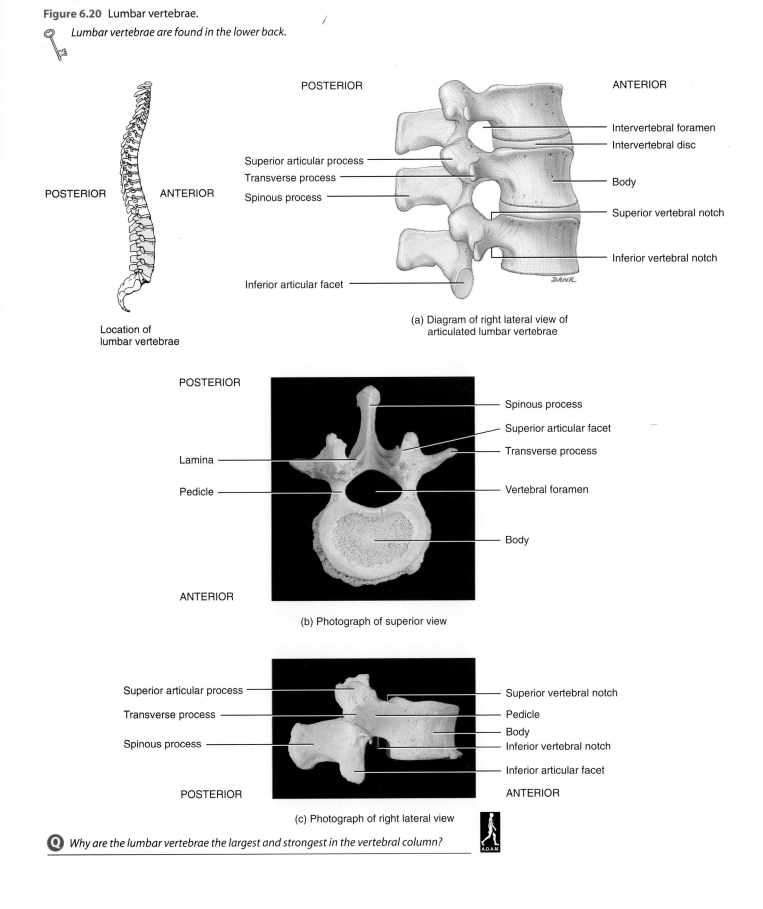

POSTERIOR ANTERIOR

Location of lumbar vertebrae

POSTERIOR

ANTERIOR

Superior articular process

Transverse process

Spinous process

Inferior articular facet

Intervertebral foramen

Intervertebral disc

Body

Superior vertebral notch

Inferior vertebral notch

DANK

(a) Diagram of right lateral view of articulated lumbar vertebrae

POSTERIOR

Lamina

Pedicle

Spinous process

Superior articular facet

Transverse process

Vertebral foramen

Body

ANTERIOR

(b) Photograph of superior view

Superior articular process

Transverse process

Spinous process

POSTERIOR

Superior vertebral notch

Pedicle

Body

Inferior vertebral notch

Inferior articular facet

ANTERIOR

(c) Photograph of right lateral view

Q *Why are the lumbar vertebrae the largest and strongest in the vertebral column?*

Table 6.6 *Comparison of Principal Structural Features of Cervical, Thoracic, and Lumbar Vertebrae*

CHARACTERISTIC	CERVICAL	THORACIC	LUMBAR
Overall structure	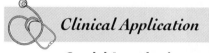		
Size	Small	Larger	Largest
Foramina	One vertebral and two transverse	One vertebral	One vertebral
Spinous processes	Slender and often bifid (C2–C6)	Long and fairly thick (most project inferiorly)	Short and blunt (project posteriorly rather than inferiorly)
Transverse processes	Small	Fairly large	Large and blunt
Articular facets for ribs	No	Yes	No
Direction of articular facets			
Superior	Posterosuperior	Posterolateral	Medial
Inferior	Anteroinferior	Anteromedial	Lateral
Size of intervertebral discs	Thick relative to size of vertebral bodies	Thin relative to vertebral bodies	Massive

Clinical Application

Caudal Anesthesia

Anesthetic agents that act on the sacral and coccygeal nerves are sometimes injected through the sacral hiatus, a procedure called **caudal anesthesia** that is used to provide anesthesia to the perianal area. Because the sacral hiatus is between the sacral cornua, the cornua are important bony landmarks for locating the hiatus. Anesthetic agents may also be injected through the posterior (dorsal) sacral foramina.

THORAX

The term **thorax** refers to the entire chest. The skeletal portion of the thorax is a bony cage formed by the sternum, costal cartilages, ribs, and the bodies of the thoracic vertebrae (Figure 6.22).

The thoracic cage is narrower at its superior end and broader at its inferior end. It is flattened from front to back. The thoracic cage encloses and protects the organs in the thoracic cavity and upper abdominal cavity. It also provides support for the bones of the shoulder girdle and upper limbs.

Sternum

The **sternum,** or breastbone, is a flat, narrow bone measuring about 15 cm (6 in.) in length. It is located in the median line of the anterior thoracic wall and may be split in the midsagittal plane to allow surgeons access to structures in the thoracic cavity such as the thymus gland, heart, and great vessels of the heart.

The sternum consists of three portions (Figure 6.22): the *manubrium* (ma-NOO-brē-um; *manubrium* = handle-like), the superior portion; the *body,* the middle, largest portion; and the *xiphoid* (ZĪ-foyd; *xipho* = swordlike) *process,* the inferior, smallest portion. The junction of the manubrium and body forms the *sternal angle.* The manubrium has a depression on its superior surface called the *suprasternal (jugular) notch.* Lateral to the suprasternal notch are *clavicular notches* that articulate with the medial ends of the clavicles to form the *sternoclavicular joints.* The manubrium also articulates with the costal cartilages of the first and second ribs to form the *sternocostal joints.*

The body of the sternum articulates directly or indirectly with the costal cartilages of the second through tenth ribs. The xiphoid process has no ribs attached to it but provides

Figure 6.21 Sacrum and coccyx.

The sacrum is formed by the union of five sacral vertebrae, and the coccyx is formed by the union of usually four coccygeal vertebrae.

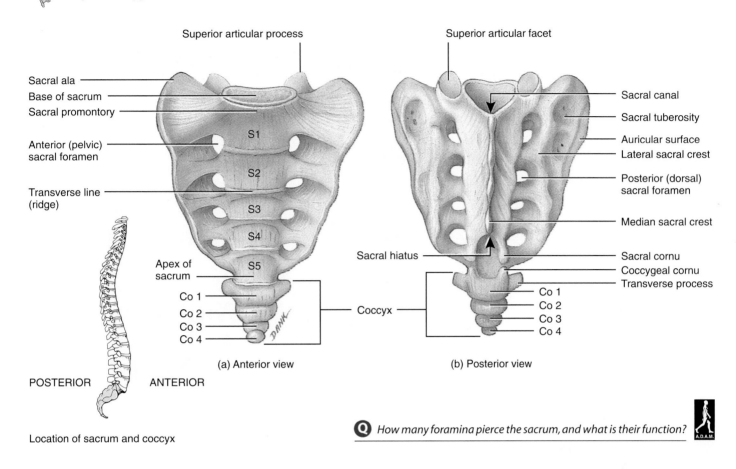

(a) Anterior view

(b) Posterior view

POSTERIOR   ANTERIOR

Location of sacrum and coccyx

Q *How many foramina pierce the sacrum, and what is their function?*

---

attachment for some abdominal muscles. The xiphoid process consists of hyaline cartilage during infancy and childhood and does not ossify completely until about age 40. If the hands of a rescuer are incorrectly positioned during cardiopulmonary resuscitation (CPR), there is danger of fracturing the xiphoid process and driving it into internal organs.

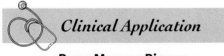

Clinical Application

Bone Marrow Biopsy

In a **bone marrow biopsy,** a sample of red bone marrow is aspirated and examined to help diagnose certain blood disorders. The sample may be taken from the posterior iliac crest (the most common site) or, less frequently, from the sternum. ■

Ribs

Twelve pairs of **ribs** make up the sides of the thoracic cavity (Figure 6.22). The ribs increase in length from the first through seventh, then decrease in length to the twelfth rib. Each articulates posteriorly with its corresponding thoracic vertebra.

The first through seventh pairs of ribs have a direct anterior attachment to the sternum by a strip of hyaline cartilage called *costal cartilage (costa* = rib). These ribs are called *true (vertebrosternal) ribs.* The remaining five pairs of ribs are referred to as *false ribs* because their costal cartilages either attach indirectly to the sternum or do not attach to the sternum at all. The cartilages of the eighth, ninth, and tenth pairs of ribs attach to each other and then to the cartilages of the seventh pair of ribs. These false ribs are called *vertebrochondral ribs.* The eleventh and twelfth pairs of ribs are false ribs designated as *floating (vertebral) ribs* because the costal

Figure 6.22 Skeleton of the thorax.

The bones of the thorax enclose and protect organs in the thoracic cavity and upper abdominal cavity.

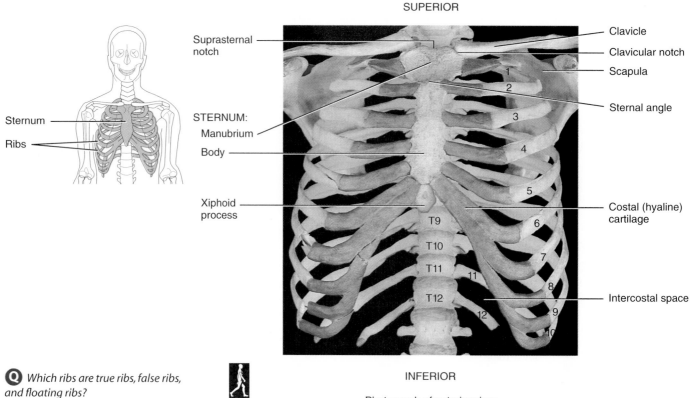

SUPERIOR

Suprasternal notch

Clavicle

Clavicular notch

Scapula

Sternal angle

Sternum

Ribs

STERNUM:

Manubrium

Body

Xiphoid process

Costal (hyaline) cartilage

T9

T10

T11

T12

Intercostal space

INFERIOR

Photograph of anterior view

Q *Which ribs are true ribs, false ribs, and floating ribs?*

cartilage at their anterior ends does not attach to the sternum at all. These ribs attach only posteriorly to the thoracic vertebrae.

We will examine the parts of a typical (third through ninth) rib (Figure 6.23). The *head* is a projection at the posterior end of the rib. It consists of one or two *facets* that articulate with facets on the bodies of adjacent thoracic vertebrae to form *vertebrocostal joints.* The *neck* is a constricted portion just lateral to the head. A knoblike structure on the posterior surface where the neck joins the body is called a *tubercle* (TOO-ber-kul). It consists of a *nonarticular part* that affords attachment to the ligament of the tubercle and an *articular part* that articulates with the facet of a transverse process of the inferior of the two vertebrae to which the head of the rib is connected. These articulations also form vertebrocostal joints. The *body (shaft)* is the main part of the rib. A short distance beyond the tubercle, an abrupt change in the curvature of the shaft occurs. This point is called the *costal angle.* The inner surface of the rib has a *costal groove* that protects blood vessels and a small nerve.

In summary, the posterior portion of the rib is connected to a thoracic vertebra by its head and articular part of a tubercle. The facet of the head fits into a facet on the body of

one vertebra or the demifacets of two adjoining vertebrae. The articular part of the tubercle articulates with the facet of the transverse process of the vertebra.

Spaces between ribs, called *intercostal* (*inter* = between) *spaces,* are occupied by intercostal muscles, blood vessels, and nerves. Surgical access to the lungs or other structures in the thoracic cavity is commonly undertaken through an intercostal space. Special rib retractors are used to create a wide separation between ribs. The costal cartilages are sufficiently elastic in younger individuals to permit considerable bending without breaking.

If you examine Figure 6.22, you will notice that the first rib is the shortest, broadest, and most sharply curved of the ribs. The first rib is an important landmark because of its close relationship to the nerves of the *brachial plexus,* which provides the entire nerve supply of the shoulder and upper limb, two major blood vessels, the subclavian artery and vein, and two skeletal muscles, the anterior and medial scalene muscles. The superior surface of the first rib has two shallow grooves, one for the subclavian vein and one for the subclavian artery and inferior trunk of the brachial plexus. The superior surface also serves as the point of attachment for the anterior and middle scalene muscles. The second rib

Figure 6.23 Typical rib. The arrow in the inset in (c) indicates the direction from which the bones are viewed (superior).

Each rib articulates posteriorly with its corresponding thoracic vertebra.

Q *How does a rib articulate with a thoracic vertebra?*

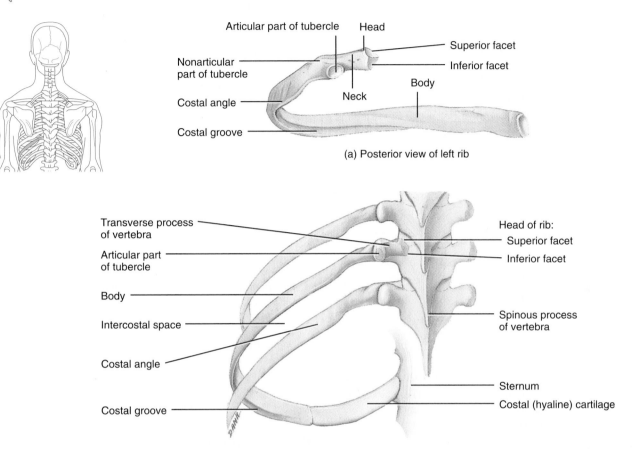

Articular part of tubercle   Head

Nonarticular part of tubercle

Superior facet

Inferior facet

Body

Neck

Costal angle

Costal groove

(a) Posterior view of left rib

Transverse process of vertebra

Articular part of tubercle

Body

Intercostal space

Costal angle

Costal groove

Head of rib:
Superior facet
Inferior facet

Spinous process of vertebra

Sternum

Costal (hyaline) cartilage

(b) Posterior view of left ribs articulated with thoracic vertebrae and the sternum

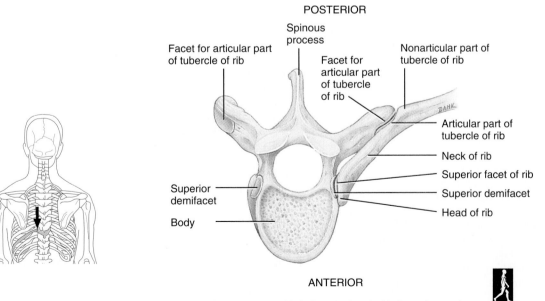

POSTERIOR

Spinous process

Facet for articular part of tubercle of rib

Facet for articular part of tubercle of rib

Nonarticular part of tubercle of rib

Articular part of tubercle of rib

Neck of rib

Superior facet of rib

Superior demifacet

Head of rib

Superior demifacet

Body

ANTERIOR

(c) Superior view of left rib articulated with thoracic vertebra is thinner, less curved, and considerably longer than the first. The tenth rib has a single articular facet on its head. The eleventh and twelfth ribs also have single articular facets on their heads, but no necks, tubercles, or costal angles.

Clinical Application

Rib Fractures

Rib fractures represent the most common chest injuries and usually result from direct blows, most frequently from steering wheel impact, falls, and crushing injuries to the chest. Ribs tend to break at the point where greatest force is applied, but may also break at the weakest point (that is, the site of greatest curvature), which is just anterior to the costal angle. In children the ribs are highly elastic, and fractures are less frequent than in adults. Because the first two ribs are protected by the clavicle and pectoralis major muscle, and the last two ribs are mobile, they are the least commonly injured. The middle ribs are the ones most commonly fractured. In some cases, fractured ribs may cause damage to the heart, great vessels of the heart, lungs, trachea, bronchi, esophagus, spleen, liver, and kidneys. ■

Study Outline

Introduction (p. 126)

1. Bones not only protect and afford movement, they also serve as landmarks for relating to other body systems.
2. The musculoskeletal system is composed of the bones, joints, and muscles working together.

Divisions of the Skeletal System (p. 126)

1. The axial skeleton consists of bones arranged along the longitudinal axis. The parts of the axial skeleton are the skull, auditory ossicles (earbones), hyoid bone, vertebral column, sternum, and ribs.
2. The appendicular skeleton consists of the bones of the girdles and the upper and lower limbs (extremities). The parts of the appendicular skeleton are the pectoral (shoulder) girdles, bones of the upper limbs, pelvic (hip) girdles, and bones of the lower limbs.

Types of Bones (p. 126)

1. On the basis of shape, bones are classified as long, short, flat, irregular, and sesamoid.
2. Sesamoid bones develop in tendons or ligaments. Sutural (Wormian) bones are found within the sutures of certain cranial bones.

Bone Surface Markings (p. 128)

1. Surface markings are structural features visible on the surfaces of bones.
2. Each marking is structured for a specific function—joint formation, muscle attachment, or passage of nerves and blood vessels (see Table 6.2).

Skull (p. 129)

1. The skull consists of the cranium and the face. It is composed of 22 bones.
2. Sutures are immovable joints between bones of the skull. Examples are coronal, sagittal, lambdoid, and squamous sutures.
3. The eight cranial bones include the frontal, parietal (2), temporal (2), occipital, sphenoid, and ethmoid.
4. The floor of the cranium contains three levels called cranial fossae.

5. The 14 facial bones are the nasal (2), maxillae (2), zygomatic (2), mandible, lacrimal (2), palatine (2), inferior nasal conchae (2), and vomer.
6. Paranasal sinuses are cavities in bones of the skull that communicate with the nasal cavity. They are lined by mucous membranes. The cranial bones containing paranasal sinuses are the frontal, sphenoid, ethmoid, and maxillae.
7. The orbits (eye sockets) are formed by seven bones of the skull.
8. Fontanels are fibrous, connective tissue membrane–filled spaces between the cranial bones of fetuses and infants. The major fontanels are the anterior, posterior, anterolaterals, and posterolaterals. They fill in with bone and become sutures.
9. The foramina of the skull bones provide passages for nerves and blood vessels.

Hyoid Bone (p. 149)

1. The hyoid bone is a U-shaped bone that does not articulate with any other bone.
2. It supports the tongue and provides attachment for some of its muscles as well as some neck muscles and muscles of the pharynx.

Vertebral Column (p. 150)

1. The vertebral column, sternum, and ribs constitute the skeleton of the trunk.
2. The bones of the adult vertebral column are the cervical vertebrae (7), thoracic vertebrae (12), lumbar vertebrae (5), sacrum (5, fused), and the coccyx (4, fused).
3. The vertebral column contains normal curves (cervical, thoracic, lumbar, and sacral) that give strength, support, and balance.
4. The vertebrae are similar in structure, each usually consisting of a body, vertebral arch, and seven processes. Vertebrae in the different regions of the column vary in size, shape, and detail.

Thorax (p. 158)

1. The thoracic skeleton consists of the sternum, ribs and costal cartilages, and thoracic vertebrae.
2. The thoracic cage protects vital organs in the chest area and upper abdomen.

Review Questions

1. Describe the importance of the skeletal system to the body. (p. 126)
2. What are the five principal types of bones on the basis of shape? Give an example of each. What is a sutural (Wormian) bone? (p.126)
3. What are bone surface markings? Give an example of each. (p. 128)
4. Distinguish between the axial and appendicular skeletons. What subdivisions and bones are contained in each? (p. 129)
5. What are the bones that compose the skull? The cranium? The face? (p. 129)
6. Define a suture. What are the four prominent sutures of the skull? Where are they located? (p. 131)
7. What is a paranasal sinus? What cranial bones contain paranasal sinuses? (p. 143)
8. Describe the components of each orbit. (p. 147)
9. What is a fontanel? Describe the location of the six fairly constant fontanels. (p. 148)
10. What bones form the skeleton of the trunk? Distinguish between the number of nonfused vertebrae found in the adult vertebral column and that of a child. (p. 150)
11. What are the normal curves in the vertebral column? What are the functions of the curves? (p. 150)
12. What are the principal distinguishing characteristics of the bones of the various regions of the vertebral column? (p. 152)
13. What bones form the skeleton of the thorax? What are the functions of the thoracic skeleton? (p. 158)
14. How are ribs classified on the basis of their attachment to the sternum? (p. 159)

Self Quiz

1. Match the following
   (a) largest facial bone; it includes the chin
   (b) the cheekbones; they contribute to the structure of the lateral walls of the orbits
   (c) the smallest facial bones; tear ducts pass through these
   (d) the bridge of the nose
   (e) auditory ossicles are located in these bones
   (f) superior to the atlas; contains the hole through which the spinal cord connects to the brain
   (g) the name means "wall"; forms most of the roof and much of the side walls of the skull
   (h) one of the jawbones; forms most of the hard palate
   (i) a light spongy bone; located in the roof and septum of the nasal cavity
   (j) the forehead; provides protection for the anterior portion of the brain
   (k) posterior portion of the hard palate
   (l) called a keystone because it articulates with all other cranial bones; also protects the pituitary gland
   (m) inferior part of the nasal septum
   (n) delicate bones; form the inferior lateral side walls of the nasal cavity

   ___ (1) zygomatic
   ___ (2) frontal
   ___ (3) inferior nasal concha
   ___ (4) ethmoid
   ___ (5) maxilla
   ___ (6) temporal
   ___ (7) lacrimal
   ___ (8) occipital
   ___ (9) palatine
   ___ (10) sphenoid
   ___ (11) vomer
   ___ (12) mandible
   ___ (13) parietal
   ___ (14) nasal

Are the following statements true or false?

2. Sesamoid bones protect tendons from wear and tear and may improve the mechanical advantage at a joint by altering the direction of pull of a tendon.
3. The ramus of the mandible has an anterior condylar process and a posterior coronoid process.
4. The tubercle of a rib articulates with demifacets on the bodies of adjacent vertebrae.
5. The opening in the base of the skull where the medulla oblongata connects with the spinal cord is the foramen magnum.
6. The coronal suture is associated with the posterior fontanel of the fetal skull.

Choose the best answer to these questions:

7. Cranial bones
   a. form the framework of the face
   b. attach to membranes that stabilize the position of the brain
   c. provide attachment for several muscles of facial expression
   d. protect and provide support for the entrances to the digestive and respiratory systems
   e. all of the above
8. Which of the following statements is/are true?
   (1) The phalanges are examples of short bones.
   (2) Flat bones have a middle layer of spongy bone between two layers of compact bone.
   (3) Irregular bones consist of compact bone only.
   a. 1 only
   b. 2 only
   c. 3 only
   d. 2 and 3
   e. all of the above

9. A foramen is
   a. a cavity within a bone
   b. a depression
   c. a hole for blood vessels and nerves
   d. a ridge
   e. a site for muscle attachment

10. Which of the following structures is part of the frontal bone?
   a. infraorbital foramen
   b. crista galli
   c. supraorbital foramen
   d. optic foramen
   e. zygomatic arch

11. The second cervical vertebra
   (1) has a characteristic feature, the dens
   (2) articulates with the first cervical vertebra in a pivot joint
   (3) is called the vertebra prominens because of its long process.
   a. 1 only
   b. 2 only
   c. 3 only
   d. both 1 and 2
   e. none of the above

12. Which of the following bones is correctly matched with the process?
   a. zygomatic bone-temporal process
   b. maxilla-pterygoid process
   c. vomer-mandibular process
   d. temporal-occipital process
   e. sphenoid-styloid process

13. All of the following articulate with the maxilla *except* the
   a. nasal bone
   b. frontal bone
   c. palatine bone
   d. mandible
   e. zygomatic bone

14. Which of the following is *not* a component of the orbit?
   a. temporal
   b. ethmoid
   c. zygomatic
   d. maxilla
   e. lacrimal

15. The sella turcica
   (1) is a depression in the ethmoid bone
   (2) is located in approximately the center of the floor of the cranium
   (3) contains the pineal gland
   (4) identifies the location of the auditory ossicles
   a. 1 only
   b. 2 only
   c. 3 only
   d. 4 only
   e. 1, 2, and 3

16. Which of the following is *not* part of the ethmoid bone?
   a. cribriform plate
   b. crista galli
   c. middle nasal concha
   d. pterygoid process
   e. superior nasal concha

17. Which of the following does *not* contain a paranasal sinus?
   a. zygomatic
   b. sphenoid
   c. ethmoid
   d. frontal
   e. maxilla

18. The skeleton of the thorax
   a. is formed by 12 pairs of ribs and costal cartilages, the sternum, and the 12 thoracic vertebrae
   b. protects the internal chest organs, as well as the liver
   c. is narrower at its superior end
   d. aids in supporting the bones of the shoulder girdle
   e. is described by all of the above

19. Ribs
   a. always articulate with the vertebrae at the sternum
   b. always articulate with the sternum by means of cartilages
   c. may be classified as true, false, or floating
   d. may be considered part of the appendicular skeleton
   e. do not support any parts of the body

20. The sternum
   a. is classified as a sesamoid bone
   b. articulates with all the true ribs directly by means of costal cartilages
   c. consists of two parts: a manubrium and a xiphoid process
   d. all of the above
   e. none of the above

21. Match the following.
   (a) small body, transverse foramina for blood vessels
   (b) atlas and axis
   (c) five in an adult
   (d) only vertebrae that articulate with ribs
   (e) massive body, blunt spinous process
   (f) usually have long spinous processes that point inferiorly, articulate with ribs
   (g) most inferior part of vertebral column
   (h) articulate with the two hipbones

   ___ (1) cervical
   ___ (2) coccyx
   ___ (3) lumbar
   ___ (4) sacral
   ___ (5) thoracic

Critical Thinking Questions

1. While investigating her new baby brother, 4-year-old Pattee finds a soft spot on the baby's skull and announces that the baby needs to go back because "it's not finished yet." Explain the presence of soft spots in the infant.
HINT: *An infant should be "soft in the head" but an adult should not.*

2. Barbara was walking down the stairs in her bunny slippers when she slipped and landed hard on her buttocks. She felt a pain sharp enough to bring tears to her eyes. Her doctor said she broke a bone but she wouldn't be given a cast. What bone did she probably break?
HINT: *Your buttocks usually keep this bone from being banged whenever you sit.*

3. The ad reads "New postureperfect mattress! Keeps spine perfectly straight—just like when you were born! A straight spine means great sleep!" Would you buy a mattress from this company? Explain.
HINT: *Take a sideways view of the vertebral column.*

4. During Pattee's new brother's well baby visit, the pediatrician said that she was checking the sutures. Pattee had sutures on her head two months ago when she fell off the bathroom counter. (Don't ask; it's a long story.) Pattee doesn't see any threads sticking out of her brother's head. What's going on?
HINT: *These sutures are under the skin.*

5. John had complaints of severe headaches and a yellow-green nasal discharge. His physician ordered x-rays of his head. John was surprised to see several fuzzy looking holes on the x-ray. Why did the doctor order the x-rays?
HINT: *No insult intended to John. These holes are in his cranium.*

6. Harry was helping some friends move. As he lifted a box (what do they have in here—rocks?!) and twisted to put it in the truck, he felt a pain in his lower back. Later that night, the pain sharpened and radiated down his thigh and into his calf. Why does Harry have pain in his leg?
HINT: *Think about what happened earlier to his back.*

Answers to Figure Questions

6.1 Axial skeleton: skull, vertebral column. Appendicular skeleton: clavicle, shoulder girdle, humerus, pelvic girdle, femur.

6.2 Flat bones.

6.3 Frontal, parietal, sphenoid, ethmoid, and temporal.

6.4 Squamous suture: parietal and temporal bones. Lambdoid suture: parietal and occipital bones. Coronal suture: parietal and frontal bones.

6.5 Parietal, temporal, sphenoid, and atlas.

6.6 Sagittal and lambdoid.

6.7 The temporal bone; the carotid foramen.

6.8 Crista galli of ethmoid bone, frontal, parietal, temporal, occipital, temporal, parietal, frontal, crista galli of ethmoid bone.

6.9 Perpendicular plate.

6.10 Maxillae and palatine bones.

6.11 They produce mucus and serve as resonating chambers for vocalization.

6.12 The mandible is the only movable skull bone, other than the auditory ossicles.

6.13 Frontal, sphenoid, zygomatic, maxilla, lacrimal, ethmoid, and palatine.

6.14 Anterolateral (sphenoid).

6.15 It does not articulate with any other bone.

6.16 The thoracic and sacral curves are concave.

6.17 The vertebral foramina enclose the spinal cord while the intervertebral foramina provide spaces for spinal nerves to exit the vertebral column.

6.18 Atlas and axis.

6.19 Facets and demifacets on the body, and facets on the transverse processes.

6.20 The body weight supported by vertebrae increases toward the inferior end of the vertebral column.

6.21 There are four pairs of foramina, for a total of eight. Each anterior sacral foramen joins a posterior sacral foramen at the intervertebral foramen. Nerves and blood vessels pass through these tunnels in the bone.

6.22 True ribs (pairs 1–7), false ribs (pairs 8–12), and floating ribs (pairs 11 and 12).

6.23 The facet of the head fits into a facet on the body of a vertebra and the articular part of the tubercle articulates with the facet of the transverse process of a vertebra.

Chapter 7

The Skeletal System: The Appendicular Skeleton

Contents at a Glance

As was noted in Chapter 6, the two main divisions of the skeletal system are the axial skeleton and the appendicular skeleton. Whereas the axial skeleton is more concerned with the protection of internal organs, the appendicular skeleton, the focus of this chapter, is more concerned with movement. The appendicular skeleton includes bones that make up the upper and lower limbs as well as bones, called girdles, that attach the limbs to the axial skeleton. As you progress through this chapter, you will see how the bones of the appendicular skeleton are coordinated with each other and with skeletal muscles and joints to bring about an array of movements. These movements are essential as we move from one activity to another during the course of a day. They permit us to walk, write, use a computer, dance, swim, and play a musical instrument.

PECTORAL (SHOULDER) GIRDLE

The **pectoral** (PEK-tō-ral) or **shoulder girdles** attach the bones of the upper limbs to the axial skeleton (Figure 7.1). Each of the two pectoral girdles consists of two bones: a clavicle and a scapula. The **clavicle** is the anterior component and articulates with the **sternum** at the *sternoclavicular joint*. The posterior component, the scapula, is the posterior component, is held in position by complex muscle attachments, and articulates with the clavicle at the *acromioclavicular joint* and with the humerus at the *glenohumeral (shoulder) joint*. The pectoral girdles have no articulation with the vertebral column. Although the joints of the shoulder girdles are not very stable, they are freely movable and thus allow movements in many directions.

Clavicle

The clavicles lie horizontally in the superior and anterior part of the thorax superior to the first rib (Figure 7.2). Each **clavicle** (KLAV-i-kul; *clavis* = key) or *collarbone*, is a

Figure 7.1 Right pectoral (shoulder) girdle.

🔑 *The clavicle is the anterior component of the pectoral girdle, and the scapula is the posterior component.*

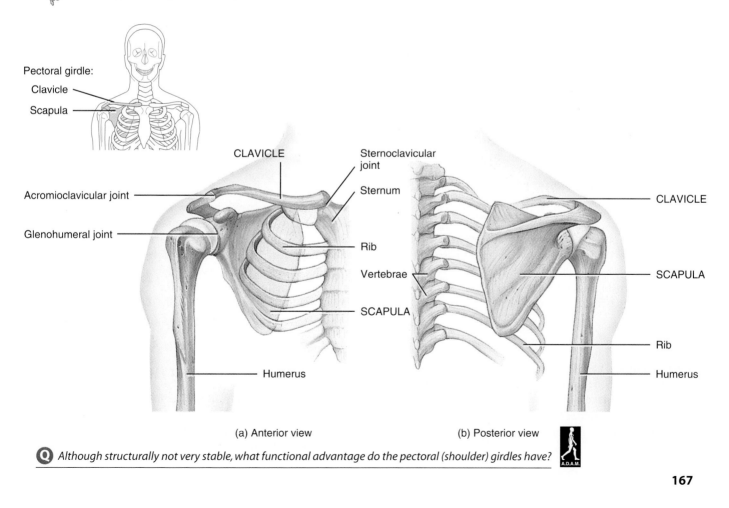

(a) Anterior view   (b) Posterior view

Q *Although structurally not very stable, what functional advantage do the pectoral (shoulder) girdles have?*

Figure 7.2 Right clavicle. The arrow in the inset indicates the direction from which the clavicle is viewed (inferior).

The clavicle articulates medially with the sternum and laterally with the acromion of the scapula.

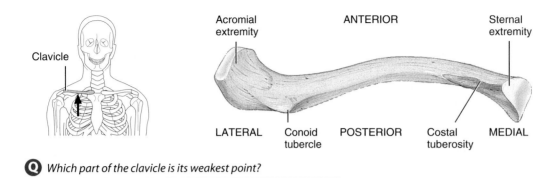

Clavicle

Acromial extremity

ANTERIOR

Sternal extremity

LATERAL   Conoid tubercle   POSTERIOR   Costal tuberosity   MEDIAL

Q *Which part of the clavicle is its weakest point?*

slender S-shaped bone with two curves, one convex and one concave. The medial one-third of the clavicle is convex anteriorly, whereas the lateral one-third is concave anteriorly. Because the junction of two curves is the weakest point of a structure, this is the most frequent site of clavicular fractures.

The medial end of the clavicle, the *sternal extremity,* is rounded and articulates with the sternum to form the *sternoclavicular joint.* The broad, flat, lateral end, the *acromial* (a-KRŌ-mē-al) *extremity,* articulates with the acromion of the scapula. This joint is called the *acromioclavicular joint.* (Refer to Figure 7.1 for a view of these articulations.) The *conoid* (KŌ-noyd; *konos* = cone) *tubercle* on the inferior surface of the lateral end of the bone serves as a point of attachment for the conoid ligament. The *costal tuberosity* on the inferior surface of the medial end also serves as a point of attachment for the costoclavicular ligament.

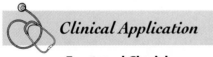

Clinical Application

Fractured Clavicle

Because of its position, the clavicle transmits mechanical force from the upper limb to the trunk. If the force transmitted to the clavicle is excessive, as in falling on one's outstretched arm, a **fractured clavicle** may result. The clavicle is one of the most frequently broken bones in the body. ■

Scapula

Each **scapula** (SCAP-yoo-la), or *shoulder blade,* is a large, triangular, flat bone situated in the posterior part of the thorax between the levels of the second and seventh ribs (Figure 7.3). The medial borders of the scapulae (plural) lie about 5 cm (2 inches) from the vertebral column.

A sharp ridge, the *spine,* runs diagonally across the posterior surface of the flattened, triangular body of the scapula. The lateral end of the spine projects as a flattened, expanded process called the *acromion* (a-KRŌ-mē-on; *acro* = top or summit), easily felt as the high point of the shoulder. (Tailors measure the length of the upper limb from the acromion.) The acromion articulates with the acromial extremity of the clavicle (*acromioclavicular joint*). Inferior to the acromion is a depression called the *glenoid cavity (fossa).* This cavity articulates with the head of the humerus (arm bone) to form the *glenohumeral (shoulder) joint* (see Figure 7.1).

The thin edge of the bone near the vertebral column is the *medial (vertebral) border.* The thick edge closer to the arm is the *lateral (axillary) border.* The medial and lateral borders join at the *inferior angle.* The superior edge of the scapula, called the *superior border,* joins the vertebral border at the *superior angle.* The *scapular notch* is a prominent indentation along the superior border through which the suprascapular nerve passes.

At the lateral end of the superior border is a projection of the anterior surface called the *coracoid* (KOR-a-koyd; *korakodes* = like a crow's beak) *process* to which muscles attach. Above and below the spine are two fossae: the *supraspinous* (soo′-pra-SPĪ-nus) *fossa* and the *infraspinous fossa,* respectively. Both serve as surfaces of attachment for shoulder muscles that have similar names (supraspinatus and infraspinatus). On the ventral (costal) surface is a slightly hollowed-out area called the *subscapular fossa,* also a surface of attachment for shoulder muscles.

UPPER LIMB (EXTREMITY)

The **upper limbs (extremities)** consist of 60 bones. Each upper limb includes a humerus in the arm, ulna and radius in the forearm, and carpals (wrist bones), metacarpals (palm bones), and phalanges (finger bones) in the hand (Figure 7.4).

Figure 7.3 Right scapula.

The glenoid cavity of the scapula articulates with the head of the humerus to form the shoulder joint.

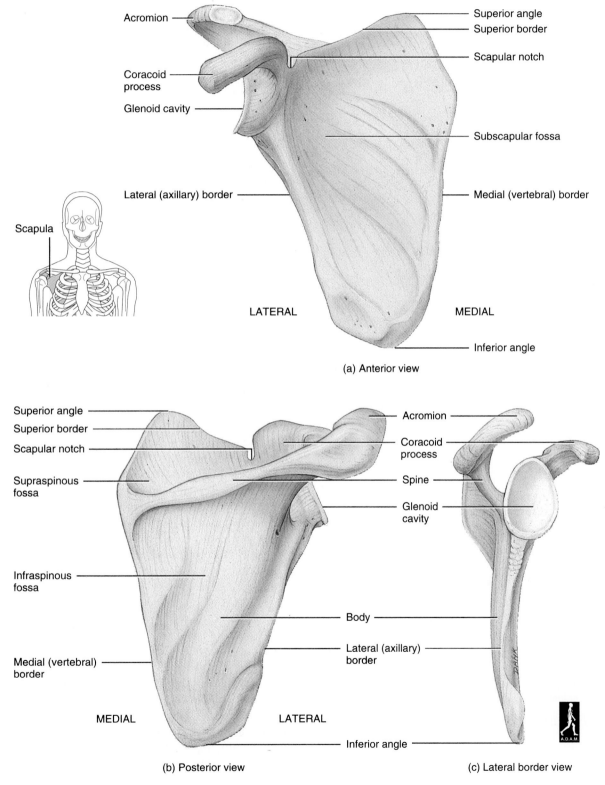

(a) Anterior view

(b) Posterior view

(c) Lateral border view

Q *Which part of the scapula forms the high point of the shoulder?*

Figure 7.4 Right upper limb.

The pectoral girdle attaches the bones of the upper limb to the axial skeleton.

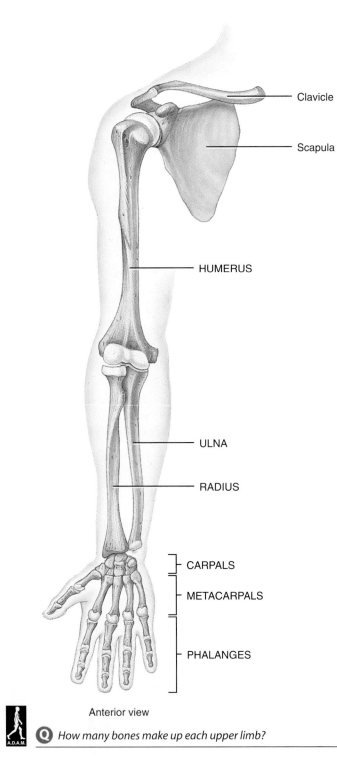

Clavicle

Scapula

HUMERUS

ULNA

RADIUS

CARPALS

METACARPALS

PHALANGES

Anterior view

Q *How many bones make up each upper limb?*

Humerus

The **humerus** (HYOO-mer-us), or arm bone, is the longest and largest bone of the upper limb (Figure 7.5). It articulates proximally with the scapula and distally at the elbow with both the ulna and radius.

The proximal end of the humerus features a *head* that articulates with the glenoid cavity of the scapula (*glenohumeral* or *shoulder joint*). It also has an *anatomical neck,* the former site of the epiphyseal plate (see Figure 5.6), which is an oblique groove just distal to the head. The *greater tubercle* is a lateral projection distal to the neck. It is the most laterally palpable bony landmark of the shoulder region. The *lesser tubercle* is an anterior projection. Between these tubercles runs an *intertubercular sulcus (bicipital groove).* The *surgical neck* is a constricted portion just distal to the tubercles and is so named because fractures often occur here.

The *body (shaft)* of the humerus is cylindrical at its proximal end. It gradually becomes triangular and is flattened and broad at its distal end. Laterally, at the middle portion of the shaft, there is a roughened, V-shaped area called the *deltoid tuberosity.* This area serves as a point of attachment for the deltoid muscle.

The following parts are found at the distal end of the humerus. The *capitulum* (ka-PIT-yoo-lum), meaning small head, is a rounded knob that articulates with the head of the radius. The *radial fossa* is an anterior depression that receives the head of the radius when the forearm is flexed (bent). The *trochlea* (TRŌK-lē-a), located medial to the capitulum, is a spool-shaped surface that articules with the ulna. The *coronoid* (KOR-o-noyd; *korne* = crown-shaped) *fossa* is an anterior depression that receives the coronoid process of the ulna when the forearm is flexed. The *olecranon* (ō-LEK-ra-non) *fossa* is a posterior depression that receives the olecranon of the ulna when the forearm is extended (straightened). The *medial epicondyle* and *lateral epicondyle* are rough projections on either side of the distal end to which most muscles of the forearm are attached. The ulnar nerve lies on the posterior surface of the medial epicondyle and may easily be rolled between the finger and the medial epicondyle.

Ulna and Radius

The **ulna** is located on the medial aspect (little finger side) of the forearm and is longer than the radius (Figure 7.6). It is connected to the radius by a broad, flat fibrous connective tissue called the *interosseous* (in′-ter-OS-ē-us; *inter* = between, *osseous* = bone) *membrane.* This membrane also provides a site of attachment for some deep skeletal muscles of the forearm. At the proximal end of the ulna (Figure 7.6a) is the *olecranon* or *olecranon process,* which forms the

Figure 7.5 Right humerus in relation to the scapula, ulna, and radius.

The humerus is the longest and largest bone of the upper limb.

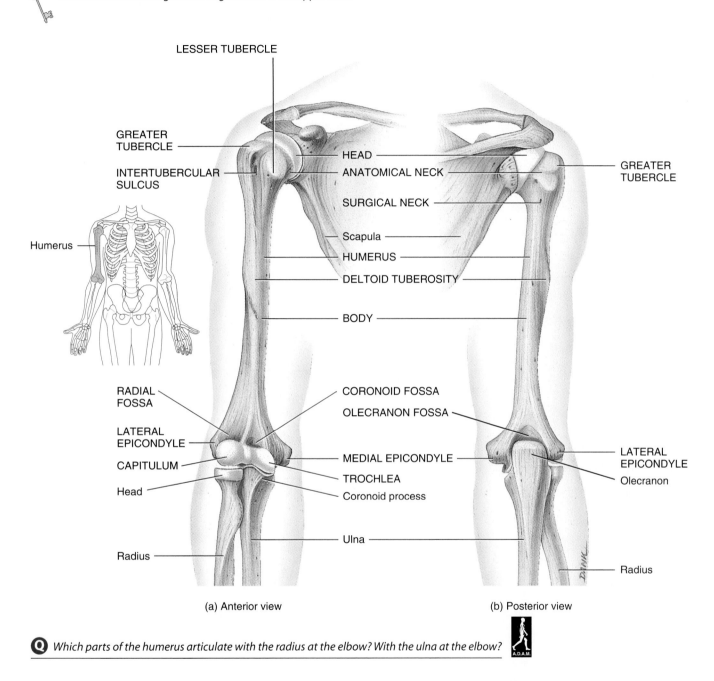

LESSER TUBERCLE

GREATER TUBERCLE

INTERTUBERCULAR SULCUS

Humerus

RADIAL FOSSA

LATERAL EPICONDYLE

CAPITULUM

Head

Radius

HEAD
ANATOMICAL NECK
SURGICAL NECK
Scapula
HUMERUS
DELTOID TUBEROSITY
BODY
CORONOID FOSSA
OLECRANON FOSSA
MEDIAL EPICONDYLE
TROCHLEA
Coronoid process
Ulna

GREATER TUBERCLE

LATERAL EPICONDYLE
Olecranon

Radius

(a) Anterior view

(b) Posterior view

Q *Which parts of the humerus articulate with the radius at the elbow? With the ulna at the elbow?*

prominence of the elbow. The *coronoid process* is an anterior projection that, together with the olecranon, receives the trochlea of the humerus. Just inferior to the coronoid process is the *ulna tuberosity.* The *trochlear (semilunar) notch* (Figure 7.6d) is a large curved area between the olecranon and the coronoid process. The trochlea of the humerus fits into this notch to form part of the *elbow joint.* The *radial notch* is a depression located laterally and inferiorly to the trochlear

notch. It receives the head of the radius to form the *proximal radioulnar joint.* The distal end of the ulna (Figure 7.6a, b) consists of a *head* that is separated from the wrist by a fibrocartilage disc. A *styloid* (*stylo* = stake or pole) *process* is on the posterior side of the distal end.

The **radius** is located on the lateral aspect (thumb side) of the forearm (Figure 7.6a, b). The proximal end of the radius has a disc-shaped *head* that articulates with the capitulum of

Figure 7.6 Right ulna and radius in relation to the humerus and carpals.

The elbow joint is formed by the articulation of the trochlea of the humerus and the trochlear notch of the ulna and by the articulation of the head of the radius with the capitulum of the humerus and radial notch of the ulna.

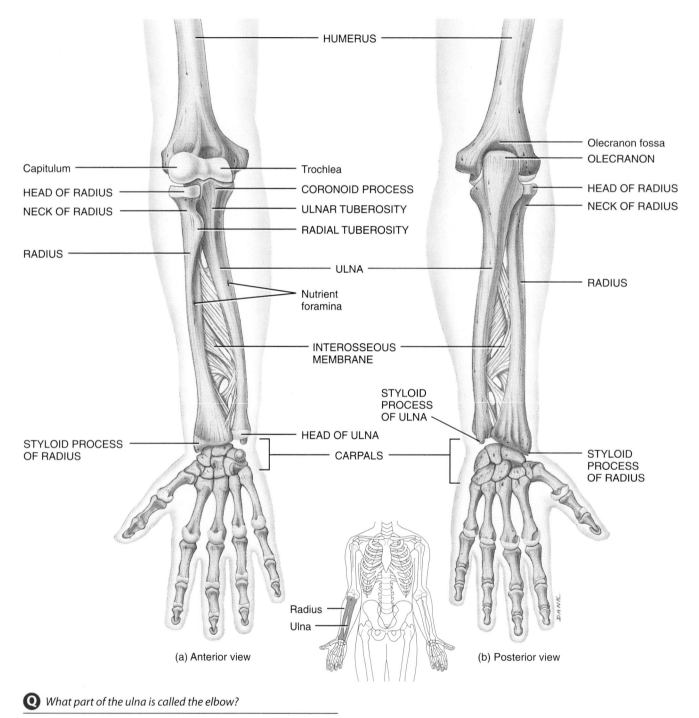

(a) Anterior view

(b) Posterior view

Q *What part of the ulna is called the elbow?*

Figure 7.6 (continued)

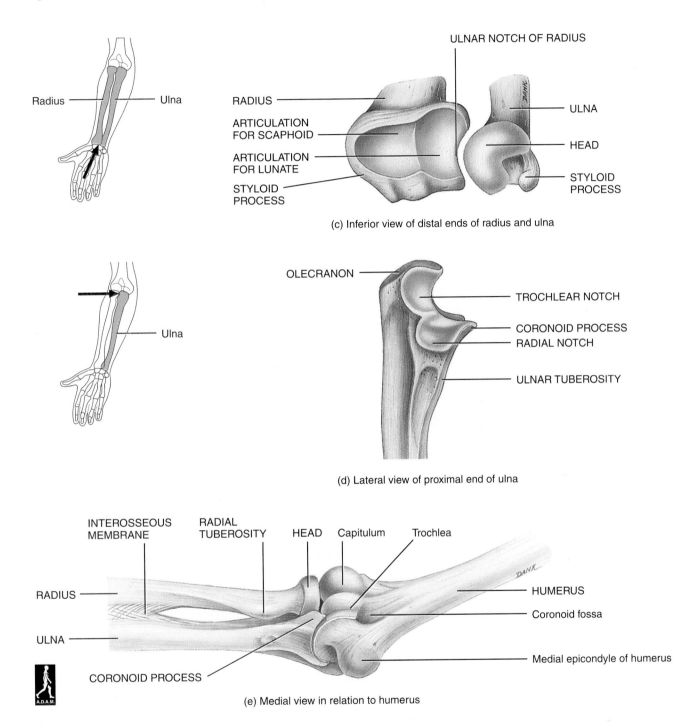

(c) Inferior view of distal ends of radius and ulna

(d) Lateral view of proximal end of ulna

(e) Medial view in relation to humerus

the humerus and radial notch of the ulna to form another part of the *elbow joint* (Figure 7.6e). It also has a raised, roughened area on the medial side called the *radial tuberosity.* This is a point of attachment for the biceps brachii muscle. Inferior to the head is the constricted *neck.* The shaft of the radius widens distally to form a concave inferior surface that articulates with three bones of the wrist called the lunate, scaphoid, and triquetrum bones to form the *radiocarpal (wrist) joint.* Also at the distal end is a *styloid process* on the lateral side and a medial, concave *ulnar notch* that articulates with the head of the ulna to form the *distal radioulnar joint* (Figure 7.6c).

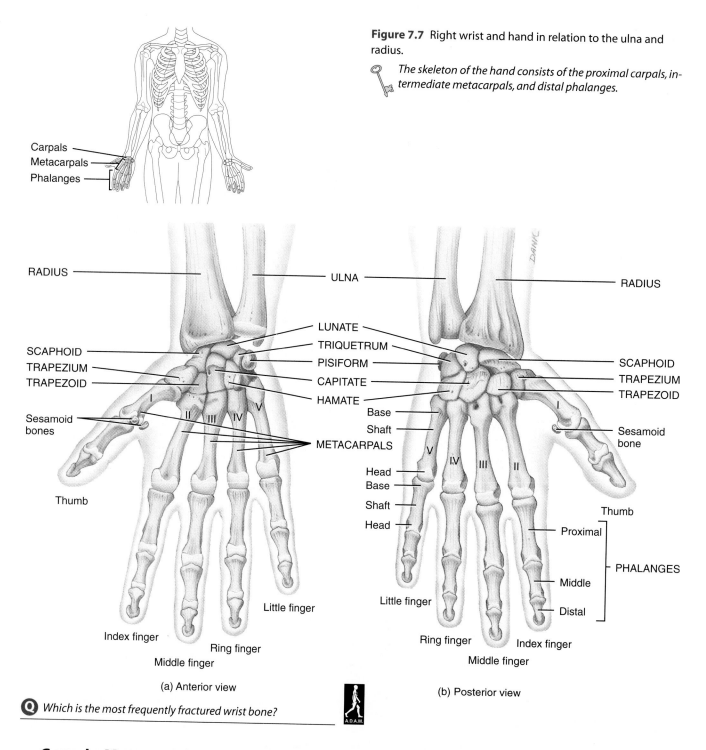

Figure 7.7 Right wrist and hand in relation to the ulna and radius.

The skeleton of the hand consists of the proximal carpals, intermediate metacarpals, and distal phalanges.

(a) Anterior view

(b) Posterior view

Q *Which is the most frequently fractured wrist bone?*

Carpals, Metacarpals, and Phalanges

The skeleton of the hand has three regions: (1) proximal carpus, (2) intermediate metacarpus, and (3) distal phalanges.

The **carpus** (wrist) consists of eight small bones, the **carpals,** joined to one another by ligaments (Figure 7.7). Articulations between carpal bones are called *intercarpal joints.* The carpals are arranged in two transverse rows, with four bones in each row, and they are named for their shapes.

In the anatomical position, the carpals in the proximal row, from the lateral to medial position, are the **scaphoid,** or **navicular** (resembles a boat), **lunate** (resembles a crescent moon in its anteroposterior aspect), **triquetrum** (has three articular surfaces), and **pisiform** (pea-shaped). In about 70% of carpal fractures, only the scaphoid is broken. The carpals in the distal row, from the lateral to medial position, are the **trapezium** (four-sided), **trapezoid** (also four-sided), **capitate** (the largest carpal bone, whose rounded projection, the

Figure 7.8 Bony pelvis. Shown here is the female bony pelvis.

The hipbones are united anteriorly at the pubic symphysis and posteriorly at the sacrum to form the bony pelvis.

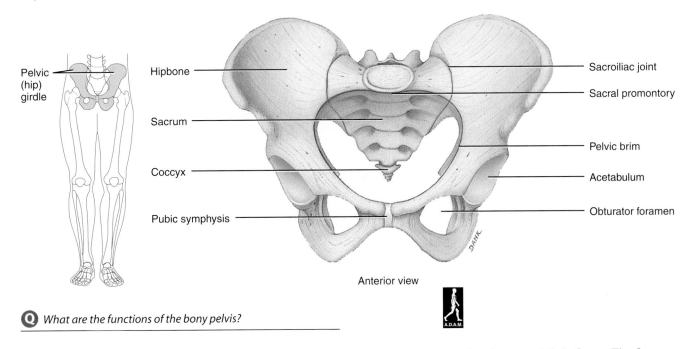

Pelvic (hip) girdle

Hipbone

Sacrum

Coccyx

Pubic symphysis

Sacroiliac joint

Sacral promontory

Pelvic brim

Acetabulum

Obturator foramen

Anterior view

Q *What are the functions of the bony pelvis?*

head, articulates with the lunate), and **hamate** (= hooked; named for a large hook-shaped projection on its anterior surface).

Together, the concavity formed by the pisiform and hamate (on the ulnar side) and the scaphoid and trapezium (on the radial side) plus the flexor retinaculum (deep fascia) constitute a space called the *carpal tunnel.* Through it pass the long flexor tendons of the digits and thumb and the median nerve. Narrowing of the carpal tunnel gives rise to a condition called carpal tunnel syndrome (see Exhibit 10.17).

The five bones of the **metacarpus** (*meta* = after or beyond), called **metacarpals,** constitute the palm of the hand. Each metacarpal bone consists of a proximal *base,* an intermediate *shaft,* and a distal *head.* The metacarpal bones are numbered I to V (or 1–5), starting with the one proximal to the thumb. The bases articulate with the distal row of carpal bones to form the *carpometacarpal joints.* The heads articulate with the proximal phalanges of the fingers to form the *metacarpophalangeal joints.* The heads of the metacarpals are commonly called the "knuckles" and are readily visible when the fist is clenched.

The **phalanges** (fa-LAN-jēz, *phalanx* = closely knit row), or bones of the digits, number 14 in each hand. Like the metacarpals, the phalanges are numbered I to V (or 1–5), beginning with the thumb. A single bone of a digit is referred to as a **phalanx** (FĀ-lanks). Each phalanx consists of a proximal *base,* an intermediate *shaft,* and a distal *head.* There are two phalanges in the thumb (*pollex*) and three phalanges in each of the other four digits. In order from the thumb, these other four digits are commonly referred to as the index fin-

ger, middle finger, ring finger, and little finger. The first row of phalanges, the *proximal row,* articulates with the metacarpal bones and second row of phalanges. The second row of phalanges, the *middle row,* articulates with the proximal row and the third row. The third row of phalanges, the *distal row,* articulates with the middle row. The thumb has no middle phalanx. Joints between phalanges are called *interphalangeal joints.*

PELVIC (HIP) GIRDLE

The **pelvic (hip) girdle** consists of the two **hipbones,** also called **coxal** (KOK-sal; *coxa* = hip) **bones** (Figure 7.8). The hipbones are united to each other anteriorly at a joint called the *pubic symphysis* (PYOO-bik SIM-fi-sis). They unite posteriorly with the sacrum. The complete ring formed by the hipbones, pubic symphysis, and sacrum is a deep, basinlike structure called the **bony pelvis** (= basin). Functionally, the bony pelvis provides a strong and stable support for the vertebral column and pelvic viscera. In addition, the pelvic girdle of the bony pelvis attaches the lower limbs to the axial skeleton.

Each of the two hipbones of a newborn consists of three bones (Figure 7.9a): a superior *ilium,* an inferior and anterior *pubis,* and an inferior and posterior *ischium* (IS-kē-um). Eventually, the three separate bones fuse into one. Although the adult hipbones are both single bones, it is common to discuss the bones as if they still consisted of three portions.

Figure 7.9 Right hipbone. In the three divisions of the hipbone shown in (a), the lines of fusion of the ilium, ischium, and pubis are not always visible in an adult.

The acetabulum is the socket formed where the three parts of the hipbone converge.

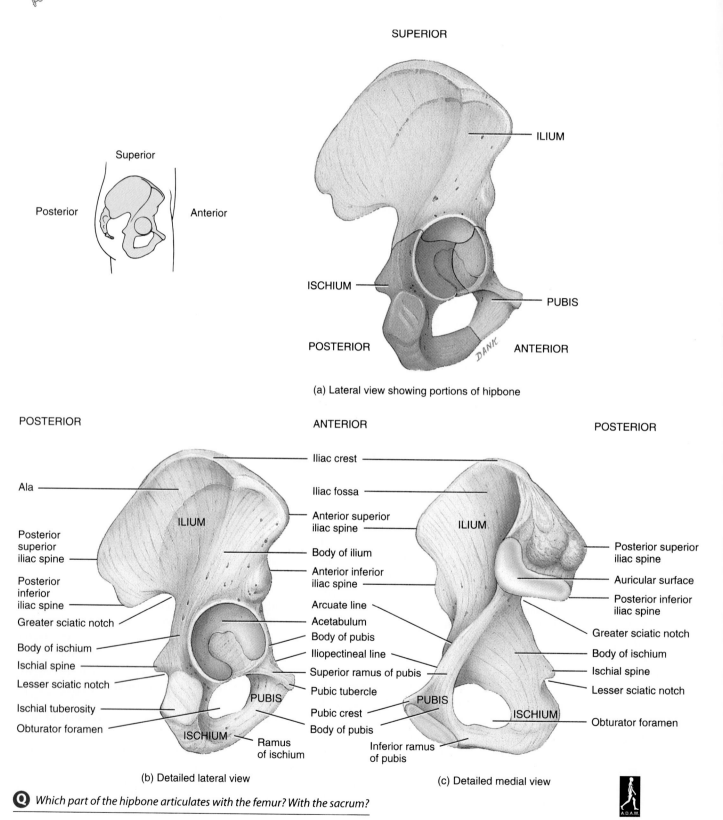

(a) Lateral view showing portions of hipbone

(b) Detailed lateral view

(c) Detailed medial view

Q *Which part of the hipbone articulates with the femur? With the sacrum?*

Ilium

The **ilium** (meaning flank) is the largest of the three subdivisions of the hipbone (Figure 7.9b, c). It is divided into a superior *ala*, meaning wing, and an inferior *body,* which enters into the formation of the acetabulum, the socket for the head of the femur. Its superior border, the *iliac crest,* ends anteriorly in a blunt *anterior superior iliac spine.* Below it is the *anterior inferior iliac spine.* Posteriorly, the iliac crest ends in a sharp *posterior superior iliac spine.* Below it is the *posterior inferior iliac spine.* The spines serve as points of attachment for muscles of the trunk, hip, and thighs. Below the posterior inferior iliac spine is the *greater sciatic* (sī-AT-ik) *notch* through which the sciatic nerve, the longest nerve in the body, passes. The medial surface of the ilium contains the *iliac fossa.* It is a concavity where the iliacus muscle attaches. Posterior to this fossa are the *iliac tuberosity,* a point of attachment for the sacroiliac ligament, and the *auricular* (*auricula* = ear-shaped) *surface,* which articulates with the sacrum to form the *sacroiliac joint* (see Figure 7.8). Projecting anteriorly and superiorly from the auricular surface is a ridge called the *arcuate* (AR-kyoo-āt; *arcuatus* = bowed) *line.* The other conspicuous markings of the ilium are three arched lines on its lateral surface called the *posterior gluteal* (*gluoutos* = buttock) *line,* the *anterior gluteral line,* and the *inferior gluteal line.* The gluteal muscles attach to the ilium between these lines.

Ischium

The **ischium** (meaning hip) is the inferior, posterior portion of the hipbone (see Figure 7.9b, c). It is composed of a superior *body* and inferior *ramus* (meaning branch), which joins the pubis. The ischium contains the prominent *ischial spine,* a *lesser sciatic notch* below the spine, and a rough and thickened *ischial tuberosity.* This prominent tuberosity may hurt someone's thigh when you sit on their lap. The ramus joins with the pubis, and together they surround the *obturator* (OB-too-rā-ter) *foramen,* the largest foramen in the skeleton. The term *obturator* means closed up. The foramen is so named because, even though blood vessels and nerves pass through it, it is nearly completely closed by a fibrous membrane (obturator membrane).

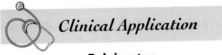

Clinical Application

Pelvimetry

Pelvimetry is the measurement of the size of the inlet and outlet of the birth canal which may be done by ultrasound or physical examination. Measurement of the pelvic cavity in pregnant females is important because the fetus must pass through the narrower opening of the pelvis at birth. ▪

Pubis

The **pubis** (meaning pubic hair) is the anterior and inferior part of the hipbone (Fig. 7.9b, c). It consists of a *superior ramus,* an *inferior ramus,* and a *body* between the rami that contributes to the formation of the pubic symphysis. The anterior border of the body is known as the *pubic crest,* and at its lateral end is a projection, the *pubic tubercle.* This tubercle is the beginning of a raised line, the *iliopectineal* (il-ē-ō-pek-TIN-ē-al) *line,* which extends superiorly and laterally along the superior ramus to merge with the arcuate line of the ilium. These lines, as you will see shortly, are important landmarks in distinguishing the bony pelvis into superior and inferior portions.

The pubic symphysis is the joint between the two hipbones (see Figure 7.8). It consists of a pad of fibrocartilage. The *pubic arch* is formed by the convergence of the inferior rami of the pubes. The *acetabulum* (as-e-TAB-yoo-lum; *acetabulum* = vinegar cup) is the fossa formed by the ilium, ischium, and pubis. It is the socket for the head of the femur, and the two together form the *hip (coxal) joint.* On the inferior portion of the acetabulum is a deep indentation, the *acetabular notch.* It forms a foramen through which nutrient vessels and nerves enter the joint and serves as a point of attachment for the ligament of the head of the femur.

True and False Pelvis

As noted previously, the two hipbones of the pelvic girdle, pubic symphysis, and sacrum together form the bony pelvis (plural is *pelvises* or *pelves*). In turn, the bony pelvis is divided into superior and inferior portions by a boundary called the *pelvic brim* (Figure 7.10a). You can trace the pelvic brim by following various landmarks around parts of the hipbones to visualize an oblique plane. Begin the plane posteriorly at the *sacral promontory* of the sacrum. Then move laterally and inferiorly to the *arcuate lines* of the ilium of the hipbones. Continue moving inferiorly to the *iliopectineal lines* of the pubis of the hipbones. Finally, move anteriorly to the superior portion of the *pubic symphysis.* Connecting these points forms an oblique plane, higher in the back then the front. The circumference of this plane is the pelvic brim.

The portion of the bony pelvis above the pelvic brim is called the *false (greater) pelvis.* This is seen best in Figure 7.10b. It is bordered by the lumbar vertebrae (posteriorly), the upper portions of the hipbones (laterally), and the abdominal wall (anteriorly). The false pelvis is actually part of the abdomen and does not contain any pelvic organs, except for the urinary bladder when it is full and the uterus during pregnancy.

The portion of the bony pelvis below the pelvic brim is called the *true (lesser) pelvis.* This is also seen best in Figure 7.10b. It is bounded by the sacrum and coccyx (posteriorly), inferior portions of the ilium and ischium (laterally),

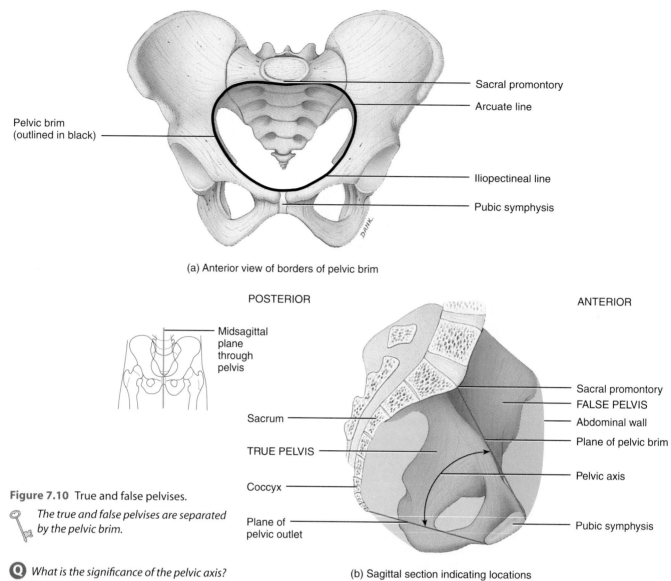

Pelvic brim
(outlined in black)

Sacral promontory

Arcuate line

Iliopectineal line

Pubic symphysis

(a) Anterior view of borders of pelvic brim

POSTERIOR

ANTERIOR

Midsagittal
plane
through
pelvis

Sacral promontory

FALSE PELVIS

Abdominal wall

Plane of pelvic brim

Sacrum

TRUE PELVIS

Pelvic axis

Coccyx

Plane of
pelvic outlet

Pubic symphysis

Figure 7.10 True and false pelvises.

*The true and false pelvises are separated
by the pelvic brim.*

Q *What is the significance of the pelvic axis?*

(b) Sagittal section indicating locations
of true and false pelvises

and pubic bones (anteriorly). The true pelvis surrounds the pelvic cavity (see Figure 1.6). The superior opening of the true pelvis is actually the pelvic brim and is called the *pelvic inlet;* the inferior opening of the true pelvis is called the *pelvic outlet* (see Figure 7.10b). The *pelvic axis* is an imaginary curved line passing through the true pelvis at right angles to the center of the planes of the pelvic inlet and outlet. During childbirth it is the course taken by the baby's head as it descends through the pelvis.

Comparison of Pectoral and Pelvic Girdles

Having studied the structure of the pectoral and pelvic girdles, we can now note some of their significant differences. Whereas the pectoral girdle has no direct articulation with the vertebral column, the pelvic girdle directly articulates with the vertebral column via the sacroiliac joint. The structure of the pectoral girdle offers more mobility than strength, whereas that of the pelvic girdle offers more strength than

mobility. Finally, the sockets (glenoid fossae) for the upper limbs in the pectoral girdle are shallow and maximize movement; the sockets (acetabula) for the lower limbs in the pelvic girdle are deep and allow less movement.

FEMALE AND MALE PELVISES

Although the bones of a male are generally larger and heavier than those of a female and generally possess larger surface features, the relative differences between skeletons of females and males are readily apparent when comparing the pelvises of both. Most of the structural differences are related to pregnancy and childbirth. Because the female's pelvis is wider and shallower than the male's pelvis, there is more space in the true pelvis of the female, especially in the pelvic inlet and pelvic outlet which accommodates the passage of the infant's head at birth. Several of the significant differences between pelvises of females and males are listed and illustrated in Table 7.1.

Table 7.1 *Comparison of Typical Female and Male Pelvises*

POINT OF COMPARISON	FEMALE	MALE
General structure	Light and thin.	Heavy and thick.
False (greater) pelvis	Shallow.	Deep.
Pelvic brim (inlet)	Larger and more oval.	Heart-shaped.
Acetabulum	Small and faces anteriorly.	Large and faces laterally.
Obturator foramen	Oval.	Round.
Pubic arch	Greater than 90° angle.	Less than 90° angle.

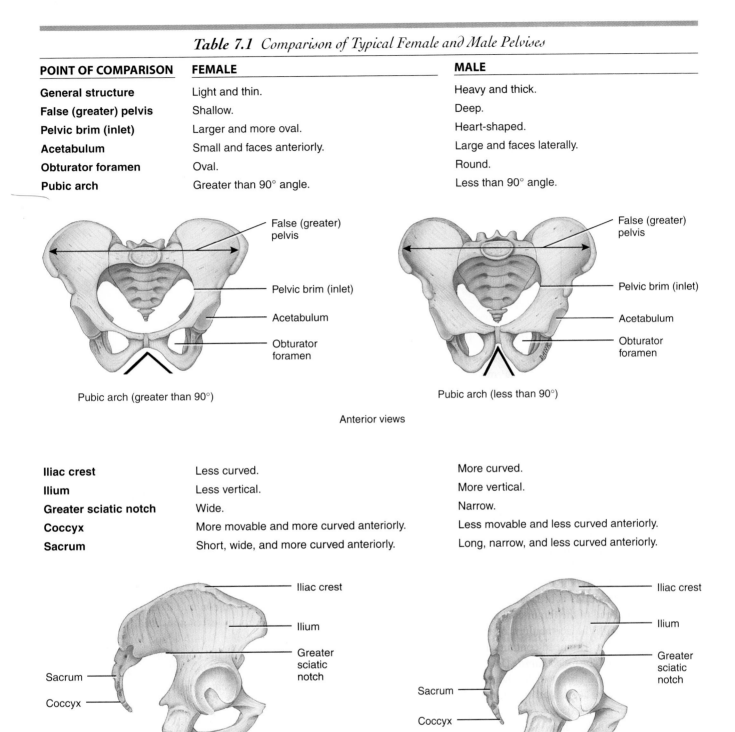

Pubic arch (greater than 90°)

Pubic arch (less than 90°)

Anterior views

Iliac crest	Less curved.	More curved.
Ilium	Less vertical.	More vertical.
Greater sciatic notch	Wide.	Narrow.
Coccyx	More movable and more curved anteriorly.	Less movable and less curved anteriorly.
Sacrum	Short, wide, and more curved anteriorly.	Long, narrow, and less curved anteriorly.

Right lateral views

LOWER LIMB (EXTREMITY)

The **lower limbs (extremities)** are composed of 60 bones. Each lower limb includes the femur in the thigh, patella (kneecap), tibia and fibula in the leg, and tarsals (ankle bones), metatarsals, and phalanges (toes) in the foot (Figure 7.11 on page 181).

The *longitudinal arch* has two parts. Both consist of tarsal and metatarsal bones arranged to form an arch from the anterior to the posterior part of the foot. The *medial* (inner) *part* of the longitudinal arch originates at the calcaneus. It rises to the talus and descends through the navicular, the three cuneiforms, and the heads of the three medial metatarsals. The *lateral* (outer) *part* of the longitudinal arch also begins at the calcaneus. It rises at the cuboid and descends to the heads of the two lateral metatarsals.

The *transverse arch* is formed by the navicular, three cuneiforms, and the bases of the five metatarsals.

Clinical Application

Flatfoot, Clawfoot, Clubfoot, and Bunions

The bones composing the arches are held in position by ligaments and tendons. If these ligaments and tendons are weakened, the height of the medial longitudinal arch may decrease or "fall." The result is **flatfoot.** Causes include excessive weight, postural abnormalities, weakened supporting tissues, and genetic predisposition. A custom-designed arch support (orthotic) often is prescribed to treat flatfoot.

Clawfoot is a condition in which the medial longitudinal arch is abnormally elevated. It is frequently caused by muscle imbalance, such as may result from poliomyelitis.

Clubfoot (**talipes equinovarus;** *talus* = ankle; *pes* = foot; *equinus* = horse) is an inherited deformity of the foot that occurs in 1 of every 1000 births. In this malformation, the foot is twisted inferiorly and medially and the arch is increased. Treatment consists of manipulating the feet to the normal position by casts or adhesive tape soon after birth. Corrective shoes or surgery may also be required.

A **bunion** (**hallux valgus;** *valgus* = bent outward) is a deformity of the great toe. Although the condition may be inherited, it is typically caused by wearing tightly fitting shoes and is characterized by lateral deviation of the proximal phalanx of the great toe and medial displacement of metatarsal I. Arthritis of the first metatarsophalangeal joint may also be a predisposing factor. The condition produces inflammation of bursae (fluid-filled sacs at the joint), bone spurs, and calluses. ▪

Study Outline

Pectoral (Shoulder) Girdle (p. 167)

1. Each pectoral (shoulder) girdle consists of a clavicle and scapula.
2. Each attaches an upper limb to the trunk.

Upper Limb (Extremity) (p. 168)

1. There are 60 bones in the upper limbs.
2. The bones of each upper limb include the humerus, ulna, radius, carpals, metacarpals, and phalanges.

Pelvic (Hip) Girdle (p. 175)

1. The pelvic (hip) girdle consists of two hipbones.
2. The hipbones, sacrum, and pubic symphysis make up the bony pelvis. It supports the vertebral column and pelvic viscera and attaches the lower limbs to the axial skeleton.
3. Each hipbone consists of three fused components: ilium, pubis, and ischium.

4. The true pelvis is separated from the false pelvis by the pelvic brim.

Female and Male Pelvises (p. 178)

1. Male's bones are generally larger and heavier than female's bones and have more prominent markings for muscle attachment.
2. The female's pelvis is adapted for pregnancy and childbirth. Differences in pelvic structure are listed in Table 7.1.

Lower Limb (Extremity) (p. 178)

1. There are 60 bones in the lower limbs.
2. The bones of each lower limb include the femur, patella, tibia, fibula, tarsals, metatarsals, and phalanges.
3. The bones of the foot are arranged in two arches, the longitudinal arch and the transverse arch, to provide support and leverage.

Review Questions

1. What is the pectoral (shoulder) girdle? Why is it important? (p. 167)
2. What are the bones of the upper limb? (p. 168)
3. What is the pelvic (hip) girdle? Why is it important? (p. 175)
4. What are the bones of the lower limb? (p. 179)
5. What is pelvimetry? What is the clinical importance of pelvimetry? (p. 177)

6. What are the principal structural differences between typical pelvises of females and males? Use Table 7.1 as a guide in formulating your response. (p. 179)
7. In what ways do the upper limb and lower limb differ structurally? (p. 179)
8. Describe the structure of the longitudinal and transverse arches of the foot. What is the function of an arch? (p. 183)

Self Quiz

Complete the following:

1. The large depression of the anterior surface of the scapula is the ___.

2. The anatomical name of the knuckle of your index finger is ___.

3. The lesser trochanter is on the ___ surface of the ___.

4. The portion of the pelvis above the pelvic brim is the ___ pelvis.

Choose the one best answer to these questions.

5. Which of the following are bones found in the axial division of the skeleton?
   a. ribs, scapula, clavicle
   b. sphenoid, sternum, coccyx
   c. parietal, ulna, carpals
   d. lacrimal, sacrum, patella
   e. pubis, ischium, ilium

6. On which bones are medial and lateral epicondyles located?
   (1) ulna
   (2) femur
   (3) radius
   (4) humerus
   (5) tibia
   a. 1, 3 and 5
   b. 2, 4, and 5
   c. 1 and 2
   d. 4 and 5
   e. 2 and 4

7. Which structures are on the posterior surface of the upper limb?
   a. radial fossa and radial notch
   b. trochlea and capitulum
   c. coronoid process and coronoid fossa
   d. olecranon process and olecranon fossa
   e. lesser tubercle and intertubercular sulcus

8. Choose the false statement.
   a. The capitulum articulates with the head of the radius.
   b. The medial and lateral condyles are located at the proximal end of the humerus.
   c. The coronoid fossa articulates with the ulna when the ulna is flexed.
   d. The trochlea articulates with the trochlear notch of the ulna.
   e. There is a styloid process on the distal end of the ulna.

9. The bones of the pectoral girdle
   a. articulate with the sternum anteriorly and the vertebrae posteriorly
   b. include both clavicles and scapulae and the manubrium of the sternum
   c. are considered to be part of the axial skeleton
   d. both articulate with the head of the humerus, forming the shoulder joint
   e. are described by none of the above.

10. The anatomical name for the socket into which the humerus fits is the
    a. acetabulum
    b. coronoid fossa
    c. glenoid cavity
    d. supraspinous fossa
    e. iliac fossa

11. The radius
    a. has a tuberosity on its distal end
    b. has an acromial process at its proximal end
    c. is longer and larger than the ulna
    d. has a styloid process on its distal end
    e. is the medial bone of the forearm

12. The greater trochanter is a large bony prominence located
    a. on the proximal part of the humerus
    b. on the proximal part of the femur
    c. near the tuberosity of the tibia
    d. on the ilium
    e. on the posterior surface of the scapula

13. Which of the following is a part of the femur?
    a. obturator foramen
    b. trochlear notch
    c. patellar surface
    d. medial malleolus
    e. lateral malleolus

14. Which of the following statements regarding the male pelvis is *not* true?
    a. The bones are heavier and thicker than in the female.
    b. The male pelvis is narrow and deep.
    c. The coccyx is less curved anteriorly.
    d. The pelvic outlet is narrower than in the female.
    e. none of the above

15. Which of the following is *not* a carpal bone?
    a. hamate
    b. cuboid
    c. pisiform
    d. trapezium
    e. scaphoid

Are the following statements true or false?

16. The metatarsals are proximal to the tarsals.
17. The phalanges are distal to the metatarsals and the tarsals.
18. The acromial extremity is located at the lateral end of the clavicle.
19. The lesser tubercle is on the anterior surface of the humerus.
20. Match the following.
    (a) sternal extremity   ___ (1) radius
    (b) acromion   ___ (2) femur
    (c) coronoid process   ___ (3) tibia
    (d) lateral malleolus   ___ (4) humerus
    (e) deltoid tuberosity   ___ (5) scapula
    (f) radial tuberosity   ___ (6) ischium
    (g) anterior inferior iliac spine   ___ (7) clavicle
    (h) linea aspera   ___ (8) fibula
    (i) ischial tuberosity   ___ (9) ilium
    (j) fibular notch   ___ (10) ulna

Critical Thinking Questions

1. The local news reported that farmer Bob Dole caught his hand in a piece of machinery on Tuesday. He lost the lateral two fingers of his left hand. Science reporter Kent Clark, really a high school junior, reports that farmer Dole has 3 remaining phalanges. Is Kent correct or is he anatomy-challenged?
   HINT: *How many phalanges in one finger?*

2. You're building a skeleton with a preschool class and singing "the thigh bone's connected to the hip bone." Leigh, a precocious 3-year-old, wanders over and demands to know the anatomical names for each bone starting with the toes. List the bones of the leg in order and don't worry if they don't rhyme.
   HINT: *The toe bones have the same name as the bones in the first question.*

3. Since the hips and shoulders are located in the trunk of the body, why aren't they part of the axial skeleton like the rest of the bones composing the trunk?
   HINT: *Have you ever seen a snake with shoulders?*

4. Drew tripped on the basement steps while running up the stairs barefoot. He landed squarely on his big toe, breaking the larger bone with a snap. Using proper anatomical terms, describe the position of the broken bone.
   HINT: *Which way are the toes pointed in the anatomical position?*

5. Amy was visiting her grandmother in the hospital. Grandmother said she had a broken hip but when Amy peeked at the chart, it said she had a fractured femur. Explain the discrepancy.
   HINT: *The area that people usually think of as the hip includes more than just the hip bone.*

6. Mary jogs on the same banked high school track at the same time and in the same direction every day. Recently she has been experiencing an ache in one knee, especially after she sits and reads the newspaper following her jog. Nothing else in her routine has changed. What could be the problem with her knee?
   HINT: *The answer is in her routine.*

Answers to Figure Questions

7.1 They afford movements in many directions.
7.2 The junction of the two curves.
7.3 Acromion.
7.4 30.
7.5 Radius: capitulum and radial fossa; ulna: trochlea, coronoid fossa, and olecranon fossa.
7.6 Olecranon.
7.7 Scaphoid.
7.8 It attaches the lower limbs to the axial skeleton and supports the backbone and pelvic viscera.

7.9 Femur: acetabulum; sacrum: auricular surface.
7.10 It is the course taken by a baby's head as it descends through the pelvis during childbirth.
7.11 30.
7.12 The angle of the neck.
7.13 Sesamoid.
7.14 Tibia.
7.15 Talus.
7.16 They are not rigid, yielding when weight is applied and springing back when weight is lifted.

Chapter 8

Joints

Student Objectives

1. Define a joint and identify the factors that determine the degree of movement at a joint.

2. Contrast the structural and functional classification of joints.

3. Describe the structure of fibrous, cartilaginous, and synovial joints.

4. Describe the nerve and blood supply of joints.

5. Discuss and compare the movements possible at various synovial joints.

6. Describe selected joints of the body with respect to the bones that enter into their formation, structural classification, anatomical components, and movements.

7. Explain the effects of aging on joints.

Bones are too rigid to bend without being damaged. Fortunately, flexible connective tissues form joints that hold bones together while still permitting some degree of movement, in most cases. Because most movements of the body occur at joints, you can appreciate their importance if you imagine how a cast over the knee joint makes walking difficult or how a splint on a finger limits the ability to manipulate small objects. The scientific study of joints is referred to as **arthrology** (ar-THROL-ō-jē; *arthro* = joint; *logos* = study of). The study of motion of the human body is called **kinesiology** (ki-nē-sē'-OL-ō-jē ; *kinesis* = movement).

An **articulation (joint)** is a point of contact between bones, between cartilage and bones, or between teeth and bones. When we say one bone *articulates* with another bone, we mean that one bone forms a joint with another bone.

Some joints permit no movement, others permit slight movement, and still others afford fairly free movement. A joint's structure determines its combination of strength and flexibility. In general, the closer the fit at the point of contact, the stronger the joint. At tightly fitted joints, however, movement is restricted. The looser the fit, the greater the movement, but loosely fitted joints are prone to dislocation (displacement). Movement at joints is also determined by (1) the structure (shape) of articulating bones, (2) the flexibility (tension or tautness) of the connective tissue that binds the bones together, and (3) the position of associated ligaments, muscles, and tendons. Joint flexibility may also be affected by hormones. For example, relaxin, a hormone produced by the placenta and ovaries, relaxes the pubic symphysis and ligaments between the sacrum, hipbone, and coccyx toward the end of pregnancy, which assists in delivery.

CLASSIFICATION OF JOINTS

Joints can be categorized into structural classes, based on anatomical characteristics, or into functional classes, based on the type of movement they permit.

Structural Classification

The structural classification of joints is based on the presence or absence of a space between the articulating bones called a synovial (joint) cavity (described shortly) and the type of connective tissue that binds the bones together. Structurally, a joint is classified as follows:

1. A **fibrous joint** has no synovial cavity, and the bones are held together by fibrous (collagenous) connective tissue.
2. A **cartilaginous joint** has no synovial cavity, and the bones are held together by cartilage.

3. A **synovial joint** has a synovial cavity, and the bones forming the joint are united by a surrounding articular capsule and frequently by accessory ligaments (described in detail later).

Functional Classification

The functional classification of joints takes into account the degree of movement they permit. Functionally, a joint is classified as follows:

1. A **synarthrosis** (sin'-ar-THRŌ-sis; *syn* = together; *arthros* = joint) is an immovable joint; *synarthroses* is plural.
2. An **amphiarthrosis** (am'-fē-ar-THRŌ-sis; *amphi* = on both sides) is a partially movable joint; *amphiarthroses* is plural.
3. A **diarthrosis** (dī-ar-THRŌ-sis; *diarthros* = movable joint) is a freely movable joint; *diarthroses* is plural.

We will discuss the joints of the body based upon their structural classification, referring to their functional classification as well.

FIBROUS JOINTS

Fibrous joints lack a synovial cavity, and the articulating bones are held very closely together by fibrous connective tissue. They permit little or no movement. The three types of fibrous joints are (1) sutures, (2) syndesmoses, and (3) gomphoses. The terms *suture, syndesmosis,* and *gomphosis* are singular.

Suture

A **suture** (SOO-cher; *sutura* = seam) is a fibrous joint composed of a thin layer of dense fibrous connective tissue that unites bones of the skull. An example is the coronal suture between the frontal and parietal bones (see Figure 6.4). The irregular, interlocking edges of sutures give them added strength and decrease their chance of fractures. Because a suture is immovable, it is functionally classified as a synarthrosis.

Some sutures, although present during childhood, are replaced by bone in the adult. Such a suture is called a **synostosis** (sin'-os-TŌ-sis; *syn* = together; *osteon* = bone), or bony joint—a joint in which there is a complete fusion of bone across the suture line. An example is the frontal suture between the left and right sides of the frontal bone that begins to fuse during infancy (see Figure 6.14a). A synostosis is also functionally classified as a synarthrosis.

Syndesmosis

A **syndesmosis** (sin′-dez-MŌ-sis; *syndesmo* = band or ligament) is a fibrous joint in which there is considerably more fibrous connective tissue than in a suture. The fit between the bones thus is not quite as tight. The fibrous connective tissue forms an interosseous membrane or ligament that permits some flexibility and movement. An example of a syndesmosis is the distal tibiofibular joint (between the tibia and fibula; see Figure 7.14a, b). Because it permits partial movement, a syndesmosis is functionally classified as an amphiarthrosis.

Gomphosis

A **gomphosis** (gom-FŌ-sis; *gomphosis* = to bolt together) is a type of fibrous joint in which a cone-shaped peg fits into a socket. The only examples of gomphoses are the articulations of the roots of the teeth with the alveoli (sockets) of the maxillae and mandible (see Figure 6.12a). The connective tissue between the two is the periodontal ligament. A gomphosis is functionally classified as a synarthrosis.

CARTILAGINOUS JOINTS

Like a fibrous joint, a **cartilaginous joint** lacks a synovial cavity. Here the articulating bones are tightly connected by cartilage. Like fibrous joints, they allow little or no movement. The two types of cartilaginous joints are (1) synchondroses and (2) symphyses. The terms *synchondrosis* and *symphysis* are singular.

Synchondrosis

A **synchondrosis** (sin′-kon-DRŌ-sis; *syn* = together; *chondro* = cartilage) is a cartilaginous joint in which the connecting material is hyaline cartilage. An example of a synchondrosis is the epiphyseal plate (see Figure 5.5) that connects the epiphysis and diaphysis of a growing bone. Because this form of synchondrosis is immovable, it is functionally a synarthrosis. When bone growth ceases, bone replaces the hyaline cartilage and the joint becomes a synostosis. Other examples of synchondroses are the sternocostal joints between true (vertebrosternal) ribs and the sternum. Functionally, these joints are amphiarthroses.

Symphysis

A **symphysis** (SIM-fi-sis; *symphysis* = growing together) is a cartilaginous joint in which the connecting material is a broad, flat disc of fibrocartilage. This type of joint is found at the intervertebral joints (between bodies of vertebrae; see Figure 6.17b). A portion of the intervertebral disc is fibrocartilaginous material. The pubic symphysis between the anterior surfaces of the hipbones is another example (see Figure 7.8). Functionally, a symphysis is an amphiarthrosis.

SYNOVIAL JOINTS

Synovial (si-NŌ-vē-al) **joints** have a space called a **synovial (joint) cavity** between the articulating bones (Figure 8.1). The surface of a synovial joint is covered by **articular cartilage,** which is hyaline cartilage that shields the surfaces of the articulating bones but does not bind them together. Articular cartilage reduces friction at the joint when the bones move and helps absorb shock.

Articular Capsule

A sleevelike **articular capsule** surrounds a synovial joint, encloses the synovial cavity, and unites the articulating bones. The articular capsule is composed of two layers. The outer layer, the *fibrous capsule,* usually consists of dense, irregular connective tissue. It attaches to the periosteum of the articulating bones at a variable distance from the edge of the articular cartilage. The flexibility of the fibrous capsule permits movement at a joint, whereas its great tensile strength (resistance to stretching) helps prevent the bones from dislocating. The fibers of some fibrous capsules are arranged in parallel bundles and are therefore highly adapted to resist recurrent strain. Such bundles of fibers are called **ligaments** (*ligare* = to bind) and are given special names. The strength of the ligaments is one of the principal factors in holding bone to bone. A synovial joint is functionally classified as a diarthrosis because of the flexibility afforded by the synovial cavity and the arrangement of the articular capsule and ligaments.

The inner layer of the articular capsule is formed by a **synovial membrane.** The synovial membrane is composed of areolar connective tissue with elastic fibers and a variable amount of adipose tissue.

Synovial Fluid

The synovial membrane secretes **synovial fluid (SF).** This fluid, which fills the synovial cavity, has several functions. It lubricates and reduces friction in the joint. It also supplies nutrients to and removes metabolic wastes from the chondrocytes of the articular cartilage. (Recall that cartilage is an avascular tissue.) Synovial fluid also contains phagocytic cells that remove microbes and debris resulting from wear and tear in the joint. Synovial fluid consists of hyaluronic acid and interstitial fluid formed from blood plasma. It is similar in appearance and consistency to uncooked egg white. When there is no joint movement, the fluid is quite

Figure 8.1 Structure of a synovial joint.

A synovial joint is distinguished by a synovial (joint) cavity between articulating bones.

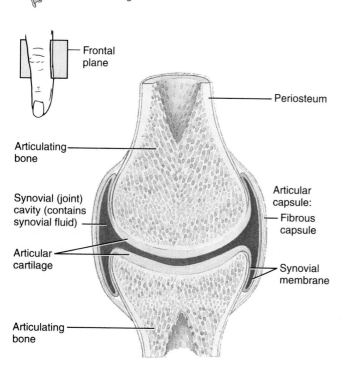

(a) Diagram of frontal section of a typical synovial joint

viscous (gel-like), but as joint movement increases, the fluid becomes less viscous. The amount of synovial fluid varies according to the size of the joint. For example, a large joint, such as the knee, may contain 3–4 ml (about ⅛ oz) of fluid. It forms a thin, viscous film over the surfaces within the articular capsule. One of the benefits of a "warm-up" prior to exercise is that it stimulates the production and secretion of synovial fluid.

Accessory Ligaments and Menisci

Many synovial joints also contain **accessory ligaments,** called extracapsular ligaments and intracapsular ligaments. *Extracapsular ligaments* lie outside the articular capsule. An example is the fibular (lateral) collateral ligament of the knee joint (see Figure 8.12a). *Intracapsular ligaments* occur within the articular capsule but are excluded from the synovial cavity by folds of the synovial membrane. Examples are the cruciate ligaments of the knee joint (see Figure 8.12d).

Inside some synovial joints, such as the knee, are pads of fibrocartilage that lie between the articular surfaces of the bones and are attached to the fibrous capsule. These pads are called **articular discs (menisci;** singular is *meniscus*). In Figure 8.12d you can see lateral and medial menisci in the knee joint. The discs usually subdivide the synovial cavity into two separate spaces. Articular discs modify the shape of

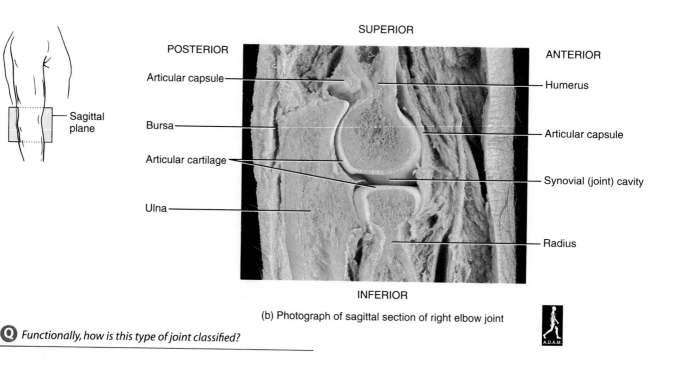

(b) Photograph of sagittal section of right elbow joint

Q *Functionally, how is this type of joint classified?*

the joint surfaces of the articulating bones. In doing so, they allow two bones of different shapes to fit tightly. Articular discs also help to maintain the stability of the joint and direct the flow of synovial fluid to areas of greatest friction.

Clinical Application

Torn Cartilage and Arthroscopy

A tearing of articular discs (menisci) in the knee, commonly called **torn cartilage,** occurs often among athletes. Such damaged cartilage requires surgical removal (meniscectomy) or it will begin to wear and may precipitate arthritis. Repair of the torn cartilage may be assisted by **arthroscopy** (ar-THROS-kō-pē; *arthro* = joint; *skopein* = to view), a procedure that involves examination of the interior of a joint, usually the knee, using an arthroscope, a lighted instrument the diameter of a pencil. It is used to determine the nature and extent of damage following knee injury; to help remove torn cartilage and repair cruciate ligaments in the knee; to help obtain tissue samples for analysis and to perform surgery on other joints, such as the shoulder, elbow, ankle, and wrist; and to monitor the progression of disease and the effects of therapy. ▪

Nerve Supply

The nerves that supply a joint are the same as those that supply the skeletal muscles that move the joint. Synovial joints contain many nerve endings that are distributed to the articular capsule and associated ligaments. Some of the nerve endings convey information about pain from the joint to the spinal cord and brain for processing. Other nerve endings are responsive to movements of joints and how much stretch occurs at a joint. This information, too, is relayed to the spinal cord and brain, which may respond by sending impulses through different nerves to the muscles to adjust body movements.

Blood Supply

Arteries in the area of a synovial joint send out numerous branches that penetrate the ligaments and articular capsule to deliver oxygen and nutrients. Veins remove carbon dioxide and wastes from the joints. It is common for arterial branches to come from several different arteries and join together around a joint before penetrating the articular capsule. This also happens with nerves, with the result that synovial joints have an abundant blood and nerve supply. This gathering of vessels and nerves around a joint is important in maintaining the vascular and nerve supply should some of the blood vessels or nerves become damaged. Except for the actual articulating portions of a synovial joint, such as the articular cartilage, all other tissues of the joint receive their nourishment either directly or indirectly from the blood. The articulating portions receive their nourishment from synovial fluid. As noted earlier, synovial fluid is formed from blood plasma. The source of this blood is blood vessels in the synovial membrane.

BURSAE AND TENDON SHEATHS

The various motions of the body create friction between moving parts. Saclike structures called **bursae** (*bursa* = pouch or purse) are strategically situated to alleviate friction in some joints such as the shoulder and knee joints. Bursae resemble joint capsules in that their walls consist of connective tissue lined by a synovial membrane. They are also filled with a fluid similar to synovial fluid. Bursae are located between the skin and bone in places where skin rubs over bone. They are also found between tendons and bones, muscles and bones, and ligaments and bones and within articular capsules. (See Figure 8.12c.) Such fluid-filled sacs cushion the movement of one part of the body over another. Inflammation of a bursa is called bursitis (see Applications to Health, p.7).

In addition to bursae, structures called tendon sheaths also reduce friction at joints. **Tendon sheaths** are tubelike bursae that wrap around tendons where there is considerable friction, such as in the wrist and ankle, where many tendons come together in a confined space (see Figure 10.18a); in the fingers and toes, where there is a great deal of movement (see Figure 10.19a); and where tendons pass through synovial cavities (see Figure 8.9c).

FACTORS AFFECTING CONTACT AND RANGE OF MOTION AT SYNOVIAL JOINTS

How the articular surfaces of synovial joints contact one another determines the type and range of motion that is possible. Several factors contribute to keeping the surfaces in contact:

1. *Structure or shape of articulating bones.* The structure or shape of the articulating bones determines how closely they can fit together. The articular surfaces of some bones interlock with one another. This spatial relationship is very obvious at the hip joint, where the head of the femur articulates with the acetabulum of the hipbone. An interlocking fit allows rotational movement. The shape of synovial joints permits a diverse range of motions, which will be described shortly.

2. *Strength and tension (tautness) of the joint ligaments.* The different components of a fibrous capsule are tense only when the joint is in certain positions. Tense ligaments not only restrict the range of motion but also direct

the movement of the articulating bones with respect to each other. In the knee joint, for example, the anterior cruciate ligament is tense and the posterior cruciate ligament is loose when the knee is straightened, and the reverse occurs when the knee is bent. These ligaments are shown in Figure 8.12d. Also, when the knee is straightened, the surfaces of the articulating bones are closest to each other.

3. *Arrangement and tension of the muscles.* Muscle tension reinforces the restraint placed on a joint by its ligaments and further seems to restrict movement. A good example of the effect of muscle tension on a joint is seen at the hip joint. When the thigh is raised with the knee straight, the movement is restricted by the tension of the hamstring muscles on the posterior surface of the thigh. But if the knee is bent, the tension on the hamstring muscles is lessened, and the thigh can be raised farther.

4. *Apposition of soft parts.* The point at which one body surface contacts another may limit mobility. For example, if you bend your arm at the elbow, it can move no farther after the anterior surface of the forearm meets with and presses against the biceps brachii muscle of the arm. Indeed, significant enlargement of the biceps brachii muscle through body building can limit mobility. Joint movement may also be restricted by the presence of adipose tissue.

5. *Hormones.* Joint flexibility may also be affected by hormones. We noted previously how the action of relaxin results in expansion of the pelvic outlet during childbirth.

6. *Disuse.* Movement at a joint may be restricted if a joint has not been used for an extended period. For example, if an elbow joint is immobilized by a cast, only a limited range of motion at the joint may be possible for a time after the cast is removed. Restricted movement due to disuse may be caused by decreased amounts of synovial fluid, by diminished flexibility of ligaments and tendons, and by muscular atrophy, a reduction in size or wasting away of a muscle.

MOVEMENTS AT SYNOVIAL JOINTS

Although you engage in a host of complex physical actions during the course of each day, your body's synovial joints produce only three general forms of motion. The axis and direction of these motions, and the combination in which they occur, are the basis of the specific types of joint movement discussed in this section. As you will see, the specific movements that a joint can produce are based on the joint's structure.

General Kinds of Motion at Synovial Joints

Three general kinds of motion occur at synovial joints: (1) gliding motion, (2) angular motion, and (3) rotational motion.

- *Gliding motion.* A gliding motion occurs when the nearly flat surfaces of articulating bones slide across one another. Gliding motions can be demonstrated in a lab by pressing a film of oil between two slides, which enables them to shift easily from side-to-side and back-and-forth.

- *Angular motion.* In this form of motion, there is a change in the angle between articulating bones. You can visualize angular motion by imagining sticking a ski pole in the ground in front of you and then moving it toward and away from you, and then to the left and to the right. Each time the ski pole moves, the angle that it forms relative to the ground changes.

- *Rotational motion.* When a bone rotates, it moves around its own length. You can simulate this form of motion with a screwdriver, which rotates as you turn a screw.

Axes of Movement at Synovial Joints

In any movement at a joint, one bone remains in a fixed position while the other moves around an axis or plane (line). If a joint permits movement in only one axis, it is said to be *monaxial* (mon-AKS-ē-al) or *uniaxial.* An example is bending your upper limb at the elbow and then straightening it. If a joint permits movement in two axes, it is referred to as *biaxial* (bī-AKS-ē-al). This can be demonstrated by bending the hand at the wrist and then straightening it (the first axis) and then moving the hand from side to side at the wrist (the second axis). A *triaxial* (trī-AKS-ē-al) joint, such as the shoulder joint, permits movement in three axes. For example, at the shoulder joint you can swing your arm forward and backward as in walking (the first axis), move your arm straight out from your side and then back again (the second axis), and turn your palm forward and backward by rotating the arm (the third axis).

Specific Movements at Synovial Joints

Anatomists and kinesiologists use specific terminology to designate exact kinds of movement that can occur at a synovial joint. These precise terms may reflect the form of motion, the direction of movement, or the relationship of one body part to another during movement. The use of these standard descriptions enables scientists and health care professionals to describe body movements with a high degree of precision.

Specific movements are grouped into four principal categories: (1) gliding, (2) angular movements, (3) rotation, and

(4) special movements. This last category includes movements that occur only at specific joints.

Gliding

Gliding is a simple movement in which relatively flat bone surfaces move back and forth and from side to side with respect to one another (see Figure 8.7a). There is no significant alteration of the angle between the bones. Gliding movements are limited in range due to the loose-fitting structure of the articular capsule and associated ligaments and bones. Gliding occurs at the intercarpal joints (between carpal bones), intertarsal joints (between tarsal joints), and sternoclavicular joints (between the sternum and clavicles).

Angular Movements

In **angular movements,** there is an increase or a decrease in the angle between articulating bones. Among the principal angular movements are flexion, extension, abduction, and adduction. Please remember that as these movements are discussed it is assumed that the body is in the anatomical position.

Flexion In **flexion** (FLEK-shun; *flexus* = to bend) there is a *decrease* in the angle between articulating bones, usually occurring in the sagittal plane (Figure 8.2a–f). Examples of flexion include bending the head toward the chest at the atlanto-occipital joint (between the atlas and occipital bone) and the cervical intervertebral joints (between the cervical vertebrae), bending the trunk forward at the intervertebral joints, moving the humerus forward at the shoulder or glenohumeral joint (between the humerus and glenoid fossa of the scapula) as in walking, moving the forearm toward the arm at the elbow joint (between the humerus, ulna, and radius), moving the palm toward the forearm at the wrist or radiocarpal joint (between the radius and carpals), bending the digits of the hand or feet at the interphalangeal joints (between phalanges), moving the femur forward at the hip or coxal joint (between the femur and hipbone) as in walking, and moving the leg toward the thigh at the knee or tibiofemoral joint (between the tibia, femur, and patella) as occurs when bending the knee.

Although flexion usually occurs in the sagittal plane, there are a few examples where this is not so. For example, flexion of the thumb involves movement of the thumb medially across the palm at the carpometacarpal joint (between the trapezium and metacarpal of the thumb) as in touching the little finger with the thumb (see Figure 10.19f). Another example is movement of the trunk sideways to the right or left at the waist. This movement, which occurs in the frontal plane and which involves the intervertebral joints, is called **lateral flexion** (Figure 8.2g).

Extension Whereas flexion decreases the angle between articulating bones, **extension** (eks-TEN-shun; *extensio* = to

stretch out) increases the angle, often to restore a part of the body to the anatomical position after it has been flexed (Figure 8.2a–f). Extension usually occurs in the sagittal plane as well, with the exceptions noted for flexion. Continuation of extension beyond the anatomical position is called **hyperextension** (*hyper* = excessive) (Figure 8.2a, b, d, e). Examples include bending the head backward at the atlanto-occipital and cervical intervertebral joints, bending the trunk backward at the intervertebral joints, moving the humerus backward at the shoulder joint as in walking, moving the palm backward at the wrist joint, and moving the femur backward at the hip joint as in walking. Hyperextension of other joints, such as the elbow, interphalangeal, and knee joints, is usually prevented by factors such as the arrangement of ligaments and the anatomical alignment of the bones, and soft tissues.

Abduction **Abduction** (ab-DUK-shun) means "taking away" and refers to the movement of a bone away from the midline, usually in the frontal plane. Examples include moving the humerus laterally at the shoulder joint, moving the palm laterally at the wrist joint, and moving the femur laterally at the hip joint (Figure 8.3a–c). With respect to the fingers and toes, the midline is not used as a point of reference. In abduction of the fingers (but not the thumb), an imaginary line is drawn through the middle (longest) finger, and the fingers move away (spread out) from the middle finger (Figure 8.3d). In abduction of the thumb, the thumb moves away from the palm in the sagittal plane (see Figure 10.19f). Abduction of the toes is relative to an imaginary line drawn through the second (usually the longest) toe.

Adduction **Adduction** (ad-DUK-shun) means "to move toward" and refers to the movement of a bone toward the midline, usually in the frontal plane (Figure 8.3a–c). As with abduction, adduction of the fingers and toes is relative to an imaginary line through the middle finger and second toe. Adduction of the fingers involves bringing the spread fingers together (Figure 8.3d). Adduction of the thumb moves the thumb toward the palm in the sagittal plane (see Figure 10.19f).

Circumduction **Circumduction** (ser-kum-DUK-shun; *circum* = around; *duco* = to draw) means "moving in a circle" and refers to movement of the distal end of a part of the body in a circle (Figure 8.4 on page 198). Circumduction occurs as a result of a continuous sequence of flexion, abduction, extension, and adduction. Because circumduction is a combined movement, it is not a separate axis of movement. Examples of circumduction are moving the humerus in a circle at the shoulder joint, moving the hand in a circle at the wrist joint, moving the thumb in a circle at the carpometacarpal joint, moving the fingers in a circle at the metacarpophalangeal joints (between the metacarpals and phalanges), and moving the femur in a circle at the hip joint.

Figure 8.2 Angular movements at synovial joints: flexion, extension, hyperextension, and lateral flexion.

In angular movements, there is an increase or decrease in the angle between articulating bones.

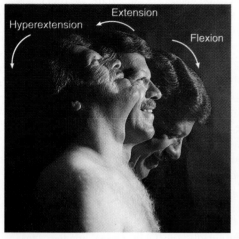

(a) Atlanto-occipital and cervical intervertebral joints

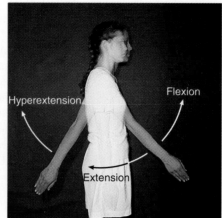

(b) Shoulder joint

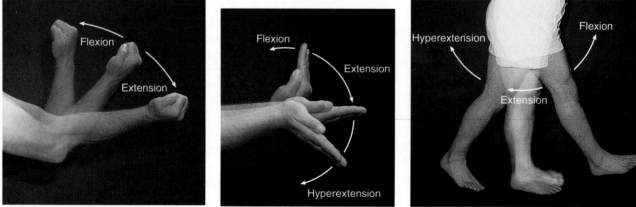

(c) Elbow joint

(d) Wrist joint

(e) Hip joint

Q *What are two examples of flexion that do not occur in the sagittal plane?*

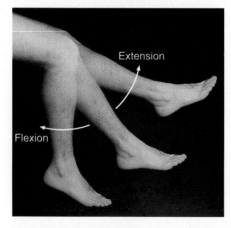

(f) Knee joint

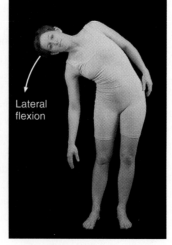

(g) Intervertebral joints

Figure 8.3 Angular movements at synovial joints: abduction and adduction.

🔑 *Abduction and adduction usually occur in the frontal plane.*

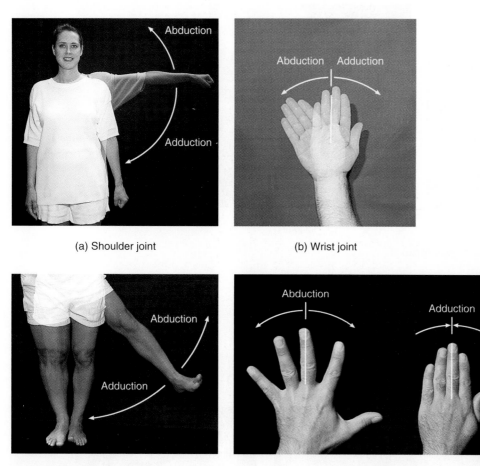

(a) Shoulder joint

(b) Wrist joint

(c) Hip joint

(d) Metacarpophalangeal joints of the fingers (not the thumb)

Q *You can remember adduction as "adding your limb to your trunk." Why is this an effective learning device?*

Rotation

Rotation (rō-TĀ-shun; *rotatus* = to revolve) is the movement of a bone around its own longitudinal axis. One example is turning the head from side to side at the atlanto-axial joint (between the atlas and axis), as in signifying "no" (Figure 8.5a). Another is turning the trunk from side to side at the intervertebral joints while keeping the hips and lower limbs in the anatomical position. In the limbs, rotation is defined relative to the midline, and specific qualifying terms are used. If the anterior surface of a bone of the limb is turned toward the midline, the movement is called *medial (internal) rotation*. You can medially rotate the humerus at the shoulder joint as follows. Starting in the anatomical position, flex your elbow and then draw your palm across the chest (Figure 8.5b). Medial rotation of the forearm at the radioulnar joints (between the radius and ulna) involves turning the palm medially from the anatomical position (see Figure 8.6h). You can medially rotate the femur at the hip joint as follows. Lie on your back, bend your knee, and then move

Figure 8.4 Movements at synovial joints: circumduction.

Circumduction occurs as a result of flexion, abduction, extension, and adduction in succession.

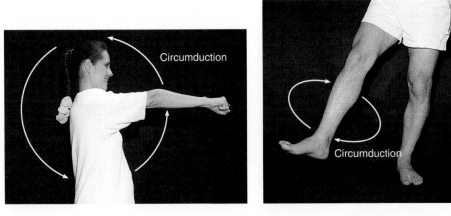

(a) Shoulder joint   (b) Hip joint

Q *Why is circumduction not considered a separate axis of movement?*

Figure 8.5 Movements at synovial joints: rotation.

In rotation, a bone moves in a single plane around its own longitudinal axis.

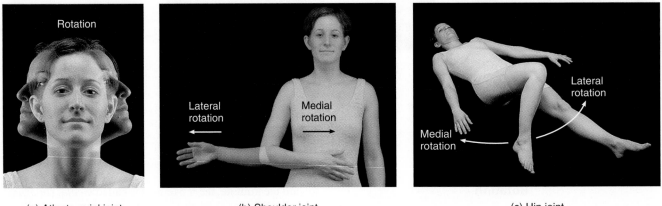

(a) Atlanto-axial joint   (b) Shoulder joint   (c) Hip joint

Q *What is medial rotation?*

your foot laterally from the midline. Although you are moving your foot laterally, the femur is rotating medially (Figure 8.5c). Medial rotation of the leg at the knee joint can be produced by sitting on a chair, bending your knee, raising your lower limb off the floor, and turning your toes medially. If the anterior surface of the bone of a limb is turned away from the midline, the movement is called *lateral (external) rotation* (Figure 8.5b, c).

Figure 8.6 Special movements at synovial joints.

Special movements occur only at specific synovial joints.

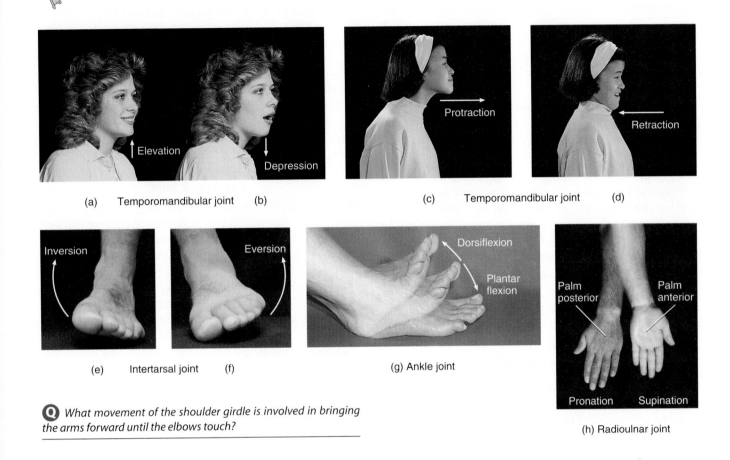

(a)   Temporomandibular joint   (b)

(c)   Temporomandibular joint   (d)

(e)   Intertarsal joint   (f)

(g) Ankle joint

(h) Radioulnar joint

Q *What movement of the shoulder girdle is involved in bringing the arms forward until the elbows touch?*

Special Movements

As noted previously, **special movements** occur only at specific joints. Among these movements are elevation, depression, protraction, retraction, inversion, eversion, dorsiflexion, plantar flexion, supination, pronation, and opposition (Figure 8.6).

- **Elevation** (el-e-VĀ-shun; *elevo* = to lift up) is an upward movement of a part of the body, such as closing the mouth at the temporomandibular joint (between the mandible and temporal bone) to elevate the mandible (Figure 8.6a) or shrugging the shoulders at the sternoclavicular joint to elevate the scapula.

- **Depression** (dē-PRESH-un; *depressio* = to press down) is a downward movement of a part of the body, such as opening the mouth to depress the mandible (Figure 8.6b) or returning shrugged shoulders to the anatomical position to depress the scapula.

- **Protraction** (prō-TRAK-shun; *protractus* = to draw forth) is a movement of a part of the body anteriorly in the transverse plane. You can protract your mandible at the temporomandibular joint by thrusting it outward (Figure 8.6c) or protract your clavicles at the sternoclavicular joint by crossing your arms.

- **Retraction** (rē-TRAK-shun; *retractio* = to draw back) is a movement of a protracted part of the body back to the anatomical position (Figure 8.6d).

- **Inversion** (in-VER-zhun) is movement of the soles medially at the intertarsal joints (between the tarsals) so that they face each other (Figure 8.6e).

- **Eversion** (ē-VER-zhun) is a movement of the soles laterally at the intertarsal joints so that they face away from each other (Figure 8.6f).

- **Dorsiflexion** (dor-si-FLEK-shun) refers to bending of the foot at the ankle or talocrural joint (between the tibia, fibula, and talus) in the direction of the dorsum (superior surface) (Figure 8.6g). Dorsiflexion occurs when you stand on your heels.

- **Plantar flexion** involves bending of the foot at the ankle joint in the direction of the plantar surface (Figure 8.6g), as when standing on your toes.

- **Supination** (soo-pi-NĀ-shun) is a movement of the forearm at the proximal and distal radioulnar joints in which the palm is turned anteriorly or superiorly (Figure 8.6h). This position of the palms is one of the defining features of the anatomical position.

- **Pronation** (prō-NĀ-shun) is a movement of the forearm at the proximal and distal radioulnar joints in which the distal end of the radius crosses over the distal end of the ulna and the palm is turned posteriorly or inferiorly (Figure 8.6h).

- **Opposition** (op-ō-ZISH-un) is the movement of the thumb at the carpometacarpal joint (between the trapezium and metacarpal of the thumb) in which the thumb moves across the palm to touch the tips of the fingers on the same hand (see Figure 10.19f). This is the single most distinctive digital movement that gives humans and other primates the ability to grasp and manipulate objects very precisely.

A summary of the movements that occur at synovial joints is presented in Table 8.1.

Types of Synovial Joints

Though all synovial joints have a generally similar structure, the shape of the articulating surfaces varies. Accordingly, synovial joints are divided into six subtypes: gliding, hinge, pivot, condyloid, saddle, and ball-and-socket joints (Figure 8.7).

Gliding Joint

The articulating surfaces of bones in a **gliding** (*planar* or *arthrodial;* ar-THRŌ-dē-al) **joint** are usually flat. A gliding movement is the simplest kind that can occur at a joint. Only side-to-side and back-and-forth movements are permitted (Figure 8.7a). Because gliding joints do not involve movement around any axis, they are said to be *nonaxial.* Some examples of gliding joints are the intercarpal, intertarsal, sternoclavicular, acromioclavicular, sternocostal, and vertebrocostal (between the heads and tubercles of ribs and transverse processes of thoracic vertebrae) joints.

Hinge Joint

In a **hinge** or *ginglymus* (JIN-gli-mus; *ginglymos* = hinge) **joint,** the convex surface of one bone fits into the concave surface of another bone (Figure 8.7b). As the name implies, hinge joints produce an opening-and-closing motion like that of a hinged door, usually flexion and extension (see Figure 8.2c, f). Hinge joints are monaxial because they typically articulate in a single axis. Examples of hinge joints are the knee, elbow, ankle, and interphalangeal joints.

Pivot Joint

In a **pivot** or *trochoid* (TRŌ-koyd; *trokhos* = wheel) **joint,** the rounded or pointed surface of one bone articulates with a ring formed partly by another bone and partly by a ligament (Figure 8.7c). The primary movement permitted at a pivot joint is rotation, movement of a bone in a single axis around its own longitudinal axis. The joint is therefore monaxial. Examples of pivot joints are the atlanto-axial joint, in which the atlas rotates around the axis and permits the head to turn from side to side as in signifying "no" (see Figure 8.5a), and the radioulnar joints that allow us to turn the palms anteriorly (or superiorly) and posteriorly (or inferiorly) (see Figure 8.6h).

Condyloid Joint

In a **condyloid** (KON-di-loyd; *kondylos* = knuckle) or *ellipsoidal* **joint,** the oval-shaped projection of one bone fits into the oval-shaped depression of another bone (Figure 8.7d). The movement permitted by such a joint is in two axes, side to side and back and forth, as when you abduct and adduct (see Figure 8.3b) and flex and extend your wrist (see Figure 8.2d). Such a joint is biaxial. Examples are the wrist and metacarpophalangeal joints for the second through fifth digits. At a condyloid joint it is possible to combine flexion, abduction, extension, and adduction in succession to produce circumduction (see Figure 8.4).

Saddle Joint

In a **saddle** or *sellaris* (sel-A-ris; *sellar* = saddle) **joint,** the articular surface of one bone is saddle-shaped, and the articular surface of the other bone fits into the "saddle" as a sitting rider would (Figure 8.7e). A saddle joint is a modified condyloid joint in which the movement is somewhat freer. Movements at a saddle joint are side to side and back and forth. Such joints are biaxial and, like condyloid joints, also permit circumduction. An example of a saddle joint is the carpometacarpal joint between the trapezium of the carpus and metacarpal of the thumb. In circumduction of the thumb, the thumb moves in a circle.

Ball-and-Socket Joint

A **ball-and-socket** or *spheroid* (SFĒ-royd; *spheroideum* = sphere-shaped) **joint** consists of the ball-like surface of one bone fitting into a cuplike depression of another bone (Figure 8.7f). Such joints permit movement in three axes: flexion–extension, abduction–adduction, and rotation. Thus they are triaxial. The only examples of ball-and-socket joints are the shoulder and hip joints. Although both joints also permit circumduction, it occurs more easily in the shoulder joints (see Figure 8.4) because flexion, abduction, extension,

Table 8.1 *Summary of Movements at Synovial Joints*

MOVEMENT	DEFINITION
Gliding	The relatively flat surfaces of bones move back and forth and from side to side over another surface. During the movement there is no significant alteration in the angle between the bones.
Angular	There is an increase or decrease in the angle between bones.
Flexion	Involves a decrease in the angle between articulating bones and usually occurs in the sagittal plane.
Lateral flexion	Movement of the trunk in the frontal plane.
Extension	Involves an increase in the angle between articulating bones and usually occurs in the sagittal plane.
Hyperextension	Continuation of extension beyond the anatomical position.
Abduction	Movement of a bone away from the midline, usually in the frontal plane.
Adduction	Movement of a bone toward the midline, usually in the frontal plane.
Circumduction	A combination of flexion, abduction, extension, and adduction in succession, in which the distal end of a part of the body moves in a circle.
Rotation	Movement of a bone around its own longitudinal axis; in the limbs, it may be medial (toward the midline) or lateral (away from the midline).
Special	Occur at specific joints.
Elevation	Movement of a part of the body superiorly.
Depression	Movement of a part of the body inferiorly.
Protraction	Movement of a part of the body anteriorly in the transverse plane.
Retraction	Movement of a part of the body posteriorly in the transverse plane.
Inversion	Movement of the soles medially so that they face each other.
Eversion	Movement of the soles laterally so that they face away from each other.
Dorsiflexion	Bending the foot in the direction of the dorsum (superior surface).
Plantar flexion	Bending the foot in the direction of the plantar surface (sole).
Supination	Movement of the forearm in which the palm is turned anteriorly or superiorly.
Pronation	Movement of the forearm in which the palm is turned posteriorly or inferiorly.
Opposition	Movement of the thumb across the palm to touch the tips of the fingers on the same hand.

and adduction are more limited in the hip joints due to the tension on certain ligaments and muscles. These are described in detail in Exhibit 8.4 on page 211.

A summary of joints based on structural classification is presented in Table 8.2.

SELECTED JOINTS OF THE BODY

In this section of the text we will examine in detail five selected joints of the body presented in a series of exhibits. Each exhibit considers a specific joint and contains (1) a definition—that is, a description of the bones that form the joint; (2) the type of joint (its structural classification); (3) its anatomical components—a description of the major connecting ligaments, articular disc, articular capsule, and other distinguishing features of the joint; and (4) movements. Each exhibit also refers you to an illustration of the joint. The joints described are the temporomandibular joint or TMJ, humeroscapular or glenohumeral joint, elbow joint, hip (coxal) joint, and knee (tibiofemoral) joint.

Text continues on page 217

Figure 8.7 Subtypes of synovial joints. For each subtype, a drawing of the actual joint and a simplified diagram are shown.

Synovial joints are classified into subtypes on the basis of the shapes of the articulating bone surfaces.

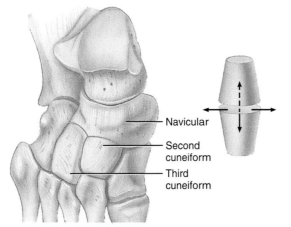

(a) Gliding joint between the navicular and second and third cuneiforms of the tarsus in the foot

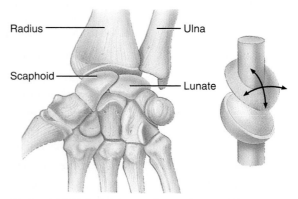

(d) Condyloid joint between radius and scaphoid and lunate bones of the carpus (wrist)

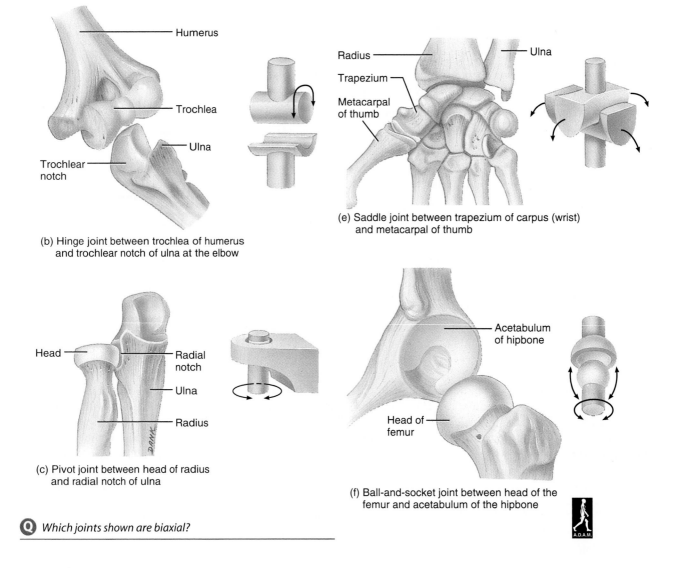

(b) Hinge joint between trochlea of humerus and trochlear notch of ulna at the elbow

(e) Saddle joint between trapezium of carpus (wrist) and metacarpal of thumb

(c) Pivot joint between head of radius and radial notch of ulna

(f) Ball-and-socket joint between head of the femur and acetabulum of the hipbone

Q *Which joints shown are biaxial?*

Table 8.2 *Summary of Joints Classified on the Basis of Structure*

FIBROUS JOINTS: Articulating bones are held together by fibrous connective tissue, and there is no synovial (joint) cavity.

Type	Description	Movement	Examples
Suture	Articulating bones are united by a thin layer of dense fibrous connective tissue; found between bones of the skull.	None (synarthrosis).	Lambdoid suture between occipital and parietal bones.
Syndesmosis	Articulating bones are united by a thick layer of dense fibrous connective tissue.	Partial (amphiarthrosis).	Distal tibiofibular joint.
Gomphosis	Articulating bones are united by a periodontal ligament; cone-shaped peg fits into a socket.	None (synarthrosis).	Roots of teeth in alveoli (sockets) of maxillae and mandible.

CARTILAGINOUS JOINTS: Articulating bones are united by cartilage, and there is no synovial (joint) cavity.

Type	Description	Movement	Examples
Synchondrosis	Connecting material is hyaline cartilage.	None (synarthrosis) or partial (amphiarthrosis).	Joint between the diaphysis and epiphysis of a long bone (epiphyseal plate), which is a synarthrosis, or sternocostal joints, which are amphiarthroses.
Symphysis	Connecting material is a broad, flat disc of fibrocartilage.	Partial (amphiarthrosis).	Intervertebral joints and pubic symphysis.

SYNOVIAL JOINTS: Characterized by a synovial cavity, articular cartilage, and an articular capsule composed of an outer fibrous capsule and an inner synovial membrane; may contain accessory ligaments, articular discs (menisci), and bursae.

Type	Description	Movement	Examples
Gliding	Articulating surfaces are usually flat.	Nonaxial.	Intercarpal and intertarsal joints.
Hinge	Convex surface fits into a concave surface.	Monaxial (flexion–extension).	Elbow, ankle, and interphalangeal joints.
Pivot	Rounded or pointed surface fits into a ring formed partly by bone and partly by a ligament.	Monaxial (rotation).	Atlanto-axial and radioulnar joints.
Condyloid	Oval-shaped projection fits into an oval-shaped depression.	Biaxial (flexion–extension, abduction–adduction). Circumduction also occurs.	Radiocarpal and metacarpophalangeal joints.
Saddle	Articular surface of one bone is saddle-shaped, and the articular surface of the other bone fits into the "saddle" like a sitting rider.	Biaxial (flexion–extension, abduction–adduction). Circumduction also occurs.	Carpometacarpal joint between trapezium and thumb.
Ball-and-socket	Ball-like surface fits into a cuplike depression.	Triaxial (flexion–extension, abduction–adduction, rotation). Circumduction also occurs.	Shoulder and hip joints.

Exhibit 8.1 *Temporomandibular Joint or TMJ (Figure 8.8)*

DEFINITION

Joint formed by the mandibular condyle of the mandible and the mandibular fossa and articular tubercle of the temporal bone. The temporomandibular joint is the only movable joint between skull bones; all other skull joints are sutures and therefore immovable.

TYPE OF JOINT

Synovial; combined hinge (ginglymus) and gliding (arthrodial) type.

Figure 8.8 Right temporomandibular joint (TMJ).

⚷ *The TMJ is the only movable joint between skull bones.*

ANATOMICAL COMPONENTS

1. *Articular disc (meniscus).* Fibrocartilage disc that separates the joint cavity into a superior and inferior compartment, each of which has a synovial membrane.

2. *Articular capsule.* Thin, fairly loose envelope around the circumference of the joint.

3. *Lateral ligament.* Two short bands on the lateral surface of the articular capsule that extend inferiorly and posteriorly from the inferior border and tubercle of the zygomatic process of the temporal bone to the lateral and posterior aspect of the neck of the mandible. It is covered by the parotid gland and helps prevent displacement of the mandible.

4. *Sphenomandibular ligament.* Thin band that extends inferiorly and anteriorly from the spine of the sphenoid bone to the ramus of the mandible.

5. *Stylomandibular ligament.* Thickened band of deep cervical fascia that extends from the styloid process of the temporal bone to the inferior and posterior border of the ramus of the mandible. The ligament separates the parotid from the submandibular gland.

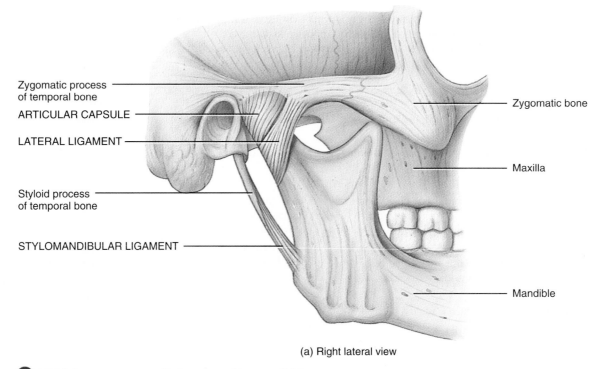

(a) Right lateral view

Q *Which ligament prevents displacement of the mandible?*

Exhibit 8.1 (continued)

MOVEMENTS

Only the mandible moves because the maxilla is firmly anchored to other bones by sutures. Accordingly, the mandible may function in depression (jaw opening), elevation (jaw closing), protraction, retraction, lateral displacement, and slight rotation (see Figure 8.6a–d).

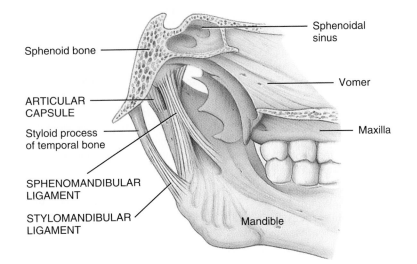

(b) Medial view

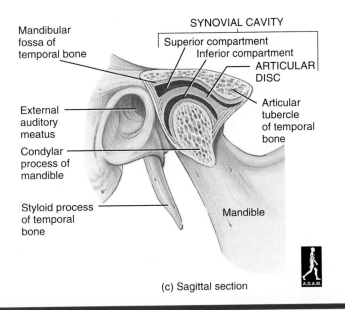

(c) Sagittal section

Exhibit 8.2 Shoulder (*Humeroscapular or Glenohumeral*) Joint (Figure 8.9)

DEFINITION

Joint formed by the head of the humerus and the glenoid cavity of the scapula.

TYPE OF JOINT

Synovial joint, ball-and-socket (spheroid) type.

ANATOMICAL COMPONENTS

1. *Articular capsule.* Thin, loose sac that completely envelops the joint. It extends from the glenoid cavity to the anatomical neck of the humerus. The inferior part of the capsule is its weakest area.

2. *Coracohumeral ligament.* Strong, broad ligament that strengthens the superior part of the articular capsule and extends from the coracoid process of the scapula to the greater tubercle of the humerus.

3. *Glenohumeral ligaments.* Three thickenings of the articular capsule over the anterior surface of the joint. They extend from the glenoid cavity to the lesser tubercle and anatomical neck of the humerus. These ligaments are often indistinct or absent and provide only minimal strength.

4. *Transverse humeral ligament.* Narrow sheet extending from the greater tubercle to the lesser tubercle of the humerus.

5. *Glenoid labrum.* Narrow rim of fibrocartilage around the edge of the glenoid cavity. It slightly deepens and enlarges the glenoid cavity.

6. Four *bursae* are associated with the shoulder joint. They are the *subscapular bursa, subdeltoid bursa, subacromial bursa,* and *subcoracoid bursa.*

MOVEMENTS

Produces flexion, extension, abduction, adduction, medial rotation, lateral rotation, and circumduction of the arm (see Figures 8.2–8.5).

The shoulder joint has more freedom of movement than any other joint of the body. This freedom results from the looseness of the articular capsule and shallowness of the glenoid cavity in relation to the large size of the head of the humerus. The cavity receives little more than one-third of the head of the humerus. The same anatomical features that lead to freedom of movement also result in instability.

Although the ligaments of the shoulder joint strengthen it to some extent, most of the strength results from the muscles that surround the joint, especially the *rotator cuff muscles.* These muscles (supraspinatus, infraspinatus, teres minor, and subscapularis) join the scapula to the humerus (see Figure 10.16). The tendons of the muscles, together called the **rotator cuff**, encircle the joint (except for the inferior portion) and fuse with the articular capsule. The rotator cuff muscles work as a group to hold the head of the humerus in the glenoid cavity. **Rotator cuff injury** is common among baseball pitchers because of shoulder movements that involve vigorous circumduction. Most often the tendon of the supraspinatus muscle is torn.

Clinical Application

Dislocated and Separated Shoulders

Dislocation, or *luxation* (luks-Ā-shun; *luxatio* = dislocation), is the displacement of a bone from a joint with tearing of ligaments, tendons, and articular capsules. It is usually caused by a blow or fall, although unusual physical effort may lead to this condition. The joint most commonly dislocated in adults is the shoulder joint because its socket is quite shallow. Usually, the head of the humerus becomes displaced inferiorly, where the articular capsule is least protected. Dislocations of the mandible, elbow, fingers, knee, or hip are less common.

A **separated shoulder** refers to an injury of the acromioclavicular joint, a joint formed by the acromion of the scapula and the acromial end of the clavicle. This condition is usually the result of forceful trauma to the joint, as when the shoulder strikes the ground. ■

Exhibit 8.2 Shoulder (continued)

Figure 8.9 Right shoulder (glenohumeral) joint. The arrow in the inset in (b) (on following page) indicates the direction from which the shoulder joint is viewed (lateral).

Most of the stability of the shoulder joints results from the arrangement of the rotator cuff muscles.

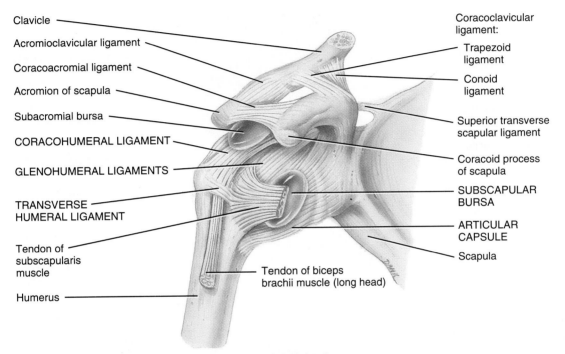

Clavicle

Acromioclavicular ligament

Coracoacromial ligament

Acromion of scapula

Subacromial bursa

CORACOHUMERAL LIGAMENT

GLENOHUMERAL LIGAMENTS

TRANSVERSE HUMERAL LIGAMENT

Tendon of subscapularis muscle

Humerus

Coracoclavicular ligament:

Trapezoid ligament

Conoid ligament

Superior transverse scapular ligament

Coracoid process of scapula

SUBSCAPULAR BURSA

ARTICULAR CAPSULE

Scapula

Tendon of biceps brachii muscle (long head)

(a) Anterior view

Exhibit 8.2 *Shoulder (Humeroscapular or Glenohumeral) Joint (continued)*

Figure 8.9 (continued)

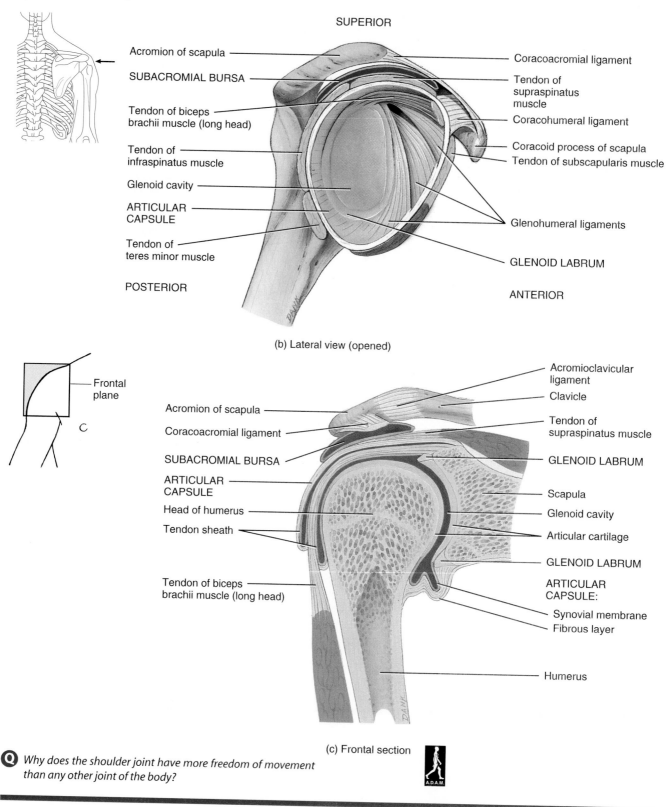

SUPERIOR

Acromion of scapula

Coracoacromial ligament

SUBACROMIAL BURSA

Tendon of supraspinatus muscle

Tendon of biceps brachii muscle (long head)

Coracohumeral ligament

Coracoid process of scapula

Tendon of infraspinatus muscle

Tendon of subscapularis muscle

Glenoid cavity

ARTICULAR CAPSULE

Glenohumeral ligaments

Tendon of teres minor muscle

GLENOID LABRUM

POSTERIOR

ANTERIOR

(b) Lateral view (opened)

Frontal plane

C

Acromioclavicular ligament

Clavicle

Acromion of scapula

Coracoacromial ligament

Tendon of supraspinatus muscle

SUBACROMIAL BURSA

GLENOID LABRUM

ARTICULAR CAPSULE

Scapula

Head of humerus

Glenoid cavity

Tendon sheath

Articular cartilage

GLENOID LABRUM

ARTICULAR CAPSULE:

Tendon of biceps brachii muscle (long head)

Synovial membrane

Fibrous layer

Humerus

(c) Frontal section

Q *Why does the shoulder joint have more freedom of movement than any other joint of the body?*

Exhibit 8.3 Elbow Joint (Figure 8.10)

DEFINITION

Joint formed by the trochlea of the humerus, the trochlear notch of the ulna, and the head of the radius.

TYPE OF JOINT

Synovial joint, hinge (ginglymus) type.

ANATOMICAL COMPONENTS

1. *Articular capsule.* The anterior part covers the anterior part of the joint from the radial and coronoid fossae of the humerus to the coronoid process of the ulna and the annular ligament of the radius. The posterior part extends from the capitulum, olecranon fossa, and lateral epicondyle of the humerus to the annular ligament of the radius, the olecranon of the ulna, and the ulna posterior to the radial notch.

2. *Ulnar collateral ligament.* Thick, triangular ligament that extends from the medial epicondyle of the humerus to the coronoid process and olecranon of the ulna.

3. *Radial collateral ligament.* Strong, triangular ligament that extends from the lateral epicondyle of the humerus to the annular ligament of the radius and the radial notch of the ulna.

MOVEMENTS

Flexion and extension of forearm (see Figure 8.2c).

Clinical Application

Tennis Elbow, Little-League Elbow, and Dislocation of the Radial Head

Tennis elbow most commonly refers to pain at or near the lateral epicondyle of the humerus, usually caused by an improperly executed backhand. The extensor muscles strain or sprain, resulting in pain.

Little-league elbow typically develops as a result of a heavy pitching schedule and/or a schedule that involves throwing curve balls, especially among youngsters. In this disorder, the elbow may enlarge, fragment, or separate.

When a strong pull is applied to the forearm while it is extended and supinated, the head of the radius may slide past or rupture the radial annular ligament, a ligament that is attached to the radial notch of the ulna and forms a collar around the head of the radius at the proximal radioulnar joint. Such a **dislocation of the radial head** is the most common upper limb dislocation in children and may occur when swinging a child around with outstretched arms. ◼

Exhibit 8.3 Elbow Joint *(continued)*

Figure 8.10 Right elbow joint. See also Figure 8.1b.

The elbow joint is formed by parts of three bones: humerus, ulna, and radius.

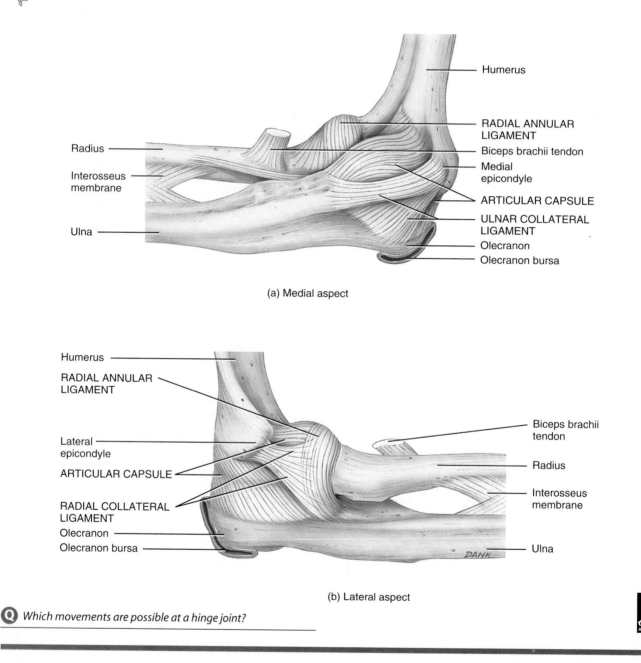

(a) Medial aspect

(b) Lateral aspect

Q *Which movements are possible at a hinge joint?*

Exhibit 8.4 Hip (Coxal) Joint (Figure 8.11)

DEFINITION

Joint formed by the head of the femur and the acetabulum of the hipbone.

TYPE OF JOINT

Synovial, ball-and-socket (spheroid) type.

ANATOMICAL COMPONENTS

1. *Articular capsule.* Very dense and strong capsule that extends from the rim of the acetabulum to the neck of the femur. One of the strongest structures of the body, the capsule consists of circular and longitudinal fibers. The circular fibers, called the *zona orbicularis,* form a collar around the neck of the femur. The longitudinal fibers are reinforced by accessory ligaments known as the iliofemoral ligament, pubofemoral ligament, and ischiofemoral ligament.

2. *Iliofemoral ligament.* Thickened portion of the articular capsule that extends from the anterior inferior iliac spine of the hipbone to the intertrochanteric line of the femur.

3. *Pubofemoral ligament.* Thickened portion of the articular capsule that extends from the pubic part of the rim of the acetabulum to the neck of the femur.

4. *Ischiofemoral ligament.* Thickened portion of the articular capsule that extends from the ischial wall of the acetabulum to the neck of the femur.

5. *Ligament of the head of the femur (capitate ligament).* Flat, triangular band that extends from the fossa of the acetabulum to the fovea capitis of the head of the femur.

6. *Acetabular labrum.* Fibrocartilage rim attached to the margin of the acetabulum that enhances the depth of the acetabulum. Because the diameter of the acetabular rim is smaller than that of the head of the femur, dislocation of the femur is rare.

7. *Transverse ligament of the acetabulum.* Strong ligament that crosses over the acetabular notch, converting it to a foramen. It supports part of the acetabular labrum and is connected with the ligament of the head of the femur and the articular capsule.

MOVEMENTS

Flexion, extension, abduction, adduction, circumduction, medial rotation, and lateral rotation of thigh (see Figures 8.2–8.5).

The extreme stability of the hip joint is related to the very strong articular capsule at its accessory ligaments, the manner in which the femur fits into the acetabulum, and the muscles surrounding the joint. Although the shoulder and hip joints are both ball-and-socket joints, the movements at the hip joints do not have a wide range of motion. Flexion is limited by the anterior surface of the thigh coming into contact with the anterior abdominal wall when the knee is flexed and by tension of the hamstring muscles when the knee is extended. Extension is limited by tension of the iliofemoral, pubofemoral, and ischiofemoral ligaments. Abduction is limited by the tension of the pubofemoral ligament and adduction is limited by contact with the opposite limb and tension in the ligament of the head of the femur. Medial rotation is limited by the tension in the ischiofemoral ligament and lateral rotation is limited by tension in the iliofemoral and pubofemoral ligaments.

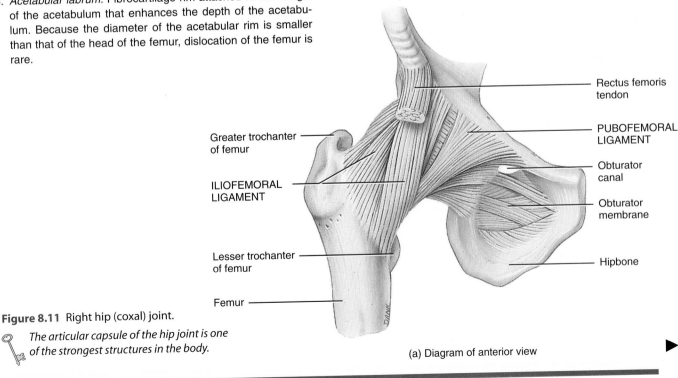

Figure 8.11 Right hip (coxal) joint.

The articular capsule of the hip joint is one of the strongest structures in the body.

Rectus femoris tendon

PUBOFEMORAL LIGAMENT

Obturator canal

Obturator membrane

Hipbone

Greater trochanter of femur

ILIOFEMORAL LIGAMENT

Lesser trochanter of femur

Femur

(a) Diagram of anterior view

Exhibit 8.4 Hip (Coxal) Joint (continued)

Figure 8.11 (continued)

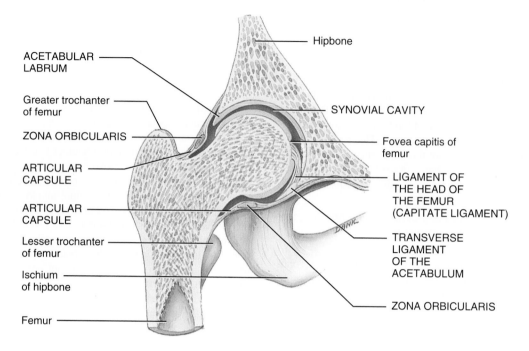

ACETABULAR
LABRUM

Greater trochanter
of femur

ZONA ORBICULARIS

ARTICULAR
CAPSULE

ARTICULAR
CAPSULE

Lesser trochanter
of femur

Ischium
of hipbone

Femur

Hipbone

SYNOVIAL CAVITY

Fovea capitis of
femur

LIGAMENT OF
THE HEAD OF
THE FEMUR
(CAPITATE LIGAMENT)

TRANSVERSE
LIGAMENT
OF THE
ACETABULUM

ZONA ORBICULARIS

(b) Diagram of frontal section

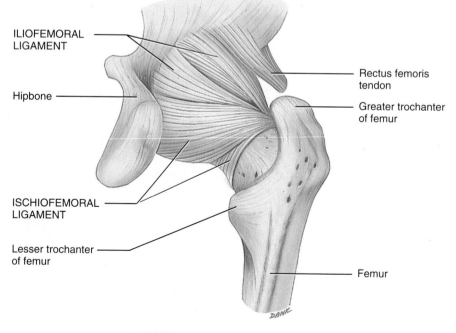

ILIOFEMORAL
LIGAMENT

Hipbone

ISCHIOFEMORAL
LIGAMENT

Lesser trochanter
of femur

Rectus femoris
tendon

Greater trochanter
of femur

Femur

(c) Diagram of posterior view

Exhibit 8.4 *(continued)*

Figure 8.11 (continued)

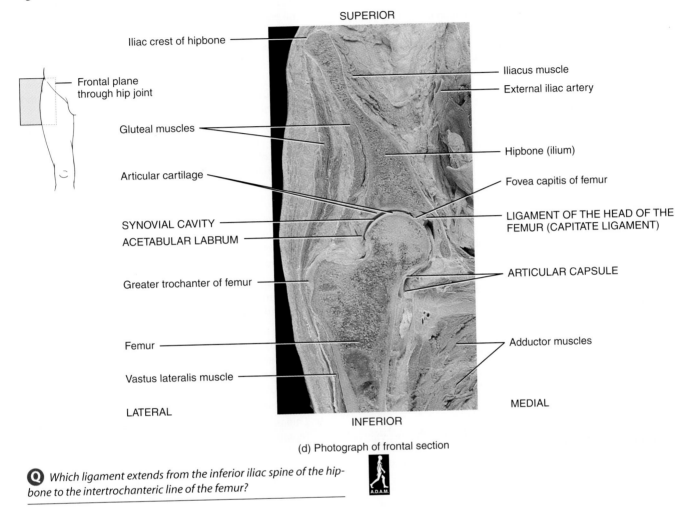

SUPERIOR

Iliac crest of hipbone

Iliacus muscle

External iliac artery

Frontal plane
through hip joint

Gluteal muscles

Hipbone (ilium)

Articular cartilage

Fovea capitis of femur

SYNOVIAL CAVITY

ACETABULAR LABRUM

LIGAMENT OF THE HEAD OF THE
FEMUR (CAPITATE LIGAMENT)

Greater trochanter of femur

ARTICULAR CAPSULE

Femur

Adductor muscles

Vastus lateralis muscle

LATERAL

MEDIAL

INFERIOR

(d) Photograph of frontal section

Q *Which ligament extends from the inferior iliac spine of the hipbone to the intertrochanteric line of the femur?*

Exhibit 8.5 Knee (Tibiofemoral) Joint (Figure 8.12)

DEFINITION

The largest joint of the body, actually consisting of three joints: (1) an intermediate patellofemoral joint between the patella and the patellar surface of the femur; (2) a lateral tibiofemoral joint between the lateral condyle of the femur, lateral meniscus, and lateral condyle of the tibia; and (3) a medial tibiofemoral joint between the medial condyle of the femur, medial meniscus, and medial condyle of the tibia.

TYPE OF JOINT

Patellofemoral joint: partly synovial, gliding (arthrodial) type. Lateral and medial tibiofemoral joints: synovial, modified hinge (ginglymus) type.

ANATOMICAL COMPONENTS

1. *Articular capsule.* No complete, independent capsule unites the bones. The ligamentous sheath surrounding the joint consists mostly of muscle tendons or expansions of them. There are, however, some capsular fibers connecting the articulating bones.

2. *Medial and lateral patellar retinacula.* Fused tendons of insertion of the quadriceps femoris muscle and the fascia lata (deep fascia of thigh) that strengthen the anterior surface of the joint.

3. *Patellar ligament.* Continuation of the common tendon of insertion of the quadriceps femoris muscle that extends from the patella to the tibial tuberosity. This ligament also strengthens the anterior surface of the joint. The posterior surface of the ligament is separated from the synovial membrane of the joint by an *infrapatellar fat pad.*

4. *Oblique popliteal ligament.* Broad, flat ligament that extends from the intercondylar fossa of the femur to the head of the tibia. The tendon of the semimembranosus muscle is superficial to the ligament and passes from the medial condyle of the tibia to the lateral condyle of the femur. The ligament and tendon afford strength for the posterior surface of the joint.

5. *Arcuate popliteal ligament.* Extends from the lateral condyle of the femur to the styloid process of the head of the fibula. It strengthens the lower lateral part of the posterior surface of the joint.

6. *Tibial (medial) collateral ligament.* Broad, flat ligament on the medial surface of the joint that extends from the medial condyle of the femur to the medial condyle of the tibia. The ligament is crossed by tendons of the sartorius, gracilis, and semitendinosus muscles, all of which strengthen the medial aspect of the joint. Because the tibial collateral ligament is firmly attached to the medial meniscus (see 9a), tearing of the ligament frequently results in tearing of the meniscus and damage to the anterior cruciate ligament, described under 8a.

7. *Fibular (lateral) collateral ligament.* Strong, rounded ligament on the lateral surface of the joint that extends from the lateral condyle of the femur to the lateral side of the head of the fibula. It strengthens the lateral aspect of the joint. The ligament is covered by the tendon of the biceps femoris muscle. The tendon of the popliteal muscle is deep to the ligament.

8. *Intracapsular ligaments.* Ligaments within the capsule that connect the tibia and femur.

    a. *Anterior cruciate ligament (ACL).* Extends posteriorly and laterally from the area anterior to the intercondylar eminence of the tibia to the posterior part of the medial surface of the lateral condyle of the femur. This ligament is stretched or torn in about 70 percent of all serious knee injuries.

    b. *Posterior cruciate ligament (PCL).* Extends anteriorly and medially from a depression on the posterior intercondylar area of the tibia and lateral meniscus to the anterior part of the medial surface of the medial condyle of the femur.

9. *Articular discs (menisci).* Two fibrocartilage discs between the tibial and femoral condyles that help compensate for the irregular shapes of the bones and circulate synovial fluid.

    a. *Medial meniscus.* Semicircular piece of fibrocartilage (C-shaped). Its anterior end is attached to the anterior intercondylar fossa of the tibia, anterior to the anterior cruciate ligament. Its posterior end is attached to the posterior intercondylar fossa of the tibia between the attachments of the posterior cruciate ligament and lateral meniscus.

    b. *Lateral meniscus.* Nearly circular piece of fibrocartilage (approaches an incomplete O in shape). Its anterior end is attached anterior to the intercondylar eminence of the tibia and lateral and posterior to the anterior cruciate ligament. Its posterior end is attached posterior to the intercondylar eminence of the tibia and anterior to the posterior end of the medial meniscus. The medial and lateral menisci are connected to each other by the *transverse ligament* and to the margins of the head of the tibia by the *coronary ligaments* (not illustrated).

10. The more important *bursae* of the knee include the following:

    a. *Prepatellar bursa* between the patella and skin.

    b. *Infrapatellar bursa* between superior part of tibia and patellar ligament.

    c. *Suprapatellar bursa* between inferior part of femur and deep surface of quadriceps femoris muscle.

MOVEMENTS

Flexion, extension, slight medial rotation, and lateral rotation of leg in flexed position (see Figures 8.2f and 8.5c).

Exhibit 8.5 (continued)

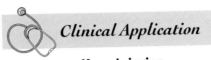

Clinical Application

Knee Injuries

The knee joint is the joint most vulnerable to damage because of the stresses to which it is subjected and because there is no interlocking of the articulating bones; reinforcement is strictly by ligaments and tendons. The most common type of **knee injury** in football is rupture of the tibial (medial) collateral ligaments, often associated with tearing of the anterior cruciate ligament and medial meniscus (torn cartilage). Usually, a hard blow to the lateral side of the knee while the foot is fixed on the ground causes the injury. When a knee is examined for such an injury, the three Cs are kept in mind: collateral ligament, cruciate ligament, and cartilage, the "unhappy triad."

A **swollen knee** may occur immediately or hours after an injury. Immediate swelling is due to escape of blood from dam-aged blood vessels adjacent to areas involving rupture of the anterior cruciate ligament, damage to synovial membranes, torn menisci, fractures, or collateral ligament sprains. Delayed swelling is due to excessive production of synovial fluid as a result of irritation of the synovial membrane, a condition commonly referred to as "water on the knee."

A **dislocated knee** refers to the displacement of the tibia relative to the femur. Accordingly, such dislocations are classified as anterior, posterior, medial, lateral, or rotatory. The most common type is anterior dislocation, resulting from hyperextension of the knee. A frequent consequence of a dislocated knee is damage to the popliteal artery.

Figure 8.12 Right knee (tibiofemoral) joint.

The knee joint is the largest and most complex joint in the body.

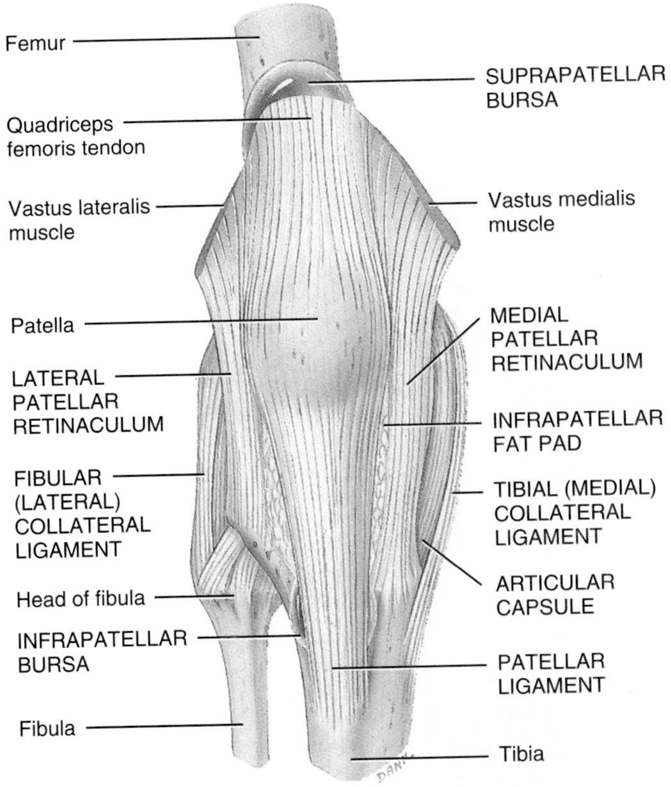

Femur

Quadriceps femoris tendon

Vastus lateralis muscle

Patella

LATERAL PATELLAR RETINACULUM

FIBULAR (LATERAL) COLLATERAL LIGAMENT

Head of fibula

INFRAPATELLAR BURSA

Fibula

SUPRAPATELLAR BURSA

Vastus medialis muscle

MEDIAL PATELLAR RETINACULUM

INFRAPATELLAR FAT PAD

TIBIAL (MEDIAL) COLLATERAL LIGAMENT

ARTICULAR CAPSULE

PATELLAR LIGAMENT

Tibia

(a) Diagram of anterior superficial view

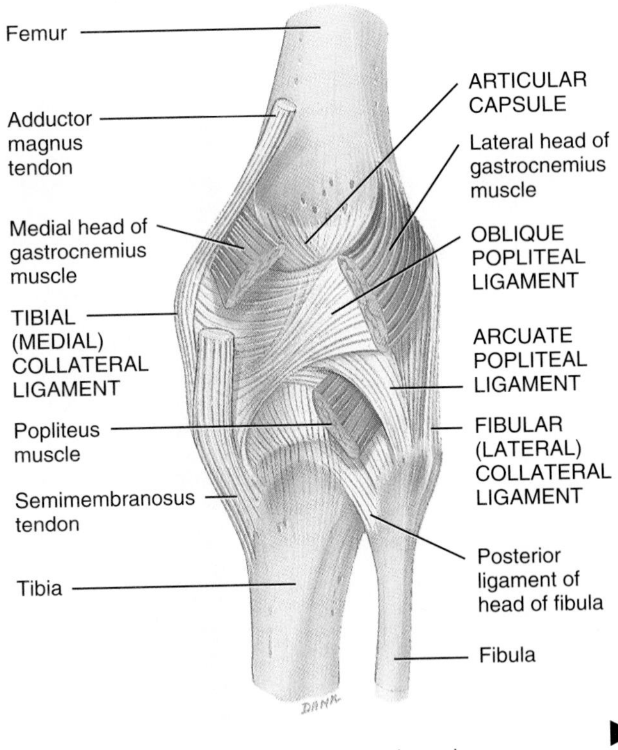

Femur

Adductor magnus tendon

Medial head of gastrocnemius muscle

TIBIAL (MEDIAL) COLLATERAL LIGAMENT

Popliteus muscle

Semimembranosus tendon

Tibia

ARTICULAR CAPSULE

Lateral head of gastrocnemius muscle

OBLIQUE POPLITEAL LIGAMENT

ARCUATE POPLITEAL LIGAMENT

FIBULAR (LATERAL) COLLATERAL LIGAMENT

Posterior ligament of head of fibula

Fibula

(b) Diagram of posterior deep view

Q *What structures are involved in the knee injury called "torn cartilage"?*

Exhibit 8.5 Knee (Tibiofemoral) Joint (continued)

Figure 8.12 (continued)

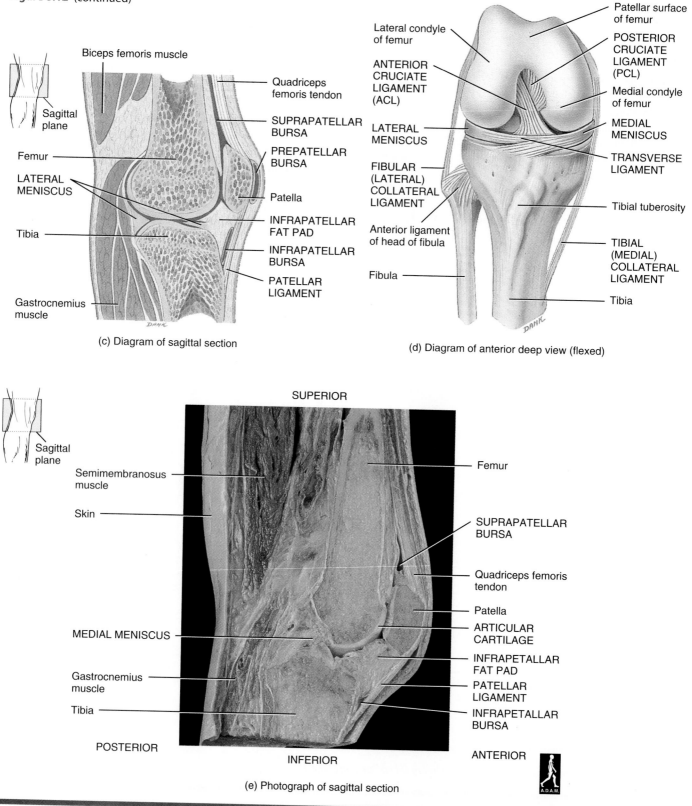

(c) Diagram of sagittal section

(d) Diagram of anterior deep view (flexed)

(e) Photograph of sagittal section

Table 8.3 *Types of Joints According to Articular Components, Structural and Functional Classification, and Movements*

JOINT	ARTICULAR COMPONENT	CLASSIFICATION	MOVEMENTS
Axial Skeleton			
Suture	Between skull bones.	*Structural:* fibrous. *Functional:* synarthrosis.	None.
Temporomandibular (TMJ)	Between condylar process of mandible and mandibular fossa and articular tubercle of temporal bone.	*Structural:* synovial (combined hinge and gliding joint). *Functional:* diarthrosis.	Depression, elevation, protraction, retraction, lateral displacement, and slight rotation of mandible.
Atlanto-occipital	Between superior articular facets of atlas and occipital condyles of occipital bone.	*Structural:* synovial (condyloid). *Functional:* diarthrosis.	Flexion and extension of head and slight lateral flexion of head to either side.
Atlanto-axial	(1) Between dens of axis and anterior arch of atlas and (2) between lateral masses of atlas and axis.	*Structural:* synovial (pivot) between dens and anterior arch and synovial (gliding) between lateral masses. *Functional:* diarthrosis.	Rotation of head.
Intervertebral	(1) Between vertebral bodies and (2) between vertebral arches.	*Structural:* cartilaginous (symphysis) between vertebral bodies and synovial (gliding) between vertebral arches. *Functional:* amphiarthrosis between vertebral bodies and diarthrosis between vertebral arches.	Flexion, extension, lateral flexion, and rotation of vertebral column.
Vertebrocostal	(1) Between facets of heads of ribs and facets of bodies of adjacent thoracic vertebrae and intervertebral discs between them and (2) between articular part of tubercles of ribs and facets of transverse processes of thoracic vertebrae.	*Structural:* synovial (gliding). *Functional:* diarthrosis.	Slight gliding.
Sternocostal	Between sternum and first seven pairs of ribs.	*Structural:* cartilaginous (synchondrosis) between sternum and first pair of ribs and synovial (gliding) between sternum and second through seventh pairs of ribs. *Functional:* synarthrosis between sternum and first pair of ribs and diarthrosis between sternum and second through seventh pairs of ribs.	None between sternum and first pair of ribs; slight gliding between sternum and second through seventh pairs of ribs.
Lumbosacral	(1) Between body of fifth lumbar vertebra and base of sacrum and (2) between inferior articular facets of fifth lumbar vertebra and superior articular facets of first vertebra of sacrum.	*Structural:* cartilaginous (symphysis) between body and base and synovial (gliding) between articular facets. *Functional:* amphiarthrosis between body and base and diarthrosis between articular facets.	Flexion, extension, lateral flexion, and rotation of vertebral column.

Now that you have learned the major bones and their markings (Chapters 6 and 7) and the classification of joints according to structure and function and movements that occur at joints (this chapter), refer to Table 8.3. It contains a list of joints according to articular components (bones that enter into their formation), structural and functional classifications, and movements. Table 8.3 will help you to apply the information you learned about bones by integrating it with information about joint classification and movements.

Table 8.3 *Types of Joints According to Articular Components, Structural and Functional Classification, and Movements (continued)*

JOINT	ARTICULAR COMPONENT	CLASSIFICATION	MOVEMENTS
Pectoral Girdles and Upper Limbs			
Sternoclavicular	Between sternal end of clavicle, manubrium of sternum, and first costal cartilage.	*Structural:* synovial (gliding and pivot). *Functional:* diarthrosis.	Gliding with limited movements in nearly every direction.
Acromioclavicular	Between acromion of scapula and acromial end of clavicle.	*Structural:* synovial (gliding). *Functional:* diarthrosis.	Gliding and rotation of scapula on clavicle.
Shoulder (gleno-humeral)	Between head of humerus and glenoid cavity of scapula.	*Structural:* synovial (ball-and-socket). *Functional:* diarthrosis.	Flexion, extension, hyper-extension, abduction, adduction, medial and lateral rotation, and circumduction of arm.
Elbow	Between trochlea of humerus, trochlear notch of ulna, and head of radius.	*Structural:* synovial (hinge). *Functional:* diarthrosis.	Flexion and extension of forearm.
Radioulnar	Proximal radioulnar joint between head of radius and radial notch of ulna; distal radioulnar joint between ulnar notch of radius and head of ulna.	*Structural:* synovial (pivot). *Functional:* diarthrosis.	Rotation of forearm.
Wrist (radiocarpal)	Between distal end of radius and scaphoid, lunate, and triquetrum of carpus.	*Structural:* synovial (condyloid). *Functional:* diarthrosis.	Flexion, extension, abduction, adduction, and circumduction of wrist.
Intercarpal	Between proximal row of carpal bones, distal row of carpal bones, and between both rows of carpal bones (midcarpal joints).	*Structural:* synovial (gliding), except for hamate, scaphoid, and lunate (midcarpal) joint, which is synovial (saddle). *Functional:* diarthrosis.	Gliding plus flexion and abduction at midcarpal joints.
Carpometacarpal	Carpometacarpal joint of thumb between trapezium of carpus and first metacarpal. Carpometacarpal joints of remaining digits formed between carpus and second through fifth metacarpals.	*Structural:* synovial (saddle) at thumb and synovial (gliding) at remaining digits. *Functional:* diarthrosis.	Flexion, extension, abduction, adduction, and circumduction at thumb and gliding at remaining digits.
Metacarpophalangeal	Between heads of metacarpals and bases of proximal phalanges.	*Structural:* synovial (condyloid). *Functional:* diarthrosis.	Flexion, extension, abduction, adduction, and circumduction of phalanges.
Interphalangeal	Between heads of phalanges and bases of more distal phalanges.	*Structural:* synovial (hinge). *Functional:* diarthrosis.	Flexion and extension of phalanges.

Table 8.3 *(continued)*

JOINT	ARTICULAR COMPONENT	CLASSIFICATION	MOVEMENTS
Pelvic Girdle and Lower Limbs			
Sacroiliac	Between auricular surfaces of sacrum and ilia of hipbones.	*Structural:* synovial (gliding). *Functional:* diarthrosis.	Slight gliding (even more so during pregnancy).
Pubic symphysis	Between anterior surfaces of hipbones.	*Structural:* cartilaginous (symphysis). *Functional:* amphiarthrosis.	Slight movements (even more so during pregnancy).
Hip (coxal)	Between head of femur and acetabulum of hipbone.	*Structural:* synovial (ball-and-socket). *Functional:* diarthrosis.	Flexion, extension, abduction, adduction, circumduction, and rotation of thigh.
Knee (tibiofemoral)	Formed by (1) lateral tibiofemoral joint between lateral condyle of femur, lateral meniscus, and lateral condyle of tibia; (2) intermediate patellofemoral joint between patella and patellar surface of femur; and (3) medial tibiofemoral joint between medial condyle of femur, medial meniscus, and medial condyle of tibia.	*Structural:* synovial (hinge) at lateral and medial tibiofemoral joints; partly synovial (hinge) and synovial (gliding) at patellofemoral joint. *Functional:* diarthrosis.	Flexion, extension, slight medial rotation, lateral rotation of leg in flexed position.
Tibiofibular	Proximal tibiofibular joint between lateral condyle of tibia and head of fibula; distal tibiofibular joint between distal end of fibula and fibular notch of tibia.	*Structural:* synovial (gliding) at proximal joint and fibrous (syndesmosis) at distal joint. *Functional:* diarthrosis at proximal joint and amphiarthrosis at distal joint.	Slight gliding at proximal joint and slight rotation of fibula during dorsiflexion of foot.
Ankle (talocrural)	(1) Between distal end of tibia and its medial malleolus and talus and (2) between lateral malleolus of fibula and talus.	*Structural:* synovial (hinge). *Functional:* diarthrosis.	Dorsiflexion and plantar flexion of foot.
Intertarsal	Subtalar joint between talus and calcaneus of tarsus; talocalcaneonavicular joint between talus and calcaneus and navicular of tarsus; calcaneocuboid joint between calcaneus and cuboid of tarsus.	*Structural:* synovial (gliding) at subtalar and calcaneocuboid joints and synovial (spheroid) at talocalcaneonavicular joint. *Functional:* diarthrosis.	Inversion and eversion of foot.
Tarsometatarsal	Between three cuneiforms of tarsus and bases of five metatarsal bones.	*Structural:* synovial (gliding). *Functional:* diarthrosis.	Slight gliding.
Metatarsophalangeal	Between heads of metatarsals and bases of proximal phalanges.	*Structural:* synovial (condyloid). *Functional:* diarthrosis.	Flexion, extension, abduction, adduction, and circumduction of phalanges.
Interphalangeal	Between heads of phalanges and bases of more distal phalanges.	*Structural:* synovial (hinge). *Functional:* diarthrosis.	Flexion and extension of phalanges.

Aging and Joints

The aging process usually results in decreased production of synovial fluid in joints. In addition, the articular cartilage becomes thinner with age, and ligaments shorten and lose some of their flexibility. The effects of aging on joints vary considerably from one person to another and are affected by genetic factors as well as wear and tear at joints. Although degenerative changes in joints may start as early as 20 years of age, most do not occur until much later. In fact, by age 80, almost everyone develops some type of degeneration in the knees and elbows, as well as in the hips and shoulders. Males also commonly develop degenerative changes in the vertebral column, resulting in a hunched-over posture and pressure on nerve roots. One type of arthritis, called osteoarthritis, is at least partially age-related (see Applications to Health, p. 6).

Key Medical Terms
Associated with Joints

Arthralgia (ar-THRAL-jē-a; *arthros* = joint; *algia* = pain) Pain in a joint.

Arthrosis (ar-THRŌ-sis) Refers to an articulation; also a disease of a joint.

Bursectomy (bur-SEK-tō-mē-; *ectomy* = removal of) Removal of a bursa.

Chondritis (kon-DRĪ-tis; *chondros* = cartilage) Inflammation of cartilage.

Luxation (luks-Ā-shun; *luxatio* = dislocation) The dislocation (displacement) of a bone from a joint with tearing of ligaments, tendons, and articular capsules; usually caused by a blow or fall.

Subluxation (sub-luks-Ā-shun) A partial or incomplete dislocation.

Synovitis (sin′-ō-VĪ-tis) Inflammation of a synovial membrane in a joint.

Study Outline

Classification of Joints (p. 190)

1. An articulation (joint) is a point of contact between two bones, cartilage and bone, or teeth and bone.
2. Structural classification is based on the presence or absence of a synovial (joint) cavity and type of connecting tissue. Structurally, joints are classified as fibrous, cartilaginous, or synovial.
3. Functional classification of joints is based on the degree of movement permitted. Joints may be synarthroses (immovable), amphiarthroses (partially movable), or diarthroses (freely movable).

Fibrous Joints (p. 190)

1. Bones held by fibrous connective tissue, with no synovial cavity, are fibrous joints.
2. These joints include immovable sutures (found between skull bones), partially movable syndesmoses (such as the distal tibiofibular articulation), and immovable gomphoses (roots of teeth in alveoli of mandible and maxilla).

Cartilaginous Joints (p. 191)

1. Bones held together by cartilage, with no synovial cavity, are cartilaginous joints.
2. These joints include immovable synchondroses united by hyaline cartilage (epiphyseal plates between diaphyses and epiphyses), partially movable synchondroses united by hyaline cartilage (sternocostal joints), and partially movable symphyses united by fibrocartilage (pubic symphysis).

Synovial Joints (p. 191)

1. A synovial joint contains a space between bones called the synovial (joint) cavity. All synovial joints are diarthroses.
2. Other characteristics of a synovial joint are the presence of articular cartilage and an articular capsule (fibrous capsule and synovial membrane).
3. Many synovial joints also contain accessory ligaments (extracapsular and intracapsular), articular discs (menisci), bursae, and tendon sheaths.
4. Synovial joints contain an extensive nerve and blood supply.
5. The nerves convey information about pain, joint movements, and the degree of stretch at a joint.
6. Blood vessels penetrate the articular capsule and ligaments. Articular portions of joints are nourished by synovial fluid.

Bursae and Tendon Sheaths (p. 193)

1. Bursae are saclike structures, similar in structure to joint capsules, that alleviate friction in joints such as the shoulder and knee joints.
2. Tendon sheaths are tubelike bursae that wrap around tendons where there is considerable friction.

Factors Affecting Contact and Range of Motion at Synovial Joints (p. 193)

1. Among the factors that keep the articular surfaces of synovial joints in contact are the shape of the articulating bones, strength and tension of ligaments, arrangement and tension of muscles, and apposition of soft parts.

2. Joint flexibility may also be affected by hormones, such as relaxin, which relaxes the pubic symphysis during childbirth, and by lack of use.

Movements at Synovial Joints (p. 194)

Axes of Movement at Synovial Joints (p. 194)

1. Axes (planes) of movement include monaxial (one axis), biaxial (two axes), and triaxial (three axes).

Specific Movements at Synovial Joints (p. 194)

1. The four principal movements are gliding, angular movements, rotation, and special movements.
2. In a gliding movement, the nearly flat surfaces of bones move back and forth and from side to side.
3. In angular movements, a change in the angle between bones occurs. Examples are flexion–extension, lateral flexion, hyperextension, and abduction–adduction. Circumduction refers to flexion, abduction, extension, and adduction in succession.
4. In rotation, a bone moves around its own longitudinal axis.
5. Special movements occur at specific joints. Examples are elevation–depression, protraction–retraction, inversion–eversion, dorsiflexion–plantar flexion, supination–pronation, and opposition.

Types of Synovial Joints (p. 201)

1. Gliding joint: articulating surfaces are flat; bone glides back and forth and side to side (nonaxial); found between carpals and tarsals.
2. Hinge joint: convex surface of one bone fits into concave surface of another; movement is flexion–extension (monaxial); example is the elbow.
3. Pivot joint: a round or pointed surface of one bone fits into a ring formed by another bone and a ligament; movement is rotation (monaxial); example is the atlanto-axial joint.
4. Condyloid joint: an oval-shaped projection of one bone fits into an oval cavity of another; movements are flexion–extension and abduction–adduction (biaxial), and circumduction; example is the wrist joint.

5. Saddle joint: articular surface of one bone is shaped like a saddle and the other bone fits into the "saddle" like a sitting rider; movements are flexion–extension and abduction–adduction (biaxial), and circumduction; example is the carpometacarpal joint between trapezium and metacarpal of thumb.
6. Ball-and-socket joint: ball-shaped surface of one bone fits into cuplike depression of another; movements are flexion–extension, abduction–adduction, and rotation (triaxial), and circumduction; example is the shoulder.

Selected Joints of the Body (p. 201)

1. Temporomandibular joint is between the condyle of the mandible and mandibular fossa and articular tubercle of the temporal bone.
2. Shoulder (humeroscapular) joint is between the head of the humerus and glenoid cavity of the scapula.
3. Elbow joint is between the trochlea of the humerus, the trochlear notch of the ulna, and the head of the radius.
4. Hip (coxal) joint is between the head of the femur and acetabulum of the hipbone.
5. Knee (tibiofemoral) joint is between the patella and patellar surface of femur; lateral condyle of femur, lateral meniscus, and lateral condyle of tibia; and medial condyle of femur, medial meniscus, and medial condyle of tibia.
6. A summary of selected joints of the body, including articular components, structural and functional classifications, and movements, is presented in Table 8.3 on page 217.

Aging and Joints (p. 220)

1. With aging, a decrease in synovial fluid, thinning of articular cartilage, and decreased flexibility of ligaments occur.
2. Most individuals experience some degeneration in the knees, elbows, hips, and shoulders due to the aging process.

Review Questions

1. Define an articulation. What factors determine the degree of movement at joints? (p. 190)
2. Classify joints on the basis of structure and function. (p. 190)
3. Set up an outline for the structural classification of joints, list and define the various subtypes, and give an example of each. (p. 190)
4. Describe the structure of a synovial joint. (p. 191)
5. What is an accessory ligament? Define the two principal types. (p. 192)
6. What are articular discs? Why are they important? (p. 192)
7. Describe the nerve and blood supply of a joint. (p. 193)
8. Explain how the articulating bones in a synovial joint are held together. (p. 193)

9. Have another person assume the anatomical position and execute each of the movements at joints discussed in the text. Reverse roles and see if you can execute the same movements. (p. 200)
10. List the six subtypes of synovial joints, define each, describe the movements at each, and give an example of each. (p. 200)
11. Refer to Table 8.3 on page 217 and for each joint listed, be sure that you can identify the bones that form the joint, define the joint according to structural and functional classification, and indicate the movements possible at the joint.
12. Define the effects of aging on joints. (p. 220)
13. Refer to the glossary of key medical terms associated with joints. Be sure that you can define each term. (p. 220)

Self Quiz

Complete the following:

1. The functional classification of the pubic symphysis is ___.
2. The sagittal suture is an example of a immovable joint in the body, a ___.
3. The functional classification of the temporomandibular joint is ___.
4. The ends of bones at synovial joints are covered with a protective layer of ___.
5. Saclike structures that reduce friction between body parts at joints are called ___.
6. Match the following terms with their descriptions:
   ___ (1) synovial, saddle
   ___ (2) gomphosis
   ___ (3) syndesmosis
   ___ (4) synovial, pivot
   ___ (5) cartilaginous (fibrocartilage)
   ___ (6) cartilaginous (hyaline cartilage)
   ___ (7) suture
   ___ (8) synovial, gliding
   ___ (9) synovial, hinge

   (a) distal tibiofibular joint
   (b) pubic symphysis
   (c) coronal suture
   (d) tooth in socket
   (e) atlanto-axial joint
   (f) intercarpal joint
   (g) elbow joint
   (h) carpometacarpal joint of thumb
   (i) epiphyseal plate

Are the following statements true or false?

7. A shoulder dislocation is an injury to the acromioclavicular joint.
8. A tendon sheath is a tubelike bursa found at the wrist and ankle joints.
9. All fibrous joints are synarthrotic and all cartilaginous joints are amphiarthrotic.
10. Sutures, syndesmoses, and symphyses are kinds of fibrous joints.
11. The nerves that supply a joint are different from those that supply the skeletal muscles that move that joint.
12. The shoulder joint has more freedom of movement than any other joint in the body.
13. Match the following joints with their descriptions:
    ___ (1) ball-and-socket
    ___ (2) condyloid
    ___ (3) gliding
    ___ (4) hinge
    ___ (5) pivot

    (a) monaxial joint; only rotation is possible
    (b) sternoclavicular and intercarpal joints
    (c) shoulder and hip joints
    (d) monaxial joint; only flexion and extension are possible
    (e) biaxial joint

Choose the one best answer to the following questions:

14. Choose the pair of terms that is most closely associated or matched.
    a. wrist joint, pronation
    b. intertarsal joints, inversion
    c. elbow joint, hyperextension
    d. ankle joint, eversion
    e. interphalangeal joints, circumduction

15. Which of the following structures is extracapsular in location?
    a. acetabular labrum
    b. posterior cruciate ligament
    c. ligament of the head of the femur
    d. fibular collateral ligament
    e. articular cartilage
16. Which of the following structures is *not* associated with the knee joint?
    a. glenoid labrum
    b. patellar ligament
    c. infrapatellar bursa
    d. cruciate ligaments
    e. tibial collateral ligament
17. Which of the following four statements is (are) correct?
    (1) An articulation is a point of contact between bones.
    (2) An articulation is a point of contact between bone and cartilage.
    (3) An articulation is a point of contact between bone and muscle.
    (4) An articulation is a point of contact between bones and teeth.
    a. 1 only
    b. 1 and 2
    c. 1, 2, and 3
    d. 1, 2, and 4
    e. 1, 2, 3, and 4
18. Which of the following three statements is (are) true about the joint between the epiphysis and the diaphysis of a long bone?
    (1) It is one example of a synchondrosis.
    (2) It is a synarthrosis.
    (3) It is temporary and becomes a synostosis.
    a. 1 only
    b. 2 only
    c. 3 only
    d. 1 and 3
    e. 1, 2, and 3
19. Which of the following three statements is (are) true?
    (1) Synovial joints are enclosed in a synovial-lined capsule and have ligaments uniting the bones of the joint.
    (2) Synovial joints constitute relatively few of the joints of the body.
    (3) Synovial joints always permit flexion, extension, abduction, and adduction.
    a. 1 only
    b. 2 only
    c. 3 only
    d. all of the above
    e. none of the above
20. Which of the following three statements is (are) true?
    (1) The synovial membrane secretes synovial fluid into a joint cavity.
    (2) The synovial membrane lines the fibrous capsule of movable joints.
    (3) The synovial membrane is an epithelial membrane.

a. 1 only
b. 2 only
c. 3 only
d. 1 and 2
e. all of the above

21. Which of the following pairs of terms is most closely associated or matched?
    a. elbow joint, coracohumeral ligament
    b. knee joint, transverse humeral ligament
    c. symphysis, epiphyseal plate
    d. condyloid joint, interphalangeal joints
    e. shoulder joint, glenoid labrum

22. Which of the following pairs of terms is most closely associated or matched?
    a. knee joint, lateral and medial menisci
    b. shoulder joint, acetabulum
    c. elbow joint, radioulnar joint
    d. cartilaginous joint, joint cavity
    e. wrist joint, radioulnar joint

23. Match the following movements with their descriptions:
    ___ (1) abduction
    ___ (2) dorsiflexion
    ___ (3) extension
    ___ (4) flexion
    ___ (5) gliding
    ___ (6) rotation

    (a) decrease in angle between bones
    (b) simplest kind of movement that can occur at a joint; no angular or rotary motion involved; for example, ribs moving against vertebrae
    (c) state of entire body when it is in the anatomical position
    (d) movement away from the midline of the body in the frontal plane
    (e) movement of a bone around its own axis
    (f) position of foot when heel is on the floor and rest of foot is raised

Critical Thinking Questions

1. Lin and Juan have been golf partners for 50 years. Lin's golf game improved by five strokes this spring, and he credits the hip replacement he had last year. Juan's knee has been bothering him for years, but when he asked his orthopedist about a new knee joint, he was told, "It's not that simple." Juan told Lin, "One joint's like any other joint" and wants a second opinion. What will the new doctor say?
   HINT: *Juan knows more about golf than he does about joints.*

2. When Laura gets bored during anatomy class, she rocks her lower jaw back and forth, making a clicking sound, which drives Fran (the student in the next seat) crazy! Fran was so distracted that she started writing everything she knew about that particular joint in her notebook. Fortunately, Fran is in her first semester of anatomy. What has she written?
   HINT: *The rocking motion is related to the joint structure.*

3. Rasheed tripped over a chair leg. He twisted his knee joint during his fall and then hit the tiled floor directly on his patella. Rasheed now has a bad case of "water on the knee." What is the cause of his swollen knee?
   HINT: *Fluid is involved, but it's not really water.*

4. After your second Human Anatomy exam, you dropped to one knee, raised one arm over your head with your hand clenched into a fist, pumped the arm up and down, bent your head back, looked straight up, and yell, "YES!" Use the proper terms to describe the movements at the various joints.
   HINT: *Think about how the position of each joint varies from the anatomical position.*

5. Tish was just getting the hang of bodysurfing during her first trip to the shore when she got caught in the break of a wave. While she was being violently rolled in the surf, she felt her shoulder "pop." When Tish finally got back to her towel she was out of breath and in pain, and her arm was hanging at an odd angle. What's the prognosis for the rest of Tish's vacation at the shore?
   HINT: *The "popped" joint is the most frequently dislocated joint in adults.*

6. Periodontal disease is a major cause of dental problems and tooth loss in adults. Describe the location and structure of the affected joints.
   HINT: *If you don't floss daily, your dentist may have reason to show you pictures of periodontal disease.*

Answers to Figure Questions

8.1 Diarthrosis.
8.2 Flexion of the thumb and lateral flexion of the trunk.
8.3 When you adduct your arm or leg, you bring it closer to the midline of the body, thus "adding" it to the trunk.
8.4 Because it is a combined movement.
8.5 Movement of the anterior surface of a bone or limb toward the midline.
8.6 Protraction.
8.7 Condyloid and saddle joints.
8.8 Lateral ligament.
8.9 Because of the looseness of the articular capsule and the shallowness of the glenoid cavity in relation to the size of the head of the humerus.
8.10 Flexion and extension.
8.11 Iliofemoral ligament.
8.12 Articular discs (menisci).

the same. At other times, there is no visible shortening and thickening of the muscle, but the tension increases greatly.

3. **Extensibility** means that muscle tissue can be stretched without being damaged. Most skeletal muscles are arranged in opposing pairs. While one is contracting, the other not only is relaxed but usually is being stretched.

4. **Elasticity** means that muscle tissue tends to return to its original shape after contraction or extension.

Skeletal muscle is the focus of much of this chapter. Cardiac muscle and smooth muscle are described briefly here and are discussed in more detail later (with discussions of the heart in Chapter 13, the various organs containing smooth muscle, and the autonomic nervous system in Chapter 21).

SKELETAL MUSCLE TISSUE

To understand how skeletal muscle contraction produces movement, one needs some knowledge of its connective tissue components, nerve and blood supply, and microscopic anatomy.

Connective Tissue Components

Connective tissue surrounds and protects muscle tissue. The **fascia** (FASH-ē-a; *fascia* = bandage) is a sheet or broad band of fibrous connective tissue deep to the skin or surrounding muscles and other organs of the body. **Superficial fascia (subcutaneous layer)** is immediately deep to the skin (see Figure 10.22). It is composed of areolar connective tissue and adipose tissue and has four important functions: (1) It stores water and fat. Much of the fat of an overweight person is in the superficial fascia. (2) It reduces the rate of heat loss. (3) It provides mechanical protection against traumatic blows. (4) It provides a pathway for nerves and blood vessels to enter and exit muscles.

Deep fascia is dense, irregular connective tissue that lines the body wall and limbs and holds muscles together, separating them into functional groups (see Figure 10.22). Deep fascia allows free movement of muscles, carries nerves and blood and lymphatic vessels, and fills spaces between muscles.

Three layers of dense, irregular connective tissue extend from the deep fascia to further protect and strengthen skeletal muscle. These are readily seen in the sectioned skeletal muscle in Figure 9.1. The outermost layer, encircling the whole muscle, is the **epimysium** (ep-i-MĪZ-ē-um; *epi* = upon). **Perimysium** (per-i-MĪZ-ē-um; *peri* = around) surrounds bundles of 10 to 100 or more individual muscle fibers called **fascicles** (FAS-i-kuls) or **fasciculi** (fa-SIK-you-lī).

Penetrating into the interior of each fascicle and separating individual muscle fibers from one another is **endomysium** (en'-do-MĪZ-ē-um; *endo* = within).

Epimysium, perimysium, and endomysium are all continuous with and contribute collagen fibers to the connective tissue that attaches the muscle to other structures, such as bone or other muscle. All three may extend beyond the muscle fibers as a **tendon** (*tendere* = to stretch out)—a cord of dense connective tissue that attaches a muscle to the periosteum of a bone. An example is the calcaneal (Achilles) tendon of the gastrocnemius muscle (see Figure 10.23a). When the connective tissue elements extend as a broad, flat layer, the tendon is called an **aponeurosis** (*apo* = from; *neuron* = sinew). This structure also attaches to the coverings of a bone, another muscle, or the skin. An example of an aponeurosis is the galea aponeurotica on top of the skull (see Figure 10.4c). Certain tendons, especially those of the wrist and ankle, are enclosed by tubes of fibrous connective tissue called **tendon (synovial) sheaths,** which are similar in structure to bursae. The inner layer of a tendon sheath, the *visceral layer,* is attached to the surface of the tendon. The outer layer is known as the *parietal layer* (see Figure 10.19a). Between the layers is a cavity that contains a film of synovial fluid. Tendon sheaths reduce friction as tendons slide back and forth.

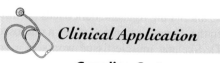

Clinical Application

Ganglion Cyst

A **ganglion (synovial) cyst** is a swelling, usually on the dorsum of the wrist and hand, that appears periodically, especially following activities that involve repetitive hand motions. The cyst contains fluid within fibrous connective tissue and is an extension of a tendon sheath that encloses a long extensor tendon in the wrist. Although they are usually not painful, ganglion cysts do cause some discomfort when they are inadvertently hit or when the wrist is flexed or extended. They result from an accumulation of fluid that has leaked from a tendon sheath or joint. As a rule they are harmless, generally requiring no treatment and having little effect on activities. ▪

Nerve and Blood Supply

Skeletal muscles are well supplied with nerves and blood vessels. Those neurons that stimulate skeletal muscle to contract are called **motor neurons.** When muscle contracts, it uses a good deal of adenosine triphosphate (ATP) and therefore needs large amounts of nutrients and oxygen for ATP

Figure 9.1 Relationship of connective tissue to a skeletal muscle, showing the relative positions of the epimysium, perimysium, endomysium, and tendons.

The connective tissue layers associated with skeletal muscle are extensions of deep fascia.

OVERVIEW OF FUNCTIONS OF MUSCLE TISSUE
1. Motion. Movements such as walking and running.
2. Movement of substances within the body. All three types of muscle tissue help move substances such as blood, food, sperm and ova, and urine.
3. Body position stability and organ volume regulation. Skeletal muscle contractions maintain stable body positions and posture. Sustained contractions of smooth muscles block exit of food from the stomach and urine from the urinary bladder for temporary storage.
4. Heat production. Skeletal muscle contractions may generate 85 percent of all body heat to help maintain normal body temperature.

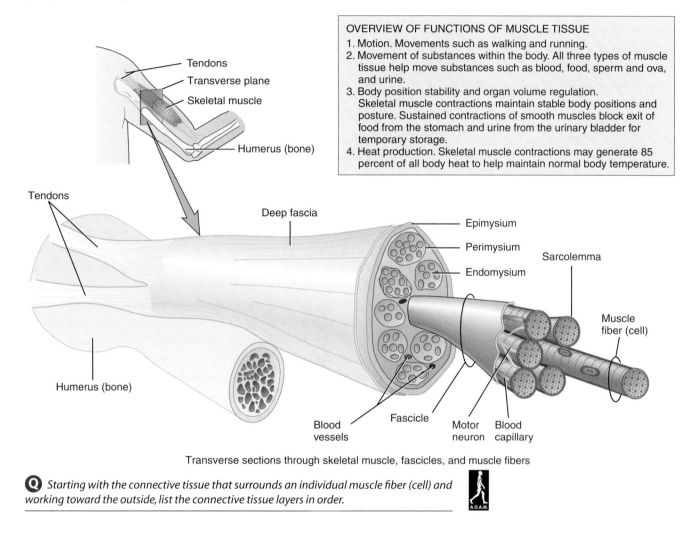

Transverse sections through skeletal muscle, fascicles, and muscle fibers

Q *Starting with the connective tissue that surrounds an individual muscle fiber (cell) and working toward the outside, list the connective tissue layers in order.*

production during cellular respiration. ATP stores energy for cellular use. Moreover, the waste products of the reactions that produce ATP must be eliminated. Thus, prolonged muscle action depends on a rich blood supply to deliver nutrients and oxygen and remove wastes and heat. Microscopic blood vessels called capillaries are plentiful in muscle tissue; each muscle fiber (cell) is in close contact with one or more capillaries (Figure 9.1).

Generally, an artery and one or two veins accompany each nerve that penetrates a skeletal muscle (Figure 9.2). The larger branches of the blood vessels accompany the nerve branches through the connective tissue of the muscle. Capillaries are plentiful within the endomysium surrounding the muscle fibers. Each muscle fiber is thus in close contact with one or more capillaries.. Each skeletal muscle fiber also makes contact with a portion of a nerve cell (see Figures 9.2 and 9.4a).

The Motor Unit

A motor neuron delivers the stimulus that ultimately causes a muscle fiber to contract. A motor neuron plus all the skeletal muscle fibers it stimulates is called a **motor unit** (Figure 9.3). A single motor neuron makes contact with an average of 150 muscle fibers. This means that when one neuron activates, about 150 muscle fibers simultaneously contract.

Muscles that control precise movements consist of many small motor units. For example, muscles of the larynx (voice box) that control voice production have as few as two or three muscle fibers per motor unit, and muscles controlling eye movements may have 10–20 muscle fibers per motor unit. On the other hand, muscles of the body that are responsible for powerful gross movements, such as the biceps brachii in the arm and gastrocnemius in the leg, have some

Figure 9.2 Relationship of blood vessels and nerves to skeletal muscle. The muscles, blood vessels, and nerves shown will be discussed in later chapters.

Generally, an artery and one or two veins accompany each nerve that penetrates a skeletal muscle.

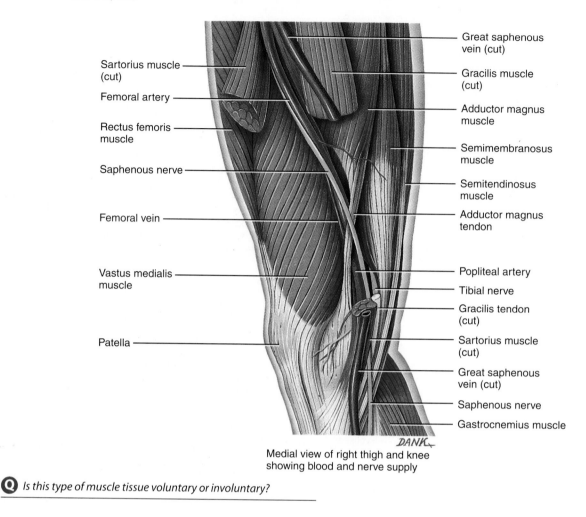

Sartorius muscle (cut)

Femoral artery

Rectus femoris muscle

Saphenous nerve

Femoral vein

Vastus medialis muscle

Patella

Great saphenous vein (cut)

Gracilis muscle (cut)

Adductor magnus muscle

Semimembranosus muscle

Semitendinosus muscle

Adductor magnus tendon

Popliteal artery

Tibial nerve

Gracilis tendon (cut)

Sartorius muscle (cut)

Great saphenous vein (cut)

Saphenous nerve

Gastrocnemius muscle

DANK

Medial view of right thigh and knee showing blood and nerve supply

Q *Is this type of muscle tissue voluntary or involuntary?*

motor units with as many as 2000 muscle fibers each. Remember, all the muscle fibers of a motor unit contract and relax together. Accordingly, the total strength of a contraction depends, in part, on how many motor units are activated.

The Neuromuscular Junction

Excitable cells (neurons and muscle fibers) communicate with one another and with other target cells—for example, endocrine or exocrine gland cells—at specialized regions called **synapses** (*synapsis* = connection). At most synapses a small gap, called the **synaptic cleft,** separates the two cells. Because the cells do not physically touch, the action potential from one cell cannot "jump the gap" to excite the next cell. Rather, the first cell communicates with the second by releasing a chemical called a **neurotransmitter.** The particular type of synapse formed between a motor neuron and

a skeletal muscle fiber is called the **neuromuscular junction** (NMJ) or *myoneural junction* (Figure 9.4).

Each motor neuron has a threadlike **axon** that extends from the spinal cord to a group of skeletal muscle fibers. Close to its target skeletal muscle fibers, the axon branches into several **axon terminals** (Figure 9.4a). In motor neurons, the distal tip of each axon terminal expands into a cluster of **synaptic end bulbs** that contain many membrane-enclosed sacs called **synaptic vesicles** (Figure 9.4b, c). Inside each synaptic vesicle are thousands of neurotransmitter molecules. Although there are many kinds of neurotransmitters, the kind released at the NMJ is **acetylcholine** (as'-ē-til-KŌ-lēn), abbreviated **ACh.** The region of the muscle fiber plasma membrane that is adjacent to the axon terminals is called the **motor end plate.** It contains *acetylcholine recep-*

Figure 9.3 Motor units. Shown are two motor neurons, one in red and one in green, each supplying the muscle fibers of its motor unit.

🔑 *A motor unit consists of a motor neuron plus the muscle fibers it stimulates.*

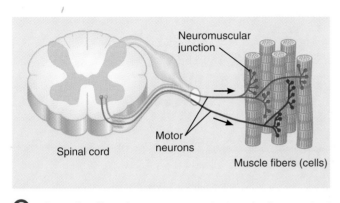

Q *What is the effect of motor unit size on strength of contraction?*

---

tors (Figure 9.4c), which are integral membrane proteins that recognize and bind specifically to ACh. The neuromuscular junction includes all motor neuron axon terminals and their synaptic end bulbs, plus the motor end plate of the muscle fiber, which typically contains 30–40 million ACh receptors.

When a nerve impulse (nerve action potential) reaches the synaptic end bulbs, the synaptic vesicles fuse with the plasma membrane and liberate ACh, which diffuses into the synaptic cleft between the motor neuron and the motor end plate. When ACh binds to its receptor, a sequence of events is initiated: Sodium ions (Na^+) enter the muscle fibers and trigger a muscle action potential that travels along the muscle cell plasma membrane (its sarcolemma), resulting in muscle contraction.

Because skeletal muscle fibers often are very long cells, the neuromuscular junction usually is located near the midpoint of the fiber. Muscle action potentials then spread from the center of the fiber toward both ends. This arrangement permits nearly simultaneous contraction of all parts of the fiber.

Clinical Application

Electromyogram

The recording of electrical activity in resting and contracting muscles is called **electromyography** (e-lek′-trō-mī-OG-ra-fē; *electro* = electricity; *myo* = muscle; *graph* = to write). An **elec-**tromyogram (EMG) is a record of skeletal muscle electrical activity. It is similar to the more familiar electrocardiogram (ECG), which is a record of the heart's electrical signals. EMGs may be used to determine the cause of muscular weakness or paralysis, to evaluate involuntary muscle twitching, to determine why abnormal levels of muscle enzymes (such as creatine phosphokinase) appear in blood, and to serve as a component of biofeedback studies.

Nerve conduction studies are sometimes done together with EMG testing. A nerve is stimulated electrically through the skin while a recording device detects the muscle's response. Such studies can help determine whether nerve damage is the cause of muscle weakness. ▪

Microscopic Anatomy

Microscopic examination of a typical skeletal muscle reveals hundreds or thousands of very long, cylindrical cells called **muscle fibers** or **myofibers** (*myo* = muscle) (Figure 9.5b on page 232). The muscle fibers lie parallel to one another and range from 10 to 100 μm* in diameter. Although a typical length is 100 μm, some are up to 30 cm (12 in.) long. The **sarcolemma** (*sarco* = flesh; *lemma* = sheath) is a muscle fiber's plasma membrane, and it surrounds the muscle fiber's cytoplasm, called **sarcoplasm** (Figure 9.5c). Because skeletal muscle fibers arise from the fusion of many smaller cells during embryonic development, each fiber has many nuclei to direct synthesis of new proteins. The nuclei are at the periphery of the cell, next to the sarcolemma, conveniently out of the way of the contractile elements (described shortly). The mitochondria lie in rows throughout the muscle fiber, strategically close to the muscle proteins that use ATP during contraction.

At high magnification the sarcoplasm appears stuffed with little threads. These small structures are the **myofibrils** (Figure 9.5d). Although myofibrils extend lengthwise within the muscle fiber, their prominent alternating light and dark bands make the whole muscle cell look striped. The bands are called *striations* (Figure 9.5f).

Myofibrils

Myofibrils are the contractile elements of skeletal muscle. They are 1–2 μm in diameter and contain three types of even smaller structures called **filaments (myofilaments).** The diameter of the *thin filaments* is about 8 nm, whereas that of the *thick filaments* is about 16 nm** (Figure 9.5e). The thick and thin filaments overlap one another, to a greater or lesser extent, depending on whether the muscle is con-

---

*One micrometer (μm) = 1/10,000 cm or 1/25,000 inch.
**One nanometer (nm) is 10^{-9} meter (0.001 μm).

Figure 9.4 Neuromuscular junction (NMJ).

A neuromuscular junction is the synapse between a motor neuron and a skeletal muscle fiber.

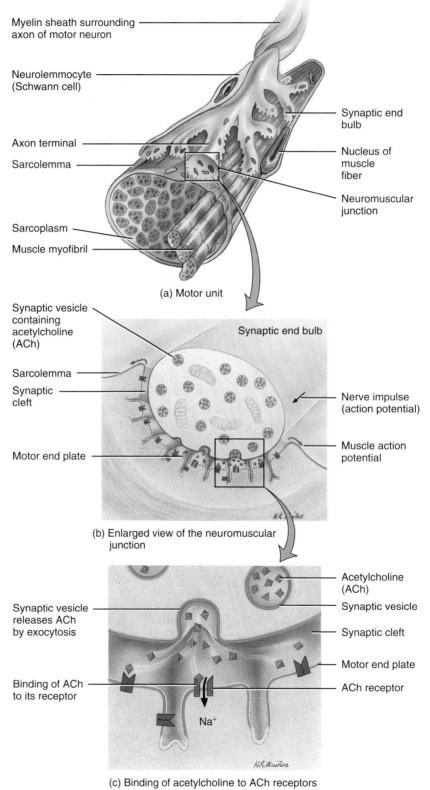

Myelin sheath surrounding axon of motor neuron

Neurolemmocyte (Schwann cell)

Axon terminal

Sarcolemma

Sarcoplasm

Muscle myofibril

Synaptic end bulb

Nucleus of muscle fiber

Neuromuscular junction

(a) Motor unit

Synaptic vesicle containing acetylcholine (ACh)

Sarcolemma

Synaptic cleft

Motor end plate

Synaptic end bulb

Nerve impulse (action potential)

Muscle action potential

(b) Enlarged view of the neuromuscular junction

Synaptic vesicle releases ACh by exocytosis

Binding of ACh to its receptor

Acetylcholine (ACh)

Synaptic vesicle

Synaptic cleft

Motor end plate

ACh receptor

Na⁺

(c) Binding of acetylcholine to ACh receptors in the motor end plate

Figure 9.4 (continued)

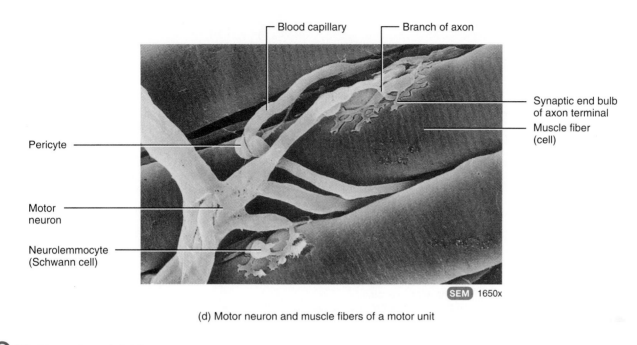

(d) Motor neuron and muscle fibers of a motor unit

Q *What is a motor end plate?*

tracting or relaxing. The pattern of their overlap causes the striations seen both in single myofibrils and in whole muscle fibers. A third type of filament, the elastic filament, will be described shortly.

The filaments inside a myofibril do not extend the entire length of a muscle fiber. They are arranged in compartments called **sarcomeres** (*meros* = part), which are the basic functional units of striated muscle tissue (Figures 9.6 and 9.7a on page 233). Narrow plate-shaped regions of dense material called *Z discs (lines)* separate one sarcomere from the next. Within a sarcomere, the darker area is called the *A (anisotropic) band,* which consists of mostly thick filaments and includes portions of the thin filaments where they overlap the thick filaments. A lighter, less dense area is called the *I (isotropic) band,* which contains the rest of the thin filaments but no thick filaments. The Z disc passes through the center of each I band. The alternating dark A bands and light I bands give the muscle fiber its striated appearance. A narrow *H (helle* = bright) *zone* in the center of each A band contains thick but not thin filaments. Dividing the H zone is the *M*

line, formed by protein molecules that connect adjacent thick filaments.

The two *contractile* proteins in muscle are myosin and actin. About 200 molecules of the protein **myosin** form a single thick filament (Figure 9.7b). Each myosin molecule is shaped like two golf clubs twisted together. The *myosin tails* (golf club handles) point toward the M line in the center of the sarcomere. The projections, called *myosin heads* or *cross bridges,* extend out toward the thin filaments. Tails of neighboring myosin molecules lie parallel to one another, forming the shaft of the thick filament. The heads project from all around the shaft in a spiraling fashion.

Thin filaments extend from anchoring points within the Z discs. Their main component is **actin.** Also present in the thin filament are smaller amounts of two *regulatory* proteins, **tropomyosin** and **troponin** (Figure 9.7c). Individual actin molecules join to form an actin filament that is twisted into a helix. On each actin molecule is a *myosin-binding site,* a site where a myosin head can attach.

Figure 9.5 Organization of skeletal muscle from macroscopic to microscopic levels.

The structural organization of a skeletal muscle from macroscopic to microscopic is: skeletal muscle, fascicle (bundle of muscle fibers), muscle fiber, myofibril, thin and thick filaments.

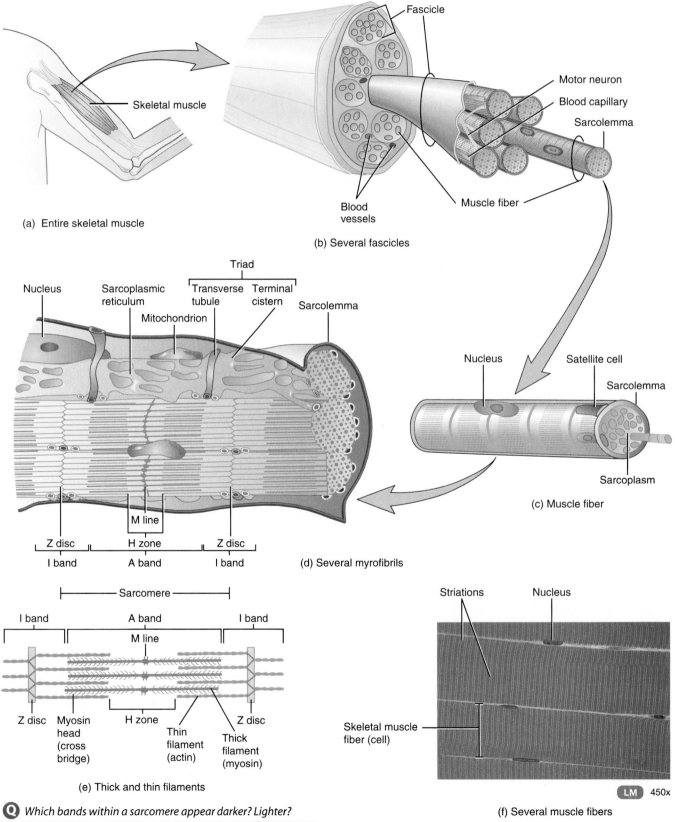

(a) Entire skeletal muscle

(b) Several fascicles

(c) Muscle fiber

(d) Several myofibrils

(e) Thick and thin filaments

(f) Several muscle fibers

Q *Which bands within a sarcomere appear darker? Lighter?*

LM 450x

Figure 9.6 Transmission electron micrograph of mammalian skeletal muscle. The Z discs separate one sarcomere from another.

The overlap of thick and thin filaments produces an alternating pattern of lighter and darker bands (striations).

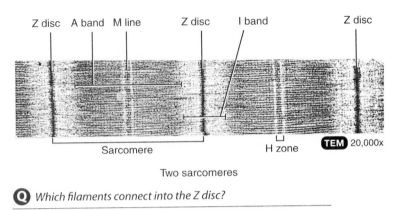

Z disc   A band   M line   Z disc   I band   Z disc

Sarcomere

H zone   **TEM** 20,000x

Two sarcomeres

Q *Which filaments connect into the Z disc?*

Figure 9.7 Detailed structure of muscle filaments. (a) The relationship of thick (myosin), thin (actin), and elastic (titin) filaments in a sarcomere. Note that actin filaments are anchored directly at the Z discs, whereas myosin filaments are connected to the Z discs by titin (also known as connectin). (b) About 200 myosin molecules make up a thick filament. The myosin tails all point toward the center of the sarcomere. (c) Thin filaments contain actin, troponin, and tropomyosin.

Myofibrils contain three types of filaments: thick, thin, and elastic.

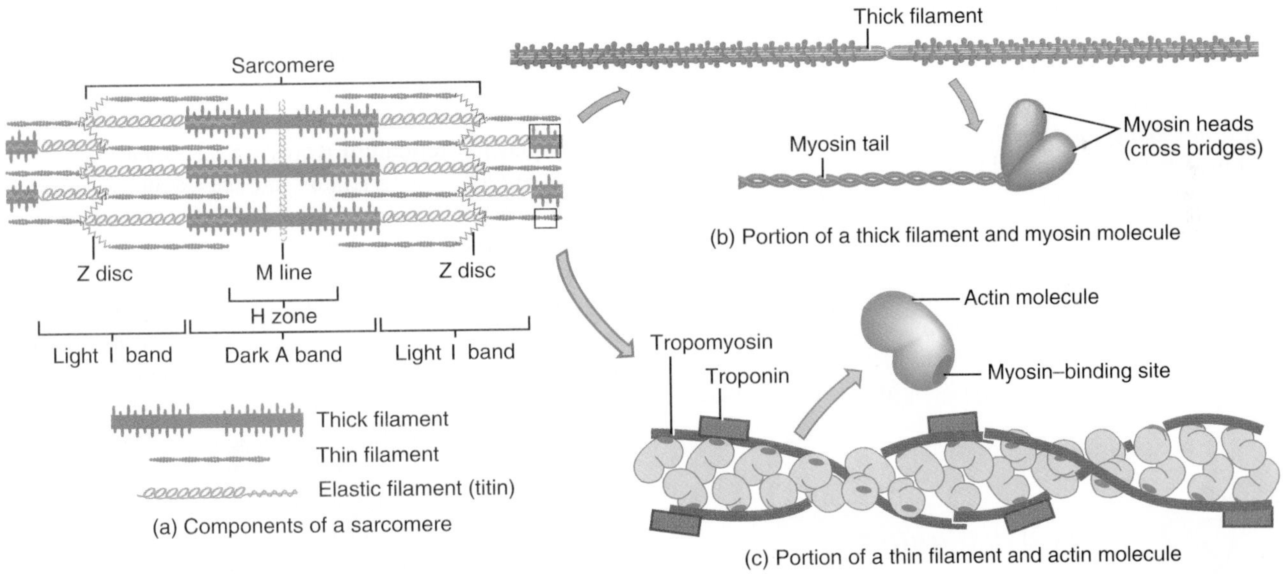

Thick filament

Sarcomere

Myosin tail

Myosin heads (cross bridges)

(b) Portion of a thick filament and myosin molecule

Z disc   M line   Z disc

H zone

Light I band   Dark A band   Light I band

Thick filament
Thin filament
Elastic filament (titin)

(a) Components of a sarcomere

Actin molecule

Tropomyosin
Troponin

Myosin–binding site

(c) Portion of a thin filament and actin molecule

Q *Which proteins are present in the A band? In the I band?*

Table 9.1 *Types of Filaments in Skeletal Muscle Fibers*

FILAMENT	SIZE AND PROTEIN COMPOSITION	FUNCTIONS
Thick filament	16 nm diameter; contains myosin.	Myosin heads (cross bridges) move thin filaments toward center of sarcomere during contraction.
Thin filament	8 nm diameter; contains actin, troponin, and tropomyosin.	Contains myosin head binding sites; slides along thick filament during contraction.
Elastic filament	Less than 1 nm diameter; contains titin, also called connectin.	Anchors thick filaments to Z discs and stabilizes them during contraction and relaxation.

The concentration of calcium ions (Ca^{2+}) in sarcoplasm is a major factor in muscle relaxation or contraction. For example, in relaxed muscle, the concentration of Ca^{2+} in the sarcoplasm is low. As a result, tropomyosin covers the myosin-binding sites on actin and thus blocks attachment of myosin heads to actin. Troponin holds the tropomyosin strand in place. When the Ca^{2+} concentration in the sarcoplasm increases, it induces a change in the position of troponin. This removes tropomyosin from the myosin binding sites so that myosin heads can attach to actin, enabling contraction to occur.

A more recently recognized component of the sarcomere is the *elastic filament* (Figure 9.7a). It is composed of the protein *titin (connectin)*, the third most plentiful protein in skeletal muscle (after actin and myosin). Titin anchors thick filaments to the Z discs and thereby helps stabilize the position of the thick filaments. It may also play a role in recovery of the resting sarcomere length when a muscle is stretched or during relaxation. The protein was named titin because it has a huge (titanic) molecular weight (or, alternatively, connectin because of its connecting function).

Table 9.1 above reviews the types of filaments in skeletal muscle fibers.

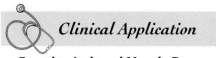

Clinical Application

Exercise-Induced Muscle Damage

Electron micrographs of muscle tissue taken from athletes before and after a marathon race or other types of severe exercise reveal considerable damage, including torn sarcolemmas in some muscle fibers, damaged myofibrils, and disrupted Z discs. Microscopic muscle damage after exercise also is indicated by increases in blood levels of myoglobin (a pigment) and certain enzymes (lactic acid dehydrogenase and creatine phosphokinase, for example) that normally are confined inside muscle fibers. From 12 to 48 hours after a period of strenuous exercise, skeletal muscles often become sore. Such **delayed onset muscle soreness (DOMS)** is accompanied by stiffness, tenderness, and swelling. Although the causes of DOMS are not completely understood, microscopic muscle damage appears to be a major contributing factor. ■

Sarcoplasmic Reticulum and Transverse Tubules

A fluid-filled system of cisterns called the **sarcoplasmic reticulum (SR)** encircles each myofibril (see Figure 9.5d). This elaborate tubular system is similar to smooth endoplasmic reticulum in nonmuscle cells. In a relaxed muscle fiber, the sarcoplasmic reticulum stores Ca^{2+}. Release of Ca^{2+} from the sarcoplasmic reticulum into the sarcoplasm around the thick and thin filaments triggers muscle contraction. The calcium ions pass out through special pores in the sarcoplasmic reticulum membrane.

The **transverse tubules (T tubules)** are deep tunnel-like infoldings of the sarcolemma. They penetrate the muscle fiber at right angles to the sarcoplasmic reticulum and the myofilaments. In mammalian skeletal muscle, there are two T tubules in each sarcomere, one at each A–I band junction. T tubules are open to the outside of the fiber and are filled with extracellular fluid. On both sides of a T tubule are dilated end sacs of the sarcoplasmic reticulum called *terminal cisterns (cistern = reservoir or cavity)* that store Ca^{2+}.

The muscle action potential generated along the sarcolemma also moves along the T tubules to penetrate the deepest portions of the muscles fibers. This arrangement ensures that each myofibril will respond to the action potential at the same time. The term **triad** refers to a T tubule and the terminal cisterns on either side of it at the A–I band.

Figure 9.8 Sliding-filament mechanism of muscle contraction shown in two sarcomeres. For simplicity, the elastic filaments are not illustrated.

🔑 *During muscle contraction, thin filaments move inward toward the H zone of a sarcomere.*

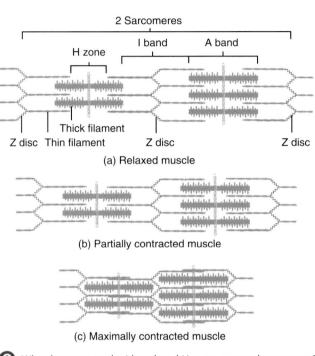

(a) Relaxed muscle

(b) Partially contracted muscle

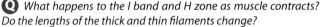

(c) Maximally contracted muscle

Q *What happens to the I band and H zone as muscle contracts? Do the lengths of the thick and thin filaments change?*

CONTRACTION OF MUSCLE
Sliding-Filament Mechanism

The physical changes that occur during muscle contraction result in the sliding of thick and thin filaments past one another. The model of how this occurs is known as the **sliding-filament mechanism** of muscle contraction.

Muscle contraction occurs because myosin heads attach to thin filaments at both ends of a sarcomere and pull the thin filaments toward the center (Figure 9.8b). During this process, each myosin head pivots inward, which generates the mechanical force that pulls on the thin filaments. Then, each myosin head detaches from the thin filament and straightens to its original position so that it can attach to another myosin-binding site further along the thin filament. This cycle of myosin head attachment, pivoting, detachment, and return to its original position repeats itself over and over again as part of a single muscle contraction. As a result of these events, the thin filaments slide inward and meet at the center of a sarcomere. They may even move so far inward that their ends overlap (Figure 9.8c). As the thin filaments slide inward, the Z discs come closer together, and

the sarcomere shortens. However, the lengths of the thick and thin filaments do not change. Shortening of the sarcomeres causes shortening of the whole muscle fiber and ultimately shortening of the entire muscle.

The contractile process is very much dependent on calcium ions, as well as on ATP as an energy source. Calcium ions alter the position of troponin to remove it from the myosin-binding sites on the thin filaments so that the myosin heads can attach to the thin filaments. ATP provides the energy for the various movements of the myosin head.

In the next chapter you will learn that while some muscles are contracting, other muscles are extending. For example, when you flex your forearm at the elbow joint, the muscle on the anterior surface of the arm, called the biceps brachii, contracts. At the same time, the muscle on the posterior surface of the arm, the triceps brachii, is extended or stretched. When you return your arm to its extended position, the triceps brachii contracts, moving the forearm and stretching the fibers of the biceps brachii back into their fully extended position. This coordinated action of opposing muscles is essential for us to perform opposing movements, such as flexion and extension or abduction and adduction, at the same joint. A contracted muscle cannot return to its fully extended length on its own; an opposing muscle must stretch it back to its full length.

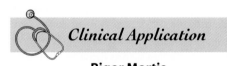

Clinical Application

Rigor Mortis

After death, chemical deterioration in muscle fibers permits Ca^{2+} to leak out of the sarcoplasmic reticulum. The Ca^{2+} binds to troponin and triggers attachment of myosin heads to actin. ATP synthesis has ceased, however, so the myosin heads cannot detach from actin. The resulting condition, in which muscles are in a state of rigidity (cannot contract or stretch), is called **rigor mortis** (rigidity of death). Rigor mortis lasts about 24 hours but disappears as tissues begin to disintegrate. ▪

Muscle Tone

In a skeletal muscle, a small number of motor units are involuntarily activated to produce a sustained contraction of their muscle fibers while the majority of the motor units are not activated and their muscle fibers remain relaxed. This process gives rise to **muscle tone** (*tonos* = tension). In order to sustain muscle tone, other small groups of motor units activate and then become inactive, always in a constantly shifting pattern. Muscle tone keeps skeletal muscles firm, but it does not result in a contraction strong enough to produce movement. For example, when the muscles in the back of the neck are in normal tonic contraction, they keep the head upright and prevent it from slumping forward on the chest, but they do not generate enough force to draw the head backward into hyperextension. Muscle tone is important in smooth muscle such as that found in the gastrointestinal tract, where the walls of the digestive organs maintain a steady pressure on their contents. The tone of smooth muscles surrounding the walls of blood vessels plays a crucial role in maintaining a steady blood pressure.

Clinical Application

Hypotonia and Hypertonia

Hypotonia (*hypo* = below) refers to decreased or lost muscle tone. Such muscles are said to be **flaccid** (FLAK-sid or FLAS-sid). Flaccid muscles are loose and appear flattened rather than rounded; the affected limbs are hyperextended. Certain disorders of the nervous system and electrolyte disturbances (especially sodium and calcium and to a lesser extent magnesium) may result in *flaccid paralysis,* which is characterized by loss of muscle tone, loss or reduction of tendon reflexes, and atrophy (wasting away) and degeneration of muscles.

Hypertonia (*hyper* = above) refers to increased muscle tone and is expressed in two ways: spasticity or rigidity. **Spasticity** is characterized by increased muscle tone (stiffness) associated with an increase in tendon reflexes and pathological reflexes (such as the Babinski sign, in which the great toe extends with or without fanning of the other toes in response to stroking the outer margin of the sole). Certain disorders of the nervous system and electrolyte disturbances such as those previously noted may result in *spastic paralysis,* partial paralysis in which the muscles exhibit spasticity. **Rigidity** refers to increased muscle tone, although reflexes are not affected. ■

Types of Skeletal Muscle Fibers

Skeletal muscle fibers are not all alike in structure or function. For example, they vary in color depending on their content of **myoglobin** (mī-ō-GLŌB-in), a red, oxygen-storing protein. Myoglobin, which has a greater affinity for oxygen than hemoglobin, may serve as an oxygen reservoir and facilitate the transport of oxygen in a muscle fiber. Myoglobin loads and unloads its oxygen very rapidly. Skeletal muscle fibers that have a high myoglobin content are termed *red muscle fibers.* Those that have a low content of myoglobin, on the other hand, are called *white muscle fibers.* Red muscle fibers also have more mitochondria and more blood capillaries than white muscle fibers.

Skeletal muscle fibers contract and relax with different velocities. A fiber is categorized as slow, fast, or intermediate depending on how rapidly it uses ATP. Fast fibers use ATP more quickly. In addition, skeletal muscle fibers vary in the metabolic reactions they use to generate ATP and in how quickly they fatigue. Based on these structural and functional characteristics, skeletal muscle fibers are classified into three types (Table 9.2):

1. *Slow fibers.* These fibers, which are smallest in diameter, contain large amounts of myoglobin and many blood capillaries. They thus look dark red. Slow fibers have the fewest myofibrils among the three types. Because slow fibers generate ATP by aerobic (oxygen-requiring) reactions, they have large numbers of mitochondria, the organelles where the aerobic generation of ATP occurs. They use ATP at a slow rate and, as a result, have a slow contraction velocity but are capable of sustained contractions. These fibers are very resistant to fatigue and are adapted for maintaining posture and endurance-type activities, such as running a marathon.

2. *Fast fibers.* These fibers, which are largest in diameter, contain more myofibrils than slow fibers and thus can generate more powerful contractions. They have a low myoglobin content, relatively few blood capillaries, and appear white in color. Fast fibers generate ATP by anaerobic (oxygen-free) reactions and thus have relatively few mitochondria. Anaerobic reactions cannot supply these fibers continuously with sufficient ATP, and they fatigue quickly. Fast fibers contain large numbers of glycogen granules, which store glucose to provide the fuel for ATP generation. Owing to their large size and their ability to use ATP at a high rate, fast fibers contract strongly and rapidly. They are adapted for intense movements of short duration, such as weight lifting or throwing a ball.

3. *Intermediate fibers.* These fibers are intermediate in diameter between slow and fast fibers. Like slow fibers, they contain large amounts of myoglobin, many blood capillaries, and thus look dark red. Also like slow fibers, they generate ATP by aerobic reactions and have many mitochondria. However, because they use ATP at a rapid rate like fast fibers, their contraction velocity is fast. Intermediate fibers are fatigue-resistant, but not quite as much as slow fibers. They are adapted for activities such as walking and sprinting.

Most skeletal muscles of the body are a mixture of all three types of skeletal muscle fibers, but their proportions vary depending on the usual action of the muscle. For example, postural muscles of the neck, back, and legs have a high proportion of slow fibers. Muscles of the shoulders and arms are not constantly active but are used intermittently, usually for short periods of time, to produce large amounts of tension such as in lifting and throwing. These muscles have a high proportion of fast fibers. Leg muscles not only support the body but are also used for walking and running. Such muscles have large numbers of both slow and intermediate fibers.

Even though most skeletal muscles are a mixture of all three types of skeletal muscle fibers, the skeletal muscle fibers of any one motor unit are all the same type. But the different motor units in a muscle may be used in various

Table 9.2 Comparison of Structural and Functional Characteristics of Skeletal Muscle Fibers

Characteristic	SLOW FIBERS	INTERMEDIATE FIBERS	FAST FIBERS
Fiber diameter	Smallest	Intermediate	Largest
Myoglobin content	Large amount	Large amount	Small amount
Mitochondria	Many	Many	Few
Capillaries	Many	Many	Few
Color	Red	Red	White
Capacity to generate ATP	High capacity by aerobic (oxygen-requiring) reactions	High capacity by aerobic (oxygen-requiring) reactions	Low capacity by anaerobic (nonoxygen-requiring) reactions
Rate of ATP use	Slow	Fast	Fast
Contraction velocity	Slow	Fast	Fast
Resistant to fatigue	High	High, but not as resistant as slow fibers	Low
Location where fibers are abundant	Postural muscles such as those of the neck	Leg muscles	Arm muscles
Primary functions of fibers	Maintaining posture and endurance-type activities	Walking, sprinting	Intense movements of short duration

Slow fiber

Fast fiber

Intermediate fiber

LM 440x

Three types of skeletal muscle fibers in transverse section

ways, depending on need. For example, if only a weak contraction is needed to perform a task, only slow motor units are activated. If a stronger contraction is needed, the motor units of fast fibers are recruited. And if a maximal contraction is required, motor units of intermediate fibers are also called into action. Activation of various motor units is determined in the brain and spinal cord.

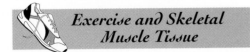

Exercise and Skeletal Muscle Tissue

The relative percentages of fast and slow fibers in each muscle are genetically determined, and the relative proportions of each account for individual differences in physical perfor-

mance among people. For example, a person with a higher proportion of fast fibers would do very well in activities that require periods of intense activity, such as weight lifting. People with higher percentages of slow fibers are better able to perform activities that require endurance, such as running a marathon.

Although the total number of skeletal muscle fibers usually does not change, the characteristics of those present can be altered to some extent. Various types of exercises can induce changes in the fibers in a skeletal muscle. Endurance-type exercises cause a gradual transformation of some fast fibers into intermediate fibers. The transformed muscle fibers show slight increases in diameter, mitochondria, blood

Figure 9.9 Histology of cardiac muscle tissue.

Cardiac muscle tissue is striated and involuntary.

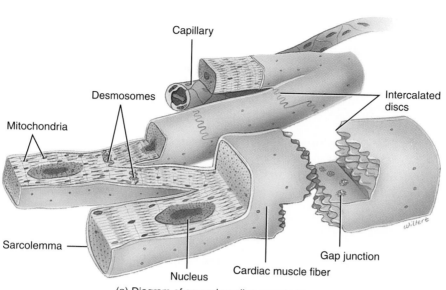

(a) Diagram of several cardiac muscle fibers

capillaries, and strength. Endurance exercises result in cardiovascular and respiratory changes that cause skeletal muscles to receive better supplies of oxygen and nutrients but do not increase muscle mass. On the other hand, exercises that require great strength for short periods of time produce an increase in the size and strength of fast fibers. The increase in size is due to increased synthesis of thin and thick filaments. The overall result is muscle enlargement.

The exact way that muscle grows in size with training is not clear. Some studies show increases in size of individual muscle fibers, whereas others claim increases in the overall number of muscle fibers. However, the number of muscle fibers is not thought to increase significantly after birth. During childhood, the increase in the *size* of muscle fibers appears to be at least partially under the control of human growth hormone, which is produced by the anterior pituitary gland. In males, a further increase in the size of muscle fibers is due to the hormone testosterone, produced by the testes.

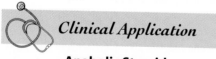

Clinical Application

Anabolic Steroids

Steroids are lipids derived from cholesterol that serve many useful functions in the body (discussed in later chapters). In recent years, the illegal use of **anabolic steroids** by athletes has received widespread attention. These steroids, similar to the male sex hormone testosterone, are taken to increase muscle size and therefore to increase strength and endurance during athletic events.

With an optimal training program, however, there is little evidence that anabolic steroids confer significant additional strength or endurance when taken in *moderate* doses. The *large* doses needed to produce an effect can cause damaging and devastating side effects. These include liver cancer, kidney damage, increased risk of heart disease, stunted growth in young people, increased irritability and aggressive behavior, and wide mood swings. Females may become sterile, develop facial hair, experience deepening of the voice and atrophy of the breasts and uterus, and suffer menstrual irregularities. Males may suffer atrophy of the testes, diminished hormone secretion and sperm production by the testes, and baldness. ■

CARDIAC MUSCLE TISSUE

The principal tissue in the heart wall is **cardiac muscle** tissue. Although it is striated like skeletal muscle, it cannot be controlled voluntarily. Also, certain cardiac muscle fibers (and some smooth muscle fibers and nerve cells in the brain and spinal cord) display autorhythmicity (spontaneous, rhythmical self-excitation) for alternating contraction and relaxation.

In contrast to skeletal muscle fibers, cardiac muscle fibers are shorter in length, larger in diameter, and squarish rather than circular in transverse section. They also exhibit branching, which gives an individual fiber a Y-shaped appearance (Figure 9.9a). The endomysium (areolar connective tissue)

Figure 9.9 (continued)

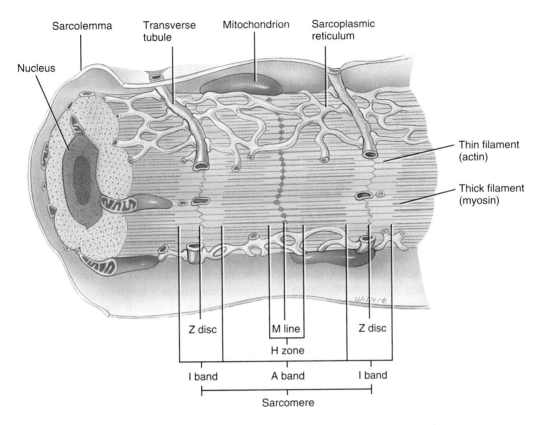

(b) Diagram of several cardiac myofibrils based on an electron micrograph

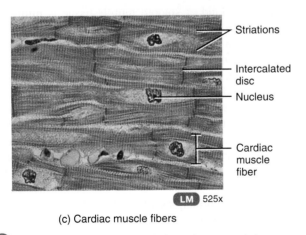

(c) Cardiac muscle fibers

Q *Which structures permit muscle action potentials to spread from one muscle fiber to another?*

and numerous capillaries occupy the spaces between cardiac muscle fibers. A typical cardiac muscle fiber is 50–100 μm long and has a diameter of about 14 μm. Usually there is only one centrally located nucleus, although an occasional cell may have two nuclei. The sarcolemma of cardiac muscle fibers is similar to that of skeletal muscle, but the sarcoplasm is more abundant and the mitochondria are larger and more numerous. Cardiac muscle fibers have the same arrangement of actin and myosin and the same bands, zones, and Z discs as skeletal muscle fibers (Figure 9.9b). The transverse tubules of mammalian cardiac muscle are wider but less abundant than those of skeletal muscle: there is only one transverse tubule per sarcomere, located at the Z disc. Also, the sarcoplasmic reticulum of cardiac muscle fibers is scanty in comparison with the SR of skeletal muscle fibers. As a result, cardiac muscle has a limited intracellular reserve of Ca^{2+}. During contraction a substantial amount of Ca^{2+} enters cardiac muscle fibers from extracellular fluid.

Although cardiac muscle fibers branch and interconnect with each other, they form two separate networks. The muscular walls and the partition of the superior chambers of the heart

14. Describe how the muscular system develops. (p. 243)
15. Describe the effects of aging on muscle tissue. (p. 243)
16. Define the following: ganglion cyst (p. 226), electromyography (p. 229), delayed onset muscle soreness (p. 234), rigor mortis (p. 235), hypotonia (p. 236), flaccidity (p. 236), hyper-tonia (p. 236), spasticity (p. 236), rigidity (p. 236), and anabolic steroids. (p. 238)
17. Refer to the glossary of key medical terms associated with the muscular system. Be sure that you can define each term. (p. 244)

Self Quiz

Complete the following:

1. Red skeletal muscle fibers have a high ___ content.
2. A transverse tubule, along with sarcoplasmic reticulum (terminal cisterns) on either side, is called a ___.
3. Each muscle cell (also known as a ___) is enclosed by a cell membrane called the ___ and contains cytoplasm called ___.
4. To describe a sarcomere during contraction, complete each statement with one of the following:
   L (lengthens)
   S (shortens)
   U (is unchanged in length)
   a. The sarcomere ___.
   b. Each thick myofilament (A band) ___.
   c. Each thin myofilament ___.
   d. The I band ___.
   e. The H zone ___.

Are the following statements true or false?

5. Muscle tone does not produce movement.
6. In adult life, growth of skeletal muscle is due to an increase in the number of muscle cells rather than an increase in the size of existing cells.
7. Elasticity is the property of muscle tissue that allows it to be stretched.
8. The study of motion of the human body is called myology.
9. Intermediate skeletal muscle fibers are adapted for walking and sprinting.
10. Match the following terms with their descriptions:
    (1) cardiac muscle tissue
    (2) skeletal muscle tissue
    (3) smooth muscle tissue
    ___ (a) involuntary muscle found in blood vessels and intestine
    ___ (b) involuntary striated muscle
    ___ (c) striated voluntary muscle attached to bones
    ___ (d) spindle-shaped cells
    ___ (e) intercalated discs
    ___ (f) no transverse tubules
    ___ (g) several nuclei, peripherally arranged

Choose the one best answer to each of the following questions:

11. Which of the following is *not* a part of thin filaments?
    a. actin
    b. myosin
    c. tropomyosin
    d. troponin
    e. myosin-binding site
12. The ability of muscle tissue to respond to a stimulus by producing action potentials is referred to as
    a. contractility
    b. excitability
    c. elasticity
    d. extensibility
    e. conductivity
13. Which of the following is *not* performed by muscles?
    a. motion
    b. regulation of organ volume
    c. maintenance of posture
    d. heat production
    e. production of hormones
14. Which of the following pairs is *not* appropriately matched?
    a. slow fibers, small diameter
    b. fast fibers, low myoglobin content
    c. intermediate fibers, low myoglobin content
    d. fast fibers, large diameter
    e. slow fibers, high myoglobin content
15. Which of the following statements is false?
    a. Cardiac muscle has a relatively long refractory period.
    b. Cardiac muscle usually has one centrally located nucleus per fiber.
    c. Because cardiac muscle cells have less sarcoplasmic reticulum than skeletal muscle cells, they store very little calcium.
    d. Cardiac muscle fibers are separated at their ends by intercalated discs.
    e. Cardiac muscle fibers are spindle-shaped and lack striations.
16. A motor unit is defined as a
    a. nerve and a muscle
    b. single neuron and a single muscle fiber
    c. neuron and all the muscle fibers it stimulates
    d. single muscle fiber and the nerves that innervate it
    e. muscle and the motor and sensory nerves that innervate it

Exhibit 10.1 *Muscles of Facial Expression (Figure 10.4)*

The muscles in this group provide humans with the ability to express a wide variety of emotions, including surprise, fear, and happiness. The muscles themselves lie within the layers of superficial fascia. They usually originate in the fascia or bones of the skull and insert into the skin. Because of their insertions, the muscles of facial expression move the skin rather than a joint when they contract.

Among the noteworthy muscles in this group are those surrounding the orifices (openings) of the head such as the eyes, nose, and mouth. These muscles function as *sphincters* (SFINGK-ters), which close the orifices, and *dilators,* which open the orifices. For example, the **orbicularis oculi** muscle closes the eye, whereas the **levator palpebrae superioris** muscle opens the eye. As another example, the **orbicularis oris** muscle closes the mouth, whereas several other muscles **(zygomaticus major, levator labii superioris, depressor labii inferioris, mentalis,** and **risorius)** radiate out from the lips and open the mouth. The **epicranius** is another unusual muscle in this group because it is made up of two parts: an anterior part called the **frontalis,** which is superficial to the frontal bone, and a posterior part called the **occipitalis,** which is superficial to the occipital bone. The two muscular portions are held together by a strong aponeurosis (sheetlike tendon), the **galea aponeurotica** (GĀ-lē-a ap-ō-noo′-RŌ-ti-ka; *galea* = helmet) or *epicranial aponeurosis,* which covers the superior and lateral surfaces of the skull. The frontalis and occipitalis are treated as separate muscles in this exhibit. The **buccinator** muscle forms the major muscular portion of the cheek. It functions in whistling, blowing, and sucking and also assists in chewing. The duct of the parotid gland (salivary gland) pierces the buccinator muscle to reach the oral cavity. The buccinator muscle is so named because it compresses the cheeks (*bucc* = cheek) during blowing—for example, when a musician plays a wind instrument such as a trumpet. Some trumpeters have stretched their buccinator muscles so much that their cheeks bulge considerably when they blow forcibly.

Clinical Application

Bell's Palsy

Bell's palsy, also known as **facial paralysis,** is a unilateral paralysis of the muscles of facial expression as a result of damage or disease of the facial (VII) nerve. Although the cause is unknown, inflammation of the facial nerve and a relationship to the herpes simplex virus have been suggested. The paralysis causes the entire side of the face to droop in severe cases, and the person cannot wrinkle the forehead, close the eye, or pucker the lips on the affected side. Difficulty in swallowing and drooling also occur. Complete recovery after total paralysis varies from 20–90%. ▪

RELATING MUSCLES TO MOVEMENTS

After you have studied the muscles in this exhibit, arrange them into two groups: those that act on the mouth and those that act on the eyes.

INNERVATION

All muscles are innervated by the facial (VII) nerve, except for the levator palpebrae superioris muscle, which is innervated by the oculomotor (III) nerve.

MUSCLE	ORIGIN	INSERTION	ACTION
Epicranius (ep-i-KRĀ-nē-us; *epi* = over; *crani* = skull)			
Frontalis (fron-TA-lis; *front* = forehead)	Galea aponeurotica.	Skin superior to supraorbital margin.	Draws scalp anteriorly, elevates (raises) eyebrows, and wrinkles skin of forehead horizontally.
Occipitalis (ok-si′-pi-TA-lis; *occipito* = base of skull)	Occipital bone and mastoid process of temporal bone.	Galea aponeurotica.	Draws scalp posteriorly.
Orbicularis oris (or-bi′-kyoo-LAR-is OR-is; *orb* = circular; *or* = mouth)	Muscle fibers surrounding opening of mouth.	Skin at corner of mouth.	Closes lips, compresses lips against teeth, protrudes lips, and shapes lips during speech.

▶

Exhibit 10.1 Muscles of Facial Expression (continued)

MUSCLE	ORIGIN	INSERTION	ACTION
Zygomaticus (zī-gō-MA-ti-kus) **major** (*zygomatic* = cheek bone; *major* = greater)	Zygomatic bone.	Skin at angle of mouth and orbicularis oris.	Draws angle of mouth superiorly and laterally as in smiling or laughing.
Zygomaticus minor	Zygomatic bone.	Upper lip.	Draws angle of mouth superiorly to form nasolabial furrow.
Levator labii superioris (le-VĀ-tor LA-bē-ī soo-per'-ē-OR-is; *levator* = raises or elevates; *labii* = lip; *superioris* = upper)	Superior to infraorbital foramen of maxilla.	Skin at angle of mouth and orbicularis oris.	Elevates upper lip.
Depressor labii inferioris (de-PRE-sor LĀ-bē-ī in-fer'-ē-OR-is; *depressor* = depresses or lowers; *inferioris* = lower)	Mandible.	Skin of lower lip.	Depresses (lowers) lower lip.
Depressor anguli oris (*angul* = angle or corner)	Mandible.	Angle of mouth.	Depresses angle of mouth.
Buccinator (BUK-si-nā'-tor; *bucc* = cheek)	Alveolar processes of maxilla and mandible and pterygomandibular raphe (fibrous band extending from the pterygoid process to the mandible).	Orbicularis oris.	Major cheek muscle; presses cheeks against teeth and lips as in whistling, blowing, and sucking; draws corner of mouth laterally; and assists in mastication (chewing) by keeping food between the teeth (and not between teeth and cheeks).
Mentalis (men-TA-lis; *mentum* = chin)	Mandible.	Skin of chin.	Elevates and protrudes lower lip and pulls skin of chin up as in pouting.
Platysma (pla-TIZ-ma; *platy* = flat, broad)	Fascia over deltoid and pectoralis major muscles.	Mandible, muscles around angle of mouth, and skin of lower face.	Draws outer part of lower lip inferiorly and posteriorly as in pouting; depresses mandible.
Risorius (ri-ZOR-ē-us; *risor* = laughter)	Fascia over parotid (salivary) gland.	Skin at angle of mouth.	Draws angle of mouth laterally as in tenseness.
Orbicularis oculi (or-bi'-kyoo-LAR-is OK-yoo-li; *oculus* = eye)	Medial wall of orbit.	Circular path around orbit.	Closes eye.
Corrugator supercilii (KOR-a-gā'-tor soo-per-SI-lē-i; *corrugo* = wrinkle; *supercilium* = eyebrow)	Medial end of superciliary arch of frontal bone.	Skin of eyebrow.	Draws eyebrow inferiorly as in frowning.
Levator palpebrae superioris (le-VĀ-tor PAL-pe-brē soo-per'-ē-OR-is; *palpebrae* = eyelids) (see also Figure 10.5a)	Roof of orbit (lesser wing of sphenoid bone).	Skin of upper eyelid.	Elevates upper eyelid (opens eye).

Exhibit 10.1 (continued)

Figure 10.4 Muscles of facial expression.

When they contract, muscles of facial expression move the skin rather than a joint.

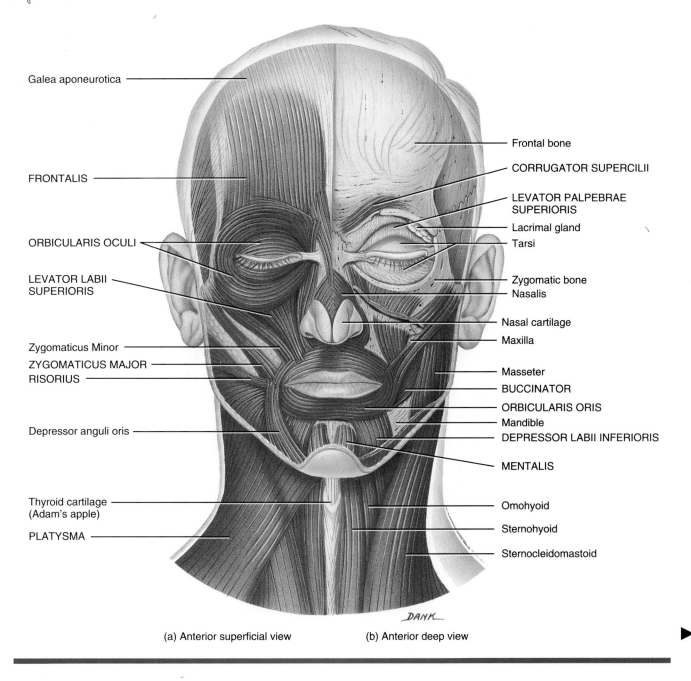

Galea aponeurotica

FRONTALIS

ORBICULARIS OCULI

LEVATOR LABII
SUPERIORIS

Zygomaticus Minor

ZYGOMATICUS MAJOR

RISORIUS

Depressor anguli oris

Thyroid cartilage
(Adam's apple)

PLATYSMA

Frontal bone

CORRUGATOR SUPERCILII

LEVATOR PALPEBRAE
SUPERIORIS

Lacrimal gland

Tarsi

Zygomatic bone

Nasalis

Nasal cartilage

Maxilla

Masseter

BUCCINATOR

ORBICULARIS ORIS

Mandible

DEPRESSOR LABII INFERIORIS

MENTALIS

Omohyoid

Sternohyoid

Sternocleidomastoid

DANK

(a) Anterior superficial view   (b) Anterior deep view

Exhibit 10.1 *Muscles of Facial Expression (continued)*

Figure 10.4 (continued)

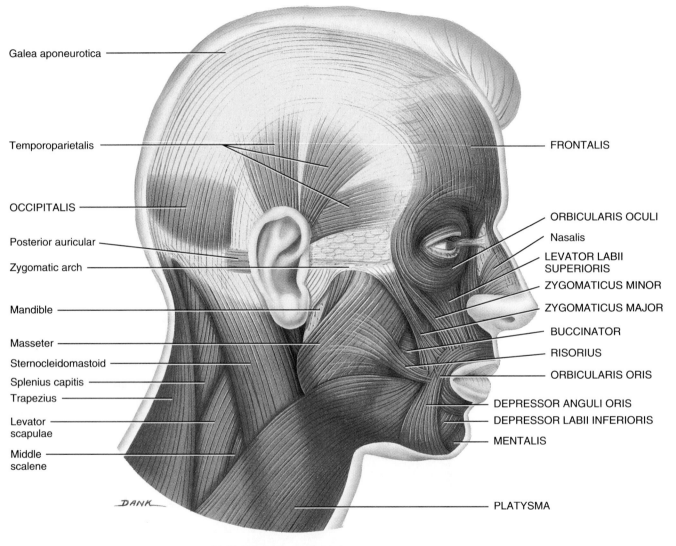

Galea aponeurotica

Temporoparietalis

OCCIPITALIS

Posterior auricular

Zygomatic arch

Mandible

Masseter

Sternocleidomastoid

Splenius capitis

Trapezius

Levator scapulae

Middle scalene

DANK

FRONTALIS

ORBICULARIS OCULI

Nasalis

LEVATOR LABII SUPERIORIS

ZYGOMATICUS MINOR

ZYGOMATICUS MAJOR

BUCCINATOR

RISORIUS

ORBICULARIS ORIS

DEPRESSOR ANGULI ORIS

DEPRESSOR LABII INFERIORIS

MENTALIS

PLATYSMA

(c) Right lateral superficial view

Exhibit 10.1 (*continued*)

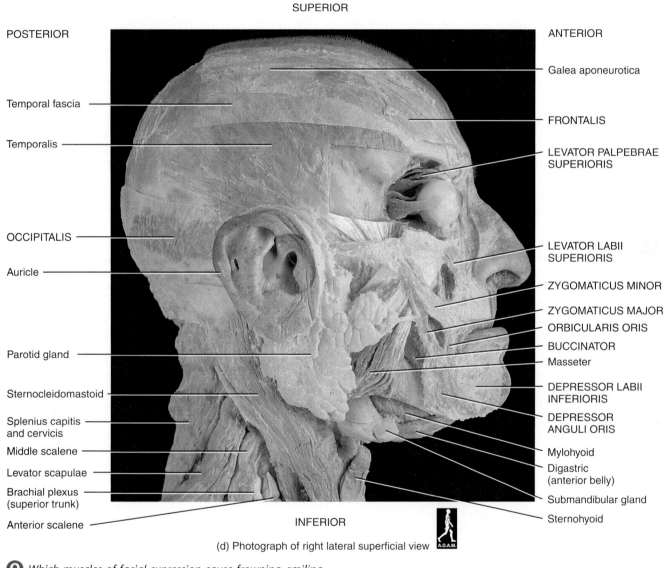

SUPERIOR

POSTERIOR

ANTERIOR

Galea aponeurotica

Temporal fascia

FRONTALIS

Temporalis

LEVATOR PALPEBRAE
SUPERIORIS

OCCIPITALIS

LEVATOR LABII
SUPERIORIS

Auricle

ZYGOMATICUS MINOR

ZYGOMATICUS MAJOR

ORBICULARIS ORIS

BUCCINATOR

Parotid gland

Masseter

Sternocleidomastoid

DEPRESSOR LABII
INFERIORIS

Splenius capitis
and cervicis

DEPRESSOR
ANGULI ORIS

Middle scalene

Mylohyoid

Levator scapulae

Digastric
(anterior belly)

Brachial plexus
(superior trunk)

Submandibular gland

Anterior scalene

Sternohyoid

INFERIOR

(d) Photograph of right lateral superficial view

Q *Which muscles of facial expression cause frowning, smiling, pouting, and squinting?*

Exhibit 10.2 *Muscles that Move the Eyeballs—Extrinsic Muscles (Figure 10.5)*

Two types of muscles are associated with the eyeballs, extrinsic and intrinsic. **Extrinsic muscles** originate outside the eyeballs (in the orbit) and are inserted on the outer surface of the sclera ("white of the eye"). They move the eyeballs in various directions. The extrinsic muscles of the eyeballs are among the fastest contracting and most precisely controlled skeletal muscles in the body. **Intrinsic muscles** originate and insert entirely within the eyeballs. They move structures within the eyeballs.

Movements of the eyeballs are controlled by three pairs of extrinsic muscles: (1) superior and inferior recti, (2) lateral and medial recti, and (3) superior and inferior oblique. The four recti muscles (superior, inferior, lateral, and medial) arise from a tendinous ring in the orbit and insert into the sclera of the eye. Whereas the **superior** and **inferior recti** lie in the same vertical plane, the **medial** and **lateral recti** lie in the same horizontal plane. The actions of the recti muscles can be deduced from their insertions on the sclera. The superior and inferior recti move the eyeballs superiorly and inferiorly, respectively; the lateral and medial recti move the eyeballs laterally and medially, respectively. It should be noted that neither the superior nor the inferior rectus muscle pulls directly parallel to the long axis of the eyeballs, and as a result both muscles also move the eyeballs medially.

It is not as easy to deduce the actions of the oblique muscles (superior and inferior) because of their paths through the orbits. For example, the **superior oblique** muscle originates posteriorly near the tendinous ring and then passes anteriorly and ends in a round tendon, which runs through a pulleylike loop called the *trochlea* in the anterior and medial part of the roof of the orbit. The tendon then turns and inserts on the posterolateral aspect of the eyeballs. Accordingly, the superior oblique muscle moves the eyeballs inferiorly and laterally. The **inferior oblique** muscle originates on the maxilla at the anteromedial aspect of the floor of the orbit. It then passes posteriorly and laterally and inserts on the posterolateral aspect of the eyeballs. Because of this arrangement, the inferior oblique muscle moves the eyeballs superiorly and laterally.

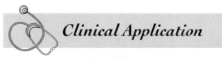

Clinical Application

Strabismus

Strabismus is a condition in which the two eyes are not properly aligned. This could be due to heredity, birth injuries, poor attachments of the muscles, or localized disease. In strabismus, the two eyes do not work as a team. Each eye sends an image to a different area of the brain. This confuses the brain, which usually ignores the messages sent by one of the eyes. As a result, the ignored eye becomes weaker, hence "lazy eye" or *amblyopia* develops. *External strabismus* results when a lesion on the oculomotor nerve (III), which controls the superior, inferior, and medial recti, and inferior oblique muscles of the eye, causes the eyeball to move laterally when at rest, and results in an inability to move the eyeball medially and inferiorly. A lesion in the abducens nerve (VI), which innervates the lateral rectus muscle, results in *internal strabismus,* a condition where the eyeball moves medially when at rest and cannot move laterally.

RELATING MUSCLES TO MOVEMENTS

After you have studied the muscles in this exhibit, arrange them according to their actions on the eyeballs: (1) elevation, (2) depression, (3) abduction, (4) adduction, (5) medial rotation, and (6) lateral rotation. The same muscle may be mentioned more than once.

INNERVATION

The superior rectus, inferior rectus, medial rectus, and inferior oblique muscles are innervated by the oculomotor (III) nerve; the lateral rectus is innervated by the abducens (VI) nerve; and the superior oblique is innervated by the trochlear (IV) nerve.

Exhibit 10.2 (continued)

MUSCLE	ORIGIN	INSERTION	ACTION
Superior rectus	Tendinous ring attached to orbit around optic foramen.	Superior and central part of eyeball.	Moves eyeball superiorly (elevation), medially (adduction), and rotates it medially (clockwise).
Inferior rectus	Same as above.	Inferior and central part of eyeball.	Moves eyeball inferiorly (depression), medially (adduction), and rotates it laterally (counterclockwise).
Lateral rectus	Same as above.	Lateral side of eyeball.	Moves eyeball laterally (abduction).
Medial rectus	Same as above.	Medial side of eyeball.	Moves eyeball medially (adduction).
Superior oblique	Sphenoid bone, superior and medial to the tendinous ring in the orbit.	Eyeball between superior and lateral recti. The muscle inserts into the superior and lateral surfaces of the eyeball via a tendon that passes through a ring of fibrocartilaginous tissue called the trochlea (*trochlea* = pulley).	Moves eyeball inferiorly (depression), laterally (abduction), and rotates it medially (clockwise).
Inferior oblique	Maxilla in floor of orbit.	Eyeball between inferior and lateral recti.	Moves eyeball superiorly (elevation), laterally (abduction), and rotates it laterally (counterclockwise).

Figure 10.5 Extrinsic muscles of the eyeball.

The extrinsic muscles of the eyeball are among the fastest contracting and most precisely controlled skeletal muscles in the body.

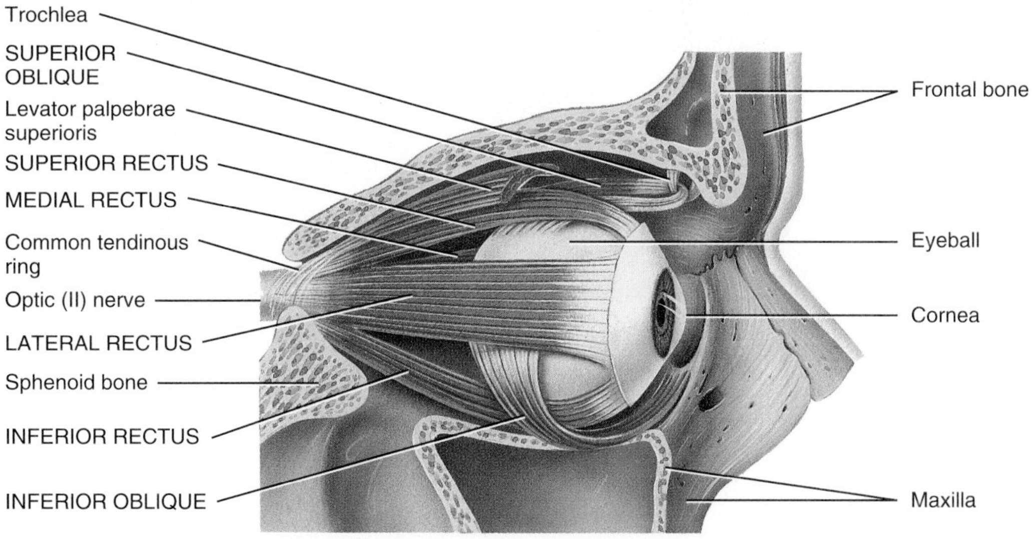

(a) Diagram of lateral view of right eyeball

Exhibit 10.2 *Muscles that Move the Eyeballs—Extrinsic Muscles (continued)*

Figure 10.5 (continued)

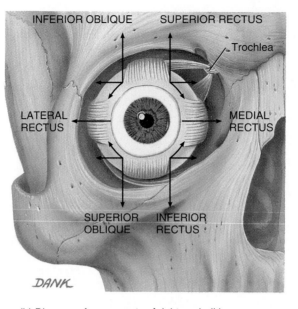

INFERIOR OBLIQUE   SUPERIOR RECTUS

Trochlea

LATERAL RECTUS

MEDIAL RECTUS

SUPERIOR OBLIQUE   INFERIOR RECTUS

DANK

(b) Diagram of movements of right eyeball in response to contraction of its extrinsic muscles

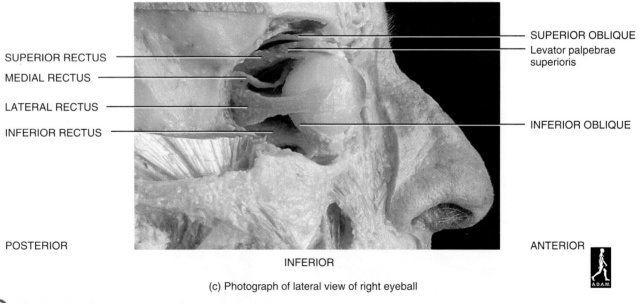

SUPERIOR

SUPERIOR OBLIQUE

Levator palpebrae superioris

SUPERIOR RECTUS

MEDIAL RECTUS

LATERAL RECTUS

INFERIOR RECTUS

INFERIOR OBLIQUE

POSTERIOR

ANTERIOR

INFERIOR

(c) Photograph of lateral view of right eyeball

Q *Why does the inferior oblique muscle move the eyeball superiorly and laterally?*

Exhibit 10.3 *Muscles that Move the Mandible (Lower Jaw) (Figure 10.6)*

The muscles that move the mandible (lower jaw) at the temporomandibular joint (TMJ) are known as the muscles of mastication because they are involved in chewing (mastication). Of the four pairs of muscles involved in mastication, three are powerful closers of the jaw and account for the strength of the bite: **masseter, temporalis,** and **medial pterygoid.** Of these, the masseter is the strongest muscle of mastication. The medial and **lateral pterygoid** muscles assist in mastication by moving the mandible from side to side to help grind food. Additionally, these muscles protrude the mandible.

In 1996, researchers at the University of Maryland reported that they have identified a new muscle in the skull, tentatively named the *sphenomandibularis muscle.* It extends from the sphenoid bone to the mandible. The muscle is believed to be ei-

ther a fifth muscle of mastication or a previously unidentified component of an already identified muscle (temporalis).

RELATING MUSCLES TO MOVEMENTS

After you have studied the muscles in this exhibit, arrange them according to their actions on the mandible: (1) elevation, (2) depression, (3) retraction, (4) protraction, and (5) side-to-side movement. The same muscle may be mentioned more than once.

INNERVATION

All muscles are innervated by the mandibular division at the trigeminal (V) nerve.

MUSCLE	ORIGIN	INSERTION	ACTION
Masseter (MA-se-ter; *maseter* = chewer)	Maxilla and zygomatic arch.	Angle and ramus of mandible.	Elevates mandible, as in closing mouth, and retracts (draws back) mandible.
Temporalis (tem′-por-A-lis; *tempora* = temples)	Temporal and frontal bones.	Coronoid process and ramus of mandible.	Elevates and retracts mandible.
Medial pterygoid (TER-i-goid; *medial* = closer to midline; *pterygoid* = like a wing; pterygoid process of sphenoid bone)	Medial surface of lateral portion of pterygoid process of sphenoid; maxilla.	Angle and ramus of mandible.	Elevates and protracts (protrudes) mandible and moves mandible from side to side.
Lateral pterygoid (TER-i-goid; *lateral* = farther from midline)	Greater wing and lateral surface of lateral portion of pterygoid process of sphenoid.	Condyle of mandible; temporomandibular articulation.	Protracts mandible, depresses mandible as in opening mouth, and moves mandible from side to side.

▶

Exhibit 10.3 *Muscles that Move the Mandible (Lower Jaw) (continued)*

Figure 10.6 Muscles that move the mandible (lower jaw).

The muscles that move the mandible are also known as muscles of mastication.

Parietal bone

TEMPORALIS

Occipital bone

Zygomatic arch

Condylar process of mandible

Coronoid process of mandible

MASSETER

Frontal bone

Nasal bone

Zygomatic bone

Maxilla

Levator labii superioris

Buccinator

Orbicularis oris

Body of mandible

DANK

(a) Right lateral superficial view

Exhibit 10.3   *(continued)*

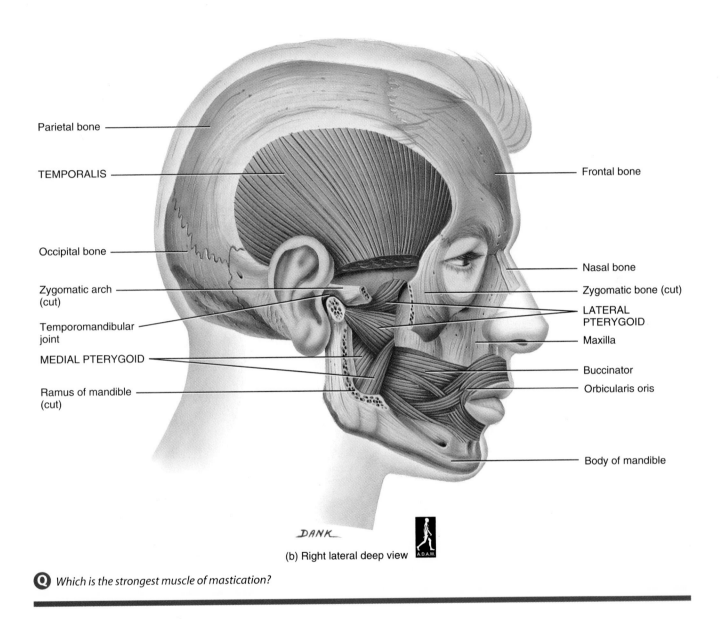

Parietal bone

TEMPORALIS

Occipital bone

Zygomatic arch (cut)

Temporomandibular joint

MEDIAL PTERYGOID

Ramus of mandible (cut)

Frontal bone

Nasal bone

Zygomatic bone (cut)

LATERAL PTERYGOID

Maxilla

Buccinator

Orbicularis oris

Body of mandible

DANK

(b) Right lateral deep view

Q *Which is the strongest muscle of mastication?*

Exhibit 10.4 *Muscles that Move the Tongue—Extrinsic Muscles (Figure 10.7)*

The tongue is a highly mobile structure that is vital to digestive functions such as mastication, perception of taste, and deglutition (swallowing). It is also important in speech. The tongue's mobility is greatly aided by its suspension from the mandible, styloid process of the temporal bone, and hyoid bone.

The tongue is divided into lateral halves by a median fibrous septum. The septum extends throughout the length of the tongue and is attached inferiorly to the hyoid bone. Like the muscles of the eyeballs, muscles of the tongue are of two principal types—extrinsic and intrinsic. **Extrinsic muscles** originate outside the tongue and insert into it. They move the entire tongue in various directions, such as anteriorly, posteriorly, and laterally. **Intrinsic muscles** originate and insert within the tongue. These muscles alter the shape of the tongue rather than moving the entire tongue. The extrinsic and intrinsic muscles of the tongue are arranged in both lateral halves of the tongue.

When you study the extrinsic muscles of the tongue, you will notice that all of the names end in *glossus,* meaning tongue. You will also notice that the actions of the muscles are obvious considering the position of the mandible, styloid process, hyoid bone, and soft palate, which serve as origins for these muscles. For example, the **genioglossus** (originates on the mandible) pulls the tongue downward and forward, the **styloglossus** (originates on the styloid process) pulls the tongue upward and backward, the **hyoglossus** (originates on the hyoid bone) pulls the tongue downward and flattens it, and the **palatoglossus** (originates on the soft palate) raises the back portion of the tongue.

Clinical Application

Intubation During Anesthesia

When general anesthesia is administered during surgery, a total relaxation of the genioglossus muscle results. This will cause the tongue to fall posteriorly and may lead to obstructing the airway to the lungs. To avoid this, the mandible is either manually thrust forward and held in place, or a tube is inserted from the lips through the laryngopharynx (inferior portion of the throat) into the trachea (endotracheal intubation). ■

RELATING MUSCLES TO MOVEMENTS

After you have studied the muscles in this exhibit, arrange them according to the following actions on the tongue: (1) depression, (2) elevation, (3) protraction, and (4) retraction. The same muscle may be mentioned more than once.

INNERVATION

All muscles are innervated by the hypoglossal (XII) nerve, except the palatoglossus, which is innervated by the pharyngeal plexus.

MUSCLE	ORIGIN	INSERTION	ACTION
Genioglossus (jē′-nē-ō-GLOS-us; *geneion* = chin; *glossus* = tongue)	Mandible.	Undersurface of tongue and hyoid bone.	Depresses tongue and thrusts it anteriorly (protraction).
Styloglossus (stī′-lō-GLOS-us; *stylo* = stake or pole; styloid process of temporal bone)	Styloid process of temporal bone.	Side and undersurface of tongue.	Elevates tongue and draws it posteriorly (retraction).
Palatoglossus (pal′-a-tō-GLOS-us; *palato* = palate)	Anterior surface of soft palate.	Side of tongue.	Elevates posterior portion of tongue and draws soft palate down on tongue.
Hyoglossus (hī′-ō-GLOS-us)	Greater horn and body of hyoid bone.	Side of tongue.	Depresses tongue and draws down its sides.

Exhibit 10.4 (continued)

Figure 10.7 Muscles that move the tongue.

The extrinsic and intrinsic muscles of the tongue are arranged in both lateral halves of the tongue.

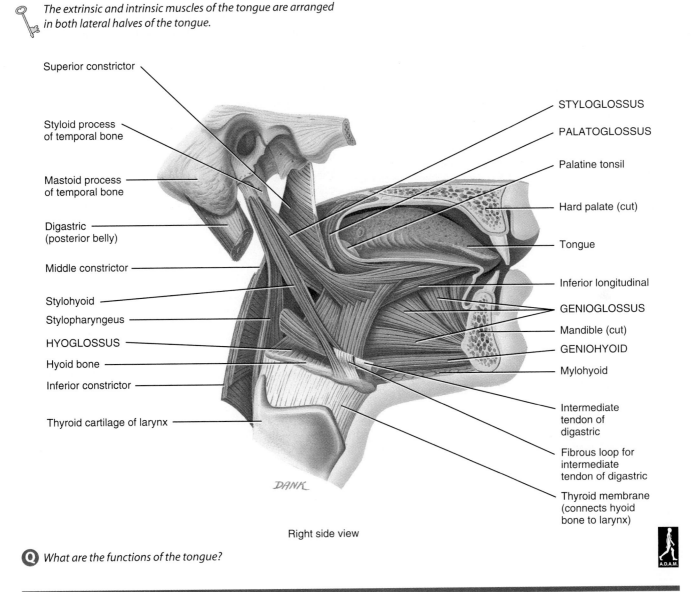

Superior constrictor

Styloid process of temporal bone

Mastoid process of temporal bone

Digastric (posterior belly)

Middle constrictor

Stylohyoid

Stylopharyngeus

HYOGLOSSUS

Hyoid bone

Inferior constrictor

Thyroid cartilage of larynx

DANK

STYLOGLOSSUS

PALATOGLOSSUS

Palatine tonsil

Hard palate (cut)

Tongue

Inferior longitudinal

GENIOGLOSSUS

Mandible (cut)

GENIOHYOID

Mylohyoid

Intermediate tendon of digastric

Fibrous loop for intermediate tendon of digastric

Thyroid membrane (connects hyoid bone to larynx)

Right side view

Q *What are the functions of the tongue?*

Exhibit 10.5 Muscles of the Floor of the Oral Cavity (Mouth) (Figure 10.8)

Two groups of muscles are associated with the anterior aspect of the neck: (1) **suprahyoid muscles,** so called because they are located superior to the hyoid bone, and (2) **infrahyoid muscles,** so named because they lie inferior to the hyoid bone. Acting with the infrahyoid muscles, the suprahyoid muscles fix the hyoid bone, thus serving as a firm base upon which the tongue can move. In this exhibit we will consider the suprahyoid muscles, which are associated with the floor of the oral cavity; Exhibit 10.6 describes the infrahyoid muscles.

As a group, the suprahyoid muscles elevate the hyoid bone, floor of the oral cavity, and tongue during swallowing. As its name suggests, the **digastric** muscle has two bellies, an anterior and a posterior, united by an intermediate tendon that is held in position by a fibrous loop (see Figure 10.7). This muscle elevates the hyoid bone and larynx (voice box) during swallowing and speech and also depresses the mandible. The **stylohyoid** muscle elevates and draws the hyoid bone posteriorly, thus elongating the floor of the oral cavity during swallowing. The **mylohyoid** muscle elevates the hyoid bone and helps press the tongue against the roof of the oral cavity during swallowing to move food from the oral cavity into the throat. The **geniohyoid** muscle (see Figure 10.7) elevates and draws the hyoid bone anteriorly to shorten the floor of the oral cavity and to widen the throat to receive food that is being swallowed. It also depresses the mandible.

RELATING MUSCLES TO MOVEMENTS

After you have studied the muscles in this exhibit, arrange them according to the following actions on the hyoid bone: (1) elevation, (2) drawing it anteriorly, and (3) drawing it posteriorly. The same muscle may be mentioned more than once.

INNERVATION

The anterior belly of the digastric and the mylohyoid are innervated by the mandibular division of the trigeminal (V) nerve; the posterior belly of the digastric and the stylohyoid are innervated by the facial (VII) nerve; and the geniohyoid is innervated by the first cervical nerve.

MUSCLE	ORIGIN	INSERTION	ACTION
Digastric (dī′-GAS-trik; *di* = two; *gaster* = belly)	Anterior belly from inner side of inferior border of mandible; posterior belly from mastoid process of temporal bone.	Body of hyoid bone via an intermediate tendon.	Elevates hyoid bone and depresses mandible as in opening the mouth.
Stylohyoid (stī′-lō-HĪ-oid; *stylo* = stake or pole, styloid process of temporal bone; *hyoedes* = U-shaped, pertaining to hyoid bone)	Styloid process of temporal bone.	Body of hyoid bone.	Elevates hyoid bone and draws it posteriorly.
Mylohyoid (mī′-lō-HĪ-oid)	Inner surface of mandible.	Body of hyoid bone.	Elevates hyoid bone and floor of mouth and depresses mandible.
Geniohyoid (jē′-nē-ō-HĪ-oid; *geneion* = chin) (see also Figure 10.7)	Inner surface of mandible.	Body of hyoid bone.	Elevates hyoid bone, draws hyoid bone and tongue anteriorly, and depresses mandible.

Exhibit 10.5 (continued)

Figure 10.8 Muscles of the floor of the oral cavity and front of the neck.

The suprahyoid muscles elevate the hyoid bone, floor of the oral cavity, and tongue during swallowing.

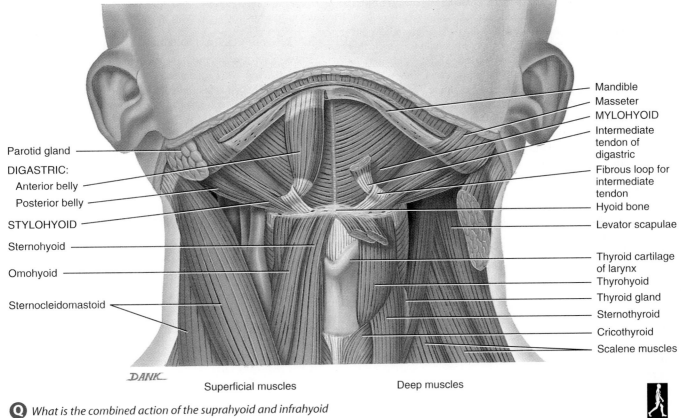

Superficial muscles   Deep muscles

Q *What is the combined action of the suprahyoid and infrahyoid muscles?*

Exhibit 10.6 *Muscles of the Larynx (Voice Box) (Figure 10.9)*

The muscles of the larynx (voice box), like those of the eyeballs and tongue, are grouped into **extrinsic and intrinsic muscles.** The extrinsic muscles of the larynx, which are associated with the anterior aspect of the neck, are called **infrahyoid muscles** because they lie inferior to the hyoid bone. (These muscles are sometimes called "strap" muscles because of their ribbonlike appearance.) As a group, the extrinsic muscles depress the hyoid bone and larynx during swallowing and speech.

The **omohyoid** muscle, like the digastric muscle, is composed of two bellies connected by an intermediate tendon. In this case, however, the two bellies are referred to as *superior* and *inferior,* rather than anterior and posterior. Together, the omohyoid, **sternohyoid,** and **thyrohyoid** muscles depress the hyoid bone. In addition, the **sternothyroid** muscle depresses the thyroid cartilage (Adam's apple) of the larynx, whereas the thyrohyoid muscle elevates the thyroid cartilage. These actions are necessary during the production of low and high tones, respectively, during phonation.

Whereas the extrinsic muscles move the larynx as a whole, the intrinsic muscles only move parts of the larynx. Based upon their actions, the intrinsic muscles may be grouped into three functional sets according to their actions. The first set includes the **cricothyroid** and **thyroarytenoid** muscles, which regulate the tension of the vocal folds. The second set varies the size of the *rima glottidis* (space between the vocal folds). These muscles in-clude the **lateral cricoarytenoid,** which brings the vocal folds together (adduction), thus closing the rima glottidis, and the **posterior cricoarytenoid,** which moves the vocal folds apart (abduction), thus opening the rima glottidis. The **transverse arytenoid** closes the posterior portion of the rima glottidis. The last intrinsic muscle functions as a sphincter to control the size of the inlet of the larynx, which is the opening anteriorly from the pharynx (throat) into the larynx. This muscle is the **oblique arytenoid.**

RELATING MUSCLES TO MOVEMENTS

After you have studied the muscles in this exhibit, arrange them according to the following actions on the thyroid cartilage: (1) depression and (2) elevation and the following actions on the vocal cords: (1) increasing tension, (2) moving them apart, and (3) moving them together. The same muscle may be mentioned more than once.

INNERVATION

The omohyoid, sternohyoid, and sternothyroid are innervated by branches of the ansa cervicalis (C1–C3); the thyrohyoid is innervated by branches of the ansa cervicalis (C1–C2) and the descending hypoglossal (XII) nerve; and the intrinsic muscles are innervated by the recurrent laryngeal branch of the vagus (X) nerve.

Exhibit 10.6 *(continued)*

MUSCLE	ORIGIN	INSERTION	ACTION
Extrinsic Muscles of the Larynx (Infrahyoid Muscles)			
Omohyoid (ō-mō-HĪ-oid; *omo* = relationship to the shoulder; *hyoedes* = U-shaped, pertaining to hyoid bone)	Superior border of scapula and superior transverse ligament.	Body of hyoid bone.	Depresses hyoid bone.
Sternohyoid ster′-nō-HĪ-oid; *sterno* = sternum)	Medial end of clavicle and manubrium of sternum.	Body of hyoid bone.	Depresses hyoid bone.
Sternothyroid ster′-nō-THĪ-roid; *thyro* = thyroid gland)	Manubrium of sternum.	Thyroid cartilage of larynx.	Depresses thyroid cartilage of larynx.
Thyrohyoid (thī′-rō-HĪ-oid)	Thyroid cartilage of larynx.	Greater horn of hyoid bone.	Elevates thyroid cartilage and depresses hyoid bone.
Intrinsic Muscles of the Larynx			
Cricothyroid (kri-kō-THĪ-roid, *crico* = cricoid cartilage of larynx)	Anterior and lateral portion of cricoid cartilage of larynx.	Anterior border of thyroid cartilage of larynx and posterior part of inferior border of thyroid cartilage of larynx.	Elongates and places tension on vocal folds.
Thyroarytenoid (thī′-rō-ar′-i-TĒ-noid)	Inferior portion of thyroid cartilage of larynx and middle of cricothyroid ligament.	Base and anterior surface of arytenoid cartilage of larynx.	Shortens and relaxes vocal folds.
Lateral cricoarytenoid (kri′-kō-ar′-i-TĒ-noid)	Superior border of cricoid cartilage of larynx.	Anterior surface of arytenoid cartilage of larynx.	Brings vocal folds together (adduction), thus closing the rima glottidis.
Posterior cricoarytenoid (kri′-kō-ar′-i-TĒ-noid; *arytaina* = shaped like a jug)	Posterior surface of cricoid cartilage of larynx.	Posterior surface of arytenoid cartilage of larynx.	Moves the vocal folds apart (abduction), thus opening the rima glottidis.
Transverse and **Oblique arytenoid** (ar′-i-TĒ-noid)	Posterior surface and lateral border of one arytenoid cartilage of larynx.	Corresponding parts of opposite arytenoid cartilage of larynx.	The transverse arytenoid closes the posterior portion of the rima glottidis; the oblique arytenoid regulates the size of the inlet of the larynx.

▶

Exhibit 10.6 *Muscles of the Larynx (Voice Box) (continued)*

Figure 10.9 Muscles of the larynx (voice box).

Intrinsic muscles of the larynx adjust the tension of the vocal folds and open or close the rima glottidis.

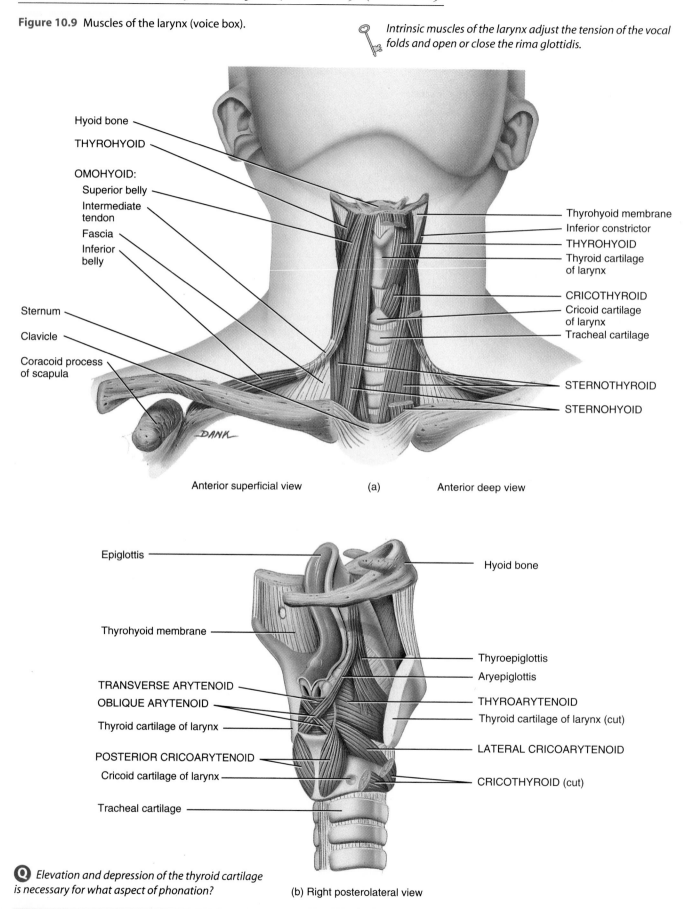

Hyoid bone
THYROHYOID

OMOHYOID:
Superior belly
Intermediate tendon
Fascia
Inferior belly

Sternum
Clavicle
Coracoid process of scapula

Thyrohyoid membrane
Inferior constrictor
THYROHYOID
Thyroid cartilage of larynx
CRICOTHYROID
Cricoid cartilage of larynx
Tracheal cartilage
STERNOTHYROID
STERNOHYOID

DANK

Anterior superficial view   (a)   Anterior deep view

Epiglottis
Hyoid bone

Thyrohyoid membrane

Thyroepiglottis
Aryepiglottis

TRANSVERSE ARYTENOID
OBLIQUE ARYTENOID
Thyroid cartilage of larynx

THYROARYTENOID
Thyroid cartilage of larynx (cut)

POSTERIOR CRICOARYTENOID
Cricoid cartilage of larynx

LATERAL CRICOARYTENOID

CRICOTHYROID (cut)

Tracheal cartilage

Q *Elevation and depression of the thyroid cartilage is necessary for what aspect of phonation?*

(b) Right posterolateral view

Exhibit 10.7 Muscles of the Pharynx (Throat) (Figure 10.10)

The pharynx (throat) is a somewhat funnel-shaped tube posterior to the nasal and oral cavities. It is a common chamber for the respiratory and digestive systems, opening into the anterior larynx and posterior esophagus.

The muscles of the pharynx are arranged in two layers, an outer circular layer and an inner longitudinal layer. The **circular layer** is composed of three constrictor muscles, each overlapping the muscle above it, an arrangement that resembles stacked flowerpots. Their names are the inferior, middle, and superior constrictor muscles. The **longitudinal layer** is composed of three muscles that descend from the styloid process of the temporal bone, auditory (Eustachian) tube, and soft palate. Their names are the stylopharyngeus, salpingopharyngeus, and palatopharyngeus, respectively.

The **inferior constrictor** muscle is the thickest of the constrictor muscles. Its inferior fibers are continuous with the musculature of the esophagus, whereas its superior fibers overlap the middle constrictor. The **middle constrictor** is fan-shaped and smaller than the inferior constrictor, and it overlaps the superior constrictor. The **superior constrictor** is quadrilateral and thinner than the other constrictors. As a group, the constrictor muscles constrict the pharynx during swallowing. The sequential contraction of the muscles moves food and drink from the mouth into the esophagus.

The **stylopharyngeus** muscle is a long, thin muscle that enters the pharynx between the middle and superior constrictors and inserts along with the palatopharyngeus. The **salpingopharyngeus** muscle is a thin muscle that descends in the lateral wall of the pharynx and also inserts along with the palatopharyngeus muscle. The **palatopharyngeus** muscle also descends in the lateral wall of the pharynx and inserts along with the stylopharyngeus. All three muscles of the longitudinal layer elevate the pharynx and larynx during swallowing and speaking. Elevation of the pharynx widens it to receive food and liquids, and elevation of the larynx causes a structure called the epiglottis to close over the rima glottidis (space between the vocal cords) and seal the respiratory passageway. Food and drink are further kept out of the respiratory tract by the suprahyoid muscles of the larynx, which elevate the hyoid bone, and thus the larynx, which is attached to the hyoid bone. Additionally, the respiratory passageway is sealed by the action of the intrinsic muscles of the larynx, which bring the vocal folds together to close off the rima glottidis. After swallowing, the infrahyoid muscles of the larynx depress the hyoid bone and larynx.

RELATING MUSCLES TO MOVEMENTS

After you have studied the muscles in this exhibit, arrange them according to the following actions on the pharynx: (1) constriction and (2) elevation, and according to the following action on the larynx: elevation. The same muscle may be mentioned more than once.

INNERVATION

All muscles, except the stylopharyngeus, are innervated by the pharyngeal plexus, branches of the vagus (X) nerve. The stylopharyngeus is innervated by the glossopharyngeal (IX) nerve.

MUSCLE	ORIGIN	INSERTION	ACTION
Circular Layer			
Inferior constrictor (*inferior* = below; *constrictor* = decreases diameter of a lumen)	Cricoid and thyroid cartilages of larynx.	Posterior median raphe (slender band of collagen fibers) of pharynx.	Constricts inferior portion of pharynx to propel food and drink into esophagus.
Middle constrictor	Greater and lesser horns of hyoid bone and stylohyoid ligament.	Posterior median raphe of pharynx.	Constricts middle portion of pharynx to propel food and drink into esophagus.
Superior constrictor (*superior* = above)	Pterygoid process of sphenoid, pterygomandibular raphe, and medial surface of mandible.	Posterior median raphe of pharynx.	Constricts superior portion of pharynx to propel food and drink into esophagus.
Longitudinal layer			
Stylopharyngeus (stī′-lō-far-IN-jē-us; *stylo* = stake or pole; styloid process of temporal bone; *pharyngo* = pharynx) (see also Figure 10.7)	Styloid process of temporal bone.	Thyroid cartilage with the palatopharyngeus.	Elevates larynx and pharynx.
Salpingopharyngeus (sal-pin′-gō-far-IN-jē-us; *salping* = pertaining to the auditory or uterine tube)	Inferior portion of auditory (Eustachian) tube.	Thyroid cartilage with the palatopharyngeus.	Elevates larynx and pharynx and opens orifice of auditory (Eustachian) tube.
Palatopharyngeus (pal′-a-tō-far-IN-jē-us; *palato* = palate)	Soft palate.	Thyroid cartilage with the stylopharyngeus.	Elevates larynx and pharynx and helps close nasopharynx during swallowing.

▶

Exhibit 10.7 *Muscles of the Pharynx (Throat) (continued)*

Figure 10.10 Muscles of the pharynx.

The muscles of the pharynx assist in swallowing and in closing off the respiratory passageway.

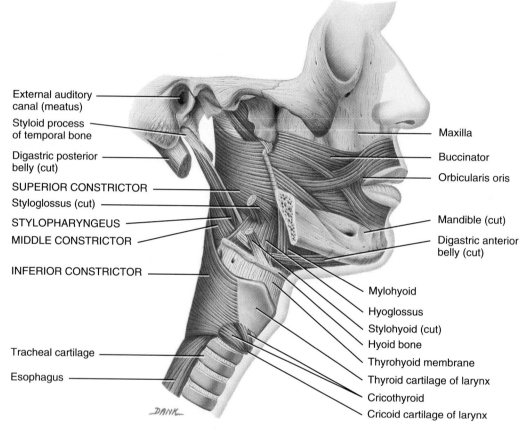

External auditory canal (meatus)
Styloid process of temporal bone
Digastric posterior belly (cut)
SUPERIOR CONSTRICTOR
Styloglossus (cut)
STYLOPHARYNGEUS
MIDDLE CONSTRICTOR
INFERIOR CONSTRICTOR
Tracheal cartilage
Esophagus

Maxilla
Buccinator
Orbicularis oris
Mandible (cut)
Digastric anterior belly (cut)
Mylohyoid
Hyoglossus
Stylohyoid (cut)
Hyoid bone
Thyrohyoid membrane
Thyroid cartilage of larynx
Cricothyroid
Cricoid cartilage of larynx

DANK

(a) Right lateral view

Exhibit 10.7 *(continued)*

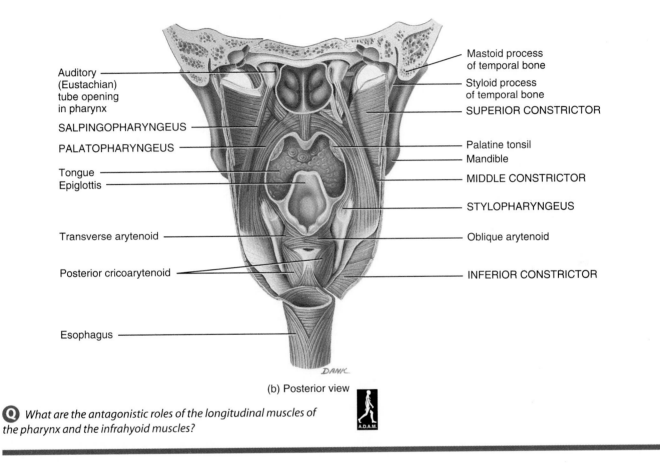

Auditory (Eustachian) tube opening in pharynx

SALPINGOPHARYNGEUS

PALATOPHARYNGEUS

Tongue
Epiglottis

Transverse arytenoid

Posterior cricoarytenoid

Esophagus

Mastoid process of temporal bone

Styloid process of temporal bone

SUPERIOR CONSTRICTOR

Palatine tonsil
Mandible

MIDDLE CONSTRICTOR

STYLOPHARYNGEUS

Oblique arytenoid

INFERIOR CONSTRICTOR

(b) Posterior view

Q *What are the antagonistic roles of the longitudinal muscles of the pharynx and the infrahyoid muscles?*

Exhibit 10.8 *Muscles that Move the Head*

The head is attached to the vertebral column at the atlanto-occipital joint formed by the atlas and occipital bone. Balance and movement of the head on the vertebral column involves the action of several neck muscles. For example, contraction of the two **sternocleidomastoid** muscles together (bilaterally) flexes the cervical portion of the vertebral column and head. Acting singly (unilaterally), the sternocleidomastoid muscle laterally flexes and rotates the head. Bilateral contraction of the **semispinalis capitis, splenius capitis,** and **longissimus capitis** muscles extends the head. However, when these same muscles contract unilaterally, their actions are quite different, involving primarily rotation of the head.

RELATING MUSCLES TO MOVEMENT

After you have studied the muscles in this exhibit, arrange them according to the following actions on the head: (1) flexion, (2) lateral flexion, (3) extension, (4) rotation to side opposite contracting muscle, and (5) rotation to same side as contracting muscle. The same muscle may be mentioned more than once.

INNERVATION

The sternocleidomastoid is innervated by the accessory (XI) nerve; the semispinalis capitis is innervated by the dorsal rami of cervical nerves; and the splenius capitis and longissimus capitis are innervated by the dorsal rami of middle and inferior cervical nerves.

MUSCLE	ORIGIN	INSERTION	ACTION
Sternocleidomastoid (ster'-nō-klī'-dō-MAS-toid; *sternum* = breastbone; *cleido* = clavicle; *mastoid* = mastoid process of temporal bone) (see Figure 10.15c)	Sternum and clavicle.	Mastoid process of temporal bone.	Acting together (bilaterally), flex cervical portion of vertebral column and head; acting singly (unilaterally) laterally flex and rotate head to side opposite contracting muscle.
Semispinalis capitis (se'-mē-spi-NA-lis KAP-i-tis; *semi* = half; *spine* = spinous process; *caput* = head) (see Figure 10.20a)	Transverse processes of first six or seven thoracic vertebrae and seventh cervical vertebra and articular processes of fourth, fifth, and sixth cervical vertebrae.	Occipital bone between superior and inferior nuchal lines.	Acting together, extend head; acting singly, rotate head to side opposite contracting muscle.
Splenius capitis (SPLĒ-nē-us KAP-i-tis; *splenion* = bandage) (see Figure 10.20a)	Ligamentum nuchae and spinous processes of seventh cervical vertebra and first three or four thoracic vertebrae.	Occipital bone and mastoid process of temporal bone.	Acting together, extend head; acting singly, laterally flex and rotate head to same side as contracting muscle.
Longissimus capitis (lon-JIS-i-mus KAP-i-tis; *longissimus* = longest) (see Figure 10.20a)	Transverse processes of upper four thoracic vertebrae and articular processes of last four cervical vertebrae.	Mastoid process of temporal bone.	Acting together, extend head; acting singly, laterally flex and rotate head to same side as contracting muscle.

Exhibit 10.9 *Muscles that Act on the Abdominal Wall (Figure 10.11)*

The anterolateral abdominal wall is composed of skin, fascia, and four pairs of muscles: the external oblique, internal oblique, transversus abdominis, and rectus abdominis. The first three muscles are flat muscles; the last is a straplike vertical muscle. The **external oblique** is the external flat muscle with its fibers directed inferiorly and medially. The **internal oblique** is the intermediate flat muscle with its fibers directed at right angles to those of the external oblique. The **transversus abdominis** is the deepest of the flat muscles, with most of its fibers directed horizontally around the abdominal wall. Together, the external oblique, internal oblique, and transversus abdominus form three layers of muscle around the abdomen. The muscle fibers of each layer run cross-directionally to one another, a structural arrangement that affords considerable protection to the abdominal viscera, especially when the muscles have good tone.

The **rectus abdominis** muscle is a long flat muscle that extends the entire length of the anterior abdominal wall, from the pubic crest and pubic symphysis to the cartilages of ribs 5–7 and the xiphoid process of the sternum. The anterior surface of the muscle is interrupted by three transverse fibrous bands of tissue called **tendinous intersections,** believed to be remnants of septa that separated myotomes during embryological development.

As a group the muscles of the anterolateral abdominal wall help contain and protect the abdominal viscera; flex, laterally flex, and rotate the vertebral column at the intervertebral joints; compress the abdomen during forced expiration; and produce the force required for defecation, urination, and childbirth.

The aponeuroses of the external oblique, internal oblique, and transversus abdominis muscles form the **rectus sheath,** which encloses the rectus abdominis muscles and meet at the midline to form the **linea alba** (white line), a tough, fibrous band that extends from the xiphoid process of the sternum to the pubic symphysis. In the latter stages of pregnancy, the linea alba stretches to increase the distance between the rectus abdominis muscles. The inferior free border of the external oblique aponeurosis, plus some collagen fibers, forms the **inguinal ligament,** which runs from the anterior superior iliac spine to the pubic tubercle (see Figure 10.21a). Just superior to the medial end of the inguinal ligament is a triangular slit in the aponeurosis referred to as the **superficial inguinal ring,** the outer opening of the **inguinal canal** (see Figure 26.2). The canal contains the spermatic cord and ilioinguinal nerve in males and round ligament of the uterus and ilioinguinal nerve in females.

The posterior abdominal wall is formed by the lumbar vertebrae, parts of the ilia of the hipbones, psoas major and iliacus muscles (described in Exhibit 10.19), and quadratus lumborum muscle. Whereas the anterolateral abdominal wall can contract and distend, the posterior abdominal wall is bulky and stable by comparison.

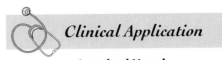

Clinical Application

Inguinal Hernia

The inguinal region is a weak area in the abdominal wall. It is often the site of an **inguinal hernia,** a rupture or separation of a portion of the inguinal area of the abdominal wall resulting in the protrusion of a part of the small intestine. These conditions are much more common in males than in females because the inguinal canals in males are larger and represent especially weak points in the abdominal wall.

RELATING MUSCLES TO MOVEMENTS

After you have studied the muscles in this exhibit, arrange them according to the following actions on the vertebral column: (1) flexion, (2) lateral flexion, (3) extension, and (4) rotation. The same muscle may be mentioned more than once.

INNERVATION

The rectus abdominis is innervated by branches of thoracic nerves T7–T12; the external oblique is innervated by branches of thoracic nerves T7–T12 and the iliohypogastric nerve; the internal oblique and transversus abdominis are innervated by branches of thoracic nerves T8–T12 and the iliohypogastric and ilioinguinal nerves; the quadratus lumborum is innervated by branches of thoracic nerve T12 and lumbar nerves L1–L3 or L1–L4.

▶

Exhibit 10.9 *Muscles that Act on the Abdominal Wall (continued)*

MUSCLE	ORIGIN	INSERTION	ACTION
Rectus abdominis (REK-tus-ab-DOM-in-us; *rectus* = fibers parallel to midline; *abdomino* = abdomen)	Pubic crest and pubic symphysis.	Cartilage of fifth to seventh ribs and xiphoid process.	Flexes vertebral column, especially lumbar portion, and compresses abdomen to aid in defecation, urination, forced expiration, and parturition (childbirth).
External oblique (ō-BLĒK; *external* = closer to surface; *oblique* = fibers diagonal to midline)	Inferior eight ribs.	Iliac crest and linea alba.	Acting together (bilaterally), compress abdomen and flex vertebral column; acting singly (unilaterally), laterally flex vertebral column, especially lumbar portion, and rotate vertebral column.
Internal oblique (ō-BLĒK; *internal* = farther from surface)	Iliac crest, inguinal ligament, and thoracolumbar fascia.	Cartilage of last three or four ribs and linea alba.	Acting together, compress abdomen and flex vertebral column; acting singly, laterally flex vertebral column, especially lumbar portion, and rotate vertebral column.
Transversus abdominis (tranz-VER-sus ab-DOM-in-us; *transverse* = fibers perpendicular to midline)	Iliac crest, inguinal ligament, lumbar fascia, and cartilages of inferior six ribs.	Xiphoid process, linea alba, and pubis.	Compresses abdomen.
Quadratus lumborum (kwod-RĀ-tus lum-BOR-um; *quad* = four; *lumbo* = lumbar region) (see Figure 10.12)	Iliac crest and iliolumbar ligament.	Inferior border of twelfth rib and transverse processes of first four lumbar vertebrae.	Acting together, pull twelfth ribs inferiorly during forced expiration, fix twelfth ribs to prevent their elevation during deep inspiration, and help extend lumbar portion of vertebral column; acting singly, laterally flex vertebral column, especially lumbar portion.

Exhibit 10.9 (continued)

Figure 10.11 Muscles of the male anterolateral abdominal wall.

The anterolateral abdominal muscles protect the abdominal viscera, move the vertebral column, and assist in forced expiration, defecation, urination, and childbirth.

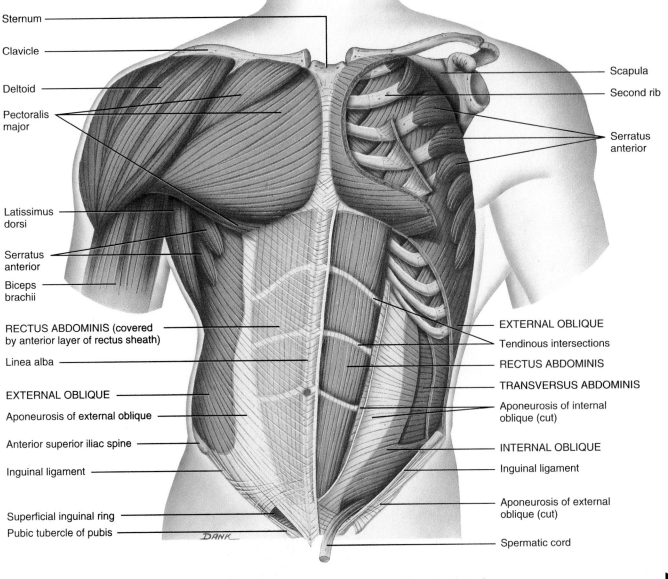

(a) Diagram of anterior superficial view   (b) Diagram of anterior deep view

Exhibit 10.9 *Muscles that Act on the Abdominal Wall (continued)*

Figure 10.11 (continued)

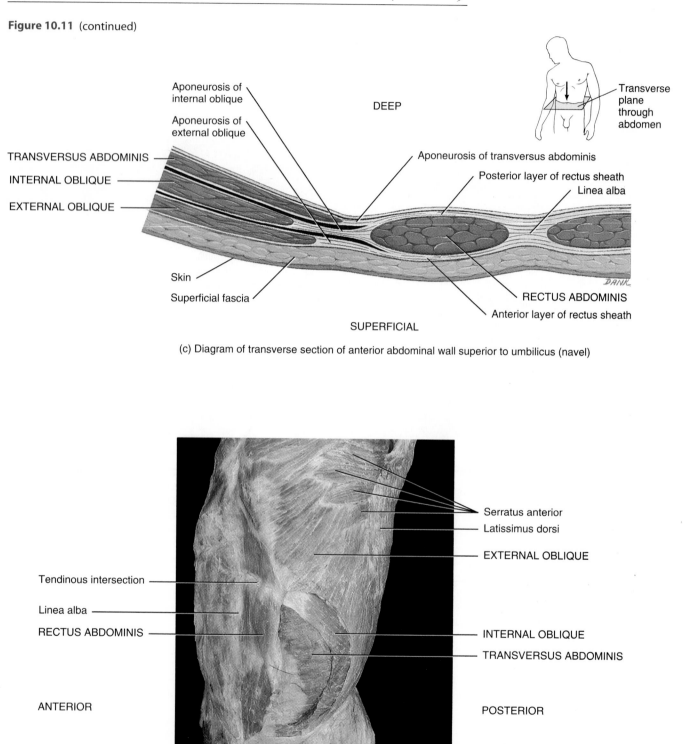

Aponeurosis of internal oblique

Aponeurosis of external oblique

DEEP

Transverse plane through abdomen

TRANSVERSUS ABDOMINIS

INTERNAL OBLIQUE

EXTERNAL OBLIQUE

Aponeurosis of transversus abdominis

Posterior layer of rectus sheath

Linea alba

Skin

Superficial fascia

RECTUS ABDOMINIS

Anterior layer of rectus sheath

SUPERFICIAL

(c) Diagram of transverse section of anterior abdominal wall superior to umbilicus (navel)

Serratus anterior

Latissimus dorsi

EXTERNAL OBLIQUE

Tendinous intersection

Linea alba

RECTUS ABDOMINIS

INTERNAL OBLIQUE

TRANSVERSUS ABDOMINIS

ANTERIOR

POSTERIOR

INFERIOR

(d) Photograph of right anterolateral view

Q *Which abdominal muscle aids in urination?*

Exhibit 10.10 *Muscles Used in Ventilation (Breathing) (Figure 10.12)*

The muscles described here alter the size of the thoracic cavity so that ventilation (breathing) can occur. Ventilation consists of two phases, inspiration (inhalation) and expiration (exhalation). Essentially, inspiration occurs when the thoracic cavity increases in size, and expiration occurs when the thoracic cavity decreases in size.

The **diaphragm** is the most important muscle in ventilation. It is a dome-shaped musculotendinous partition that separates the thoracic and abdominal cavities. Although the diaphragm descends during inspiration, only its central dome region moves downward because the peripheral portions are attached to the sternum, costal cartilages, and lumbar vertebrae. The diaphragm is composed of two parts: a peripheral muscular portion and a central portion called the central tendon. The **central tendon** is a strong aponeurosis that serves as the tendon of insertion for all the peripheral muscular fibers of the diaphragm. It fuses with the inferior surface of the fibrous pericardium, the external covering of the heart, and the parietal pleurae, the external coverings of the lungs.

In addition to its function in inspiration, movements of the diaphragm help to return to the heart venous blood passing through the abdomen. Together with the anterolateral abdominal muscles, the diaphragm helps to increase intra-abdominal pressure to evacuate the pelvic contents during defecation, urination, and childbirth. This mechanism is further assisted when you take a deep breath and close the rima glottidis. The trapped air in the respiratory system prevents the diaphragm from elevating. The increase in intra-abdominal pressure as just described will also help support the vertebral column and prevent flexion during weight lifting. This greatly assists the back muscles in lifting a heavy weight.

The diaphragm has three major openings through which various structures pass between the thorax and abdomen. These structures include the aorta along with the thoracic duct and azygos vein, which pass through the **aortic hiatus;** the esophagus with accompanying vagus (X) nerves, which pass through the **esophageal hiatus;** and the inferior vena cava, which passes through the **foramen for the vena cava.** In a condition called a hiatus hernia, the stomach protrudes superiorly through the esophageal hiatus.

The other muscles involved in ventilation are called **intercostal** muscles and occupy the intercostal spaces, the spaces between ribs. These muscles are arranged in three layers. The 11 **external intercostal** muscles occupy the superficial layer, and their fibers run obliquely inferiorly and anteriorly from the rib above to the rib below. Their role in respiration is to elevate the ribs during inspiration to help expand the thoracic cavity. The **internal intercostal** muscles, also 11 in number, occupy the intermediate layer of the intercostal spaces. The fibers of these muscles run obliquely inferiorly and posteriorly from the inferior border of the rib above to the superior border of the rib below. Their function is to draw adjacent ribs together during forced expiration to help decrease the size of the thoracic cavity. The deepest muscle layer is made up of the **transversus thoracis** muscles. These poorly developed muscles (not described in the exhibit) run in the same direction as the internal intercostals, and they may have the same role.

RELATING MUSCLES TO MOVEMENTS

After you have studied the muscles in this exhibit, arrange them according to the following actions on the size of the thorax: (1) increase in vertical length, (2) increase in lateral and anteroposterior dimensions, and (3) decrease in lateral and anteroposterior dimensions.

INNERVATION

The diaphragm is innervated by the phrenic nerve (contains axons from cervical nerves C3–C5), and the external and internal intercostals are innervated by the intercostal nerves.

MUSCLE	ORIGIN	INSERTION	ACTION
Diaphragm (DĪ-a-fram; *dia* = across; *phragma* = wall)	Xiphoid process of the sternum, costal cartilages of inferior six ribs, and lumbar vertebrae.	Central tendon.	Forms floor of thoracic cavity; pulls central tendon inferiorly during inspiration, and as dome of diaphragm flattens increases vertical length of thorax.
External intercostals (in'-ter-KOS-tals; *external* = closer to surface; *inter* = between; *costa* = rib)	Inferior border of rib above.	Superior border of rib below.	Elevate ribs during inspiration and thus increase lateral and anteroposterior dimensions of thorax.
Internal intercostals (in'-ter-KOS-tals; *internal* = farther from surface)	Superior border of rib below.	Inferior border of rib above.	Draw adjacent ribs together during forced expiration and thus decrease lateral and anteroposterior dimensions of thorax.

Exhibit 10.10 *Muscles Used in Ventilation (Breathing) (continued)*

Figure 10.12 Muscles used in breathing, as seen in a male.

Openings in the diaphragm permit the passage of the aorta, esophagus, and inferior vena cava.

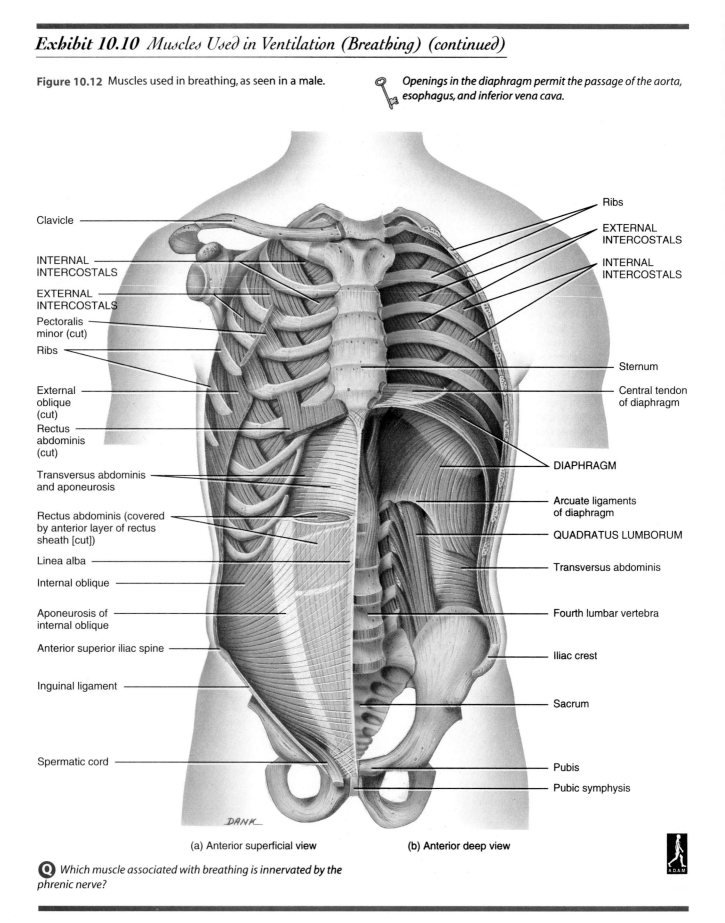

(a) Anterior superficial view   (b) Anterior deep view

Q *Which muscle associated with breathing is innervated by the phrenic nerve?*

Exhibit 10.11 *Muscles of the Pelvic Floor (Figure 10.13)*

The muscles of the pelvic floor are the levator ani and coccygeus. Together with the fascia covering their internal and external surfaces, these muscles are referred to as the **pelvic diaphragm,** which stretches from the pubis anteriorly to the coccyx posteriorly, and from one lateral wall of the pelvis to the other. This arrangement gives the pelvic diaphragm the appearance of a funnel suspended from its attachments. The pelvic diaphragm is pierced by the anal canal and urethra in both sexes and also by the vagina in females.

The **levator ani** muscle is divisible into two muscles called the **pubococcygeus** and **iliococcygeus.** These muscles in the female are shown in Figure 10.13 and in the male in Figure 10.14. The levator ani is the largest and most important muscle of the pelvic floor. It supports the pelvic viscera and resists the inferior thrust that accompanies increases in intra-abdominal pressure during functions such as forced expiration, coughing, vomiting, urination, and defecation. The muscle also functions as a sphincter at the anorectal junction, urethra, and vagina. During childbirth, the levator ani muscle supports the head of the fetus, and the muscle may be injured during a difficult childbirth or traumatized during an episiotomy. This may cause urinary stress incontinence in which there is a leakage of urine whenever intra-abdominal pressure is increased—for example, during coughing. In addition to assisting the levator ani, the **coccygeus** muscle pulls the coccyx anteriorly after it has been pushed posteriorly following defecation or childbirth. In order to treat urinary stress incontinence, it may be necessary to strengthen and tighten the muscles that support the pelvic viscera. This is accomplished by *Kegel exercises,* the alternate contraction and relaxation of muscles of the pelvic floor.

RELATING MUSCLES TO MOVEMENTS

After you have studied the muscles in this exhibit, arrange them according to the following actions on the pelvic viscera: supporting and maintaining their position; resisting an increase in intra-abdominal pressure; and according to the following action on the anus, urethra, and vagina: constriction. The same muscle may be mentioned more than once.

INNERVATION

The pubococcygeus and iliococcygeus are innervated by sacral nerves S3–S4 or S4 and perineal branch of pudendal nerve; the coccygeus is innervated by sacral nerves S4–S5.

MUSCLE	ORIGIN	INSERTION	ACTION
Levator ani (le-VĀ-tor Ā-nē; *levator* = raises; *ani* = anus)	This muscle is divisible into two parts, the pubococcygeus muscle and the iliococcygeus muscle.		
Pubococcygeus (pu'-bō-kok-SIJ-ē-us; *pubo* = pubis; *coccygeus* = coccyx)	Pubis.	Coccyx, urethra, anal canal, central tendon of perineum, and anococcygeal raphe (narrow fibrous band that extends from anus to coccyx).	Supports and maintains position of pelvic viscera; resists increase in intra-abdominal pressure during forced expiration, coughing, vomiting, urination, and defecation; constricts anus, urethra, and vagina; and supports fetal head during childbirth.
Iliococcygeus (il'-ē-ō-kok-SIJ-ē-us; *ilio* = ilium)	Ischial spine.	Coccyx.	As above.
Coccygeus (kok-SIJ-ē-us)	Ischial spine.	Lower sacrum and upper coccyx.	Supports and maintains position of pelvic viscera; resists increase in intra-abdominal pressure during forced expiration, coughing, vomiting, urination, and defecation; and pulls coccyx anteriorly following defecation or childbirth.

▶

Exhibit 10.11 *Muscles of the Pelvic Floor (continued)*

Figure 10.13 Muscles of the pelvic floor, as seen in the female perineum.

The pelvic diaphragm supports the pelvic viscera.

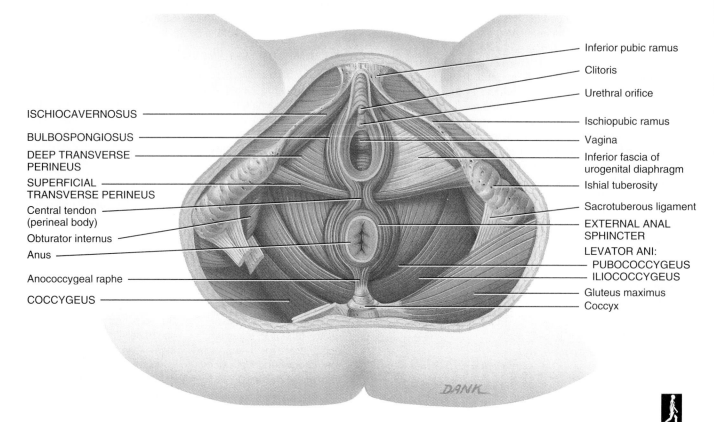

ISCHIOCAVERNOSUS

BULBOSPONGIOSUS

DEEP TRANSVERSE PERINEUS

SUPERFICIAL TRANSVERSE PERINEUS

Central tendon (perineal body)

Obturator internus

Anus

Anococcygeal raphe

COCCYGEUS

Inferior pubic ramus

Clitoris

Urethral orifice

Ischiopubic ramus

Vagina

Inferior fascia of urogenital diaphragm

Ishial tuberosity

Sacrotuberous ligament

EXTERNAL ANAL SPHINCTER

LEVATOR ANI:
  PUBOCOCCYGEUS
  ILIOCOCCYGEUS

Gluteus maximus

Coccyx

Superficial muscles

Q *What are the borders of the pelvic diaphragm?*

Exhibit 10.12 *Muscles of the Perineum (Figures 10.13 and 10.14)*

The **perineum,** the region of the trunk inferior to the pelvic diaphragm, is a diamond-shaped area that extends from the pubic symphysis anteriorly, to the coccyx posteriorly, and to the ischial tuberosities laterally. The female and the male perineums may be compared in Figures 10.13 and 10.14, respectively. A transverse line drawn between the ischial tuberosities divides the perineum into an anterior **urogenital triangle** that contains the external genitals and a posterior **anal triangle** that contains the anus (see Figure 26.20). In the center of the perineum is a wedge-shaped mass of fibrous tissue called the **central tendon (perineal body).** It is a strong tendon into which several perineal muscles insert. Clinically, the perineum is very important to physicians who care for women during pregnancy and treat disorders related to the female genital tract, urogenital organs, and the anorectal region.

The muscles of the perineum are arranged in two layers: **superficial** and **deep.** The muscles of the superficial layer are the superficial transverse perineus, bulbospongiosus, and ischiocavernosus. The **superficial transverse perineus** muscle is a narrow muscle that passes more or less transversely across the perineum anterior to the anus. It joins with the external sphincter posteriorly. The **bulbospongiosus** muscle assists in urination and erection of the penis in males and in decreasing the vaginal orifice and erection of the clitoris in females. The **ischiocaver-**nosus muscle assists in maintaining erection of the penis and clitoris. The deep muscles of the perineum are the **deep transverse perineus** muscle and **external urethral spinchter** (see Figure 25.12a). The deep transverse perineus, external urethral spinchter, and their fascia are known as the **urogenital diaphragm.** The muscles of this diaphragm assist in urination and ejaculation in males and urination in females. The **external anal sphincter** closely adheres to the skin around the margin of the anus and assists in defecation.

RELATING MUSCLES TO MOVEMENTS

After you have studied the muscles in this exhibit, arrange them according to the following actions: (1) expulsion of urine and semen, (2) erection of the clitoris and penis, (3) closing the anal orifice, and (4) closing the vaginal orifice. The same muscle may be mentioned more than once.

INNERVATION

All muscles are innervated by the perineal branch of the pudendal nerve, except for the external anal sphincter, which is innervated by sacral nerve S4 and the inferior rectal branch of the pudendal nerve. The pudendal nerve is derived from the sacral plexus, which is described and illustrated in Exhibit 17.4 on pages 538–539.

MUSCLE	ORIGIN	INSERTION	ACTION
Superficial Perineal Muscles			
Superficial transverse perineus (per-i-NĒ-us; *superficial* = closer to surface; *transverse* = across; *perineus* = perineum)	Ischial tuberosity.	Central tendon of perineum.	Helps to stabilize central tendon of perineum.
Bulbospongiosus (bul'-bō-spon'-jē-Ō-sus; *bulbus* = bulb; *spongia* = sponge)	Central tendon of perineum.	Inferior fascia of urogenital diaphragm, corpus spongiosum of penis, and deep fascia on dorsum of penis in male; pubic arch and root and dorsum of clitoris in female.	Helps expel last drops of urine during micturition (urination), helps propel semen along urethra, and assists in erection of the penis in male; decreases vaginal orifice and assists in erection of clitoris in female.
Ischiocavernosus (is'-kē-ō-ka'-ver-NŌ-sus; *ischion* = hip)	Ischial tuberosity and ischial and pubic rami.	Corpus cavernosum of penis in male and clitoris in female.	Maintains erection of penis in male and clitoris in female.
Deep Perineal Muscles			
Deep transverse perineus (per-i-NĒ-us; *deep* = farther from surface)	Ischial rami.	Central tendon of perineum.	Helps expel last drops of urine and semen in male and urine in female.
External urethral sphincter (yoo-RĒ-thral SFINGK-ter; *urethrae* = urethra; *sphincter* = circular muscle that decreases size of an opening) (see Figure 25.12a)	Ischial and pubic rami.	Median raphe in male and vaginal wall in female.	Helps expel last drops of urine and semen in male and urine in female.
External anal sphincter (Ā-nal)	Anococcygeal raphe.	Central tendon of perineum.	Keeps anal canal and orifice closed.

▶

Exhibit 10.12 Muscles of the Perineum (continued)

Figure 10.14 Muscles of the male perineum.

The urogenital diaphragm surrounds the urogenital duct and helps strengthen the pelvic floor.

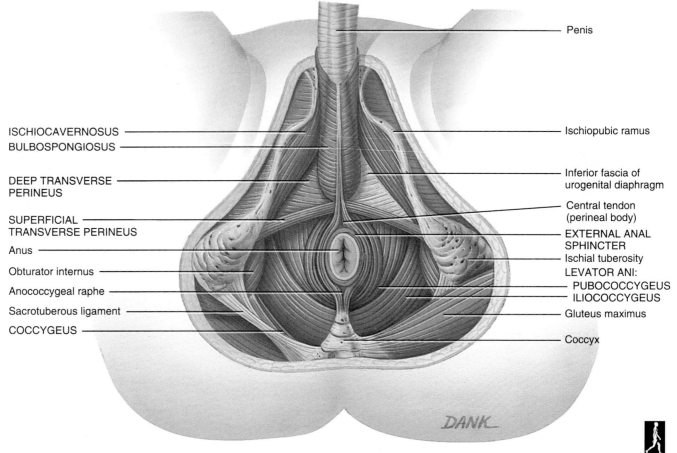

ISCHIOCAVERNOSUS

BULBOSPONGIOSUS

DEEP TRANSVERSE PERINEUS

SUPERFICIAL TRANSVERSE PERINEUS

Anus

Obturator internus

Anococcygeal raphe

Sacrotuberous ligament

COCCYGEUS

Penis

Ischiopubic ramus

Inferior fascia of urogenital diaphragm

Central tendon (perineal body)

EXTERNAL ANAL SPHINCTER

Ischial tuberosity

LEVATOR ANI:
PUBOCOCCYGEUS
ILIOCOCCYGEUS

Gluteus maximus

Coccyx

DANK

Superficial muscles

Q *What are the borders of the perineum?*

Exhibit 10.13 *Muscles that Move the Pectoral (Shoulder) Girdle (Figure 10.15)*

Muscles that move the pectoral girdle can be classified into two groups based on their location in the thorax: **anterior** and **posterior thoracic** muscles. The principal action of the muscles that move the pectoral girdle is to stabilize the scapula so it can function as a stable point of origin for most of the muscles that move the humerus. Because scapular movements usually accompany humeral movements in the same direction, the muscles also move the scapula to increase the range of motion of the humerus. For example, it would not be possible to abduct the humerus past the horizontal if the scapula did not move with the humerus. During abduction, the scapula follows the humerus by rotating upward.

The anterior thoracic muscles are the subclavius, pectoralis minor, and serratus anterior. The **subclavius** is a small, cylindrical muscle under the clavicle that extends from the clavicle to the first rib. It steadies the clavicle during movements of the pectoral girdle. The **pectoralis minor** is a thin, flat, triangular muscle that is deep to the pectoralis major. In addition to its role in movements of the scapula, the pectoralis minor muscle assists in forced inspiration. The **serratus anterior** is a large, flat, fan-shaped muscle between the ribs and scapula. It is named because of the saw-toothed appearance of its origins on the ribs.

The posterior thoracic muscles are the trapezius, levator scapulae, rhomboideus major, and rhomboideus minor. The **trapezius** is a large, flat, triangular sheet of muscle extending from the skull and vertebral column medially to the pectoral girdle laterally. It is the most superficial back muscle and covers the posterior neck region and superior portion of the trunk. The two trapezius muscles form a trapezium (diamond-shaped quadrangle)—hence its name. The **levator scapulae** is a narrow, elongated muscle in the posterior portion of the neck. It is deep to the sternocleidomastoid and trapezius muscles. As its name suggests, one of its actions is to elevate the scapula. The **rhomboideus major** and **rhomboideus minor** lie deep to the trapezius and are not always distinct from each other. They appear as parallel bands that pass inferiolaterally from the vertebrae to the scapula. They are named on the basis of their shape—that is, a rhomboid (an oblique parallelogram). The rhomboideus major is about two times wider than the rhomboideus minor. Both muscles are used when forcibly lowering the raised upper limbs, as in driving a stake with a sledgehammer.

In order to understand the actions of muscles that move the scapula, it is first helpful to describe the various movements of the scapula:

- **Elevation.** Superior movement of the scapula, such as shrugging the shoulders or lifting a weight over the head.

- **Depression.** Inferior movement of the scapula, as in doing a "pull-up."

- **Abduction (protraction).** Movement of the scapula laterally and anteriorly, as in doing a "push-up" or punching.

- **Adduction (retraction).** Movement of the scapula medially and posteriorly, as in pulling the oars in a rowboat.

- **Upward rotation.** Movement of the inferior angle of the scapula laterally so that the glenoid cavity is moved upward. This movement is required to abduct the humerus past the horizontal.

- **Downward rotation.** Movement of the inferior angle of the scapula medially so that the glenoid cavity is moved downward. This movement is seen when the weight of the body is supported on the hands by a gymnast on parallel bars.

RELATING MUSCLES TO MOVEMENTS

After you have studied the muscles in this exhibit, arrange them according to the following actions on the scapula: (1) depression, (2) elevation, (3) abduction, (4) adduction, (5) upward rotation, and (6) downward rotation. The same muscle may be mentioned more than once.

INNERVATION

The subclavius is innervated by the subclavian nerve; the pectoralis minor is innervated by the medial pectoral nerve; the serratus anterior is innervated by the long thoracic nerve; the trapezius is innervated by the accessory (XI) nerve and cervical nerves C3–C5; the levator scapulae is innervated by the dorsal scapular nerve and cervical nerves C3–C5; and the rhomboideus major and rhomboideus minor are innervated by the dorsal scapular nerve.

▶

Exhibit 10.13 *Muscles that Move the Pectoral (Shoulder) Girdle (continued)*

MUSCLE	ORIGIN	INSERTION	ACTION
Anterior Thoracic Muscles			
Subclavius (sub-KLĀ-vē-us; *sub* = under; *clavius* = clavicle)	First rib.	Clavicle.	Depresses and moves clavicle anteriorly and helps stabilize pectoral girdle.
Pectoralis (pek′-tor-A-lis) **minor** (*pectus* = breast, chest, thorax; *minor* = lesser)	Third through fifth ribs.	Coracoid process of scapula.	Depresses and abducts scapula and rotates it downward; elevates third through fifth ribs during forced inspiration when scapula is fixed.
Serratus (ser-Ā-tus) **anterior** (*serratus* = sawtoothed; *anterior* = front)	Superior eight or nine ribs.	Vertebral border and inferior angle of scapula.	Abducts scapula and rotates it upward; elevates ribs when scapula is fixed; known as "boxer's muscle."
Posterior Thoracic Muscles			
Trapezius (tra-PĒ-zē-us; *trapezoides* = trapezoid-shaped)	Superior nuchal line of occipital bone, ligamentum nuchae, and spines of seventh cervical and all thoracic vertebrae.	Clavicle and acromion and spine of scapula.	Superior fibers elevate scapula and can help extend head; middle fibers adduct scapula; inferior fibers depress scapula; superior and inferior fibers together rotate scapula upward; stabilizes scapula.
Levator scapulae (le-VĀ-tor SKA-pyoo-lē; *levator* = raises; *scapulae* = scapula)	Superior four or five cervical vertebrae.	Superior vertebral border of scapula.	Elevates scapula and rotates it downward.
Rhomboideus (rom-BOID-ē-us) **major** (*rhomboides* = rhomboid or diamond-shaped)	Spines of second to fifth thoracic vertebrae.	Vertebrae border of scapula inferior to spine.	Elevates and adducts scapula and rotates it downward; stabilizes scapula.
Rhomboideus (rom-BOID-ē-us) **minor**	Spines of seventh cervical and first thoracic vertebrae.	Vertebral border of scapula superior to spine.	Elevates and adducts scapula and rotates it downward; stabilizes scapula.

Exhibit 10.13 (continued)

Figure 10.15 Muscles that move the pectoral (shoulder) girdle.

Muscles that move the pectoral girdle originate on the axial skeleton and insert on the clavicle or scapula.

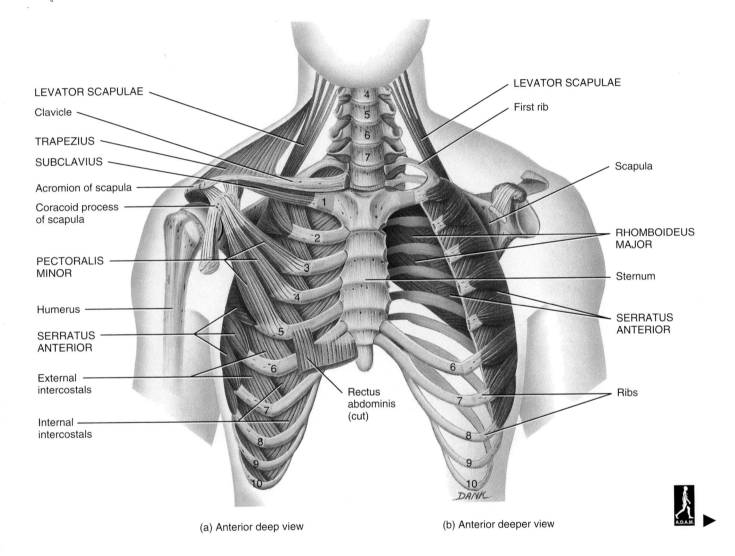

LEVATOR SCAPULAE

Clavicle

TRAPEZIUS

SUBCLAVIUS

Acromion of scapula

Coracoid process of scapula

PECTORALIS MINOR

Humerus

SERRATUS ANTERIOR

External intercostals

Internal intercostals

LEVATOR SCAPULAE

First rib

Scapula

RHOMBOIDEUS MAJOR

Sternum

SERRATUS ANTERIOR

Ribs

Rectus abdominis (cut)

(a) Anterior deep view

(b) Anterior deeper view

Exhibit 10.13 *Muscles that Move the Pectoral (Shoulder) Girdle (continued)*

Figure 10.15 (continued)

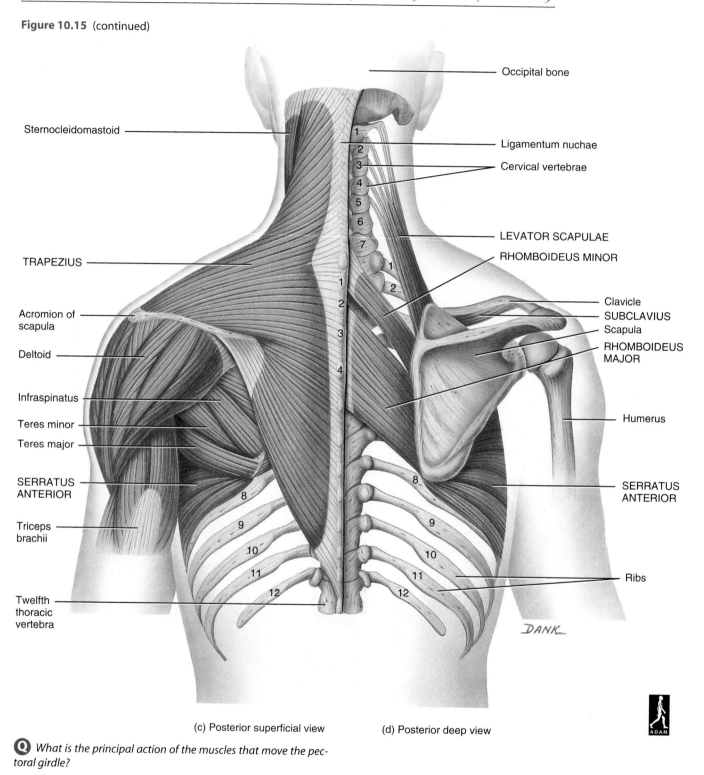

Occipital bone

Sternocleidomastoid

Ligamentum nuchae

Cervical vertebrae

LEVATOR SCAPULAE

RHOMBOIDEUS MINOR

TRAPEZIUS

Clavicle

SUBCLAVIUS

Acromion of scapula

Scapula

RHOMBOIDEUS MAJOR

Deltoid

Infraspinatus

Teres minor

Teres major

Humerus

SERRATUS ANTERIOR

SERRATUS ANTERIOR

Triceps brachii

Ribs

Twelfth thoracic vertebra

DANK

(c) Posterior superficial view   (d) Posterior deep view

Q *What is the principal action of the muscles that move the pectoral girdle?*

Exhibit 10.14 *Muscles that Move the Humerus (Arm) (Figure 10.16)*

Of the nine muscles that cross the shoulder joint, only two of them (pectoralis major and latissimus dorsi) do not originate on the scapula. These two muscles are thus designated as **axial muscles,** because they originate on the axial skeleton. The remaining seven muscles, the **scapular muscles,** arise from the scapula.

Of the two axial muscles that move the humerus, the **pectoralis major** is a large, thick, fan-shaped muscle that covers the superior part of the thorax. It has two origins: a smaller clavicular head and a larger sternocostal head. The **latissimus dorsi** is a broad, triangular muscle located on the inferior part of the back. It is commonly called the "swimmer's muscle" because its many actions are used while swimming.

Among the scapular muscles, the **deltoid** is a thick, powerful shoulder muscle that covers the shoulder joint and forms the rounded contour of the shoulder. This muscle is a frequent site of intramuscular injections. As you study the deltoid, note that its fibers originate from three different points and that each group of fibers moves the humerus differently. The **subscapularis** is a large triangular muscle that fills the subscapular fossa of the scapula and forms part of the posterior wall of the axilla. The **supraspinatus** is a rounded muscle, named for its location in the supraspinous fossa of the scapula. It lies deep to the trapezius. The **infraspinatus** is a triangular muscle, also named for its location in the infraspinous fossa of the scapula. The **teres major** is a thick, flattened muscle inferior to the teres minor that also helps form part of the posterior wall of the axilla. The **teres minor** is a cylindrical, elongated muscle, often inseparable from the infraspinatus, which lies along its superior border. The **coracobrachialis** is an elongated, narrow muscle in the arm that is pierced by the musculocutaneous nerve.

The strength and stability of the shoulder joint are not provided by the shape of the articulating bones or its ligaments. Instead, four deep muscles of the shoulder—subscapularis, supraspinatus, infraspinatus, and teres minor—strengthen and stabilize the shoulder joint. These muscles join the scapula to the humerus. Their flat tendons fuse together to form a nearly complete circle around the shoulder joint, like the cuff on a shirt sleeve. This arrangement is referred to as the **rotator (musculo-tendinous) cuff.** The supraspinatus muscle is especially predisposed to wear and tear because of its location between the head of the humerus and acromion of the scapula, which compress the tendon during shoulder movements.

Clinical Application

Impingement Syndrome

One of the most common causes of shoulder pain and dysfunction in the athlete is known as **impingement syndrome.** The repetitive overhead motion that is common in baseball, overhead racquet sports, and swimming, places these athletes at risk for developing this syndrome. Continual pinching of the supraspinatus tendon causes it to become inflamed and results in pain. If the condition persists, the tendon may degenerate near the attachment to the humerus and ultimately may tear away from the bone (rotator cuff injury). ■

RELATING MUSCLES TO MOVEMENTS

After you have studied the muscles in this exhibit, arrange them according to the following actions on the humerus at the shoulder joint: (1) flexion, (2) extension, (3) abduction, (4) adduction, (5) medial rotation, and (6) lateral rotation. The same muscle may be mentioned more than once.

INNERVATION

The pectoralis major is innervated by the medial and lateral pectoral nerve; the latissimus dorsi is innervated by the thoracodorsal nerve; the deltoid and teres minor are innervated by the axillary nerve; the subscapularis is innervated by the upper and lower subscapular nerves; the supraspinatus and infraspinatus are innervated by the suprascapular nerve; the teres major is innervated by the lower subscapular nerve; and the coracobrachialis is innervated by the musculocutaneous nerve.

▶

Exhibit 10.14 *Muscles that Move the Humerus (Arm) (continued)*

Figure 10.16 (continued)

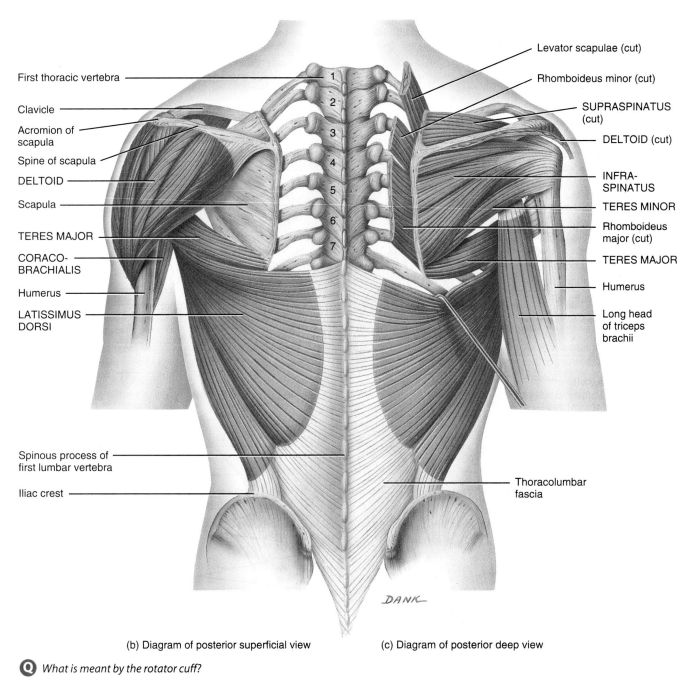

Levator scapulae (cut)

Rhomboideus minor (cut)

SUPRASPINATUS (cut)

DELTOID (cut)

INFRA-SPINATUS

TERES MINOR

Rhomboideus major (cut)

TERES MAJOR

Humerus

Long head of triceps brachii

First thoracic vertebra

Clavicle

Acromion of scapula

Spine of scapula

DELTOID

Scapula

TERES MAJOR

CORACO-BRACHIALIS

Humerus

LATISSIMUS DORSI

Spinous process of first lumbar vertebra

Iliac crest

Thoracolumbar fascia

DANK

(b) Diagram of posterior superficial view   (c) Diagram of posterior deep view

Q *What is meant by the rotator cuff?*

Exhibit 10.14 (*continued*)

SUPERIOR

DELTOID

Pectoralis minor

PECTORALIS MAJOR

LATISSIMUS DORSI

Biceps brachii

Serratus anterior

External oblique

POSTERIOR

ANTERIOR

INFERIOR

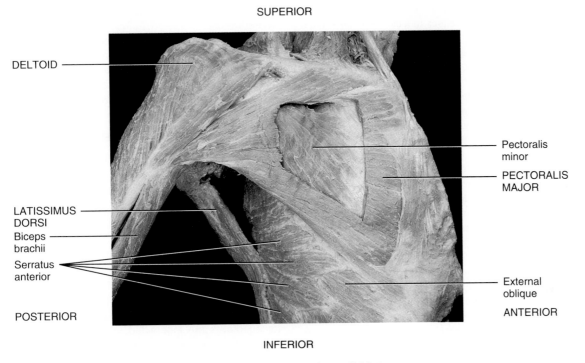

(d) Photograph of right lateral superficial view

Exhibit 10.15 *Muscles that Move the Radius and Ulna (Forearm)* *(Figure 10.17)*

Most of the muscles that move the radius and ulna (forearm) are involved in flexion and extension at the elbow, which is a hinge joint. The biceps brachii, brachialis, and brachioradialis muscles are the flexor muscles. The extensor muscles are the triceps brachii and the **anconeus.**

The **biceps brachii** is the large muscle located on the anterior surface of the arm. As indicated by its name, it has two heads of origin (long and short), both from the scapula. The muscle spans both the shoulder and elbow joints. In addition to its role in flexing the forearm at the elbow joint, it also supinates the forearm at the radioulnar joints and flexes the arm at the shoulder joint. The **brachialis** is deep to the biceps brachii muscle. It is the most powerful flexor of the forearm at the elbow joint. For this reason, it is called the "workhorse" of the elbow flexors. The **brachioradialis** flexes the forearm at the elbow joint, especially when a quick movement is required or when a weight is lifted slowly during flexion of the forearm.

The **triceps brachii** is the large muscle located on the posterior surface of the arm. It is the more powerful of the extensors of the forearm at the elbow joint. As its name implies, it has three heads of origin, one from the scapula (long head) and two from the humerus (lateral and medial heads). The long head crosses the shoulder joint; the other heads do not. The **anconeus** is a small muscle located on the lateral part of the posterior aspect of the elbow that assists the triceps brachii in extending the forearm at the elbow joint.

Some muscles that move the radius and ulna are involved in pronation and supination at the radioulnar joints. The pronators, as suggested by their names, are the **pronator teres** and **prona-** **tor quadratus** muscles. The supinator of the forearm is the aptly named the **supinator** muscle. You use the powerful action of the supinator when you twist a corkscrew or turn a screw with a screwdriver.

In the limbs, functionally related skeletal muscles and their associated blood vessels and nerves are grouped together by fascia into regions called **compartments.** In the arm, the biceps brachii, brachialis, and coracobrachialis muscles constitute the *flexor compartment;* the triceps brachii muscle forms the *extensor compartment.*

RELATING MUSCLES TO MOVEMENTS

After you have studied the muscles in this exhibit, arrange them according to the following actions on the elbow joint: (1) flexion and (2) extension; the following actions on the forearm at the radioulnar joints: (1) supination and (2) pronation; and the following actions on the humerus at the shoulder joint: (1) flexion and (2) extension. The same muscle may be mentioned more than once.

INNERVATION

The biceps brachii is innervated by the musculocutaneous nerve; the brachialis is innervated by the musculocutaneous and radial nerves; the brachioradialis, triceps brachii, and anconeus are innervated by the radial nerve; the pronator teres and pronator quadratus are innervated by the median nerve; and the supinator is innervated by the deep radial nerve. All the nerves listed here are derived from the brachial plexus, which is described and illustrated in Exhibit 17.2 on pages 530–535.

MUSCLE	ORIGIN	INSERTION	ACTION
Forearm Flexors			
Biceps brachii (BĪ-ceps BRĀ-kē-ī; *biceps* = two heads of origin; *brachion* = arm)	*Long head* originates from tubercle above glenoid cavity of scapula (supraglenoid tubercle); *short head* originates from coracoid process of scapula.	Radial tuberosity and bicipital aponeurosis (broad aponeurosis from tendon of insertion of biceps brachii muscle that descends medially across brachial artery and fuses with deep fascia over forearm flexor muscles).	Flexes forearm at elbow joint, supinates forearm at radioulnar joints, and flexes arm at shoulder joint.
Brachialis (brā-kē-A-lis)	Distal, anterior surface of humerus.	Ulnar tuberosity and coronoid process of ulna.	Flexes forearm at elbow joint.
Brachioradialis (bra′-kē-ō-rā-dē-A-lis; *radialis* = radius) (see also Figure 10.18a,c)	Medial and lateral borders of distal end of humerus.	Superior to styloid process of radius.	Flexes forearm at elbow joint and supinates and pronates forearm at radioulnar joints to neutral position.

Exhibit 10.15 (continued)

MUSCLE	ORIGIN	INSERTION	ACTION
Forearm Extensors			
Triceps brachii (TRĪ-ceps BRĀ-kē-ī; *triceps* = three heads of origin)	*Long head* originates from a projection inferior to glenoid cavity (infraglenoid tubercle) of scapula; *lateral head* originates from lateral and posterior surface of humerus superior to radial groove; *medial head* originates from entire posterior surface of humerus inferior to a groove for the radial nerve.	Olecranon of ulna.	Extends forearm at elbow joint and extends arm at shoulder joint.
Anconeus (an-KŌ-nē-us; *anconeal* = pertaining to elbow) (see Figure 10.18c)	Lateral epicondyle of humerus.	Olecranon and superior portion of shaft of ulna.	Extends forearm at elbow joint.
Forearm Pronators			
Pronator teres (PRŌ-nā-tor TE-rēz; *pronation* = turning palm downward or posteriorly) (see Figure 10.18a)	Medial epicondyle of humerus and coronoid process of ulna.	Midlateral surface of radius.	Pronates forearm at radioulnar joints and weakly flexes forearm at elbow joint.
Pronator quadratus (PRŌ-nā-tor kwod-RĀ-tus; *quadratus* = squared, four-sided) (see Figure 10.18 a,b)	Distal portion of shaft of ulna.	Distal portion of shaft of radius.	Pronates forearm at radioulnar joints.
Forearm Supinator			
Supinator (SOO-pi-nā-tor; *supination* = turning palm upward or anteriorly) (see Figure 10.18b)	Lateral epicondyle of humerus and ridge near radial notch of ulna (supinator crest).	Lateral surface of proximal one-third of radius.	Supinates forearm at radioulnar joints.

▶

Exhibit 10.15 *Muscles that Move the Radius and Ulna (Forearm) (continued)*

Figure 10.17 Muscles that move the radius and ulna (forearm).

Whereas the anterior arm muscles flex the forearm, the posterior arm muscles extend it.

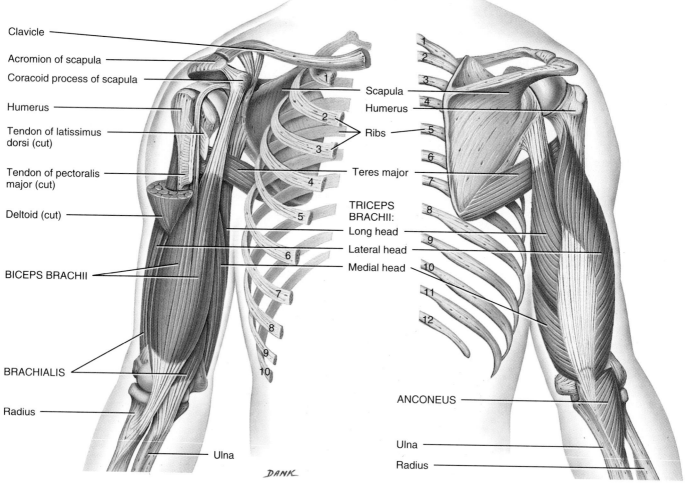

(a) Diagram of anterior view

(b) Diagram of posterior view

Exhibit 10.15 *(continued)*

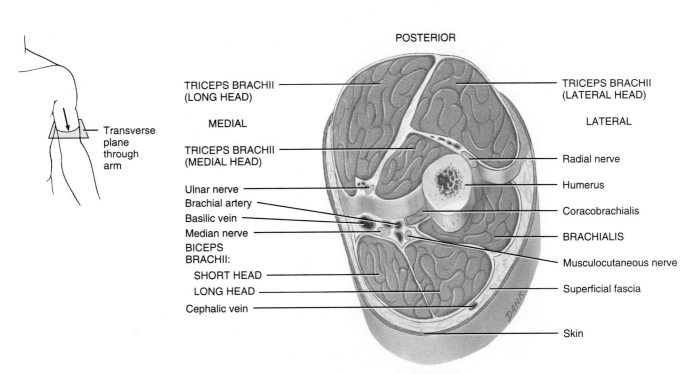

Transverse plane through arm

POSTERIOR

TRICEPS BRACHII (LONG HEAD)

TRICEPS BRACHII (LATERAL HEAD)

MEDIAL

LATERAL

TRICEPS BRACHII (MEDIAL HEAD)

Radial nerve

Ulnar nerve

Humerus

Brachial artery

Coracobrachialis

Basilic vein

Median nerve

BRACHIALIS

BICEPS BRACHII:

Musculocutaneous nerve

SHORT HEAD

LONG HEAD

Superficial fascia

Cephalic vein

Skin

ANTERIOR

(c) Diagram of superior view of transverse section of arm

Exhibit 10.15 *Muscles that Move the Radius and Ulna (Forearm) (continued)*

Figure 10.17 (continued)

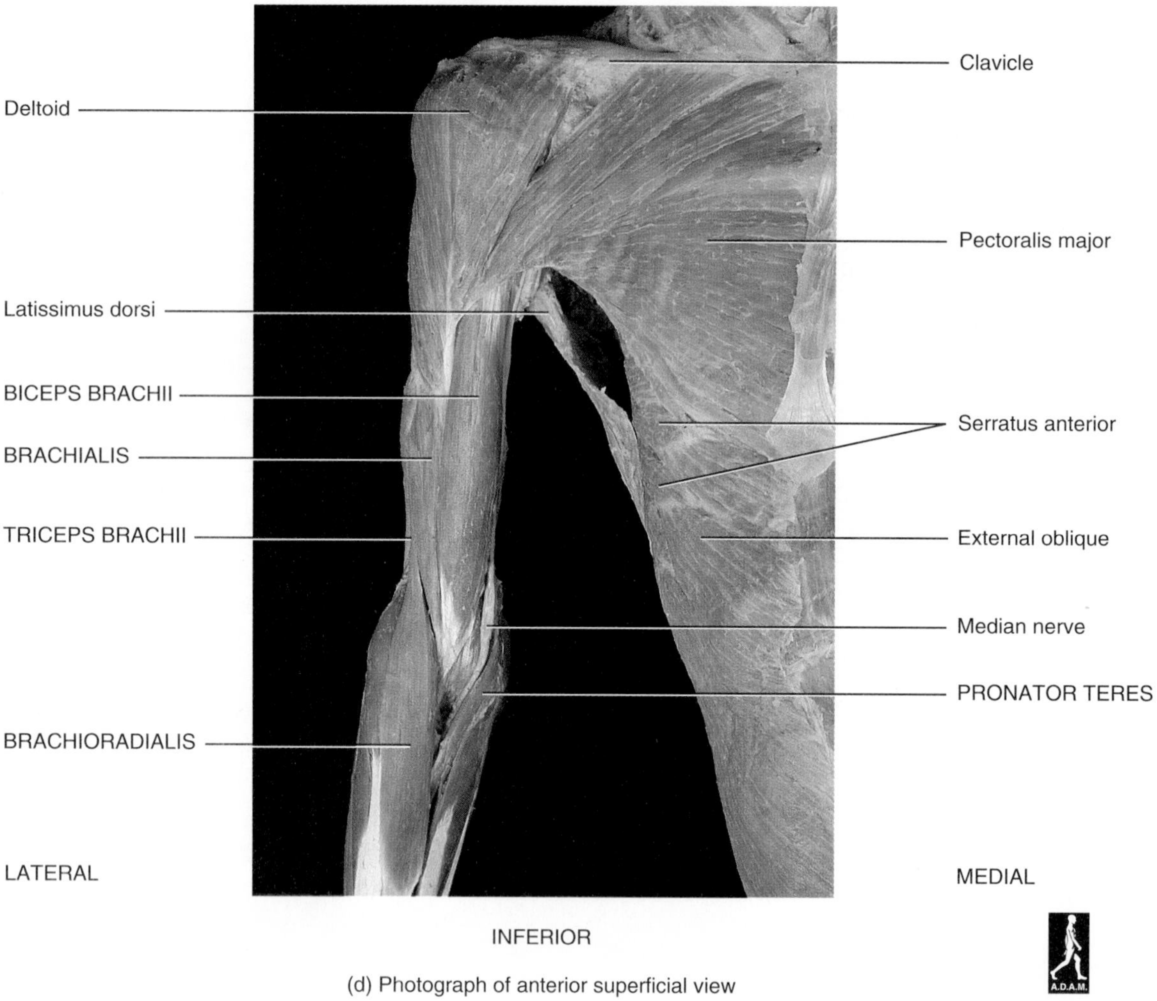

Deltoid

Latissimus dorsi

BICEPS BRACHII

BRACHIALIS

TRICEPS BRACHII

BRACHIORADIALIS

LATERAL

Clavicle

Pectoralis major

Serratus anterior

External oblique

Median nerve

PRONATOR TERES

MEDIAL

INFERIOR

(d) Photograph of anterior superficial view

Q *Which muscles are the most powerful flexor and extensor of the forearm?*

Exhibit 10.16 *Muscles that Move the Wrist, Hand, and Digits (Figure 10.18)*

Muscles that move the wrist, hand, and digits are many and varied. Those in this group that act on the digits are known as **extrinsic muscles** because they originate *outside* the hand and insert within it. As you will see, the names for the muscles that move the wrist, hand, and digits give some indication of their origin, insertion, or action. On the basis of location and function, the muscles are divided into two groups: (1) anterior compartment muscles and (2) posterior compartment muscles. The **anterior compartment** muscles originate on the humerus; typically insert on the carpals, metacarpals, and phalanges; and function as flexors. The bellies of these muscles form the bulk of the forearm. The **posterior compartment** muscles originate on the humerus, insert on the metacarpals and phalanges; and function as extensors. Within each compartment, the muscles are grouped as superficial and deep.

The **superficial anterior compartment** muscles are arranged in the following order from lateral to medial: **flexor carpi radialis, palmaris longus** (absent in about 10% of the population), and **flexor carpi ulnaris** (the ulnar nerve and artery are just lateral to the tendon of this muscle at the wrist). The **flexor digitorum superficialis** muscle is actually deep to the other three muscles and is the largest superficial muscle in the forearm.

The **deep anterior compartment** muscles are arranged in the following order from lateral to medial: **flexor pollicis longus** (the only flexor of the distal phalanx of the thumb) and **flexor digitorum profundus** (ends in four tendons that insert into the distal phalanges of the fingers).

The **superficial posterior compartment** muscles are arranged in the following order from lateral to medial: **extensor carpi radialis longus, extensor carpi radialis brevis, extensor digitorum** (occupies most of the posterior surface of the forearm and divides into four tendons that insert into the middle and distal phalanges of the fingers), **extensor digiti minimi** (a slender muscle generally connected to the extensor digitorum), and **extensor carpi ulnaris.**

The **deep posterior compartment** muscles are arranged in the following order from lateral to medial: **abductor pollicis longus, extensor pollicis brevis, extensor pollicis longus,** and **extensor indicis.**

The tendons of the muscles of the forearm that attach to the wrist or continue into the hand, along with blood vessels and nerves, are held close to bones by strong fascial structures. The tendons are also surrounded by tendon sheaths. At the wrist, the deep fascia is thickened into fibrous bands called **retinacula** (*retinere* = retain). The **flexor retinaculum** *(transverse carpal ligament)* is located over the palmar surface of the carpal bones. Through it pass the long flexor tendons of the digits and wrist and the median nerve. The **extensor retinaculum** *(dorsal carpal ligament)* is located over the dorsal surface of the carpal bones. Through it pass the extensor tendons of the wrist and digits.

RELATING MUSCLES TO MOVEMENTS

After you have studied the muscles in this exhibit, arrange them according to the following actions on the wrist joint: (1) flexion, (2) extension, (3) abduction, and (4) adduction; the following actions on the fingers at the metacarpophalangeal joints: (1) flexion and (2) extension; the following actions on the fingers at the interphalangeal joints: (1) flexion and (2) extension; the following actions on the thumb at the carpometacarpal, metacarpophalangeal, and interphalangeal joints: (1) extension and (2) abduction; and the following action on the thumb at the interphalangeal joint: flexion. The same muscle may be mentioned more than once.

INNERVATION

The flexor carpi radialis, palmaris longus, flexor digitorum superficialis, and flexor pollicis longus are innervated by the median nerve; the flexor carpi ulnaris is innervated by the ulnar nerve; the flexor digitorum profundus is innervated by the median and ulnar nerves; the extensor carpi radialis longus, extensor carpi radialis brevis, and extensor digitorum are innervated by the radial nerve; and the extensor digiti minimi, extensor carpi ulnaris, and all of the deep muscles of the posterior compartment are innervated by the deep radial nerve. All the nerves listed here are derived from the brachial plexus, which is described and illustrated in Exhibit 17.2 on pages 530–535.

▶

Exhibit 10.16 *Muscles that Move the Wrist, Hand, and Digits (continued)*

MUSCLE	ORIGIN	INSERTION	ACTION
Superficial Anterior Compartment (Flexors)			
Flexor carpi radialis (FLEK-sor KAR-pē rā'-dē-A-lis; *flexor* = decreases angle at joint; *carpus* = wrist; *radialis* = radius)	Medial epicondyle of humerus.	Second and third metacarpals.	Flexes and abducts hand (radial deviation) at wrist joint.
Palmaris longus (pal-MA-ris LON-gus; *palma* = palm; *longus* = long)	Medial epicondyle of humerus.	Flexor retinaculum and palmar aponeurosis (deep fascia in center of palm).	Weakly flexes hand at wrist joint.
Flexor carpi ulnaris (FLEK-sor KAR-pē ul-NAR-is; *ulnaris* = ulna)	Medial epicondyle of humerus and superior posterior border of ulna.	Pisiform, hamate, and fifth metacarpal.	Flexes and adducts hand (ulnar deviation) at wrist joint.
Flexor digitorum superficialis (FLEK-sor di'-ji-TOR-um soo'-per-fish'-ē-A-lis; *digit* = finger or toe; *superficialis* = closer to surface)	Medial epicondyle of humerus, coronoid process of ulna, and a ridge along lateral margin of anterior surface (anterior oblique line) of radius.	Middle phalanges of each finger*.	Flexes middle phalanx of each finger at proximal interphalangeal joint, proximal phalanx of each finger at metacarpophalangeal joint, and hand at wrist.
Deep Anterior Compartment (Flexors)			
Flexor pollicis longus (FLEK-sor POL-li-kis LON-gus; *pollex* = thumb)	Anterior surface of radius and interosseous membrane (sheet of fibrous tissue that holds shafts of ulna and radius together).	Base of distal phalanx of thumb.	Flexes distal phalanx of thumb at interphalangeal joint.
Flexor digitorum profundus (FLEK-sor di'-ji-TOR-um prō-FUN-dus; *profundus* = deep)	Anterior medial surface of body of ulna.	Bases of distal phalanges of each finger.	Flexes distal and middle phalanges of each finger at interphalangeal joints, proximal phalanx of each finger at metacarpophalangeal joint, and hand at wrist.

*Each finger (little, ring, middle, index) has three phalanges: proximal, middle, and distal. The thumb has two phalanges: proximal and distal.

Exhibit 10.16 *(continued)*

MUSCLE	ORIGIN	INSERTION	ACTION
Superficial Posterior Compartment (Extensors)			
Extensor carpi radialis longus (eks-TEN-sor KAR-pē rā′-dē-A-lis LON-gus; *extensor* = increases angle at joint)	Lateral supracondylar ridge of humerus.	Second metacarpal.	Extends and abducts hand at wrist joint.
Extensor carpi radialis brevis (eks-TEN-sor KAR-pē rā′-dē-A-lis BREV-is; *brevis* = short)	Lateral epicondyle of humerus.	Third metacarpal.	Extends and abducts hand at wrist joint.
Extensor digitorum (eks-TEN-sor di′-ji-TOR-um)	Lateral epicondyle of humerus.	Distal and middle phalanges of each finger.	Extends distal and middle phalanges of each finger at interphalangeal joints, proximal phalanx of each finger at metacarpophalangeal joint, and hand at wrist.
Extensor digiti minimi (eks-TEN-sor DIJ-i-tē MIN-i-mē; *minimi* = little finger)	Lateral epicondyle of humerus.	Tendon of extensor digitorum on fifth phalanx.	Extends proximal phalanx of little finger at metacarpophalangeal joint and hand at wrist joint.
Extensor carpi ulnaris (eks-TEN-sor KAR-pē ul-NAR-is)	Lateral epicondyle of humerus and posterior border of ulna.	Fifth metacarpal.	Extends and adducts hand at wrist joint.
Deep Posterior Compartment (Extensors)			
Abductor pollicis longus (ab-DUK-tor POL-li-kis LON-gus; *abductor* = moves part away from midline)	Posterior surface of middle of radius and ulna and interosseous membrane.	First metacarpal.	Abducts and extends thumb at carpometacarpal joint and abducts hand at wrist.
Extensor pollicis brevis (eks-TEN-sor POL-li-kis BREV-is)	Posterior surface of middle of radius and interosseous membrane.	Base of proximal phalanx of thumb.	Extends proximal phalanx of thumb at metacarpophalangeal joint, first metacarpal of thumb at carpometacarpal joint, and hand at wrist.
Extensor pollicis longus (eks-TEN-sor POL-li-kis LON-gus)	Posterior surface of middle of ulna and interosseous membrane.	Base of distal phalanx of thumb.	Extends distal phalanx of thumb at interphalangeal joint, first metacarpal of thumb at carpometacarpal joint, and abducts hand at wrist joint.
Extensor indicis (eks-TEN-sor IN-di-kis; *indicis* = index)	Posterior surface of ulna.	Tendon of extensor digitorum of index finger.	Extends distal and middle phalanges of index finger at interphalangeal joints, proximal phalanx of index finger at metacarpophalangeal joint, and hand at wrist.

▶

Exhibit 10.16 *Muscles that Move the Wrist, Hand, and Digits (continued)*

Figure 10.18 Muscles that move the wrist, hand, and digits.

The anterior compartment muscles function as flexors, and the posterior compartment muscles function as extensors.

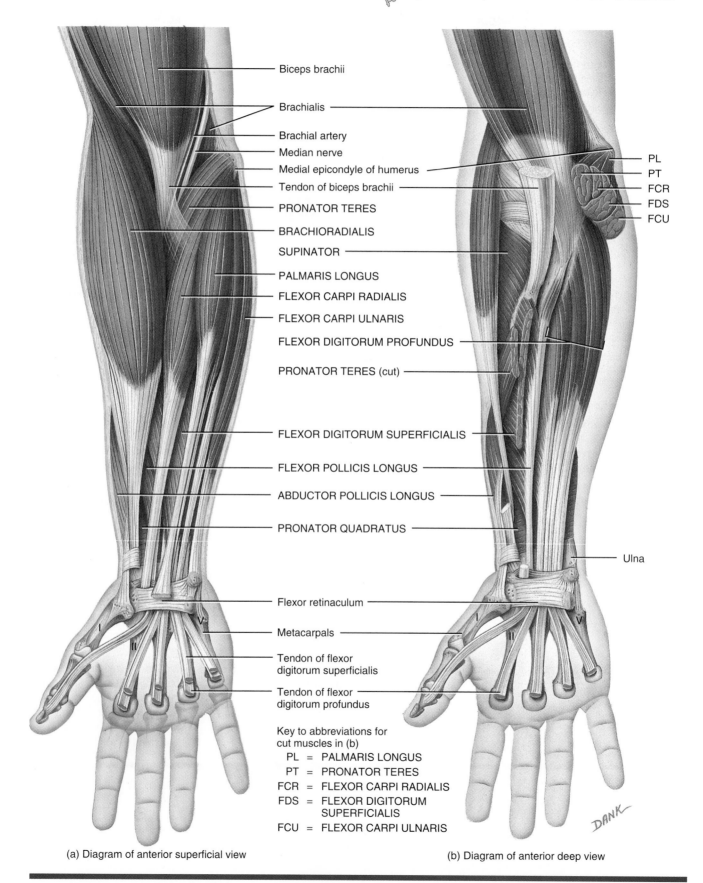

Biceps brachii

Brachialis

Brachial artery

Median nerve

Medial epicondyle of humerus

Tendon of biceps brachii

PRONATOR TERES

BRACHIORADIALIS

SUPINATOR

PALMARIS LONGUS

FLEXOR CARPI RADIALIS

FLEXOR CARPI ULNARIS

FLEXOR DIGITORUM PROFUNDUS

PRONATOR TERES (cut)

FLEXOR DIGITORUM SUPERFICIALIS

FLEXOR POLLICIS LONGUS

ABDUCTOR POLLICIS LONGUS

PRONATOR QUADRATUS

Flexor retinaculum

Metacarpals

Tendon of flexor digitorum superficialis

Tendon of flexor digitorum profundus

PL
PT
FCR
FDS
FCU

Ulna

Key to abbreviations for cut muscles in (b)

PL = PALMARIS LONGUS
PT = PRONATOR TERES
FCR = FLEXOR CARPI RADIALIS
FDS = FLEXOR DIGITORUM SUPERFICIALIS
FCU = FLEXOR CARPI ULNARIS

DANK

(a) Diagram of anterior superficial view

(b) Diagram of anterior deep view

Exhibit 10.16 (*continued*)

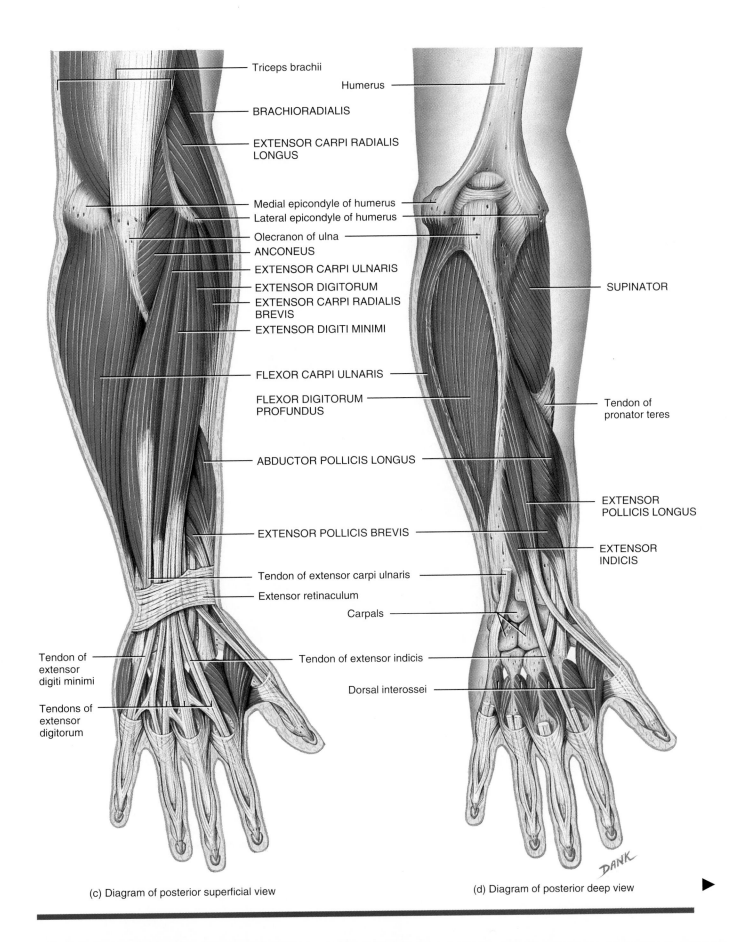

Triceps brachii

Humerus

BRACHIORADIALIS

EXTENSOR CARPI RADIALIS
LONGUS

Medial epicondyle of humerus
Lateral epicondyle of humerus
Olecranon of ulna
ANCONEUS
EXTENSOR CARPI ULNARIS
EXTENSOR DIGITORUM
EXTENSOR CARPI RADIALIS
BREVIS
EXTENSOR DIGITI MINIMI

SUPINATOR

FLEXOR CARPI ULNARIS

FLEXOR DIGITORUM
PROFUNDUS

Tendon of
pronator teres

ABDUCTOR POLLICIS LONGUS

EXTENSOR
POLLICIS LONGUS

EXTENSOR POLLICIS BREVIS

EXTENSOR
INDICIS

Tendon of extensor carpi ulnaris

Extensor retinaculum

Carpals

Tendon of
extensor
digiti minimi

Tendon of extensor indicis

Tendons of
extensor
digitorum

Dorsal interossei

(c) Diagram of posterior superficial view

(d) Diagram of posterior deep view

DANK

Exhibit 10.16 *Muscles that Move the Wrist, Hand, and Digits (continued)*

Figure 10.18 (continued)

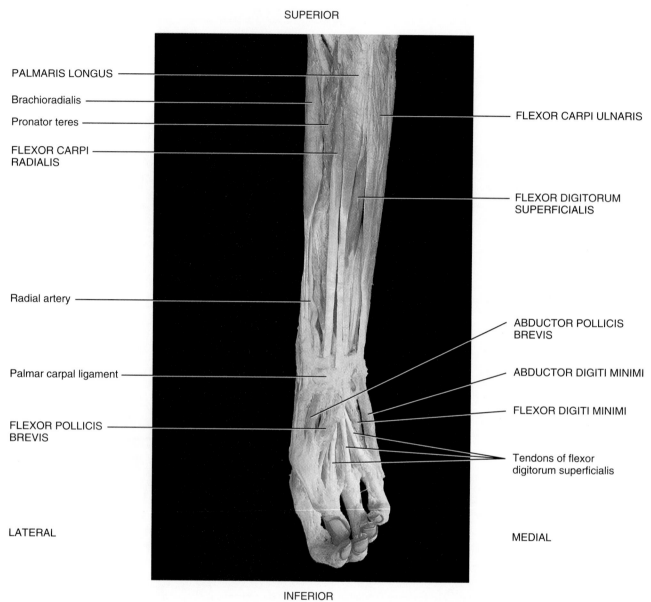

SUPERIOR

PALMARIS LONGUS

Brachioradialis

Pronator teres

FLEXOR CARPI RADIALIS

Radial artery

Palmar carpal ligament

FLEXOR POLLICIS BREVIS

FLEXOR CARPI ULNARIS

FLEXOR DIGITORUM SUPERFICIALIS

ABDUCTOR POLLICIS BREVIS

ABDUCTOR DIGITI MINIMI

FLEXOR DIGITI MINIMI

Tendons of flexor digitorum superficialis

LATERAL

MEDIAL

INFERIOR

(e) Photograph of anterior superficial view

Exhibit 10.16 (continued)

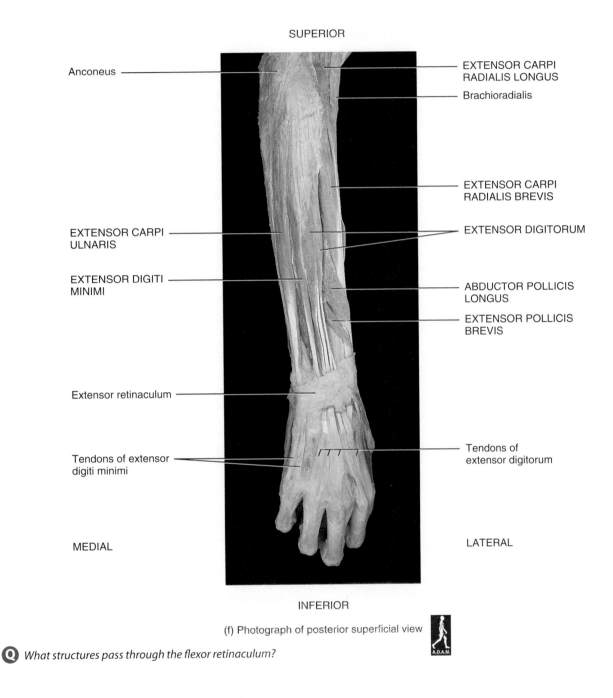

SUPERIOR

Anconeus

EXTENSOR CARPI
RADIALIS LONGUS

Brachioradialis

EXTENSOR CARPI
RADIALIS BREVIS

EXTENSOR CARPI
ULNARIS

EXTENSOR DIGITORUM

EXTENSOR DIGITI
MINIMI

ABDUCTOR POLLICIS
LONGUS

EXTENSOR POLLICIS
BREVIS

Extensor retinaculum

Tendons of
extensor digitorum

Tendons of extensor
digiti minimi

MEDIAL

LATERAL

INFERIOR

(f) Photograph of posterior superficial view

Q *What structures pass through the flexor retinaculum?*

Exhibit 10.17 *Intrinsic Muscles of the Hand (Figure 10.19)*

Several of the muscles discussed in Exhibit 10.16 move the digits in various ways and are known as extrinsic muscles. They produce the powerful but crude movements of the digits. The **intrinsic muscles** in the palm produce weak but intricate and precise movements of the digits that characterize the human hand. The muscles in this group are so named because their origins and insertions are *within* the hands.

The intrinsic muscles of the hand are divided into three groups: (1) **thenar,** (2) **hypothenar,** and (3) **intermediate.** The four thenar muscles act on the thumb and form the **thenar eminence,** the lateral rounded contour on the palm that is also called the ball of the thumb. The thenar muscles include the abductor pollicis brevis, opponens pollicis, flexor pollicis brevis, and adductor pollicis. The **abductor pollicis brevis** is a thin, short, relatively broad superficial muscle on the lateral side of the thenar eminence. The **opponens pollicus** is a small, triangular muscle that is deep to the abductor pollicis brevis muscle. The **flexor pollicis brevis** is a short, wide muscle that is medial to the abductor pollicis brevis muscle. The **adductor pollicis** muscle is fan-shaped and has two heads (oblique and transverse) separated by a gap through which the radial artery passes.

The three hypothenar muscles act on the little finger and form the **hypothenar eminence,** the medial rounded contour on the palm that is also called the ball of the little finger. The hypothenar muscles are the abductor digiti minimi, flexor digiti minimi brevis, and opponens digiti minimi. The **abductor digiti minimi** is a short, wide muscle and is the most superficial of the hypothenar muscles. It is a powerful muscle that plays an important role in grasping an object with outspread fingers. The **flexor digiti minimi brevis** muscle is also short and wide and is lateral to the abductor digiti minimi muscle. The **opponens digiti minimi** muscle is triangular and deep to the other two hypothenar muscles.

The 12 intermediate (midpalmar) muscles act on all the digits except the thumb. The intermediate muscles include the lumbricals, palmar interossei, and dorsal interossei. The **lumbricals,** as their name indicates, are worm-shaped. They originate from and insert into the tendons of other muscles (flexor digitorum profundus and extensor digitorum). The **palmar interossei** are the smaller and most superficial of the interossei muscles. The **dorsal interossei** are the deep interossei muscles. Both sets of interossei muscles are located between the metacarpals and are important in abduction, adduction, flexion, and extension of the fingers, and in movements in skilled activities such as writing, typing, and playing a piano.

The functional importance of the hand is readily apparent when one considers that certain hand injuries can result in permanent disability. Most of the dexterity of the hand depends on movements of the thumb. The general activities of the hand are free motion, power grip (forcible movement of the fingers and thumb against the palm, as in squeezing), precision handling (a change in position of a handled object that requires exact control of finger and thumb positions, as in winding a watch or threading a needle), and pinch (compression between the thumb and index finger or between the thumb and first two fingers).

Movements of the thumb are very important in the precise activities of the hand, and they are defined in different planes from comparable movements of other digits because the thumb is positioned at a right angle to the other digits. The five principal movements of the thumb are illustrated in Figure 10.19f and include *flexion* (movement of the thumb medially across the palm), *extension* (movement of the thumb laterally away from the palm), *abduction* (movement of the thumb in an anteroposterior plane away from the palm), *adduction* (movement of the thumb in an anteroposterior plane toward the palm), and *opposition* (movement of the thumb across the palm so that the tip of the thumb meets the tip of a finger). Opposition is the single most distinctive digital movement that gives humans and other primates the ability to precisely grasp and manipulate objects.

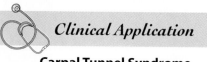

Clinical Application

Carpal Tunnel Syndrome

The **carpal tunnel** is a narrow passageway formed anteriorly by the flexor retinaculum and posteriorly by the carpal bones. Through this tunnel pass the median nerve, the most superficial structure, and the long flexor tendons for the digits. Structures within the carpal tunnel, especially the median nerve, are vulnerable to nerve compression and the resulting condition is called **carpal tunnel syndrome.** Compression of the median nerve leads to sensory changes over the lateral side of the hand and muscle weakness in the thenar eminence. The condition may be caused by inflammation of the digital tendon sheaths, fluid retention, excessive exercise, infection, trauma, and repetitive activities that involve flexion of the wrist, such as keyboarding, typing, cutting hair, and playing a piano.

RELATING MUSCLES TO MOVEMENTS

After you have studied the muscles in this exhibit, arrange them according to the following actions on the thumb at the carpometacarpal and metaphalangeal joints: (1) abduction, (2) adduction, (3) flexion, and (4) opposition, and the following actions on the fingers at the metacarpophalangeal and interphalangeal joints: (1) abduction, (2) adduction, (3) flexion, and (4) extension. The same muscle may be mentioned more than once.

INNERVATION

The abductor pollicis brevis and opponens pollicis are innervated by the median nerve. The adductor pollicis, abductor digiti minimi, flexor digit minimi brevis, opponens digiti minimi, dorsal interossei, and palmar interossei are innervated by the ulnar nerve. The flexor pollicis brevis and lumbricals are innervated by the median and ulnar nerves. The nerves listed here are derived from the brachial plexus, which is described and illustrated in Exhibit 17.2 on pages 530–535.

Exhibit 10.17 (*continued*)

MUSCLE	ORIGIN	INSERTION	ACTION
Thenar			
Abductor pollicis brevis (*abductor* = moves part away from middle; *pollex* = thumb; *brevis* = short)	Flexor retinaculum, scaphoid, and trapezium.	Lateral side of proximal phalanx of thumb.	Abducts thumb at carpometacarpal and metacarpophalangeal joints.
Opponens pollicis (*opponens* = opposes)	Flexor retinaculum and trapezium.	Lateral side of first metacarpal (thumb).	Moves thumb across palm to meet little finger (opposition) at the carpometacarpal joint.
Flexor pollicis brevis (*flexor* = decreases angle at joint)	Flexor retinaculum, trapezium, capitate, and trapezoid.	Lateral side of proximal phalanx of thumb.	Flexes thumb at carpometacarpal and metacarpophalangeal joints.
Adductor pollicis (*adductor* = moves part toward midline)	Oblique head; capitate and second and third metacarpals; transverse head: third metacarpal.	Medial side of proximal phalanx of thumb by a tendon containing a sesamoid bone.	Adducts thumb at carpometacarpal and metacarpophalangeal joints.
Hypothenar			
Abductor digiti minimi (*digit* = finger or toe; *minimi* = little finger)	Pisiform and tendon of flexor carpi ulnaris.	Medial side of proximal phalanx of little finger.	Abducts and flexes little finger at metacarpophalangeal joint.
Flexor digiti minimi brevis	Flexor retinaculum and hamate.	Medial side of proximal phalanx of little finger.	Flexes little finger at carpometacarpal and metacarpophalangeal joints.
Opponens digiti minimi	Flexor retinaculum and hamate.	Medial side of fifth metacarpal (little finger).	Moves little finger across palm to meet thumb (opposition) at the carpometacarpal joint.
Intermediate (Midpalmar)			
Lumbricals (LUM-bri-kals; *lumbricus* = earthworm) (four muscles)	Lateral sides of tendons and flexor digitorum profundus of all four fingers.	Lateral sides of tendons of extensor digitorum on proximal phalanges of all four fingers.	Flex fingers at metacarpophalangeal joints and extend fingers at interphalangeal joints.
Palmar interossei (in′-ter-OS-ē-ī (*palmar* = palm; *inter* = between; *ossei* = bones)	Sides of shafts of metacarpals of all digits (except the middle one)	Sides of bases of proximal phalanges of all digits (except the middle one).	Abduct fingers at metacarpophalangeal joints; flex fingers at metacarpophalangeal joints.
Dorsal interossei (in′-ter-OS-ē-ī; *dorsal* = back surface) (four muscles)	Adjacent sides of metacarpals.	Proximal phalanx of all four fingers.	Abduct fingers at metacarpophalangeal joints; flex fingers at metacarpophalangeal joints; and extend fingers at interphalangeal joints.

▶

Exhibit 10.17 *Intrinsic Muscles of the Hand (continued)*

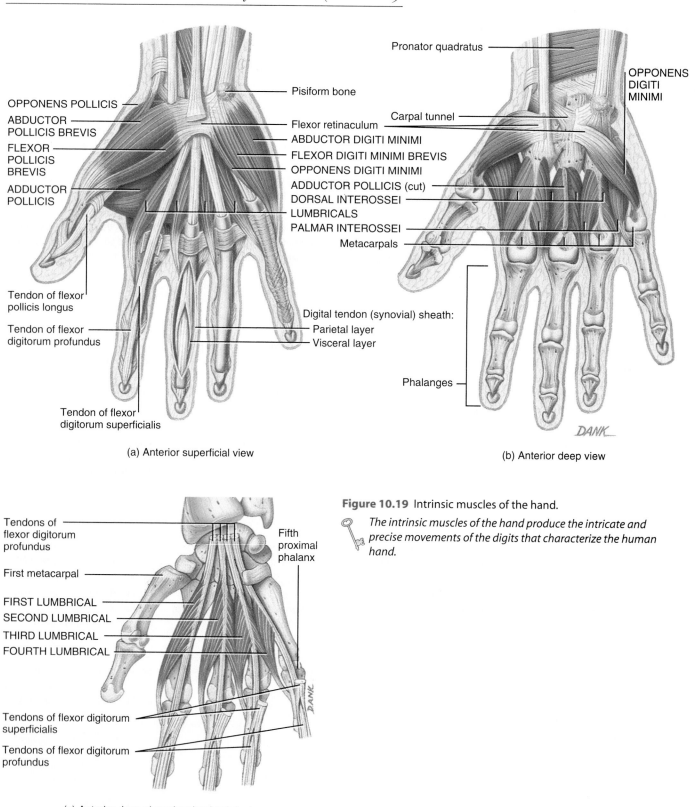

OPPONENS POLLICIS
ABDUCTOR POLLICIS BREVIS
FLEXOR POLLICIS BREVIS
ADDUCTOR POLLICIS

Pisiform bone

Flexor retinaculum
ABDUCTOR DIGITI MINIMI
FLEXOR DIGITI MINIMI BREVIS
OPPONENS DIGITI MINIMI
ADDUCTOR POLLICIS (cut)
DORSAL INTEROSSEI
LUMBRICALS
PALMAR INTEROSSEI
Metacarpals

Tendon of flexor pollicis longus

Tendon of flexor digitorum profundus

Digital tendon (synovial) sheath:
Parietal layer
Visceral layer

Tendon of flexor digitorum superficialis

(a) Anterior superficial view

Pronator quadratus

OPPONENS DIGITI MINIMI

Carpal tunnel

Phalanges

DANK

(b) Anterior deep view

Tendons of flexor digitorum profundus

First metacarpal

FIRST LUMBRICAL
SECOND LUMBRICAL
THIRD LUMBRICAL
FOURTH LUMBRICAL

Fifth proximal phalanx

Tendons of flexor digitorum superficialis

Tendons of flexor digitorum profundus

DANK

(c) Anterior deep view showing lumbricals

Figure 10.19 Intrinsic muscles of the hand.

The intrinsic muscles of the hand produce the intricate and precise movements of the digits that characterize the human hand.

Exhibit 10.17 *(continued)*

Figure 10.19 (continued)

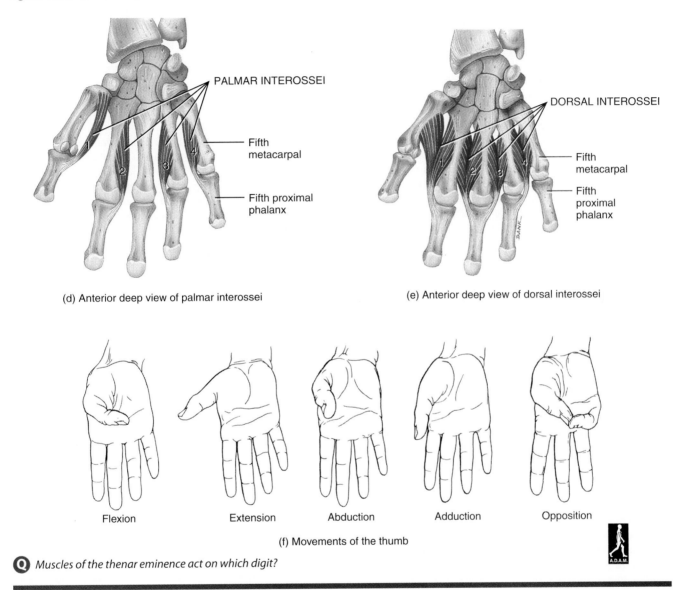

PALMAR INTEROSSEI

Fifth metacarpal

Fifth proximal phalanx

(d) Anterior deep view of palmar interossei

DORSAL INTEROSSEI

Fifth metacarpal

Fifth proximal phalanx

(e) Anterior deep view of dorsal interossei

Flexion   Extension   Abduction   Adduction   Opposition

(f) Movements of the thumb

Q *Muscles of the thenar eminence act on which digit?*

Exhibit 10.18 *Muscles that Move the Vertebral Column (Backbone)* *(Figure 10.20)*

The muscles that move the vertebral column (backbone) are quite complex because they have multiple origins and insertions and there is considerable overlap among them. One way to group the muscles is on the basis of the general direction of the muscle bundles and their approximate lengths. For example, the splenius muscles arise from the midline and extend laterally and superiorly to their insertions. The erector spinae (sacrospinalis) muscle arises from either the midline or more laterally but usually runs almost longitudinally, with neither a significant lateral nor medial direction as it is traced superiorly. The transversospinalis muscles arise laterally but extend toward the midline as they are traced superiorly. Deep to these three muscle groups are small segmental muscles that extend between spinous processes or transverse processes of vertebrae. Because the scalene muscles also assist in moving the vertebral column, they are included in this exhibit. Note in Exhibit 10.9 that the rectus abdominis, external oblique, internal oblique, and quadratus lumborum muscles also play a role in moving the vertebral column.

The bandage-like **splenius** muscles are attached to the sides and back of the neck. The two muscles in this group are named on the basis of their superior attachments (insertions): **splenius capitis** (head region) and **splenius cervicis** (cervical region). They extend the head and laterally flex and rotate the head.

The **erector spinae (sacrospinalis)** is the largest muscle mass of the back, forming a prominent bulge on either side of the vertebral column. It is the chief extensor of the vertebral column. It is also important in controlling flexion, lateral flexion, and rotation of the vertebral column and in maintaining the lumbar curve, because the main mass of the muscle is in the lumbar region. It consists of three groups: iliocostalis (laterally placed), longissimus (intermediately placed), and spinalis (medially placed). These groups, in turn, consist of a series of overlapping muscles, and the muscles within the groups are named according to the regions of the body with which they are associated. The **iliocostalis group** consists of three muscles: the **iliocostalis cervicis** (cervical region), **iliocostalis thoracis** (thoracic region), and **iliocostalis lumborum** (lumbar region). The **longissimus group** resembles a herring bone and consists of three muscles: the **longissimus capitis** (head region), **longissimus cervicis** (cervical region), and **longissimus thoracis** (thoracic region). The **spinalis group** also consists of three muscles: the **spinalis capitis, spinalis cervicis,** and **spinalis thoracis.**

The **transversospinalis** muscles are named because their fibers run from the transverse processes to the spinous processes of the vertebrae. The semispinalis muscles in this group are also named according to the region of the body with which they are associated: **semispinalis capitis** (head region), **semispinalis cervicis** (cervical region), and **semispinalis thoracis** (thoracic region). These muscles extend the vertebral column and rotate the head. The **multifidus** muscle in this group, as

its name implies, is split into several bundles. It extends and laterally flexes the vertebral column and rotates the head. The **rotatores** muscles of this group are short and run the entire length of the vertebral column. They extend and rotate the vertebral column.

Within the **segmental** muscle group are the **interspinales** and **intertransversarii** muscles, which unite the spinous and transverse processes of consecutive vertebrae. They function primarily in stabilizing the vertebral column during its movements.

Within the **scalene** group, the **anterior scalene** muscle is anterior to the middle scalene muscle, the **middle scalene** muscle is intermediate in placement and is the longest and largest of the scalene muscles, and the **posterior scalene** muscle is posterior to the middle scalene muscle and is the smallest of the scalene muscles. These muscles flex, laterally flex, and rotate the head and assist in deep inspiration.

RELATING MUSCLES TO MOVEMENTS

After you have studied the muscles in this exhibit, arrange them according to the following actions on the head at the atlanto-occipital and intervertebral joints: (1) flexion, (2) extension, (3) lateral flexion, (4) rotation to same side as contracting muscle, and (5) rotation to opposite side as contracting muscle; the following actions on the vertebral column at the intervertebral joints: (1) flexion, (2) extension, (3) lateral flexion, (4) rotation, and (5) stabilization; and the following action on the ribs: elevation during deep inspiration. The same muscle may be mentioned more than once.

INNERVATION

The splenius capitis is innervated by the dorsal rami of the middle cervical nerves; the splenius cervicis is innervated by the dorsal rami of the inferior cervical nerves; the iliocostalis cervicis and semispinalis capitis are innervated by the dorsal rami of cervical nerves; the iliocostalis thoracis is innervated by the dorsal rami of the thoracic (intercostal) nerves; the iliocostalis lumborum is innervated by the dorsal rami of the lumbar nerves; the longissimus capitis is innervated by the middle and inferior cervical nerves; the longissimus cervicis, longissimus thoracis, all spinalis muscles, multifidus, rotatores, and interspinales are innervated by the dorsal rami of spinal nerves; the semispinalis cervicis and semispinalis thoracis are innervated by the dorsal rami of cervical and thoracic nerves; the intertransversarii are innervated by the dorsal and ventral rami of spinal nerves; the anterior scalene is innervated by the ventral rami of cervical nerves C5–C6; the middle scalene is innervated by the ventral rami of cervical nerves C3–C8; and the posterior scalene is innervated by the ventral rami of cervical nerves C6–C8.

Exhibit 10.18 *(continued)*

MUSCLE	ORIGIN	INSERTION	ACTION
Splenius (SPLĒ-nē-us)			
Splenius capitis (KAP-i-tis; *splenium* = bandage; *caput* = head)	Ligamentum nuchae and spinous processes of seventh cervical vertebra and first three or four thoracic vertebrae.	Occipital bone and mastoid process of temporal bone.	Acting together (bilaterally), extend head; acting singly (unilaterally), laterally flex and rotate head to same side as contracting muscle.
Splenius cervicis (SER-vi-kis; *cervix* = neck)	Spinous processes of third through sixth thoracic verte-brae.	Transverse processes of first two or four cervical vertebrae.	Acting together, extend head; acting singly, laterally flex and rotate head to same side as contracting muscle.
Erector Spinae (e-REK-tor SPI-nē) **(Sacrospinalis)**			
Iliocostalis (Lateral) Group			
Iliocostalis cervicis (il′-ē-ō-kos-TAL-is SER-vi-kis; *ilium* = flank; *costa* = rib)	Superior six ribs.	Transverse processes of fourth to sixth cervical vertebrae.	Acting together, muscles of each region (cervical, thoracic, and lumbar) extend and maintain erect posture of vertebral column of their respective regions; acting singly, laterally flex vertebral column of their respective regions.
Iliocostalis thoracis (il′-ē-ō-kos-TAL-is thō-RA-kis; *thorax* = chest)	Inferior six ribs.	Superior six ribs.	
Iliocostalis lumborum (il′-ē-ō-kos-TAL-is lum-BOR-um)	Iliac crest.	Inferior six ribs.	
Longissimus (Intermediate) Group			
Longissimus capitis (lon-JIS-i-mus KAP-i-tis; *longis-simus* = longest)	Transverse processes of su-perior four thoracic vertebrae and articular processes of in-ferior four cervical vertebrae.	Mastoid process of temporal bone.	Acting together, both longis-simus capitis muscles extend head; acting singly, rotate head to same side as contracting muscle. Acting to-gether, longissimus cervicis and both longissimus thoracis muscles extend vertebral col-umn of their respective re-gions; acting singly, laterally flex vertebral column of their respective regions.
Longissimus cervicis (lon-JIS-i-mus SER-vi-kis)	Transverse processes of fourth and fifth thoracic vertebrae.	Transverse processes of second to sixth cervical vertebrae.	
Longissimus thoracis (lon-JIS-i-mus thō-RA-kis)	Transverse processes of lum-bar vertebrae.	Transverse processes of all thoracic and superior lumbar vertebrae and ninth and tenth ribs.	

▶

Exhibit 10.18 *Muscles that Move the Vertebral Column (Backbone) (continued)*

MUSCLE	ORIGIN	INSERTION	ACTION
Spinalis (Medial) Group			
Spinalis capitis (spi-NA-lis KAP-i-tis; *spinalis* = vertebral column)	Arises with semispinalis capitis.	Occipital bone.	Acting together, muscles of each region (cervical, thoracic, and lumbar) extend vertebral column of their respective regions.
Spinalis cervicis (spi-NA-lis SER-vi-kis)	Ligamentum nuchae and spinous process of seventh cervical vertebra.	Spinous process of axis.	
Spinalis thoracis (spi-NA-lis thō-RA-kis)	Spinous processes of superior lumbar and inferior thoracic vertebrae.	Spinous processes of superior thoracic vertebrae.	
Transversospinalis (trans-ver'-sō-spi-NA-lis)			
Semispinalis capitis (sem'-ē-spi-NA-lis KAP-i-tis; *semi* = partially or one-half)	Transverse processes of first six or seven thoracic vertebrae and seventh cervical vertebra, and articular processes of fourth, fifth, and sixth cervical vertebrae.	Occipital bone.	Acting together, extend head; acting singly, rotate head to side opposite contracting muscle.
Semispinalis cervicis (sem'-ē-spi-NA-lis SER-vi-kis)	Transverse processes of superior five or six thoracic vertebrae.	Spinous processes of first to fifth cervical vertebrae.	Acting together, both semispinalis cervicis and both semispinalis thoracis muscles extend vertebral column of their respective regions; acting singly, rotate head to side opposite contracting muscle.
Semispinalis thoracis (sem'-ē-spi-NA-lis thō-RA-kis)	Transverse processes of sixth to tenth thoracic vertebrae.	Spinous processes of superior four thoracic and last two cervical vertebrae.	
Multifidus (mul-TIF-i-dus; *multi* = many; *findere* = to split)	Sacrum, ilium, transverse processes of lumbar, thoracic, and inferior four cervical vertebrae.	Spinous process of a more superior vertebra.	Acting together, extend vertebral column; acting singly, laterally flex vertebral column and rotate head to side opposite contracting muscle.
Rotatores (rō'-ta-TŌ-rez; *rotare* = to turn)	Transverse processes of all vertebrae.	Spinous process of vertebra superior to the one of origin.	Acting together, extend vertebral column; acting singly, rotate vertebral column to side opposite contracting muscle.

Exhibit 10.18 (continued)

MUSCLE	ORIGIN	INSERTION	ACTION
Segmental (seg-MEN-tal)			
Interspinales (in-ter-SPĪ-nāl-ez; *inter* = between)	Superior surface of all spinous processes.	Inferior surface of spinous process of vertebra superior to the one of origin.	Acting together, extend vertebral column; acting singly, stabilize vertebral column during movement.
Intertransversarii (in'-ter-trans-vers-AR-ē-ī; *inter* = between)	Transverse processes of all vertebrae.	Transverse process of vertebra superior to the one of origin.	Acting together, extend vertebral column; acting singly, laterally flex vertebral column and stabilize it during movements.
Scalene (SKĀ-lēn)			
Anterior scalene (SKĀ-lēn; *anterior* = front; *skalenos* = uneven)	Transverse processes of third through sixth cervical vertebrae.	First rib.	Acting together, both anterior scalene and middle scalene muscles flex head and elevate first ribs during deep inspiration; acting singly, laterally flex head and rotate head to side opposite contracting muscle.
Middle scalene (SKĀ-lēn)	Transverse processes of inferior six cervical vertebrae.	First rib.	
Posterior scalene (SKĀ-lēn)	Transverse processes of fourth through sixth cervical vertebrae.	Second rib.	Acting together, flex head and elevate second ribs during deep inspiration; acting singly, laterally flex head and rotate head to side opposite contracting muscle.

►

Exhibit 10.18 *Muscles that Move the Vertebral Column (Backbone) (continued)*

Figure 10.20 Muscles that move the vertebral column (backbone).

The erector spinae group is the largest muscular mass of the body and is the chief extensor of the vertebral column.

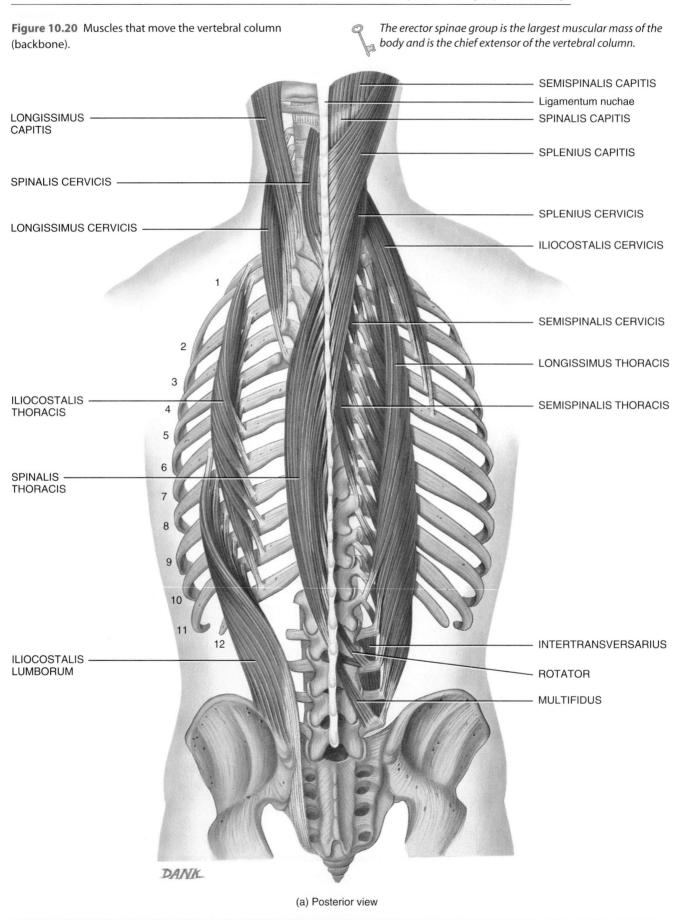

LONGISSIMUS CAPITIS

SPINALIS CERVICIS

LONGISSIMUS CERVICIS

ILIOCOSTALIS THORACIS

SPINALIS THORACIS

ILIOCOSTALIS LUMBORUM

SEMISPINALIS CAPITIS

Ligamentum nuchae

SPINALIS CAPITIS

SPLENIUS CAPITIS

SPLENIUS CERVICIS

ILIOCOSTALIS CERVICIS

SEMISPINALIS CERVICIS

LONGISSIMUS THORACIS

SEMISPINALIS THORACIS

INTERTRANSVERSARIUS

ROTATOR

MULTIFIDUS

DANK

(a) Posterior view

Exhibit 10.18 *(continued)*

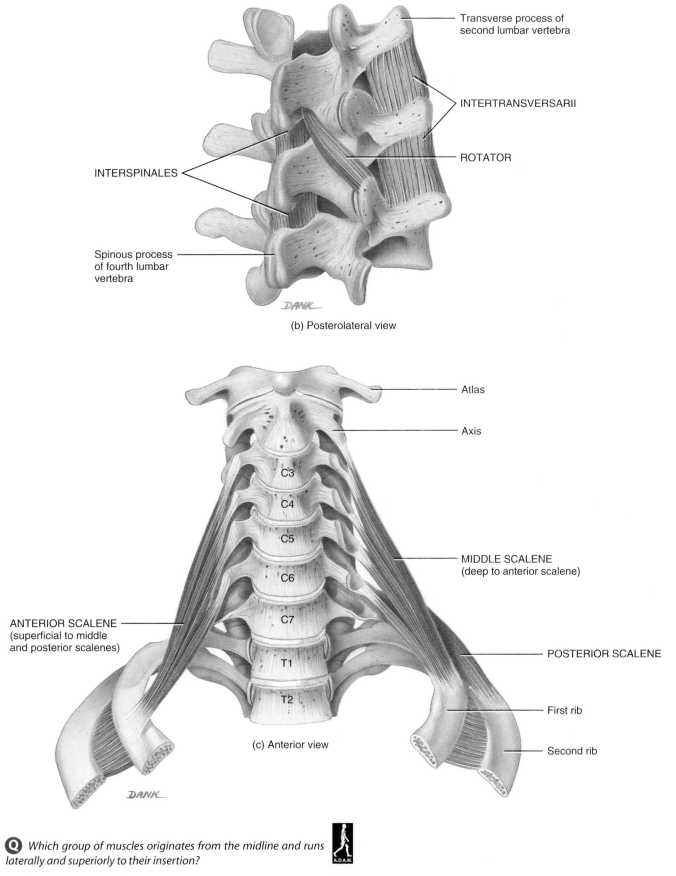

Transverse process of
second lumbar vertebra

INTERTRANSVERSARII

ROTATOR

INTERSPINALES

Spinous process
of fourth lumbar
vertebra

DANK

(b) Posterolateral view

Atlas

Axis

C3

C4

C5

C6

MIDDLE SCALENE
(deep to anterior scalene)

C7

ANTERIOR SCALENE
(superficial to middle
and posterior scalenes)

T1

POSTERIOR SCALENE

T2

First rib

(c) Anterior view

Second rib

DANK

Q *Which group of muscles originates from the midline and runs
laterally and superiorly to their insertion?*

Exhibit 10.19 *Muscles that Move the Femur (Thigh) (Figure 10.21)*

As you will see, muscles of the lower limbs are larger and more powerful than those of the upper limbs because lower limb muscles function in stability, locomotion, and maintenance of posture. Upper limb muscles are characterized by versatility of movement. In addition, muscles of the lower limbs often cross two joints and act equally on both.

The majority of muscles that move the femur originate on the pelvic girdle and insert on the femur. The **psoas major** and **iliacus** muscles are together referred to as the **iliopsoas** (il′-ē-ō-SŌ-as) muscle because they share a common insertion (lesser trochanter of femur). There are three gluteal muscles: gluteus maximus, gluteus medius, and gluteus minimus. The **gluteus maximus** is the largest and heaviest of the three muscles and is actually one of the largest muscles in the body. It is the chief extensor of the femur. The **gluteus medius** is mostly deep to the gluteus maximus and is a powerful abductor of the femur at the hip joint. It is a common site for an intramuscular injection. The **gluteus minimus** is the smallest of the gluteal muscles and lies deep to the gluteus medius.

The **tensor fasciae latae** muscle is located on the lateral surface of the thigh. A layer of deep fascia composed of dense connective tissue encircles the entire thigh and is referred to as the **fascia lata.** It is well-developed laterally where, together with the tendons of the tensor fasciae and gluteus maximus muscles, it forms a structure called the **iliotibial tract.** The tract inserts into the lateral condyle of the tibia.

The **piriformis, obturator internus, obturator externus, superior gemellus, inferior gemellus,** and **quadratus femoris** muscles are all deep to the gluteus maximus muscle and function as lateral rotators of the femur at the hip joint.

Three muscles on the medial aspect of the thigh are the **adductor longus, adductor brevis,** and **adductor magnus.** They originate on the pubic bone and insert on the femur. All three muscles adduct, flex, and medially rotate the femur at the hip joint. The **pectineus** muscle also adducts and flexes the femur at the hip joint.

Technically, the adductor muscles and pectineus muscles are components of the medial compartment of the thigh and could also be included in Exhibit 10.20. However, they are included here because they act on the femur.

Clinical Application

Pulled Groin

Certain athletic and other activities may cause a strain, stretching, or tearing of the distal attachments of the medial muscles of the thigh, iliopsoas and/or adductor muscles. It generally results from activities that involve quick sprints, such as may occur in soccer, tennis, football, and running. When this occurs, the condition is called a **pulled groin.**

RELATING MUSCLES TO MOVEMENTS

After you have studied the muscles in this exhibit, arrange them according to the following actions on the thigh at the hip joint: (1) flexion, (2) extension, (3) abduction, (4) adduction, (5) medial rotation, and (6) lateral rotation. The same muscle may be mentioned more than once.

INNERVATION

The psoas major is innervated by lumbar nerves L2–L3; the iliacus and pectineus are innervated by the femoral nerve; the gluteus maximus is innervated by the inferior gluteal nerve; the gluteus medius and minimus and tensor fasciae latae are innervated by the superior gluteal nerve; the piriformis is innervated by sacral nerves S1 or S2, mainly S1; the obturator internus and superior gemellus are innervated by the nerve to obturator internus; the obturator externus, adductor longus, and adductor brevis are innervated by the obturator nerve; the inferior gemellus and quadratus femoris are innervated by the nerve to quadratus femoris; and the adductor magnus is innervated by the obturator and sciatic nerves. The nerves listed here are derived from the lumbar and sacral plexuses, which are described and illustrated in Exhibits 17.3 and 17.4 on pages 536–539.

Exhibit 10.19 *(continued)*

MUSCLE	ORIGIN	INSERTION	ACTION
Psoas (SŌ-as) **major** (*psoa* = muscle of loin)	Transverse processes and bodies of lumbar vertebrae.	With iliacus into lesser trochanter of femur.	Both psoas major and iliacus muscles acting together flex thigh at hip joint, rotate thigh laterally, and flex trunk on the hip as in sitting up from the supine position.
Iliacus (il'-ē-AK-us; *iliac* = ilium)	Iliac fossa.	With psoas major into lesser trochanter of femur.	
Gluteus maximus (GLOO-tē-us MAK-si-mus; *glutos* = buttock; *maximus* = largest; strongest single muscle in body)	Iliac crest, sacrum, coccyx, and aponeurosis of sacrospinalis.	Iliotibial tract of fascia lata and lateral part of linea aspera under greater trochanter (gluteal tuberosity) of femur.	Extends thigh at hip joint and laterally rotates thigh.
Gluteus medius (GLOO-tē-us MĒ-dē-us; *media* = middle)	Ilium.	Greater trochanter of femur.	Abducts thigh at hip joint and medially rotates thigh.
Gluteus minimus (GLOO-tē-us MIN-i-mus; *minimus* = smallest)	Ilium.	Greater trochanter of femur.	Abducts thigh at hip joint and medially rotates thigh.
Tensor fasciae latae (TEN-sor FA-shē-ē LĀ-tē; *tensor* = makes tense; *fascia* = band; *latus* = wide)	Iliac crest.	Tibia by way of the iliotibial tract.	Flexes and abducts thigh at hip joint.
Piriformis (pir-i-FOR-mis; *pirum* = pear; *forma* = shape)	Anterior sacrum.	Superior border of greater trochanter of femur.	Laterally rotates and abducts thigh at hip joint.
Obturator internus (OB-too-rā'-tor in-TER-nus; *obturator* = obturator foramen; *internus* = inside)	Inner surface of obturator foramen, pubis, and ischium.	Greater trochanter of femur.	Laterally rotates and abducts thigh at hip joint.
Obturator externus (OB-too-rā'-tor ex-TER-nus; *externus* = outside)	Outer surface of obturator membrane.	Deep depression inferior to greater trochanter (trochanteric fossa) of femur.	Laterally rotates and abducts thigh at hip joint.

▶

Exhibit 10.19 *Muscles that Move the Femur (Thigh) (continued)*

MUSCLE	ORIGIN	INSERTION	ACTION
Superior gemellus (jem-EL-lus; *superior* = above; *gemellus* = twins)	Ischial spine.	Greater trochanter of femur.	Laterally rotates and abducts thigh at hip joint.
Inferior gemellus (jem-EL-lus; *inferior* = below)	Ischial tuberosity.	Greater trochanter of femur.	Laterally rotates and abducts thigh at hip joint.
Quadratus femoris (kwod-RĀ-tus FEM-or-is; *quad* = four; *femoris* = femur)	Ischial tuberosity.	Elevation superior to mid-portion of intertrochanteric crest (quadrate tubercle) on posterior femur.	Laterally rotates and abducts thigh at hip joint.
Adductor longus (LONG-us; *adductor* = moves part closer to midline; *longus* = long)	Pubic crest and pubic symphysis.	Linea aspera of femur.	Adducts and flexes thigh at hip joint and medially rotates thigh.
Adductor brevis (BREV-is; *brevis* = short)	Inferior ramus of pubis.	Superior half of linea aspera of femur.	Adducts and flexes thigh at hip joint and medially rotates thigh.
Adductor magnus (MAG-nus; *magnus* = large)	Inferior ramus of pubis and ischium to ischial tuberosity.	Linea aspera of femur.	Adducts thigh at hip joint and medially rotates thigh; anterior part flexes thigh at hip joint, posterior part extends thigh at hip joint.
Pectineus (pek-TIN-ē-us; *pecten* = comb-shaped)	Superior ramus of pubis.	Pectineal line of femur, between lesser trochanter and linea aspera.	Flexes and adducts thigh at hip joint.

Exhibit 10.19 (*continued*)

Figure 10.21 Muscles that move the femur (thigh)

Most muscles that move the femur originate on the pelvic (hip) girdle and insert on the femur.

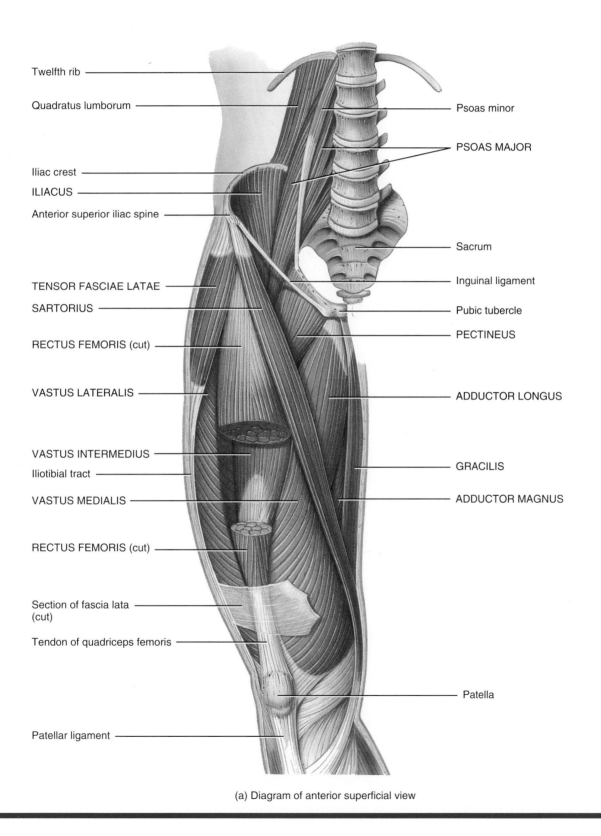

Twelfth rib

Quadratus lumborum

Iliac crest

ILIACUS

Anterior superior iliac spine

TENSOR FASCIAE LATAE

SARTORIUS

RECTUS FEMORIS (cut)

VASTUS LATERALIS

VASTUS INTERMEDIUS

Iliotibial tract

VASTUS MEDIALIS

RECTUS FEMORIS (cut)

Section of fascia lata (cut)

Tendon of quadriceps femoris

Patellar ligament

Psoas minor

PSOAS MAJOR

Sacrum

Inguinal ligament

Pubic tubercle

PECTINEUS

ADDUCTOR LONGUS

GRACILIS

ADDUCTOR MAGNUS

Patella

(a) Diagram of anterior superficial view

Exhibit 10.19 *Muscles that Move the Femur (Thigh) (continued)*

Figure 10.21 (continued)

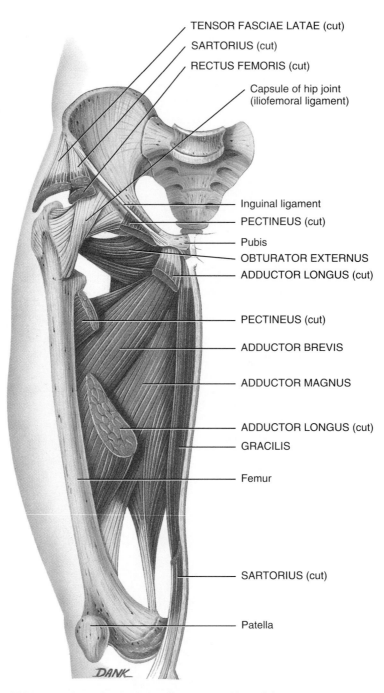

- TENSOR FASCIAE LATAE (cut)
- SARTORIUS (cut)
- RECTUS FEMORIS (cut)
- Capsule of hip joint (iliofemoral ligament)
- Inguinal ligament
- PECTINEUS (cut)
- Pubis
- OBTURATOR EXTERNUS
- ADDUCTOR LONGUS (cut)
- PECTINEUS (cut)
- ADDUCTOR BREVIS
- ADDUCTOR MAGNUS
- ADDUCTOR LONGUS (cut)
- GRACILIS
- Femur
- SARTORIUS (cut)
- Patella

DANK

(b) Diagram of anterior deep view (femur rotated laterally)

Exhibit 10.19 *(continued)*

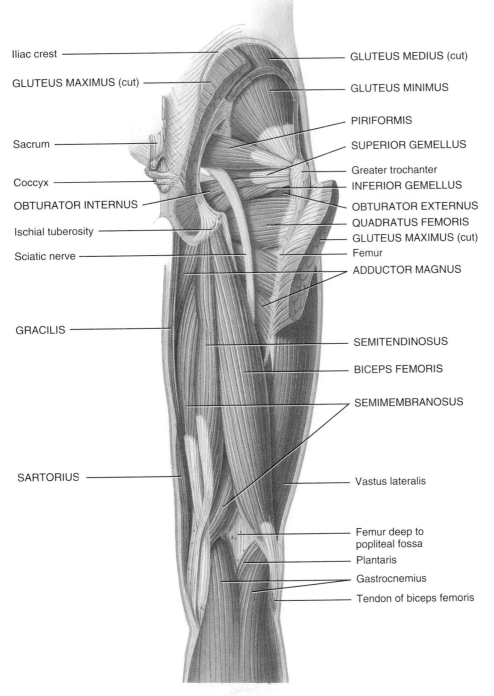

Iliac crest

GLUTEUS MAXIMUS (cut)

Sacrum

Coccyx

OBTURATOR INTERNUS

Ischial tuberosity

Sciatic nerve

GRACILIS

SARTORIUS

GLUTEUS MEDIUS (cut)

GLUTEUS MINIMUS

PIRIFORMIS

SUPERIOR GEMELLUS

Greater trochanter

INFERIOR GEMELLUS

OBTURATOR EXTERNUS

QUADRATUS FEMORIS

GLUTEUS MAXIMUS (cut)

Femur

ADDUCTOR MAGNUS

SEMITENDINOSUS

BICEPS FEMORIS

SEMIMEMBRANOSUS

Vastus lateralis

Femur deep to
popliteal fossa

Plantaris

Gastrocnemius

Tendon of biceps femoris

(c) Diagram of posterior superficial view

Exhibit 10.19 *Muscles that Move the Femur (Thigh) (continued)*

Figure 10.21 (continued)

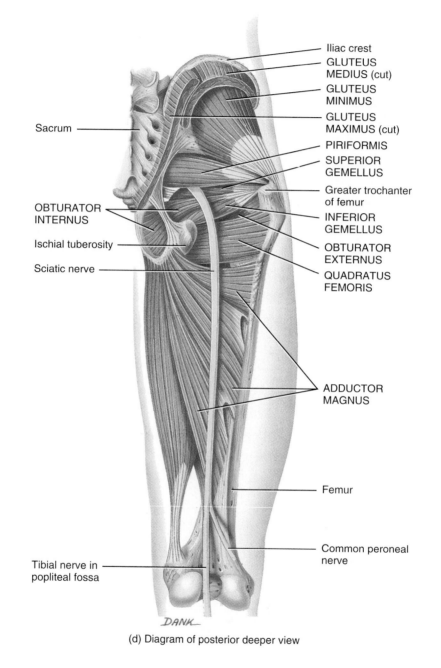

Iliac crest
GLUTEUS MEDIUS (cut)
GLUTEUS MINIMUS
GLUTEUS MAXIMUS (cut)
PIRIFORMIS
SUPERIOR GEMELLUS
Greater trochanter of femur
INFERIOR GEMELLUS
OBTURATOR EXTERNUS
QUADRATUS FEMORIS

Sacrum
OBTURATOR INTERNUS
Ischial tuberosity
Sciatic nerve

ADDUCTOR MAGNUS

Femur

Common peroneal nerve

Tibial nerve in popliteal fossa

DANK

(d) Diagram of posterior deeper view

Exhibit 10.19 *(continued)*

Figure 10.21 (continued)

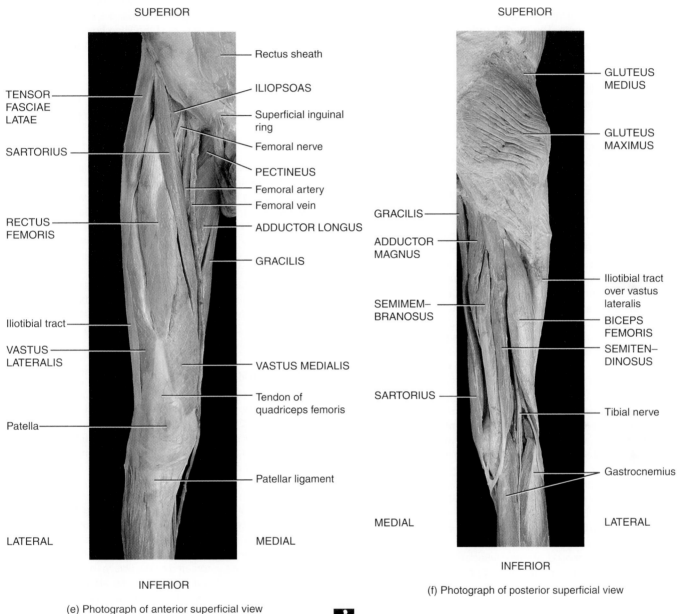

(e) Photograph of anterior superficial view

(f) Photograph of posterior superficial view

Q *What are the principal differences between the muscles of the upper and lower limbs?*

Exhibit 10.20 *Muscles that Act on the Femur (Thigh) and Tibia and Fibula (Leg) (Figures 10.21 and 10.22)*

The muscles that act on the femur (thigh) and tibia and fibula (leg) are separated by deep fascia into medial, anterior, and posterior compartments. The **medial (adductor) compartment** is so named because its muscles adduct the femur at the hip joint. (See the adductor magnus, adductor longus, adductor brevis, and pectineus, which are components of the medial compartment, in Exhibit 10.19.) The **gracilis,** the other muscle in the medial compartment, not only adducts the thigh, but also flexes the leg at the knee joint. For this reason, it is discussed in this exhibit. The gracilis is a long, straplike muscle that lies on the medial aspect of the thigh and knee.

The **anterior (extensor) compartment** is so designated because its muscles extend the leg (and also flex the thigh). This compartment is composed of the quadriceps femoris and sartorius muscles. The **quadriceps femoris** muscle is the biggest muscle in the body, covering almost all of the anterior surface and sides of the thigh. The muscle is actually a composite muscle that includes four distinct parts, usually described as four separate muscles: (1) **rectus femoris,** on the anterior aspect of the thigh; (2) **vastus lateralis,** on the lateral aspect of the thigh; (3) **vastus medialis,** on the medial aspect of the thigh; and (4) **vastus intermedius,** located deep to the rectus femoris between the vastus lateralis and vastus medialis. The common tendon for the four muscles is known as the **quadriceps tendon,** which inserts into the patella. The tendon continues inferior to the patella as the **patellar ligament,** which attaches to the tibial tuberosity. The quadriceps femoris muscle is the great extensor muscle of the leg. The **sartorius** is a long, narrow muscle that forms a band across the thigh from the ilium of the hipbone to the medial side of the tibia. The various movements it produces help effect the cross-legged sitting position in which the heel of one limb is placed on the knee of the opposite limb. It is known as the tailor's muscle because tailors frequently assume this cross-legged sitting position. (Because the major action of the sartorius muscle is to move the thigh rather than the leg, it could also have been included in Exhibit 10.19.)

The **posterior (flexor) compartment** is so named because its muscles flex the leg (and also extend the thigh). This compartment is composed of three muscles collectively called the **hamstrings:** (1) **biceps femoris,** (2) **semitendinosus,** and (3) **semimembranosus.** The hamstrings are so named because their tendons are long and stringlike in the popliteal area, and from an old practice of butchers in which they hung hams for smoking by these long tendons. Because the hamstrings span

two joints (hip and knee), they are both extensors of the thigh and flexors of the leg. The **popliteal fossa** is a diamond-shaped space on the posterior aspect of the knee bordered laterally by the tendons of the biceps femoris muscle and medially by the tendons of the semitendinosus and semimembranosus muscles.

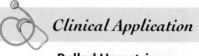

Clinical Application

Pulled Hamstrings

A strain or partial tear of the proximal hamstring muscles is referred to as **pulled hamstrings** or **hamstring strains.** They are common sports injuries in individuals who run very hard and/or are required to perform quick starts and stops. Sometimes the violent muscular exertion required to perform a feat tears off part of the tendinous origins of the hamstrings, especially the biceps femoris, from the ischial tuberosity. This is usually accompanied by a contusion (bruising), tearing of some of the muscle fibers, and rupture of blood vessels, producing a hematoma (collection of blood) and pain. Adequate training with good balance between the quadriceps femoris and hamstrings and stretching exercises before running or competing are important in preventing this injury. ■

RELATING MUSCLES TO MOVEMENTS

After you have studied the muscles in this exhibit, arrange them according to the following actions on the thigh at the hip joint: (1) abduction, (2) adduction, (3) lateral rotation, (4) flexion, and (5) extension, and according to the following actions on the leg at the knee joint: (1) flexion and (2) extension. The same muscle may be mentioned more than once.

INNERVATION

The gracilis is innervated by the obturator nerve; the quadriceps femoris and sartorius are innervated by the femoral nerve; the biceps femoris is innervated by the tibial and common peroneal nerves from the sciatic nerve; and the semimembranosus and semitendinosus are innervated by the tibial nerve from the sciatic nerve. The obturator and femoral nerves are derived from the lumbar plexus (see Exhibit 17.3 on pages 536–537), whereas the sciatic nerve branches from the sacral plexus (see Exhibit 17.4 on pages 538–539).

Exhibit 10.20 (*continued*)

MUSCLE	ORIGIN	INSERTION	ACTION
Medial (adductor) compartment			
Adductor magnus (MAG-nus)			
Adductor longus (LONG-us)	See Exhibit 10.19.		
Adductor brevis (BREV-is)			
Pectineus (pek-TIN-ē-us)			
Gracilis (gra-SIL-is; *gracilis* = slender)	Pubic symphysis and pubic arch.	Medial surface of body of tibia.	Adducts thigh at hip joint, medially rotates thigh, and flexes leg at knee joint.
Anterior (extensor) compartment			
Quadriceps femoris (KWOD-ri-ceps FEM-or-is; *quadriceps* = four heads of origin; *femoris* = femur)			
Rectus femoris (REK-tus FEM-or-is; *rectus* = fibers parallel to midline)	Anterior inferior iliac spine.	Patella via quadriceps tendon and then tibial tuberosity via patellar ligament.	All four heads extend leg at knee joint; rectus femoris muscle acting alone also flexes thigh at hip joint.
Vastus lateralis (VAS-tus lat'-er-A-lis; *vastus* = large; *lateralis* = lateral)	Greater trochanter and linea aspera of femur.		
Vastus medialis (VAS-tus mē'-dē-A-lis; *medialis* = medial)	Linea aspera of femur.		
Vastus intermedius (VAS-tus in'-ter-MĒ-dē-us; *intermedius* = middle)	Anterior and lateral surfaces of body of femur.		
Sartorius (sar-TOR-ē-us; *sartor* = tailor; longest muscle in body)	Anterior superior iliac spine.	Medial surface of body of tibia.	Flexes leg at knee joint; flexes, abducts, and laterally rotates thigh at hip joint.
Posterior (flexor) compartment			
Hamstrings A collective designation for three separate muscles.			
Biceps femoris (BĪ-ceps FEM-or-is; *biceps* = two heads of origin)	Long head arises from ischial tuberosity; short head arises from linea aspera of femur.	Head of fibula and lateral condyle of tibia.	Flexes leg at knee joint and extends thigh at hip joint.
Semitendinosus (sem'-ē-TEN-di-nō-sus; *semi* = half; *tendo* = tendon)	Ischial tuberosity.	Proximal part of medial surface of shaft of tibia.	Flexes leg at knee joint and extends thigh at hip joint.
Semimembranosus (sem'-ē-MEM-bra-nō-sus; *membran* = membrane)	Ischial tuberosity.	Medial condyle of tibia.	Flexes leg at knee joint and extends thigh at hip joint.

▶

Exhibit 10.20 *Muscles that Act on the Femur (Thigh) and Tibia and Fibula (Leg) (continued)*

Figure 10.22 Muscles that act on the femur (thigh) and tibia and fibula (leg).

Muscles that act on the leg originate in the hip and thigh and are separated into compartments by deep fascia.

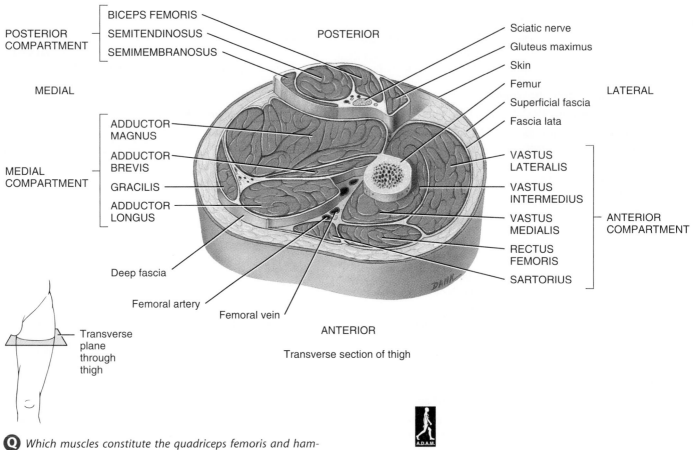

Transverse section of thigh

Q *Which muscles constitute the quadriceps femoris and hamstring muscles?*

Exhibit 10.21 *Muscles that Move the Foot and Toes (Figure 10.23)*

Muscles that move the foot and toes are located in the leg. The musculature of the leg, like that of the thigh, is divided by deep fascia into three compartments: anterior, lateral, and posterior. The **anterior compartment** consists of muscles that dorsiflex the foot. In a situation analogous to the wrist, the tendons of the muscles of the anterior compartment are held firmly to the ankle by thickenings of deep fascia called the **superior extensor retinaculum** *(transverse ligament of the ankle)* and **inferior extensor retinaculum** *(cruciate ligament of the ankle).*

Within the anterior compartment, the **tibialis anterior** is a long, thick muscle against the lateral surface of the tibia, where it is easy to palpate. The **extensor hallucis longus** is a thin muscle between and partly deep to the **tibialis anterior** and **extensor digitorum longus** muscles. This latter muscle is featherlike and lies lateral to the tibialis anterior muscle, where it can easily be palpated. The **peroneus tertius** muscle is actually part of the extensor digitorum longus, with which it shares a common origin.

The **lateral (peroneal) compartment** contains two muscles that plantar flex and evert the foot: **peroneus longus** and **peroneus brevis.**

The **posterior compartment** consists of muscles that are divisible into superficial and deep groups. The superficial muscles share a common tendon of insertion, the **calcaneal (Achilles) tendon,** the strongest tendon of the body, that inserts into the calcaneal bone of the ankle. The superficial and most of the deep muscles plantar flex the foot at the ankle joint. The superficial muscles of the posterior compartment are the gastrocnemius, soleus, and plantaris—the so-called calf muscles. The large size of these muscles is directly related to our upright stance, a characteristic of humans. The **gastrocnemius** is the most superficial muscle and forms the prominence of the calf. The **soleus** is broad, flat, and lies deep to the gastrocnemius. It derives its name from its resemblance to a flat fish (sole). The **plantaris** is a small muscle that may be absent; sometimes there are two of them in each leg. It runs obliquely between the gastrocnemius and soleus muscles.

The deep muscles of the posterior compartment are the popliteus, tibialis posterior, flexor digitorum longus, and flexor hallucis longus. The **popliteus** is a triangular muscle that forms the floor of the popliteal fossa. The **tibialis posterior** is the deep-est muscle in the posterior compartment. It lies between the flexor digitorum longus and flexor hallucis longus muscles. The **flexor digitorum longus** is smaller than the **flexor hallucis longus,** even though the former flexes four toes, whereas the latter flexes only the great toe at the interphalangeal joint.

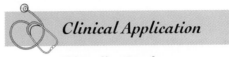

Clinical Application

Shinsplint Syndrome

Shinsplint syndrome, or simply **shinsplints,** refers to pain or soreness along the tibia, specifically the medial, distal two-thirds. It may be caused by tendinitis of the anterior compartment muscles, especially the tibialis anterior muscle, inflammation of the periosteum (periostitis) around the tibia, or stress fractures of the tibia. The tendinitis usually occurs when poorly conditioned runners run on hard or banked surfaces with poorly supportive running shoes. The condition may also occur as a result of vigorous activity of the legs following a period of relative inactivity. The muscles in the anterior compartment (mainly the tibialis anterior) can be strengthened to balance the stronger posterior compartment muscles. ▨

RELATING MUSCLES TO MOVEMENTS

After you have studied the muscles in this exhibit, arrange them according to the following actions on the foot at the ankle joint: (1) dorsiflexion and (2) plantar flexion; according to the following actions on the foot at the intertarsal joints: (1) inversion and (2) eversion; and according to the following actions on the toes at the metatarsophalangeal and intertarsal joints: (1) flexion and (2) extension. The same muscle may be mentioned more than once.

INNERVATION

Muscles of the anterior compartment are innervated by the deep peroneal nerve; muscles of the lateral compartment are innervated by the superficial peroneal nerve; and muscles of the posterior compartment are innervated by the tibial nerve. The nerves listed here are derived from the sacral plexus, which is described and illustrated in Exhibit 17.4 on pages 538–539.

►

Exhibit 10.21 *Muscles that Move the Foot and Toes (continued)*

MUSCLE	ORIGIN	INSERTION	ACTION
Anterior Compartment			
Tibialis anterior (tib'-ē-A-lis; *tibialis* = tibia; *anterior* = front)	Lateral condyle and body of tibia and interosseous membrane (sheet of fibrous tissue that holds shafts of tibia and fibula together).	First metatarsal and first (medial) cuneiform.	Dorsiflexes foot at ankle joint and inverts foot at intertarsal joints.
Extensor hallucis longus (HAL-a-kis LON-gus; *extensor* = increases angle at joint; *hallucis* = hallux or great toe; *longus* = long)	Anterior surface of fibula and interosseous membrane.	Distal phalanx of great toe.	Dorsiflexes foot at ankle joint, inverts foot at intertarsal joints, and extends proximal phalanx of great toe at metatarsophalangeal joint.
Extensor digitorum longus (di'-ji-TOR-um LON-gus)	Lateral condyle of tibia, anterior surface of fibula, and interosseous membrane.	Middle and distal phalanges of toes 2–5.*	Dorsiflexes foot at ankle joint, everts foot at intertarsal joints, and extends distal and middle phalanges of each toe at interphalangeal joints and proximal phalanx of each toe at metatarsophalangeal joint.
Peroneus tertius (per'-ō-NĒ-us TER-shus; *perone* = fibula; *tertius* = third)	Distal third of fibula and interosseous membrane.	Fifth metatarsal.	Dorsiflexes foot at ankle joint and everts foot at intertarsal joints.
Lateral (Peroneal) Compartment			
Peroneus longus (per'-ō-NĒ-us LON-gus)	Head and body of fibula and lateral condyle of tibia.	First metatarsal and first cuneiform.	Plantar flexes foot at ankle joint and everts foot at intertarsal joints.
Peroneus brevis (per'-ō-NĒ-us BREV-is; *brevis* = short)	Body of fibula.	Fifth metatarsal.	Plantar flexes foot at ankle joint and everts foot at intertarsal joints.

*The great toe or hallux is the first toe and has two phalanges: proximal and distal. The remaining toes, toes 2–5, have three phalanges: proximal, middle, and distal.

Exhibit 10.21 *(continued)*

MUSCLE	ORIGIN	INSERTION	ACTION
Superficial Posterior Compartment			
Gastrocnemius (gas'-trok-NĒ-mē-us; *gaster* = belly; *kneme* = leg)	Lateral and medial condyles of femur and capsule of knee.	Calcaneus by way of calcaneal (Achilles) tendon.	Plantar flexes foot at ankle joint and flexes leg at knee joint.
Soleus (SŌ-lē-us; *soleus* = sole of foot)	Head of fibula and medial border of tibia.	Calcaneus by way of calcaneal (Achilles) tendon.	Plantar flexes foot at ankle joint.
Plantaris (plan-TA-ris; *plantar* = sole of foot)	Femur superior to lateral condyle.	Calcaneus by way of calcaneal (Achilles) tendon.	Plantar flexes foot at ankle joint and flexes leg at knee joint.
Deep Posterior Compartment of the Leg			
Popliteus (pop-LIT-ē-us; *poples* = posterior surface of knee)	Lateral condyle of femur.	Proximal tibia.	Flexes leg at knee joint and medially rotates leg.
Tibialis (tib'-ē-A-lis) **posterior** (*posterior* = back)	Tibia, fibula, and interosseous membrane.	Second, third, and fourth metatarsals; navicular; all three cuneiforms; and cuboid.	Plantar flexes foot at ankle joint and inverts foot at intertarsal joints.
Flexor digitorum longus (di'-ji-TOR-um LON-gus; *digitorum* = finger or toe)	Posterior surface of tibia.	Distal phalanges of toes 2–5.	Plantar flexes foot at ankle joint; flexes distal and middle phalanges of each toe at interphalangeal joints and proximal phalanx of each toe at metatarsophalangeal joint.
Flexor hallucis longus (HAL-a-kis LON-gus; *flexor* = decreases angle at joint)	Inferior two-thirds of fibula.	Distal phalanx of great toe.	Plantar flexes foot at ankle joint; flexes distal phalanx of great toe at interphalangeal joint and proximal phalanx of great toe at metatarsophalangeal joint.

▶

Exhibit 10.21 Muscles that Move the Foot and Toes (continued)

Figure 10.23 Muscles that move the foot and toes.

The superficial muscles of the posterior compartment share a common tendon of insertion, the calcaneal (Achilles) tendon, that inserts into the calcaneal bone of the ankle.

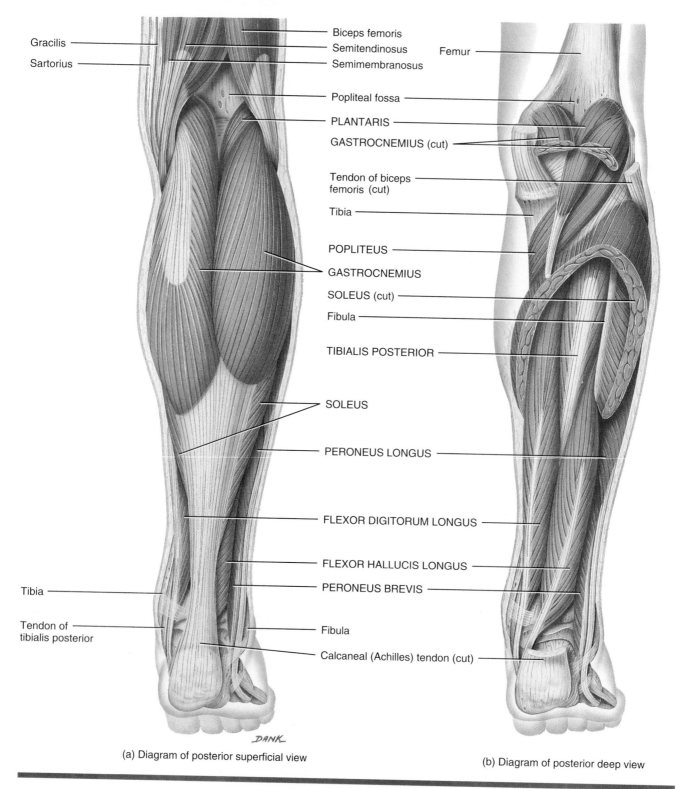

Gracilis
Sartorius
Biceps femoris
Semitendinosus
Semimembranosus
Femur
Popliteal fossa
PLANTARIS
GASTROCNEMIUS (cut)
Tendon of biceps femoris (cut)
Tibia
POPLITEUS
GASTROCNEMIUS
SOLEUS (cut)
Fibula
TIBIALIS POSTERIOR
SOLEUS
PERONEUS LONGUS
FLEXOR DIGITORUM LONGUS
FLEXOR HALLUCIS LONGUS
PERONEUS BREVIS
Tibia
Tendon of tibialis posterior
Fibula
Calcaneal (Achilles) tendon (cut)

DANK

(a) Diagram of posterior superficial view

(b) Diagram of posterior deep view

Exhibit 10.21 *(continued)*

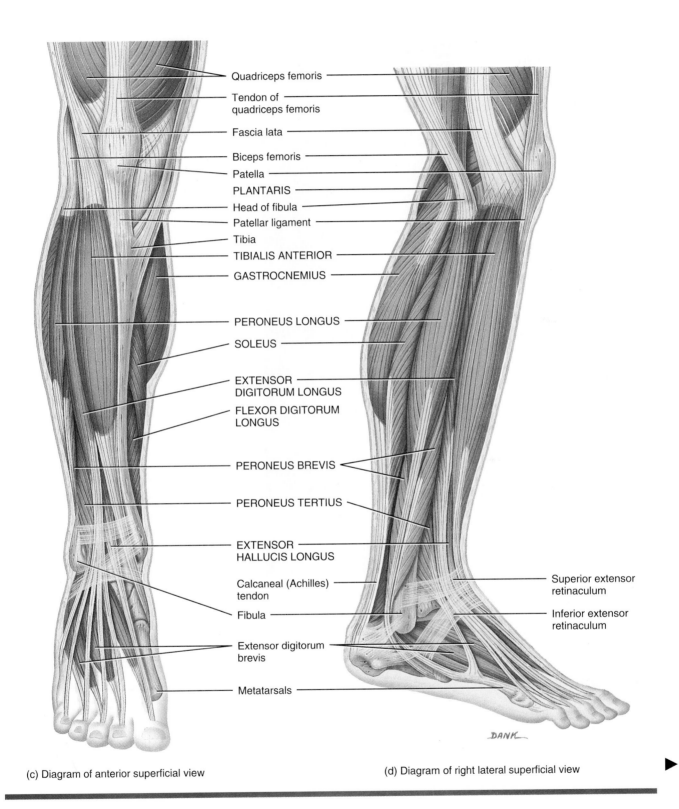

Quadriceps femoris

Tendon of
quadriceps femoris

Fascia lata

Biceps femoris

Patella

PLANTARIS

Head of fibula

Patellar ligament

Tibia

TIBIALIS ANTERIOR

GASTROCNEMIUS

PERONEUS LONGUS

SOLEUS

EXTENSOR
DIGITORUM LONGUS

FLEXOR DIGITORUM
LONGUS

PERONEUS BREVIS

PERONEUS TERTIUS

EXTENSOR
HALLUCIS LONGUS

Calcaneal (Achilles)
tendon

Fibula

Extensor digitorum
brevis

Metatarsals

Superior extensor
retinaculum

Inferior extensor
retinaculum

DANK

(c) Diagram of anterior superficial view

(d) Diagram of right lateral superficial view

Exhibit 10.21 *Muscles that Move the Foot and Toes (continued)*

Figure 10.23 (continued)

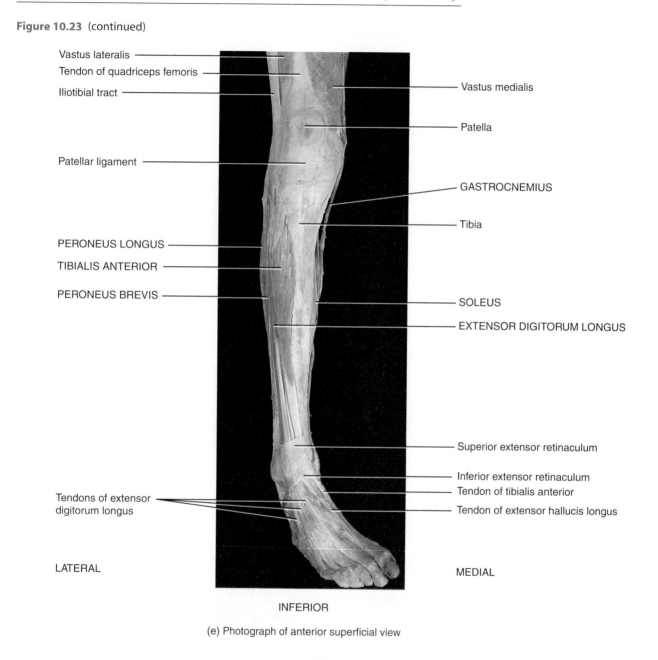

Vastus lateralis

Tendon of quadriceps femoris

Iliotibial tract

Patellar ligament

PERONEUS LONGUS

TIBIALIS ANTERIOR

PERONEUS BREVIS

Tendons of extensor digitorum longus

Vastus medialis

Patella

GASTROCNEMIUS

Tibia

SOLEUS

EXTENSOR DIGITORUM LONGUS

Superior extensor retinaculum

Inferior extensor retinaculum

Tendon of tibialis anterior

Tendon of extensor hallucis longus

LATERAL

MEDIAL

INFERIOR

(e) Photograph of anterior superficial view

Exhibit 10.21 *(continued)*

SUPERIOR

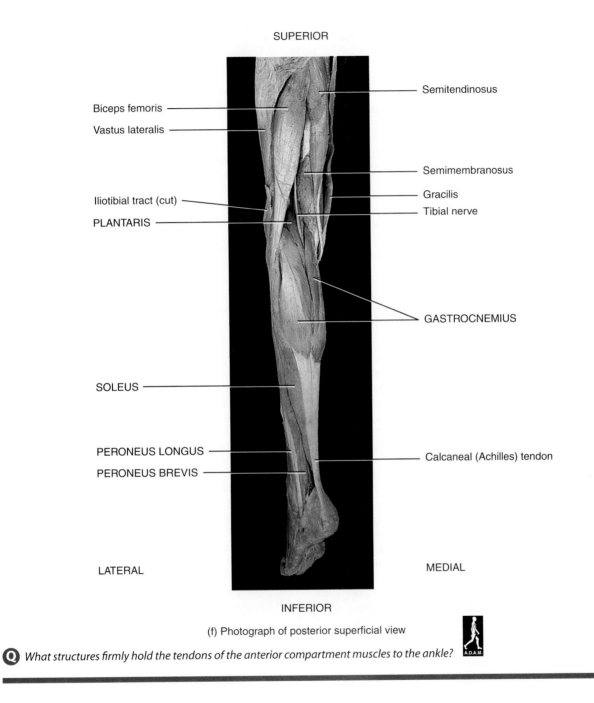

Semitendinosus

Biceps femoris

Vastus lateralis

Semimembranosus

Gracilis

Iliotibial tract (cut)

Tibial nerve

PLANTARIS

GASTROCNEMIUS

SOLEUS

PERONEUS LONGUS

Calcaneal (Achilles) tendon

PERONEUS BREVIS

LATERAL

MEDIAL

INFERIOR

(f) Photograph of posterior superficial view

Q *What structures firmly hold the tendons of the anterior compartment muscles to the ankle?*

Exhibit 10.22 *Intrinsic Muscles of the Foot (Figure 10.24)*

Several of the muscles discussed in Exhibit 10.21 move the toes in various ways and are known as extrinsic muscles. The muscles in this exhibit are called **intrinsic muscles** because they originate and insert *within* the foot. The intrinsic muscles of the foot are, for the most part, comparable to those in the hand. Whereas the muscles of the hand are specialized for precise and intricate movements, those of the foot are limited to support and locomotion. The deep fascia of the foot forms the **plantar aponeurosis (fascia)** that extends from the calcaneus bone to the phalanges of the toes. The aponeurosis supports the longitudinal arch of the foot and encloses the flexor tendons of the foot.

The intrinsic musculature of the foot is divided into two groups: **dorsal** and **plantar.** There is only one dorsal muscle, the **extensor digitorum brevis,** a four-part muscle deep to the tendons of the extensor digitorum longus muscle, which extends toes 2–5 at the metatarsophalangeal joints.

The plantar muscles are arranged in four layers, the more superficial layer being referred to as the first layer. The muscles in the **first (superficial) layer** are the **abductor hallucis,** which lies along the medial border of the sole, is comparable to the abductor pollicis brevis in the hand, and abducts the great toe at the metatarsophalangeal joint; the **flexor digitorum brevis,** which lies in the middle of the sole and flexes toes 2–5 at the interphalangeal and metatarsophalangeal joints; and the **abductor digiti minimi,** which lies along the lateral border of the sole, is comparable to the same muscle in the hand, and abducts the small toe.

The **second layer** consists of the **quadratus plantae,** a rectangular muscle that arises by two heads and flexes toes 2–5 at the metatarsophalangeal joints, and the **lumbricals,** four small muscles that are similar to the lumbricals in the hands. They flex the proximal phalanges, and extend the distal phalanges of toes 2–5.

The **third layer** is composed of the **flexor hallucis brevis,** which lies adjacent to the plantar surface of the metatarsal of the great toe, is comparable to the same muscle in the hand, and flexes the great toe; the **adductor hallucis,** which has an oblique and transverse head like the adductor pollicis in the hand and adducts the great toe; and the **flexor digiti minimi brevis,** which lies superficial to the metatarsal of the small toe, is comparable to the same muscle in the hand, and flexes the small toe.

The **fourth (deep) layer** consists of the **dorsal interossei,** four muscles that adduct toes 2–4, flex the proximal phalanges, and extend the distal phalanges, and the three **plantar interos-** **sei,** which adduct toes 3–5, flex the proximal phalanges, and extend the distal phalanges. The interossei of the feet are similar to those of the hand, except that their actions are relative to the midline of the second digit rather than to the third as in the hand.

Clinical Application

Plantar Fasciitis

Plantar fasciitis (fas-ē-Ī-tis) or **painful heel syndrome** is an inflammatory reaction due to chronic irritation of the plantar aponeurosis (fascia) at its origin on the calcaneus (heel bone). This condition is related to age (the aponeurosis becomes less elastic with age), weight-bearing activities (walking, jogging, lifting heavy objects), improperly constructed or fitting shoes, excess weight (puts pressure on the feet), and poor biomechanics (flat feet, high arches, and abnormalities in gait may cause uneven distribution of weight on the feet). It is the most common cause of heel pain in runners and arises in response to the repeated impact of running. ▪

RELATING MUSCLES TO MOVEMENTS

After you have studied the muscles in this exhibit, arrange them according to the following actions on the great toe at the metatarsophalangeal joint: (1) flexion, (2) extension, (3) abduction, and (4) adduction, and according to the following actions on toes 2–5 at the metatarsophalangeal and interphalangeal joints: (1) flexion, (2) extension, (3) abduction, and (4) adduction. The same muscle may be mentioned more than once.

INNERVATION

The extensor digitorum brevis is innervated by the deep peroneal nerve; the abductor hallucis and flexor digitorum brevis are innervated by the medial plantar nerve; the abductor digiti minimi, quadratus plantae, adductor hallucis, flexor digiti minimi brevis, and dorsal and plantar interossei are innervated by the lateral plantar nerve; and the lumbricals are innervated by the medial and lateral plantar nerves. The nerves listed here are derived from the sacral plexus, which is described and illustrated in Exhibit 17.4 on pages 538–539.

Exhibit 10.22 *(continued)*

MUSCLE	ORIGIN	INSERTION	ACTION
Dorsal			
Extensor digitorum brevis (*extensor* = increases angle at joint; *digit* = finger or toe; *brevis* = short) (see Figure 10.23c, d)	Calcaneus and inferior extensor retinaculum.	Tendons of extensor digitorum longus on toes 2–4 and proximal phalanx of great toe.	Extends great toe at metatarsophalangeal joint and toes 2–4 at interphalangeal joints.
Plantar			
First (Superficial) Layer	Calcaneus, plantar aponeurosis, and flexor retinaculum.	Medial side of proximal phalanx of great toe with the tendon of the flexor hallucis brevis.	Abducts and flexes great toe at metatarsophalangeal joint.
Abductor hallucis (*abductor* = moves part away from midline; *hallucis* = hallux or great toe)			
Flexor digitorum brevis (*flexor* = decreases angle at joint)	Calcaneus and plantar aponeurosis.	Sides of middle phalanx of toes 2–5.	Flexes toes 2–5 at proximal interphalangeal and metatarsophalangeal joints.
Abductor digiti minimi (*minimi* = small toe)	Calcaneus and plantar aponeurosis.	Lateral side of proximal phalanx of small toe with the tendon of the flexor digiti minimi brevis.	Abducts and flexes small toe at metatarsophalangeal joint.
Second layer			
Quadratus plantae (*quad* = four; *planta* = sole)	Calcaneus.	Tendon of flexor digitorum longus.	Assists flexor digitorum longus to flex toes 2–5 at interphalangeal and metatarsophalangeal joints.
Lumbricals (LUM-bri-kals; *lumbricus* = earthworm)	Tendons of flexor digitorum longus.	Tendons of extensor digitorum longus on proximal phalanges of toes 2–5.	Extend toes 2–5 at interphalangeal joints and flex toes 2–5 at metatarsophalangeal joints.
Third Layer			
Flexor hallucis brevis	Cuboid and third (lateral) cuneiform.	Medial and lateral sides of proximal phalanx of great toe via a tendon containing a sesamoid bone.	Flexes great toe at metatarsophalangeal joint.
Adductor hallucis	Metatarsals 2–4, ligaments of 3–5 metatarsophalangeal joints, and tendon of peroneus longus.	Lateral side of proximal phalanx of great toe.	Adducts and flexes great toe at metatarsophalangeal joint.
Flexor digiti minimi brevis	Metatarsal 5 and tendon of peroneus longus.	Lateral side of proximal phalanx of small toe.	Flexes small toe at metatarsophalangeal joint.
Fourth (Deep) Layer			
Dorsal interossei	Adjacent side of metatarsals.	Proximal phalanges: both sides of toe 2 and lateral side of toes 3 and 4.	Abduct and flex toes 2–4 at metatarsophalangeal joints and extend toes at interphalangeal joints.
Plantar interossei	Metatarsals 3–5.	Medial side of proximal phalanges of toes 3–5.	Adduct and flex proximal metatarsophalangeal joints and extend toes at interphalangeal joints.

▶

Exhibit 10.22 *Intrinsic Muscles of the Foot (continued)*

Figure 10.24 Intrinsic muscles of the foot.

Whereas the muscles of the hand are specialized for precise and intricate movements, those of the foot are limited to support and movement.

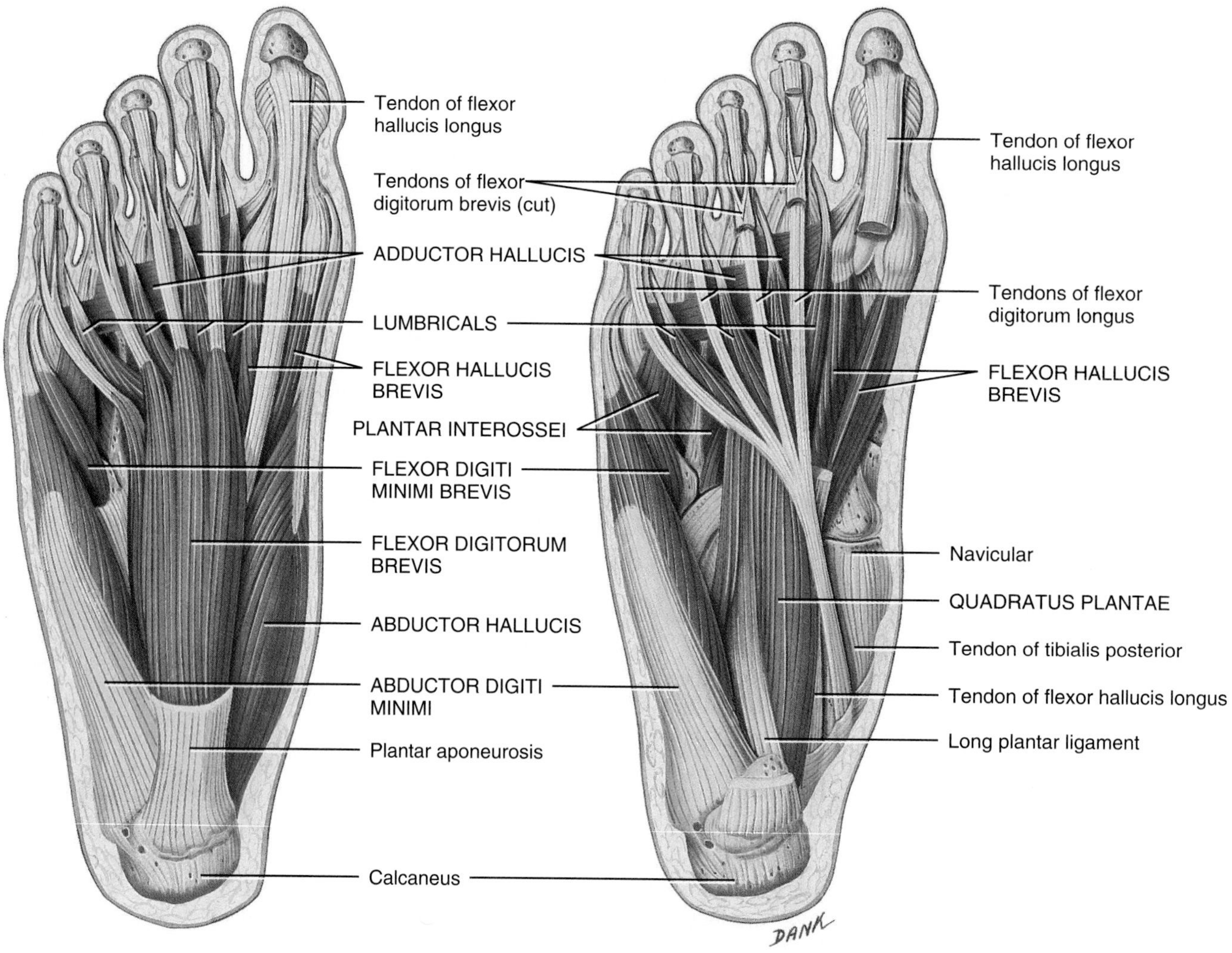

Tendon of flexor hallucis longus

Tendons of flexor digitorum brevis (cut)

ADDUCTOR HALLUCIS

LUMBRICALS

FLEXOR HALLUCIS BREVIS

PLANTAR INTEROSSEI

FLEXOR DIGITI MINIMI BREVIS

FLEXOR DIGITORUM BREVIS

ABDUCTOR HALLUCIS

ABDUCTOR DIGITI MINIMI

Plantar aponeurosis

Calcaneus

Tendon of flexor hallucis longus

Tendons of flexor digitorum longus

FLEXOR HALLUCIS BREVIS

Navicular

QUADRATUS PLANTAE

Tendon of tibialis posterior

Tendon of flexor hallucis longus

Long plantar ligament

DANK

(a) Plantar superficial and deep view

(b) Plantar deep view

Exhibit 10.22 (continued)

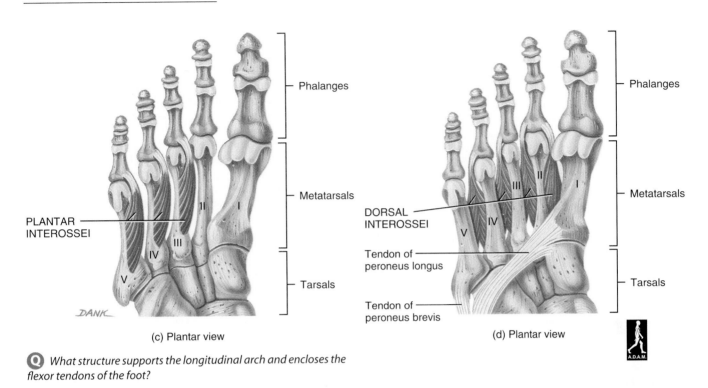

PLANTAR INTEROSSEI

Phalanges

Metatarsals

Tarsals

(c) Plantar view

DANK

DORSAL INTEROSSEI

Tendon of peroneus longus

Tendon of peroneus brevis

Phalanges

Metatarsals

Tarsals

(d) Plantar view

A.D.A.M.

Q What structure supports the longitudinal arch and encloses the flexor tendons of the foot?

Study Outline

How Skeletal Muscles Produce Movement (p. 249)

1. Skeletal muscles produce movement by pulling on bones.
2. The attachment to the stationary bone is the origin. The attachment to the movable bone is the insertion.
3. Bones serve as levers, and joints serve as fulcrums. The lever is acted on by two different forces: resistance and effort.
4. Levers are categorized into three types—first-class, second-class, and third-class (most common)—according to the position of the fulcrum, effort, and resistance on the lever.
5. Fascicular arrangements include parallel, fusiform, pennate, circular, and triangular. Fascicular arrangement is correlated with the power of a muscle and the range of motion.
6. The prime mover produces the desired action. The antagonist produces an opposite action. The synergist assists the prime mover by reducing unnecessary movement. The fixator stabilizes the origin of the prime mover so that it can act more efficiently.

Naming Skeletal Muscles (p. 254)

1. Skeletal muscles are named on the basis of distinctive criteria: direction of fibers, size, number of origins (or heads), shape, action, location, and origin and insertion.

Principal Skeletal Muscles (p. 254)

1. Muscles of facial expression move the skin rather than a joint when they contract and permit us to express a wide variety of emotions.
2. The extrinsic muscles that move the eyeballs are among the fastest contracting and most precisely controlled skeletal muscles in the body. They permit us to elevate, depress, abduct, adduct, and medially and laterally rotate the eyeballs.
3. Muscles that move the mandible (lower jaw) are also known as the muscles of mastication because they are involved in chewing.
4. The extrinsic muscles that move the tongue are important in chewing, swallowing, and speech.
5. Muscles of the floor of the oral cavity (mouth) are called suprahyoid muscles because they are located above the hyoid bone. They elevate the hyoid bone, oral cavity, and tongue during swallowing.
6. Muscles of the larynx (voice box) depress the hyoid bone and larynx during swallowing and speech and move the internal portion of the larynx during speech.
7. Muscles of the pharynx are arranged into a circular layer, which functions in swallowing, and a longitudinal layer, which functions in swallowing and speaking.
8. Muscles that move the head alter the position of the head and help balance the head on the vertebral column.
9. Muscles that act on the abdominal wall help contain and protect the abdominal viscera, move the vertebral column, compress the abdomen, and produce the force required for defecation, urination, vomiting, and childbirth.
10. Muscles used in breathing alter the size of the thoracic cavity so that ventilation can occur and assist in venous return of blood to the heart.
11. Muscles of the pelvic floor support the pelvic viscera, resist the thrust that accompanies increases in intra-abdominal pressure, and function as sphincters at the anorectal junction, urethra, and vagina.
12. Muscles of the perineum assist in urination, erection of the penis and clitoris, ejaculation, and defecation.
13. Muscles that move the pectoral (shoulder) girdle stabilize the scapula so it can function as a stable point of origin for most of the muscles that move the humerus.
14. Muscles that move the humerus (arm) originate for the most part on the scapula (scapular muscles); the remaining muscles originate on the axial skeleton (axial muscles).
15. Muscles that move the radius and ulna (forearm) are involved in flexion and extension at the elbow joint and are organized into flexor and extensor compartments.
16. Muscles that move the wrist, hand, and digits are many and varied, and those muscles that act on the digits are called extrinsic muscles.
17. The intrinsic muscles of the hand are important in skilled activities and provide humans with the ability to grasp precisely and manipulate objects.
18. Muscles that move the vertebral column are quite complex because they have multiple origins and insertions and because there is considerable overlap among them.
19. Muscles that move the femur (thigh) originate for the most part on the pelvic girdle and insert on the femur, and these muscles are larger and more powerful than comparable muscles in the upper limb.
20. Muscles that move the femur (thigh) and tibia and fibula (leg) are separated into medial (adductor), anterior (extensor), and posterior (flexor) compartments.
21. Muscles that move the foot and toes are divided into anterior, lateral, and posterior compartments.
22. Intrinsic muscles of the foot, unlike those of the hand, are limited to support and locomotion functions.

Review Questions

1. What is meant by the muscular system? (p. 249)
2. Using the terms origin, insertion, and belly in your discussion, describe how skeletal muscles produce body movements by pulling on bones. (p. 249)
3. What is a lever? Fulcrum? Mechanical advantage? Mechanical disadvantage? Describe the three classes of levers and provide one example for each in the body. (p. 249)
4. Describe the various arrangements of fascicles. How is fascicular arrangement correlated with the strength of a muscle and its range of motion? (p. 252)
5. Define the role of the prime mover (agonist), antagonist, synergist, and fixator in producing body movements. (p. 252)
6. Select at random several muscles presented in Exhibits 10.1 through 10.22 and see if you can determine the criterion or criteria used for naming each. In addition, refer to the prefixes, suffixes, roots, and definitions in each exhibit as a guide. Select as many muscles as you wish, as long as you feel you understand the concept involved.
7. In what respect are insertions of muscles of facial expression different from those of other skeletal muscles? What muscles would you use to do the following: (a) frown, (b) pout, (c) show surprise, (d) show your upper teeth, (e) pucker your lips, (f) squint, (g) blow up a balloon, (h) smile? (p. 259)
8. What extrinsic muscles move the eyeballs? In which direction does each muscle move the eyeball? (p. 264)
9. What are the principal muscles that move the mandible (lower jaw)? Give the function of each. (p. 267)
10. What would happen if you lost tone in the masseter and temporalis muscles? (p. 267)
11. Distinguish between extrinsic and intrinsic muscles of the tongue. Describe the action of each of the muscles acting on the tongue. (p. 270)
12. What tongue, facial, and mandibular muscles would you use when chewing food? (p. 270)
13. How do suprahyoid and infrahyoid muscles differ in function? (p. 272)
14. Describe the actions of the extrinsic and intrinsic muscles of the larynx (voice box). (p. 274)
15. How do the circular and longitudinal muscles of the pharynx differ in function? (p. 277)
16. What muscles are responsible for moving the head? And how do they move the head? Distinguish the anterior from the posterior triangle of the neck. (p. 280)
17. What muscles would you use to signify "yes" and "no" by moving your head? (p. 281)
18. Describe the composition of the anterolateral and posterior abdominal wall. (p. 281)
19. What muscles accomplish compression of the abdominal wall? (p. 281)
20. What are the principal muscles involved in breathing? What are their actions? (p. 285)
21. Describe the actions of the muscles of the pelvic floor. What is the pelvic diaphragm? (p. 287)
22. Describe the actions of the muscles of the perineum. What is the urogenital diaphragm? (p. 289)
23. In what directions can the scapula be drawn? What muscles accomplish these movements? Give an example of each. (p. 291)
24. What muscles are used to raise your shoulders, lower your shoulders, join your hands behind your back, and join your hands in front of your chest? (p. 291)
25. What movements are possible at the shoulder joint? What muscles accomplish these movements? (p. 291)
26. Distinguish axial and scapular muscles involved in moving the humerus (arm). (p. 295)
27. What muscles move the humerus (arm)? In which directions do these movements occur? (p. 295)
28. Organize the muscles that move the radius and ulna (forearm) into flexors and extensors. What muscles move the radius and ulna (forearm), and what actions are used when striking a match? (p. 300)
29. Discuss the various movements possible at the wrist, hand, and fingers. What muscles accomplish these movements? (p. 305)
30. How many muscles and actions of the wrist, hand, and digits that are used when writing can you list? (p. 305)
31. What is the flexor retinaculum? Extensor retinaculum? (p. 305)
32. Distinguish among the roles of the thenar, hypothenar, and intermediate intrinsic muscles of the hand. (p. 312)
33. Discuss the various muscles and movements of the vertebral column (backbone). How are the muscles grouped? (p. 316)
34. What muscles accomplish movements of the femur (thigh)? What actions are produced by these muscles? What is the iliotibial tract? (p. 322)
35. Organize the muscles that act on the femur (thigh) and tibia and fibula (leg) into medial, anterior, and posterior compartments. What is the popliteal fossa? (p. 322)
36. What muscles act at the knee joint? What kinds of movements do these muscles perform? (p. 330)
37. Name the muscles that plantar flex, evert, pronate, and dorsiflex the foot. What is the superior extensor retinaculum? Inferior extensor retinaculum? (p. 333)
38. How do the intrinsic muscles of the hand and foot differ in function? (p. 340)
39. Define the following: Bell's palsy (p. 259), strabismus (p. 264), inguinal hernia (p. 281), impingement syndrome (p. 295), carpal tunnel syndrome (p. 312), pulled groin (p. 322), "pulled hamstrings" (p. 330), shinsplint syndrome (p. 333), and plantar fasciitis (p. 340).

Self Quiz

1. Match the following terms with their descriptions:
   (1) first-class lever   ___ (a) FER
   (2) second-class lever   ___ (b) FRE
   (3) third-class lever   ___ (c) EFR
                              ___ (d) the elbow joint, the bones of the forearm, and the biceps brachii
                              ___ (e) the atlanto-occipital joint, the weight of the anterior skull, and the muscles of the posterior neck
                              ___ (f) the ball of the foot, the tarsal bones, and the calf muscles
                              ___ (g) the most common lever system in the body

2. Match each of the muscles with the clue(s) given by the parts of its name:
   (1) action   ___ (a) rectus abdominis
   (2) direction of fibers   ___ (b) gluteus maximus
   (3) location   ___ (c) biceps brachii
   (4) number of tendons of origin   ___ (d) sternocleidomastoid
   (5) points of attachment and insertion ___ (e) adductor longus
   (6) size or shape

Complete the following:

3. In flexion, the forearm serves as a rigid rod, or ___, that moves about a fixed point, called a ___ (the elbow joint, in this case).

4. The attachment of a muscle to the bone that moves is called the ___ of the muscle.

5. A muscle that contracts to cause the desired action is called the ___. Muscles that assist or cooperate with the muscle that causes the desired action are known as ___.

6. Fascicular arrangement may take one of five characteristic patterns: ___, ___, ___, ___, or ___.

7. Muscles having their fascicles arranged in a circular pattern around an orifice are called ___.

8. The skeletal muscle that:
   (a) produces a frowning expression is the ___.
   (b) closes the eye is the ___.
   (c) assists in mastication by pressing the cheeks against the teeth so that food is kept between the teeth is the ___.
   (d) elevates the eyebrows and wrinkles the forehead to show surprise is the ___.

9. The ___ muscle is responsible for lateral rotation of the eyeball.

10. The ___ muscle, which assists in speech and swallowing, has two bellies united by an intermediate tendon held by a fibrous loop to the hyoid bone.

11. The sternocleidomastoid and longissimus capitis muscles insert on the ___ of the temporal bone.

12. The ___ muscle divides the neck into a posterior triangle and an anterior triangle.

13. Three muscles that help to form the posterior abdominal wall are the ___, ___, and ___.

14. The ___ tendon is the tendon of insertion for the diaphragm.

15. The deep transverse perineus, urethral sphincter, and their fascia constitute the ___ diaphragm.

16. Three posterior thoracic muscles that adduct the scapula are ___, ___, and ___.

17. The extensor compartment of the arm consists of the ___ muscle.

18. Whereas the ___ muscles act on the thumb, the ___ muscles act on the little finger.

19. The largest muscle mass of the back, the ___, is the chief extensor of the vertebral column and helps to maintain the ___ curve.

20. The ___ muscle, a common site for intramuscular injection, is a powerful abductor of the femur at the hip joint.

21. The quadriceps femoris is the main muscle mass on the ___ surface of the thigh. The name of this muscle mass indicates it has ___ heads of origin. All converge to insert on the ___ bone by means of the patellar ligament. All therefore act to ___ the leg.

22. The hamstring muscles, which share a common origin on the ___, span two joints and therefore ___ the thigh and ___ the leg.

23. The two muscles that utilize the calcaneal tendon to insert on the calcaneus are the ___ and ___.

24. The deep fascia of the foot forms the ___ aponeurosis that supports the longitudinal arch and encloses the flexor tendons of the foot.

25. Match the following terms with their descriptions:
   (1) inferior, middle, and superior constrictor muscles   ___ (a) extrinsic muscles of the larynx
   (2) thenar, hypothenar, and intermediate muscles   ___ (b) muscles of the floor of the oral cavity
   (3) stylopharyngeus and palatopharyngeus muscles   ___ (c) outer circular muscle layer of the pharynx
   (4) infrahyoid muscles   ___ (d) inner longitudinal muscle layer of the pharynx
   (5) subclavius, pectoralis minor, and serratus anterior muscles ___ (e) extrinsic muscles of the tongue
   (6) styloglossus and genioglossus muscles   ___ (f) anterior thoracic muscles
   (7) suprahyoid muscles   ___ (g) intrinsic muscles of the hand

Are the following statements true or false?

26. When the fingers are flexed, the flexor digitorum muscles act as the prime movers while the wrist extensor muscles act as synergists and the extensor digitorum muscle acts as an antagonist.

27. Each of the four recti muscles of the eye moves the eyeball in a direction the same as that given in the muscle name. For example, the lateral rectus moves the eyeball laterally.
28. The superior oblique muscle originates from the orbit, passes through the trochlea, and then re-inserts on the orbit.
29. Relaxation of the diaphragm flattens the dome, causing the size of the thorax to increase, as occurs during inspiration.
30. The muscles that dorsiflex the foot at the ankle joint are located in the anterior compartment of the leg.

Choose the one best answer to the following questions:

31. Which of the following statements is false concerning hyper-extending your head as if to look up at the sky?
    a. The weight of the face and jaw serves as the R (resistance).
    b. The posterior neck muscles provide the E (effort).
    c. The F (fulcrum) is the atlanto-occipital joint.
    d. This is an example of a second-class lever.
    e. All of the above are true.
32. Which of the following generalizations concerning movement by skeletal muscles is false?
    a. Muscles produce movements by pulling on bones.
    b. During contractions, the two articulating bones move equally.
    c. The tendon attachment to the stationary bone is the origin.
    d. Bones serve as levers and joints as fulcrums of the levers.
    e. In the limbs, the insertion is usually distal.
33. Which of the following muscles of facial expression does *not* act on the mouth?
    a. zygomaticus major
    b. levator palpebrae superioris
    c. levator labii superioris
    d. orbicularis oris
    e. risorius
34. Which of the following muscles depress(es) the mandible?
    a. lateral pterygoid
    b. digastric
    c. medial pterygoid
    d. both a and b
    e. all of the above
35. Which of the following statements is true?
    a. The trapezius is used to elevate the scapula and extend the head.
    b. The latissimus dorsi is considered part of the abdominal muscle group.
    c. The sartorius is found on the lateral and posterior part of the thigh.
    d. The rectus femoris is considered one of the hamstring muscles.
    e. The triceps brachii is used to extend the elbow and abduct the arm.
36. The action of the pectoralis major muscle is to
    a. abduct the arm and rotate the arm laterally
    b. flex, adduct, and rotate the arm medially
    c. adduct the arm and rotate the arm laterally
    d. abduct the arm and rotate the arm medially
    e. abduct and raise the arm

37. Which of the following statements are true about the muscles that form the anterior compartment of the forearm?
    (1) They originate on the medial epicondyle of the humerus or portions of the radius and ulna.
    (2) They originate on the lateral epicondyle of the humerus or portions of the radius and ulna.
    (3) They flex the forearm at the elbow.
    (4) They flex the wrist and fingers.
    (5) They extend the forearm at the elbow.
    (6) They extend the wrist and fingers.
        a. 1, 3, and 4
        b. 2, 5, and 6
        c. 1 and 6
        d. 2 and 4
        e. 1 and 4
38. Which of the following is *not* a flexor of the femur?
    a. pectineus
    b. psoas major
    c. adductor longus
    d. superior gemellus
    e. iliacus
39. Which of the following muscles can cause inversion of the foot at the intertarsal joints?
    a. peroneus longus
    b. tibialis posterior
    c. extensor digitorum longus
    d. flexor digitorum longus
    e. plantaris
40. Match the following muscles with the appropriate nerves:
    ___ (a) most of the muscles of facial expression
    ___ (b) muscles that move the mandible
    ___ (c) most muscles that move the tongue
    ___ (d) most muscles of the pharynx
    ___ (e) muscles that move the head
    ___ (f) muscles of the anterolateral abdominal wall
    ___ (g) diaphragm
    ___ (h) quadriceps femoris muscles
    ___ (i) muscles of the anterior compartment of the leg
    ___ (j) muscles of the lateral compartment of the leg
    ___ (k) muscles of the posterior compartment of the leg

    (1) hypoglossal (XII) nerve
    (2) thoracic nerves 7–12
    (3) facial (VII) nerve
    (4) tibial nerve
    (5) femoral nerve
    (6) trigeminal (V) nerve, mandibular division
    (7) deep peroneal nerve
    (8) accessory (XI) and cervical spinal nerves
    (9) superficial peroneal nerve
    (10) phrenic nerve
    (11) vagus (X) nerve

Critical Thinking Questions

1. Three-year-old Keesha likes to form her tongue into a cylindrical shape and use it as a straw when she drinks her milk. Name the muscles that Keesha uses to protrude her lips and tongue and to suck up the milk.
   HINT: *The tongue muscles have "glossus" as part of their names.*

2. Baby Eddie weighed 13 lb. 4 oz. at birth, and he's been gaining ever since. His mom has noticed a sharp pain between her shoulder blades when she picks Eddie up to put him in his high chair (again). Which muscle may Eddie's mother have strained?
   HINT: *She feels like she's going to pull her arms and shoulders right off her back if Eddie gets any bigger.*

3. A group of dental students was working out while singing "We might, we might, we might improve our bite!" Which muscles work to elevate and depress the jaw?
   HINT: *These are also muscles of mastication.*

4. Three major muscles make up the bulk of the buttocks. List the muscles, their origins, and their insertions.
   HINT: *These muscles are named for their location and their sizes.*

5. While you sleep, your muscles take turns being the agonist or antagonist—that is, being contracted or relaxed. Name some antagonistic pairs of muscles in the arm, thigh, torso, and leg.
   HINT: *Antagonistic pairs are often located on opposite sides of a limb or body region.*

6. Taylor felt something "pop" in her shoulder during swim team practice. The coach thinks that she may have injured her rotator cuff. What is the rotator cuff and where is it located?
   HINT: *The location is a joint with a wide range of movements, including rotation.*

Answers to Figure Questions

10.1 Prime mover (agonist); antagonist; synergist.

10.2 Second-class levers.

10.3 Possible response: direction of fibers—external oblique; shape—deltoid; action—extensor digitorum; size—gluteus maximus; origin and insertion—sternocleidomastoid; location—tibialis anterior; number of tendons of origin—biceps brachii.

10.4 Frowning—frontalis; smiling—zygomaticus major; pouting—platysma; squinting—orbicularis oculi.

10.5 It originates at the anteromedial aspect of the floor of the orbit and inserts on the posterolateral aspect of the eyeball.

10.6 Masseter.

10.7 Mastication, perception of taste, deglutition, and speech.

10.8 They fix the hyoid bone to assist in tongue movements.

10.9 Producing high and low tones.

10.10 The longitudinal muscles elevate the larynx, whereas the infrahyoid muscles depress the larynx.

10.11 Rectus abdominis.

10.12 Diaphragm.

10.13 Pubic symphysis anteriorly, coccyx posteriorly, and lateral walls of the pelvis.

10.14 Pubic symphysis anteriorly, coccyx posteriorly, and ischial tuberosities laterally.

10.15 To stabilize the scapula to assist in movements of the humerus.

10.16 The term refers to the flat tendons of the subscapularis, supraspinatus, infraspinatus, and teres minor muscles that form a nearly complete circle around the shoulder joint.

10.17 Flexor—brachialis: extensor—triceps brachii.

10.18 Flexor tendons of the digits and wrist and median nerve.

10.19 Thumb.

10.20 Splenius.

10.21 Upper limb muscles exhibit diversity of movement; lower limb muscles function in stability, locomotion, and maintenance of posture. Lower limb muscles also usually cross two joints and act on both.

10.22 Quadriceps femoris—rectus femoris, vastus lateralis, vastus medialis, and vastus intermedialius; hamstrings—biceps femoris, semitendinosus, and semimembranosus.

10.23 Superior and inferior extensor retinacula.

10.24 Plantar aponeurosis.

Chapter 11

Surface Anatomy

Student Objectives

1. Define surface anatomy and describe its importance in anatomical studies and in clinical settings.

2. Name the five principal regions of the body that are examined in surface anatomy.

3. Describe the surface anatomy features of the head (cranium and face).

4. Describe the surface anatomy features of the neck, and explain how the neck is divided into triangles.

5. Describe the surface anatomy features of the various regions of the trunk—back, chest, abdomen, and pelvis.

6. Describe the surface anatomy features of the various regions of the upper limbs—axilla, shoulder, arm, elbow, forearm, wrist, and hand.

7. Describe the surface anatomy features of the various regions of the lower limbs—buttock, thigh, knee, leg, ankle, and foot.

In Chapter 1 we introduced several branches of anatomy and noted their relationship to how we understand the body's structure. In this chapter we will take a closer look at surface anatomy. Recall that **surface anatomy** is the study of the anatomical landmarks on the exterior of the body. A knowledge of surface anatomy will not only help you identify structures on the body's exterior, but will also assist you in locating the positions of various internal structures.

The study of surface anatomy involves two related, although distinct, activities: visualization and palpation. **Visualization** involves looking in a very selective and purposeful manner at a specific part of the body. **Palpation** (pal-PĀ-shun) means using the sense of touch to determine the location of an internal part of the body through the skin. Like visualization, palpation is performed selectively and purposefully, and it supplements information already gained through other methods, including visualization. Because of the varying depth of different structures from the surface of the body and due to differences in the thickness of the subcutaneous layer over different parts of the body between females (thicker) and males (thinner), palpations may range from light, to moderate, to deep.

Knowledge of surface anatomy has many applications, both anatomically and clinically. From an anatomical viewpoint, a knowledge of surface anatomy can provide valuable information regarding the location of structures such as bones, muscles, blood vessels, nerves, lymph nodes, and internal organs. Clinically, a knowledge of surface anatomy is the basis for conducting a physical examination and performing certain diagnostic tests. Health care professionals use their understanding of surface anatomy to learn where to take the pulse, measure blood pressure, draw blood, stop bleeding, insert needles and tubes, make surgical incisions, reduce (straighten) fractured bones, and listen to sounds made by the heart, lungs, and intestines. Clinicians also draw on a knowledge of surface anatomy to assess the status of lymph nodes and to identify the presence of tumors or other unusual masses in the body.

Recall that there are five principal regions of the body: (1) head, (2) neck, (3) trunk, (4) upper limbs, and (5) lower limbs. Using these regions as our guide, we will first study the surface anatomy of the head and then discuss the other regions, concluding with the lower limbs. Because living bodies are best suited for studying surface anatomy, you will find it helpful to visualize and/or palpate the various structures described on your own body as you study each region. Following a brief introduction to a region of the body, you will be presented with a list of the prominent structures to locate. This directed approach will help to organize your learning efforts. Labeled photographs illustrate most of the structures listed in each region.

SURFACE ANATOMY OF THE HEAD

The **head** (*cephalic region,* or *caput*) contains the brain and sense organs (eyes, ears, nose, and tongue) and is divided into the cranium and face. The **cranium** (*skull,* or *brain case*) is that portion of the head that surrounds and protects the brain; the **face** is the anterior portion of the head.

Regions of the Cranium and Face

The cranium and face are divided into several regions, which are described here and illustrated in Figure 11.1.

- **Frontal region:** forms the front of the skull and includes the frontal bone.
- **Parietal region:** forms the crown of the skull and includes the parietal bones.
- **Temporal region:** forms the sides of the skull and includes the temporal bones.
- **Occipital region:** forms the base of the skull and includes the occipital bone.
- **Orbital (ocular) region:** includes the eyeballs, eyebrows, and eyelids.
- **Infraorbital region:** the region inferior to the orbit.
- **Zygomatic region:** the region inferolateral to the orbit that includes the zygomatic bone.
- **Nasal region:** region of the nose.
- **Oral region:** region of the mouth.
- **Mental region:** anterior portion of the mandible.
- **Buccal region:** region of the cheek.
- **Auricular region:** region of the external ear.

In addition to locating the various regions of the head, it is also possible to visualize and/or palpate certain structures within each region. We will examine various bone and muscle structures of the head and then look at specific areas such as the eyes, ears, and nose.

Many of the bony landmarks of the head and muscles of facial expression are labeled in Figure 11.2. As you read about each feature, refer to the figure, and try to visualize and/or palpate each feature on your own head or on the head of a study partner.

Bony Landmarks of the Head

Several bony structures of the skull can be detected through palpation, including the following:

- **Sagittal suture.** Move the fingers from side to side over the superior aspect of the scalp in order to palpate this suture.
- **Coronal** and **lambdoid sutures.** Use the same side-to-side technique to palpate these structures located on the frontal and occipital regions of the skull.

Figure 11.1 Principal regions of the cranium and face.

The head consists of a cranium that surrounds the brain and an anterior face.

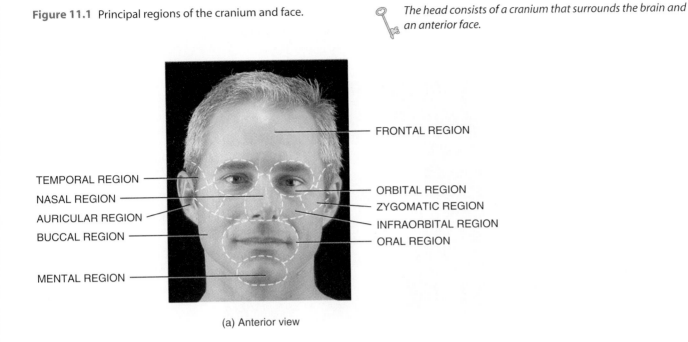

FRONTAL REGION

TEMPORAL REGION

NASAL REGION

AURICULAR REGION

BUCCAL REGION

MENTAL REGION

ORBITAL REGION

ZYGOMATIC REGION

INFRAORBITAL REGION

ORAL REGION

(a) Anterior view

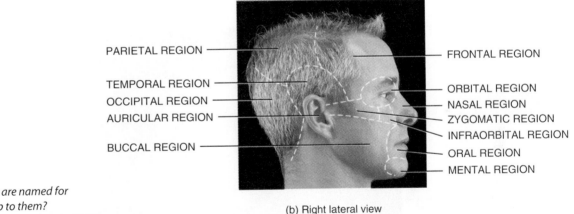

PARIETAL REGION

TEMPORAL REGION

OCCIPITAL REGION

AURICULAR REGION

BUCCAL REGION

FRONTAL REGION

ORBITAL REGION

NASAL REGION

ZYGOMATIC REGION

INFRAORBITAL REGION

ORAL REGION

MENTAL REGION

Q *Which regions are named for bones that are deep to them?*

(b) Right lateral view

- **External occipital protuberance.** This is the most prominent bony landmark on the occipital region of the skull.
- **Orbit.** Deep to the eyebrow, the superior aspect of the orbit can be palpated. Actually, the entire circumference of the orbit can be palpated.
- **Nasal bones.** These can be palpated in the nasal region between the orbits on either side of the midline.

- **Mandible.** The *ramus* (vertical portion) and *body* (horizontal portion) of the mandible can be palpated easily at the mental and buccal regions of the head.
- **Zygomatic arch.** This can be palpated in the zygomatic region.

Figure 11.2 Surface anatomy of the head.

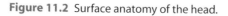 *Several muscles of facial expression can be palpated while they are contracting.*

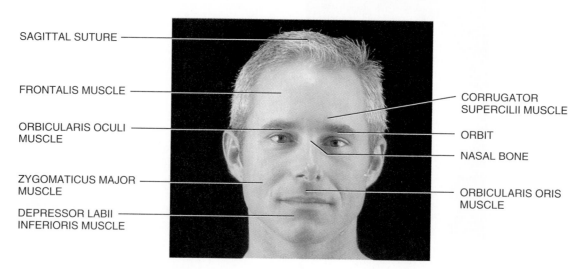

SAGITTAL SUTURE

FRONTALIS MUSCLE

ORBICULARIS OCULI MUSCLE

ZYGOMATICUS MAJOR MUSCLE

DEPRESSOR LABII INFERIORIS MUSCLE

CORRUGATOR SUPERCILII MUSCLE

ORBIT

NASAL BONE

ORBICULARIS ORIS MUSCLE

(a) Anterior view

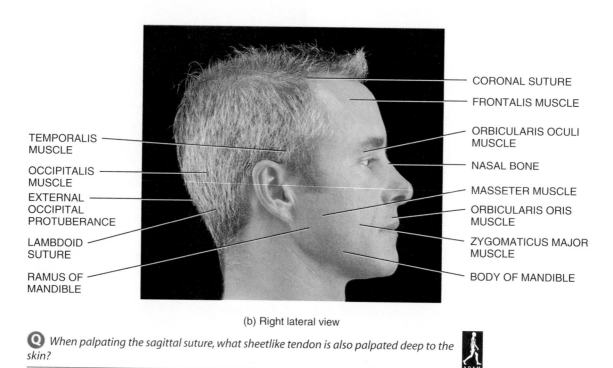

TEMPORALIS MUSCLE

OCCIPITALIS MUSCLE

EXTERNAL OCCIPITAL PROTUBERANCE

LAMBDOID SUTURE

RAMUS OF MANDIBLE

CORONAL SUTURE

FRONTALIS MUSCLE

ORBICULARIS OCULI MUSCLE

NASAL BONE

MASSETER MUSCLE

ORBICULARIS ORIS MUSCLE

ZYGOMATICUS MAJOR MUSCLE

BODY OF MANDIBLE

(b) Right lateral view

Q *When palpating the sagittal suture, what sheetlike tendon is also palpated deep to the skin?*

Figure 11.3 Surface anatomy of the right eye.

The eyeball is protected by the eyebrow, eyelashes, and eyelids.

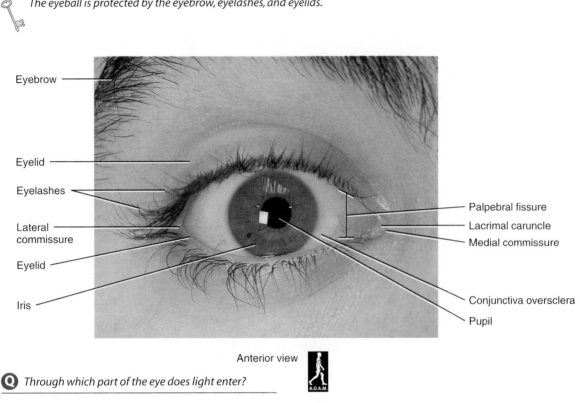

Eyebrow

Eyelid

Eyelashes

Lateral commissure

Eyelid

Iris

Palpebral fissure

Lacrimal caruncle

Medial commissure

Conjunctiva oversclera

Pupil

Anterior view

Q *Through which part of the eye does light enter?*

Muscles of Facial Expression

Several muscles of facial expression can be palpated while they are contracting:

- **Frontalis** and **occipitalis.** By raising and lowering the eyebrows, it is possible to palpate the two muscles in the frontal and occipital regions as they alternately contract and the scalp moves forward and backward.

- **Orbicularis oculi.** By closing the eyes and placing the fingers on the eyelids, the muscle can be palpated in the orbital region by tightly squeezing the eyes shut.

- **Corrugator supercilii.** By frowning, this muscle can be felt above the nose near the medial end of the eyebrow.

- **Zygomaticus major.** By smiling, this muscle can be palpated between the corner of the mouth and zygomatic (cheek) bone.

- **Depressor labii inferioris.** This muscle can be palpated in the mental region between the lower lip and chin when moving the lower lip inferiorly to expose the lower teeth.

- **Orbicularis oris.** By closing the lips tightly, this muscle can be palpated in the oral region around the margin of the lips.

- **Temporalis.** By alternately clenching the teeth and then opening the mouth, it is possible to palpate this muscle in the temporal region just superior to the zygomatic arch.

- **Masseter.** Again, by alternately clenching the teeth and then opening the mouth, this muscle can be palpated just superior and anterior to the angle of the mandible.

Surface Features of the Eyes

Now we will examine the surface features of the eyes, illustrated in Figure 11.3.

- **Iris.** Circular pigmented muscular structure behind the cornea (transparent covering over the iris).

- **Pupil.** Opening in the center of the iris through which light travels.

- **Sclera.** "White" of the eye; a coat of fibrous tissue that covers the entire eyeball except for the cornea.

- **Conjunctiva.** Membrane that covers the exposed surface of the eyeball and lines the eyelids.

- **Eyelids (palpebrae).** Folds of skin and muscle lined on their inner aspect by conjunctiva.

Figure 11.4 Surface anatomy of the right ear.

The most conspicuous surface feature of the ear is the auricle.

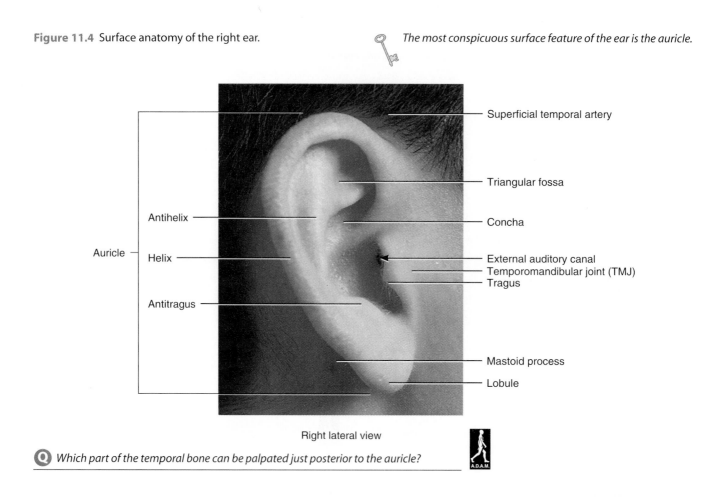

Right lateral view

Q *Which part of the temporal bone can be palpated just posterior to the auricle?*

- **Palpebral fissure.** Space between the eyelids when they are open; when the eye is closed, it lies just inferior to the level of the pupil.
- **Medial commissure.** Medial (near the nose) site of the union of the upper and lower eyelids.
- **Lateral commissure.** Lateral (away from the nose) site of the union of the upper and lower eyelids.
- **Lacrimal caruncle.** Fleshy, yellowish projection of the medial commissure that contains modified sudoriferous (sweat) and sebaceous (oil) glands.
- **Eyelashes.** Hairs on the margin of the eyelids, usually arranged in two or three rows.
- **Eyebrows (supercilia).** Several rows of hairs superior to the upper eyelids.

Surface Features of the Ears

Next, locate the surface features of the ears in Figure 11.4.

- **Auricle.** Shell-shaped portion of the external ear; also called the *pinna.* It funnels sound waves into the external auditory canal. Just posterior to it, the **mastoid process** of the temporal bone can be palpated.

- **Tragus** (*tragos* = goat). Cartilaginous projection anterior to the external auditory canal. Just anterior to the tragus and between it and the neck of the mandible you can palpate the **superficial temporal artery** and feel the pulse in the vessel. Inferior to this point, directly anterior to the tragus, it is possible to feel the **temporomandibular joint (TMJ).** As you open and close your jaw, you can feel the mandibular condyle moving.
- **Antitragus.** Cartilaginous projection opposite the tragus.
- **Concha.** Hollow of the auricle.
- **Helix.** Superior and posterior free margin of the auricle.
- **Antihelix.** Semicircular ridge superior and posterior to the tragus.
- **Triangular fossa.** Depression in the superior portion of the antihelix.
- **Lobule.** Inferior portion of the auricle; it does not contain cartilage.
- **External auditory canal (meatus).** Canal about 3 cm (1 in.) long extending from the external ear to the eardrum (tympanic membrane). It contains ceruminous glands that secrete cerumen (earwax). The **condylar process** of the mandible can be palpated by placing your little finger in the canal and opening and closing your mouth.

Figure 11.5 Surface anatomy of the nose and lips.

The external nares permit air to move into and out of the nose during breathing.

Anterior view

Q *What is the name of the anterior border of the nose that connects the root and apex?*

Surface Features of the Nose and Mouth

To complete the discussion of the surface anatomy of the head, locate the following features of the nose and mouth (Figure 11.5):

- **Root.** Superior attachment of the nose at the forehead, between the eyes.

- **Apex.** Tip of the nose.

- **Dorsum nasi.** Rounded anterior border connecting the root and apex; in profile, it may be straight, convex, concave, or wavy.

- **External naris.** External opening into the nose.

- **Ala.** Convex flared portion of the inferior lateral surfaces of the nose.

- **Bridge.** Superior portion of the dorsum nasi, superficial to the nasal bones.

- **Philtrum** (FIL-trum; *phileo* = to love). The vertical groove on the upper lip that extends along the midline to the inferior portion of the nose.

- **Lips (labia).** Superior and inferior fleshy borders of the oral cavity.

SURFACE ANATOMY OF THE NECK

The **neck** (*collum*) connects the head to the trunk and is divided into an *anterior cervical region,* two *lateral cervical regions,* and a *posterior cervical region (nucha).*

The following are the major surface features of the neck, most of which are illustrated in Figure 11.6a and b:

- **Thyroid cartilage (Adam's apple).** The largest of the cartilages that compose the larynx (voice box). It is the most prominent structure in the midline of the anterior cervical region.

- **Hyoid bone.** Located just superior to the thyroid cartilage. It is the first structure palpated in the midline inferior to the chin.

- **Cricoid cartilage.** An inferior laryngeal cartilage located just inferior to the thyroid cartilage, it attaches the larynx to the trachea (windpipe). After you pass your finger over the cricoid cartilage moving inferiorly, your fingertip sinks in. The cricoid cartilage is used as a landmark for performing a tracheostomy, which is described on page 688.

- **Thyroid gland.** Two-lobed gland with one lobe on either side of the trachea.

Figure 11.6 Surface anatomy of the neck.

The anatomical subdivisions of the neck are the anterior cervical region, lateral cervical regions, and posterior cervical region.

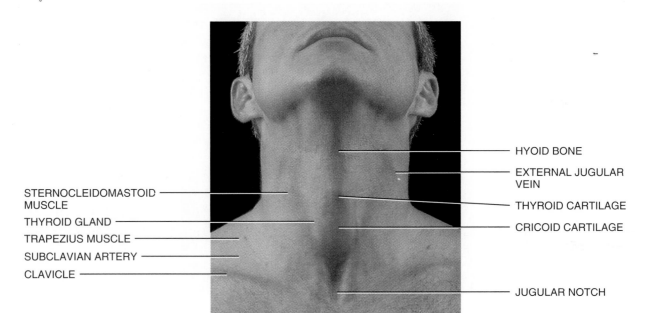

STERNOCLEIDOMASTOID MUSCLE

THYROID GLAND

TRAPEZIUS MUSCLE

SUBCLAVIAN ARTERY

CLAVICLE

HYOID BONE

EXTERNAL JUGULAR VEIN

THYROID CARTILAGE

CRICOID CARTILAGE

JUGULAR NOTCH

(a) Anterior view of surface anatomy features

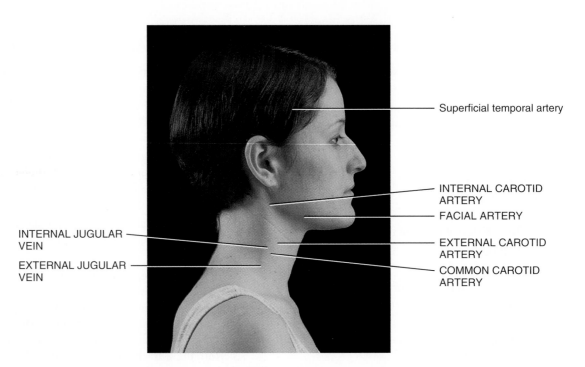

Superficial temporal artery

INTERNAL CAROTID ARTERY

FACIAL ARTERY

EXTERNAL CAROTID ARTERY

COMMON CAROTID ARTERY

INTERNAL JUGULAR VEIN

EXTERNAL JUGULAR VEIN

(b) Right lateral view

Figure 11.6 (continued)

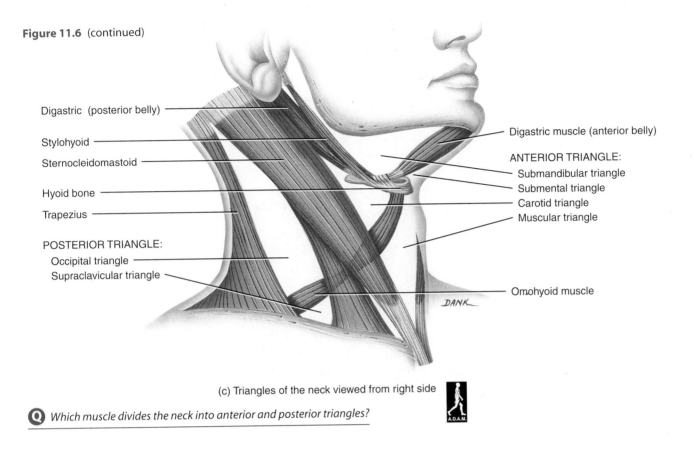

Digastric (posterior belly)

Stylohyoid

Sternocleidomastoid

Hyoid bone

Trapezius

POSTERIOR TRIANGLE:
Occipital triangle
Supraclavicular triangle

Digastric muscle (anterior belly)

ANTERIOR TRIANGLE:
Submandibular triangle
Submental triangle
Carotid triangle
Muscular triangle

Omohyoid muscle

DANK

(c) Triangles of the neck viewed from right side

Q *Which muscle divides the neck into anterior and posterior triangles?*

- **Sternocleidomastoid muscle.** Forms the major portion of the lateral aspect of the neck.
- **Common carotid artery.** Lies just deep to the sternocleidomastoid muscle along its anterior border.
- **Internal jugular vein.** Located lateral to the common carotid artery.
- **Subclavian artery.** Located just lateral to the inferior portion of the sternocleidomastoid muscle. Pressure on this artery can stop bleeding in the upper limb because this vessel supplies blood to the limb.
- **External carotid artery.** Located superior to the larynx, just anterior to the sternocleidomastoid muscle, this artery is the site of the carotid (neck) pulse.
- **External jugular vein.** Located superficial to the sternocleidomastoid muscle, this vessel is readily seen if you are angry or your collar is too tight.
- **Trapezius muscle.** Extends inferiorly and laterally from the base of the skull and occupies a portion of the lateral cervical region. A "stiff neck" is frequently associated with inflammation of this muscle.
- **Vertebral spines.** The spinous processes of the cervical vertebrae may be felt along the midline of the posterior region of the neck. Especially prominent at the base of the neck is the spinous process of the seventh cervical vertebra (vertebra prominens). See Figure 11.7.

The sternocleidomastoid muscle is not only a landmark for several arteries and veins, it is also the muscle that divides the neck into two major triangles: anterior and posterior (Figure 11.6c). The triangles are important because of the structures that lie within their boundaries.

The **anterior triangle** is bordered superiorly by the mandible, inferiorly by the sternum, medially by the cervical midline, and laterally by the anterior border of the sternocleidomastoid muscle. The anterior triangle is subdivided into an unpaired submental triangle and three paired triangles: submandibular, carotid, and muscular. The *submental triangle* is inferior to the chin and contains the submental lymph nodes and the beginning of the anterior jugular vein. The *submandibular triangle* is inferior to the mandible and contains the submandibular salivary gland, a portion of the parotid salivary gland, submandibular lymph nodes, the facial artery and vein, the carotid arteries, internal jugular vein, and the glossopharyngeal (IX), vagus (X), and hypoglossal (XII) cranial nerves. The *carotid triangle* is posterior to the hyoid bone. It contains the common carotid artery dividing into internal and external carotid arteries, the internal jugular vein, deep cervical lymph nodes, and the vagus (X), accessory (XI), and hypoglossal (XII) cranial nerves. The *muscular triangle* lies inferior to the hyoid bone. Beneath its floor are located the thyroid gland, larynx, trachea, and esophagus.

Figure 11.7 Surface anatomy of the back.

The posterior boundary of the axilla, the posterior axillary fold, is formed mainly by the latissimus dorsi and teres major muscles.

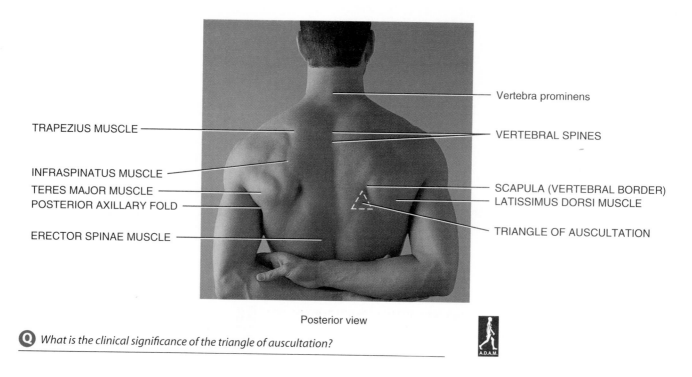

TRAPEZIUS MUSCLE

INFRASPINATUS MUSCLE
TERES MAJOR MUSCLE
POSTERIOR AXILLARY FOLD

ERECTOR SPINAE MUSCLE

Vertebra prominens

VERTEBRAL SPINES

SCAPULA (VERTEBRAL BORDER)
LATISSIMUS DORSI MUSCLE

TRIANGLE OF AUSCULTATION

Posterior view

Q *What is the clinical significance of the triangle of auscultation?*

The **posterior triangle** is bordered inferiorly by the clavicle, anteriorly by the posterior border of the sternocleidomastoid muscle, and posteriorly by the anterior border of the trapezius muscle. The posterior triangle is subdivided into two triangles, occipital and supraclavicular, by the inferior belly of the omohyoid muscle. The *occipital triangle* lies inferior to the occipital bone. The *supraclavicular triangle* lies superior to the middle third of the clavicle. The posterior triangle as a whole contains part of the subclavian artery, external jugular vein, cervical lymph nodes, brachial plexus, and accessory (XI) cranial nerve.

SURFACE ANATOMY OF THE TRUNK

The **trunk** is divided into the back, chest, and abdomen and pelvis. We will consider each region separately.

Surface Features of the Back

Among the prominent features of the **back** or *dorsum* are several superficial bones and muscles (Figure 11.7).

- **Vertebral spines.** The spinous processes of vertebrae, especially the thoracic and lumbar vertebrae, are quite prominent when the vertebral column is flexed.

- **Scapulae.** These easily identifiable surface landmarks on the back lie between ribs 2–7. In fact, it is also possible to palpate some ribs on the back. Depending on how lean a person is, it might be possible to palpate various parts of the scapula, such as the **vertebral border, axillary border, inferior angle, spine,** and **acromion.** The **spinous process of T3** is at about the same level as the spine of the scapula, and the *spinous process of T7* is about opposite the inferior angle of the scapula.

- **Latissimus dorsi muscle.** A broad, flat, triangular muscle of the lumbar region that extends superiorly to the axilla.

- **Erector spinae (sacrospinalis) muscle.** Located on either side of the vertebral column between the twelfth ribs and iliac crests.

- **Infraspinatus muscle.** Located inferior to the spine of the scapula.

- **Trapezius muscle.** Extends from the cervical and thoracic vertebrae to the spine and acromion of the scapula and the lateral end of the clavicle. It also occupies a portion of the lateral region of the neck and forms the posterior border of the posterior triangle of the neck.

- **Teres major muscle.** Located inferior to the infraspinatus muscle, and together with the latissimus dorsi muscle forms the inferior border of the posterior axillary fold (posterior wall of the axilla).

Figure 11.8 Surface anatomy of the chest.

Thoracic viscera are protected by the sternum, ribs, and thoracic vertebrae, which form the skeleton of the thorax.

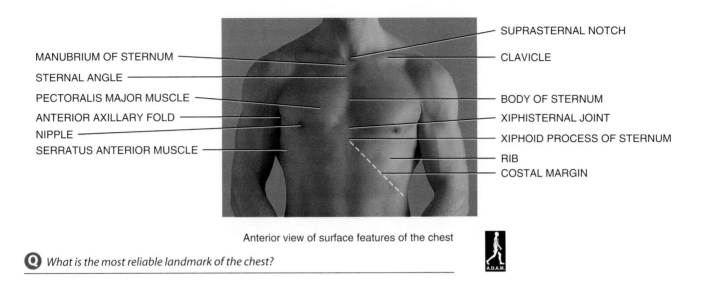

MANUBRIUM OF STERNUM

STERNAL ANGLE

PECTORALIS MAJOR MUSCLE

ANTERIOR AXILLARY FOLD

NIPPLE

SERRATUS ANTERIOR MUSCLE

SUPRASTERNAL NOTCH

CLAVICLE

BODY OF STERNUM

XIPHISTERNAL JOINT

XIPHOID PROCESS OF STERNUM

RIB

COSTAL MARGIN

Anterior view of surface features of the chest

Q *What is the most reliable landmark of the chest?*

- **Posterior axillary fold.** Formed by the latissimus dorsi and teres major muscles, it can be palpated between the fingers and thumb at the posterior aspect of the axilla; forms the posterior wall of the axilla.

- **Triangle of auscultation** (aw-skul-TĀ-shun; *ausculto* = listening). A triangular region of the back just medial to the inferior part of the scapula, where the rib cage is not covered by superficial muscles. It is bounded by the latissimus dorsi and trapezius muscles and vertebral border of the scapula.

Clinical Application

Triangle of Auscultation and Lung Sounds

The triangle of auscultation is a landmark of clinical significance because in this area respiratory sounds can be heard clearly through a stethoscope pressed against the skin. If a patient folds the arms across the chest and bends forward, the lung sounds can be heard clearly in the intercostal space between ribs 6 and 7.

Surface Features of the Chest

Our discussion of the surface anatomy of the **chest** or *thorax* will consider several distinguishing surface landmarks (Figure 11.8), as well as some surface markings used to identify the location of the heart and lungs within the chest.

- **Clavicle.** Visible at the junction of the neck and thorax. Inferior to the clavicle, especially where it articulates with the manubrium (superior portion) of the sternum, the **first**

rib can be palpated. In a depression superior to the medial end of the clavicle, just lateral to the sternocleidomastoid muscle, the **trunks of the brachial plexus** can be palpated (see Figure 17.9a).

- **Suprasternal (jugular) notch of sternum.** A depression on the superior border of the manubrium of the sternum between the medial ends of the clavicles; the **trachea** can be palpated posterior to the notch.

- **Manubrium of sternum.** Superior portion of the sternum at the same levels as the bodies of the third and fourth thoracic vertebrae and anterior to the arch of the aorta.

- **Sternal angle of sternum.** Formed at the junction between the manubrium and body of the sternum, located about 4 cm (1.5 in.) inferior to the suprasternal notch. It is palpable deep to the skin and locates the costal cartilage of the second rib. It is the most reliable landmark of the chest and the starting point from which ribs are counted. At or inferior to the sternal angle and slightly to the right, the trachea divides into right and left primary bronchi.

- **Body of sternum.** Midportion of the sternum anterior to the heart and the vertebral bodies of T5–T8.

- **Xiphoid process of sternum.** Inferior portion of the sternum, medial to the seventh costal cartilages. The joint between the xiphoid process and body of the sternum is called the **xiphisternal joint**. The heart lies on the diaphragm deep to this joint.

- **Costal margin.** Inferior edges of the costal cartilages of ribs 7 through 10. The first costal cartilage lies inferior to the medial end of the clavicle; the seventh costal cartilage

Figure 11.9 Surface anatomy of the abdomen and pelvis.

The linea alba is a frequent site for an abdominal incision because cutting through it severs no muscles and few blood vessels and nerves.

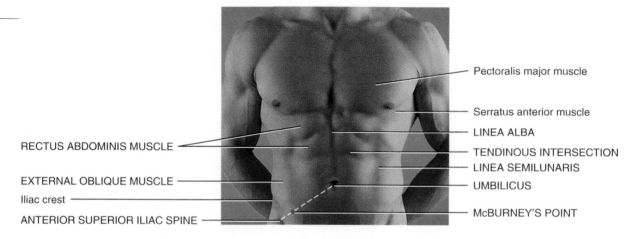

RECTUS ABDOMINIS MUSCLE

EXTERNAL OBLIQUE MUSCLE

Iliac crest

ANTERIOR SUPERIOR ILIAC SPINE

Pectoralis major muscle

Serratus anterior muscle

LINEA ALBA

TENDINOUS INTERSECTION

LINEA SEMILUNARIS

UMBILICUS

McBURNEY'S POINT

(a) Anterior view of abdomen

is the most inferior costal cartilage to articulate directly with sternum; the tenth costal cartilage forms the most inferior part of the costal margin when viewed anteriorly. At the superior end of the costal margin is the xiphisternal joint.

- **Serratus anterior muscle.** Inferior and lateral to the pectoralis major muscle.

- **Ribs.** Twelve pairs of ribs help to form the bony cage of the thoracic cavity. Depending on body leanness, they may or may not be visible. One site for listening to the heartbeat in adults is the left fifth intercostal space, just medial to the left midclavicular line.

- **Mammary glands.** Accessory organs of the female reproductive system located inside the breasts. These glands overlie the pectoralis major muscle (two-thirds) and serratus anterior muscle (one-third). After puberty, the mammary glands enlarge to their hemispherical shape, and in young adult females they extend from the second through the sixth ribs and from the lateral margin of the sternum to the midaxillary line. The **midaxillary line** (see Figure 23.9e) is a vertical line that extends downward on the lateral thoracic wall from the center of the axilla.

- **Nipples.** Superficial to the fourth intercostal space or fifth rib, about 10 cm (4 in.) from the midline in males and most females. The position of the nipples in females is variable depending on the size and pendulousness of the breasts. The right dome of the diaphragm is just inferior to the right nipple, the left dome is about 2–3 cm (1 in.) inferior to the left nipple, and the central tendon is at the level of the xiphisternal joint.

- **Anterior axillary fold.** Formed by the lateral border of the pectoralis major muscle; can be palpated between the fingers and thumb; forms the anterior wall of the axilla.

- **Pectoralis major muscle.** Principal upper chest muscle; in the male, the inferior border of the muscle forms a curved line leading to the anterior wall of the axilla and serves as a guide to the fifth rib. In the female, the inferior border is mostly covered by the breast.

Although structures within the chest are almost totally hidden by the sternum, ribs, and thoracic vertebrae, it is important to indicate the surface markings of the heart and lungs. The projection (shape of an organ on the surface of the body) of the heart on the anterior surface of the chest is indicated by four points as follows (see Figure 13.2c): The *inferior left point* is the apex (inferior, pointed end) of the heart, which projects downward and to the left and can be palpated in the fifth intercostal space, about 9 cm (3.5 in.) to the left of the midline. The *inferior right point* is at the lower border of the costal cartilage of the right sixth rib, about 3 cm to the right of the midline. The *superior right point* is located at the superior border of the costal cartilage of the right third rib, about 3 cm to the right of the midline. The *superior left point* is located at the inferior border of the costal cartilage of the left second rib, about 3 cm to the left of the midline. If you connect the four points, you can determine the heart's location and size (about the size of a person's closed fist).

The lungs almost totally fill the thorax. The *apex* (superior, pointed end) of the lungs lies just superior to the medial third of the clavicles and is the only area that can be palpated (see Figure 23.9e). The anterior, lateral, and posterior surfaces of the lung lie against the ribs. The *base* (inferior,

Figure 11.9 (continued)

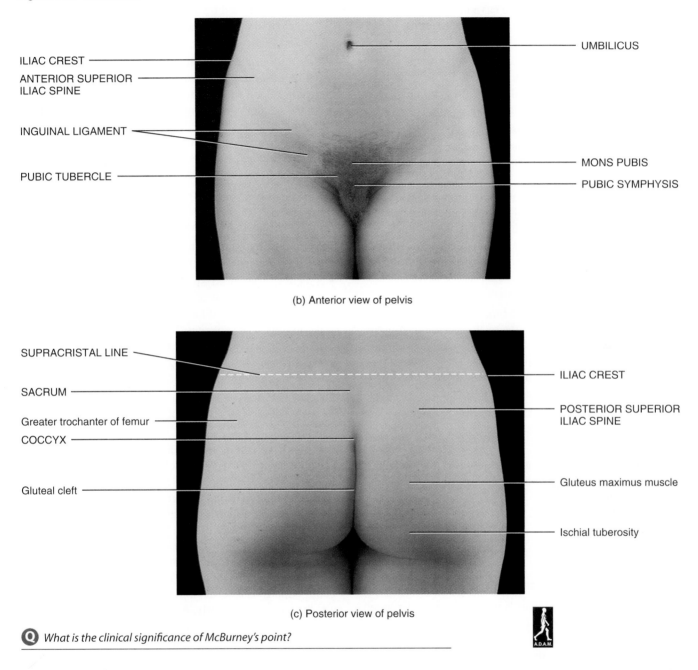

(b) Anterior view of pelvis

(c) Posterior view of pelvis

Q *What is the clinical significance of McBurney's point?*

broad end) of the lungs is concave and fits over the convex area of the diaphragm. It extends from the sixth costal cartilage anteriorly to the spinous process of the tenth thoracic vertebra posteriorly. (The base of the right lung is usually slightly higher than that of the left due to the position of the liver below it.) The covering around the lungs is called the pleura. At the base of each lung, the pleura extends about 5 cm below the base from the sixth costal cartilage anteriorly to the twelfth rib posteriorly. Thus, the lungs do not completely fill the pleural cavity in this area. A needle is typically inserted into this space in the pleural cavity to withdraw fluid from the chest.

Surface Features of the Abdomen and Pelvis

Following are some of the prominent surface anatomy features of the **abdomen** and **pelvis** (Figure 11.9):

- **Umbilicus.** Also called the *navel;* it is the previous site of the attachment of the umbilical cord in the fetus. It is level with the intervertebral disc between the bodies of vertebrae L3 and L4. The **abdominal aorta,** which branches into the right and left common iliac arteries anterior to the body of vertebra L4, can easily be palpated through the upper part of the anterior abdominal wall just to the left of the midline. The **inferior vena cava** lies to the right of the abdominal aorta and is wider; it arises anterior to the body of vertebra L5.

- **External oblique muscle.** Located inferior to the serratus anterior muscle. The aponeurosis of the muscle on its inferior border is the **inguinal ligament,** a structure along which hernias frequently occur.

- **Rectus abdominis muscles.** Located just lateral to the midline of the abdomen. They can be seen by raising the shoulders while in the supine position without using the arms.

- **Linea alba.** Flat, tendinous raphe forming a furrow along the midline between the rectus abdominis muscles. The furrow extends from the xiphoid process to the pubic symphysis. It is broad above the umbilicus and narrow below it. The linea alba is a frequently selected site for abdominal surgery because an incision through it severs no muscles and only a few blood vessels and nerves.

- **Tendinous intersection.** A fibrous band that runs transversely or obliquely across the rectus abdominis muscles. One intersection is at the level of the umbilicus, one is at the level of the xiphoid process, and one is midway between. The rectus abdominis muscles thus contain four prominent bulges and three tendinous intersections.

- **Linea semilunaris.** The lateral edge of the rectus abdominis muscle that crosses the costal margin at the top of the ninth costal cartilage.

- **McBurney's point.** Located two-thirds of the way down an imaginary line drawn between the umbilicus and anterior superior iliac spine.

Clinical Application

McBurney's Point and the Appendix

McBurney's point is an important landmark related to the appendix. When the appendix must be removed in an operation called an *appendectomy,* an oblique incision is made through McBurney's point. Moreover, pressure of the finger on McBurney's point produces tenderness in acute *appendicitis,* inflammation of the appendix. ▪

- **Iliac crest.** Superior margin of the ilium of the hipbone. It forms the outline of the superior border of the buttock. When you rest your hands on your hips, they rest on the iliac crests. A horizontal line drawn across the highest point of each iliac crest is called the **supracristal line,** which intersects the spinous process of the fourth lumbar vertebra. This vertebra is a landmark for performing a lumbar puncture (see page 517).

- **Anterior superior iliac spine.** The anterior end of the iliac crest that lies at the upper lateral end of the fold of the groin.

- **Posterior superior iliac spine.** The posterior end of the iliac crest, indicated by a dimple in the skin that coincides with the middle of the sacroiliac joint, where the hipbone attaches to the sacrum.

- **Pubic tubercle.** Projection on the superior border of the pubis of the hipbone. Attached to it is the medial end of the inguinal ligament. The lateral end of the ligament is attached to the anterior superior iliac spine. The **inguinal ligament** is the inferior free edge of the external oblique muscle aponeurosis that forms the **inguinal canal.** The spermatic cord in males and the round ligament of the uterus in females pass through the inguinal canal.

- **Pubic symphysis.** Anterior joint of the hipbones; palpated as a firm resistance in the midline at the inferior portion of the anterior abdominal wall,

- **Mons pubis.** An elevation of adipose tissue covered by skin and pubic hair that is anterior to the pubic symphysis.

- **Sacrum.** The fused spinous processes of the sacrum, called the median sacral crest, can be palpated beneath the skin in the superior of the **gluteal cleft,** a depression along the midline that separates the buttocks (and is considered part of the discussion on the buttocks on page 368).

- **Coccyx.** The inferior surface of the tip of the coccyx can be palpated in the gluteal cleft, about 2.5 cm (1 in.) posterior to the anus.

Chapter 1 discussed the division of the abdomen and pelvis into nine regions (see Figure 1.9). The nine-region designation subdivided the abdomen and pelvis by drawing two vertical lines (left and right midclavian), an upper horizontal line (transpyloric), and a lower horizontal line (transtubercular). Take time to examine Figure 1.9 carefully, so that you can see which organs or parts of organs lie within each region.

In Chapter 10, as part of your study of skeletal muscles, you were introduced to the **perineum,** the diamond-shaped region medial to the thigh and buttocks. It is bounded by the pubic symphysis anteriorly, the ischial tuberosities laterally, and the coccyx posteriorly. A transverse line drawn between the ischial tuberosities divides the perineum into an anterior *urogenital triangle* that contains the external genitals, and a posterior *anal triangle* that contains the anus. Details concerning the perineum are provided in Chapter 26. At this point keep in mind that the musculature of the perineum constitutes the floor of the pelvic cavity.

SURFACE ANATOMY OF THE UPPER LIMB (EXTREMITY)

The **upper limb** consists of the shoulder, armpit, arm, elbow, forearm, wrist, and hand.

Figure 11.10 Surface anatomy of the shoulder.

The deltoid muscle gives the shoulder its rounded prominence.

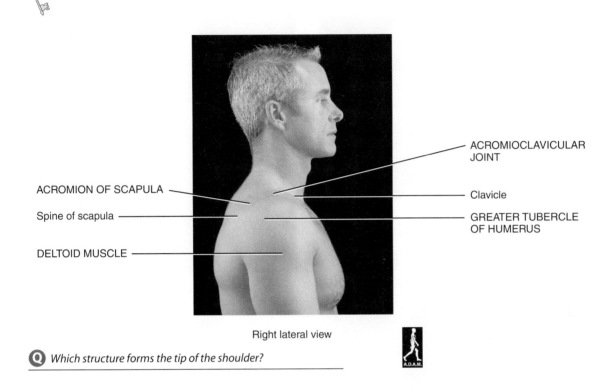

ACROMIOCLAVICULAR JOINT

ACROMION OF SCAPULA

Clavicle

Spine of scapula

GREATER TUBERCLE OF HUMERUS

DELTOID MUSCLE

Right lateral view

Q *Which structure forms the tip of the shoulder?*

Surface Features of the Shoulder

The **shoulder** or *acromial region* is located at the lateral aspect of the clavicle, where the clavicle joins the scapula, and the scapula joins the humerus. The region presents several conspicuous surface anatomy features (Figure 11.10).

- **Acromioclavicular joint.** A slight elevation at the lateral end of the clavicle. It is the joint between the acromion of the scapula and the clavicle.
- **Acromion.** The expanded lateral end of the spine of the scapula that forms the top of the shoulder. It can be palpated about 2.5 cm (1 in.) distal to the acromioclavicular joint.
- **Humerus.** The **greater tubercle** of the humerus may be palpated on the superior aspect of the shoulder. It is the most laterally palpable bony structure.
- **Deltoid muscle.** Triangular muscle that forms the rounded prominence of the shoulder.

Clinical Application

Shoulder Intramuscular Injection

The *deltoid muscle* is a frequent site for an intramuscular injection. To avoid injury to major blood vessels and nerves, the injection is given in the midportion of the muscle about 2–3 finger-breadths inferior to the acromion of the scapula and lateral to the axilla.

Surface Features of the Armpit

The **armpit** or *axilla* is a pyramid-shaped area at the junction of the arm and the chest that enables blood vessels and nerves to pass between the neck and the upper limbs (see Figure 11.11a). The *apex* of the axilla is surrounded by the clavicle, scapula, and first rib. The *base* of the axilla is formed by the concave skin and fascia that extends from the arm to the chest wall. It contains hair, and deep to the base the axillary lymph nodes can be palpated. The *anterior wall* of the axilla is composed mainly of the pectoralis major muscle (anterior axillary fold; see also Figure 11.8a). The *posterior wall* of the axilla is formed mainly by the teres major and latissimus dorsi muscles (posterior axillary fold; see also Figure 11.7. The *medial wall* of the axilla is formed by ribs 1–4 and their corresponding intercostal muscles, plus the overlying serratus anterior muscle. Finally, the *lateral wall* of the axilla is formed by the coracobrachialis muscle and the superior portion of the shaft of the humerus. Passing through the axilla are the axillary artery and vein, branches of the brachial plexus, and axillary lymph nodes.

Figure 11.11 Surface anatomy of the arm and elbow.

The biceps brachii and triceps brachii muscles form the bulk of the musculature of the arm.

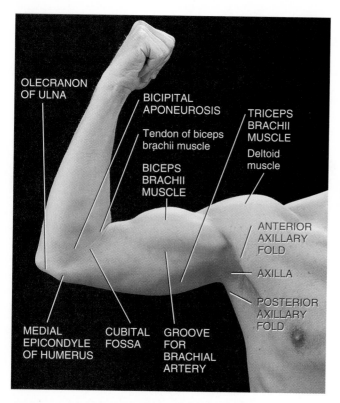

(a) Medial view of arm

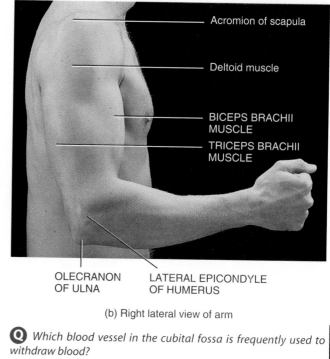

(b) Right lateral view of arm

Q *Which blood vessel in the cubital fossa is frequently used to withdraw blood?*

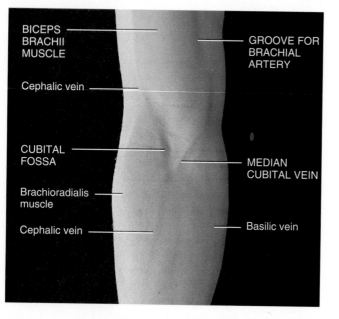

(c) Anterior view of cubital fossa

Surface Features of the Arm and Elbow

The **arm** or *brachium* is the region between the shoulder and elbow. The **elbow** or *cubitus* is the region where the arm and forearm join. The arm and elbow present several surface anatomy features (Figure 11.11).

- **Humerus.** This may be palpated along its entire length, especially at the elbow.
- **Biceps brachii muscle.** Forms the bulk of the anterior surface of the arm. On the medial side of the muscle is a groove that contains the **brachial artery.**
- **Triceps brachii muscle.** Forms the bulk of the posterior surface of the arm.
- **Medial epicondyle.** Medial projection of the humerus near the elbow.
- **Lateral epicondyle.** Lateral projection of the humerus near the elbow.

- **Olecranon.** Projection of the proximal end of the ulna between and slightly superior to the epicondyles when the forearm is extended; it forms the elbow.

- **Ulnar nerve.** Can be palpated in a groove posterior to the medial epicondyle. The "funny bone" is the region where the ulnar nerve rests against the medial epicondyle. Hitting the nerve at this point produces a sharp pain along the medial side of the forearm.

- **Cubital fossa.** Triangular space in the anterior region of the elbow bounded proximally by an imaginary line between the humeral epicondyles, laterally by the medial border of the brachioradialis muscle, and medially by the lateral border of the pronator teres muscle; contains the tendon of the biceps brachii muscle, the median cubital vein, brachial artery and its terminal branches (radial and ulnar arteries), and parts of the median and radial nerves.

- **Median cubital vein.** Crosses the cubital fossa obliquely and connects the lateral cephalic with the medial basilic veins of the arm. The median cubital vein is frequently used to withdraw blood from a vein for diagnostic purposes or to introduce substances into blood, such as medications, contrast media for radiographic procedures, nutrients, and blood cells and/or plasma for transfusion.

- **Brachial artery.** Continuation of the axillary artery that passes posterior to the coracobrachialis muscle and then medial to the biceps brachii muscle. It enters the middle of the cubital fossa and passes deep to the bicipital aponeurosis, which separates it from the median cubital vein.

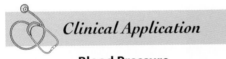

Clinical Application

Blood Pressure

Blood pressure is usually measured in the brachial artery, when the cuff of a *sphygmomanometer* is wrapped around the arm and a stethoscope is placed over the brachial artery in the cubital fossa. Pulse can also be detected in the artery in the cubital fossa. In cases of severe hemorrhage in the forearm and hand, pressure may be applied to the brachial artery to help stop the bleeding. ■

- **Bicipital aponeurosis.** An aponeurotic band that inserts the biceps brachii muscle into the deep fascia in the medial aspect of the forearm. It can be felt when the muscle contracts.

Surface Features of the Forearm and Wrist

The **forearm** or *antebrachium* is the region between the elbow and wrist. The **wrist** or carpus is between the forearm and palm. Following are some prominent surface anatomy features of the forearm and wrist (Figure 11.12).

- **Ulna.** The medial bone of the forearm. It can be palpated along its entire length from the olecranon (see Figure 11.11a, b) to the **styloid process,** a projection on the distal end of the bone at the medial side of the wrist. The **head of the ulna** is a conspicuous enlargement just proximal to the styloid process.

- **Radius.** When the forearm is rotated, the distal half of the radius can be palpated; the proximal half is covered by muscles. The **styloid process** of the radius is a projection on the distal end of the bone at the lateral side of the wrist.

- **Muscles.** Because of their close proximity, it is difficult to identify individual muscles of the forearm. Instead, it is much easier to identify tendons as they approach the wrist and then trace them proximally to the following muscles:

  Brachioradialis muscle. Located at the superior and lateral aspect of the forearm.

  Flexor carpi radialis muscle. The tendon of this muscle is on the lateral side of the forearm about 1 cm medial to the styloid process of the radius.

  Palmaris longus muscle. The tendon of this muscle is medial to the flexor carpi radialis tendon and can be seen if the wrist is slightly flexed and the base of the thumb and little finger are drawn together.

  Flexor digitorum superficialis muscle. The tendon of this muscle is medial to the palmaris longus tendon and can be palpated by flexing the fingers at the metacarpophalangeal and proximal interphalangeal joints.

  Flexor carpi ulnaris muscle. The tendon of this muscle is on the medial aspect of the forearm.

- **Radial artery.** Located on the lateral aspect of the wrist between the flexor carpi radialis tendon and styloid process of the radius. It is frequently used to take a pulse.

- **Pisiform bone.** Medial bone of the proximal row of carpals that can be palpated as a projection distal and anterior to the styloid process of the ulna.

- **"Anatomical snuffbox."** A triangular depression between tendons of extensor pollicis brevis and extensor pollicis longus muscles. It derives its name from a habit of previous centuries in which a person would take a pinch of snuff (powdered tobacco or scented powders) and place it in the depression before sniffing it into the nose. The styloid process of the radius, the base of the first metacarpal, trapezium, scaphoid, and radial artery can all be palpated in the depression.

- **Wrist creases.** Three more or less constant lines on the anterior aspect of the wrist (named *proximal, middle,* and *distal*) where the skin is firmly attached to underlying deep fascia.

Figure 11.12 Surface anatomy of the forearm and wrist.

Muscles of the forearm are most easily identified by locating their tendons near the wrist and tracing them proximally.

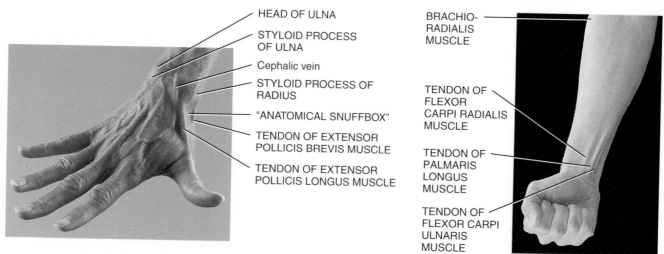

HEAD OF ULNA

STYLOID PROCESS OF ULNA

Cephalic vein

STYLOID PROCESS OF RADIUS

"ANATOMICAL SNUFFBOX"

TENDON OF EXTENSOR POLLICIS BREVIS MUSCLE

TENDON OF EXTENSOR POLLICIS LONGUS MUSCLE

BRACHIO-RADIALIS MUSCLE

TENDON OF FLEXOR CARPI RADIALIS MUSCLE

TENDON OF PALMARIS LONGUS MUSCLE

TENDON OF FLEXOR CARPI ULNARIS MUSCLE

(a) Dorsum of wrist

(b) Anterior aspect of forearm and wrist

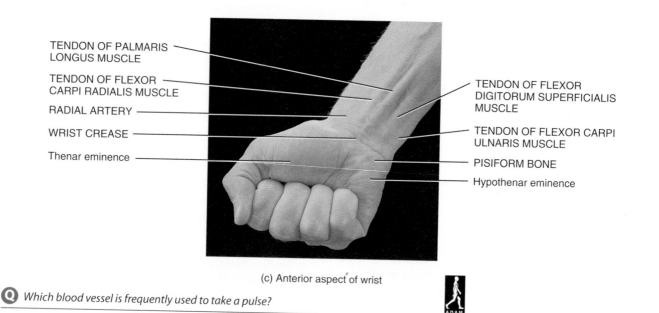

TENDON OF PALMARIS LONGUS MUSCLE

TENDON OF FLEXOR CARPI RADIALIS MUSCLE

RADIAL ARTERY

WRIST CREASE

Thenar eminence

TENDON OF FLEXOR DIGITORUM SUPERFICIALIS MUSCLE

TENDON OF FLEXOR CARPI ULNARIS MUSCLE

PISIFORM BONE

Hypothenar eminence

(c) Anterior aspect of wrist

Q *Which blood vessel is frequently used to take a pulse?*

Figure 11.13 Surface anatomy of the hand.

🔑 *Several tendons on the dorsum of the hand can be identified by their alignment with phalanges of the digits.*

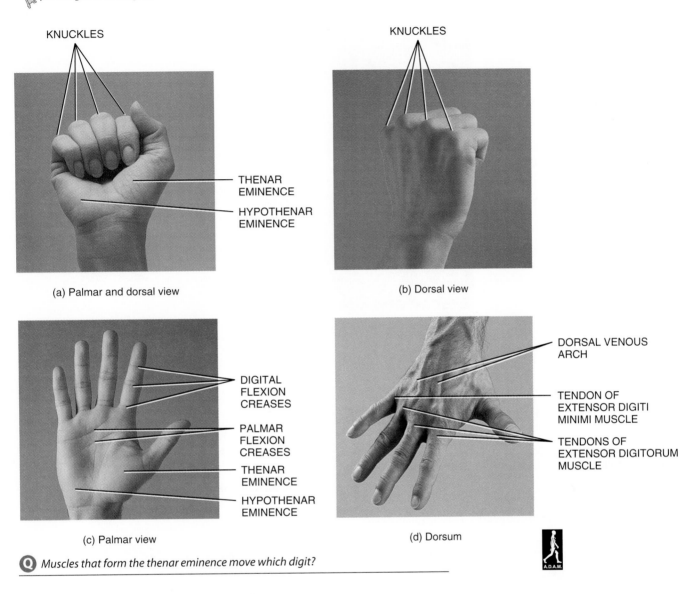

(a) Palmar and dorsal view

(b) Dorsal view

(c) Palmar view

(d) Dorsum

Q *Muscles that form the thenar eminence move which digit?*

Surface Features of the Hand

The **hand** or *manus* is the region from the wrist to the termination of the upper limb, and in which several conspicuous surface anatomy features are visible (Figure 11.13).

- **Knuckles.** Commonly refers to the dorsal aspect of the distal ends of metacarpals II–V (or 2–5), but also includes the dorsal aspects of the metacarpophalangeal and interphalangeal joints.

- **Dorsal venous arch.** Superficial veins on the dorsum of the hand that drain blood into the cephalic vein. It can be displayed by compressing the blood vessels at the wrist for a few moments as the hand is opened and closed.

- **Tendon of extensor digiti minimi muscle.** This can be seen on the dorsum of the hand in line with the phalanx of the little finger.

- **Tendons of extensor digitorum muscle.** These can be seen on the dorsum of the hand in line with the phalanges of the ring, middle, and index fingers.

Figure 11.14 Surface anatomy of the buttocks.

The gluteus medius muscle is a frequent site for an intramuscular injection.

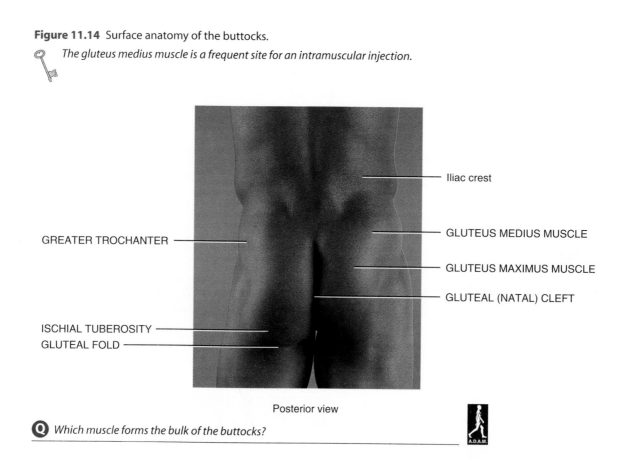

Iliac crest

GLUTEUS MEDIUS MUSCLE

GLUTEUS MAXIMUS MUSCLE

GLUTEAL (NATAL) CLEFT

GREATER TROCHANTER

ISCHIAL TUBEROSITY
GLUTEAL FOLD

Posterior view

Q *Which muscle forms the bulk of the buttocks?*

- **Tendon of the extensor pollicis brevis muscle.** This tendon (described previously) is in line with the phalanx of the thumb (see Figure 11.12a).

- **Thenar eminence.** Larger, rounded contour on the lateral aspect of the palm formed by muscles that move the thumb.

- **Hypothenar eminence.** Smaller, rounded contour on the medial aspect of the palm formed by muscles that move the little finger.

- **Palmar flexion creases.** Skin creases on the palm.

- **Digital flexion creases.** Skin creases on the anterior surface of the fingers.

SURFACE ANATOMY OF THE LOWER LIMB (EXTREMITY)

The **lower limb** consists of the buttock, thigh, knee, leg, ankle, and foot.

Surface Features of the Buttock

The **buttock** or *gluteal region* is formed mainly by the gluteus maximus muscle. The outline of the superior border of the buttock is formed by the iliac crests (see Figure 11.9c).

Following are some surface anatomy features of the buttock (Figure 11.14).

- **Gluteus maximus muscle.** Forms the major portion of the prominence of the buttocks. The sciatic nerve is deep to this muscle.

- **Gluteus medius muscle.** Superior and lateral to the gluteus maximus muscle.

Clinical Application

Gluteal Intramuscular Injection

Another common site for an intramuscular injection is the *gluteus medius muscle*. In order to give this injection, the buttock is divided into quadrants and the upper outer quadrant is used as an injection site. The iliac crest serves as the landmark for this quadrant. This site is chosen because the gluteus medius muscle in this area is quite thick, and there is less chance of injury to the sciatic nerve or major blood vessels. ■

- **Gluteal (natal) cleft.** Depression along the midline that separates the left and right buttocks.

- **Gluteal fold.** Inferior limit of the buttock that roughly corresponds to the inferior margin of the gluteus maximus muscle.

- **Ischial tuberosity.** Just superior to the medial side of the gluteal fold, the tuberosity bears the weight of the body when seated.
- **Greater trochanter.** A projection of the proximal end of the femur on the lateral side of the thigh. It is about 20 cm (8 in.) inferior to the highest point of the iliac crest.

Surface Features of the Thigh and Knee

The **thigh** or *femoral region* is the region from the hip to the knee. The **knee** or *genu* is the region where the thigh and leg join. Several muscles are clearly visible in the thigh. Following are several surface anatomy features of the thigh and knee (Figure 11.15).

- **Sartorius muscle.** Superficial anterior muscle that can be traced from the lateral aspect of the thigh to the medial aspect of the knee.
- **Quadriceps femoris muscle.** Three of the four components of the muscle can be seen: **rectus femoris** at the midpoint of the anterior aspect of the thigh; **vastus medialis** at the anteromedial aspect of the thigh; and **vastus lateralis** at the anterolateral aspect of the thigh.

Clinical Application

Thigh Intramuscular Injection

The *vastus lateralis muscle* of the quadriceps femoris group is another site that may be used for an intramuscular injection. This site is located at a point midway between the greater trochanter (see Figure 11.14) of the femur and the patella (knee cap). Injection in this area reduces the chance of injury to major blood vessels and nerves.

- **Adductor longus muscle.** Located at the superior aspect of the thigh, medial to the sartorius.
- **Hamstring muscles.** Superficial, posterior thigh muscles located below the gluteal folds. They are the **biceps femoris,** which lies more laterally as it passes inferiorly to the knee, and the **semitendinosus** and **semimembranosus,** which lie medially as they pass inferiorly to the knee.
- **Femoral triangle.** A large space formed by the inguinal ligament superiorly, the sartorius muscle laterally, and the adductor longus muscle medially. The triangle contains the femoral artery, vein, and nerve; inguinal lymph nodes; and the terminal portion of the great saphenous vein. The triangle is an important arterial pressure point in cases of severe hemorrhage of the lower limb. Hernias frequently occur in this area.

The knee presents several distinguishing surface features.

- **Patella.** Kneecap. A large sesamoid bone located within the tendon of the quadriceps femoris muscle on the anterior surface of the knee along the midline.

- **Patellar ligament.** Continuation of the quadriceps femoris tendon inferior to the patella. Infrapatellar fat pads are found on both sides of it.
- **Medial condyle of femur.** Medial projection on the distal end of the femur.
- **Medial condyle of tibia.** Medial projection on the proximal end of the tibia.
- **Lateral condyle of femur.** Lateral projection on the distal end of the femur.
- **Lateral condyle of tibia.** Lateral projection on the proximal end of the tibia. All four condyles can be palpated just inferior to the patella on either side of the patellar ligament.
- **Popliteal fossa.** A diamond-shaped area on the posterior aspect of the knee that is clearly visible when the knee is flexed. The fossa is bordered superolaterally by the biceps femoris muscle, superomedially by the semimembranosus and semitendinosus muscles, and inferolaterally and inferomedially by the lateral and medial heads of the gastrocnemius muscle, respectively. The **head of the fibula** can easily be palpated on the lateral side of the popliteal fossa. The fossa also contains the popliteal artery and vein. It is sometimes possible to detect a pulse in the popliteal artery.

Surface Features of the Leg, Ankle, and Foot

The **leg** or *crus* is the region between the knee and ankle. The **ankle** or *tarsus* is between the leg and foot and the foot is the region from the ankle to the termination of the lower limb. Following are several surface anatomy features of the leg, ankle, and foot (Figure 11.16).

- **Tibial tuberosity.** Bony prominence on the superior, anterior surface of the tibia into which the patellar ligament inserts.
- **Tibialis anterior muscle.** Lies against the lateral surface of the tibia, where it is easy to palpate, particularly when the foot is dorsiflexed.
- **Tibia.** The medial surface and anterior border (shin) of the tibia are subcutaneous and can be palpated throughout the length of the bone.
- **Peroneus longus muscle.** A superficial lateral muscle that overlies the fibula.
- **Gastrocnemius muscle.** Forms the bulk of the midportion and superior portion of the posterior aspect of the leg. The medial and lateral heads can be clearly seen in a person standing on tiptoe.
- **Soleus muscle.** Located deep to the gastrocnemius muscle; it and the gastrocnemius together are referred to as the calf muscles.

Figure 11.15 Surface anatomy of the thigh and knee.

The quadriceps femoris and hamstrings form the bulk of the musculature of the thigh.

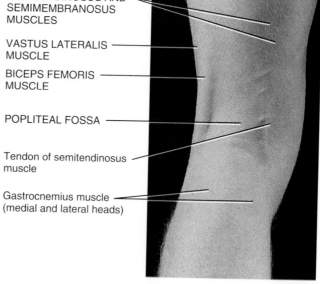

SEMITENDINOSUS AND SEMIMEMBRANOSUS MUSCLES

VASTUS LATERALIS MUSCLE

BICEPS FEMORIS MUSCLE

POPLITEAL FOSSA

Tendon of semitendinosus muscle

Gastrocnemius muscle (medial and lateral heads)

(b) Posterior view of popliteal fossa

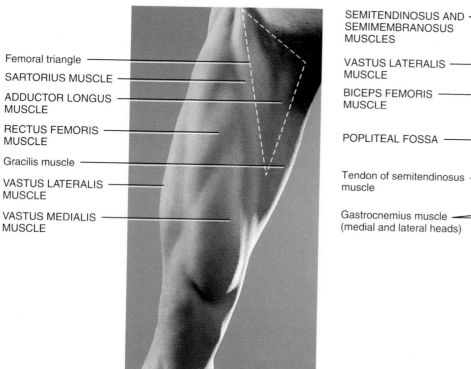

Femoral triangle

SARTORIUS MUSCLE

ADDUCTOR LONGUS MUSCLE

RECTUS FEMORIS MUSCLE

Gracilis muscle

VASTUS LATERALIS MUSCLE

VASTUS MEDIALIS MUSCLE

(a) Anterior view of thigh

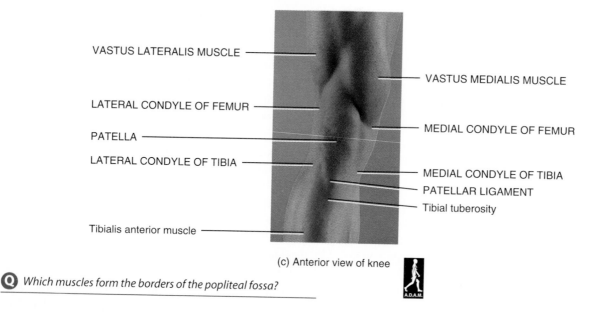

VASTUS LATERALIS MUSCLE

LATERAL CONDYLE OF FEMUR

PATELLA

LATERAL CONDYLE OF TIBIA

Tibialis anterior muscle

VASTUS MEDIALIS MUSCLE

MEDIAL CONDYLE OF FEMUR

MEDIAL CONDYLE OF TIBIA

PATELLAR LIGAMENT

Tibial tuberosity

(c) Anterior view of knee

Q *Which muscles form the borders of the popliteal fossa?*

Figure 11.16 Surface anatomy of the leg, ankle, and foot.

The calcaneal tendon is a common tendon for the gastrocnemius and soleus (calf muscles) and inserts into the calcaneus (heel bone).

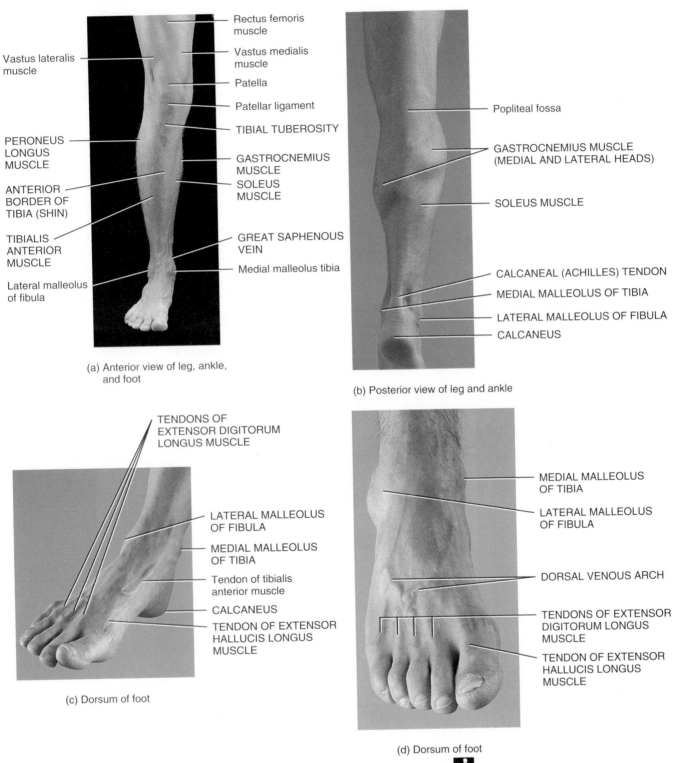

Rectus femoris muscle

Vastus medialis muscle

Patella

Patellar ligament

TIBIAL TUBEROSITY

GASTROCNEMIUS MUSCLE

SOLEUS MUSCLE

GREAT SAPHENOUS VEIN

Medial malleolus tibia

Vastus lateralis muscle

PERONEUS LONGUS MUSCLE

ANTERIOR BORDER OF TIBIA (SHIN)

TIBIALIS ANTERIOR MUSCLE

Lateral malleolus of fibula

(a) Anterior view of leg, ankle, and foot

Popliteal fossa

GASTROCNEMIUS MUSCLE (MEDIAL AND LATERAL HEADS)

SOLEUS MUSCLE

CALCANEAL (ACHILLES) TENDON

MEDIAL MALLEOLUS OF TIBIA

LATERAL MALLEOLUS OF FIBULA

CALCANEUS

(b) Posterior view of leg and ankle

TENDONS OF EXTENSOR DIGITORUM LONGUS MUSCLE

LATERAL MALLEOLUS OF FIBULA

MEDIAL MALLEOLUS OF TIBIA

Tendon of tibialis anterior muscle

CALCANEUS

TENDON OF EXTENSOR HALLUCIS LONGUS MUSCLE

(c) Dorsum of foot

MEDIAL MALLEOLUS OF TIBIA

LATERAL MALLEOLUS OF FIBULA

DORSAL VENOUS ARCH

TENDONS OF EXTENSOR DIGITORUM LONGUS MUSCLE

TENDON OF EXTENSOR HALLUCIS LONGUS MUSCLE

(d) Dorsum of foot

Q *Which artery is just lateral to the tendon of the extensor digitorum muscle and is a site where a pulse may be detected?*

- **Calcaneal (Achilles) tendon.** Prominent tendon of the gastrocnemius and soleus muscles on the posterior aspect of the ankle; inserts into the **calcaneus** (heel) bone of the foot.

- **Lateral malleolus of fibula.** Projection of the distal end of the fibula that forms the lateral prominence of the ankle. The head of the fibula, at the proximal end of the bone, lies at the same level as the tibial tuberosity.

- **Medial malleolus of tibia.** Projection of the distal end of the tibia that forms the medial prominence of the ankle.

- **Dorsal venous arch.** Superficial veins on the dorsum of the foot that unite to form the small and great saphenous veins. The great saphenous vein is the longest vein of the body.

- **Tendons of extensor digitorum longus muscle.** Visible in line with phalanges II–V (or 2–5).

- **Tendon of extensor hallucis longus muscle.** Visible in line with phalanx I (great toe). Pulsations in the dorsalis pedis artery may be felt in most people just lateral to this tendon where the blood vessel passes over the navicular and cuneiform bones of the tarsus.

Now that you have studied some of the surface anatomy features of the principal regions of the body, it is hoped you can appreciate their importance in locating the positions of many internal structures, especially bones, joints, and muscles. In later chapters you will learn to relate surface anatomy features to various blood vessels, lymph nodes, nerves, and organs in the chest, abdomen, and pelvis. Once you have visualized and palpated as many of these structures as possible on your own body, you can likely better appreciate the application of surface anatomy to your anatomical and clinical studies.

Study Outline

Introduction (p. 350)

1. Surface anatomy is the study of anatomical landmarks on the exterior of the body.
2. Surface features may be noted by visualization and/or palpation.
3. Knowledge of surface anatomy has many applications, both anatomically and clinically.
4. The regions of the body include the head, neck, trunk, upper limbs, and lower limbs.

Surface Anatomy of the Head (p. 350)

1. The head is divided into a cranium and face, each of which is divided into distinct regions.
2. Many muscles of facial expression and skull bones are easily palpated.
3. The eyes, ears, and nose present numerous readily identifiable surface features.

Surface Anatomy of the Neck (p. 355)

1. The neck connects the head to the trunk.
2. It contains several important arteries and veins.
3. The neck is divisible into several triangles by specific muscles and bones.

Surface Anatomy of the Trunk (p. 358)

1. The trunk is divided into the back, chest, abdomen, and pelvis.
2. The back presents several muscles and superficial bones that serve as landmarks.
3. The triangle of auscultation is a region of the back not covered by superficial muscles where respiratory sounds can be heard clearly.
4. The skeleton of the chest protects the organs within and provides several surface landmarks for identifying the position of the heart and lungs.

5. The abdomen and pelvis contain several important landmarks, such as the linea alba, linea semilunaris, McBurney's point, and supracristal line.
6. The perineum forms the floor of the pelvis.

Surface Anatomy of the Upper Limb (Extremity) (p. 362)

1. The upper limb consists of the shoulder, armpit, arm, elbow, forearm, wrist, and hand.
2. The prominence of the shoulder is formed by the deltoid muscle, a frequent site of an intramuscular injection.
3. The axilla is a pyramid-shaped area at the junction of the arm and chest that contains blood vessels and nerves that pass between the neck and upper limbs.
4. The arm presents several blood vessels for withdrawing blood, introducing fluids, taking the pulse, and measuring blood pressure.
5. Several muscles of the forearm are more easily identifiable by their tendons. The radial artery is a principal blood vessel for detecting a pulse.
6. The hand contains the origin of the cephalic vein and extensor tendons on the dorsum, and the thenar and hypothenar eminences on the palm.

Surface Anatomy of the Lower Limb (Extremity) (p. 368)

1. The lower limb consists of the buttock, thigh, knee, leg, ankle, and foot.
2. The buttock is formed mainly by the gluteus maximus muscle.
3. The major muscles of the thigh are the anterior quadriceps femoris and the posterior hamstrings. An important landmark in the thigh is the femoral triangle.
4. The popliteal fossa is a diamond-shaped area on the posterior aspect of the knee.
5. The gastrocnemius muscle forms the bulk of the midportion of the posterior aspect of the leg and together with the soleus muscle forms the calf of the leg.
6. The foot contains the origin of the great saphenous vein, the largest vein in the body.

Review Questions

1. What is surface anatomy? Why are visualization and palpation important in learning surface anatomy? (p. 350)
2. What are some of the applications of a knowledge of surface anatomy? (p. 350)
3. What are the divisions of the head? Describe the regions of the head and name one structure in each. (p. 350)
4. List and define five surface features of the eyes, ears, and nose. (p. 353)
5. How would you locate the superficial temporal artery, temporomandibular joint, zygomatic arch, mastoid process, and condylar process of the mandible? (p. 354)
6. What is the neck? Why is it important? (p. 355)
7. How would you palpate the hyoid bone, cricoid cartilage, thyroid gland, common carotid artery, external carotid artery, and external jugular vein? (p. 355)
8. How is the neck divided into triangles? Describe the major structures in each triangle. (p. 357)
9. What are the divisions of the trunk? (p. 358)
10. Describe the location of the scapulae relative to the ribs and vertebrae. (p. 358)
11. What is the triangle of auscultation? Why is it important clinically? (p. 359)
12. Describe the location of the sternum relative to the ribs and vertebrae. (p. 359)
13. What is the significance of the xiphisternal joint? (p. 359)
14. Where would you palpate the first rib and brachial plexus? (p. 359)
15. Why is the midaxillary line an important anatomical landmark? (p. 360)
16. How would you indicate the surface markings of the heart? (p. 360)
17. Describe the position of the lungs in the thorax. (p. 360)
18. Why do the lungs not completely fill the pleural cavity? Why is this important clinically? (p. 361)
19. Why are the linea alba, linea semilunaris, McBurney's point, and supracristal line important landmarks? (p. 362)
20. What is the inguinal ligament, and why is it important? (p. 362)
21. How would you palpate the sacrum and coccyx? (p. 362)
22. What is the perineum? (p. 362)
23. What are the major divisions of the upper limb? (p. 362)
24. Define the axilla. What are its borders? Why is it important? (p. 363)
25. What three bones are present at the shoulder? How would you palpate each? (p. 363)
26. What is the significance of the deltoid muscle? (p. 363)
27. In which arteries of the upper limb can a pulse be detected? (p. 365)
28. What is the "funny bone"? Why is it so designated? (p. 365)
29. Define the borders and significance of the cubital fossa. (p. 365)
30. What is the clinical importance of the median cubital vein? (p. 365)
31. What artery is normally used to take blood pressure? (p. 365)
32. How would you palpate the ulna, styloid process of the ulna, head of the ulna, and styloid process of the radius? (p. 365)
33. Explain the easiest way to identify the muscles of the forearm. (p. 365)
34. What is the "anatomical snuffbox"? (p. 365)
35. What are the knuckles? Why are the thenar and hypothenar eminences important? (p. 367)
36. What are the divisions of the lower limb? (p. 368)
37. Define the buttock. What is the gluteal cleft? Gluteal fold? (p. 368)
38. Why is the greater trochanter of the femur important? (p. 369)
39. What are the major anterior and posterior muscles of the thigh? (p. 369)
40. Define the femoral triangle. Why is it important? (p. 369)
41. What are the borders of the popliteal fossa? What structures pass through it? (p. 369)
42. Define the calf muscles. Where do they insert? (p. 369)

Self Quiz

Complete the following:

1. The head is divided into the ___ region and the ___ region.
2. The ___ muscle is a useful landmark for locating the carotid artery.
3. The ___ artery is commonly used to measure blood pressure.
4. Use these directional terms (anterior, posterior, superior, inferior, proximal, distal, medial, lateral) to complete the following: (a) The frontal region is ___ to the parietal region. (b) The orbital region is ___ to the mental region. (c) The antebrachium is ___ to the brachium. (d) The sternal angle is ___ to the xiphoid process. (e) The external auditory meatus is ___ to the helix. (f) The iris of the eye is ___ to the medial commissure.

5. Match the following terms with their descriptions:
   (1) olecranon
   (2) iliac crest
   (3) ischial tuberosity
   (4) acromion
   (5) tibial tuberosity
   (6) calcaneus
   (7) styloid process

   ___ (a) bony prominence inferior to the patella
   ___ (b) protuberance at distal end of the ulna
   ___ (c) superior border of the buttock
   ___ (d) forms the tip of the shoulder
   ___ (e) projection that forms the dorsum of the elbow
   ___ (f) bears most of the weight of a person when sitting
   ___ (g) attachment point for calcaneal (Achilles) tendon

Are the following statements true or false?

6. The ulnar nerve lies in a groove posterior to the medial epicondyle.

7. All of the following surface features are located on the lower limbs: tibial tuberosity, calcaneal (Achilles) tendon, lateral malleolus.

8. The trapezius muscle forms the posterior border of the posterior triangle of the lateral cervical region.

9. The linea alba is located on the anterior surface of the abdomen.

10. A landmark for performing a tracheostomy is the cricoid cartilage.

11. The triangle of auscultation is formed by the latissimus dorsi, trapezius, and scapula.

12. The femoral triangle is bordered by the inguinal ligament, sartorius muscle, and adductor longus muscle.

13. A vein of the upper limb often used for intravenous therapy is located in the popliteal fossa.

14. Match the following common and anatomical terms:
    (1) supercilia   ___ (a) skull
    (2) palpebrae   ___ (b) eyebrows
    (3) auricular region   ___ (c) cheek
                            ___ (d) hand
    (4) buccal region   ___ (e) ear
    (5) cranium   ___ (f) eyelids
    (6) gluteal region   ___ (g) buttock
    (7) manus

Choose the one best answer to the following questions:

15. The sternal angle is
    a. the site of articulation of the second costal cartilage
    b. the starting point from which the ribs are counted
    c. the most reliable surface landmark of the chest
    d. the junction line between the manubrium and the body of the sternum
    e. all of the above

16. Of the regions into which the face is subdivided, which of the following does *not* belong?
    a. temporal
    b. nasal
    c. mental
    d. oral
    e. auricular

17. Match the following terms with their descriptions:
    (1) palmaris longus
    (2) triceps brachii
    (3) trapezius
    (4) deltoid
    (5) vastus lateralis
    (6) extensor pollicus longus
    (7) sternocleidomastoid
    (8) latissimus dorsi

    ___ (a) frequent site for intramuscular injections in the upper limb
    ___ (b) a "stiff neck" results from an inflammation of this muscle
    ___ (c) divides the neck into anterior and posterior (lateral) triangles
    ___ (d) contributes to the posterior wall of the axilla
    ___ (e) occupies most of the posterior surface of the arm
    ___ (f) tendon visible on the anterior surface of the wrist when making a fist
    ___ (g) tendon that forms one border of the "anatomical snuffbox"
    ___ (h) frequently used as a site for insulin injection

Critical Thinking Questions

1. Fred the football fanatic ran headfirst into the goalpost and then landed on his back. He told the coach that he felt all right, but when he reached behind his head he felt a bump at the base of his neck. The team physician examined Fred and told him that the bump was nothing to worry about; it had always been there. What is the bump? Do you feel one on your neck?
   HINT: *Your vertebral column is also called your spine.*

2. Great-aunt Edith heard that you're bringing your new boyfriend—the one with the tattooed head and six nose rings—to Thanksgiving dinner. She told your mother that she's having "palpations." Your boyfriend, a second-year medical student, said she probably meant "palpitations." What's the difference?
   HINT: *Would you palpitate someone's pulse point?*

3. Tongun arrived at the health center complaining of vomiting, fever, and severe pain in his abdomen. The nurse on duty pushed on an area in his lower abdomen, which made Tongun "go through the roof." She called an ambulance for a possible case of appendicitis. Where exactly was the tender area on Tongun, and why did the nurse push on that spot?
   HINT: *Refer to Chapter 1 to locate the appendix.*

4. The instructor in CPR class was telling the students how to position their hands on the chest to compress the heart. One of the landmarks used was the xiphoid process. Describe the location of the xiphoid process.
   HINT: *The hands will be positioned on the region superior to the xiphoid to do the "cardio" part of CPR.*

5. Sara donated a pint of blood at the Red Cross blood drive. The phlebotomist prepared the commonly used site on her arm to insert the needle. Where did Sara get stuck?
   HINT: *If she flexes her arm, she can hold the gauze pad in place without using her hand.*

6. While the new mom and dad were diapering their baby, the mom said that the baby had the cutest little dimples in her cheeks. The dad agreed that their baby was the cutest in the world but said he hadn't noticed any dimples on the baby's face. Can you help Mom clear up Dad's misunderstanding?
   HINT: *She's talking about the OTHER cheeks.*

Answers to Figure Questions

11.1 Frontal, parietal, temporal, occipital, zygomatic, and nasal.
11.2 Galea aponeurotica.
11.3 Pupil.
11.4 Mastoid process.
11.5 Dorsum nasi.
11.6 Sternocleidomastoid.
11.7 With the aid of a stethoscope, respiratory sounds can be heard clearly in the triangle.
11.8 Sternal angle.
11.9 Tenderness produced by pressure of the finger on it indicates acute appendicitis; the incision for an appendectomy is made through it.

11.10 Acromion of the scapula.
11.11 Median cubital vein.
11.12 Radial artery.
11.13 The thumb.
11.14 Gluteus maximus.
11.15 The fossa is bordered superolaterally by the biceps femoris muscle, superomedially by the semimembranosus and semitendinosus muscle, and inferolaterally and inferomedially by the lateral and medial heads of the gastrocnemius muscle.
11.16 Dorsalis pedis artery.

Chapter 12

The Cardiovascular System: Blood

Student Objectives

1. Describe the transportation, regulation, and protective functions of blood.
2. Describe the physical characteristics of blood.
3. Explain the divisions of blood into plasma and formed elements.
4. Describe the developmental sequence by which blood cells are formed.
5. Describe the structure and functions of red blood cells.
6. Explain the basis for the ABO and Rh blood grouping systems.
7. Describe the structure and functions of white blood cells.
8. Describe the structure and functions of platelets.

Contents at a Glance

Introduction

The **cardiovascular** (*cardio* = heart; *vascular* = blood or blood vessels) **system** consists of three interrelated components: blood, the heart, and blood vessels. The focus of this chapter is blood, and the next two chapters will cover the heart and blood vessels, respectively.

Most cells of a multicellular organism cannot move around to obtain oxygen and nutrients and get rid of carbon dioxide and other wastes. Instead, to accomplish these needs the cells must be serviced by two fluids: blood and interstitial fluid. **Blood** is a connective tissue composed of a liquid portion called plasma and a cellular portion consisting of various cells and cell fragments. **Interstitial fluid** is the fluid that bathes body cells. Oxygen brought into the lungs and nutrients brought into the gastrointestinal tract are transported by the blood to the cells of the body. The oxygen and nutrients diffuse from the blood into the interstitial fluid and then into body cells. Carbon dioxide and other wastes move in the reverse direction from cells into the interstitial fluid and then into the blood. Blood then transports the wastes to various organs—the lungs, kidneys, skin, and digestive system—for elimination from the body. As you will see in this chapter, blood not only transports various substances throughout the body, it also helps regulate several life processes and affords protection against disease. The branch of science concerned with the study of blood, blood-forming tissues, and the disorders associated with them is **hematology** (hēm-a-TOL-ō-jē; *heme* = blood; *logos* = study of).

In order for blood to reach the cells of the body, it must be moved throughout the body. The *heart,* the center of the cardiovascular system, is the pump that circulates the blood. Blood vessels provide the routes that convey blood from the heart to body cells and from body cells back to the heart. The largest blood vessels that take blood away from the heart are called *arteries.* These branch into smaller vessels called *arterioles.* As an arteriole enters a tissue, it divides into numerous microscopic vessels called *capillaries.* Substances exchanged between the blood and interstitial fluid pass through the thin walls of capillaries. Before leaving a tissue, capillaries unite to form small veins called *venules.* These merge to form progressively larger vessels called *veins,* which convey blood back to the heart.

For all of its similarities in origin, composition, and functions, blood is as unique from one person to another as are skin, bone, and hair. Health care professionals routinely examine and analyze its differences through various blood tests when attempting to determine the cause of different diseases. Yet, despite these differences, blood is the most easily and widely shared of human tissues, saving many thousands of lives every year through blood transfusions.

FUNCTIONS OF BLOOD

Blood has three general functions: transportation, regulation, and protection.

1. **Transportation.** Blood transports oxygen from the lungs to the cells of the body, and carbon dioxide from the cells to the lungs. It also carries nutrients from the gastrointestinal tract to body cells, heat and waste products away from cells, and hormones from endocrine glands to body cells.

2. **Regulation.** Blood contains dissolved chemicals called buffers that help regulate the body's pH. Blood also adjusts body temperature due to the heat-absorbing properties of its water content and because of its variable rate of flow through the skin, where excess heat can be lost to the environment. Blood exerts an osmotic pressure that influences the water content within cells, principally through the action of dissolved ions and proteins in the plasma (discussed shortly).

3. **Protection.** Blood can clot, which protects against excessive loss of blood from the cardiovascular system following an injury. In addition, white blood cells protect against disease by carrying on phagocytosis and producing proteins called antibodies. Blood also contains additional proteins, called interferons and complement, that protect us against disease.

PHYSICAL CHARACTERISTICS OF BLOOD

Blood is a viscous (sticky) fluid; it is heavier, thicker, and more viscous than water. Its temperature is about 38°C (100.4°F), its pH range is 7.35 to 7.45 (slightly alkaline), and its salt (NaCl) concentration is 0.90%. The blood volume of an average-sized adult male is 5–6 liters (5–6 qt); an average-sized adult female has 4–5 liters. Blood constitutes about 8% of total body weight.

Whole blood is composed of two portions: (1) blood plasma, a watery liquid that contains dissolved substances, and (2) formed elements, which are cells and cell fragments. If a sample of blood is centrifuged (spun) in a small glass tube, the cells sink to the bottom of the tube while the lighter-weight plasma forms a layer on top (Figure 12.1a). Figure 12.1b shows the composition of blood plasma and the numbers of the various types of formed elements in blood.

Clinical Application

Withdrawing Blood

Blood samples for laboratory testing may be obtained in several ways. The most often used procedure is **venipuncture,** withdrawal of blood from a vein. A commonly used vein is the median

Figure 12.1 Components of blood in a normal adult.

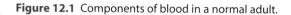

Blood is a connective tissue that consists of plasma (liquid) plus formed elements (erythrocytes, leukocytes, and platelets).

Plasma (55%)

Buffy coat, composed of leukocytes (white blood cells) and platelets

Erythrocytes (red blood cells) (45%)

(a) Appearance of centrifuged blood

OVERVIEW OF FUNCTIONS OF BLOOD
1. Transportation. Blood transports oxygen, carbon dioxide, nutrients, hormones, heat, and wastes.
2. Regulation. Blood helps regulate pH, body temperature, and water content of cells.
3. Protection. Blood protects against blood loss through clotting, and against foreign microbes through the activities of phagocytic white blood cells and certain plasma proteins.

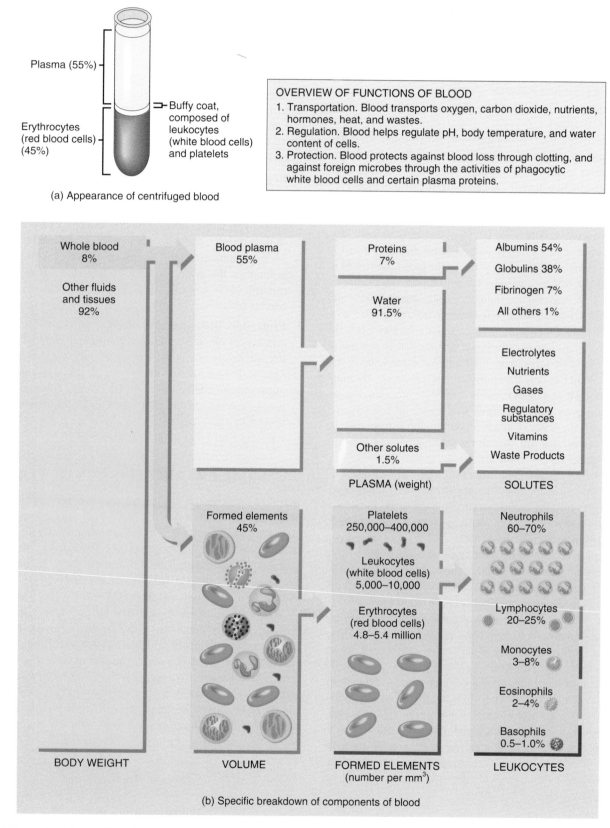

Whole blood 8%

Other fluids and tissues 92%

Blood plasma 55%

Proteins 7%

Albumins 54%

Globulins 38%

Fibrinogen 7%

All others 1%

Water 91.5%

Electrolytes

Nutrients

Gases

Regulatory substances

Vitamins

Other solutes 1.5%

Waste Products

PLASMA (weight)

SOLUTES

Formed elements 45%

Platelets 250,000–400,000

Neutrophils 60–70%

Leukocytes (white blood cells) 5,000–10,000

Lymphocytes 20–25%

Erythrocytes (red blood cells) 4.8–5.4 million

Monocytes 3–8%

Eosinophils 2–4%

Basophils 0.5–1.0%

BODY WEIGHT

VOLUME

FORMED ELEMENTS (number per mm³)

LEUKOCYTES

(b) Specific breakdown of components of blood

Q What is the volume of blood in the body?

cubital vein anterior to the elbow (see Figure 11.11c). A tourniquet is wrapped around the arm to stop blood flow through the veins. This makes the veins distal to the tourniquet stand out. Opening and closing the fist has the same effect.

Another procedure used to withdraw blood is the **finger-stick.** A drop or two of capillary blood is taken from a finger, earlobe, or heel of the foot for evaluation. Diabetics may monitor their blood glucose levels in this way.

Finally, an **arterial stick** may be used to withdraw blood. The sample usually is taken from the radial artery in the wrist (see Figure 11.12c) or the femoral artery in the groin (see Figure 9.2).

BLOOD PLASMA

When the formed elements are removed from blood, a straw-colored liquid called **blood plasma** or simply **plasma** is left. Plasma is about 91.5% water and 8.5% solutes, most of which by weight (7%) are proteins. Some of the proteins in plasma are also found elsewhere in the body, but those confined to blood are called *plasma proteins.* These proteins play a role in maintaining proper blood osmotic pressure, which is important in total body fluid balance. Most plasma proteins are synthesized by the liver, including the *albumins* (54% of plasma proteins), *globulins* (38%), and *fibrinogen* (7%). Other solutes in plasma include electrolytes, nutrients, gases, regulatory substances such as enzymes and hormones, and waste products such as urea, uric acid, creatine, creatinine, bilirubin, and ammonium salts. Table 12.1 summarizes the chemical composition of the plasma portion of blood.

FORMED ELEMENTS

The formed elements of the blood include three principal components: **erythrocytes** (red blood cells), **leukocytes** (white blood cells), and **platelets** (thrombocytes). Although erythrocytes and leukocytes are living cells, platelets are only cell fragments.

Unlike erythrocytes and platelets, each of which performs limited roles, leukocytes play a number of specialized roles. For this reason, leukocytes develop into several distinct forms, each having a unique appearance under the microscope. First, leukocytes are classified as either granular or agranular, depending on whether they contain chemical-filled vesicles (granules) in their cytoplasm. *Granular leukocytes* include neutrophils, eosinophils, and basophils. *Agranular leukocytes* include lymphocytes—such as T cells, B cells, or natural killer cells—and cells called monocytes. The role of each type of leukocyte will be discussed later in this chapter.

Formation of Blood Cells

The process by which the formed elements of blood develop is called **hemopoiesis** (hē-mō-poy-Ē-sis; *hemo* = blood; *poiem* = to make) or *hematopoiesis.* About 0.05–0.1% of red bone marrow cells are cells derived from mesenchyme, called **hemopoietic stem cells** (Figure 12.2). These cells replenish themselves and give rise to *all* the formed elements of the blood. For this reason, hemopoietic stem cells are also called **pluripotential stem cells.**

Originating from the pluripotential stem cells are two types of **multipotential stem cells,** which also have the ability to replenish themselves but that differentiate to give rise to more specific formed elements of blood. For example, *myeloid stem cells* give rise to erythrocytes, platelets, granular leukocytes, and monocytes. *Lymphoid stem cells* give rise to lymphocytes. Whereas myeloid stem cells begin and complete their development in red bone marrow, lymphoid stem cells begin their development in red bone marrow but complete it in lymphatic tissue.

As the development of the formed elements of blood continues, multipotential stem cells form **progenitor cells**—cells that no longer are capable of replenishing themselves and can only give rise to more specific formed elements of blood. The individual forms of progenitor cells are called *colony-forming cells* and are named on the basis of the mature elements in blood that they will ultimately produce. Accordingly, *erythrocyte-colony-forming cells (ECFCs)* develop into erythrocytes, *basophil-colony-forming cells (BCFCs)* develop into basophils, and so on. Stem cells and progenitor cells cannot be distinguished morphologically (on the basis of their appearance under the microscope) and resemble lymphocytes. Progenitor cells develop into **precursor cells,** known as **blasts,** which will develop over several cell divisions into the actual formed elements of blood. For example, *proerythroblasts* develop into erythrocytes, *lymphoblasts* develop into lymphocytes, and so on. Precursor cells exhibit recognizable morphological characteristics.

During embryonic and fetal life, the yolk sac, liver, spleen, thymus gland, lymph nodes, and red bone marrow all participate at various times in producing the formed elements. After birth, however, hemopoiesis takes place only in red bone marrow, although small numbers of stem cells circulate in the bloodstream. Red bone marrow is found in the epiphyses (ends) of long bones such as the humerus and femur; flat bones such as the sternum, ribs, and cranial bones; the vertebrae; and the pelvis. Once they leave the bone marrow, most of the formed elements do not divide and survive only a few hours or days. The exceptions are the immature lymphocytes, which may live for years and can be stimulated to undergo further proliferation by mitosis within lymphatic tissues.

Table 12.1 Substances in Plasma

CONSTITUENT	DESCRIPTION
Water	Liquid portion of blood; constitutes about 91.5% of plasma. Acts as solvent and suspending medium for solid components of blood and absorbs, transports, and releases heat.
Proteins	Constitute about 7.0% (by weight) of plasma.
Albumins	Smallest plasma proteins; produced by liver. Exert osmotic pressure, which helps maintain water balance between blood and tissues, and regulates blood volume. Also function as transport proteins for several steroid hormones.
Globulins	Protein group to which antibodies (immunoglobulins) belong. Produced by liver and plasma cells that develop from B lymphocytes. Antibodies help to attack viruses and bacteria. Alpha and beta globulins transport iron, fats, and fat-soluble vitamins.
Fibrinogen	Produced by liver. Plays essential role in blood clotting.
Other Solutes	Constitute about 1.5% (by weight) of plasma.
Electrolytes	Inorganic salts. Positively charged ions (cations) include Na^+, K^+, Ca^{2+}, Mg^{2+}; negatively charged ions (anions) include Cl^-, HPO_4^{2-}, SO_4^{2-}, and HCO_3^-. Help maintain osmotic pressure and play an essential role in the function of cells.
Nutrients	Products of digestion passed into blood for distribution to all body cells. Include amino acids (from proteins), glucose (from carbohydrates), and fatty acids and glycerol (from triglycerides).
Gases	Oxygen (O_2), carbon dioxide (CO_2), and nitrogen (N_2). Whereas there is more O_2 associated with hemoglobin inside red blood cells, there is more CO_2 dissolved in plasma. N_2 has no known function in the body.
Regulatory substances	Enzymes, produced by body cells, catalyze chemical reactions. Hormones, produced by endocrine glands, regulate growth and development in body.
Wastes	Most are breakdown products of protein metabolism and are carried by blood to organs of excretion. Include urea, uric acid, creatine, creatinine, bilirubin, and ammonium salts.

Erythrocytes (Red Blood Cells)

More than 99% of the formed elements in blood are **erythrocytes** (e-RITH-rō-sīts; *erythros* = red; *cyte* = cell) or *red blood cells (RBCs)*. They contain the oxygen-carrying protein **hemoglobin,** which is a pigment that gives whole blood its red color. To maintain normal quantities of erythrocytes, new mature cells must enter the circulation at the astonishing rate of at least 2 million per second. This pace balances the equally high rate of RBC destruction.

RBC Anatomy

Under the microscope, RBCs appear as biconcave discs that have a diameter of 7–8 μm (Figure 12.3a). Mature red blood cells have a simple structure. They lack a nucleus and other organelles and can neither reproduce nor carry on extensive metabolic activities. The plasma membrane encloses hemoglobin, which was synthesized before loss of the nucleus and which constitutes about 33% of the cell weight, dissolved in the cytosol. As you will see later, certain proteins (antigens) on the surfaces of red blood cells are responsible for the various blood groups. The ABO and Rh groups are examples.

RBC Functions

As blood passes through the lungs, hemoglobin inside RBCs combines with oxygen to form oxyhemoglobin. A hemoglobin molecule consists of a protein called *globin*, composed of four polypeptide chains (two called alpha and two called beta), plus four nonprotein pigments called *hemes* (Figure 12.3b). Each heme is associated with one polypeptide chain and contains an iron ion (Fe^{2+}) that can combine reversibly with one oxygen molecule. The oxygen is transported in this state to other tissues of the body. In the tissues, the iron–oxygen reaction reverses. Hemoglobin releases oxygen, which diffuses into the interstitial fluid and from there into cells.

Red blood cells are highly specialized for their oxygen transport function. Each one contains about 280 million hemoglobin molecules. Because RBCs have no nucleus, all their internal space is available for oxygen transport. Moreover, because they lack mitochondria and generate ATP anaerobically (without oxygen), RBCs do not consume any of the oxygen they transport. Even the shape of a RBC facilitates its function. A biconcave disk has a much greater surface area for its volume than, say, a sphere or a cube having the same volume. This shape confers two advantages.

Figure 12.2 Origin, development, and structure of blood cells.

Blood cell production is called hemopoiesis and occurs only in red bone marrow after birth.

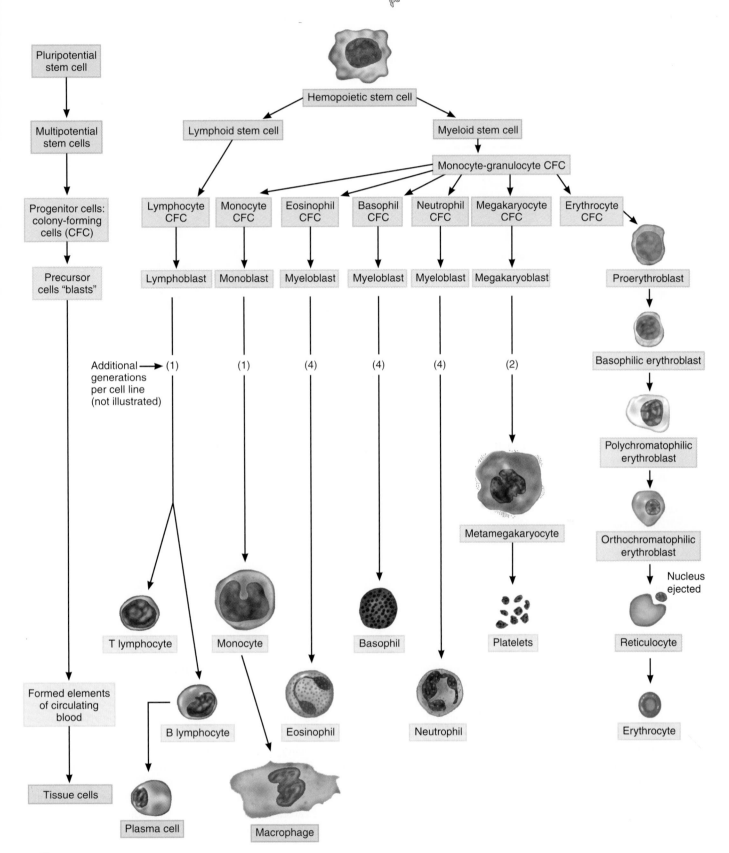

Q *List the following physical characteristics of blood: temperature, pH, and percentage of total body weight.*

Figure 12.3 Shape of red blood cells (RBCs) and the hemoglobin molecule. In (b), the four globin (protein) portions of the molecule are shown in blue; the heme (iron-containing) portions are shown in red. (Modified from R. E. Dickerson and I. Geis.)

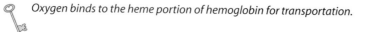

Oxygen binds to the heme portion of hemoglobin for transportation.

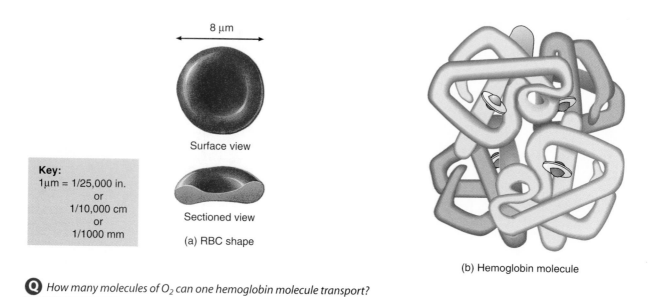

8 μm

Surface view

Key:
1μm = 1/25,000 in.
or
1/10,000 cm
or
1/1000 mm

Sectioned view

(a) RBC shape

(b) Hemoglobin molecule

Q *How many molecules of O_2 can one hemoglobin molecule transport?*

First, there is a large surface area for the diffusion of gas molecules into or out of the RBC. Second, the biconcave disk is a very flexible shape, which permits RBCs to squeeze through smaller, narrow capillaries, some as narrow as 3 μm in diameter.

Hemoglobin also transports about 23% of the body's total carbon dioxide, a waste product of metabolism. Blood flowing through tissue capillaries picks up carbon dioxide, some of which combines with amino acids in the globin portion of hemoglobin to form carbaminohemoglobin. This complex is transported to the lungs, where the carbon dioxide is released and then exhaled.

In addition to its key role in transporting oxygen and carbon dioxide, hemoglobin also assumes a function in blood pressure regulation, a recent and surprising discovery. *Nitric oxide (NO)* is a gas produced by endothelial (lining) cells of blood vessels. The iron ions in the heme portion of hemoglobin have a strong affinity for NO. Hemoglobin that has already circulated through the body and given up its oxygen to tissues passes into the lungs, where it releases both carbon dioxide to be exhaled and NO. Then the hemoglobin picks up a fresh supply of oxygen and a different form of NO called *super nitric oxide (SNO)*, probably produced by lung cells. In body tissues, hemoglobin releases its oxygen and SNO and picks up carbon dioxide. The released SNO can make blood vessels relax, if necessary. At the tissue level, hemoglobin can also pick up any local excess NO, thus making blood vessels contract, if necessary. By ferrying NO and SNO throughout the body, hemoglobin functions as a blood pressure regulator by adjusting the amount of NO or SNO to which blood vessels are exposed. This new role of hemoglobin could have implications for treating blood pressure and for the development of artificial blood.

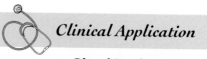

Clinical Application

Blood Doping

In recent years, some athletes have been tempted to try **blood doping.** In this procedure, blood cells are removed from the body, stored for a month or so, and then reinjected a few days before an athletic event. Because delivery of oxygen to muscle is a limiting factor in muscular feats, and red blood cells carry oxygen, it was predicted that increasing the oxygen-carrying capacity of the blood could increase muscular performance. Indeed, blood doping can improve athletic performance in endurance events. The practice is dangerous, however, because it increases the workload of the heart. With increased numbers of RBCs, the viscosity of the blood rises, which makes the blood more difficult for the heart to pump. Moreover, the procedure is banned by the International Olympic Committee. ■

RBC Life Span and Numbers

Red blood cells live only about 120 days because of the wear and tear inflicted on their plasma membranes as they squeeze through blood capillaries. Without a nucleus and other organelles, RBCs cannot synthesize new components to replace damaged ones. The plasma membrane thus becomes progressively fragile with age and the cell is more likely to burst, especially as it squeezes through narrow channels in the spleen. Worn-out red blood cells are removed from circulation and destroyed by fixed phagocytic macrophages in the spleen and liver, and the breakdown products are recycled.

A healthy adult male has about 5.4 million red blood cells per cubic millimeter (mm^3) of blood, and a healthy adult female has about 4.8 million. (There are about 50 mm^3 in a drop of blood.) The higher value in males is due to their higher levels of testosterone, which stimulates the synthesis of red blood cells.

Production of RBCs

The process of erythrocyte formation, **erythropoiesis** (e-rith′-rō-poy-Ē-sis), starts in the red bone marrow with a precursor cell called a proerythroblast (see Figure 12.2). The *proerythroblast (rubriblast)* gives rise to a *basophilic erythroblast (prorubricyte)*, which then develops into a *polychromatophilic erythroblast (rubricyte)*, the first cell in the sequence that begins to synthesize hemoglobin. The polychromatophilic erythroblast next develops into an *orthochromatophilic erythroblast (normoblast)*, in which hemoglobin synthesis is at maximum. In the next stage, the orthochromatophilic erythroblast ejects its nucleus and becomes a *reticulocyte*. Loss of the nucleus causes the center of the cell to indent, giving it a distinctive biconcave shape. Reticulocytes contain about 34% hemoglobin and retain some mitochondria, ribosomes, and endoplasmic reticulum. They pass from red bone marrow into the bloodstream by squeezing between the endothelial cells of blood capillaries. Reticulocytes usually develop into *erythrocytes,* or mature red blood cells, within one to two days after their release from red bone marrow.

Normally, erythropoiesis and red blood cell destruction proceed at the same pace. If the oxygen-carrying capacity of the blood falls because erythropoiesis is not keeping up with RBC destruction, erythrocyte production is increased. Oxygen delivery may fall due to **anemia,** a lower than normal number of RBCs or quantity of hemoglobin, or because of circulatory problems that reduce blood flow to tissues. Cellular oxygen deficiency, called **hypoxia** (hī-POKS-ē-a), may also occur if not enough oxygen enters the blood—for example, when you do not breathe in enough oxygen. This situation commonly occurs at high altitudes, where the air contains less oxygen. Whatever the cause, hypoxia stimulates the kidneys to step up release of the hormone **erythro-poietin** (e-rith′-rō-POI-e-tin). This hormone circulates through the blood to the red bone marrow, where it speeds the development of proerythroblasts into reticulocytes.

Blood Group Systems

More than 100 kinds of antigens have been detected on the surface of red blood cells. These antigens are genetically determined and are referred to as **agglutinogens** (ag′-loo-TIN-ō-jens). Many of these antigens appear in characteristic patterns, a fact that enables scientists or health care professionals to identify a person's blood as belonging to one or more blood grouping systems, of which at least 14 are currently recognized. Each system is characterized by the presence or absence of specific antigens on the surface of a red blood cell's plasma membrane. The two major blood groups distinguished on the basis of these antigens are called the ABO and Rh blood groups.

ABO blood grouping system is based on two antigens, symbolized as *A* and *B*. Individuals whose erythrocytes manufacture only antigen *A* are said to have blood type A. Those who manufacture only antigen *B* are type B. Individuals who manufacture both *A* and *B* are type AB. Those who manufacture neither are type O.

The *Rh blood grouping system* is so named because it was first worked out using the blood of the *Rh*esus monkey. Individuals whose erythrocytes have the Rh antigens (D antigens) are designated Rh^+. Those who lack Rh antigens are designated Rh^-.

Leukocytes (White Blood Cells)

WBC Anatomy and Types

Unlike red blood cells, **leukocytes** (LOO-kō-sīts; *leukos* = white; *cyte* = cell), or *white blood cells (WBCs),* have a nucleus and do not contain hemoglobin (Figure 12.4). As noted, the two major groups of WBCs are granular leukocytes and agranular leukocytes.

Granular Leukocytes (Granulocytes) Granular leukocytes have conspicuous granules in the cytoplasm that can be seen under a light microscope. The three types of granular leukocytes are eosinophils, basophils, and neutrophils. Each class of cells displays a distinctive coloration under a light microscope after staining.

The nucleus of an **eosinophil** (ē-ō-SIN-ō-fil) usually has two lobes connected by a thin or thick strand (Figure 12.4a). Large, uniform-sized granules pack the cytoplasm but usually do not cover or obscure the nucleus. These eosinophilic (= eosin-loving) granules stain red-orange with acidic dyes. The nucleus of a **basophil** (BĀ-sō-fil) is bilobed or irregular in shape, often in the form of the letter S (Figure 12.4b). The cytoplasmic basophilic granules are round, variable in size, stain blue-purple with basic dyes, and commonly obscure the nucleus.

Figure 12.4 Structure of blood cells.

White blood cells are distinguished on the basis of the shape of their nuclei and the color of their cytoplasm after staining.

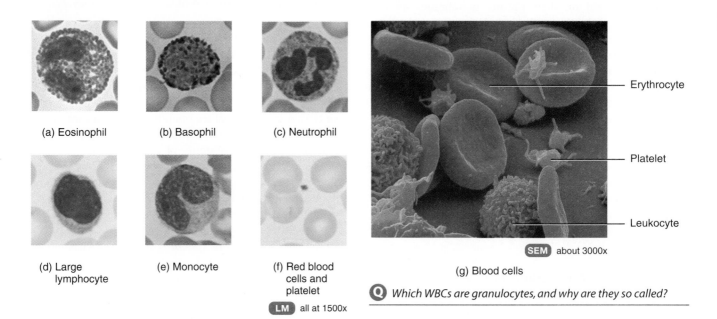

(a) Eosinophil

(b) Basophil

(c) Neutrophil

(d) Large lymphocyte

(e) Monocyte

(f) Red blood cells and platelet

LM all at 1500x

Erythrocyte

Platelet

Leukocyte

SEM about 3000x

(g) Blood cells

Q *Which WBCs are granulocytes, and why are they so called?*

The nucleus of a **neutrophil** (NOO-tro-fil) has two to five lobes, connected by very thin strands of chromatin (Figure 12.4c). As the cells age, the extent of nuclear lobulation increases. Because older neutrophils appear to have many differently shaped nuclei, they are often called *polymorphonuclear leukocytes (PMNs),* polymorphs, or "polys." Younger neutrophils are often called *bands* because their nucleus is more rod-shaped. When stained, the cytoplasm of neutrophils includes fine, evenly distributed pale lilac-colored granules, which may be difficult to see.

Agranular Leukocytes (Agranulocytes) Agranular leukocytes do not have cytoplasmic granules that can be seen under a light microscope, owing to their small size and poor staining qualities. The two kinds of agranular leukocytes are **lymphocytes** (LIM-fō-sīts) and **monocytes** (MON-ō-sīts).

Small lymphocytes are 6–9 μm in diameter; large lymphocytes are 10–14 μm in diameter (Figure 12.4d). (Although the functional significance of the size difference between small and large lymphocytes is unclear, the distinction is still useful clinically because an increase in the number of large lymphocytes has diagnostic significance in acute viral infections and some immunodeficiency diseases). The nuclei of lymphocytes are darkly stained and round, or slightly indented. The cytoplasm stains sky blue and forms a rim around the nucleus. The larger the cell, the more cytoplasm is visible.

Monocytes are 12–20 μm in diameter (Figure 12.4e). The nucleus of a monocyte is usually kidney-shaped or horseshoe-shaped, and the cytoplasm is blue-gray and has a

foamy appearance. The blood is merely a conduit for monocytes, which migrate out into the tissues, enlarge, and differentiate into **macrophages** (*macro* = large; *phagein* = to eat). Some are **fixed macrophages,** which means they reside in a particular tissue—for example, alveolar macrophages, spleen macrophages, or stellate reticuloendothelial (Kupffer) cells in the liver. Others are **wandering (free) macrophages,** which roam the tissues and gather at sites of infection or inflammation.

Just as red blood cells have surface proteins, so do white blood cells and all other nucleated cells in the body. Some of these proteins, called *major histocompatibility (MHC) antigens,* are unique for each person (except for identical twins).

Clinical Application

Histocompatibility Testing

The success of a proposed organ or tissue transplant depends on *histocompatibility* (his'-tō-kom-pat-i-BIL-i-tē)—that is, the tissue compatibility between the donor and the recipient. The more similar the MHC antigens, the greater the histocompatibility, and the higher the probability the transplant will not be rejected. **Histocompatibility testing (tissue typing)** is done before any organ transplant. A nationwide computerized registry helps physicians select the most histocompatible and neediest organ transplant recipients whenever donor organs become available. Also, in cases of disputed parentage, tissue typing can be used to identify the biological parents. ■

WBC Functions

The skin and mucous membranes of the body are continuously exposed to microbes and their toxins. Some of these microbes can invade deeper tissues to cause disease. Once pathogens enter the body, the general function of white blood cells is to combat them via phagocytosis or by engaging in other immune responses. To accomplish these tasks, many WBCs leave the bloodstream and collect at points of pathogen invasion or inflammation. When granulocytes and monocytes leave the bloodstream to fight injury or infection, they never return. Lymphocytes, on the other hand, continually recirculate from the blood, to interstitial spaces of tissues, to lymphatic fluid, and back to the blood. Only 2% of the total lymphocyte population is in the blood at any given time. The rest are in lymphatic fluid and organs such as skin, lungs, lymph nodes, and spleen.

WBCs leave the bloodstream by a process termed **emigration** (em′-i-GRĀ-shun; *e* = out; *migrare* = to move), also called *migration* or *extravasation*. In this process, white blood cells squeeze between the membrane junctions of endothelial cells that line the capillaries. The precise signals that stimulate emigration through a particular capillary vary for the different types of WBCs.

Neutrophils and macrophages (derived from monocytes) are **phagocytes,** cells that can ingest bacteria and dead matter. Toxins released by microbes and substances called *kinins* that are released by inflamed tissues attract phagocytes, a process called **chemotaxis.**

Among the WBCs, neutrophils respond most quickly to tissue destruction by bacteria. After engulfing a pathogen during phagocytosis, a neutrophil unleashes several destructive chemicals held within its granules. These include the enzyme *lysozyme,* which destroys certain bacteria, and *strong oxidants,* such as the superoxide anion (O_2^-), hydrogen peroxide (H_2O_2), and hypochlorite anion (OCl^-), which is similar to household bleach. Neutrophils also contain *defensins,* proteins that exhibit a broad range of antibiotic activity against bacteria, fungi, and viruses. Defensins form peptide spears that poke holes in microbial membranes. The resulting leakiness kills the invader.

Monocytes take longer to reach a site of infection than do neutrophils, but they arrive in larger numbers and destroy more microbes. Upon arrival they enlarge and differentiate into wandering macrophages, which clean up cellular debris and microbes following an infection.

Eosinophils leave the capillaries and enter tissue fluid. They are believed to release enzymes, such as *histaminase,* that combat the effects of histamine and other mediators of inflammation in allergic reactions. Eosinophils also phagocytize antigen–antibody complexes (described shortly) and are effective against certain parasitic worms. Consequently, a high eosinophil count may indicate an allergic condition or a parasitic infection.

Basophils are also involved in inflammatory and allergic reactions. They leave capillaries, enter tissues, and liberate heparin, histamine, and serotonin. These substances intensify the inflammatory reaction and are involved in hypersensitivity (allergic) reactions.

The two major types of lymphocytes are B cells and T cells. These cells are the major combatants in immune responses (described in detail in Chapter 15). Briefly, a substance that stimulates an immune response is called an **antigen** (AN-ti-jen). Most antigens are foreign proteins, ones not synthesized by the body. They include bacterial enzymes, toxins, and structural proteins. In response to certain antigens, **B cells** develop into **plasma cells** (see Figure 12.2) that produce antibodies. **Antibodies** (AN-ti-bod′-ēz) are proteins of the immunoglobulin family. A specific antibody will generally bind only to a certain antigen. The two together are called an *antigen–antibody complex.* However, unlike enzymes, which enhance the reactivity of the substrate, antibodies "cover" their antigens so the antigens cannot come in contact with other chemicals in the body. In this way, bacterial poisons are inactivated, and the bacteria themselves are destroyed. This process is called the *antigen–antibody response.*

Antigens also stimulate a variety of **T cells** to take part in immune responses. One group of T cells, the *cytotoxic (killer) T cells,* also called *T8 cells,* react by destroying foreign invaders directly. Another group, called *helper T cells (T4 cells),* assists both B cells and cytotoxic T cells. T cells are especially effective against viruses, fungi, transplanted cells, cancer cells, and some bacteria. You will see later that one type of T cell is destroyed by the AIDS virus, and this greatly weakens the immune system.

Immune responses mediated by B and T cells help combat infection and provide protection against some diseases. They also are responsible for transfusion reactions, allergies, and the body's rejection of organs transplanted from a person with different MHC antigens.

An increase in the number of circulating WBCs usually indicates inflammation or infection. Because each type of white blood cell plays a different role, determining the *percentage* of each type in the blood assists in diagnosing the condition.

Clinical Application

Differential White Blood Cell Count

To evaluate infection or inflammation, determine the effects of possible poisoning by chemicals or drugs, monitor blood disorders (for example, leukemia) and effects of chemotherapy, or detect allergic reactions and parasitic infections, a physician may order a **differential white blood cell count.** The percentage of each type of white blood cell in a normal differential white blood cell count and possible causes of increased or decreased counts are as follows:

	Percentage		Percentage
Neutrophils	60–70	Eosinophils	2–4
Lymphocytes	20–25	Basophils	0.5–1
Monocytes	3–8	Total	100

A high neutrophil count might result from bacterial infections, burns, stress, or inflammation; a low count might be caused by radiation, certain drugs, vitamin B_{12} deficiency, or systemic lupus erythematosus (SLE). High lymphocyte counts could indicate viral infections and some leukemias; low counts might occur as a result of prolonged severe illness, high steroid levels, and immunosuppression. A high monocyte count could result from certain viral or fungal infections, tuberculosis, some leukemias, and chronic diseases; below normal monocyte levels rarely occur. A high eosinophil count could indicate allergic reactions, parasitic infections, or autoimmune disease; a low count could be caused by certain drugs or stress. Basophils could be elevated in some types of allergic responses, leukemias, cancers, and due to an underactive thyroid gland; decreases could occur during pregnancy, ovulation, stress, and due to an overactive thyroid gland. ∎

WBC Life Span and Numbers

Bacteria exist everywhere in the environment and have continuous access to the body through the mouth, nose, and pores of the skin. Furthermore, many cells, especially those of epithelial tissue, age and die daily, and their remains must be disposed of by phagocytes that actively ingest bacteria and debris. However, the engulfed debris interferes with normal metabolic activities and sooner or later causes death of the phagocyte, which then is consumed by another phagocyte. In a healthy body, some WBCs, especially lymphocytes, can live for several months or years, but most live only a few days. During a period of infection, phagocytic WBCs may live only a few hours.

WBCs are far less numerous than red blood cells, averaging from 5000 to 10,000 cells per cubic millimeter (mm^3) of blood. RBCs therefore outnumber white blood cells about 700:1. The term **leukocytosis** (loo′-kō-sī-TŌ-sis) refers to an increase in the number of WBCs. An abnormally low level of white blood cells (below 5000/mm^3) is termed **leukopenia** (loo-kō-PĒ-nē-a).

Production of WBCs

We will first consider the origin of the granular leukocytes: eosinophils, basophils, and neutrophils. Note in Figure 12.2 that eosinophils, basophils, and neutrophils develop from a monocyte-granulocyte-colony-forming cell derived from a myeloid stem cell.

Agranular leukocytes—lymphocytes and monocytes—develop from different cells. Whereas lymphocytes develop from a lymphoid stem cell, monocytes develop from a monocyte-granulocyte-colony-forming cell derived from a myeloid stem cell.

Clinical Application

Bone Marrow Transplant

A **bone marrow transplant** is the intravenous transfer of red bone marrow from a healthy donor to a recipient to establish normal hematopoiesis in a patient with defective or damaged red bone marrow. The donor marrow must be very closely matched to that of the recipient through histocompatibility testing. In the procedure, donor marrow is withdrawn from the hipbones, mixed with heparin (to prevent blood clotting), and then passed through screens. The suspension of red bone marrow cells is treated to remove T cells and injected into a vein. Stem cells in the suspension pass through the lungs, enter the general circulation, and reseed and grow in the red bone marrow cavities of the recipient's bones. In patients with cancer or certain genetic diseases, the recipient's faulty red bone marrow first must be destroyed by irradiation before transplantation of the healthy marrow.

Bone marrow transplants have been used to treat aplastic anemia, certain types of leukemia, severe combined immunodeficiency disease (SCID), Hodgkin's disease, non-Hodgkin's lymphoma, multiple myeloma, thalassemia, sickle-cell disease, breast cancer, ovarian cancer, testicular cancer, and hemolytic anemia. ∎

Platelets

Besides the immature cell types that develop into erythrocytes and leukocytes, hemopoietic stem cells also differentiate into cells that produce platelets. Myeloid stem cells develop into megakaryocyte-colony-forming cells that, in turn, develop into precursor cells called megakaryoblasts (see Figure 12.2). Megakaryoblasts transform into metamegakaryocytes, huge cells that break apart into 2000–3000 fragments. Each fragment, enclosed by a piece of the cell membrane, is a **platelet** (*thrombocyte*). Platelets break off from the metamegakaryocytes in red bone marrow and then enter the blood circulation. Between 250,000 and 400,000 platelets are present in each cubic millimeter (mm^3) of blood. They are disc-shaped, 2–4 μm in diameter, and exhibit many granules but no nucleus. Platelets help stop blood loss from damaged blood vessels by forming a platelet plug. Their granules also contain chemicals that upon release promote blood clotting. Platelets have a short life span, normally just 5–9 days. Aged and dead platelets are removed by fixed macrophages in the spleen and liver.

A summary of the formed elements in blood is presented in Table 12.2.

Table 12.2 *Summary of the Formed Elements in Blood*

NAME AND APPEARANCE	NUMBER	CHARACTERISTICS[a]	FUNCTIONS
Red blood cells (RBCs) or **erythrocytes**	4.8 million/mm^3 in females; 5.4 million/mm^3 in males.	7–8 μm diameter, biconcave discs, without a nucleus; live for about 120 days.	Hemoglobin within RBCs transports most of the oxygen and part of the carbon dioxide in the blood. It also transports NO and SNO which help adjust blood pressure.
White blood cells (WBCs) or **leukocytes**	5000–10,000/mm^3	Most live for a few hours to a few days.[b]	Combat pathogens and other foreign substances that enter the body.
Granular leukocytes			
Neutrophils Chromatin	60–70% of all WBCs.	10–12 μm diameter; nucleus has 2–5 lobes connected by thin strands of chromatin; cytoplasm has very fine, pale lilac granules.	Phagocytosis. Destruction of bacteria with lysozyme, defensins, and strong oxidants, such as superoxide anion, hydrogen peroxide, and hypochlorite anion.
Eosinophils	2–4% of all WBCs.	10–12 μm diameter; nucleus usually has 2 lobes; large, red-orange granules fill the cytoplasm.	Combat the effects of histamine in allergic reactions, phagocytize antigen–antibody complexes, and destroy certain parasitic worms.
Basophils	0.5–1% of all WBCs.	8–10 μm diameter; nucleus is bilobed or irregular in shape; large cytoplasmic granules appear deep blue-purple.	Liberate heparin, histamine, and serotonin in allergic reactions that intensify the overall inflammatory response.
Agranular leukocytes			
Lymphocytes (T cells, B cells, and natural killer cells)	20–25% of all WBCs.	Small lymphocytes are 6–9 μm in diameter; large lymphocytes are 10–14 μm in diameter; nucleus is round or slightly indented; cytoplasm forms a rim around the nucleus that looks sky blue; the larger the cell, the more cytoplasm is visible.	Mediate immune responses, including antigen–antibody reactions. B cells develop into plasma cells, which secrete antibodies. T cells attack invading viruses, cancer cells, and transplanted tissue cells. Natural killer cells attack a wide variety of infectious microbes and certain spontaneously arising tumor cells.
Monocytes	3–8% of all WBCs.	12–20 μm diameter; nucleus is oval or kidney-shaped or horseshoe-shaped; cytoplasm is blue-gray and has foamy appearance.	Phagocytosis (after transforming into fixed or wandering macrophages).
Platelets (thrombocytes)	250,000–400,000/mm^3.	2–4 μm diameter cell fragments that live for 5–9 days; contain many granules but no nucleus.	Blood clotting and other mechanisms that prevent excessive loss of blood.

[a]Colors are those seen when using Wright's stain.
[b]Some lymphocytes, called T and B memory cells, can live for many years once they are established.

Clinical Application

Complete Blood Count

A **complete blood count (CBC)** is a valuable test that screens for anemia and various infections. Usually included are a determination of red blood cell count (the number of RBCs per mm^3 of blood), amount of hemoglobin (grams per ml of blood), hematocrit (volume percentage of RBCs in whole blood), white blood cell count (number of WBCs per mm^3 of blood), differential white blood cell count, and platelet count (number of platelets per mm^3 of blood). ▪

Key Medical Terms
Associated with Blood

Acute normovolemic (nor-mō-vō-LĒ-mik) **hemodilution** (hē-mō-di-LOO-shun) Removal of blood immediately before surgery and replacing it with a cell-free solution to maintain normal blood volume for adequate circulation. At the end of surgery, when bleeding has been controlled, the collected blood is returned to the body.

Autologous (aw-TOL-o-gus; *auto* = self) **intraoperative transfusion** (trans-FYOO-zhun) **(AIT)** Procedure in which blood lost during surgery is suctioned from the patient, treated with an anticoagulant, and reinfused into the patient.

Autologous preoperative transfusion Donating one's own blood; can be done up to 6 weeks before elective surgery. Also called *predonation.*

Blood bank A stored supply of blood for future use by the donor or others. Because blood banks have now assumed additional and diverse functions (immunohematology reference work, continuing medical education, bone and tissue storage, and clinical consultation), they are more appropriately referred to as **centers of transfusion medicine.**

Cyanosis (sī-a-NŌ-sis; *cyano* = blue) Slightly bluish, dark purple skin coloration, most easily seen in the nail beds and mucous membranes, due to increased quantity of reduced hemoglobin (hemoglobin not combined with oxygen) in systemic blood.

Gamma globulin (GLOB-yoo-lin) Solution of immunoglobulins from blood consisting of antibodies that react with specific pathogens, such as viruses. It is prepared by injecting the specific virus into animals, removing blood from the animals after antibodies have accumulated, isolating the antibodies, and injecting them into a human to provide short-term immunity.

Hemochromatosis (hē-mō-krō-ma-TŌ-sis; *heme* = iron; *chroma* = color) Disorder of iron metabolism characterized by excess deposits of iron in tissues, especially the liver, heart, pituitary gland, gonads, and pancreas, that result in bronze coloration of the skin, cirrhosis, diabetes mellitus, and bone and joint abnormalities.

Hemorrhage (HEM-or-ij; *rhegnynai* = bursting forth) Loss of a large amount of blood, either internal (from blood vessels into tissues) or external (from blood vessels directly to the surface of the body).

Multiple myeloma (mī-e-LŌ-ma) Malignant disorder of plasma cells in red bone marrow; symptoms (pain, osteoporosis, hypercalcemia, thrombocytopenia, kidney damage) are caused by the growing tumor cell mass or antibodies produced by malignant cells.

Phlebotomist (fle-BOT-ō-mist; *phlebo* = vein; *tome* = to cut) A technician who specializes in withdrawing blood.

Septicemia (sep′-ti-SĒ-mē-a; *septicus* = decay; *emia* = condition of blood) Toxins or disease-causing bacteria growing in the blood. Also called *"blood poisoning."*

Thrombocytopenia (throm′-bō-sī′-tō-PĒ-nē-a; *thrombo* = clot; *penia* = poverty) Very low platelet count that results in a tendency to bleed from capillaries.

Transfusion (trans-FYOO-shun) Transfer of whole blood, blood components (red blood cells only or plasma only), or red bone marrow directly into the bloodstream.

Venesection (vē′-ne-SEK-shun; *veno* = vein) Opening of a vein for withdrawal of blood. Although **phlebotomy** (fle-BOT-ō-me) is a synonym for venesection, in clinical practice, phlebotomy refers to therapeutic bloodletting, such as removing some blood to lower the viscosity of blood of a patient with polycythemia.

Whole blood Blood containing all formed elements, plasma, and plasma solutes in natural concentrations.

Study Outline

Functions of Blood (p. 377)

1. Blood transports oxygen, carbon dioxide, nutrients, wastes, and hormones.
2. It helps to regulate pH, body temperature, and water content of cells.
3. It prevents blood loss through clotting and combats toxins and microbes through certain phagocytic white blood cells or specialized plasma proteins.

Physical Characteristics of Blood (p. 377)

1. Physical characteristics of blood include a viscosity greater than that of water, a temperature of 38°C (100.4°F), and a pH of 7.35 to 7.45. Blood constitutes about 8% of body weight and volume ranges from 4 to 6 liters in an adult.

Blood Plasma (p. 379)

1. Blood consists of 55% plasma and 45% formed elements.
2. Plasma consists of 91.5% water and 8.5% solutes.
3. Principal solutes include proteins (albumins, globulins, fibrinogen), nutrients, hormones, respiratory gases, electrolytes, and waste products.

Formed Elements (p. 379)

1. The formed elements in blood include erythrocytes (red blood cells), leukocytes (white blood cells), and platelets.

Formation of Blood Cells (p. 379)

1. Blood cells are formed from hematopoietic stem cells in red bone marrow.
2. The process is called hemopoiesis.

3. Myeloid stem cells form erythrocytes, platelets, granular leukocytes, and monocytes. Lymphoid stem cells give rise to lymphocytes.

Erythrocytes (Red Blood Cells) (p. 380)

1. Mature erythrocytes are biconcave discs without nuclei that contain hemoglobin.
2. The function of the hemoglobin in red blood cells is to transport oxygen and some carbon dioxide. It also transports nitric oxide and super nitric oxide, which participate in blood pressure regulation.
3. After phagocytosis of aged red blood cells by macrophages, hemoglobin is recycled.
4. Erythrocyte formation, called erythropoiesis, occurs in adult red bone marrow of certain bones. It starts with a precursor cell called a proerythroblast and is stimulated by hypoxia.
5. Red blood cells live about 120 days. A healthy adult male has about 5.4 million/mm^3 of blood; a healthy adult female, about 4.8 million/mm^3.

Leukocytes (White Blood Cells) (p. 383)

1. Leukocytes are nucleated cells. Two principal types are granular (neutrophils, eosinophils, basophils) and agranular (lymphocytes and monocytes).
2. Eosinophils, basophils, and neutrophils develop from a monocyte-granulocyte-colony-forming cell derived from a myeloid stem cell.

3. Whereas lymphocytes develop from a lymphoid stem cell, monocytes develop from a monocyte-granulocyte-colony-forming cell derived from a myeloid stem cell.
4. The general function of leukocytes is to combat inflammation and infection. Neutrophils and macrophages (which develop from monocytes) do so through phagocytosis.
5. Eosinophils combat the effects of histamine in allergic reactions, phagocytize antigen–antibody complexes, and combat parasitic worms; basophils develop into mast cells that liberate heparin, histamine, and serotonin in allergic reactions that intensify the inflammatory response.
6. B lymphocytes, in response to the presence of foreign substances called antigens, differentiate into plasma cells that produce antibodies. Antibodies attach to the antigens and render them harmless. This antigen–antibody response combats infection and provides immunity. T lymphocytes destroy foreign invaders directly.
7. White blood cells usually live for only a few hours or a few days. Normal blood contains 5000 to 10,000/mm^3.

Platelets (p. 386)

1. Platelets are disc-shaped structures without nuclei.
2. They are formed from a myeloid stem cell that develops into a megakaryocyte-colony-forming cell that, in turn, develops into a precursor cell called a megakaryoblast. Platelets are involved in clotting.
3. Normal blood contains 250,000 to 400,000/mm^3.

Review Questions

1. What is the cardiovascular system? (p. 377)
2. List the functions of blood and their relationships to other systems of the body. (p. 377)
3. List the principal physical characteristics of blood. (p. 377)
4. Distinguish between plasma and formed elements. (p. 379)
5. What are the major constituents in plasma? What do they do? (p. 380)
6. Describe the origin of blood cells. (p. 379)
7. Describe the microscopic appearance of erythrocytes. What is the function of erythrocytes? (p. 380)
8. Define erythropoiesis. What factors accelerate and decelerate erythropoiesis? (p. 383)

9. Describe the classification of leukocytes and describe their microscopic appearance. What are their functions? (p. 383)
10. What is the importance of emigration, chemotaxis, and phagocytosis in fighting bacterial invasion? (p. 385)
11. Describe the antigen–antibody response. How is it protective? (p. 385)
12. Distinguish between leukocytosis and leukopenia. (p. 386)
13. Describe the structure and function of platelets. (p. 386)
14. Compare erythrocytes, leukocytes, and platelets with respect to size, number per mm^3, and life span. (p. 387)
15. Refer to the glossary of key medical terms associated with blood. Be sure that you can define each term. (p. 388)

Self Quiz

Complete the following:

1. The three types of plasma proteins are ___, ___, and ___.
2. A test to determine the percentage of each type of white blood cell is known as a ___ count.
3. Blood is a connective tissue that consists of about ___% extracellular material (plasma) and about ___% formed elements.
4. The process of red blood cell formation is called ___.
5. The study of blood and blood-forming tissues is called ___.

Choose the one best answer to the following questions:

6. Which of the following is *not* a site of hemopoiesis in the body?
   a. sternum
   b. liver
   c. vertebrae
   d. head of the humerus
   e. parietal bones
7. Which of the following statements is false?
   a. Blood is thicker than water.
   b. Blood normally has a pH of 6.5–7.0.
   c. The adult human male body normally contains about 5–6 liters of blood.
   d. Normal blood temperature is 38°C.
   e. Blood constitutes about 8% of total body weight.
8. The normal red blood cell count in healthy adult males is
   a. 5.4 million RBCs/mm^3
   b. 2 million RBCs/mm^3
   c. 0.5 million RBCs/mm^3
   d. 250,000 RBCs/mm^3
   e. 8000 RBCs/mm^3
9. The ability of white blood cells to crawl through capillary walls and reach an injured or infected body tissue is
   a. emigration
   b. phagocytosis
   c. leukocytosis
   d. leukopenia
   e. pinocytosis
10. Erythrocytes
    a. lack nuclei
    b. are disc-shaped
    c. bear ABO and Rh antigens on their cell membranes
    d. all of the above
    e. none of the above
11. Which of the following is *not* a function of blood?
    a. protection against fluid loss
    b. transportation of gases and nutrients
    c. regulation of body temperature
    d. protection against microbes and toxins
    e. synthesis of plasma proteins

12. Arrange the following four stages of erythropoiesis in chronological order.
    ___ (1) Basophilic erythroblast is formed.   a. 1, 2, 3, 4
    ___ (2) Erythroblast loses its nucleus.   b. 1, 3, 2, 4
    ___ (3) Polychromatophilic erythroblast   c. 2, 1, 3, 4
          starts to synthesize hemoglobin.   d. 2, 4, 1, 3
    ___ (4) Reticulocyte moves out of bone   e. 3, 2, 1, 4
          marrow and into general circulation.
13. Which of the following statements about neutrophils is false?
    a. They are actively phagocytic.
    b. They are the most abundant type of leukocyte.
    c. Their numbers decrease during most infections.
    d. They combat bacteria with lysozyme, oxidants, and defensins.
    e. They are granular leukocytes.
14. Granular leukocytes develop from which precursor cell?
    a. megakaryoblast
    b. lymphoblast
    c. monoblast
    d. proerythroblast
    e. myeloblast
15. The parent (stem) cell of all blood cells is the
    a. myeloid stem cell
    b. megakaryoblast
    c. pluripotential stem cell
    d. lymphoblast
    e. lymphoid stem cell
16. The precursor cells that form mature red blood cells are
    a. proerythroblasts
    b. lymphoblasts
    c. myeloblasts
    d. megakaryoblasts
    e. monoblasts
17. Lymphocytes may develop into
    a. B cells
    b. T cells
    c. killer cells
    d. helper cells
    e. all of the above
18. Lymphoid stem cells
    a. develop into myeloid stem cells
    b. develop into lymphocytes
    c. develop into platelets and granular leukocytes
    d. complete their development in red bone marrow
    e. all of the above
19. One function of platelets is to aid in the
    a. transport of carbon dioxide
    b. production of vitamin K
    c. utilization of calcium and phosphorus
    d. destruction of bacteria
    e. clotting of blood

Are the following statements true or false?

20. All nucleated cells (including white blood cells) have surface MHC antigens.

21. Blood has a salt (NaCl) concentration of 0.9%.

22. The pigment hemoglobin is responsible for the color of blood and the transport of respiratory gases.

23. The three kinds of granular leukocytes are neutrophils, monocytes, and eosinophils.

24. Match the following terms with their descriptions:

___ (a) constitute the largest percentage (60–70%) of leukocytes

___ (b) make up 20–25% of leukocytes

___ (c) are involved in immunity; form plasma cells for antibody production

___ (d) are involved in allergic response; release serotonin, heparin, and histamine

___ (e) are involved in allergic response; release antihistamines

___ (f) develop into macrophages that clean up sites of infection

___ (g) are important in phagocytosis (two answers)

(1) basophils
(2) eosinophils
(3) lymphocytes
(4) monocytes
(5) neutrophils

Critical Thinking Questions

1. Felicia was telling her roommate some facts she had learned about blood in her Human Anatomy class. Her roommate is a math major and thought it would be interesting to calculate how much Felicia's blood weighed. Felicia weighs 146 pounds. How much of her weight is from her blood?
HINT: *Blood makes up what percentage of body weight?*

2. A professional cyclist finished third in a grueling cross-country bicycle race. Although he appeared to be in great physical condition, he suffered a heart attack a few hours after the race. The media speculated that his condition was caused by blood doping. Explain.
HINT: *Blood "dope"-ing doesn't refer to his intelligence in this case, but it ought to.*

3. Geordy has been sniffling and sneezing his way through Human Anatomy class all spring, and his eyes are watery and itchy. While checking a smear of his own blood, he thought his blood had a lot more of the bluish-black granular cells than his lab partner's blood, so he asked his instructor to confirm his finding. What are these cells, and why does Geordy have a higher number of them than usual?
HINT: *Geordy used to take antihistamines to relieve his symptoms, but he quit because they made him sleepy.*

4. Becky is on a perpetual diet of diet soda, pizza, and chocolate. Recently, she visited the clinic complaining of a lack of energy and constant fatigue. She felt chilled in the overheated clinic, even though she wore a sweater. The nurse noticed that her skin appeared pale and colorless. A Complete Blood Count (CBC) was ordered. What condition does the nurse suspect, and what will the CBC reveal?
HINT: *Think back to what you learned about skin color in the chapter on the integumentary system.*

5. While looking at slides of their own blood under the microscope, Antonio teased Renee, "My blood's redder than your blood. My blood's redder than yours!" In a way, he's right. How does the composition of blood differ in males versus females?
HINT: *What makes blood red?*

6. A college student visited the bloodmobile to make a donation. He overheard a technician say something about how his blood was "a positive." (Of course, what she said was, "A pos.") What does "A pos." mean when referring to blood?
HINT: *"A pos." has nothing to do with usefulness or advantage and everything to do with red blood cells.*

Answers to Figure Questions

12.1 About 6 liters (1.5 gal) in an average-sized adult male and 4–5 liters (1.2 gal) in an average-sized adult female.

12.2 Temperature, 38°C (100.4°F); pH, 7.35–7.45; proportion of body weight, 8%.

12.3 Four—one bound to each heme group.

12.4 Neutrophils, eosinophils, and basophils are called granular leukocytes because all have granules in their cytoplasm that after staining can be seen using a light microscope.

Chapter 13

The Cardiovascular System: The Heart

Student Objectives

1. Describe the general flow of blood through the systemic and pulmonary circulation.

2. Describe the location of the heart and its surface projection.

3. Describe the structure of the pericardium and heart wall.

4. Discuss the external and internal anatomy of the chambers of the heart.

5. Explain the flow of blood through the heart.

6. Describe the structure and function of the valves of the heart.

7. Discuss the blood supply of the heart.

8. Explain the structural and functional features of the conduction system of the heart.

9. Explain the relationship between exercise and the heart.

10. Describe the development of the heart.

As noted in Chapter 12, the cardiovascular system consists of the blood, heart, and blood vessels. Having considered the origin, structural features, and functions of blood, we will now examine the heart, the center of the cardiovascular system. This chapter explores the design of the heart and its unique properties that permit a lifetime of pumping with never a minute's rest.

The heart is magnificently designed for its task of propelling blood through an estimated 100,000 km (60,000 mi) of blood vessels. Even while you are sleeping, your heart pumps 30 times its own weight each minute, about 5 liters (5.3 qt) to the lungs and the same volume to the rest of the body. At this rate, the heart would pump more than 14,000 liters (3,600 gal) of blood in a day or 10 million liters (2.6 million gal) in a year. You don't spend all your time sleeping, however, and your heart pumps more vigorously when you are active. The actual flow thus is much larger. The study of the normal heart and diseases associated with it is known as **cardiology** (kar-dēOL-ō-jē).

OVERVIEW OF CIRCULATION

With each beat, the heart pumps blood into two closed circuits—**systemic circulation** and **pulmonary** (*pulmonis* = lung) **circulation** (Figure 13.1). The left side of the heart is the pump for systemic circulation. It receives freshly oxygenated (oxygen-rich) blood from the lungs and propels it into the largest artery in the body, the **aorta** (*aorte* = to suspend, because the aorta once was believed to suspend the heart). From the aorta, the blood divides into separate streams, entering progressively smaller and smaller **systemic arteries** that carry it to organs throughout the body except the air sacs (alveoli) of the lungs.

In the tissues, arteries give rise to smaller diameter vessels called **arterioles,** which finally lead into extensive microscopic **capillaries.** Exchange of nutrients and gases occurs across the thin capillary walls. Here, blood unloads oxygen (O_2) and picks up carbon dioxide (CO_2). In most cases, blood flows through only one capillary and then enters a small vein called a **venule.** Venules carry deoxygenated (oxygen-poor) blood away from tissues and merge to form larger vessels called **veins,** and ultimately the blood flows back to the right side of the heart.

The right side of the heart is the pump for pulmonary circulation. It receives all deoxygenated blood returning from the systemic circulation. Blood ejected from the right side of the heart flows into the **pulmonary trunk,** a large vessel that branches into **pulmonary arteries,** which carry blood to the right and left lungs. In pulmonary capillaries, then, blood unloads carbon dioxide, which is exhaled, and picks up oxygen. The freshly oxygenated blood then returns to the left side of the heart to begin another pass through systemic circulation.

The upper chambers of the heart receive blood from both systemic and pulmonary circulations and they are called atria. The lower chambers of the heart, which pump blood into both systemic and pulmonary circulations, are called ventricles.

LOCATION AND SURFACE PROJECTION OF THE HEART

For all its might, the hollow, cone-shaped heart is relatively small, about the same size as a person's closed fist, and weighs only about 300 g (10 oz) in an adult. The heart is about 12 cm (5 in.) long, 9 cm (3.5 in.) wide at its broadest point, and 6 cm (2.5 in.) thick.

The heart rests on the diaphragm, near the midline of the thoracic cavity in the **mediastinum** (mē′-dē-a-STĪ-num), a mass of tissue that extends from the sternum to the vertebral column and between the coverings (pleurae) of the lungs (Figure 13.2a). About two-thirds of the mass of the heart lies to the left of the body's midline. The position of the heart in the mediastinum is more readily appreciated by examining its ends, surfaces, and borders (Figure 13.2b). First, visualize the heart as a cone lying on its side. The pointed end of the heart is the **apex,** which is directed anteriorly, inferiorly, and to the left. The end opposite the apex is the **base,** the broad portion of the heart that points posteriorly, superiorly, and to the right. In addition to the apex and base, there are several surfaces and borders (margins) that are useful in describing anatomical relationships and pathological conditions that relate to deviations from normal shape and contour. The **anterior (sternocostal) surface** is deep to the sternum and ribs and is presented in the view in Figure 13.2b. The **inferior (diaphragmatic) surface** is the portion of the heart that rests mostly on the diaphragm. It is found between the apex and right border. The **right border** faces the right lung and extends from the inferior surface to the base. Finally, the **left border,** also called the pulmonary border, faces the left lung and extends from the base to the apex.

Surface projection refers to outlining the shape of an organ on the surface of the body. This practice is useful for diagnostic procedures (for example, a lumbar puncture), auscultation (for example, listening to heart and lung sounds), and anatomical studies. We can project the heart on the anterior surface of the chest by locating the following landmarks. The **superior right point** is located at the superior border of the third right costal cartilage, about 3 cm (1 in.) to the right of the midline

Figure 13.1 Systemic and pulmonary circulation. Throughout this book, blood vessels that carry oxygenated blood are colored red, whereas those that carry deoxygenated blood are colored blue.

Whereas oxygenated blood is oxygen-rich, deoxygenated blood is oxygen-poor.

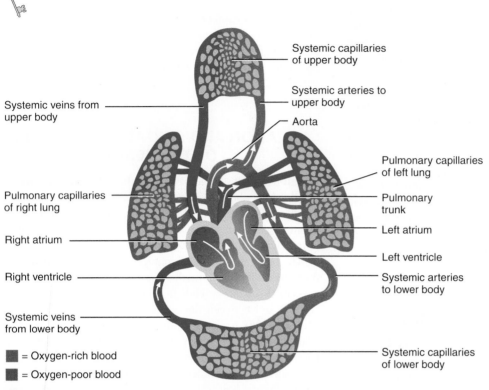

Systemic capillaries of upper body

Systemic arteries to upper body

Aorta

Systemic veins from upper body

Pulmonary capillaries of left lung

Pulmonary trunk

Left atrium

Left ventricle

Systemic arteries to lower body

Pulmonary capillaries of right lung

Right atrium

Right ventricle

Systemic veins from lower body

Systemic capillaries of lower body

■ = Oxygen-rich blood

■ = Oxygen-poor blood

OVERVIEW OF FUNCTIONS OF THE HEART

1. The heart pumps blood into two closed circuits—the systemic and the pulmonary circulations.
2. The left side of the heart pumps freshly oxygenated blood into the systemic circulation to all tissues of the body except the air sacs (alveoli) of the lungs. In systemic capillaries, blood unloads oxygen and picks up carbon dioxide.
3. At the same time, the right side of the heart pumps deoxygenated (oxygen-poor) blood into the pulmonary circulation to the lungs. In pulmonary capillaries, blood picks up oxygen and unloads carbon dioxide, which is exhaled.
4. Freshly oxygenated blood returns to the left side of the heart to begin another pass through the systemic circulation.

Q *Which heart chambers receive blood from systemic and pulmonary circulation? Which chambers pump blood into systemic and pulmonary circulation?*

(Figure 13.2c). The **superior left point** is located at the inferior border of the second left costal cartilage, about 3 cm to the left of the midline. A line connecting these two points corresponds to the base of the heart.

The **inferior left point** is located at the apex of the heart in the fifth left intercostal space, about 9 cm (3.5 in.) to the left of the midline. A line connecting the superior and inferior left points corresponds to the left border of the heart. The **inferior right point** is located at the superior border of the sixth right costal cartilage, about 3 cm to the right of the midline. A line connecting the inferior left and right points corresponds to the inferior surface of the heart, and a line connecting the inferior and superior right points corresponds to the right border of the heart. When all four points are connected, they form a shape that roughly reveals the size and shape of the heart.

Clinical Application

Cardiopulmonary Resuscitation

Because the heart lies between two rigid structures, the vertebral column and the sternum (Figure 13.2a), external pressure (compression) on the chest can be used to force blood out of the heart into the circulation. In cases in which the heart suddenly stops beating, **cardiopulmonary resuscitation (CPR)**—properly applied cardiac compressions, performed in conjunction with artificial ventilation of the lungs—saves lives by keeping oxygenated blood circulating until the heart can be restarted. ▧

PERICARDIUM

The **pericardium** (*peri* = around) is a triple-layered bag that surrounds and protects the heart. It confines the heart to its position in the mediastinum, yet allows it sufficient freedom of movement for vigorous and rapid contraction.

Figure 13.2 Position of the heart and associated structures in the mediastinum, and the points of the heart that correspond to its surface projection. The arrow in (a) indicates the direction from which the section is viewed (superior).

The heart is located in the mediastinum; two-thirds of its mass is to the left of the midline.

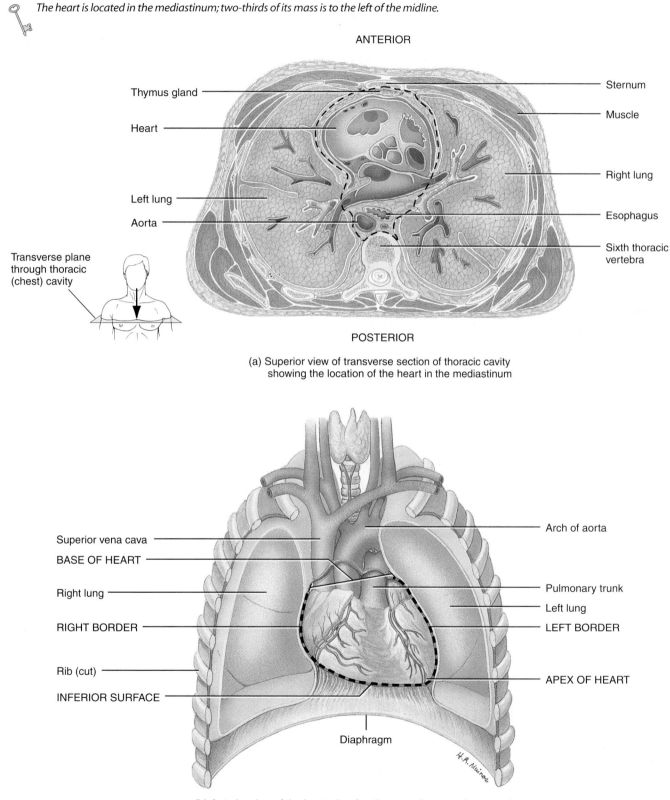

ANTERIOR

Thymus gland

Heart

Left lung

Aorta

Sternum

Muscle

Right lung

Esophagus

Sixth thoracic vertebra

Transverse plane through thoracic (chest) cavity

POSTERIOR

(a) Superior view of transverse section of thoracic cavity showing the location of the heart in the mediastinum

Superior vena cava

BASE OF HEART

Right lung

RIGHT BORDER

Rib (cut)

INFERIOR SURFACE

Arch of aorta

Pulmonary trunk

Left lung

LEFT BORDER

APEX OF HEART

Diaphragm

(b) Anterior view of the heart showing the apex, base, surfaces, and borders in the mediastinum

Figure continues

Figure 13.2 (continued)

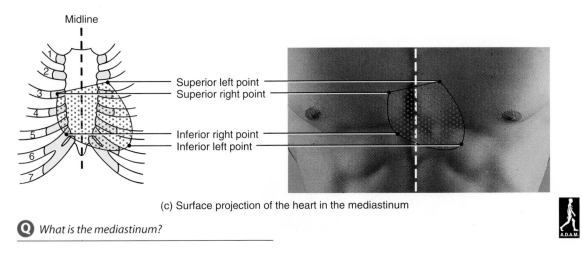

(c) Surface projection of the heart in the mediastinum

Q *What is the mediastinum?*

The pericardium consists of two principal portions: the fibrous pericardium and the serous pericardium (Figure 13.3a). The superficial **fibrous pericardium** is a tough, inelastic, dense irregular fibrous connective tissue membrane. It resembles a sac that rests on and attaches to the diaphragm and sternum, with its open end fused to the connective tissues of the blood vessels entering and leaving the heart. Its lateral surfaces lie against the parietal pleurae, the membranes of the lungs that line the thoracic cavity. The fibrous pericardium prevents overstretching of the heart, provides protection, and anchors the heart in the mediastinum.

The deeper **serous pericardium** is a thinner, more delicate membrane that forms a double layer around the heart. The superficial **parietal layer** of the serous pericardium is fused to the fibrous pericardium. The deeper **visceral layer** of the serous pericardium, also called the **epicardium** (*epi* = on top), adheres tightly to the muscle of the heart. Between the parietal and visceral layers of the serous pericardium is a thin film of serous fluid known as **pericardial fluid.** It is a slippery secretion of the pericardial cells that reduces friction between the membranes as the heart moves. The potential space that houses the pericardial fluid is called the **pericardial cavity.**

HEART WALL

Three layers form the wall of the heart (Figure 13.3a): the epicardium (external layer), myocardium (middle layer), and endocardium (inner layer). The outermost **epicardium,** also called the **visceral layer** of the serous pericardium, is the thin, transparent outer layer of the wall. It is composed of mesothelium and delicate connective tissue that imparts a smooth, slippery texture to the outermost surface of the heart.

The middle **myocardium** (*myo* = muscle), which is cardiac muscle tissue, makes up the bulk of the heart and is responsible for its pumping action. Cardiac muscle fibers (cells) are involuntary, striated, and branched (Figure 13.3b). They swirl diagonally around the heart in interlacing bundles and form two large networks—one atrial and one ventricular (Figure 13.3c). At its ends, each fiber physically contacts neighboring fibers in its network by transverse thickenings of the sarcolemma called **intercalated discs.** Within the discs are **gap junctions** that allow muscle action potentials to spread from one fiber to another. As a result, the whole atrial network contracts as one unit, and the ventricular network contracts as another. The intercalated discs also contain **desmosomes,** which act as reinforcing spot welds. They prevent adjacent cardiac fibers from pulling apart during their vigorous contractions.

The innermost **endocardium** (*endo* = within) is a thin layer of endothelium overlying a thin layer of connective tissue. It provides a smooth lining for the inside of the heart and covers the valves of the heart. The endocardium is continuous with the endothelial lining of the large blood vessels associated with the heart and the rest of the cardiovascular system.

Figure 13.3 Pericardium and heart wall.

The pericardium is a triple-layered sac that surrounds and protects the heart.

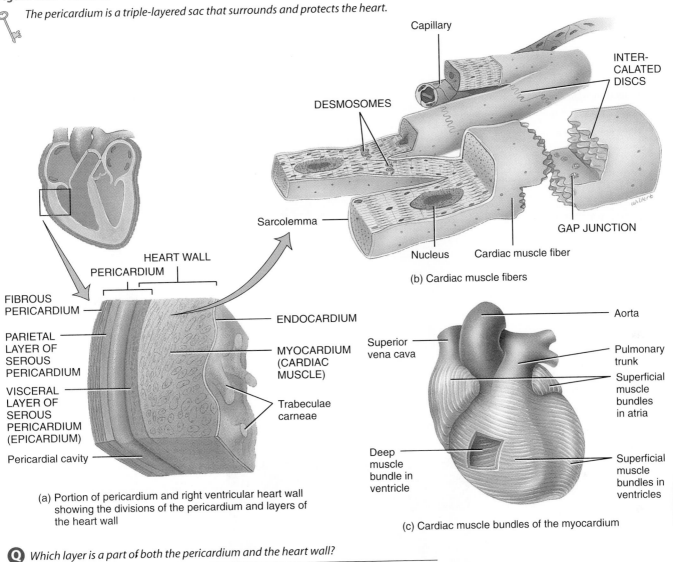

Capillary

INTER-
CALATED
DISCS

DESMOSOMES

Sarcolemma

HEART WALL
PERICARDIUM

FIBROUS
PERICARDIUM

PARIETAL
LAYER OF
SEROUS
PERICARDIUM

VISCERAL
LAYER OF
SEROUS
PERICARDIUM
(EPICARDIUM)

Pericardial cavity

ENDOCARDIUM

MYOCARDIUM
(CARDIAC
MUSCLE)

Trabeculae
carneae

Nucleus   Cardiac muscle fiber

GAP JUNCTION

(b) Cardiac muscle fibers

Aorta

Superior
vena cava

Pulmonary
trunk

Superficial
muscle
bundles
in atria

Deep
muscle
bundle in
ventricle

Superficial
muscle
bundles in
ventricles

(a) Portion of pericardium and right ventricular heart wall
showing the divisions of the pericardium and layers of
the heart wall

(c) Cardiac muscle bundles of the myocardium

Q *Which layer is a part of both the pericardium and the heart wall?*

CHAMBERS OF THE HEART

As indicated earlier, the heart contains four chambers. Each of the two superior chambers is called an **atrium** (*atrium* = court). The **right atrium** collects deoxygenated blood from systemic circulation, and the **left atrium** collects oxygenated blood from pulmonary circulation. The two inferior chambers are called **ventricles** (*ventricle* = little belly). The **right ventricle** receives deoxygenated blood from the right atrium and pumps it to the lungs to become oxygenated. The **left ventricle** receives oxygenated blood from the left atrium and pumps it to all body cells, except the air sacs of the lungs.

Before we discuss the internal anatomy of the chambers of the heart, we will first examine the relationship of the chambers to the external features of the heart.

External Features of the Heart

During one phase of a heartbeat, the atria empty of blood and become smaller. This produces the appearance of a wrinkled pouch on the anterior surface of each atrium called an **auricle** (OR-i-kul; *auris* = ear), so named because of its resemblance to a dog's ear (Figure 13.4). Each auricle slightly increases the surface area of an atrium so that it can hold a greater volume of blood. On the surface of the heart are a series of grooves, each called a **sulcus** (SUL-kus), which contain coronary blood vessels and a variable amount of fat. The sulci (SUL-kē) also mark the external boundaries between chambers of the heart. The deep **coronary** (*corona* = crown) **sulcus** encircles most of the heart and marks the boundary between the superior atria and inferior ventricles. The **anterior interventricular sulcus** is a shallow groove on the anterior surface of the heart that

Figure 13.4 Structure of the heart: surface features.

Sulci are grooves that contain blood vessels and fat and mark the boundaries between the various chambers.

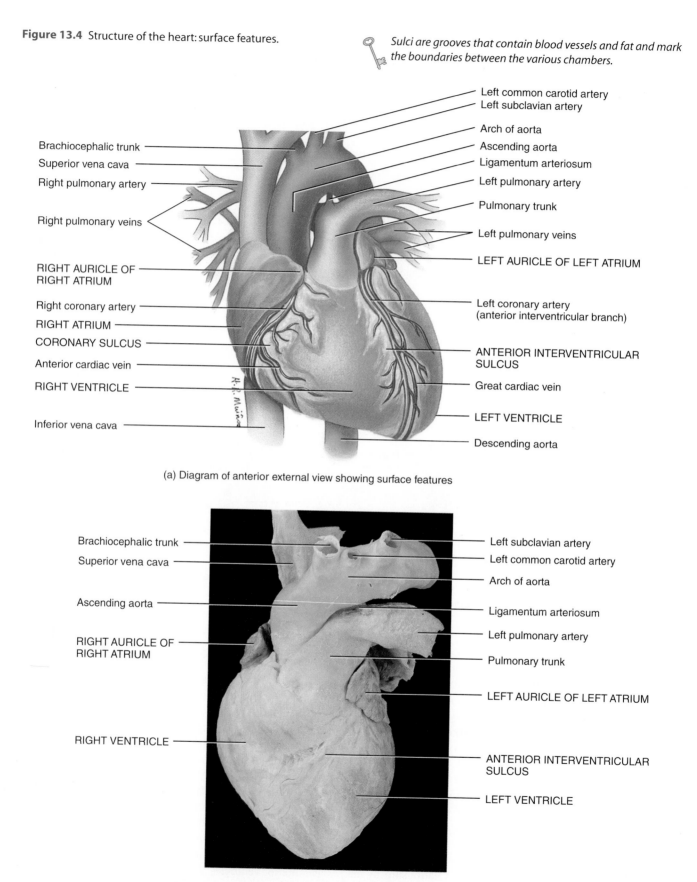

(a) Diagram of anterior external view showing surface features

(b) Photograph of anterior external view showing surface features

Figure continues

Figure 13.4 (continued)

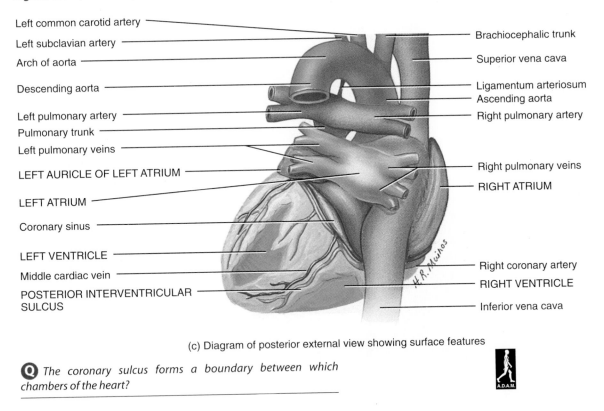

Left common carotid artery
Left subclavian artery
Arch of aorta
Descending aorta
Left pulmonary artery
Pulmonary trunk
Left pulmonary veins
LEFT AURICLE OF LEFT ATRIUM
LEFT ATRIUM
Coronary sinus
LEFT VENTRICLE
Middle cardiac vein
POSTERIOR INTERVENTRICULAR SULCUS

Brachiocephalic trunk
Superior vena cava
Ligamentum arteriosum
Ascending aorta
Right pulmonary artery
Right pulmonary veins
RIGHT ATRIUM
Right coronary artery
RIGHT VENTRICLE
Inferior vena cava

(c) Diagram of posterior external view showing surface features

Q *The coronary sulcus forms a boundary between which chambers of the heart?*

marks the boundary between the right and left ventricles. This sulcus continues around to the posterior surface of the heart as the **posterior interventricular sulcus,** which marks the boundary between the ventricles of the posterior aspect of the heart.

Internal Anatomy of the Heart Chambers

Right Atrium

The **right atrium** forms the right border of the heart (see Figure 13.2b). It receives deoxygenated blood from systemic circulation through three veins: superior vena cava, inferior vena cava, and coronary sinus (Figure 13.5a).

The anterior and posterior walls of the right atrium differ considerably. Whereas the posterior wall is smooth, the anterior wall is rough due to the presence of muscular ridges called **pectinate** (*pectin* = comb) **muscles** (Figure 13.5b). Between the right atrium and left atrium is a connective tissue partition called the **interatrial septum.** A prominent feature of this septum is an oval depression called the **fossa ovalis.** It is the remnant of the foramen ovale, an opening in the interatrial septum of the fetal heart that normally closes soon after birth (see Figure 14.22).

Right Ventricle

The **right ventricle** forms most of the anterior surface of the heart. Deoxygenated blood passes from the right atrium into the right ventricle through a valve called the *tricuspid valve* (Figure 13.5a). The cusps of the valve are connected to tendonlike cords, the **chordae tendineae** (KOR-dē ten-DIN-ē-ē; *chorda* = cord; *tendo* = tendon), which, in turn, are connected to cone-shaped muscular columns in the inner surface of the ventricle called **papillary** (*papilla* = nipple) **muscles.** (Details about the tricuspid and other valves mentioned in this discussion of the internal anatomy of the heart chambers will follow shortly.) Another prominent feature of the right ventricle is a series of ridges and folds called **trabeculae carneae** (tra-BEK-yoo-lē KAR-nē-ē; *trabecula* = little beam; *carneous* = fleshy). The right ventricle is separated from the left ventricle by a partition called the **interventricular septum.**

Deoxygenated blood passes from the right ventricle through a valve called the *pulmonary semilunar valve* into the **pulmonary** (*pulmo* = lung) **trunk** to enter pulmonary circulation. The pulmonary trunk divides into a right and a left **pulmonary artery,** each of which carries blood to their respective lungs. In the lungs, the deoxygenated blood releases carbon dioxide and takes on oxygen. The blood is then oxygenated blood.

Figure 13.5 Structure of the heart: internal anatomy. The arrow in the inset in (c) indicates the direction from which the heart is viewed (inferior).

The thickness of the four chambers varies according to their functions.

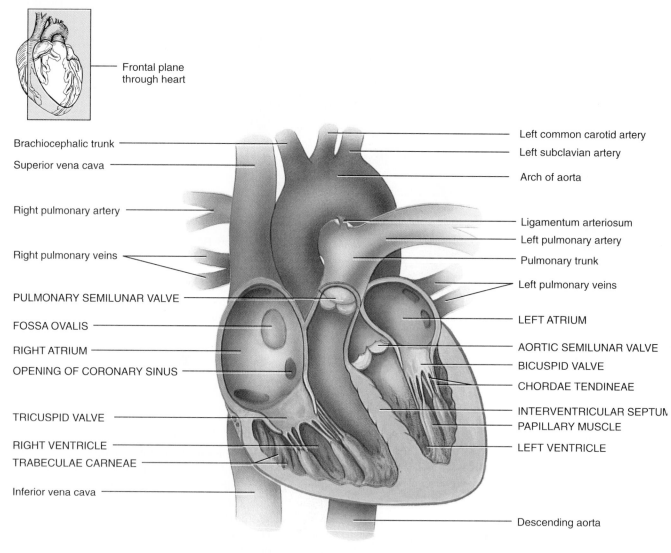

Frontal plane through heart

Brachiocephalic trunk

Superior vena cava

Right pulmonary artery

Right pulmonary veins

PULMONARY SEMILUNAR VALVE

FOSSA OVALIS

RIGHT ATRIUM

OPENING OF CORONARY SINUS

TRICUSPID VALVE

RIGHT VENTRICLE

TRABECULAE CARNEAE

Inferior vena cava

Left common carotid artery

Left subclavian artery

Arch of aorta

Ligamentum arteriosum

Left pulmonary artery

Pulmonary trunk

Left pulmonary veins

LEFT ATRIUM

AORTIC SEMILUNAR VALVE

BICUSPID VALVE

CHORDAE TENDINEAE

INTERVENTRICULAR SEPTUM

PAPILLARY MUSCLE

LEFT VENTRICLE

Descending aorta

(a) Diagram of anterior view of frontal section showing internal anatomy

Figure continues

Left Atrium

The **left atrium** forms most of the base of the heart (see Figure 13.2b). It receives oxygenated blood from pulmonary circulation in the lungs through four *pulmonary veins*. Like the right atrium, the left atrium is characterized by a smooth posterior wall and a rough anterior wall due to the presence of pectinate muscles.

Left Ventricle

The **left ventricle** forms the apex of the heart (see Figure 13.2b). Oxygenated blood passes from the left atrium into the left ventricle through a valve called the *bicuspid (mitral)*

valve. Like the tricuspid valve, the bicuspid valve has chordae tendineae attached to papillary muscles. However, as the name indicates, the bicuspid valve has only two cusps (not three as in the tricuspid valve). The left ventricle also contains trabeculae carneae.

Oxygenated blood passes into systemic circulation from the left ventricle through a valve called the *aortic semilunar valve* into the **ascending aorta.** From here some of the blood flows into the left and right **coronary arteries,** which branch from the ascending aorta and carry the blood to the heart wall. The remainder of the blood passes into the **arch of the aorta** and **descending aorta (thoracic aorta** and **abdominal aorta).** Branches of the arch of the aorta and descending aorta carry the blood throughout the systemic circulation.

Figure 13.5 (continued)

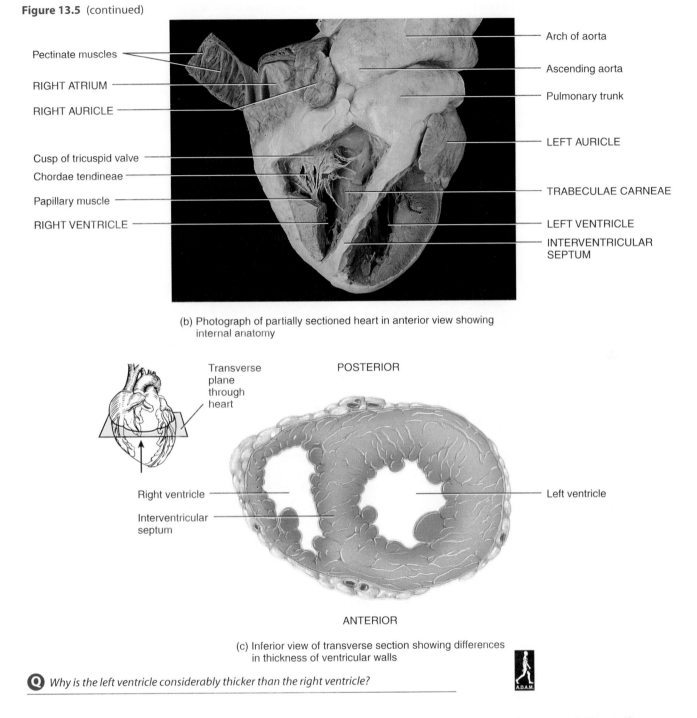

Pectinate muscles

RIGHT ATRIUM

RIGHT AURICLE

Cusp of tricuspid valve

Chordae tendineae

Papillary muscle

RIGHT VENTRICLE

Arch of aorta

Ascending aorta

Pulmonary trunk

LEFT AURICLE

TRABECULAE CARNEAE

LEFT VENTRICLE

INTERVENTRICULAR SEPTUM

(b) Photograph of partially sectioned heart in anterior view showing internal anatomy

Transverse plane through heart

POSTERIOR

Right ventricle

Interventricular septum

Left ventricle

ANTERIOR

(c) Inferior view of transverse section showing differences in thickness of ventricular walls

Q *Why is the left ventricle considerably thicker than the right ventricle?*

During fetal life, a temporary blood vessel, called the ductus arteriosus, connects the pulmonary trunk with the aorta (see Figure 14.22). It redirects blood so that only a small amount enters the nonfunctioning fetal lungs. The ductus arteriosus normally closes shortly after birth, leaving a remnant known as the **ligamentum arteriosum,** which interconnects the arch of the aorta and pulmonary trunk (Figure 13.5a).

Chamber Thickness and Function

The thickness of the walls of the four chambers varies according to the chambers' functions. The atria are thin-walled because they only have to deliver blood into the ventricles (Figure 13.5a). Although the right and left ventricles act as two separate pumps and pump the same volume of blood at the same time, the left side has a much larger workload. This is because the right ventricle pumps blood only to the lungs (pulmonary circulation), whereas the left ventricle pumps

Figure 13.6 Blood flow through the heart.

The right side of the heart pumps deoxygenated blood and the left side pumps oxygenated blood.

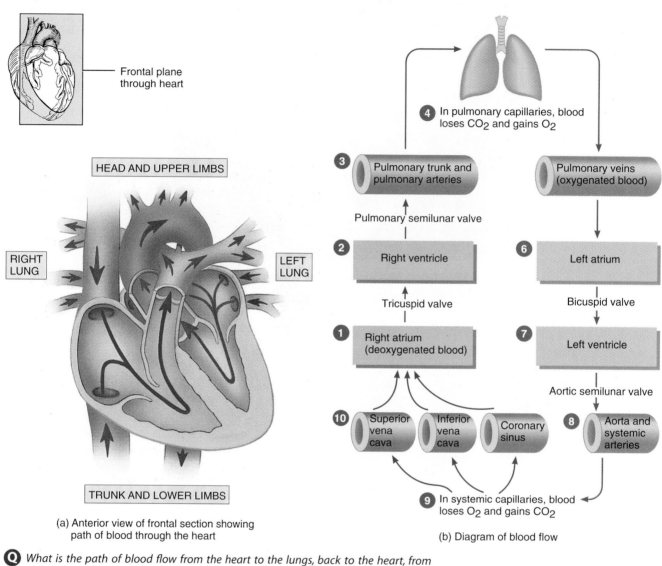

Frontal plane through heart

HEAD AND UPPER LIMBS

RIGHT LUNG

LEFT LUNG

TRUNK AND LOWER LIMBS

(a) Anterior view of frontal section showing path of blood through the heart

4 In pulmonary capillaries, blood loses CO_2 and gains O_2

3 Pulmonary trunk and pulmonary arteries

Pulmonary veins (oxygenated blood)

Pulmonary semilunar valve

2 Right ventricle

6 Left atrium

Tricuspid valve

Bicuspid valve

1 Right atrium (deoxygenated blood)

7 Left ventricle

Aortic semilunar valve

10 Superior vena cava | Inferior vena cava | Coronary sinus

8 Aorta and systemic arteries

9 In systemic capillaries, blood loses O_2 and gains CO_2

(b) Diagram of blood flow

Q *What is the path of blood flow from the heart to the lungs, back to the heart, from the heart to the systemic circulation, and back to the heart again? (Be sure to name all the vessels, chambers, and valves of the heart, starting in the right atrium and ending in the superior vena cava.)*

blood to all other parts of the body (systemic circulation). Thus the left ventricle must work harder than the right ventricle to maintain the same rate of blood flow. The anatomy of the two ventricles confirms this functional difference: the muscular wall of the left ventricle is considerably thicker than that of the right ventricle (Figure 13.5c). Note also that the shape of the lumen (space) of the left ventricle is circular, whereas that of the right ventricle is crescent-shaped.

Blood Flow Through the Chambers

Next is a summary of blood flow through the chambers of the heart. The right atrium receives deoxygenated blood from various parts of the body. Figure 13.6 shows the route of blood flow through the pulmonary and systemic circulations.

From the right atrium ❶, blood flows into the right ventricle ❷, which pumps it into the **pulmonary trunk** ❸. The pulmonary trunk divides into a **right** and **left pulmonary artery,** each of which carries blood to one lung. As blood flows through **pulmonary capillaries,** it loses CO_2 and takes on O_2 ❹. The oxygenated blood returns to the heart via the **pulmonary veins** ❺ that empty into the left atrium ❻. The blood then passes into the **left ventricle** ❼ , which pumps the blood into the **ascending aorta** ❽. Branches of the arch of the aorta and descending aorta (thoracic aorta and abdominal aorta) deliver blood to **systemic arteries,** which lead into **systemic capillaries** ❾. In systemic capillaries, blood loses O_2 and gains CO_2. The deoxygenated blood returns to the right side of the heart through three veins ❿: the **superior vena cava (SVC)** brings blood from most parts of the body superior to the heart, the **inferior vena cava (IVC)** brings blood from all parts of the body inferior to the diaphragm, and the **coronary sinus** drains blood from most of the vessels serving the heart itself.

VALVES OF THE HEART

As each chamber of the heart contracts, it pushes a volume of blood into a ventricle or out of the heart through an artery. To prevent backflow of blood, the heart has **valves.** These structures are composed of dense connective tissue covered by endocardium. Valves open and close in response to pressure changes as the heart contracts and relaxes.

Atrioventricular Valves

Atrioventricular (AV) valves lie between the atria and ventricles (see Figure 13.5a). The right AV valve between the right atrium and right ventricle is also called the **tricuspid (trī-KUS-pid) valve** because it consists of three cusps (flaps). The left AV valve between the left atrium and left ventricle has two cusps and is called the **bicuspid (mitral) valve.** When an AV valve is open, the pointed ends of the cusps project into a ventricle. Tendonlike cords called **chordae tendineae** connect the pointed ends and undersurfaces to **papillary muscles** (muscular columns) that are located on the inner surface of the ventricles.

Blood moves from the atria into the ventricles through open AV valves when ventricular pressure is lower than atrial pressure (Figure 13.7a). At this time the papillary muscles are relaxed, and the chordae tendineae are slack. When the ventricles contract, the pressure of the blood drives the cusps upward until their edges meet and close the opening (Figure 13.7b). At the same time, the papillary muscles are also contracting, which pulls on and tightens the chordae tendineae. This prevents the valve cusps from everting (swinging upward into the atria). If the AV valves or chordae tendineae are damaged, blood may regurgitate (flow back) into the atria when the ventricles contract.

Semilunar Valves

Near the origin of both arteries that emerge from the heart, there are heart valves that allow ejection of blood from the heart but prevent blood from flowing backward into the heart. These are the **semilunar (SL) valves** (see Figure 13.5a). The **pulmonary semilunar valve** lies in the opening where the pulmonary trunk leaves the right ventricle. The **aortic semilunar valve** is situated at the opening between the left ventricle and the aorta.

Both valves consist of three semilunar (half-moon, or crescent-shaped) cusps (Figure 13.7c). Each cusp is attached by its convex margin to the artery wall. The free borders of the cusps curve outward and project into the opening inside the blood vessel. When blood starts to flow backward toward the heart as the ventricles relax, it fills the cusps and tightly closes the semilunar valves.

Fibrous Skeleton of the Heart

In addition to cardiac muscle tissue, the heart wall also contains dense connective tissue that forms the **fibrous skeleton of the heart.** The skeleton forms the foundation to which the heart valves attach, serves as a point of insertion for cardiac muscle bundles (see Figure 13.3c), prevents overstretching of the valves as blood passes through them, and acts as an electrical insulator that prevents the direct spread of action potentials from the atria to the ventricles.

Essentially, the fibrous skeleton consists of dense connective tissue rings that surround the valves of the heart, fuse with one another, and merge with the interventricular septum. The components of the fibrous skeleton of the heart are as follows (Figure 13.8):

1. Four **fibrous rings** that support the four valves of the heart and are fused to each other: **right atrioventricular fibrous ring, left atrioventricular fibrous ring, pulmonary fibrous ring,** and **aortic fibrous ring.**

2. **Right fibrous trigone** (TRĪ-gōn), a relatively large triangular mass that is formed by the fusion of the fibrous connective tissue of the left atrioventricular, aortic, and right atrioventricular fibrous rings.

3. **Left fibrous trigone,** a smaller mass that is formed by the fusion of the fibrous connective tissue of the left atrioventricular and aortic fibrous rings.

4. **Conus tendon,** a mass that is formed by the fusion of the fibrous connective tissue of the pulmonary and aortic fibrous rings.

Heart Sounds

The act of listening to sounds within the body is called **auscultation** (aws-kul-TĀ-shun; *auscultare* = to listen). Auscultation is usually done with a stethoscope. The sound of the

Figure 13.7 Valves of the heart. The arrow in (c) indicates the direction from which the valves are viewed (superior).

Heart valves prevent the backflow of blood.

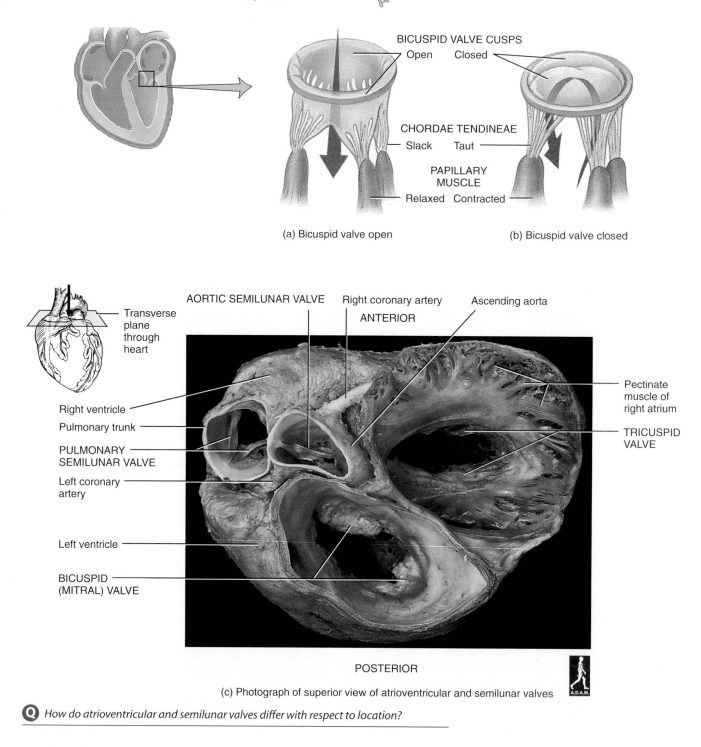

BICUSPID VALVE CUSPS
Open   Closed

CHORDAE TENDINEAE
Slack   Taut

PAPILLARY MUSCLE
Relaxed   Contracted

(a) Bicuspid valve open

(b) Bicuspid valve closed

Transverse plane through heart

AORTIC SEMILUNAR VALVE   Right coronary artery   Ascending aorta
ANTERIOR

Right ventricle

Pulmonary trunk

PULMONARY SEMILUNAR VALVE

Left coronary artery

Left ventricle

BICUSPID (MITRAL) VALVE

Pectinate muscle of right atrium

TRICUSPID VALVE

POSTERIOR

(c) Photograph of superior view of atrioventricular and semilunar valves

Q *How do atrioventricular and semilunar valves differ with respect to location?*

heartbeat comes primarily from blood turbulence caused by closing of the heart valves. During each heartbeat four **heart sounds** are generated. In a normal heart, however, only the first two (first and second heart sounds) are loud enough to be heard by listening through a stethoscope.

The first sound, which can be described as a **lubb** sound, is louder and a bit longer than the second sound. The lubb is the sound created by blood turbulence associated with closure of the AV valves soon after ventricular contraction begins. The second sound, which is shorter and not as loud

Figure 13.8 Fibrous skeleton of the heart (illustrated in blue).

The fibrous skeleton provides a base for the attachment of heart valves, prevents over-stretching of the valves, serves as a point of insertion for cardiac muscle bundles, and prevents the direct spread of action potentials from the atria to the ventricles.

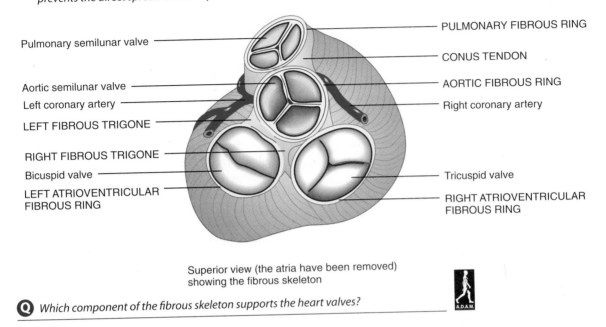

Pulmonary semilunar valve

Aortic semilunar valve

Left coronary artery

LEFT FIBROUS TRIGONE

RIGHT FIBROUS TRIGONE

Bicuspid valve

LEFT ATRIOVENTRICULAR FIBROUS RING

PULMONARY FIBROUS RING

CONUS TENDON

AORTIC FIBROUS RING

Right coronary artery

Tricuspid valve

RIGHT ATRIOVENTRICULAR FIBROUS RING

Superior view (the atria have been removed) showing the fibrous skeleton

Q *Which component of the fibrous skeleton supports the heart valves?*

Figure 13.9 Surface projections for heart valves and optimal sites for auscultation of heart sounds (red circles).

Heart sounds are best heard on the surface of the chest in locations that differ slightly from the actual locations of the valves.

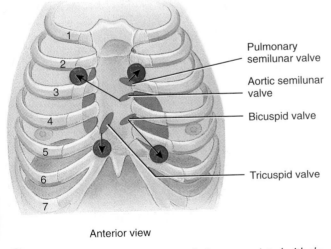

Pulmonary semilunar valve

Aortic semilunar valve

Bicuspid valve

Tricuspid valve

Anterior view

Q *Which heart sound is related to turbulence associated with closure of the atrioventricular valves?*

as the first, can be described as a **dupp** sound. Dupp is the sound created by blood turbulence associated with closure of the semilunar valves at the beginning of ventricular relaxation. A pause between the second sound of one heartbeat and the first sound of the next heartbeat is about two times longer than the pause between the first and second sound of each heartbeat. Thus, the heartbeat can be heard as a lubb, dupp, pause; lubb, dupp, pause; lubb, dupp, pause.

As noted previously, it is possible to project the heart onto the anterior surface of the chest by locating its superior right, superior left, inferior left, and inferior right points (see Figure 13.2c). The location of the valves of the heart may also be projected to the surface (Figure 13.9). The pulmonary and aortic semilunar valves are represented on the surface by a line about 2.5 cm (1 in.) in length. The pulmonary semilunar valve lies horizontally posterior to the medial end of the left third costal cartilage and the adjoining part of the sternum. The aortic semilunar valve is placed obliquely posterior to the left side of the sternum, at the level of the third intercostal space. The tricuspid valve lies posterior to the sternum, extending from the midline at the level of the fourth costal cartilage down toward the right sixth chondrosternal junction. The bicuspid valve lies posterior to the left side of the sternum obliquely at the level of the fourth costal cartilage. It is represented by a line about 3 cm in length.

Although heart sounds are produced in part by the closure of valves, they are not necessarily heard best over these valves. Because of the paths they travel, heart sounds are best heard at the surface of the chest in locations that differ slightly from the actual locations of the valves. The aortic semilunar valve sound is best heard near the superior right point, the pulmonary semilunar valve sound near the superior left point, the bicuspid valve sound near the inferior left point, and the tricuspid valve sound near the inferior right point.

Clinical Application

Heart Murmurs

A **heart murmur** is an abnormal sound that consists of a flow noise that is heard before, between, or after the normal heart sounds or that may mask the normal heart sounds. Although some heart murmurs are "innocent," meaning they do not suggest a heart problem, most often a murmur indicates a valve disorder.

Among the valvular abnormalities that may contribute to murmurs are **mitral stenosis** (narrowing of the mitral valve by scar formation or a congenital defect), **mitral insufficiency** (backflow or regurgitation of blood from the left ventricle into the left atrium due to a damaged mitral valve or ruptured chordae tendineae), **aortic stenosis** (narrowing of the aortic semilunar valve), and **aortic insufficiency** (backflow of blood from the aorta into the left ventricle). Another cause of a heart murmur is **mitral valve prolapse (MVP),** an inherited disorder in which one or both cusps of a mitral valve bulge superiorly and is pushed back too far (prolapsed) during ventricular contraction. Although a small volume of blood may flow back into the left atrium during ventricular contraction, and result in a murmur, mitral valve prolapse does not always pose a serious threat. MVP occurs in 10–15% of the population, more often in females (65%). ▪

HEART BLOOD SUPPLY

Nutrients could not possibly diffuse from the chambers of the heart through all the layers of cells that make up the heart tissue. For this reason, the wall of the heart has its own blood vessels. The flow of blood through the many vessels that pierce the myocardium is called the **coronary (cardiac) circulation.** The arteries of the heart encircle it like a crown encircles the head (*corona* = crown). While it is contracting, the heart receives little flow of oxygenated blood by way of the **coronary arteries,** which branch from the ascending aorta. When the heart relaxes, however, the high pressure of blood in the aorta propels blood through the coronary arteries, into capillaries, and then into **coronary veins.**

Coronary Arteries

Two coronary arteries, the right and left coronary arteries, branch from the ascending aorta and supply oxygenated blood to the myocardium (Figure 13.10a). The **left coronary artery** passes inferior to the left auricle and divides into the anterior interventricular and circumflex branches. The **anterior interventricular branch** or **left anterior descending (LAD) artery** is in the anterior interventricular sulcus and supplies oxygenated blood to the walls of both ventricles. The **circumflex branch** lies in the coronary sulcus and distributes oxygenated blood to the walls of the left ventricle and left atrium.

The **right coronary artery** supplies small branches (atrial branches) to the right atrium. It continues inferior to the right auricle and divides into the posterior interventricular and marginal branches. The **posterior interventricular branch** follows the posterior interventricular sulcus, supplying walls of the two ventricles with oxygenated blood. The **marginal branch** in the coronary sulcus transports oxygenated blood to the myocardium of the right ventricle.

Most parts of the body receive blood from branches of more than one artery, and where two or more arteries supply the same region, they usually connect. The connections, called **anastomoses** (a-nas'-tō-MŌ-sēs), provide alternate routes for blood to reach a particular organ or tissue. The myocardium contains many anastomoses, connecting branches of one coronary artery or extending between branches of different coronary arteries.

Coronary Veins

After blood passes through the arteries of coronary circulation, where it delivers oxygen and nutrients, it passes into veins, where it collects carbon dioxide and wastes. The deoxygenated blood then drains into a large vascular sinus on the posterior surface of the heart, called the **coronary sinus** (Figure 13.10b), which empties into the right atrium. A vascular sinus is a vein with a thin wall that has no smooth muscle to alter its diameter. The principal tributaries carrying blood into the coronary sinus are the **great cardiac vein,** which drains the anterior aspect of the heart, and the **middle cardiac vein,** which drains the posterior aspect of the heart.

CONDUCTION SYSTEM OF THE HEART

Recall that cardiac muscle fibers form two separate networks—one atrial and one ventricular. Each fiber is in physical contact with other fibers in the network by transverse thickenings of the sarcolemma called intercalated discs (see Figure 13.3b). Within the discs are gap junctions that aid in the conduction of muscle action potentials (nerve impulses)

Figure 13.10 Coronary (cardiac) circulation. The heart is viewed from the anterior aspect, but is drawn as if it were transparent to reveal blood vessels on the posterior aspect.

The right and left coronary arteries deliver blood to the myocardium; the coronary veins drain blood from the myocardium into the coronary sinus.

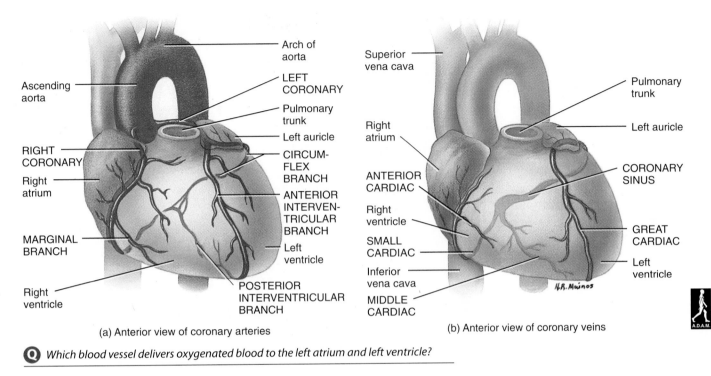

(a) Anterior view of coronary arteries

(b) Anterior view of coronary veins

Q *Which blood vessel delivers oxygenated blood to the left atrium and left ventricle?*

between cardiac muscle fibers. The gap junctions provide low-resistance bridges for the spread of muscle action potentials from one fiber to another. Thus, the spread of muscle action potentials between fibers is rapid, and the atria contract as one unit and the ventricles as another. The intercalated discs also function in adhesion of cardiac muscle fibers so that they do not pull apart. Each network contracts as a functional unit.

The heart is innervated by the autonomic nervous system (Chapter 21), but the autonomic neurons only increase or decrease the time it takes to complete a cardiac cycle (heartbeat); that is, they do not initiate contraction. The chamber walls can repeatedly contract and relax without any direct stimulus from the nervous system.

During embryonic development, about 1% of the cardiac muscle fibers become **autorhythmic** (self-excitable)—that is, able to repeatedly and rhythmically generate impulses. Autorhythmic fibers have two important functions. They act as a **pacemaker,** setting the rhythm for the entire heart, and they form the **conduction system,** the route for conducting action potentials throughout the heart muscle. The conduction system assures that cardiac chambers contract in a coordinated manner, which makes the heart an effective pump. Figure 13.11a shows the components of the conduction system: ❶ the sinoatrial (SA) node, ❷ the atrioventricular (AV) node, ❸ the atrioventricular (AV) bundle (bundle of His), ❹ the right and left bundle branches, and ❺ the conduction myofibers (Purkinje fibers).

The **sinoatrial (SA) node** is located in the right atrial wall just inferior to the opening to the superior vena cava. The SA node initiates each heartbeat and sets the basic pace for the heart rate by spontaneously generating action potentials more rapidly than other components of the conduction system. As a result, action potentials from the SA node spread to the other cells of the conduction system and stimulate them so frequently that they are unable to generate action potentials at their own inherently slower rates. Thus the faster SA node sets the rhythm for the rest of the heart. The rate set by the SA node may be altered by action potentials from the autonomic nervous system or by chemicals such as thyroid hormones and epinephrine.

When an action potential is initiated by the SA node, it spreads out over both atria, causing them to contract and at the same time causing depolarization (spread of the action potential) of the slowly conducting **atrioventricular (AV) node.** This node is located in the interatrial septum and is one of the last portions of the atria to be depolarized, which allows time for the atria to empty their blood into the ventricles before the ventricles begin to contract; the atria finish their contraction before the ventricles begin theirs.

Figure 13.11 Conduction system of the heart. The arrows represent the conduction of action potentials (impulses) through the atria.

🔑 *The conduction system ensures that the heart chambers contract in a coordinated manner.*

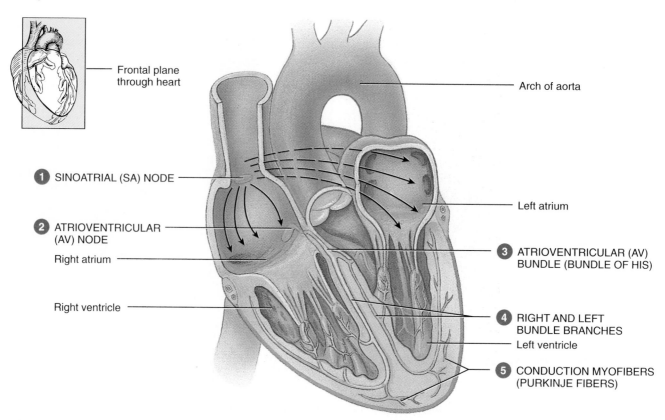

Frontal plane through heart

Arch of aorta

1 SINOATRIAL (SA) NODE

Left atrium

2 ATRIOVENTRICULAR (AV) NODE

Right atrium

3 ATRIOVENTRICULAR (AV) BUNDLE (BUNDLE OF HIS)

Right ventricle

4 RIGHT AND LEFT BUNDLE BRANCHES

Left ventricle

5 CONDUCTION MYOFIBERS (PURKINJE FIBERS)

(a) Anterior view of frontal section

Q *Which component of the conduction system provides the only electrical connection between the atria and the ventricles?*

From the AV node, the action potential passes to a tract of conducting fibers at the superior portion of the interventricular septum called the **atrioventricular (AV) bundle** or **bundle of His** (HISS). The AV bundle is the only electrical connection between the atria and ventricles. From here the action potential passes toward the heart's apex through the right and left **bundle branches** in the interventricular septum. The AV bundle distributes the action potential over the ventricles. Actual contraction of the ventricles is stimulated by the **conduction myofibers (Purkinje fibers)** that emerge from the bundle branches and distribute the action potential first to the apex of the ventricular myocardium and then superiorly to the remaining ventricular myocardial cells.

Transmission of action potentials through the conduction system generates electric currents that can be detected on the body's surface. A recording of the electrical changes that

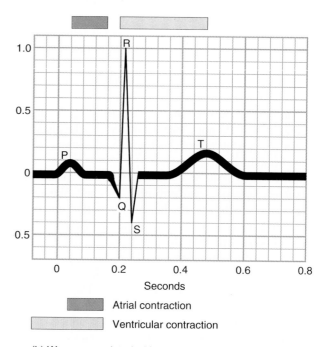

Atrial contraction

Ventricular contraction

(b) Waves associated with a normal electrocardiogram of a single heartbeat

Figure 13.12 The cardiac cycle (heartbeat).

A cardiac cycle comprises all the events associated with a single heartbeat.

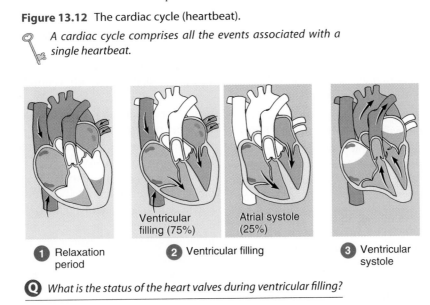

Ventricular filling (75%)

Atrial systole (25%)

1 Relaxation period

2 Ventricular filling

3 Ventricular systole

Q *What is the status of the heart valves during ventricular filling?*

accompany the heartbeat is called an **electrocardiogram** (e-lek′-trō-KAR-dē-ō-gram), which is abbreviated as either *ECG* or *EKG*. The action potentials are graphed as a series of up-and-down waves during an ECG. Three clearly recognizable waves normally accompany each cardiac cycle (Figure 13.11b). The first, called the *P wave*, is the spread of an action potential from the SA node through the two atria. A fraction of a second after the P wave begins, the atria contract. The second wave, called the *QRS wave*, is the spread of the action potential through the ventricles. Shortly after the QRS wave begins, the ventricles contract. The third wave, the *T wave*, indicates ventricular relaxation. There is no wave to show atrial relaxation because the stronger QRS wave masks this event.

Variations in the size and duration of deflection waves of an ECG are useful in diagnosing abnormal cardiac rhythms and conduction patterns and in following the course of recovery from a heart attack. It can also detect the presence of fetal life.

In major disruptions of the conduction system, an irregular heart rhythm may occur. Normal heart rhythm can be restored and maintained with an **artificial pacemaker,** a device that sends out small electrical charges to the right atrium, right ventricle, or both in order to stimulate the heart.

Clinical Application

Arrhythmias

Arrhythmia (a-RITH-mē-a) or **dysrhythmia** is a general term referring to an irregularity in heart rhythm. It results when there is a disturbance in the conduction system of the heart. Arrhythmias

are caused by such factors as caffeine, nicotine, alcohol, anxiety, certain drugs, hyperthyroidism, potassium deficiency, and certain heart diseases. One serious arrhythmia is called a **heart block,** a slowing or interruption of an action potential from one part of the heart's conduction system to an adjoining structure. The most common blockage is in the atrioventricular (AV) node, which conducts impulses from the atria to the ventricles. This disturbance is called **atrioventricular (AV) block.**

In **atrial flutter,** the atrial rhythm is about 300 beats per minute. The condition is essentially rapid atrial contractions accompanied by AV block. **Atrial fibrillation** is an uncoordinated contraction of the atrial muscles that causes the atria to contract irregularly and still faster. When the muscle fibrillates, the muscle fibers of the atrium quiver individually instead of contracting together. The quivering cancels out the pumping of the atrium. **Ventricular fibrillation (VF)** is characterized by asynchronous, haphazard, ventricular muscle contractions. Ventricular contraction becomes ineffective and circulatory failure and death occur. ■

CARDIAC CYCLE (HEARTBEAT)

The **cardiac cycle** comprises all the events associated with one heartbeat. In a normal cardiac cycle, the two atria contract while the two ventricles relax. Then while the two ventricles contract, the two atria relax. **Systole** (SIS-tō-lē) refers to the phase of contraction of a chamber of the heart; **diastole** (dī′-AS-tō-lē) is the phase of relaxation. For the purposes of our discussion, we will divide the cardiac cycle into the following phases illustrated in Figure 13.12:

1 **Relaxation period.** At the end of a cardiac cycle when the ventricles start to relax, all four chambers are in diastole. This is the beginning of the *relaxation* period. As the ventricles relax, pressure within the chambers drops, and blood starts to flow from the pulmonary trunk and aorta back toward the ventricles. As this blood becomes trapped in the semilunar cusps, however, the semilunar valves close. As the ventricles continue to relax, the space inside expands, and the pressure falls. When ventricular pressure drops below atrial pressure, the AV valves open and ventricular filling begins.

❷ **Ventricular filling.** The major part of ventricular filling (75 percent) occurs just after the AV valves open. This occurs *without atrial systole*. Atrial contraction accounts for the remaining 25 percent of the blood that fills the ventricles. Throughout the period of ventricular filling, the AV valves are open and the semilunar valves are closed.

❸ **Ventricular systole (contraction).** Ventricular contraction pushes blood up against the AV valves, forcing them shut. For a very brief period of time, all four valves are closed again. As ventricular contraction continues, pressure inside the chambers rises sharply. When left ventricular pressure rises above the pressure in the arteries, both semilunar valves open, and ejection of blood from the heart begins. This lasts until the ventricles start to relax. Then, the semilunar valves close and another relaxation period begins.

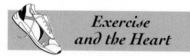

Exercise and the Heart

No matter what a person's level of fitness, it can be improved at any age with regular exercise. Of the various types of exercise, some (such as aerobics) are more effective than others for improving the health of the cardiovascular system. **Aerobics,** or any activity that works large body muscles for at least 20 minutes, elevates cardiac output (the amount of blood pumped by the left ventricle each minute) and accelerates metabolic rate. Three to five such sessions a week are usually recommended for improving the health of the cardiovascular system. Brisk walking, running, bicycling, cross-country skiing, and swimming are examples of aerobic exercises.

Sustained exercise increases the oxygen demand of the muscles. Whether the demand is met depends primarily on the adequacy of cardiac output and proper functioning of the respiratory system. After several weeks of training, a healthy person increases maximal cardiac output and thereby increases the maximal rate of oxygen delivery to the tissues. Oxygen delivery also rises because hemoglobin level increases and skeletal muscles develop more capillary networks in response to long-term training. More forceful contractions of skeletal muscles also help to return blood to the heart. In addition, regular exercise also strengthens the smooth muscle of blood vessels enabling them to assist overall circulation.

During strenuous activity, a well-trained athlete can achieve a cardiac output double or triple that of a sedentary person because training causes hypertrophy (enlargement) of the heart. Even though the heart of a well-trained athlete is larger, *resting* cardiac output is about the same as in a healthy untrained person. This is because the amount of blood pumped by the left ventricle with each ventricular contraction is increased while heart rate is decreased. The resting heart rate of a trained athlete often is only 40–60 beats per minute.

Other benefits of physical conditioning are an increase in high-density lipoprotein (HDL), a decrease in triglyceride levels, and improved lung function. Exercise also helps to reduce blood pressure, anxiety, and depression; control weight; and increase the body's ability to dissolve blood clots. Intense exercise increases levels of endorphins, the body's natural painkillers. This may explain the psychological "high" that athletes experience with strenuous training and the "low" they feel when they miss regular workouts. Exercise also helps make bones stronger and thus may be a factor in inhibiting and treating osteoporosis. Some research indicates that exercise may even offer protection against cancer and diabetes.

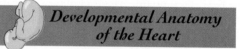

Developmental Anatomy of the Heart

The heart, a derivative of **mesoderm,** begins to develop before the end of the third week of gestation. It begins its development in the ventral region of the embryo beneath the foregut (see Figure 24.20b). The first step is the formation of a pair of tubes, the **endothelial (endocardial) tubes,** from mesodermal cells (Figure 13.13). These tubes then unite to form a common tube, the **primitive heart tube.** Next, the primitive heart tube develops into five regions: (1) **truncus arteriosus,** (2) **bulbus cordis,** (3) **ventricle,** (4) **atrium,** and (5) **sinus venosus.** Because the bulbus cordis and ventricle grow more rapidly than the others and the heart enlarges more rapidly than its superior and inferior attachments, the heart assumes a U-shape and later an S-shape. The flexures of the heart reorient the regions so that the atrium and sinus venosus eventually come to lie superior to the bulbus cordis, ventricle, and truncus arteriosus. Contractions of the primitive heart begin by day 22. They begin in the sinus venosus and force blood through the tubular heart.

At about the seventh week, a partition, the **interatrial septum,** forms in the atrial region, dividing it into a *right atrium* and *left atrium.* The opening in the partition is the **foramen ovale,** which normally closes at birth and later forms a depression called the *fossa ovalis.* An **interventricular septum** also develops and partitions the ventricular region into a *right ventricle* and *left ventricle.* The bulbus cordis and truncus arteriosus divide into two vessels, the *aorta* (arising from the left ventricle) and the *pulmonary trunk* (arising from the right ventricle). The great veins of the heart, *superior vena cava* and *inferior vena cava,* develop from the venous end of the primitive heart tube. ■

Figure 13.13 Development of the heart. Arrows within the structures show the direction of blood flow.

The heart develops from mesoderm and begins its development before the end of the third week of gestation.

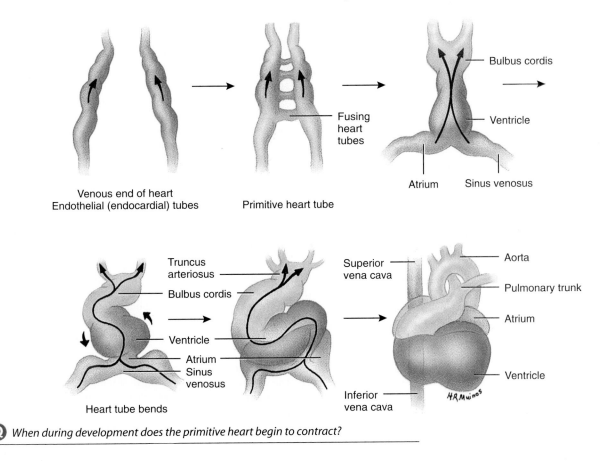

Venous end of heart
Endothelial (endocardial) tubes

Primitive heart tube

Bulbus cordis

Ventricle

Atrium   Sinus venosus

Truncus arteriosus

Bulbus cordis

Ventricle

Atrium

Sinus venosus

Heart tube bends

Superior vena cava

Inferior vena cava

Aorta

Pulmonary trunk

Atrium

Ventricle

Q *When during development does the primitive heart begin to contract?*

Key Medical Terms Associated with the Heart

Cardiac angiography (an′-jē-OG-ra-fē; *angio* = vessel; *cardio* = heart; *graph* = writing) Procedure in which a cardiac catheter (plastic tube) is used to inject a radiopaque contrast medium into blood vessels or heart chambers. The procedure may be used to visualize blood vessels and the ventricles to assess structural abnormalities. Angiography can also be used to inject clot-dissolving drugs, such as streptokinase or tissue plasminogen activator (t-PA), into a coronary artery to dissolve an obstructing thrombus.

Cardiac arrest (KAR-dē-ak a-REST) A clinical term meaning cessation of an effective heartbeat. The heart may be completely stopped or quivering ineffectively (ventricular fibrillation).

Cardiac catheterization (kath′-e-ter-i-ZĀ-shun) Procedure that is used to visualize the heart's coronary arteries, chambers, valves, and great vessels. It may also be used to measure pressure in the heart and blood vessels; to assess cardiac output; and to measure the flow of blood through the heart and blood vessels, the oxygen content of blood, and the status of heart valves and conduction system. The basic procedure involves inserting a catheter (plastic tube) into a peripheral vein (for right heart catheterization) or

artery (for left heart catheterization) and guiding it under fluoroscopy (x-ray observation).

Cardiomegaly (kar′-dē-ō-MEG-a-lē; *mega* = large) Heart enlargement.

Cor pulmonale (CP) (kor pul-mōn-ALE; *cor* = heart; *pulmon* = lung) This term refers to right ventricular hypertrophy from disorders that bring about hypertension (high blood pressure) in the pulmonary circulation.

Incompetent valve (in-KOM-pe-tent) Any valve that does not close properly, thus permitting a backflow of blood; also called **valvular insufficiency.**

Palpitation (pal′-pi-TĀ-shun) A fluttering of the heart or abnormal rate or rhythm of the heart.

Paroxysmal tachycardia (par′-ok-SIZ-mal tak′-e-KAR-dē-ā) A period of rapid heartbeats that begins and ends suddenly.

Sudden cardiac death The unexpected cessation of circulation and breathing due to an underlying heart disease such as ischemia, myocardial infarction, or a disturbance in cardiac rhythm.

Study Outline

Overview of Circulation (p. 393)

1. The left side of the heart is the pump for systemic circulation, the circulation of blood throughout the body except for the air sacs of the lungs.
2. The right side of the heart is the pump for pulmonary circulation, the circulation of blood through the lungs.

Location and Surface Projection of the Heart (p. 393)

1. The heart is located in the mediastinum, with about two-thirds of its mass to the left of the midline.
2. The heart is basically a cone lying on its side; it consists of an apex, a base, anterior and inferior surfaces, and right and left borders.
3. The heart contains four points that are used to project its location to the surface of the chest.

Pericardium (p. 394)

1. The pericardium consists of a superficial fibrous layer and a deep serous pericardium.
2. The serous pericardium is composed of a parietal and a visceral layer.
3. Between the parietal and visceral layers of the serous pericardium is the pericardial cavity, a potential space filled with pericardial fluid that reduces friction between the two membranes.

Heart Wall (p. 396)

1. The wall of the heart has three layers: epicardium (also called the visceral layer of the serous pericardium), myocardium, and endocardium.
2. The epicardium consists of mesothelium and connective tissue, the myocardium is composed of cardiac muscle tissue, and the endocardium consists of endothelium and connective tissue.

Chambers of the Heart (p. 397)

1. The chambers include two superior chambers, the right and left atria, and two inferior chambers, the right and left ventricles.
2. External features of the heart include the auricles (flaps on each atrium that increase their size), the coronary sulcus between the atria and ventricles, and the anterior and posterior sulci between the ventricles on the anterior and posterior surfaces of the heart, respectively.
3. The right atrium receives deoxygenated blood from the superior vena cava, inferior vena cava, and coronary sinus. It is separated from the left atrium by the interatrial septum, which contains the fossa ovalis.
4. The right ventricle receives deoxygenated blood from the right atrium through the tricuspid valve. It is separated from the left ventricle by the interventricular septum and pumps blood to the lungs through the pulmonary semilunar valve and pulmonary trunk.
5. The left atrium receives oxygenated blood from the pulmonary veins.
6. The left ventricle receives oxygenated blood from the left atrium through the bicuspid (mitral) valve. It pumps oxy-

genated blood into systemic circulation through the aortic semilunar valve and aorta.
7. The thickness of the walls of the four chambers varies according to the chambers' functions. The left ventricle has the thickest wall because of its high workload.
8. Deoxygenated blood flows through the heart from the superior and inferior venae cavae and the coronary sinus to the right atrium, through the tricuspid valve to the right ventricle, through the pulmonary semilunar valve, through the pulmonary trunk and pulmonary arteries to the lungs. Returning oxygenated blood flows through the pulmonary veins into the left atrium, through the bicuspid valve to the left ventricle, through the aortic semilunar valve, and out through the aorta.
9. Divisions of the aorta are the ascending aorta, arch of the aorta, and descending aorta (thoracic aorta and abdominal aorta).

Valves of the Heart (p. 403)

1. Valves prevent backflow of blood in the heart.
2. Atrioventricular (AV) valves, between the atria and their ventricles, are the tricuspid valve on the right side of the heart and the bicuspid (mitral) valve on the left.
3. The chordae tendineae and papillary muscles stabilize the flaps of the AV valves and stop blood from backing into the atria.
4. Each of the two arteries that leaves the heart has a semilunar valve (aortic and pulmonary).

Fibrous Skeleton of the Heart (p. 403)

1. The skeleton of the heart consists of fibrous connective tissue around the valves of the heart that fuse with each other.
2. The skeleton anchors the valves, prevents the valves from overstretching, serves as a point of insertion for the muscle bundles, and acts as an electrical insulator between atria and ventricles.

Heart Sounds (p. 403)

1. Heart sounds result from blood turbulence caused by closing of the heart's valves.
2. The first heart sound (lubb) is created by blood turbulence associated with the closing of the atrioventricular valves.
3. The second sound (dupp) is created by blood turbulence associated with the closing of semilunar valves.
4. The valves may be located by surface projection on the anterior chest wall.
5. Heart sounds are best heard in locations that are slightly different from the actual locations of the valves.

Heart Blood Supply (p. 406)

1. The flow of blood through the heart is called coronary (cardiac) circulation.
2. The principal arteries that supply blood to the myocardium are the left and right coronary; the principal veins that drain blood from the myocardium are the cardiac and coronary sinus.

Conduction System of the Heart (p. 406)

1. The conduction system consists of tissue specialized for generation and conduction of action potentials.

2. Components of this system are the sinoatrial (SA) node (pacemaker), atrioventricular (AV) node, atrioventricular (AV) bundle (bundle of His), bundle branches, and conduction myofibers (Purkinje fibers).
3. A recording of the electrical changes that accompany the heartbeat is called an electrocardiogram.

Cardiac Cycle (Heartbeat) (p. 409)

1. A cardiac cycle consists of systole (contraction) and diastole (relaxation) of the chambers of the heart.
2. The phases of the cardiac cycle are (a) the relaxation period, in which all chambers are in diastole; (b) ventricular filling, most of which occurs without atrial systole; and (c) ventricular systole.

Exercise and the Heart (p. 410)

1. The most effective way to improve the health of the cardiovascular system is through aerobic exercise.
2. Sustained and frequent exercise increases cardiac output and oxygen delivery to cells, helps develop more capillaries, decreases heart rate, increases high-density lipoprotein (HDL) levels, decreases triglyceride levels, and reduces blood pressure and blood clot formation.

Developmental Anatomy of the Heart (p. 410)

1. The heart develops from mesoderm.
2. The endothelial tubes develop into the four-chambered heart and great vessels of the heart.

Review Questions

1. Distinguish between system and pulmonary circulation and describe the role of the heart in each. (p. 393)
2. Describe the position of the heart in the mediastinum by defining its apex, base, anterior and posterior surfaces, and right and left borders. (p. 393)
3. Explain the location of the superior right point, superior left point, inferior left point, and inferior right point. Why are these points significant? (p. 393)
4. Distinguish among the layers of the pericardium. Why is the pericardium important? (p. 396)
5. Compare the three layers of the heart wall with respect to structure, location, and function. (p. 396)
6. Name the four chambers of the heart and identify their locations relative to each other. (p. 397)
7. Define each of the following external features of the heart: auricle, coronary sulcus, anterior interventricular sulcus, and posterior interventricular sulcus. (p. 397)
8. For each chamber of the heart, describe its unique internal features and which blood vessels enter or exit it. (p. 399)
9. Describe the relationship between wall thickness of heart chambers and each chamber's function. (p. 401)
10. Describe the flow of blood through the heart beginning with the right atrium and ending with the aorta. (p. 402)
11. Describe the location, structure, and function of the valves of the heart. (p. 403)
12. What is the fibrous skeleton of the heart? What are its functions? (p. 403)
13. How are heart sounds produced? Where are they best heard? What is a heart murmur? (p. 403)
14. Explain how the locations of the heart valves can be projected onto the surface of the chest. (p. 405)
15. What is coronary circulation? Which arteries deliver oxygenated blood to which chambers of the heart? Which veins drain deoxygenated blood from which aspects of the heart? (p. 406)
16. Describe the structure and function of the conduction system of the heart. What is an electrocardiogram? (p. 406)
17. Briefly describe the major events of the cardiac cycle. (p. 409)
18. What are the benefits of aerobic exercise? (p. 410)
19. Describe how the heart develops. (p. 410)
20. Refer to the glossary of key medical terms associated with the heart. Be sure that you can define each term. (p. 411)

Self Quiz

Complete the following statements:

1. The heart is derived from ___-derm. The heart begins to develop during the ___ month. Its initial formation consists of two endothelial tubes that unite to form the ___ tube.
2. Two types of junctions found in intercalated discs are ___. and ___.
3. The first heart sound is created by turbulence of blood at the closing of the ___ valves.
4. The pacemaker of the heart is the ___ of the conduction system.
5. The heart is divided into superior chambers called ___ and inferior chambers called ___.
6. The first two vessels to branch from the ascending aorta are ___ and ___.
7. The ___ valve is posterior to the left side of the sternum, near the fourth costal cartilage.

8. During the period of ventricular filling, the AV valves are in the ___ position, and the semilunar valves are in the ___ position.

9. Match the following terms to their descriptions:
   ___ (a) also called the mitral valve
   ___ (b) prevents backflow of blood from right ventricle to right atrium
   ___ (c) prevents backflow of blood from pulmonary trunk to right ventricle
   ___ (d) prevents backflow of blood into left ventricle

   (1) aortic semilunar valve
   (2) pulmonary semilunar valve
   (3) bicuspid valve
   (4) tricuspid valve

Are the following statements true or false?

10. The inferior surface of the heart rests mainly on the diaphragm.
11. The visceral pericardium is also called the epicardium.
12. Chordae tendineae attach AV valve cusps to pectinate muscles.
13. A semilunar valve consists of three half-moon shaped cusps.
14. The fibrous skeleton of the heart is a layer of dense connective tissue that anchors the heart to the surrounding structures of the mediastinum.
15. Most of the ventricular filling occurs during ventricular systole.
16. The conduction myofibers (Purkinje fibers) distribute the action potentials first to the apex of the ventricles, then superiorly to the remainder of the ventricles.
17. The left side of the heart is the pump for systemic circulation.
18. The right border of the heart extends from the base to the apex.
19. The inferior left point of the heart is located in the left fifth intercostal space.
20. The boundary between the atria and ventricles is marked by the coronary sulcus.
21. The chamber of the heart with the thickest wall is the right ventricle.
22. The sound of the tricuspid valves is best heard near the inferior right point.

23. A very positive effect of aerobic exercise is to increase cardiac output.

Choose the one best answer to each of the following questions:

24. Which of the following arteries directly supplies oxygenated blood to the walls of the atria?
    a. right coronary
    b. pulmonary
    c. marginal branch
    d. aorta
    e. anterior interventricular branch

25. Which of the following statements is true about the right atrium of the heart?
    (1) It contains the pacemaker.
    (2) It receives blood from the superior and inferior vena cavae.
    (3) It receives blood directly from the lungs.
    (4) It empties into the aorta.
    a. 1 only
    b. 2 only
    c. 3 only
    d. 4 only
    e. 1 and 2

26. Which of the following vessels carries oxygenated blood?
    a. superior vena cava
    b. coronary sinus
    c. inferior vena cava
    d. pulmonary vein
    e. none of the above

27. Which of the following vessels is *not* associated with the right side of the heart?
    a. superior vena cava
    b. pulmonary trunk
    c. inferior vena cava
    d. aorta
    e. coronary sinus

28. Which of the following sequences correctly represents the conduction of an impulse through the heart?
    a. SA node, AV node, AV bundle, bundle branches
    b. SA node, AV bundle, AV node, bundle branches
    c. AV node, SA node, AV bundle, bundle branches
    d. SA node, bundle branches, AV node, AV bundle
    e. AV node, AV bundle, SA node, bundle branches

29. Which of the following structures is/are located in the ventricles?
    a. trabeculae carneae
    b. fossa ovalis
    c. ligamentum arteriosum
    d. pectinate muscles
    e. orifice of coronary sinus

Critical Thinking Questions

1. Compare the structure and function of the AV valves and their adjacent structures to that of a parachute.
   HINT: *Why doesn't the parachute turn inside out?*
2. Your uncle had an artificial pacemaker put in after his last bout with heart trouble. For what heart structure does the pacemaker substitute? Where is it located?
   HINT: *The autorhythmic fibers that depolarize the fastest will lead the rest.*

3. An adult male was diagnosed with blockages in his anterior interventricular and circumflex branches. What regions of the heart may be affected by these blockages?
   HINT: *Think of which sulci contain these vessels.*
4. The heart is constantly beating, which means constantly moving. Why doesn't the heart beat its way out of position? Why don't the muscle contractions pull the fibers apart?

HINT: *Think about how the heart is anchored in the mediastinum and how cardiac muscle fibers are anchored to each other.*

5. Compare the causes and effects of angina pectoris to MI (myocardial infarction).

HINT: *What do the words* pectoris *and* myocardial *mean?*

6. The physician moved the stethoscope around to various locations on J.J.'s chest during his precamp physical. The doctor explained that she was listening to the closure of the heart valves. Where would she position her stethoscope to best hear the aortic semilunar valve? How does this compare to the valve's actual position in the thorax?

HINT: *Think about the position of the heart relative to the sternum.*

Answers to Figure Questions

13.1 Atria; ventricles.

13.2 The mass of tissue that extends from the sternum to the vertebral column and between the coverings (pleurae) of the lungs.

13.3 Visceral layer of the serous pericardium (epicardium).

13.4 Atria and ventricles.

13.5 The left ventricle has a greater workload than the right ventricle.

13.6 Right atrium, tricuspid valve, right ventricle, pulmonary semilunar valve, pulmonary trunk, pulmonary arteries, pulmonary capillaries, pulmonary veins, left atrium, bicuspid valve, left ventricle, aortic semilunar valve, aorta, systemic capillaries, systemic veins, superior vena cava.

13.7 Atrioventricular valves are located between the atria and ventricles; semilunar valves are located between the ventricles and arteries.

13.8 The four fibrous rings.

13.9 First sound (lubb).

13.10 Circumflex branch.

13.11 Atrioventricular (AV) bundle.

13.12 The atrioventricular valves are open and the semilunar valves are closed.

13.13 By the 22nd day of gestation.

Chapter 14

The Cardiovascular System: Blood Vessels

Student Objectives

1. Contrast the structure and functions of arteries, arterioles, capillaries, venules, and veins.

2. Identify the principal circulatory routes.

3. Describe the principal parts of the aorta and the general distribution of blood from each part.

4. Describe the distribution of blood to the head and neck, upper limbs, thorax, abdomen and pelvis, and lower limbs.

5. Explain the role of the superior vena cava, inferior vena cava, and coronary sinus in returning venous blood to the heart.

6. Describe the venous drainage of the head and neck, upper limbs, thorax, abdomen and pelvis, and lower limbs.

7. Describe the development of blood vessels and blood.

8. Describe the effects of aging on the cardiovascular system.

Blood vessels form a closed system of tubes that carries blood away from the heart, transports it to the tissues of the body, and then returns it to the heart. **Arteries** (AR-ter-ēz) are vessels that carry blood from the heart to the tissues. Arteries then divide into smaller vessels called **arterioles** (ar-TER-ē-ōls). As the arterioles enter a tissue, they branch into countless microscopic vessels called **capillaries** (KAP-i-lar'-ēs). Substances are exchanged between the blood and body tissues through the thin walls of capillaries. Before leaving the tissue, groups of capillaries unite to form small veins called **venules** (VEN-yools). These, in turn, merge to form progressively larger blood vessels called veins. **Veins** (VĀNZ) then convey blood from the tissues back to the heart.

ANATOMY OF BLOOD VESSELS

Arteries

In ancient times, **arteries** (*aer* = air; *tereo* = to carry), found empty at death, were thought to contain only air. The hollow center through which blood flows is called the **lumen** (Figure 14.1). The surrounding arterial wall has three coats or tunics. The inner coat, the **tunica interna (intima),** is composed of a lining of *endothelium* (simple squamous epithelium) that is in contact with the blood, a *basement membrane,* and a layer of elastic tissue called the *internal elastic lamina.* The endothelium is a continuous layer of cells that lines the inner surface of the entire cardiovascular system (heart and blood vessels). The middle coat, or **tunica media,** is usually the thickest layer. It consists of elastic fibers and smooth muscle fibers (cells). The outer coat, the **tunica externa (adventitia),** is composed principally of elastic and collagen fibers. An *external elastic lamina* may separate the tunica externa from the tunica media.

The structure of arteries, especially of the tunica media, gives them two important functional properties: elasticity and contractility. As the ventricles of the heart contract and eject blood from the heart, the large arteries stretch to accommodate the extra blood. Then, as the ventricles relax, the elastic recoil of the arteries forces the blood onward. The contractility of an artery comes from its smooth muscle, which is arranged circularly (in rings) around the lumen. Sympathetic fibers of the autonomic nervous system innervate vascular smooth muscle. Usually when there is an increase in sympathetic stimulation, the smooth muscle contracts, squeezes the wall around the lumen, and narrows the vessel. Such a decrease in the size of the lumen of a blood vessel is called **vasoconstriction.** Conversely, when sympathetic stimulation decreases, certain chemicals (nitric oxide,

K^+, H^+, and lactic acid) cause smooth muscle fibers to relax and the size of the lumen to increase. This increase is called **vasodilation.**

The smooth muscle layer of blood vessels, especially of arteries and arterioles (described shortly), also helps limit bleeding from wounds. When an artery or arteriole is cut, the smooth muscle contracts, producing vascular spasm of the vessel. However, there is a limit to how much vascular spasm can prevent hemorrhaging because the heart's pumping action causes blood to flow through arteries under great pressure.

Elastic (Conducting) Arteries

Large arteries are referred to as **elastic (conducting) arteries.** They include the aorta and the brachiocephalic, common carotid, subclavian, vertebral, and common iliac arteries. The walls of elastic arteries are thin in proportion to their diameters, and their tunica media contains more elastic fibers and less smooth muscle. Elastic arteries are called conducting arteries because they *conduct* blood from the heart to medium-sized muscular arteries.

As the heart alternately contracts and relaxes, blood flow tends to be intermittent. When the heart contracts and forces blood into the aorta, the walls of the elastic arteries stretch to accommodate the surge of blood and store the pressure energy. During relaxation of the heart, the walls of the elastic arteries recoil to create pressure, moving the blood forward in a more continuous flow.

Muscular (Distributing) Arteries

Medium-sized arteries are called **muscular (distributing) arteries.** They include the axillary, brachial, radial, intercostal, splenic, mesenteric, femoral, popliteal, and tibial arteries. In comparison with elastic arteries, their tunica media contains more smooth muscle and fewer elastic fibers. Thus they are capable of greater vasoconstriction and vasodilation to adjust the rate of blood flow to suit the needs of the structure supplied. The walls of muscular arteries are relatively thick, due mainly to the large amount of smooth muscle. Muscular arteries are called distributing arteries because they *distribute* blood to various parts of the body.

Anastomoses

Most tissues of the body receive blood from more than one artery. The union of the branches of two or more adjacent but separate arteries supplying the same body region is called an **anastomosis** (a-nas-tō-MŌ-sis; = coming together). See Figure 14.12b, which illustrates anastomosis between branches of the superior mesenteric artery as they approach the jejunum of the small intestine. See also the cerebral arterial circle (circle of Willis) in Figure 14.10c. Anastomoses may also occur between veins and between arterioles and

Figure 14.1 Comparative structure of blood vessels. The relative size of the capillary in (c) is enlarged.

Arteries carry blood away from the heart to tissue cells; veins carry blood from tissue cells to the heart.

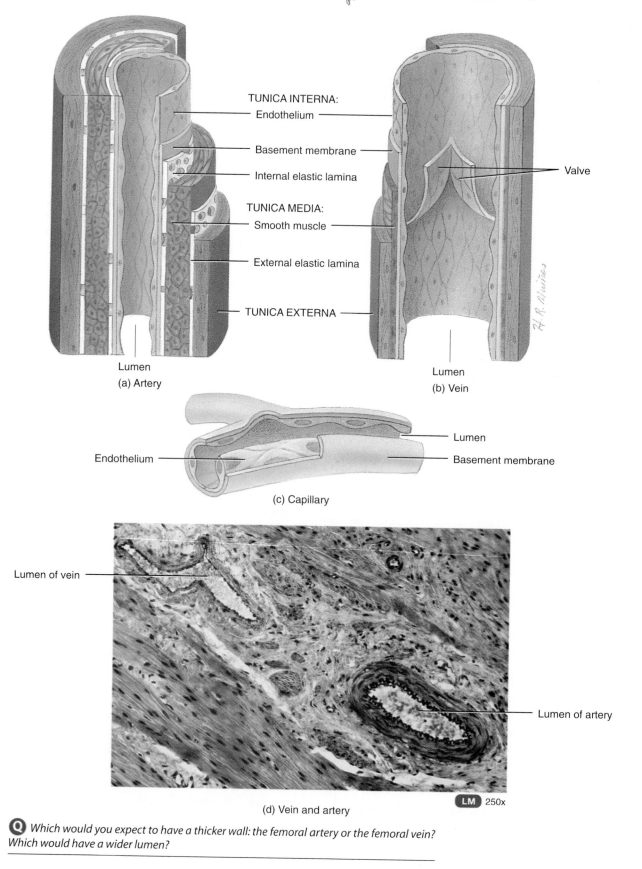

TUNICA INTERNA:
Endothelium
Basement membrane
Internal elastic lamina

TUNICA MEDIA:
Smooth muscle

External elastic lamina

TUNICA EXTERNA

Valve

Lumen
(a) Artery

Lumen
(b) Vein

Endothelium

Lumen
Basement membrane

(c) Capillary

Lumen of vein

Lumen of artery

LM 250x

(d) Vein and artery

Q *Which would you expect to have a thicker wall: the femoral artery or the femoral vein? Which would have a wider lumen?*

Figure 14.2 Structure of an arteriole.

Arterioles are almost-microscopic arteries that deliver blood to the capillaries.

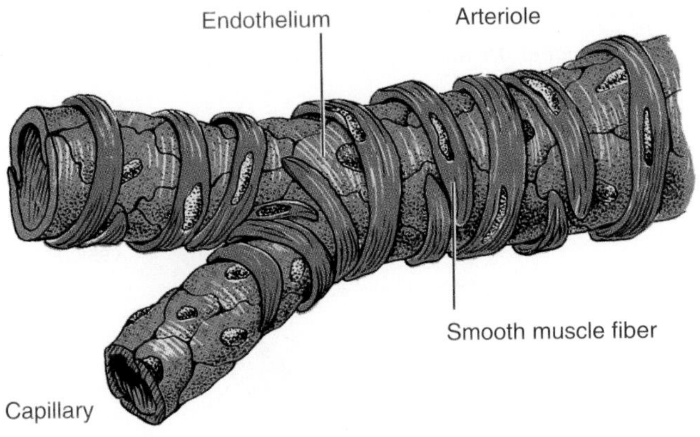

Endothelium   Arteriole

Smooth muscle fiber

Capillary

Q *What causes vasoconstriction?*

venules. Anastomoses between arteries provide alternate routes for blood to reach a tissue or organ. Thus if a vessel is blocked by disease, injury, or surgery, circulation to a part of the body is not necessarily stopped. The alternate route of blood flow to a body part through an anastomosis is known as **collateral circulation.** An alternate blood route may also be from nonanastomosing vessels that supply the same region of the body.

Arteries that do not anastomose are known as **end arteries.** Obstruction of an end artery interrupts the blood supply to a whole segment of an organ, producing necrosis (death) of that segment.

Arterioles

An **arteriole** (*arteriola* = small artery) is a very small, almost-microscopic artery that delivers blood to capillaries. Arterioles closer to the arteries from which they branch have a tunica interna like that of arteries, a tunica media composed of smooth muscle and very few elastic fibers, and a tunica externa composed mostly of elastic and collagen fibers. In the smallest-diameter arterioles, which are closest to capillaries, the tunics consist of little more than a ring of endothelium surrounded by a few scattered smooth muscle fibers (Figure 14.2).

Arterioles play a key role in regulating blood flow from arteries into capillaries. When the smooth muscle of arterioles contracts, causing vasoconstriction, blood flow into capillaries decreases. When the smooth muscle relaxes, the arteriole vasodilates, and blood flow into capillaries increases. A change in diameter of arterioles can also significantly affect blood pressure.

Capillaries

Capillaries (*capillaris* = hairlike) are microscopic vessels that usually connect arterioles and venules. The flow of blood from arterioles to venules through capillaries is called **microcirculation.** Capillaries are found near almost every cell in the body, but their distribution varies with the metabolic activity of the tissue. Body tissues with high metabolic activity—for example, muscles, the liver, kidneys, lungs, and nervous system—require more oxygen and nutrients. Accordingly, they have extensive capillary networks. In areas where activity is lower, such as tendons and ligaments, there are fewer capillaries. A few tissues have no capillaries—all epithelia, the cornea and lens of the eye, and cartilage.

The primary function of capillaries is to permit the exchange of nutrients and wastes between the blood and tissue cells through interstitial fluid. The structure of capillaries is admirably suited to this purpose. Capillary walls are composed of only a single layer of endothelial cells and a basement membrane (see Figure 14.1c). They have no tunica media or tunica externa. Thus a substance in the blood passes through just one cell layer into interstitial fluid before reaching tissue cells. Exchange of materials occurs only through capillary walls: the walls of arteries and veins present too thick a barrier.

Although capillaries directly link arterioles to venules in some places in the body, in other places they form extensive branching networks. These networks increase the surface area for diffusion and filtration and thereby allow rapid exchange of large quantities of materials. In most tissues, blood flows through only a small portion of the capillary network when metabolic needs are low. But when a tissue is active, such as when muscle contracts, the entire capillary network fills with blood.

The flow of blood through capillaries is regulated by vessels with smooth muscle in their walls. A **metarteriole** (*met* = beyond) is a vessel that emerges from an arteriole, passes through the capillary network, and empties into a venule (Figure 14.3). The proximal portion of a metarteriole is surrounded by scattered smooth muscle fibers whose contraction and relaxation help regulate blood flow. The distal portion of a metarteriole, which empties into a venule, has no smooth muscle fibers and is called a **thoroughfare channel.** It serves as a low-resistance pathway that opens when constriction of precapillary sphincters reduces blood flow through the capillary network. Thoroughfare channels thus bypass the capillary bed and sustain blood flow through a region when the capillaries are not being utilized.

True capillaries emerge from arterioles or metarterioles and are not on the direct route from arteriole to venule. At their sites of origin, there is a ring of smooth muscle fibers called a **precapillary sphincter** that controls the flow of blood entering a true capillary. Blood usually does not flow in a continuous manner through capillary networks. Rather, it flows intermittently, because of contraction and relaxation of the smooth muscle of metarterioles and the precapillary sphincters of true capillaries.

Many capillaries of the body are said to be **continuous capillaries.** Except for **intercellular clefts,** which are gaps between neighboring endothelial cells, the plasma membranes of the cells form a continuous, uninterrupted ring

around the capillary (Figure 14.4a). Continuous capillaries are found in skeletal and smooth muscle, connective tissues, and the lungs. Other capillaries of the body are called **fenes-**

Figure 14.3 Arteriole, capillaries, and venule.

Arterioles regulate blood flow into capillaries, where nutrients, gases, and wastes are exchanged between blood and tissue cells.

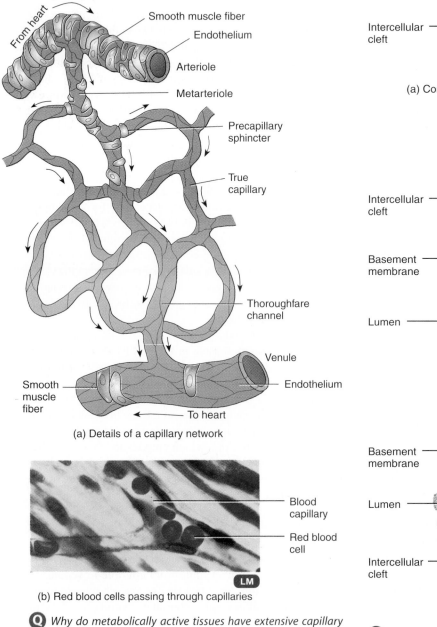

(a) Details of a capillary network

(b) Red blood cells passing through capillaries

Q *Why do metabolically active tissues have extensive capillary networks?*

Figure 14.4 Types of capillaries.

Capillaries are microscopic blood vessels that connect arterioles and venules.

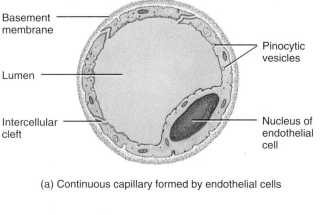

(a) Continuous capillary formed by endothelial cells

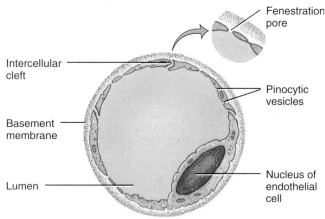

(b) Fenestrated capillary

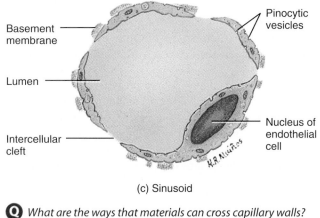

(c) Sinusoid

Q *What are the ways that materials can cross capillary walls?*

trated (*fenestra* = window) **capillaries.** They differ from continuous capillaries in that their endothelial cells have many fenestrations (pores) in the plasma membrane (Figure 14.4b). These range from 70 to 100 nm in diameter. Fenestrated capillaries are found in the kidneys, villi of the small intestine, choroid plexuses of the ventricles in the brain, ciliary processes of the eyes, and endocrine glands.

Blood capillaries in certain parts of the body, such as the liver, are termed **sinusoids.** They are wider than other capillaries and more tortuous (with many twists and turns). Instead of the usual endothelial lining, sinusoids contain spaces between endothelial cells, and the basement membrane is incomplete or absent (Figure 14.4c). In addition, sinusoids contain specialized lining cells that are adapted to the function of the tissue. For example, sinusoids in the liver contain phagocytic cells called **stellate reticuloendothelial (Kupffer's) cells.** Like other capillaries, sinusoids convey blood from arterioles to venules. Other regions containing sinusoids include the spleen, anterior pituitary gland, parathyroid glands, and red bone marrow.

Materials can cross the blood capillary walls through four basic routes: through intercellular clefts, via pinocytic vesicles, directly across endothelial membranes, and through fenestrations.

Venules

When several capillaries unite, they form small veins called **venules** (*venula* = little vein). Venules collect blood from capillaries and drain it into veins. The venules closest to the capillaries consist of a tunica interna of endothelium and a tunica media that has only a few scattered smooth muscle fibers (see Figure 14.3a). As the venules become larger as they approach the veins, they also contain a tunica externa characteristic of veins.

Veins

Veins are composed of essentially the same three coats as arteries, but there are variations in their relative thickness. The tunica interna of veins is thinner than that of their companion arteries, and the tunica media of veins is much thinner, with relatively little smooth muscle and few elastic fibers (see Figure 14.1b). The tunica externa of veins is their thickest layer, consisting of collagen and elastic fibers. The tunica externa of the inferior vena cava also contains longitudinal bundles of smooth muscle. Veins do not contain the internal or external elastic laminae found in arteries. Despite these differences, veins are still distensible enough to adapt to variations in the volume and pressure of blood passing through them. Also, the lumen of a vein is larger than that of a comparable artery, and a vein frequently appears collapsed (flattened) when sectioned (see Figure 14.1d).

The pumping action of the heart is a major factor in moving venous blood back to the heart. Valves in veins, described shortly, also assume a key function. The average blood pressure in veins is considerably lower than in arteries. The difference in pressure can be noticed when blood flows from a cut vessel. Blood leaves a cut vein in an even, slow flow but spurts rapidly from a cut artery. Most of the structural differences between arteries and veins reflect this pressure difference. For example, the walls of veins are not as strong as those of arteries. Many veins also feature **valves** (see Figure 14.1b), which are needed because venous blood pressure is so low. When you stand, the pressure pushing blood up the veins in your lower limbs is barely enough to overcome the force of gravity pulling it back down.

Each valve is composed of two or more thin folds of tunica interna that form flaplike cusps; the cusps project into the lumen of the veins pointing toward the heart (Figure 14.5). Veins pass between groups of skeletal muscles, and when these muscles contract venous pressure is increased; the valve opens as the cusps are pushed against the wall of the vein by the blood as it flows through the valve toward the heart. When the muscles relax, blood would tend to move back toward the feet, but is stopped from doing so as the cusps come together and close the lumen of the vein. As a result, blood is prevented from moving away from the heart. In this way, valves prevent backflow of blood and aid in moving blood in one direction only—toward the heart.

A **vascular (venous) sinus** is a vein with a thin endothelial wall that has no smooth muscle to alter its diameter. The surrounding dense connective tissue replaces the tunica media and tunica externa to provide support. Dural venous sinuses, which are supported by the dura mater, return cerebrospinal fluid and deoxygenated blood from the brain to the heart. Another example of a vascular sinus is the coronary sinus of the heart.

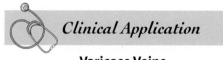

Clinical Application

Varicose Veins

In people with weak venous valves in their lower limbs, gravity forces large quantities of blood back down into the distal parts of the vein. The resulting back-pressure overloads the vein and pushes its wall outward. After repeated overloading, the walls lose their elasticity and become stretched and flabby. Such veins are called **varicose** (*varicosus* = swollen, knotted) **veins.** They may be due to heredity, mechanical factors (prolonged standing and pregnancy), or aging. Because a varicosed wall is not able to exert a firm resistance against the blood, blood tends to accumulate in the pouched-out area of the vein. This causes it to bulge and also forces fluid into the surrounding tissue. Veins close to the surface of the legs, especially the saphenous vein, are highly susceptible to varicosities, whereas deeper veins are not as vulnerable because surrounding skeletal muscles prevent their walls from excessive stretching. ▪

Figure 14.5 Role of skeletal muscle contractions and venous valves in returning blood to the heart. (a) When skeletal muscles contract, the proximal valve opens, and blood is forced toward the heart. (b) Sections of a venous valve.

🔑 *Venous return depends on the pumping action of the heart, skeletal muscle contractions, and valves in veins.*

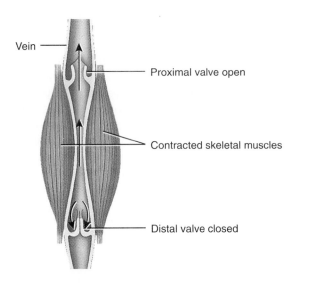

Vein

Proximal valve open

Contracted skeletal muscles

Distal valve closed

(a) Diagram of contracted skeletal muscles

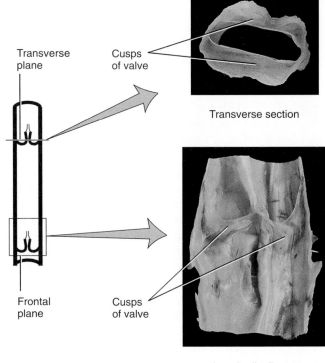

Transverse plane

Cusps of valve

Transverse section

Frontal plane

Cusps of valve

Longitudinally cut

(b) Photograph of a valve in a vein

Q *Why is it more important for leg veins to have valves than it is for neck veins to have valves?*

Blood Distribution

The largest portion of your blood volume at rest, about 60%, is in systemic veins and venules (Figure 14.6). Systemic capillaries hold only about 5% of the blood volume, and arteries and arterioles about 15%. Because systemic veins and venules contain so much of the blood, they are called **blood reservoirs.** They serve as storage depots for blood, which can be diverted quickly to other vessels if the need arises. For example, when there is increased muscular activity, an area in the medulla oblongata of the brain called the vasomotor center sends increasing sympathetic impulses to veins serving as blood reservoirs. The result is vasoconstriction, which reduces the volume of blood in venous reservoirs. A greater blood volume then can flow to skeletal muscles, where it is needed most. A similar mechanism operates in cases of hemorrhage, when blood volume and pressure decrease. Vasoconstriction of veins in venous reservoirs helps to compensate for the blood loss. Among the principal blood reservoirs are the veins of the abdominal organs (especially the liver and spleen) and the veins of the skin.

Figure 14.6 Blood distribution in the heart and blood vessels in a person at rest.

🔑 *Because systemic veins and venules contain so much blood, they are called blood reservoirs.*

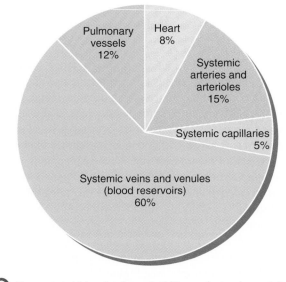

Pulmonary vessels 12%

Heart 8%

Systemic arteries and arterioles 15%

Systemic capillaries 5%

Systemic veins and venules (blood reservoirs) 60%

Q *If your total blood volume is 5 liters, what volume is in your venules and veins right now? In your capillaries?*

CIRCULATORY ROUTES

Arteries, arterioles, capillaries, venules, and veins are organized into routes that deliver blood throughout the body. We can now look at the basic routes the blood takes as it is transported through its vessels.

Figure 14.7 shows the **circulatory routes** for blood flow. The routes are parallel; that is, in most cases a portion of the cardiac output flows separately to each tissue of the body. Thus each organ receives its own supply of freshly oxygenated blood. The two basic postnatal (after birth) routes for blood flow are the systemic and pulmonary circulations. The **systemic circulation** includes all the arteries and arterioles that carry oxygenated blood from the left ventricle to systemic capillaries plus the veins and venules that carry deoxygenated blood returning to the right atrium after flowing through body organs. The nutrient arteries to the lungs, such as the bronchial arteries, also are part of the systemic circulation. Some subdivisions of the systemic circulation are the **coronary (cardiac) circulation** (see Figure 13.10), which supplies the myocardium of the heart; **cerebral circulation,** which supplies the brain (see Figure 14.10c); and the **hepatic portal circulation,** which extends from the gastrointestinal tract to the liver (see Figure 14.20). Blood leaving the aorta and flowing through the systemic arteries is a bright red color. As it moves through capillaries, it loses some of its oxygen and picks up carbon dioxide, so that blood in systemic veins is a dark red color.

When blood returns to the heart from the systemic route, it is pumped out of the right ventricle through the **pulmonary circulation** to the lungs (see Figure 14.21). In pulmonary capillaries of the air sacs (alveoli) of the lungs, it loses some of its carbon dioxide and takes on oxygen. Bright red again, it returns to the left atrium of the heart and reenters the systemic circulation as it is pumped out by the left ventricle.

Another major route—**fetal circulation**—exists only in the fetus and contains special structures that allow the developing fetus to exchange materials with its mother (see Figure 14.22).

Systemic Circulation

The systemic circulation carries oxygen and nutrients to body tissues and removes carbon dioxide and other wastes and heat from the tissues. All systemic arteries branch from the aorta. The portion of the aorta that passes superiorly and then posteriorly to the pulmonary trunk as it emerges from the left ventricle is called the **ascending aorta.** It gives off two coronary artery branches to the myocardium of the heart. Then it turns to the left, forming the **arch of the aorta,** which descends to the level of the fourth thoracic vertebra. The **descending aorta** begins at this point. It lies close to the vertebral bodies, passes through the diaphragm, and divides

at the level of the fourth lumbar vertebra into two **common iliac arteries,** which carry blood to the lower limbs. The section of the descending aorta between the arch of the aorta and the diaphragm is called the **thoracic aorta.** The section between the diaphragm and the common iliac arteries is termed the **abdominal aorta.** Each section of the aorta gives off arteries that continue to branch into distributing arteries leading to organs and finally into the arterioles and capillaries that service the systemic tissues (all tissues except the air sacs of the lungs).

Deoxygenated blood returns to the heart through the systemic veins. All the veins of the systemic circulation drain into the **superior vena cava, inferior vena cava,** or **coronary sinus,** which in turn empty into the right atrium.

The principal arteries and veins of systemic circulation are described and illustrated in Exhibits 14.1–14.12 and Figures 14.8–14.19 to assist you in learning their names. The blood vessels are organized in the exhibits according to regions of the body. Figure 14.8a shows an overview of the major arteries, and Figure 14.14 shows an overview of the major veins. As you study the various blood vessels in the exhibits, refer to these two figures to see the relationships of the blood vessels under consideration to other regions of the body.

Each of the exhibits contains the following information:

- **An overview.** This information provides a general orientation to the blood vessels under consideration, with emphasis on how the blood vessels are organized into various regions and on distinguishing and/or interesting features about the blood vessels.

- **Blood vessel names.** Students often have difficulty with the pronunciations and meanings of blood vessels' names. To learn them more easily, study the phonetic pronunciations and word derivations that indicate how blood vessels get their names.

- **Region supplied or drained.** For each artery listed, there is a description of the parts of the body that receive blood from the vessel. For each vein listed, there is a description of the parts of the body that are drained by the vessel.

- **Illustrations and photographs.** The figures that accompany the exhibits contain several elements. There is always a drawing of the blood vessels under consideration. In many cases, flow diagrams are provided to indicate the patterns of blood distribution or drainage. Cadaver photographs have also been included to provide more realistic views of the blood vessels.

Text continues on page 466

Figure 14.7 Circulatory routes. Heavy black arrows indicate systemic circulation, thin black arrows pulmonary circulation, and thin red arrows hepatic portal circulation. Refer to Figure 13.10 for the details of coronary (cardiac) circulation and to Figure 14.22 for the details of fetal circulation.

Blood vessels are organized into routes that deliver blood to various tissues of the body.

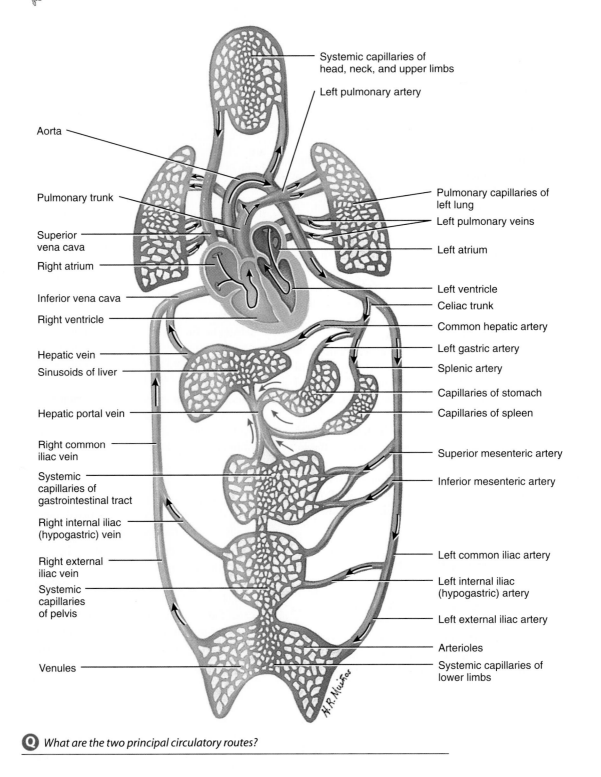

Q *What are the two principal circulatory routes?*

Exhibit 14.1 *The Aorta and Its Branches (Figure 14.8)*

The **aorta** (*aorte*-to lift up) is the largest artery of the body and has a diameter of about 2–3 cm. (about 1 inch). It begins at the left ventricle and contains a valve at its origin, called the aortic semilunar valve (see Figure 13.5a), which prevents backflow of blood into the left ventricle during diastole (relaxation). The principal divisions of the aorta are the **ascending aorta, arch of the aorta, thoracic aorta,** and **abdominal aorta.**

ASCENDING AORTA

BRANCH	REGION SUPPLIED
Right and left coronary arteries	Heart.

ARCH OF THE AORTA

BRANCH	REGION SUPPLIED
Brachiocephalic (brā′-kē-ō-se-FAL-ik) **trunk**	
Right common carotid (ka-ROT-id) **artery**	Right side of head and neck.
Right subclavian (sub-KLĀ-vē-an) **artery**	Right upper limb.
Left common carotid artery	Left side of head and neck.
Left subclavian artery	Left upper limb.

THORACIC (*THORAX*-CHEST) AORTA

BRANCH	REGION SUPPLIED
Intercostal (in′-ter-KOS-tal) **arteries**	Intercostal and chest muscles and plurae.
Superior phrenic (FREN-ik) **arteries**	Posterior and superior surfaces of diaphragm.
Bronchial (BRONG-kē-al) **arteries**	Bronchi of lungs.
Esophageal (e-sof′-a-JĒ-al) **arteries**	Esophagus.

ABDOMINAL AORTA

BRANCH	REGION SUPPLIED
Inferior phrenic (FREN-ik) **arteries**	Inferior surface of diaphragm.
Celiac (SĒ-lē-ak) **trunk**	
Common hepatic (he-PAT-ik) **artery**	Liver.
Left gastric (GAS-trik) **artery**	Stomach and esophagus.
Splenic (SPLĒN-ik) **artery**	Spleen, pancreas, and stomach.
Superior mesenteric (MES-en-ter′-ik) **artery**	Small intestine, cecum, ascending and transverse colons, and pancreas.
Suprarenal (soo′-pra-RĒ-nal) **arteries**	Adrenal (suprarenal) glands.
Renal (RĒ-nal) **arteries**	Kidneys.
Gonadal (gō-NAD-al) **arteries**	
Testicular (tes-TIK-yoo-lar) **arteries**	Testes (male).
Ovarian (ō-VAR-ē-an) **arteries**	Ovaries (female).
Inferior mesenteric artery	Transverse, descending, and sigmoid colons; rectum.
Common iliac (IL-ē-ak) **arteries**	
External iliac arteries	Lower limbs.
Internal iliac (hypogastric) arteries	Uterus (female), prostate gland (male), muscles of buttocks, and urinary bladder.

Exhibit 14.1 *The Aorta and Its Branches*

Figure 14.8 Aorta and its principal branches.

All systemic arteries branch from the aorta.

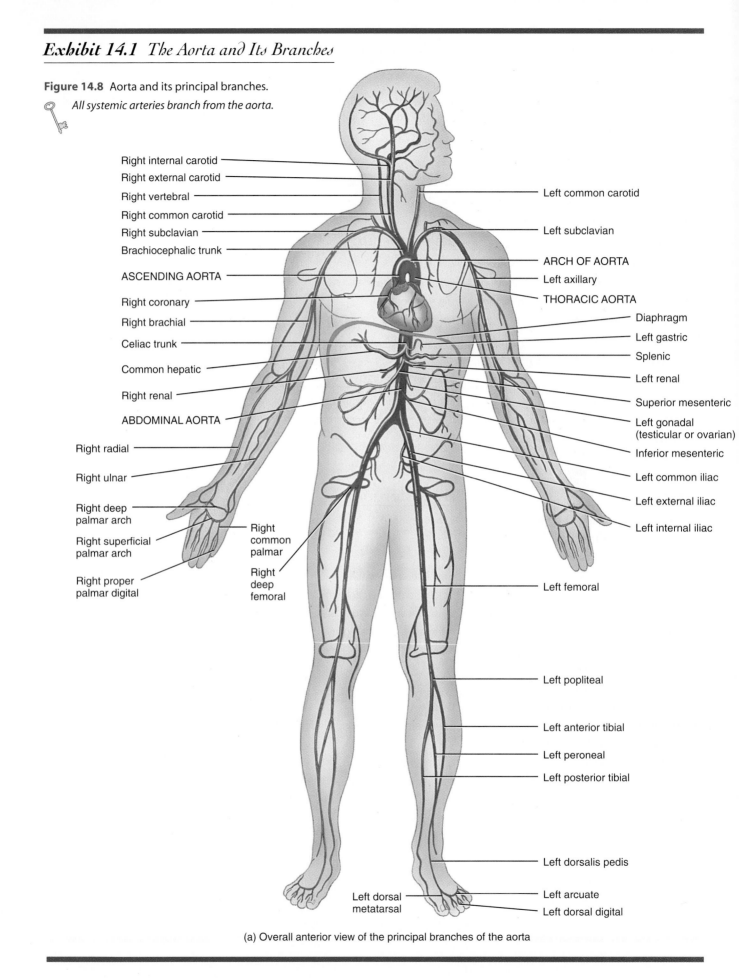

Right internal carotid
Right external carotid
Right vertebral
Right common carotid
Right subclavian
Brachiocephalic trunk
ASCENDING AORTA
Right coronary
Right brachial
Celiac trunk
Common hepatic
Right renal
ABDOMINAL AORTA
Right radial
Right ulnar
Right deep palmar arch
Right superficial palmar arch
Right proper palmar digital
Right common palmar
Right deep femoral

Left common carotid
Left subclavian
ARCH OF AORTA
Left axillary
THORACIC AORTA
Diaphragm
Left gastric
Splenic
Left renal
Superior mesenteric
Left gonadal (testicular or ovarian)
Inferior mesenteric
Left common iliac
Left external iliac
Left internal iliac
Left femoral
Left popliteal
Left anterior tibial
Left peroneal
Left posterior tibial
Left dorsalis pedis
Left dorsal metatarsal
Left arcuate
Left dorsal digital

(a) Overall anterior view of the principal branches of the aorta

Exhibit 14.1 (*continued*)

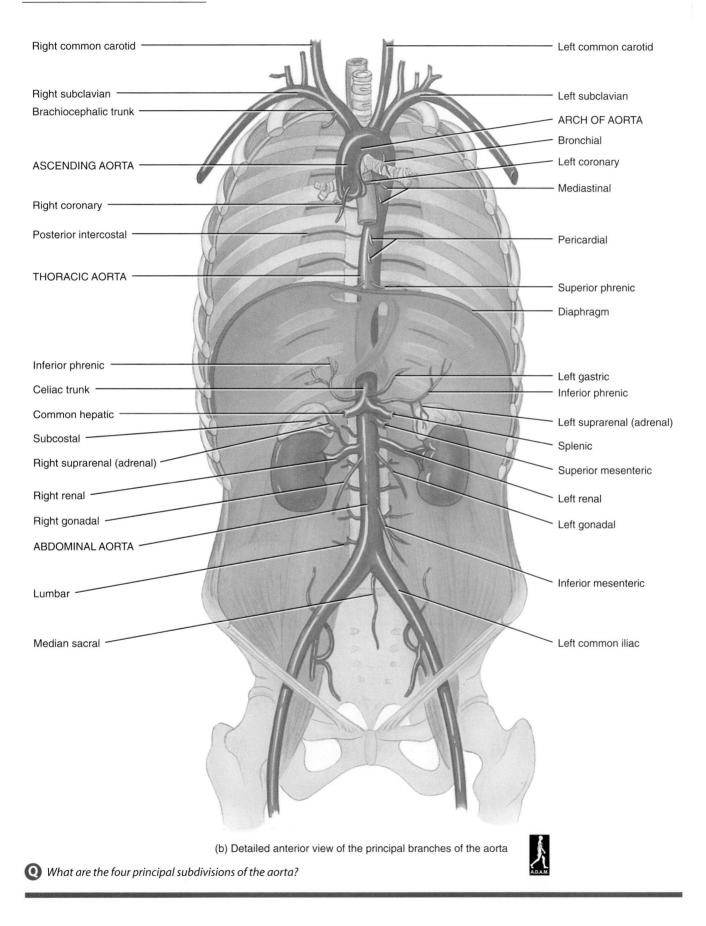

Right common carotid

Right subclavian
Brachiocephalic trunk

ASCENDING AORTA

Right coronary

Posterior intercostal

THORACIC AORTA

Inferior phrenic

Celiac trunk

Common hepatic

Subcostal

Right suprarenal (adrenal)

Right renal

Right gonadal

ABDOMINAL AORTA

Lumbar

Median sacral

Left common carotid

Left subclavian

ARCH OF AORTA

Bronchial

Left coronary

Mediastinal

Pericardial

Superior phrenic

Diaphragm

Left gastric

Inferior phrenic

Left suprarenal (adrenal)

Splenic

Superior mesenteric

Left renal

Left gonadal

Inferior mesenteric

Left common iliac

(b) Detailed anterior view of the principal branches of the aorta

Q *What are the four principal subdivisions of the aorta?*

Exhibit 14.2 *Ascending Aorta (Figure 14.9)*

The **ascending aorta** is the first division of the aorta, about 5 cm (2 in.) in length. It begins at the aortic semilunar valve in the left ventricle and is directed superiorly, slightly anteriorly, and to the right and ends at the level of the sternal angle, where it becomes the arch of the aorta. The beginning of the ascending aorta is covered by the pulmonary trunk and right auricle; the right pulmonary artery is posterior to it. At its origin, the ascending aorta contains three dilations, called aortic sinuses. Two of these, the right and left sinuses, give rise to the right and left coronary arteries, respectively. The ascending aorta is more susceptible to aneurysms than succeeding parts of the aorta because of the higher pressure of blood from the left ventricle during ventricular systole (contraction).

BRANCH	DESCRIPTION AND REGION SUPPLIED
Coronary (KOR-ō-nar-ē; *corona* = crown) **arteries**	The right and left **coronary arteries** arise from the ascending aorta just superior to the aortic semilunar valve. They form a crownlike ring around the heart, giving off branches to the atrial and ventricular myocardium. The **posterior interventricular** (in'-ter-ven-TRIK-yoo-lar; *inter* = between) **branch** of the right coronary artery supplies both ventricles, and the **marginal branch** supplies the right ventricle. The **anterior interventricular branch (left anterior descending)** of the left coronary artery supplies both ventricles, and the **circumflex** (SER-kum-flex; *circum* = around; *flex* = to bend) **branch** supplies the left atrium and left ventricle.

Figure 14.9 Ascending aorta and its branches.

🔑 *The ascending aorta is the first division of the aorta.*

Q *Which arteries arise from the ascending aorta?*

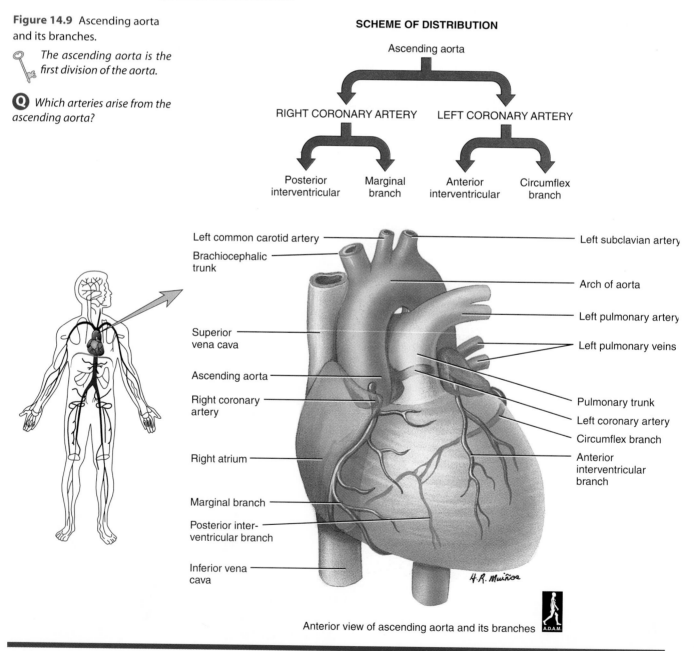

SCHEME OF DISTRIBUTION

Ascending aorta

RIGHT CORONARY ARTERY   LEFT CORONARY ARTERY

Posterior interventricular · Marginal branch · Anterior interventricular · Circumflex branch

Left common carotid artery
Brachiocephalic trunk
Superior vena cava
Ascending aorta
Right coronary artery
Right atrium
Marginal branch
Posterior interventricular branch
Inferior vena cava

Left subclavian artery
Arch of aorta
Left pulmonary artery
Left pulmonary veins
Pulmonary trunk
Left coronary artery
Circumflex branch
Anterior interventricular branch

H.R. Muñoz

Anterior view of ascending aorta and its branches

Exhibit 14.3 *Arch of the Aorta (Figure 14.10)*

The **arch of the aorta** is about 4½ cm (almost 2 in.) in length and is the continuation of the ascending aorta. It emerges from the pericardium posterior to the sternum at the level of the sternal angle. Initially, the arch is directed superiorly, posteriorly, and to the left, and then inferiorly on the left side of the body of the fourth thoracic vertebra. Actually, the arch is directed not only from right to left, but from anterior to posterior as well. The arch of the aorta ends at the level of the intervertebral disc between the fourth and fifth thoracic vertebrae, where it becomes the thoracic aorta. The thymus gland lies anterior to the arch of the aorta, whereas the trachea lies posterior to it.

Interconnecting the arch of the aorta and the pulmonary trunk is a structure called the **ligamentum arteriosum.** It is a remnant of a fetal blood vessel called the ductus arteriosus (see Figure 14.22). Within the arch of the aorta are receptors (nerve cells) called baroreceptors that are sensitive to changes in blood pressure. Situated near the arch of the aorta are other receptors, referred to as **aortic bodies,** that are sensitive to changes in blood concentrations of chemicals such as O_2, CO_2, and H^+.

Three major arteries branch from the superior aspect of the arch of the aorta. In order of their origination, they are the **brachiocephalic trunk, left common carotid artery,** and **left subclavian artery.**

BRANCH	DESCRIPTION AND REGION SUPPLIED
Brachiocephalic (brā′-kē-ō-se-FAL-ik; *brachium* = arm; *cephalic* = head) **trunk**	The **brachiocephalic trunk,** which is found only on the right side, is the first and largest branch off the arch of the aorta (Figure 14.10a). There is no left brachiocephalic artery. It bifurcates (divides) at the right sternoclavicular joint to form the right subclavian artery and right common carotid artery.

The **right subclavian** (sub-KLĀ-vē-an) **artery** extends from the brachiocephalic trunk to the first rib and then passes into the armpit (axilla). The general distribution of the artery is to the brain and spinal cord, neck, shoulder, thoracic viscera and wall, and scapular muscles.

Continuation of the right subclavian artery into the axilla is called the **axillary** (AK-si-ler-ē) **artery.** (Note that the right subclavian artery, which passes deep to the clavicle, is a good example of the practice of giving the same vessel different names as it passes through different regions.) Its general distribution is the shoulder, thoracic and scapular muscles, and humerus.

The **brachial** (BRĀ-kē-al) **artery** is the continuation of the axillary artery into the arm. The brachial artery provides the main blood supply to the arm and is superficial and palpable along its course. It begins at the tendon of the teres major muscle and ends just distal to the bend of the elbow. At first, the brachial artery is medial to the humerus, but as it descends it gradually curves laterally and passes through the cubital fossa, a triangular depression anterior to the elbow (see Figure 11.11c). It is here that you can easily detect the pulse of the brachial artery and listen to the various sounds when taking a person's blood pressure. Just distal to the bend in the elbow, the brachial artery divides into the radial artery and ulnar artery.

The **radial** (RĀ-dē-al = radius) **artery** is the smaller branch and appears to be a direct continuation of the brachial artery. It passes along the lateral (radial) aspect of the forearm and then through the wrist and hand, supplying these structures with blood. At the wrist, the radial artery comes into contact with the distal end of the radius, where it is covered only by fascia and skin. Because of its superficial location at this point, it is a common site for measuring radial pulse.

The **ulnar** (UL-nar = ulna) **artery,** the larger branch of the brachial artery, passes along the medial (ulnar) aspect of the forearm and then into the wrist and hand, also supplying these structures with blood. In the palm, branches of the radial and ulnar arteries anastomose to form the superficial palmar arch and the deep palmar arch.

The **superficial palmar** (*palma* = palm) **arch** is formed mainly by the ulnar artery, with a contribution from a branch of the radial artery. The arch is superficial to the long flexor tendons of the fingers and extends across the palm at the bases of the metacarpals. It gives rise to **common palmar digital arteries,** which supply the palm. Each divides into a pair of **proper palmar digital arteries,** which supply the fingers.

The **deep palmar arch** is formed mainly by the radial artery, with a contribution from a branch of the ulnar artery. The arch is deep to the long flexor tendons of the fingers and extends across the palm, just distal to the bases of the metacarpals. Arising from the deep palmar arch are **palmar metacarpal arteries,** which supply the palm and anastomose with the common palmar digital arteries of the superficial palmar arch. |

Exhibit 14.3 *Arch of the Aorta (continued)*

BRANCH	DESCRIPTION AND REGION SUPPLIED
	Before passing into the axilla, the right subclavian artery gives off a major branch to the brain called the **right vertebral** (VER-te-bral) **artery** (Figure 14.10b). The right vertebral artery passes through the foramina of the transverse processes of the sixth through first cervical vertebrae and enters the skull through the foramen magnum to reach the inferior surface of the brain. Here it unites with the left vertebral artery to form the **basilar** (BAS-i-lar) **artery.** The vertebral artery supplies the posterior portion of the brain with blood. The basilar artery passes along the midline of the anterior aspect of the brain stem and supplies the cerebellum and pons of the brain and the inner ear.
	The **right common carotid** (ka-ROT-id) **artery** begins at the bifurcation of the brachiocephalic trunk, posterior to the right sternoclavicular joint, and passes superiorly in the neck to supply structures in the head (Figure 14.10b). At the superior border of the larynx (voice box), it divides into the right external and right internal carotid arteries.
	The **external carotid artery** begins at the superior border of the larynx and terminates near the temporomandibular joint in the substance of the parotid gland, where it divides into two branches: the superficial temporal and maxillary arteries. The carotid pulse can be detected in the external carotid artery just anterior to the sternocleidomastoid muscle at the superior border of the larynx. The general distribution of the external carotid artery is to structures *external* to the skull.
	The **internal carotid artery** has no branches in the neck and supplies structures *internal* to the skull. It enters the cranial cavity through the carotid foramen in the temporal bone. At the proximal portion of the internal carotid artery there is a slight dilation called the **carotid sinus,** which contains receptors (nerve cells) called **baroreceptors** that monitor changes in blood pressure. Near the baroreceptors is a small mass of tissue at the bifurcation of the common carotid artery called the **carotid body.** It contains receptors that monitor changes in concentrations of chemicals in blood such as O_2, CO_2, and H^+. The internal carotid artery supplies blood to the eyeball and other orbital structures, ear, most of the cerebrum of the brain, pituitary gland, and external nose. The terminal branches of the internal carotid artery are the **anterior cerebral artery,** which supplies most of the medial surface of the cerebrum, and the **middle cerebral artery,** which supplies most of the lateral surface of the cerebrum (Figure 14.10c).
	Inside the cranium, anastomoses of the left and right internal carotid arteries along with the basilar artery form an arrangement of blood vessels at the base of the brain near the sella turcica called the **cerebral** (se-RĒ-bral) **arterial circle (circle of Willis).** From this circle (see Figure 14.10c) arise arteries supplying most of the brain. Essentially the cerebral arterial circle is formed by the union of the **anterior cerebral arteries** (branches of internal carotids) and **posterior cerebral arteries** (branches of basilar artery). The posterior cerebral arteries are connected with the internal carotid arteries by the **posterior communicating** (ko-MYOO-ni-kā′-ting) **arteries.** The anterior cerebral arteries are connected by the **anterior communicating arteries.** The **internal carotid arteries** are also considered part of the cerebral arterial circle. The functions of the cerebral arterial circle are to equalize blood pressure to the brain and provide alternate routes for blood to the brain, should the arteries become damaged.
Left common carotid (ka-ROT-id) **artery**	The **left common carotid artery** is the second branch off the arch of the aorta (Figure 14.10a). Corresponding to the right common carotid, it divides into basically the same branches with the same names, except that the arteries are now labeled "left" instead of "right."
Left subclavian (sub-KLĀ-vē-an) **artery**	The **left subclavian artery** is the third branch off the arch of the aorta (Figure 14.10a). It distributes blood to the left vertebral artery and vessels of the left upper limb. Arteries branching from the left subclavian artery are named like those of the right subclavian artery.

Exhibit 14.3 (*continued*)

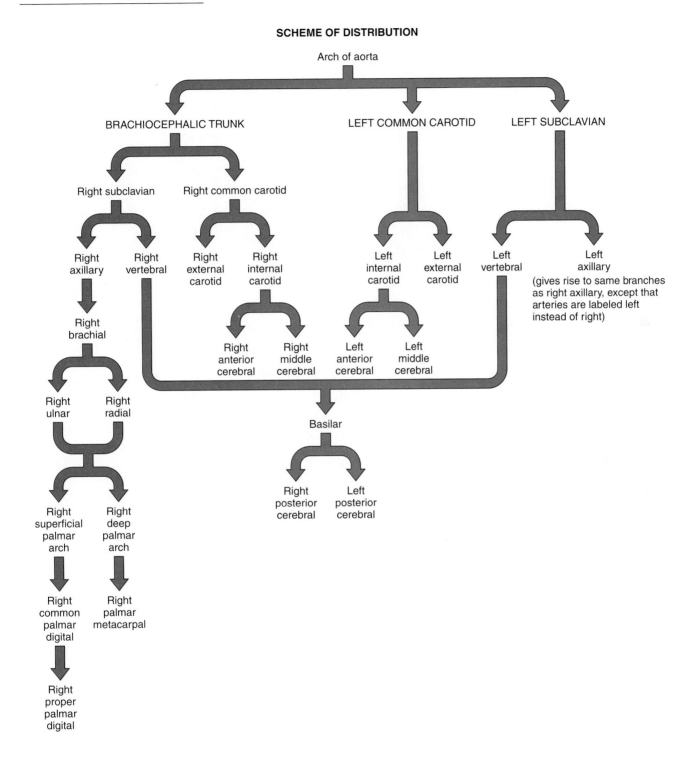

SCHEME OF DISTRIBUTION

Arch of aorta

Exhibit 14.3 Arch of the Aorta (continued)

Figure 14.10 Arch of the aorta and its branches. Note in (c) the arteries that constitute the cerebral arterial circle (circle of Willis).

🗝 *The arch of the aorta is the continuation of the ascending aorta.*

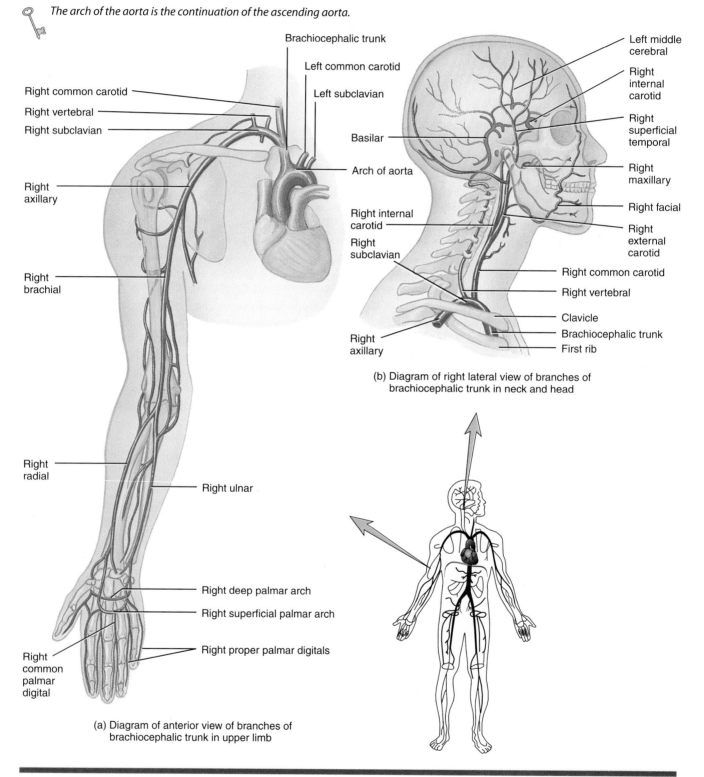

(a) Diagram of anterior view of branches of brachiocephalic trunk in upper limb

(b) Diagram of right lateral view of branches of brachiocephalic trunk in neck and head

Exhibit 14.3 (continued)

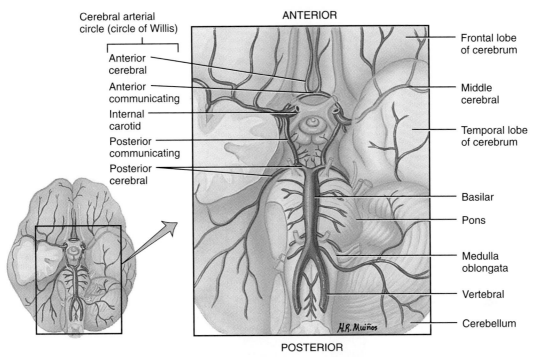

Cerebral arterial
circle (circle of Willis)

Anterior
cerebral

Anterior
communicating

Internal
carotid

Posterior
communicating

Posterior
cerebral

ANTERIOR

Frontal lobe
of cerebrum

Middle
cerebral

Temporal lobe
of cerebrum

Basilar

Pons

Medulla
oblongata

Vertebral

Cerebellum

H.R. Muiños

POSTERIOR

(c) Diagram of inferior view of base of brain showing cerebral anterial circle

Q *What are the three major branches of the arch of the aorta in order of their origination?*

Exhibit 14.3 *Arch of the Aorta (continued)*

SUPERIOR

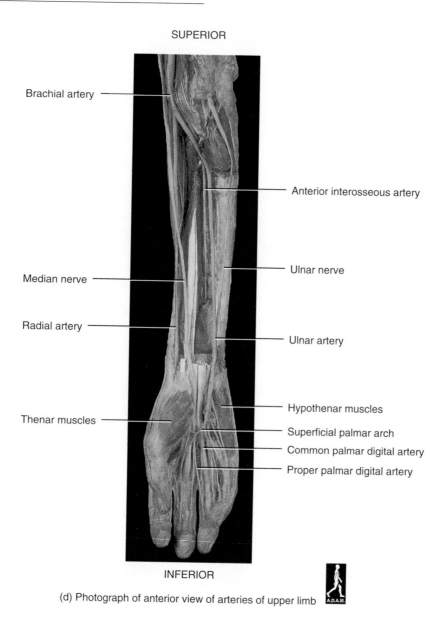

Brachial artery

Anterior interosseous artery

Median nerve

Ulnar nerve

Radial artery

Ulnar artery

Thenar muscles

Hypothenar muscles

Superficial palmar arch

Common palmar digital artery

Proper palmar digital artery

INFERIOR

(d) Photograph of anterior view of arteries of upper limb

Exhibit 14.4 *Thoracic Aorta (Figure 14.11)*

The **thoracic aorta** is about 20 cm (8 in.) long and is a continuation of the arch of the aorta. It begins at the level of the intervertebral disc between the fourth and fifth thoracic vertebrae, where it lies to the left of the vertebral column. As it descends, it moves closer to the midline and ends at an opening in the diaphragm (aortic hiatus) anterior to the vertebral column at the level of the intervertebral disc between the twelfth thoracic and first lumbar vertebrae.

Along its course, the thoracic aorta sends off numerous small arteries to viscera (**visceral branches**) and body wall structures (**parietal branches**).

BRANCH	DESCRIPTION AND REGION SUPPLIED
Visceral	
Pericardial (per'-i-KAR-dē-al); *peri* = around; *cardia* = heart) **arteries**	Two or three minute **pericardial arteries** supply blood to the pericardium.
Bronchial (BRONG-kē-al; *bronchus* = windpipe) **arteries**	One right and two left **bronchial arteries** supply the bronchial tubes, pleurae, bronchial lymph nodes, and esophagus. (Whereas the right bronchial artery arises from the third posterior intercostal artery, the two left bronchial arteries arise from the thoracic aorta.)
Esophageal (e-sof'-a-JĒ-al; *oisein* = to carry; *phagema* = food) **arteries**	Four or five **esophageal arteries** supply the esophagus.
Mediastinal (mē'-dē-as-TĪ-nal) **arteries**	Numerous small **mediastinal arteries** supply blood to structures in the mediastinum.
Parietal	
Posterior intercostal (in'-ter-KOS-tal; *inter* = between; *costa* = rib) **arteries**	Nine pairs of **posterior intercostal arteries** supply the intercostal, pectoralis major and minor, and serratus anterior muscles; overlying subcutaneous tissue and skin; mammary glands; and vertebrae, meninges, and spinal cord.
Subcostal (sub-KOS-tal; *sub* = under) **arteries**	The left and right **subcostal arteries** have a distribution similar to that of the posterior intercostals.
Superior phrenic (FREN-ik; *phren* = diaphragm) **arteries**	Small **superior phrenic arteries** supply the superior and posterior surfaces of the diaphragm.

SCHEME OF DISTRIBUTION

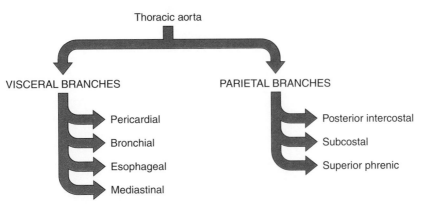

Exhibit 14.4 *Thoracic Aorta (continued)*

Figure 14.11 Thoracic and abdominal aorta and their principal branches.

🔑 *The thoracic aorta ends at the level of the intervertebral disc between the twelfth thoracic and first lumbar vertebrae.*

Right common carotid

Right vertebral

Right subclavian

Brachiocephalic trunk

Right posterior intercostal

Celiac trunk

Right inferior phrenic

Common hepatic

Right middle suprarenal (adrenal)

Right renal

Right gonadal (testicular or ovarian)

Right lumbars

Left common carotid

Left vertebral

Left subclavian

Arch of aorta

Left axillary

Thoracic aorta

Left brachial

Left gastric

Left inferior phrenic

Left middle suprarenal (adrenal)

Splenic

Superior mesenteric

Left renal

Left gonadal (testicular or ovarian)

Inferior mesenteric

Abdominal aorta

Median sacral

Left common iliac

Inguinal ligament

Left internal iliac (hypogastric)

Left external iliac

Left femoral

Thoracic and abdominal aorta in anterior view

Q *Where does the thoracic aorta begin?*

Exhibit 14.5 *Abdominal Aorta (Figure 14.12)*

The **abdominal** (ab-DOM-i-nal) **aorta** is the continuation of the thoracic aorta. It begins at the aortic hiatus in the diaphragm and ends at about the level of the fourth lumbar vertebra, where it divides into right and left common iliac arteries. The abdominal aorta lies anterior to the vertebral column.

As with the thoracic aorta, the abdominal aorta gives off visceral and parietal branches. The unpaired visceral branches arise from the anterior surface of the aorta and include the **celiac, superior mesenteric,** and **inferior mesenteric arteries** (see Figure 14.11). The paired visceral branches arise from the lateral surfaces of the aorta and include the **suprarenal, renal,** and **gonadal arteries.** The paired parietal branches arise from the posterolateral surfaces of the aorta and include the **inferior phrenic** and **lumbar arteries.** The unpaired parietal branch is the **median sacral artery.**

BRANCH	DESCRIPTION AND REGION SUPPLIED
Visceral	
Celiac (SĒ-lē-ak; *koilia=* abdominal cavity) **trunk**	The **celiac trunk (artery)** is the first visceral branch off the aorta, inferior to the diaphragm at about the level of the twelfth thoracic vertebra (Figure 14.12a). Its general distribution is to the esophagus, stomach, pancreas, liver, gallbladder, part of the small intestine, and greater omentum (a portion of the peritoneum that drapes over the intestines). Almost immediately following its origination, the celiac trunk divides into three branches: left gastric, splenic, and common hepatic arteries (Figure 14.12a).
	The **left gastric** (GAS-trik; *gastro* = stomach) **artery** is the smallest of the three branches. It passes superiorly to the left toward the esophagus and then turns to follow the lesser curvature of the stomach. It supplies the stomach and esophagus.
	The **splenic** (SPLEN-ik) **artery** is the largest branch of the celiac trunk. It arises from the left side of the celiac trunk distal to the left gastric artery. It passes horizontally to the left along the pancreas to reach the spleen. Its branches are the (1) **pancreatic** (pan-krē-AT-ik) **artery,** which supplies the pancreas; (2) **left gastroepiploic** (gas′-trō-ep′-i-PLŌ-ik; *epiplo* = omentum) **artery,** which supplies the stomach and greater omentum; and (3) **short gastric artery,** which supplies the stomach.
	The **common hepatic** (he-PAT-ik; *hepato* = liver) **artery** is intermediate in size between the left gastric and splenic arteries. Unlike the other two branches of the celiac trunk, the common hepatic artery arises from the right side. Its branches are the (1) **hepatic artery proper,** which supplies the liver, gallbladder, and stomach; (2) **right gastric artery,** which supplies the stomach; and (3) **gastroduodenal** (gas′-trō-doo′-ō-DĒ-nal) **artery,** which supplies the stomach, duodenum of the small intestine, pancreas, and greater omentum.
Superior mesenteric (MES-en-ter′-ik; *meso* = middle; *enteron* = intestine) **artery**	The **superior mesenteric artery** (Figure 14.12b) arises from the anterior surface of the abdominal aorta about 1.25 cm inferior to the celiac trunk at the level of the first lumbar vertebra. It runs inferiorly and anteriorly and between the layers of mesentery, a portion of the peritoneum that attaches the small intestine to the posterior abdominal wall. The general distribution of the superior mesenteric artery is to the pancreas, most of the small intestine, and part of the large intestine. It anastomoses extensively and has the following branches: (1) **inferior pancreaticoduodenal** (pan′-krē-at′-i-kō-doo-ō-DĒ-nal) **artery,** which supplies the pancreas and duodenum; (2) **jejunal** (je-JOO-nal) and **ileal** (IL-ē-al) **arteries,** which supply the jejunum and ileum of the small intestine, respectively; (3) **ileocolic** (il′-ē-ō-KOL-ik) **artery,** which supplies the ileum and ascending colon of the large intestine; (4) **right colic** (KOL-ik) **artery,** which supplies the ascending colon; and (5) **middle colic artery,** which supplies the transverse colon of the large intestine.

Exhibit 14.5 *Abdominal Aorta (continued)*

BRANCH	DESCRIPTION AND REGION SUPPLIED
Suprarenal (soo′-pra-RĒ-nal; *supra* = above; *renal* = kidney) **arteries**	Although there are three pairs of **suprarenal (adrenal) arteries** that supply the adrenal (suprarenal) glands (superior, middle, and inferior), only the middle pair originates directly from the abdominal aorta (see Figure 14.11). The middle suprarenal arteries arise at the level of the first lumbar vertebra at or superior to the renal arteries. The superior suprarenal arteries arise from the inferior phrenic artery, and the inferior suprarenal arteries originate from the renal arteries.
Renal (RĒ-nal; *renal* = kidney) **arteries**	The right and left **renal arteries** usually arise from the lateral aspects of the abdominal aorta at the superior border of the second lumbar vertebra, about 1 cm inferior to the superior mesenteric artery (see Figure 14.11). The right renal vein, which is longer than the left, arises slightly lower than the left and passes posterior to the right renal vein and inferior vena cava. The left renal artery is posterior to the left renal vein and is crossed by the inferior mesenteric vein. The renal arteries carry blood to the kidneys, adrenal (suprarenal) glands, and ureters. Their distribution within the kidneys is discussed in Chapter 25.
Gonadal (gō-NAD-al; *gonos* = seed) **[testicular** tes-TIK-yoo-lar) **or ovarian** (ō-VA-rē-an)**] arteries**	The **gonadal arteries** arise from the abdominal aorta at the level of the second lumbar vertebra just inferior to the renal arteries (see Figure 14.11). In the male, the gonadal arteries are specifically referred to as the **testicular arteries.** They pass through the inguinal canal and supply the testes, epididymis, and ureters. In the female, the gonadal arteries are called the **ovarian arteries.** They are much shorter than the testicular arteries and supply the ovaries, uterine (Fallopian) tubes, and ureters.
Inferior mesenteric (MES-en-ter′-ik) **artery**	The **inferior mesenteric artery** (Figure 14.12c) arises from the anterior aspect of the abdominal aorta at the level of the third lumbar vertebra. It passes inferiorly to the left of the aorta and anastomoses extensively. The principal branches of the inferior mesenteric artery are the (1) **left colic** (KŌL-ik) **artery,** which supplies the transverse and descending colons of the large intestine; (2) **sigmoid** (SIG-moyd) **arteries,** which supply the descending and sigmoid colons of the large intestine; and (3) **superior rectal** (REK-tal) **artery,** which supplies the rectum of the large intestine.
Parietal	
Inferior phrenic (FREN-ik; *phrenic* = diaphragm) **arteries**	The **inferior phrenic arteries** are the first paired branches of the abdominal aorta, immediately superior to the origin of the celiac trunk (see Figure 14.11). (They may also arise from the renal arteries.) The inferior phrenic arteries are distributed to the inferior surface of the diaphragm and adrenal (suprarenal) glands.
Lumbar (LUM-bar; *lumbar* = loin) **arteries**	The four pairs of **lumbar arteries** arise from the posterolateral surface of the abdominal aorta (see Figure 14.11). They supply the lumbar vertebrae, spinal cord and its meninges, and the muscles and skin of the lumbar region of the back.
Median sacral (SĀ-kral; *sacral* = sacrum) **artery**	The **median sacral artery** arises from the posterior surface of the abdominal aorta about 1 cm superior to the bifurcation of the aorta into the right and left common iliac arteries (see Figure 14.11). The median sacral artery supplies the sacrum and coccyx.

Exhibit 14.5 *(continued)*

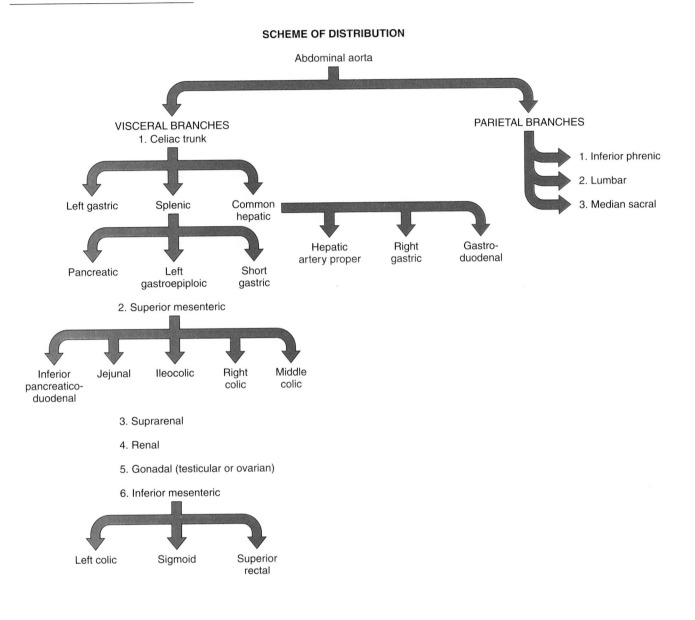

SCHEME OF DISTRIBUTION

Exhibit 14.5 *Abdominal Aorta (continued)*

Figure 14.12 (continued)

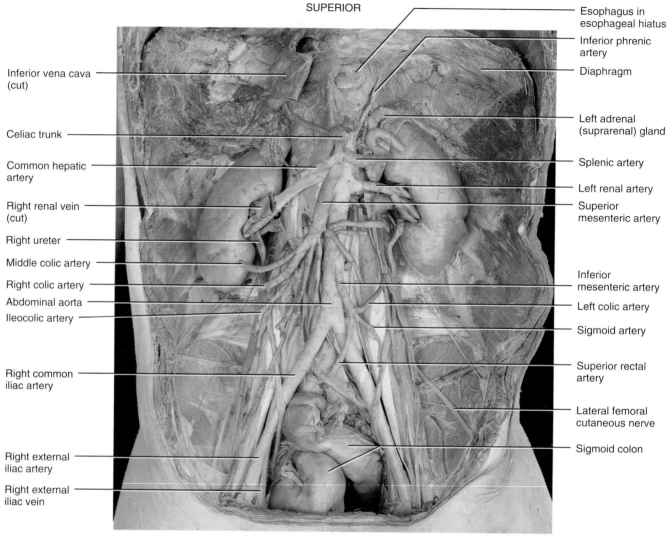

SUPERIOR

Inferior vena cava (cut)

Celiac trunk

Common hepatic artery

Right renal vein (cut)

Right ureter

Middle colic artery

Right colic artery

Abdominal aorta

Ileocolic artery

Right common iliac artery

Right external iliac artery

Right external iliac vein

Esophagus in esophageal hiatus

Inferior phrenic artery

Diaphragm

Left adrenal (suprarenal) gland

Splenic artery

Left renal artery

Superior mesenteric artery

Inferior mesenteric artery

Left colic artery

Sigmoid artery

Superior rectal artery

Lateral femoral cutaneous nerve

Sigmoid colon

INFERIOR

(d) Photograph of anterior view of arteries of the abdomen and pelvis

Q *Where does the abdominal aorta begin?*

Exhibit 14.6 *Arteries of the Pelvis and Lower Limbs (Figure 14.13)*

The abdominal aorta terminates by dividing into the right and left **common iliac arteries.** These, in turn, divide into the **internal** and **external iliac arteries.** Upon entering the thighs, the external iliac arteries become the **femoral arteries,** then the **popliteal arteries** posterior to the knee, and then the **anterior** and **posterior tibial arteries** at the knees.

BRANCH	DESCRIPTION AND REGION SUPPLIED
Common iliac (IL-ē-ak; *iliac* = ilium) **arteries**	At about the level of the fourth lumbar vertebra, the abdominal aorta divides into the right and left **common iliac arteries,** the terminal branches of the abdominal aorta. Each passes inferiorly about 5 cm (2 in.) and gives rise to two branches: internal iliac and external iliac arteries. The general distribution of the common iliac arteries is to the pelvis, external genitals, and lower limbs.
Internal iliac arteries	The **internal iliac (hypogastric) arteries** are the primary arteries of the pelvis. They begin at the bifurcation of the common iliac arteries anterior to the sacroiliac joint at the level of the lumbosacral intervertebral disc. They pass posteromedially as they descend in the pelvis and divide into anterior and posterior divisions. The general distribution of the internal iliac arteries is to the pelvis, buttocks, external genitals, and thigh.
External iliac arteries	The **external iliac arteries** are larger than the internal iliac arteries. Like the internal iliac arteries, they begin at the bifurcation of the common iliac arteries. They descend along the medial border of the psoas major muscles following the pelvic brim, pass posterior to the midportion of the inguinal ligaments, and become the femoral arteries. The general distribution of the external iliac arteries is to the lower limbs. Specifically, branches of the external iliac arteries supply the muscles of the anterior abdominal wall, the cremasteric muscle in males and the round ligament of the uterus in females, and the lower limbs.
	The **femoral** (FEM-ō-ral; *femur* = thigh) **arteries** descend along the anteriomedial aspects of the thighs to the junction of the middle and lower third of the thighs. Here they pass through an opening in the tendon of the adductor magnus muscle, where they emerge posterior to the femurs as the popliteal arteries. A pulse may be felt in the femoral artery just inferior to the inguinal ligament. The general distribution of the femoral arteries is to the lower abdominal wall, groin, external genitals, and muscles of the thigh.
	The **popliteal** (pop'-li-TĒ-al; *poples* = posterior surface of knee) **arteries** are the continuation of the femoral arteries through the popliteal fossa (space behind the knee). They descend to the inferior border of the popliteus muscles, where they divide into the anterior and posterior tibial arteries. A pulse may be detected in the popliteal arteries. In addition to supplying the adductor magnus and hamstring muscles and the skin on the posterior aspect of the legs, branches of the popliteal arteries also supply the gastrocnemius, soleus, and plantaris muscles of the calf, knee joint, femur, patella, and fibula.

Exhibit 14.6 *Arteries of the Pelvis and Lower Limbs (continued)*

Figure 14.13 Arteries of the pelvis and right lower limb.

The internal iliac arteries carry most of the blood supplied to the pelvic viscera and wall.

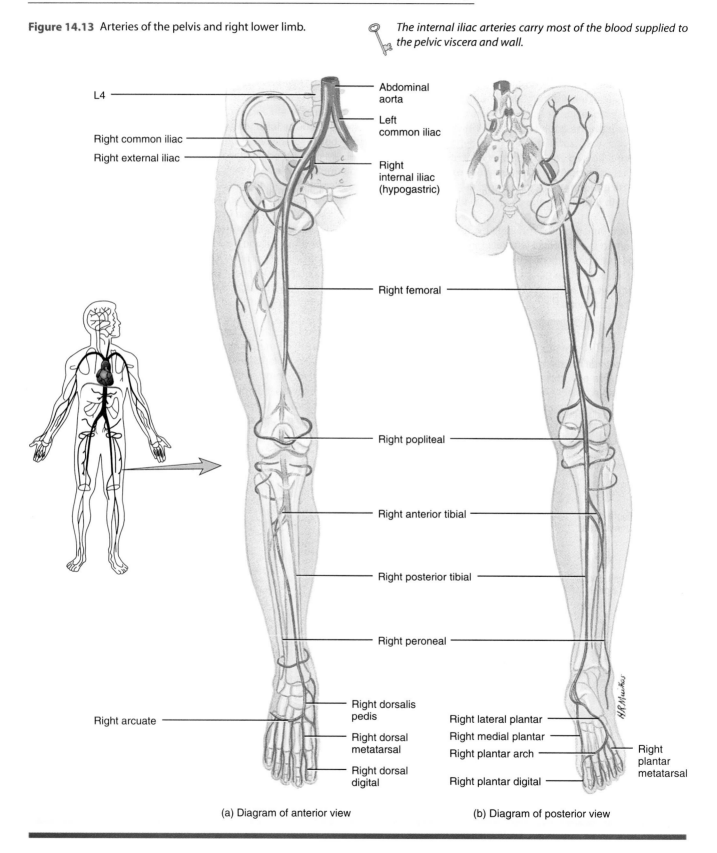

L4

Abdominal aorta

Left common iliac

Right common iliac

Right external iliac

Right internal iliac (hypogastric)

Right femoral

Right popliteal

Right anterior tibial

Right posterior tibial

Right peroneal

Right arcuate

Right dorsalis pedis

Right dorsal metatarsal

Right dorsal digital

Right lateral plantar

Right medial plantar

Right plantar arch

Right plantar digital

Right plantar metatarsal

(a) Diagram of anterior view

(b) Diagram of posterior view

Exhibit 14.6 (*continued*)

SUPERIOR

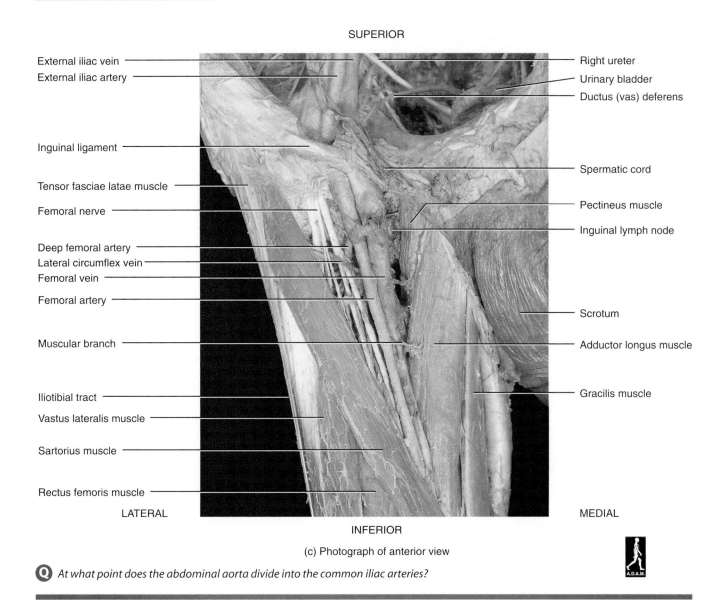

External iliac vein

External iliac artery

Inguinal ligament

Tensor fasciae latae muscle

Femoral nerve

Deep femoral artery

Lateral circumflex vein

Femoral vein

Femoral artery

Muscular branch

Iliotibial tract

Vastus lateralis muscle

Sartorius muscle

Rectus femoris muscle

Right ureter

Urinary bladder

Ductus (vas) deferens

Spermatic cord

Pectineus muscle

Inguinal lymph node

Scrotum

Adductor longus muscle

Gracilis muscle

LATERAL

MEDIAL

INFERIOR

(c) Photograph of anterior view

Q *At what point does the abdominal aorta divide into the common iliac arteries?*

Exhibit 14.7 *Veins of Systemic Circulation (Figure 14.14)*

Whereas arteries distribute blood to various parts of the body, veins drain blood away from them. For the most part, arteries are deep, whereas veins may be superficial or deep. Superficial veins are located just beneath the skin and can easily be seen. Because there are no large superficial veins, the names of superficial veins do not correspond to those of arteries. Superficial veins are clinically important as sites for withdrawing blood or giving injections. Deep veins generally travel alongside arteries and usually bear the same name. Arteries usually follow definite pathways; veins are more difficult to follow because they connect in irregular networks in which many tributaries merge to form a large vein. Although only one systemic artery, the aorta, takes oxygenated blood away from the heart (left ventricle), three systemic veins, the **coronary sinus, superior vena cava,** and **inferior vena cava** deliver deoxygenated blood to the heart (right atrium). The coronary sinus receives blood from the cardiac veins; the superior vena cava receives blood from other veins superior to the diaphragm, except the air sacs (alveoli) of the lungs. The inferior vena cava receives blood from veins inferior to the diaphragm.

VEIN	DESCRIPTION AND REGION SUPPLIED
Coronary (KOR-ō-nar-ē; *corona* = crown) **sinus**	The **coronary sinus** is the main vein of the heart; it receives almost all venous blood from the myocardium. It is located in the coronary sulcus (see Figure 13.4a) and opens into the right atrium between the orifice of the inferior vena cava and the tricuspid valve. It is a wide venous channel into which three veins drain. It receives the **great cardiac vein** (in the anterior interventricular sulcus) into its left end and the **middle cardiac vein** (in the posterior interventricular sulcus) and the **small cardiac vein** into its right end. Several **anterior cardiac veins** drain directly into the right atrium.
Superior vena cava (VĒ-na CA-va; *vena* = vein; *cava* = cavelike) **(SVC)**	The **SVC** is about 7½ cm (3 in.) long and 2 cm (1 in.) in diameter and empties its blood into the superior part of the right atrium. It begins posterior to the right first costal cartilage by the union of the right and left brachiocephalic veins and ends at the level of the right third costal cartilage where it enters the right atrium. The SVC drains the head, neck, chest, and upper limbs.
Inferior vena cava (IVC)	The **IVC** is the largest vein in the body, about 3½ cm (1½ in.) in diameter. It begins anterior to the fifth lumbar vertebra by the union of the common iliac veins, ascends behind the peritoneum to the right of the midline, pierces the costal tendon of the diaphragm at the level of the eighth thoracic vertebra, and enters the inferior part of the right atrium. The IVC drains the abdomen, pelvis, and lower limbs. The inferior vena cava is commonly compressed during the later stages of pregnancy by the enlarging uterus. This produces edema of the ankles and feet and temporary varicose veins.

Exhibit 14.7 *(continued)*

Figure 14.14 Principal veins.

🔑 *Deoxygenated blood returns to the heart via the superior and inferior venae cavae and the coronary sinus.*

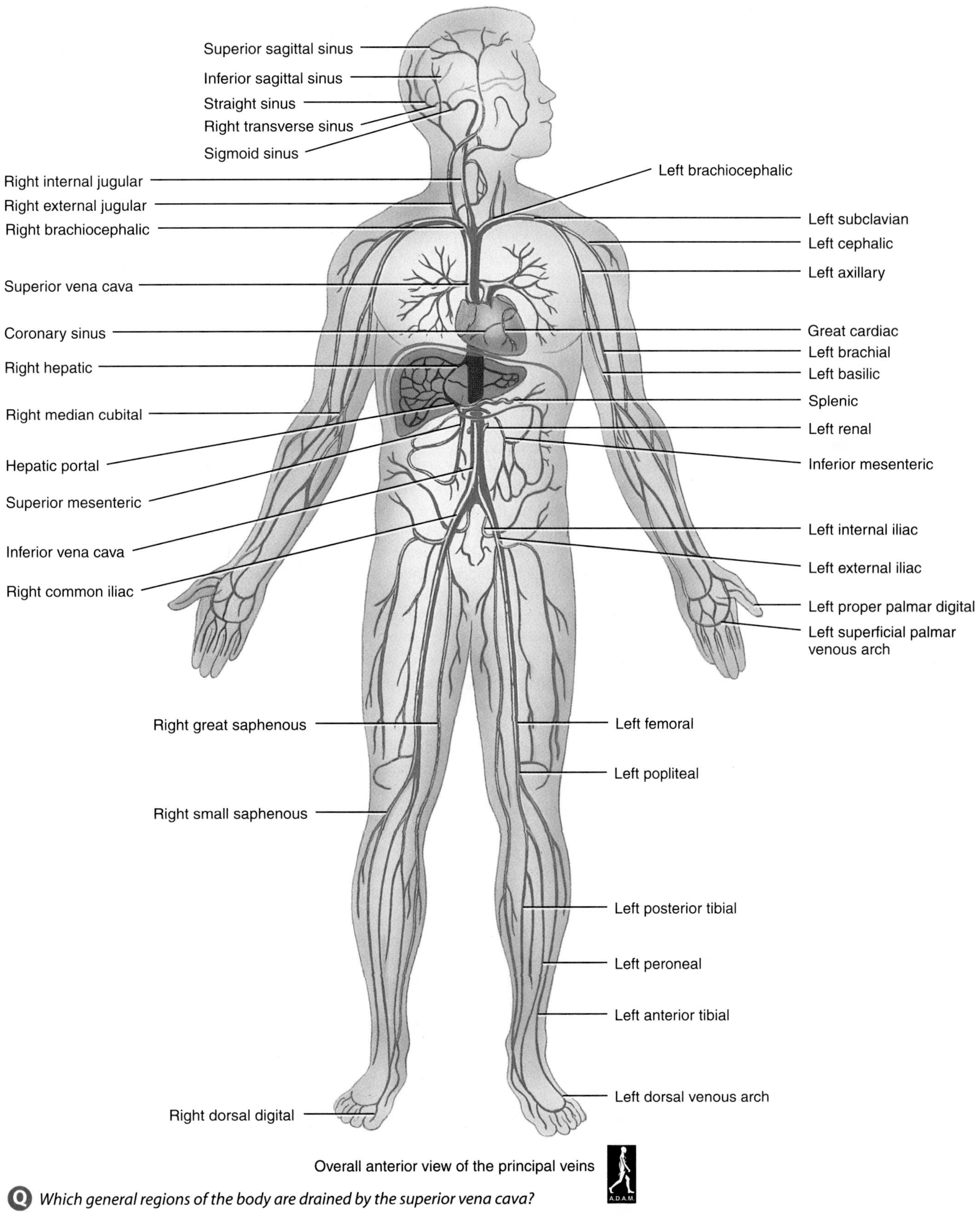

Superior sagittal sinus

Inferior sagittal sinus

Straight sinus

Right transverse sinus

Sigmoid sinus

Right internal jugular

Right external jugular

Right brachiocephalic

Superior vena cava

Coronary sinus

Right hepatic

Right median cubital

Hepatic portal

Superior mesenteric

Inferior vena cava

Right common iliac

Left brachiocephalic

Left subclavian

Left cephalic

Left axillary

Great cardiac

Left brachial

Left basilic

Splenic

Left renal

Inferior mesenteric

Left internal iliac

Left external iliac

Left proper palmar digital

Left superficial palmar venous arch

Right great saphenous

Left femoral

Left popliteal

Right small saphenous

Left posterior tibial

Left peroneal

Left anterior tibial

Left dorsal venous arch

Right dorsal digital

Overall anterior view of the principal veins

❓ *Which general regions of the body are drained by the superior vena cava?*

Exhibit 14.8 *Veins of the Head and Neck (Figure 14.15)*

The majority of blood draining from the head passes into three pairs of veins: **internal jugular, external jugular,** and **vertebral veins.** Within the brain, all veins drain into dural venous sinuses and then into the internal jugular veins. **Dural venous sinuses** are endothelial-lined venous channels between layers of the cranial dura mater.

VEIN	DESCRIPTION AND REGION DRAINED
Internal jugular (JUG-yoo-lar; *jugular* = throat) **veins**	The flow of blood from the dural venous sinuses into the internal jugular veins is as follows (Figure 14.15): The **superior sagittal** (SAJ-i-tal; *sagittalis* = straight) **sinus** begins at the frontal bone, where it receives a vein from the nasal cavity, and passes posteriorly to the occipital bone. Along its course, it receives blood from the superior, medial, and lateral aspects of the cerebral hemispheres, meninges, and cranial bones. The superior sagittal sinus usually turns to the right and drains into the right transverse sinus.
	The **inferior sagittal sinus** is much smaller than the superior sagittal sinus and begins posterior to the attachment of the falx cerebri and receives the great cerebral vein to become the straight sinus. The great cerebral vein drains the deeper parts of the brain. Along its course, the inferior sagittal sinus also receives tributaries from the superior and medial aspects of the cerebral hemispheres.
	The **straight sinus** runs in the tentorium cerebelli and is formed by the union of the inferior sagittal sinus and the great cerebral vein. The straight sinus also receives blood from the cerebellum and usually drains into the left transverse sinus.
	The **transverse (lateral) sinuses** begin near the occipital bone, pass laterally and anteriorly, and become the sigmoid sinuses near the temporal bone. The transverse sinuses receive blood from the cerebrum, cerebellum, and cranial bones.
	The **sigmoid** (SIG-moid = S-shaped) **sinuses** are located along the temporal bone. They pass through the jugular foramina where they terminate in the internal jugular veins. The sigmoid sinuses drain the transverse sinuses.
	The right and left **internal jugular veins** pass inferiorly on either side of the neck lateral to the internal carotid and common carotid arteries. They then unite with the subclavian veins posterior to the clavicles at the sternoclavicular joints to form the right and left **brachiocephalic** (brā-kē-ō-se-FAL-ik; *brachium* = arm; *cephalic* = head) **veins.** From here blood flows into the superior vena cava. The general structures drained by the internal jugular veins are the brain (through the dural venous sinuses), face, and neck.
External jugular veins	The right and left **external jugular veins** begin in the parotid glands near the angle of the mandible. They are superficial veins that descend through the neck across the sternocleidomastoid muscles. They terminate at a point opposite the middle of the clavicle, where they empty into the subclavian veins. The general structures drained by the external jugular veins are external to the cranium, such as the scalp and superficial and deep regions of the face.
	In cases of heart failure, the venous pressure in the right atrium may rise. In such patients the pressure in the column of blood in the external jugular vein rises so that, even with the patient at rest and sitting in a chair, the external jugular vein will be visibly distended. Temporary distention of the vein is often seen in healthy adults when the intrathoracic pressure is raised during coughing and physical exertion.
Vertebral (VER-te-bral; *vertebra* = vertebrae) **veins**	The right and left **vertebral veins** originate inferior to the occipital condyles. They descend through successive transverse foramina of the first six cervical vertebrae and emerge from the foramina of the sixth cervical vertebra to enter the brachiocephalic veins in the root of the neck. The vertebral veins drain deep structures in the neck such as the cervical vertebrae, cervical spinal cord, and some neck muscles.

Exhibit 14.8 (continued)

SCHEME OF DRAINAGE

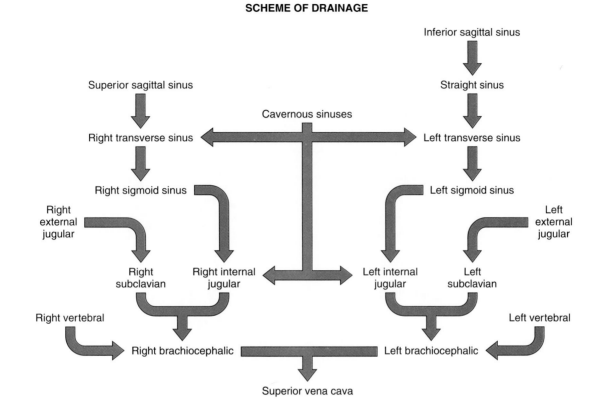

Exhibit 14.8 *Veins of the Head and Neck (continued)*

Figure 14.15 Principal veins of the head and neck.

Blood draining from the head passes into the internal jugular, external jugular, and vertebral veins.

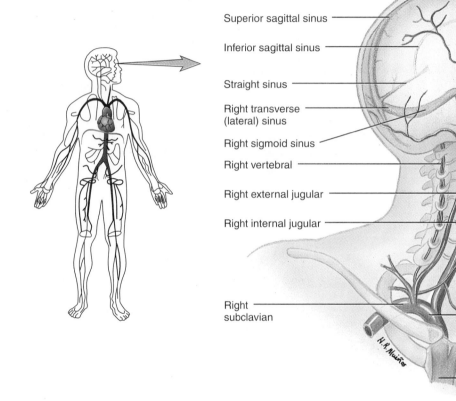

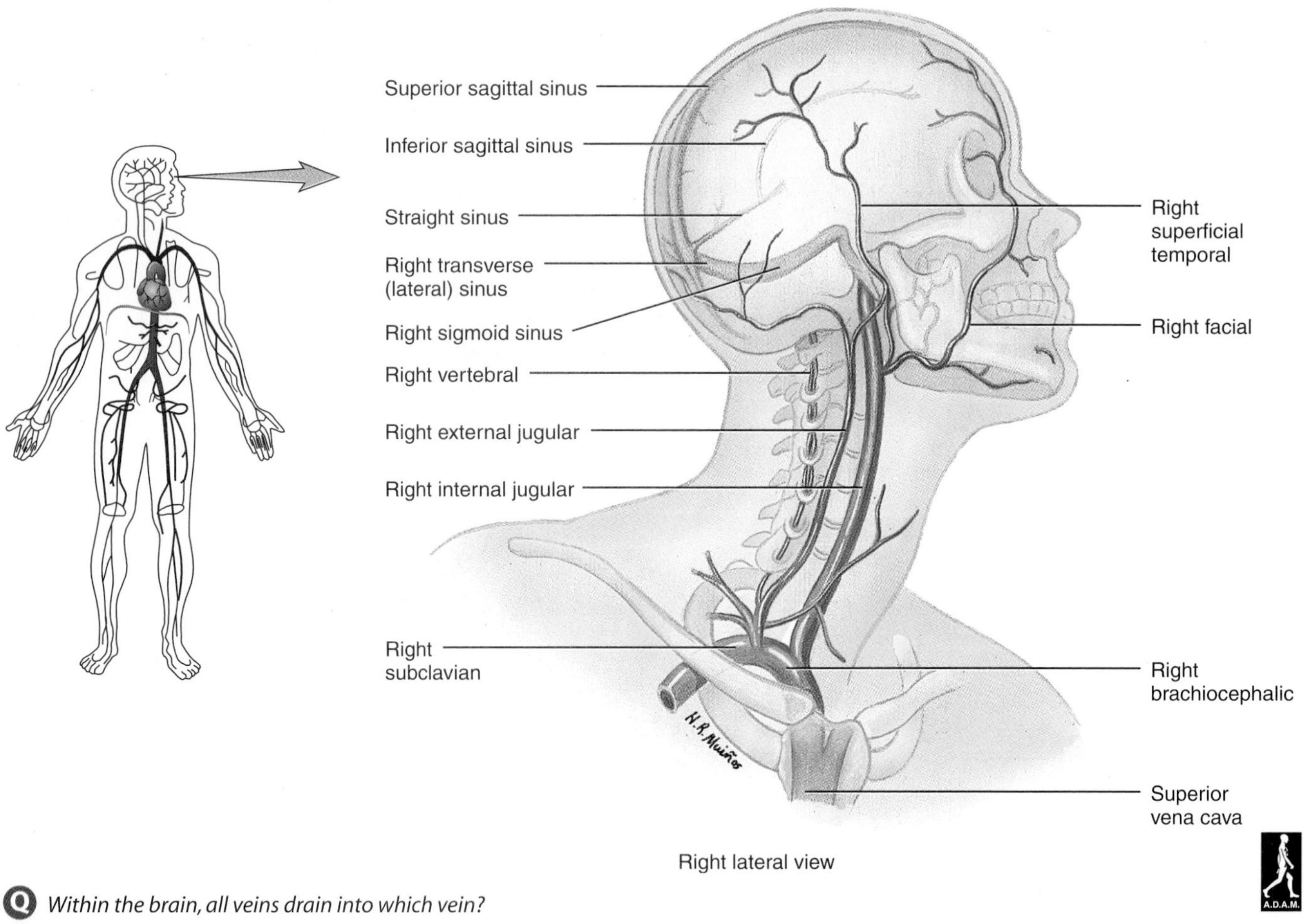

Superior sagittal sinus

Inferior sagittal sinus

Straight sinus

Right transverse (lateral) sinus

Right sigmoid sinus

Right vertebral

Right external jugular

Right internal jugular

Right subclavian

Right superficial temporal

Right facial

Right brachiocephalic

Superior vena cava

Right lateral view

Q *Within the brain, all veins drain into which vein?*

Exhibit 14.9 *Veins of the Upper Limbs (Figure 14.16)*

Blood from the upper limbs is returned to the heart by both superficial and deep veins. Both sets of veins have valves, which are more numerous in the deep veins. **Superficial veins** are located just deep to the skin and are often visible. They anastomose extensively with one another and with deep veins, and they do not accompany arteries. Superficial veins are larger than deep veins and return most of the blood from the upper limbs. **Deep veins** are located deep in the body. They usually accompany arteries and have the same names as the corresponding arteries.

VEIN	DESCRIPTION AND REGION DRAINED
Superficial	
Cephalic (se-FAL-ik; *ceph* = head) **veins**	The principal superficial veins that drain the upper limbs are the cephalic and basilic veins. They originate in the hand and convey blood from the smaller superficial veins into the axillary veins. The **cephalic veins** begin on the lateral aspect of the **dorsal venous** (VĒ-nus) **arches,** networks of veins on the dorsum of the hands formed by the **dorsal metacarpal veins.** These veins, in turn, drain the **dorsal digital veins,** which pass along the sides of the fingers. Following their formation from the dorsal venous arches, the cephalic veins arch around the radial side of the forearms to the anterior surface and ascend through the entire limbs along the anterolateral surface. The cephalic veins end where they join the axillary veins, just inferior to the clavicles. **Accessory cephalic veins** originate either from a venous plexus on the dorsum of the forearms or from the medial aspects of the dorsal venous arches and unite with the cephalic veins just inferior to the elbow. The cephalic veins drain blood from the lateral aspect of the upper limbs.
Basilic (ba-SIL-ik; *basilikos* = royal) **veins**	The **basilic veins** begin on the medial aspects of the dorsal venous arches and ascend along the posteromedial surface of the forearm and anteromedial surface of the arm. They drain blood from the medial aspects of the upper limbs. Anterior to the elbow, the basilic veins are connected to the cephalic veins by the **median cubital** (*cubitus* = elbow) **veins,** which drain the forearm. If veins must be punctured for an injection, transfusion, or removal of a blood sample, the medial cubital veins are preferred. After receiving the median cubital veins, the basilic veins continue ascending until they reach the middle of the arm. There they penetrate the tissues deeply and run alongside the brachial arteries until they join the brachial veins. As the basilic and brachial veins merge in the axillary area, they form the axillary veins.
Median antebrachial (an'-tē-BRĀ-kē-al; *ante* = before, in front of; *brachium* = arm) **veins**	The **median antebrachial veins (median veins of the forearm)** begin in the **palmar venous plexuses,** networks of veins on the palms. The median antebrachial veins ascend anteriorly in the forearms to join the basilic or median cubital veins, sometimes both. They drain the palms and forearms.
Deep	
Radial RĀ-dē-al = radius) **veins**	The paired **radial veins** begin at the **deep venous palmar arches.** These arches drain the **palmar metacarpal veins** in the palms. The radial veins drain the lateral aspects of the forearms and pass alongside the radial arteries. Just inferior to the elbow joint, the radial veins unite with the ulnar veins to form the brachial veins.
Ulnar (UL-nar = ulna) **veins**	The paired **ulnar veins,** which are larger than the radial veins, begin at the **superficial palmar venous arches.** These arches drain the **common palmar digital veins** and the **proper palmar digital veins** in the fingers. The ulnar veins drain the medial aspect of the forearms, pass alongside the ulnar arteries, and join with the radial veins to form the brachial veins.
Brachial (BRĀ-kē-al; *brachium* = arm) **veins**	The paired **brachial veins** accompany the brachial arteries. They drain the forearms, elbow joints, arms, and humerus. They pass superiorly and join with the basilic veins to form the axillary veins.
Axillary (AK-si-ler'-ē; *axilla* = armpit) **veins**	The **axillary veins** ascend to the outer borders of the first ribs, where they become the subclavian veins. The axillary veins receive tributaries that correspond to the branches of the axillary arteries. The axillary veins drain the arms, axillas, and superolateral chest wall.
Subclavian (sub-KLĀ-vē-an; *sub* = under; *clavicula* = clavicle) **veins**	The **subclavian veins** are continuations of the axillary veins that terminate at the sternal end of the clavicles, where they unite with the internal jugular veins to form the brachiocephalic veins. The subclavian veins drain the arms, neck, and thoracic wall. The thoracic duct of the lymphatic system delivers lymph into the left subclavian vein at the junction with the internal jugular. The right lymphatic duct delivers lymph into the right subclavian vein at the corresponding junction (see Figure 15.4).

Exhibit 14.9 *Veins of the Upper Limbs (continued)*

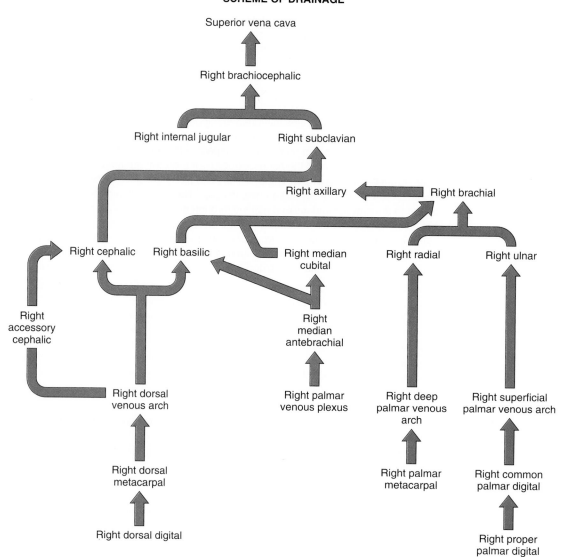

SCHEME OF DRAINAGE

Exhibit 14.9 *(continued)*

Figure 14.16 Principal veins of the right upper limb.

Usually, deep veins accompany arteries that have similar names.

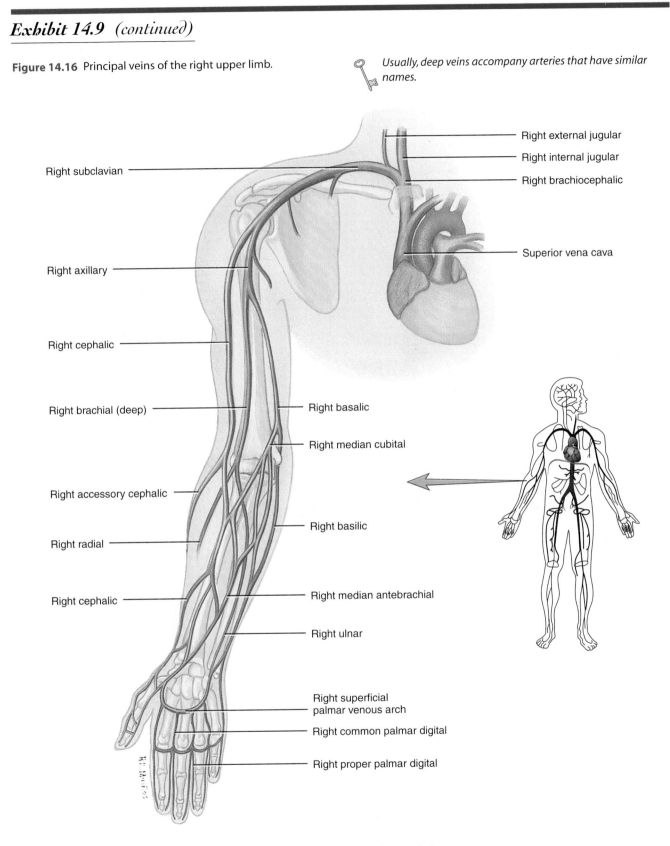

Right external jugular

Right internal jugular

Right brachiocephalic

Right subclavian

Superior vena cava

Right axillary

Right cephalic

Right brachial (deep)

Right basalic

Right median cubital

Right accessory cephalic

Right basilic

Right radial

Right cephalic

Right median antebrachial

Right ulnar

Right superficial palmar venous arch

Right common palmar digital

Right proper palmar digital

(a) Diagram of anterior view of veins of upper limb

Figure continues

Exhibit 14.9 *Veins of the Upper Limbs (continued)*

Figure 14.12 (continued)

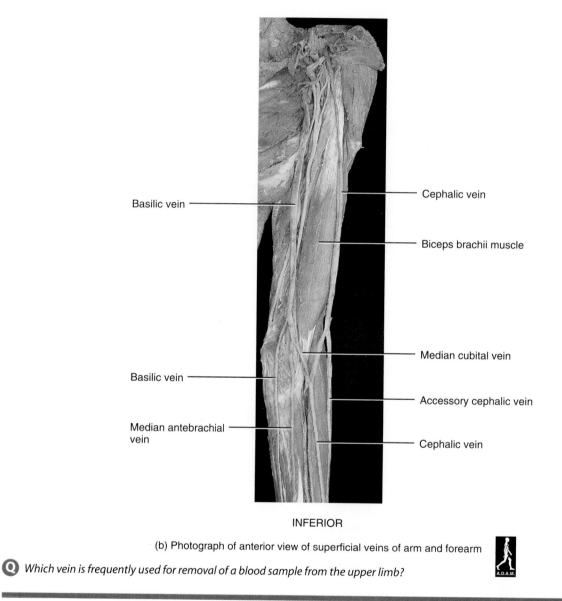

INFERIOR

(b) Photograph of anterior view of superficial veins of arm and forearm

Q *Which vein is frequently used for removal of a blood sample from the upper limb?*

Exhibit 14.10 *Veins of the Thorax (Figure 14.17)*

Although the brachiocephalic veins drain some portions of the thorax, most thoracic structures are drained by a network of veins called the **azygos system**. This is a network of veins on each side of the vertebral column: **azygos, hemiazygos,** and **accessory hemiazygos.** They show considerable variation in origin, course, tributaries, anastomoses, and termination. Ultimately, they empty into the superior vena cava.

VEIN	DESCRIPTION AND REGION DRAINED
Brachiocephalic (brā′-kē-ō-se-FAL-ik; *brachium* = arm; *cephalic* = head) **vein**	The right and left **brachiocephalic veins,** formed by the union of the subclavian veins and internal jugular veins, drain blood from the head, neck, upper limbs, mammary glands, and superior thorax. The brachiocephalic veins unite to form the superior vena cava. Because the superior vena cava is to the right, the left brachiocephalic vein is longer than the right. The right brachiocephalic vein is anterior and to the right of the brachiocephalic trunk. The left brachiocephalic vein is anterior to the brachiocephalic trunk, the left common carotid and left subclavian arteries, the trachea, and the left vagus (X) and phrenic nerves.
Azygos (az-Ī-gos = unpaired) **system**	The **azygos system,** besides collecting blood from the thorax and abdominal wall, may serve as a bypass for the inferior vena cava that drains blood from the lower body. Several small veins directly link the azygos system with the inferior vena cava. Large veins that drain the lower limbs and abdomen pass blood into the azygos system. If the inferior vena cava or hepatic portal vein becomes obstructed, the azygos system can return blood from the lower body to the superior vena cava.
Azygos vein	The **azygos vein** lies anterior to the vertebral column, slightly to the right of the midline. It usually begins at the junction of the right ascending lumbar and right subcostal veins near the diaphragm. At the level of the fourth thoracic vertebra, it arches over the root of the right lung to end in the superior vena cava. Generally, the azygos vein drains the right side of the thoracic wall, thoracic viscera, and abdominal wall. Specifically, the azygos vein receives blood from most of the **right posterior intercostal, hemiazygos, accessory hemiazygos, esophageal, mediastinal, pericardial, and bronchial veins.**
Hemiazygos (HEM-ē-az-Ī-gos; *hemi* = half) **vein**	The **hemiazygos vein** is anterior to the vertebral column and slightly to the left of the midline. It begins at the junction of the left ascending lumbar and left subcostal veins. It terminates by joining the azygos vein at about the level of the ninth thoracic vertebra. Generally, the hemiazygos vein drains the left side of the thoracic wall, thoracic viscera, and abdominal wall. Specifically, the hemiazygos vein receives blood from the ninth through eleventh **left posterior intercostal veins, esophageal, mediastinal,** and sometimes the accessory **hemizygos veins.**
Accessory hemiazygos vein	The **accessory hemiazygos vein** is also anterior to the vertebral column and to the left of the midline. It begins at the fourth or fifth intercostal space and descends from the fifth to the eighth thoracic vertebra. It terminates by joining the azygos vein at about the level of the eighth thoracic vertebra. The accessory hemiazygos vein drains the left side of the thoracic wall. It receives blood from the fourth through eighth **left posterior intercostal veins,** (the first through third left intercostal veins open into the left brachiocephalic vein), **left bronchiol, and mediastinal veins.**

SCHEME OF DRAINAGE

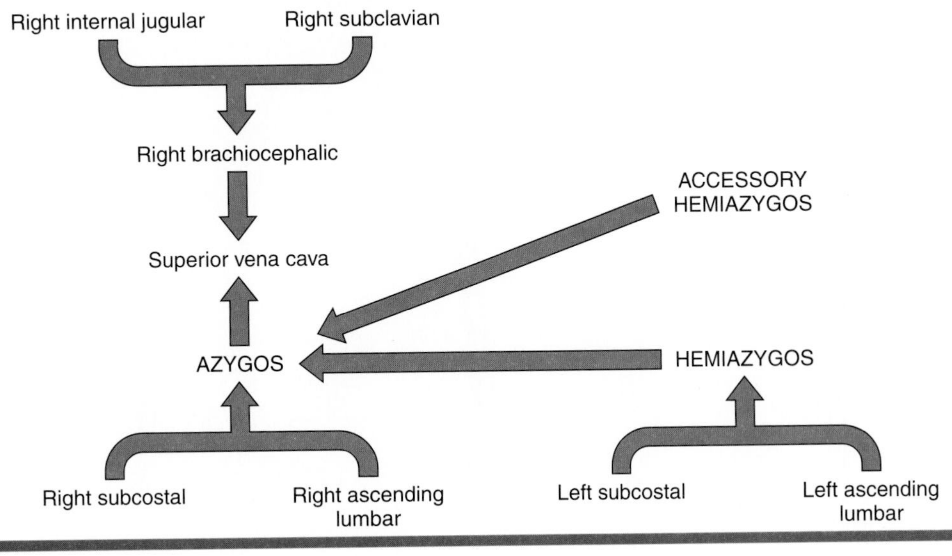

Exhibit 14.10 *Veins of the Thorax (continued)*

Figure 14.17 Principal veins of the thorax, abdomen, and pelvis

Most thoracic structures are drained by the azygos system of veins.

Q *Which vein returns blood from the abdominopelvic viscera to the heart?*

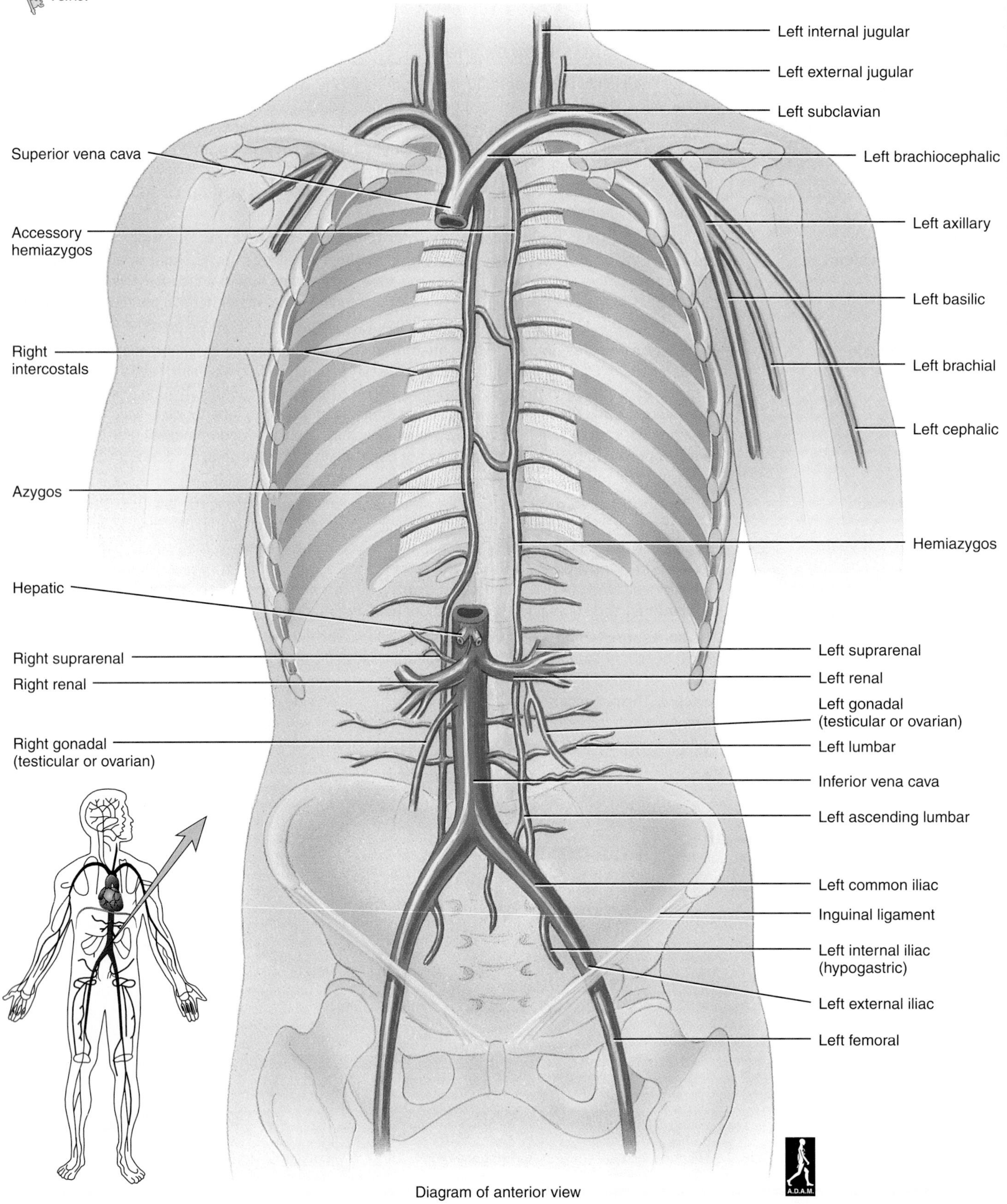

Left internal jugular
Left external jugular
Left subclavian
Left brachiocephalic
Left axillary
Left basilic
Left brachial
Left cephalic
Hemiazygos
Left suprarenal
Left renal
Left gonadal (testicular or ovarian)
Left lumbar
Inferior vena cava
Left ascending lumbar
Left common iliac
Inguinal ligament
Left internal iliac (hypogastric)
Left external iliac
Left femoral

Superior vena cava
Accessory hemiazygos
Right intercostals
Azygos
Hepatic
Right suprarenal
Right renal
Right gonadal (testicular or ovarian)

Diagram of anterior view

Exhibit 14.11 Veins of the Abdomen and Pelvis (Figure 14.18. See also Figure 14.17)

Blood from the abdominopelvic viscera and abdominal wall returns to the heart via the **inferior vena cava.** Many small veins enter the inferior vena cava. Most carry return flow from parietal branches of the abdominal aorta, and their names correspond to the names of the arteries.

The inferior vena cava does not receive veins directly from the gastrointestinal tract, spleen, pancreas, and gallbladder. These

organs pass their blood into a common vein, the **hepatic portal vein,** which delivers the blood to the liver. The hepatic portal vein is formed by the union of the superior mesenteric and splenic veins (see Figure 14.20). This special flow of venous blood is called **hepatic portal circulation,** which is described shortly. After passing through the liver for processing, blood drains into the hepatic veins, which empty into the inferior vena cava.

VEIN	DESCRIPTION AND REGION DRAINED
Inferior vena cava (VĒ-na CA-va; *vena* = vein; *cava* = cavelike)	The **inferior vena cava** is formed by the union of two common iliac veins that drain the lower limbs, pelvis, and abdomen. The inferior vena cava extends superiorly through the abdomen and thorax to the right atrium.
Common iliac (IL-ē-ak; *iliac* = pertaining to the ilium) **veins**	The **common iliac veins** are formed by the union of the internal (hypogastric) and external iliac veins anterior to the sacroiliac joint and represent the distal continuation of the inferior vena cava at their bifurcation. The right common iliac vein is much shorter than the left and is also more vertical. Generally, the common iliac veins drain the pelvis, external genitals, and lower limbs.
Internal iliac veins	The **internal iliac veins** begin near the superior portion of the greater sciatic notch and run medial to their corresponding arteries. Generally, the veins drain the thigh, buttocks, external genitals, and pelvis.
External iliac veins	The **external iliac veins** are companions of the internal iliac arteries and begin at the inguinal ligaments as continuations of the femoral veins. They end anterior to the sacroiliac joint where they join with the internal iliac veins to form the common iliac veins. The external iliac veins drain the lower limbs, cremasteric muscle in males, and the abdominal wall.
Lumbar (LUM-bar; *lumbar* = loin) **veins**	A series of parallel **lumbar veins,** usually four on each side, drain blood from both sides of the posterior abdominal wall, vertebral canal, spinal cord, and meninges (see Figure 14.17). The lumbar veins run horizontally with the lumbar arteries. The lumbar veins connect at right angles with the right and left **ascending lumbar veins,** which form the origin of the corresponding azygos or hemiazygos vein. The lumbar veins drain blood into the ascending lumbars and then run to the inferior vena cava, where they release the remainder of the flow.
Gonadal (gō-NAD-al; *gono* = seed) **[testicular** (tes-TIK-yoo-lor) or **ovarian]** (ō-VA-rē-an) **veins**	The **gonadal veins** ascend with the gonadal arteries along the posterior abdominal wall (see Figure 14.17). In the male, the gonadal veins are called the testicular veins. The **testicular veins** drain the testes (the left testicular vein empties into the left renal vein, and the right testicular vein drains into the inferior vena cava). In the female, the gonadal veins are called the ovarian veins. The **ovarian veins** drain the ovaries (the left ovarian vein empties into the left renal vein, and the right ovarian vein drains into the inferior vena cava).
Renal (RĒ-nal; *renal* = kidney) **veins**	The **renal veins** are large and pass anterior to the renal arteries (see Figure 14.17). The left renal vein is longer than the right renal vein and passes anterior to the abdominal aorta. It receives the left testicular (or ovarian), left inferior phrenic, and usually left suprarenal veins. The right renal vein empties into the inferior vena cava posterior to the duodenum. The renal veins drain the kidneys.
Suprarenal (soo′-pra-RĒ-nal; *supra* = above) **veins**	The **suprarenal veins** drain the adrenal (suprarenal) glands (the left suprarenal vein empties into the left renal vein, and the right suprarenal vein empties into the inferior vena cava). See Figure 14.17.
Inferior phrenic (FREN-ik; *phrenic* = diaphragm) **veins**	The **inferior phrenic veins** drain the diaphragm (the left inferior phrenic vein usually sends one tributary to the left suprarenal vein, which empties into the left renal vein, and another tributary that empties into the inferior vena cava; the right inferior phrenic vein empties into the inferior vena cava). See figure 14.17.
Hepatic (he-PAT-ik; *hepatic* = liver) **veins**	The **hepatic veins** drain the liver. (See Figure 14.17).

Exhibit 14.11 *Veins of the Abdomen and Pelvis continued)*

SCHEME OF DRAINAGE

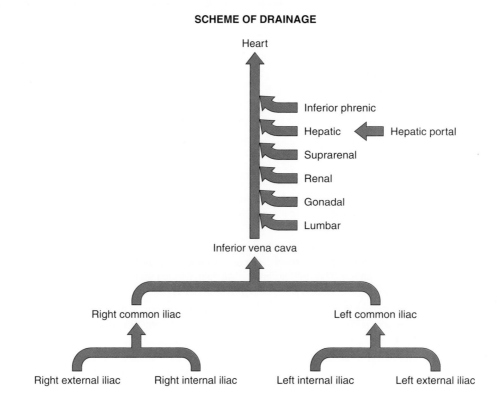

Exhibit 14.11 (continued)

Figure 14.18 Principal veins of the abdomen and pelvis.

Blood from the abdominopelvic viscera and wall returns to the heart via the inferior vena cava.

SUPERIOR

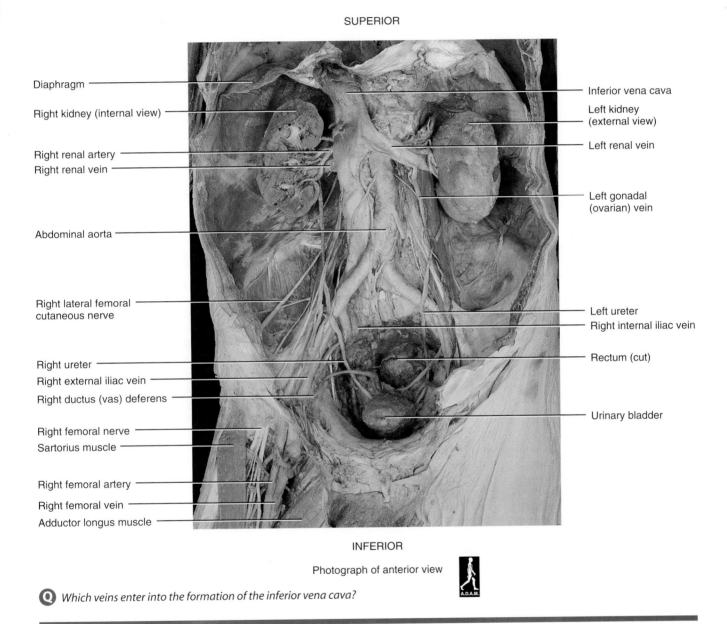

Diaphragm

Right kidney (internal view)

Right renal artery

Right renal vein

Abdominal aorta

Right lateral femoral cutaneous nerve

Right ureter

Right external iliac vein

Right ductus (vas) deferens

Right femoral nerve

Sartorius muscle

Right femoral artery

Right femoral vein

Adductor longus muscle

Inferior vena cava

Left kidney (external view)

Left renal vein

Left gonadal (ovarian) vein

Left ureter

Right internal iliac vein

Rectum (cut)

Urinary bladder

INFERIOR

Photograph of anterior view

Q *Which veins enter into the formation of the inferior vena cava?*

Exhibit 14.12 *Veins of the Lower Limbs (Figure 14.19)*

As with the upper limbs, blood from the lower limbs is drained by both **superficial** and **deep veins.** The superficial veins often anastomose with each other and with deep veins along their length. Deep veins, for the most part, have the same names as corresponding arteries. All veins of the lower limbs have valves, which are more numerous than in veins of the upper limbs.

VEIN	DESCRIPTION AND REGION DRAINED
Superficial Veins	
Great saphenous (sa-FĒ-nus; *saphenes* = clearly visible) **veins**	The **great (long) saphenous veins,** the longest veins in the body, ascend from the foot to the groin in the subcutaneous layer. They begin at the medial end of the dorsal venous arches of the foot. The **dorsal venous** (VĒ-nus) **arches** are networks of veins on the dorsum of the foot formed by the **dorsal digital veins,** which collect blood from the toes, and then unite in pairs to form the **dorsal metatarsal veins,** which parallel the metatarsals. As the dorsal metatarsal veins approach the foot, they combine to form the dorsal venous arches. The great saphenous veins pass anterior to the medial malleolus of the tibia and then superiorly along the medial aspect of the leg and thigh just deep to the skin. They receive tributaries from superficial tissues and connect with the deep veins as well. They empty into the femoral veins at the groin. Generally, the great saphenous veins drain mainly the medial side of the leg and thigh, the groin, external genitals, and abdominal wall.
	Along their length, the great saphenous veins have from 10 to 20 valves, with more located in the leg than the thigh. These veins are more likely to be subject to varicosities than other veins in the lower limbs because they must support a long column of blood and are not well supported by skeletal muscles.
	The great saphenous veins are often used for prolonged administration of intravenous fluids. This is particularly important in very young children and in patients of any age who are in shock and whose veins are collapsed. The great saphenous veins are also often used as a source of vascular grafts, especially for coronary bypass surgery. In the procedure, the vein is removed and then reversed so that the valves do not obstruct the flow of blood.
Small saphenous veins	The **small (short) saphenous veins** begin at the lateral aspect of the dorsal venous arches of the foot. They pass posterior to the lateral malleolus of the fibula and ascend deep to the skin along the posterior aspect of the leg. They empty into the popliteal veins in the popliteal fossa, posterior to the knee. Along their length, the small saphenous veins have from 9 to 12 valves. The small saphenous veins drain the foot and posterior aspect of the leg. They may communicate with the great saphenous veins in the proximal thigh.

Exhibit 14.12 *(continued)*

VEIN	DESCRIPTION AND REGION DRAINED
Deep Veins	
Posterior tibial (TIB-ē-al) **veins**	The **plantar digital veins** on the plantar surfaces of the toes unite to form the **plantar metatarsal veins,** which parallel the metatarsals. They unite to form the **deep plantar venous arches.** From each arch emerges the **medial** and **lateral plantar veins.**
	The paired **posterior tibial veins,** which sometimes unite to form a single vessel, are formed by the medial and lateral plantar veins, posterior to the medial malleolus of the tibia. They accompany the posterior tibial artery through the leg. They ascend deep to the muscles in the posterior aspect of the leg and drain the foot and posterior compartment muscles. About two-thirds the way up the leg, the posterior tibial veins drain blood from the **peroneal** (per′-ō-NĒ-al; *perone* =fibula) **veins,** which drain the lateral and posterior leg muscles. The posterior tibial veins unite with the anterior tibial veins just inferior to the popliteal fossa to form the popliteal veins.
Anterior tibial veins	The **paired anterior tibial veins** arise in the dorsal venous arch and accompany the anterior tibial artery. They ascend in the interosseous membrane between the tibia and fibula and unite with the posterior tibial veins to form the popliteal vein. The anterior tibial veins drain the ankle joint, knee joint, tibiofibular joint, and anterior portion of leg.
Popliteal (pop′-li-TĒ-al; *popliteus* = hollow behind knee) **veins**	The **popliteal veins** are formed by the union of the anterior and posterior tibial veins. The popliteal veins also receive blood from the small saphenous veins and tributaries that correspond to branches of the popliteal artery. The popliteal veins drain the knee joint and the skin, muscles, and bones of portions of the calf and thigh around the knee joint.
Femoral (FEM-o-ral) **veins**	The **femoral veins** accompany the femoral arteries and are the continuations of the popliteal veins just superior to the knee. The femoral veins extend up the posterior surface of the thighs and drain the muscles of the thighs, femurs, external genitals, and superficial lymph nodes. The largest tributaries of the femoral veins are the **deep femoral veins.** Just before penetrating the abdominal wall, the femoral veins receive the deep femoral veins and the great saphenous veins. The veins formed from their union penetrate the body wall and enter the pelvic cavity. Here they are known as the **external iliac veins.**

Figure 14.22 Fetal circulation and changes at birth. The boxes indicate the fate of certain fetal structures once postnatal circulation is established.

The lungs, kidneys, and gastrointestinal organs begin to function at birth.

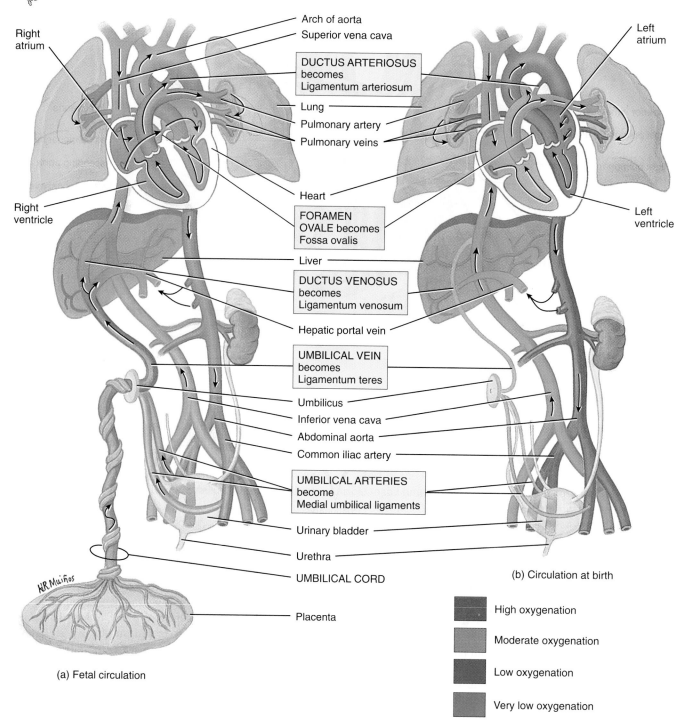

Arch of aorta
Superior vena cava

DUCTUS ARTERIOSUS
becomes
Ligamentum arteriosum

Lung
Pulmonary artery
Pulmonary veins

Right atrium

Left atrium

Heart

Right ventricle

FORAMEN
OVALE becomes
Fossa ovalis

Left ventricle

Liver

DUCTUS VENOSUS
becomes
Ligamentum venosum

Hepatic portal vein

UMBILICAL VEIN
becomes
Ligamentum teres

Umbilicus
Inferior vena cava
Abdominal aorta
Common iliac artery

UMBILICAL ARTERIES
become
Medial umbilical ligaments

Urinary bladder

Urethra

UMBILICAL CORD

Placenta

(a) Fetal circulation

(b) Circulation at birth

High oxygenation

Moderate oxygenation

Low oxygenation

Very low oxygenation

H.R. Muiños

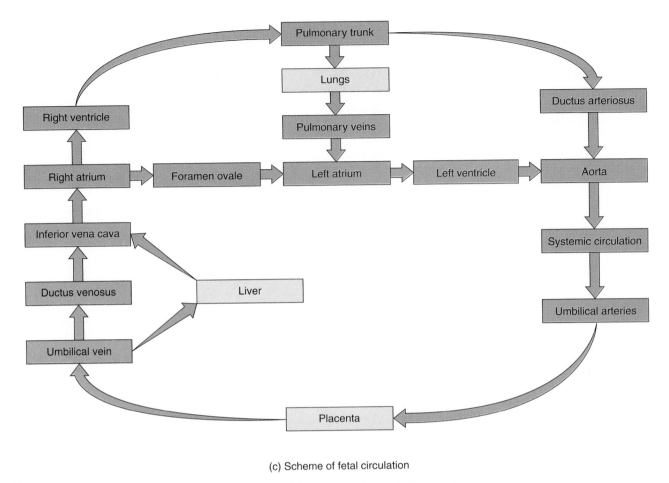

(c) Scheme of fetal circulation

Q *Through which structure does the exchange of materials between mother and fetus occur?*

At birth, when pulmonary (lung), renal, and digestive functions are established, the following vascular changes occur (Figure 14.22b):

1. When the umbilical cord is tied off, no blood flows through the umbilical arteries, they fill with connective tissue, and the distal portions of the umbilical arteries become the **medial umbilical ligaments.**

2. Tying off the umbilical cord results in the conversion of the umbilical vein into the **ligamentum teres (round ligament),** a structure that attaches the umbilicus to the liver.

3. The placenta is expelled from the mother's birth canal as the so-called **afterbirth.**

4. The ductus venosus collapses as blood stops flowing through the umbilical vein and becomes the **ligamentum venosum,** a fibrous cord on the inferior surface of the liver.

5. The foramen ovale normally closes shortly after birth to become the **fossa ovalis,** a depression in the interatrial septum. When an infant takes its first breath, the lungs expand and blood flow to the lungs increases. Blood returning from the lungs to the heart increases pressure in the left atrium. This closes the foramen ovale by pushing the valve that guards it against the interatrial septum.

6. The ductus arteriosus closes by vasoconstriction, atrophies, and becomes the **ligamentum arteriosum.**

Developmental Anatomy of Blood Vessels and Blood

The human yolk sac has little yolk to nourish the developing embryo, and blood vessel and red blood cell formation starts as early as 15 to 16 days. The development begins in the **mesoderm** of the yolk sac, chorion, and body stalk.

Blood vessels develop from isolated masses and cords of mesenchyme in the mesoderm called **blood islands** (Figure 14.23). Spaces soon appear in the islands and become the lumens of the blood vessels. Some of the mesenchymal cells immediately around the spaces give rise to the *endothelial lining of the blood vessels.* Mesenchyme around the endothelium forms the *tunics* (intima, media, externa) of the larger blood vessels. Growth and fusion of blood islands form an extensive network of blood vessels throughout the embryo.

Blood plasma and *blood cells* are produced by the endothelial cells and appear in the blood vessels of the yolk sac and allantois quite early. Blood formation in the embryo itself begins at about the second month in the liver and spleen, a little later in red bone marrow, and much later in lymph nodes. ■

Aging and the Cardiovascular System

General changes associated with aging and the cardiovascular system include loss of compliance (extensibility) of the aorta, reduction in cardiac muscle fiber size, progressive loss of cardiac muscular strength, reduced cardiac output, a decline in maximum heart rate, and an increase in blood pressure. Total blood cholesterol tends to increase with age. The heart loses about 1% of its reserve pumping capacity each year after age 30. There is an increase in the incidence of coronary artery disease (CAD), the major cause of heart disease and death in older Americans. Congestive heart failure (CHF), a set of symptoms associated with impaired pumping of the heart, also occurs. Changes in blood vessels that serve brain tissue, such as atherosclerosis, reduce nourishment to the brain and result in the malfunction or death of brain cells. By age 80, cerebral blood flow is 20% less and renal blood flow is 50% less than in the same person at age 30.

Figure 14.23 Development of blood vessels and blood cells from blood islands.

Blood vessel development begins in the embryo on about the 15th or 16th day.

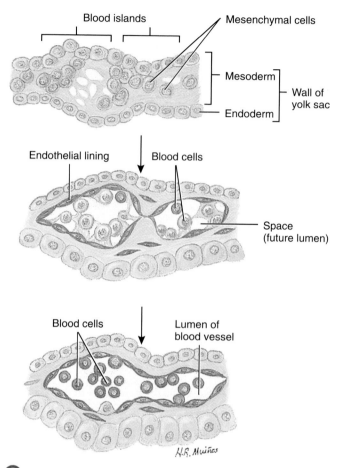

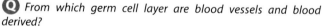

Q *From which germ cell layer are blood vessels and blood derived?*

Key Medical Terms
Associated with Blood Vessels

Angiogenesis (an′-jē-ō-JEN-e-sis) Formation of new blood vessels.

Aortography (ā′-or-TOG-ra-fē) X-ray examination of the aorta and its main branches after injection of radiopaque dye.

Arteritis (ar′-te-RĪ-tis; *itis* = inflammation of) Inflammation of an artery, probably due to an autoimmune response.

Carotid endarterectomy (ka-ROT-id end′-ar-ter-EK-tō-mē) The removal of atherosclerotic plaque from the carotid artery to restore greater blood flow to the brain.

Claudication (klaw′-di-KĀ-shun) Pain and lameness or limping caused by defective circulation of the blood in the vessels of the limbs.

Hypotension (hī′-pō-TEN-shun; *hypo* = below; *tension* = pressure) Low blood pressure; most commonly used to describe an acute drop in blood pressure, as occurs during excessive blood loss.

Normotensive (nor′-mō-TEN-siv) Characterized by normal blood pressure.

Occlusion (o-KLOO-shun) The closure or obstruction of the lumen of a structure such as a blood vessel. An example is an atherosclerotic plaque in an artery.

Orthostatic (or′-thō-STAT-ik) **hypotension** (*ortho* = straight; *statikos* = causing to stand) An excessive lowering of systemic blood pressure with the assumption of an erect or semierect posture; it is usually a sign of a disease. May be caused by excessive fluid loss, certain drugs (antihypertensives), and cardiovascular or neurogenic factors. Also called *postural hypotension*.

Phlebitis (fle-BĪ-tis; *phleb* = vein) Inflammation of a vein, often in a leg.

Raynaud's (rā-NOZ) **disease** A vascular disorder, primarily of females, characterized by bilateral attacks of ischemia, usually of the fingers and toes, in which the skin becomes pale and exhibits burning and pain. It is brought on by cold temperatures or emotional stimuli.

Thrombectomy (throm-BEK-tō-mē; *thrombo* = clot) An operation to remove a blood clot from a blood vessel.

Thrombophlebitis (throm′-bō-fle-BĪ-tis) Inflammation of a vein with clot formation. Superficial thrombophlebitis occurs in veins under the skin, especially the calf.

White coat (office) hypertension A stress-induced syndrome found in patients who have elevated blood pressure while being examined by healthcare personnel, but otherwise have normal blood pressure.

Study Outline

Anatomy of Blood Vessels (p. 417)

Arteries (p. 417)
1. Arteries carry blood away from the heart. The wall of an artery consists of a tunica interna, tunica media (which maintains elasticity and contractility), and tunica externa.
2. Large arteries are referred to as elastic (conducting) arteries, and medium-sized arteries are called muscular (distributing) arteries.
3. Many arteries anastomose: the distal ends of two or more vessels unite. An alternate blood route from an anastomosis is called collateral circulation. Arteries that do not anastomose are called end arteries.

Arterioles (p. 419)
1. Arterioles are small arteries that deliver blood to capillaries.
2. Through constriction and dilation, they assume a key role in regulating blood flow from arteries into capillaries and in altering arterial blood pressure.

Capillaries (p. 419)
1. Capillaries are microscopic blood vessels through which materials are exchanged between blood and tissue cells; some capillaries are continuous, whereas others are fenestrated.
2. Capillaries branch to form an extensive capillary network throughout the tissue. This network increases the surface area, allowing a rapid exchange of large quantities of materials.
3. Precapillary sphincters regulate blood flow through capillaries.
4. Microscopic blood vessels in the liver are called sinusoids.

Venules (p. 421)
1. Venules are small vessels that continue from capillaries and merge to form veins.
2. They drain blood from capillaries into veins.

Veins (p. 421)
1. Veins consist of the same three tunics as arteries but have a thinner tunica interna and media. The lumen of a vein is also larger than that of an accompanying artery.
2. They contain valves to prevent backflow of blood.
3. Weak valves can lead to varicose veins.
4. Vascular (venous) sinuses are veins with very thin walls.

Blood Distribution (p. 422)
1. Systemic veins are collectively called blood reservoirs.
2. They store blood, which through vasoconstriction can move to other parts of the body, if the need arises.
3. The principal reservoirs are the veins of the abdominal organs (liver and spleen) and skin.

Circulatory Routes (p. 423)
1. The two basic postnatal circulatory routes are the systemic and pulmonary circulations.
2. Among the subdivisions of the systemic circulation are coronary (cardiac), cerebral, and hepatic portal circulation.
3. Fetal circulation exists only in the fetus.

Systemic Circulation (p. 423)
1. The systemic circulation takes oxygenated blood from the left ventricle through the aorta to all parts of the body, including some lung tissue (but does *not* supply the air sacs or alveoli of the lungs) and returns the deoxygenated blood to the right atrium.
2. The aorta is divided into the ascending aorta, the arch of the aorta, and the descending aorta. Each section gives off arteries that branch to supply the whole body.
3. Blood returns to the heart through the systemic veins. All the veins of the systemic circulation flow into the superior or inferior venae cavae or the coronary sinus, which in turn empty into the right atrium.
4. The principal blood vessels of systemic circulation may be reviewed in Exhibits 14.1–14.12.

Hepatic Portal Circulation (p. 466)
1. The hepatic portal circulation detours venous blood from the gastrointestinal organs and spleen and directs it into the hepatic portal vein of the liver before it is returned to the heart. This circulation enables the liver to utilize nutrients and detoxify harmful substances in the blood.

Pulmonary Circulation (p. 466)
1. The pulmonary circulation takes deoxygenated blood from the right ventricle to the air sacs (alveoli) of the lungs and returns

oxygenated blood from the air sacs of the lungs to the left atrium.

2. It allows blood to be oxygenated for systemic circulation.

Fetal Circulation (p. 466)

1. The fetal circulation involves the exchange of materials by diffusion between fetus and mother.

2. The fetus derives its oxygen and nutrients and eliminates its carbon dioxide and wastes through the maternal blood supply by means of a structure called the placenta.

3. At birth, when pulmonary (lung), digestive, and liver functions are established, the special structures of fetal circulation are no longer needed.

Developmental Anatomy of Blood Vessels and Blood (p. 471)

1. Blood vessels develop from isolated masses of mesenchyme in mesoderm called blood islands.

2. Blood is produced by the endothelium of blood vessels.

Aging and the Cardiovascular System (p. 472)

1. General changes include loss of elasticity of blood vessels, reduction in cardiac muscle size, reduced cardiac output, and increased blood pressure.

2. The incidence of coronary artery disease (CAD), congestive heart failure (CHF), and atherosclerosis increases with age.

Review Questions

1. Describe the structural and functional differences among arteries, arterioles, capillaries, venules, and veins. (p. 417)

2. Discuss the importance of the elasticity and contractility of arteries. (p.417)

3. Distinguish between elastic (conducting) and muscular (distributing) arteries in terms of location, histology, and function. What is an anastomosis? What is collateral circulation? (p. 417)

4. Describe how capillaries are structurally adapted for exchanging materials between blood and body cells. (p. 419)

5. Define varicose veins. (p. 421)

6. What are blood reservoirs? Why are they important? (p. 422)

7. What is meant by a circulatory route? Define systemic circulation. (p. 423)

8. Diagram the major divisions of the aorta, their principal arterial branches, and the regions supplied. (p. 423)

9. Trace a drop of blood from the arch of the aorta through its systemic circulatory route to the tip of the big toe on your left foot and back to the heart again. Remember that the major branches of the arch are the brachiocephalic artery, left common carotid artery, and left subclavian artery. Be sure to also indicate which veins return the blood to the heart. (pp. 429–463)

10. What is the cerebral arterial circle (circle of Willis)? Why is it important? (p. 430)

11. Distinguish between visceral and parietal branches of an artery. What major organs are supplied by branches of the thoracic aorta? How is blood returned from these organs to the heart? (p. 435)

12. What organs are supplied by the celiac, superior mesenteric, renal, inferior mesenteric, inferior phrenic, and median sacral arteries? How is blood returned to the heart? (p. 437)

13. Trace a drop of blood from the brachiocephalic artery into the digits of the right upper limb and back again to the right atrium. (p. 429)

14. What are the three major groups of systemic veins? (p. 448)

15. What is hepatic portal circulation? Trace the route by means of a diagram. Why is this route significant? (p. 466)

16. Define pulmonary circulation. Prepare a diagram to indicate the route. What is the purpose of the route? (p. 466)

17. Discuss in detail the anatomy and physiology of fetal circulation. Be sure to indicate the function of the umbilical arteries, umbilical vein, ductus venosus, foramen ovale, and ductus arteriosus. (p. 466)

18. Describe the development of blood vessels and blood. (p. 471)

19. Describe the effects of aging on the cardiovascular system. (p. 472)

20. Refer to the glossary of key medical terms associated with blood vessels. Be sure that you can define each term. (p. 472)

Self Quiz

Complete the following:

1. The major artery from which all systemic arteries branch is the ___, which exits from the ___ ventricle of the heart.

2. The abdominal aorta ends at about the level of the ___ vertebra, where it divides into right and left ___ arteries.

3. Blood from gastrointestinal organs and the spleen is transported to the liver via the ___ vein.

4. The union of branches of two or more arteries supplying the same body region is called a(n) ___.

5. Most of the smooth muscle of arteries is in the tunica ___.

6. Decrease in the size of the lumen of a blood vessel by contraction of smooth muscle is called ___.
7. The three branches of the celiac trunk are the ___ , ___, and ___.
8. Blood from all the dural venous sinuses eventually drains into the ___ veins, which descend through the neck.
9. Match the following terms with their definitions:

(1) left external iliac artery	___ (a) travels through cervical transverse foramina
(2) esophageal artery	___ (b) continues as the left axillary artery
(3) cephalic vein	
(4) sigmoid sinus	___ (c) gives rise to the left common carotid artery
(5) coronary sinus	
(6) posterior tibial vein	___ (d) is a branch of the thoracic aorta
(7) left subclavian artery	
(8) arch of the aorta	___ (e) abdominal vessel that branches to supply blood to the pancreas, small intestine, and part of the large intestine
(9) great saphenous vein	
(10) vertebral artery	
(11) superior mesenteric artery	___ (f) continues as the left femoral artery
	___ (g) receives venous blood from the myocardium
	___ (h) is located in the arm
	___ (i) is the longest vein in the body
	___ (j) is completely located in the leg
	___ (k) is completely located in the head

Are the following statements true or false?

10. The wall of the femoral artery is thicker than the wall of the femoral vein.
11. End arteries are vessels that anastomose.
12. In order for blood to flow from the left brachial vein to the right arm, it must pass through the heart and capillaries in the lungs.
13. Elastic arteries, such as the aorta and common carotids, have walls that are thin in proportion to their diameters.
14. Fenestrated capillaries are found in skeletal and smooth muscle.

Choose the one best answer to the following questions:

15. A small vessel connecting the pulmonary trunk with the aorta and bypassing the fetal lungs is the
    a. foramen ovale
    b. ductus venosus
    c. ductus arteriosus
    d. fossa ovalis
    e. vasa vasorum
16. Which of the following statements best describes arteries?
    a. All carry oxygenated blood to the heart.
    b. All contain valves to prevent the backflow of blood.
    c. All carry blood away from the heart.
    d. Only large arteries are lined with endothelium.
    e. All branch from the descending aorta.

17. Which of the following statements is true of veins?
    a. Their tunica media is thinner than that in arteries.
    b. Their tunica externa is thicker than that in arteries.
    c. Most veins in the limbs have valves.
    d. Veins always carry deoxygenated blood.
    e. All veins ultimately empty into the inferior vena cava.
18. All of the following are vessels in the leg or foot *except* the
    a. saphenous vein
    b. azygos vein
    c. peroneal artery
    d. dorsalis pedis artery
    e. popliteal artery
19. Which of the following is a muscular (distributing) artery?
    a. radial artery
    b. vertebral artery
    c. axillary artery
    d. common iliac artery
    e. aorta
20. In fetal circulation, blood passes from the right atrium to the left atrium through the
    a. ductus venosus
    b. ductus arteriosus
    c. fossa ovalis
    d. coronary sulcus
    e. none of the above
21. Which of the following groups of three vessels is involved in pulmonary circulation?
    a. superior vena cava, right atrium, and left ventricle
    b. inferior vena cava, right atrium, and left ventricle
    c. right ventricle, pulmonary trunk, and left atrium
    d. left ventricle, aorta, and inferior vena cava
    e. superior vena cava, right atrium, and right ventricle
22. For blood to flow from the left brachial veins to the left brachial artery, it must pass through
    a. the heart
    b. the lungs
    c. a brachial capillary
    d. the superior vena cava
    e. all of the above
23. Which of the following vessels carry blood into the cerebral arterial circle (circle of Willis)?
    (1) internal carotid artery
    (2) external carotid artery
    (3) posterior cerebral artery
    (4) anterior cerebral artery
    (5) basilar artery
    a. 3 and 4
    b. 1 and 5
    c. 1, 2, and 5
    d. 2, 3, and 4
    e. 1, 4, and 5
24. Put the following vessels in the correct order to trace the route of a drop of blood moving from the right side of the heart to the left side of the heart.
    (1) pulmonary vein
    (2) pulmonary artery

(3) arterioles
(4) venules
(5) capillaries
    a. 1, 4, 5, 3, 2
    b. 1, 3, 5, 4, 2
    c. 2, 4, 5, 3, 1
    d. 2, 5, 4, 1, 3
    e. 2, 3, 5, 4, 1

25. Blood is supplied to the pelvic viscera by way of the
    a. inferior vena cava
    b. superior mesenteric artery
    c. external iliac artery
    d. internal iliac artery
    e. femoral artery

26. Put the following vessels in the correct order to trace the route of a drop of blood moving from the small intestine to the heart.
    (1) hepatic portal vein
    (2) hepatic vein
    (3) inferior vena cava
    (4) superior mesenteric vein
    (5) small vessels within the liver
    a. 4, 1, 5, 2, 3
    b. 3, 1, 2, 5, 4
    c. 1, 4, 5, 3, 2
    d. 4, 2, 1, 5, 3
    e. 4, 3, 2, 5, 1

Critical Thinking Questions

1. In Chapter 13 you learned that the right ventricle is smaller than the left. Relate the difference in ventricular size to the differences between pulmonary and systemic circulation.
HINT: *Greater ventricular size imparts a more forceful "push" to the blood.*

2. Which structures in the fetal circulation are absent in an adult's circulatory system? Why do these changes occur?
HINT: *A mother may want to do everything for her newborn, but she can't breathe for her baby.*

3. Use the cerebral arterial circle (circle of Willis) to illustrate the structure and function of an anastomosis.
HINT: *Several arteries converge to form this structure.*

4. Geneeva was studying blood vessels for the next anatomy exam when she realized that the arterial circulation was not entirely bilaterally symmetrical. Name the vessels that are unpaired.
HINT: *Trace the aorta until it splits at the iliac level.*

5. Khalil spent the week before the anatomy exam lying on a beach studying the back of his eyelids. "What was that question about sinuses doing on the cardiovascular test?" he complained afterwards. "We covered that when we studied bones!" See if you can enlighten him.
HINT: *In bone, sinuses are filled with air; these sinuses aren't.*

6. You've read about varicose veins. Why aren't there varicose arteries?
HINT: *What do veins have that arteries don't?*

Answers to Figure Questions

14.1 The artery, the vein.
14.2 Contraction of smooth muscle in the tunica media.
14.3 They use oxygen and produce wastes more rapidly than inactive tissues.
14.4 Through intercellular clefts, via pinocytic vesicles, directly across endothelial membranes, and through fenestrations.
14.5 Due to the effect of gravity when you are standing, blood tends to pool in your leg veins. The valves prevent backflow as the blood is pushed back toward the right atrium after each beat of the left ventricle. Gravity aids the flow of blood in neck veins back toward the heart when you are erect.
14.6 Volume in veins is about 60% $\times$ 5 liters = 3 liters; volume in capillaries is about 5% $\times$ 5 liters = 250 ml.
14.7 Systemic and pulmonary circulations.
14.8 Ascending aorta, arch of the aorta, thoracic aorta, and abdominal aorta.

14.9 Left and right coronary arteries.
14.10 Brachiocephalic trunk, left common carotid artery, and left subclavian artery.
14.11 At the level of the intervertebral disc between T4 and T5.
14.12 At the aortic hiatus in the diaphragm.
14.13 About the level of L4.
14.14 Regions above the diaphragm.
14.15 Internal jugular vein.
14.16 Median cubital vein.
14.17 Inferior vena cava.
14.18 Left and right common iliac veins.
14.19 Dorsal venous arch, dorsal metatarsal, dorsal digital, great saphenous, and small saphenous veins.
14.20 Hepatic veins.
14.21 Pulmonary arteries.
14.22 Placenta.
14.23 Mesoderm

Chapter 15

The Lymphatic System

Student Objectives

1. Describe the components of the lymphatic system, and list their functions.

2. Describe the structure and origin of lymphatic vessels, and contrast them with veins.

3. Trace the general plan of lymph circulation from lymphatic vessels into the thoracic duct or right lymphatic duct.

4. Describe the structure and function of the thymus gland, lymph nodes, spleen, and lymphatic nodules.

5. Describe the principal lymph nodes of the head and neck, limbs, and trunk, their location, and the areas they drain.

6. Describe the development of the lymphatic system.

7. Describe the effects of aging on the lymphatic system.

of the viscera generally follow arteries, forming plexuses (networks) around them.

Lymphatic Capillaries

Lymphatic capillaries are found throughout the body, except in avascular tissues (e.g., cartilage, the epidermis, and the cornea of the eye), the central nervous system, portions of the spleen, and red bone marrow. Lymphatic capillaries have a slightly larger diameter than blood capillaries and have a unique structure that permits interstitial fluid to flow into them but not out. The ends of endothelial cells that make up the wall of a lymphatic capillary overlap. When pressure is greater in the interstitial fluid than in lymph, the cells separate slightly, like a one-way valve opening, and fluid enters the lymphatic capillary. When pressure is greater inside the lymphatic capillary, the cells adhere more closely so lymph cannot flow back into interstitial fluid.

At right angles to the lymphatic capillary are structures called **anchoring filaments** that attach lymphatic endothelial cells to surrounding tissues (Figure 15.2b). During edema (described shortly), excess interstitial fluid accumulates and causes tissue swelling. This swelling pulls on the anchoring filaments, making the openings between cells even larger so that more fluid can flow into the lymphatic capillary.

As you will see in Chapter 24, the lining of the small intestine contains finger-like projections called **villi.** Each contains blood capillaries and a specialized lymphatic capillary called a **lacteal** (LAK-tē-al; *lacteus* = milky). Lacteals transport digested triglycerides (fats) from the small intestine into lymphatic vessels and ultimately into blood. By following this route, most digested triglycerides bypass the liver. Because of the presence of triglycerides in the lymph, the lymph draining the small intestine is creamy white and is referred to as **chyle** (KĪL; *chylus* = juice). In other parts of the body, lymph is a clean, transparent, sometimes faintly yellow fluid.

Formation and Flow of Lymph

Most components of blood plasma freely move through the capillary walls to form interstitial fluid. More fluid moves out of blood capillaries than returns to them. The excess fluid, about 3 liters per day, drains into lymphatic vessels and becomes lymph. Ultimately, lymph drains into venous blood through the right lymphatic duct and thoracic duct (left lymphatic duct) at the junction of the internal jugular and subclavian veins (see Figure 15.4). Thus the sequence of fluid flow (Figure 15.3) is: arteries (blood) → blood capillaries (blood) → interstitial spaces (interstitial fluid) → lymphatic capillaries (lymph) → lymphatic vessels (lymph) → lymphatic ducts (lymph) → subclavian veins (blood).

The **lymphatic** (lim-FAT-ik) **system** (Figure 15.1) consists of a fluid called lymph flowing within lymphatic vessels (lymphatics), several structures and organs that contain lymphatic tissue, and red bone marrow, which is the site of lymphocyte production. Interstitial (tissue) fluid and lymph are basically the same. The major difference between the two is location. After fluid passes from interstitial spaces into lymphatic vessels, it is called **lymph** (*lympha* = clear water). Lymphatic tissue is a specialized form of reticular connective tissue that contains large numbers of lymphocytes.

The lymphatic system has several functions:

1. **Draining interstitial fluid.** Lymphatic vessels drain tissue spaces of excess interstitial fluid.

2. **Transporting dietary lipids.** Lymphatic vessels carry to the blood the lipids and lipid-soluble vitamins absorbed by the gastrointestinal tract.

3. **Protecting against invasion.** Lymphatic tissue carries out **immune responses.** These are highly specific responses targeted to particular invaders or abnormal cells. Lymphocytes and macrophages, previously discussed in Chapter 12, recognize foreign cells and substances, microbes (bacteria, viruses, and so on), and cancer cells and respond to them in two basic ways. Some lymphocytes (T cells) destroy the intruders directly or indirectly by releasing cytotoxic (cell-killing) substances. Other lymphocytes (B cells) differentiate into plasma cells that secrete antibodies. These are proteins that combine with and cause destruction of specific foreign substances. In carrying out specific immune responses, the lymphatic system concentrates foreign substances in certain lymphatic organs, circulates lymphocytes through the organs to make contact with the foreign substances, and destroys the foreign substances and eliminates them from the body.

LYMPHATIC VESSELS AND LYMPH CIRCULATION

Lymphatic vessels begin as closed-ended vessels called **lymphatic capillaries** in spaces between cells (Figure 15.2a). Just as blood capillaries converge to form venules and veins, lymphatic capillaries unite to form larger tubes called **lymphatic vessels** (see Figure 15.1). Lymphatic vessels resemble veins in structure but have thinner walls and more valves. At intervals along the lymphatic vessels, lymph flows through lymphatic tissue structures called **lymph nodes.** In the skin, lymphatic vessels lie in subcutaneous tissue and generally follow veins. Lymphatic vessels

Figure 15.1 The lymphatic system.

The lymphatic system consists of lymph, lymphatic vessels, lymphatic tissue, and red bone marrow.

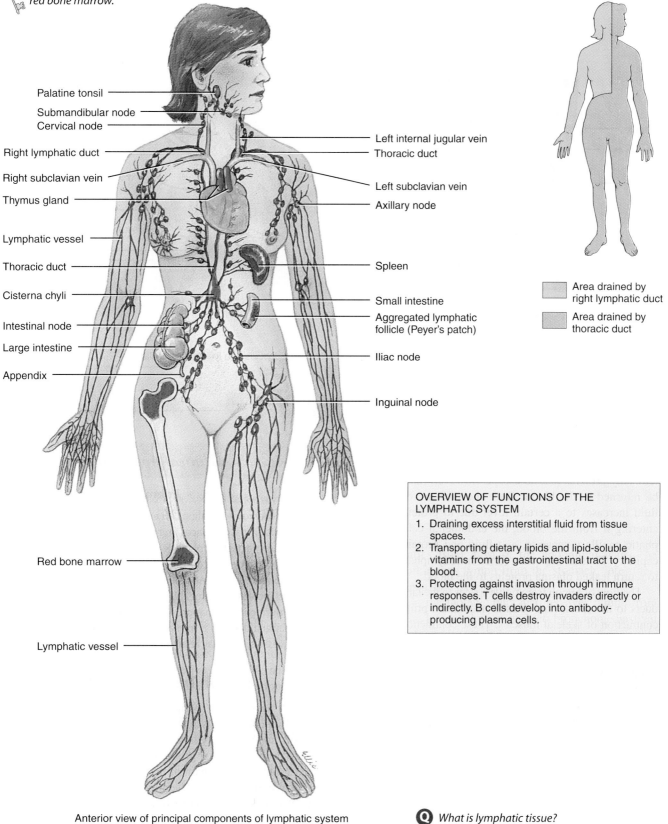

Palatine tonsil

Submandibular node

Cervical node

Right lymphatic duct

Right subclavian vein

Thymus gland

Lymphatic vessel

Thoracic duct

Cisterna chyli

Intestinal node

Large intestine

Appendix

Left internal jugular vein

Thoracic duct

Left subclavian vein

Axillary node

Spleen

Small intestine

Aggregated lymphatic follicle (Peyer's patch)

Iliac node

Inguinal node

Red bone marrow

Lymphatic vessel

Area drained by right lymphatic duct

Area drained by thoracic duct

Anterior view of principal components of lymphatic system

OVERVIEW OF FUNCTIONS OF THE LYMPHATIC SYSTEM

1. Draining excess interstitial fluid from tissue spaces.
2. Transporting dietary lipids and lipid-soluble vitamins from the gastrointestinal tract to the blood.
3. Protecting against invasion through immune responses. T cells destroy invaders directly or indirectly. B cells develop into antibody-producing plasma cells.

Q *What is lymphatic tissue?*

Figure 15.4

Routes for drainage of lymph from lymph trunks into the thoracic and right lymphatic ducts.

All lymph returns to the bloodstream through the thoracic (left lymphatic) duct and the right lymphatic duct.

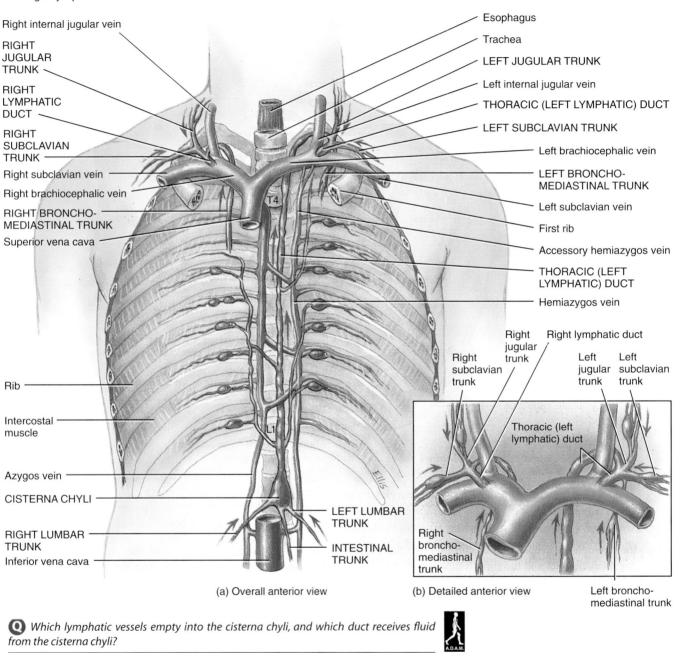

(a) Overall anterior view

(b) Detailed anterior view

Which lymphatic vessels empty into the cisterna chyli, and which duct receives fluid from the cisterna chyli?

lymph from the right jugular trunk, which drains the right side of the head and neck; the right subclavian trunk, which drains the right upper limb; and the right bronchomediastinal trunk, which drains the right side of the thorax, right lung, right side of the heart, and part of the liver. The right lymphatic duct drains lymph into venous blood via the **right subclavian vein.**

LYMPHATIC TISSUES

The **primary lymphatic (lymphoid) organs** of the body are the **red bone marrow** in flat bones and the epiphyses of long bones and the **thymus gland** because they produce B and T cells, the lymphocytes that carry out immune responses. Hematopoietic stem cells in red bone marrow give rise to B cells and pre-T cells. The pre-T cells then populate the

Figure 15.5 Thymus gland.

The bilobed thymus gland is largest at puberty; with age it atrophies.

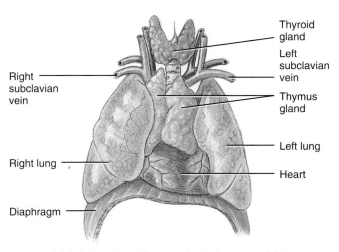

(a) Anterior view of thymus gland of a young child

Right subclavian vein

Right lung

Diaphragm

Thyroid gland

Left subclavian vein

Thymus gland

Left lung

Heart

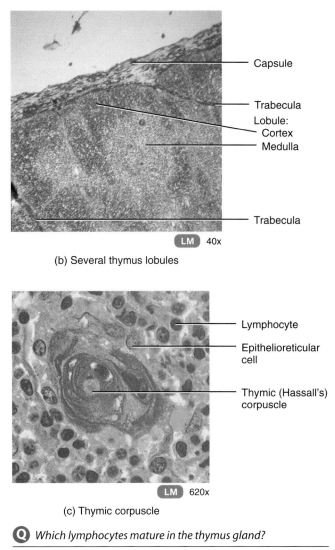

(b) Several thymus lobules

Capsule

Trabecula

Lobule:
Cortex
Medulla

Trabecula

LM 40x

(c) Thymic corpuscle

Lymphocyte

Epithelioreticular cell

Thymic (Hassall's) corpuscle

LM 620x

Q *Which lymphocytes mature in the thymus gland?*

thymus gland. The major **secondary lymphatic organs** are the **lymph nodes** and **spleen.** Also included among the secondary lymphatic organs are the **lymphatic nodules** (although, strictly speaking, they are not discrete organs because they are not surrounded by a capsule). Lymphatic nodules are clusters of lymphocytes that populate all mucous membranes, where microorganisms might attempt to enter the body. Mucous membranes line the gastrointestinal tract, respiratory passageways, urinary tract, and reproductive tract. Most immune responses occur in secondary lymphatic organs.

Thymus Gland

Usually a bilobed lymphatic organ, the **thymus gland** is located in the superior mediastinum, posterior to the sternum and medial to the lungs (Figure 15.5a). An enveloping layer of connective tissue holds the two **thymic lobes** closely together, but a connective tissue **capsule** encloses each lobe. The capsule gives off extensions into the lobes called **trabeculae,** which divide the lobes into **lobules** (Figure 15.5b).

Each lobule consists of a deeply staining outer **cortex** and a lighter-staining central **medulla.** The cortex is composed almost entirely of tightly packed lymphocytes held in place by reticular fibers. Pre-T cells migrate (via the blood) from red bone marrow to the thymus gland, where they proliferate and develop into mature T cells. The medulla consists mostly of epithelial cells and more widely scattered lymphocytes. The epithelial cells produce thymic hormones, which are thought to aid in maturation of T cells. Exactly what the hormones do, however, is not known. In addition, the

medulla contains characteristic **thymic (Hassall's) corpuscles,** concentric layers of epithelial cells (Figure 15.5c). Possibly, they are remnants of dying cells.

The thymus gland is large in the infant, and it reaches its maximum size of about 40 g (about 1.4 oz) at 10 to 12 years of age. After puberty, fat and connective tissue begin to replace the thymic tissue. By the time a person reaches maturity, the gland has largely atrophied (wasting with age). Most T cells arise before puberty, however, and some continue to mature throughout life. Lymphatic vessels that drain the thymus gland terminate in lymph nodes in the thorax (see Exhibit 15.5).

Lymph Nodes

The oval or bean-shaped structures located along lymphatic vessels are called **lymph nodes.** They range from 1 to 25 mm (0.04 to 1 in.) in length. Lymph nodes are scattered throughout the body, usually in groups (see Figure 15.1). Typically, the groups are arranged in two sets: **superficial** and **deep.**

Each lymph node (Figure 15.6) is covered by a **capsule** of dense connective tissue that extends strands into the node. The extensions, called **trabeculae** (tra-BEK-yoo-lē; *trabecula* = little beam), divide the node into compartments, provide support, and convey blood vessels into the interior of a node. The superficial region of a lymph node, the **cortex,** contains many **follicles,** which are regions of densely packed lymphocytes. The outer rim of each follicle contains **T cells (T lymphocytes)** as well as **macrophages** and **follicular dendritic cells,** which participate in the activation of T cells. When an immune response is occurring, the follicles surround lighter-staining central areas, the **germinal centers,** where **B cells (B lymphocytes)** are proliferating into antibody-secreting plasma cells. The deeper region of a lymph node, the **medulla,** contains lymphocytes in tightly packed strands called **medullary cords.** These cords also contain macrophages and plasma cells.

Lymph flows through a node in one direction. It enters through **afferent** (*afferre* = bring to) **lymphatic vessels,** which contain valves that open toward the node so that the lymph is directed *inward.* Inside the node, lymph enters the **sinuses,** which are a series of irregular channels. Lymph flows through sinuses in the cortex and then in the medulla and exits the lymph node via one or two **efferent** (*efferre* = to carry outward) **lymphatic vessels.** Efferent lymphatic vessels are wider than the afferent vessels and contain valves that open away from the node to convey lymph *out* of the node. Efferent lymphatic vessels emerge from one side of the lymph node at a slight depression called a **hilus** (HĪ-lus). Blood vessels also pass through the hilus.

Among lymphatic tissues, only lymph nodes filter lymph by having it enter at one end and exit at another. They filter foreign substances from lymph as it passes back toward the bloodstream. These substances are trapped by the reticular fibers within the node. Then macrophages destroy some foreign substances by phagocytosis, and lymphocytes bring about destruction of others by immune responses. Plasma cells and T cells that have proliferated within a lymph node also can leave and circulate to other parts of the body.

Clinical Application

Metastasis Through the Lymphatic System

Knowledge of the location of the lymph nodes and the direction of lymph flow is important in the diagnosis and prognosis of the spread of cancer by **metastasis** (me-TAS-ta-sis). Cancer cells may travel via the lymphatic system and later via the cardiovascular system and produce clusters of tumor cells where they lodge. Such secondary tumor sites are predictable by the direction of lymph flow from the organ primarily involved. (Cancer may also spread by extending locally or being carried by the cardiovascular system.) Cancerous lymph nodes feel enlarged, firm, and nontender and may be generalized or widespread. Most lymph nodes that enlarge during an infection, by contrast, are not firm and are very tender and are often localized. ■

Spleen

The oval **spleen** (Figure 15.7a) is the largest single mass of lymphatic tissue in the body, measuring about 12 cm (5 in.) in length. It is situated in the left hypochondriac region between the stomach and diaphragm (see Figure 1.9c). The superior surface of the spleen is smooth and convex and conforms to the concave surface of the diaphragm. Neighboring organs make indentations in the spleen—the gastric impression (stomach), renal impression (left kidney), and colic impression (left flexure of colon). Like lymph nodes, the spleen has a hilus. The splenic artery and vein and the efferent lymphatic vessels pass through the hilus.

A capsule of dense connective tissue surrounds the spleen. Trabeculae extend inward from the capsule. The capsule, in turn, is covered by a serous membrane, the visceral peritoneum. The capsule plus trabeculae, reticular fibers, and fibroblasts constitute the stroma of the spleen.

The parenchyma of the spleen consists of two different kinds of tissue called white pulp and red pulp (Figure 15.7b). **White pulp** is lymphatic tissue, mostly lymphocytes and macrophages, arranged around central arteries, branches of the splenic artery. The **red pulp** consists of **venous sinuses** filled with blood and cords of splenic tissue called **splenic (Billroth's) cords.** Splenic cords consist of red blood cells, macrophages, lymphocytes, plasma cells, and granulocytes. Veins are closely associated with the red pulp.

Blood flowing into the spleen through the splenic artery enters the central arteries of the white pulp. Within the white pulp, the spleen carries out its immune functions: (1) T lymphocytes, in response to antigens (such as microbes), directly attack and destroy the antigens in blood; (2) B lymphocytes, in response to antigens, develop into antibody-producing plasma cells, and the antibodies inactivate the antigens in blood; and (3) macrophages destroy antigens in blood by phagocytosis. Within the red pulp, the spleen carries out several functions related to blood cells; (1) removal by macrophages of worn out or defective red blood cells, white blood cells, and platelets; (2) storage of blood platelets, perhaps up to one-third of the body's supply; and (3) production of blood cells (hemopoiesis) during the second trimester of pregnancy (after birth, blood cell formation usually occurs only in the red bone marrow, but the spleen can also participate if necessary).

Figure 15.6 Structure of a lymph node. Arrows indicate direction of lymph flow.

🔑 *Lymph nodes are present throughout the body, usually clustered in groups.*

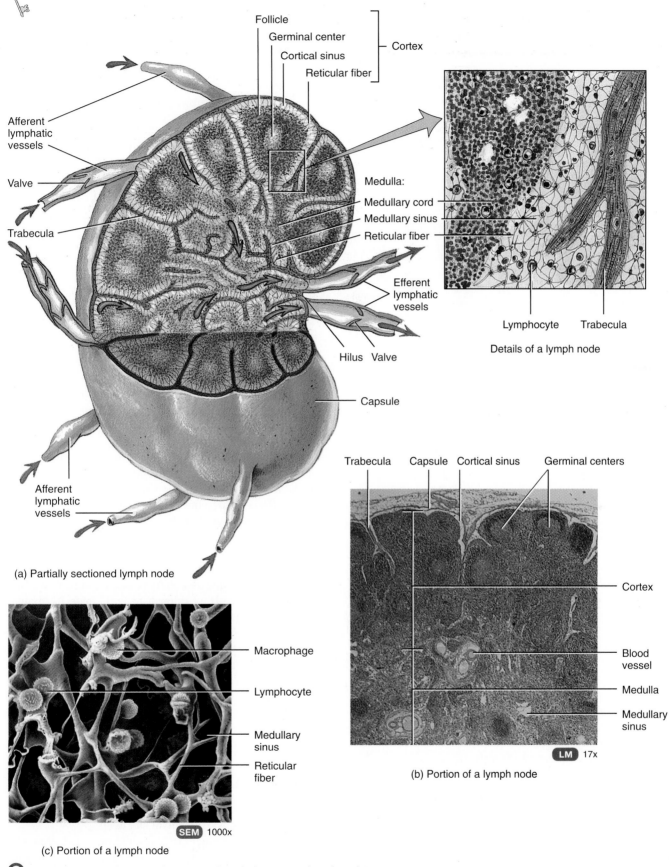

Follicle
Germinal center
Cortical sinus — Cortex
Reticular fiber

Afferent lymphatic vessels

Valve

Trabecula

Medulla:
Medullary cord
Medullary sinus
Reticular fiber

Efferent lymphatic vessels

Hilus   Valve

Capsule

Afferent lymphatic vessels

Lymphocyte   Trabecula

Details of a lymph node

(a) Partially sectioned lymph node

Macrophage

Lymphocyte

Medullary sinus

Reticular fiber

SEM 1000x

(c) Portion of a lymph node

Trabecula   Capsule   Cortical sinus   Germinal centers

Cortex

Blood vessel

Medulla

Medullary sinus

LM 17x

(b) Portion of a lymph node

Q *What happens to foreign substances in lymph that enter a lymph node?*

Figure 15.7 Structure of the spleen.

The spleen is the largest single mass of lymphatic tissue in the body.

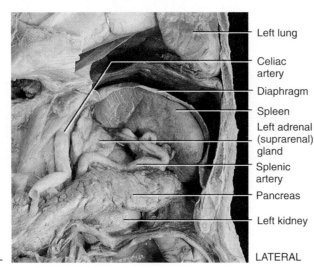

SUPERIOR

MEDIAL

LATERAL

INFERIOR

(a) Photograph of anterior view of a portion of the abdominal cavity

- Left lung
- Celiac artery
- Diaphragm
- Spleen
- Left adrenal (suprarenal) gland
- Splenic artery
- Pancreas
- Left kidney

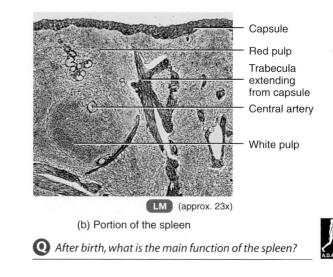

- Capsule
- Red pulp
- Trabecula extending from capsule
- Central artery
- White pulp

LM (approx. 23x)

(b) Portion of the spleen

Q *After birth, what is the main function of the spleen?*

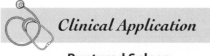

Clinical Application

Ruptured Spleen

The spleen is the organ most often damaged in cases of abdominal trauma. Severe blows over the inferior left chest or superior abdomen may fracture the protecting ribs. Such a crushing injury may rupture the spleen, which causes severe intraperitoneal hemorrhage and shock. Prompt removal of the spleen, called a **splenectomy,** is needed to prevent the patient from bleeding to death. Other structures, particularly red bone marrow and the liver, can take over functions normally carried out by the spleen. The spleen's absence can place the patient at higher risk for sepsis, a blood infection resulting from the loss of the filtering and phagocytic function of the spleen. Patients with a history of a splenectomy take prophylactic antibiotics before any invasive procedures to reduce the risk of sepsis. ■

Lymphatic Nodules

Lymphatic nodules are oval-shaped concentrations of lymphatic tissue. Although they are not surrounded by a capsule, most lymphatic nodules are solitary, small, and discrete. They are scattered throughout the lamina propria (connective tissue) of mucous membranes lining the gastrointestinal tract, respiratory airways, urinary tract, and reproductive tract. This lymphatic tissue is referred to as **mucosa-associated lymphoid tissue (MALT).**

Some lymphatic nodules occur in multiple, large aggregations in specific parts of the body. Among these are the tonsils in the pharyngeal region and **aggregated lymphatic follicles (Peyer's patches)** in the ileum of the small intestine. Aggregations of lymphatic nodules also occur in the appendix.

Usually there are five **tonsils,** which form a ring at the junction of the oral cavity and oropharynx and the junction of the nasal cavity and nasopharynx (see Figure 23.2b). Thus they are strategically positioned to participate in immune responses against foreign substances that are inhaled or ingested. The single **pharyngeal** (fa-RIN-jē-al) **tonsil** or **adenoid** is embedded in the posterior wall of the nasopharynx. The two **palatine** (PAL-a-tīn) **tonsils** lie at the lateral, posterior regions of the oral cavity. These are the ones commonly removed by a tonsillectomy. The paired **lingual** (LIN-gwal) **tonsils** are located at the base of the tongue and may also require removal by a tonsillectomy.

PRINCIPAL GROUPS OF LYMPH NODES

Lymph nodes usually appear in groups and typically are arranged in two sets: **superficial** and **deep.**

Exhibits 15.1 through 15.5 list the principal groups of lymph nodes of the body by region and the general areas of the body they drain.

Exhibit 15.1 *Principal Lymph Nodes of the Head and Neck (Figure 15.8)*

LYMPH NODES OF THE HEAD	LOCATION	DRAINAGE
Occipital nodes	Near trapezius and semispinalis capitis muscles.	Occipital portion of scalp and upper neck.
Retroauricular nodes	Posterior to ear.	Skin of ear and posterior parietal region of scalp.
Preauricular nodes	Anterior to ear.	Auricle of ear and temporal region of scalp.
Parotid nodes	Embedded in and inferior to parotid gland.	Root of nose, eyelids, anterior temporal region, external auditory meatus, tympanic cavity, nasopharynx, and posterior portions of nasal cavity.
Facial nodes	Consist of three groups: infraorbital, buccal, and mandibular.	
Infraorbital nodes	Inferior to the orbit.	Eyelids and conjunctiva.
Buccal nodes	At angle of mouth.	Skin and mucous membrane of nose and cheek.
Mandibular nodes	Over mandible.	Skin and mucous membrane of nose and cheek.

LYMPH NODES OF THE NECK	LOCATION	DRAINAGE
Submandibular nodes	Along inferior border of mandible.	Chin, lips, nose, nasal cavity, cheeks, gums, inferior surface of palate, and anterior portion of tongue.
Submental nodes	Between digastric muscles.	Chin, lower lip, cheeks, tip of tongue, and floor of mouth.
Superficial cervical nodes	Along external jugular vein.	Inferior part of ear and parotid region.
Deep cervical nodes	Largest group of nodes in neck, consisting of numerous large nodes forming a chain extending from base of skull to root of neck. They are arbitrarily divided into superior deep cervical nodes and inferior deep cervical nodes.	
Superior deep cervical nodes	Deep to sternocleidomastoid muscle.	Posterior head and neck, auricle, tongue, larynx, esophagus, thyroid gland, nasopharynx, nasal cavity, palate, and tonsils.
Inferior deep cervical nodes	Near subclavian vein.	Posterior scalp and neck, superficial pectoral region, and part of arm.

Exhibit 15.1 Principal Lymph Nodes of the Head and Neck (continued)

Figure 15.8 Principal lymph nodes of the head and neck.

🔑 *The facial lymph nodes include the infraorbital, buccal, and mandibular lymph nodes.*

PREAURICULAR LYMPH NODE

RETROAURICULAR LYMPH NODE

OCCIPITAL LYMPH NODE

SUPERFICIAL CERVICAL LYMPH NODE

DEEP CERVICAL LYMPH NODE

Orbicularis oculi muscle

INFRAORBITAL LYMPH NODE

PAROTID LYMPH NODE

Parotid gland

BUCCAL LYMPH NODE

MANDIBULAR LYMPH NODE

SUBMENTAL LYMPH NODE

Submandibular gland

SUBMANDIBULAR LYMPH NODE

Sternocleidomastoid muscle

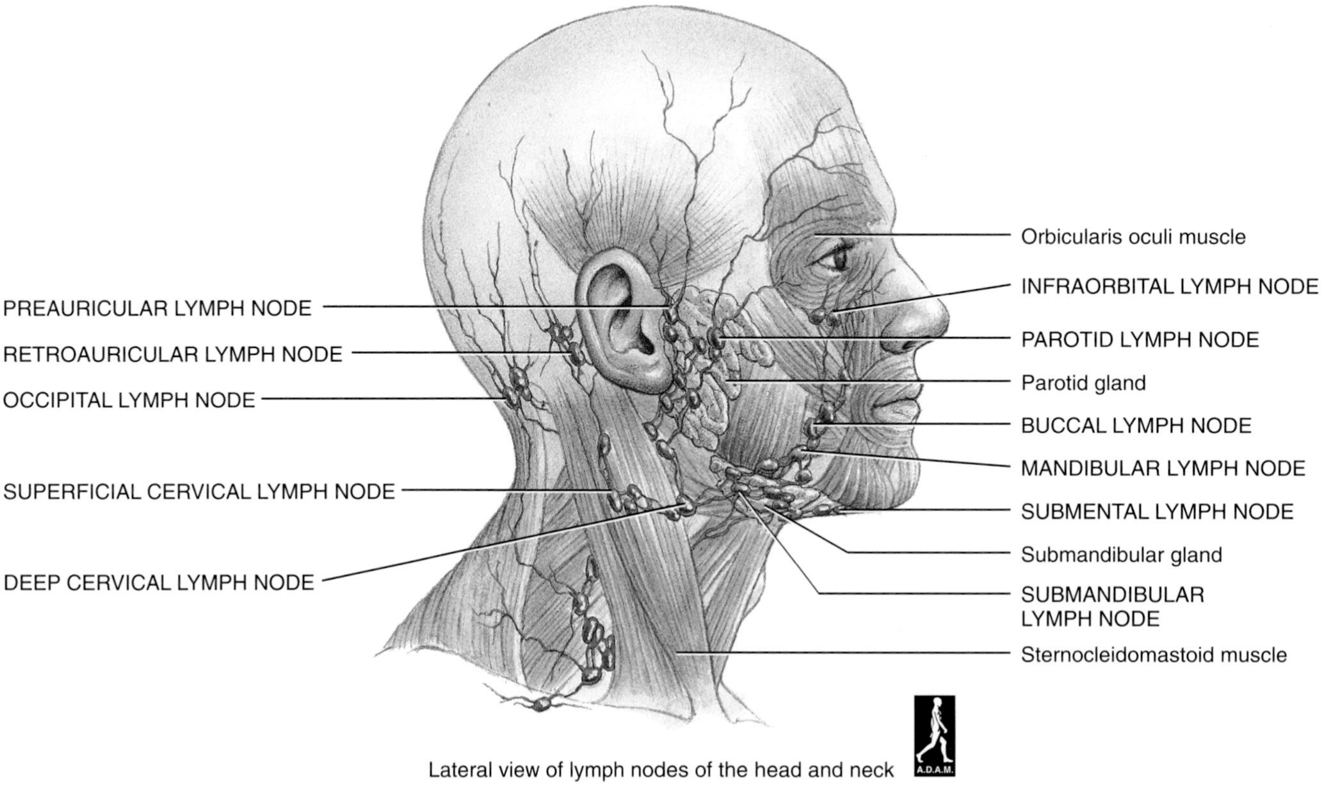

Lateral view of lymph nodes of the head and neck

Q *Which is the largest group of lymph nodes in the neck?*

Exhibit 15.2 *Principal Lymph Nodes of the Upper Limbs (Figure 15.9)*

LYMPH NODES	LOCATION	DRAINAGE
Supratrochlear nodes	Superior to medial epicondyle of humerus.	Medial fingers, palm, and forearm.
Deltopectoral nodes	Inferior to clavicle.	Lymphatic vessels on radial side of upper limb.
Axillary nodes	Most deep lymph nodes of the upper limbs are in the axilla and are called the axillary nodes. They are large.	
Lateral nodes	Medial and posterior aspects of axillary artery.	Most of entire upper limb.*
Pectoral (anterior) nodes	Along inferior border of the pectoralis minor muscle.	Skin and muscles of anterior and lateral thoracic walls and central and lateral portions of mammary gland.
Subscapular (posterior) nodes	Along subscapular artery.	Skin and muscles of posterior part of neck and thoracic wall.
Central (intermediate) nodes	Base of axilla embedded in adipose tissue.	Lateral, pectoral (anterior), and subscapular (posterior) nodes.
Subclavicular (apical) nodes	Posterior and superior to pectoralis minor muscle.	Deltopectoral nodes.

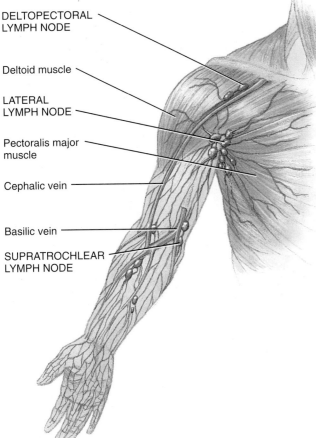

DELTOPECTORAL LYMPH NODE

Deltoid muscle

LATERAL LYMPH NODE

Pectoralis major muscle

Cephalic vein

Basilic vein

SUPRATROCHLEAR LYMPH NODE

SUBCLAVICULAR LYMPH NODE

LATERAL LYMPH NODE

CENTRAL LYMPH NODE
SUBSCAPULAR LYMPH NODE

Pectoralis major muscle
STERNAL LYMPH NODE
PECTORAL LYMPH NODE

Mammary gland

(b) Anterior view of mostly axillary lymph nodes

(a) Anterior view of lymph nodes of the upper limb

*Because infection or malignancy of the upper limbs may cause tenderness and swelling in the axilla, the axillary nodes, especially the lateral group, are clinically important in that they filter lymph from much of the upper limb.

Figure 15.9 Principal lymph nodes of the upper limbs. In (b) the direction of drainage is indicated by arrows.

Most of the lymph drainage of the breast is to the pectoral group of axillary lymph nodes.

Q *Which lymph nodes drain most of the upper limb?*

Exhibit 15.3 Principal Lymph Nodes of the Lower Limbs (Figure 15.10)

LYMPH NODES	LOCATION	DRAINAGE
Popliteal nodes	In adipose tissue in popliteal fossa.	Knee and portions of leg and foot, especially heel.
Superficial inguinal nodes	Parallel to saphenous vein.	Anterior and lateral abdominal wall to level of umbilicus, gluteal region, external genitals, perineal region, and entire superficial lymphatics of lower limb.
Deep inguinal nodes	Medial to femoral vein.	Deep lymphatics of lower limb, penis, and clitoris.

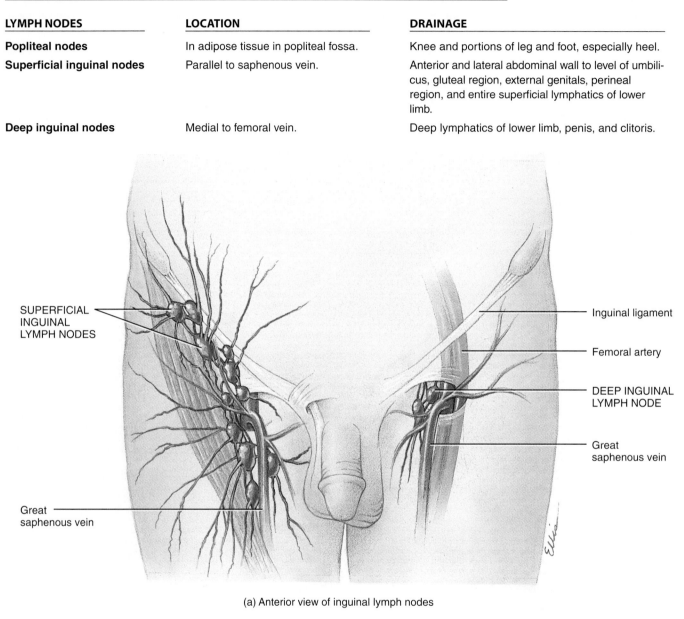

(a) Anterior view of inguinal lymph nodes

Figure 15.10 Principal lymph nodes of the lower limbs.

The inguinal lymph nodes drain the lymphatics of the lower limbs.

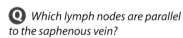 *Which lymph nodes are parallel to the saphenous vein?*

Exhibit 15.4 Principal Lymph Nodes of the Abdomen and Pelvis (Figure 15.11)

PARIETAL LYMPH NODES	LOCATION	DRAINAGE
Parietal nodes are located retroperitoneally (behind the parietal peritoneum) and in close association with larger blood vessels.		
External iliac nodes	Arranged about external iliac vessels.	Deep lymphatics of abdominal wall inferior to umbilicus, adductor region of thigh, urinary bladder, prostate gland, ductus (vas) deferens, seminal vesicles, prostatic and membranous urethra, uterine (Fallopian) tubes, uterus, and vagina.
Common iliac nodes	Arranged along course of common iliac vessels.	Pelvic viscera.
Internal iliac nodes	Near internal iliac artery.	Pelvic viscera, perineum, gluteal region, and posterior surface of thigh.
Sacral nodes	In hollow of sacrum.	Rectum, prostate gland, and posterior pelvic wall.
Lumbar nodes	From aortic bifurcation to diaphragm; arranged around aorta and designated as *right lateral aortic nodes, left lateral aortic nodes, preaortic nodes,* and *retroaortic nodes.*	Efferents from testes, ovaries, uterine (Fallopian) tubes, uterus, kidneys, adrenal (suprarenal) glands, abdominal surface of diaphragm, and lateral abdominal wall.

VISCERAL LYMPH NODES	LOCATION	DRAINAGE
Visceral nodes are found in association with visceral arteries.		
Celiac nodes	Consist of three groups: gastric, hepatic, and pancreaticosplenic.	
Gastric nodes	Lie along lesser curvature of stomach.	Lesser curvature of stomach; inferior, anterior, and posterior aspects of stomach; esophagus.
Hepatic nodes	Along hepatic artery.	Stomach, duodenum, liver, gallbladder, and pancreas.
Pancreaticosplenic nodes	Along splenic artery.	Stomach, spleen, and pancreas.
Superior mesenteric nodes	Consist of three groups: mesenteric, ileocolic, and transverse mesocolic.	
Mesenteric nodes	Along superior mesenteric artery.	Jejunum and all parts of ileum, except for terminal portion.
Ileocolic nodes	Along ileocolic artery.	Terminal portion of ileum, appendix, cecum, and ascending colon.
Transverse mesocolic nodes	Between layers of transverse mesocolon.	Descending iliac and sigmoid parts of colon.
Inferior mesenteric nodes	Near left colic, sigmoid, and superior rectal arteries.	Descending, iliac, and sigmoid parts of colon; superior part of rectum; superior anal canal.

Exhibit 15.4 *Principal Lymph Nodes of the Abdomen and Pelvis (continued)*

Figure 15.11 Principal lymph nodes of the abdomen and pelvis.

The parietal lymph nodes are retroperitoneal and in close association with larger blood vessels.

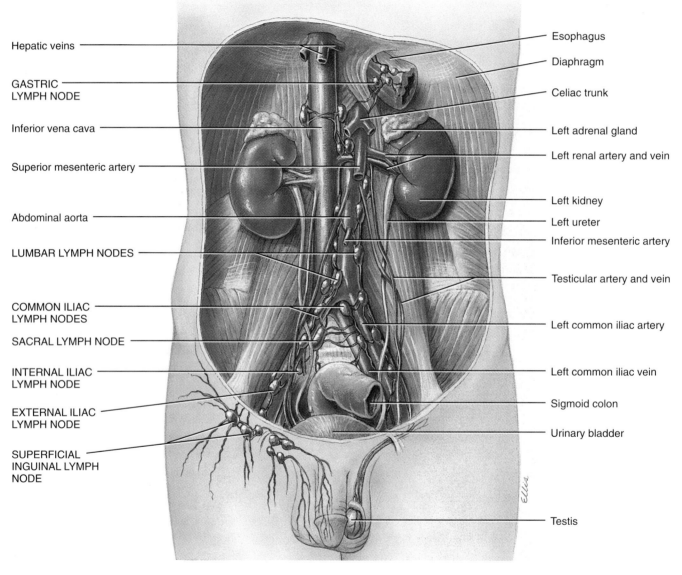

(a) Anterior view of abdominal and pelvic lymph nodes

Exhibit 15.4 *(continued)*

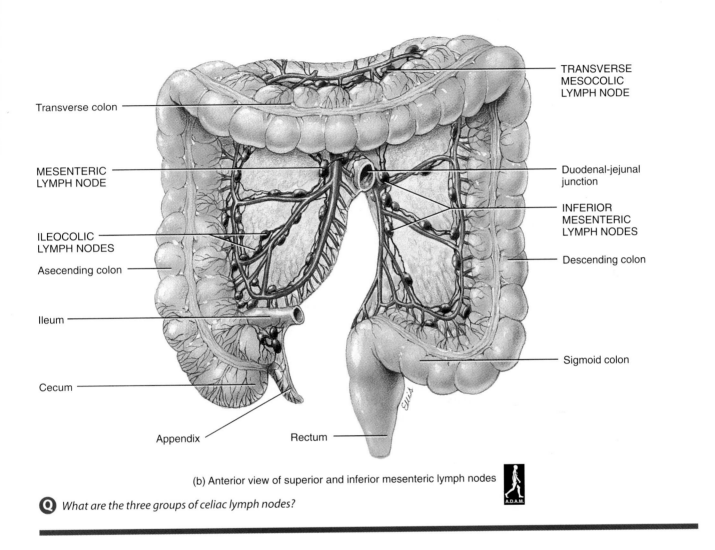

Transverse colon

TRANSVERSE
MESOCOLIC
LYMPH NODE

MESENTERIC
LYMPH NODE

Duodenal-jejunal
junction

INFERIOR
MESENTERIC
LYMPH NODES

ILEOCOLIC
LYMPH NODES

Descending colon

Asecending colon

Ileum

Cecum

Sigmoid colon

Appendix

Rectum

(b) Anterior view of superior and inferior mesenteric lymph nodes

Q *What are the three groups of celiac lymph nodes?*

Exhibit 15.5 Principal Lymph Nodes of the Thorax (Figure 15.12)

PARIETAL LYMPH NODES	LOCATION	DRAINAGE
Parietal nodes drain the wall of the thorax.		
Sternal (parasternal) nodes	Alongside internal thoracic artery.	Central and lateral parts of mammary gland, deeper structures of anterior abdominal wall superior to umbilicus, diaphragmatic surface of liver, and deeper parts of anterior portion of thoracic wall.
Intercostal nodes	Near heads of ribs at posterior parts of intercostal spaces.	Posterolateral aspect of thoracic wall.
Phrenic (diaphragmatic) nodes	On thoracic aspect of diaphragm and divisible into three sets called anterior phrenic, middle phrenic, and posterior phrenic.	
Anterior phrenic nodes	Posterior to base of xiphoid process.	Convex surface of liver, diaphragm, and anterior abdominal wall.
Middle phrenic nodes	Close to phrenic nerves where they pierce diaphragm.	Medial part of diaphragm and convex surface of liver.
Posterior phrenic nodes	Posterior surface of diaphragm near aorta.	Posterior part of diaphragm.

VISCERAL LYMPH NODES	LOCATION	DRAINAGE
Visceral nodes drain the viscera in the thorax.		
Anterior mediastinal nodes	Anterior part of superior mediastinum anterior to arch of aorta.	Thymus gland and pericardium.
Posterior mediastinal nodes	Posterior to pericardium.	Esophagus, posterior aspect of the pericardium, diaphragm, and convex surface of liver.
Tracheobronchial nodes	Are divided into five groups: tracheal, superior and inferior tracheobronchial, bronchopulmonary, and pulmonary nodes.	
Tracheal nodes	Either side of trachea.	Trachea and upper esophagus.
Superior tracheobronchial nodes	Between trachea and bronchi.	Trachea and bronchi.
Inferior tracheobronchial nodes	Between bronchi.	Trachea and bronchi.
Bronchopulmonary nodes	In hilus of each lung.	Lungs and bronchi.
Pulmonary nodes	Within lungs on larger bronchial tube branches.	Lungs and bronchi.

Exhibit 15.5 *(continued)*

Figure 15.12 Principal lymph nodes of the thorax.

Whereas parietal lymph nodes drain the thoracic wall, visceral lymph nodes drain the viscera in the thorax.

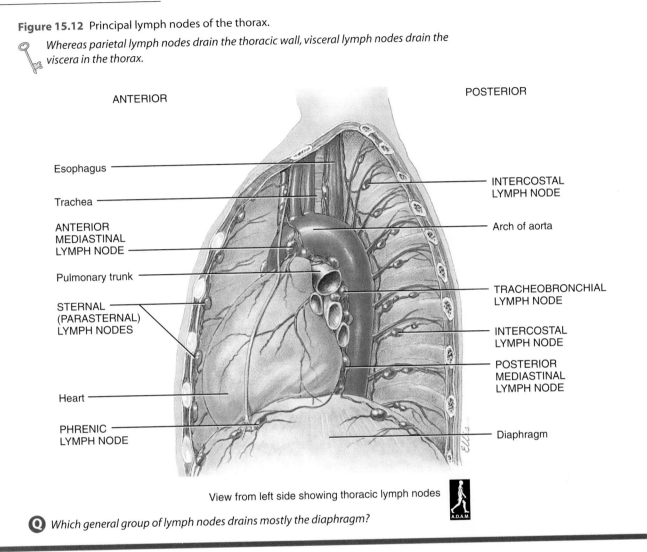

ANTERIOR

POSTERIOR

Esophagus

INTERCOSTAL LYMPH NODE

Trachea

Arch of aorta

ANTERIOR MEDIASTINAL LYMPH NODE

Pulmonary trunk

TRACHEOBRONCHIAL LYMPH NODE

STERNAL (PARASTERNAL) LYMPH NODES

INTERCOSTAL LYMPH NODE

POSTERIOR MEDIASTINAL LYMPH NODE

Heart

PHRENIC LYMPH NODE

Diaphragm

View from left side showing thoracic lymph nodes

Q *Which general group of lymph nodes drains mostly the diaphragm?*

Clinical Application

Breast Cancer and Lymphatic Draining

More than 75% of the lymph drainage of the breast is to the pectoral (anterior) group of axillary lymph nodes. In breast cancer, it is possible that cancer cells leave the breast and lodge in the pectoral nodes. From here, metastasis may develop in other axillary nodes. Most of the remaining lymphatic drainage is to the sternal (parasternal) lymph nodes (see also Figure 15.12). ■

Developmental Anatomy of the Lymphatic System

The lymphatic system begins to develop by the end of the fifth week. *Lymphatic vessels* develop from **lymph sacs** that arise from developing veins. Thus the lymphatic system is also derived from **mesoderm.**

The first lymph sacs to appear are the paired **jugular lymph sacs** at the junction of the internal jugular and subclavian veins (Figure 15.13). From the jugular lymph sacs, capillary plexuses spread to the *thorax, upper limbs, neck,* and *head.* Some of the plexuses enlarge and form lymphatic vessels in their respective regions. Each jugular lymph sac retains at least one connection with its jugular vein, the left one developing into the superior portion of the thoracic duct (left lymphatic duct).

The next lymph sac to appear is the unpaired **retroperitoneal lymph sac** at the root of the mesentery of the intestine. It develops from the primitive vena cava and mesonephric (primitive kidney) veins. Capillary plexuses and lymphatic vessels spread from the retroperitoneal lymph sac to the *abdominal viscera* and *diaphragm.* The sac establishes connections with the cisterna chyli but loses its connections with neighboring veins.

At about the time the retroperitoneal lymph sac is developing, another lymph sac, the **cisterna chyli,** develops inferior to the diaphragm on the posterior abdominal wall. It gives rise to the inferior portion of the *thoracic duct* and the *cisterna chyli* of the thoracic duct. Like the retroperitoneal lymph sac, the cisterna chyli also loses its connections with surrounding veins.

The last of the lymph sacs, the paired **posterior lymph sacs,** develop from the iliac veins at their union with the posterior cardinal veins. The posterior lymph sacs produce capillary plexuses and lymphatic vessels of the *abdominal wall, pelvic region,* and *lower limb.* The posterior lymph sacs join the cisterna chyli and lose their connections with adjacent veins.

Figure 15.13 Development of the lymphatic system.

The lymphatic system is derived from mesoderm.

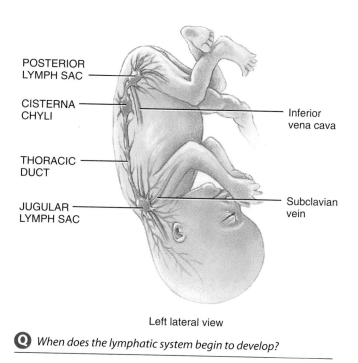

POSTERIOR LYMPH SAC

CISTERNA CHYLI

THORACIC DUCT

JUGULAR LYMPH SAC

Inferior vena cava

Subclavian vein

Left lateral view

Q *When does the lymphatic system begin to develop?*

With the exception of the anterior part of the sac from which the cisterna chyli develops, all lymph sacs become invaded by **mesenchymal cells** and are converted into groups of *lymph nodes.*

The *spleen* develops from **mesenchymal cells** between layers of the dorsal mesentery of the stomach. The *thymus gland* arises as an outgrowth of the **third pharyngeal pouch** (see Figure 22.9a). ■

Aging and the Lymphatic System

As the immune system ages, elderly individuals become more susceptible to all types of infections and malignancies. Their response to vaccines is decreased, and they tend to produce more autoantibodies (antibodies against their body's own antigens). In addition, the immune system exhibits lowered levels of function. There is a decrease in the production of T lymphocytes that attach to and destroy antigens and in the production of B lymphocytes that develop into antibody-producing plasma cells.

Key Medical Terms Associated
with the Lymphatic System

Adenitis (ad′-e-NĪ-tis; *adeno* = gland; *itis* = inflammation of) Enlarged, tender, and inflamed lymph nodes resulting from an infection.

Hypersplenism (hī′-per-SPLĒN-izm; *hyper* = over) Abnormal splenic activity due to splenic enlargement and associated with an increased rate of destruction of normal blood cells.

Lymphadenectomy (lim-fad′-e-NEK-tō-mē; *ectomy* = removal) Removal of a lymph node.

Lymphadenopathy (lim-fad′-e-NOP-a-thē; *patho* = disease) Enlarged, sometimes tender lymph glands.

Lymphangioma (lim-fan′-jē-Ō-ma; *angio* = vessel; *oma* = tumor) A benign tumor of the lymphatic vessels.

Lymphangitis (lim′-fan-JĪ-tis) Inflammation of the lymphatic vessels.

Lymphedema (lim′-fe-DĒ-ma; *edema* = swelling) Accumulation of lymph producing subcutaneous tissue swelling.

Lymphostasis (lim-FŌ-stā-sis; *stasis* = halt) A lymph flow stoppage.

Splenomegaly (splē′-no-MEG-a-lē; *mega* = large) Enlarged spleen.

Tonsillectomy (ton′-si-LEK-tō-mē; *tonsilla* = a stake; *ektome* = excision) Removal of a tonsil.

Study Outline

Introduction (p. 478)

1. The lymphatic system consists of lymph, lymphatic vessels, and structures and organs that contain lymphatic tissue (specialized reticular tissue containing many lymphocytes).
2. The lymphatic system functions to drain interstitial fluid, transport dietary lipids, and protect against invasion through immune responses.

Lymphatic Vessels and Lymph Circulation (p. 478)

1. Lymphatic vessels begin as closed-ended lymph capillaries in tissue spaces between cells.
2. Interstitial fluid drains into lymphatic capillaries, thus forming lymph.
3. Lymph capillaries merge to form larger vessels, called lymphatic vessels, which convey lymph into and out of structures called lymph nodes.
4. The passage of lymph is from lymph capillaries to lymphatic vessels to lymph trunks to the thoracic duct or right lymphatic duct to the subclavian veins.
5. Lymph flows as a result of skeletal muscle contractions and respiratory movements. It is also aided by valves in the lymphatic vessels.

Lymphatic Tissues (p. 482)

1. The primary lymphatic organs are red bone marrow and the thymus gland. Secondary lymphatic organs are the lymph nodes, spleen, and lymphatic nodules.
2. The thymus gland lies superior to the sternum and the large blood vessels superior to the heart. It is the site of T cell maturation.
3. Lymph nodes are encapsulated, oval structures located along lymphatic vessels.

4. Lymph enters nodes through afferent lymphatic vessels, is filtered, and exits through efferent lymphatic vessels.
5. Lymph nodes are the site of proliferation of plasma cells and T cells.
6. The spleen is the largest mass of lymphatic tissue in the body. Its white pulp carries out immune functions, and its red pulp has several functions related to blood cells.
7. Lymphatic nodules are scattered throughout the mucosa of the gastrointestinal, respiratory, urinary, and reproductive tracts. This lymphatic tissue is referred to as mucosa-associated lymphoid tissue (MALT).

Principal Groups of Lymph Nodes (p. 486)

1. Lymph nodes are scattered throughout the body as superficial and deep groups.
2. The principal groups of lymph nodes are found in the head and neck, upper limbs, lower limbs, abdomen and pelvis, and thorax.

Developmental Anatomy of the Lymphatic System (p. 496)

1. Lymphatic vessels develop from lymph sacs, which develop from veins. Thus, they are derived from mesoderm.
2. Lymph nodes develop from lymph sacs that become invaded by mesenchymal cells.

Aging and the Lymphatic System (p. 496)

1. With advancing age, individuals become more susceptible to infections and malignancies, response to vaccines is decreased, and more autoantibodies are produced.
2. Immune responses also diminish with age.

Review Questions

1. Identify the components and functions of the lymphatic system. (p. 478)
2. Compare veins and lymphatic vessels with regard to structure. (p. 478)
3. Construct a diagram to show the route of lymph circulation. (p. 478)
4. Describe the role of the thymus gland in immunity. (p. 483)
5. Describe the structure of a lymph node. What functions do lymph nodes serve? (p. 484)
6. Describe the location, gross anatomy, histology, and functions of the spleen. (p. 484)
7. Identify the tonsils by location. (p. 486)
8. For each of the following regions of the body, list the major lymph nodes and the areas of the body they drain: head, neck, upper limbs, lower limbs, abdomen and pelvis, and thorax. (p. 487)
9. Describe how the lymphatic system develops. (p. 496)
10. What are the effects of aging on the lymphatic system? (p. 496)

Self Quiz

Complete the following:

1. The ___ and the ___ are the primary lymphatic organs because they produce the lymphocytes for the immune system.
2. The thoracic duct starts in the lumbar region as a dilation known as the ___.
3. The lymphatic system derives from the ___-derm.
4. Lymphatic nodules in the lamina propria of mucous membranes are collectively referred to as ___.
5. The parenchyma of the spleen consists of two types of tissue, ___ and ___.

Are the following statements true or false?

6. Lymph is conveyed into a node at several points by afferent lymphatic vessels.
7. The main difference between lymph and interstitial fluid is location.
8. Most of the lymph nodes that drain viscera bear the same names as arteries that supply those viscera.
9. Blood capillaries are more permeable than lymph capillaries.
10. The palatine tonsils are the ones most often removed in a tonsillectomy.
11. Lymph may pass through several lymph nodes in a number of regions before returning to blood.
12. Match the following lymph nodes with their descriptions:
   ___ (a) facial nodes
   ___ (b) axillary nodes
   ___ (c) popliteal nodes
   ___ (d) common iliac nodes
   ___ (e) celiac nodes
   ___ (f) phrenic nodes
   ___ (g) sacral nodes
   ___ (h) deep cervical nodes

   (1) drain the knee, leg, and heel
   (2) three groups of nodes that drain the stomach, liver, pancreas, and spleen
   (3) largest group of nodes in the neck
   (4) three group of nodes that drain the mucous mem--branes of the cheeks, eyelids, and nose
   (5) anterior to sacrum; drain the rectum, prostate gland, and pelvic wall
   (6) three groups of nodes that drain parts of the liver, diaphragm, and abdominal wall
   (7) deep nodes; drain the upper limbs, skin and muscles of the thorax, and part of the neck
   (8) arranged along blood vessels of the same name; drain the pelvic viscera

Choose the one best answer to the following questions:

13. Put the following items in the correct order for the pathway of lymph from the stomach to blood.
   (1) thoracic duct
   (2) lymphatic vessels and lymph nodes
   (3) lymph capillaries
   (4) interstitial fluid
   (5) left subclavian vein
   a. 2, 3, 4, 5, 1
   b. 3, 2, 4, 1, 5
   c. 1, 5, 3, 2, 4
   d. 4, 2, 1, 5, 3
   e. 4, 3 2, 1, 5
14. Which of the following is *not* a major site of lymphatic tissue?
   a. tonsils
   b. thymus gland
   c. kidneys
   d. spleen
   e. lymph nodes
15. Which of the following is an important function of the lymphatic system?
   a. controlling body temperature by evaporation of sweat
   b. manufacturing all white blood cells
   c. transporting fluids out to, and back from, body tissues
   d. returning fluid and proteins to the cardiovascular system
   e. all of the above
16. The spleen
   a. serves as a storage site for blood platelets
   b. is an organ in which phagocytosis of aged red blood cells occurs
   c. is an organ in which phagocytosis of aged or damaged platelets occurs
   d. is a site of blood formation in the fetus
   e. all of the above

Answers to Figure Questions

17. A major difference between the spleen and lymph nodes is that
    a. lymph nodes are not enclosed in a capsule, whereas the spleen is
    b. the spleen is located in the abdomen; lymph nodes are not
    c. lymph nodes have afferent and efferent lymphatic vessels; the spleen has neither
    d. lymph nodes have afferent and efferent lymphatic vessels; the spleen has only efferent lymphatic vessels
    e. lymph flows through sinuses in the spleen, but not through sinuses in lymph nodes

18. Which of the following lymph nodes drains the leg?
    a. axillary
    b. anterior mediastinal
    c. deltopectoral
    d. deep inguinal
    e. superior mesenteric

19. Both the thoracic duct and the right lymphatic duct empty directly into the
    a. axillary lymph nodes
    b. superior vena cava
    c. cisterna chyli
    d. subclavian arteries
    e. junction of the internal jugular and subclavian veins

20. Which of the following statements is true about the flow of lymph?
    (1) It is aided in its movement toward the thoracic region by a pressure gradient.
    (2) It is maintained primarily by the milking action of muscles.
    (3) It is possible because of valves in lymphatic vessels.
    a. 1 only
    b. 2 only
    c. 3 only
    d. all of the above
    e. 2 and 3

21. Which of the following statements about HIV transmission is true?
    a. Standard chlorination effectively eliminates HIV from swimming pools and hot tubs.
    b. The transmission of HIV requires the transfer of, or direct contact with, infected body fluids.
    c. Organ transplants and artificial insemination are routes for HIV transmission.
    d. HIV may be present in semen, blood, and breast milk.
    e. All of the above are true.

Critical Thinking Questions

1. Jack got a splinter in his right heel at the beach. He neglected to clean it properly and it became infected. Trace the route of the microbes through the lymphatic system.
   HINT: *The right lymphatic duct does not drain the right foot.*

2. Three-year-old Kelsey has chronic sore throats, averaging 10 infections per year. She breathes loudly through her mouth and even snores. The pediatrician recommended that she have her tonsils removed. Where are these located and what is their function?
   HINT: *The traditional diet to ease the pain following a tonsillectomy is ice chips and ice cream.*

3. Kate is miserable every spring. She has itchy, watery eyes and nasal congestion. What is her problem? Why is her body reacting this way?
   HINT: *Antihistamine would probably help relieve her symptoms.*

4. Can you live without a spleen? What happens if you need to have your spleen removed?
   HINT: *Some organs are more essential than others.*

5. Following a woman's diagnosis with breast cancer, the surgeon removed the tumor and some lymph nodes from her breast. Why?
   HINT: *What lymph tissue drains the breast?*

Answers to Figure Questions

15.1 Reticular connective tissue that contains large numbers of lymphocytes.
15.2 Interstitial fluid, because the protein content is low.
15.3 Capillaries.
15.4 Left and right lumbar trunks and intestinal trunk; thoracic duct.
15.5 T cells.
15.6 Macrophages may phagocytize them or lymphocytes may attack them via immune responses.
15.7 White pulp carries out immune functions; red pulp carries out functions related to blood.
15.8 Deep cervical lymph nodes.
15.9 Lateral lymph nodes.
15.10 Superficial inguinal lymph nodes.
15.11 Gastric, hepatic, and pancreaticosplenic nodes.
15.12 Phrenic lymph nodes.
15.13 By the end of the fifth week of gestation.

Chapter 16

Nervous Tissue

Student Objectives

1. **Classify the organs of the nervous system into central and peripheral divisions.**

2. **Contrast the histological characteristics and functions of neuroglia and neurons.**

3. **Describe the parts of a neuron and the functions of each.**

4. **Classify neurons by structure and function.**

5. **Distinguish between gray matter and white matter.**

6. **Describe the various types of neuronal circuits in the nervous system.**

The body has two control centers, one for rapid response—the nervous system—and one that makes slower but no less important adjustments for maintaining homeostasis—the endocrine system. The nervous system has three basic functions. First, it senses changes (stimuli) within the body (internal environment) and in the outside environment (external environment); this is its *sensory* function. Second, it analyzes the sensory information, stores some of it, and makes decisions regarding appropriate behaviors; this is its *integrative* function. Third, it responds to stimuli by initiating action in the form of muscular contractions or glandular secretions; this is its *motor* function.

Because the nervous system is quite complex, we will consider its parts in several related chapters. This chapter focuses on nervous tissue. We will discuss the basic structure and functions of nerve cells and supporting cells called neuroglia. In chapters that follow, we will examine the structure and functions of the spinal cord and spinal nerves, as well as that of the brain and cranial nerves. Then we will discuss the general senses and the motor and sensory pathways to understand how nerve impulses pass from receptors to the spinal cord and brain or from the spinal cord and brain to muscles to glands. Our journey continues with a discussion of the special senses: smell, taste, vision, hearing, and equilibrium. We conclude our study of the nervous system by examining the autonomic nervous system, the part of the nervous system that operates without conscious control.

The branch of medical science that deals with the normal functioning and disorders of the nervous system is called **neurology** (noo-ROL-ō-jē; *neuro* = nerve or nervous system; *logos* = study of).

NERVOUS SYSTEM DIVISIONS

The two principal divisions of the nervous system are the **central nervous system (CNS)** and the **peripheral** (pe-RIF-er-al) **nervous system (PNS)** (Figure 16.1).

The CNS consists of the **brain** and **spinal cord.** Within the CNS, various sorts of incoming sensory information are integrated and correlated, thoughts and emotions are generated, and memories are formed and stored. Most nerve impulses that stimulate muscles to contract and glands to secrete originate in the CNS.

The CNS is connected to sensory receptors, muscles, and glands in peripheral parts of the body by the PNS. The PNS consists of **cranial nerves** that arise from the brain and **spinal nerves** that emerge from the spinal cord. Portions of these nerves carry nerve impulses into the CNS, whereas other portions carry impulses out of the CNS.

The input component of the PNS consists of nerve cells called **sensory** or **afferent neurons.** They conduct nerve impulses from sensory receptors located in various parts of the body *to the CNS* and end within the CNS. The output component consists of nerve cells, called **motor** or **efferent neurons.** They originate within the CNS and conduct nerve impulses *from the CNS* to muscles and glands.

Based primarily on the part of the body that responds, the PNS may be further subdivided into a **somatic** (*soma* = body) **nervous system (SNS)** and an **autonomic** (*auto* = self; *nomos* = law) **nervous system (ANS).** The SNS consists of sensory neurons that convey information from cutaneous and special sense receptors primarily in the head, body wall, and limbs to the CNS and motor neurons from the CNS that conduct impulses to *skeletal muscles* only. Because these motor responses can be consciously controlled, this portion of the SNS is *voluntary.*

The ANS consists of sensory neurons that convey information from receptors primarily in the viscera to the CNS and motor neurons from the CNS that conduct impulses to *smooth muscle, cardiac muscle,* and *glands.* Because its motor responses are not normally under conscious control, the ANS is *involuntary.*

The motor portion of the ANS consists of two branches, the **sympathetic division** and the **parasympathetic division.** With few exceptions, the viscera receive instructions from both. Usually, the two divisions have opposing actions. For example, sympathetic neurons speed the heartbeat, whereas parasympathetic neurons slow it down. Processes promoted by sympathetic neurons often involve expenditure of body energy, whereas those promoted by parasympathetic neurons restore and conserve body energy.

HISTOLOGY OF THE NERVOUS SYSTEM

Despite the complexity of the nervous system, it consists of only two principal kinds of cells: neuroglia and neurons. Neuroglia support, nurture, and protect the neurons and maintain the chemical balance of the fluid that bathes neurons. Neurons are responsible for most of the activities attributed to the nervous system: sensing, thinking, remembering, controlling muscle activity, and regulating glandular secretions. When injured, mature neurons have little capability for repair because they do not normally undergo mitosis.

Neuroglia

Neuroglia (noo-rō-GLĒ-a; *neuro* = nerve; *glia* = glue), or *glia,* are generally smaller than neurons, outnumber them by 5 to 50 times, and take up more than half the volume of the brain. Unlike neurons, neuroglia can multiply and divide in

Figure 16.1 Organization of the nervous system.

The two principal subdivisions of the nervous system are (1) the central nervous system (CNS), consisting of the brain and spinal cord, and (2) the peripheral nervous system (PNS), consisting of cranial and spinal nerves; the PNS has somatic and autonomic components.

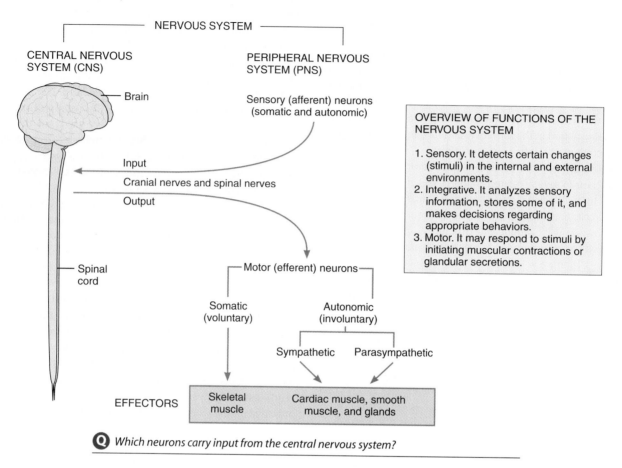

OVERVIEW OF FUNCTIONS OF THE NERVOUS SYSTEM

1. Sensory. It detects certain changes (stimuli) in the internal and external environments.
2. Integrative. It analyzes sensory information, stores some of it, and makes decisions regarding appropriate behaviors.
3. Motor. It may respond to stimuli by initiating muscular contractions or glandular secretions.

Q *Which neurons carry input from the central nervous system?*

the mature nervous system. In cases of traumatic injury, neuroglia multiply to fill in the spaces formerly occupied by neurons. Brain tumors commonly involve neuroglia. Such tumors, called *gliomas,* tend to be highly malignant and grow rapidly. Neuroglia are active participants in the physiology of the brain.

Of the six types of neuroglia, four—astrocytes, oligodendrocytes, microglia, and ependymal cells—are found in the CNS.

Astrocytes (AS-trō-sīts; *astro* = star; *cyte* = cell) are star-shaped cells with many processes. *Protoplasmic astrocytes* are found in the gray matter of the CNS, and *fibrous astrocytes* are found in the white matter of the CNS. (The distinction between gray matter and white matter will be described shortly.) Astrocytes participate in the metabolism of neurotransmitters, maintain the proper balance of potassium ions (K^+) for generation of nerve impulses, participate in brain development, help form the blood–brain barrier,

which regulates entry of substances into the brain, and provide a link between neurons and blood vessels.

Oligodendrocytes (OL-i-gō-den′-drō-sīts; *oligo* = few; *dendro* = tree) are the most common glial cells in the CNS. They have few processes and are smaller than astrocytes. They form a supporting network by twining around neurons and produce a lipid and protein wrapping called a myelin sheath (described shortly).

Microglia (mī-KROG-lē-a; *micro* = small; singular is *microglial cell*) are small, phagocytic neuroglia derived from monocytes. They protect the central nervous system from disease by engulfing invading microbes and clearing away debris from dead cells.

Ependymal (ep-EN-di-mal; *epi* = above; *dyma* = garment) **cells** are epithelial cells. They range in shape from cuboidal to columnar, and many are ciliated. Ependymal cells line the fluid-filled ventricles, cavities within the brain, and the central canal, a narrow passageway in the spinal cord. They form the fluid—termed cerebrospinal fluid (CSF)—and assist its circulation.

The remaining two types of neuroglia or supporting cells are found in the PNS: **neurolemmocytes** (noo′-rō-LE-mō-sīts) or **Schwann cells,** which produce myelin sheaths around PNS neurons, and **satellite** (SAT-i-līt) **cells,** which support neurons in ganglia (clusters of neuron cell bodies) of the PNS.

Table 16.1 summarizes the types of neuroglia.

Neurons

Neurons have the ability to respond to stimuli and convert them into nerve impulses. Very simply, a **nerve impulse (nerve action potential)** is a sudden change in electrochemical force that propagates itself along the surface of the membrane of a neuron. Among other things, a nerve impulse is sustained by the movement of sodium, potassium, and other ions between interstitial fluid and the inside of a neuron. For a nerve impulse to begin, a stimulus of adequate strength must be applied to the neuron. A **stimulus** is any change in the environment that may be of sufficient strength to initiate a nerve impulse. The ability of a neuron to respond to a stimulus and convert it into a nerve impulse is known as **excitability.**

Some neurons are tiny and relay signals over a short distance (less than 1 mm) within the CNS. Others are the longest cells in your body. Motor neurons that cause muscles to wiggle your toes, for example, extend from the lumbar region of your spinal cord (just above waist level) to your foot. Some sensory neurons are even longer. Those that allow you to feel the position of your wiggling toes stretch all the way from your foot to the lower portion of your brain. Nerve impulses travel these great distances at speeds ranging from 0.5 to 130 meters per second (1 to 280 mi/hr).

The functional contact between two neurons or between a neuron and an effector (muscle or gland) cell is called a **synapse.** The synapse between a motor neuron and a muscle fiber is called a **neuromuscular junction** (see Figure 9.4c), whereas the synapse between a neuron and glandular cells is called a **neuroglandular junction.**

Parts of a Neuron

Most neurons have three parts: (1) cell body, (2) dendrites, and (3) axon (Figure 16.2). The **cell body** (**soma** or **perikaryon**) contains a nucleus surrounded by cytoplasm that includes typical organelles such as lysosomes, mitochondria, and a Golgi complex. Many neurons also contain cytoplasmic inclusions such as **lipofuscin** pigment that occurs as clumps of yellowish brown granules. Lipofuscin, probably an end-product of lysosomal activity, collects as one ages but does not seem to harm the neuron.

Other structures in the cytoplasm are characteristic of neurons: chromatophilic substance and neurofibrils. The chromatophilic substance (**Nissl bodies**) is an orderly arrangement of rough endoplasmic reticulum, the site of protein synthesis. Newly synthesized proteins replace those being recycled and are used for growth of neurons and regeneration of damaged peripheral nerve axons. **Neurofibrils,** composed of intermediate filaments, form the cytoskeleton, which provides support and shape for the cell.

Neurons have two kinds of processes that differ in structure and function: dendrites and axons. **Dendrites** (*dendro* = tree) are the receiving or input portion of a neuron. They are usually short, tapering, and highly branched. Often, the dendrites form a tree-shaped array of processes that emerge from the cell body. Usually, dendrites are not myelinated. Their cytoplasm contains chromatophilic substance, mitochondria, and other organelles.

The second type of process, the **axon** (*axon* = axis), is a long, thin, cylindrical projection that may be myelinated and that conducts nerve impulses toward another neuron, muscle fiber, or gland cell. It joins the cell body at a cone-shaped elevation called the **axon hillock** (*hilloc* = small hill). The first portion of the axon is called the **initial segment.** Except in sensory neurons, nerve impulses arise at the junction of the axon hillock and initial segment, which is called the **trigger zone,** and then conduct along the axon. An axon contains mitochondria, neurofibrils, and microtubules but no rough endoplasmic reticulum; thus it does not carry on protein synthesis. Its cytoplasm, called **axoplasm,** is surrounded by a plasma membrane known as the **axolemma** (*lemma* = sheath or husk). Along the length of an axon, side branches called **axon collaterals** may branch off, typically at a right angle to the axon. The axon and its collaterals end by dividing into many fine processes called **axon terminals.**

The tips of the axon terminals swell into bulb-shaped structures called **synaptic end bulbs.** They contain many minute membrane-enclosed sacs called **synaptic vesicles** that store a chemical substance known as a **neurotransmitter.** Neurons influence the activity of other neurons, muscle fibers, or gland cells by releasing neurotransmitter from synaptic vesicles at synapses.

The axons of most mammalian neurons are surrounded by a multilayered lipid and protein covering produced by neuroglia that is called the **myelin sheath.** The sheath electrically insulates the axon of a neuron and increases the speed of nerve impulse conduction. Axons with such a covering are said to be *myelinated,* whereas those without it are *unmyelinated.* However, electron micrographs reveal that even unmyelinated axons are surrounded by a thin coat of neuroglial plasma membrane.

Two types of neuroglia produce myelin sheaths: neurolemmocytes (in the PNS) and oligodendrocytes (in the CNS). In the PNS, neurolemmocytes (Schwann cells) begin to form myelin sheaths around axons during fetal development. Each neurolemmocyte wraps about 1 millimeter (1 mm = 0.04 in.) of a single axon's length by spiraling

Table 16.1 *Neuroglia*

CENTRAL NERVOUS SYSTEM

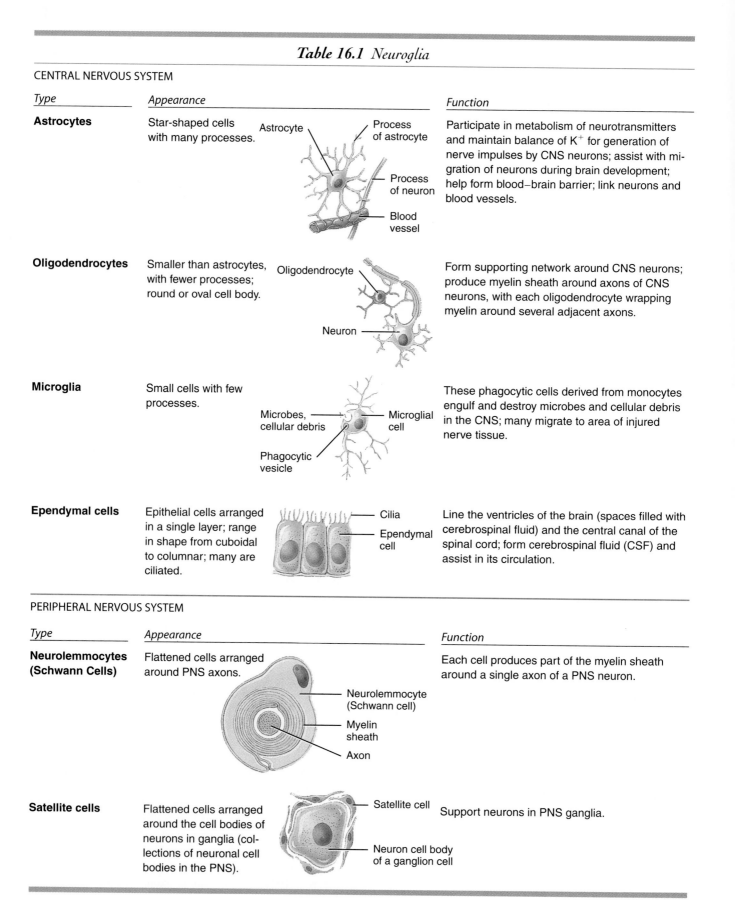

Type	Appearance	Function
Astrocytes	Star-shaped cells with many processes.	Participate in metabolism of neurotransmitters and maintain balance of K^+ for generation of nerve impulses by CNS neurons; assist with migration of neurons during brain development; help form blood–brain barrier; link neurons and blood vessels.
Oligodendrocytes	Smaller than astrocytes, with fewer processes; round or oval cell body.	Form supporting network around CNS neurons; produce myelin sheath around axons of CNS neurons, with each oligodendrocyte wrapping myelin around several adjacent axons.
Microglia	Small cells with few processes.	These phagocytic cells derived from monocytes engulf and destroy microbes and cellular debris in the CNS; many migrate to area of injured nerve tissue.
Ependymal cells	Epithelial cells arranged in a single layer; range in shape from cuboidal to columnar; many are ciliated.	Line the ventricles of the brain (spaces filled with cerebrospinal fluid) and the central canal of the spinal cord; form cerebrospinal fluid (CSF) and assist in its circulation.

Labels in Astrocyte figure: Astrocyte, Process of astrocyte, Process of neuron, Blood vessel

Labels in Oligodendrocyte figure: Oligodendrocyte, Neuron

Labels in Microglia figure: Microbes, cellular debris, Microglial cell, Phagocytic vesicle

Labels in Ependymal figure: Cilia, Ependymal cell

PERIPHERAL NERVOUS SYSTEM

Type	Appearance	Function
Neurolemmocytes (Schwann Cells)	Flattened cells arranged around PNS axons.	Each cell produces part of the myelin sheath around a single axon of a PNS neuron.
Satellite cells	Flattened cells arranged around the cell bodies of neurons in ganglia (collections of neuronal cell bodies in the PNS).	Support neurons in PNS ganglia.

Labels in Neurolemmocyte figure: Neurolemmocyte (Schwann cell), Myelin sheath, Axon

Labels in Satellite cell figure: Satellite cell, Neuron cell body of a ganglion cell

Figure 16.2 Structure of a motor (efferent) neuron. Arrows indicate the direction of information flow. The break indicates that the axon is actually longer than shown.

The basic parts of a neuron are (1) dendrites, (2) cell body, and (3) axon.

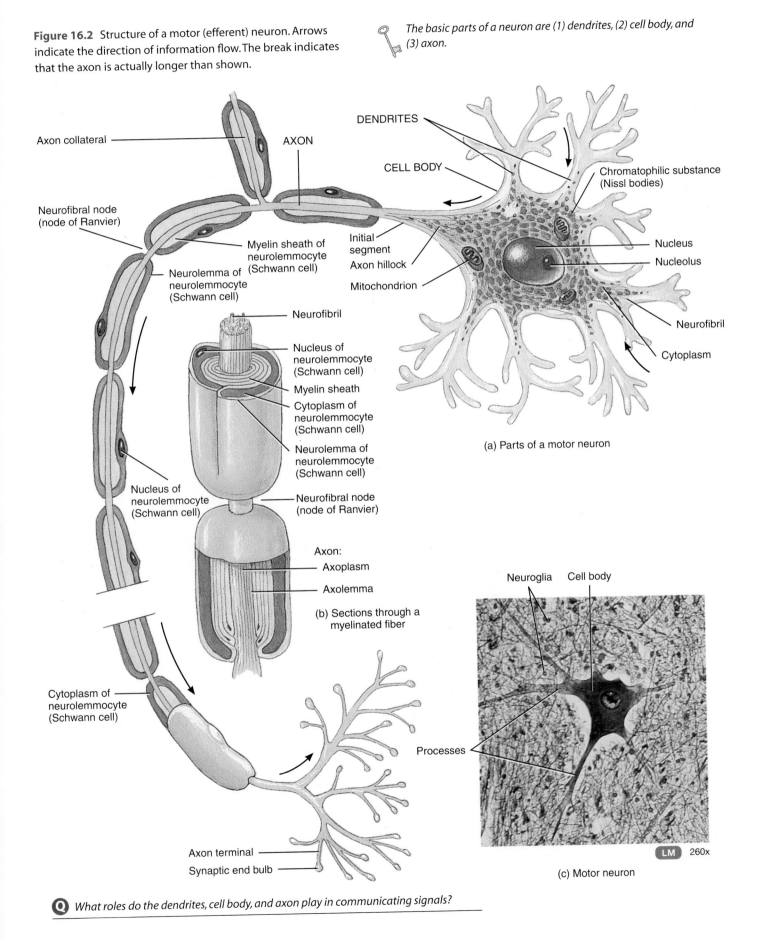

Axon collateral

AXON

DENDRITES

CELL BODY

Chromatophilic substance (Nissl bodies)

Neurofibral node (node of Ranvier)

Myelin sheath of neurolemmocyte (Schwann cell)

Neurolemma of neurolemmocyte (Schwann cell)

Initial segment

Axon hillock

Mitochondrion

Nucleus

Nucleolus

Neurofibril

Cytoplasm

Neurofibril

Nucleus of neurolemmocyte (Schwann cell)

Myelin sheath

Cytoplasm of neurolemmocyte (Schwann cell)

Neurolemma of neurolemmocyte (Schwann cell)

Nucleus of neurolemmocyte (Schwann cell)

Neurofibral node (node of Ranvier)

Axon:
- Axoplasm
- Axolemma

(a) Parts of a motor neuron

(b) Sections through a myelinated fiber

Cytoplasm of neurolemmocyte (Schwann cell)

Neuroglia   Cell body

Processes

Axon terminal

Synaptic end bulb

LM 260x

(c) Motor neuron

Q *What roles do the dendrites, cell body, and axon play in communicating signals?*

Figure 16.3 Myelinated and unmyelinated axons.

Axons of most mammalian neurons are surrounded by a myelin sheath that is produced by neurolemmocytes in the peripheral nervous system and by oligodendrocytes in the central nervous system.

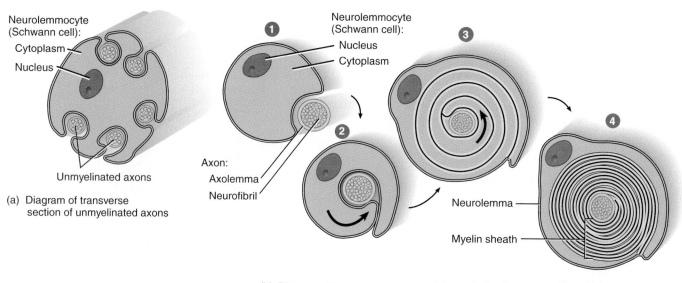

(a) Diagram of transverse section of unmyelinated axons

(b) Diagram of transverse sections of stages in the formation of a myelin sheath

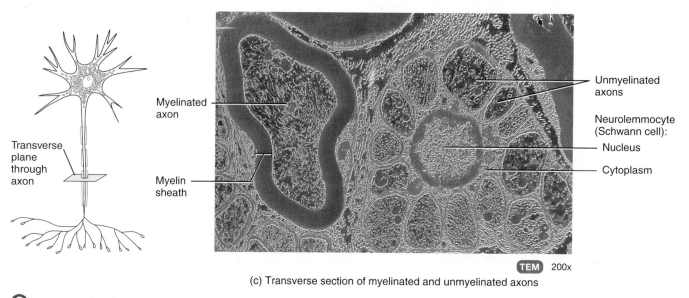

(c) Transverse section of myelinated and unmyelinated axons

Q *What is the functional advantage of myelination?*

many times around the axon (Figure 16.3b). Up to 500 neurolemmocytes participate in forming a myelin sheath around the longest axons in your body. Eventually, multiple layers of glial plasma membrane surround the axon, with the neurolemmocyte cytoplasm and nucleus forming the outermost layer. The inner portion, consisting of up to 100 layers of neurolemmocyte membrane, is the myelin sheath. The outer nucleated cytoplasmic layer of the neurolemmocyte, which encloses the myelin sheath, is called the **neurolemma (sheath of Schwann).** A neurolemma is found only around axons in the PNS. When an axon is injured, the neurolemma aids regeneration by forming a regeneration tube that guides and stimulates regrowth of the axon. At intervals along an axon, the myelin sheath has gaps called **neurofibral nodes** or **nodes of Ranvier** (RON-vē-ā; see Figure 16.2). Each neurolemmocyte wraps the axon segment between two nodes.

In the CNS an **oligodendrocyte** myelinates parts of many axons in somewhat the same manner as a neurolemmocyte myelinates part of a single PNS axon (see Table 16.1). It puts forth an average of 15 broad, flat processes that spiral about CNS axons and deposit a myelin sheath. A neurolemma is not formed, however, because the oligodendrocyte cell body and nucleus do not envelop the axon. Neurofibral nodes are present, but they are fewer in number. Axons in the CNS display little regrowth after injury. This is thought to be due, in part, to the absence of a neurolemma and in part to an inhibitory influence exerted by CNS neuroglia.

The amount of myelin increases from birth to maturity, and its presence greatly increases the speed of nerve impulse conduction. Because myelination is still in progress during infancy, an infant's responses to stimuli are not as rapid or coordinated as those of an older child or an adult. Certain diseases such as multiple sclerosis and Tay–Sachs disease cause destruction of myelin sheaths.

Clinical Application

Multiple Sclerosis (MS)

Multiple sclerosis (MS) is a progressive destruction of myelin sheaths of neurons in the CNS. It is a chronic, often disabling disease that usually occurs between ages 20–40. It is two times more common in females than males and affects 2.6 million people worldwide. The name multiple sclerosis describes the anatomical pathology: myelin sheaths deteriorate to *scleroses,* which are hardened scars or plaques, in *multiple* regions. The destruction of myelin sheaths slows and short-circuits conduction of nerve impulses. Usually, the first symptoms, such as muscular weakness, abnormal sensations, or double vision, occur in early adult life. After an attack there is a period of remission during which the symptoms temporarily disappear. Sometime later new plaques develop, and the victim suffers a second attack. One attack follows another over the years, usually every year or two. The result is a progressive loss of function interspersed with remission periods during which neurons that are undamaged regain their ability to conduct nerve impulses. The symptoms worsen dramatically when body temperature is raised. Although the precise cause of MS is unclear, some evidence implicates a viral infection that precipitates an autoimmune response, specifically the destruction of myelin-producing oligodendrocytes by the body's cytotoxic (killer) cells. Initiating factors include infection, pregnancy, injury, stress, and fatigue. In 1993 the Food and Drug Administration licensed treatment of certain patients with MS with injections of Betaseron (a form of interferon beta). This is the first drug approved specifically for treatment of MS, and its effectiveness in slowing or preventing progression of MS is still being evaluated. ▪

Nerve fiber is a general term for any neuronal process (dendrite or axon). Most often, it refers to an axon and its sheaths. A **nerve** is a bundle of many nerve fibers that course along the same path in the PNS, such as the ulnar nerve in the arm or the sciatic nerve in the thigh. Most nerves include bundles of both sensory and motor fibers and are surrounded by three connective tissue coats (Figure 16.4). Individual axons, whether myelinated or unmyelinated, are wrapped in **endoneurium** (en′-dō-NOO-rē-um). Groups of axons with their endoneurium are arranged in bundles called **fascicles,** and each fascicle is wrapped in **perineurium** (per′-i-NOO-rē-um). The superficial covering around the entire nerve is the **epineurium** (ep′-i-NOO-rē-um). You may recall from Chapter 10 that the connective tissue coverings of skeletal muscles—endomysium, perimysium, and epimysium—are similar in organization to those of nerves. This organization is common to excitable tissues.

Nerve cell bodies in the PNS generally are clustered together to form **ganglia** (GANG-lē-a; *ganglion* = knot). A **tract** is a bundle of nerve fibers, without connective tissue elements, in the CNS. Tracts may interconnect different regions of the brain or extend long distances up or down the spinal cord and connect with special regions of the brain. A **nucleus** is a collection of nerve cell bodies in the CNS.

Structural Variation in Neurons

Neurons display great diversity in size and shape. For example, their cell bodies range in diameter from 5 μm (smaller than a red blood cell) up to 135 μm (barely large enough to see with the unaided eye). The pattern of dendritic branching is varied and distinctive for neurons in different parts of the nervous system. A few small neurons lack an axon and many others have very short axons. Among the longest neurons, however, are ones that have axons that extend for a meter (3.2 ft) or more.

Classification of Neurons

Both structural and functional features are used to classify the different neurons in the body. *Structural classification* is based on the number of processes extending from the cell body (Figure 16.5). **Multipolar neurons** usually have several dendrites and one axon (see also Figure 16.2). Most neurons in the brain and spinal cord are of this type.

Bipolar neurons have one main dendrite and one axon and are found in the retina of the eye, inner ear, and olfactory area of the brain. **Unipolar neurons** have just one process extending from the cell body and are always sensory neurons. They originate in the embryo as bipolar neurons, but during development the axon and dendrite fuse into a single process. The single process divides into two branches a short distance from the cell body. Both branches have the structure and function of an axon. They are long, cylindrical processes that may be myelinated, and they propagate action potentials.

Figure 16.4 Connective tissue coverings of a spinal nerve.

Three layers of connective tissue wrappings protect axons: endoneurium surrounds individual axons, perineurium surrounds bundles of axons (fascicles), and epineurium surrounds an entire nerve.

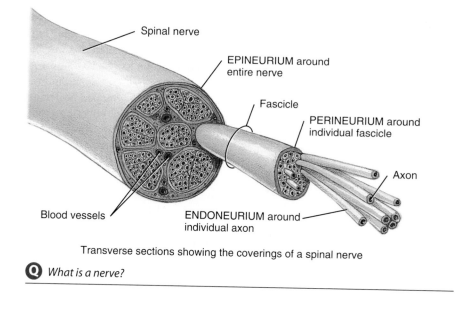

Transverse sections showing the coverings of a spinal nerve

Q *What is a nerve?*

Figure 16.5 Structural classification of neurons. The breaks indicate that axons are actually longer than shown.

A multipolar neuron has many processes extending from the cell body, whereas a bipolar neuron has two and a unipolar neuron has one.

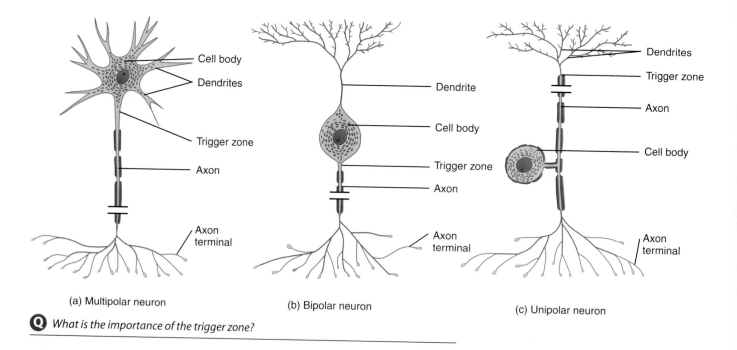

(a) Multipolar neuron

(b) Bipolar neuron

(c) Unipolar neuron

Q *What is the importance of the trigger zone?*

Figure 16.6 Structure of a typical sensory (afferent) neuron. Arrows indicate the direction of information flow. The break indicates that the process is actually longer than shown.

🔑 *Sensory neurons transmit nerve impulses from receptors to the central nervous system.*

Figure 16.7 Two examples of association neurons (interneurons). Arrows indicate the direction in which information flows.

🔑 *Association neurons carry nerve impulses from one neuron to another.*

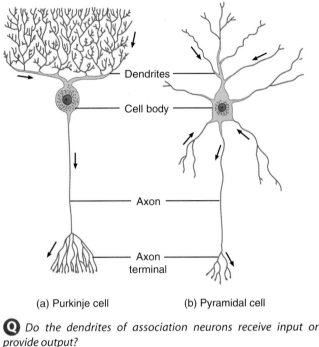

(a) Purkinje cell   (b) Pyramidal cell

Q *Do the dendrites of association neurons receive input or provide output?*

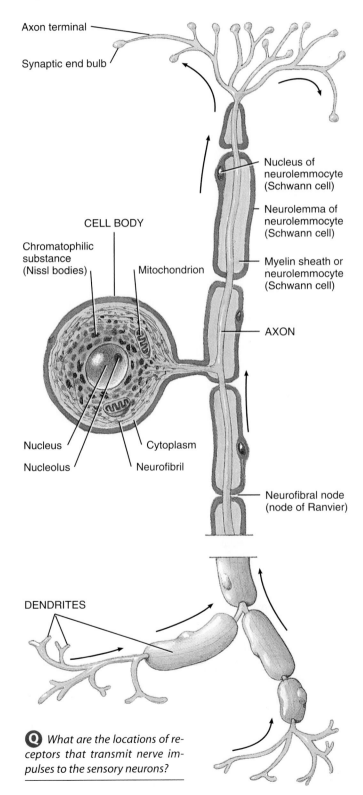

Q *What are the locations of receptors that transmit nerve impulses to the sensory neurons?*

Dendrites, located at the distal tip of the axon, monitor a sensory stimulus such as touch or pain. The trigger zone for nerve impulses in a unipolar neuron is at the junction of the dendrites and axon. The impulses then propagate toward the axon terminals, which are in the CNS.

The direction in which they transmit nerve impulses is the basis for *functional classification* of neurons. **Sensory (afferent) neurons** (Figure 16.6) transmit sensory nerve impulses from receptors in the skin, sense organs, muscles, joints, and viscera *toward* the brain and spinal cord. **Motor (efferent) neurons** convey motor nerve impulses *from* the brain and spinal cord to effectors, which may be either muscles or glands. All other neurons that are not specifically sensory or motor neurons are called **association neurons** or **interneurons.** Most neurons in the body, perhaps 90%, are association neurons.

There are thousands of different types of association neurons. Neurons often are named for the histologist who first described them, such as **Purkinje** (pur-KIN-jē) **cells** in the cerebellum (Figure 16.7a) or **Renshaw cells** in the spinal cord. Or, they are named for some aspect of their shape or appearance. For example, **pyramidal** (pi-RAM-i-dal) **cells,** found in the cerebral cortex (the superficial layer of the cerebrum), have a cell body shaped like a pyramid (Figure 16.7b).

Figure 16.8 Distribution of gray and white matter in the spinal cord and brain.

White matter consists of myelinated processes of many neurons. Gray matter consists of neuron cell bodies, dendrites, axon terminals, bundles of unmyelinated axons, and neuroglia.

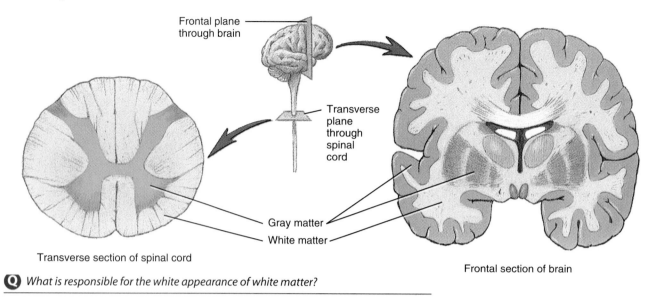

Frontal plane through brain

Transverse plane through spinal cord

Gray matter

White matter

Transverse section of spinal cord

Frontal section of brain

Q *What is responsible for the white appearance of white matter?*

Gray Matter and White Matter

In a freshly dissected section of the brain or spinal cord, some regions look white and glistening, whereas others appear gray. **White matter** refers to aggregations of myelinated processes from many neurons. The whitish color of myelin gives white matter its name. The **gray matter** of the nervous system contains either nerve cell bodies, dendrites, and axon terminals or bundles of unmyelinated axons and neuroglia. It looks grayish, rather than white, because there is no myelin in these areas.

In the spinal cord, the white matter surrounds an inner core of gray matter shaped like a butterfly or the letter **H** (Figure 16.8). In the brain, a thin superficial shell of gray matter covers the surface of the largest portion of the brain, the two cerebral hemispheres (Figure 16.8). There also are many nuclei of gray matter lying deep within the brain. A **nucleus** is a mass of nerve cell bodies and dendrites inside the CNS. Because they consist mainly of nerve cell bodies, nuclei are masses of gray matter. Most nerves in the PNS and all tracts in the CNS are white matter. The arrangement of gray and white matter in the spinal cord and brain is described more extensively in Chapters 17 and 18.

Neuronal Circuits

The CNS contains billions of neurons organized into complicated patterns called **neuronal pools.** Each pool differs from all others and has its own role in regulating the activities of the body. A neuronal pool may contain thousands or even millions of neurons. Between neurons, the cell that conducts a message toward the synapse is called the **presynaptic neuron;** the cell that transmits a message away from a synapse is called a **postsynaptic neuron.**

Neuronal pools in the CNS are arranged in patterns called **circuits** over which the nerve impulses are conducted. In **simple series circuits** a presynaptic neuron stimulates only a single neuron in a pool. The single neuron then stimulates another, and so on. But most circuits are more complex.

A single presynaptic neuron may synapse with several postsynaptic neurons. Such an arrangement, called **divergence,** permits one presynaptic neuron to influence several postsynaptic neurons or several muscle fibers or gland cells at the same time. In a **diverging circuit,** the nerve impulse from a single presynaptic neuron causes the stimulation of increasing numbers of cells along the circuit (Figure 16.9a). For example, a small number of neurons in the brain that govern a particular body movement stimulate a much larger number of neurons in the spinal cord. Sensory signals also feed into diverging circuits and are often relayed to several regions of the brain.

In another arrangement, called **convergence,** several presynaptic neurons synapse with a single postsynaptic neuron. This arrangement permits more effective stimulation or inhibition of the postsynaptic neuron. In one type of **converging circuit** (Figure 16.9b), the postsynaptic neuron receives nerve impulses from several different sources. For example, a single motor neuron that synapses with skeletal muscle fibers at neuromuscular junctions (see Figure 9.4c) receives input from several pathways that originate in different brain regions.

Some circuits in your body are constructed so that once the presynaptic cell is stimulated, it will cause the postsynaptic cell to transmit a series of nerve impulses. One such circuit is called a **reverberating (oscillatory) circuit** (Figure 16.9c). In this pattern, the incoming impulse stimulates the first neuron, which stimulates the second, which stimulates the third,

Figure 16.9 Examples of neuronal circuits.

Neuronal circuits are groups of neurons arranged in patterns over which impulses are conducted.

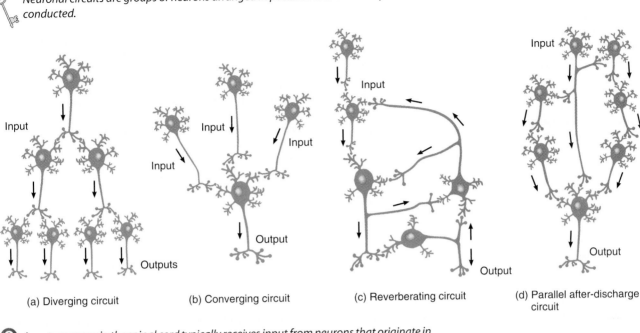

(a) Diverging circuit (b) Converging circuit (c) Reverberating circuit (d) Parallel after-discharge circuit

Q *A motor neuron in the spinal cord typically receives input from neurons that originate in several different regions of the brain. Is this an example of convergence or divergence?*

and so on. Branches from later neurons synapse with earlier ones, however, sending the impulse back through the circuit again and again. The output signal may last from a few seconds to many hours, depending on the number of synapses and arrangement of neurons in the circuit. Inhibitory neurons may turn off a reverberating circuit after a period of time. Among the body responses thought to be the result of output signals from reverberating circuits are breathing, coordinated muscular activities, waking up, sleeping (when reverberation stops), and short-term memory. One form of epilepsy (grand mal) is probably caused by abnormal reverberating circuits.

A fourth type of circuit is the **parallel after-discharge circuit** (Figure 16.9d). In this circuit, a single presynaptic cell stimulates a group of neurons, each of which synapses with a common postsynaptic cell. A signal entering such a circuit causes a prolonged output of discharge that may last from a few milliseconds to as much as several minutes after the input signal is over. If the input is excitatory, the postsynaptic neuron then can send out a stream of impulses in quick succession. It is thought that parallel after-discharge circuits may be involved in precise activities such as mathematical calculations.

Study Outline

Nervous System Divisions (p. 501)

1. The nervous system helps regulate and integrate all body activities by sensing changes (sensory function), interpreting them (integrative function), and reacting to them (motor function).
2. The central nervous system (CNS) consists of the brain and spinal cord.
3. The peripheral nervous system (PNS) consists of cranial and spinal nerves. It has sensory (afferent) and motor (efferent) components.
4. The PNS also is subdivided into somatic (voluntary) and autonomic (involuntary) nervous systems.
5. The somatic nervous system (SNS) consists of neurons that conduct impulses from cutaneous and special sense receptors to the CNS and motor neurons from the CNS that conduct impulses to skeletal muscle tissue.

6. The autonomic nervous system (ANS) contains sensory neurons from visceral organs and motor neurons that convey impulses from the CNS to smooth muscle tissue, cardiac muscle tissue, and glands.

Histology of the Nervous System (p. 501)

Neuroglia (p. 501)

1. Neuroglia are specialized tissue cells that support neurons, attach neurons to blood vessels, produce the myelin sheath around axons, and carry out phagocytosis.
2. Neuroglia include astrocytes, oligodendrocytes, microglia, ependymal cells, neurolemmocytes (Schwann cells), and satellite cells.
3. Two types of neuroglia produce myelin sheaths: oligodendrocytes myelinate axons in the CNS, and neurolemmocytes myelinate axons in the PNS.

Neurons (p. 503)

1. Most neurons, or nerve cells, consist of a cell body (soma), many dendrites, and usually a single axon. The axon conducts nerve impulses from one neuron to the dendrites or cell body of another neuron or to an effector organ of the body (muscle or gland).
2. On the basis of structure, neurons are multipolar, bipolar, or unipolar.
3. On the basis of function, neurons are sensory (afferent) neurons, which conduct impulses from receptors to the CNS; motor (efferent) neurons, which conduct impulses to effectors; or association neurons (interneurons).

Gray Matter and White Matter (p. 510)

1. White matter is aggregations of myelinated processes, whereas gray matter contains nerve cell bodies, dendrites, and axon terminals or bundles of unmyelinated axons and neuroglia.
2. In the spinal cord, gray matter forms an H-shaped inner core surrounded by white matter. In the brain, a thin outer shell of gray matter covers the cerebral hemispheres.

Neuronal Circuits (p. 510)

1. Neurons in the central nervous system are organized into patterns called neuronal pools. Each pool differs from all others and has its own role in regulating body processes.
2. Neuronal pools are organized into circuits. These include simple series, diverging, converging, reverberating (oscillatory), and parallel after-discharge circuits.

Review Questions

1. Describe the three basic functions of the nervous system. (p. 501)
2. Distinguish between the central and peripheral nervous systems and describe the functions of each subdivision. (p. 501)
3. Relate the terms *voluntary* and *involuntary* to the nervous system. (p. 501)
4. What are neuroglia? List the principal types and their functions. Why are they important clinically? (p. 501)
5. Define a neuron. Diagram and label a neuron. Next to each part, list its function. (p. 503)
6. Contrast myelination between neurolemmocytes (Schwann cells) and oligodendrocytes. (p. 503)
7. Describe the three coverings around nerve fibers. (p. 507)
8. Define the neurolemma. Why is it important? (p. 506)
9. Discuss the structural classification of neurons. Give an example of each. (p. 507)
10. What are the functional differences between a sensory and a motor neuron? Define association neuron. (p. 509)
11. Distinguish between gray matter and white matter. Where is each found? (p. 510)
12. What is a neuronal circuit? Distinguish among simple series, diverging, converging, reverberating (oscillatory), and parallel after-discharge circuits. (p. 510)

Self Quiz

Complete the following:

1. The parts of the peripheral nervous system (PNS) that connect to the central nervous system are the ___ nerves and the ___ nerves.
2. A tract is a bundle of nerve fibers located in the ___ nervous system.
3. The part of a neuron that conducts nerve impulses toward the cell body is the ___.
4. The area of communication between neurons or between a neuron and an effector is called a ___.
5. Sacs in the axon terminals that store neurotransmitters are called ___.
6. In a ___ circuit, a nerve impulse from a single presynaptic neuron causes stimulation of several postsynaptic neurons.
7. Match the following terms with their definitions:

   (1) axon
   (2) axon terminal
   (3) cell body
   (4) chromatophilic substance (Nissl bodies)
   (5) dendrite
   (6) lipofuscin
   (7) initial segment
   (8) neurofibrils

   ___ (a) contains a nucleus; cannot regenerate because it lacks mitotic apparatus
   ___ (b) yellowish pigment that increases with age; appears to be a by-product of lysosomes
   ___ (c) first portion of an axon
   ___ (d) long, thin fibrils composed of intermediate filaments
   ___ (e) orderly arrangement of rough endoplasmic reticulum; site of protein synthesis
   ___ (f) conducts nerve impulses toward cell body
   ___ (g) conducts nerve impulses away from cell body; has synaptic end bulbs that contain neurotransmitters
   ___ (h) fine filaments that are branching ends of axons and axon collaterals

Choose the one best answer to the following questions:

8. Which of the following statements is true?
   a. A neuron usually has many axons that are connected to other neurons.
   b. Most unipolar neurons are found in the central nervous system.
   c. Multipolar neurons possess numerous dendrites and only one axon.
   d. Sensory neurons are usually bipolar.
   e. All of the above are true.

9. The myelin sheath of central nervous system neurons is produced by
   a. neurolemmocytes (Schwann cells)
   b. astrocytes
   c. oligodendrocytes
   d. microglia
   e. ependymal cells
10. The term *nerve fiber* usually refers to
    a. a synaptic end bulb
    b. an axon terminal
    c. an axon and its sheaths
    d. a cell body
    e. a bundle of axons
11. The point of contact between a nerve fiber and a muscle is called the
    a. axon hillock
    b. neurofibral node
    c. trigger zone
    d. neuromuscular junction
    e. ganglion
12. Which of the following functions does *not* correctly pertain to any of the various types of specialized cells called neuroglia?
    a. regulation of entry of substances into the brain
    b. production of a myelin sheath around central nervous system neurons
    c. engulfment and destruction of microbes
    d. assistance in the production and circulation of cerebrospinal fluid
    e. replacement of damaged neurons with new ones

13. Match the following terms with their descriptions:
    (1) afferent neurons
    (2) efferent neurons
    (3) interneurons
    ___ (a) association neurons
    ___ (b) unipolar neurons
    ___ (c) sensory neurons
    ___ (d) motor neurons
    ___ (e) most central nervous system neurons are of this type
    ___ (f) these cells conduct impulses away from the CNS
    ___ (g) these cells conduct impulses toward the CNS

Are the following statements true or false?

14. The two principal divisions of the nervous system are the brain and the spinal cord.
15. Motor responses of the autonomic nervous system are under subconscious control.
16. White matter refers to aggregations of myelinated processes from many axons.
17. A nucleus is a mass of nerve cell bodies and dendrites inside the central nervous system.
18. *Nerve* is a general term for any neuronal process.
19. A neuron with several dendrites and one axon is classified as a sensory neuron.

Critical Thinking Questions

1. Mei-Ling was stopped at an intersection at which five different roads came together. Explain why she might have been reminded of a neuronal circuit.
   HINT: *Five people coming together would have a conversation.*
2. Mei-Ling had heard a catchy advertising jingle on the radio that morning and found herself still singing it hours later, while she waited at the five-way intersection. Her mind was still on anatomy because the song in her head also reminded her of a type of neuronal circuit. Name the type of circuit and explain what use the body has for it.
   HINT: *If the temperature of your oven oscillates, it goes up and down, up and down, and so on.*
3. When asked to define "gray matter," Kenny answered that it's white matter from a really old person. How would you define gray matter if you wanted to please your anatomy teacher?

HINT: *The difference between gray and white matter has to do with its structure, not with age.*
4. The nervous system has several major divisions. Trace an impulse along its route through the nervous system, starting with the division that allows you to smell breakfast cooking and ending with the division that would cause you to salivate.
   HINT: *Remember Pavlov's dogs? Salivation is automatic.*
5. How does myelin sheath production differ in the central and peripheral nervous systems?
   HINT: *The neuroglia are more important than you might think.*
6. A typical neuron has many of the same structures that are seen in other types of cells. However, centrosomes are not commonly found in adult neurons. Why is this significant?
   HINT: *Return to Chapter 2 to review the function of these missing organelles.*

Answers to Figure Questions

16.1 Motor (efferent) neurons.
16.2 Dendrites receive (motor or association neurons) or generate (sensory neurons) inputs; the cell body also receives input signals; the axon conducts nerve impulses and transmits the message to another neuron or effector cell by releasing neurotransmitter at its synaptic end bulbs.
16.3 It increases the speed of nerve impulse conduction.
16.4 A bundle of many nerve fibers that travel along the same path in the PNS.
16.5 It is the site where nerve impulses arise.
16.6 The skin, sense organs, muscles, joints, and viscera.
16.7 Receive input.
16.8 Myelin.
16.9 Convergence.

Chapter 17

The Spinal Cord and the Spinal Nerves

Student Objectives

1. Describe the protective coverings of the spinal cord.

2. Describe the external and internal anatomy of the spinal cord.

3. Describe the functions of the spinal cord.

4. Explain the origin and composition of a spinal nerve.

5. Describe the branches of a spinal nerve.

6. Define a plexus and describe the origin and distribution of nerves of the cervical, brachial, lumbar, and sacral plexuses.

7. Define dermatomes and describe their importance.

Together, the spinal cord and spinal nerves contain neuronal circuits that mediate some of your quickest reactions to environmental changes. If you pick up something very hot, for example, the grasping muscles may relax and you may drop it even before the sensation of extreme heat or pain reaches your conscious perception. This is an example of a spinal cord reflex, a quick, automatic response to certain kinds of stimuli that involves neurons (nerve cells) only in the spinal nerves and spinal cord. Besides processing reflexes, the spinal cord also is the site for integration (summing) of nerve impulses that arise locally or arrive from the periphery and brain. Moreover, the spinal cord is the highway traveled by sensory nerve impulses headed for the brain and by motor nerve impulses destined for spinal nerves. Keep in mind that the spinal cord is continuous with the brain and that together they constitute the central nervous system (CNS).

SPINAL CORD ANATOMY
Protection and Coverings

Two types of connective tissue coverings, the bony vertebrae and the tough meninges, plus a cushion of cerebrospinal fluid (produced by the brain) surround and protect the delicate nervous tissue of the spinal cord and brain.

Vertebral Column

The spinal cord is located within the vertebral (spinal) canal of the vertebral column. The vertebral foramina of all the vertebrae, stacked one on top of the other, form the canal. The surrounding vertebrae provide a sturdy shelter for the enclosed spinal cord. The vertebral ligaments, meninges, and cerebrospinal fluid provide additional protection.

Meninges

The **meninges** (me-NIN-jēz; singular, *meninx* [MĒ-ninks]) are connective tissue coverings that encircle the spinal cord and brain. They are called, respectively, the *spinal meninges* (Figure 17.1a) and the *cranial meninges* (see Figure 18.4). The most superficial of the three spinal meninges is the **dura mater** (DYOO-ra MĀ-ter; *dura* = tough, *mater* = mother), composed of dense, irregular connective tissue. It forms a sac from the level of the foramen magnum of the occipital bone, where it is continuous with the dura mater of the brain, to the second sacral vertebra, where it is close-ended. The

spinal cord is also protected by a cushion of fat and connective tissue located in the **epidural space,** a space between the dura mater and the wall of the vertebral canal.

The middle meninx is an avascular covering called the **arachnoid** (a-RAK-noyd; *arachne* = spider) because of its spider's web arrangement of delicate collagen fibers and some elastic fibers. it lies inside the dura mater and is also continuous with the arachnoid of the brain. Between the dura mater and the arachnoid is a thin **subdural space,** which contains interstitial fluid.

The innermost meninx is the **pia mater** (PĪ-a MĀ-ter; *pia* = delicate), a thin transparent connective tissue layer that adheres to the surface of the spinal cord and brain. It consists of interlacing bundles of collagen fibers and some fine elastic fibers and contains many blood vessels that supply nutrients and oxygen to the spinal cord. Between the arachnoid and the pia mater is the **subarachnoid space,** which contains cerebrospinal fluid. Inflammation of the meninges is known as **meningitis.**

All three spinal meninges cover the spinal nerves up to the point of exit from the spinal column through the intervertebral foramina. Triangular-shaped membranous extensions of the pia mater suspend the spinal cord in the middle of its dural sheath. These extensions, called *denticulate* (den-TIK-yoo-lāt; *denticulatus* = small-toothed) *ligaments,* are thickenings of the pia mater. They project laterally and fuse with the arachnoid and inner surface of the dura mater along the length of the cord between the nerve roots of spinal nerves on either side. They protect the spinal cord against shock and sudden displacement.

External Anatomy of the Spinal Cord

The **spinal cord** is roughly cylindrical but flattened slightly in its anterior–posterior dimension. In the adult, it extends from the medulla oblongata, the most inferior part of the brain, to the superior border of the second lumbar vertebra (Figure 17.2). In a newborn infant, it extends to the third or fourth lumbar vertebra. During early childhood, both the spinal cord and the vertebral column grow longer as part of the overall body growth. Around age 4 or 5, however, elongation of the spinal cord stops. Because the vertebral column continues to elongate, the spinal cord does not extend the entire length of the vertebral column. The length of the adult spinal cord ranges from 42 to 45 cm (16 to 18 in.). Its diameter is about 2 cm (3/4 in.) in the midthoracic region, somewhat larger in the lower cervical and midlumbar regions, and smallest at the inferior tip.

When the spinal cord is viewed externally, two conspicuous enlargements can be seen. The superior enlargement, the **cervical enlargement,** extends from the fourth cervical to the first thoracic vertebrae. Nerves to and from the upper limbs arise from the cervical enlargement. The inferior

Text continues on page 519.

Figure 17.1 Spinal meninges. The arrow in the inset in (b) indicates the direction from which the spinal cord is viewed (superior).

Meninges are connective tissue coverings that surround the brain and spinal cord.

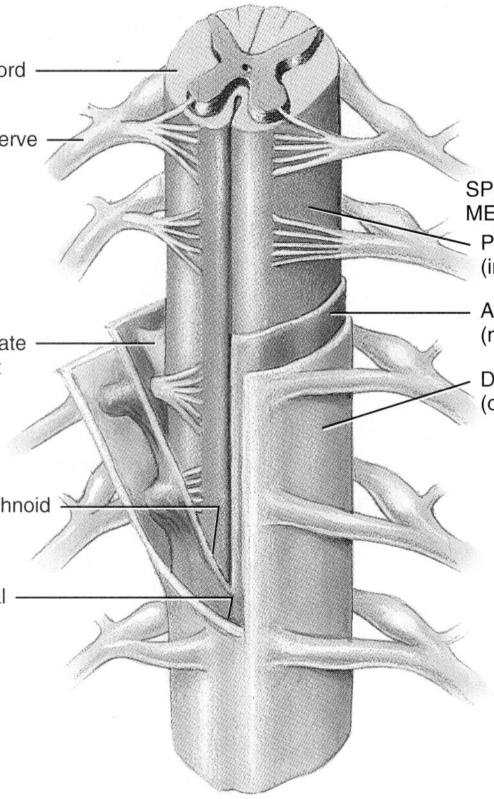

Spinal cord

Spinal nerve

SPINAL MENINGES:
Pia mater (inner)

Arachnoid (middle)

Dura mater (outer)

Denticulate ligament

Subarachnoid space

Subdural space

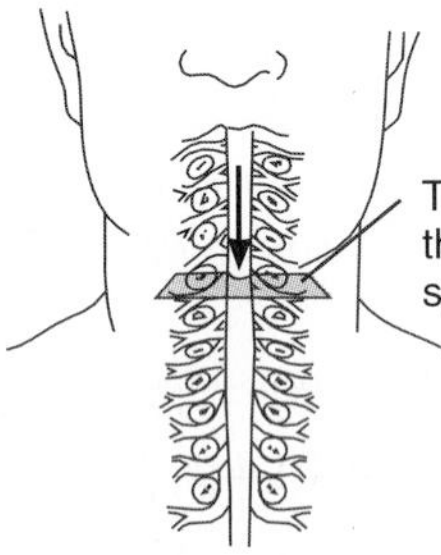

Transverse plane through cervical spinal cord

(a) Diagram of sections through spinal cord

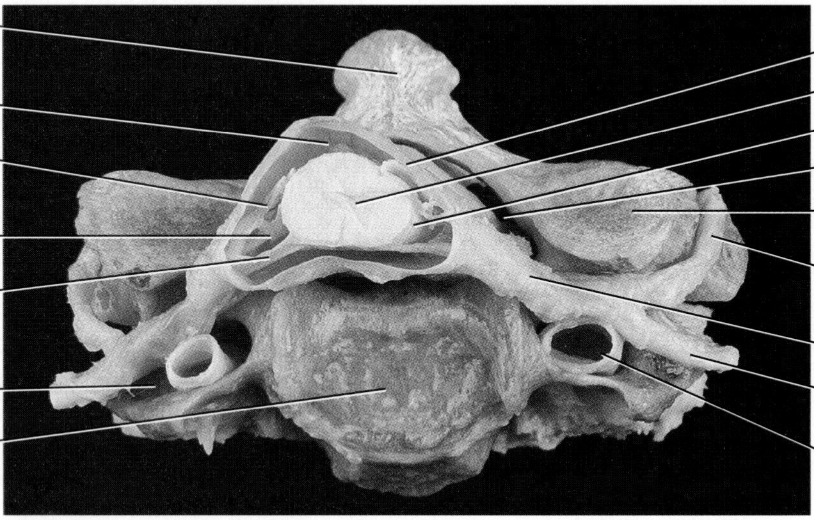

POSTERIOR

Spinous process of vertebra

Subarachnoid space

Posterior (dorsal) root of spinal nerve

Denticulate ligament

Anterior (ventral) root of spinal nerve

Transverse foramen

Body of vertebra

Dura mater and arachnoid

Spinal cord

Pia mater

Epidural space

Superior articular facet of vertebra

Posterior (dorsal) ramus of spinal nerve

Spinal nerve

Anterior (ventral) ramus of spinal nerve

Vertebral artery in transverse foramen

ANTERIOR

(a) Photograph of a transverse section of the spinal cord within a cervical vertebra

Q *What are the superior and inferior boundaries of the spinal dura mater?*

Figure 17.2 Spinal cord and spinal nerves.

🔑 *The spinal cord extends from the medulla oblongata of the brain to the superior border of the second lumbar vertebra.*

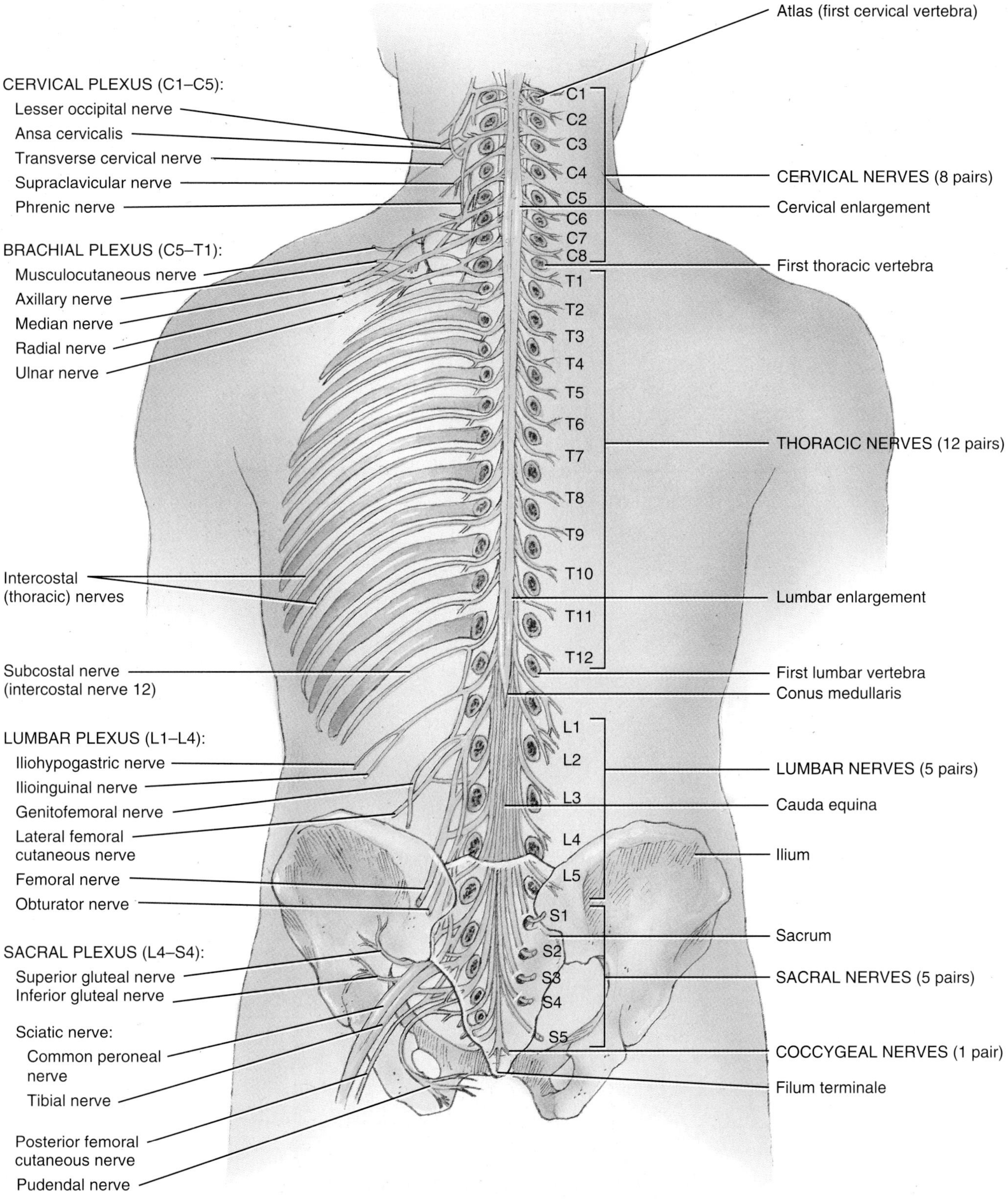

CERVICAL PLEXUS (C1–C5):
- Lesser occipital nerve
- Ansa cervicalis
- Transverse cervical nerve
- Supraclavicular nerve
- Phrenic nerve

BRACHIAL PLEXUS (C5–T1):
- Musculocutaneous nerve
- Axillary nerve
- Median nerve
- Radial nerve
- Ulnar nerve

Intercostal (thoracic) nerves

Subcostal nerve (intercostal nerve 12)

LUMBAR PLEXUS (L1–L4):
- Iliohypogastric nerve
- Ilioinguinal nerve
- Genitofemoral nerve
- Lateral femoral cutaneous nerve
- Femoral nerve
- Obturator nerve

SACRAL PLEXUS (L4–S4):
- Superior gluteal nerve
- Inferior gluteal nerve

Sciatic nerve:
- Common peroneal nerve
- Tibial nerve

Posterior femoral cutaneous nerve

Pudendal nerve

Atlas (first cervical vertebra)

C1
C2
C3
C4
C5
C6
C7
C8

CERVICAL NERVES (8 pairs)

Cervical enlargement

First thoracic vertebra

T1
T2
T3
T4
T5
T6
T7
T8
T9
T10
T11
T12

THORACIC NERVES (12 pairs)

Lumbar enlargement

First lumbar vertebra

Conus medullaris

L1
L2
L3
L4
L5

LUMBAR NERVES (5 pairs)

Cauda equina

Ilium

S1
S2
S3
S4
S5

Sacrum

SACRAL NERVES (5 pairs)

COCCYGEAL NERVES (1 pair)

Filum terminale

(a) Diagram of posterior view of entire spinal cord and portions of spinal nerves

Q *What portion of the spinal cord attaches to sensory and motor nerves of the upper limb?*

Figure 17.2 (continued)

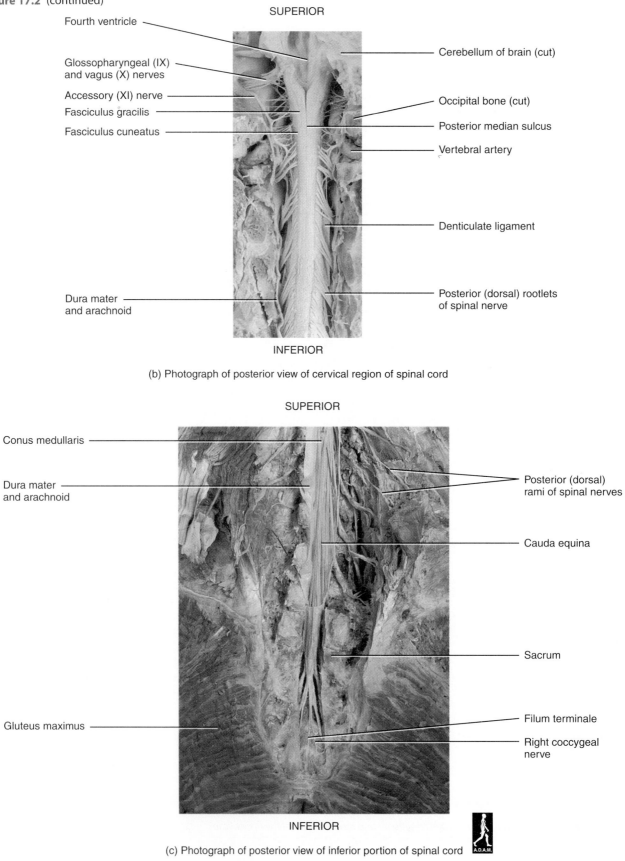

SUPERIOR

Fourth ventricle

Glossopharyngeal (IX) and vagus (X) nerves

Accessory (XI) nerve

Fasciculus gracilis

Fasciculus cuneatus

Cerebellum of brain (cut)

Occipital bone (cut)

Posterior median sulcus

Vertebral artery

Denticulate ligament

Dura mater and arachnoid

Posterior (dorsal) rootlets of spinal nerve

INFERIOR

(b) Photograph of posterior view of cervical region of spinal cord

SUPERIOR

Conus medullaris

Dura mater and arachnoid

Posterior (dorsal) rami of spinal nerves

Cauda equina

Sacrum

Gluteus maximus

Filum terminale

Right coccygeal nerve

INFERIOR

(c) Photograph of posterior view of inferior portion of spinal cord enlargement, called the **lumbar enlargement,** extends from the ninth to the twelfth thoracic vertebrae. Nerves to and from the lower limbs arise from the lumbar enlargement. Two grooves divide the cord into right and left sides (see Figure 17.3a). The **anterior median fissure** is a deep, wide groove on the anterior (ventral) side, and the **posterior median sulcus** is a shallower, narrow groove on the posterior (dorsal) surface.

Inferior to the lumbar enlargement, the spinal cord tapers to a conical portion known as the **conus medullaris** (KŌ-nus med-yoo-LAR-is; *konos* = cone), which ends at the level of the intervertebral disc between the first and second lumbar vertebrae in an adult. Arising from the conus medullaris is the **filum terminale** (FĪ-lum ter-mi-NAL-ē; *filum* = filament; *terminale* = terminal), an extension of the pia mater that extends inferiorly and anchors the spinal cord to the coccyx.

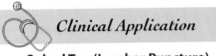

Clinical Application

Spinal Tap (Lumbar Puncture)

A **spinal tap (lumbar puncture)** is used to withdraw cerebrospinal fluid (CSF) for diagnostic purposes; introduce antibiotics, contrast media for myelography, and anesthetics; administer chemotherapy; measure CSF pressure; and evaluate the effects of treatment. In an adult, a spinal tap is normally performed between the third and fourth or fourth and fifth lumbar vertebrae. This region is inferior to the lowest portion of the spinal cord and provides relatively safe access. (A line drawn across the highest points of the iliac crests, called the supracristal line (see Figure 11.9c), passes through the spinous process of the fourth lumbar vertebra.) A local anesthetic is given, and a long needle is inserted into the subarachnoid space. ■

Some nerves that arise from the inferior part of the cord do not leave the vertebral column at the same level as they exit from the spinal cord. The roots (points of attachment to the spinal cord) of these nerves angle inferiorly in the vertebral canal from the end of the cord like wisps of hair. Appropriately, the roots of these nerves are collectively named the **cauda equina** (KAW-da ē-KWĪ-na), meaning "horse's tail."

Although the spinal cord is not actually segmented, it appears that way due to the presence of 31 pairs of spinal nerves that emerge from it. Each pair of spinal nerves is said to arise from a spinal segment.

If you examine Figure 17.2a carefully, you will notice that the right side of the illustration indicates the locations of the 31 pairs of spinal nerves. The left side of the illustration shows how most of the spinal nerves are organized into networks called plexuses. Later in the chapter, the plexuses will be discussed in more detail, and reference will be made to this illustration so that you can see how the plexuses relate to each other.

Internal Anatomy of the Spinal Cord

The gray matter of the spinal cord is shaped like the letter H or a butterfly and is surrounded by white matter. The gray matter consists primarily of cell bodies of neurons and neuroglia and unmyelinated axons and dendrites of association and motor neurons. The white matter consists of bundles of myelinated axons of motor and sensory neurons. The **gray commissure** (KOM-mi-shur) forms the crossbar of the H. In the center of the gray commissure is a small space called the **central canal.** This canal extends the entire length of the spinal cord. At its superior end, the central canal is continuous with the fourth ventricle (a space that contains cerebrospinal fluid) in the medulla oblongata of the brain. Anterior to the gray commissure is the **anterior (ventral) white commissure,** which connects the white matter of the right and left sides of the spinal cord.

The gray matter on each side of the spinal cord is subdivided into regions called **horns.** Those closer to the front of the spinal cord are called **anterior (ventral) gray horns,** whereas those closer to the back are **posterior (dorsal) gray horns.** Between the anterior and posterior gray horns are the **lateral gray horns.** Lateral gray horns are present only in the thoracic, upper lumbar, and sacral segments of the cord. The cell bodies of motor neurons are located in the gray matter of the cord. If a motor neuron supplies a skeletal muscle, its cell body is located in the anterior gray horn. If, however, a motor neuron supplies smooth muscle, cardiac muscle, or a gland through the autonomic nervous system, its cell body is located in the lateral gray horn. The gray matter of the cord also contains several nuclei that serve as neural processing centers for nerve impulses and origins for certain nerves. Nuclei are clusters of neuron cell bodies in the spinal cord and brain.

The white matter, like the gray matter, also is organized into regions. The anterior and posterior gray horns divide the white matter on each side into three broad areas called **columns** *(funiculi):* **anterior (ventral) white columns, posterior (dorsal) white columns,** and **lateral white columns.** Each column, in turn, contains distinct bundles of nerve fibers having a common origin or destination and carrying similar information. These bundles are called **tracts** *(fasciculi).* Tracts may extend long distances up or down the spinal cord. **Ascending (sensory) tracts** consist of axons that conduct nerve impulses upward to the brain. Tracts consisting of axons that carry nerve impulses down the cord are called **descending (motor) tracts.** Sensory and motor tracts of the spinal cord are continuous with sensory and motor tracts in the brain.

The various spinal cord segments vary in size, shape, relative amounts of gray and white matter, and distribution and shape of gray matter. These features are summarized in Table 17.1.

Figure 17.3 The organization of gray matter and white matter in the spinal cord. For simplicity, in this and several other illustrations of transverse sections of the spinal cord, dendrites are not shown. Blue and red arrows indicate the direction of nerve impulse propagation.

In the spinal cord, white matter surrounds the gray matter.

> **OVERVIEW OF FUNCTIONS OF THE SPINAL CORD**
> 1. Tracts in the white matter propagate sensory impulses from the periphery to the brain and motor impulses from the brain to the periphery.
> 2. The gray matter receives and integrates incoming and outgoing information.

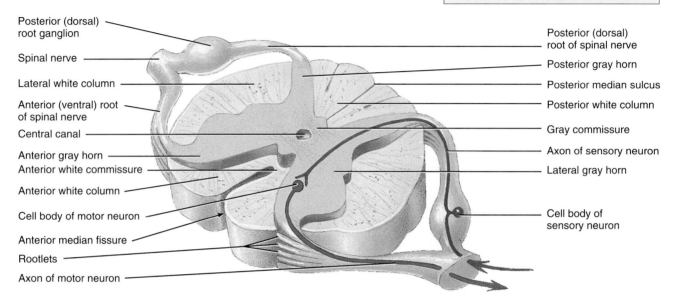

Posterior (dorsal) root ganglion
Spinal nerve
Lateral white column
Anterior (ventral) root of spinal nerve
Central canal
Anterior gray horn
Anterior white commissure
Anterior white column
Cell body of motor neuron
Anterior median fissure
Rootlets
Axon of motor neuron

Posterior (dorsal) root of spinal nerve
Posterior gray horn
Posterior median sulcus
Posterior white column
Gray commissure
Axon of sensory neuron
Lateral gray horn
Cell body of sensory neuron

(a) Diagram of a transverse section of the thoracic spinal cord

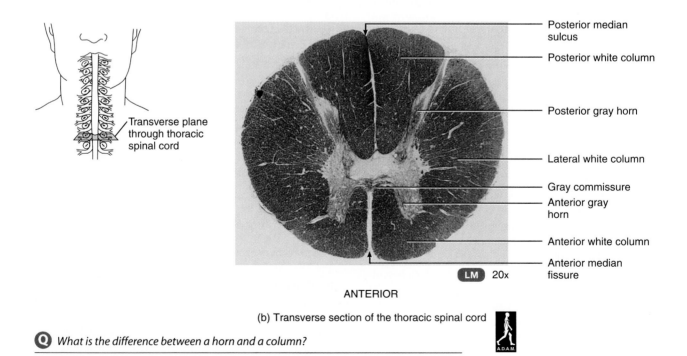

Transverse plane through thoracic spinal cord

Posterior median sulcus
Posterior white column
Posterior gray horn
Lateral white column
Gray commissure
Anterior gray horn
Anterior white column
Anterior median fissure

LM 20x

ANTERIOR

(b) Transverse section of the thoracic spinal cord

Q *What is the difference between a horn and a column?*

Table 17.1 *Comparison of Various Spinal Cord Segments*

SEGMENT	DISTINGUISHING CHARACTERISTICS
Cervical 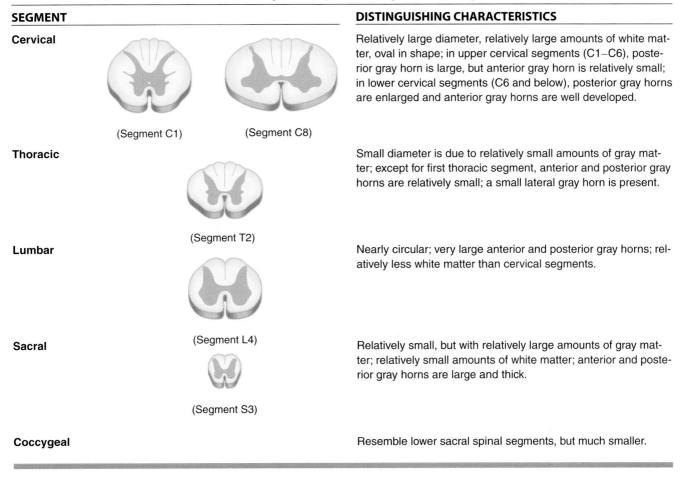 (Segment C1) (Segment C8)	Relatively large diameter, relatively large amounts of white matter, oval in shape; in upper cervical segments (C1–C6), posterior gray horn is large, but anterior gray horn is relatively small; in lower cervical segments (C6 and below), posterior gray horns are enlarged and anterior gray horns are well developed.
Thoracic (Segment T2)	Small diameter is due to relatively small amounts of gray matter; except for first thoracic segment, anterior and posterior gray horns are relatively small; a small lateral gray horn is present.
Lumbar (Segment L4)	Nearly circular; very large anterior and posterior gray horns; relatively less white matter than cervical segments.
Sacral (Segment S3)	Relatively small, but with relatively large amounts of gray matter; relatively small amounts of white matter; anterior and posterior gray horns are large and thick.
Coccygeal	Resemble lower sacral spinal segments, but much smaller.

SPINAL CORD FUNCTIONS

The spinal cord has two principal functions. (1) The *white matter tracts* in the spinal cord are highways for nerve impulse propagation. Along these highways sensory impulses flow from the periphery to the brain and motor impulses flow from the brain to the periphery. (2) The *gray matter* of the spinal cord receives and integrates all of the incoming and outgoing information.

Sensory and Motor Tracts

Often, the name of a tract indicates its position in the white matter, where it begins and ends, and the direction of nerve impulse conduction. For example, the anterior spinothalamic tract is located in the *anterior* white column, it begins in the *spinal cord,* and it ends in the *thalamus* (a region of the brain). Because it conveys nerve impulses from the cord upward to the brain, it is a sensory (ascending) tract. Figure 17.4 shows the principal sensory and motor tracts in the

spinal cord. These tracts are described in detail in Chapter 19 in Table 19.1 on page 602.

Sensory information from receptors travels up the spinal cord to the brain along two main routes on each side of the cord: the spinothalamic tracts and the posterior column tracts. The **spinothalamic tracts** convey impulses for sensing pain, temperature, crude (poorly localized) touch, and deep pressure. The **posterior column tracts** (the fasciculus gracilis and fasciculus cuneatus) carry nerve impulses for sensing: (1) proprioception, which is awareness of the movements of muscles, tendons, and joints; (2) discriminative touch, the ability to feel exactly what part of the body is touched; (3) two-point discrimination, the ability to distinguish that two different points on the skin are touched even though they are close together; (4) pressure; and (5) vibrations.

The sensory systems keep the CNS informed of changes in the external and internal environments. Responses to this information are brought about by motor systems, which enable us to move about and change our relationship to the

Figure 17.4 Selected sensory and motor tracts of the spinal cord. Both types of tracts actually are present on both sides of the spinal cord but are depicted here separately for simplicity. These functions are discussed in Chapter 19.

The name of a tract often indicates its location in the white matter and where it begins and ends.

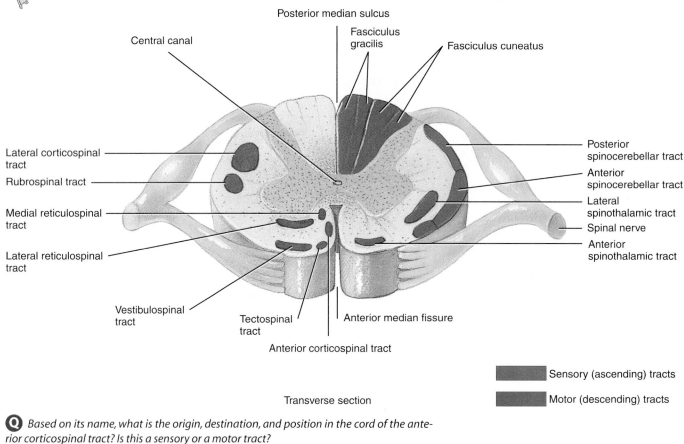

Transverse section

Sensory (ascending) tracts

Motor (descending) tracts

Q *Based on its name, what is the origin, destination, and position in the cord of the anterior corticospinal tract? Is this a sensory or a motor tract?*

world around us. As sensory information is conveyed to the CNS, it becomes part of a large pool of sensory input. Each piece of incoming information is integrated with all the other information arriving from activated sensory neurons.

With the help of association neurons, the integration process occurs not just once but in many regions of the CNS. It occurs within the spinal cord, the brain stem (which is continuous with the spinal cord), the cerebellum, and the cerebrum of the brain. As a result, motor responses to make a muscle contract or a gland secrete can be initiated at any of these levels. Most regulation of involuntary activities of smooth muscle, cardiac muscle, and glands by the autonomic nervous system originates in the brain stem and a nearby brain region called the hypothalamus. Details are given in Chapter 18.

The cerebral cortex (superficial gray matter of the cerebrum) plays a major role in controlling precise, voluntary muscular movements, whereas other brain regions provide important integration for regulation of automatic movements, such as arm swinging during walking. Motor output to skeletal muscles travels down the spinal cord in two types of descending

pathways: direct and indirect. **Direct (pyramidal) pathways** (the lateral corticospinal, anterior corticospinal, and corticobulbar tracts) convey nerve impulses that originate in the cerebral cortex and are destined to cause precise, *voluntary* movements of skeletal muscles. The term *pyramidal* refers to neurons in the cerebral cortex called pyramidal cells (see Figure 16.7b) that have pyramid-shaped cell bodies. **Indirect pathways** (rubrospinal, tectospinal, and vestibulospinal tracts) convey nerve impulses that originate in the cerebral cortex and program *automatic* movements that help coordinate body movements with visual stimuli, maintain skeletal muscle tone and posture, and play a major role in equilibrium by regulating muscle tone in response to movements of the head.

Reflexes

Reflexes are fast, predictable, automatic responses to changes in the environment that help maintain normal physiological activities. The second principal function of the spinal cord is to serve as an integrating center for reflexes that occur in the gray matter of the cord. These are called *spinal reflexes.* The integrating center for a few reflexes is the brain stem. Such reflexes involve cranial nerves and are called *cranial reflexes.* You are most aware of *somatic reflexes,* which involve contraction of skeletal muscles. Equally important, however, are the *autonomic (visceral) reflexes,* which generally are not consciously perceived. They involve responses of smooth muscle, cardiac muscle, and glands. Body functions such as heart rate, respiration, digestion, urination, and defecation are controlled by the autonomic nervous system through autonomic reflexes.

Spinal nerves are the paths of communication between the spinal cord and most of the body. Figure 17.3a shows the two separate points of attachment called **roots** that connect each spinal nerve to a segment of the cord. The **posterior** or **dorsal (sensory) root** contains sensory nerve fibers and conducts nerve impulses from the periphery into the spinal cord. As each posterior root approaches the spinal cord, it divides into a series of small branches called *rootlets* that attach the root of the spinal cord. Each posterior root also has a swelling, the **posterior** or **dorsal (sensory) root ganglion,** which contains the cell bodies of the sensory neurons from the periphery.

The other point of attachment of a spinal nerve to the cord is the **anterior** or **ventral (motor) root.** It contains motor neuron axons and conducts impulses from the spinal cord to the periphery. Each anterior root is formed by the uniting of a series of rootlets that emerge from the spinal cord. The cell bodies of motor neurons are located in the gray matter of the cord. If a motor neuron supplies a skeletal muscle, its cell body is located in the anterior (ventral) gray horn. If, however, a motor neuron supplies smooth muscle, cardiac muscle, or a gland through the autonomic nervous system, its cell body is located in the lateral gray horn.

The route followed by a series of nerve impulses from their origin in the dendrites or cell body of a neuron in one part of the body to their arrival elsewhere in the body is called a *pathway.* Pathways are specific neuronal circuits and thus include at least one synapse. The simplest kind of pathway is known as a **reflex arc.** A reflex arc includes five functional components, as follows (Figure 17.5):

❶ **Receptor.** The distal end of a sensory neuron (dendrite) or an associated sensory structure serves as a receptor. It responds to a specific stimulus, a change in the internal or external environment, by producing a change that will ultimately trigger one or more nerve impulses.

❷ **Sensory neuron.** The nerve impulses propagate from the receptor to the axon terminals of the sensory neuron, which are located in the gray matter of the spinal cord or brain stem.

❸ **Integrating center.** This is one or more regions of gray matter within the CNS. The integrating center for the sim-

plest type of reflex is a single synapse, between a sensory neuron and a motor neuron. A reflex pathway having only one synapse in the CNS is termed a *monosynaptic reflex arc.* More often, however, the integrating center consists of one or more association neurons, which may relay the impulse to other association neurons as well as to a motor neuron. A *polysynaptic reflex arc* is one that involves more than two types of neurons and more than one CNS synapse.

❹ **Motor neuron.** Impulses triggered by the integrating center propagate out of the CNS along a motor neuron to the part of the body that will respond.

❺ **Effector.** The part of the body that responds to the motor nerve impulse, such as a muscle or gland, is the effector. Its action is called a **reflex.** If the effector is skeletal muscle, the reflex is a *somatic reflex.* If the effector is smooth muscle, cardiac muscle, or a gland, the reflex is an *autonomic (visceral) reflex.*

Reflexes are often used for diagnosing disorders of the nervous system and locating injured tissue. If a reflex is absent or abnormal, the damage may be somewhere along a particular conduction pathway. Autonomic reflexes, however, are usually not practical tools for diagnosis. It is difficult to stimulate visceral receptors, because they are deep in the body. In contrast, many somatic reflexes can be tested simply by tapping or stroking the body surface.

SPINAL NERVES

Spinal nerves connect the CNS to receptors, muscles, and glands and are a part of the peripheral nervous system (PNS). The 31 pairs of spinal nerves are named and numbered according to the region and level of the spinal cord from which they emerge (see the right side of Figure 17.2a). The first cervical pair emerges between the atlas (first cervical vertebra) and the occipital bone. All other spinal nerves emerge from the vertebral column through the intervertebral foramina between adjoining vertebrae. Thus, there are 8 pairs of cervical nerves, 12 pairs of thoracic, 5 pairs of lumbar, 5 pairs of sacral, and 1 pair of coccygeal nerves.

Not all spinal cord segments are in line with their corresponding vertebrae. Recall that the spinal cord ends near the level of the superior border of the second lumbar vertebra. Thus the roots of the lower lumbar, sacral, and coccygeal nerves descend at an angle to reach their respective foramina before emerging from the vertebral column. This arrangement constitutes the cauda equina (horse's tail). See Figure 17.2a.

As noted earlier, a typical spinal nerve has two separate points of attachment to the cord: a posterior root and an anterior root (see Figure 17.3a). The posterior and anterior

Figure 17.5 General components of a reflex arc. The arrows show the direction of nerve impulse propagation.

🔑 *Reflexes are fast, predictable, automatic responses to changes in the environment.*

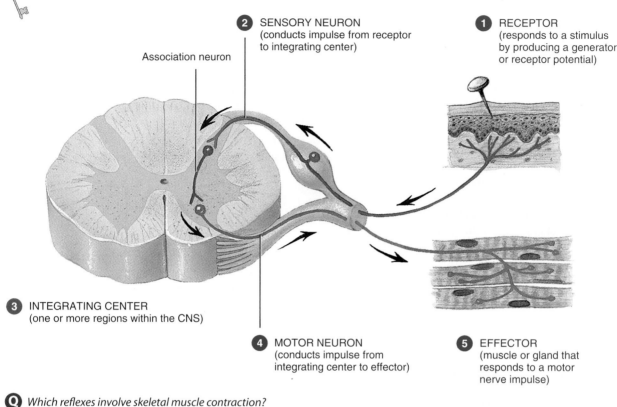

2 SENSORY NEURON
(conducts impulse from receptor to integrating center)

1 RECEPTOR
(responds to a stimulus by producing a generator or receptor potential)

Association neuron

3 INTEGRATING CENTER
(one or more regions within the CNS)

4 MOTOR NEURON
(conducts impulse from integrating center to effector)

5 EFFECTOR
(muscle or gland that responds to a motor nerve impulse)

Q *Which reflexes involve skeletal muscle contraction?*

roots unite to form a spinal nerve at the intervertebral foramen. Because the posterior root contains sensory fibers and the anterior root contains motor fibers, a spinal nerve is a *mixed nerve,* at least at its origin. Recall from Chapter 16 that three connective tissue coverings—from outside to inside, the epineurium, perineurium, and endoneurium—surround and protect spinal nerves (see Figures 16.4 and 17.6).

Distribution of Spinal Nerves

Branches

A short distance after passing through its intervertebral foramen, a spinal nerve divides into several branches (Figure 17.7). These branches are known as **rami** (RĀ-mī; singular, **ramus**). The *posterior (dorsal) ramus* (RĀ-mus) serves the deep muscles and skin of the posterior surface of the trunk. The *anterior (ventral) ramus* serves the muscles and structures of the upper and lower limbs and the lateral and anterior trunk. In addition to posterior and anterior rami, spinal nerves also give off a *meningeal branch.* This branch reenters the spinal canal through the intervertebral foramen and supplies the vertebrae, vertebral ligaments, blood vessels of the spinal cord, and meninges. Other branches of a spinal nerve are the *rami communicantes* (kō-myoo-ni-KAN-tēz),

components of the autonomic nervous system whose structure and function are discussed in Chapter 21.

Plexuses

The anterior (ventral) rami of spinal nerves, except for thoracic nerves T2–T12, do not go directly to the body structures they supply. Instead, they form networks on both the left and right sides of the body by joining with various numbers of nerve fibers from anterior rami of adjacent nerves. Such a network is called a **plexus** (*plexus* = braid). The principal plexuses are the **cervical plexus, brachial plexus, lumbar plexus,** and **sacral plexus.** Refer to Figure 17.2a to see their relationships to one another. Emerging from the plexuses are nerves bearing names that are often descriptive of the general regions they serve or the course they take. Each of the nerves, in turn, may have several branches named for the specific structures they innervate.

The principal plexuses are summarized in Exhibits 17.1 through 17.4.

Recall that plexuses are networks of nerves formed by the anterior rami of spinal nerves (except for T2–T12) and the rami do not go directly to the body structures they innervate. The anterior rami of spinal nerves T2–T12 are called intercostal nerves and will be discussed now.

Figure 17.6 Coverings of a spinal nerve. See also Figure 16.4.

Three layers of connective tissue wrappings protect axons: endoneurium surrounds individual axons, perineurium surrounds bundles of axons (fascicles), and epineurium surrounds an entire nerve.

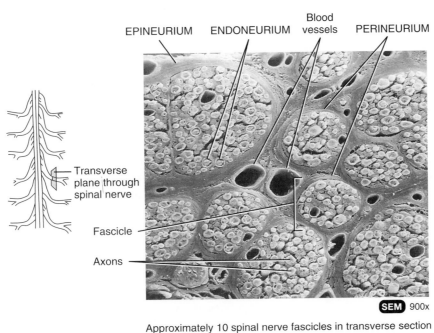

Approximately 10 spinal nerve fascicles in transverse section

Q *Why are all spinal nerves classified as mixed nerves?*

Intercostal (Thoracic) Nerves

The anterior rami of spinal nerves T2–T12 do not enter into the formation of plexuses and are known as **intercostal (thoracic) nerves.** These nerves directly innervate the structures they supply in the intercostal spaces (see Figure 17.2a). After leaving its intervertebral foramen, the anterior ramus of nerve T2 supplies the intercostal muscles of the second intercostal space, the skin of the axilla, and posteromedial aspect of the arm. Nerves T3–T6 pass in the costal grooves of the ribs and then to the intercostal muscles and skin of the anterior and lateral chest wall. Nerves T7–T12 supply the intercostal muscles and the abdominal muscles and overlying skin. The posterior rami of the intercostal nerves supply the deep back muscles and skin of the dorsal aspect of the thorax.

Dermatomes

The skin over the entire body is supplied by somatic sensory neurons that carry impulses from the skin into the spinal cord and brain stem. Likewise, the underlying skeletal muscles receive signals from somatic motor neurons that carry impulses out of the spinal cord. Each spinal nerve contains both sensory and motor neurons and serves a specific, constant segment of the body. Most of the skin of the face and scalp is served by cranial nerve V (the trigeminal nerve). The area of the skin that provides sensory input to one pair of spinal nerves or to cranial nerve V (for the face) is called a **dermatome** (*derma* = skin; *temnein* = to cut), as seen in Figure 17.8 on page 527. The nerve supply in adjacent dermatomes overlaps somewhat; in some cases the overlap is considerable. Thus there may be little loss of sensation if only the single nerve supplying one dermatome is damaged. Skeletal muscles receive their innervation from motor neurons within spinal nerves.

Because physicians know which spinal cord segments supply each dermatome and myotome, it is possible to locate damaged regions of the spinal cord. If the skin in a particular region is stimulated and the sensation is not perceived, the nerves supplying that dermatome are probably damaged. Because of overlap, however, deliberate production of a region of complete anesthesia requires that at least three adjacent spinal nerves be cut or blocked by an anesthetic drug.

Figure 17.7 Branches of a typical spinal nerve. See also Figure 17.1b.

🔑 *The branches of a spinal nerve are the dorsal ramus, ventral ramus, meningeal branch, and rami communicantes.*

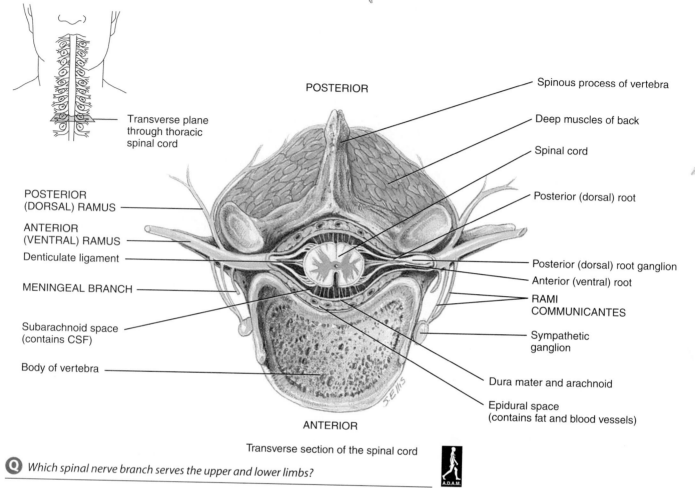

POSTERIOR

Transverse plane through thoracic spinal cord

Spinous process of vertebra

Deep muscles of back

Spinal cord

POSTERIOR (DORSAL) RAMUS

ANTERIOR (VENTRAL) RAMUS

Denticulate ligament

Posterior (dorsal) root

MENINGEAL BRANCH

Posterior (dorsal) root ganglion

Anterior (ventral) root

RAMI COMMUNICANTES

Subarachnoid space (contains CSF)

Sympathetic ganglion

Body of vertebra

Dura mater and arachnoid

Epidural space (contains fat and blood vessels)

ANTERIOR

Transverse section of the spinal cord

Q *Which spinal nerve branch serves the upper and lower limbs?*

Figure 17.8 Distribution of spinal nerves to dermatomes. An illustration can only imprecisely approximate the degree of overlap of innervation by cutaneous nerves.

A dermatome is an area of skin that provides sensory input to the dorsal roots of one pair of spinal nerves or to cranial nerve V.

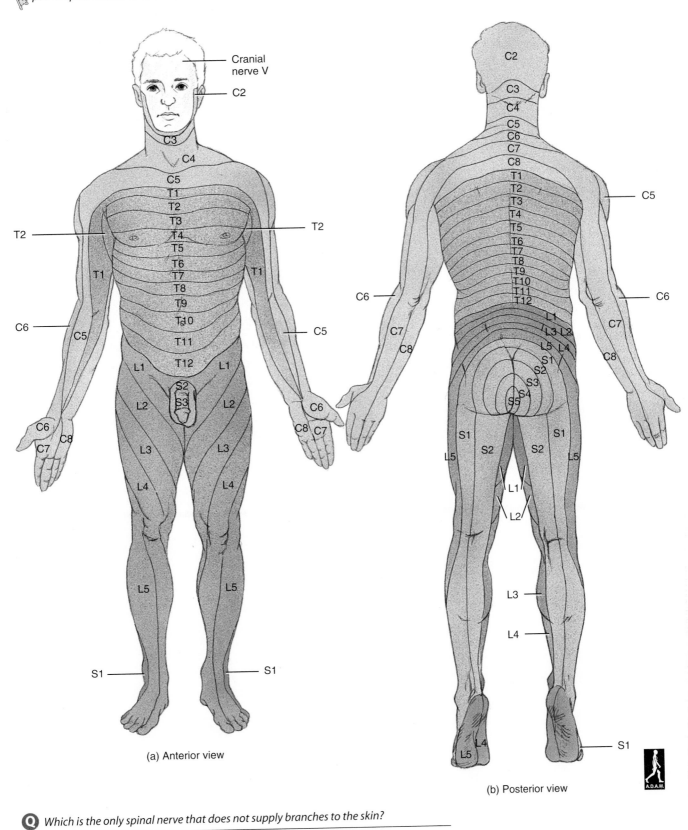

(a) Anterior view

(b) Posterior view

Q *Which is the only spinal nerve that does not supply branches to the skin?*

Exhibit 17.1 *Cervical Plexus (Figure 17.9)*

The **cervical** (SER-vi-kul) **plexus** is formed by the anterior (ventral) rami of the first four cervical nerves (C1–C4), with contributions from C5. There is one on each side of the neck alongside the first four cervical vertebrae. The roots of the plexus indicated in the diagram are the anterior rami.

The cervical plexus supplies the skin and muscles of the head, neck, and superior part of the shoulders and chest. The phrenic nerves arise from the cervical plexuses and supply motor fibers to the diaphragm. Branches of the cervical plexus also run parallel to the accessory (XI) and hypoglossal (XII) cranial nerves.

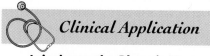

Clinical Application

Injuries to the Phrenic Nerves

Complete severing of the spinal cord above the origin of the phrenic nerves (C3, C4, and C5) causes respiratory arrest. Breathing stops because the phrenic nerves no longer send impulses to the diaphragm. ▨

SUPERFICIAL (SENSORY) BRANCHES

NERVE	ORIGIN	DISTRIBUTION
Lesser occipital	C2	Skin of scalp posterior and superior to ear.
Great auricular (aw-RIK-yoo-lar)	C2–C3	Skin anterior, inferior, and over ear and over parotid glands.
Transverse cervical	C2–C3	Skin over anterior aspect of neck.
Supraclavicular	C3–C4	Skin over superior portion of chest and shoulder.

DEEP (LARGELY MOTOR) BRANCHES

NERVE	ORIGIN	DISTRIBUTION
Ansa cervicalis (AN-sa ser-vi-KAL-is)		This nerve divides into superior and inferior roots.
Superior root	C1	Infrahyoid and geniohyoid muscles of neck.
Inferior root	C2–C3	Infrahyoid muscles of neck.
Phrenic (FREN-ik)	C3–C5	Diaphragm between thorax and abdomen.
Segmental branches	C1–C5	Prevertebral (deep) muscles of neck, levator scapulae, and middle scalene muscles.

Exhibit 17.1 *(continued)*

Figure 17.9 Cervical plexus.

The cervical plexus supplies the skin and muscles of the head, neck, superior portion of the shoulders and chest, and diaphragm.

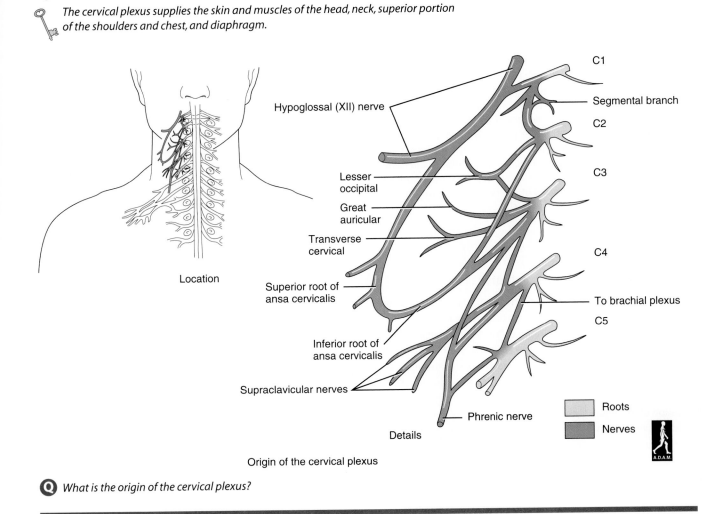

Location

Hypoglossal (XII) nerve

Lesser occipital

Great auricular

Transverse cervical

Superior root of ansa cervicalis

Inferior root of ansa cervicalis

Supraclavicular nerves

Phrenic nerve

Details

C1

Segmental branch

C2

C3

C4

To brachial plexus

C5

Roots

Nerves

Origin of the cervical plexus

Q *What is the origin of the cervical plexus?*

Exhibit 17.2 Brachial Plexus (Figures 17.10 and 17.11)

The anterior (ventral) rami of spinal nerves C5–C8 and T1 form the **brachial** (BRĀ-kē-al) **plexus,** which extends inferiorly and laterally on either side of the last four cervical and first thoracic vertebrae (Figure 17.10a). It passes superior to the first rib posterior to the clavicle and then enters the axilla.

The brachial plexus provides the entire nerve supply of the shoulders and upper limbs (Figure 17.10b and c). Five important nerves arise from the brachial plexus. (1) The **axillary nerve** supplies the deltoid and teres minor muscles. (2) The **musculo-cutaneous nerve** supplies the flexors of the arm. (3) The **radial nerve** supplies the muscles on the posterior aspect of the arm and forearm. (4) The **median nerve** supplies most of the muscles of the anterior forearm and some of the muscles of the hand. (5) The **ulnar nerve** supplies the anteromedial muscles of the forearm and most of the muscles of the hand.

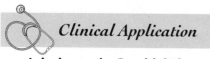

Clinical Application

Injuries to the Brachial Plexus

Injuries to the brachial plexus affect movements and cutaneous sensations of the upper limbs. The plexus may be injured by disease, stretching, and traumatic injury to the neck or axilla.

Injury to the superior roots of the plexus (C5–C6) may result from forceful pulling away of the head from the shoulder, as might occur from a heavy fall on the shoulder or during childbirth in which the infant's head is excessively stretched. The presentation of this injury is characterized by an upper limb in which the shoulder is adducted, the arm is medially rotated, the elbow is extended, the forearm is pronated, and the wrist is flexed (Figure 17.11a). This condition is called **Erb-Duchene palsy** or **waiter's tip position.** There is also loss of sensation along the lateral side of the arm.

Injury to the inferior roots of the plexus (C8–T1) results from trauma that forcefully abducts an infant's arm as may occur during difficult childbirth, or in a person clutching an object to break a fall. This form of nerve injury mostly affects the intrinsic muscles of the hand, causing a condition called **Klumpke's palsy,** in which there is hyperextension of the metacarpophalangeal joints and flexion of the interphalangeal joints.

Radial Nerve Damage

The radial (and axillary) nerves can be damaged by improperly administered intramuscular injections into the deltoid muscle. The radial nerve may also be injured when a cast is applied too tightly around the mid-humerus. **Radial nerve damage** is indicated by **wrist drop,** the inability to extend the wrist and fingers (Figure 17.11b). Sensory loss is minimal due to the overlap of sensory innervation by adjacent nerves.

Median Nerve Damage

Median nerve damage is indicated by numbness, tingling, and pain in the palm and fingers. There is also inability to pronate the forearm and flex the proximal interphalangeal joints of all digits and the distal interphalangeal joints of the second and third digits (Figure 17.11c). In addition, wrist flexion is weak and is accompanied by adduction, and thumb movements are weak.

Carpal tunnel syndrome results from compression of the median nerve inside the carpal tunnel. This narrow passageway is formed anteriorly by the flexor retinaculum (transverse carpal ligament) and posteriorly by the carpal bones (see Figure 10.19). It may be caused by any condition that aggravates compression of the contents of the carpal tunnel, such as trauma, edema, and repetitive flexion of the wrist as a result of activities such as playing video games, keyboarding at a computer terminal or typing, driving a car, cutting hair, or playing a piano.

Ulnar Nerve Damage

Ulnar nerve damage is indicated by an inability to abduct or adduct the fingers, atrophy of the interosseus muscles of the hand, hyperextension of the metacarpophalangeal joints, and flexion of the interphalangeal joints, a condition called **claw-hand** (Figure 17.11d). There is also loss of sensation over the little finger.

Long Thoracic Nerve Damage

Long thoracic nerve damage results in paralysis of the serratus anterior muscle. The medial border of the scapula protrudes, giving it the appearance of a wing. When the arm is raised the vertebral border and inferior angle of the scapula pull away from the thoracic wall, a condition called **winged scapula** (Figure 17.11e). The arm cannot be abducted beyond the horizontal position.

Exhibit 17.2 *(continued)*

NERVE	ORIGIN	DISTRIBUTION
Dorsal scapular (SKAP-yoo-lar)	C5	Levator scapulae, rhomboideus major, and rhomboideus minor muscles.
Long thoracic (tho-RAS-ik)	C5–C7	Serratus anterior muscle.
Nerve to subclavius (sub-KLĀ-ve-us)	C5–C6	Subclavius muscle.
Suprascapular (soo′-pra-SKAP-yoo-lar)	C5–C6	Supraspinatus and infraspinatus muscles.
Musculocutaneous (mus′-kyoo-lo-kyoo-TĀN-ē-us)	C5–C7	Coracobrachialis, biceps brachii, and brachialis muscles.
Lateral pectoral (PEK-to-ral)	C5–C7	Pectoralis major muscle.
Upper subscapular (sub-SKAP-yoo-lar)	C5–C6	Subscapularis muscle.
Thoracodorsal (tho-RA-ko-dor-sal)	C6–C8	Latissimus dorsi muscle.
Lower subscapular (sub-SKAP-yoo-lar)	C5–C6	Subscapularis and teres major muscles.
Axillary (AK-si-lar-ē) or **circumflex** (SER-kum-fleks)	C5–C6	Deltoid and teres minor muscles; skin over deltoid and superior posterior aspect of arm.
Median	C5–T1	Flexors of forearm, except flexor carpi ulnaris and some muscles of the hand (lateral palm); skin of lateral two-thirds of palm of hand and fingers.
Radial (RĀ-dē-al)		Triceps brachii and other extensor muscles of arm and extensor muscles of forearm; skin of posterior arm and forearm, lateral two-thirds of dorsum of hand, and fingers over proximal and middle phalanges.
Medial pectoral (PEK-to-ral)	C8–T1	Pectoralis major and pectoralis minor muscles.
Medial brachial (BRĀ-kē-al) **cutaneous** (kyoo′-TĀ-nē-us)	C8–T1	Skin of medial and posterior aspects of distal third of arm.
Medial antebrachial (an′-tē-BRĀ-kē-el) **cutaneous** (kyoo′-TĀ-nē-us)	C8–T1	Skin of medial and posterior aspects of forearm.
Ulnar (UL-nar)	C8–T1	Flexor carpi ulnaris, flexor digitorum profundus, and most muscles of the hand; skin of medial side of hand, little finger, and medial half of ring finger.

▶

Exhibit 17.2 *Brachial Plexus (continued)*

Figure 17.10 Brachial plexus.

The brachial plexus supplies the shoulders and upper limbs.

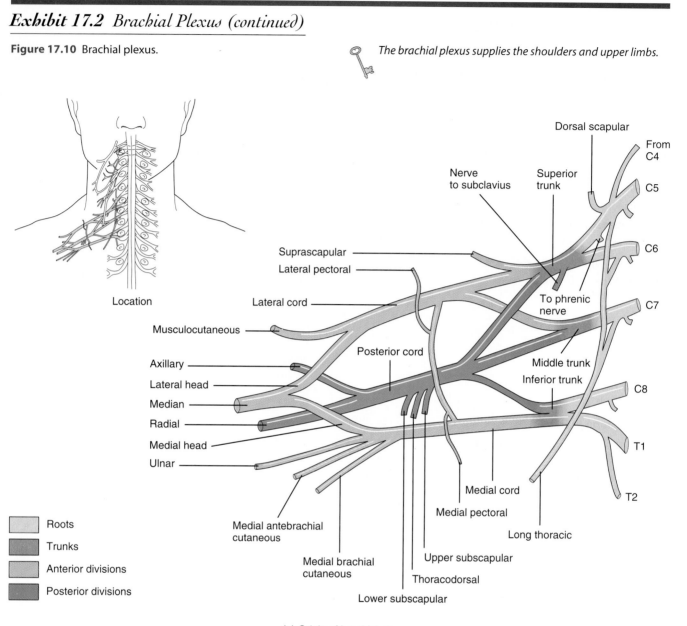

Roots

Trunks

Anterior divisions

Posterior divisions

(a) Origin of brachial plexus

Exhibit 17.2 (continued)

Figure 17.10 (continued)

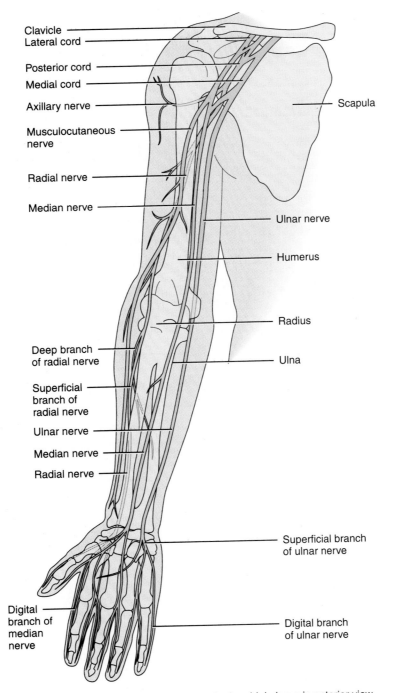

Clavicle
Lateral cord
Posterior cord
Medial cord
Axillary nerve
Musculocutaneous nerve
Radial nerve
Median nerve
Deep branch of radial nerve
Superficial branch of radial nerve
Ulnar nerve
Median nerve
Radial nerve
Digital branch of median nerve

Scapula
Ulnar nerve
Humerus
Radius
Ulna
Superficial branch of ulnar nerve
Digital branch of ulnar nerve

(b) Diagram of distribution of nerves from the brachial plexus in anterior view

Figure continues

Exhibit 17.2 Brachial Plexus (continued)

Figure 17.10 (continued)

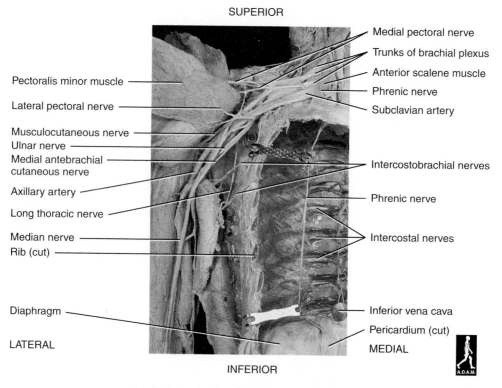

SUPERIOR

Medial pectoral nerve

Trunks of brachial plexus

Anterior scalene muscle

Pectoralis minor muscle

Phrenic nerve

Lateral pectoral nerve

Subclavian artery

Musculocutaneous nerve

Ulnar nerve

Medial antebrachial
cutaneous nerve

Intercostobrachial nerves

Axillary artery

Phrenic nerve

Long thoracic nerve

Median nerve

Intercostal nerves

Rib (cut)

Diaphragm

Inferior vena cava

Pericardium (cut)

LATERAL

MEDIAL

INFERIOR

(c) Photograph of brachial plexus in anterior view

Q *What is the origin of the brachial plexus?*

Exhibit 17.2 *(continued)*

Figure 17.11 Injuries to the brachial plexus.

Injuries to the brachial plexus affect the movements and sensations of the upper limbs.

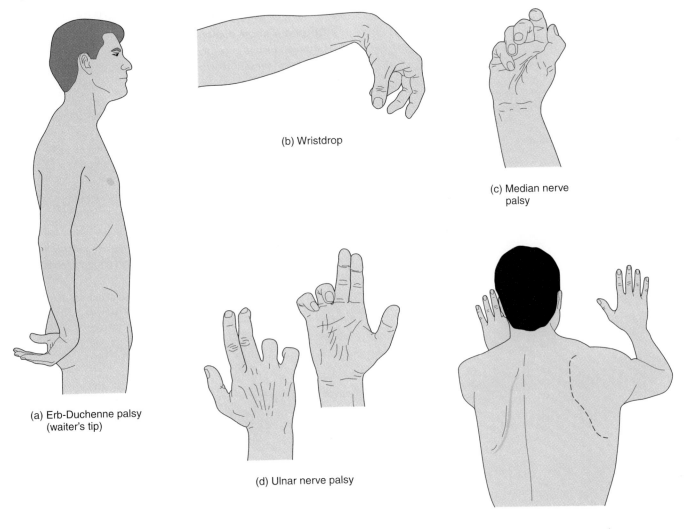

(b) Wristdrop

(c) Median nerve palsy

(a) Erb-Duchenne palsy (waiter's tip)

(d) Ulnar nerve palsy

(e) Winging of left scapula

Q *Damage to which nerve of the brachial plexus affects sensations on the palm and fingers?*

Exhibit 17.3 *Lumbar Plexus* *(Figure 17.12)*

Anterior (ventral) rami of spinal nerves L1–L4 form the **lumbar** (LUM-bar) **plexus.** It differs from the brachial plexus in that there is no intricate intermingling of fibers. On either side of the first four lumbar vertebrae, the lumbar plexus passes obliquely outward, posterior to the psoas major muscle and anterior to the quadratus lumborum muscle. It then gives rise to its peripheral nerves.

The lumbar plexus supplies the anterolateral abdominal wall, external genitals, and part of the lower limbs.

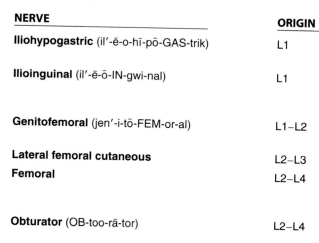

Clinical Application

Femoral Nerve Injury

The largest nerve arising from the lumbar plexus is the femoral nerve. **Femoral nerve injury,** as in stab or gunshot wounds, is indicated by an inability to extend the leg and by loss of sensation in the skin over the anteromedial aspect of the thigh.

Obturator Nerve Injury

Obturator nerve damage is a common complication of childbirth and results in paralysis of the adductor muscles of the leg and loss of sensation over the medial aspect of the thigh. ▪

NERVE	ORIGIN	DISTRIBUTION
Iliohypogastric (il'-ē-o-hī-pō-GAS-trik)	L1	Muscles of anterolateral abdominal wall; skin of inferior abdomen and buttock.
Ilioinguinal (il'-ē-ō-IN-gwi-nal)	L1	Muscles of anterolateral abdominal wall; skin of superior medial aspect of thigh, root of penis and scrotum in male, and labia majora and mons pubis in female.
Genitofemoral (jen'-i-tō-FEM-or-al)	L1–L2	Cremaster muscle; skin over middle anterior surface of thigh, scrotum in male, and labia majora in female.
Lateral femoral cutaneous	L2–L3	Skin over lateral, anterior, and posterior aspects of thigh.
Femoral	L2–L4	Flexor muscles of thigh and extensor muscles of leg; skin over anterior and medial aspect of thigh and medial side of leg and foot.
Obturator (OB-too-rā-tor)	L2–L4	Adductor muscles of leg; skin over medial aspect of thigh.

Figure 17.12 Lumbar plexus.

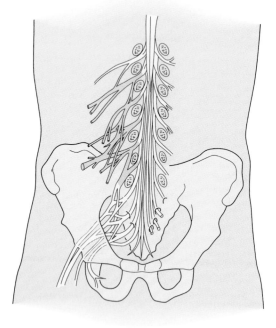

Location

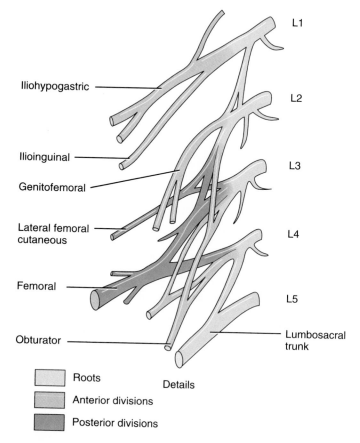

Iliohypogastric
Ilioinguinal
Genitofemoral
Lateral femoral cutaneous
Femoral
Obturator

L1
L2
L3
L4
L5
Lumbosacral trunk

 Roots
Anterior divisions
Posterior divisions

Details

(a) Origin of lumbar plexus

Exhibit 17.3 *Lumbar Plexus (continued)*

Figure 17.12 (continued)

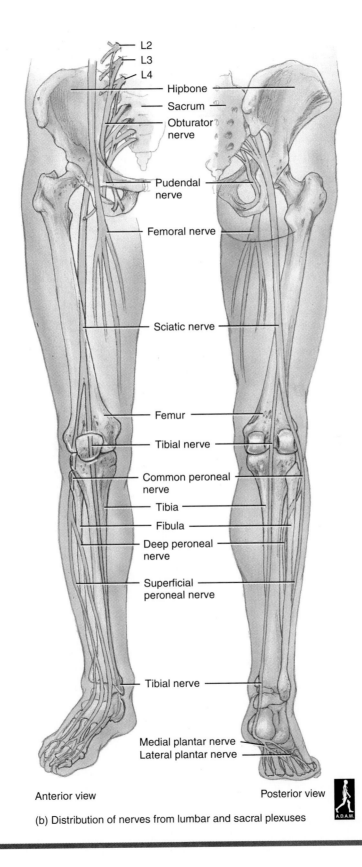

— L2
— L3
— L4

Hipbone
Sacrum
Obturator nerve
Pudendal nerve
Femoral nerve
Sciatic nerve
Femur
Tibial nerve
Common peroneal nerve
Tibia
Fibula
Deep peroneal nerve
Superficial peroneal nerve
Tibial nerve
Medial plantar nerve
Lateral plantar nerve

Anterior view   Posterior view

Q *What is the origin of the lumbar plexus?*

(b) Distribution of nerves from lumbar and sacral plexuses

Exhibit 17.4 Sacral Plexus *(Figure 17.13)*

The anterior (ventral) rami of spinal nerves L4–L5 and S1–S4 form the **sacral** (SĀ-kral) **plexus.** It is situated largely anterior to the sacrum. The sacral plexus supplies the buttocks, perineum, and lower limbs. The largest nerve in the body—the sciatic nerve—arises from the sacral plexus.

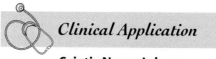

Clinical Application

Sciatic Nerve Injury

Injury to the sciatic nerve and its branches results in **sciatica,** pain that may extend from the buttock down the posterior and lateral aspect of the leg and the lateral aspect of the foot. The sciatic nerve may be injured because of a herniated (slipped) disc,

dislocated hip, osteoarthritis of the lumbosacral spine, pressure from the uterus during pregnancy, or an improperly administered gluteal intramuscular injection.

In the majority of sciatic nerve injuries, the common peroneal portion is the most affected, frequently from fractures of the fibula or by pressure from casts or splints. Damage to the common peroneal nerve causes the foot to be plantar flexed, a condition called **footdrop,** and inverted, a position called **equinovarus.** There is also loss of function along the anterolateral aspects of the leg and dorsum of the foot and toes. Injury to the tibial portion of the sciatic nerve results in dorsiflexion of the foot plus eversion, a position called **calcaneovalgus.** Loss of sensation on the sole also occurs. ▪

Figure 17.13 Sacral plexus.

The sacral plexus supplies the buttocks, perineum, and lower limbs. The distribution of the nerves of the sacral plexus is shown in Figure 17.12b.

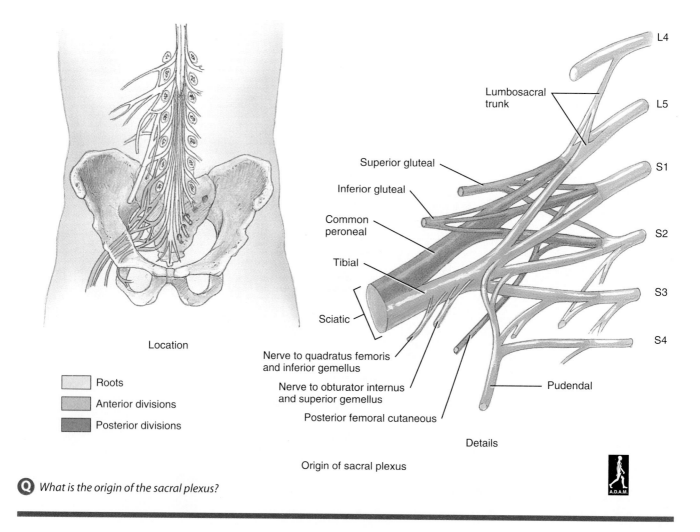

Location

☐ Roots
☐ Anterior divisions
☐ Posterior divisions

Origin of sacral plexus

Details

Q *What is the origin of the sacral plexus?*

Exhibit 17.4 Sacral Plexus (continued)

NERVE	ORIGIN	DISTRIBUTION
Superior gluteal (GLOO-tē-al)	L4–L5 and S1	Gluteus minimus and gluteus medius muscles and tensor fasciae latae. Gluteus maximus muscle.
Inferior gluteal	L5–S2	Piriformis muscle.
Nerve to piriformis (pir-i-FORM-is)	S1–S2	
Nerve to quadratus femoris (quod-RĀ-tus FEM-or-is) **and** **inferior gemellus** (jem-EL-us)	L4–L5 and S1	Quadratus femoris and inferior gemellus muscles.
Nerve to obturator internus (OB-too-rā'-tor in-TER-nus) **and** **superior gemellus**	L5–S2	Obturator internus and superior gemellus muscles.
Perforating cutaneous (PER-fō-rā-ting kyoo'-TĀ-ne-us)	S2–S3	Skin over inferior medial aspect of buttock.
Posterior femoral cutaneous	S1–S3	Skin over anal region, inferior lateral aspect of buttock, superior posterior aspect of thigh, superior part of calf, scrotum in male, and labia majora in female.
Sciatic (sī-AT-ik)	L4–S3	Actually two nerves—tibial and common peroneal—bound together by common sheath of connective tissue. It splits into its two divisions, usually at knee. (See below for distributions.) As sciatic nerve descends through the thigh, it sends branches to hamstring muscles and adductor magnus.
Tibial (TIB-ē-al)	L4–S3	Gastrocnemius, plantaris, soleus, popliteus, tibialis posterior, flexor digitorum longus, and flexor hallucis longus muscles. Branches of tibial nerve in foot are medial plantar nerve and lateral plantar nerve.
Medial plantar (PLAN-tar)		Abductor hallucis, flexor digitorum brevis, and flexor hallucis brevis muscles; skin over medial two-thirds of plantar surface of foot.
Lateral plantar		Remaining muscles of foot not supplied by medial plantar nerve; skin over lateral third of plantar surface of foot.
Common peroneal (per'-ō-NĒ-al)	L4–S2	Divides into a superficial peroneal and a deep peroneal branch.
Superficial peroneal		Peroneus longus and peroneus brevis muscles; skin over distal third of anterior aspect of leg and dorsum of foot.
Deep peroneal		Tibialis anterior, extensor hallucis longus, peroneus tertius, and extensor digitorum longus and brevis muscles; skin on adjacent sides of great and second toes.
Pudendal (pyoo-DEN-dal)	S2–S4	Muscles of perineum; skin of penis and scrotum in male and clitoris, labia majora, labia minora, and vagina in female.

Study Outline

Spinal Cord Anatomy (p. 515)

Protection and Coverings (p. 515)

1. The spinal cord is protected by the vertebral column, meninges, cerebrospinal fluid, and vertebral ligaments.
2. The meninges are three coverings that run continuously around the spinal cord and brain: dura mater, arachnoid, and pia mater.

External Anatomy of the Spinal Cord (p. 515)

1. The spinal cord begins as a continuation of the medulla oblongata and ends at about the second lumbar vertebra in an adult.
2. It contains cervical and lumbar enlargements that serve as points of origin for nerves to the limbs.
3. The tapered inferior portion of the spinal cord is the conus medullaris, from which arise the filum terminale and cauda equina.
4. The spinal cord is partially divided into right and left sides by the anterior median fissure and posterior median sulcus.
5. The gray matter in the spinal cord is divided into horns and the white matter into columns.
6. In the center of the spinal cord is the central canal, which runs the length of the spinal cord.

Internal Anatomy of the Spinal Cord (p. 519)

1. Parts of the spinal cord observed in transverse section are the gray commissure; central canal; anterior, posterior, and lateral gray horns; anterior, posterior, and lateral white columns; and ascending and descending tracts.
2. The spinal cord conveys sensory and motor information by way of the ascending and descending tracts, respectively.

Spinal Cord Functions (p. 521)

Sensory and Motor Tracts (p. 521)

1. A major function of the spinal cord is to convey nerve impulses from the periphery to the brain (sensory tracts) and to conduct motor impulses from the brain to the periphery (motor tracts).
2. Sensory information travels along two main routes: posterior column tracts and spinothalamic tracts.
3. Motor information travels along two main routes: direct and indirect tracts.

Reflexes (p. 522)

1. Another function of the spinal cord is to serve as an integrating center for spinal reflexes. This occurs in the gray matter of the spinal cord.
2. A reflex is a fast, predictable, automatic response to changes in the environment that helps to maintain normal physiological activities.

3. Reflexes may be spinal or cranial and somatic or autonomic (visceral).
4. A reflex arc is the simplest type of pathway.
5. Its components are a receptor, sensory neuron, integrating center, motor neuron, and effector.

Spinal Nerves (p. 523)

1. The 31 pairs of spinal nerves are named and numbered according to the region and level of the spinal cord from which they emerge.
2. There are 8 pairs of cervical, 12 pairs of thoracic, 5 pairs of lumbar, 5 pairs of sacral, and 1 pair of coccygeal nerves.
3. Spinal nerves typically are attached to the spinal cord by a posterior (dorsal) root and an anterior (ventral) root. All spinal nerves include both sensory and motor fibers (mixed).
4. Spinal nerves are covered by endoneurium, perineurium, and epineurium.

Distribution of Spinal Nerves (p. 524)

1. Branches of a spinal nerve include the posterior (dorsal) ramus, anterior (ventral) ramus, meningeal branch, and rami communicantes.
2. The anterior rami of spinal nerves, except for T2–T12, form networks of nerves called plexuses.
3. Emerging from the plexuses are nerves bearing names that are often descriptive of the general regions they supply or the course they take.
4. The cervical plexus supplies the skin and muscles of the head, neck, upper part of the shoulders and chest, and diaphragm, and connects with some cranial nerves.
5. The brachial plexus constitutes the nerve supply for the upper limbs and several neck and shoulder muscles.
6. The lumbar plexus supplies the anterolateral abdominal wall, external genitals, and part of the lower limbs.
7. The sacral plexus supplies the buttocks, perineum, and part of the lower limbs.
8. Anterior rami of nerves T2–T12 do not form plexuses and are called intercostal (thoracic) nerves. They are distributed directly to the structures they supply in intercostal spaces.

Dermatomes (p. 525)

1. All spinal nerves except C1 innervate specific, constant segments of the skin called dermatomes.
2. Knowledge of dermatomes helps a physician to determine which segment of the spinal cord or a spinal nerve is malfunctioning.

Review Questions

1. Describe the bony covering of the spinal cord. (p. 515)
2. Explain the location and composition of the spinal meninges. Describe the location of the epidural, subdural, and subarachnoid spaces. Define meningitis. (p. 515)
3. Describe the location of the spinal cord. What are the cervical and lumbar enlargements? (p. 515–519)
4. Define conus medullaris, filum terminale, and cauda equina.

What is a spinal segment? How is the spinal cord partially divided into a right and left side? (p. 519)
5. Based on your knowledge of the structure of the spinal cord in transverse section, define the following: gray commissure, central canal, anterior gray horn, lateral gray horn, posterior gray horn, anterior white column, lateral white column, posterior white column, ascending tract, and descending tract. (p. 519)

6. Describe the function of the spinal cord as a conduction pathway. (p. 521)
7. Describe how the spinal cord serves as an integrating center for reflexes. (p. 522)
8. What is a reflex arc? List and define the components of a reflex arc. (p. 523)
9. Define a spinal nerve. How are spinal nerves named and numbered? Why are all spinal nerves classified as mixed nerves? (p. 523–524)
10. Describe how a spinal nerve is attached to the spinal cord. (p. 523)
11. Explain how a spinal nerve is enveloped by its connective tissue coverings. (p. 524)
12. Describe the branches and innervations of a typical spinal nerve. (p. 524)
13. What is a plexus? Describe the principal plexuses and the regions they supply. (p. 524)
14. What are intercostal (thoracic) nerves? (p. 525)
15. Define a dermatome. Why is a knowledge of dermatomes important? (p. 525)
16. Describe nerve injuries to the cervical plexus (p. 528), brachial plexus (p. 530), lumbar plexus (p. 536), and sacral plexus (p. 538).

Self Quiz

Complete the following:

1. The spinal cord extends from the ___ of the brain to the level of the ___ of the vertebral column.
2. The order of the meningeal layers from deep to superficial is ___ , ___ , and ___ .
3. Individual nerve fibers are wrapped in a connective tissue covering known as ___-neurium. Groups of nerve fibers are held in bundles by ___-neurium. The entire nerve is wrapped with ___-neurium.
4. The arachnoid layer of the meninges lies between two spaces, the ___ space and the ___ space.
5. Cell bodies of motor neurons leading to skeletal muscles are located in the ___ gray horns of the spinal cord, whereas cell bodies of neurons supplying smooth muscle, cardiac muscle, or glands lie in the ___ gray horns of the spinal cord.
6. There are ___ pairs of spinal nerves that emerge from the spinal cord, named and numbered as follows: ___ pair(s) of cervical nerves, ___ pair(s) of thoracic nerves, ___ pair(s) of lumbar nerves, ___ pair(s) of sacral nerves, ___ pair(s) of coccygeal nerves.
7. Match the following terms with their descriptions:
   (1) cauda equina
   (2) conus medullaris
   (3) cervical enlargement
   (4) lumbar enlargement
   (5) filum terminale
   (6) spinal segment

   ___ (a) tapering inferior end of spinal cord
   ___ (b) any region of spinal cord from which one pair of spinal nerves arises
   ___ (c) nonnervous inferior extension of pia mater; anchors spinal cord in place
   ___ (d) "horse's tail"; extension of spinal nerve roots in lumbar and sacral regions within subarachnoid space
   ___ (e) gives rise to nerves that serve the upper limbs
   ___ (f) gives rise to nerves that serve the lower limbs

Choose the one best answer to the following questions:

8. Which of the following is true?
   a. Ascending tracts are in the white matter; descending tracts are in the gray matter.
   b. Tracts appear white because they consist of bundles of unmyelinated nerve fibers.
   c. Ascending tracts are all motor tracts.
   d. All tracts are located in gray horns.
   e. None of the above is true.
9. The branch of a spinal nerve that innervates the vertebrae, vertebral ligaments, and blood vessels of the spinal cord is known as the
   a. meningeal branch
   b. posterior (dorsal) ramus
   c. ramus communicantes
   d. white ramus
   e. anterior (ventral) ramus
10. Which of the following statements is true about a reflex arc?
    (1) It always includes at least a sensory and a motor neuron.
    (2) It always has its integrating center in the brain or spinal cord.
    (3) It always terminates in a muscle or gland.

    a. 1 only
    b. 2 only
    c. 3 only
    d. all of the above
    e. none of the above
11. The posterior (dorsal) root ganglion contains
    a. cell bodies of motor neurons
    b. cell bodies of sensory neurons
    c. cranial nerve axons
    d. synapses
    e. all of the above
12. Which of the following statements about the spinal cord is *false*?
    a. It has enlargements in the cervical and lumbar areas.
    b. It lies in the vertebral foramen.
    c. In transverse section, it contains an H-shaped area of white matter surrounded by gray matter.
    d. It is surrounded by meninges, which in turn are surrounded by an epidural space.
    e. None of the above is false.

13. The spinothalamic tracts
    a. convey motor input to skeletal muscles
    b. convey impulses for sensing pain, temperature, crude touch, and deep pressure
    c. convey impulses for sensing proprioception, discriminative touch, and vibrations
    d. play a major role in equilibrium by regulating muscle tone in response to movements of the head
    e. contain integrating centers for reflexes

14. Which of the following sequences best represents the course of a nerve impulse over a reflex arc?
    a. receptor, integrating center, sensory neuron, motor neuron, effector
    b. effector, sensory neuron, integrating center, motor neuron, receptor
    c. receptor, sensory neuron, integrating center, motor neuron, effector
    d. receptor, motor neuron, integrating center, sensory neuron, effector
    e. effector, integrating center, sensory neuron, motor neuron, receptor

Are the following statements true or false?

15. Cell bodies of sensory neurons are located in the posterior (dorsal) root ganglia.

16. The posterior median sulcus is the deeper, wider groove along the outer surface of the spinal cord.

17. The spinal cord is located in the central canal of the vertebral column.

18. The two main functions of the spinal cord are to serve as a reflex center and to be the site where sensations are felt.

19. Lateral extensions of the pia mater that help to protect the spinal cord against displacement are called denticulate ligaments.

20. A tract is a bundle of nerve fibers inside the central nervous system.

21. Networks called plexuses are formed by the anterior (ventral) rami of spinal nerves, except for T2–T12.

22. Match the following terms with their descriptions:
    (1) brachial plexus
    (2) cervical plexus
    (3) sacral plexus
    (4) lumbar plexus
    (5) intercostal nerves

    ___ (a) provides the entire nerve supply for the upper limb.
    ___ (b) contains the origin of the phrenic nerve that supplies the diaphragm
    ___ (c) forms the median, radial, and axillary nerves
    ___ (d) not a plexus at all, but rather segmentally arranged nerves
    ___ (e) supplies fibers to the scalp, neck, and part of the shoulder and chest
    ___ (f) supplies fibers to the femoral nerve that innervates the flexor muscles of the thigh and extensor muscles of the leg
    ___ (g) forms the largest nerve in the body, the sciatic nerve, that supplies the posterior of the thigh and the leg

Critical Thinking Questions

1. A high school senior dove headfirst into a murky pond. Unfortunately, he hit a submerged log with his head and is now a quadriplegic. Can you deduce the location of the injury? What is the likelihood of recovery from his injury?
   HINT: *He didn't just break bones.*

2. A woman was in labor with her first child. Her birthing coach was urging her to try for an "all-natural" delivery. She vetoed the idea and asked for an epidural anesthetic. List the tissues that the needle will pierce in order, from superficial to deep.
   HINT: Epi*dural means* above *what structure?*

3. The spinal cord is covered by protective layers and enclosed by the vertebral column. How is it able to send and receive messages from the periphery of the body?
   HINT: *The vertebral column is a stack of bones separated by intervertebral discs.*

4. Why doesn't the spinal cord "creep" up toward the head every time you bend over? Why doesn't it get all twisted out of position when you exercise?
   HINT: *How would you prevent a boat from floating away from a dock?*

5. Mohammed had a spinal tap performed at the hospital. The technician had to make several attempts, so several needle marks were clustered in the same location. Where would the needle marks be located? Why are they all in the same spot?
   HINT: *A spinal tap is also called a lumbar puncture.*

6. Trace the route taken by nerve impulses along a simple reflex arc such as would occur after you burn your finger with a match.
   HINT: *Be sure to list five components in the arc.*

Answers to Figure Questions

17.1 The foramen magnum of the occipital bone and the second sacral vertebra.

17.2 Cervical enlargement.

17.3 A horn is an area of gray matter, whereas a column is a region of white matter.

17.4 The anterior corticospinal tract originates in the cortex of the cerebrum, ends in the spinal cord, and is located on the anterior side of the cord. It contains descending fibers and is a motor tract.

17.5 Somatic reflexes.

17.6 The poster (dorsal) root contains sensory fibers, and the anterior (ventral) root contains motor fibers. The two roots unite to form the spinal nerve.

17.7 Anterior (ventral) ramus.

17.8 C1.

17.9 Anterior (ventral) rami of cervical nerves C1–C4, with contributions from C5.

17.10 Anterior (ventral) rami of spinal nerves C5–C8 and T1.

17.11 Median nerve.

17.12 Anterior (ventral) rami of spinal nerves L1–L4.

17.13 Anterior (ventral) rami of spinal nerves L4–L5 and S1–S4.

BRAIN

Principal Parts

The brain is divided into four principal parts: brain stem, diencephalon, cerebrum, and cerebellum, which are shown in sagittal section in Figure 18.1. The **brain stem** consists of the medulla oblongata, pons, and midbrain and is continuous with the spinal cord. Superior to the brain stem is the **diencephalon** (di-en-SEF-a-lon), consisting primarily of the thalamus and hypothalamus. Like the cap of a mushroom, the **cerebrum** spreads over the diencephalon. The cerebrum, which occupies most of the cranium, has right and left halves called *cerebral hemispheres.* Inferior to the cerebrum and posterior to the brain stem is the **cerebellum,** which has right and left halves called *cerebellar hemispheres.*

The brain grows very rapidly during the first few years of life. Growth is due mainly to an increase in the size of neurons already present, proliferation and growth of neuroglia, development of synaptic contacts and dendritic branching, and myelination of the various fiber tracts.

Protection and Coverings

The brain is protected by the cranial bones and cranial meninges. The **cranial meninges** shown in Figures 18.2 and 18.4 surround the brain. They are continuous with the spinal meninges, have the same basic structure, and bear the same names: the outer **dura mater,** middle **arachnoid,** and inner **pia mater.** The cranial dura mater consists of two layers. The thicker, outer *periosteal layer* tightly adheres to the cranial bones and functions as an inner periosteum of the cranial bones. The thinner, inner *meningeal layer* includes a mesothelial layer on its smooth surface. The spinal dura mater corresponds to the meningeal layer of the cranial dura mater. Between the dural layers of the brain are found the dural venous sinuses (modified veins), which drain blood from the brain into the internal jugular veins (see Figure 14.15).

The two hemispheres (sides) of the cerebrum are separated by an extension of the dura mater called the **falx cerebri.** In addition, the two hemispheres of the cerebellum are separated by an extension of the dura mater called the **falx cerebelli.** The cerebrum is separated from the cerebellum by an extension of the dura mater called the **tentorium cerebelli.**

Cerebrospinal Fluid (CSF)

The brain and spinal cord are nourished and protected against chemical or physical injury by **cerebrospinal fluid (CSF).** This fluid continuously circulates through the subarachnoid space (between the arachnoid and pia mater) around the brain and spinal cord and through cavities within the brain and spinal cord.

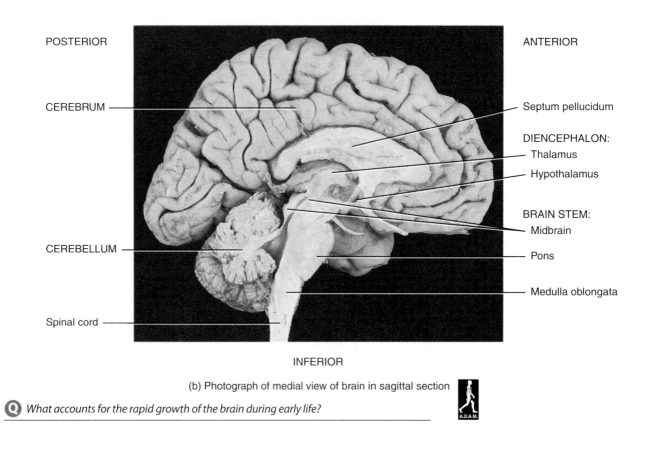

POSTERIOR   ANTERIOR

CEREBRUM

Septum pellucidum

DIENCEPHALON:
Thalamus
Hypothalamus

BRAIN STEM:
Midbrain

CEREBELLUM

Pons

Medulla oblongata

Spinal cord

INFERIOR

(b) Photograph of medial view of brain in sagittal section

Q *What accounts for the rapid growth of the brain during early life?*

Figure 18.2 Superior portion of the skull showing the protective coverings of the brain.

Cranial bones and cranial meninges protect the brain.

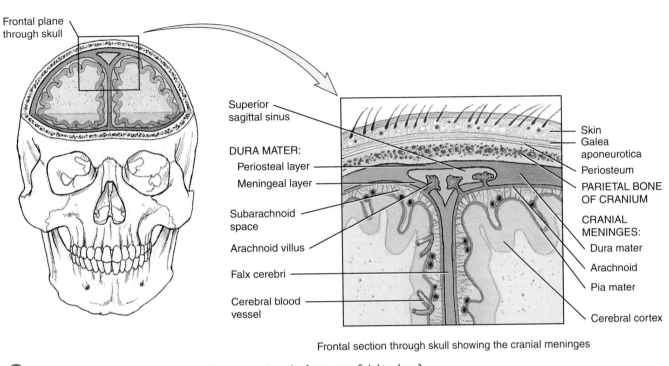

Frontal plane through skull

Superior sagittal sinus

DURA MATER:
Periosteal layer
Meningeal layer

Subarachnoid space

Arachnoid villus

Falx cerebri

Cerebral blood vessel

Skin
Galea aponeurotica
Periosteum
PARIETAL BONE OF CRANIUM

CRANIAL MENINGES:
Dura mater
Arachnoid
Pia mater

Cerebral cortex

Frontal section through skull showing the cranial meninges

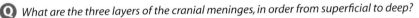

Q *What are the three layers of the cranial meninges, in order from superficial to deep?*

Figure 18.3 shows in lateral and anterior views the four CSF-filled cavities within the brain, which are called **ventricles** (VEN-tri-kuls; *ventriculus* = little belly or cavity). Each of the two **lateral ventricles** is located in a hemisphere of the cerebrum. Anteriorly, the lateral ventricles are separated by a thin membrane, the **septum pellucidum** (SEP-tum pe-LOO-si-dum; *pellucidus* = allowing passage of light). See also Figure 18.12. The **third ventricle** is a vertical slit at the midline medial to and inferior to the right and left halves of the thalamus and between the lateral ventricles. The **fourth ventricle** lies between the brain stem and the cerebellum.

The entire central nervous system contains between 80 and 150 ml (3 to 5 oz) of CSF. It is a clear, colorless liquid that contains glucose, proteins, lactic acid, urea, cations (Na$^+$, K$^+$, Ca^{2+}, Mg^{2+}), and anions (Cl$^-$ and HCO$_3^-$). It also contains some lymphocytes. CSF has three basic functions:

1. **Mechanical protection.** The fluid serves as a shock-absorbing medium to protect the delicate tissue of the brain and spinal cord from jolts that would otherwise cause them to crash against the bony walls of the cranial and vertebral cavities. The fluid also buoys the brain so that it "floats" in the cranial cavity.

2. **Chemical protection.** CSF provides an optimal chemical environment for accurate neuronal signaling. Even slight changes in the ionic composition of CSF within the brain could seriously disrupt production of action potentials.

3. **Circulation.** CSF is a medium for exchange of nutrients and waste products between the blood and nervous tissue.

We will begin the discussion of the circulation of CSF by first examining where the fluid is formed. The **choroid** (KŌ-royd; *chorion* = membrane) **plexuses** (Figure 18.4) are networks of capillaries in the walls of the ventricles. The capillaries are covered by ependymal cells that form cerebrospinal fluid from blood plasma by filtration and secretion. Because the ependymal cells are joined by tight junctions, substances entering the CSF from choroid capillaries cannot leak between these cells. Rather, they must diffuse through the ependymal cells. This **blood–cerebrospinal fluid barrier** permits certain substances to enter the CSF but excludes others. Such a barrier protects the brain and spinal cord from potentially harmful substances in the blood.

The CSF formed in the choroid plexuses of the lateral ventricles flows into the third ventricle through a narrow, oval opening, the **interventricular foramen (foramen of Monro).** See also Figure 18.3. More fluid is added by the choroid plexus in the roof of the third ventricle. The fluid then flows through the **cerebral aqueduct (aqueduct of Sylvius),** which passes through the midbrain, into the fourth ventricle. The choroid plexus of the fourth ventricle

Figure 18.3 Structures associated with the formation, circulation, and reabsorption of cerebrospinal fluid.

Ventricles are cerebrospinal fluid–filled cavities within the brain.

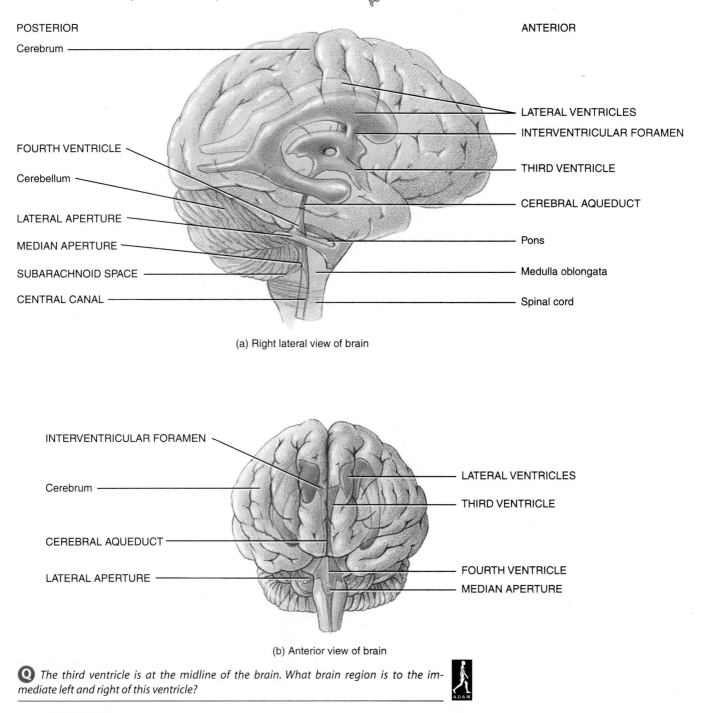

POSTERIOR

Cerebrum

FOURTH VENTRICLE

Cerebellum

LATERAL APERTURE

MEDIAN APERTURE

SUBARACHNOID SPACE

CENTRAL CANAL

ANTERIOR

LATERAL VENTRICLES

INTERVENTRICULAR FORAMEN

THIRD VENTRICLE

CEREBRAL AQUEDUCT

Pons

Medulla oblongata

Spinal cord

(a) Right lateral view of brain

INTERVENTRICULAR FORAMEN

Cerebrum

CEREBRAL AQUEDUCT

LATERAL APERTURE

LATERAL VENTRICLES

THIRD VENTRICLE

FOURTH VENTRICLE

MEDIAN APERTURE

(b) Anterior view of brain

Q *The third ventricle is at the midline of the brain. What brain region is to the immediate left and right of this ventricle?*

contributes more fluid. CSF enters the subarachnoid space through three openings in the roof of the fourth ventricle: a **median aperture (of Magendie)** and two **lateral apertures (of Luschka).** The fluid then circulates in the subarachnoid space around the posterior surface of the brain. It also passes inferiorly in the central canal of the spinal cord and to the subarachnoid space around the posterior surface of the

spinal cord, up the anterior surface of the spinal cord, and around the anterior part of the brain. From there it is gradually reabsorbed into a blood vascular sinus called the **superior sagittal sinus.** The actual reabsorption occurs through **arachnoid villi**—fingerlike extensions of the arachnoid that project into the dural venous sinuses, especially the superior sagittal sinus (see Figure 18.2). Normally, cerebrospinal

Figure 18.4 Pathways of circulating cerebrospinal fluid.

Cerebrospinal fluid is formed by ependymal cells that cover the choroid plexus of the ventricles.

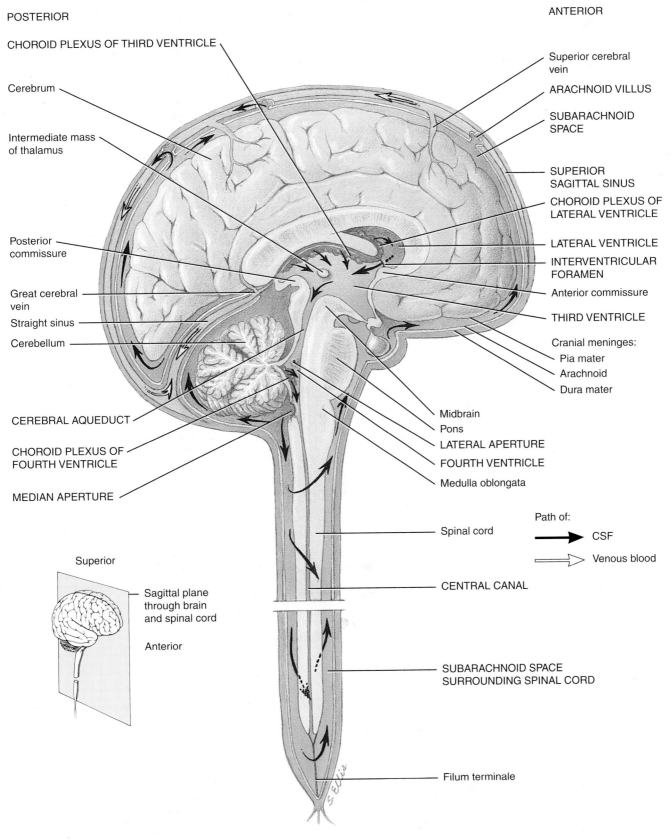

POSTERIOR

ANTERIOR

CHOROID PLEXUS OF THIRD VENTRICLE

Cerebrum

Intermediate mass of thalamus

Posterior commissure

Great cerebral vein

Straight sinus

Cerebellum

CEREBRAL AQUEDUCT

CHOROID PLEXUS OF FOURTH VENTRICLE

MEDIAN APERTURE

Superior cerebral vein

ARACHNOID VILLUS

SUBARACHNOID SPACE

SUPERIOR SAGITTAL SINUS

CHOROID PLEXUS OF LATERAL VENTRICLE

LATERAL VENTRICLE

INTERVENTRICULAR FORAMEN

Anterior commissure

THIRD VENTRICLE

Cranial meninges:
Pia mater
Arachnoid
Dura mater

Midbrain
Pons
LATERAL APERTURE
FOURTH VENTRICLE
Medulla oblongata

Spinal cord

CENTRAL CANAL

SUBARACHNOID SPACE SURROUNDING SPINAL CORD

Filum terminale

Path of:

CSF

Venous blood

Superior

Sagittal plane through brain and spinal cord

Anterior

Sagittal section of brain and spinal cord

Figure 18.4 (continued)

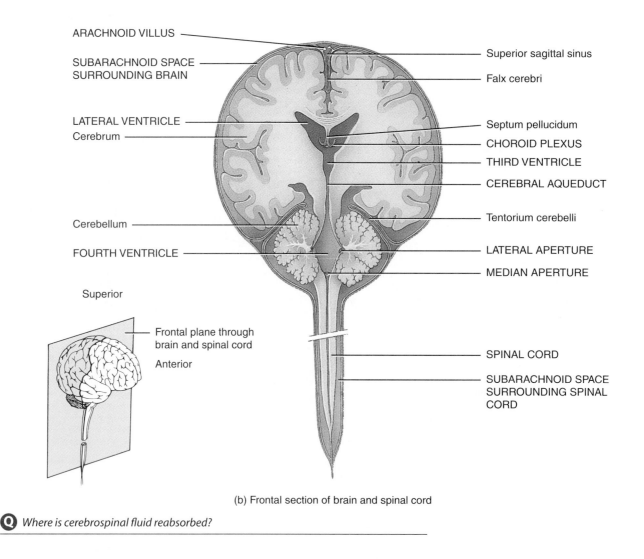

ARACHNOID VILLUS

SUBARACHNOID SPACE
SURROUNDING BRAIN

LATERAL VENTRICLE
Cerebrum

Cerebellum

FOURTH VENTRICLE

Superior

Frontal plane through
brain and spinal cord

Anterior

Superior sagittal sinus

Falx cerebri

Septum pellucidum

CHOROID PLEXUS

THIRD VENTRICLE

CEREBRAL AQUEDUCT

Tentorium cerebelli

LATERAL APERTURE

MEDIAN APERTURE

SPINAL CORD

SUBARACHNOID SPACE
SURROUNDING SPINAL
CORD

(b) Frontal section of brain and spinal cord

Q *Where is cerebrospinal fluid reabsorbed?*

fluid is reabsorbed as rapidly as it is formed. This occurs at a rate of about 20 ml/hr (about 480 ml/day).

Clinical Application

Hydrocephalus

An obstruction in the brain, such as from a tumor or a congenital blockage, or due to brain inflammation, can interfere with the drainage of CSF from the ventricles into the subarachnoid space. As excess fluid accumulates in the ventricles, the CSF pressure rises. This condition is called **hydrocephalus** (*hydro* = water; *kephale* = head). In a baby, if the fontanels have not yet closed, the head bulges in response to the increased pressure. In time, however, the fluid buildup compresses and damages the delicate nervous tissue. Hydrocephalus is relieved by ventricular drainage of CSF. A neurosurgeon may implant a drain, called a shunt, into the lateral ventricle to divert CSF into the superior vena cava or abdominal cavity. In adults, hydrocephalus may occur following head injury, meningitis, or subarachnoid hemorrhage. ■

Blood Supply

The brain is well supplied with oxygen and nutrients, mainly by blood vessels that form the cerebral arterial circle (circle of Willis) at the base of the brain. Cerebral circulation is outlined in Exhibits 14.3 and 14.8 and Figures 14.10c and 14.15. Blood vessels of the brain pass along its surface, and as they penetrate inward, they are surrounded by a loose-fitting layer of pia mater. The space between the penetrating blood vessel and pia mater is called a **perivascular space.**

Glucose, oxygen, carbon dioxide, water, and most lipid-soluble substances (such as alcohol, caffeine, nicotine, heroin, and most anesthetics) pass rapidly from the circulating blood into brain cells. Other substances, such as creatinine, urea, and most ions (for example, Na$^+$, K$^+$, and Cl$^-$), enter quite slowly. Still other substances—proteins and most antibiotics—do not pass at all from the blood into brain cells. The different rates of passage of certain materials from the blood into most parts of the brain depend on the **blood–brain barrier (BBB).** Brain capillaries are much less leaky than most other body capillaries. Tight junctions seal together the endothelial cells of brain capillaries, which also are surrounded by a continuous basement membrane. Also, processes of large numbers of astrocytes (one type of neuroglia) press up against the capillaries. Astrocytes are thought to pass selectively some substances from the blood but inhibit the passage of others. As mentioned, substances that cross the BBB are either soluble in lipids or water-soluble substances that receive the assistance of a transporter (carrier) to cross by active transport.

Clinical Application

Breaching the Blood–Brain Barrier

The blood–brain barrier functions as a selective anatomical and physiological barrier to protect brain cells from harmful substances and pathogens. An injury to the brain due to trauma, inflammation, or toxins may cause a breakdown of the BBB, permitting the passage of substances into brain tissue that normally are kept out. On the other hand, the BBB may prevent entry of drugs that could be used as therapy for brain cancer or other CNS disorders. Researchers are working on ways to slip drugs past the BBB. In one method, the drug is injected together with a concentrated sugar solution. Temporarily, the endothelial cells shrink, due to the osmotic effect of the sugar solution, and pull apart to open up gaps in the tight junctions. The result is entry of the drug into the brain tissue. ▓

Several small brain regions lying in the walls of the third and fourth ventricles and called **circumventricular organs (CVOs)** can monitor chemical changes in the blood because they lack the blood–brain barrier. CVOs include part of the hypothalamus, the pineal gland, the pituitary gland, and a few other nearby structures. Functionally, these regions coordinate activities of the endocrine and nervous systems, such as regulation of blood pressure, fluid balance, hunger, and thirst. CVOs are also thought to be the site of entry for the AIDS virus into the brain, where the virus may cause dementia (irreversible deterioration of mental state) and other neurologic disorders.

Brain Stem

The brain stem connects the spinal cord to the diencephalon and consists of the medulla oblongata, pons, midbrain, and reticular formation.

Medulla Oblongata

The **medulla oblongata** (me-DULL-la ob'-long-GA-ta), or more simply the **medulla** is a continuation of the superior portion of the spinal cord and forms the inferior part of the brain stem (Figure 18.5). The medulla begins at the foramen magnum and extends upward to the inferior border of the pons, a distance of about 3 cm (1.2 in.)

Within the medulla are all ascending (sensory) and descending (motor) white matter tracts that connect the spinal cord with the brain and many nuclei (gray matter masses of cell bodies and dendrites in the CNS) that regulate various vital body functions. It also contains the nuclei that receive sensory input from or provide motor output to five of the twelve cranial nerves (see Table 18.2 on page 579). Major structural and functional regions of the medulla are as follows.

On the anterior aspect of the medulla are two conspicuous external bulges called the **pyramids** (Figures 18.5 and 18.6). They contain the largest motor tracts that pass from the cerebrum to the spinal cord. Just superior to the junction of the medulla with the spinal cord, most of the axons in the left pyramid cross to the right side, and most of the axons in the right pyramid cross over to the left. This crossing is called the **decussation** (dē'-ku-SA-shun) **of pyramids** (Figure 18.6). Thus neurons in the left cerebral cortex control muscles on the right side of the body, and neurons in the right cerebral cortex control muscles on the left side. Just lateral to each pyramid is an oval-shaped swelling called an **olive** (Figures 18.5 and 18.6). The swelling is caused mostly by the **inferior olivary nucleus** within the medulla. Axons of neurons in the inferior olivary nuclei communicate with the cerebellum by paired tracts called the **inferior cerebellar peduncles** (pe-DUNG-kulz; *peduncles* = stemlike part; see Figure 18.7a). The peduncles carry from the inferior olivary nuclei signals that ensure the efficiency of precise, voluntary movements and maintain equilibrium and posture.

In addition to the olivary nuclei, the medulla also contains the nuclei of origin for five pairs of cranial nerves (Figures 18.5 and 18.6). These are: (1) the vestibulocochlear (VIII) nerves, which are concerned with hearing and equilibrium (there are also nuclei for the vestibular branches in the pons); (2) the glossopharyngeal (IX) nerves, which relay nerve impulses for swallowing, salivation, and taste; (3) the vagus (X) nerves, which relay nerve impulses to and from many thoracic and abdominal viscera; (4) the cranial portion of the accessory (XI) nerves, which conveys nerve impulses related to head and shoulder movements (a part of these

Figure 18.5 Medulla oblongata in relation to the rest of the brain stem. The arrow in the inset indicates the direction from which the brain is viewed (inferior).

The brain stem consists of the medulla oblongata, pons, midbrain, and reticular formation.

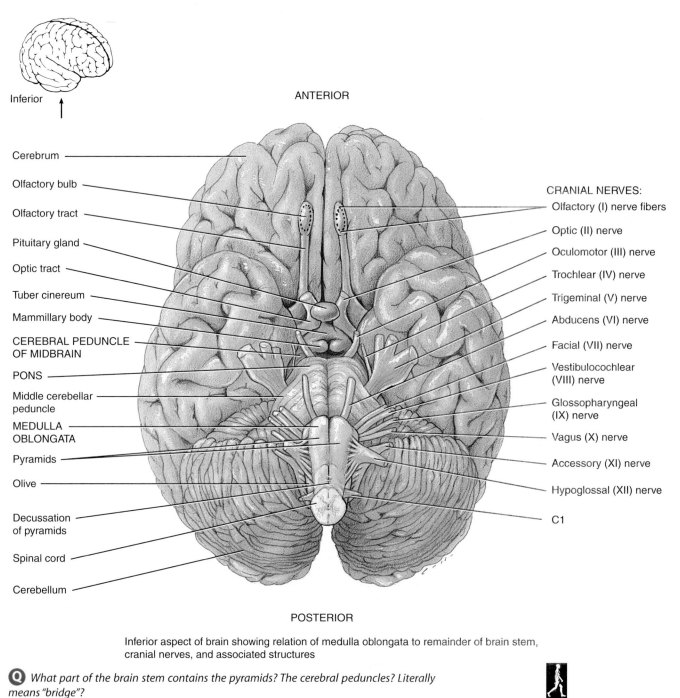

Inferior

ANTERIOR

Cerebrum

Olfactory bulb

Olfactory tract

Pituitary gland

Optic tract

Tuber cinereum

Mammillary body

CEREBRAL PEDUNCLE OF MIDBRAIN

PONS

Middle cerebellar peduncle

MEDULLA OBLONGATA

Pyramids

Olive

Decussation of pyramids

Spinal cord

Cerebellum

CRANIAL NERVES:

Olfactory (I) nerve fibers

Optic (II) nerve

Oculomotor (III) nerve

Trochlear (IV) nerve

Trigeminal (V) nerve

Abducens (VI) nerve

Facial (VII) nerve

Vestibulocochlear (VIII) nerve

Glossopharyngeal (IX) nerve

Vagus (X) nerve

Accessory (XI) nerve

Hypoglossal (XII) nerve

C1

POSTERIOR

Inferior aspect of brain showing relation of medulla oblongata to remainder of brain stem, cranial nerves, and associated structures

Q *What part of the brain stem contains the pyramids? The cerebral peduncles? Literally means "bridge"?*

nerves, the spinal portion, originates in the superior five cervical segments of the spinal cord); and (5) the hypoglossal (XII) nerves, which convey nerve impulses that involve tongue movements.

The nuclei associated with sensory information are located on the dorsal aspect of the medulla. They are the right and left **nucleus gracilis** (gras-I-lis; *gracilis* = slender)

Figure 18.6 Internal anatomy of the medulla oblongata.

The pyramids of the medulla contain the largest motor tracts from the cerebrum to the spinal cord.

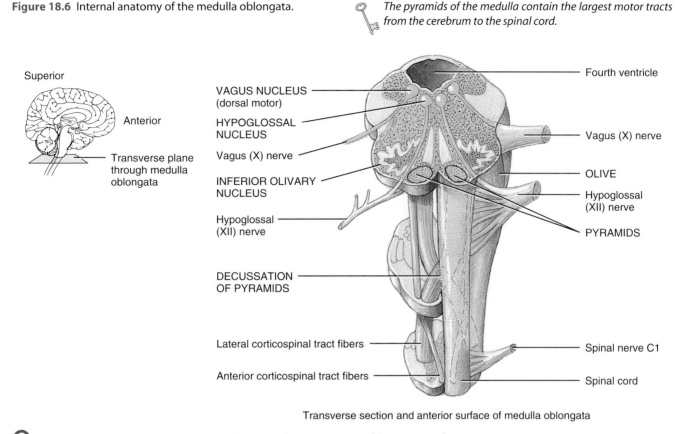

Superior

Anterior

Transverse plane through medulla oblongata

VAGUS NUCLEUS (dorsal motor)

HYPOGLOSSAL NUCLEUS

Vagus (X) nerve

INFERIOR OLIVARY NUCLEUS

Hypoglossal (XII) nerve

DECUSSATION OF PYRAMIDS

Lateral corticospinal tract fibers

Anterior corticospinal tract fibers

Fourth ventricle

Vagus (X) nerve

OLIVE

Hypoglossal (XII) nerve

PYRAMIDS

Spinal nerve C1

Spinal cord

Transverse section and anterior surface of medulla oblongata

Q *What does decussation mean? Functionally, what is the consequence of decussation of the pyramids?*

and **nucleus cuneatus** (kyoo-nē-Ā-tus; *cuneus* = wedge). Some ascending sensory axon terminals form synapses in these nuclei, and postsynaptic neurons then relay the sensory information to the thalamus on the opposite side (see Figure 19.4). Most sensory nerve impulses initiated on one side of the body cross to the opposite side either in the spinal cord or in these medullary nuclei and finally are received in the cerebral cortex on the opposite side by way of the thalamus.

Finally, the medulla contains nuclei related to several autonomic functions. These include the **cardiovascular center,** which regulates the rate and force of heartbeat and blood vessel diameter; the **medullary rhythmicity area** of the respiratory center, which helps regulate the basic rhythm of respiration (see Figure 23.14); and other centers in the medulla that coordinate activities such as swallowing, vomiting, coughing, sneezing, and hiccuping.

In view of the many vital activities controlled by the medulla, it is not surprising that a hard blow to the back of the head or upper neck can be fatal. Damage of the respiratory center is particularly serious and can rapidly lead to death. Before the advent of vaccines for polio (see *Applications to Health,* p. 16) many victims of this disease were confined to respirators ("iron lungs") because the respiratory

center is a primary target of the polio virus. Symptoms of nonfatal medullary injury may include cranial nerve malfunctions on the same side of the body as the injury, paralysis and loss of sensation on the opposite side of the body, and irregularities in breathing or heart function.

Pons

The relationship of the **pons** (*pons* = bridge) to other parts of the brain can be seen in Figures 18.1 and 18.5. The pons lies directly superior to the medulla and anterior to the cerebellum and is about 2½ cm (1 in.) long. Like the medulla, the pons consists of both nuclei and tracts of white matter. As its name implies, the pons is a bridge connecting the spinal cord with the brain and parts of the brain with each other. These connections are provided by axons that extend in two principal directions. The transverse axons are within paired tracts that connect the right and left sides of the cerebellum. They form the **middle cerebellar peduncles** (Figure 18.7a). The longitudinal axons of the pons belong to the motor and sensory tracts that connect the medulla with the midbrain.

Figure 18.7 Midbrain.

The midbrain connects the pons to the diencephalon.

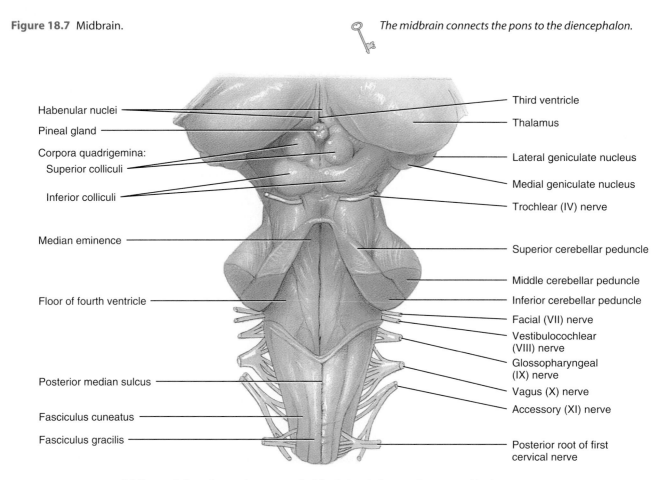

Habenular nuclei

Pineal gland

Corpora quadrigemina:
Superior colliculi

Inferior colliculi

Median eminence

Floor of fourth ventricle

Posterior median sulcus

Fasciculus cuneatus

Fasciculus gracilis

Third ventricle

Thalamus

Lateral geniculate nucleus

Medial geniculate nucleus

Trochlear (IV) nerve

Superior cerebellar peduncle

Middle cerebellar peduncle

Inferior cerebellar peduncle

Facial (VII) nerve

Vestibulocochlear (VIII) nerve

Glossopharyngeal (IX) nerve

Vagus (X) nerve

Accessory (XI) nerve

Posterior root of first cervical nerve

(a) External view of posterior aspect of midbrain in relation to other parts of brain stem

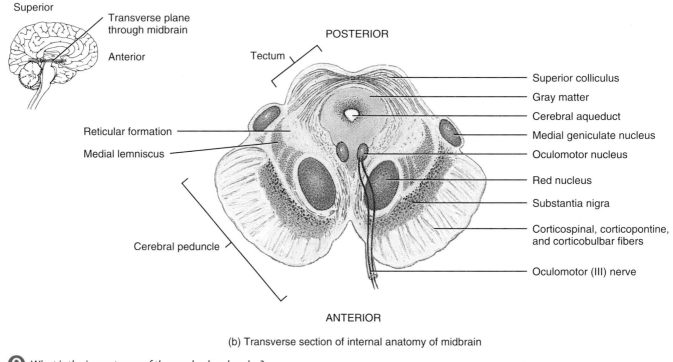

Superior

Transverse plane through midbrain

Anterior

POSTERIOR

Tectum

Reticular formation

Medial lemniscus

Cerebral peduncle

ANTERIOR

Superior colliculus

Gray matter

Cerebral aqueduct

Medial geniculate nucleus

Oculomotor nucleus

Red nucleus

Substantia nigra

Corticospinal, corticopontine, and corticobulbar fibers

Oculomotor (III) nerve

(b) Transverse section of internal anatomy of midbrain

Q *What is the importance of the cerebral peduncles?*

The nuclei for four pairs of cranial nerves are also contained in the pons (see Figure 18.5). These include: (1) the trigeminal (V) nerves, which relay nerve impulses for chewing and for sensations of the head and face; (2) the abducens (VI) nerves, which regulate certain eyeball movements; (3) the facial (VII) nerves, which conduct impulses related to taste, salivation, and facial expression; and (4) some of the vestibular branches of the vestibulocochlear (VIII) nerves, which are concerned with equilibrium.

Other important nuclei in the pons are the **pneumotaxic** (noo-mō-TAK-sik) **area** and the **apneustic** (ap-NOO-stik) **area** (see Figure 23.14). Together with the medullary rhythmicity area, they help control respiration.

Midbrain

The **midbrain, or mesencephalon,** extends from the pons to the diencephalon (Figures 18.1 and 18.5). It is about 2½ cm (1 in.) in length. The cerebral aqueduct passes through the midbrain, connecting the third ventricle above with the fourth ventricle below. Like the medulla and the pons, the midbrain contains both white matter tracts and gray matter nuclei. Main regions are as follows.

The anterior portion of the midbrain contains a pair of tracts called **cerebral peduncles** (Figure 18.7b). They contain some of the motor fibers (corticospinal, corticopontine, and corticobulbar) that convey nerve impulses from the cerebral cortex to the pons, medulla, and spinal cord and sensory fibers passing from the medulla to the thalamus. Another pair of tracts are called the **superior cerebellar peduncles.** They connect the midbrain with the cerebellum (Figure 18.7a).

The posterior portion of the midbrain is called the **tectum** (*tectum* = roof) and contains four rounded elevations, the **corpora quadrigemina** (KOR-po-ra kawd-ri-JEM-in-a; *corpus* = body; *quadrigeminus* = group of four). The two superior elevations are known as the **superior colliculi** (ko-LIK-yoo-lī; singular is **colliculus** = small mound). These serve as reflex centers for movements of the eyes, head, and neck in response to visual and other stimuli. The two inferior elevations, the **inferior colliculi,** are reflex centers for movements of the head and trunk in response to auditory stimuli (Figure 18.7a).

The midbrain contains several nuclei. Among these are the left and right **substantia nigra** (sub-STAN-shē-a NĪ-gra = black substance). See Figure 18.7b. These large, darkly pigmented nuclei that control subconscious muscle activities are near the cerebral peduncles. In Parkinson's disease (see *Applications to Health*, p.17), dopamine-containing neurons in the substantia nigra degenerate. Also present are the left and right **red nuclei** (see Figure 18.7b). The name derives from their rich blood supply and an iron-containing pigment in their neuronal cell bodies. Fibers from the cerebellum and cerebral cortex form synapses in the red nu-

clei, which function with the basal ganglia and cerebellum to coordinate muscular movements. Two nuclei in the midbrain are associated with cranial nerves (see Figure 18.5); (1) the oculomotor (III) nerves, which mediate some movements of the eyeballs and changes in pupil size and lens shape; and (2) the trochlear (IV) nerves, which conduct impulses that coordinate movements of the eyeballs.

Finally, a band of white matter called the **medial lemniscus** (*lemniskos* = ribbon or band) extends through the medulla, pons, and midbrain (see Figure 18.7b). The medial lemniscus contains axons that convey nerve impulses from the medulla to the thalamus for discriminative touch, proprioception (sensations of joint and muscle position), pressure, and vibration sensations.

Reticular Formation

A large portion of the brain stem (medulla, pons, and midbrain) consists of small areas of gray matter interspersed among threads of white matter. This region is called the **reticular formation** (Figure 18.8). It also extends into the spinal cord and diencephalon. The reticular formation has both sensory and motor functions. It receives input from higher brain regions that control skeletal muscles and makes its own important contribution to regulating muscle tone (the slight degree of contraction that characterizes muscles at rest). It also alerts the cerebral cortex to incoming sensory signals. The portion of the reticular formation that consists of fibers that project to the cerebral cortex and has powerful influences on broad regions of the cerebral cortex is called the **reticular activating system (RAS);** it is responsible for maintaining consciousness and awakening from sleep. Incoming impulses from the ears, eyes, and skin are effective stimulators of the RAS. For example, we awaken to the sound of an alarm clock, to a bright light flash, or to a painful pinch because of activity in the RAS that arouses the cerebral cortex.

Diencephalon

The **diencephalon** begins where the midbrain ends and surrounds the third ventricle. From superior to inferior it consists of the **epithalamus, thalamus, subthalamus,** and **hypothalamus** (see Figure 18.1 and Table 18.1). These structures form the walls of the third ventricle as follows: (1) the roof of the third ventricle consists of the epithalamus, which includes a single **pineal** (PĪN-ē-al) **gland** at the midline (see Figure 18.7a), paired right and left **habenular** (ha-BEN-yoo-lar) **nuclei** (see Figure 18.7a), and the choroid plexus of the third ventricle; (2) the two lobes of the thalamus form the superior part of the lateral walls of the third ventricle (Figure 18.4); and (3) the paired right and left subthalamus and hypothalamus form the inferior part of the lateral walls and the floor of the third ventricle (Figure 18.4).

Figure 18.8 Reticular formation. The reticular activating system (RAS) consists of fibers that project to the cerebral cortex through the thalamus from the reticular formation.

The reticular activating system is responsible for maintaining consciousness and awakening from sleep.

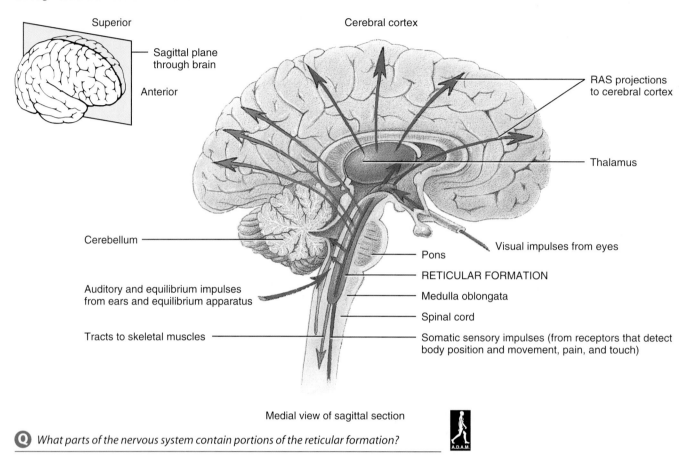

Medial view of sagittal section

Q *What parts of the nervous system contain portions of the reticular formation?*

Details of the most prominent regions—the pineal gland, thalamus, and hypothalamus—are described next. The habenular nucleus is involved in olfaction (smell), especially emotional responses to odors. The small subthalamus includes both tracts and the paired **subthalamic nuclei,** which connect to motor areas of the cerebrum. Portions of two pairs of midbrain nuclei, the red nucleus and the substantia nigra, also extend into the subthalamus. The subthalamic nuclei, red nuclei, and substantia nigra all work together with the basal ganglia, cerebellum, and cerebrum in control of body movements.

Pineal Gland

The **pineal gland** (named for its resemblance to a pine cone) is about the size of a lentil or small pea and protrudes from the dorsal midline of the third ventricle. Although its physiological role is not completely clear, the pineal gland secretes the hormone **melatonin** and thus is an endocrine gland. More melatonin is liberated during darkness and it is thought to promote sleepiness. Melatonin also seems to contribute to setting the timing of the body's biological clock, which is controlled from the suprachiasmatic nucleus of the hypothalamus (see Figure 18.10). In some mammals, such as sheep (which have a very large pineal gland), it also contributes to regulation of seasonal breeding capability. The pineal gland is discussed in more detail in Chapter 22.

Thalamus

The **thalamus** (THAL-a-mus; *thalamos* = inner chamber) measures about 3 cm (1.2 in.) in length and comprises 80% of the diencephalon. It consists of paired oval masses of gray matter organized into nuclei with interspersed tracts of white matter (Figure 18.9). Usually, a bridge of gray matter called the **intermediate mass** crosses the third ventricle to join the right and left portions of the thalamus (see Fig. 18.10).

The thalamus is the principal relay station for sensory impulses that reach the cerebral cortex from the spinal cord, brain stem, cerebellum, and other parts of the cerebrum. It also allows crude appreciation of some sensations, such as pain, temperature, and pressure. Precise localization of such

Figure 18.9 Thalamus. In (a), the red arrows indicate some of the connections between the thalamus and cerebral cortex.

The thalamus is the principal relay station for sensory impulses that reach the cerebral cortex from other parts of the brain and the spinal cord.

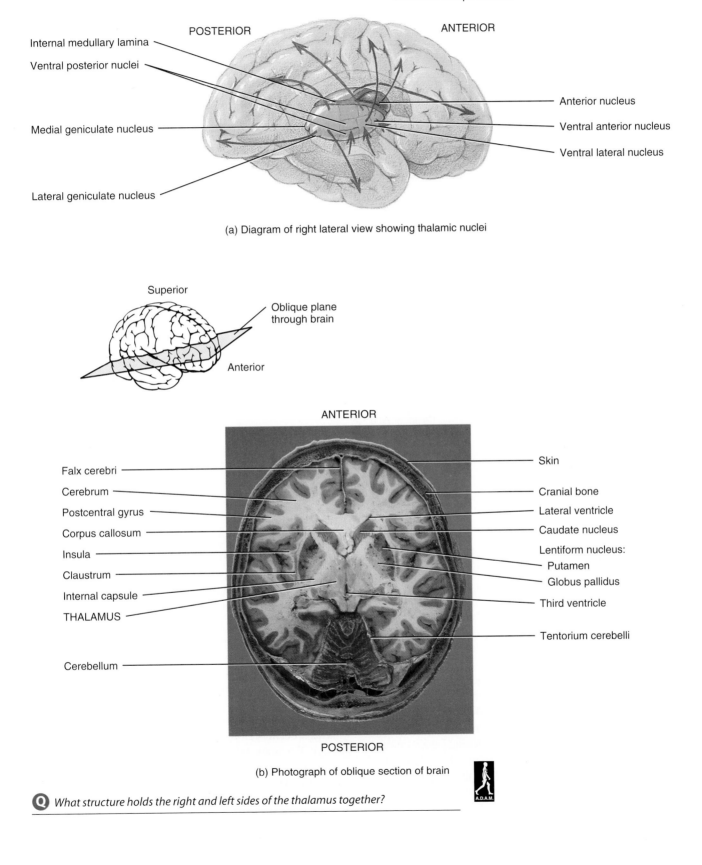

POSTERIOR

ANTERIOR

Internal medullary lamina

Ventral posterior nuclei

Medial geniculate nucleus

Lateral geniculate nucleus

Anterior nucleus

Ventral anterior nucleus

Ventral lateral nucleus

(a) Diagram of right lateral view showing thalamic nuclei

Superior

Oblique plane through brain

Anterior

ANTERIOR

Falx cerebri

Cerebrum

Postcentral gyrus

Corpus callosum

Insula

Claustrum

Internal capsule

THALAMUS

Cerebellum

Skin

Cranial bone

Lateral ventricle

Caudate nucleus

Lentiform nucleus:
Putamen
Globus pallidus

Third ventricle

Tentorium cerebelli

POSTERIOR

(b) Photograph of oblique section of brain

Q *What structure holds the right and left sides of the thalamus together?*

Figure 18.10 Hypothalamus. Shown are selected areas of the hypothalamus and a three-dimensional representation of hypothalamic nuclei (after Netter).

Despite its small size, the hypothalamus controls many body activities.

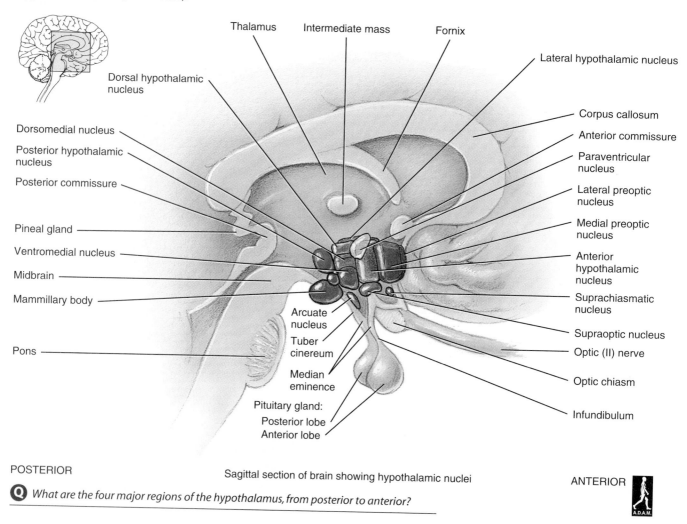

Sagittal section of brain showing hypothalamic nuclei

POSTERIOR   ANTERIOR

Q *What are the four major regions of the hypothalamus, from posterior to anterior?*

sensations depends on nerve impulses being relayed from the thalamus to the cerebral cortex. Within each half of the thalamus are nuclei that have various roles. Certain nuclei in the thalamus relay all sensory input to the cerebral cortex. These include the **medial geniculate** (je-NIK-yoo-lāt; *geniculatus* = bent like a knee) **nucleus** (hearing), the **lateral geniculate nucleus** (vision), and the **ventral posterior nucleus** (taste and somatic sensations such as touch, pressure, vibration, heat, cold and pain). Other nuclei are centers for synapses in the somatic motor system. These include the **ventral lateral nucleus** and **ventral anterior nucleus** (voluntary motor actions and arousal). The **anterior nucleus** in the floor of the lateral ventricle is concerned with certain emotions and memory. The thalamus is also thought to play an essential role in awareness and in the acquisition of knowledge, which is termed **cognition** (*cognitus* = to get to know).

Hypothalamus

The **hypothalamus** (*hypo* = under) is a small portion of the diencephalon located inferior to the thalamus (Figures 18.1 and 18.13b). It is divided into a dozen or so nuclei in four major regions (Figure. 18.10). The **mammillary** (*mammilla* = nipple-shaped structure) **region,** adjacent to the midbrain, is the most posterior portion of the hypothalamus. It includes the mammillary bodies and posterior hypothalamic nucleus. The **mammillary bodies** are two, small, rounded projections that serve as relay stations in reflexes related to the sense of smell.

The **tuberal region,** in the middle, is the widest portion of the hypothalamus. It includes the dorsomedial, ventromedial, and arcuate nuclei. On its ventral surface are the **tuber cinereum** (TOO-ber si-NE-rē-um; *tuber* = swelling; *cinerum* = ash-colored), an elevated mass of gray matter, and the **infundibulum.** The stalklike infundibulum attaches the pituitary gland to the hypothalamus. The **median eminence** of the tuber cinereum is a slightly raised region that encircles the site where the infundibulum becomes the stalk of the pituitary gland. The median eminence contains neurons that synthesize hypothalamic regulating hormones, which regulate hormonal secretions of the anterior lobe of the pituitary gland.

The **supraoptic region** lies superior to the optic chiasm (point of crossing of optic nerves) and contains the paraventricular nucleus, supraoptic nucleus, anterior hypothalamic nucleus, and suprachiasmatic nucleus. Axons from the paraventricular and supraoptic nuclei form the supraopticohypophyseal tract, which extends through the infundibulum to the posterior pituitary gland (see Figure 22.4).

The **preoptic region** anterior to the supraoptic region is usually considered part of the hypothalamus because it regulates certain autonomic activities together with the hypothalamus. The preoptic region contains the preoptic periventricular nucleus, medial preoptic nucleus, and lateral preoptic nucleus.

The hypothalamus controls many body activities. Sensory input from the external and internal environments ultimately comes to the hypothalamus via afferent pathways originating in somatic and visceral sense organs. Impulses from sound, taste, and smell receptors all reach the hypothalamus. Other receptors within the hypothalamus itself continually monitor osmotic pressure, certain hormone concentrations, and the temperature of blood. The hypothalamus has several very important connections with the pituitary gland and is itself able to produce a variety of hormones, which will be described in more detail in Chapter 22. Although some functions can be attributed to specific nuclei, others are not so precisely localized and may overlap nuclear boundaries. The chief functions of the hypothalamus are as follows:

1. It controls and integrates activities of the autonomic nervous system (ANS), which regulates contraction of smooth muscle and cardiac muscle and secretions of many glands. Through the ANS, the hypothalamus is the main regulator of visceral activities, such as heart rate, movement of food through the gastrointestinal tract, and contraction of the urinary bladder.

2. It is associated with feelings of rage and aggression.

3. It regulates body temperature.

4. It regulates food intake through two centers. The **feeding (hunger) center** is responsible for hunger sensations. When sufficient food has been ingested, the **satiety** (sa-TĪ-e-tē; *satio* = full, satisfied) **center** is stimulated and sends out nerve impulses that inhibit the feeding center.

5. It contains a **thirst center.** When certain cells in the hypothalamus are stimulated by rising osmotic pressure of the extracellular fluid, they produce the sensation of thirst. If thirst is satisfied, osmotic pressure returns to normal.

6. It is one of the centers that maintains the waking state and sleep patterns.

Epithalamus

The **epithalamus** (*epi* = above) helps to form the superior margin (roof) of the diencephalon. Among the structures that constitute the epithalamus is the pineal gland (see Figure 18.1a and Table 18.1), whose structure and functions will be discussed in Chapter 22.

Subthalamus

The **subthalamus** lies lateral to the thalamus (see Table 18.1). It contains the subthalamic nuclei, components of the basal ganglia to be described shortly. Two nuclei, the substantia nigra and red nucleus, extend from the midbrain into the subthalamus. The subthalamus may assume a role in influencing the output of the basal ganglia, which control skeletal muscle movements and muscle tone.

Cerebrum

Supported on the diencephalon and brain stem and forming the bulk of the brain is the **cerebrum** (see Figure 18.1). The surface of the cerebrum is composed of gray matter (Figure 18.11a) and is called the **cerebral cortex** (*cortex* = rind or bark). It is only 2–4mm (0.08–0.16 in.) thick and contains billions of neurons. Deep to the cortex lies the cerebral white matter. The cerebrum is the seat of intelligence—the ability to read, write, and speak; make calculations and compose music; remember the past and plan for the future; and create works that have never existed before.

During embryonic development, when brain size increases rapidly, the gray matter of the cortex enlarges much faster than the deeper white matter. As a result, the cortical region rolls and folds upon itself. The folds are called **gyri** (JĪ-rī; singular is **gyrus;** *gyros* = circle) or **convolutions.** The deepest grooves between folds are called **fissures;** the shallower grooves between folds are termed **sulci** (SUL-kē; singular is **sulcus;** *sulcus* = groove). These are illustrated in Figure 18.11. The most prominent fissure, the **longitudinal fissure,** nearly separates the cerebrum into right and left halves called **cerebral hemispheres.** The hemispheres, however, are connected internally by white matter that forms a large bundle of transverse fibers called the **corpus callosum** (kal-LŌ-sum; *corpus* = body; *callous* = hard; see Figure 18.12). Between the hemispheres is an extension of the cranial dura mater called the **falx** (FALKS; *falx* = sickle-shaped) **cerebri (cerebral fold).** It encloses the superior and inferior sagittal sinuses (see Figure 18.2).

Figure 18.13 Basal ganglia. In (a), the basal ganglia have been projected to the surface and are shown in dark green. In (b) they are shown in dark green and blue.

The basal ganglia control large automatic movements of skeletal muscles and muscle tone.

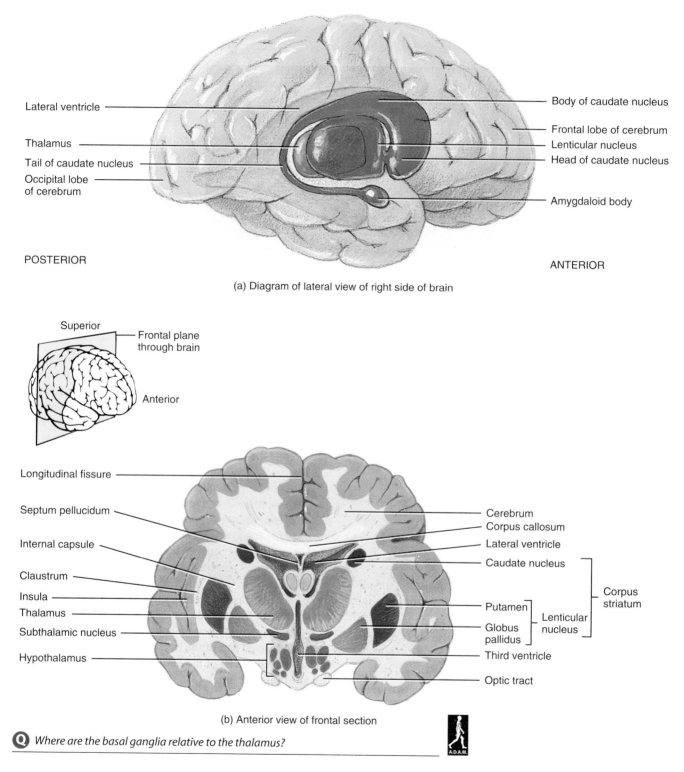

(a) Diagram of lateral view of right side of brain

(b) Anterior view of frontal section

Q *Where are the basal ganglia relative to the thalamus?*

Figure 18.14 Components of the limbic system and surrounding structures.

The limbic system governs emotional aspects of behavior.

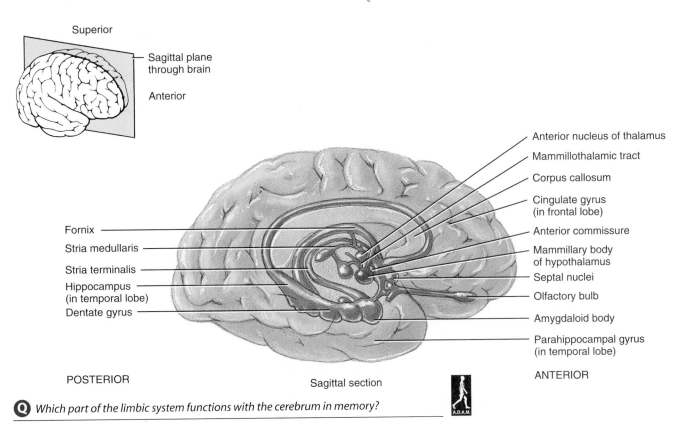

Superior

Sagittal plane through brain

Anterior

Anterior nucleus of thalamus

Mammillothalamic tract

Corpus callosum

Cingulate gyrus (in frontal lobe)

Anterior commissure

Mammillary body of hypothalamus

Septal nuclei

Olfactory bulb

Amygdaloid body

Parahippocampal gyrus (in temporal lobe)

Fornix

Stria medullaris

Stria terminalis

Hippocampus (in temporal lobe)

Dentate gyrus

POSTERIOR

Sagittal section

ANTERIOR

Q *Which part of the limbic system functions with the cerebrum in memory?*

of skeletal muscles, such as swinging the arms while walking. The globus pallidus is concerned with the regulation of muscle tone required for specific body movements.

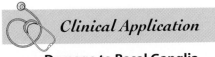

Clinical Application

Damage to Basal Ganglia

Damage to the basal ganglia results in abnormal body movements, such as uncontrollable shaking, called **tremor,** and **involuntary movements of skeletal muscles.** Moreover, destruction of a substantial portion of the caudate nucleus results in almost total **paralysis** of the side of the body opposite to the damage. The caudate nucleus is an area often affected by a stroke.

A lesion in the subthalamic nucleus results in a motor disturbance on the opposite side of the body called **hemiballismus** (*hemi* = half; *ballismos* = jumping), which is characterized by involuntary movements occurring suddenly with great force and rapidity. The movements are purposeless and of a throwing or flailing type, although they may be jerky. The spontaneous movements affect the proximal portions of the limbs most severely, especially the arms.

Limbic System

Encircling the brain stem is a ring of structures on the inner border of the cerebrum and floor of the diencephalon that forms the **limbic** (*limbus* = border) **system.** Among its components are the following regions (Figure 18.14):

1. The **limbic lobe** consists of the **parahippocampal** and **cingulate** (*cingo* = to surround) **gyri,** both gyri of the cerebral hemispheres, and the **hippocampus** (*hippokampos* = seahorse), a portion of the parahippocampal gyrus that extends into the floor of the lateral ventricle.

2. The **dentate** (*dentatus* = toothed) **gyrus** is between the hippocampus and parahippocampal gyrus.

3. The **amygdaloid** (*amygdale* = almond-shaped) **body (amygdala)** is several groups of neurons located at the tail end of the caudate nucleus.

4. The **septal nuclei** are within the septal area, formed by the regions under the corpus callosum and a cerebral gyrus (paraterminal).

5. The **mammillary bodies of the hypothalamus** are two round masses close to the midline near the cerebral peduncles.

6. The **anterior nucleus of the thalamus** is located in the floor of the lateral ventricle.

7. The **olfactory bulbs** are flattened bodies of the olfactory pathway that rest on the cribriform plate (see Figure 20.1b).

8. Bundles of **interconnecting myelinated axons** link various components of the limbic system. They include the **fornix** (white matter tract that connects the hippocampus to the hypothalamus), **stria terminalis, stria medullaris,** and **mammillothalamic tract.**

The limbic system functions in mediating emotional behavior related to survival. The hippocampus, together with portions of the cerebrum, also functions in memory. Because of this relationship, events that cause a strong emotional response are committed to memory much more efficiently than those that do not. Memory impairment results from lesions in parts of the limbic system. People with such damage forget recent events and cannot commit anything to memory. Specifically how the limbic system functions in memory is not clear. Although behavior is a function of the entire nervous system, the limbic system controls most of its involuntary aspects.

Experiments have shown that the limbic system is associated with pleasure and pain. When certain areas of the limbic system are stimulated in animals, their reactions indicate they are experiencing intense punishment. When other areas are stimulated, the animals' reactions indicate they are experiencing extreme pleasure. Stimulation of the amygdaloid body or certain nuclei of the hypothalamus in a cat results in a behavioral pattern called rage. The cat assumes a defensive posture—extending its claws, raising its tail, hissing, spitting, and opening its eyes wide. Stimulating other areas of the limbic system results in an opposite behavioral pattern: docility, tameness, and affection. Because the limbic system has a primary function in emotions such as pain, pleasure, anger, rage, fear, sorrow, sexual feelings, docility, and affection, it is sometimes called the "emotional" brain.

Clinical Application

Brain Injuries

Brain injuries are commonly associated with head injuries and result in part from displacement and distortion of neuronal tissue at the moment of impact. Secondary effects include decreased blood pressure, sustained increased intracranial pressure, infections, and respiratory complications. In addition, some damage occurs when normal blood flow is restored after a period of ischemia (reduced blood flow). The sudden increase in oxygen level produces large numbers of oxygen free radicals (charged oxygen molecules with an unpaired electron). Brain cells recovering from the effects of a stroke or cardiac arrest also release free radicals. Free radicals cause damage by disrupting cellular DNA and enzymes and altering plasma membrane permeability. Various degrees of brain injury are described by the following terms:

Concussion. An abrupt but temporary loss of consciousness (from seconds to hours) following a blow to the head or the sudden stopping of a moving head. It is the most common brain injury. A concussion produces no obvious bruising of the brain, but headache, drowsiness, lack of concentration, confusion, and post-traumatic amnesia (memory loss) may occur.

Contusion. A visible bruising of the brain due to trauma and blood leaking from microscopic vessels. It is usually associated with a concussion. In a contusion, the pia mater may be torn, allowing blood to enter the subarachnoid space. The most common area affected is the frontal lobe. A contusion usually results in an immediate loss of consciousness (generally no longer than 5 minutes), loss of reflexes, transient cessation of respirations, and decreased blood pressure. Vital signs typically stabilize in a few seconds.

Laceration. Tearing of the brain, usually from a skull fracture or gunshot wound. A laceration results in rupture of large blood vessels with bleeding into the brain and subarachnoid space. Consequences include cerebral hematoma (localized pool of blood, usually clotted, that swells against the brain tissue), edema, and increased intracranial pressure. ■

Functional Areas of the Cerebral Cortex

Specific types of sensory, motor, and integrative signals are processed in certain cerebral regions (Figure 18.15). Generally, the **sensory areas** receive and interpret sensory impulses, the **motor areas** control muscular movement, and the **association areas** deal with more complex integrative functions such as memory, emotions, reasoning, will, judgment, personality traits, and intelligence.

Sensory Areas Sensory input to the cerebral cortex flows mainly to the posterior half of the hemispheres, to regions posterior to the central sulci. In the cortex, primary sensory areas have the most direct connections with peripheral sensory receptors. Secondary sensory areas and sensory association areas often are adjacent to the primary areas. Usually, they receive input from the primary areas and from diverse other regions of the brain. They participate in the interpretation of sensory experiences into meaningful patterns of recognition and awareness. For example, a person with damage in the primary visual cortex would be blind in at least part of this visual field. A person with damage to a visual association area, on the other hand, might see normally yet be unable to recognize a friend.

1. **Primary somatosensory area** or **general sensory area.** Located directly posterior to the central sulcus of each cerebral hemisphere in the postcentral gyrus of each parietal lobe just posterior to the primary motor area.

Figure 18.15 Functional areas of the cerebrum. Although Broca's area is in the left cerebral hemisphere of most people, it is shown here in the right hemisphere to indicate its relative location.

🔑 *Particular areas of the cerebral cortex process sensory, motor, and integrative signals.*

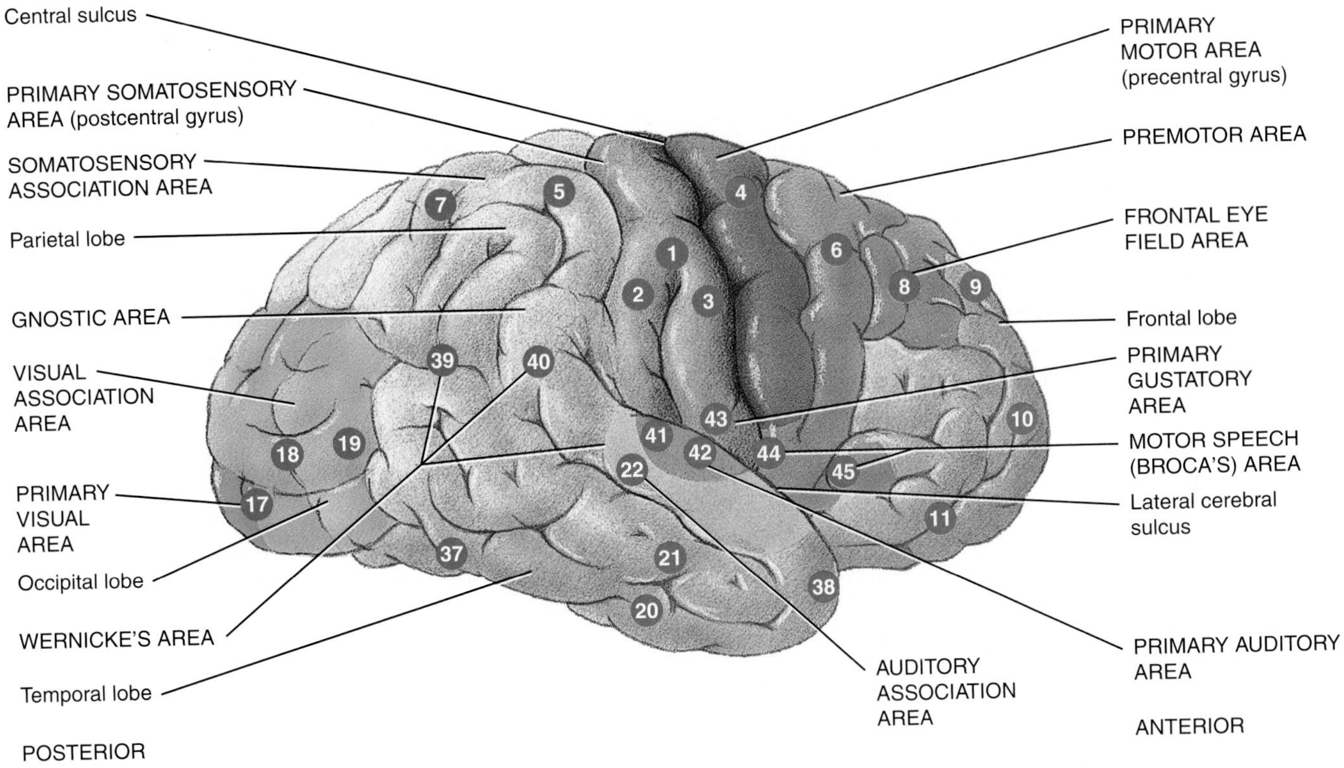

Lateral view of right cerebral hemisphere

Q *What area of the cerebrum integrates interpretation of visual, auditory, and somatic sensations? Translates thoughts into speech? Controls skilled muscular movements? Interprets sensations related to taste? Interprets pitch and rhythm? Interprets shape, color, and movement of objects? Controls voluntary scanning movements of the eyes?*

It extends from the longitudinal fissure on the superior aspect of the cerebrum to the lateral cerebral sulcus. In Figure 18.15, the somatosensory area is designated by the areas numbered 1, 2, and 3.*

The primary somatosensory area receives nerve impulses from somatic sensory receptors for touch, proprioception (joint and muscle position), pain, and temperature. Each point within the area receives sensations from a specific part of the body, and essentially the entire body is spatially represented in it. The size of the cortical area receiving impulses from a particular body part depends on the number of receptors present there rather than on the size of the part. For example, a larger portion of the sensory area receives impulses from the lips and fingertips than from the thorax or hip (see Figure 19.6a). The major function of the primary somatosensory area is to localize exactly the points of the body where the sensations originate. The thalamus is capable of registering sensations in a general way. It receives sensations from large areas of the body but cannot distinguish precisely the specific area of stimulation. This capability depends on the primary somatosensory area of the cortex.

2. **Primary visual area (area 17).** Located on the medial surface of the occipital lobe and occasionally extending around to the lateral surface; axons of neurons with cell bodies in the eye form the optic nerves (cranial nerve II), which terminate in the thalamus. From the thalamus, neurons carrying visual input project to the primary

---

*These numbers, as well as most of the others shown, are based on K. Brodmann's cytoarchitectural map of the cerebral cortex. His map, first published in 1909, was a brilliant and successful attempt to correlate specific brain regions with particular functions.

visual area, with information concerning shape, color, and movement of visual stimuli.

3. **Primary auditory area (areas 41 and 42).** Located in the superior part of the temporal lobe near the lateral cerebral sulcus, it interprets the basic characteristics of sound such as pitch and rhythm.

4. **Primary gustatory area (area 43).** Located at the base of the postcentral gyrus superior to the lateral cerebral sulcus in the parietal cortex, it receives impulses related to taste.

5. **Primary olfactory area (area 28).** Located in the temporal lobe on the medial aspect, it receives impulses related to smell.

Motor Areas Motor output from the cerebral cortex flows mainly from the anterior portion of each hemisphere.

1. **Primary motor area (area 4).** Located in the precentral gyrus of the frontal lobe (Figure 18.15). Each region in the primary motor area controls voluntary contractions of specific muscles or groups of muscles (see Figure 19.6b). Electrical stimulation of any point in the primary motor area results in contraction of specific skeletal muscle fibers on the opposite side of the body. As is true for the primary somatosensory area, body parts are represented unequally here. More cortical area is devoted to those muscles where skilled, complex, or delicate movement is required—for example, finger maneuvers.

2. **Motor speech area (areas 44 and 45).** The translation of speech or written words into thought involves both sensory and association areas—primary auditory, auditory association, primary visual, visual association, and gnostic (described shortly). The production of speech occurs in the **motor speech area**, also called **Broca's (BRŌ-kaz) area.** It is located in one frontal lobe, usually the *left* frontal lobe, just superior to the lateral cerebral sulcus.

Association Areas The association areas of the cerebrum consist of motor and sensory areas and large parts of the cortex on the lateral surfaces of the occipital, parietal, and temporal lobes and the frontal lobes anterior to the motor areas. They are connected with one another by association tracts and include the following:

1. **Somatosensory association area (areas 5 and 7).** Just posterior to the primary somatosensory area, it receives input from the thalamus, other inferior portions of the brain, and the primary somatosensory area. Its role is to integrate and interpret sensations. This area permits you to determine the exact shape and texture of an object without looking at it, to determine the orientation of one object to another as they are felt, and to sense the relationship of one body part to another. Another role of the somatosensory association area is the storage of memories of past sensory experiences. Thus you can compare sensations with previous experiences.

2. **Visual association area (areas 18 and 19).** Located in the occipital lobe, it receives sensory impulses from the primary visual area and the thalamus. It relates present to past visual experiences and is essential for recognizing and evaluating what is seen.

3. **Auditory association (area 22).** Located inferior and posterior to the primary auditory area in the temporal cortex, it determines if a sound is speech, music, or noise.

4. **Wernicke's (posterior language) area.** This is located in the posterior portion of area 22. Some authors also include areas 39 and 40. Wernicke's area interprets the meaning of speech by recognizing spoken words; it translates words into thoughts.

5. **Gnostic (NOS-tik; *gnosis* = knowledge) area (areas 5, 7, 39, and 40).** This **common integrative area** is located among the somatosensory, visual, and auditory association areas. The gnostic area receives nerve impulses from these areas, as well as from the taste and smell areas, the thalamus, and inferior portions of the brain stem. It integrates sensory interpretations from the association areas and impulses from other areas so that a common thought can be formed from the various sensory inputs. It then transmits signals to other parts of the brain to cause the appropriate response to the interpretation of the sensory signals.

6. **Premotor area (area 6).** Immediately anterior to the primary motor area is a motor association area. Neurons in this region communicate with the primary motor cortex, sensory association areas in the parietal lobe, the basal ganglia, and the thalamus. The premotor area deals with learned motor activities of a complex and sequential nature. It generates nerve impulses that cause a specific group of muscles to contract in a specific sequence—for example, to write a word. The premotor area controls learned skilled movements and serves as a memory bank for such movements.

7. **Frontal eye field area (area 8).** This area in the frontal cortex is sometimes included in the premotor area. It controls voluntary scanning movements of the eyes—searching for a word in a dictionary, for instance.

8. **Language areas.** From Broca's area, nerve impulses pass to the premotor regions that control the muscles of the larynx, pharynx, and mouth. The impulses from the premotor area to the muscles result in specific, coordinated contractions that enable you to speak. Simultaneously, impulses are sent from the motor speech area to the primary motor area. From here, impulses reach your breathing muscles to regulate the proper flow of air past the vocal cords. The coordinated contractions of your speech and breathing muscles enable you to speak your thoughts.

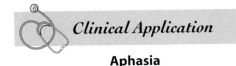

Clinical Application

Aphasia

Much of what we know about language areas comes from studies of patients with language or speech disturbances that have resulted from brain damage. The motor speech (Broca's) area, auditory association area, and other language areas are located in the left cerebral hemisphere of most people, regardless of whether they are left-handed or right-handed. Injury to the association or motor speech areas results in **aphasia** (a-FĀ-zē-a; *a* = without; *phasis* = speech), an inability to speak. Damage to the motor speech area results in nonfluent aphasia, an inability to properly articulate or form words. The person knows what she wishes to say but cannot speak. Damage to the gnostic area or auditory association area (areas 39 and 22) results in fluent aphasia, faulty understanding of spoken or written words. Such a patient may fluently produce strings of words that have no meaning. The deficiency may be **word deafness,** an inability to understand spoken words, or **word blindness,** an inability to understand written words, or both.

Brain Waves

At any instant, brain cells are generating millions of nerve impulses (nerve action potentials) and other changes in membrane potentials in individual neurons. These electrical potentials added together are called **brain waves** and indicate electrical activity of the cerebral cortex. Brain waves pass through the skull and can be detected by sensors called electrodes. A record of such waves is called an **electroencephalogram** (**EEG**; e-lek′-trō-en-SEF-a-lō-gram′).

Clinical Application

Electroencephalogram (EEG)

An **electroencephalogram (EEG or brain wave test)** is used primarily to diagnose epilepsy and other seizure disorders. It may also be used to help diagnose infectious diseases, tumors, trauma, hematomas, metabolic abnormalities, degenerative diseases, and periods of unconsciousness and confusion. In some cases, an EEG provides useful information regarding sleep and wakefulness. An EEG may also be one criterion in establishing brain death (complete absence of brain waves in two EEGs taken 24 hours apart).

Lateralization of the Cerebral Cortex

On gross examination, the brain appears the same on both sides. However, detailed examination reveals subtle anatomical differences between the two hemispheres. For example,

Figure 18.16 Summary of the principal functional differences between the left and right cerebral hemispheres.

The two cerebral hemispheres differ slightly in structure and function.

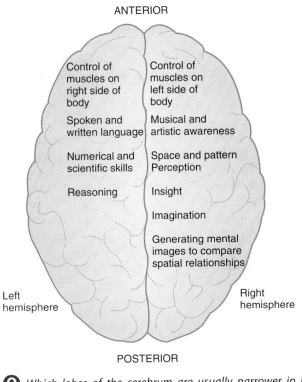

Q *Which lobes of the cerebrum are usually narrower in left-handed people?*

in left-handed people the parietal and occipital lobes of the right hemisphere are usually narrower than the corresponding lobes of the left hemisphere. In addition, the frontal lobe of the left hemisphere of such individuals is typically narrower than that of the right hemisphere.

Besides the structural differences between the hemispheres, there are important functional differences (Figure 18.16). The left hemisphere receives sensory signals from and controls the right side of the body, whereas the right hemisphere receives sensory signals from and controls the left side of the body. Also, in most people the left hemisphere is more important for spoken and written language, numerical and scientific skills, ability to use and understand sign language, and reasoning. Conversely, it has been shown that the right hemisphere is more important for musical and artistic awareness, space and pattern perception, insight, imagination, and generating mental images of sight, sound, touch, taste, and smell to compare relationships.

Cerebellum

The **cerebellum** is the second largest portion of the brain and occupies the inferior and posterior aspects of the cranial cavity. Specifically, it is posterior to the medulla oblongata and pons and inferior to the occipital lobes of the cerebrum (see Figure 18.1). It is separated from the cerebrum by the **transverse fissure** and by an extension of the cranial dura mater called the **tentorium** (*tentorium* = tent) **cerebelli** (see Figure 18.3c). The tentorium cerebelli partially encloses the transverse sinuses (veins) and supports the occipital lobes of the cerebrum.

The cerebellum is shaped somewhat like a butterfly. The central constricted area is the **vermis,** which means "worm-shaped," and the lateral "wings" or lobes are termed **cerebellar hemispheres** (Figure 18.17). Each hemisphere consists of lobes that are separated by deep and distinct fissures. The **anterior lobe** and **posterior (middle) lobe** are concerned with subconscious movements of skeletal muscles. The **flocculonodular** (*flocculus* = little tuft) **lobe** on the inferior surface is concerned with the sense of equilibrium (see Chapter 20). Between the hemispheres is a sickle-shaped extension of the cranial dura mater, the **falx** (*falx* = sickle-shaped) **cerebelli.** It passes only a short distance between the cerebellar hemispheres and contains the occipital sinus (vein).

The surface of the cerebellum, called the **cerebellar cortex,** consists of gray matter in a series of slender, parallel ridges called **folia** (*folia* = leaf). They are less prominent than the convolutions of the cerebral cortex. Deep to the gray matter are **white matter tracks** called **arbor vitae** (*arbor* = tree; *vita* = life) that resemble branches of a tree. Deep within the white matter are masses of gray matter, the **cerebellar nuclei.** The nuclei give rise to nerve fibers that convey information out of the cerebellum to other parts of the brain and the spinal cord.

The cerebellum is attached to the brain stem by three paired cerebellar peduncles (see Figure 18.7a). The **inferior cerebellar peduncles** connect the cerebellum with the medulla oblongata and the spinal cord. These peduncles contain both sensory and motor axons that carry information into and out of the cerebellum. The **middle cerebellar peduncles** contain only sensory axons, which conduct input from the pons into the cerebellum. The **superior cerebellar peduncles** contain mostly motor axons that conduct output from the cerebellum into the midbrain.

The cerebellum compares the intended movement programmed by motor areas in the cerebrum with what is actually happening. It constantly receives sensory input from proprioceptors in muscles, tendons, and joints, receptors for equilibrium, and visual receptors of the eyes. If the intent of the cerebral motor areas is not being attained by skeletal muscles, the cerebellum detects the variation and sends feedback signals to the motor areas to either stimulate or inhibit the activity of skeletal muscles. This interaction helps to smooth and coordinate complex sequences of skeletal muscle contractions. Besides coordinating skilled movements, the cerebellum is the main brain region that regulates posture and balance. These aspects of cerebellar function make possible all skilled motor activities, from catching a baseball to dancing.

Clinical Application

Damage to Cerebellum

Damage to the cerebellum through trauma or disease is characterized by certain symptoms involving skeletal muscles on the same side of the body as the damage. The effects are on the same side of the body as the damaged side of the cerebellum because of a double crossing of tracts within the cerebellum. There may be lack of muscle coordination, called **ataxia** (*a* = without; *taxes* = order). Blindfolded people with ataxia cannot touch the tip of their nose with a finger because they cannot coordinate movement with their sense of where a body part is located. Additionally, the person cannot walk heel to toe along a line. (If there is cerebellar damage, the person will fall toward the side of the lesion.) Another sign of ataxia is a change in the speech pattern due to a lack of coordination of speech muscles. Cerebellar damage may also result in **disturbances of gait,** in which the subject staggers or cannot coordinate normal walking movements, and **severe dizziness.** ■

A summary of the functions of the various parts of the brain is presented in Table 18.1.

CRANIAL NERVES

Cranial nerves, like spinal nerves, are part of the peripheral nervous system (PNS). Of the 12 pairs of **cranial nerves,** 10 originate from the brain stem (see Figure 18.5a), but all pass through foramina of the skull. The cranial nerves are designated with roman numerals and with names. The roman numerals indicate the order in which the nerves arise from the brain from anterior to posterior. The names indicate the distribution or function.

Some cranial nerves contain only sensory fibers and thus are called **sensory nerves.** The remainder contain both sensory and motor fibers and are referred to as **mixed nerves,** though a few of them are predominantly motor in function. The cell bodies of sensory neurons are found outside the brain, whereas the cell bodies of motor neurons lie in nuclei within the brain.

Although the cranial nerves are mentioned singly in the following description of their type, location, and function, remember that they are paired structures.

Figure 18.17 Cerebellum. The arrows in the insets in (a) and (b) indicate the directions from which the cerebellum is viewed (superior and inferior, respectively).

The cerebellum coordinates skilled movements and regulates posture and balance.

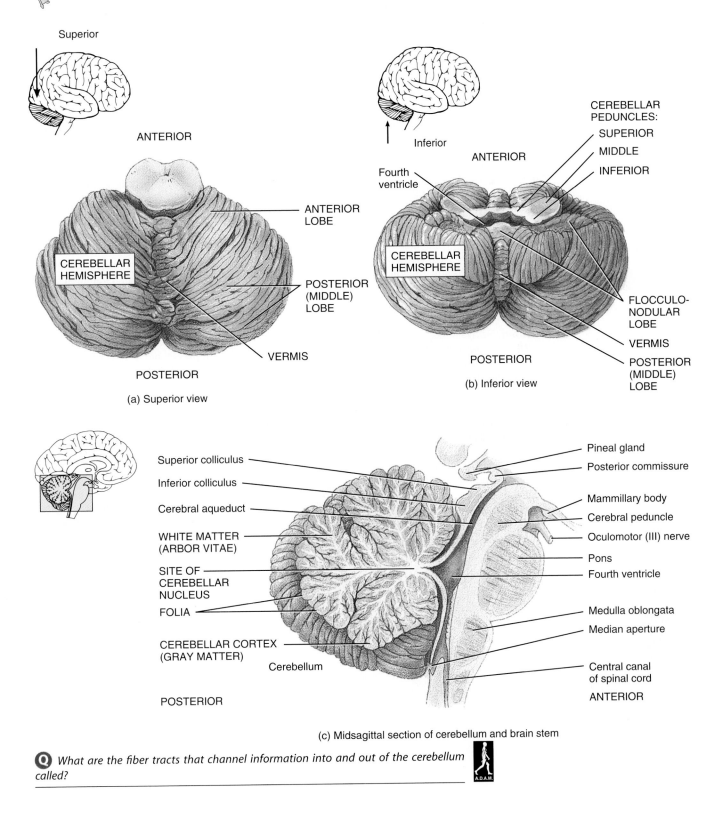

Superior

ANTERIOR

ANTERIOR LOBE

CEREBELLAR HEMISPHERE

POSTERIOR (MIDDLE) LOBE

VERMIS

POSTERIOR

(a) Superior view

Inferior

ANTERIOR

Fourth ventricle

CEREBELLAR PEDUNCLES:
SUPERIOR
MIDDLE
INFERIOR

CEREBELLAR HEMISPHERE

FLOCCULO-NODULAR LOBE

VERMIS

POSTERIOR

POSTERIOR (MIDDLE) LOBE

(b) Inferior view

Superior colliculus

Inferior colliculus

Cerebral aqueduct

WHITE MATTER (ARBOR VITAE)

SITE OF CEREBELLAR NUCLEUS

FOLIA

CEREBELLAR CORTEX (GRAY MATTER)

Cerebellum

POSTERIOR

Pineal gland

Posterior commissure

Mammillary body

Cerebral peduncle

Oculomotor (III) nerve

Pons

Fourth ventricle

Medulla oblongata

Median aperture

Central canal of spinal cord

ANTERIOR

(c) Midsagittal section of cerebellum and brain stem

Q *What are the fiber tracts that channel information into and out of the cerebellum called?*

Table 18.1 *Summary of Functions of Principal Parts of the Brain*

PART	FUNCTION	PART	FUNCTION
Brain Stem 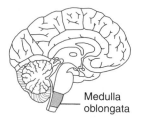 Medulla oblongata	**Medulla oblongata:** Relays motor and sensory impulses between other parts of the brain and the spinal cord. Reticular formation (also in pons, midbrain, and diencephalon) functions in consciousness and arousal. Vital centers regulate heartbeat, breathing (together with pons), and blood vessel diameter. Other centers coordinate swallowing, vomiting, coughing, sneezing, and hiccuping. Contains nuclei of origin for cranial nerves VIII, IX, X, XI, and XII. **Pons:** Relays impulses from one side of the cerebellum to the other and between the medulla and midbrain. Contains nuclei of origin for cranial nerves V, VI, VII, and VIII. Pneumotaxic area and apneustic area, together with the medulla, help control breathing. **Midbrain:** Relays motor impulses from the cerebral cortex to the pons and sensory impulses from the spinal cord to the thalamus. Superior colliculi coordinate movements of the eyeballs in response to visual and other stimuli, and the inferior colliculi coordinate movements of the head and trunk in response to auditory stimuli. Substantia nigra and red nucleus contribute to control of movement. Contains nuclei of origin for cranial nerves III and IV. **Epithalamus:** Contains pineal gland (which secretes melatonin), habenular nuclei, and the choroid plexus of the third ventricle. **Thalamus:** Relays all sensory input to the cerebral cortex. Provides crude appreciation	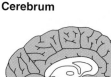Subthalamus Hypothalamus **Cerebrum** Cerebrum **Cerebellum** 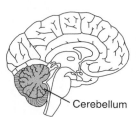 Cerebellum	of touch, pressure, pain, and temperature. Includes nuclei involved in voluntary motor actions and arousal; anterior nucleus functions in emotions and memory. Also functions in cognition and awareness. **Subthalamus:** Contains the subthalamic nuclei and portions of the red nucleus and the substantia nigra, which are positioned mostly lateral to the midline. These regions communicate with the basal ganglia to help control muscle movements. **Hypothalamus:** Controls and integrates activities of the autonomic nervous system and pituitary gland. Regulates emotional and behavioral patterns and diurnal rhythms. Controls body temperature and regulates eating and drinking behavior. Helps maintain the waking state and establishes patterns of sleep. Sensory areas interpret sensory impulses, motor areas control muscular movement, and association areas function in emotional and intellectual processes. Basal ganglia coordinate gross, automatic muscle movements and regulate muscle tone. Limbic system functions in emotional aspects of behavior related to survival. Compares intended movements with what is actually happening to smooth and coordinate complex, skilled movements. Regulates posture and balance.

Brain Stem

Medulla oblongata

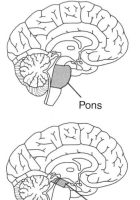
Pons

Midbrain

Diencephalon

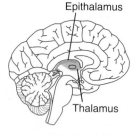
Epithalamus

Thalamus

Olfactory (I)

The **olfactory (I) nerve** (*olfacere* = to smell) is entirely sensory and conveys nerve impulses related to smell. It arises as bipolar neurons from the olfactory mucosa of the nasal cavity (see Figure 20.1). The dendrites and cell bodies of these neurons are generally limited to the mucosa covering the superior nasal conchae and the adjacent nasal septum. Axons from the neurons pass through the cribriform plate of the ethmoid bone and synapse with other olfactory neurons in the **olfactory bulb,** an extension of the brain lying superior to the cribriform plate. The axons of these neurons make up the **olfactory tract.** The fibers from the tract terminate in the primary olfactory area in the temporal lobe of the cerebral cortex.

Optic (II)

The **optic (II) nerve** (*optikos* = vision, eye, or optics) is entirely sensory and conveys nerve impulses related to vision. Signals initiated by rods and cones of the retina are relayed by bipolar neurons to ganglion cells (see Figure 20.6). Axons of the ganglion cells, the optic nerve fibers, exit the optic foramen, after which the two optic nerves unite to form the **optic chiasm** (kī-AZM). Within the chiasma, fibers from the medial half of each retina cross to the opposite side; those from the lateral half remain on the same side. From the chiasm the regrouped fibers pass posteriorly to the **optic tracts.** From the optic tracts the majority of fibers terminate in a nucleus (lateral geniculate) in the thalamus. They then synapse with neurons that pass to the visual areas of the cerebral cortex (see Figure 20.8). Some fibers from the optic chiasm terminate in the superior colliculi of the midbrain. They synapse with neurons whose fibers terminate in the nuclei that convey impulses to the oculomotor (III), trochlear (IV), and abducens (VI) nerves—nerves that control the extrinsic (external) and intrinsic (internal) eye muscles. Through this relay there are widespread motor responses to light stimuli.

Oculomotor (III)

The **oculomotor (III) nerve** (*oculus* = eye; *motor* = mover) is a mixed cranial nerve. Although the oculomotor (III), trochlear (IV), abducens (VI), accessory (XI), and hypoglossal (XII) cranial nerves are referred to as mixed nerves because they contain fibers from proprioceptors, they are primarily motor, serving to stimulate skeletal muscle contractions.

The oculomotor nerve originates from neurons in a nucleus in the ventral portion of the midbrain (Figure 18.18a). It runs anteriorly and divides into superior and inferior divisions, both of which pass through the superior orbital fissure into the orbit. The superior branch is distributed to the superior rectus (an extrinsic eyeball muscle) and the levator palpebrae superioris (the muscle of the upper eyelid). The inferior branch is distributed to the medial rectus, inferior rectus, and inferior oblique muscles—all extrinsic eyeball muscles. These distributions to the levator palpebrae superioris and extrinsic eyeball muscles constitute the motor portion of the oculomotor nerve. Through these distributions are sent nerve impulses that control movements of the eyeball and upper eyelids.

The inferior branch of the oculomotor nerve also provides parasympathetic innervation to the **ciliary ganglion,** a relay center of the autonomic nervous system that connects a nucleus in the midbrain with the intrinsic eyeball muscles. These intrinsic muscles include the ciliary muscle of the eyeball and the sphincter muscle of the iris. Through the ciliary ganglion the oculomotor nerve controls the smooth muscle (ciliary muscle) responsible for adjustment of the lens for near vision and the smooth muscle (sphincter muscle of iris) responsible for constriction of the pupil.

The sensory portion of the oculomotor nerve consists of fibers from proprioceptors in the eyeball muscles supplied by the nerve to the midbrain. These fibers convey nerve impulses related to muscle sense (proprioception).

Trochlear (IV)

The **trochlear** (TROK-lē-ar) **(IV) nerve** (*trokhileia* = pulley) is a mixed cranial nerve. It is the smallest of the 12 cranial nerves and is also unique because it is the only cranial nerve to arise from the posterior aspect of the brain stem. The motor portion originates in a nucleus in the midbrain, and axons from the nucleus pass through the superior orbital fissure of the orbit (Figure 18.18b). The motor fibers innervate the superior oblique muscle of the eyeball, another extrinsic eyeball muscle. The trochlear nerve controls movement of the eyeball.

The sensory portion of the trochlear nerve consists of fibers that run from proprioceptors in the superior oblique muscle to a nucleus of the nerve in the midbrain. The sensory portion is responsible for muscle sense, the awareness of the activity of muscles.

Trigeminal (V)

The **trigeminal (V) nerve** (*tri;* = three; *geminus* = twin) is a mixed cranial nerve and the largest of the cranial nerves. As indicated by its name, the trigeminal nerve has three branches: ophthalmic (of-THAL-mik), maxillary, and mandibular (Figure 18.19). The trigeminal nerve contains two roots on the ventrolateral surface of the pons. The large sensory root has a swelling called the **semilunar (Gasserian) ganglion** located in a fossa on the inner surface of the petrous portion of the temporal bone. The ganglion contains

Figure 18.18 Oculomotor (III), trochlear (IV), and abducens (VI) nerves.

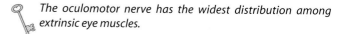

The oculomotor nerve has the widest distribution among extrinsic eye muscles.

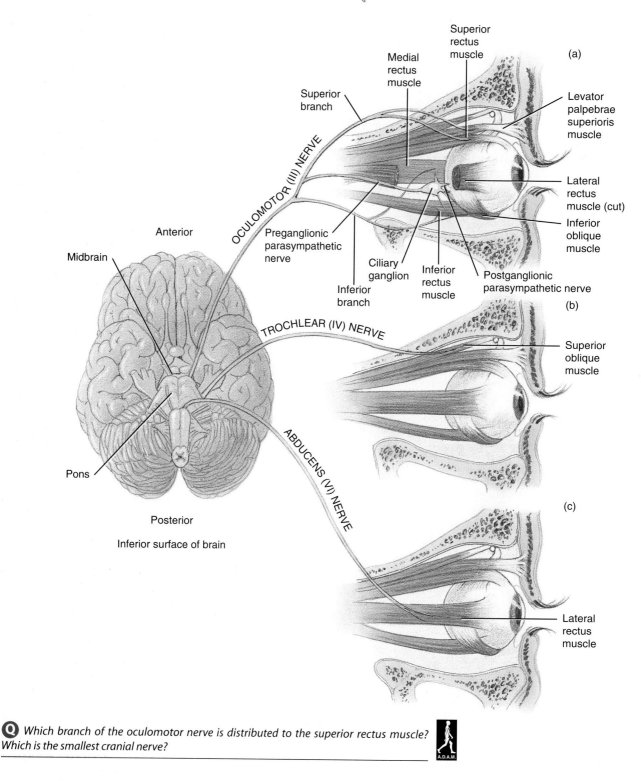

Q *Which branch of the oculomotor nerve is distributed to the superior rectus muscle? Which is the smallest cranial nerve?*

Figure 18.19 Trigeminal (V) nerve.

The three branches of the trigeminal nerve leave the cranium through the superior orbital fissure, foramen rotundum, and foramen ovale.

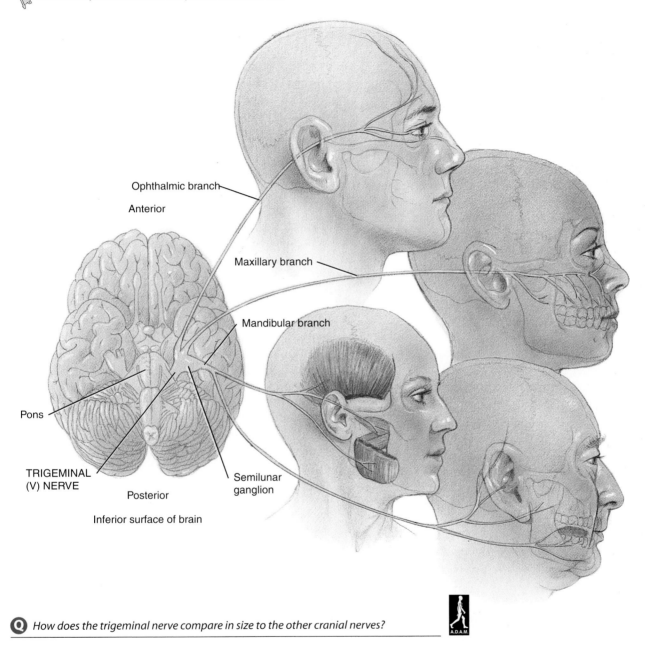

Ophthalmic branch

Anterior

Maxillary branch

Mandibular branch

Pons

TRIGEMINAL (V) NERVE

Posterior

Semilunar ganglion

Inferior surface of brain

Q *How does the trigeminal nerve compare in size to the other cranial nerves?*

cell bodies of most of the primary sensory neurons. From this ganglion, the **ophthalmic branch** (*ophthalmos* = eye), the smallest branch, enters the orbit via the superior orbital fissure; the **maxillary branch** (*maxilla* = upper jaw bone), intermediate in size between the ophthalmic and mandibular branches, enters the foramen rotundum; and the **mandibular branch** (*mandible* = lower jaw bone), the largest branch, exits through the foramen ovale. The smaller motor root originates in a nucleus in the pons. The motor fibers join the mandibular branch and supply the muscles of mastication. These motor fibers, which control chewing movements, constitute the motor portion of the trigeminal nerve and nerve to the mylohyoid and anterior belly of the digastric muscles.

The sensory portion of the trigeminal nerve delivers nerve impulses related to touch, pain, and temperature and consists of the ophthalmic, maxillary, and mandibular branches. The ophthalmic branch receives sensory fibers from the skin over the upper eyelid, eyeball, lacrimal glands, upper part of the nasal cavity, side of the nose, forehead, and anterior half of the scalp. The maxillary branch receives sensory fibers from the mucosa of the nose, palate, parts of the pharynx, upper teeth, upper lip, and lower eyelid. The mandibular branch transmits sensory fibers from the anterior two-thirds of the tongue (not taste), cheek and mucosa deep to it, lower teeth, skin over the mandible and side of the head anterior to the ear, and mucosa of the floor of the mouth. Sensory fibers from the three branches of the trigeminal nerve enter the semilunar ganglion and terminate in nuclei in the pons. There are also sensory fibers from proprioceptors in the muscles of mastication.

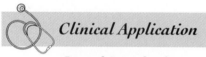

Clinical Application

Dental Anesthesia

The inferior alveolar nerve, a branch of the mandibular nerve, supplies all the teeth in one-half of the mandible and is frequently anesthetized in dental procedures. The same procedure will anesthetize the lower lip because the mental nerve is a branch of the inferior alveolar nerve. Because the lingual nerve runs very close to the inferior alveolar near the mental foramen, it too is often anesthetized at the same time. For anesthesia to the upper teeth, the superior alveolar nerve endings, which are branches of the maxillary branch, are blocked by inserting the needle beneath the mucous membrane; the anesthetic solution is then infiltrated slowly throughout the area of the roots of the teeth to be treated. ▪

Abducens (VI)

The **abducens** (ab-DOO-sens) **(VI) nerve** (*abducere* = to lead away) is a mixed cranial nerve that originates from a nucleus in the pons (see Figure 18.18c). The motor fibers extend from the nucleus to the lateral rectus muscle of the eyeball, an extrinsic eyeball muscle. Nerve impulses over the fibers bring about abduction of the eyeball (turn it laterally), the reason for the name of the nerve. The sensory fibers run from proprioceptors in the lateral rectus muscle to the pons and mediate muscle sense. The abducens nerve reaches the lateral rectus muscle through the superior orbital fissure of the orbit.

Facial (VII)

The **facial (VII) nerve** (*facies* = face) is a mixed cranial nerve. Its motor fibers originate from a nucleus in the pons, enter the petrous portion of the temporal bone, and are distributed to facial, scalp, and neck muscles (Figure 18.20). Nerve impulses along these fibers cause contraction of the muscles of facial expression and posterior belly of the digastric and stylohyoid muscles. Some motor fibers are also distributed via parasympathetics to the lacrimal, nasal, and palatine glands, and to saliva-producing sublingual and submandibular glands.

The sensory fibers extend from the taste buds of the anterior two-thirds of the tongue to the **geniculate ganglion,** a swelling of the facial nerve. From here, the fibers pass to a nucleus in the pons, which sends fibers to the thalamus for relay to the gustatory area of the cerebral cortex. The sensory portion of the facial nerve also conveys deep general sensations from the face. There are also sensory fibers from proprioceptors in the muscles of the face and scalp.

Vestibulocochlear (VIII)

The **vestibulocochlear** (ves-tib′-yoo-lō-KŌK-lē-ar) **(VIII) nerve** (*vestibulum* = vestibule; *kokhlos* = land snail), formerly known as the **acoustic** or **auditory nerve,** is another sensory cranial nerve. It consists of two branches: the cochlear (auditory) branch and the vestibular branch (see Figure 20.12b). The **cochlear branch,** which conveys nerve impulses associated with hearing, arises in the spiral organ (organ of Corti) in the cochlea of the internal ear. The cell bodies of the cochlear branch are located in the **spiral ganglion** of the cochlea. From here the axons synapse in nuclei in the medulla oblongata. The next neurons synapse in the inferior colliculus and pass on to the medial geniculate nucleus in the thalamus. Ultimately, the fibers synapse with neurons that relay the nerve impulses to the auditory areas of the cerebral cortex.

The **vestibular branch** arises in the semicircular canals, the saccule, and the utricle of the inner ear. Fibers from these structures extend to the **vestibular ganglion,** where the cell bodies are contained (see Figure 20.12b). The fibers of the neurons with cell bodies in the vestibular ganglion terminate in nuclei in the thalamus. Some fibers also enter the cerebellum. The vestibular branch transmits impulses related to equilibrium.

Glossopharyngeal (IX)

The **glossopharyngeal** (glos′-ō-fa-RIN-jē-al) **(IX) nerve** (*glossa* = tongue; *pharyngeal* = pharynx) is a mixed cranial nerve. Its motor fibers originate in nuclei in the medulla

Figure 18.20 Facial (VII) nerve.

The facial nerve causes contraction of the muscles of facial expression.

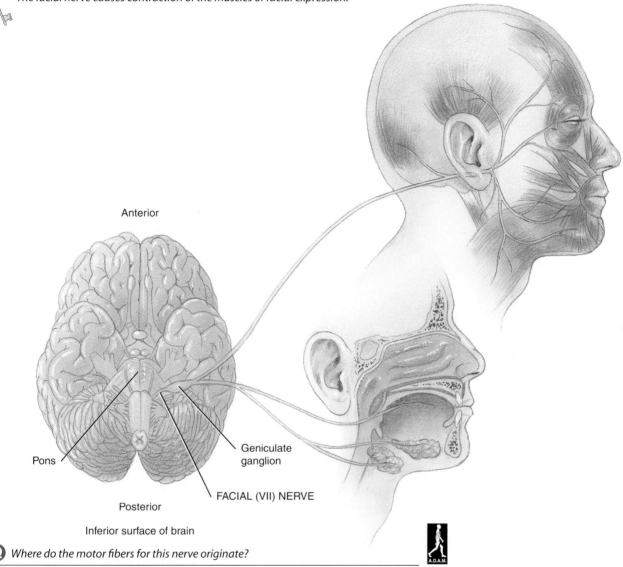

Anterior

Pons

Geniculate
ganglion

FACIAL (VII) NERVE

Posterior

Inferior surface of brain

Q *Where do the motor fibers for this nerve originate?*

(Figure 18.21). The nerve exits the skull through the jugular foramen. The motor fibers are distributed to the stylopharyngeus muscle and parotid gland (parasympathetic) to mediate the secretion of saliva. Some of the cell bodies of parasympathetic motor neurons are located in the otic ganglion.

The sensory fibers of the glossopharyngeal nerve supply the pharynx and taste buds of the posterior third of the tongue. Some sensory fibers also originate from receptors in the carotid sinus, which assumes a major role in blood pressure regulation. The sensory fibers terminate in a nucleus in the thalamus. There are also sensory fibers from proprioceptors in the muscles innervated by this nerve. Sensory neuron cell bodies are located in the superior and inferior ganglia.

Vagus (X)

The **vagus (X) nerve** (*vagus* = vagrant or wandering) is a mixed cranial nerve that is widely distributed from the head and neck into the thorax and abdomen (Figure 18.22). The nerve derives its name from its wide distribution. In the neck, it is positioned medial and posterior to the internal jugular vein and common carotid artery. Motor fibers of the vagus nerve originate in nuclei of the medulla and terminate in the muscles of the respiratory passageways, lungs, heart, esophagus, stomach, small intestine, most of the large intestine, and gallbladder. Nerve impulses along the motor fibers generate visceral, cardiac, and skeletal muscle movement.

Figure 18.21 Glossopharyngeal (IX) nerve.

Sensory fibers of the glossopharyngeal nerve supply the taste buds.

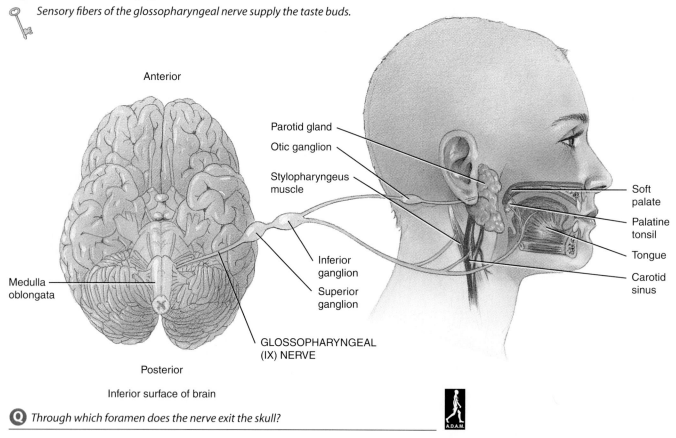

Anterior

Parotid gland

Otic ganglion

Stylopharyngeus muscle

Soft palate

Palatine tonsil

Tongue

Carotid sinus

Inferior ganglion

Superior ganglion

Medulla oblongata

GLOSSOPHARYNGEAL (IX) NERVE

Posterior

Inferior surface of brain

Q *Through which foramen does the nerve exit the skull?*

Parasympathetic fibers innervate involuntary muscles and glands of the gastrointestinal tract.

Sensory fibers of the vagus nerve supply essentially the same structures as the motor fibers. They convey nerve impulses for various sensations from the larynx, the viscera, and the ear. The fibers terminate in the medulla and pons. There are also sensory fibers from proprioceptors in the muscles supplied by this nerve.

Accessory (XI)

The **accessory (XI) nerve** (*accessorius* = assisting; formerly the **spinal accessory nerve**) is a mixed cranial nerve. It differs from all other cranial nerves in that it originates from *both* the brain stem and the spinal cord (Figure 18.23). The **cranial portion** originates from nuclei in the medulla, passes through the jugular foramen, and supplies the voluntary muscles of the pharynx, larynx, and soft palate that are used in swallowing. The **spinal portion** originates in the anterior gray horn of the first five segments of the cervical portion of the spinal cord. The fibers from the segments join, enter the foramen magnum, and exit through the jugular foramen along with the cranial portion.

The spinal portion conveys motor impulses to the sternocleidomastoid and trapezius muscles to coordinate head movements. The sensory fibers originate from proprioceptors in the muscles supplied by its motor neurons and terminate in upper cervical posterior root ganglia.

Hypoglossal (XII)

The **hypoglossal (XII) nerve** (*hypo* = below; *glossa* = tongue) is a mixed cranial nerve. The motor fibers originate in a nucleus in the medulla, pass through the hypoglossal canal, and supply the muscles of the tongue (Figure 18.24). These fibers conduct nerve impulses related to speech and swallowing.

The sensory portion of the hypoglossal nerve consists of fibers originating from proprioceptors in the tongue muscles and terminating in the medulla. The sensory fibers conduct nerve impulses for muscle sense.

A summary of cranial nerves and clinical applications related to dysfunction is presented in Table 18.2 on page 579.

Text continues on page 583

Figure 18.22 Vagus (X) nerve.

The vagus nerve is widely distributed in the head, neck, thorax, and abdomen.

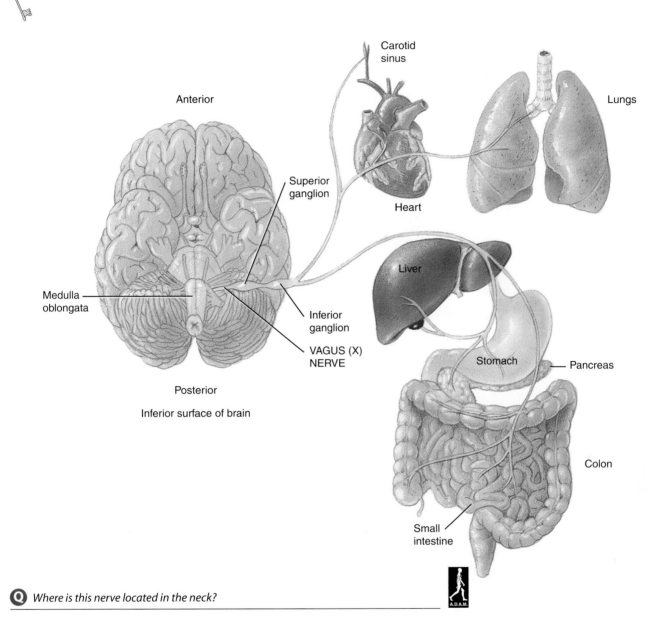

Anterior

Carotid
sinus

Lungs

Superior
ganglion

Heart

Medulla
oblongata

Liver

Inferior
ganglion

VAGUS (X)
NERVE

Stomach

Pancreas

Posterior

Colon

Inferior surface of brain

Small
intestine

Q *Where is this nerve located in the neck?*

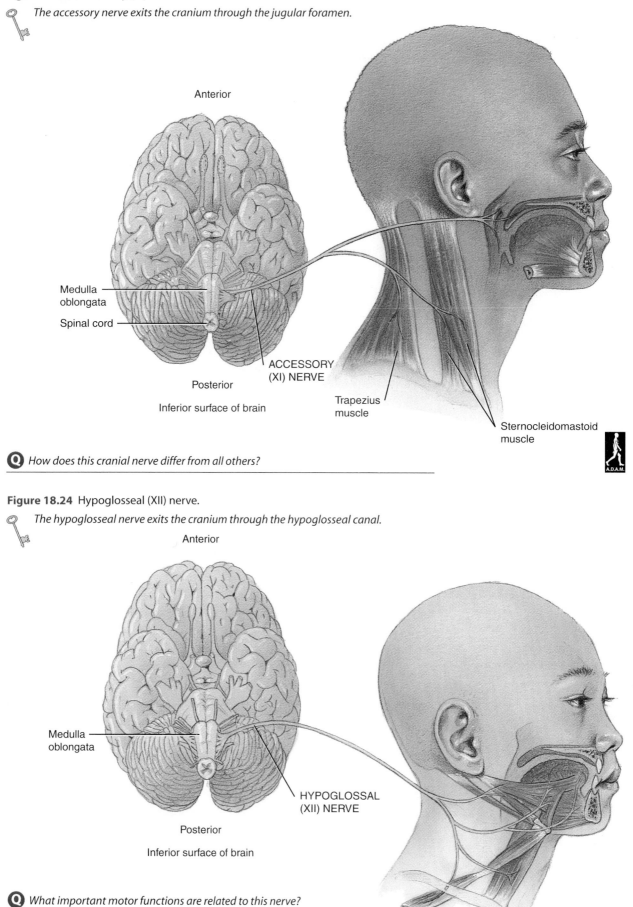

Figure 18.23 Accessory (XI) nerve.

The accessory nerve exits the cranium through the jugular foramen.

Anterior

Medulla oblongata

Spinal cord

ACCESSORY (XI) NERVE

Posterior

Inferior surface of brain

Trapezius muscle

Sternocleidomastoid muscle

Q *How does this cranial nerve differ from all others?*

Figure 18.24 Hypoglosseal (XII) nerve.

The hypoglosseal nerve exits the cranium through the hypoglosseal canal.

Anterior

Medulla oblongata

HYPOGLOSSAL (XII) NERVE

Posterior

Inferior surface of brain

Q *What important motor functions are related to this nerve?*

Table 18.2 *Summary of Cranial Nerves*

NUMBER, NAME, TYPE (SENSORY, MOTOR, OR MIXED)	LOCATION	FUNCTION AND CLINICAL APPLICATION
Cranial nerve I: **olfactory** *(olfacere = to smell)* Sensory 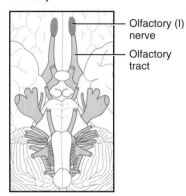	Arises in olfactory mucosa, passes through olfactory foramina in the cribriform plate of ethmoid bone, and ends in the olfactory bulb. The olfactory tract extends via two pathways to olfactory areas in the temporal lobe of the cerebral cortex.	*Function:* Smell. *Clinical application:* Loss of the sense of smell, called *anosmia,* may result from head injuries in which the cribriform plate of the ethmoid bone is fractured and from lesions along the olfactory pathway.
Cranial nerve II: **optic** *(optikos = vision, eye, or optics)* Sensory 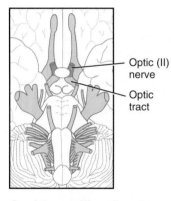	Arises in retina of the eye, passes through optic foramen, forms optic chiasm, passes through optic tracts, and terminates in lateral geniculate nuclei of thalamus. From thalamus, projections extend to visual areas in the occipital lobe of the cerebral cortex.	*Function:* Vision. *Clinical application:* Fractures in the orbit, lesions along the visual pathway, and diseases of the nervous system may result in visual field defects and loss of visual acuity. Loss of vision is called *anopsia.*
Cranial nerve III: **oculomotor** *(oculus = eye; motor = mover)* Mixed, primarily motor 	*Motor portion:* Originates in midbrain, passes through superior orbital fissure, and is distributed to levator palpebrae superioris of upper eyelid and four extrinsic eyeball muscles (superior rectus, medial rectus, inferior rectus, and inferior oblique); parasympathetic innervation to ciliary muscle of eyeball and sphincter muscle of iris. *Sensory portion:* Consists of fibers from proprioceptors in eyeball muscles that pass through superior orbital fissure and terminate in midbrain.	*Motor function:* Movement of eyelid and eyeball, accommodation of lens for near vision, and constriction of pupil. *Sensory function:* Muscle sense (proprioception). *Clinical application:* A lesion in the nerve causes *strabismus* (a deviation of the eye in which both eyes do not fix on the same object), *ptosis* (drooping) of the upper eyelid, pupil dilation, the movement of the eyeball downward and outward on the damaged side, a loss of accommodation for near vision, and *diplopia* (double vision).

Olfactory (I) nerve
Olfactory tract

Optic (II) nerve
Optic tract

Oculomotor (III) nerve

Table 18.2 Summary of Cranial Nerves (continued)

NUMBER, NAME, TYPE (SENSORY, MOTOR, OR MIXED)	LOCATION	FUNCTION AND CLINICAL APPLICATION
Cranial nerve IV: **trochlear** *(trokhileia = pulley)* Mixed, primarily motor  — Trochlear (IV) nerve	**Motor portion:** Originates in midbrain, passes through superior orbital fissure, and is distributed to superior oblique muscle, an extrinsic eyeball muscle. **Sensory portion:** Consists of fibers from proprioceptors in superior oblique muscles that pass through superior orbital fissure and terminate in midbrain.	**Motor function:** Movement of eyeball. **Sensory function:** Muscle sense (proprioception). **Clinical application:** In trochlear nerve paralysis, diplopia and strabismus occur.
Cranial nerve V: **trigeminal** *(tri = three; geminus = twin; trigeminus = threefold, for its three branches)* Mixed 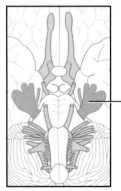 — Trigeminal (V) nerve	**Motor portion:** Is part of the mandibular branch; originates in pons, passes through foramen ovale, and ends in muscles of mastication (anterior belly of digastric and mylohyoid muscles). **Sensory portion:** Consists of three branches: **ophthalmic** *(ophthalmos = eye)*, which contains fibers from skin over upper eyelid, eyeball, lacrimal glands, nasal cavity, side of nose, forehead, and anterior half of scalp that pass through superior orbit fissure; **maxillary** *(maxilla = upper jaw bone)*, which contains fibers from mucosa of nose, palate, parts of pharynx, upper teeth, upper lip, and lower eyelid that pass through foramen rotundum; **mandibular** *(mandibula = lower jaw bone)*, which contains somatic sensory fibers (but not special sense of taste) from anterior two-thirds of tongue, lower teeth, skin over mandible, cheek and mucosa deep to it, and side of head in front of ear that pass through foramen ovale. The three branches end in pons. Sensory portion also consists of fibers from proprioceptors in muscles of mastication.	**Motor function:** Chewing. **Sensory function:** Conveys sensations for touch, pain, and temperature from structures supplied; muscle sense (proprioception). **Clinical application:** Injury results in paralysis of the muscles of mastication and a loss of sensation of touch and temperature. *Neuralgia* (pain) of one or more branches of trigeminal nerve is called *trigeminal neuralgia (tic douloureux).*

Table 18.2 (continued)

NUMBER, NAME, TYPE (SENSORY, MOTOR, OR MIXED)	LOCATION	FUNCTION AND CLINICAL APPLICATION
Cranial nerve VI: **abducens** (*ab = away; ducere = to lead*) Mixed, primarily motor 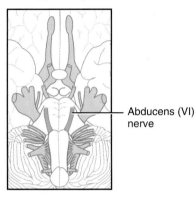 — Abducens (VI) nerve	***Motor portion:*** Originates in pons, passes through superior orbital fissure, and is distributed to lateral rectus muscle, an extrinsic eyeball muscle. ***Sensory portion:*** Consists of fibers from proprioceptors in lateral rectus muscle that pass through superior orbital fissure and end in pons.	***Motor function:*** Movement of eyeball. ***Sensory function:*** Muscle sense (proprioception). ***Clinical application:*** With damage to this nerve, the affected eyeball cannot move laterally beyond the midpoint and the eye is usually directed medially.
Cranial nerve VII: **facial** (*facies = face*) Mixed 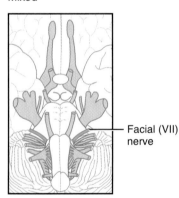 — Facial (VII) nerve	***Motor portion:*** Originates in pons, passes through stylomastoid foramen, and is distributed to facial, scalp, and neck muscles; parasympathetic fibers are distributed to lacrimal, sublingual, submandibular, nasal, and palatine glands. ***Sensory portion:*** Arises from taste buds on anterior two-thirds of tongue, passes through stylomastoid foramen, and ends in geniculate ganglion, a nucleus in pons that sends fibers to the thalamus for relay to gustatory areas in the parietal lobe of the cerebral cortex. Also contains fibers from proprioceptors in muscles of face and scalp.	***Motor function:*** Facial expression and secretion of saliva and tears. ***Sensory function:*** Muscle sense (proprioception) and taste. ***Clinical application:*** This is the most frequently paralyzed cranial nerve. Injury produces paralysis of the facial muscles, called *Bell's palsy,* loss of taste, and loss of ability to close the eyes, even during sleep.
Cranial nerve VIII: **vestibulocochlear** (*vestibulum = vestibule; kokhlos = land snail*) Sensory  — Vestibulocochlear nerve	***Cochlear branch:*** Arises in spiral organ (organ of Corti), forms spiral ganglion, passes through internal auditory meatus, nuclei in the medulla, and ends in thalamus. Fibers synapse with neurons that relay impulses to auditory areas in the temporal lobe of the cerebral cortex. ***Vestibular branch:*** Arises in semicircular canals, saccule, and utricle and forms vestibular ganglion; fibers end in pons and cerebellum.	***Cochlear branch function:*** Conveys impulses associated with hearing. ***Vestibular branch function:*** Conveys impulses associated with equilibrium. ***Clinical application:*** Injury to the cochlear branch may cause *tinnitus* (ringing) or deafness. Injury to the vestibular branch may cause *vertigo* (a subjective feeling of rotation), *ataxia,* and *nystagmus* (involuntary rapid movement of the eyeball).

Table 18.2 *Summary of Cranial Nerves (continued)*

NUMBER, NAME, TYPE (SENSORY, MOTOR, OR MIXED)	LOCATION	FUNCTION AND CLINICAL APPLICATION
Cranial nerve IX: **glossopharyngeal** (*glossa = tongue; pharynx = throat*) Mixed 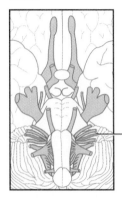 — Glossopharyngeal nerve	***Motor portion:*** Originates in medulla, passes through jugular foramen, and is distributed to stylopharyngeus muscle; parasympathetic fibers extend to parotid gland. ***Sensory portion:*** Arises from taste buds on posterior one-third of tongue and from carotid sinus, passes through jugular foramen, and ends in medulla. Also contains fibers from somatic sensory receptors on posterior one-third of tongue and proprioceptors in swallowing muscles supplied by motor portion.	***Motor function:*** Secretion of saliva. ***Sensory function:*** Taste, regulation of blood pressure, and muscle sense (proprioception). ***Clinical application:*** Injury results in difficulty during swallowing, reduced secretion of saliva, loss of sensation in the throat, and loss of taste.
Cranial nerve X: **vagus** (*vagus = vagrant or wandering*) Mixed 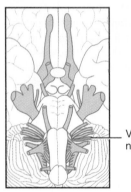 — Vagus (X) nerve	***Motor portion:*** Originates in medulla, passes through jugular foramen, and terminates in muscles of airways, lungs, esophagus, heart, stomach, small intestine portions of large intestine, and gallbladder; parasympathetic fibers innervate involuntary muscles and most glands of the gastrointestinal (GI) tract. ***Sensory portion:*** Arises from essentially same structures supplied by motor fibers, passes through jugular foramen, and ends in medulla and pons.	***Motor function:*** Smooth muscle contraction and relaxation; secretion of digestive fluids. ***Sensory function:*** Sensations from visceral organs supplied; muscle sense (proprioception). ***Clinical application:*** Severing of both nerves in the upper body interferes with swallowing, paralyzes vocal cords, and interrupts sensations from many organs.

Table 18.2 (continued)

NUMBER, NAME, TYPE (SENSORY, MOTOR, OR MIXED)	LOCATION	FUNCTION AND CLINICAL APPLICATION
Cranial nerve XI: **accessory** (*accessorius* = *assisting*) Mixed, primarily motor  — Accessory (XI) nerve	**Motor portion:** Consists of a cranial portion and a spinal portion. *Cranial portion* originates from medulla, passes through jugular foramen, and supplies voluntary muscles of pharynx, larynx, and soft palate. *Spinal portion* originates from anterior gray horn of first five cervical segments of spinal cord, passes through jugular foramen, and supplies sternocleidomastoid and trapezius muscles. **Sensory portion:** Consists of fibers from proprioceptors in muscles supplied by motor portion and passes through jugular foramen.	**Motor function:** Cranial portion mediates swallowing movements; spinal portion mediates movement of head. **Sensory function:** Muscle sense (proprioception). **Clinical application:** If nerves are damaged, the sternocleidomastoid and trapezius muscles become paralyzed, with resulting inability to raise the shoulders and difficulty in turning the head.
Cranial nerve XII: **hypoglossal** (*hypo* = *below; glossa* = *tongue*) Mixed, primarily motor  — Hypoglossal (XII) nerve	**Motor portion:** Originates in medulla, passes through hypoglossal canal, and supplies muscles of tongue. **Sensory portion:** Consists of fibers from proprioceptors in tongue muscles that pass through hypoglossal canal and end in medulla.	**Motor function:** Movement of tongue during speech and swallowing. **Sensory function:** Muscle sense (proprioception). **Clinical application:** Injury results in difficulty in chewing, speaking, and swallowing. The tongue, when protruded, curls toward the affected side and the affected side becomes atrophied, shrunken, and deeply furrowed.

*A mnemonic device used to remember the names of the nerves is "**O**h, **o**h, **o**h, **t**o **t**ouch **a**nd **f**eel **v**ery **g**reen **v**egetables–**AH**!" The initial letter of each word corresponds to the initial letter of each pair of cranial nerves.

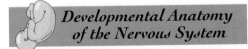 ## Developmental Anatomy of the Nervous System

The development of the nervous system begins early in the third week with a thickening of the **ectoderm** called the **neural plate** (Figure 18.25 a and b). The plate folds inward and forms a longitudinal groove, the **neural groove.** The raised edges of the neural plate are called **neural folds.** As development continues, the neural folds increase in height and meet to form a tube called the **neural tube.**

Three types of cells differentiate from the wall that encloses the neural tube. The outer or **marginal layer** develops into the *white matter* of the nervous system; the middle or **mantle layer** develops into the *gray matter,* and the inner

Figure 18.25 Origin of the nervous system. (a) Embryo in which the neural folds have partially united, forming the early neural tube. (b) Transverse sections through the embryo showing the formation of the neural tube.

🔑 *The nervous system begins developing in the third week from a thickening of ectoderm called the neural plate.*

Q *What is the origin of the gray matter of the nervous system?*

or **ependymal layer** eventually forms the *lining of the central canal of the spinal cord* and *ventricles of the brain.*

The **neural crest** is a mass of tissue between the neural tube and the skin ectoderm (Figure 18.25b). It differentiates and eventually forms the *posterior (dorsal) root ganglia of spinal nerves, spinal nerves, ganglia of cranial nerves, cranial nerves, ganglia of the autonomic nervous system, adrenal medulla,* and *meninges.*

When the neural tube forms from the neural plate, its anterior portion develops at the end of the fourth week of embryonic development into three fluid-filled enlargements called **primary brain vesicles** (Figure 18.26a). These are the **prosencephalon** (prōs'-en-SEF-a-lon; *proso* = before; *enkephalos* = brain) or forebrain, **mesencephalon** (mes'-en-SEF-a-lon; *mesos* = middle) or midbrain, and **rhombencephalon** (rom'-ben-SEF-a-lon; *rhombos* = diamond-shaped) or hindbrain. During the fifth week of development, the prosencephalon develops into two **secondary brain vesicles** called the **telencephalon** (tel-en-SEF-a-lon; *tele* = distant) and **diencephalon** (dī-en-SEF-a-lon; *dia* = through). The rhombencephalon also develops into two secondary brain vesicles called the **myelencephalon** (mī-el-en-SEF-a-lon; *myelos* = marrow) and **metencephalon** (met-en-SEF-a-lon; *meta* = after) (Figure 18.26b). Because the mesencephalon remains unchanged, there are five secondary brain vesicles.

Ultimately, the telencephalon develops into the *cerebral hemispheres* and *basal ganglia* and houses the paired *lateral ventricles;* the diencephalon develops into the *thalamus, hypothalamus,* and *pineal gland* and houses the *third ventricle;* the mesencephalon develops into the *midbrain* and houses the *cerebral aqueduct;* the metencephalon develops into the *pons* and *cerebellum;* and the myelencephalon develops into the *medulla oblongata,* each housing a portion of the *fourth ventricle.* The area of the neural tube inferior to the myelencephalon gives rise to the *spinal cord.*

Two neural tube defects—spina bifida and anencephaly (absence of the skull and cerebral hemispheres)—are associated with low levels of folic acid, a B vitamin. The incidence of both disorders is greatly decreased when women who may become pregnant take folic acid supplements. ■

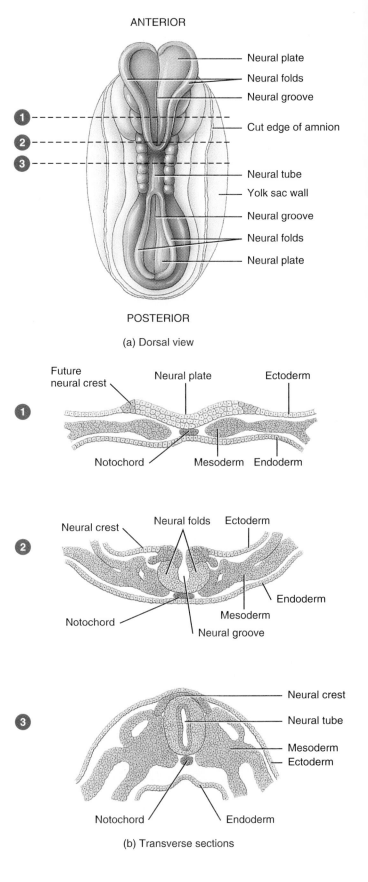

ANTERIOR

Neural plate
Neural folds
Neural groove
Cut edge of amnion
Neural tube
Yolk sac wall
Neural groove
Neural folds
Neural plate

POSTERIOR

(a) Dorsal view

Future neural crest
Neural plate
Ectoderm
Notochord
Mesoderm
Endoderm

Neural crest
Neural folds
Ectoderm
Notochord
Mesoderm
Neural groove
Endoderm

Neural crest
Neural tube
Mesoderm
Ectoderm
Notochord
Endoderm

(b) Transverse sections

Figure 18.26 Development of the brain and spinal cord.

🔑 *The various portions of the brain develop from the primary brain vesicles.*

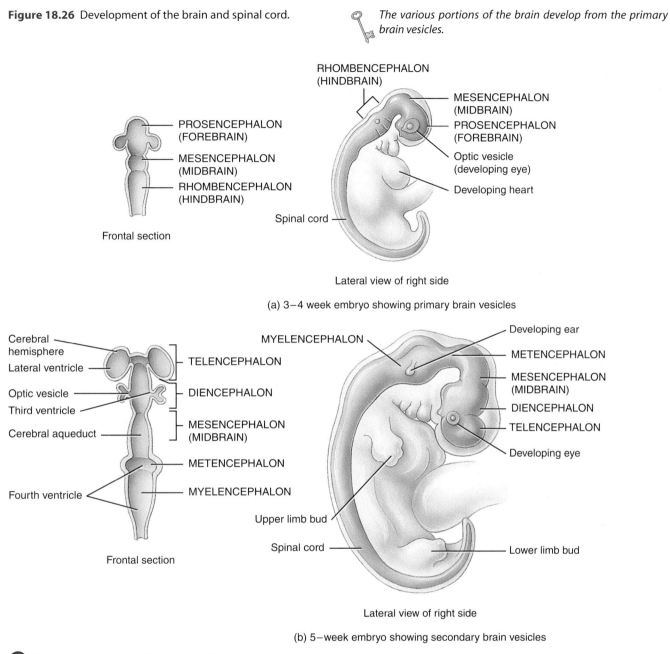

(a) 3–4 week embryo showing primary brain vesicles

(b) 5–week embryo showing secondary brain vesicles

Q *Which primary brain vesicle does not develop into a secondary brain vesicle?*

Aging and the Nervous System

One of the effects of aging on the nervous system is loss of neurons. This is a consequence of the aging process, not necessarily because of a disease state. By the age of 30, the brain begins to lose thousands of neurons each day. By the time a person reaches 80, the brain weighs about 7% less than it did in the prime of life. The cerebral cortex may lose as much as 45% of its cells. Associated with this decline in neurons is a decreased capacity for sending nerve impulses to and from the brain so that processing of information diminishes. Conduction velocity decreases, voluntary motor movements slow down, and reflex times increase. Parkinson's disease is the most common movement disorder of the CNS. Degenerative changes and disease states involving the sense organs can alter vision, hearing, taste, smell, and touch. Impaired hearing associated with aging, known as presbycusis, is usually the result of changes in important structures of the inner ear.

Key Medical Terms
Associated with the Central Nervous System

Agnosia (ag-NŌ-zē-a; *a* = without; *gnosis* = knowledge) Inability to recognize the significance of sensory stimuli such as auditory, visual, olfactory, gustatory, and tactile stimuli.

Apraxia (a-PRAK-sē-a; *pratto* = to do) Inability to carry out purposeful movements in the absence of paralysis.

Coma (KŌ-ma) Abnormally deep unconsciousness with an absence of voluntary response to stimuli and with varying degrees of reflex activity. It may be due to illness or to an injury.

Delirium (de-LIR-ē-um; *deliria* = off the track) Also called **acute confusional state (ACS).** A transient disorder of abnormal cognition (perception, thinking, and memory) and disordered attention that is accompanied by disturbances of the sleep–wake cycle and psychomotor behavior (hyperactivity or hypoactivity of movements and speech).

Dementia (de-MEN-shē-a; *de* = away from; *mens* = mind) An organic mental disorder that results in permanent or progressive general loss of intellectual abilities such as impairment of memory, judgment, and abstract thinking and changes in personality.

Electroconvulsive therapy (ECT) (e-lek′-trō-con-VUL-siv THER-a-pē) A form of shock therapy in which convulsions are induced by the passage of a brief electric current through the brain. A patient undergoing ECT is properly anesthetized and given a muscle relaxant to minimize the convulsions. ECT is regarded as an important therapeutic option in the treatment of severe depression and acute mania. Side effects include acute confusional states and memory deficits.

Huntington's chorea (HUNT-ing-tunz kō-RĒ-a; *choreia* = dance) A rare hereditary disease characterized by involuntary jerky movements and mental deterioration that terminates in dementia.

Lethargy (LETH-ar-jē) A condition of functional torpor or sluggishness.

Nerve block Loss of sensation in a region, such as in local dental anesthesia, due to injection of a local anesthetic.

Neuralgia (noo-RAL-jē-a; *neur* = nerve) Attacks of pain along the entire course or branch of a peripheral sensory nerve.

Paralysis (pa-RAL-a-sis) Diminished or total loss of motor function resulting from damage to nervous tissue or a muscle.

Spastic (SPAS-tik; *spas* = draw or pull) An increase in muscle tone (stiffness) associated with an increase in tendon reflexes and abnormal reflexes (Babinski sign).

Stupor (STOO-por) Unresponsiveness from which a patient can be aroused only briefly and by vigorous and repeated stimulation.

Torpor (TOR-por) State of lethargy and sluggishness that precedes stupor, which precedes semicoma, which precedes coma.

Viral encephalitis (VĪ-ral en′-sef-a-LĪ-tis) An acute inflammation of the brain caused by a direct attack by various viruses or by an allergic reaction to any of the many viruses that are normally harmless to the central nervous system. If the virus affects the spinal cord as well, it is called **encephalomyelitis.**

Study Outline

Brain (p. 546)

Principal Parts (p. 546)

1. The principal parts of the brain are the brain stem, diencephalon, cerebrum, and cerebellum.
2. The brain stem consists of the medulla oblongata, pons, and midbrain.
3. The diencephalon is composed of the thalamus, hypothalamus, epithalamus, and subthalamus.

Protection and Coverings (p. 546)

1. The brain is protected by cranial bones, cranial meninges, and cerebrospinal fluid.
2. The cranial meninges are continuous with the spinal meninges and are named dura mater, arachnoid, and pia mater.

Cerebrospinal Fluid (CSF) (p. 546)

1. Cerebrospinal fluid is formed by the choroid plexuses and circulates through the subarachnoid space, ventricles, and central

canal. Most of the fluid is absorbed by the arachnoid villi of the superior sagittal blood sinus.

2. Cerebrospinal fluid provides mechanical protection, chemical protection, and circulation.

3. If cerebrospinal fluid accumulates in the ventricles or subarachnoid space, the resulting condition is called hydrocephalus.

Blood Supply (p. 550)

1. The blood supply to the brain is via the cerebral arterial circle (circle of Willis).

2. The blood–brain barrier (BBB) is a concept that explains the differential rates of passage of certain material from the blood into the brain.

Brain Stem (p. 551)

1. The medulla oblongata is continuous with the superior part of the spinal cord and contains portions of both motor and sensory tracts. It contains nuclei that are reflex centers for regulation of heart rate, respiratory rate, vasoconstriction, swallowing, coughing, vomiting, sneezing, and hiccuping. It also contains the nuclei of origin for cranial nerves VII (cochlear and vestibular branches) through XII.

2. The pons is superior to the medulla. It connects the spinal cord with the brain and links parts of the brain with one another by way of tracts. It relays nerve impulses related to voluntary skeletal movements from the cerebral cortex to the cerebellum. It contains the nuclei for cranial nerves V through VII and the vestibular branch of VIII. The pons contains the pneumotaxic and apneustic centers, which help control respiration.

3. The midbrain connects the pons and diencephalon. It conveys motor impulses from the cerebrum to the cerebellum and spinal cord, sends sensory impulses from the spinal cord to the thalamus, and regulates auditory and visual reflexes. It also contains the nuclei of origin for cranial nerves III and IV.

4. A large part of the brain stem consists of small areas of gray matter called the reticular formation. It contributes to regulating muscle tone and helps maintain consciousness and awakening from sleep.

Diencephalon (p. 555)

1. The diencephalon consists of the thalamus, hypothalamus, epithalamus, and subthalamus.

2. The thalamus is superior to the midbrain and contains nuclei that serve as relay stations for all sensory impulses to the cerebral cortex. It also registers conscious recognition of pain and temperature and some awareness of touch and pressure.

3. The hypothalamus is inferior to the thalamus. It controls and integrates the autonomic nervous system, receives sensory impulses from viscera, connects the nervous and endocrine systems, secretes a variety of regulating hormones, functions in rage and aggression, controls body temperature, regulates food and fluid intake, and maintains the waking state and sleep patterns.

4. The epithalamus contains the pineal gland, and the subthalamus contains the substantia nigra and red nucleus.

Cerebrum (p. 559)

1. The cerebrum is the largest part of the brain. Its cortex contains convolutions, fissures, and sulci.

2. The cerebral lobes are named the frontal, parietal, temporal, and occipital.

3. The white matter is deep to the cortex and consists of mostly myelinated axons running in three principal directions.

4. The basal ganglia are paired masses of gray matter in the cerebral hemispheres. They help to control muscular movements.

5. The limbic system is found in the cerebral hemispheres and diencephalon. It functions in emotional aspects of behavior and memory.

6. The sensory areas of the cerebral cortex are concerned with the interpretation of sensory impulses. The motor areas are the regions that govern muscular movement. The association areas are concerned with more complex integrative functions.

7. Brain waves generated by the cerebral cortex are recorded as an electroencephalogram (EEG). It may be used to diagnose epilepsy, infections, and tumors.

Lateralization of the Cerebral Cortex (p. 567)

1. The two hemispheres of the brain are not bilaterally symmetrical, either anatomically or functionally.

2. The left hemisphere is more important for right-handed control, spoken and written language, numerical and scientific skills, and reasoning.

3. The right hemisphere is more important for left-handed control, musical and artistic awareness, space and pattern perception, insight, imagination, and generating mental images of sight, sound, touch, taste, and smell.

Cerebellum (p. 568)

1. The cerebellum occupies the inferior and posterior aspects of the cranial cavity. It consists of two lateral hemispheres and a medial, constricted vermis.

2. It is attached to the brain stem by three pairs of cerebellar peduncles.

3. The cerebellum functions in the coordination of skeletal muscles and the maintenance of normal muscle tone and body equilibrium.

Cranial Nerves (p. 568)

1. Twelve pairs of cranial nerves originate from the brain.

2. They are named primarily on the basis of distribution and numbered by order of attachment to the brain. (See Table 18.2 on page 579 for a summary of cranial nerves.)

Developmental Anatomy of the Nervous System (p. 583)

1. The development of the nervous system begins with a thickening of ectoderm called the neural plate.

2. During embryological development, primary brain vesicles are formed and serve as forerunners of various parts of the brain.

3. The telencephalon forms the cerebrum, the diencephalon develops into the thalamus and hypothalamus, the mesencephalon develops into the midbrain, the metencephalon develops into the pons and cerebellum, and the myelencephalon forms the medulla.

Aging and the Nervous System (p. 586)

1. Age-related effects involve loss of neurons and decreased capacity for sending nerve impulses.

2. Degenerative changes also affect the sense organs.

Review Questions

1. Identify the four principal parts of the brain and the components of each, where applicable. (p. 546)
2. Describe the location of the cranial meninges. (p. 546)
3. Where is cerebrospinal fluid (CSF) formed? Describe its circulation and functions. Where is CSF absorbed? (p. 546)
4. What is hydrocephalus? (p. 550)
5. What is the blood–brain barrier (BBB)? Describe the passage of several substances with respect to the BBB. (p. 551)
6. Describe the location and structure of the medulla oblongata. Define decussation of pyramids. Why is it important? List the principal functions of the medulla. (p. 551)
7. Describe the location and structure of the pons. What are its functions? (p. 553)
8. Describe the location and structure of the midbrain. What are its functions? (p. 555)
9. Describe the location and structure of the thalamus. List its functions. (p. 556)
10. Where is the hypothalamus located? Explain some of its major functions. (p. 558)
11. Where is the cerebrum located? Describe the cortex, convolutions, fissures, and sulci of the cerebrum. (p. 559)
12. List and locate the lobes of the cerebrum. How are they separated from one another? What is the insula? (p. 560)
13. Describe the organization of cerebral white matter. Be sure to indicate the function of each group of fibers. (p. 561)
14. What are basal ganglia? Name the important basal ganglia and list the function of each. (p. 561)
15. Define the limbic system. Explain several of its functions. (p. 563)
16. What is meant by a sensory area of the cerebral cortex? List, locate, and give the function of each sensory area. (p. 564)
17. What is meant by a motor area of the cerebral cortex? List, locate, and give the function of each motor area. (p. 568)
18. What conditions may result from damage to sensory or motor speech areas? (p. 569)
19. What is an association area of the cerebral cortex? What are its functions? (p. 566)
20. Define an electroencephalogram (EEG). What is the diagnostic value of an EEG? (p. 567)
21. Describe lateralization of the cerebral cortex. (p. 567)
22. Describe the location of the cerebellum. List the principal parts of the cerebellum. (p. 568)
23. What are cerebellar peduncles? List and explain the function of each. (p. 568)
24. Explain the functions of the cerebellum. Describe some effects of cerebellar damage. (p. 568)
25. Define a cranial nerve. How are cranial nerves named and numbered? Distinguish between a mixed and a sensory cranial nerve. (p. 568)
26. For each of the 12 pairs of cranial nerves, list (a) its name, number, and type; (b) its location; and (c) its function. In addition, list the effects of damage, where applicable. (p. 568)
27. Describe the development of the nervous system. (p. 583)
28. Describe the effects of aging on the nervous system. (p. 586)
29. Refer to the glossary of key medical terms associated with the nervous system. Be sure that you can define each term. (p. 586)

Self Quiz

1. Match the regions of the brain with the brain vesicles from which they develop.
   ___ (a) thalamus
   ___ (b) pons
   ___ (c) midbrain
   ___ (d) hypothalamus
   ___ (e) cerebellum
   ___ (f) medulla oblongata
   ___ (g) cerebrum

   (1) telencephalon
   (2) diencephalon
   (3) myelencephalon
   (4) mesencephalon
   (5) metencephalon

Complete the following sentences:

2. The nervous system begins to develop during the third week of gestation when the ___-derm forms a thickening called the ___ plate.
3. The dural extension between the cerebrum and cerebellum is the ___.
4. The two nuclei of the medulla oblongata that relay sensory information from the spinal cord to the opposite side of the brain are the nucleus gracilis and nucleus ___.
5. The optic (II) nerve exits from the eye and joins its partner from the other eye at the optic ___, where some fibers cross to the opposite side of the brain.
6. The ___ colliculi of the corpora quadrigemina are reflex centers for head and trunk movements in response to auditory stimuli.
7. The caudate and lentiform nuclei, together with several other nuclei, are called ___. They are islands of gray matter embedded deep within the ___.
8. White matter fibers that transmit nerve impulses between gyri in the same hemisphere are called ___ fibers.
9. The outer layer of the cerebrum is called the ___. It is composed of ___ matter, which means that it contains mainly neuron cell bodies. The surface of the cerebrum is a series of tightly packed ridges, called ___, with shallow grooves between them called ___.

For the following questions, fill in the blanks as requested.

10. Write "w" for white matter or "g" for gray matter:
(a) corpus callosum: ___ ; (b) olive: ___ ; (c) cerebral cortex: ___ ; (d) corpora quadrigemina: ___ ; (e) arbor vitae: ___ ; (f) caudate and lenticular nuclei: ___ .

11. Write the name of the cranial nerve that fits each description: (a) the only cranial nerve that originates partly from the spinal cord: ___ ; (b) eighth (VIII) cranial nerve: ___ ; (c) is widely distributed into neck, thorax, and abdomen: ___ ; (d) senses toothache, pain under a contact lens, wind on the face: ___ ; (e) the largest cranial nerve, with ophthalmic, maxillary, and mandibular parts: ___ ; (f) controls contraction of muscle of the iris, causing constriction of the pupil: ___ ; (g) innervates muscles of facial expression: ___ ; (h) two nerves that contain taste fibers and autonomic fibers to salivary glands: ___ and ___ ; (i) three purely sensory cranial nerves: ___, ___, and ___ .

12. Number the following in the correct sequence from anterior to posterior:
(a) fourth ventricle: ___ ; (b) pons and medulla oblongata: ___ ; (c) cerebellum: ___ .

13. Number the following in the correct sequence from superior to inferior:
(a) thalamus: ___ ; (b) hypothalamus: ___ ; (c) corpus callosum: ___ .

14. Number the following cranial nerves in the order in which they attach to the brain:
(a) olfactory: ___ ; (b) trochlear: ___ ; (c) optic: ___ ; (d) trigeminal: ___ ; (e) facial: ___ ; (f) oculomotor: ___ .

15. Number the following in the correct order for the circulation of CSF, from its production to its reabsorption:
(a) arachnoid villi: ___ ; (b) median and lateral apertures: ___ ; (c) cerebral aqueduct: ___ ; (d) lateral ventricle: ___ ; (e) third ventricle: ___ ; (f) fourth ventricle: ___ ; (g) interventricular foramen: ___ ; (h) subarachnoid space: ___ ; (i) choroid plexuses: ___ ; (j) superior sagittal sinus: ___ .

Are the following statements true or false?

16. The blood–brain barrier helps protect the brain from harmful substances in the blood, and the blood–cerebrospinal fluid barrier helps protect the brain from harmful substances in the cerebrospinal fluid.

17. The language areas are located in the cerebellar cortex.
18. The limbic system functions in the control of behavior.
19. The three parts of the brain stem are medulla oblongata, pons, and cerebellum.
20. The dural extension between cerebral hemispheres is known as the falx cerebri.
21. The cerebral nuclei that control large subconscious movements, such as swinging the arms while walking, are the caudate and putamen.
22. Match the following terms with their definitions:

(1) hypothalamus
(2) medulla oblongata
(3) midbrain
(4) pons
(5) thalamus
(6) cerebellum
(7) cerebrum

___ (a) has two main parts, separated by a longitudinal fissure, but joined internally by the corpus callosum

___ (b) responsible for coordination of skilled movements and regulation of posture and balance

___ (c) cranial nerves III and IV attach to this part of the brain
___ (d) cranial nerves V–VIII attach to this part of the brain
___ (e) cranial nerves VIII–XII attach to this part of the brain
___ (f) regulates food and fluid intake and body temperature
___ (g) all sensations are relayed through here
___ (h) centers for control of heart rate and respiration are located here
___ (i) constitutes four-fifths of the diencephalon
___ (j) connects to the pituitary gland via the infundibulum
___ (k) part of the brain stem; contains tracts that connect the cerebellum, midbrain, and medulla oblongata
___ (l) surrounds the cerebral aqueduct and contains nuclei that serve as reflex centers for head and eye movements

Choose the one best answer to the following questions:

23. The meninx that adheres to the surface of the brain and spinal cord and contains blood vessels is the
a. arachnoid
b. dura mater
c. pia mater
d. falx cerebri
e. denticulate ligament

24. The motor areas of the right cerebral cortex control voluntary movements on the left side of the body because
a. the cerebrum contains projection fibers
b. the cerebellum controls voluntary movements
c. the medulla contains decussating pyramids
d. the pons connects the spinal cord with the brain
e. the midbrain reroutes all motor impulses

25. The primary motor area is located in the
a. precentral gyrus
b. basal ganglia
c. corpus callosum
d. hypothalamus
e. postcentral gyrus

26. Which of the following signs would you expect to observe in a patient with a tumor of the cerebellum?
(1) loss of general sensation
(2) inability to execute any voluntary movements
(3) inability to execute smooth, steady movement
a. 1 only
b. 2 only
c. 3 only
d. none of the above
e. all of the above

27. Damage to the occipital lobe of the cerebrum would most likely cause
a. loss of hearing
b. loss of vision
c. loss of ability to smell
d. paralysis
e. loss of muscle sense (proprioception)

Critical Thinking Questions

1. An elderly relative suffered a stroke and now has both difficulty moving her right arm and speech problems. What areas of the brain were damaged by the stroke?
   HINT: *What results from decussation of the pyramids of the medulla?*

2. Casey complained to her swim coach that "my bathing cap is so tight, it'll squeeze my brains out through my ears!" Her coach is majoring in physical therapy and tells her that's anatomically impossible. Explain the coach's position.
   HINT: *Put your hands on your head and squeeze. What do you feel?*

3. Guy was having a rough morning. He couldn't do anything right. When his anatomy teacher jokingly asked him, "What do you have between your ears?" Guy made the mistake of answering, "I don't know. What?" Now he has to find out. Name the structures from ear to ear.
   HINT: *Your answer to Casey is a good start.*

4. A drunk driver had a car accident that caused severe damage to the brain stem, crushing the medulla oblongata. Predict the outcome of these injuries.
   HINT: *Your cerebrum houses your intelligence, but what part of your brain controls your vital functions?*

5. Beatriz just figured out that her Human Anatomy class actually starts at 9 A.M. and not at 9:15, which had been her arrival time since the beginning of the term. One of the other students remarked that Beatriz's "gray matter is pretty thin." Should Beatriz thank him?
   HINT: *Gray matter is something that you DON'T want to lose when you diet.*

6. Dennis's first trip to the dentist after a 10-year absence resulted in extensive dental work. He received numbing injections of anesthetic in several locations during the session. While having lunch right after the appointment, soup dribbles out of Dennis's mouth because he has no feeling in his left upper lip, right lower lip, and the tip of his tongue. What happened to Dennis?
   HINT: *The dentist wanted to block the nerves' ability to transmit the sensation of pain from the teeth.*

Answers to Figure Questions

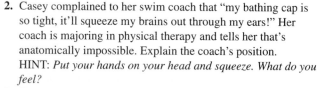

18.1 Increase in the size of neurons, proliferation of neuroglia, development of synapses and dendritic branching, and myelination of fiber tracts.

18.2 Dura mater, arachnoid, pia mater.

18.3 The thalamus (see Figure 18.13b).

18.4 CSF is reabsorbed by the arachnoid villi that project into the dural venous sinuses.

18.5 Medulla oblongata; midbrain; pons.

18.6 Decussation means crossing to the opposite side. The pyramids contain motor tracts that extend from the cortex into the spinal cord. Because they convey impulses for contraction of skeletal muscle, the functional consequence of decussation of the pyramids is that one side of the cerebrum controls muscles on the opposite side of the body.

18.7 They are the main connections for tracts between upper parts of the brain and lower parts of the brain and spinal cord.

18.8 Brain stem, diencephalon, and spinal cord.

18.9 Intermediate mass.

18.10 Mammillary region, tuberal region, supraoptic region, preoptic region.

18.11 The gray matter enlarges more rapidly, producing convolutions or gyri (folds), sulci (shallow grooves), and fissures (deep grooves).

18.12 Association fibers.

18.13 Lateral, superior, and inferior to the thalamus.

18.14 Hippocampus.

18.15 Gnostic area; motor speech area; premotor area; gustatory areas; auditory areas; visual areas; frontal eye field area.

18.16 Parietal and occipital.

18.17 Cerebellar peduncles.

18.18 Superior branch; trochlear (IV) nerve.

18.19 It's the largest.

18.20 Pons.

18.21 Jugular foramen.

18.22 Between and behind the internal jugular vein and common carotid artery.

18.23 It originates from both the brain and spinal cord.

18.24 Speech and swallowing.

18.25 The mantle layer of the neural tube.

18.26 Mesencephalon.

Chapter 19

General Senses and Sensory and Motor Pathways

Student Objectives

1. Define where in the central nervous system sensations occur.

2. Describe the events that must occur for a sensation to occur.

3. Describe the classification of receptors.

4. List the locations and functions of the receptors for tactile sensations (touch, pressure, vibration), thermal sensations (warmth and cold), pain, and proprioception.

5. Describe the role of first-order, second-order, and third-order neurons in sensory pathways.

6. Discuss the neuronal components and functions of the posterior column-medial lemniscus, anterolateral, and spinocerebellar pathways.

7. Distinguish between the somatosensory cortex and the motor cortex in terms of location and function.

8. Compare the locations and functions of the direct and indirect motor pathways.

9. Explain how sensory input and motor output are integrated.

The previous three chapters described the organization of the nervous system. Now we will see how certain parts cooperate to carry out some aspects of its three basic functions: (1) receiving sensory input; (2) integrating, associating, and storing information; and (3) transmitting motor impulses that result in movement or secretion.

In this chapter we will explore the pathways that convey somatic sensory input from the body to the brain and the pathways that carry motor commands for control of movements from the brain to the skeletal muscles. Chapter 20 deals with input from the special senses of smell, taste, vision, hearing, and equilibrium, and Chapter 21 covers output via the autonomic nervous system to smooth muscle, cardiac muscle, and glands.

SENSATION

Consider what would happen if you could not feel pain, or if you could not see, hear, smell, taste, or maintain your balance. If pain could not be sensed, burns would be common, and an inflamed appendix might burst without advance warning. A lack of sight would make negotiating your environment difficult, and a lack of hearing would enable danger to approach undetected. It would be impossible to initiate proper responses to changes in your internal and external environments. Responding to these changes requires a continual flow of sensory input into the CNS, and at any given time, your brain receives and responds to many aspects of this input. You don't notice some sensory input because it never reaches the thalamus and cerebral cortex. For example, usually one is unaware of blood pressure, yet it is constantly monitored; the input flows to the cardiovascular center in the medulla oblongata. Even sensory input that is relayed to the thalamus and cerebral cortex is noticed only some of the time. There is no question that we would quickly exhibit "sensory overload" if our conscious mind had to deal with all the information arriving at once. Filtering of sensory input assures that only a small portion reaches conscious perception at any given time.

Levels of Sensation

In its broadest context, **sensation** is the conscious or subconscious awareness of external or internal stimuli. If the stimulus is strong enough, one or more impulses arise in sensory nerve fibers. After the nerve impulses have been conducted to a region of the spinal cord or brain, they are then translated into a sensation.

The nature of the sensation and the type of reaction generated vary with the level of the central nervous system at which the sensation is translated. Sensory fibers terminating in the spinal cord can generate spinal reflexes without immediate action by the brain. Sensory fibers terminating in the lower brain stem bring about far more complex motor reactions. When sensory impulses reach the lower brain stem, they cause subconscious motor reactions—for example, changes in heart or breathing rate. Sensory impulses that reach the thalamus can be localized crudely in the body. At the thalamic level, there is crude identification of the type of sensation—that is, as a *specific* sensation such as touch, pressure, pain, position, hearing, or taste. When sensory information reaches the cerebral cortex, we experience precise identification and localization. It is at this level that memories of previous sensory information are stored and the perception of sensation occurs based on past experience. **Perception** is the conscious awareness and interpretation of sensations.

Components of Sensation

For a sensation to arise, four events typically occur:

1. **Stimulation.** A **stimulus,** or change in the environment, capable of activating certain sensory neurons must occur.

2. **Transduction.** A **sensory receptor** or **sense organ** must respond to the stimulus and ultimately **transduce** (convert) it to a nerve impulse. A sensory receptor or sense organ is a specialized cell or type of neuron or group of neurons that is selectively sensitive to certain types of changes (stimuli) in internal or external conditions.

3. **Conduction.** Nerve impulses are conducted into the CNS by sensory neurons. Such neurons are called **first-order neurons.**

4. **Translation.** A region of the CNS must receive and translate the nerve impulses into a sensation. Most conscious sensations or perceptions occur in the cerebral cortex. You seem to see with your eyes, hear with your ears, and feel pain in an injured part of your body because sensory impulses from each part of the body arrive in specific regions of the cerebral cortex, which interpret the sensation as coming from the stimulated sensory receptors.

Sensory Receptors

The process of sensation begins in a large variety of different types of sensory receptors, each of which is sensitive to a particular type of stimulus.

Selectivity of Receptors

Sensory receptors respond vigorously to one particular kind of stimulus and weakly or not at all to others. This characteristic is termed **selectivity.** The stimulus may be in the form of electromagnetic energy, such as light or heat; mechanical energy, such as pressure; or a chemical, such as carbon dioxide in body fluids. For instance, auditory receptors in the ears respond to sound waves but not to light.

Classification of Receptors

Simple receptors are associated with the **general** (or **somatic**) **senses,** which include touch, pressure, vibration, temperature, pain, and proprioception (body position and movement). Anatomically, the simplest receptors are called **free nerve endings** because they have no apparent structural specializations. Examples are receptors for pain, temperature, tickle, and itch sensations (some of which are shown in Figure 19.1). Other receptors for general sensations such as touch, pressure, and vibration have distinctive structures.

Complex receptors are associated with **special senses.** The receptors for each special sense are located in only one or two specific areas of the body such as the eye and ear. The special senses include smell, taste, vision, hearing, and equilibrium.

Two widely used classifications of receptors are based on the location of the receptors and the type of stimuli they detect.

By Location One method of classifying receptors is by their location. **Exteroceptors** (eks′-ter-ō-SEP-tors) are located at or near the surface of the body and provide information about the *external* environment. They are sensitive to stimuli outside the body and transmit sensations of hearing, vision, smell, taste, touch, pressure, vibration, temperature, and pain.

Interoceptors or **visceroceptors** (vis′-er-ō-SEP-tors) are located in blood vessels and viscera and provide information about the *internal* environment. These sensations arise from within the body and often do not reach conscious perception. Occasionally, they may be felt as pain or pressure.

Proprioceptors (prō′-prē-ō-SEP-tors) are located in muscles, tendons, joints, and the internal ear. They provide information about body position and movement. Such sensations provide information about muscle tension, the position and activity of our joints, and equilibrium.

By Type of Stimulus Another method of classifying receptors is by the type of stimuli they detect. **Mechanoreceptors** detect mechanical pressure or stretching. Stimuli so detected include those related to touch, pressure, vibration, proprioception, hearing, equilibrium, and blood pressure. **Thermoreceptors** detect changes in temperature. **Nociceptors** detect pain, usually as a result of physical or chemical damage to tissues. **Photoreceptors** detect light that strikes the retina of the eye. **Chemoreceptors** detect chemicals in the mouth (taste), nose (smell), and body fluids.

GENERAL SENSES

General or somatic senses arise in receptors located in the skin (cutaneous sensations) or embedded in muscles, tendons, joints, and the inner ear (proprioceptive sensations).

Cutaneous Sensations

Cutaneous (*cuta* = skin) **sensations** include tactile sensations (touch, pressure, vibration), thermal sensations (cold and warmth), and pain. The receptors for these sensations are in the skin, connective tissue deep to the skin, mucous membranes, mouth, and anus.

Cutaneous receptors are distributed over the body surface in such a way that some parts of the body are densely populated with receptors and other parts contain only a few. Areas of the body that have few cutaneous receptors are not very sensitive; those containing many are highly sensitive. Areas with high sensitivity include the tip of the tongue, lips, and fingertips.

Cutaneous receptors consist of the dendrites of sensory neurons that may be enclosed in a capsule of epithelial or connective tissue or have no apparent structural specialization (free nerve endings). Nerve impulses generated by cutaneous receptors pass along somatic sensory neurons in spinal and cranial nerves, through the thalamus, to the somatosensory area of the parietal lobe of the cortex (see Figure 19.4).

Tactile Sensations

Even though **tactile** (*tact* = touch) **sensations** are divided into touch, pressure, and vibration plus itch and tickle, they are all detected by mechanoreceptors.

Touch Sensations of **touch** generally result from stimulation of tactile receptors in the skin or tissues immediately deep to the skin. **Crude touch** is the ability to perceive that something has touched the skin, although its exact location, shape, size, or texture cannot be determined. **Discriminative touch** is the ability to recognize exactly what point of the body is touched. Tactile receptors include corpuscles of touch, hair root plexuses, and type I and II cutaneous mechanoreceptors (Figure 19.1).

Corpuscles of touch, or **Meissner's** (MĪS-ners) **corpuscles,** are egg-shaped receptors for discriminative touch. They are a mass of dendrites enclosed by connective tissue and are located in the dermal papillae of the skin. Corpuscles of touch are most plentiful in the fingertips, palms, and soles. They are also abundant in the eyelids, tip of the tongue, lips, nipples, clitoris, and tip of the penis. Other touch receptors are the **hair root plexuses,** which are dendrites arranged in networks around hair follicles. Movement of the hair shaft

Figure 19.1 Structure and location of cutaneous receptors.

Cutaneous senses include touch, pressure, vibration, warmth, cold, pain, and proprioception.

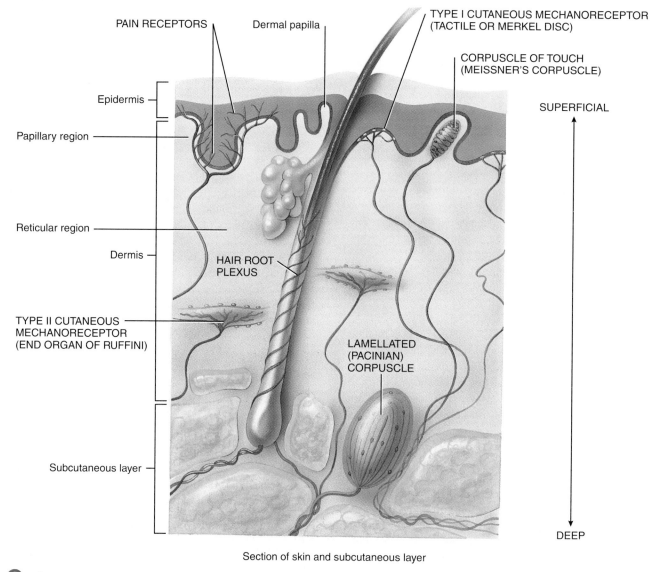

PAIN RECEPTORS

Dermal papilla

TYPE I CUTANEOUS MECHANORECEPTOR
(TACTILE OR MERKEL DISC)

CORPUSCLE OF TOUCH
(MEISSNER'S CORPUSCLE)

Epidermis

SUPERFICIAL

Papillary region

Reticular region

Dermis

HAIR ROOT
PLEXUS

TYPE II CUTANEOUS
MECHANORECEPTOR
(END ORGAN OF RUFFINI)

LAMELLATED
(PACINIAN)
CORPUSCLE

Subcutaneous layer

DEEP

Section of skin and subcutaneous layer

Q *What receptors detect pressure?*

stimulates the dendrites. Hair root plexuses detect movements mainly on the skin when hairs are disturbed.

Type I cutaneous mechanoreceptors, also called **tactile** or **Merkel** (MER-kel) **discs,** are the flattened portions of dendrites of sensory neurons that make contact with epidermal cells of the stratum basale called Merkel cells (see Figure 4.2). They are distributed in many of the same locations as corpuscles of touch and also function in discriminative touch. **Type II cutaneous mechanoreceptors,** or **end organs of Ruffini,** are embedded deeply in the dermis and in deeper tissues of the body. They detect heavy and continuous touch sensations.

Pressure Sensations of pressure generally result from stimulation of tactile receptors in deeper tissues. **Pressure** is a sustained sensation that is felt over a larger area than is touch.

Pressure receptors are type II cutaneous mechanoreceptors and lamellated corpuscles. **Lamellated,** or **Pacinian** (pa-SIN-ē-an), **corpuscles** (Figure 19.1) are oval structures composed of a connective tissue capsule, layered like an onion, that enclose a dendrite. They are located in subcutaneous tissues, deep submucosal tissues that lie under mucous membranes, and serous membranes; around joints, tendons, and muscles; and in the mammary glands, external genitalia, and certain viscera, such as the pancreas and urinary bladder.

Vibration Sensations of **vibration** result from rapidly repetitive sensory signals from tactile receptors. The receptors for vibration sensations are corpuscles of touch and lamellated corpuscles. Whereas corpuscles of touch detect low-frequency vibrations, lamellated corpuscles detect higher-frequency vibrations.

Itch and Tickle The **itch** sensation results from stimulation of free nerve endings by certain chemicals, such as bradykinin, often as a result of a local inflammatory response. Free nerve endings also are thought to mediate the **tickle** sensation. This unusual sensation is the only one that you may not be able to elicit on yourself. Why the tickle sensation may arise only when someone else touches you is not understood.

Thermal Sensations

The **thermal** (*therm* = heat) **sensations** are warmth and cold. For a long time, the receptors detecting thermal sensations were thought to be the end organs of Ruffini and similar encapsulated structures called Krause's corpuscles. Experiments now confirm, however, that **thermoreceptors** are free nerve endings. Separate thermoreceptors respond to warmth and cold stimuli.

Pain Sensations

Pain is indispensable for a normal life. It provides us with information about noxious, tissue-damaging stimuli and thus often enables us to protect ourselves from greater damage. From a medical standpoint, the subjective description and indication of the location of pain may help pinpoint the underlying cause of disease.

The receptors for **pain,** called **nociceptors** (NŌ-sē-septors; *noci* = harmful), are free nerve endings (Figure 19.1). Pain receptors are found in almost every tissue of the body. They may respond to any type of stimulus if it is strong enough to cause tissue damage. When stimuli for other sensations, such as touch, pressure, warmth, and cold, reach a certain intensity, they stimulate the sensation of pain as well. Excessive stimulation of most sensory receptors causes pain. Other stimuli that elicit pain include excessive distension or dilation of a structure, prolonged muscular contractions, muscle spasms, inadequate blood flow to an organ, or the presence of certain chemical substances. Tissue irritation or injury releases chemicals—for example, prostaglandins and kinins—that stimulate nociceptors. Pain persists even after the initial trauma occurs because these substances linger. Pain receptors, because of their sensitivity to all stimuli, perform a protective function by identifying changes that may endanger the body.

Based on the location of the stimulated receptors, pain may be divided into two types: somatic and visceral. **Somatic pain** that arises from stimulation of receptors in the skin is called **superficial somatic pain,** whereas stimulation of receptors in skeletal muscles, joints, tendons, and fascia causes **deep somatic pain. Visceral pain** results from stimulation of receptors in the viscera.

In most instances of somatic pain and in some instances of visceral pain, the cerebral cortex accurately localizes the pain to the stimulated area. If you burn your finger, you feel the pain in your finger. If the pleural membranes of the lungs are inflamed, you experience pain in the chest. In most instances of visceral pain, however, the pain is felt in or just deep to the skin that overlies the stimulated organ. The pain may also be felt in a surface area far from the stimulated organ. This phenomenon is called **referred pain.** In general, the area to which the pain is referred and the visceral organ involved are served by the same segment of the spinal cord. For example, sensory fibers from the heart and skin over the heart and along the medial aspect of the left upper extremity enter spinal cord segments T1–T4. Thus the pain of a heart attack is typically felt in the skin over the heart and along the left arm. Figure 19.2 illustrates skin regions to which visceral pain may be referred.

The pain often experienced by patients who have had a limb amputated is called **phantom pain (phantom limb sensation).** They still experience sensations such as itching, pressure, tingling, or pain in the limb as if the limb were still there. This occurs in part because nerve impulses arise in the

Figure 19.2 Referred pain. The colored parts of the diagrams indicate skin areas to which visceral pain is referred.

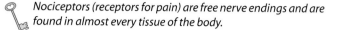 *Nociceptors (receptors for pain) are free nerve endings and are found in almost every tissue of the body.*

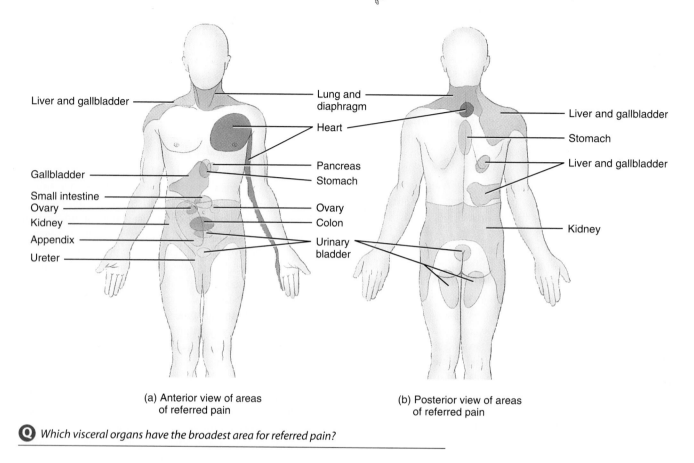

(a) Anterior view of areas of referred pain

(b) Posterior view of areas of referred pain

 *Which visceral organs have the broadest area for referred pain?*

remaining proximal portions of the sensory nerves that previously received impulses from the limb. The brain interprets impulses from these nerves as coming from the nonexistent (phantom) limb.

Clinical Application

Anesthesia

During certain surgical or diagnostic procedures, **anesthesia** (an'-es-THĒ-zē-a; *an* = without; *aisthesis* = sensation) is used to block sensations. Two commonly used forms of anesthesia are general and spinal. **General anesthesia** removes all sensations, including pain, and also produces unconsciousness and sometimes muscular relaxation. **Spinal anesthesia** involves injection of a drug into the subarachnoid space to block pain and other somatic sensations from that point downward. The procedure is widely used for surgery inferior to the diaphragm such as hernia repair, procedures on the hips and lower limbs, and operations involving the rectum, urinary bladder, prostate gland, and other pelvic structures. ▮

Proprioceptive Sensations

An awareness of body positions and movements of parts of the body is provided by the **proprioceptive** (*proprio* = one's own), or **kinesthetic** (kin'-es-THET-ik; *kinesis* = motion), **sense.** It informs us of the degree to which muscles are contracted, the amount of tension created in the tendons, the change of position of a joint, and the orientation of the head relative to the ground and in response to movements. Proprioception tells us the location and rate of movement of one body part in relation to others, so we can walk, type, or dress without using our eyes. It also allows us to estimate weight and determine the muscular work necessary to perform a task. For example, recall when you have "sensed" the weight of an object before throwing it, such as when you crumpled a piece of paper in your hands before tossing it into the wastebasket, or when you moved a basketball back and forth before attempting a free throw. Both actions involve proprioception.

Proprioceptors, the receptors for proprioception, continually inform the brain of the position of different parts of the body. The brain uses this sensory information to coordinate body movements or to stabilize the body's position. Proprioceptors include muscle spindles, tendon organs, and joint kinesthetic receptors. They are located within skeletal muscles, tendons, and joint capsules. Also classified as proprioceptors are hair cells of the internal ear that provide information for maintaining balance (see Figure 20.12d).

Impulses for conscious proprioception pass along ascending tracts in the spinal cord to the thalamus and from there to the cerebral cortex. The sensation is perceived in the somatosensory area in the parietal lobe of the cerebral cortex posterior to the central sulcus. At the same time, proprioceptive impulses also pass to the cerebellum along the spinocerebellar tracts.

Muscle Spindles

Muscle spindles are specialized groupings of muscle fibers interspersed among regular skeletal muscle fibers and oriented parallel to them (Figure 19.3a). The ends of the spindles are anchored to endomysium and perimysium. A muscle spindle consists of 3–10 specialized muscle fibers called **intrafusal muscle fibers** that are partially enclosed in a spindle-shaped connective tissue capsule. The central region of each intrafusal fiber contains several nuclei but has few or no actin and myosin filaments. Both ends of the intrafusal muscle fibers do contain actin and myosin filaments. They contract when stimulated by small-diameter motor neurons called **gamma motor neurons.** Surrounding the muscle spindle are the regular skeletal muscle fibers, which are called **extrafusal muscle fibers.** Large **alpha motor neurons** innervate the extrafusal fibers. Both types of motor neurons arise in the anterior gray horn of the spinal cord.

The central area of an intrafusal fiber cannot contract because it lacks actin and myosin, but it does contain two types of sensory fibers. The first are large diameter, rapidly conducting sensory fibers, called **type Ia fibers.** The dendrites of the Ia fiber wrap in a spiral manner around the central area of each intrafusal fiber. Stretching the central part of the spindle stimulates the dendrites, and nerve impulses propagate toward the spinal cord. The central receptive area of some muscle spindles is also served by smaller diameter sensory fibers called **type II fibers.** Their dendrites are located on either side of the type Ia dendrites. Type II dendrites are also stimulated when the central part of the spindle is stretched, and they too send impulses to the spinal cord.

Either sudden or prolonged stretch on the central areas of the intrafusal muscle fibers stimulates the type Ia and type II dendrites. Muscle spindles monitor changes in the length of a skeletal muscle by responding to the rate and degree of change in length. This information is relayed to the cerebrum, which allows conscious perception of limb position (see Figure 19.4). It also passes to the cerebellum to aid in the coordination and efficiency of muscle contraction.

Tendon Organs

Tendon organs (Golgi tendon organs) are proprioceptors found at the junction of a tendon with a muscle. Each tendon organ consists of a thin capsule of connective tissue that encloses a few collagen fibers (Figure 19.3b). Penetrating the capsule are one or more sensory **type Ib fibers** whose dendrites entwine among and around the collagen fibers. When tension is applied to a tendon, tendon organs are stimulated, and nerve impulses are conducted into the CNS (see Figure 19.4). Tendon organs help protect tendons and their associated muscles from damage due to excessive tension. Also, they function as contraction receptors; that is, they monitor the force of contraction of associated muscles.

Joint Kinesthetic Receptors

There are several types of **joint kinesthetic receptors** within and around the articular capsules of synovial joints. Encapsulated receptors, similar to type II cutaneous mechanoreceptors (end organs of Ruffini), are present in the capsules of joints and respond to pressure. Small lamellated (Pacinian) corpuscles in the connective tissue outside articular capsules are receptors that respond to acceleration and deceleration of joint movement. Articular ligaments contain receptors similar to tendon organs that protect against excessive tension on adjacent muscles when excessive strain is placed on the joint.

SENSORY (ASCENDING) PATHWAYS

Sensory pathways from receptors to the cerebral cortex involve three-neuron sets (see Figure 19.4). Axon collaterals (branches) of sensory neurons simultaneously carry signals into the cerebellum and the reticular formation of the brain stem.

1. **First-order neurons** carry signals from the somatic receptors into either the brain stem or spinal cord. From the face, mouth, teeth, and eyes, somatic sensory impulses propagate along *cranial nerves* into the brain stem. From the posterior aspect of the head, neck, and body, somatic sensory impulses propagate along *spinal nerves* into the spinal cord.

Figure 19.3 Proprioceptors.

Proprioceptors are located mainly in muscles, tendons, and the inner ear; they provide information about movements and position of the body.

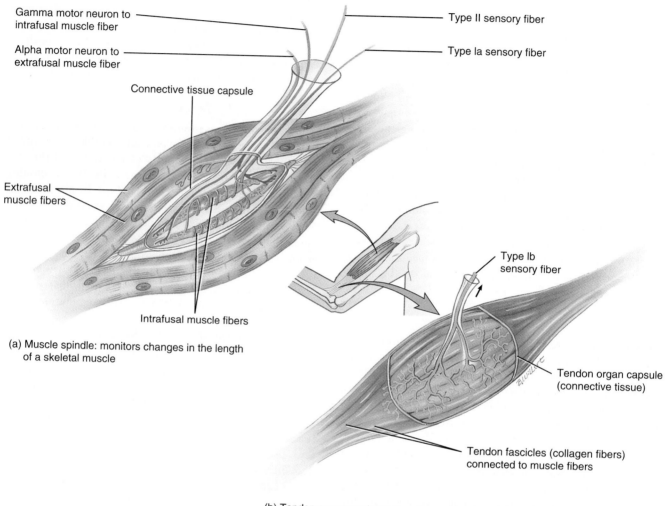

Gamma motor neuron to intrafusal muscle fiber

Alpha motor neuron to extrafusal muscle fiber

Connective tissue capsule

Type II sensory fiber

Type Ia sensory fiber

Extrafusal muscle fibers

Intrafusal muscle fibers

Type Ib sensory fiber

Tendon organ capsule (connective tissue)

Tendon fascicles (collagen fibers) connected to muscle fibers

(a) Muscle spindle: monitors changes in the length of a skeletal muscle

(b) Tendon organ: protects muscles and tendons from damage due to excess tension and monitors force of muscle contraction

Q *Which parts of a muscle spindle are associated with the sensory and motor neurons?*

2. **Second-order neurons** carry signals from the spinal cord and brain stem to the thalamus. Axons of second-order neurons cross over (decussate) to the opposite side in the spinal cord or brain stem before ascending to the thalamus.

3. **Third-order neurons** project from the thalamus to the primary somatosensory area of the cortex (postcentral gyrus), where conscious perception of the sensations results.

There are two general pathways by which somatic sensory signals entering the spinal cord ascend to the cerebral cortex: the posterior column–medial lemniscus pathway and the anterolateral (spinothalamic) pathways.

Posterior Column–Medial Lemniscus Pathway to the Cortex

Nerve impulses for conscious proprioception and most tactile sensations ascend to the cortex along a common pathway formed by three-neuron sets (Figure 19.4). First-order neurons extend from sensory receptors into the spinal cord and up to the medulla oblongata on the same side of the body. The cell bodies of these *first-order neurons* are in the posterior (dorsal) root ganglia of spinal nerves. Their axons form the **posterior column (fasciculus gracilis** and **fasciculus cuneatus)** in the spinal cord (see Table 19.1 on page 602).

Figure 19.4 Posterior column–medial lemniscus pathway.

Nerve impulses for conscious proprioception and most tactile sensations are conducted along sets of first-order, second-order, and third-order neurons to the somatosensory area of the cerebral cortex.

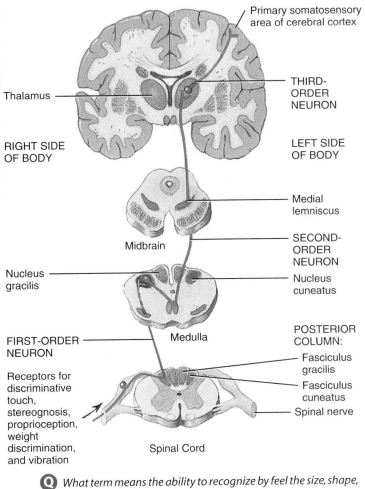

Q *What term means the ability to recognize by feel the size, shape, and texture of an object?*

The axon terminals form synapses with *second-order neurons* in the medulla. The cell body of a second-order neuron is located in the nucleus cuneatus (which receives input conducted along axons in the fasciculus cuneatus from the neck, upper limbs, and upper chest) or nucleus gracilis (which receives input conducted along axons in the fasciculus gracilis from the trunk and lower limbs). The axon of the second-order neuron crosses to the opposite side of the

medulla and enters the **medial lemniscus,** a projection tract that extends from the medulla to the thalamus. In the thalamus, the axon terminals of second-order neurons synapse with *third-order neurons,* which project their axons to the somatosensory area of the cerebral cortex.

Impulses conducted along the posterior column–medial lemniscus pathway give rise to several highly evolved and refined sensations:

1. **Discriminative touch,** the ability to recognize the exact location of a light touch and to make two-point discriminations.

2. **Stereognosis,** the ability to recognize by feel the size, shape, and texture of an object. Examples are identifying (with closed eyes) a paperclip put into your hand or reading braille.

3. **Proprioception,** the awareness of the precise position of body parts, and **kinesthesia,** the awareness of directions of movement.

4. **Weight discrimination,** the ability to assess the weight of an object.

5. **Vibratory sensations,** the ability to sense rapidly fluctuating touch.

Anterolateral (Spinothalamic) Pathways to the Cortex

The **anterolateral (spinothalamic) pathways** carry mainly pain and temperature impulses. In addition, they relay the sensations of tickle and itch and some tactile impulses, which give rise to a very crude, not well-localized touch or pressure sensation. Like the posterior column–medial lemniscus pathway, the anterolateral pathways are also composed of three-neuron sets (Figure 19.5). The first-order neuron connects a receptor of the neck, trunk, or limbs with the spinal cord. The cell body of the first-order neuron is in the posterior root ganglion. The axon of the first-order neuron synapses with the second-order neuron, which is located in the posterior gray horn of the spinal cord. The axon of the second-order neuron continues to the opposite side of the spinal cord and passes superiorly to the brain stem in either the **lateral spinothalamic tract** or **anterior spinothalamic tract** (see Table 19.1 on page 602). The axon from the second-order neuron ends in the thalamus. There, it synapses with the third-order neuron. The axon of the third-order neuron projects to the somatosensory area of the cerebral cortex. The lateral spinothalamic tract conveys sensory impulses for pain and temperature, whereas the anterior spinothalamic tract conveys impulses for tickle, itch, crude touch, and pressure.

Somatosensory Cortex

Regions of the primary somatosensory area (postcentral gyrus) that receive sensory information from different parts of the body have been mapped out. Figure 19.6a shows the location and areas of representation of the somatosensory cortex of the right cerebral hemisphere. The left cerebral hemisphere has a similar somatosensory cortex.

Note that some parts of the body are represented by large areas in the primary somatosensory area. These include the lips, face, tongue, and thumb. Other parts of the body, such as the trunk and lower limbs, are represented by much smaller areas. The relative sizes of the areas in the primary somatosensory area are directly proportional to the number of specialized sensory receptors in each respective part of the body. Thus there are many receptors in the skin of the lips but few in the skin of the trunk. The size of the cortical area for a particular part of the body relates directly to the functional importance of high sensitivity for that body part.

Pathways to the Cerebellum

Two tracts in the spinal cord, the **posterior spinocerebellar tract** and the **anterior spinocerebellar tract** (see Table 19.1 on page 602), are the major routes whereby proprioceptive input reaches the cerebellum. Sensory input conveyed to the cerebellum along these two pathways is critical for cerebellar regulation of posture and balance and coordination of skilled movements. With respect to the posterior spinocerebellar tract, impulses originating in proprioceptors of muscles, tendons, and joints of the trunk and lower limbs on one side of the body enter the posterior gray horn of the spinal cord. Here the axon terminals synapse with second-order neurons. Axons of the second-order neurons pass into the *posterior* part of the lateral white column on the same side of the cord and ascend as the posterior spinocerebellar tract. The tract enters the inferior cerebellar peduncle from the medulla and terminates in the cerebellar cortex.

With respect to the anterior spinocerebellar tract, impulses also originate from proprioceptors of muscles, tendons, and joints of the trunk and lower limbs on the same side of the body and enter the posterior gray horn of the spinal cord. Here the axon terminals synapse with the second-order neurons. Axons of the second-order neurons pass into the *anterior* part of the lateral white column on the opposite side of the spinal cord and ascend as the anterior spinocerebellar tract. The tract enters the superior cerebellar peduncle from the medulla but, before terminating in the cerebellar cortex, axons in the tract recross within the cerebellum. Thus, the anterior spinocerebellar tract, like the posterior cerebellar tract, conveys information about proprioception from one side of the body to the same side of the cerebellum.

Before leaving our discussion of somatic sensory pathways to the cerebellum, it should be noted that two other

Figure 19.5 Anterolateral (spinothalamic) pathways.

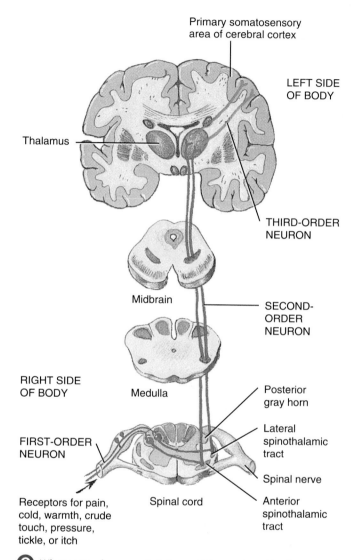

Nerve impulses for pain, temperature, crude touch, pressure, tickle, and itch are conducted along sets of first-order, second-order, and third-order neurons to the somatosensory area of the cerebral cortex.

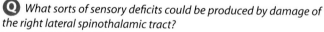

Q *What sorts of sensory deficits could be produced by damage of the right lateral spinothalamic tract?*

pathways also convey proprioceptive input to the cerebellum. These are the **cuneocerebellar tract** and the **rostral spinocerebellar tract.** Both tracts convey impulses from proprioceptors of the trunk and upper limbs.

A summary of the major sensory (ascending) tracts in the spinal cord and pathways in the brain is presented in Table 19.1.

Figure 19.6 Primary somatosensory area (postcentral gyrus) and primary motor area (precentral gyrus) of the right cerebral hemisphere. The left hemisphere has similar representation. (after Penfield and Rasmussen).

Each point on the body surface maps to a specific region in the sensory cortex and motor cortex.

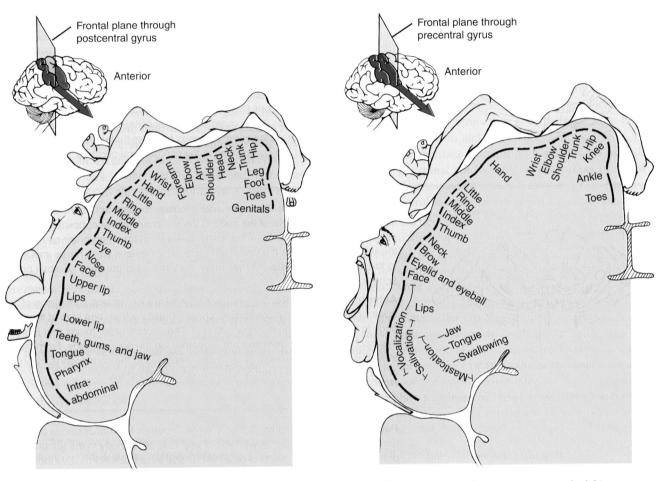

(a) Frontal section of primary somatosensory area in right cerebral hemisphere

(b) Frontal section of primary motor area in right cerebral hemisphere

Q *How do the somatosensory and motor representations in this figure compare for the hand, and what does this difference imply?*

MOTOR (DESCENDING) PATHWAYS

The most direct motor pathways extend from the cortex of the brain to skeletal muscles. Other pathways are less direct and include synapses in the basal ganglia, thalamus, reticular formation, and cerebellum.

Motor Cortex

The **primary motor area (precentral gyrus)** of the cerebral cortex is the major control region for initiation of voluntary movements. The adjacent **premotor area** and even the somatosensory area in the postcentral gyrus (see Figure 18.15) also contribute fibers to the descending motor pathways. As is true for sensory representation in the somatosensory area, different muscles are not represented equally in the primary motor area (Figure 19.6b). The degree of representation is proportional to the number of motor units in a particular muscle of the body. For example, the muscles in the thumb, fingers, lips, tongue, and vocal cords have large representations, whereas the trunk has a much smaller representation. Figures 19.6a and 19.6b show that the primary somatosensory and primary motor representations are similar but not identical for the same part of the body.

7. Why is the concept of referred pain useful to a physician in diagnosing internal disorders? (p. 595)
8. What is the proprioceptive sense? Where are the receptors for this sense located? (p. 596)
9. Describe the structure of muscle spindles, tendon organs (Golgi tendon organs), and joint kinesthetic receptors. (p. 597)
10. Describe the locations of first-order, second-order, and third-order neurons in a sensory pathway. (p. 597)
11. Distinguish between the posterior column–medial lemniscus pathway and anterolateral (spinothalamic) pathways in terms of location and function. (p. 598)
12. Describe how various parts of the body are represented in the primary somatosensory area. (p. 600)

13. What are the functions of the cerebellar tracts? (p. 600)
14. Describe how various parts of the body are represented in the primary motor area. (p. 601)
15. How do upper and lower motor neurons differ in location and function? (p. 603)
16. Distinguish between direct and indirect motor pathways in terms of location and function. (p. 603)
17. Describe how sensory input and motor output are linked in the central nervous system. (p. 606)
18. How do the basal ganglia and cerebellum function in body movements? (p. 606)

Self Quiz

Complete the following:

1. Visual receptors in the eye respond to light but not to sound. This is an example of ___ .
2. Sensory receptors may be classified by location into three groups: ___ , ___ , and ___ .
3. The sensations of touch, pressure, vibration, itch, and tickle are all ___-receptors.
4. Upper motor neurons of the indirect pathways begin in the ___ and synapse with lower motor or ___ neurons.
5. Fill in the blanks with 1 for first-order, 2 for second-order, and 3 for third-order sensory neurons: (a) ___ extend from the thalamus to the postcentral gyrus of the cerebral cortex; (b) ___ extend from the facial region along cranial nerves into the brain stem; (c) ___ extend from the spinal cord to the thalamus, or from the brain stem to the thalamus; (d) ___ extend along spinal nerves into the spinal cord.
6. Number the following in the correct sequence to indicate the route of nerve impulses along the direct (pyramidal) pathway: (a) anterior gray horn (lower motor neuron): ___ ; (b) midbrain and pons: ___ ; (c) effector (skeletal muscle): ___ ; (d) internal capsule: ___ ; (e) lateral corticospinal tract: ___ ; (f) medulla oblongata decussation site: ___ ; (g) precentral gyrus (upper motor neuron): ___ ; (h) anterior root of spinal nerve: ___ .
7. Number the following to indicate the correct sequence of events: (a) conduction: ___ ; (b) stimulation: ___ ; (c) transduction: ___ ; (d) translation: ___ .
8. Match the following terms to their definitions:
   (1) corpuscles of touch (Meissner's corpuscles)
   (2) tactile (Merkel) discs
   (3) lamellated (Pacinian) corpuscles
   (4) hair root plexuses

   ___ (a) sensitive to movement of hair shaft
   ___ (b) egg-shaped masses located in dermal papillae; plentiful in fingertips, palms, and soles; receptors for discriminative touch
   ___ (c) onion-shaped structures sensitive to pressure in skin, membranes, joints, and some viscera
   ___ (d) dendrites that contact epidermal cells in the stratum basale of skin; function in discriminative touch

Are the following statements true or false?

9. Stimulation of pain receptors in skeletal muscles, joints, and tendons results in visceral pain.
10. A sensory pathway, from receptor to the cerebral cortex, involves three neurons.
11. The precentral gyrus of the cerebral cortex is the premotor area.
12. The postcentral gyrus of the cerebral cortex is the primary somatosensory area.
13. Match the following terms to their definitions:
   (1) tendon organ (Golgi tendon organ)
   (2) muscle spindle
   (3) joint kinesthetic receptor

   ___ (a) located in skeletal muscles
   ___ (b) located within and around articular capsules of synovial joints
   ___ (c) located at the junction of a tendon and a muscle
   ___ (d) helps protect tendons and their associated muscles from damage due to excessive tension
   ___ (e) responds to acceleration and deceleration of joint movement; monitors pressure at joints
   ___ (f) monitors changes in the length of a skeletal muscle

Choose the one best answer to the following questions:

14. Which of the following is the receptor for pain?
   a. chemoreceptor
   b. photoreceptor
   c. nociceptor
   d. thermoreceptor
   e. mechanoreceptor

15. All of the following sensations are conveyed by the posterior column–medial lemniscus pathway *except*
   a. pain and temperature   d. stereognosis
   b. proprioception   e. kinesthesia
   c. discriminative touch

16. The incorrect interpretation of pain as having come from regions far from the actual site of pain is known as
   a. visceral pain   d. somatosensory pain
   b. phantom pain   e. somatic pain
   c. referred pain

17. The two major routes whereby proprioceptive input reaches the cerebellum are the
   a. anterior and posterior spinocerebellar tracts
   b. anterior and lateral spinothalamic tracts
   c. anterior and lateral corticospinal tracts
   d. direct and indirect pathways
   e. fasciculus gracilis and fasciculus cuneatus tracts

18. Which of the following receptors does *not* belong with the others?
   a. muscle spindles
   b. tactile (Merkel) discs
   c. tendon (Golgi tendon organs) organs
   d. joint kinesthetic receptors

19. Match the following terms to their definitions:
   (1) anterior spinothalamic   (4) lateral spinothalamic
   (2) anterior and lateral corticospinal   (5) tectospinal
   (3) posterior column– medial lemniscus   (6) vestibulospinal

___ (a) conveys nerve impulses that tell you that you touched a hot object

___ (b) allows you to be conscious of the position of your body parts; helps you assess weight, texture, size, and shape of objects

___ (c) located in anterior white column; starts in spinal cord and ends in thalamus

___ (d) sends nerve impulses from your brain to enable you to coordinate precise, discrete, voluntary movements

___ (e) conveys nerve impulses from one side of the medulla oblongata to the same side of your head to regulate head movements—for example, when you are trying to maintain your balance

___ (f) controls head movements on opposite side, so that the right tract causes you to turn to your left when you hear or see something to your left

Critical Thinking Questions

1. When you first enter a chemistry lab, the odors are quite strong. After several minutes, the odors are barely noticeable. Why?
HINT: *Has something happened to the odors or has something happened to you?*

2. Wouldn't it be nice to be free from pain? Well, not really. What is the purpose of pain? A few rare individuals do not feel pain. What kind of problems could they experience due to their inability to feel pain?
HINT: *If the shoe doesn't fit, don't wear it.*

3. Liv was laughing uncontrollably. Her brother had her by the leg and was tickling her foot. How does her brain know that her foot is being tickled?
HINT: *If you blocked the pathway for itch, you'd still be able to feel pain and temperature.*

4. Compare the roles of the cerebrum and the cerebellum in directing the body's movements while riding a bicycle.
HINT: *Police sobriety checks also test these functions.*

5. Very young children are usually given only spoons (no forks or knives) and cups with lids at the dinner table. Why?
HINT: *This is the same age group that's still in diapers.*

6. Ahmad used to predict the weather by how much the bunion on his left foot was bothering him. Unfortunately, his left foot was amputated last year due to complications from diabetes. Still, sometimes he thinks he feels that bunion. Explain Ahmad's experience.
HINT: *Even if his bunion had been right about the weather sometimes, it cannot be involved anymore.*

Answers to Figure Questions

19.1 Lamellated (Pacinian) corpuscles.
19.2 The kidneys.
19.3 Type Ia and type II sensory fibers wrap around the central region of intrafusal muscle fibers. Gamma motor neurons innervate intrafusal muscle fibers.
19.4 Stereognosis.

19.5 Loss of pain and temperature sensations on the left side of the body below the level of the damage.
19.6 The hand has a larger representation in the primary motor area than in the primary somatosensory area, which implies a greater precision in its movement control than discriminative ability in its sensation.
19.7 Corticobulbar.

Chapter 20

Special Senses

Student Objectives

1. Locate the receptors for olfaction and describe the neural pathway for smell.

2. Identify the gustatory receptors and describe the neural pathway for taste.

3. List and describe the accessory structures of the eye and the structural divisions of the eyeball.

4. Describe how the retina processes visual information, and describe the neural pathway between the retina and the brain.

5. Describe the anatomical subdivisions of the ear.

6. Identify the receptor organs for equilibrium and explain how they function.

Contents at a Glance

Receptors for the general senses are scattered throughout the body and are relatively simple in structure. For the most part, receptors for the general senses are modified dendrites of sensory neurons. Receptors for the special senses—smell, taste, vision, hearing, and equilibrium—are anatomically distinct from one another and are concentrated in specific locations in the head. They are housed in complex sensory organs such as the eyes and ears. Moreover, receptors for special senses are receptor cells that are usually embedded in epithelial tissue within special sense organs. Neural pathways for the special senses are also more complex than those for the general senses.

This chapter will examine the anatomy of the special sense organs and the pathways involved in conveying information from them to the central nervous system. **Ophthalmology** (op-thal-MOL-ō-jē; *ophthalmo* = eye) is the science that deals with the eye and its disorders. The other special senses are, in large part, the concern of **otolaryngology** (ō-tō-lar-in-GOL-ō-jē; *ot* = ear; *laryngo* = larynx).

OLFACTORY SENSATIONS: SMELL

Smell is a chemical sense; that is, the sensation arises from the interaction of airborne molecules and chemically sensitive smell receptors. Among all sensations, only smell and taste nerve impulses project both to higher cortical areas and the limbic system. This is probably the reason that certain odors and tastes can evoke strong emotional responses or a flood of memories.

Anatomy of Olfactory Receptors

Between 10 and 100 million receptors for the **olfactory** (ol-FAK-tō-rē; *olfactus* = smell) **sense,** or sense of smell, lie in the nasal epithelium in the superior portion of the nasal cavity. The total area of the olfactory epithelium is 5 cm^2 (a little less than 1 in.^{2}). It occupies the superior portion of the nasal septum and extends along the superior and upper part of the middle nasal conchae (Figure 20.1a). The olfactory epithelium consists of three principal kinds of cells: olfactory receptors, supporting cells, and basal cells.

Olfactory receptors are bipolar neurons whose distal (apical) end is a knob-shaped dendrite. Several cilia, called **olfactory hairs,** project from the dendrite. The cilia are the sites where olfactory stimuli are converted to signals that will ultimately produce nerve impulses. Olfactory receptors respond to the chemical stimulation of an odorant molecule and ultimately initiate the generation of nerve impulses that result in olfaction. From the proximal (basal) part of each olfactory receptor, a single axon projects to the olfactory bulb (Figure 20.1b).

Supporting (sustentacular) cells are columnar epithelial cells of the mucous membrane lining the nose. **Basal cells** lie between the bases of the supporting cells. They are stem cells that continually produce new olfactory receptors, which live for only a month or so before being replaced. This process is remarkable because olfactory receptors are neurons. It is an exception to the general rule that mature neurons are not replaced in the nervous system.

Within the connective tissue that supports the olfactory epithelium are **olfactory (Bowman's) glands.** They produce mucus, which is carried to the surface of the epithelium by ducts. The secretion moistens the surface of the olfactory epithelium and dissolves odorant gases.

The supporting cells of the nasal epithelium and the olfactory glands are innervated by branches of the facial (VII) nerve. For this reason, some odors can irritate the facial (VII) nerves that innervate the olfactory epithelium. Impulses generated in the nerves, in turn, stimulate the lacrimal glands in the eyes and nasal mucous glands, which accounts for the tears and runny nose that may occur when you accidentally inhale substances such as pepper or the vapors of household ammonia.

Olfactory Pathway

On each side of the nose, bundles of the slender, unmyelinated axons of the olfactory receptors extend through about 20 olfactory foramina in the cribriform plate of the ethmoid bone (Figure 20.1b). Collectively, these 40 or so bundles of axons are termed the **olfactory (I) nerves.** They terminate in paired masses of gray matter in the brain called the **olfactory bulbs,** which lie inferior to the frontal lobes of the cerebrum and lateral to the crista galli of the ethmoid bone. The first synapse of the olfactory pathway occurs in the olfactory bulbs between the axon terminals of olfactory receptors and dendrites of second-order neurons inside the olfactory bulbs.

Axons of olfactory bulb neurons extend posteriorly and form the **olfactory tract** (Figure 20.1a). They project to a region called the lateral olfactory area, which is located at the inferior and medial surface of the temporal lobe. This primitive cortical region is a part of the limbic system, and includes some of the amygdaloid body. Because many olfactory tract axons terminate there, the lateral olfactory area is considered the primary olfactory area, where conscious awareness of smells begins. From the lateral olfactory area, pathways also extend to the frontal lobe both directly and indirectly via the thalamus.

Figure 20.1 Olfactory receptors.

🔑 *The olfactory epithelium consists of olfactory receptors, supporting cells, and basal cells.*

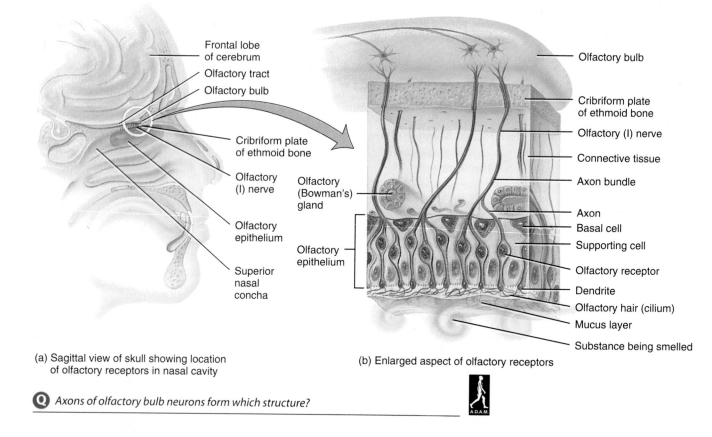

(a) Sagittal view of skull showing location of olfactory receptors in nasal cavity

(b) Enlarged aspect of olfactory receptors

Q *Axons of olfactory bulb neurons form which structure?*

GUSTATORY SENSATIONS: TASTE

Like olfaction, taste is a chemical sense; that is, the sensation arises due to the interaction of molecules of a liquid and chemically sensitive taste (gustatory) receptors. Solid substances must be dissolved by saliva in the mouth before they can be tasted. Individuals with colds or allergies sometimes complain that they cannot taste food. Although their taste sensations may be operating normally, their olfactory sensations are not. This shows that much of the experience we think of as taste is actually smell. Odors from foods pass upward into the nasopharynx (portion of the throat behind the nose) and nasal cavity to stimulate olfactory receptors. A given concentration of a substance will stimulate the olfactory sense thousands of times more strongly than it stimulates the gustatory sense.

Anatomy of Gustatory Receptors

The receptors for **gustatory** (GUS-ta-tō′-rē; *gusto* = taste) **sensations,** or sensations of taste, are located in the taste buds (Figure 20.2). The nearly 10,000 taste buds of a young adult are located mainly on the tongue, but they are also found on the soft palate (back portion of roof of mouth), larynx (voice box), and pharynx (throat). The number of taste buds declines with age. Each **taste bud** is an oval body consisting of three kinds of epithelial cells: supporting cells, gustatory receptor cells, and basal cells (Figure 20.2c). The **supporting (sustentacular) cells** form a capsule; inside are about 50 **gustatory (taste) receptor cells.** A single, hairlike **gustatory hair (microvillus)** projects from each gustatory receptor cell to the external surface through an opening in the taste bud called the **taste pore.** The gustatory hairs make contact with taste stimuli through the taste pore. **Basal cells** are found at the periphery of the taste bud near the connective tissue layer. These epithelial cells produce supporting cells, which then develop into gustatory receptor cells that have a life span of about 10 days. At their base, the receptor cells synapse with dendrites of sensory nerve fibers that form the first part of the gustatory pathway. The dendrites of a single fiber branch profusely and contact many receptors in several taste buds.

Figure 20.2 Gustatory receptors and tongue papillae.

Gustatory (taste) receptors are located in taste buds.

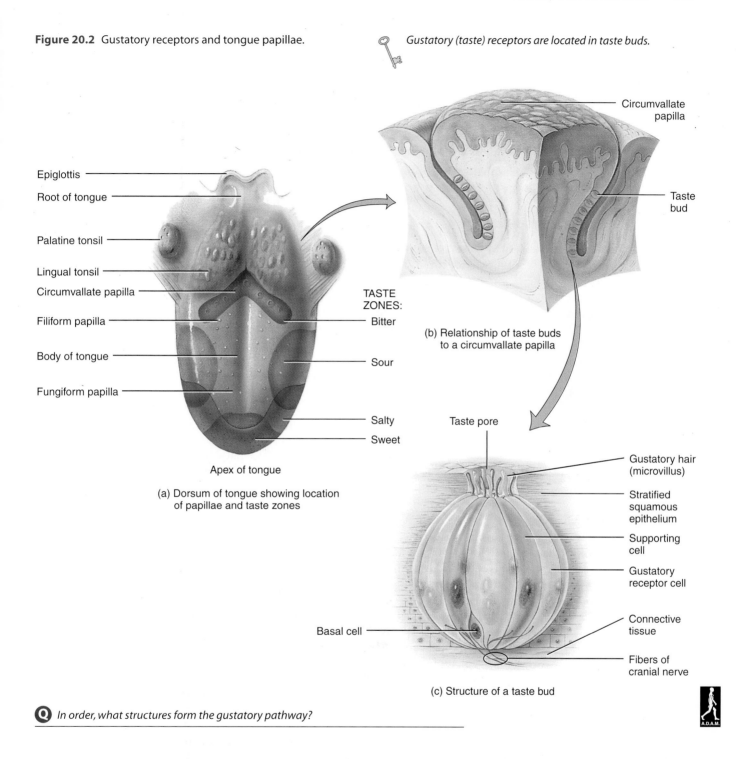

Circumvallate papilla

Taste bud

(b) Relationship of taste buds to a circumvallate papilla

Epiglottis

Root of tongue

Palatine tonsil

Lingual tonsil

Circumvallate papilla

Filiform papilla

Body of tongue

Fungiform papilla

Apex of tongue

(a) Dorsum of tongue showing location of papillae and taste zones

TASTE ZONES:

Bitter

Sour

Salty

Sweet

Taste pore

Gustatory hair (microvillus)

Stratified squamous epithelium

Supporting cell

Gustatory receptor cell

Basal cell

Connective tissue

Fibers of cranial nerve

(c) Structure of a taste bud

Q In order, what structures form the gustatory pathway?

Taste buds are found in elevations on the tongue called **papillae** (pa-PIL-ē). The papillae give the upper surface of the tongue its rough appearance (Figure 20.2a and b). **Circumvallate** (ser-kum-VAL-āt) **papillae,** the largest type, are circular and form an inverted V-shaped row at the posterior portion of the tongue. **Fungiform** (FUN-ji-form; meaning mushroom-shaped) **papillae** are knoblike elevations scattered over the entire surface of the tongue. All circumvallate and most fungiform papillae contain taste buds. **Filiform** (FIL-i-form) **papillae** are pointed threadlike structures that are also distributed over the entire surface of the tongue. They rarely contain taste buds.

Despite the many substances we seem to taste, there are only four primary taste sensations: sour, salty, bitter, and sweet (Figure 20.2a). All other "tastes," such as chocolate, pepper, and coffee, are combinations of these four, modified by accompanying olfactory sensations. Individual gustatory receptor cells may respond to more than one of the four primary tastes, but receptors in certain regions of the tongue react more strongly than others to the primary taste sensations. Although the tip of the tongue reacts to all four, it is highly sensitive to sweet and salty substances. The posterior portion of the tongue is highly sensitive to bitter substances. The lateral edges of the tongue are more sensitive to sour substances.

Gustatory Pathway

Three cranial nerves include sensory fibers that synapse with taste buds: the facial (VII) serves the anterior two-thirds of the tongue; the glossopharyngeal (IX) serves the posterior one-third of the tongue; and the vagus (X) serves the throat and epiglottis (cartilage lid over the voice box). Taste impulses conduct from the taste buds along these cranial nerves to the medulla oblongata. From the medulla, some taste fibers project to the limbic system and the hypothalamus, whereas others project to the thalamus. Fibers that extend from the thalamus to the primary gustatory area in the parietal lobe of the cerebral cortex (see Figure 18.15) are responsible for the conscious perception of taste.

VISUAL SENSATIONS

More than half the sensory receptors in the human body are located in the eyes, and a major portion of the cerebral cortex is reserved exclusively for processing visual information. Both facts demonstrate that vision is by far the most complex of the special senses, and it is perhaps the sense we most cherish. Vision is a complex sensory experience that is the result of nerve impulses that arise in two kinds of receptors in the eyes and pass via the optic (II) nerves to the brain, where the impulses undergo extensive integration and highly selective processing. This discussion will focus on the accessory structure of the eye, the eyeball, and the visual pathway.

Accessory Structures of the Eye

The **accessory structures** of the eye are the eyelids, eyelashes, eyebrows, lacrimal (tearing) apparatus, and extrinsic eye muscles.

Eyelids

The upper and lower **eyelids,** or **palpebrae** (PAL-pe-brē), shade the eyes during sleep, protect the eyes from excessive light and foreign objects, and spread lubricating secretions over the eyeballs. The upper eyelid is more movable than the lower and contains in its superior region a special levator muscle known as the **levator palpebrae superioris.** The space between the upper and lower eyelids that exposes the eyeball is called the **palpebral fissure.** Its angles are known as the **lateral commissure** (KOM-i-shūr), which is narrower and closer to the temporal bone, and the **medial commissure,** which is broader and nearer the nasal bone. In the medial commissure, there is a small, reddish elevation, the **lacrimal caruncle** (KAR-ung-kul), containing sebaceous (oil) and sudoriferous (sweat) glands. A whitish material secreted by the caruncle collects in the medial commissure.

From superficial to deep, each eyelid consists of epidermis, dermis, subcutaneous tissue, fibers of the orbicularis oculi muscle, a tarsal plate, tarsal glands, and conjunctiva (Figure 20.3a). The **tarsal plate** is a thick fold of connective tissue that gives form and support to the eyelids. Embedded in each tarsal plate is a row of elongated modified sebaceous glands known as **tarsal** or **Meibomian** (mī-BŌ-mē-an) **glands.** Their oily secretion helps keep the eyelids from adhering to each other. Infection of the tarsal glands produces a tumor or cyst on the eyelid called a **chalazion** (ka-LĀ-zē-on). The **conjunctiva** (kon′-junk-TĪ-va) is a thin, protective mucous membrane composed of stratified columnar epithelium with numerous goblet cells and areolar connective tissue. The **palpebral conjunctiva** lines the inner aspect of the eyelids, and the **bulbar (ocular) conjunctiva** passes from the eyelids onto the anterior surface of the eyeball. When the blood vessels of the bulbar conjunctiva are dilated and congested due to local irritation or infection, the person has bloodshot eyes.

Eyelashes and Eyebrows

The **eyelashes** project from the border of each eyelid and together with the **eyebrows,** which arch transversely above the upper eyelids, help protect the eyeballs from foreign objects, perspiration, and the direct rays of the sun. Sebaceous glands at the base of the hair follicles of the eyelashes, called **sebaceous ciliary glands** or **glands of Zeis** (ZĪS), pour a lubricating fluid into the follicles. Infection of these glands is called a **sty.**

Lacrimal Apparatus

The **lacrimal** (*lacrima* = tear) **apparatus** is a group of structures that produces and drains lacrimal fluid or tears (Figure 20.3b). The **lacrimal glands,** each about the size and shape of an almond, secrete lacrimal fluid that drains into 6 to 12 **excretory lacrimal ducts** that empty tears onto the surface of the conjunctiva of the upper lid. From here the tears pass medially over the anterior surface of the eyeball to enter two small openings called **lacrimal puncta.** Tears then pass into two ducts, the **lacrimal canals,** which lead into the **nasolacrimal duct,** a canal that carries the lacrimal fluid into the nasal cavity just inferior to the inferior nasal concha.

Figure 20.3 Accessory structures of the eye.

🗝 *Accessory structures of the eye are the eyelids, eyelashes, eyebrows, lacrimal apparatus, and extrinsic eye muscles.*

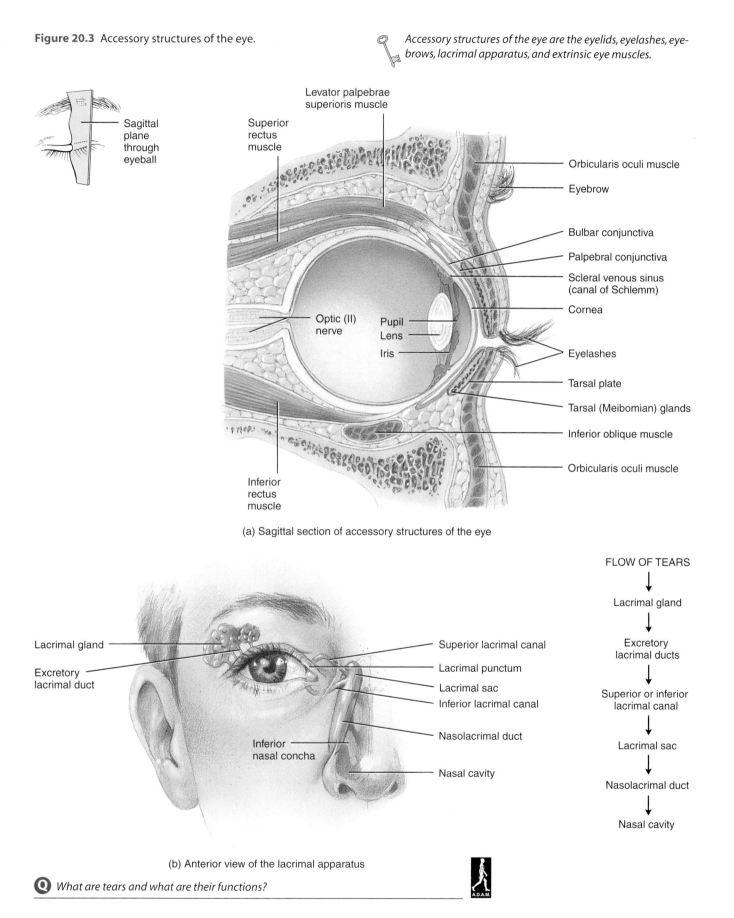

Sagittal plane through eyeball

Levator palpebrae superioris muscle

Superior rectus muscle

Orbicularis oculi muscle

Eyebrow

Bulbar conjunctiva

Palpebral conjunctiva

Scleral venous sinus (canal of Schlemm)

Cornea

Optic (II) nerve

Pupil
Lens
Iris

Eyelashes

Tarsal plate

Tarsal (Meibomian) glands

Inferior oblique muscle

Orbicularis oculi muscle

Inferior rectus muscle

(a) Sagittal section of accessory structures of the eye

Lacrimal gland

Excretory lacrimal duct

Superior lacrimal canal

Lacrimal punctum

Lacrimal sac

Inferior lacrimal canal

Nasolacrimal duct

Inferior nasal concha

Nasal cavity

FLOW OF TEARS
↓
Lacrimal gland
↓
Excretory lacrimal ducts
↓
Superior or inferior lacrimal canal
↓
Lacrimal sac
↓
Nasolacrimal duct
↓
Nasal cavity

(b) Anterior view of the lacrimal apparatus

Q *What are tears and what are their functions?*

Lacrimal fluid is a watery solution containing salts, some mucus, and a bactericidal enzyme called **lysozyme.** The fluid cleans, lubricates, and moistens the eyeball. After being secreted by the lacrimal glands, it is spread medially over the surface of the eyeball by the blinking of the eyelids. Each gland produces about 1 ml per day.

Normally, tears are cleared away by evaporation or by passing into the lacrimal canals and then into the nasal cavity as fast as they are produced. If an irritating substance contacts the conjunctiva, however, the lacrimal glands are stimulated to oversecrete, and tears accumulate (watery eyes). This is a protective mechanism, because the tears dilute and wash away the irritating substance. Watery eyes also occur when an inflammation of the nasal mucosa, such as a cold, obstructs the nasolacrimal ducts and blocks drainage of tears. Humans are unique in expressing emotions, both happiness and sadness, by **crying.** In response to parasympathetic stimulation, the lacrimal glands produce excessive tears that may spill over the edges of the eyelids and even fill the nasal cavity with fluid.

Extrinsic Eye Muscles

The six extrinsic eye muscles that move each eye receive their innervation from cranial nerves III, IV, and VI. These muscles are the **superior rectus, inferior rectus, lateral rectus, medial rectus, superior oblique,** and **inferior oblique.** In general, the size of the motor units is small in these muscles. Some motor neurons serve only two or three muscles fibers, fewer than in any other part of the body except the larynx (voice box). This permits smooth, precise, and rapid movement of eyes. As indicated in Exhibit 10.5, the extrinsic eye muscles move the eyeball in various directions (laterally, medially, superiorly, or inferiorly). For example, looking to the right requires simultaneous contraction of the right lateral rectus and left medial rectus muscles and relaxation of the left lateral rectus and right medial rectus. The oblique muscles function to preserve rotational stability of the eyeball. Coordinated movement of the two eyes involves circuits in both the brain stem and the cerebellum.

The surface anatomy of the accessory structures of the eye and the eyeball may be reviewed in Figure 11.3.

Anatomy of the Eyeball

The adult **eyeball** measures about 2.5 cm (1 in.) in diameter. Of its total surface area, only the anterior one-sixth is exposed. The remainder is recessed and protected by the orbit into which it fits. Anatomically, the wall of the eyeball can be divided into three layers: fibrous tunic, vascular tunic, and retina or nervous tunic.

Fibrous Tunic

The **fibrous tunic** is the outer coat of the eyeball consisting of the anterior cornea and posterior sclera (Figure 20.4a). The **cornea** (KOR-nē-a) is a nonvascular, transparent coat that covers the colored iris. Because it is curved, the cornea helps focus light. Its outer surface consists of nonkeratinized stratified squamous epithelium that is continuous with the epithelium of the bulbar conjunctiva. The middle coats of the cornea consist of collagen fibers and fibroblasts, and the inner surface is simple squamous epithelium. The **sclera** (SKLE-ra; *skleros* = hard), the "white" of the eye, is a coat of dense connective tissue made up mostly of collagen fibers and a few fibroblasts. The sclera covers all the eyeball except the cornea. The sclera gives shape to the eyeball, makes it more rigid, and protects its inner parts. At the junction of the sclera and cornea is an opening known as the **scleral venous sinus,** or **canal of Schlemm.**

Clinical Application

Corneal Transplants

Corneal transplants are the most common organ transplant operation and the most successful type of transplant because rejection rarely occurs. Because the cornea is avascular, blood-borne antibodies that might cause rejection do not enter the transplanted tissue. The defective cornea is removed and a donor cornea of similar diameter is sewn in. The shortage of donated corneas has partially been overcome by the development of artificial corneas made of plastic. ■

Vascular Tunic

The **vascular tunic** or **uvea** (YOO-vē-a) is the middle layer of the eyeball. It has three portions: choroid, ciliary body, and iris (Figure 20.4a). The highly vascularized **choroid** (KŌ-royd) is the posterior portion of the vascular tunic and lines most of the internal surface of the sclera. It provides nutrients to the posterior surface of the retina. Melanocytes, which produce the dark pigment melanin, give the choroid a brown-black appearance.

In the anterior portion of the vascular tunic, the choroid becomes the **ciliary** (SIL-ē-ar′-ē) **body.** This is the thickest portion of the vascular tunic. It extends from the **ora serrata** (Ō-ra ser-RĀ-ta), the jagged anterior margin of the retina, to a point just posterior to the sclerocorneal junction. The ciliary body consists of the ciliary processes and ciliary muscle. The **ciliary processes** are protrusions or folds on the internal surface of the ciliary body that contain blood capillaries that secrete a watery fluid called aqueous humor. The **ciliary muscle** is a circular band of smooth muscle that alters the shape of the lens for near or far vision.

Figure 20.4 Gross structure of the eyeball and responses of the pupil to light. The arrow in the inset in (a) indicates the direction from which the eyeball is viewed (superior).

🔑 *The wall of the eyeball consists of the fibrous tunic, vascular tunic, and retina (nervous tunic).*

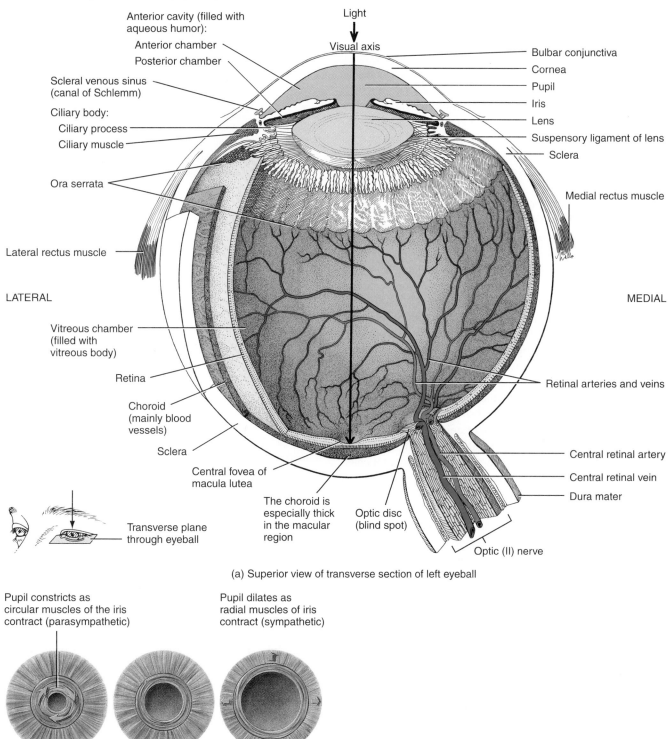

(a) Superior view of transverse section of left eyeball

(b) Anterior view of responses of the pupil to varying brightness of light

Pupil constricts as circular muscles of the iris contract (parasympathetic)

Pupil dilates as radial muscles of iris contract (sympathetic)

Bright light　　Normal light　　Dim light

Q *Which structures in the eye do not have their own blood supply and therefore rely on diffusion of oxygen and nutrients from adjacent tissues or fluid? What advantage does this avascularity confer?*

The **iris** (*irid* = colored circle) is the colored portion of the eyeball and is shaped like a flattened donut. It is suspended between the cornea and the lens and is attached at its outer margin to the ciliary processes. It consists of circular and radial smooth muscle fibers. The hole in the center of the iris is the **pupil.** A principal function of the iris is to regulate the amount of light entering the vitreous chamber of the eyeball through the pupil (Figure 20.4b). When bright light stimulates the eye, parasympathetic nerve fibers stimulate the **circular muscles (constrictor** or **sphincter pupillae)** of the iris to contract and decrease the size of the pupil (constriction). In dim light, sympathetic nerve fibers stimulate the **radial muscles (dilator pupillae)** of the iris to contract and increase the pupil's size (dilation). These responses are visceral reflexes.

Retina (Nervous Tunic)

The third and inner coat of the eye, the **retina (nervous tunic),** lines the posterior three-quarters of the eyeball and is the beginning of the visual pathway (Figure 20.4a). By using an ophthalmoscope to peer through the pupil, one can see a magnified image of the retina and the blood vessels that course across its anterior surface. The retina is the only place in the body where blood vessels can be viewed directly and examined for pathological changes such as occur with hypertension or diabetes. Several landmarks are visible (Figure 20.5). The **optic disc** is the site where the optic nerve exits the eyeball. Bundled together with the optic nerve are the retinal blood vessels (described shortly).

The retina consists of a pigment epithelium (nonvisual portion) and a neural portion (visual portion). The **pigment epithelium** is a sheet of melanin-containing epithelial cells (see Figure 20.6) that lies between the choroid and the neural portion of the retina. Some histologists classify it as part of the choroid rather than the retina. Melanin in the choroid and the pigment epithelium absorbs stray light rays, which prevents reflection and scattering of light within the eyeball. This ensures that the image cast on the retina by the cornea and lens remains sharp and clear. Albinos lack melanin in all parts of the body, including the eye. They often wear sunglasses, even indoors, because moderately bright light is perceived as nothing but glare.

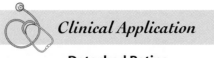

Clinical Application

Detached Retina

Detachment of the retina may occur in trauma, such as a blow to the head, or in various eye disorders. The detachment occurs between the neural portion of the retina and the pigment epithelium. Fluid accumulates between these layers, forcing the thin, pliable retina to billow outward. The result is distorted vision

Figure 20.5 Normal retina as seen through an ophthalmoscope.

In the retina, blood vessels can be viewed directly and examined for pathological changes.

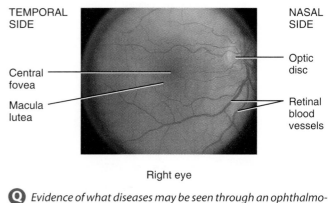

Right eye

Evidence of what diseases may be seen through an ophthalmoscope?

and blindness in the corresponding field of vision. The retina may be reattached by laser surgery or cryosurgery (which involves the local application of extreme cold). ▪

The **neural portion** of the retina is a multilayered outgrowth of the brain. It processes visual data extensively before transmitting nerve impulses to the thalamus, which then relays nerve impulses to the primary visual cortex. Three distinct layers of retinal neurons are separated by two zones where synaptic contacts are made, the inner and outer synaptic layers. The three layers of retinal neurons, named in the order in which they process visual input, are the **photoreceptor layer, bipolar cell layer,** and **ganglion cell layer** (Figure 20.6). Note that light passes through the ganglion and bipolar cell layers before reaching the photoreceptor layer. There also are two other types of cells present in the retina called **horizontal cells** and **amacrine cells.** These cells form laterally directed pathways that modify the signals being transmitted along the pathway from photoreceptors to bipolar cells to ganglion cells.

Photoreceptors are specialized to begin the process in which light rays are ultimately converted into nerve impulses. The two types of photoreceptors are rods and cones, named for the differing shapes of their ends, which nestle among fingerlike extensions of the pigment epithelium cells. Each retina has about 6 million cones and 120 million rods. **Rods** are most important for seeing shades of gray in dim light. They also allow us to see shapes and movement. **Cones** provide color vision in bright light. In moonlight we cannot see colors because only the rods are functioning. Due to the low light level cones are not functioning.

The **macula lutea** (MAK-yoo-la LOO-tē-a; *macula* = spot; *lutea* = yellow) is in the exact center of the posterior

Figure 20.6 Microscopic structure of the retina. The downward (blue) arrow indicates the direction of the signals passing through the neural portion of the retina. Ultimately, nerve impulses arise in ganglion cells and pass into the optic (II) nerve.

The neural portion of the retina processes visual data before axons of ganglion cells propagate nerve impulses to the thalamus via the optic nerve.

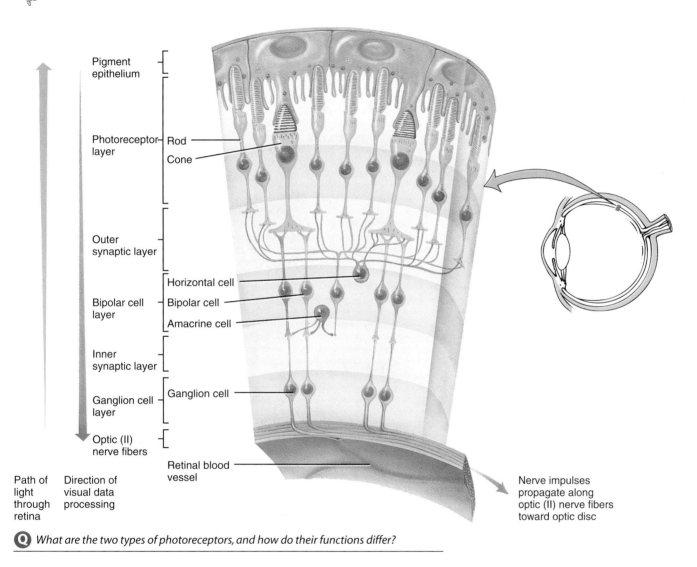

Pigment epithelium

Photoreceptor layer — Rod / Cone

Outer synaptic layer

Bipolar cell layer — Horizontal cell / Bipolar cell / Amacrine cell

Inner synaptic layer

Ganglion cell layer — Ganglion cell

Optic (II) nerve fibers

Retinal blood vessel

Path of light through retina

Direction of visual data processing

Nerve impulses propagate along optic (II) nerve fibers toward optic disc

Q *What are the two types of photoreceptors, and how do their functions differ?*

portion of the retina, at the visual axis of the eye (see Figure 20.4a). The **central fovea,** a small depression in the center of the macula lutea, contains only cone photoreceptors. In addition, the layers of bipolar and ganglion cells, which scatter light to some extent, do not cover the cones here; these layers are displaced to the periphery of the fovea. As a result, the central fovea is the area of highest **visual acuity** (sharpness of vision). Rods are absent from the fovea and macula and increase in number toward the periphery of the retina. Because rod vision is more sensitive than cone vision, you can see a faint star on a moonless night better if you gaze to one side of the star rather than directly at it.

The principal blood supply of the retina is from the **central retinal artery,** a branch of the ophthalmic artery. The central retinal artery enters the retina at about the middle of the optic disc, an area where the optic (II) nerve exits the eyeball (see Figure 20.4a). After emerging through the disc, the central retinal artery divides into superior and inferior branches, each of which subdivides into nasal and temporal branches. They fan out to nourish the anterior surface of the retina. The **central retinal vein** drains blood from the retina through the optic disc.

Figure 20.7 Section through the anterior portion of the eyeball at the sclerocorneal junction. Arrows indicate the flow of aqueous humor.

The lens separates the anterior cavity from the vitreous chamber.

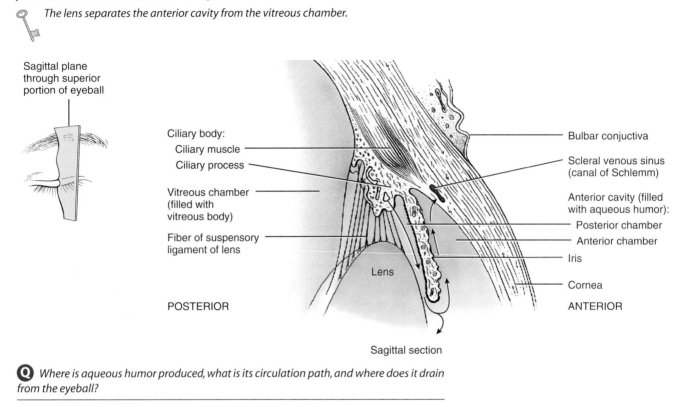

Sagittal plane through superior portion of eyeball

Ciliary body:
Ciliary muscle
Ciliary process

Vitreous chamber (filled with vitreous body)

Fiber of suspensory ligament of lens

Lens

POSTERIOR

Bulbar conjuctiva

Scleral venous sinus (canal of Schlemm)

Anterior cavity (filled with aqueous humor):
Posterior chamber
Anterior chamber

Iris

Cornea

ANTERIOR

Sagittal section

Q *Where is aqueous humor produced, what is its circulation path, and where does it drain from the eyeball?*

From photoreceptors, information flows to bipolar cells and then to ganglion cells (see Figure 20.6). The axons of ganglion cells extend posteriorly to the optic disc and exit the eyeball as the optic nerve. The optic disc is also called the **blind spot.** Because it contains no rods or cones, we cannot see an image that strikes the blind spot. Normally, you are not aware of having a blind spot, but you can easily demonstrate its presence. Cover your left eye and gaze directly at the cross below. Then increase or decrease the distance between the book and your eye. At some point the square will disappear as its image falls on the blind spot.

$+$ ■

Lens

Just posterior to the pupil and iris, within the cavity of the eyeball, is the nonvascular **lens** (see Figure 20.4a). Proteins called **crystallins,** arranged like the layers of an onion, make up the lens. Normally, the lens is perfectly transparent. It is enclosed by a clear connective tissue capsule and held in position by encircling **suspensory ligaments.** The lens fine-tunes focusing of light rays for clear vision.

Clinical Application

Cataracts

The leading cause of blindness is a loss of transparency of the lens known as a **cataract.** This problem often occurs with aging but may also be caused by injury, exposure to ultraviolet rays, certain medications (such as long-term use of steroids), or complications of other diseases (for example, diabetes). People who smoke also have increased risk of developing cataracts. The lens becomes cloudy or less transparent due to changes in the structure of the lens proteins. Fortunately, sight can usually be restored by surgical removal of the old lens and implantation of an artificial one. ■

Interior of the Eyeball

The interior of the eyeball is a large space divided by the lens into two cavities: the anterior cavity and vitreous chamber. The **anterior cavity,** the space anterior to the lens, is further divided into the **anterior chamber,** which lies behind the cornea and in front of the iris, and the **posterior chamber,** which lies behind the iris and in front of the suspensory ligaments and lens (Figure 20.7). The anterior cavity is filled with a watery fluid called the **aqueous** (*aqua* = water) **humor** that is continually secreted by the blood capillaries

Table 20.1 *Summary of Structures Associated with the Eyeball*

STRUCTURE	FUNCTION	STRUCTURE	FUNCTION
Fibrous tunic	**Cornea:** Admits and refracts (bends) light. **Sclera:** Provides shape and protects inner parts.	**Lens**	Refracts light.
Vascular tunic	**Iris:** Regulates amount of light that enters eyeball. **Ciliary body:** Secretes aqueous humor and alters shape of lens for near or far vision (accommodation). **Choroid:** Provides blood supply and absorbs scattered light.	**Anterior cavity**	Contains aqueous humor that helps maintain shape of eyeball and supplies oxygen and nutrients to lens and cornea.
Retina (nervous tunic)	Receives light and converts light into receptor potentials and nerve impulses. Output to brain is via axons of ganglion cells, which form the optic (II) nerve.	**Vitreous chamber**	Contains vitreous body that helps maintain shape of eyeball and keeps retina evenly applied to choroid.

in the ciliary processes posterior to the iris. Once the fluid is formed, it flows into the posterior chamber and then forward between the iris and the lens, through the pupil, and into the anterior chamber. Aqueous humor helps nourish the lens and cornea. From the anterior chamber, aqueous humor drains into the scleral venous sinus (canal of Schlemm) and then into the blood. Normally, aqueous humor is completely replaced about every 90 minutes.

Functionally, aqueous humor helps nourish the lens and cornea. Also, the pressure in the eye, called **intraocular pressure,** is produced mainly by the aqueous humor. The intraocular pressure, along with the vitreous body (described shortly), maintains the shape of the eyeball and keeps the retina smoothly pressed against the choroid so the retina will maintain an even surface for the reception of clear images. Because there is a balance between production and outflow

of the aqueous humor, intraocular pressure normally stays about 16 mm Hg. Excessive intraocular pressure is called **glaucoma** (glaw-KŌ-ma). It leads to degeneration of the retina, causing blindness.

The second, and larger, cavity of the eyeball is the **vitreous chamber (posterior cavity).** It lies between the lens and the retina and contains a jellylike substance called the **vitreous body.** This substance contributes to intraocular pressure, helps to prevent the eyeball from collapsing, and holds the retina flush against the internal portions of the eyeball. The vitreous body, unlike the aqueous humor, does not undergo constant replacement. It is formed during embryonic life and is not replaced thereafter. The vitreous body may partially liquefy and create the appearance of specs, hairs, or strings that dart in and out of the field of vision. These are called **floaters.** Those that appear gradually and

Figure 20.8 Photograph of the visual pathway. The brain is partially dissected to reveal the optic radiations (axons extending from the thalamus to the occipital lobe). The arrow in the inset indicates the direction from which the visual pathway is viewed (inferior).

The optic chiasm is the crossing point of the optic nerves.

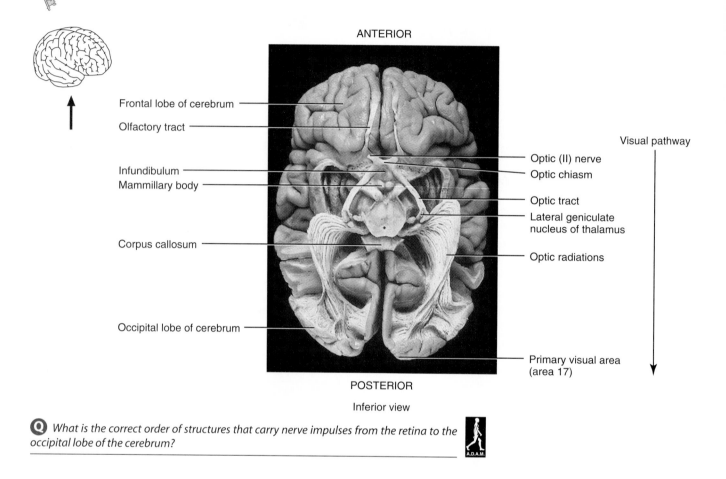

ANTERIOR

- Frontal lobe of cerebrum
- Olfactory tract
- Infundibulum
- Mammillary body
- Corpus callosum
- Occipital lobe of cerebrum

Visual pathway

- Optic (II) nerve
- Optic chiasm
- Optic tract
- Lateral geniculate nucleus of thalamus
- Optic radiations
- Primary visual area (area 17)

POSTERIOR

Inferior view

Q *What is the correct order of structures that carry nerve impulses from the retina to the occipital lobe of the cerebrum?*

then become less noticeable are usually harmless and do not require treatment.

A summary of structures associated with the eyeball is presented in Table 20.1 on page 621.

Visual Pathway

Considerable processing of visual input occurs in the retina, at synapses among the various types of cells (see Figure 20.6). The axons of retinal ganglion cells provide output from the retina to the brain. They exit the eyeball via the **optic (II) nerve.**

Retinal Processing of Visual Input

Within the retina, certain features of visual input are enhanced while other features may be discarded. Input from several cells may converge upon a smaller number of post-synaptic neurons or may diverge to a large number. On the whole, however, convergence predominates because there

are only one million ganglion cells that serve about 126 million photoreceptors.

Chemicals released by rods and cones induce changes in both bipolar cells and horizontal cells that will ultimately lead to the generation of nerve impulses (see Figure 20.6). Amacrine cells synapse with ganglion cells and transmit information to them. When bipolar or amacrine cells transmit signals to ganglion cells, the ganglion cells initiate nerve impulses.

Brain Pathway and Visual Fields

The axons of the optic (II) nerve pass through the **optic chiasm** (kī-AZM; *chiasma* = letter X), a crossing point of the optic nerves (Figure 20.8). Some fibers cross to the opposite side. Others remain uncrossed. After passing through the optic chiasm, the fibers, now part of the optic tract, enter the brain and terminate in the lateral geniculate

Figure 20.9 Structure of the ear; illustrated is the right ear.

The ear is divided into three principal regions: external (outer), middle, and internal (inner).

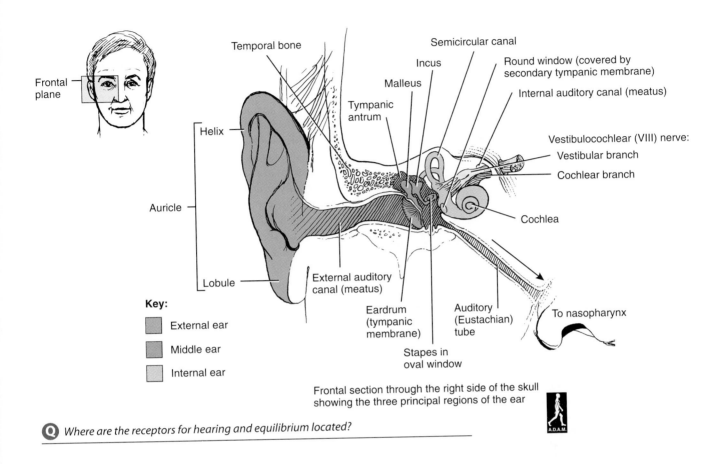

Frontal section through the right side of the skull showing the three principal regions of the ear

Q *Where are the receptors for hearing and equilibrium located?*

nucleus of the thalamus. Here the fibers synapse with neurons whose axons form the **optic radiations.** These fibers project to the visual areas of the cerebral cortex, which are located in the occipital lobes (see area 17 in Figure 18.15).

AUDITORY SENSATIONS AND EQUILIBRIUM

Besides receptors for sound waves, the ear also contains receptors for equilibrium. Anatomically, the ear is divided into three principal regions: the external (outer) ear, the middle ear, and the internal (inner) ear.

External (Outer) Ear

The **external (outer) ear** collects sound waves and passes them inward (Figure 20.9). It consists of the auricle, external auditory canal, and eardrum. The **auricle (pinna)** is a flap of elastic cartilage shaped like the flared end of a trumpet and covered by skin. The rim of the auricle is called the **helix;**

the inferior portion is the **lobule.** The auricle is attached to the head by ligaments and muscles. The **external auditory** (*audire* = hearing) **canal (meatus)** is a curved tube about 2.5 cm (1 in.) long that lies in the temporal bone and leads from the auricle to the eardrum. The **eardrum** or **tympanic** (tim-PAN-ik; *tympano* = drum) **membrane** is a thin, semi-transparent partition between the external auditory canal and middle ear. The eardrum is covered by epidermis and lined by simple cuboidal epithelium. Between the epithelial layers is connective tissue composed of collagen and elastic fibers and fibroblasts.

Near the exterior opening, the external auditory canal contains a few hairs and specialized sebaceous (oil) glands called **ceruminous** (se-ROO-mi-nus) **glands** that secrete **cerumen** (se-ROO-min) (earwax). The combination of hairs and cerumen helps prevent dust and foreign objects from entering the ear. Usually, cerumen dries up and falls out of the ear canal. Some people, however, produce an abnormal amount of cerumen. It then becomes impacted and muffles incoming sounds. The treatment for **impacted cerumen** is usually periodic ear irrigation or removal of wax with a blunt instrument by trained medical personnel.

Middle Ear

The **middle ear (tympanic cavity)** is a small, air-filled cavity in the temporal bone that is lined by epithelium (Figure 20.10a). It is separated from the external ear by the eardrum and from the internal ear by a thin bony partition that contains two small membrane-covered openings: the oval window and the round window.

The posterior wall of the middle ear communicates with the mastoid air "cells" of the temporal bone through a chamber called the **tympanic antrum.** This anatomical feature explains why a middle ear infection may spread to the temporal bone, causing mastoiditis, or even to the brain.

The anterior wall of the middle ear contains an opening that leads directly into the **auditory (Eustachian) tube.** The auditory tube consists of both bone and hyaline cartilage and connects the middle ear with the nasopharynx (upper portion of the throat). It is normally closed at its medial (pharyngeal) end; during swallowing and yawning, it opens. The atmospheric air from the throat enters or leaves the middle ear until the internal pressure equals the external pressure. When the pressures are balanced, the eardrum vibrates freely as sound waves strike it. If the pressure is not equalized, intense pain, hearing impairment, ringing in the ears, and vertigo could develop. Sudden pressure changes against the eardrum may be equalized by yawning, swallowing or pinching the nose closed, closing the mouth, and gently forcing air from the lungs into the nasopharynx. The auditory tube also is a route whereby pathogens may travel from the nose and throat to the middle ear. The fact that the auditory tubes of babies are shorter and more horizontal than those of adults may account for the higher incidence of middle ear infections in children.

Extending across the middle ear and attached to it by ligaments are three tiny bones called **auditory ossicles** (OS-si-kuls). The bones, named for their shape, are the malleus, incus, and stapes, commonly called the hammer, anvil, and stirrup, respectively. They are connected by synovial joints. The "handle" of the **malleus** is attached to the internal surface of the tympanic membrane. Its head articulates with the body of the incus. The **incus** is the intermediate bone in the series and articulates with the head of the stapes. The base or footplate of the **stapes** fits into a membrane-covered opening in the thin bony partition between the middle and inner ear. The opening is called the **oval window.** Directly below the oval window is another opening, the **round window.** This opening is enclosed by a membrane called the **secondary tympanic membrane.**

Three ligaments are associated with the malleus (Figure 20.10b). The **anterior ligament of the malleus** is attached by one end to the anterior process of the malleus and by the other to the anterior wall of the middle ear. The **superior ligament of the malleus** extends from the head of the malleus to the roof of the middle ear. The **lateral ligament of the malleus** extends from the neck of the malleus to the lateral wall of the middle ear. Two ligaments are associated with the incus. The **posterior ligament of the incus** extends from the short crus (a projection) of the incus to the posterior wall of the middle ear, and the **superior ligament of the incus** extends from the body of the incus to the roof of the middle ear. A single ligament, the **annular ligament of the base of the stapes,** is associated with the stapes. It extends from the base of the stapes to the fenestra vestibuli.

Besides the ligaments, two skeletal muscles also attach to the ossicles (Figure 20.10b and c). The **tensor tympani muscle,** which is innervated by the mandibular branch of the trigeminal (V) nerve, limits movement and increases tension on the eardrum to prevent damage to the inner ear from loud noises. It only protects the inner ear from prolonged loud noises, not brief ones, such as a gunshot. The **stapedius muscle,** which is innervated by the facial (VII) nerve, is the smallest of all skeletal muscles, and it also has a protective function in that it dampens (checks) large vibrations that result from loud noises. It is for this reason that paralysis of the stapedius muscle is associated with **hyperacusia** (abnormally sensitive hearing).

Internal (Inner) Ear

The **internal (inner) ear** is also called the **labyrinth** (LAB-i-rinth) because of its complicated series of canals (Figure 20.11). Structurally, it consists of two main divisions: an outer bony labyrinth that encloses an inner membranous labyrinth. The **bony labyrinth** is a series of cavities in the petrous portion of the temporal bone. It can be divided into three areas: (1) the semicircular canals and (2) vestibule, both of which contain receptors for equilibrium, and (3) the cochlea, which contains receptors for hearing. The bony labyrinth is lined with periosteum and contains a fluid called **perilymph.** This fluid, which is chemically similar to cerebrospinal fluid, surrounds the **membranous labyrinth,** a series of sacs and tubes floating inside and having the same general form as the bony labyrinth. The membranous labyrinth is lined with epithelium and contains a fluid called **endolymph,** which is chemically similar to intracellular fluid.

The **vestibule** is the oval central portion of the bony labyrinth. The membranous labyrinth in the vestibule consists of two sacs called the **utricle** (YOO-tri-kul; = little bag) and **saccule** (SAK-yool; = little sac). These structures are connected to each other by a small duct. Both the utricle and saccule give off a branch to form the **endolymphatic duct,** which enlarges to form the **endolymphatic sac.** Projecting superiorly and posteriorly from the vestibule are the three bony **semicircular canals.** Each lies at approximately right angles to the other two. Based on their positions, they are called the anterior, posterior, and lateral canals. The

Figure 20.10 Auditory ossicles in the middle ear of the right ear.

Common names for the malleus, incus, and stapes are the hammer, anvil, and stirrup, respectively.

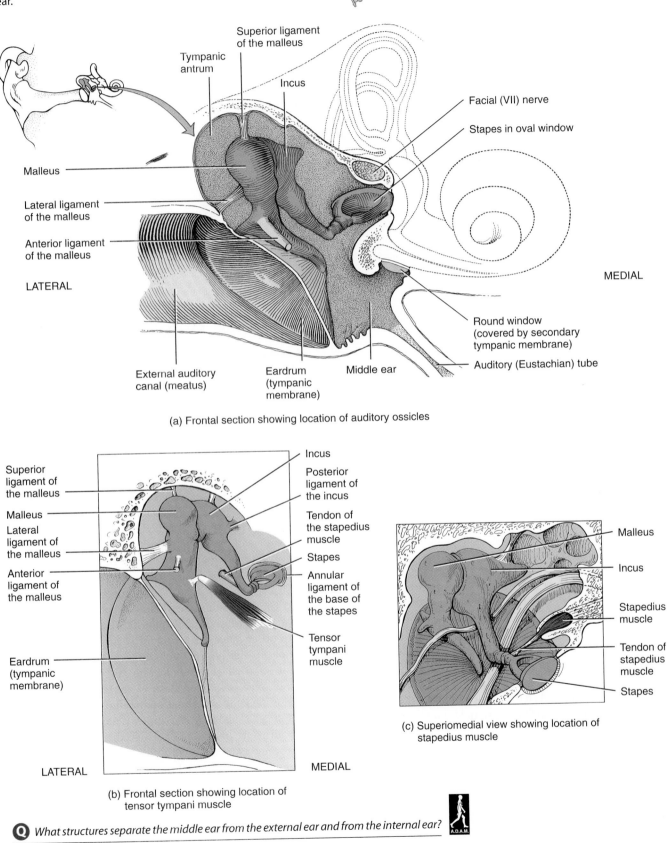

(a) Frontal section showing location of auditory ossicles

(b) Frontal section showing location of tensor tympani muscle

(c) Superiomedial view showing location of stapedius muscle

What structures separate the middle ear from the external ear and from the internal ear?

Figure 20.11 The internal ear of the right ear. The outer, blue-colored area is part of the bony labyrinth. The inner, pink-colored area belongs to the membranous labyrinth.

The bony labyrinth contains perilymph, and the membranous labyrinth contains endolymph.

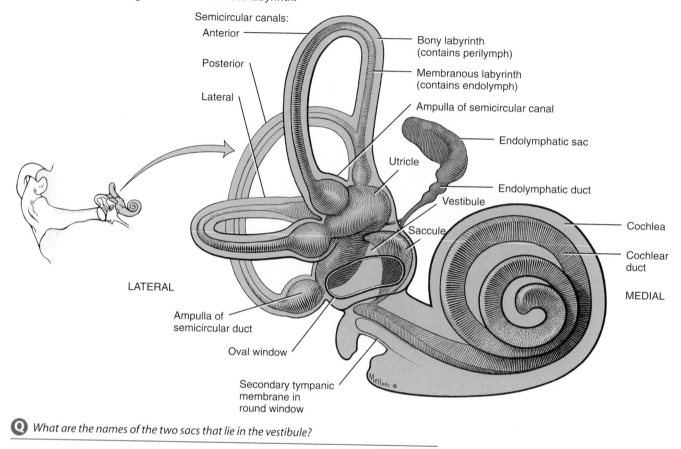

Semicircular canals:
Anterior
Posterior
Lateral

Bony labyrinth (contains perilymph)
Membranous labyrinth (contains endolymph)
Ampulla of semicircular canal
Endolymphatic sac
Utricle
Endolymphatic duct
Vestibule
Cochlea
Saccule
Cochlear duct
MEDIAL
Cochlea

LATERAL

Ampulla of semicircular duct
Oval window
Secondary tympanic membrane in round window

Q *What are the names of the two sacs that lie in the vestibule?*

anterior and posterior semicircular canals are oriented vertically; the lateral one is oriented horizontally. One end of each canal enlarges into a swelling called the **ampulla** (am-POOL-la; = little jar). The portions of the membranous labyrinth that lie inside the bony semicircular canals are called the **semicircular ducts (membranous semicircular canals).** These structures communicate with the utricle of the vestibule.

Anterior to the vestibule is the **cochlea** (KŌK-lē-a; = snail's shell). This bony spiral canal (Figure 20.12) resembles a snail's shell and makes almost three turns around a central bony core called the **modiolus** (Figure 20.12b). Cross sections through the cochlea (Figure 20.12a–c) show that it is divided into three channels. Together, the partitions that separate the channels have a shape like the letter Y. The stem of the Y is a bony shelf that protrudes into the canal; the wings of the Y are composed mainly of membranous labyrinth. The channel above the bony partition is the **scala vestibuli,** which ends at the oval window; the channel below is the **scala tympani,** which ends at the round window.

The scala vestibuli and scala tympani both contain perilymph and are completely separated except for an opening at the apex of the cochlea called the **helicotrema.** The cochlea adjoins the wall of the vestibule, into which the scala vestibuli opens. The perilymph of the vestibule is continuous with that of the scala vestibuli. The third channel (between the wings of the Y) is the **cochlear duct (scala media).** The **vestibular membrane** separates the cochlear duct from the scala vestibuli, and the **basilar membrane** separates the cochlear duct from the scala tympani.

Resting on the basilar membrane is the **spiral organ (organ of Corti),** the organ of hearing (Figure 20.12c and d). The spiral organ is a coiled sheet of epithelial cells, including supporting cells and about 16,000 **hair cells,** which are the receptors for auditory sensations. There are two groups of hair cells. The *inner hair cells* are medially placed in a single row and extend the entire length of the cochlea. The *outer hair cells* are arranged in several rows. The hair cells

Figure 20.12 Views of the semicircular canals, vestibule, and cochlea of the right ear. Note that the cochlea makes almost three complete turns.

The three channels in the cochlea are the (1) scala vestibuli, (2) scala tympani, and (3) cochlear duct.

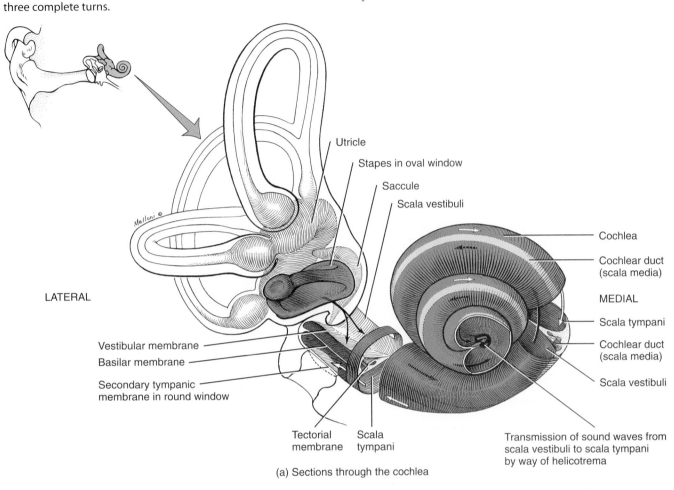

Utricle

Stapes in oval window

Saccule

Scala vestibuli

Cochlea

Cochlear duct (scala media)

MEDIAL

Scala tympani

Cochlear duct (scala media)

Scala vestibuli

LATERAL

Vestibular membrane

Basilar membrane

Secondary tympanic membrane in round window

Tectorial membrane

Scala tympani

Transmission of sound waves from scala vestibuli to scala tympani by way of helicotrema

(a) Sections through the cochlea

Figure continues

have long hairlike processes (specialized microvilli) at their apical ends that extend into the endolymph of the cochlear duct. The basal ends of the hair cells synapse with fibers of the cochlear branch of the vestibulocochlear nerve (cranial nerve VIII). Projecting over and in contact with the hair cells of the spiral organ is the **tectorial** (*tectum* = cover) **membrane,** a delicate and flexible gelatinous membrane.

Clinical Application

High-Intensity Sounds and Deafness

Hair cells of the spiral organ are easily damaged by exposure to high-intensity sounds such as loud music or the engine roar of jet planes, revved-up motorcycles, pneumatic drills, lawn mowers, and vacuum cleaners. As a result of the noises, hair cells become arranged in disorganized patterns, or they and their supporting cells may degenerate. Continued exposure to high-intensity sounds permanently damages the hair cells. Usually, deafness (significant or total hearing loss) begins with loss of sensitivity for high-pitched sounds. The deficiency may not be noticed until destruction is extensive, which is why ear protectors should be worn by those exposed to high-intensity noises. ■

Mechanism of Hearing

Sound waves result from the alternate compression and decompression of air molecules. They originate from a vibrating object, much the same way that ripples travel over the surface of water when you toss a stone into it. The events involved in hearing are as follows (Figure 20.13 on page 630):

❶ The auricle directs sound waves into the external auditory canal.

Figure 20.12 (continued)

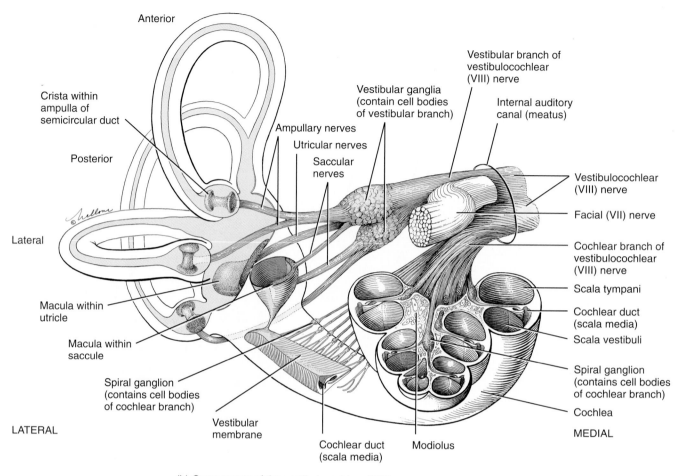

(b) Components of the vestibulocochlear (VIII) nerve

❷ When sound waves strike the eardrum, it vibrates back and forth.

❸ The central area of the eardrum connects to the malleus, which also starts to vibrate. The vibration is transmitted from the malleus to the incus and then to the stapes.

❹ As the stapes moves back and forth, it pushes the membrane of the oval window in and out.

❺ The movement of the oval window sets up fluid pressure waves in the perilymph of the cochlea.

❻ As the oval window bulges inward, it pushes on the perilymph of the scala vestibuli. Pressure waves are transmitted from the scala vestibuli to the scala tympani and eventually to the round window, causing it to bulge outward into the middle ear. (See 9 in the illustration.)

❼ As the pressure waves push the walls of the scala vestibuli and scala tympani, they also push the vestibular membrane back and forth. As a result, the pressure of the endolymph inside the cochlear duct increases and decreases.

❽ The pressure fluctuations of the endolymph move the basilar membrane, and the hair cells of the spiral organ move against the tectorial membrane. The bending of the microvilli ultimately leads to the generation of nerve impulses in cochlear nerve fibers.

Auditory Pathway

The cochlear branch of the vestibulocochlear (VIII) nerve conducts auditory impulses to the cochlear nuclei in the medulla oblongata. Here, most fibers cross to the opposite side, extend through the midbrain, and terminate in the thalamus. From the thalamus, auditory signals project to the primary auditory area in the temporal lobe of the cerebral cortex (areas 41 and 42 in Figure 18.15).

Figure 20.12 (continued)

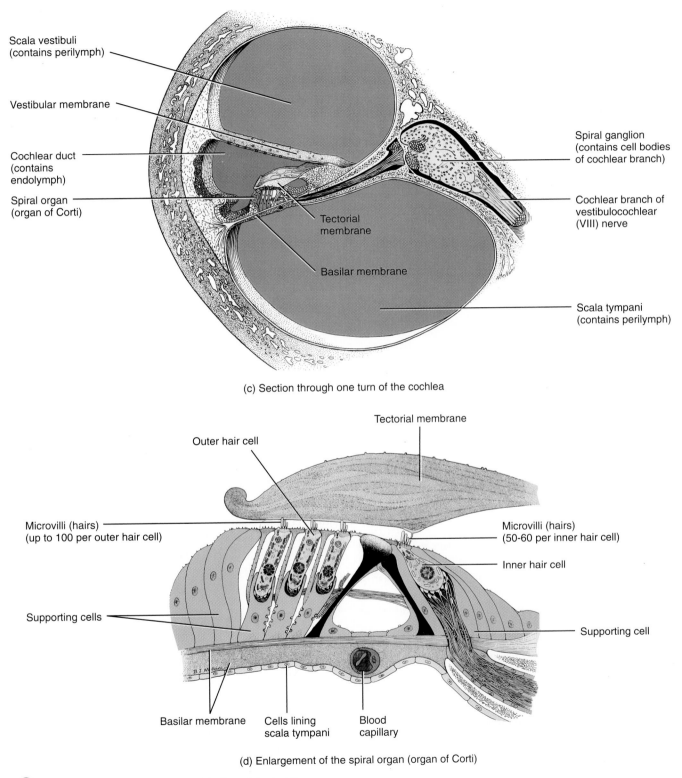

(c) Section through one turn of the cochlea

(d) Enlargement of the spiral organ (organ of Corti)

Q *What are the three subdivisions of the bony labyrinth?*

Figure 20.13 Events in the stimulation of auditory receptors in the right ear. The cochlea has been uncoiled to visualize more easily the transmission of sound waves and their distortion of the vestibular and basilar membranes of the cochlear duct.

The function of hair cells of the spiral organ (organ of Corti) is to ultimately convert a mechanical force (stimulus) into an electrical signal (nerve impulse).

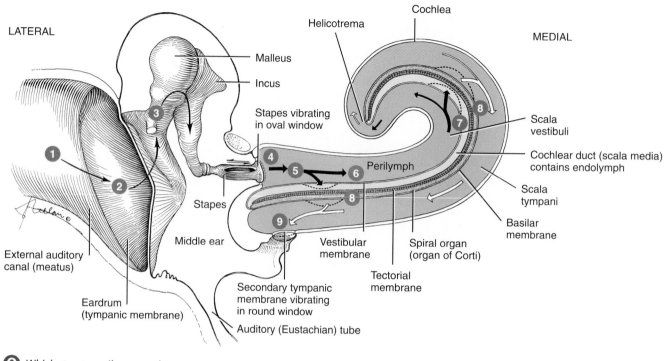

Q *Which structure vibrates and sets up pressure waves in the perilymph?*

Clinical Application

Cochlear Implants

Cochlear implants (artificial ears) are devices that translate sounds into electronic signals that can be interpreted by the brain. They take the place of hair cells of the spiral organ (organ of Corti), which normally convert sound waves into nerve impulses. The implants are used for people with deafness due to disease or injury that has destroyed hair cells of the spiral organ.

Sound waves enter a tiny microphone in the ear and travel to a microprocessor pack, where they are converted into electrical signals. The signals then travel to electrodes implanted in the cochlea, where they trigger nerve impulses in fibers of the cochlear branch of the vestibulocochlear nerve. These artificially induced nerve impulses then conduct over their normal pathways to the brain. Sounds perceived are crude compared to normal hearing, but they provide a sense of rhythm and loudness as well as information about noises such as telephones, automobiles, and the pitch and cadence of speech. ■

Mechanism of Equilibrium

There are two kinds of **equilibrium** (balance). One, called **static equilibrium,** refers to the maintenance of the position of the body (mainly the head) relative to the force of gravity. The second kind, **dynamic equilibrium,** is the maintenance of body position (mainly the head) in response to sudden movements such as rotation, acceleration, and deceleration. Collectively, the receptor organs for equilibrium are called the **vestibular apparatus,** which includes the saccule, utricle, and semicircular ducts.

Otolithic Organs: Saccule and Utricle

The walls of both the utricle and saccule contain a small, thickened region called a **macula** (plural is **maculae,** shown in Figure 20.14a). The maculae are the receptors for static equilibrium and also contribute to some aspects of dynamic equilibrium. For static equilibrium, they provide sensory information on the position of the head in space and are essential for maintaining appropriate posture and balance. For dynamic equilibrium, they detect linear acceleration and deceleration—for example, the sensations you feel while in an elevator or a car that is speeding up or slowing down.

Text continues on page 633

Figure 20.14 Location and structure of receptors in the maculae of the right ear.

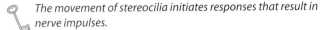 *The movement of stereocilia initiates responses that result in nerve impulses.*

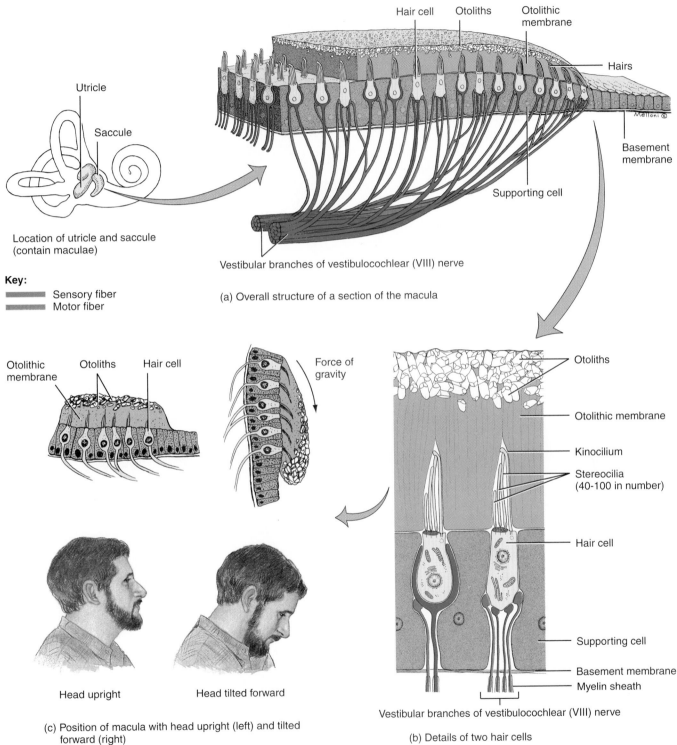

Hair cell   Otoliths   Otolithic membrane

Hairs

Utricle

Saccule

Basement membrane

Supporting cell

Location of utricle and saccule (contain maculae)

Vestibular branches of vestibulocochlear (VIII) nerve

Key:
Sensory fiber
Motor fiber

(a) Overall structure of a section of the macula

Otolithic membrane   Otoliths   Hair cell

Force of gravity

Otoliths

Otolithic membrane

Kinocilium

Stereocilia (40-100 in number)

Hair cell

Supporting cell

Basement membrane
Myelin sheath

Vestibular branches of vestibulocochlear (VIII) nerve

(b) Details of two hair cells

Head upright   Head tilted forward

(c) Position of macula with head upright (left) and tilted forward (right)

Q *With which type of equilibrium are the maculae mainly concerned?*

Figure 20.15 Semicircular ducts of the right ear. The ampullary nerves are branches of the vestibular division of the vestibulo-cochlear (VIII) nerve.

The positions of the semicircular ducts permit detection of rotational movements.

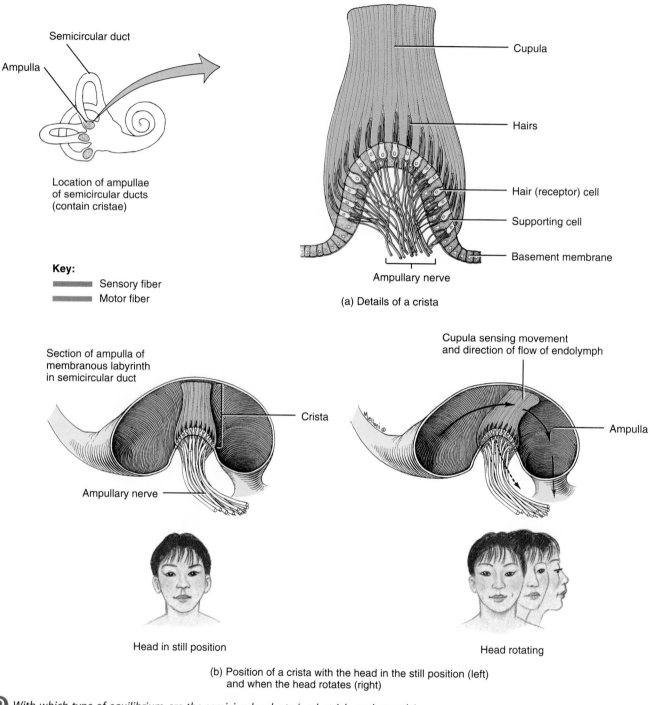

Semicircular duct

Ampulla

Location of ampullae of semicircular ducts (contain cristae)

Key:
Sensory fiber
Motor fiber

Cupula

Hairs

Hair (receptor) cell

Supporting cell

Basement membrane

Ampullary nerve

(a) Details of a crista

Section of ampulla of membranous labyrinth in semicircular duct

Crista

Ampullary nerve

Head in still position

Cupula sensing movement and direction of flow of endolymph

Ampulla

Head rotating

(b) Position of a crista with the head in the still position (left) and when the head rotates (right)

Q With which type of equilibrium are the semicircular ducts (and utricle and saccule) associated?

The two maculae are perpendicular to one another. They consist of two kinds of cells: **hair (receptor) cells** and **supporting cells** (Figure 20.14b). Hair cells have long extensions of the cell membrane consisting of 70 or more **stereocilia** (they are actually microvilli) and one **kinocilium** (a conventional cilium) anchored firmly to its basal body and extending beyond the longest microvilli. Scattered among the hair cells are columnar supporting cells. They probably secrete the thick, gelatinous, glycoprotein layer, called the **otolithic membrane,** that rests on the hair cells. A layer of heavy calcium carbonate crystals, called **otoliths** (*oto* = ear; *lithos* = stone), extends over the entire surface of the otolithic membrane.

Because the heavy otolithic membrane sits on top of the macula, if you tilt your head forward, the otolithic membrane along with the otoliths is pulled by gravity, slides downhill over the hair cells in the direction of the tilt, and stimulates the hair cells (Figure 20.14c). Similarly, if you are sitting upright in a car that suddenly jerks forward, the otolithic membrane, due to its inertia, slides backward and stimulates the hair cells. As the otoliths move, they pull on the gelatinous layer, which pulls on the stereocilia and makes them bend. The movement of the stereocilia initiates responses that will ultimately lead to generation of a nerve impulse. The hair cells synapse with first-order sensory neurons in the vestibular branch of the vestibulocochlear (VIII) nerve.

Semicircular Ducts

The three semicircular ducts, together with the saccule and utricle, maintain dynamic equilibrium. The ducts lie at right angles to one another in three planes: the two vertical ones are the anterior and posterior semicircular ducts, and the horizontal one is the lateral semicircular duct. This positioning permits detection of rotational acceleration or deceleration (see Figure 20.11). In the ampulla, the dilated portion of each duct, there is a small elevation called the **crista** (Figure 20.15a). Each crista contains a group of **hair (receptor) cells** and supporting cells covered by a mass of gelatinous material called the **cupula.** When the head moves, the endolymph in the semicircular ducts flows over the hairs and bends them (Figure 20.15b). The movement of the hairs stimulates sensory neurons, and the resulting nerve impulses pass along the vestibular branch of the vestibulocochlear (VIII) nerve.

Equilibrium Pathway

Most of the vestibular branch fibers of the vestibulocochlear (VIII) nerve enter the brain stem and terminate in the vestibular nuclei in the pons. The remaining fibers enter the cerebellum through the inferior cerebellar peduncle. Fibers from the vestibular nuclei extend to tracts that send nerve impulses to the nuclei of cranial nerves that control eye movements—oculomotor (III), trochlear (IV), and abducens (VI)—and to the accessory (XI) nerve nucleus that helps control head and neck movements. In addition, fibers from the lateral vestibular nucleus form the vestibulospinal tract, which conveys impulses to skeletal muscles that regulate muscle tone in response to head movements. Various pathways among the vestibular nuclei, cerebellum, and cerebrum enable the cerebellum to play a key role in maintaining static and dynamic equilibrium. The cerebellum continuously receives updated sensory information from the utricle and saccule and sends continuous nerve impulses to the motor areas of the cerebrum, in response to input from the utricle, saccule, and semicircular ducts, causing the motor system to increase or decrease its impulses to specific skeletal muscles to maintain equilibrium.

A summary of the structures of the ear related to hearing and equilibrium is presented in Table 20.2 on page 634.

Key Medical Terms
Associated with Sensory Structures

Achromatopsia (a-krō′-ma-TOP-sē-a; *a* = without; *chrom* = color) Complete color blindness.

Amblyopia (am′-blē-Ō-pē-a) is the term used to describe the loss of vision in an otherwise normal eye that, because of muscle imbalance, cannot focus in synchrony with the other eye.

Anopsia (an-OP-sē-a; *opsia* = vision) A defect of vision.

Blepharitis (blef-a-RĪ-tis; *blepharo* = eyelid; *itis* = inflammation of) An inflammation of the eyelid.

Conjunctivitis (pinkeye) An inflammation of the conjunctiva, caused by bacteria such as pneumococci, staphylococci, or *He-* *mophilus influenzae,* that is very contagious and more common in children. Conjunctivitis may also be caused by irritants, such as dust, smoke, or pollutants in the air, in which case it is not contagious.

Eustachitis (yoo′-stā-KĪ-tis) An inflammation or infection of the auditory (Eustachian) tube.

Exotropia (ek′-sō-TRŌ-pē-a; *ex* = out; *tropia* = turning) Turning outward of the eyes.

Keratitis (ker′-a-TĪ-tis; *kerato* = cornea) An inflammation or infection of the cornea.

Table 20.2 *Summary of Structures of the Ear Related to Hearing and Equilibrium*

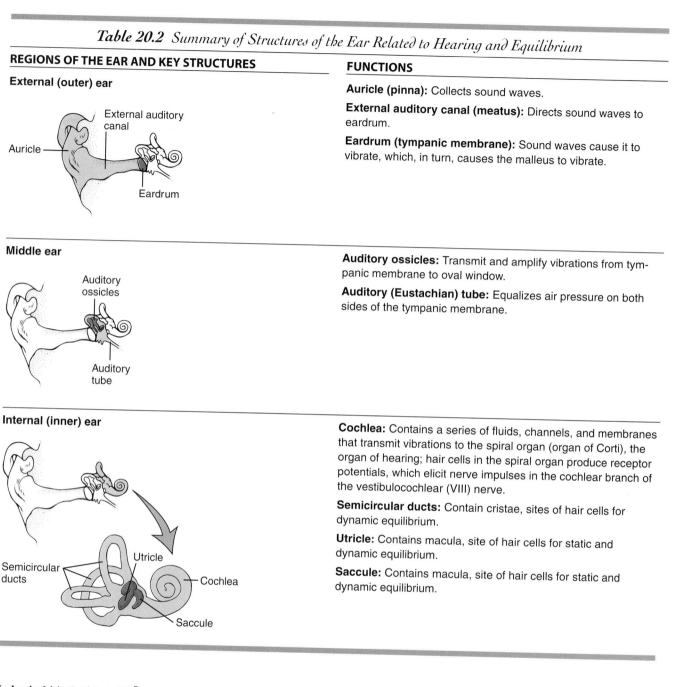

REGIONS OF THE EAR AND KEY STRUCTURES	FUNCTIONS
External (outer) ear	**Auricle (pinna):** Collects sound waves.
	External auditory canal (meatus): Directs sound waves to eardrum.
	Eardrum (tympanic membrane): Sound waves cause it to vibrate, which, in turn, causes the malleus to vibrate.
Middle ear	**Auditory ossicles:** Transmit and amplify vibrations from tympanic membrane to oval window.
	Auditory (Eustachian) tube: Equalizes air pressure on both sides of the tympanic membrane.
Internal (inner) ear	**Cochlea:** Contains a series of fluids, channels, and membranes that transmit vibrations to the spiral organ (organ of Corti), the organ of hearing; hair cells in the spiral organ produce receptor potentials, which elicit nerve impulses in the cochlear branch of the vestibulocochlear (VIII) nerve.
	Semicircular ducts: Contain cristae, sites of hair cells for dynamic equilibrium.
	Utricle: Contains macula, site of hair cells for static and dynamic equilibrium.
	Saccule: Contains macula, site of hair cells for static and dynamic equilibrium.

Labyrinthitis (lab′-i-rin-THĪ-tis) An inflammation of the labyrinth (inner ear).

Mydriasis (mi-DRĪ-a-sis) Dilated pupil.

Myringitis (mir′-in-JĪ-tis; *myringa* = eardrum) An inflammation of the eardrum; also called *tympanitis.*

Nystagmus (nis-TAG-mus; *nystazein* = to nod) A rapid involuntary movement of the eyeballs, possibly caused by a disease of the central nervous system. It is associated with conditions that cause vertigo.

Ophthalmologist (of-thal-MOL-ō-jist′; *ophthalmo* = eye) A physician who specializes in the treatment of eye disorders using drugs, surgery, and corrective lenses.

Optician (op-TISH-an; *optico* = eye) A technician who fits, adjusts, and dispenses corrective lenses prescribed by an ophthalmologist or optometrist.

Optometrist (op-TOM-e-trist′) A specialist with a doctorate in optometry who is licensed to examine and test the eyes and treat visual defects by prescribing corrective lenses

Otalgia (ō-TAL-jē-a; *oto* = ear; *algia* = pain) Earache.

Otosclerosis (ō'-tō-skle-RŌ-sis; *oto* = ear; *sclerosis* = hardening) Pathological process that may be hereditary in which new bone is deposited around the oval window. The result may be immobilization of the stapes, leading to conduction deafness.

Photophobia (fō'-tō-FŌ-bē-a; *photo* = light; *phobia* = fear) Abnormal visual intolerance to light.

Ptosis (TŌ-sis; *ptosis* = fall) Falling or drooping of the eyelid. (This term is also used for the slippage of any organ below its normal position.)

Retinoblastoma (ret'-i-nō-blas-TŌ-ma; *blast* = bud; *oma* = tumor) A tumor arising from immature retinal cells and accounting for 2% of childhood malignancies.

Scotoma (skō-TŌ-ma; *scotoma* = darkness) An area of reduced or lost vision in the visual field. Also called a *blind spot* (other than the normal blind spot at the optic disc).

Strabismus (stra-BIZ-mus) An imbalance in the extrinsic eye muscles that produces a squint (formerly called "cross-eyes").

Tinnitus (ti-NĪ-tus) A ringing, roaring, or clicking in the ears.

Trachoma (tra-KŌ-ma) A serious form of conjunctivitis, the greatest single cause of blindness in the world. It is caused by a bacterium called *Chlamydia trachomatis*. The disease produces an excessive growth of subconjunctival tissue and invasion of blood vessels into the cornea, which progresses until the entire cornea is opaque, causing blindness.

Vertigo (VER-ti-gō; *vertex* = whorl) A sensation of spinning or movement in which the world is revolving or the person is revolving in space.

Study Outline

Olfactory Sensations: Smell (p. 611)

1. The receptors for olfaction, which are bipolar neurons, are in the nasal epithelium.
2. Axons of olfactory receptors form the olfactory (I) nerves, which convey nerve impulses to the olfactory bulbs, olfactory tracts, limbic system, and cerebral cortex (temporal and frontal lobes).

Gustatory Sensations: Taste (p. 612)

1. The receptors for gustation, the gustatory receptor cells, are located in taste buds.
2. Gustatory receptor cells trigger nerve impulses in cranial nerves VII, IX, and X. Taste signals then pass to the medulla oblongata, thalamus, and cerebral cortex (parietal lobe).

Visual Sensations (p. 614)

1. Accessory structures of the eyes include the eyebrows, eyelids, eyelashes, and the lacrimal apparatus.
2. The lacrimal apparatus consists of structures that produce and drain tears.
3. The eye is constructed of three coats: (a) fibrous tunic (sclera and cornea), (b) vascular tunic (choroid, ciliary body, and iris), and (c) retina (nervous tunic).
4. The retina consists of pigment epithelium and a neural portion (photoreceptor layer, bipolar cell layer, ganglion cell layer, horizontal cells, and amacrine cells).
5. The anterior cavity contains aqueous humor; the vitreous chamber contains the vitreous body.
6. Chemicals released by rods and cones induce changes in bipolar and horizontal cells that ultimately lead to the generation of nerve impulses.

7. Impulses from ganglion cells are conveyed into the optic (II) nerve, and through the optic chiasm, the optic tract, and to the thalamus. From the thalamus visual signals pass to the cerebral cortex (occipital lobe).

Auditory Sensations and Equilibrium (p. 623)

1. The external or outer ear consists of the auricle, external auditory canal, and eardrum (tympanic membrane).
2. The middle ear consists of the auditory (Eustachian) tube, ossicles, oval window, and round window.
3. The internal or inner ear consists of the bony labyrinth and membranous labyrinth. The internal ear contains the spiral organ (organ of Corti), the organ of hearing.
4. Sound waves enter the external auditory canal, strike the eardrum, pass through the ossicles, strike the oval window, set up waves in the perilymph, strike the vestibular membrane and scala tympani, increase pressure in the endolymph, strike the basilar membrane, and stimulate hairs on the spiral organ (organ of Corti).
5. Hair cells release a chemical that initiates nerve impulses.
6. The cochlear branch of the vestibulocochlear (VIII) nerve terminates in the thalamus. From the thalamus auditory signals pass to the temporal lobes of the cerebral cortex.
7. Static equilibrium is the orientation of the body relative to the pull of gravity. The maculae of the utricle and saccule are the sense organs of static equilibrium.
8. Dynamic equilibrium is the maintenance of body position in response to movement. The cristae in the semicircular ducts are the sense organs of dynamic equilibrium.
9. Most vestibular branch fibers of the vestibulocochlear (VIII) nerve enter the brain stem and terminate in the pons; the remaining fibers enter the cerebellum.

Review Questions

1. Describe the structure of olfactory receptors. (p. 611)
2. Discuss the origin and path of a nerve impulse that results in olfaction. (p. 611)
3. Describe the structure of gustatory receptors. (p. 612)
4. Discuss how a nerve impulse for gustation travels from a taste bud to the brain. (p. 614)
5. Describe the structure and importance of the following accessory structures of the eye: eyelids, eyelashes, and eyebrows. (p. 614)
6. What is the function of the lacrimal apparatus? Explain how it operates. (p. 614)
7. By means of a labeled diagram, indicate the principal anatomical structures of the eye. (p. 617)
8. Describe the location and contents of the chambers of the eye. (p. 620)
9. Describe the histology of the neural portion of the retina. (p. 618)
10. Describe the path of a visual impulse from the optic (II) nerve to the brain. (p. 622)
11. Diagram and label the principal parts of the external, middle, and internal ear. Describe the function of each part labeled. (p. 623)
12. Explain the events involved in the transmission of sound from the auricle to the spiral organ (organ of Corti). (p. 627)
13. What is the sensory pathway for sound impulses from the cochlear branch of the vestibulocochlear (VIII) nerve to the brain? (p. 628)
14. Compare the function of the maculae in the saccule and utricle in maintaining static equilibrium with the role of the cristae in the semicircular ducts in maintaining dynamic equilibrium. (p. 630)
15. Describe the path of a nerve impulse that results in static and dynamic equilibrium. (p. 633)
16. Define each of the following: corneal transplant (p. 616), detached retina (p. 618), and cataract. (p. 620)
17. Refer to the glossary of key medical terms associated with the sensory structures. Be sure that you can define each term. (p. 633)

Self Quiz

Complete the following:

1. Three types of cells make up the olfactory epithelium: ___, ___, and ___.
2. Taste buds are located on elevated projections of the tongue called ___. The ___ type is scattered over the entire surface of the tongue.
3. The study of the structure, function, and diseases of the eye is known as ___.
4. In the macula, the gelatinous membrane is embedded with calcium carbonate crystals called ___. Collectively, the receptor organs for equilibrium are called the ___.
5. Aqueous humor is produced by the ___ body, and is reabsorbed into the ___ sinus.
6. The three layers of the eye, from superficial to deep are ___, ___, and ___.
7. The mucous membrane that lines the eyelids and covers the anterior surface of the eyeball is called the ___.
8. The ___ epithelium is a layer of melanin-containing epithelial cells between the choroid and neural portion of the retina.
9. Number the following in the correct sequence for the conduction pathway for the sense of smell: (a) olfactory bulbs: ___; (b) olfactory receptors: ___; (c) olfactory (I) nerves: ___; (d) olfactory tracts: ___; (e) cortical region of the temporal lobe: ___.
10. Number the following in the correct sequence for the pathway of sound waves: (a) external auditory canal: ___; (b) stapes: ___; (c) malleus: ___; (d) incus: ___; (e) oval window: ___; (f) eardrum (tympanic membrane): ___.

11. Number the following in the correct sequence from anterior to posterior: (a) anterior chamber: ___; (b) iris: ___; (c) cornea: ___; (d) vitreous chamber: ___; (e) lens: ___; (f) optic (II) nerve: ___: (g) posterior chamber: ___.
12. Match the following terms to their descriptions:
    (1) temporal lobe of cerebral cortex
    (2) occipital lobe of cerebral cortex
    (3) parietal lobe of cerebral cortex

    ___ (a) primary olfactory area
    ___ (b) primary gustatory area
    ___ (c) primary auditory area
    ___ (d) primary visual area

Choose the one best answer to the following questions:

13. The first synapse in the olfactory pathway occurs in the olfactory
    a. glands
    b. nerves
    c. bulbs
    d. cortex
    e. tract
14. Receptors for hearing are located in the
    a. middle ear
    b. cochlea
    c. semicircular canals
    d. tympanic membrane
    e. vestibule

15. Which of the following is *not* part of the vascular tunic of the eyeball?
    a. cornea
    b. choroid
    c. ciliary body
    d. iris
    e. all of the above

16. Choose the true statement:
    a. Olfactory receptors are specialized columnar epithelial cells.
    b. Basal cells are stem cells that continually produce new olfactory receptors.
    c. Dendrites of olfactory cells make up the olfactory (I) nerves.
    d. Olfactory (Bowman's) glands contain the olfactory receptors.
    e. Olfactory receptors are evenly distributed over the surface of the nasal mucosa.

17. The three layers of retinal neurons are the (1) ganglion cell layer, (2) bipolar cell layer, and (3) photoreceptor layer. The order in which they process visual input is
    a. 1, 2, 3
    b. 3, 2, 1
    c. 2, 1, 3
    d. 3, 1, 2
    e. 2, 3, 1

18. The bony labyrinth of the inner ear
    a. is lined with ceruminous glands
    b. consists of semicircular canals, vestibule, and tympanic antrum
    c. contains a fluid called perilymph
    d. houses hearing and equilibrium receptors
    e. both c and d

19. Cones in the eye
    a. function best in low light levels
    b. are more plentiful in the peripheral region of the retina
    c. receive sensory impulses from ganglion cells
    d. are concentrated in the central fovea of the macula lutea
    e. form part of the optic disc

20. Match the following terms to their descriptions:
    ___ (a) "white of the eye"
    ___ (b) a clear structure composed of protein layers arranged like an onion
    ___ (c) blind spot; area in which there are no cones or rods
    ___ (d) exact center of the posterior portion of the retina
    ___ (e) nonvascular, transparent, fibrous coat; most anterior eye structure
    ___ (f) layer containing neurons; if detached, causes blindness
    ___ (g) dark brown layer; prevents reflection of light rays; also nourishes eyeball because it is vascular
    ___ (h) a hole; appears black, like a circular doorway leading into a dark room
    ___ (i) regulates the amount of light entering the eye; colored part of the eye
    ___ (j) attaches to the lens by means of radially arranged fibers called the suspensory ligaments
    ___ (k) serrated margin of the retina
    ___ (l) located at the junction of sclera and cornea; drains aqueous humor

    (1) macula lutea
    (2) choroid
    (3) cornea
    (4) ciliary muscle
    (5) iris
    (6) lens
    (7) optic disc
    (8) ora serrata
    (9) pupil
    (10) retina
    (11) sclera
    (12) scleral venous sinus (canal of Schlemm)

Are the following statements true or false?

21. Aqueous humor passes from the posterior chamber of the eye into the anterior chamber, from which it is reabsorbed.
22. A receptor cell in a taste bud is a first-order sensory neuron in the gustatory pathway.
23. Tears pass medially over the anterior surface of the eyeball.
24. Contraction of the circular muscles of the iris enlarges the pupil of the eye.

Critical Thinking Questions

1. Mother held a steaming dish under Mariko's nose. "How can you say it tastes terrible? You haven't even tried it!" Mariko's lips remained clamped tight. Why is she so sure that she won't like the food?
   HINT: *How would she know if she had tried it before?*

2. An optometrist puts a few drops of fluorescent dye onto the surface of the eye to check the drainage of lacrimal fluid. The dye soon appears at the nostrils. Trace the route of the dye.
   HINT: *Why does your nose run when you cry?*

3. Reuven noticed that the colored rings in his mother's eyes were different shapes. His mother explained that getting poked in the eye with a stick when she was a little girl had caused the damage to her eye. List the structures of the eye that the stick had penetrated.
   HINT: *Corrective contact lenses float over this region.*

4. Harriet's mother always complains that "anything I say to you goes in one ear and out the other!" Follow the path of sound from the external ear to the organ of hearing (organ of Corti).
   HINT: *The inner ear on one side of the head does not connect to the inner ear on the other side.*

5. Compare the effects of damage to the left cranial nerve II to damage to the left occipital lobe.
   HINT: *Trace the route of the nerve fibers from the eye to the brain.*

6. Vikram's mother said, "You scream so loud, you'll burst my eardrums!" So now he asks, "Where's this drum, and how do you whack it?" Explain eardrums to Vikram.
   HINT: *In the orchestra, the person who plays the kettle drum is the tympanist.*

Answers to Figure Questions

20.1 Olfactory tract.

20.2 Gustatory receptors → cranial nerves VII, IX, or X → medulla → (1) limbic system and hypothalamus or (2) thalamus → primary gustatory area in the parietal lobe of the cerebral cortex.

20.3 Tears or lacrimal fluid is a watery solution containing salts, some mucus, and lysozyme. Tears clean, lubricate, and moisten the eyeballs.

20.4 Avascular structures in the eye include the cornea, lens, and retina. Because these structures have no blood vessels, they are more transparent to light, an advantage because the photoreceptors lie at the posterior aspect of the eye.

20.5 Hypertension and diabetes; also cataract and macular degeneration.

20.6 Two types of photoreceptors are rods and cones. Rods provide black-and-white vision in dim light, whereas cones provide color vision in bright light.

20.7 Aqueous humor is secreted by the ciliary process, then flows into the posterior chamber, around the iris, into the anterior chamber, and out of the eyeball through the scleral venous sinus.

20.8 Retinal ganglion cell axons form the optic nerve (cranial nerve II) → optic chiasm → optic tract → lateral geniculate nucleus of the thalamus → optic radiations → occipital lobe of cerebrum.

20.9 Inner ear: cochlea (hearing) and semicircular ducts (equilibrium).

20.10 The eardrum (tympanic membrane) separates the external ear from the middle ear. The oval and round windows separate the middle ear from the internal ear.

20.11 The utricle and the saccule.

20.12 Semicircular canals, vestibule, and cochlea.

20.13 Membrane of the oval window.

20.14 Static equilibrium.

20.15 Dynamic equilibrium.

Chapter 21

The Autonomic Nervous System

Student Objectives

1. Compare the structural and functional differences between the somatic and autonomic portions of the nervous system.

2. Identify the principal structural features of the autonomic nervous system.

3. Compare the sympathetic and parasympathetic divisions of the autonomic nervous system in terms of structure, physiology, and neurotransmitters released.

4. Describe the various postsynaptic receptors involved in autonomic responses.

5. Describe the components of an autonomic reflex arc.

The **autonomic nervous system (ANS)** regulates the activity of smooth muscle, cardiac muscle, and certain glands. Traditionally, the ANS has been described as a specific *motor output* portion of the peripheral nervous system. In order to operate, however, the ANS depends on a continual flow of *sensory input* from visceral organs and blood vessels into the CNS. Thus it is reasonable to include these sensory neurons as part of the ANS. Structurally, then, the ANS includes two main components: general visceral sensory neurons and general visceral motor neurons.

Functionally, the ANS usually operates without conscious control. The system was originally named *autonomic* because it was thought to function autonomously or in a self-governing manner, without control by the central nervous system (CNS). The ANS, however, is regulated by centers in the brain, mainly the hypothalamus and medulla oblongata, which receive input from the limbic system and other regions of the cerebrum.

In this chapter we will compare the structural and functional features of the somatic and autonomic nervous systems. Then we will discuss the anatomy of the ANS and compare the organization and actions of its sympathetic and parasympathetic divisions.

COMPARISON OF SOMATIC AND AUTONOMIC NERVOUS SYSTEMS

The somatic nervous system includes both sensory and motor neurons. The sensory neurons convey input from receptors for the special senses (vision, hearing, taste, smell, and equilibrium), proprioceptors (muscle and joint position), and general somatic receptors (pain, temperature, and tactile sensations). All these sensations normally are consciously perceived. In turn, somatic motor neurons innervate skeletal muscle, the effector tissue of the somatic nervous system, and produce conscious, voluntary movements. In the somatic nervous system, the effect of a motor neuron always is *excitation*. When a somatic motor neuron stimulates a skeletal muscle, the muscle contracts. When the neuron ceases to stimulate the muscle, contraction stops.

The input component of the ANS consists of **general visceral sensory neurons.** Mostly, these are associated with interoceptors, such as chemoreceptors that monitor blood CO_2 level and mechanoreceptors that detect the degree of stretch of organs or blood vessels. These motor signals are not consciously perceived most of the time, although intense activation of interoceptors may give rise to conscious sensations. Examples of this are the sensations of pain or nausea from damaged viscera, fullness of the urinary bladder, and angina pectoris (chest pain) from inadequate blood flow to the heart.

General visceral motor neurons regulate visceral activities by either *exciting* or *inhibiting* their effector tissues, which are cardiac muscle, smooth muscle, and glands. Responses include changes in the size of the pupil, accommodation for near vision, dilation of blood vessels, adjustment of the rate and force of the heartbeat, movements of the gastrointestinal tract, and secretion by most glands. These activities usually lie beyond conscious control. They are automatic. Input from the general somatic and special senses, acting via the limbic system, also may modify responses of autonomic motor neurons. Seeing a bike about to hit you or hearing squealing brakes of a nearby car, or being grabbed by an attacker, for example, would increase the rate and force of your heartbeat.

Autonomic motor pathways consist of sets of two motor neurons in series (one following the other) (Figure 21.1a). The first has its cell body in the CNS; its axon extends from the CNS to an **autonomic ganglion.** (Recall that a ganglion is a collection of neuronal cell bodies outside the CNS.) The cell body of the second neuron is in that autonomic ganglion; its unmyelinated axon extends directly from the ganglion to the effector (smooth muscle, cardiac muscle, or gland). In contrast, a single myelinated somatic motor neuron extends from the CNS to skeletal muscle (Figure 21.1b). Also, whereas somatic motor neurons release acetylcholine (ACh) as their neurotransmitter, autonomic motor neurons release either ACh or norepinephrine (NE).

The output (motor) part of the ANS has two principal divisions: **sympathetic** and **parasympathetic.** Many organs receive autonomic motor fibers from both divisions. In general, nerve impulses from one division stimulate the organ to start or increase activity (excitation), whereas impulses from the other division decrease the organ's activity (inhibition). For example, whereas the sympathetic division increases heart rate and force of contraction, the parasympathetic division decreases heart rate and force of contraction. Organs that receive impulses from both sympathetic and parasympathetic fibers are said to have **dual innervation.** The rest of the chapter focuses on the anatomy and functions of the sympathetic and parasympathetic divisions and the CNS centers that regulate outflow along these two divisions. A summary of the similarities and differences between the somatic and autonomic nervous systems is presented in Table 21.1.

Figure 21.1 Comparison of the autonomic and somatic nervous systems.

Stimulation by the autonomic nervous system either excites or inhibits visceral effectors; somatic nervous system stimulation always excites its effectors.

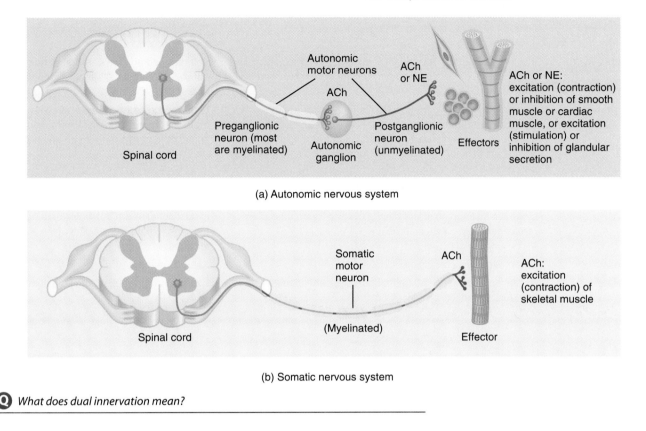

(a) Autonomic nervous system

(b) Somatic nervous system

Q *What does dual innervation mean?*

ANATOMY OF AUTONOMIC MOTOR PATHWAYS

Overview

The first of the two autonomic motor neurons is called a **preganglionic neuron** (Figure 21.1a). Its cell body is in the brain or spinal cord. Its myelinated axon, called a **preganglionic fiber,** passes out of the CNS as part of a cranial or spinal nerve. At some point, the fiber separates from the nerve and passes to an autonomic ganglion. There it synapses with the postganglionic neuron, the second neuron in the autonomic motor pathway. The **postganglionic neuron** lies entirely outside the CNS. Its cell body and dendrites are located in an autonomic ganglion, where it synapses with one or more preganglionic fibers. The axon of a postganglionic neuron, called a **postganglionic fiber,** is unmyelinated and terminates in a visceral effector. Thus preganglionic neurons convey motor impulses from the CNS to autonomic ganglia, and postganglionic neurons relay the motor impulses from autonomic ganglia to visceral effectors. Note a major difference between autonomic ganglia and posterior root ganglia: both contain cell bodies, but only autonomic ganglia contain synapses.

Preganglionic Neurons

In the sympathetic division, the preganglionic neurons have their cell bodies in the lateral gray horns of the 12 thoracic segments and first two or three lumbar segments of the spinal cord (Figure 21.2). For this reason, the sympathetic division is also called the **thoracolumbar** (thō′-ra-kō-LUM-bar) **division,** and the fibers of the sympathetic preganglionic neurons are known as the **thoracolumbar outflow.**

The cell bodies of the preganglionic neurons of the parasympathetic division are located in the nuclei of the oculomotor (III), facial (VII), glossopharyngeal (IX), and vagus (X) cranial nerves in the brain stem and in the lateral gray horns of the second through fourth sacral segments of the spinal cord. Hence the parasympathetic division is also known as the **craniosacral division,** and the fibers of the parasympathetic preganglionic neurons are referred to as the **craniosacral outflow.**

Table 21.1 Comparison of Autonomic and Somatic Nervous Systems

	AUTONOMIC	SOMATIC
Sensory input	General visceral senses (mainly from interoceptors).	Special senses; general somatic senses; proprioceptors.
Control	Involuntary control via limbic system, hypothalamus, medulla oblongata, pons, and spinal cord; (limited control via cerebral cortex).	Voluntary control via cerebral cortex (other active regions: basal ganglia, cerebellum, brain stem, and spinal cord).
Motor output	Two-neuron pathway: preganglionic (first-order) neuron extending from brain stem or spinal cord synapses with post-ganglionic (second-order) neuron in an autonomic ganglion. Postganglionic neuron extending from ganglion synapses with a visceral effector.	One-neuron pathway: axon of somatic motor neuron extending from brain stem or spinal cord synapses with an effector.
Neurotransmitters	Preganglionic axons release acetylcholine (ACh); post-ganglionic axons release ACh (parasympathetic division and sympathetic fibers to sweat glands) or norepinephrine (NE; remainder of sympathetic division).	At the synapse with effector (the neuro-muscular junction), the neurotransmitter is acetylcholine (ACh), which is released from the synaptic end bulbs of the axon terminals.
Effectors	Smooth muscle; cardiac muscle; glands.	Skeletal muscle.
Response	Excitation or inhibition of smooth muscle, cardiac muscle, or glandular secretions.	Excitation (contraction) of skeletal muscle.

Autonomic Ganglia

The autonomic ganglia may be divided into three general groups: two of the groups are components of the sympathetic division and one group is a component of the parasympathetic division.

Sympathetic Ganglia The **sympathetic trunk ganglia** (also called *vertebral chain ganglia,* the *paravertebral ganglia,* or the *lateral ganglia*) lie in a vertical row on either side of the vertebral column, extending from the base of the skull to the coccyx (Figures 21.2 and 21.4). They receive preganglionic fibers only from the *sympathetic division.* Because the sympathetic trunk ganglia are so near the spinal cord, sympathetic preganglionic fibers tend to be short.

The second kind of autonomic ganglion also receives preganglionic fibers from the *sympathetic division.* It is called a **prevertebral** *(collateral)* **ganglion** (Figure 21.2). The ganglia of this group lie anterior to the spinal column and close to the large abdominal arteries. Examples of prevertebral ganglia are the **celiac ganglion,** on either side of the celiac artery just inferior to the diaphragm; the **superior mesenteric ganglion,** near the beginning of the superior mesenteric artery in the upper abdomen; and the **inferior mesenteric ganglion,** located near the beginning of the inferior mesenteric artery in the middle of the abdomen.

Parasympathetic Ganglia Preganglionic fibers from the *parasympathetic division* make synapses in **terminal** *(intramural)* **ganglia.** These ganglia are located at the end of an autonomic motor pathway very close to or actually within the wall of a visceral organ. Examples of terminal ganglia include the **ciliary ganglion, pterygopalatine ganglion, submandibular ganglion,** and **otic ganglion** (Figure 21.2). Because parasympathetic preganglionic fibers extend from the CNS to the terminal ganglion in an innervated organ, they tend to be long.

Autonomic Plexuses

In the beginning of the chapter we indicated that many organs receive dual innervation; that is, they receive nerve impulses from both the sympathetic and parasympathetic divisions of the ANS. Here we will consider the anatomical basis for dual innervation. Later in the chapter we will discuss the physiological basis.

You will notice in Figure 21.2 that sympathetic postganglionic fibers from the superior cervical ganglion (a sympathetic trunk ganglion) innervate certain structures in the head. Also note that the same structures are innervated by parasympathetic postganglionic fibers from the ciliary, pterygopalatine, submandibular, and otic ganglia (all terminal ganglia of the parasympathetic division of the ANS), which will be described later.

Figure 21.2 Structure of the sympathetic and parasympathetic divisions of the autonomic nervous system (ANS). Although the parasympathetic division is shown only on one side and the sympathetic division is shown only on the other side, keep in mind that each division innervates tissues on both sides of the body.

Sympathetic and parasympathetic stimulation have opposing effects on organs that receive dual innervation.

Q *Which division has longer preganglionic fibers? Why?*

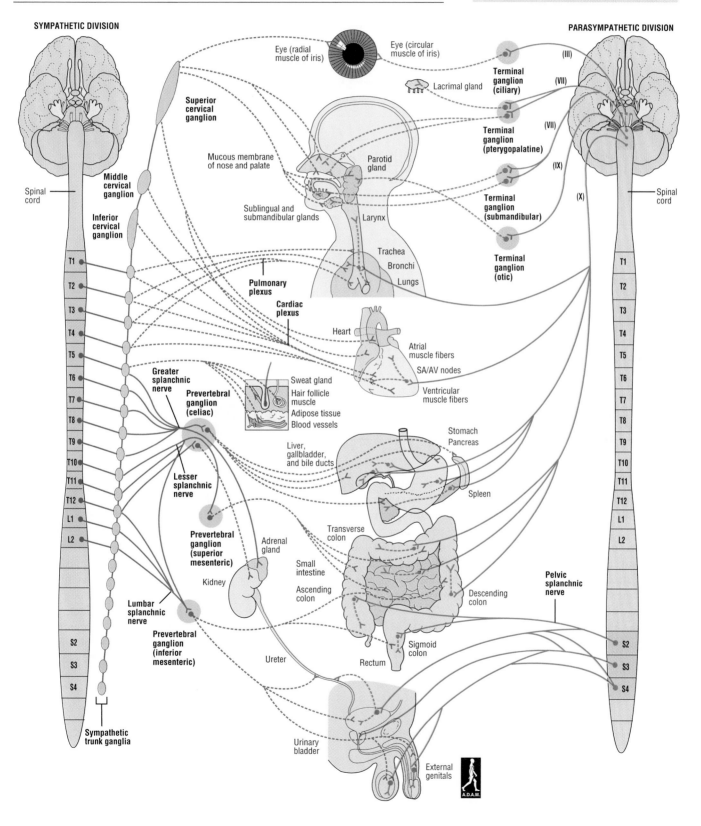

KEY: Sympathetic preganglionic fibers
Sympathetic postganglionic fibers
Parasympathetic preganglionic fibers
Parasympathetic postganglionic fibers

Figure 21.3 Autonomic plexuses in the abdomen and pelvis.

An autonomic plexus is an aggregate of sympathetic and parasympathetic fibers and autonomic ganglia.

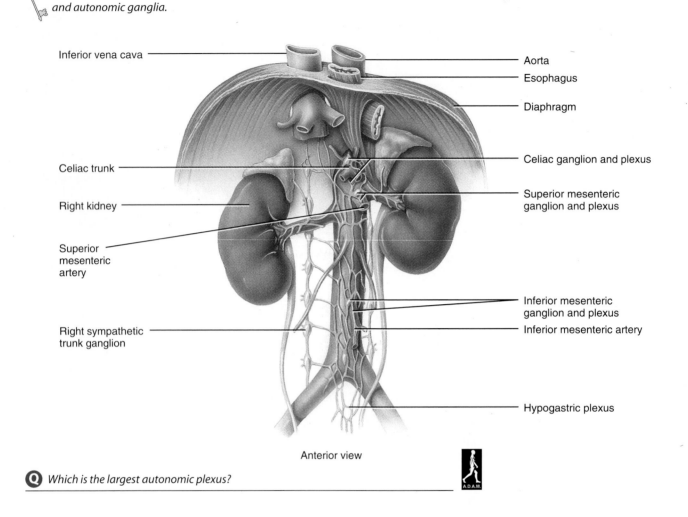

Inferior vena cava

Aorta

Esophagus

Diaphragm

Celiac trunk

Celiac ganglion and plexus

Right kidney

Superior mesenteric ganglion and plexus

Superior mesenteric artery

Inferior mesenteric ganglion and plexus

Inferior mesenteric artery

Right sympathetic trunk ganglion

Hypogastric plexus

Anterior view

Q *Which is the largest autonomic plexus?*

In the thorax, abdomen, and pelvis, nerve fibers of the sympathetic and parasympathetic divisions of the ANS and autonomic ganglia aggregate to form **autonomic plexuses** (Figure 21.3). Nerves from the plexuses pass along most branches of large blood vessels, which they surround on their way to supplying viscera. The plexuses are usually named after the artery along which they are distributed. The major plexuses are the cardiac, pulmonary, celiac, and hypogastric.

In the thorax, the **cardiac plexus** is at the base of the heart, surrounding the large blood vessels emerging from the heart (see Figure 21.2). The **pulmonary plexus** is located mostly posterior to each lung.

In the abdomen and pelvis, the **celiac** or **solar plexus** is found at the level of the last thoracic and first lumbar vertebrae. It is the largest autonomic plexus and surrounds the celiac and superior mesenteric arteries. It contains two large celiac ganglia and a dense network of autonomic fibers. Secondary plexuses from or connected to the celiac plexus are distributed to the diaphragm, liver, gallbladder, stomach, pancreas, spleen, kidneys, medulla (inner region) of the suprarenal gland, testes, and ovaries. The **superior mesenteric plexus,** a secondary plexus of the celiac plexus, contains the superior mesenteric ganglion and supplies the small and large intestine. The **inferior mesenteric plexus,** also a secondary plexus of the celiac plexus, contains the inferior mesenteric ganglion and supplies the large intestine. The **hypogastric plexus** is located anterior to the fifth lumbar vertebra and superior portion of the sacrum. It supplies pelvic viscera.

Postganglionic Neurons

Axons of preganglionic neurons of the sympathetic division pass to ganglia of the sympathetic trunk. Some synapse with postganglionic neurons in the sympathetic trunk ganglia (Figure 21.4). Others continue, without synapsing, through the sympathetic trunk ganglia to end at a prevertebral ganglion and synapse with the postganglionic neurons there. In either case, a single sympathetic preganglionic fiber may synapse with 20 or more postganglionic fibers. This is an example of *divergence* (see Figure 16.9a) and is part of the reason why sympathetic responses tend to be widespread throughout the body. After exiting their ganglia, the postganglionic fibers typically innervate several visceral effectors (see Figure 21.2).

Axons of preganglionic neurons of the parasympathetic division pass to terminal ganglia near or within a visceral effector (see Figure 21.2). In the ganglion, the presynaptic neuron usually synapses with only four or five postsynaptic neurons, all of which supply a single visceral effector. Thus parasympathetic effects tend to be localized. With this background in mind, we can now examine some specific structural features of the sympathetic and parasympathetic divisions of the ANS.

Sympathetic Division

Cell bodies of sympathetic preganglionic neurons are part of the lateral gray horns of all thoracic segments and the first two or three lumbar segments of the spinal cord (see Figure 21.2). The preganglionic axons are myelinated and leave the spinal cord through the anterior root of a spinal nerve along with the somatic motor fibers at the same segmental level. After exiting through the intervertebral foramina, the myelinated preganglionic sympathetic fibers enter a short pathway called a **white ramus** before passing to the nearest sympathetic trunk ganglion on the same side.

Collectively, the white rami are called the **white rami communicantes** (kō-myoo-ni-KAN-tēz; singular is **ramus communicans;** see Figure 21.4). Their name ("white") indicates that they contain myelinated fibers. Only the thoracic and first lumbar nerves have white rami communicantes. The white rami communicantes connect the anterior ramus of the spinal nerve with the ganglia of the sympathetic trunk. The paired sympathetic trunk ganglia are arranged anterior

and lateral to the vertebral column, one on either side. Typically, there are 3 cervical, 11 or 12 thoracic, 4 or 5 lumbar, and 4 or 5 sacral sympathetic trunk ganglia. Although the sympathetic trunk ganglia extend inferiorly from the neck, chest, and abdomen to the coccyx, they receive preganglionic fibers only from the thoracic and lumbar segments of the spinal cord (see Figure 21.2).

The cervical portion of each sympathetic trunk is located in the neck anterior to the prevertebral muscles. It is subdivided into superior, middle, and inferior ganglia (see Figure 21.2). The **superior cervical ganglion** is posterior to an internal carotid artery and anterior to a transverse process of the second cervical vertebra. Postganglionic fibers leaving the ganglion serve the head. They are distributed to sweat glands, smooth muscle of the eye and blood vessels of the face, nasal mucosa, and the submandibular, sublingual, and parotid salivary glands. Gray rami communicantes (described shortly) from the ganglion also pass to the upper two to four cervical spinal nerves. The **middle cervical ganglion** lies near the sixth cervical vertebra at the level of the cricoid cartilage. Postganglionic fibers from it innervate the heart. The **inferior cervical ganglion** is located near the first rib, anterior to the transverse processes of the seventh cervical vertebra. Its postganglionic fibers also supply the heart.

Clinical Application

Horner's Syndrome

If the cervical sympathetic trunk is cut or damaged on one side, the sympathetic supply to that side of the head is removed, resulting in **Horner's syndrome,** in which the patient exhibits (on the affected side): ptosis (drooping of the upper eyelid), slight elevation of the lower eyelid, narrowing of the space between eyelids, enophthalmos (the eye appears sunken), miosis (constricted pupil), and anhidrosis (lack of sweating). ▨

The thoracic portion of each sympathetic trunk lies anterior to the necks of the corresponding ribs. This portion of the sympathetic trunk receives most of the sympathetic preganglionic fibers. Postganglionic fibers from the thoracic sympathetic trunk innervate the heart, lungs, bronchi, and other thoracic viscera. In the skin, they also innervate sweat glands, blood vessels, and arrector pili muscles of hair follicles.

The lumbar portion of each sympathetic trunk lies lateral to the corresponding lumbar vertebrae. The sacral portion of the sympathetic trunk lies in the pelvic cavity on the medial side of the sacral foramina. Unmyelinated postganglionic fibers from the lumbar and sacral sympathetic chain ganglia enter a short pathway called a **gray ramus** and then merge

Figure 21.4 Ganglia and rami communicantes of the sympathetic division of the ANS.

Sympathetic ganglia lie in two chains on either side of the vertebral column (sympathetic trunk ganglia) and near large abdominal arteries anterior to the vertebral column (prevertebral ganglia).

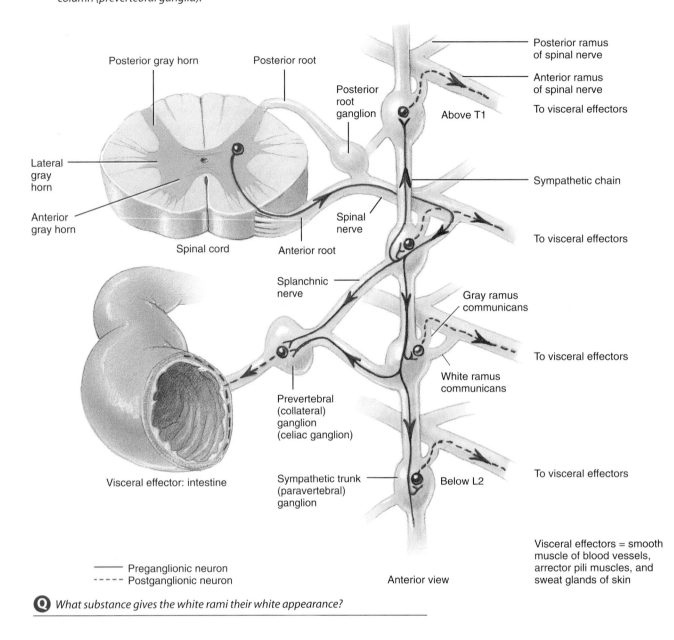

Preganglionic neuron
Postganglionic neuron

Anterior view

Visceral effectors = smooth muscle of blood vessels, arrector pili muscles, and sweat glands of skin

Q *What substance gives the white rami their white appearance?*

with a spinal nerve or join the hypogastric plexus via direct visceral branches. The **gray ramus communicans** (kō-MYOO-ni-kanz) is the structure containing the postganglionic fibers that connect the ganglion of the sympathetic trunk to the spinal nerve (Figure 21.4). The fibers are unmyelinated. Gray rami communicantes outnumber the white rami because there is a gray ramus leading to each of the 31 pairs of spinal nerves.

As preganglionic fibers extend from a white ramus communicans into the sympathetic trunk, they give off several axon collaterals (branches). These collateral fibers terminate and synapse in several ways. Some synapse in the first ganglion at the level of entry. Others pass up or down the sympathetic trunk for a variable distance to form the fibers on which the ganglia are strung. These fibers, known as **sympathetic chains** (Figure 21.4), may not synapse until they reach a ganglion in the cervical or sacral area. Many postganglionic fibers rejoin the spinal nerves through gray rami and supply peripheral visceral effectors such as sweat glands and the smooth muscle in blood vessels and around hair follicles.

Some preganglionic fibers pass through the sympathetic trunk without terminating in the trunk. Beyond the trunk, they form nerves known as **splanchnic** (SPLANK-nik) **nerves** (Figure 21.4), which extend to and terminate in the outlying prevertebral ganglia. Splanchnic nerves from the thoracic area terminate in the **celiac** (SĒ-lē-ak) **ganglion.** Here, the preganglionic fibers synapse with postganglionic cell bodies. The *greater splanchnic nerve* enters the celiac ganglion of the celiac plexus. From here, postganglionic fibers extend to the stomach, spleen, liver, kidney, and small intestine (Figure 21.2). The *lesser splanchnic nerve* passes through the celiac plexus to enter the superior mesenteric ganglion of the superior mesenteric plexus. Postganglionic fibers from this ganglion innervate the small intestine and colon. The *lowest splanchnic nerve,* not always present, enters the renal plexus near the kidney. Postganglionic fibers supply the renal arterioles and ureter.

The *lumbar splanchnic nerve* enters the inferior mesenteric plexus, which is inferior to the superior mesenteric plexus. In the plexus, the preganglionic fibers synapse with postganglionic fibers in the inferior mesenteric ganglion. These fibers pass through the hypogastric plexus and supply the distal colon and rectum, urinary bladder, and genital organs. Postganglionic fibers leaving the prevertebral ganglia follow the course of various arteries to abdominal and pelvic visceral effectors.

Sympathetic *preganglionic fibers* also extend to the medullae (inner layers) of the adrenal glands. Developmentally, the adrenal medulla is a modified sympathetic ganglion. Its cells are like sympathetic postganglionic neurons. Rather than extending to another organ, however, these cells release the hormones norepinephrine (20%) and epinephrine (80%) into the blood. This is one exception to the usual pattern of two motor neurons in an autonomic motor pathway.

Parasympathetic Division

Cell bodies of parasympathetic preganglionic neurons are found in nuclei in the brain stem and the lateral gray horn of the second through fourth sacral segments of the spinal cord (see Figure 21.2). Their fibers emerge as part of a cranial nerve or as part of the anterior root of a spinal nerve. The **cranial parasympathetic outflow** consists of preganglionic fibers that extend from the brain stem in four cranial nerves. The **sacral parasympathetic outflow** consists of preganglionic fibers in anterior roots of the second through fourth sacral nerves. The preganglionic fibers of both the cranial and sacral outflows end in terminal ganglia, where they

synapse with postganglionic neurons. We will look first at the cranial outflow.

The cranial outflow has five components: four pairs of ganglia and the plexuses associated with the vagus (X) nerve. The four pairs of cranial parasympathetic ganglia innervate structures in the head and are located close to the organs they innervate (see Figure 21.2). The **ciliary ganglia** lie lateral to each optic (II) nerve near the posterior aspect of the orbit. Preganglionic fibers pass with the oculomotor (III) nerve to the ciliary ganglion. Postganglionic fibers from the ganglion innervate smooth muscle cells in the eyeball. The **pterygopalatine** (ter′-i-gō-PAL-a-tin) **ganglia** are lateral to the sphenopalatine foramen. They receive preganglionic fibers from the facial (VII) nerve and send postganglionic fibers to the nasal mucosa, palate, pharynx, and lacrimal glands. The **submandibular ganglia** are found near the ducts of the submandibular salivary glands. They receive preganglionic fibers from the facial (VII) nerve and send postganglionic fibers to the submandibular and sublingual salivary glands. The **otic ganglia** are situated just inferior to each foramen ovale. Each otic ganglion receives preganglionic fibers from the glossopharyngeal (IX) nerve and sends postganglionic fibers to the parotid salivary gland.

Preganglionic fibers that leave the brain as part of the vagus (X) nerves are the last components of the cranial outflow. The vagus nerves carry nearly 80% of the total craniosacral outflow. Vagal fibers extend to many terminal ganglia in the thorax and abdomen. Because the terminal ganglia are close to or in the walls of their visceral effectors, postganglionic parasympathetic fibers are very short. (By comparison, postganglionic sympathetic fibers are long.) As it passes through the thorax, the vagus (X) nerve sends fibers to the heart and the airways of the lungs. In the abdomen, it supplies the liver, gallbladder, stomach, pancreas, small intestine, and part of the large intestine.

The sacral parasympathetic outflow consists of preganglionic fibers from the anterior roots of the second through fourth sacral nerves. Collectively, they form the **pelvic splanchnic nerves** (see Figure 21.2). These nerves synapse with parasympathetic postganglionic neurons located in terminal ganglia in the walls of the innervated viscera. From the ganglia, parasympathetic postganglionic fibers innervate smooth muscle and glands in the walls of the colon, ureters, urinary bladder, and reproductive organs.

Anatomical features of the sympathetic and parasympathetic divisions are compared in Table 21.2.

ACTIVITIES OF THE ANS

As noted earlier, most body structures receive dual innervation, and the anatomical basis for it was described. This may be reviewed in Figure 21.2. Structures that receive only

Table 21.2 *Comparative Anatomy of Sympathetic and Parasympathetic Divisions*

	SYMPATHETIC	PARASYMPATHETIC
Distribution	Bodywide: skin, sweat glands, arrector pili muscles of hair follicles, adipose tissue, smooth muscle of blood vessels.	Limited mainly to head and viscera of thorax, abdomen, and pelvis; some blood vessels.
Outflow	Thoracolumbar.	Craniosacral.
Associated ganglia	Two: sympathetic trunk and prevertebral ganglia.	One: terminal ganglia.
Ganglia locations	Close to CNS and distant from visceral effectors.	Near or within wall of visceral effectors.
Fiber length and divergence	Short preganglionic fibers synapse with many long postganglionic neurons that pass to many visceral effectors.	Long preganglionic fibers usually synapse with four to five short postganglionic neurons that pass to a single visceral effector.
Rami communicantes	Both present; white rami communicantes contain myelinated preganglionic fibers, and gray rami communicantes contain unmyelinated postganglionic fibers.	Neither present.

sympathetic innervation include sweat glands, arrector pili muscles attached to hair follicles in the skin, adipose (fat) cells, the kidneys, and most blood vessels. Lacrimal (tear) glands, on the other hand, receive only parasympathetic fibers. Elsewhere in the body, the two divisions generally have opposing effects on a given organ, one causing excitation and the other inhibition. This is possible because they use different neurotransmitters and have different neurotransmitter receptors. This is the physiological basis for dual innervation. The hypothalamus regulates the balance of sympathetic versus parasympathetic activity or tone. This balance can change from one moment to the next according to demands imposed by the internal and external environments.

ANS Neurotransmitters

Like other neurons, autonomic neurons release neurotransmitters at synapses. Based on the neurotransmitter they produce and liberate, autonomic neurons are classified as either cholinergic or adrenergic (Figure 21.5).

Cholinergic (kō'-lin-ER-jik) **neurons** release **acetylcholine (ACh)** and include the following: (1) all sympathetic and parasympathetic preganglionic neurons, (2) all parasympathetic postganglionic neurons, and (3) a few sympathetic postganglionic neurons. The cholinergic sympathetic postganglionic fibers include those to most sweat glands and a few blood vessels in skeletal muscles.

ACh diffuses the short distance across the synaptic cleft to bind with specific receptors on the postsynaptic membrane. The membrane—of an autonomic postganglionic neuron, a smooth or cardiac muscle fiber, or a glandular cell—becomes either excited or inhibited. (Recall from Chapter 9 that ACh always causes depolarization and excitation of skeletal muscle membranes.) Because acetylcholine is quickly inactivated by the enzyme **acetylcholinesterase (AChE),** effects triggered by short-lived activity of cholinergic fibers are brief.

Adrenergic (ad'-ren-ER-jik) **neurons** release **norepinephrine (noradrenalin)** or **epinephrine (adrenalin).** Most sympathetic postganglionic axons are adrenergic; they secrete norepinephrine (NE). NE diffuses across the synaptic cleft and combines with specific receptors on the postsynaptic membrane to bring about either excitation or inhibition.

The effects of sympathetic stimulation are longer lasting and more widespread than the effects of parasympathetic stimulation for three reasons. First, there is much more divergence of sympathetic postganglionic fibers. Thus, many tissues may be activated simultaneously. Second, NE lingers in the synaptic cleft for a longer time than does ACh. Third, NE and epinephrine secreted into the blood by the adrenal medullae of the two adrenal glands intensify the action of NE liberated from sympathetic postganglionic axons. These hormones in the blood circulate throughout the body, affecting whichever tissues have the appropriate receptors. In time, liver enzymes degrade blood-borne NE and epinephrine.

Cholinergic and Adrenergic Receptors

There are two main categories of both cholinergic and adrenergic receptors plus several subcategories of adrenergic receptors (Figure 21.5).

Figure 21.5 Neurotransmitters and receptors. Cholinergic neurons release acetylcholine, whereas adrenergic neurons release norepinephrine.

Most sympathetic postganglionic neurons are adrenergic (orange); other autonomic neurons and somatic motor neurons are cholinergic (blue).

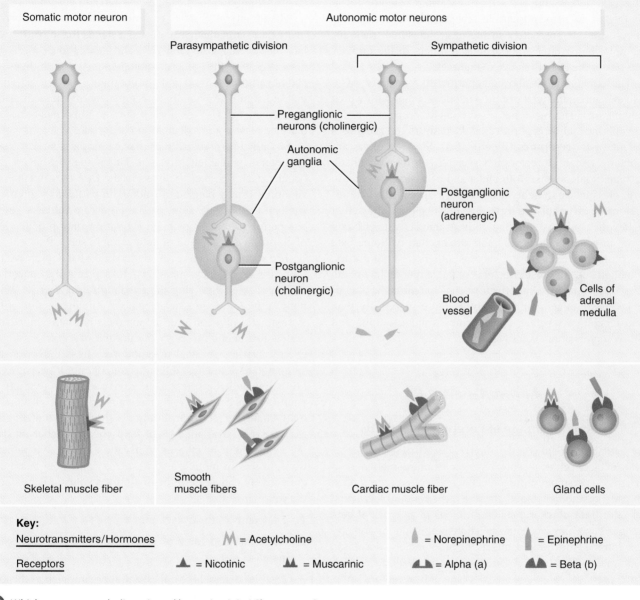

Q *Which neurons are cholinergic and have nicotinic ACh receptors?*

The two types of cholinergic (ACh) receptors are nicotinic receptors and muscarinic receptors. **Nicotinic receptors** are found on the dendrites and cell bodies of both sympathetic and parasympathetic postganglionic neurons. They are so named because nicotine mimics the action of ACh on such receptors. **Muscarinic receptors** are present on all effectors (muscles and glands) innervated by parasympathetic postganglionic axons. Most sweat glands and the smooth muscle in certain blood vessels, which receive their innervation from *cholinergic* sympathetic postganglionic fibers, also possess muscarinic receptors. These postsynaptic receptors are so named because a mushroom poison called muscarine mimics the actions of ACh on them. Nicotine does not activate muscarinic receptors, nor does muscarine stimulate nicotinic receptors, but ACh activates both receptor types.

Activation of nicotinic receptors leads to excitation of the postsynaptic cell, be it a postganglionic neuron or a visceral effector. Activation of muscarinic receptors sometimes causes excitation and sometimes inhibition, depending on which particular cell bears the muscarinic receptors. The binding of ACh to muscarinic receptors inhibits (relaxes) smooth muscle sphincters in the gastrointestinal tract, for example, but excites smooth muscle fibers in the circular muscles of the iris of the eye to contract.

The effects of NE and epinephrine, like those of ACh, also depend on the type of postsynaptic receptor with which they interact. The two types of adrenergic receptors for norepinephrine and epinephrine are called **alpha (α) receptors** and **beta (β) receptors** (Figure 21.5). Such receptors are found on visceral effectors innervated by most sympathetic postganglionic axons. Both alpha and beta receptors are, in turn, classified into subtypes—α_1, α_2, β_1, and β_2. The receptors are distinguished by the specific responses they elicit and by their selective binding of drugs that activate or block them. In general, alpha receptors are excitatory. Some beta receptors are excitatory, whereas others are inhibitory. Although cells of most effectors contain either alpha or beta receptors, some visceral effector cells contain both. In general, NE stimulates alpha receptors more vigorously than beta receptors, whereas epinephrine stimulates alpha and beta receptors about equally.

Clinical Application

Taking Advantage of Receptor Selectivity

The availability of drugs and natural products that selectively activate or block certain receptors is a cornerstone of modern drug therapy. For example, propranolol (Inderal®) often is prescribed for patients with high blood pressure (hypertension). It is a nonselective **beta blocker,** meaning it binds to all types of beta receptors and prevents their activation by epinephrine and norepinephrine. The desired effects of propranolol—decreased heart rate and contraction force and a consequent decrease in blood pressure—are due to its blockage of β_1 receptors. Undesired effects due to blockage of β_2 receptors may include hypoglycemia (low blood glucose) due to decreased glycogen breakdown and decreased gluconeogenesis (the conversion of a noncarbohydrate into glucose in the liver) and mild bronchoconstriction (narrowing of the airways). If these side effects pose a threat to the patient, a selective β_1 blocker such as metoprolol (Lopressor®) can be prescribed instead of propranolol. ▪

Parasympathetic and Sympathetic Responses

The parasympathetic division regulates primarily those activities that conserve and restore body energy during times of rest or recovery. It is an **energy conservation–restorative system.** Normally, parasympathetic impulses to the digestive glands and the smooth muscle of the gastrointestinal tract dominate over sympathetic impulses. Thus energy-supplying food can be digested and absorbed.

The acronym "SLUD" is a mental key for remembering many parasympathetic responses. It stands for salivation (S), lacrimation (L), urination (U), and defecation (D). These responses, except for lacrimation, have to do with digestion and absorption of food and elimination of wastes, activities that are stimulated mainly by the parasympathetic division. Besides the "SLUD" responses, another important parasympathetic response is a decrease in heart rate.

So-called paradoxical fear may cause massive activation of the parasympathetic division. This is the sort of fear that occurs when one feels helpless in the face of a threatening situation. It may happen to soldiers in battle, to unprepared students taking an exam, or sometimes to athletes before competition. Parasympathetic activation may produce loss of control over urination or defecation in such situations.

The sympathetic division, on the other hand, prepares the body for emergency situations. It is primarily concerned with processes involving the expenditure of energy. When body activities are in a steady state, the main function of the sympathetic division is to counteract the parasympathetic effects just enough to carry out normal processes requiring energy. During physical or emotional stress, however, the sympathetic dominates the parasympathetic. Physical exertion stimulates the sympathetic division, as do a variety of emotions (such as fear, embarrassment, or rage). Visualizing body changes that occur during "E situations" (emergency, exercise, embarrassment) will help you remember most of the sympathetic responses. Activation of the sympathetic division sets in motion a series of physiological responses collectively called the **fight-or-flight response.** It produces the following effects:

1. The pupils of the eyes dilate.

2. Heart rate and force of contraction and blood pressure increase.

3. The blood vessels of nonessential organs such as the kidneys and gastrointestinal tract constrict.

4. Blood vessels of organs involved in exercise or fighting off danger—skeletal muscles, cardiac muscle, liver, and adipose tissue—dilate to allow faster flow of blood. (The liver converts glycogen to glucose, and adipose tissue converts triglycerides to fatty acids, both of which are used by muscle fibers to generate ATP.)

5. The rate and depth of breathing increase and the airways dilate, which allow faster movement of air in and out of the lungs.

Table 21.3 Activities of Sympathetic and Parasympathetic Divisions

Glands

VISCERAL EFFECTOR	EFFECT OF SYMPATHETIC STIMULATION (α OR β RECEPTORS, EXCEPT AS NOTED)[a]	EFFECT OF PARASYMPATHETIC STIMULATION (MUSCARINIC RECEPTORS)
Sweat	Increases secretion locally on palms and soles (α); increases secretion in most body regions (muscarinic ACh receptors).	None.
Lacrimal (tear)	None.	Stimulates secretion.
Adrenal medulla	Promotes epinephrine and norepinephrine secretion (nicotinic ACh receptors).	None.
Liver	Promotes conversion of glycogen in the liver into glucose (glycogenolysis), stimulates conversion of noncarbohydrates in the liver into glucose (gluconeogenesis), and decreases bile secretion (α and β_2).	Promotes glycogen synthesis; increases bile secretion.
Adipose (fat) cells[b]	Promotes the breakdown of triglycerides into fatty acids and glycerol (lipolysis) (β_1) and release of fatty acids and glycerol into blood (β_1 and β_3). Promotes thermogenesis (heat production) in brown fat (β_3).	None.
Kidney, juxta-glomerular cells	Stimulates secretion of renin (β_1).	None.
Pancreas	Inhibits secretion of digestive enzymes and the hormone insulin (α); promotes secretion of the hormone glucagon (β_2).	Promotes secretion of digestive enzymes and the hormone insulin.

Smooth Muscle

VISCERAL EFFECTOR	EFFECT OF SYMPATHETIC STIMULATION (α OR β RECEPTORS, EXCEPT AS NOTED)[a]	EFFECT OF PARASYMPATHETIC STIMULATION (MUSCARINIC RECEPTORS)
Iris, radial muscle	Contraction → dilation of pupil (α).	None.
Iris, circular muscle	None.	Contraction → constriction of pupil.
Ciliary muscle of eye	Relaxation for far vision (β).	Contraction for near vision.
Salivary gland arterioles	Vasoconstriction, which decreases secretion of saliva (β_2).	Vasodilation, which increases K^+ and water secretion.
Gastric gland arterioles	Vasoconstriction, which inhibits secretion (?).	Promotes secretion.
Intestinal gland arterioles	Vasoconstriction, which inhibits secretion (α).	Promotes secretion.
Lungs, bronchial muscle	Relaxation → airway dilation (β_2).	Contraction → airway constriction.

[a]Subcategories of β receptors are listed where known.
[b]Grouped with glands because they release substances into the blood.

6. Blood glucose level rises as liver glycogen is converted to glucose.

7. The medullae of the adrenal glands are stimulated to release epinephrine and norepinephrine. These hormones intensify and prolong the sympathetic effects just described.

8. Processes that are not essential for meeting the stress situation are inhibited. For example, muscular movements of the gastrointestinal tract and digestive secretions slow down or even stop.

Table 21.3 summarizes the responses of glands, smooth muscle, and cardiac muscle to stimulation by the sympathetic and parasympathetic branches of the ANS.

Table 21.3 *Activities of Sympathetic and Parasympathetic Divisions (continued)*

Smooth Muscle

VISCERAL EFFECTOR	EFFECT OF SYMPATHETIC STIMULATION (α OR β RECEPTORS, EXCEPT AS NOTED)[a]	EFFECT OF PARASYMPATHETIC STIMULATION (MUSCARINIC RECEPTORS)
Heart arterioles	Relaxation → dilation of coronary blood vessels (β_1).	Contraction → constriction.
Skin and mucosal arterioles	Contraction → constriction (α).	Dilation, which may not be physiologically significant.
Skeletal muscle arterioles	Contraction → constriction (α); relaxation → dilation (β_2). Relaxation → dilation (muscarinic).	None.
Abdominal viscera arterioles	Contraction → constriction (α, β).	None.
Brain arterioles	Slight contraction → constriction (α).	None.
Systemic veins	Contraction → constriction (α); relaxation → dilation (β_2).	None.
Gallbladder and ducts	Relaxation (β_2).	Contraction → increased release of bile into small intestine.
Stomach and intestines	Decreases motility and tone (α, β_2); contracts sphincters (α).	Increases motility and tone; relaxes sphincters → enhanced digestive activities and defecation.
Kidney	Constriction of blood vessels → decreased rate of urine production (α).	None.
Ureter	Increases motility.	Decreases motility.
Spleen	Contraction and discharge of stored blood into general circulation (α).	None.
Urinary bladder	Relaxation of muscular wall (β); contraction of trigone and sphincter (α).	Contraction of muscular wall; relaxation of trigone and sphincter → urination.
Uterus	Inhibits contraction in nonpregnant woman (β_2); promotes contraction in pregnant woman (α).	Minimal effect.
Sex organs	In male: contraction of smooth muscle of ductus (vas) deferens, seminal vesicle, prostate → ejaculation.	Vasodilation and erection in both sexes.
Hair follicles, arrector pili muscle	Contraction → erection of hairs in skin.	None.
Cardiac Muscle (Heart)		
	Increases heart rate and force of atrial and ventricular contractions (β_1).	Decreases heart rate; decreases force of atrial contraction.

AUTONOMIC REFLEXES

Autonomic (visceral) reflexes are responses that occur when nerve impulses pass over an autonomic reflex arc. Autonomic reflexes play a key role in activities such as *blood pressure* by adjusting heart rate, force of ventricular contraction, and blood vessel diameter; *respiration* by regulating the diameter of bronchial tubes; *digestion* by adjusting the motility (movement) and muscle tone of the gastrointestinal tract; and *defecation* and *urination* by regulating the opening and closing of sphincters. Other examples are shown in Table 21.3.

The components of an autonomic reflex arc are described below:

1. **Receptor.** Like the receptor in a somatic reflex arc (see Figure 17.5), the receptor in an autonomic reflex arc is the distal end of a sensory neuron. It responds to a stimulus and produces a change that will ultimately trigger nerve impulses. Whereas somatic sensory receptors are associated with special senses, general somatic senses, and proprioceptors, autonomic sensory receptors are mostly associated with interoceptors.

2. **Sensory neuron.** This neuron conducts nerve impulses from receptors to the CNS.

3. **Integrating center.** This is a region in the CNS composed of gray matter.

4. **Motor neuron.** Nerve impulses triggered by the integrating center propagate out of the CNS along a motor neuron to an effector. In a somatic reflex arc, there is only one set of motor neurons connecting the CNS to the effector, and the link is direct. In an autonomic reflex arc, there are two sets of motor neurons connecting the CNS to an effector. The first set, the *preganglionic neurons,* conduct motor impulses from the CNS to an autonomic ganglion (see Figure 21.4). The second set of neurons, the *postganglionic neurons,* conduct motor impulses from an autonomic ganglion to an effector.

5. **Effector.** This is the part of the body that responds to a motor impulse, and the action resulting is called a reflex. In a somatic reflex arc, the effector is a skeletal muscle, and the reflex is called a somatic reflex. In an autonomic reflex arc, the effectors, called *visceral effectors,* are smooth muscle, cardiac muscle, and glands. The reflex is called an autonomic (visceral) reflex.

Visceral sensations often do not reach the cerebral cortex to give rise to conscious perceptions. Under normal conditions, you are not aware of muscular contractions of the digestive organs, heartbeat, changes in the diameter of blood vessels, and pupil dilation and constriction. Your body adjusts such visceral activities by visceral reflex arcs whose integrating centers are in the spinal cord or lower regions of the brain. Among such integrating centers are the cardiac, respiratory, vasomotor, swallowing, and vomiting centers in the medulla and the temperature control center in the hypothalamus. Somatic or visceral sensory neurons deliver input to these centers, and autonomic motor neurons provide output that adjusts activity in the visceral effector, usually without conscious recognition.

CONTROL BY HIGHER CENTERS

Axons from many parts of the CNS connect to both the sympathetic and the parasympathetic divisions of the autonomic nervous system and thus exert considerable control over them. The hypothalamus is the major control and integration center of the ANS. Output from the hypothalamus influences autonomic centers in the medulla oblongata and spinal cord.

The hypothalamus receives input from areas of the nervous system concerned with emotions, visceral functions, olfaction (smell), and gustation (taste), as well as changes in temperature, osmolarity, and levels of various substances in blood. Anatomically, the hypothalamus is connected to both the sympathetic and the parasympathetic divisions of the ANS by axons of neurons whose dendrites and cell bodies are in various hypothalamic nuclei. The axons form tracts from the hypothalamus to sympathetic and parasympathetic nuclei in the brain stem and spinal cord through relays in the reticular formation.

The posterior and lateral portions of the hypothalamus control the sympathetic division. When these areas are stimulated, there is an increase in heart rate and force of contraction, a rise in blood pressure due to constriction (narrowing) of blood vessels, an increase in body temperature, an increase in the rate and depth of respiration, dilation of the pupils, and inhibition of the gastrointestinal tract. On the other hand, the anterior and medial portions of the hypothalamus control the parasympathetic division. Stimulation of these areas results in a decrease in heart rate, lowering of blood pressure, constriction of the pupils, and increased secretion and motility of the gastrointestinal tract.

Control of the ANS by the cerebral cortex occurs primarily during emotional stress. In extreme anxiety, the cortex can stimulate the hypothalamus as part of the limbic system. This, in turn, stimulates the cardiovascular center of the medulla oblongata, which increases the rate and force of the heartbeat and blood pressure. Stimulation of the cortex by hearing bad news or seeing an extremely unpleasant sight may cause vasodilation of blood vessels, a lowering of blood pressure, and fainting. Even though most ANS responses are involuntary, with practice it may be possible to achieve some voluntary control. Biofeedback experiments and the feats of accomplished yogis provide evidence that learning can control autonomic motor neurons to some degree.

Study Outline

Comparison of Somatic and Autonomic Nervous Systems (p. 640)

1. The somatic nervous system receives input from the special senses, general somatic senses, and proprioceptors; the ANS receives input from the special senses, general visceral senses, and general somatic senses.
2. The somatic nervous system operates under conscious control; the ANS operates without conscious control.
3. The axons of the motor neurons of the somatic nervous system extend from the CNS, synapse directly with an effector, and release acetylcholine; the axon of the first motor neuron of the ANS (preganglionic) extends from the CNS and synapses in a ganglion with the second motor neuron; the second neuron (postganglionic) synapses on an effector. Preganglionic fibers release acetylcholine, and postganglionic fibers release acetylcholine or norepinephrine.
4. Somatic nervous system effectors are skeletal muscles; ANS effectors include cardiac muscle, smooth muscle, and glands.
5. Neurotransmitters released by somatic motor neurons cause excitation; neurotransmitters released by autonomic motor neurons cause excitation or inhibition.

Anatomy of Autonomic Motor Pathways (p. 641)

1. Preganglionic neurons are myelinated; postganglionic neurons are unmyelinated.
2. The cell bodies of sympathetic preganglionic neurons are in the lateral gray horns of the 12 thoracic and first two or three lumbar segments; the cell bodies of parasympathetic preganglionic neurons are in cranial nerve nuclei (III, VII, IX, and X) in the brain stem and lateral gray horns of the second through fourth sacral segments of the cord.
3. Autonomic ganglia are classified as sympathetic trunk ganglia (on both sides of spinal column), prevertebral ganglia (anterior to spinal column), and terminal ganglia (near or inside visceral effectors.)

4. Autonomic plexuses in the thorax, abdomen, and pelvis consist of nerve fibers and ganglia. Examples are the celiac, superior mesenteric, inferior mesenteric, and hypogastric plexuses.
5. Sympathetic preganglionic neurons synapse with postganglionic neurons in ganglia of the sympathetic trunk or prevertebral ganglia; parasympathetic preganglionic neurons synapse with postganglionic neurons in terminal ganglia.

Activities of the ANS (p. 648)

1. Most body structures receive dual innervation, and usually one division causes excitation and the other causes inhibition.
2. Cholinergic neurons release acetylcholine; adrenergic neurons release norepinephrine or epinephrine.
3. The effects of sympathetic stimulation are longer lasting and more widespread than the parasympathetic effects.
4. Two types of cholinergic receptors are nicotinic and muscarinic; two types of adrenergic receptors are alpha and beta.
5. The parasympathetic division regulates activities that conserve and restore body energy; the sympathetic division prepares the body for emergency situations (fight-or-flight response).

Autonomic Reflexes (p. 652)

1. An autonomic (visceral) reflex adjusts the activities of smooth muscle, cardiac muscle, and glands.
2. An autonomic reflex arc consists of a receptor, sensory neuron, association neuron, autonomic motor neurons, and visceral effector.

Control by Higher Centers (p. 653)

1. The hypothalamus controls and integrates activities of the autonomic nervous system. It is connected to both the sympathetic and the parasympathetic divisions.
2. Control of the ANS by the cerebral cortex occurs mostly during emotional stress.

Review Questions

1. What are the principal components of the autonomic nervous system? What is its general function? Why is it called involuntary? (p. 640)
2. What are the principal differences between the somatic nervous system and the autonomic nervous system? (p. 640)
3. Distinguish between preganglionic neurons and postganglionic neurons with respect to location and function. (p. 641)
4. What is an autonomic ganglion? Describe the location and function of the three types of autonomic ganglia. Define white and gray rami communicantes. (p. 642)
5. On what basis are the sympathetic and parasympathetic divisions of the autonomic nervous system differentiated anatomically and functionally? (p. 648)
6. Discuss the distinction between cholinergic and adrenergic fibers of the autonomic nervous system. (p. 648)
7. How is acetylcholine (ACh) related to nicotinic and muscarinic receptors? (p. 649)

8. How are alpha and beta receptors related to norepinephrine (NE) and epinephrine? (p. 650)
9. Give examples of the antagonistic effects of the sympathetic and parasympathetic divisions of the autonomic nervous system. (p. 650)
10. Why is the parasympathetic division of the ANS called an energy conservation–restorative system? (p. 650)
11. Describe what happens during the fight-or-flight response. (p. 650)
12. Give the *sympathetic response* in a frightening situation for each of the following body parts: hair follicles, iris of eye, lungs, spleen, adrenal medullae, urinary bladder, stomach, intestines, gallbladder, liver, heart, arterioles of the abdominal viscera, skeletal muscles, and skin and mucosa. (p. 651)
13. Define an autonomic reflex and give three examples. (p. 652)
14. Describe an autonomic reflex in proper sequence. (p. 653)
15. Describe how the hypothalamus controls and integrates the autonomic nervous system. (p. 653)

Self Quiz

Complete the following sentences:

1. The input component of the ANS consists of ___ neurons, and the output component consists of ___ neurons.
2. Organs such as the heart or stomach that receive impulses from both divisions of the ANS are said to have ___ innervation.
3. Adrenergic neurons release the neurotransmitters ___ and ___.
4. The two types of adrenergic receptors are ___ and ___ receptors.
5. The two types of cholinergic receptors are ___ and ___ receptors.
6. The pterygopalatine, otic, and submandibular ganglia are part of the ___ division of the ANS.
7. Match the following:
   ___ (a) thoracolumbar outflow
   ___ (b) has long preganglionic fibers and short postganglionic fibers
   ___ (c) has short preganglionic fibers and long postganglionic fibers
   ___ (d) sends some preganglionic fibers through cranial nerves
   ___ (e) has some preganglionic fibers synapsing in sympathetic trunk ganglia
   ___ (f) has more widespread and longer lasting effect in the body
   ___ (g) has some fibers running in gray rami to supply sweat glands, hair follicle muscles, and blood vessels
   ___ (h) has fibers in white rami (connecting spinal nerve with sympathetic trunk ganglia)
   ___ (i) contains fibers that supply viscera with motor impulses
   ___ (j) celiac and superior mesenteric ganglia are sites of postganglionic neuron cell bodies

   (1) applies to the sympathetic division of the ANS
   (2) applies to the parasympathetic division of the ANS
   (3) applies to both divisions of the ANS

Choose the one best answer to the following questions:

8. Which of the following statements is true for the parasympathetic system?
   a. Ganglia are close to the central nervous system and distant from the effector.
   b. It forms the craniosacral outflow.
   c. It supplies nerves to blood vessels, sweat glands, and adrenal (suprarenal) glands.
   d. It has some of its nerve fibers passing through paravertebral ganglia.
   e. All of the above are true.
9. Which of the following statements about gray rami is false?
   a. They carry preganglionic neurons from the anterior rami of spinal nerves to sympathetic trunk ganglia.
   b. They contain only sympathetic nerve fibers.
   c. They carry impulses from sympathetic trunk ganglia to spinal nerves.
   d. They are located at all levels of the vertebral column.
   e. None of the above are false.
10. Which of the following statements regarding the parasympathetic ganglia of the ANS is true?
   a. They contain adrenergic neurons only.
   b. They include the cervical and celiac ganglia.
   c. They lie within the spinal cord or brain.
   d. They consist of a double chain of structures on either side of the spinal column.
   e. None of the above is true.
11. The ANS provides the chief nervous control in which of the following activities?
   a. following a moving object with the eyes
   b. removing a hand reflexively from a hot object
   c. typing
   d. digesting food
   e. writing an essay
12. Which of the following four statements apply to the preganglionic nerve fibers that arise from the thoracic and lumbar parts of the spinal cord?
   (1) They are called preganglionic sympathetic fibers.
   (2) They synapse with postganglionic fibers of the parasympathetic nervous system.
   (3) They form the posterior roots of the thoracic and lumbar spinal nerves.
   (4) They are also called preganglionic parasympathetic fibers.
   a. 1 only
   b. 2 only
   c. 3 only
   d. 4 only
   e. 2 and 4
13. Cholinergic fibers are thought to include
   a. all preganglionic axons
   b. all postganglionic parasympathetic axons
   c. a few postganglionic sympathetic axons
   d. all axons of somatic motor neurons
   e. all of the above
14. Which of the following statements would not be included in a description of the sympathetic trunks?
   a. The trunks lie anterior and lateral to the spinal cord.
   b. Each trunk contains 30 ganglia.
   c. Each trunk receives preganglionic fibers from the thoracic and sacral regions of the spinal cord.
   d. The ganglia of the trunks are called paravertebral ganglia.
   e. White rami direct nerve fibers into the trunk; gray rami direct nerve fibers out of the trunk.

15. Which of the following would indicate increased parasympathetic activity?
    a. "cotton mouth" from reduced salivation
    b. increased gastric secretion
    c. rise in blood pressure
    d. decreased flow of blood through the skin
    e. elevated blood glucose level
16. The axons of neurons that lie within the central nervous system and connect with the ANS are
    a. myelinated fibers
    b. postganglionic fibers
    c. preganglionic fibers
    d. cranial nerve fibers
    e. none of the above

Are the following statements true or false?

17. The ganglia that make up the sympathetic trunks are also called prevertebral or collateral ganglia.
18. Synapsing occurs in both sympathetic and parasympathetic ganglia.
19. Parasympathetic preganglionic neurons originate in the nuclei of the oculomotor (III), facial (VII), glossopharyngeal (IX), or vagus (X) nerves, or in lateral gray horns of the sacral region of the spinal cord.
20. An event occurring during the fight-or-flight response is increased urine production due to blood being shunted to the kidneys.
21. In a visceral autonomic reflex arc, only one motor neuron is involved.

Critical Thinking Questions

1. Skydiving, hang gliding, and bungee jumping can all give you a great rush (or get you killed). How do these activities cause this type of physiological effect?
   HINT: *A bungee jumper might say she was "pumped with adrenaline" after a jump.*
2. It's Thanksgiving and you've just eaten a huge turkey dinner with all the trimmings. Now you're heading to watch the football game on television. Which division of the nervous system will be handling your body's post-dinner activities? List several organs and the effects of the nervous system on their functions.
   HINT: *Couch potatoes know no fear.*
3. The arrangement of the sympathetic chain looks a bit like a highway planner's "spaghetti bowl" nightmare. Using this highway analogy, relate the spinal nerve, the white ramus, the gray ramus, the sympathetic trunk ganglion, and the sympathetic chain to the main east-west highway, its on and off ramps, the north-south highway, and the intersection.
   HINT: *If your body was a globe, your head would be the north pole and your feet the south pole.*

4. After lunch, baby Bobby was looking uncomfortable, squirming, and fretting. Suddenly, a pungent odor wafted from his diaper and he looked very content. Trace the reflex arc that is responsible for baby Bobby's present condition.
   HINT: *Baby Bobby's body responds to the stimuli automatically (say this three times fast).*
5. Your textbook suggested the acronym "SLUD" to help you remember the effects of the parasympathetic nervous system. Come up with your own learning aid for the parasympathetic division of the ANS, the sympathetic division of the ANS, or both divisions.
   HINT: *Familiar abbreviations or rhymes help; for example, finish this one: "feed and . . ."*
6. Cranial nerve X makes up the majority of the cranial outflow for the parasympathetic division of the ANS. Relate cranial nerve X's name to its structure.
   HINT: *There's nothing vague about this answer.*

Answers to Figure Questions

21.1 An organ of the body receives both sympathetic and parasympathetic fibers.
21.2 Most parasympathetic preganglianic fibers are longer than most sympathetic preganglionic fibers because most parasympathetic ganglia are in the walls of visceral organs, whereas most sympathetic ganglia are close to the spinal cord in the sympathetic trunk.
21.3 Celiac plexus.
21.4 Myelin.
21.5 Sympathetic and parasympathetic postganglionic neurons.

Chapter 22

The Endocrine System

Student Objectives

1. Define endocrine gland and exocrine gland, and list the endocrine glands of the body.

2. Describe the structural divisions of the pituitary gland.

3. Describe the location, histology, and blood and nerve supplies of the pituitary gland, thyroid gland, parathyroid glands, adrenal (suprarenal) glands, pancreas, pineal gland, and thymus gland.

4. Explain the structural subdivisions of the adrenal (suprarenal) glands.

5. Describe the development of the endocrine system.

6. Describe the effects of aging on the endocrine system.

7. Describe the names and actions of hormones secreted by organs other than endocrine glands.

Together, the nervous and endocrine systems coordinate functions of all body systems. The nervous system controls body activities through nerve impulses conducted along axons of neurons. At axon terminals, impulses trigger release of neurotransmitter molecules. The result is either excitation or inhibition of specific other neurons, muscle fibers (cells), or gland cells. In contrast, the glands of the endocrine system release messenger molecules, called **hormones** (*hormon* = to urge on), into the bloodstream. The circulating blood then delivers hormones to virtually all cells throughout the body.

The nervous and endocrine systems are coordinated as an interlocking supersystem, often referred to as the **neuroendocrine system.** Certain parts of the nervous system stimulate or inhibit the release of hormones. Hormones, in turn, may promote or inhibit the generation of nerve impulses. The nervous system causes muscles to contract and glands to secrete either more or less of their product. The endocrine system alters metabolic activities, regulates growth and development, and guides reproductive processes. Thus it not only helps regulate the activity of smooth and cardiac muscle and some glands, it affects virtually all other tissues as well.

The onset and duration of action and the durations of effects produced differ between the nervous and endocrine systems. Nerve impulses most often produce their effects within a few milliseconds. Although some hormones can act within seconds, others can take several hours or more to bring about their responses. Also, the effects of activating the nervous system are generally briefer than effects produced by the endocrine system. A comparison between the nervous and endocrine systems is presented in Table 22.1.

The focus of this chapter is the endocrine system. We will examine the principal endocrine glands and hormone-producing tissues and their roles in coordinating body activities. The science concerned with the structure and functions of the endocrine glands and the diagnosis and treatment of disorders of the endocrine system is called **endocrinology** (en'-dō-kri-NOL-ō-jē).

Table 22.1 *Comparison of the Nervous and Endocrine Systems*

CHARACTERISTIC	NERVOUS SYSTEM	ENDOCRINE SYSTEM
Mechanism of control	Neurotransmitters released in response to nerve impulses.	Hormones delivered to tissues throughout the body by the blood.
Cells affected	Muscle cells, gland cells, other neurons.	Virtually all body cells.
Type of action that results	Muscular contraction or glandular secretion.	Changes in metabolic activities.
Time to onset of action	Typically within milliseconds.	Seconds to hours or days.
Duration of action	Generally briefer.	Generally longer.

ENDOCRINE GLANDS

The body contains two kinds of glands: exocrine and endocrine. **Exocrine** (*ex* = out; *krinein* = to secrete) **glands** secrete their products into ducts, and the ducts carry the secretions into body cavities, into the lumen of an organ, or to the outer surface of the body. Exocrine glands include sudoriferous (sweat), sebaceous (oil), mucous, and digestive glands. **Endocrine** (*endo* = within) **glands,** by contrast, secrete their products, called hormones, into the extracellular space around the secretory cells, rather than into ducts. The secretion then diffuses into capillaries and is carried away by the blood. The endocrine glands of the body (Figure 22.1) constitute the **endocrine system** and include the pituitary (hypophysis), thyroid, parathyroids, adrenals (suprarenals), and pineal (epiphysis cerebri) glands. In addition, several organs of the body contain endocrine tissue but are not endocrine glands exclusively. These include the hypothalamus, thymus gland, pancreas, ovaries, testes, kidneys, stomach, liver, small intestine, skin, heart, and placenta.

Figure 22.1 Location of many endocrine glands, other organs that contain endocrine tissue, and associated structures.

Endocrine glands secrete hormones that diffuse into the blood for transport to target cells throughout the body.

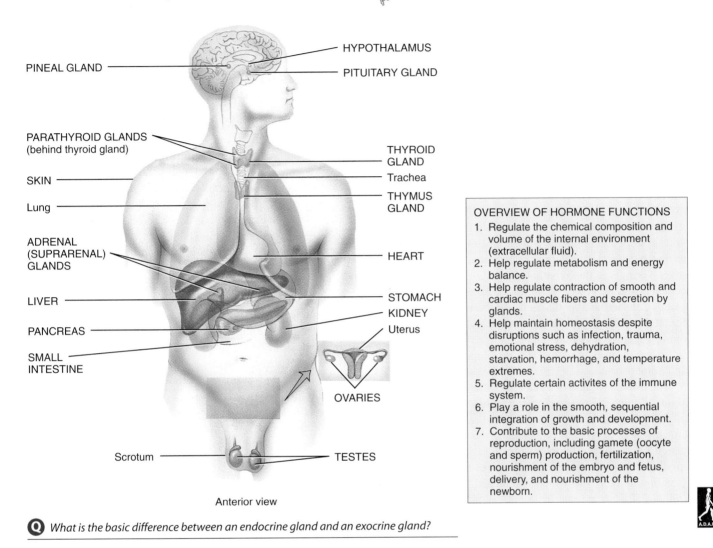

PINEAL GLAND

HYPOTHALAMUS

PITUITARY GLAND

PARATHYROID GLANDS
(behind thyroid gland)

THYROID GLAND

Trachea

SKIN

THYMUS GLAND

Lung

HEART

ADRENAL (SUPRARENAL) GLANDS

LIVER

STOMACH

KIDNEY

PANCREAS

Uterus

SMALL INTESTINE

OVARIES

Scrotum

TESTES

Anterior view

OVERVIEW OF HORMONE FUNCTIONS
1. Regulate the chemical composition and volume of the internal environment (extracellular fluid).
2. Help regulate metabolism and energy balance.
3. Help regulate contraction of smooth and cardiac muscle fibers and secretion by glands.
4. Help maintain homeostasis despite disruptions such as infection, trauma, emotional stress, dehydration, starvation, hemorrhage, and temperature extremes.
5. Regulate certain activites of the immune system.
6. Play a role in the smooth, sequential integration of growth and development.
7. Contribute to the basic processes of reproduction, including gamete (oocyte and sperm) production, fertilization, nourishment of the embryo and fetus, delivery, and nourishment of the newborn.

Q What is the basic difference between an endocrine gland and an exocrine gland?

HORMONES

Hormones produce powerful effects when present in very low concentration. As a rule, most of the over 50 or so hormones affect only a few types of cells. Why is it that some cells respond to a particular hormone and others do not?

Although a given hormone travels throughout the body in the blood, it affects only specific cells called **target cells.** Hormones influence their target cells by chemically binding to large protein or glycoprotein molecules called **receptors.** Only the target cells for a certain hormone have receptors that bind and recognize that hormone. For example, thyroid-stimulating hormone (TSH) binds to receptors on the surface of cells of the thyroid gland, but it does not bind to cells of the ovaries because ovarian cells do not have TSH receptors.

Receptors, like other cellular proteins, are constantly synthesized and broken down. Generally, a target cell has 2000 to 100,000 receptors for a particular hormone. When a hormone is present in excess, the number of target cell receptors may decrease. This decreases the responsiveness of target cells to the hormone. On the other hand, when a hormone is deficient, the number of receptors may increase. This makes a target tissue more sensitive to a hormone. The amount of hormone released by an endocrine gland or tissue is determined by the body's *need* for the hormone at any given time. This need is the basis on which the endocrine system operates. Endocrine secretion is regulated so that there is no overproduction or underproduction of a particular hormone.

HYPOTHALAMUS AND PITUITARY GLAND (HYPOPHYSIS)

For many years the **pituitary gland** or **hypophysis** (hī-POF-i-sis) was called the "master" endocrine gland because it secretes several hormones that control other endocrine glands. We now know that the pituitary gland itself has a master—the **hypothalamus.** This small region of the brain, inferior to the two lobes of the thalamus, is the major integrating link between the nervous and endocrine systems (see Figure 18.1). The hypothalamus receives impulses from several other regions of the brain, including the limbic system, cerebral cortex, thalamus, and reticular activating system. Also, it receives sensory signals from internal organs and perhaps from the visual system. Painful, stressful, and emotional experiences all cause changes in hypothalamic activity. In turn, the hypothalamus controls the autonomic nervous system and regulates body temperature, thirst, hunger, sexual behavior, and defensive reactions such as fear and rage. Not only is the hypothalamus an important regulatory center in the nervous system, it is also a crucial endocrine gland. Cells in the hypothalamus synthesize at least nine different hormones, and the pituitary gland secretes seven more. Together, they play important roles in the regulation of virtually all aspects of growth, development, metabolism, and homeostasis.

The pituitary gland is a pea-sized structure that measures about 1.3 cm (0.5 in.) in diameter. It lies in the sella turcica of the sphenoid bone (Figure 22.2). Posterior to the optic chiasm is a grayish protuberance, the **tuber cinereum,** which is part of the hypothalamus (see Figure 18.10). The **median eminence** is a raised portion of the tuber cinereum (see Figure 18.10) to which is attached the **infundibulum** (*infundibulum* = a funnel), a stalk that attaches the pituitary gland to the hypothalamus.

The pituitary gland has two anatomically and functionally separate portions. The **anterior pituitary gland (anterior lobe)** accounts for about 75% of the total weight of the gland. It derives from an outgrowth of ectoderm called the hypophyseal (Rathke's) pouch in the roof of the mouth (see Figure 22.9b). The anterior pituitary gland contains many glandular epithelial cells and forms the glandular part of the pituitary.

The **posterior pituitary gland (posterior lobe)** also derives from the ectoderm, but from an outgrowth called the neurohypophyseal bud (see Figure 22.9b). It contains axons and axon terminals of more than 10,000 neurons whose cell bodies are located in the supraoptic and paraventricular nuclei of the hypothalamus (see Figure 18.10). The axon terminals in the posterior pituitary gland are associated with specialized neuroglia called **pituicytes** (pi-TOO-i-sītz).

A third region called the **pars intermedia (intermediate lobe)** atrophies during fetal development and ceases to exist as a separate lobe in adults (see Figure 22.9b). However, its cells migrate into adjacent parts of the anterior pituitary gland when present.

Anterior Pituitary Gland (Adenohypophysis)

The **anterior pituitary gland (anterior lobe** or **adenohypophysis)** (ad'-e-nō-hī-POF-i-sis; *adeno* = glandular) secretes hormones that regulate a wide range of bodily activities from growth to reproduction. Release of anterior pituitary gland hormones is stimulated by **releasing hormones** and suppressed by **inhibiting hormones** from the hypothalamus. These hypothalamic hormones are an important link between the nervous and endocrine systems.

Hypothalamic hormones reach the anterior pituitary gland through a system of blood vessels that directly connects the two regions. Blood flows from the median eminence of the hypothalamus into the infundibulum and anterior pituitary gland principally from several **superior hypophyseal** (hī-pō-FIZ-ē-al) **arteries** (Figure 22.2). These arteries are branches of the internal carotid and posterior communicating arteries. The superior hypophyseal arteries form the **primary plexus,** a capillary network at the base of the hypothalamus. Near the median eminence and superior to the optic chiasm are two groups of specialized neurons, called **neurosecretory cells,** that secrete releasing and inhibiting hormones into the primary plexus. These hormones are synthesized in the neuronal cell bodies and released at the axon terminals. The hormones then diffuse into the capillaries of the primary plexus.

From the primary plexus, blood drains into the **hypophyseal portal veins** that pass down the outside of the infundibulum. In the anterior pituitary gland, the hypophyseal portal veins redivide into a **secondary plexus** of capillaries. This direct route permits hypothalamic hormones to act quickly on the anterior pituitary gland cells before the hormones are diluted or destroyed in the systemic circulation. Hormones secreted by anterior pituitary gland cells pass into the secondary plexus and then into the anterior hypophyseal veins for distribution to target tissues throughout the body.

Five principal types of anterior pituitary gland cells secrete seven major hormones (Figure 22.3):

1. **Somatotrophs** (*soma* = body; *tropos* = changing) produce **human growth hormone (hGH)** or **somatotropin,** which stimulates general body growth and regulates aspects of metabolism.

2. **Thyrotrophs** (*thyreos* = shield) synthesize **thyroid-stimulating hormone (TSH),** which controls the secretions and other activities of the thyroid gland.

3. **Gonadotrophs** (*gonas* = seed) produce two major hormones: **follicle-stimulating hormone (FSH)** and **luteinizing hormone (LH).** Together FSH and LH stimulate

Figure 22.2 Hypothalamus and pituitary gland and their blood supply. Note in the small figure to the right that releasing and inhibiting hormones synthesized by hypothalamic neurons are transported within their axons and released from the axon terminals. They diffuse into capillaries of the primary plexus and are carried by the hypophyseal portal veins to the secondary plexus for distribution to target cells in the anterior pituitary gland.

🔑 *Hypothalamic hormones are an important link between the nervous and endocrine systems.*

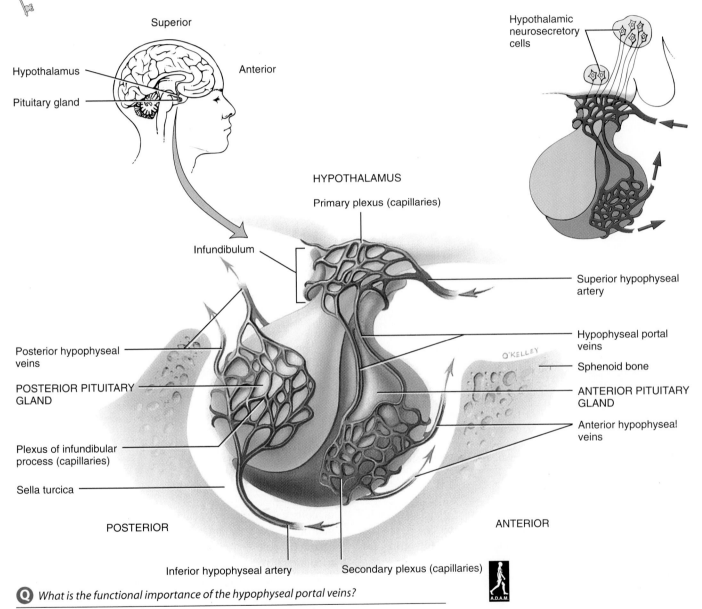

Superior

Hypothalamic neurosecretory cells

Hypothalamus

Anterior

Pituitary gland

HYPOTHALAMUS

Primary plexus (capillaries)

Infundibulum

Superior hypophyseal artery

Hypophyseal portal veins

Posterior hypophyseal veins

O'KELLEY

Sphenoid bone

POSTERIOR PITUITARY GLAND

ANTERIOR PITUITARY GLAND

Anterior hypophyseal veins

Plexus of infundibular process (capillaries)

Sella turcica

POSTERIOR

ANTERIOR

Inferior hypophyseal artery

Secondary plexus (capillaries)

Q *What is the functional importance of the hypophyseal portal veins?*

secretion of estrogens and progesterone and maturation of oocytes in the ovaries, and secretion of testosterone and production of sperm in the testes.

4. **Lactotrophs** (*lact* = milk) synthesize **prolactin (PRL),** which initiates milk production in suitably prepared mammary glands.

5. **Corticotrophs** (*cortex* = rind or bark) synthesize **adrenocorticotropic hormone (ACTH),** which stimulates the adrenal cortex to secrete glucocorticoids. Some corticotrophs, remnants of the pars intermedia, also secrete **melanocyte-stimulating hormone (MSH),** which affects skin pigmentation.

Figure 22.3 Hormone-secreting cells in the anterior pituitary gland and the target tissues and general functions of the hormones they produce.

Hormones that influence other endocrine glands are called tropins or trophic hormones.

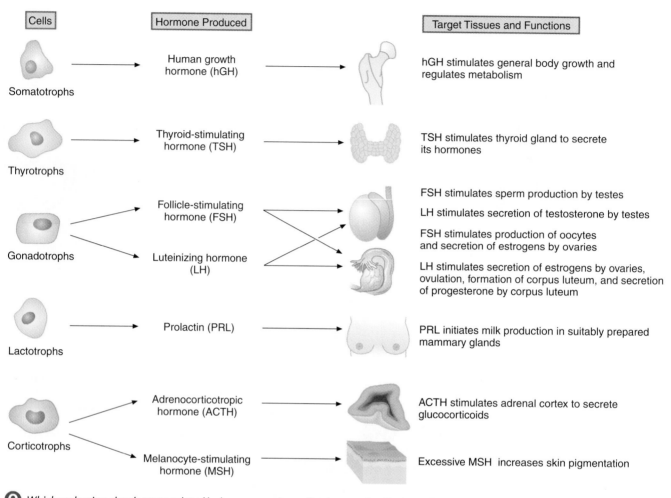

Cells	Hormone Produced	Target Tissues and Functions
Somatotrophs	Human growth hormone (hGH)	hGH stimulates general body growth and regulates metabolism
Thyrotrophs	Thyroid-stimulating hormone (TSH)	TSH stimulates thyroid gland to secrete its hormones
Gonadotrophs	Follicle-stimulating hormone (FSH) / Luteinizing hormone (LH)	FSH stimulates sperm production by testes. LH stimulates secretion of testosterone by testes. FSH stimulates production of oocytes and secretion of estrogens by ovaries. LH stimulates secretion of estrogens by ovaries, ovulation, formation of corpus luteum, and secretion of progesterone by corpus luteum
Lactotrophs	Prolactin (PRL)	PRL initiates milk production in suitably prepared mammary glands
Corticotrophs	Adrenocorticotropic hormone (ACTH) / Melanocyte-stimulating hormone (MSH)	ACTH stimulates adrenal cortex to secrete glucocorticoids. Excessive MSH increases skin pigmentation

Q *Which endocrine glands are regulated by hormones released by the anterior pituitary gland?*

Hormones that influence other endocrine glands are called **tropins** or **tropic hormones.** The anterior pituitary gland secretes several tropins. For example, two **gonadotropins,** follicle-stimulating hormone (FSH) and luteinizing hormone (LH), regulate the functions of the gonads (ovaries and testes).

Posterior Pituitary Gland (Neurohypophysis)

Although the **posterior pituitary gland (posterior lobe),** or **neurohypophysis,** does not *synthesize* hormones, it does *store* and *release* two hormones. As noted earlier, it consists of cells called pituicytes and axon terminals of secretory neurons of hypothalamic neurosecretory cells (Figure 22.4). The cell bodies of neurosecretory cells are in the paraventricular and supraoptic nuclei of the hypothalamus. Their axons form the **supraopticohypophyseal** (soo-pra-op′-ti-kō-hī-pō-FIZ-ē-al) **tract,** which extends from the hypothalamus to the posterior pituitary gland and terminates near blood capillaries in the posterior pituitary gland. The cell bodies of the neurosecretory cells produce two hormones: **oxytocin** (*oxytocia* = childbirth) or **OT,** which is involved in labor and milk secretion, and **antidiuretic hormone (ADH),** which decreases urine volume and raises blood pressure.

Figure 22.4 Axons of neurosecretory cells form the supraopticohypophyseal tract. Note in the small figure to the right that hormones synthesized by neurosecretory cells in the hypothalamus pass in their axons down to the axon terminals in the posterior pituitary gland. Nerve impulses discharge the hormones, which diffuse into the plexus of the infundibular process and posterior hypophyseal veins for distribution to target cells.

The posterior pituitary gland stores and releases oxytocin and antidiuretic hormone.

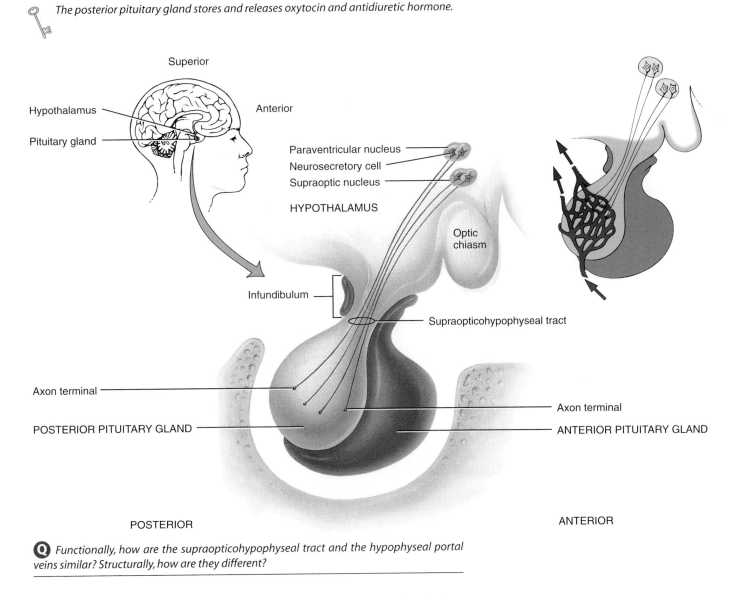

Q Functionally, how are the supraopticohypophyseal tract and the hypophyseal portal veins similar? Structurally, how are they different?

After their production in the cell bodies of neurosecretory cells, the hormones are packed into secretory vesicles, which move to the axon terminals in the posterior pituitary gland. Nerve impulses that propagate along the axon and reach the axon terminals trigger exocytosis of the secretory vesicles. The released OT and ADH then diffuse into nearby capillaries.

Clinical Application

Oxytocin and Childbirth

Years before oxytocin was discovered, it was common practice in midwifery to let a first-born twin nurse at the mother's breast to speed the birth of the second child. Now we know why this practice is helpful—it stimulates release of oxytocin. Even after a single birth, nursing promotes expulsion of the placenta (afterbirth) and

Table 22.2 Summary of Pituitary Gland Hormones

Anterior Pituitary Gland

HORMONE	PRINCIPAL ACTIONS
Human growth hormone (hGH)	Growth of body cells; protein anabolism; elevation of blood glucose concentration.
Thyroid-stimulating hormone (TSH)	Stimulates secretion of thyroid hormones by thyroid gland.
Follicle-stimulating hormone (FSH)	In females, initiates development of ova and induces ovarian secretion of estrogens; in males, stimulates testes to produce sperm.
Luteinizing hormone (LH) [also called interstitial cell-stimulating hormone (ICSH) in males]	In females, stimulates secretion of estrogens and progesterone; in males, stimulates interstitial cells in testes to develop and produce testosterone.
Prolactin (PRL)	In females, together with other hormones, stimulates milk secretion by the mammary glands.
Adrenocorticotropic hormone (ACTH)	Controls secretion of glucocorticoids (mainly cortisol) by adrenal cortex.
Melanocyte-stimulating hormone (MSH)	Role in humans is unknown; stimulates dispersion of melanin granules in melanocytes of amphibians.

Posterior Pituitary Gland

HORMONE	PRINCIPAL ACTIONS
Oxytocin (OT)	Stimulates contraction of smooth muscle cells of uterus during labor and stimulates contraction of smooth muscle cells around glandular cells of mammary glands to cause milk ejection.
Antidiuretic hormone (ADH)	Principal effect is to conserve body water by decreasing urine volume; also raises blood pressure by constricting arterioles.

helps the uterus regain its smaller size. Synthetic OT often is given to induce labor or to increase uterine tone and control hemorrhage just after giving birth. ▪

The blood supply to the posterior pituitary gland is from the **inferior hypophyseal arteries** (see Figure 22.2), derived from the internal carotid arteries. In the posterior pituitary gland, the inferior hypophyseal arteries form a plexus of capillaries called the **plexus of the infundibular process.** From this plexus, hormones pass into the **posterior hypophyseal veins** for distribution to tissue cells.

A summary of pituitary gland hormones and their actions is presented in Table 22.2.

THYROID GLAND

The **thyroid gland** is located just inferior to the larynx (voice box). The right and left **lateral lobes** lie one on either side of the trachea. Connecting the lobes is a mass of tissue called an **isthmus** (IS-mus) that lies on the anterior surface of the trachea (Figure 22.5a). A small, pyramidal-shaped

lobe sometimes extends upward from the isthmus. The gland usually weighs about 30 g (about 1 oz) and has a rich blood supply, receiving about 80 to 120 ml of blood per minute.

One of the thyroid gland's unique features is its ability to store hormones and release them in a steady flow over a long period of time.

Microscopic spherical sacs called **thyroid follicles** (Figure 22.5b and c) constitute most of the thyroid gland. The wall of each follicle consists of two types of cells. Most extend to the lumen (internal space) of the follicle and are called **follicular cells.** When the follicular cells are inactive, their shape is low cuboidal to squamous. Under the influence of TSH, they become cuboidal or low columnar and actively secretory. The follicular cells manufacture **thyroxine** (thī-ROK-sēn), which is also called **tetraiodothyronine** (tet-ra-ī′-ōd-ō-THĪ-rō-nēn, or T_4, because it contains four atoms of iodine, and **triiodothyronine** (trī-ī′-ōd-ō-THĪ-rō-nēn), or T_3, which contains three atoms of iodine. Together these hormones are referred to as **thyroid hormones.** A few cells do not reach the follicle lumen or lie between follicles and are called **parafollicular cells,** or **C (clear) cells.** They produce the hormone **calcitonin** (kal-si-TŌ-nin), which helps regulate calcium blood levels.

Figure 22.5 Location, blood supply, and histology of the thyroid gland.

Thyroid hormones regulate (1) oxygen use and basal metabolic rate, (2) cellular metabolism, and (3) growth and development.

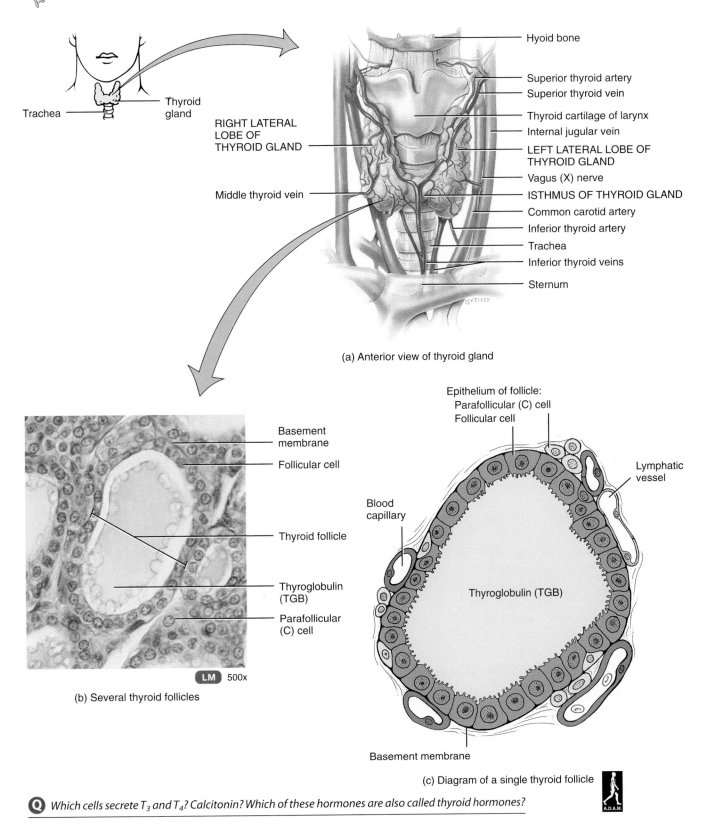

(a) Anterior view of thyroid gland

(b) Several thyroid follicles

LM 500x

(c) Diagram of a single thyroid follicle

Q *Which cells secrete T₃ and T₄? Calcitonin? Which of these hormones are also called thyroid hormones?*

The main blood supply of the thyroid gland is from the superior thyroid artery, a branch of the external carotid artery, and the inferior thyroid artery, a branch of the thyrocervical trunk from the subclavian artery. The thyroid gland is drained by the superior and middle thyroid veins, which pass into the internal jugular veins, and the inferior thyroid veins, which join the brachiocephalic veins or internal jugular veins (Figure 22.5a).

The nerve supply of the thyroid gland consists of postganglionic fibers from the superior and middle cervical sympathetic ganglia. Preganglionic fibers from the ganglia are derived from the second through seventh thoracic segments of the spinal cord.

A summary of thyroid gland hormones and their actions is presented in Table 22.3.

PARATHYROID GLANDS

Attached to the posterior surfaces of the thyroid gland are small, round masses of tissue called the **parathyroid** (*para* = beside) **glands.** Usually, there is one superior and one inferior parathyroid gland attached to each lateral thyroid lobe (Figure 22.6a).

Microscopically, the parathyroids contain two kinds of epithelial cells (Figure 22.6b and c). The more numerous cells, called **principal (chief) cells,** probably are the major source of **parathyroid hormone (PTH),** or **parathormone.** PTH increases blood calcium (Ca^{2+}) and magnesium (Mg^{2+}) levels, decreases blood phosphate (HPO_4^{2-}) levels, increases the number and activity of osteoclasts (bone-destroying cells), increases Ca^{2+} reabsorption and HPO_4^{2-} excretion by the kidneys, and promotes the formation of calcitriol (active form of vitamin D). The function of the other kind of cell, called an **oxyphil cell,** is not known.

The parathyroids are abundantly supplied with blood from branches of the superior and inferior thyroid arteries. Blood is drained by the superior, middle, and inferior thyroid veins. The nerve supply of the parathyroids is derived from the thyroid branches of cervical sympathetic ganglia.

A summary of parathyroid hormone and its actions is presented in Table 22.4.

ADRENAL (SUPRARENAL) GLANDS

The paired **adrenal (suprarenal) glands,** one of which lies superior to each kidney (Figure 22.7a), are about 3 to 5 cm in height, 2 to 3 cm in width, and less than 1 cm in thickness. They average about 3.5 to 5 g in weight. Structurally and functionally the adrenal glands are differentiated into two regions. The outer **adrenal cortex** makes up the bulk of the gland and surrounds the inner **adrenal medulla** (Figure 22.7a). The adrenal cortex is derived from mesoderm of a

Table 22.3 *Summary of Thyroid Gland Hormones*

HORMONE	PRINCIPAL ACTIONS
Thyroid hormones: T_3 (triiodothyronine) and T_4 (thyroxine) from follicular cells	Increase basal metabolic rate, stimulate synthesis of proteins, increase use of glucose for ATP production, increase breakdown of triglycerides into fatty acids and glycerol (lipolysis), enhance cholesterol excretion in bile, accelerate body growth, and contribute to normal development of the nervous system.
Calcitonin (CT) from parafollicular cells	Lowers blood levels of calcium and phosphates by accelerating their deposition in bone matrix.

Table 22.4 *Summary of Parathyroid Gland Hormone*

HORMONE	PRINCIPAL ACTIONS
Parathyroid hormone (PTH)	Increases blood Ca^{2+} and Mg^{2+} levels and decreases HPO_4^{2-} level by increasing rate of dietary Ca^{2+} and Mg^{2+} absorption; increases number and activity of osteoclasts; increases Ca^{2+} reabsorption by kidneys; increases HPO_4^{2-} excretion by kidneys; and promotes formation of calcitriol (active form of vitamin D).

developing embryo and produces steroid hormones that are essential for life. Complete loss of adrenocortical hormones leads rapidly to death in a few days to a week due to dehydration and electrolyte imbalance unless hormone replacement therapy begins promptly. The adrenal medulla, unlike the adrenal cortex, is derived from ectoderm and produces two hormones, norepinephrine and epinephrine. Covering the gland is a connective tissue capsule. The adrenal glands, like the thyroid gland, are highly vascularized.

Adrenal Cortex

The adrenal cortex is subdivided into three zones that secrete different hormones (Figure 22.7b). The outer zone, just deep to the connective tissue capsule, is called the **zona glomerulosa** (*glomerulus* = little ball). Its cells are arranged in

Figure 22.6 Location, blood supply, and histology of the parathyroid glands.

The parathyroid glands, normally four in number, are embedded in the posterior surface of the thyroid gland.

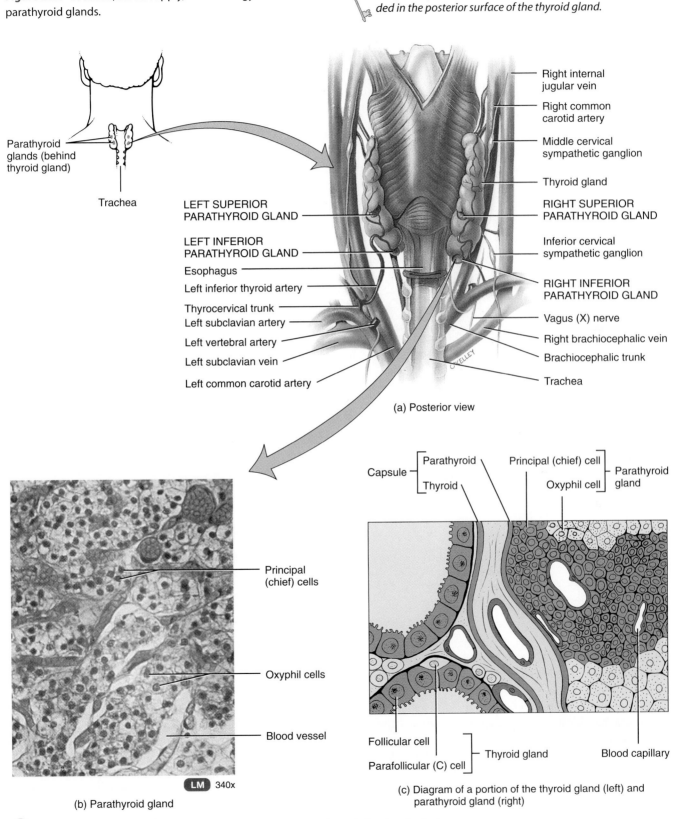

(a) Posterior view

(b) Parathyroid gland

LM   340x

(c) Diagram of a portion of the thyroid gland (left) and parathyroid gland (right)

Q *What are the secretory products of parafollicular cells of the thyroid gland and principal cells of the parathyroid glands?*

Figure 22.7 Location, blood supply, and histology of the adrenal (suprarenal) glands.

The adrenal cortex secretes steroid hormones that are essential for life; the adrenal medulla secretes norepinephrine and epinephrine.

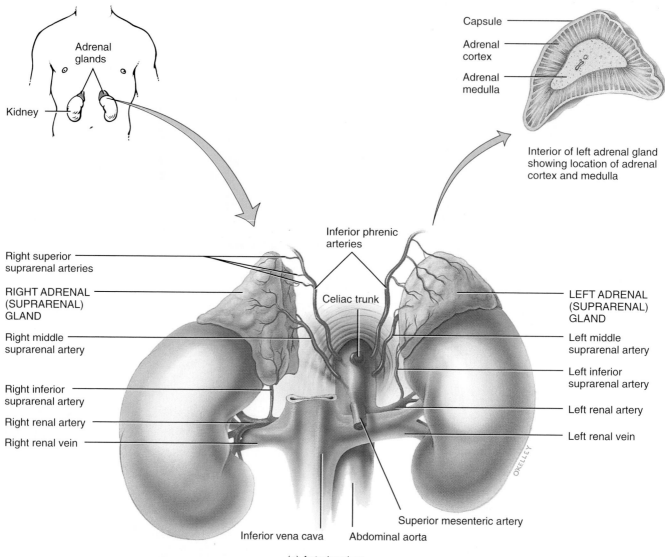

Interior of left adrenal gland showing location of adrenal cortex and medulla

Capsule

Adrenal cortex

Adrenal medulla

Inferior phrenic arteries

Right superior suprarenal arteries

RIGHT ADRENAL (SUPRARENAL) GLAND

Celiac trunk

Right middle suprarenal artery

Right inferior suprarenal artery

Right renal artery

Right renal vein

LEFT ADRENAL (SUPRARENAL) GLAND

Left middle suprarenal artery

Left inferior suprarenal artery

Left renal artery

Left renal vein

Superior mesenteric artery

Inferior vena cava   Abdominal aorta

Adrenal glands

Kidney

(a) Anterior view

arched loops or round balls. Its primary secretions are a group of hormones called **mineralocorticoids** (min′-er-al-ō-KOR-ti-koyds) because they affect mineral levels in the blood, especially sodium and potassium levels.

The middle zone, or **zona fasciculata** (*fasciculus* = little bundle), is the widest of the three cortical zones and consists of cells arranged in long, straight cords. The zona fasciculata secretes mainly **glucocorticoids** (gloo′-kō-KOR-ti-koyds), so named because they affect glucose metabolism.

The inner zone, the **zona reticularis** (*reticular* = net), contains cords of cells that branch freely. This zone synthe-

sizes minute amounts of **adrenocortical androgens** (steroid hormones that are masculinizing in their effects).

Adrenal Medulla

The adrenal medulla consists of hormone-producing cells, called **chromaffin** (krō-MAF-in; *chroma* = color; *affinis* = affinity for) **cells** (see Figure 22.7b), which surround large blood vessels. Chromaffin cells receive direct innervation from preganglionic neurons of the sympathetic division of the autonomic nervous system (ANS) and develop from the same source as all other sympathetic postganglionic cells.

SUPERFICIAL

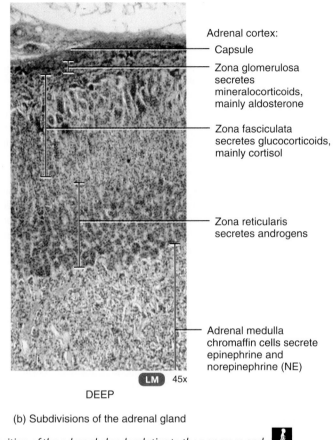

Adrenal cortex:

Capsule

Zona glomerulosa
secretes
mineralocorticoids,
mainly aldosterone

Zona fasciculata
secretes glucocorticoids,
mainly cortisol

Zona reticularis
secretes androgens

Adrenal medulla
chromaffin cells secrete
epinephrine and
norepinephrine (NE)

 45x

DEEP

(b) Subdivisions of the adrenal gland

Q *How would you describe the position of the adrenal glands relative to the pancreas and kidneys?*

Thus they are sympathetic postganglionic cells that are specialized to secrete hormones (epinephrine and norepinephrine). Because the ANS controls the chromaffin cells directly, hormone release can occur very quickly.

The two principal hormones synthesized by the adrenal medulla are **epinephrine** and **norepinephrine (NE),** also called adrenaline and noradrenaline, respectively. Epinephrine constitutes about 80% of the total secretion of the gland. Both hormones are **sympathomimetic** (sim′-pa-thō-mi-MET-ik); that is, they produce effects that mimic those brought about by the sympathetic division of the ANS. To a large extent, they are responsible for the fight-or-flight response. Like the glucocorticoids of the adrenal cortex, these hormones help the body resist stress. However, unlike the cortical hormones, the medullary hormones are not essential for life.

The main arteries that supply the adrenal glands are the several superior suprarenal arteries arising from the inferior phrenic artery, the middle suprarenal artery from the aorta, and the inferior suprarenal arteries from the renal arteries. The suprarenal vein of the right adrenal gland drains into the inferior vena cava, whereas the suprarenal vein of the left adrenal gland empties into the left renal vein (see Figure 22.7a).

The principal nerve supply to the adrenal glands is from preganglionic fibers from the thoracic splanchnic nerves, which pass through the celiac and associated sympathetic plexuses. These myelinated fibers end on the secretory cells of the gland found in a region of the medulla.

A summary of adrenal gland hormones and their actions is presented in Table 22.5.

Table 22.5 Summary of Adrenal Gland Hormones

Adrenal Cortical Hormones

HORMONE	PRINCIPAL ACTIONS
Mineralocorticoids (mainly aldosterone)	Increase blood levels of sodium and water and decrease blood levels of potassium.
Glucocorticoids (mainly cortisol)	Increase rate of protein catabolism (except in liver), stimulate conversion of other nutrients into glucose in liver (gluconeogenesis) and breakdown of triglycerides into fatty acids and glycerol (lipolysis), provide resistance to stress, dampen inflammation, and depress immune responses.
Adrenocortical androgens	Levels secreted by adult males are so low in comparison to amounts produced by testes that their effects are usually insignificant; in females, may contribute to sex drive and libido; assist in early development of axillary and pubic hair in both sexes.

Adrenal Medullary Hormones

HORMONE	PRINCIPAL ACTIONS
Epinephrine and norepinephrine from chromaffin cells	Sympathomimetic—that is, produce effects that mimic those of the sympathetic division of the autonomic nervous system (ANS) during stress.

PANCREAS

The **pancreas** (*pan* = all; *kreas* = flesh) is both an endocrine and an exocrine gland. We will treat its endocrine functions here and discuss its exocrine functions with the digestive system (Chapter 24). The pancreas is a flattened organ located posterior and slightly inferior to the stomach (Figure 22.8a). The adult pancreas consists of a head, body, and tail.

Roughly 99% of the pancreatic cells are arranged in clusters called **acini;** these cells produce digestive enzymes, which flow into the gastrointestinal tract through a network of ducts.

Scattered among the exocrine portions of the pancreas are one to two million tiny clusters of endocrine tissue called **pancreatic islets** or **islets of Langerhans** (LAHNG-er-hanz) (Figure 22.8b and c). Four types of hormone-secreting cells compose the pancreatic islets: (1) **alpha cells** secrete the hormone **glucagon** (GLOO-ka-gon), which raises blood sugar level; (2) **beta cells** secrete the hormone **insulin** (IN-soo-lin), which lowers blood sugar level; (3) **delta cells** secrete **somatostatin,** which inhibits the secretion of insulin and glucagon; and (4) **F-cells** secrete **pancreatic polypep-**

tide, which regulates release of pancreatic digestive enzymes. The islets are infiltrated by blood capillaries and surrounded by clusters of cells (acini) that form the exocrine part of the pancreas.

The arterial supply of the pancreas is from the superior and inferior pancreaticoduodenal arteries and from the splenic and superior mesenteric arteries. The veins, in general, correspond to the arteries. Venous blood reaches the hepatic portal vein by means of the splenic and superior mesenteric veins (Figure 22.8a).

The nerves to the pancreas are autonomic nerves derived from the celiac and superior mesenteric plexuses. Included are preganglionic vagal, postganglionic sympathetic, and sensory fibers. Parasympathetic vagal fibers are said to terminate at both acinar (exocrine) and islet (endocrine) cells. Although the innervation is presumed to influence enzyme formation, pancreatic secretion is controlled largely by the hormones secretin and cholecystokinin (CCK) released by the small intestine. The sympathetic fibers that enter the islets (and also end on blood vessels) are vasomotor and are accompanied by sensory fibers, especially for pain.

A summary of pancreatic hormones and their actions is presented in Table 22.6.

Figure 22.8 Location, blood supply, and histology of the pancreas.

Pancreatic hormones regulate blood glucose level.

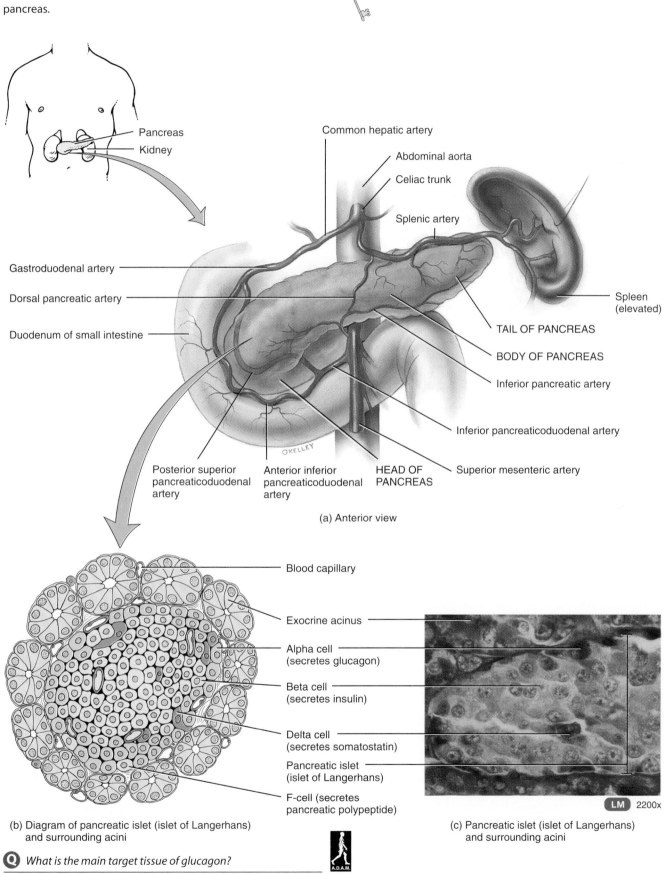

Pancreas
Kidney

Common hepatic artery
Abdominal aorta
Celiac trunk
Splenic artery

Gastroduodenal artery
Dorsal pancreatic artery
Duodenum of small intestine

Spleen (elevated)
TAIL OF PANCREAS
BODY OF PANCREAS
Inferior pancreatic artery
Inferior pancreaticoduodenal artery

OKELLEY

Posterior superior pancreaticoduodenal artery
Anterior inferior pancreaticoduodenal artery
HEAD OF PANCREAS
Superior mesenteric artery

(a) Anterior view

Blood capillary
Exocrine acinus
Alpha cell (secretes glucagon)
Beta cell (secretes insulin)
Delta cell (secretes somatostatin)
Pancreatic islet (islet of Langerhans)
F-cell (secretes pancreatic polypeptide)

LM 2200x

(b) Diagram of pancreatic islet (islet of Langerhans) and surrounding acini

(c) Pancreatic islet (islet of Langerhans) and surrounding acini

Q *What is the main target tissue of glucagon?*

Table 22.6 Summary of Hormones Produced by the Pancreas

HORMONE	PRINCIPAL ACTIONS
Glucagon from alpha cells	Raises blood glucose level by accelerating breakdown of glycogen into glucose in liver (glycogenolysis) and conversion of other nutrients into glucose in liver (gluconeogenesis), and by releasing glucose into blood.
Insulin from beta cells	Lowers blood glucose level by accelerating transport of glucose into cells, converting glucose into glycogen (glycogenesis), and decreasing glycogenolysis and gluconeogenesis; also increases lipogenesis and stimulates protein synthesis.
Somatostatin from delta cells	Inhibits secretion of insulin and glucagon.
Pancreatic polypeptide from F-cells	Regulates release of pancreatic digestive enzymes.

OVARIES AND TESTES

The female gonads, called the **ovaries,** are paired oval bodies located in the pelvic cavity. The ovaries produce female sex hormones called **estrogens** and **progesterone.** These hormones are responsible for the development and maintenance of female sexual characteristics. Along with the gonadotropic hormones of the pituitary gland, the sex hormones also regulate the female reproductive cycle, maintain pregnancy, and prepare the mammary glands for lactation. The ovaries also produce **inhibin,** a hormone that inhibits secretion of FSH (and, to a lesser extent, LH). Just before the birth of a baby, the ovaries and placenta also produce a hormone called **relaxin,** which relaxes the pubic symphysis and helps dilate the uterine cervix. These actions help ease the baby's passage through the birth canal.

The male has two oval gonads, called **testes,** that produce **testosterone,** the primary androgen (male sex hormone). Testosterone regulates production of sperm and stimulates the development and maintenance of male sexual characteristics. The testes also produce the hormone inhibin, which inhibits secretion of FSH. The detailed structure of the ovaries and testes will be discussed in Chapter 26.

Table 22.7 presents a summary of hormones produced by the ovaries and testes and their principal actions.

Table 22.7 Summary of Hormones of the Ovaries and Testes

Ovarian Hormones

HORMONE	PRINCIPAL ACTIONS
Estrogens and progesterone	Together with gonadotropic hormones of the anterior pituitary gland, regulate the female reproductive cycle, maintain pregnancy, prepare the mammary glands for lactation, regulate oogenesis, and promote development and maintenance of feminine secondary sex characteristics.
Relaxin	Increases flexibility of pubic symphysis during pregnancy and helps dilate uterine cervix during labor and delivery.
Inhibin	Inhibits secretion of FSH from anterior pituitary gland.

Testicular Hormones

HORMONE	PRINCIPAL ACTIONS
Testosterone	Stimulates descent of testes before birth, regulates spermatogenesis, and promotes development and maintenance of masculine secondary sex characteristics.
Inhibin	Inhibits secretion of FSH from the anterior pituitary gland.

PINEAL GLAND (EPIPHYSIS CEREBRI)

The endocrine gland attached to the roof of the third ventricle is known as the **pineal** (PĪN-ē-al) **gland** (named for its resemblance to a pine cone), or **epiphysis cerebri** (see Figure 22.1). The gland is covered by a capsule formed by the pia mater and consists of masses of **neuroglia** and secretory cells called **pinealocytes** (pin-ē-AL-ō-sīts). Sympathetic postganglionic fibers from the superior cervical ganglion terminate in the pineal gland.

Although many anatomical features of the pineal gland have been known for years, its physiological role is still unclear. One hormone secreted by the pineal gland is **melatonin.** It is believed to assume a primary role as the regulator of the body's internal clock, as a defense against damaging free radicals, as a promoter of sleep, and as a coordinator of the hormones involved in fertility.

The posterior cerebral artery supplies the pineal gland with blood, and the great cerebral vein drains it.

Table 22.8 Summary of Hormones Produced
by the Thymus Gland

HORMONES	PRINCIPAL ACTIONS
Thymosin Thymic humoral factor (THF) Thymic factor (TF) Thymopoietin	Promote proliferation and maturation of T cells (a type of white blood cell), which destroy microbes and foreign substances; may retard the aging process.

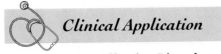

Clinical Application

Seasonal Affective Disorder

Seasonal affective disorder (SAD) is a type of depression that arises during the winter months when day-length is short. It is thought to be due, in part, to overproduction of melatonin. It is characterized by headaches, irritability, and low energy level. Bright light therapy—repeated doses of several hours of exposure to artificial light as bright as sunlight—may provide relief.

THYMUS GLAND

Because of its role in immunity, the details of the structure and functions of the **thymus gland** are discussed in Chapter 15, which deals with the lymphatic system and immunity. At this point, only its hormonal role in immunity will be discussed.

Lymphocytes are one type of white blood cell. These, in turn, are divided into two types, called T cells and B cells, based on their specific roles in immunity. Hormones produced by the thymus gland, called **thymosin, thymic humoral factor (THF), thymic factor (TF),** and **thymopoietin,** promote the proliferation and maturation of T cells, which destroy foreign microbes and substances. There is also some evidence that thymic hormones may retard the aging process.

A summary of hormones produced by the thymus gland and their actions is presented in Table 22.8.

OTHER ENDOCRINE TISSUES

Before leaving our discussion of hormones, it should be noted that body cells other than those usually classified as endocrine glands also contain endocrine cells and thus secrete hormones. These are summarized in Table 22.9.

Table 22.9 Summary of Hormones Produced
by Tissues that Contain Endocrine Cells

HORMONES	PRINCIPAL ACTIONS
Gastrointestinal Tract	
Gastrin	Promotes secretion of gastric juice and increases movements of the gastrointestinal tract.
Gastric inhibitory peptide (GIP)	Inhibits secretion of gastric juice, decreases movements of the gastrointestinal tract, and stimulates release of insulin by pancreas.
Secretin	Stimulates secretion of pancreatic juice and bile.
Cholecystokinin (CCK)	Stimulates secretion of pancreatic juice, regulates release of bile from the gallbladder, and brings about a feeling of fullness after eating.
Placenta	
Human chorionic gonadotropin (hCG)	Stimulates the corpus luteum in the ovary to continue the production of estrogens and progesterone to maintain pregnancy.
Estrogens and progesterone	Maintain pregnancy and prepare mammary glands to secrete milk.
Human chorionic somatomammotropin (hCS)	Stimulates the development of the mammary glands for lactation.
Kidneys	
Erythropoietin (EPO)	Increases rate of red blood cell production.
Calcitriol[a] (active form of vitamin D)	Aids in the absorption of dietary calcium and phosphorus.
Heart	
Atrial nutriuretic peptide (ANP)	Decreases blood pressure.

[a]Synthesis begins in the skin, continues in the liver, and ends in the kidneys.

Figure 22.9 Development of the endocrine system.

Glands of the endocrine system develop from all three primary germ layers.

Q *Why is the development of the endocrine system not as localized as that of the other body systems?*

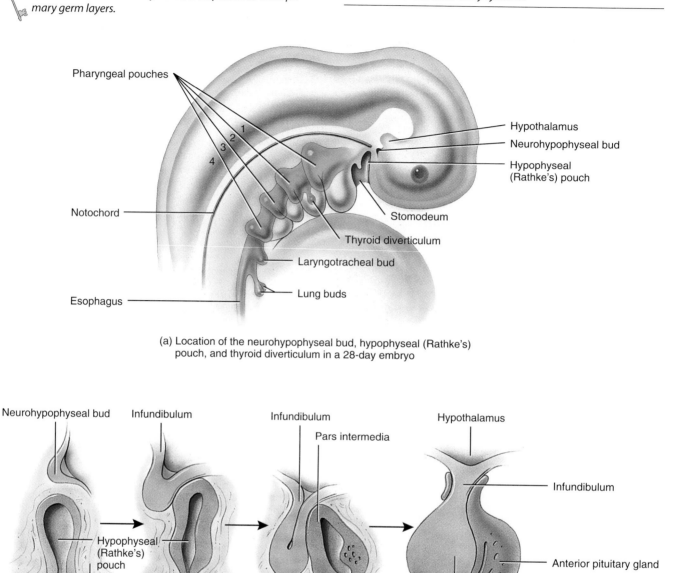

(a) Location of the neurohypophyseal bud, hypophyseal (Rathke's) pouch, and thyroid diverticulum in a 28-day embryo

(b) Development of the pituitary gland between 5 and 16 weeks

Developmental Anatomy of the Endocrine System

The development of the endocrine system is not as localized as the development of other systems because endocrine organs develop in widely separated parts of the embryo.

The *pituitary gland (hypophysis)* originates from two different regions of the **ectoderm.** The *posterior pituitary gland (neurohypophysis)* is derived from an outgrowth of ectoderm called the **neurohypophyseal bud,** located on the floor of the hypothalamus (Figure 22.9). The *infundibulum,* also an outgrowth of the neurohypophyseal bud, connects the posterior pituitary gland to the hypothalamus. The *anterior pituitary gland (adenohypophysis)* is derived from an outgrowth of **ectoderm** from the roof of the mouth called the **hypophyseal (Rathke's) pouch.** The pouch grows toward the neurohypophyseal bud and eventually loses its connection with the roof of the mouth.

The *thyroid gland* develops as a midventral outgrowth of **endoderm,** called the **thyroid diverticulum,** from the floor of the pharynx at the level of the second pair of pharyngeal pouches (Figure 22.9a). The outgrowth projects inferiorly and differentiates into the right and left lateral lobes and the isthmus of the gland. The *parathyroid glands* develop from **endoderm** as outgrowths from the third and fourth **pharyngeal pouches** (Figure 22.9b).

The adrenal cortex and adrenal medulla have completely different embryological origins. The *adrenal cortex* is derived from intermediate **mesoderm** from the same region that produces the gonads (see Figure 26.24). The *adrenal medulla* is **ectodermal** in origin and is derived from the **neural crest,** which also produces sympathetic ganglia and other structures of the nervous system (see Figure 18.25b).

The *pancreas* develops from **endoderm** from dorsal and ventral outgrowths of the part of the **foregut** that later becomes the duodenum (see Figure 24.21b). The two outgrowths eventually fuse to form the pancreas. The origin of the ovaries and testes is discussed in Chapter 25 on the reproductive systems.

The *pineal gland* arises from **ectoderm** of the **diencephalon** (see Figure 18.25b), as an outgrowth between the thalamus and colliculi.

The *thymus gland* arises from **endoderm** from the third **pharyngeal pouches** (Figure 22.9a). ■

Aging and the Endocrine System

Although endocrine glands shrink as we get older, their performance may or may not be compromised. With respect to the pituitary gland, production of human growth hormone decreases, representing one factor in muscle atrophy as aging proceeds. But production of gonadotrophins and thyroid-stimulating hormone actually increases with age. Output of adrenocorticotrophic hormone by the pituitary gland is seemingly unaffected by age. The thyroid gland decreases its output of thyroxin with age, causing a decrease in metabolic rate, an increase in body fat, and hypothyroidism, which is seen more often in older people.

The adrenal glands contain increasingly more fibrous tissue and produce less cortisol and aldosterone. However, production of epinephrine and norepinephrine remains normal. The pancreas releases insulin more slowly with age, and receptor sensitivity to glucose declines. As a result, blood glucose levels increase faster and return to normal levels more slowly than in younger individuals.

The ovaries experience a drastic decrease in size with age and no longer respond to gonadotrophins. The resultant decreased output of estrogen by the ovaries leads to conditions such as osteoporosis and arteriosclerosis. Although testosterone production by the testes decreases with age, the effects of this decrease are not usually apparent until very old age, and many elderly males can still produce active sperm in normal numbers.

Key Medical Terms
Associated with the Endocrine System

Acromegaly (ak′-rō-MEG-a-lē; *acro* = limb; *mega* = large) A disorder caused by hypersecretion of hGH during adulthood. Because the epiphyseal plates are already closed, further lengthening of bones does not occur, but the bones thicken. This is apparent in the head, hands, and feet.

Diabetes insipidus (dī-a-BĒ-tez in-SIP-i-dus; *diabetes* = overflow; *insipidus* = tasteless) A disorder caused by hyposecretion of ADH by the posterior pituitary gland that is characterized by excretion of large amounts of urine (10 liters or more per day), dehydration, and thirst. This disorder should not be confused with diabetes mellitus, which is characterized by the presence of glucose in the urine.

Gigantism (giantism) A disorder caused by hypersecretion of hGH before puberty and closure of the epiphyseal plates. It results in an abnormal increase in the length of long bones, and the person is unusually tall, typically more than seven feet.

Gynecomastia (gī-ne-kō-MAS-tē-a, *gyneca* = woman; *mast* = breast) Excessive development of male mammary glands. Sometimes a tumor of the adrenal gland may secrete sufficient quantities of estrogen to cause the condition.

Pheochromocytoma (fē-ō-krō′-mō-sī-TŌ-ma; *pheochrome* = dusky color; *cyto* = cell; *oma* = tumor) A usually benign tumor of chromaffin cells of the adrenal medulla that results in the hypersecretion of medullary hormones. It is characterized by headache, rapid heart rate, high blood pressure, high glucose levels, elevated basal metabolic rate, flushing, nervousness, sweating, and decreased gastrointestinal motility. The individual experiences a prolonged version of the sympathetic flight-or-fight response.

Pituitary dwarfism A disorder caused by the hyposecretion of hGH during the growth years that results in slow bone growth and premature closure of the epiphyseal plates before normal height is reached.

Thyroid crisis (storm) A severe state of hyperthyroidism that can be life-threatening. It is characterized by high body temperature, rapid heart rate, high blood pressure, gastrointestinal symptoms (abdominal pain, vomiting, diarrhea), agitation, tremors, confusion, seizures, and possibly coma.

Virilism (VIR-il-izm; *vivilis* = masculine) The presence of mature masculine characteristics in females or prepubescent males. In a female, virile characteristics include growth of a beard, devel-

opment of a much deeper voice, occasionally the development of baldness, development of a masculine distribution of hair on the body and on the pubis, growth of the clitoris such that it may resemble a penis, atrophy of the breasts, infrequent or absent menstruation, and increased muscularity that produces a male-like physique. In prepubescent males the syndrome causes the same characteristics as in females, plus rapid development of the male sexual organs and emergence of sexual desires.

Study Outline

Endocrine Glands (p. 658)

1. The nervous system coordinates body activities through nerve impulses; the endocrine system uses hormones.
2. The nervous system causes muscles to contract and glands to secrete; the endocrine system affects virtually all body tissues.
3. Exocrine glands (sudoriferous, sebaceous, digestive) secrete their products through ducts into body cavities or onto body surfaces.
4. Endocrine glands secrete hormones into the blood.
5. The endocrine system consists of endocrine glands and several organs that contain endocrine tissue (see Figure 22.1 on page 659).

Hormones (p. 659)

1. The amount of hormone released is determined by the body's need for the hormone.
2. Cells that respond to the effects of hormones are called target cells.
3. The combination of hormone and receptor activates a chain of events in a target cell in which the physiological effects of the hormone are expressed.

Hypothalamus and Pituitary Gland (Hypophysis) (p. 660)

1. The hypothalamus is the major integrating link between the nervous and endocrine systems.
2. The hypothalamus and pituitary gland regulate virtually all aspects of growth, development, metabolism, and other body activities.
3. The pituitary gland is located in the sella turcica and is divided into the anterior pituitary gland (glandular portion), the posterior pituitary gland (nervous portion), and pars intermedia (avascular zone in between).
4. Hormones of the anterior pituitary gland are controlled by releasing or inhibiting hormones produced by the hypothalamus.
5. The blood supply to the anterior pituitary gland is from the superior hypophyseal arteries. It carries releasing and inhibiting hormones to the anterior pituitary gland.
6. Histologically, the anterior pituitary consists of somatotrophs that produce human growth hormone (hGH); lactotrophs that produce prolactin (PRL); corticotrophs that secrete adrenocorticotropic hormone (ACTH) and melanocyte-stimulating

hormone (MSH); thyrotrophs that secrete thyroid-stimulating hormone (TSH); and gonadotrophs that synthesize follicle-stimulating hormone (FSH) and luteinizing hormone (LH).
7. hGH stimulates body growth.
8. TSH regulates thyroid gland activities.
9. FSH regulates the activities of the ovaries and testes.
10. LH regulates the activities of the ovaries and testes.
11. PRL helps initiate milk secretion.
12. MSH increases skin pigmentation.
13. ACTH regulates the activities of the adrenal cortex.
14. The neural connection between the hypothalamus and posterior pituitary gland is via the supraopticohypophyseal tract.
15. Hormones made by the hypothalamus and stored in the posterior pituitary gland are oxytocin (OT), which stimulates contraction of uterus and ejection of milk, and antidiuretic hormone (ADH), which stimulates water reabsorption by the kidneys and arteriole constriction.

Thyroid Gland (p. 664)

1. The thyroid gland is located inferior to the larynx.
2. Histologically, the thyroid gland consists of thyroid follicles composed of follicular cells, which secrete the thyroid hormones thyroxine (T_4) and triiodothyronine (T_3), and parafollicular cells, which secrete calcitonin (CT).
3. Thyroid hormones regulate the rate of metabolism, growth and development, and the reactivity of the nervous system.
4. Calcitonin (CT) lowers the blood level of calcium.

Parathyroid Glands (p. 666)

1. The parathyroid glands are embedded on the posterior surfaces of the lateral lobes of the thyroid.
2. The parathyroids consist of principal and oxyphil cells.
3. Parathyroid hormone (PTH) increases blood calcium level and decreases blood phosphate level.

Adrenal (Suprarenal) Glands (p. 666)

1. The adrenal glands are located superior to the kidneys. They consist of an outer adrenal cortex and inner adrenal medulla.
2. Histologically, the adrenal cortex is divided into a zona glomerulosa, zona fasciculata, and zona reticularis; the adrenal medulla consists of chromaffin cells and large blood vessels.
3. Cortical secretions are mineralocorticoids, glucocorticoids, and gonadocorticoids.

4. Mineralocorticoids (for example, aldosterone) increase sodium and water reabsorption and decrease potassium reabsorption.

5. Glucocorticoids (for example, cortisol) promote normal organic metabolism, help resist stress, and serve as antiinflammatory substances.

6. Androgens secreted by the adrenal cortex usually have minimal effects.

7. Medullary secretions are epinephrine and norepinephrine (NE), which produce effects similar to sympathetic responses. They are released during stress.

Pancreas (p. 670)

1. The pancreas is posterior and slightly inferior to the stomach.

2. Histologically, it consists of pancreatic islets or islets of Langerhans (endocrine cells) and clusters of enzyme-producing cells (acini). Four types of cells in the endocrine portion are alpha cells, beta cells, delta cells, and F-cells.

3. Alpha cells secrete glucagon, beta cells secrete insulin, delta cells secrete somatostatin, and F-cells secrete pancreatic polypeptide.

4. Glucagon increases blood sugar level.

5. Insulin decreases blood sugar level.

Ovaries and Testes (p. 672)

1. The ovaries are located in the pelvic cavity and produce sex hormones related to development and maintenance of female sexual characteristics, reproductive cycle, pregnancy, lactation, and normal reproductive functions.

2. The testes lie inside the scrotum and produce sex hormones related to the development and maintenance of male sexual characteristics and normal reproductive functions.

Pineal Gland (Epiphysis Cerebri) (p. 672)

1. The pineal gland is attached to the roof of the third ventricle.

2. Histologically, it consists of secretory cells called pinealo-cytes, neuroglial cells, and scattered postganglionic sympathetic fibers.

3. The pineal gland secretes melatonin in a diurnal rhythm linked to the dark–light cycle.

Thymus Gland (p. 673)

1. The thymus gland secretes several hormones related to immunity.

2. Thymosin, thymic humoral factor (THF), thymic factor (TF), and thymopoietin promote the maturation of T cells.

Other Endocrine Tissues (p. 673)

1. The gastrointestinal tract synthesizes several hormones, including gastrin, gastric inhibitory peptide (GIP), secretin, and cholecystokinin (CCK).

2. The placenta produces human chorionic gonadotropin (hCG), estrogens, progesterone, and human chorionic somatomammotropin (hCS).

3. The kidneys release erythropoietin.

4. The skin begins the synthesis of vitamin D.

5. The atria of the heart produce atrial natriuretic peptide (ANP).

Developmental Anatomy of the Endocrine System (p. 674)

1. The development of the endocrine system is not as localized as other systems.

2. The pituitary gland, adrenal medullae, and pineal gland develop from ectoderm; the adrenal cortex develops from mesoderm; and the thyroid gland, parathyroid glands, pancreas, and thymus gland develop from endoderm.

Aging and the Endocrine System (p. 675)

1. Although endocrine glands shrink with age, their performance may not necessarily be compromised.

2. Whereas the pituitary gland produces less human growth hormone, it produces more gonadotropins and thyroid-stimulating hormone.

Review Questions

1. Contrast the nervous and endocrine systems' control of body activities. (p. 658)

2. Distinguish between an endocrine gland and an exocrine gland. (p. 658)

3. What are the principal functions of hormones? (p. 659)

4. How are hormones related to receptors? (p. 659)

5. In what respect is the pituitary gland actually two glands? (p. 660)

6. Describe the histology of the anterior pituitary gland. Why does the anterior pituitary gland have such an abundant blood supply? (p. 660)

7. What hormones are produced by the anterior pituitary gland? What are their functions? (p. 664)

8. Discuss the histology of the posterior pituitary gland and the function of its hormones. Describe the structure and importance of the supraopticohypophyseal tract. (p. 662)

9. Describe the location and histology of the thyroid gland. (p. 664)

10. Discuss the physiological effects of the thyroid hormones. (p. 666)

11. Describe the function of calcitonin (CT). (p. 666)

12. Where are the parathyroid glands located? What is their histology? (p. 666)

13. What are the functions of parathyroid hormone (PTH)? (p. 666)

14. Compare the adrenal cortex and adrenal medulla with regard to location and histology. (p. 666)

15. Describe the hormones produced by the adrenal cortex in terms of type and function. (p. 670)

16. What relationship does the adrenal medulla have to the autonomic nervous system? What is the action of adrenal medullary hormones? (p. 668)

17. Describe the location of the pancreas and the histology of the pancreatic islets. (p. 670)

18. What are the actions of glucagon, insulin, somatostatin, and pancreatic polypeptide? (p. 672)

19. Why are the ovaries and testes considered to be endocrine glands? (p. 672)

20. Where is the pineal gland located? What are its presumed functions? (p. 672)

21. How are hormones of the thymus gland related to immunity? (p. 673)

22. List the hormones secreted by the gastrointestinal tract, placenta, kidneys, skin, and heart. (p. 673)

23. Describe the development of the endocrine system. (p. 674)

24. Describe the effects of aging on the endocrine system. (p. 675)

25. Refer to the glossary of key medical terms associated with the endocrine system and be sure that you can define each term. (p. 675)

Self Quiz

Complete the following:

1. Place numbers in the blanks to correctly order the vessels that supply blood to the anterior pituitary gland:
(a) hypophyseal portal veins: ___; (b) primary plexus: ___; (c) superior hypophyseal arteries: ___; (d) secondary plexus: ___.

2. Place numbers in the blanks to indicate the correct order of items involved in the route of a hormone. (Note: Not all items apply to this route.)
(a) blood capillary: ___; (b) target cells: ___; (c) duct: ___; (d) outer surface of the body or lumen of an organ: ___; (e) interstitial fluid: ___; (f) secretory cell: ___.

3. Thyroid stimulating hormone (TSH) controls the secretion of hormones by follicular cells of the thyroid gland. Therefore, TSH is a ___ hormone, and the follicular cells are the ___ cells.

4. Glucagon, produced by ___ cells of the pancreatic islets, causes blood sugar level to ___.

5. A stalk called the ___ attaches the pituitary gland to the hypothalamus. The pituitary gland lies in the ___ of the sphenoid bone.

6. Another term for posterior pituitary gland is ___.

7. The ___ cells of the thyroid gland and the ___ cells of the parathyroid glands secrete hormones that help to regulate the level of calcium in the blood.

8. Match the following terms with their definitions:

___ (a) adrenal cortex
___ (b) adrenal medulla
___ (c) anterior pituitary gland
___ (d) pineal gland
___ (e) parathyroid glands
___ (f) thymus gland
___ (g) pancreas

(1) develops from the foregut area that later becomes part of the small intestine
(2) derived from the roof of the stomodeum (mouth) called the hypophyseal (Rathke's) pouch
(3) originates from the neural crest, which also produces sympathetic ganglia
(4) derived from tissue from the same region that forms gonads
(5) arise from the third and fourth pharyngeal pouches (two answers)
(6) arises from the ectoderm of the diencephalon

Are the following statements true or false?

9. The secretory portion of the posterior pituitary gland consists of axons of neurosecretory cells; the secretory portion of the anterior pituitary gland is glandular epithelium.

10. The middle region of hormone-secreting cells of the adrenal cortex is called the zona reticularis.

11. The gland that assumes a role in the maturation of T cells is the pineal gland.

12. Chromaffin cells are the principal secreting cells of the pancreas.

13. The outer region of the adrenal (suprarenal) gland is called the adrenal cortex.

Choose the one best answer to the following questions:

14. A generalized anti-inflammatory reaction is most closely associated with
a. glucocorticoids
b. mineralocorticoids
c. parathyroid hormone (PTH)
d. insulin
e. melatonin

15. A chemical produced by the hypothalamus that causes the anterior pituitary gland to secrete a hormone is called a
a. gonadotropic hormone
b. tropic hormone
c. releasing hormone
d. target hormone
e. neurotransmitter

16. A tumor of the beta cells of the pancreatic islets would probably affect the body's ability to
a. lower blood sugar level
b. raise blood sugar level
c. lower blood calcium level
d. raise blood calcium level
e. regulate metabolism

17. Which of the following hormones is involved in milk production?
a. melanocyte-stimulating hormone (MSH)
b. follicle-stimulating hormone (FSH)
c. prolactin (PRL)
d. glucagon
e. parathyroid hormone (PTH)

18. Which of the following six glands consists of two anatomically and functionally separate portions (two endocrine glands in one structure)?
    (1) ovary
    (2) pituitary gland   a. 1, 2, 3, 4
    (3) thyroid gland   b. 2, 3, 4, 5
    (4) adrenal glands   c. 1, 3, 5, 6
    (5) pancreas   d. 2, 4
    (6) parathyroid glands   e. 3, 5
19. What the stomach, pancreas, testes, and ovaries have in common is that they
    a. are influenced by hormones from the anterior pituitary
    b. have tissues derived from embryological ectoderm
    c. form hormones that influence secondary sex characteristics
    d. receive their blood supply from the superior mesenteric artery
    e. are considered to be both exocrine and endocrine glands
20. Which of the following endocrine glands consists of lobes made up of follicles that secrete hormones?
    a. pituitary gland   d. adrenal medulla
    b. parathyroid gland   e. pancreatic islets (islets of
    c. thyroid gland   Langerhans)
21. Which of the following glands is called the "emergency gland" because it helps the body deal with sudden stress?
    a. pituitary gland   d. thymus gland
    b. pancreas   e. adrenal (suprarenal) glands
    c. thyroid gland

Critical Thinking Questions

1. The hormone melatonin is being sold as a sleep aid and a cure for jet lag. If melatonin had to be harvested from the gland that secretes it, where would you have to go?
   HINT: *Look in the brain.*
2. Mandy hates her new student ID photo. Her hair looks dry, the ten pounds she's gained shows, and her neck looks fat. In fact, there's an odd butterfly-shaped swelling across the front of her neck, under her chin. Mandy's also been feeling very tired and mentally "dull" lately but figures all new anatomy students feel that way. Should she visit the clinic or just wear turtlenecks?
   HINT: *She also constantly feels cold.*
3. As Indira walked into her Human Anatomy class (ten minutes late), she was handed an exam that she completely forgot about when she decided to stay an extra week at the beach. Predict how her body will react to this situation.
   HINT: *The stress she's feeling just wiped out the relaxed mood from the extra week at the beach.*
4. Beatrice had a large bag of double chunk chocolate chip cookies and a 16-oz. cola for lunch, but no sandwich because "she's on a diet." Explain how her endocrine system will respond to this lunch.
   HINT: *This meal turns the "food pyramid" into the "sugar cube."*
5. For several years, military pilots were given radiation treatments in their nasal cavities to reduce sinus problems, which interfere with flying. Years later, some of these former pilots began to experience problems associated with their pituitary gland. Can you propose an explanation for this relationship?
   HINT: *Refer to the structure of the skull in Chapter 6.*
6. Pei is a 15-year-old computer whiz with an interest in medicine. He likes to practice virtual surgery with a computer program he wrote himself. Today, he's transplanting an adrenal gland. Describe the location and structure of the adrenal glands.
   HINT: *It's really two glands in one.*

Answers to Figure Questions

22.1 Secretions of endocrine glands diffuse into the blood; exocrine secretions flow into ducts that lead into body cavities or to the body surface.
22.2 They carry blood from the median eminence of the hypothalamus, where hypothalamic releasing and inhibiting hormones are secreted, to the anterior pituitary gland, where these hormones act.
22.3 Thyroid gland, adrenal cortex, ovaries, testes.
22.4 Similarity: both carry hypothalamic hormones to the pituitary gland. Difference: tract is axons of neurons that extend from hypothalamus to posterior pituitary gland; portal veins extend to anterior pituitary gland.
22.5 Follicular cells secrete T_3 and T_4, also known as thyroid hormones. Parafollicular (C) cells secrete calcitonin.
22.6 Parafollicular cells secrete CT; principal cells secrete PTH.
22.7 The adrenal glands are situated superior to the kidneys in the retroperitoneal space, at about the same horizontal level as the pancreas.
22.8 Liver.
22.9 Because endocrine organs develop in widely separated parts of the embryo.

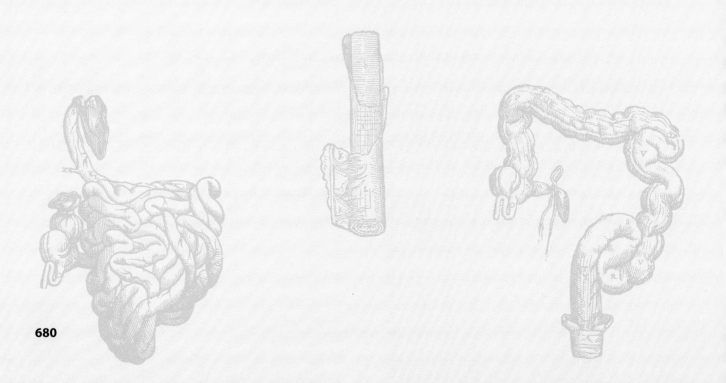

Chapter 23

The Respiratory System

Student Objectives

1. Identify the organs of the respiratory system.

2. Compare the structure and function of the external and internal nose.

3. Differentiate the three anatomical regions of the pharynx and describe their roles in respiration.

4. Describe the structure of the larynx and explain its function in respiration and voice production.

5. Describe the location and structure of the tubes that form the bronchial tree.

6. Describe the coverings of the lungs, the division of the lungs into lobes, and the composition of a lobule of the lung.

7. Explain the structure of the alveolar–capillary (respiratory) membrane and its function in the diffusion of respiratory gases.

8. Describe the development of the respiratory system.

9. Describe the effects of aging on the respiratory system.

Cells continually use oxygen (O_2) for the metabolic reactions that release energy from nutrient molecules and produce ATP. At the same time, these reactions release carbon dioxide (CO_2). Because an excessive amount of CO_2 produces acidity that is toxic to cells, the excess CO_2 must be eliminated quickly and efficiently. The two systems that supply oxygen and eliminate carbon dioxide are the cardiovascular system and the respiratory system. The respiratory system provides for gas exchange, intake of O_2, and elimination of CO_2, whereas the cardiovascular system transports the gases in the blood between the lungs and the cells. Failure of either system has the same effect on the body: rapid death of cells from oxygen starvation and buildup of waste products. In addition to functioning in gas exchange, the respiratory system also contains receptors for the sense of smell, filters inspired air, produces sounds, and helps eliminate some wastes.

The **respiratory system** consists of the nose, pharynx (throat), larynx (voice box), trachea (windpipe), bronchi, and lungs (Figure. 23.1a). Structurally, the respiratory system has two divisions. (1) The **upper respiratory system** includes the nose, pharynx, and their associated structures. (2) The **lower respiratory system** includes the larynx, trachea, bronchi, and lungs. Functionally, the respiratory system also consists of two portions. (1) The **conducting portion** consists of a series of interconnecting cavities and tubes—nose, pharynx, larynx, trachea, bronchi, bronchioles, and terminal bronchioles—that conduct air into the lungs. (2) The **respiratory portion** consists of the structures in which gas exchange occurs—respiratory bronchioles, alveolar ducts, alveolar sacs, and alveoli.

The branch of medicine that deals with the diagnosis and treatment of diseases of the ears, nose, and throat is called **otorhinolaryngology** (ō′-tō-rī′-nō-lar′-in-GOL-ō-jē; *oto* = ear; *rhino* = nose).

The first goal of this chapter is to examine the structures through which air passes between the atmosphere and lungs. Then we will discuss the mechanics of breathing and the various factors that control breathing.

ORGANS OF RESPIRATION

Nose

The **nose** has an external portion and an internal portion inside the skull. The external portion consists of a supporting framework of bone and hyaline cartilage covered with muscle and skin and lined by mucous membrane (Figure 23.2a). The bridge of the nose is formed by the nasal bones, which hold it in a fixed position. Other bones that contribute to the bony framework are the maxillae and frontal bone. The car-tilaginous framework consists of the **septal cartilage,** which forms the anterior portion of the nasal septum (see Figure 6.10a), the **lateral nasal cartilages,** inferior to the nasal bones, and the **alar cartilages,** which form a portion of the walls of the nostrils. Because it has a framework of pliable hyaline cartilage, the rest of the external nose is somewhat flexible. On the undersurface of the external nose are two openings called the **external nares** (NA-rēz; singular is **naris**), or **nostrils.** The surface anatomy of the nose is shown in Figure 11.5.

The internal portion of the nose is a large cavity in the skull that lies inferior to the anterior cranium and superior to the mouth (Figure 23.2b and c). It also consists of muscle and mucous membrane. Anteriorly, the internal nose merges with the external nose, and posteriorly it communicates with the pharynx through two openings called the **internal nares (choanae).** Ducts from the paranasal sinuses (frontal, sphenoidal, maxillary, and ethmoidal) and the nasolacrimal ducts also open into the internal nose. The lateral walls of the internal nose are formed by the ethmoid, maxillae, lacrimal, palatine, and inferior nasal conchae bones. The ethmoid also forms the roof. The floor of the internal nose is formed mostly by the palatine bones and palatine processes of the maxillae, which together constitute the hard palate.

The inside of both the external and internal nose is called the **nasal cavity.** It is divided into right and left sides by a vertical partition called the **nasal septum.** The anterior portion of the septum consists primarily of hyaline cartilage. The remainder is formed by the vomer, perpendicular plate of the ethmoid, maxillae, and palatine bones (see Figure 6.10a). The anterior portion of the nasal cavity, just inside the nostrils, is called the **vestibule** and is surrounded by cartilage. The superior part of the nasal cavity is surrounded by bone.

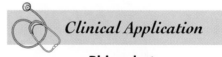

Clinical Application

Rhinoplasty

Rhinoplasty (RĪ-nō-plas′-tē; *rhino* = nose; *plassein* = to form), commonly called a "nose job," is a surgical procedure in which the structure of the external nose is altered. Although it is frequently done for cosmetic reasons, it is sometimes performed to repair a fractured nose or deviated nasal septum. In the procedure, a local or general anesthetic is given, and with instruments inserted through the nostrils, the nasal bones are fractured and repositioned to achieve the desired shape. Any extra bone fragments or cartilage are shaved down and removed through the nostrils. Both an internal packing and an external splint keep the nose in the desired position while it heals. In some cases, only the cartilage must be reshaped to achieve the desired appearance. ■

Figure 23.1 Structures of the respiratory system.

The upper respiratory system includes the nose, pharynx, and associated structures. The lower respiratory system includes the larynx, trachea, bronchi, and lungs.

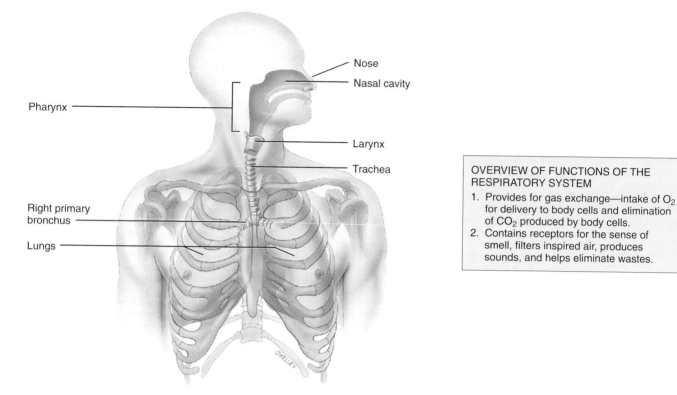

OVERVIEW OF FUNCTIONS OF THE
RESPIRATORY SYSTEM

1. Provides for gas exchange—intake of O_2 for delivery to body cells and elimination of CO_2 produced by body cells.
2. Contains receptors for the sense of smell, filters inspired air, produces sounds, and helps eliminate wastes.

(a) Diagram of anterior view showing organs of respiration

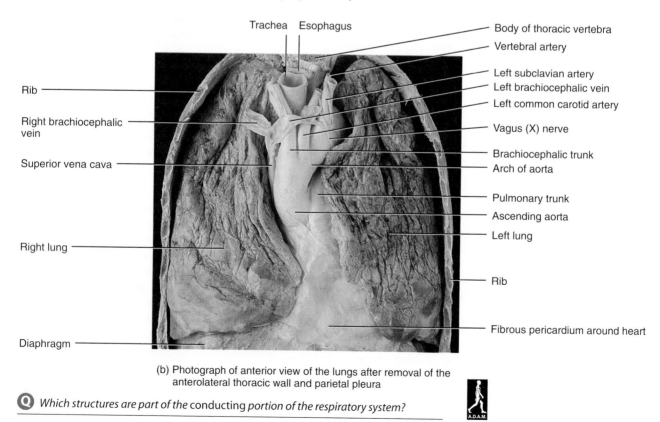

(b) Photograph of anterior view of the lungs after removal of the anterolateral thoracic wall and parietal pleura

Q *Which structures are part of the conducting portion of the respiratory system?*

Figure 23.2 Respiratory organs in the head and neck.

As air passes through the nose, it is warmed, filtered, and moistened.

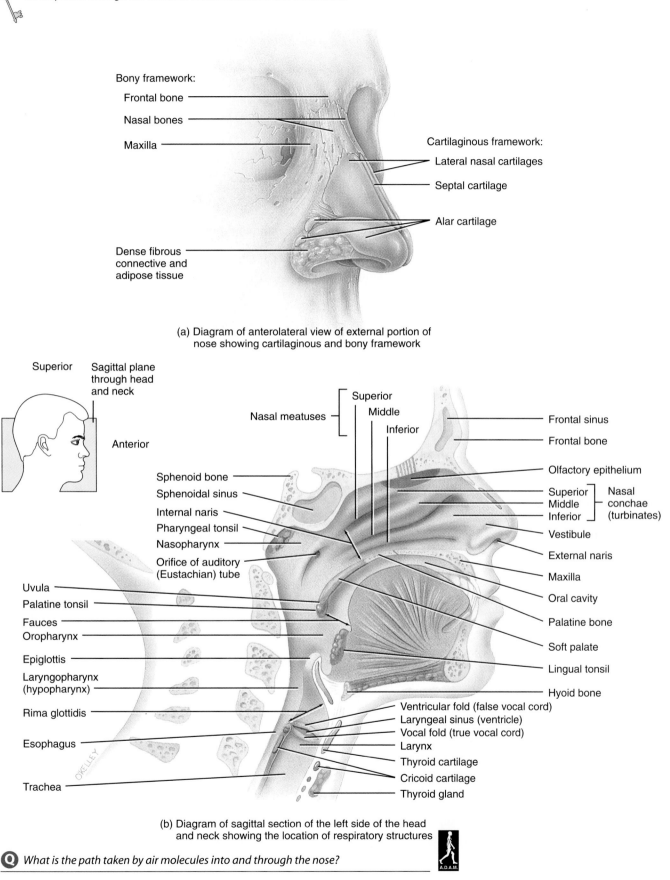

Bony framework:
Frontal bone
Nasal bones
Maxilla

Cartilaginous framework:
Lateral nasal cartilages
Septal cartilage
Alar cartilage

Dense fibrous connective and adipose tissue

(a) Diagram of anterolateral view of external portion of nose showing cartilaginous and bony framework

Superior

Sagittal plane through head and neck

Anterior

Nasal meatuses — Superior, Middle, Inferior

Superior
Middle
Inferior

Frontal sinus
Frontal bone
Olfactory epithelium

Superior
Middle
Inferior
Nasal conchae (turbinates)

Vestibule
External naris
Maxilla
Oral cavity
Palatine bone
Soft palate
Lingual tonsil
Hyoid bone

Sphenoid bone
Sphenoidal sinus
Internal naris
Pharyngeal tonsil
Nasopharynx
Orifice of auditory (Eustachian) tube

Uvula
Palatine tonsil
Fauces
Oropharynx
Epiglottis
Laryngopharynx (hypopharynx)
Rima glottidis
Esophagus
Trachea

Ventricular fold (false vocal cord)
Laryngeal sinus (ventricle)
Vocal fold (true vocal cord)
Larynx
Thyroid cartilage
Cricoid cartilage
Thyroid gland

(b) Diagram of sagittal section of the left side of the head and neck showing the location of respiratory structures

Q *What is the path taken by air molecules into and through the nose?*

The interior structures of the nose are specialized for three functions: (1) incoming air is warmed, moistened, and filtered; (2) olfactory stimuli are received; and (3) large, hollow resonating chambers modify speech sounds.

When air enters the nostrils, it passes first through the vestibule. The vestibule is lined by skin containing coarse hairs that filter out large dust particles. The air then passes into the superior nasal cavity. Three shelves formed by projections of the **superior, middle,** and **inferior nasal conchae** extend out of each lateral wall of the cavity (Figure 23.2 b–d). The conchae, almost reaching the septum, subdivide each side of the nasal cavity into a series of groovelike passageways—the **superior, middle,** and **inferior meatuses** (singular is **meatus**). Mucous membrane lines the cavity and its shelves.

The olfactory receptors lie in the membrane lining the superior nasal conchae and adjacent septum. This region is called the **olfactory epithelium.** Inferior to the olfactory epithelium, the mucous membrane contains capillaries and pseudostratified ciliated columnar epithelium with many goblet cells. As the air whirls around the conchae and meatuses, it is warmed by blood in the capillaries. Mucus secreted by the goblet cells moistens the air and traps dust particles. Drainage from the nasolacrimal ducts and perhaps secretions from the paranasal sinuses also help moisten the air. The cilia move the mucus–dust packages toward the pharynx so they can be eliminated from the respiratory tract by swallowing or expectoration (spitting).

The arterial supply to the nasal cavity is principally from the sphenopalatine branch of the maxillary artery. The remainder is supplied by the ophthalmic artery. The veins of the nasal cavity drain into the sphenopalatine vein, the facial vein, and the ophthalmic vein.

The nerve supply of the nasal cavity consists of olfactory cells in the olfactory epithelium associated with the olfactory (I) nerve and the nerves of general sensation. These nerves are branches of the ophthalmic and maxillary divisions of the trigeminal (V) nerve.

Pharynx

The **pharynx** (FAIR-inks), or throat, is a somewhat funnel-shaped tube about 13 cm (5 in.) long that starts at the internal nares and extends to the level of the cricoid cartilage, the most inferior cartilage of the larynx (voice box) (Figure 23.3). It lies just posterior to the nasal cavity, oral cavity, and larynx and just anterior to the cervical vertebrae. The pharyngeal wall is composed of skeletal muscles and lined with mucous membrane. The pharynx functions as a passageway for air and food, provides a resonating chamber for speech sounds, and houses the tonsils, which help eliminate foreign substances through immunological reactions.

The pharynx is composed of three regions: (1) nasopharynx, (2) oropharynx, and (3) laryngopharynx. The superior portion of the pharynx, called the **nasopharynx,** lies posterior to the nasal cavity and extends to the plane of the soft palate. There are five openings in its wall: two internal nares, two openings that lead into the auditory (Eustachian) tubes, and the opening into the oropharynx. The posterior wall also contains the **pharyngeal tonsil.** Through the internal nares, the nasopharynx receives air from the nasal cavity and receives packages of dust-laden mucus. It is lined with pseudostratified ciliated columnar epithelium, and the cilia move the mucus down toward the most inferior part of the pharynx. The nasopharynx also exchanges small amounts of air with the auditory (Eustachian) tubes to equalize air pressure between the pharynx and middle ear.

The intermediate portion of the pharynx, the **oropharynx,** lies posterior to the oral cavity and extends from the soft palate inferiorly to the level of the hyoid bone. It has only one opening, the **fauces** (FAW-sēz; *fauces* = throat), the opening from the mouth. This portion of the pharynx is both a respiratory and digestive passageway. Because the oropharynx is subject to abrasion by food particles, it is lined with nonkeratinized stratified squamous epithelium, which protects underlying tissues. Two pairs of tonsils, the **palatine** and **lingual tonsils,** are found in the oropharynx.

The inferior portion of the pharynx, the **laryngopharynx** (la-rin′-gō-FAIR-inks), or **hypopharynx,** begins at the level of the hyoid bone and connects the esophagus (food tube) with the larynx (voice box). Like the oropharynx, the laryngopharynx is both a respiratory and a digestive pathway and is lined by nonkeratinized stratified squamous epithelium.

The arterial supply of the pharynx includes the ascending pharyngeal artery, the ascending palatine branch of the facial artery, the descending palatine and pharyngeal branches of the maxillary artery, and the muscular branches of the superior thyroid artery. The veins of the pharynx drain into the pterygoid plexus and the internal jugular vein.

Most of the muscles of the pharynx are innervated by the pharyngeal plexus. This plexus is formed by the pharyngeal branches of the glossopharyngeal (IX), vagus (X), and cranial portion of the accessory (XI) nerves and the superior cervical sympathetic ganglion.

Larynx

The **larynx** (LAIR-inks), or voice box, is a short passageway that connects the pharynx with the trachea. It lies in the midline of the neck anterior to the fourth through sixth cervical vertebrae (C4–C6).

The wall of the larynx is composed of nine pieces of cartilage (Fig. 23.4). Three are single (thyroid cartilage, epiglottis or epiglottic cartilage, and cricoid cartilage) and three are paired (arytenoid, cuneiform, and corniculate cartilages). Of the paired cartilages, the arytenoid cartilages are

Figure 23.3 Pharynx.

The three regions of the pharynx are the (1) nasopharynx, (2) oropharynx, and (3) laryngopharynx.

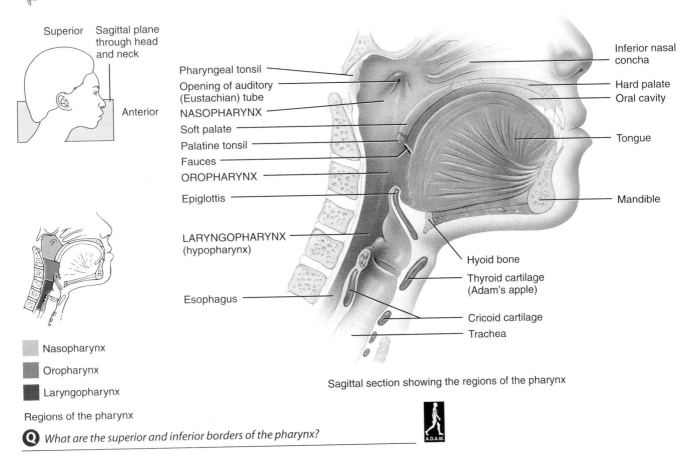

Sagittal section showing the regions of the pharynx

Regions of the pharynx

- Nasopharynx
- Oropharynx
- Laryngopharynx

Q *What are the superior and inferior borders of the pharynx?*

the most important because they influence the positions and tensions of the vocal folds (true vocal cords). Whereas the extrinsic muscles of the larynx connect the cartilages to other structures in the throat, the intrinsic muscles connect the cartilages to each other (see Figure 10.9).

The **thyroid cartilage (Adam's apple)** consists of two fused plates of hyaline cartilage that form the anterior wall of the larynx and give it its triangular shape. It is often larger in males than in females due to the influence of male sex hormones during puberty. The ligament that connects the thyroid cartilage to the hyoid bone is called the **thyrohyoid membrane.**

The **epiglottis** (*epi* = above; *glotta* = tongue) is a large, leaf-shaped piece of elastic cartilage that is covered with epithelium (see also Figure 23.3). The superior portion of the epiglottis comes into contact with whatever is swallowed, and it is covered by nonkeratinized stratified squamous epithelium. The inferior portion is part of the lining of the respiratory system and is lined with pseudostratified ciliated columnar epithelium. The "stem" of the epiglottis is attached to the anterior rim of the thyroid cartilage, but the "leaf" portion is unattached and free to move up and down like a trap door. During swallowing, the larynx rises. This causes the free edge of the epiglottis to move down and form a lid over the glottis, closing it off. The **glottis** consists of a pair of folds of mucous membrane, the vocal folds (true vocal cords) in the larynx, and the space between them called the **rima glottidis** (RĪ-ma GLOT-ti-dis; Figure 23.5). In this way, the larynx is closed off, and liquids and foods are routed into the esophagus and kept out of the larynx and airways inferior to it. When small particles, such as dust, smoke, food, or liquids, pass into the larynx, a cough reflex occurs to expel the material.

The **cricoid** (KRĪ-koyd; *krikos* = ring) **cartilage** is a ring of hyaline cartilage that forms the inferior wall of the larynx. It is attached to the first ring of cartilage of the trachea by the **cricotracheal ligament.** The thyroid cartilage is connected to the cricoid cartilage by the **cricothyroid ligament.** Clinically, the cricoid cartilage is the landmark for making an emergency airway (a tracheostomy, see page 688).

Figure 23.4 Larynx.

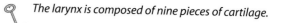

The larynx is composed of nine pieces of cartilage.

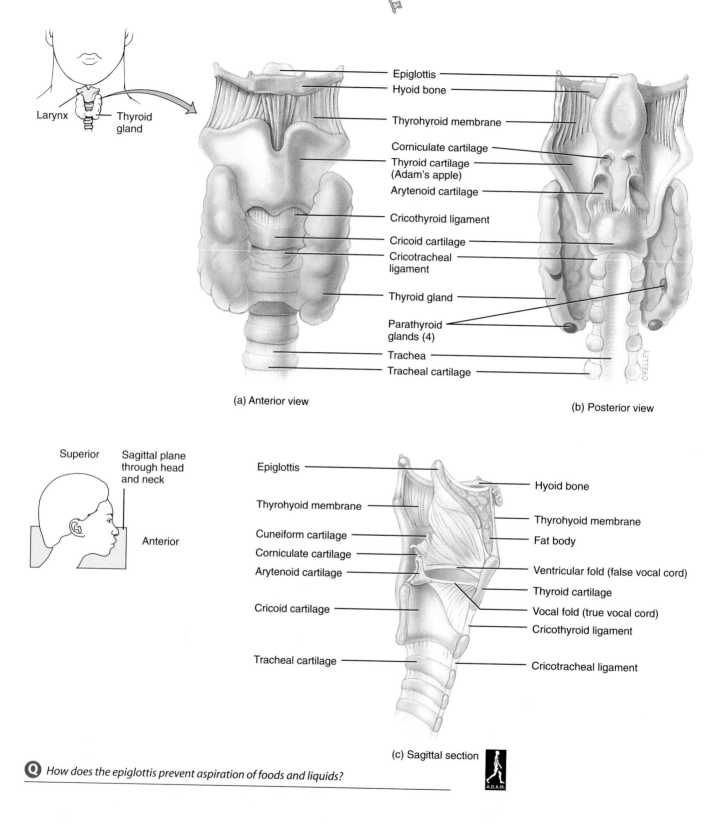

(a) Anterior view

(b) Posterior view

(c) Sagittal section

Q *How does the epiglottis prevent aspiration of foods and liquids?*

Figure 23.5 Vocal folds.

Sound originates from vibrations of the vocal folds.

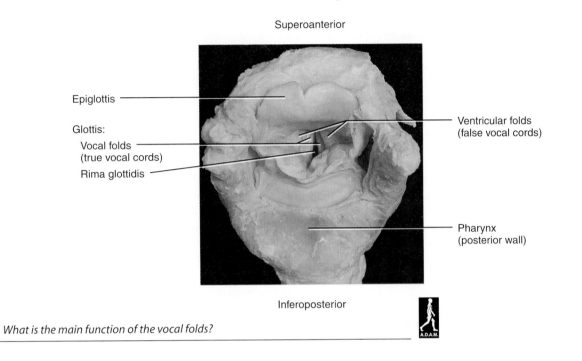

Superoanterior

Epiglottis

Glottis:
Vocal folds
(true vocal cords)
Rima glottidis

Ventricular folds
(false vocal cords)

Pharynx
(posterior wall)

Inferoposterior

Q *What is the main function of the vocal folds?*

The paired **arytenoid** (ar′-i-TĒ-noyd; *arytaina* = ladle) **cartilages** are triangular pieces of mostly hyaline cartilage located at the posterior, superior border of the cricoid cartilage. They attach to the vocal folds and intrinsic pharyngeal muscles. Supported by the arytenoid cartilages, the intrinsic pharyngeal muscles contract and thus move the vocal folds.

The paired **corniculate** (kor-NIK-yoo-lāt; *corniculate* = shaped like a small horn) **cartilages** are horn-shaped pieces of elastic cartilage. One is located at the apex of each arytenoid cartilage. The paired **cuneiform** (kyoo-NĒ-i-form; *cuneus* = wedge) **cartilages** are club-shaped elastic cartilages anterior to the corniculate cartilages (see Figure 23.4). The cuneiform cartilages support the vocal folds and lateral aspects of the epiglottis.

The lining of the larynx superior to the vocal folds is nonkeratinized stratified squamous epithelium. The lining of the larynx inferior to the vocal folds is pseudostratified ciliated columnar epithelium. It consists of ciliated columnar cells, goblet cells, and basal cells, and its mucous covering helps trap dust not removed in the upper passages. Whereas the cilia in the upper respiratory tract move mucus and trapped particles *down* toward the pharynx, the cilia in the lower respiratory tract move them *up* toward the pharynx.

The mucous membrane of the larynx forms two pairs of folds (see Figure 23.5): a superior pair called the **ventricular folds (false vocal cords)** and an inferior pair called simply the **vocal folds (true vocal cords).** The space between the ventricular folds is known as the **rima vestibuli.** The

laryngeal sinus (ventricle) is a lateral expansion of the middle portion of the laryngeal cavity between the ventricular folds above and the vocal folds below (see Figure 23.2b).

When the ventricular folds are brought together, they function in holding the breath against pressure in the thoracic cavity, such as might occur when a person strains to lift a heavy object. The mucous membrane of the vocal folds is lined by nonkeratinized stratified squamous epithelium. Deep to the membrane are bands of elastic ligaments stretched between pieces of rigid cartilage like the strings on a guitar. Skeletal muscles of the larynx, called intrinsic muscles, attach to both the rigid cartilage and the vocal folds themselves. When the muscles contract, they pull the elastic ligaments tight and stretch the vocal folds out into the airways so that the rima glottidis is narrowed. If air is directed against the vocal folds, they vibrate and set up sound waves in the column of air in the pharynx, nose, and mouth. The greater the pressure of air, the louder the sound.

Pitch is controlled by the tension on the vocal folds. If they are pulled taut by the muscles, they vibrate more rapidly, and a higher pitch results. Lower sounds are produced by decreasing the muscular tension on the vocal folds. Due to the influence of androgens (male sex hormones), vocal folds are usually thicker and longer in adult males than in females, and therefore they vibrate more slowly. Thus men generally speak or sing at a deeper pitch than women.

Sound originates from the vibration of the vocal folds, but other structures are necessary for converting the sound into recognizable speech. The pharynx, mouth, nasal cavity, and paranasal sinuses all act as resonating chambers that give the voice its human and individual quality. By constricting and relaxing the muscles in the wall of the pharynx, we produce the vowel sounds. Muscles of the face, tongue, and lips help us enunciate words.

Whispering is accomplished by closing all but the posterior portion of the rima glottidis. Because the vocal folds do not vibrate during whispering, there is no pitch to this form of speech. However, we can still produce intelligible speech while whispering by changing the shape of the oral cavity as we enunciate. As the size of the oral cavity changes, its resonance qualities change, which imparts a vowel-like pitch to the air as it rushes toward the lips. The fact that we can produce intelligible words without vocalizing only serves to demonstrate that speech is the product of many anatomical structures in the head, not just the vocal cords.

Clinical Application

Laryngitis and Cancer of the Larynx

Laryngitis is an inflammation of the larynx that is most often caused by a respiratory infection or irritants such as cigarette smoke. Inflammation of the vocal folds causes hoarseness or loss of voice by interfering with the contraction of the folds or by causing them to swell to the point at which they cannot vibrate freely. Many long-term smokers acquire a permanent hoarseness from the damage done by chronic inflammation. **Cancer of the larynx** is found almost exclusively in individuals who smoke. The condition is characterized by hoarseness, pain on swallowing, or pain radiating to an ear. Treatment consists of radiation therapy and/or surgery.

The arteries of the larynx are the superior laryngeal and inferior laryngeal arteries. The superior and inferior laryngeal veins accompany the arteries. The superior laryngeal vein empties into the superior thyroid vein, and the inferior vein empties into the inferior thyroid vein.

The nerves of the larynx are the superior and recurrent (inferior) laryngeal branches of the vagus (X) nerve.

Trachea

The **trachea** (TRĀ-kē-a; *tracheia* = sturdy), or windpipe, is a tubular passageway for air about 12 cm (5 in.) in length and 2½ cm (1 in.) in diameter. It is located anterior to the esophagus (Figure 23.6a) and extends from the larynx to the superior border of the fifth thoracic vertebra (T5), where it divides into right and left primary bronchi (see Figure 23.7).

The layers of the trachea, from deep to superficial, are (1) a mucosa, (2) submucosa, (3) hyaline cartilage, and (4) adventitia, composed of areolar connective tissue. The mucosa of the trachea consists of an epithelial layer of pseudostratified ciliated columnar epithelium (Figure 23.6b) and an underlying layer of lamina propria that contains elastic and reticular fibers. The epithelium consists of ciliated columnar cells and goblet cells that reach the luminal surface plus basal cells that do not reach the luminal surface. The epithelium provides the same protection against dust as the membrane lining the nasal cavity and larynx. The submucosa consists of areolar connective tissue and contains seromucous glands and their ducts. The 16–20 incomplete rings of hyaline cartilage look like letter Cs and are arranged horizontally and stacked one on top of another. They may be felt through the skin inferior to the larynx. The open part of each C-shaped cartilage ring faces the esophagus (Figure 23.6a). This accommodates slight expansion of the esophagus into the trachea during swallowing. Transverse smooth muscle fibers, called the **trachealis muscle,** and elastic connective tissue hold the open ends of the cartilage rings together. The solid C-shaped cartilage rings provide a semirigid support so the tracheal wall does not collapse inward (especially during inspiration) and obstruct the air passageway. The adventitia of the trachea consists of areolar connective tissue that joins the trachea to surrounding tissues.

At the point where the trachea divides into right and left primary bronchi, there is an internal ridge called the **carina** (ka-RĪ-na; *karina* = keel of a boat) (see Figure 23.7a). It is formed by a posterior and somewhat inferior projection of the last tracheal cartilage. The mucous membrane of the carina is one of the most sensitive areas of the entire larynx and trachea for triggering a cough reflex. Widening and distortion of the carina is a serious sign because it usually indicates a carcinoma of the lymph nodes around the region where the trachea divides.

Clinical Application

Tracheostomy and Intubation

In some situations—for example, a crushing injury to the chest—the rings of cartilage may not be strong enough to overcome collapse and obstruction of the trachea. Or the mucous membrane may become inflamed and swell so much that it closes off the airways; inflamed membranes secrete a great deal of mucus that may clog the lower airways. A large object may be breathed in (aspirated) while the rima glottidis is open; or an aspirated foreign object may cause spasm of the muscles of the larynx. If the obstruction is superior to the level of the larynx, a **tracheostomy** (trā-kē-OS-tō-mē) may be performed. A skin incision is made, followed by a short longitudinal incision into the trachea inferior to the cricoid cartilage. The patient breathes through a metal or plastic tracheal tube inserted through the incision. Another method is

Figure 23.6 Histology of the trachea.

🗝️ *The trachea is anterior to the esophagus and extends from the larynx to the superior border of the fifth thoracic vertebra.*

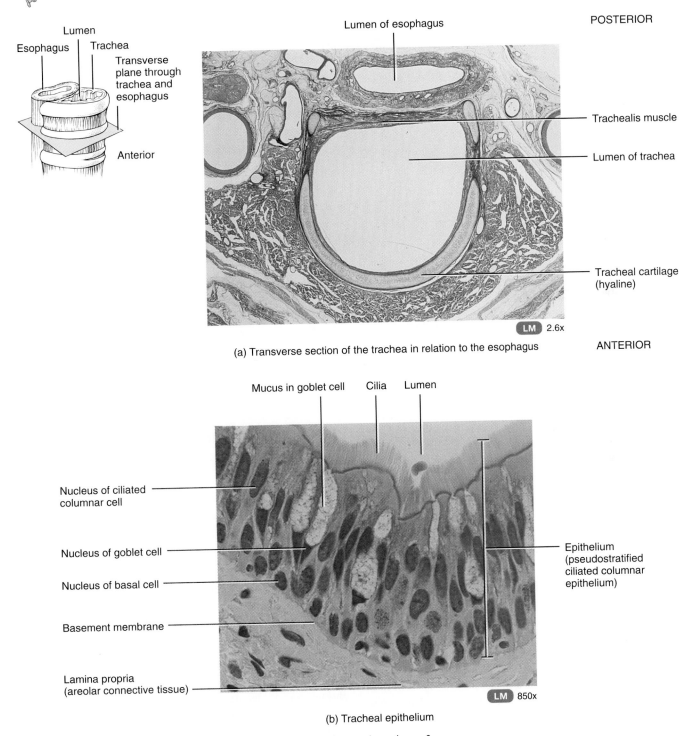

(a) Transverse section of the trachea in relation to the esophagus

(b) Tracheal epithelium

❓ *What is the benefit of not having cartilage between the trachea and esophagus?*

Figure 23.7 Bronchial tree in relation to the lungs.

The bronchial tree begins at the trachea and ends at the terminal bronchioles.

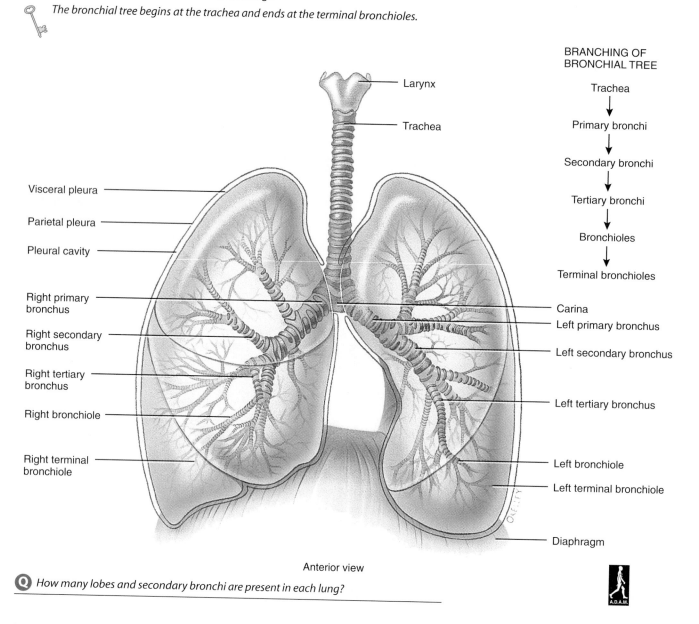

BRANCHING OF
BRONCHIAL TREE

Trachea

↓

Primary bronchi

↓

Secondary bronchi

↓

Tertiary bronchi

↓

Bronchioles

↓

Terminal bronchioles

Larynx

Trachea

Carina

Left primary bronchus

Left secondary bronchus

Left tertiary bronchus

Left bronchiole

Left terminal bronchiole

Diaphragm

Visceral pleura

Parietal pleura

Pleural cavity

Right primary bronchus

Right secondary bronchus

Right tertiary bronchus

Right bronchiole

Right terminal bronchiole

Anterior view

Q *How many lobes and secondary bronchi are present in each lung?*

intubation. A tube is inserted into the mouth or nose and passed inferiorly through the larynx and trachea. The firm wall of the tube pushes back any flexible obstruction, and the inside of the tube provides a passageway for air. If mucus is clogging the trachea, it can be suctioned out through the tube. ■

The arteries of the trachea are branches of the inferior thyroid, internal thoracic, and bronchial arteries. The veins of the trachea terminate in the inferior thyroid veins.

The smooth muscle and glands of the trachea are innervated parasympathetically via the vagus (X) nerve directly and by its recurrent laryngeal branches. Sympathetic innervation is through branches from the sympathetic trunk and its ganglia.

Bronchi

At the superior border of the fifth thoracic vertebra, the trachea divides into a **right primary bronchus** (BRON-kus; *bronchos* = windpipe), which goes into the right lung, and a **left primary bronchus,** which goes into the left lung (Figure 23.7). The right primary bronchus is more vertical, shorter, and wider than the left. As a result, an aspirated object is more likely to enter and lodge in the right primary bronchus than the left. Like the trachea, the primary bronchi (BRON-kē) contain incomplete rings of cartilage and are lined by pseudostratified ciliated columnar epithelium.

Clinical Application

Bronchoscopy

Bronchoscopy (bron-KOS-kō-pē) is the visual examination of the bronchi through a **bronchoscope,** an illuminated, tubular instrument that is passed through the trachea into the bronchi. The examiner can view the interior of the trachea and bronchi to biopsy a tumor, clear an obstructing object or secretions from an airway, take cultures or smears for microscopic examination, stop bleeding, or deliver drugs. ■

On entering the lungs, the primary bronchi divide to form smaller bronchi—the **secondary (lobar) bronchi,** one for each lobe of the lung. (The right lung has three lobes; the left lung has two.) The secondary bronchi continue to branch, forming still smaller bronchi, called **tertiary (segmental) bronchi,** that divide into **bronchioles.** Bronchioles, in turn, branch repeatedly, and the smallest bronchioles branch into even smaller tubes called **terminal bronchioles.** This extensive branching from the trachea resembles a tree trunk with its branches and is commonly referred to as the **bronchial tree.**

As the branching becomes more extensive in the bronchial tree, several structural changes may be noted. First, the epithelium gradually changes from pseudostratified ciliated columnar epithelium in the bronchi to nonciliated simple cuboidal epithelium in the terminal bronchioles. (In regions where nonciliated cuboidal epithelium is present, inhaled particles are removed by macrophages.) Second, incomplete rings of cartilage in primary bronchi are gradually replaced by plates of cartilage that finally disappear in the distal bronchioles. Third, as the cartilage decreases, the amount of smooth muscle increases. Smooth muscle encircles the lumen in spiral bands, and its contraction is affected by both the autonomic nervous system (ANS) and various chemicals.

During exercise, activity in the sympathetic division of the ANS increases, and the adrenal medulla releases the hormones epinephrine and norepinephrine. Both hormones relax smooth muscle in the bronchioles, thus dilating the airways and improving lung ventilation because more air enters the lungs. The parasympathetic division of the ANS and mediators of allergic reactions, such as histamine, cause constriction of distal bronchioles.

Clinical Application

Nebulization

Many respiratory disorders are treated by means of **nebulization** (neb-yoo-li-ZĀ-shun). This procedure consists of administering medication in the form of droplets that are suspended in air into the respiratory tract. The patient inhales the medication as a fine mist. Nebulization therapy can be used with many different types of drugs, such as chemicals that relax the smooth muscle of the airways, chemicals that reduce the thickness of mucus, and antibiotics. ■

The blood supply to the bronchi is via the left bronchial and right bronchial arteries. The veins that drain the bronchi are the right bronchial vein, which enters the azygos vein, and the left bronchial vein, which empties into the hemiazygos vein or the left superior intercostal vein.

Lungs

The **lungs** (*lunge* = lightweight, because the lungs float) are paired cone-shaped organs lying in the thoracic cavity. They are separated from each other by the heart and other structures in the mediastinum. The mediastinum separates the thoracic cavity into two anatomically distinct chambers. Accordingly, if trauma causes one lung to collapse, the other may remain expanded. Two layers of serous membrane, collectively called the **pleural** (*pleura* = side) **membrane,** enclose and protect each lung (Figure 23.8). The superficial layer lines the wall of the thoracic cavity and is called the **parietal pleura.** The deep layer, the **visceral pleura,** covers the lungs themselves. Between the visceral and parietal pleurae is a small potential space, the **pleural cavity,** which contains a lubricating fluid secreted by the membranes. This fluid reduces friction between the membranes and allows them to slide easily on one another during breathing.

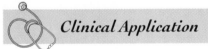

Clinical Application

Pneumothorax, Hemothorax, and Pleurisy

In certain conditions, the pleural cavity may fill with air (**pneumothorax;** *pneumo* = air or breath), blood (**hemothorax),** or pus. Air in the pleural cavity, most commonly introduced in a surgical opening of the chest or as a result of a stab or gunshot wound, may cause the lung to collapse. This condition is called **atelectasis** (at'-e-LEK-ta-sis; *ateles* = incomplete; *ectasis* = expansion). Fluid can be drained from the pleural cavity by inserting a needle, usually posteriorly through the seventh intercostal space. The needle is passed along the superior border of the lower rib to avoid damage to the intercostal nerves and blood vessels. Inferior to the seventh intercostal space there is danger of penetrating the diaphragm.

Inflammation of the pleural membrane, called **pleurisy** or **pleuritis,** may in its early stages cause pain due to friction between the parietal and visceral layers of the pleura. If the inflammation persists, fluid accumulates in the pleural space, a condition known as **pleural effusion.** One cause of pleural effusion is lung cancer. ■

Figure 23.8 Relationship of the pleural membranes to the lungs. The arrow in the inset indicates the direction from which the lungs are viewed (superior).

🔑 *The parietal pleura lines the thoracic cavity, whereas the visceral pleura covers the lungs.*

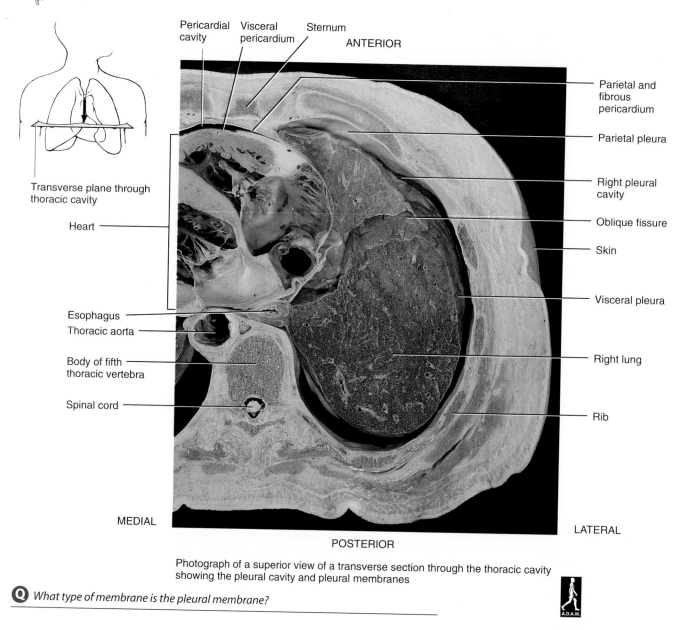

Photograph of a superior view of a transverse section through the thoracic cavity showing the pleural cavity and pleural membranes

Q *What type of membrane is the pleural membrane?*

Gross Anatomy

The lungs extend from the diaphragm to just slightly superior to the clavicles and lie against the ribs anteriorly and posteriorly. The broad inferior portion of the lung, the **base,** is concave and fits over the convex area of the diaphragm (Figure 23.9). The narrow superior portion of the lung is termed the **apex (capula).** The surface of the lung lying against the ribs, the **costal surface,** is rounded to match the curvature of the ribs. The **mediastinal (medial) surface** of each lung contains a region, the **hilus,** through which bronchi, pulmonary blood vessels, lymphatic vessels, and nerves enter and exit. These structures are held together by the pleura and connective tissue and constitute the **root** of the lung. Medially, the left lung also contains a concavity, **the cardiac notch,** in which the heart lies.

The right lung is thicker and broader than the left. It is also somewhat shorter than the left because the diaphragm is higher on the right side to accommodate the liver, which lies inferior to it.

Figure 23.9 Lungs. The arrows in the inset indicate the directions from which the lungs are viewed (lateral and medial).

The subdivisions of the lungs are lobes → bronchopulmonary segments → lobules.

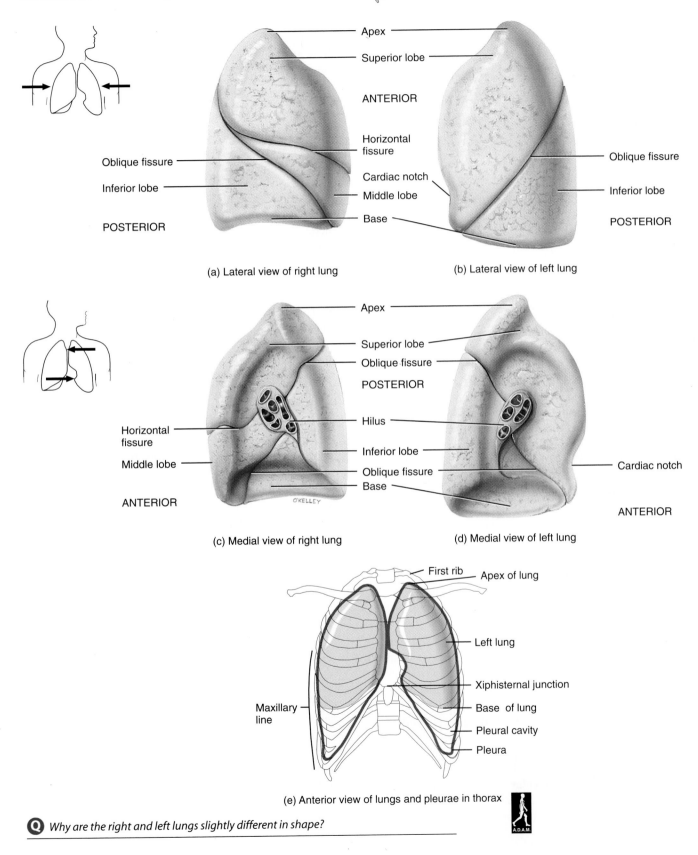

Apex
Superior lobe
ANTERIOR
Horizontal fissure
Oblique fissure
Inferior lobe
Cardiac notch
Middle lobe
POSTERIOR
Base

(a) Lateral view of right lung

Apex
Superior lobe
Oblique fissure
Inferior lobe
POSTERIOR

(b) Lateral view of left lung

Apex
Superior lobe
Oblique fissure
POSTERIOR
Horizontal fissure
Hilus
Middle lobe
Inferior lobe
Oblique fissure
Base
ANTERIOR

O'KELLEY

(c) Medial view of right lung

Apex
Superior lobe
Oblique fissure
POSTERIOR
Hilus
Inferior lobe
Oblique fissure
Cardiac notch
Base
ANTERIOR

(d) Medial view of left lung

First rib
Apex of lung
Left lung
Xiphisternal junction
Base of lung
Maxillary line
Pleural cavity
Pleura

(e) Anterior view of lungs and pleurae in thorax

Q *Why are the right and left lungs slightly different in shape?*

Position in Thorax

The lungs almost totally fill the thorax (Figure 23.9e). The apex of the lungs lies just superior to the medial third of the clavicles and is the only area that can be palpated. The anterior, lateral, and posterior surfaces of the lung lie against the ribs. The base of the lungs extends from the sixth costal cartilage anteriorly to the spinous process of the tenth thoracic vertebra posteriorly. The pleura extends about 5 cm below the base from the sixth costal cartilage anteriorly to the twelfth rib posteriorly. Thus, the lungs do not completely fill the pleural cavity in this area. A needle is typically inserted into this space in the pleural cavity to withdraw fluid from the chest.

Lobes and Fissures

Each lung is divided into lobes by one or more fissures (Figure 23.9). Both lungs have an **oblique fissure,** which extends inferiorly and anteriorly. The right lung also has a **horizontal fissure.** The oblique fissure in the left lung separates the **superior lobe** from the **inferior lobe.** The superior part of the oblique fissure of the right lung separates the superior lobe from the inferior lobe, whereas the inferior part of the oblique fissure separates the inferior lobe from the **middle lobe.** The horizontal fissure of the right lung subdivides the superior lobe, thus forming a middle lobe.

Each lobe receives its own secondary (lobar) bronchus. Thus the right primary bronchus gives rise to three secondary (lobar) bronchi called the **superior, middle,** and **inferior secondary (lobar) bronchi.** The left primary bronchus gives rise to **superior** and **inferior secondary (lobar) bronchi.** Within the substance of the lung, the secondary bronchi give rise to the **tertiary (segmental) bronchi,** which are constant in both origin and distribution. There are ten tertiary bronchi in each lung. The segment of lung tissue that each supplies is called a **bronchopulmonary segment** (Figure 23.10). Bronchial and pulmonary disorders, such as tumors or abscesses, may be localized in a bronchopulmonary segment and may be surgically removed without seriously disrupting surrounding lung tissue.

Lobules

Each bronchopulmonary segment of the lungs has many small compartments called **lobules.** Each lobule is wrapped in elastic connective tissue and contains a lymphatic vessel, an arteriole, a venule, and a branch from a terminal bronchiole (Figure 23.11a). Terminal bronchioles subdivide into microscopic branches called **respiratory bronchioles.** As the respiratory bronchioles penetrate more deeply into the lungs, the epithelial lining changes from simple cuboidal to simple squamous. Respiratory bronchioles, in turn, subdivide into several (2–11) **alveolar ducts.** From the trachea to the alveolar ducts, there are about 25 orders of branching of the respiratory passageways; that is, the trachea divides into primary bronchi (first order), the primary bronchi divide into secondary bronchi (second order), and so on.

Around the circumference of the alveolar ducts are numerous alveoli and alveolar sacs (Figure 23.11b). An **alveolus** (al-VĒ-ō-lus) is a cup-shaped outpouching lined by simple squamous epithelium and supported by a thin elastic basement membrane. **Alveolar sacs** are two or more alveoli that share a common opening (Figure 23.11a and b). The alveolar walls consist of two types of alveolar epithelial cells or **pneumocytes** (Figure 23.12 on page 697). **Type I alveolar (squamous pulmonary epithelial) cells** are simple squamous epithelial cells that form a continuous lining of the alveolar wall, interrupted by occasional **type II alveolar (septal) cells.** The thin type I alveolar cells are the main site where gas exchange takes place. Type II alveolar cells are rounded or cuboidal epithelial cells whose free surfaces contain microvilli. They secrete alveolar fluid, which contains a phospholipid and lipoprotein mixture called **surfactant** (sur-FAK-tant). The surfactant lowers the surface tension of the alveolar fluid and thus reduces the tendency of alveoli to collapse during exhalation. Associated with the alveolar wall are **alveolar macrophages (dust cells).** These wandering phagocytes remove fine dust particles and other debris in the alveolar spaces. Also present are fibroblasts that produce reticular and elastic fibers. Superficial to the layer of type I alveolar cells is an elastic basement membrane. Around the alveoli, the lobule's arteriole and venule disperse into a capillary network. The blood capillaries consist of a single layer of endothelial cells and basement membrane.

Alveolar–Capillary Membrane

The exchange of respiratory gases (O_2 and CO_2) between the lungs and blood takes place by diffusion across alveolar and capillary walls. Collectively, the layers through which the respiratory gases diffuse are known as the **alveolar–capillary (respiratory) membrane** (Figure 23.12b). It consists of:

1. A layer of type I and type II alveolar cells with wandering alveolar macrophages that constitute the **alveolar (epithelial) wall.**

2. An **epithelial basement membrane** underlying the alveolar wall.

3. A **capillary basement membrane** that is often fused to the epithelial basement membrane.

4. The **endothelial cells** of the capillary.

Despite having several layers, the alveolar–capillary membrane averages only ½ μm in thickness, about 1/16 the diameter of a red blood cell. This allows rapid diffusion of respiratory gases. Moreover, it has been estimated that the lungs contain 300 million alveoli, providing an immense surface area of 70 m^2 (750 ft^2), about the size of a handball court, for the exchange of gases.

Figure 23.10 Bronchopulmonary segments of the lungs. The bronchial branches are shown in the center of the figure. The bronchopulmonary segments are numbered and named for convenience. The arrows in the inset indicate the directions from which the lungs are viewed (lateral and medial).

There are ten tertiary (segmental) bronchi in each lung, and each tertiary bronchus supplies a bronchopulmonary segment.

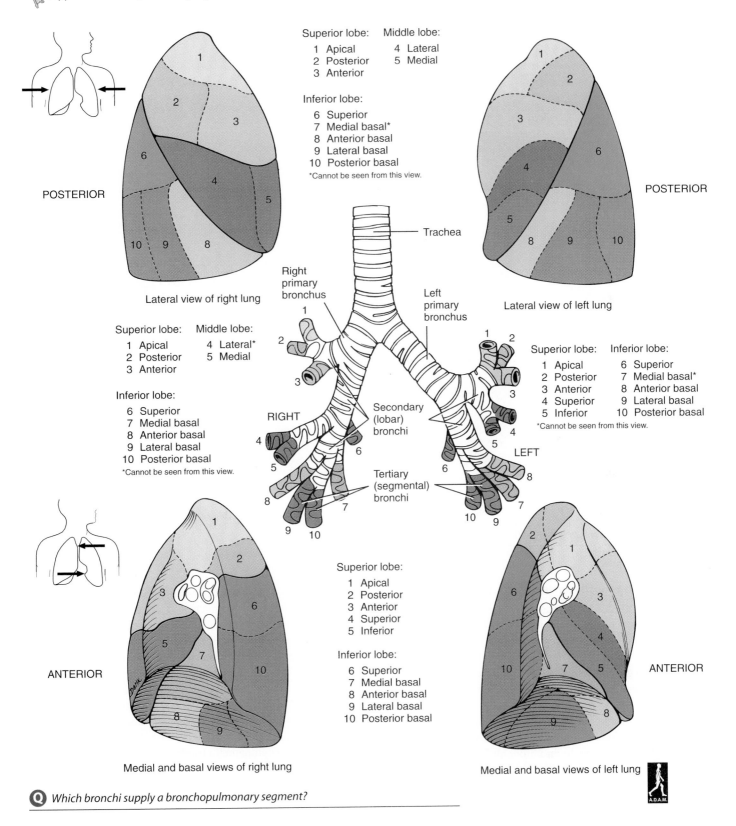

Superior lobe:
1 Apical
2 Posterior
3 Anterior

Middle lobe:
4 Lateral
5 Medial

Inferior lobe:
6 Superior
7 Medial basal*
8 Anterior basal
9 Lateral basal
10 Posterior basal
*Cannot be seen from this view.

POSTERIOR

Lateral view of right lung

Trachea

Right primary bronchus

Left primary bronchus

POSTERIOR

Lateral view of left lung

Superior lobe:
1 Apical
2 Posterior
3 Anterior

Middle lobe:
4 Lateral*
5 Medial

Inferior lobe:
6 Superior
7 Medial basal
8 Anterior basal
9 Lateral basal
10 Posterior basal
*Cannot be seen from this view.

RIGHT

Secondary (lobar) bronchi

Tertiary (segmental) bronchi

LEFT

Superior lobe:
1 Apical
2 Posterior
3 Anterior
4 Superior
5 Inferior

Inferior lobe:
6 Superior
7 Medial basal*
8 Anterior basal
9 Lateral basal
10 Posterior basal
*Cannot be seen from this view.

Superior lobe:
1 Apical
2 Posterior
3 Anterior
4 Superior
5 Inferior

Inferior lobe:
6 Superior
7 Medial basal
8 Anterior basal
9 Lateral basal
10 Posterior basal

ANTERIOR

Medial and basal views of right lung

ANTERIOR

Medial and basal views of left lung

Q *Which bronchi supply a bronchopulmonary segment?*

Figure 23.11 Structure and histology of a lobule of the lungs.

Alveolar sacs are two or more alveoli that share a common opening.

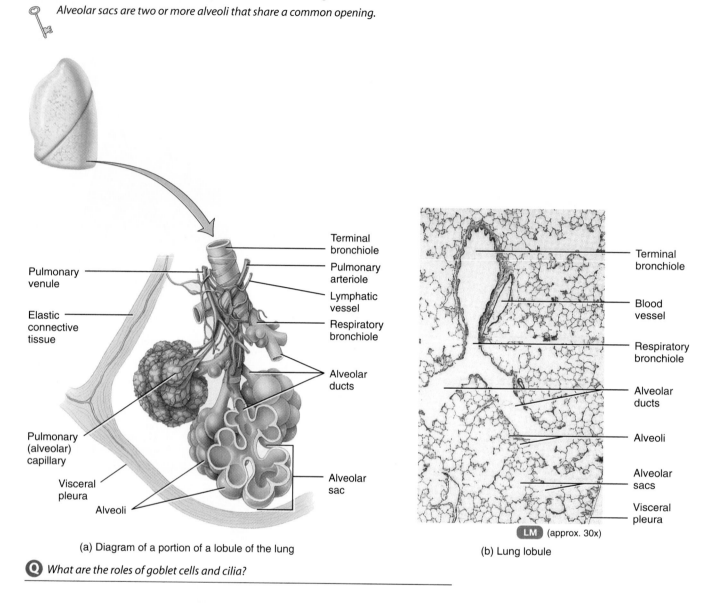

(a) Diagram of a portion of a lobule of the lung

(b) Lung lobule

Q *What are the roles of goblet cells and cilia?*

Blood and Nerve Supply to the Lungs

There is a double blood supply to the lungs. Deoxygenated blood passes through the pulmonary trunk, which divides into a left pulmonary artery that enters the left lung and a right pulmonary artery that enters the right lung. The return of the oxygenated blood to the heart is by way of the pulmonary veins, which drain into the left atrium (see Figure 14.21).

Bronchial arteries, which branch from the aorta, deliver oxygenated blood to the lungs. This blood mainly perfuses the walls of the bronchi and bronchioles. Connections exist between branches of the bronchial arteries and branches of the pulmonary arteries, however, and most blood returns to the heart via pulmonary veins. Some blood, however, drains into bronchial veins, branches of the azygos system, and returns to the heart via the superior vena cava.

The nerve supply of the lungs is derived from the pulmonary plexus, located anterior and posterior to the roots of the lungs. The pulmonary plexus is formed by branches of the vagus (X) nerves and sympathetic trunks. Motor parasympathetic fibers arise from the dorsal nucleus of the vagus (X) nerve, whereas motor sympathetic fibers are postganglionic fibers of the second to fifth thoracic paravertebral ganglia of the sympathetic trunk.

Figure 23.12 Structure of an alveolus.

The exchange of respiratory gases occurs by diffusion across the alveolar–capillary (respiratory) membrane.

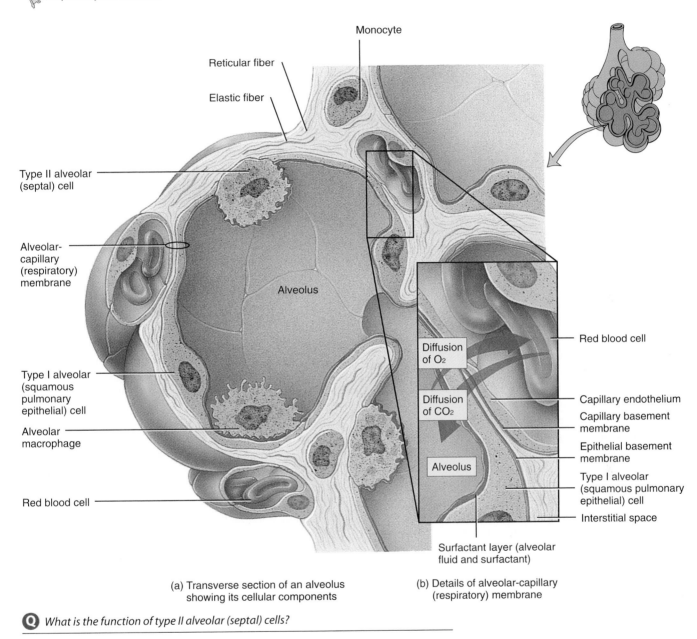

(a) Transverse section of an alveolus showing its cellular components

(b) Details of alveolar-capillary (respiratory) membrane

Q *What is the function of type II alveolar (septal) cells?*

MECHANICS OF PULMONARY VENTILATION (BREATHING)

Respiration is the exchange of gases between the atmosphere, blood, and cells. It takes place in three basic steps:

1. **Pulmonary ventilation.** The first process, pulmonary (*pulmo* = lung) ventilation or breathing, is the inspiration (inflow) and expiration (outflow) of air between the atmosphere and the lungs.

2. **External (pulmonary) respiration.** This is the exchange of gases between the air spaces of the lungs and blood in pulmonary capillaries. The blood gains O_2 and loses CO_2.

3. **Internal (tissue) respiration.** The exchange of gases between blood in systemic capillaries and tissue cells is known as internal (tissue) respiration. The blood loses O_2 and gains CO_2.

The flow of air between the atmosphere and lungs occurs for the same reason that blood flows through the body: a pressure gradient (difference) exists. Air moves into the lungs when the pressure inside the lungs is less than the air pressure in the atmosphere. Air moves out of the lungs when the pressure inside the lungs is greater than the pressure in the atmosphere.

Inspiration

Breathing in is called **inspiration (inhalation).** Just before each inspiration, the air pressure inside the lungs equals the pressure of the atmosphere, which is about 760 mm Hg, or 1 atmosphere (atm), at sea level. For air to flow into the lungs, the pressure inside the alveoli of the lungs **(alveolar pressure)** must become lower than the pressure in the atmosphere. This condition is achieved by increasing the volume (size) of the lungs.

For inspiration to occur, the lungs must expand. This increases lung volume and thus decreases the pressure in the lungs below atmospheric pressure. The first step in expanding the alveoli of the lungs involves contraction of the principal inspiratory muscles—the diaphragm and/or external intercostals (Figure 23.13).

The diaphragm, the most important muscle of inspiration, is a dome-shaped skeletal muscle that forms the floor of the thoracic cavity. It is innervated by fibers of the phrenic nerves, which emerge from both sides of the spinal cord at cervical levels 3, 4, and 5. Contraction of the diaphragm causes it to flatten, lowering its dome. This increases the vertical dimension of the thoracic cavity and accounts for the movement of about 75% of the air that enters the lungs during inspiration. The distance the diaphragm moves during inspiration ranges from 1 cm (0.4 in.) during normal quiet breathing up to about 10 cm (4 in.) during heavy breathing. Advanced pregnancy, excessive obesity, or confining abdominal clothing can prevent a complete descent of the diaphragm. At the same time the diaphragm contracts, the external intercostals contract. These skeletal muscles run obliquely downward and forward between adjacent ribs, and when these muscles contract, the ribs are pulled superiorly and the sternum is pushed anteriorly. This increases the anterior–posterior dimension of the thoracic cavity.

As the diaphragm contracts and the overall size of the thoracic cavity increases, the walls of the lungs are pulled outward. The parietal and visceral pleurae normally adhere strongly to each other because of the below-atmospheric pressure between them and because of the surface tension created by their moist adjoining surfaces. As the thoracic cavity expands, the parietal pleura lining the cavity is pulled outward in all directions, and the visceral pleura and lungs are pulled along with it.

When the volume of the lungs increases, alveolar pressure decreases from 760 to 758 mm Hg. A pressure gradient is thus established between the atmosphere and the alveoli. Air rushes from the atmosphere into the lungs due to a gas pressure difference, and inspiration takes place. Air continues to move into the lungs as long as the pressure difference exists.

During deep, labored inspiration, accessory muscles of inspiration also participate in increasing the size of the thoracic cavity (see Figure 23.13a). These include the sternocleidomastoid, which elevates the sternum; the scalenes, which elevate the superior two ribs; and the pectoralis minor, which elevates the third through fifth ribs.

Expiration

Breathing out, called **expiration (exhalation),** is also achieved by a pressure gradient, but in this case the gradient is reversed: the pressure in the lungs is greater than the pressure of the atmosphere. Normal expiration during quiet breathing depends on two factors: (1) the recoil of elastic fibers that were stretched during inspiration and (2) the inward pull of surface tension due to the film of alveolar fluid.

Expiration starts when the inspiratory muscles relax. As the external intercostals relax, the ribs move inferiorly, and as the diaphragm relaxes, its dome moves superiorly owing to its elasticity. These movements decrease the vertical and anterior–posterior dimensions of the thoracic cavity. Also, surface tension exerts an inward pull and the elastic basement membranes of the alveoli and elastic fibers in bronchioles and alveolar ducts recoil. As a result, lung volume decreases and the alveolar pressure increases to 762 mm Hg. Air then flows from the area of higher pressure in the alveoli to the area of lower pressure in the atmosphere.

During labored breathing and when air movement out of the lungs is impeded, muscles of expiration—abdominal and internal intercostals—contract. Contraction of the abdominal muscles moves the inferior ribs inferiorly and compresses the abdominal viscera, thus forcing the diaphragm superiorly. Contraction of the internal intercostals, which extend inferiorly and posteriorly between adjacent ribs, pulls the ribs inferiorly.

CONTROL OF RESPIRATION

Although breathing can be controlled voluntarily for short periods, the nervous system usually controls respirations automatically to meet the body's demand without our conscious concern.

Figure 23.13 Pulmonary ventilation: muscles of inspiration and expiration. The pectoralis minor muscle, an accessory inspiratory muscle, is not shown here but is illustrated in Figure 10.16a.

During deep, labored inspiration, accessory muscles of inspiration (sternocleidomastoids, scalenes, and pectoralis minors) participate.

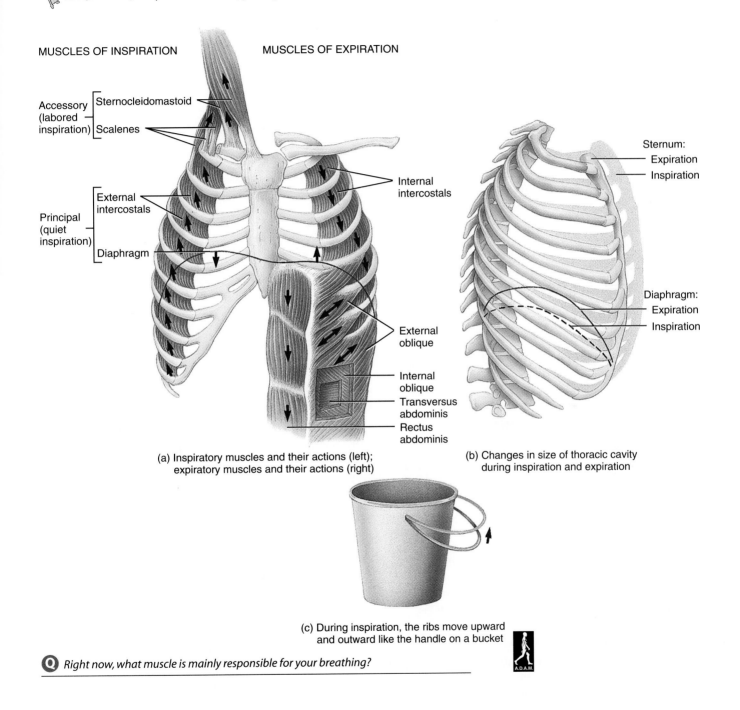

MUSCLES OF INSPIRATION   MUSCLES OF EXPIRATION

Accessory (labored inspiration)
- Sternocleidomastoid
- Scalenes

Principal (quiet inspiration)
- External intercostals
- Diaphragm

Internal intercostals

External oblique
Internal oblique
Transversus abdominis
Rectus abdominis

Sternum:
- Expiration
- Inspiration

Diaphragm:
- Expiration
- Inspiration

(a) Inspiratory muscles and their actions (left); expiratory muscles and their actions (right)

(b) Changes in size of thoracic cavity during inspiration and expiration

(c) During inspiration, the ribs move upward and outward like the handle on a bucket

Q *Right now, what muscle is mainly responsible for your breathing?*

Figure 23.14 Approximate location of areas of the respiratory center.

The respiratory center is located in the medulla oblongata and pons of the brain stem.

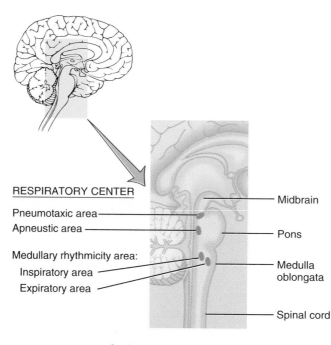

RESPIRATORY CENTER

Pneumotaxic area

Apneustic area

Medullary rhythmicity area:

Inspiratory area

Expiratory area

Midbrain

Pons

Medulla oblongata

Spinal cord

Sagittal section of brain stem showing components of the respiratory center

Q *In normal, quiet breathing, which of the areas shown contains cells that set the basic rhythm of respiration?*

Nervous Control

The size of the thorax is affected by the action of the respiratory muscles. These muscles contract and relax as a result of nerve impulses transmitted to them from centers in the brain. The area from which nerve impulses are sent to respiratory muscles is located bilaterally in the medulla oblongata and pons of the brain stem. This area is called the **respiratory center** and consists of a widely dispersed group of neurons that is functionally divided into three areas: (1) the medullary rhythmicity (rith-MIS-i-tē) area in the medulla oblongata; (2) the pneumotaxic (noo-mō-TAK-sik) area in the pons; and (3) the apneustic (ap-NOO-stik) area, also in the pons.

Medullary Rhythmicity Area

The function of the **medullary rhythmicity area** is to control the basic rhythm of respiration.

Within the medullary rhythmicity area are both inspiratory and expiratory neurons that comprise inspiratory and expiratory areas, respectively (Figure 23.14). The basic rhythm of respiration is determined by nerve impulses generated in the inspiratory area. At the beginning of expiration, the inspiratory area is inactive, but after 3 seconds it automatically becomes active. This activity seems to result from neurons that are autorhythmic cells. Even when all incoming nerve connections to the inspiratory area are cut or blocked, neurons in this area still rhythmically discharge impulses that result in inspiration. Nerve impulses from the active inspiratory area last for about 2 seconds and travel to the muscles of inspiration. The impulses reach the diaphragm by the phrenic nerves, and the external intercostal muscles by the intercostal nerves. When the impulses reach the inspiratory muscles, the muscles contract and inspiration occurs. At the end of 2 seconds the inspiratory muscles relax, and the cycle repeats itself over and over.

The expiratory neurons remain inactive during most normal, quiet respirations. During quiet respiration, inspiration is accomplished by contraction of the inspiratory muscles, and expiration results from elastic recoil of the lungs and thoracic wall as the inspiratory muscles relax. However, during high levels of ventilation, it is believed that impulses from the inspiratory area activate the expiratory area. Impulses discharged from the expiratory area cause contraction of the internal intercostals and abdominal muscles that decrease the size of the thoracic cavity to cause forced (labored) expiration.

Pneumotaxic Area

Although the medullary rhythmicity area controls the basic rhythm of respiration, other parts of the brain stem help coordinate the transition between inspiration and expiration. One of these is the **pneumotaxic** (*pneuma* = breath; *taxis* = arrangement) **area** in the superior portion of the pons (Figure 23.14). It transmits inhibitory impulses to the inspiratory area to help turn off the inspiratory area before the lungs become too full of air. In other words, the impulses limit the duration of inspiration and thus facilitate the onset of expiration. When the pneumotaxic area is more active, the breathing rate is quicker.

Apneustic Area

Another part of the brain stem that coordinates the transition between inspiration and expiration is the **apneustic area** in the inferior portion of the pons (Figure 23.14). It sends to the inspiratory area stimulatory impulses that activate it and prolong inspiration, thus inhibiting expiration. This occurs when the pneumotaxic area is inactive. When the pneumotaxic area is active, it overrides the apneustic area.

Regulation of Respiratory Center Activity

Although the basic rhythm of respiration is set and coordinated by the inspiratory area, the rhythm can be modified in response to inputs from other brain regions and receptors in the peripheral nervous system.

Cortical Influences

The cerebral cortex has connections with the respiratory center, which means we can voluntarily alter our pattern of breathing. We can even refuse to breathe at all for a short time. Voluntary control is protective because it enables us to prevent water or irritating gases from entering the lungs. The ability to not breathe is limited by the buildup of CO_2 and H^+ in the blood, however. When CO_2 and the concentration of H^+ increase to a certain level, the inspiratory area is strongly stimulated, nerve impulses are sent along the phrenic and intercostal nerves to inspiratory muscles, and breathing resumes whether or not the person wishes. It is impossible for people to kill themselves by holding their breath. Even if a person faints, breathing resumes when consciousness is lost.

Chemical Stimuli

Certain chemical stimuli determine how fast and deeply we breathe. The ultimate goal of the respiratory system is to maintain proper levels of CO_2 and O_2, and the system is highly responsive to changes in the blood levels of either.

Chemoreceptors in two locations monitor levels of CO_2 and O_2 and provide input to the respiratory center. **Central chemoreceptors** are located in the medulla oblongata (*cen-tral* nervous system), whereas **peripheral chemoreceptors** are located in the walls of systemic arteries and relay impulses to the respiratory center over two cranial nerves, the vagus (X) and glossopharyngeal (IX) nerves, of the *peripheral* nervous system. If there is even a slight increase in CO_2, central and peripheral chemoreceptors are stimulated. The chemoreceptors send nerve impulses to the brain that cause the inspiratory area to become highly active, and the rate of respiration increases. This allows the body to expel more CO_2 until the CO_2 is lowered to normal. If arterial CO_2 is lower than normal, the chemoreceptors are not stimulated, and stimulatory impulses are not sent to the inspiratory area. Consequently, the rate of respiration decreases until CO_2 accumulates and the CO_2 rises to normal.

The oxygen chemoreceptors are sensitive only to large decreases in O_2. If arterial O_2 falls moderately from normal, the chemoreceptors become stimulated and send impulses to the inspiratory area and respiration increases. But if O_2 falls significantly, the cells of the inspiratory area suffer oxygen starvation and do not respond well to stimuli from any chemical receptors. They send fewer impulses to the inspiratory muscles, and the respiration rate decreases or breathing ceases altogether.

Inflation Reflex

Located in the walls of bronchi and bronchioles within the lungs are receptors sensitive to stretch called **baroreceptors** or **stretch receptors.** When the receptors become stretched during overinflation of the lungs, nerve impulses are sent along the vagus (X) nerves to the inspiratory area and apneustic area. In response, the inspiratory area is inhibited and the apneustic area is inhibited from activating the inspiratory area. The result is that expiration begins. As air leaves the lungs during expiration, the lungs deflate and the stretch receptors are no longer stimulated. Thus the inspiratory and apneustic areas are no longer inhibited, and a new inspiration begins. This reflex is referred to as the **inflation (Hering–Breuer) reflex.** Some evidence suggests that the reflex is mainly a protective mechanism for preventing excessive inflation of the lungs.

Clinical Application

Why Smokers Have Lowered Respiratory Efficiency

It is commonly observed that smoking may cause a person to become easily "winded" with even moderate exercise. Several factors decrease respiratory efficiency. (1) Nicotine constricts terminal bronchioles, and this decreases airflow into and out of the lungs. (2) Carbon monoxide in smoke binds to hemoglobin and reduces its oxygen-carrying capability. (3) Irritants in smoke cause increased fluid secretion by the mucosa of the bronchial tree and swelling of the mucosal lining, both of which impede airflow into and out of the lungs. (4) Irritants in smoke also inhibit the movement of cilia in the lining of the respiratory system. Thus excess fluids and foreign debris are not easily removed, which further adds to the difficulty in breathing. (5) With time, smoking leads to destruction of elastic fibers in the lungs and emphysema (described on page 21 in *Applications to Health*). These changes cause collapse of small bronchioles and trapping of air in alveoli at the end of exhalation. As a result, gas exchange is less efficient. ▪

Developmental Anatomy of the Respiratory System

The development of the mouth and pharynx is considered in Chapter 24 on the digestive system. Here we consider the remainder of the respiratory system. At about 4 weeks of fetal development, the respiratory system begins as an outgrowth of the **endoderm** of the foregut (precursor of some digestive organs) just posterior to the pharynx. This outgrowth is called the **laryngotracheal bud** (see Figure 22.9a). As the bud grows, it elongates and differentiates into the future epithelial lining of the *larynx* and other structures as well. Its proximal end maintains a slitlike opening into the pharynx called the *glottis*. The middle portion of the bud gives rise to the epithelial lining of the *trachea*. The distal portion divides into two **lung buds,** which grow into the epithelial lining of the *bronchi* and *lungs* (Figure 23.15).

As the lung buds develop, they branch and rebranch and give rise to all the *bronchial tubes*. After the sixth month, the closed terminal portions of the tubes dilate and become the *alveoli*. The smooth muscle, cartilage, and connective tissues of the bronchial tubes and the pleural sacs of the lungs are contributed by **mesenchymal (mesodermal) cells.**

Aging and the Respiratory System

With advancing age, the airways and tissues of the respiratory tract, including the alveoli, become less elastic and more rigid. In addition to the lungs becoming less elastic, the chest wall also becomes more rigid. As a result, there is a decrease in lung capacity. In fact, vital capacity (the maximum amount of air that can be expired after maximal inspi-

Figure 23.15 Development of the bronchial tubes and lungs.

🔑 *The respiratory system develops from endoderm and mesoderm.*

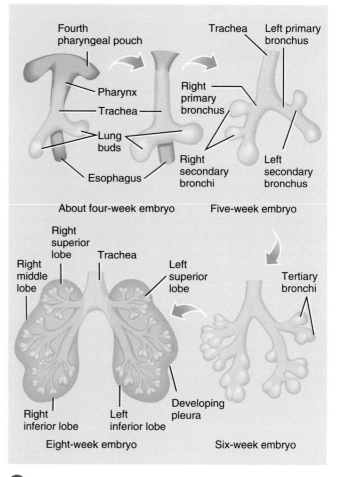

Q *When does the respiratory system begin to develop in an embryo?*

ration) can decrease as much as 35% by age 70. Also, there is a decrease in blood levels of O_2, decreased activity of alveolar macrophages, and diminished ciliary action of the epithelium lining the respiratory tract. Therefore, elderly people are more susceptible to pneumonia, bronchitis, emphysema, and other pulmonary disorders.

Key Medical Terms
Associated with the Respiratory System

Apnea (AP-nē-a; *a* = without; *pnoia* = breath) Absence of ventilatory movements.

Asphyxia (as-FIK-sē-a; *sphyxis* = pulse) Oxygen starvation due to low atmospheric oxygen or interference with ventilation, external respiration, or internal respiration.

Aspiration (as′-pi-RĀ-shun; *spirare* = breathe) Inhalation of a foreign substance such as water, food, or foreign body into the bronchial tree; drawing of a substance in or out by suction.

Atelectasis (at′-e-LAK-ta-sis; *ateles* = incomplete; *ektasis* = expansion) A collapsed lung or portion of a lung.

Bronchiectasis (bron′-kē-EK-ta-sis; *ektasis* = dilation) A chronic dilation of the bronchi or bronchioles.

Cardiopulmonary resuscitation (kar′-dē-ō-PUL-mo-ner-ē re-sus′-i-TĀ-shun) **(CPR)** The artificial establishment of normal or near-normal respiration and circulation. The **ABC's** of cardiopulmonary resuscitation are **Airway, Breathing,** and **Circulation,** meaning the rescuer must establish an airway, provide artificial ventilation if breathing has stopped, and reestablish circulation if there is inadequate cardiac action. The procedure should be performed in that order (A,B,C).

Cheyne–Stokes respiration (CHĀN STŌKS res′-pi-RĀ-shun) A repeated cycle of irregular breathing beginning with shallow breaths that increase in depth and rapidity, then decrease and cease altogether for 15 to 20 seconds. Cheyne–Stokes is normal in infants. It is also often seen just before death from pulmonary, cerebral, cardiac, and kidney disease.

Diphtheria (dif-THĒ-rē-a; *diphthera* = membrane) An acute bacterial infection that causes the mucous membranes of the oropharynx, nasopharynx, and larynx to enlarge and become leathery. Enlarged membranes may obstruct airways and cause death from asphyxiation.

Dyspnea (DISP-nē-a; *dys* = painful, difficult; *pnoia* = breath) Painful or labored breathing.

Epistaxis (ep′-i-STAK-sis) Loss of blood from the nose due to trauma, infection, allergy, neoplasms, and bleeding disorders. It can be arrested by cautery with silver nitrate, electrocautery, and firm packing. Also called **nosebleed.**

Heimlich (abdominal thrust) maneuver (HĪM-lik ma-NOO-ver) First-aid procedure designed to clear the air passageways of obstructing objects. It is preformed by applying a quick upward thrust that causes sudden elevation of the diaphragm and forceful, rapid expulsion of air in the lungs; this action forces air out the trachea to eject the obstructing object. The Heimlich maneuver is also used to expel water from the lungs of near-drowning victims before resuscitation is begun.

Hemoptysis (hē-MOP-ti-sis; *hemo* = blood; *ptein* = to spit) Spitting of blood from the respiratory tract.

Hypoxia (hī-POK-sē-a; *hypo* = below, under) Reduction in oxygen supply to cells.

Orthopnea (or′-THOP-nē-a; *ortho* = straight) Dyspnea that occurs in the horizontal position. It is an abnormal finding because a normal individual can tolerate the reduction in vital capacity that accompanies the recumbent position.

Pneumonectomy (noo′-mō-NEK-tō-mē; *pneumo* = lung; *tome* = cutting) Surgical removal of a lung.

Pulmonary edema Abnormal accumulation of interstitial fluid in the interstitial spaces and alveoli of the lungs as a result of increased pulmonary capillary permeability or pressure. Symptoms are dyspnea, wheezing, rapid respiration, feeling of suffocation, cyanosis, paleness, and excessive perspiration.

Rales (RĀLS) Sounds sometimes heard in the lungs that resemble bubbling or rattling. Rales are to the lungs what murmurs are to the heart. Different types are due to the presence of an abnormal amount or type of fluid or mucus inside the bronchi or alveoli, or to bronchoconstriction that causes turbulent air flow.

Respirator (RES-pi-rā′-tor) An apparatus fitted to a mask over the nose and mouth, or hooked directly to an endotracheal or tracheotomy tube, that is used to assist or support ventilation or to provide nebulized medication to the air passages under pressure.

Rhinitis (rī-NĪ-tis; *rhino* = nose) Chronic or acute inflammation of the mucous membrane of the nose.

Tachypnea (tak′-ip-NĒ-a; *tachy* = rapid; *pnoia* = breath) Rapid breathing.

Study Outline

Organs of Respiration (p. 681)

1. Respiratory organs include the nose, pharynx, larynx, trachea, bronchi, and lungs.
2. They act with the cardiovascular system to supply oxygen and remove carbon dioxide from the blood.

Nose (p. 681)

1. The external portion is made of cartilage and skin and is lined with mucous membrane. Openings to the exterior are the external nares.
2. The internal portion communicates with the paranasal sinuses and nasopharynx through the internal nares.
3. The nasal cavity is divided by a septum. The anterior portion of the cavity is called the vestibule.
4. The nose is adapted for warming, moistening, and filtering air; olfaction; and speech.

Pharynx (p. 684)

1. The pharynx (throat) is a muscular tube lined by a mucous membrane.

2. The anatomic regions are the nasopharynx, oropharynx, and laryngopharynx.
3. The nasopharynx functions in respiration. The oropharynx and laryngopharynx function both in digestion and in respiration.

Larynx (p. 684)

1. The larynx (voice box) is a passageway that connects the pharynx with the trachea.
2. It contains the thyroid cartilage (Adam's apple); the epiglottis, which prevents food from entering the larynx; the cricoid cartilage, which connects the larynx and trachea; and the paired arytenoid, corniculate, and cuneiform cartilages.
3. The larynx contains vocal folds, which produce sound. Taut folds produce high pitches, and relaxed ones produce low pitches.

Trachea (p. 688)

1. The trachea (windpipe) extends from the larynx to the primary bronchi.
2. It is composed of C-shaped rings of cartilage and some smooth muscle and is lined with pseudostratified ciliated columnar epithelium.

3. Two methods of bypassing obstructions from the respiratory passageways are tracheostomy and intubation.

Bronchi (p. 690)

1. The bronchial tree consists of the trachea, primary bronchi, secondary bronchi, tertiary bronchi, bronchioles, and terminal bronchioles. Walls of bronchi contain rings of cartilage; walls of bronchioles contain plates of cartilage and smooth muscle.

2. Bronchoscopy is the visual examination of the bronchus through a bronchoscope.

Lungs (p. 691)

1. Lungs are paired organs in the thoracic cavity. They are enclosed by the pleural membrane. The parietal pleura is the superficial layer; the visceral pleura is the deep layer.

2. The right lung has three lobes separated by two fissures; the left lung has two lobes separated by one fissure and a depression, the cardiac notch.

3. Secondary bronchi give rise to branches called segmental bronchi, which supply segments of lung tissue called bronchopulmonary segments.

4. Each bronchopulmonary segment consists of lobules, which contain lymphatics, arterioles, venules, terminal bronchioles, respiratory bronchioles, alveolar ducts, alveolar sacs, and alveoli.

5. Alveolar walls consist of type I alveolar cells, type II alveolar cells, and alveolar macrophages.

6. Gas exchange occurs across the alveolar–capillary (respiratory) membranes.

Mechanics of Pulmonary Ventilation (Breathing) (p. 697)

1. Pulmonary ventilation, or breathing, consists of inspiration and expiration.

2. Inspiration occurs when alveolar pressure falls below atmospheric pressure. Contraction of the diaphragm and external intercostal muscles increases the size of the thorax, thus decreasing the intrapleural pressure so that the lungs expand. Expansion of the lungs decreases alveolar pressure so that air moves along the pressure gradient from the atmosphere into the lungs.

3. During forced inspiration, accessory muscles of inspiration (sternocleidomastoids, scalenes, and pectoralis minor) are also used.

4. Expiration occurs when alveolar pressure is higher than atmospheric pressure. Relaxation of the diaphragm and external intercostal muscles results in elastic recoil of the chest wall and lungs, which increases intrapleural pressure, lung volume decreases, and alveolar pressure increases so that air moves from the lungs to the atmosphere.

5. Forced expiration employs contraction of the internal intercostals and abdominal muscles.

Control of Respiration (p. 698)

Nervous Control (p. 700)

1. The respiratory center consists of a medullary rhythmicity area (inspiratory and expiratory area), pneumotaxic area, and apneustic area.

2. The inspiratory area sets the basic rhythm of respiration.

3. The pneumotaxic and apneustic areas coordinate inspiration and expiration.

Regulation of Respiratory Center Activity (p. 701)

1. Respirations may be modified by a number of factors, both in the brain and outside.

2. Among the modifying factors are cortical influences, the inflation reflex, and chemical stimuli (CO_2 and O_2 levels).

Developmental Anatomy of the Respiratory System (p. 702)

1. The respiratory system begins as an outgrowth of endoderm called the laryngotracheal bud.

2. Smooth muscle, cartilage, and connective tissue of the bronchial tubes and pleural sacs develop from mesoderm.

Aging and the Respiratory System (p. 702)

1. Aging results in decreased vital capacity, decreased blood level of oxygen, and diminished alveolar macrophage activity.

2. Elderly people are more susceptible to pneumonia, emphysema, bronchitis, and other pulmonary disorders.

Review Questions

1. What organs make up the respiratory system? Distinguish between the upper and lower respiratory system. What functions do the respiratory and cardiovascular systems have in common? (p. 681)

2. Describe the structure of the external and internal nose, and describe their functions in filtering, warming, and moistening air. (p. 681)

3. What is the pharynx? Differentiate the three anatomical regions of the pharynx and indicate their roles in respiration. (p. 684)

4. Describe the structure of the larynx, and explain how it functions in respiration and voice production. (p. 684)

5. Describe the location and structure of the trachea. (p. 688)

6. What is the bronchial tree? Describe its structure. (p. 691)

7. Where are the lungs located? Distinguish the parietal pleura from the visceral pleura. (p. 691)

8. Define each of the following parts of a lung: base, apex, costal surface, medial surface, hilus, root, cardiac notch, and lobe. (p. 692)

9. What is a lobule of the lung? Describe its composition and function in respiration. (p. 694)

10. What is a bronchopulmonary segment? (p. 694)

11. Describe the histology and function of the alveolar–capillary (respiratory) membrane. (p. 694)

12. Discuss the basic steps involved in inspiration and expiration. (p. 698)

13. Explain how the rate of respiration is controlled. (p. 698)

14. Describe the development of the respiratory system. (p. 702)

15. Describe the effects of aging on the respiratory system. (p. 702)

16. Refer to the key medical terms associated with the respiratory system. Be sure that you can define each term. (p. 702)

Self Quiz

Complete the following:

1. The broad, inferior portion of the lung that sits on the diaphragm is called the ___. The narrow superior apex of each lung extends just superior to the ___. The costal surfaces lie against the ___.

2. The flaplike structure that prevents aspiration of solid and liquid substances into the trachea is called the ___.

3. The laryngotracheal bud is derived from ___-derm.

4. A section of lung tissue supplied by a tertiary bronchus is called a ___.

5. Place numbers in the blanks to indicate the route through which air passes as it enters a lobule enroute to alveoli.
   (a) alveolar ducts: ___; (b) respiratory bronchiole: ___;
   (c) terminal bronchiole: ___.

6. Place numbers in the blanks to indicate the pathway of inspired air.
   (a) trachea: ___; (b) internal nares: ___; (c) primary bronchi: ___; (d) meatuses: ___; (e) larynx: ___; (f) tertiary bronchi: ___; (g) alveolar ducts: ___; (h) bronchioles: ___; (i) oropharynx: ___; (j) external nares: ___; (k) vestibule: ___; (l) alveoli: ___; (m) nasopharynx: ___; (n) secondary bronchi: ___; (o) laryngopharynx: ___.

7. Match the following terms with their descriptions:
   (1) right lung   ___ (a) thicker, broader, and shorter
   (2) left lung   ___ (b) has a cardiac notch
       ___ (c) has only two secondary bronchi
       ___ (d) has a horizontal fissure
       ___ (e) has a primary bronchus that is shorter, wider, and more vertical

Choose the one best answer to the following questions:

8. Which is the correct order, from superficial to deep, of the following five structures?
   (1) parietal pleura   a. 5, 1, 3, 2, 4
   (2) visceral pleura   b. 5, 3, 1, 2, 4
   (3) pleural cavity   c. 4, 2, 3, 1, 5
   (4) lung   d. 5, 1, 2, 4, 3
   (5) wall of thoracic cavity   e. 5, 2, 3, 1, 4

9. Which of the following is *not* a part of the nasal septum?
   a. ethmoid
   b. nasal conchae
   c. hyaline cartilage
   d. vomer
   e. maxillae

10. The mucous membranes of the nose
    a. moisten, warm, and filter the inhaled air
    b. contain only alveoli
    c. contain only capillaries
    d. are not involved in the sense of smell
    e. clean, cool, and dry the inhaled air

11. Which of the following statements is/are true regarding the pharynx?
    (1) It serves as a common passageway for both the respiratory and digestive systems.
    (2) Adenoids, or pharyngeal tonsils, are located in the inferior part, the laryngopharynx.
    (3) The nasal cavity connects with it, through two internal nares, into the nasopharynx.
    a. 1 only   d. 1 and 3
    b. 2 only   e. all of the above
    c. 3 only

12. The auditory (Eustachian) tube opens into the lateral wall of the
    a. inner ear   d. laryngopharynx
    b. nasopharynx   e. nasal cavity
    c. oropharynx

13. Which of the following statements is/are true of the cartilage rings of the trachea?
    (1) They are made of tough fibrocartilage.
    (2) They are complete rings of cartilage, just like those of the bronchi.
    (3) They keep the lumen of the trachea open.
    a. 1 only   d. all of the above
    b. 2 only   e. none of the above
    c. 3 only

14. Which of the following structures is the smallest in diameter?
    a. left primary bronchus
    b. respiratory bronchioles
    c. secondary bronchi
    d. alveolar ducts
    e. right primary bronchus

15. Match the following terms to their descriptions:
    ___ (a) nasal conchae and meatuses are located here   1. nasal cavity
    ___ (b) contains pharyngeal tonsil   2. oropharynx
    ___ (c) palatine tonsils are located here   3. laryngopharynx
    ___ (d) this structure leads directly into the esophagus   4. larynx
    ___ (e) cricoid, epiglottis, and thyroid cartilages are here   5. nasopharynx
    ___ (f) vocal cords here enable voice production
    ___ (g) both a respiratory and a digestive pathway (two answers)
    ___ (h) fauces opens into this structure

Are the following statements true or false?

16. The pharynx, larynx, and trachea are parts of the upper respiratory system.

17. Phagocytic cells in the alveoli are known as alveolar macrophages.

18. The opening from the oval cavity into the oropharynx is called the fauces.

19. The cricoid cartilages attach the epiglottis to the glottis.

20. Match the following:

___ (a) superior part of nasal cavity
___ (b) alveoli
___ (c) nasopharynx
___ (d) larynx (inferior to vocal folds)

___ (e) primary bronchus
___ (f) oropharynx
___ (g) trachea
___ (h) terminal bronchioles

(1) pseudostratified ciliated columnar epithelium
(2) stratified columnar epithelium
(3) simple squamous epithelium
(4) nonciliated simple cuboidal epithelium

Critical Thinking Questions

1. Your friend Hedge wants to pierce his nose to go along with the six rings in his ear. He thinks a ring through the center would be awesome but wonders if there's any difference between piercing the center and piercing the side of the nose. Explain the structural differences.
HINT: *One of these locations is more similar in structure to the upper ear than the other.*

2. Suzanne just finished listening to one of those cute answering-machine messages recorded by a young child. She's still trying to figure out what the child said in the greeting and wonders if the child was a boy or a girl. Why can you usually tell adult females and males apart by their voices but all preteen boys and girls sound alike?
HINT: *What happens to boys and girls when they go through puberty?*

3. Dawson was riding his new mountain bike through the woods near his home when he crashed into some brush. He felt a sharp pain in his thorax, and when he rolled over, a large twig was sticking out of his side. He had trouble catching his breath and suspected a collapsed lung. List the structures that have been penetrated by the stick.
HINT: *A lung will collapse if the pleural cavity is open to the outside air.*

4. Greta watched very intently as the tenor sang an aria. At intermission she asked loudly, "What's that thing bobbing up and down on his neck?" What is the structure she's observing?
HINT: *The structure is bobbing along with the singer's tune.*

5. Gus had been a two-pack-a-day smoker for 40 years before quitting last month. A neighbor asked, "How were you able to quit cold turkey?" "Easy," Gus answered, "the hospital wouldn't let me smoke after the surgery." (Gus had the cancerous upper third of his right lung removed.) List the names and location of the structures removed.
HINT: *Gus will have no more need for one secondary bronchus.*

6. In an old "MASH" TV episode, a surgeon talks a priest through an emergency tracheostomy via radio. Grab your cellular phone, call your study partner, and use anatomical terminology to guide him/her to the correct location.
HINT: *A tracheostomy may be performed if a patient cannot breathe due to an obstruction in the throat.*

Answers to Figure Questions

22.1 Nose, pharynx, larynx, trachea, bronchi, and bronchioles, except the respiratory bronchioles.

22.2 External nares → vestibule → nasal cavity → internal nares.

22.3 Superior: internal nares; inferior: cricoid cartilage.

22.4 During swallowing, the epiglottis closes over the rima glottidis, the entrance to the trachea.

22.5 Voice production.

22.6 Because the tissues between the esophagus and trachea are soft, the esophagus can bulge into the trachea during swallowing.

22.7 Left lung has two of each; right lung has three of each.

22.8 Serous membrane.

22.9 Because two-thirds of the heart lies to the left of the midline, the left lung contains a cardiac notch to accommodate the position of the heart.

22.10 Tertiary bronchi.

22.11 Goblet cells produce and secrete mucus, and cilia move mucus laden with foreign particles toward the throat.

22.12 They secrete alveolar fluid, which includes surfactant.

22.13 If you are at rest while reading, your diaphragm.

22.14 Medullary inspiratory area.

22.15 At about 4 weeks.

Chapter 24

The Digestive System

Student Objectives

1. Define digestion, and identify the organs of the gastrointestinal (GI) tract and the accessory digestive organs.

2. Describe the structure of the wall of the gastrointestinal tract.

3. Define the mesentery, mesocolon, falciform ligament, lesser omentum, and greater omentum as extensions of the peritoneum.

4. Describe the structure of the tongue.

5. Identify the location and histology of the salivary glands.

6. Identify the parts of a typical tooth and compare deciduous and permanent dentitions.

7. Describe the anatomy, histology, and functions of the esophagus, stomach, pancreas, liver, gallbladder, small intestine, and large intestine.

8. Describe the development of the digestive system.

9. Describe the effects of aging on the digestive system.

Food contains a variety of nutrients—molecules needed for building new body tissues, repairing damaged tissues, and sustaining needed chemical reactions. Food also is vital for life because it is the source of energy that drives the chemical reactions occurring in every cell. Energy is needed for muscle contraction, conduction of nerve impulses, and secretory and absorptive activities of many cells. Food as it is consumed, however, is not in a state suitable for use as an energy source by any cell. First, it must be broken down into molecules small enough to cross the plasma (cell) membranes. The breaking down of larger food particles into molecules small enough to enter body cells is called **digestion.** The passage of these smaller molecules into blood and lymph is termed **absorption.** The organs that collectively perform these functions compose the **digestive system.**

The focus of this chapter is a discussion of the organs of digestion in terms of their structure and functions. The medical specialty that deals with the structure, function, diagnosis, and treatment of diseases of the stomach and intestines is called **gastroenterology** (gas'-trō-en'ter-OL-ō-jē; *gastro* = stomach; *enteron* = intestines; *logos* = study of). The medical specialty that deals with the diagnosis and treatment of disorders of the rectum and anus is called **proctology** (prok-TOL-ō-jē; *proct* = rectum).

ORGANIZATION

The organs of digestion are traditionally divided into two main groups. First is the **gastrointestinal (GI) tract,** or **alimentary** (*alimentum* = nourishment) **canal,** a continuous tube that extends from the mouth to the anus through the ventral body cavity (Figure 24.1). Organs composing the gastrointestinal tract include the mouth, pharynx, esophagus, stomach, small intestine, and large intestine.* The length of the GI tract taken from a cadaver is about 9 m (30 ft). In a living person it is shorter because the muscles in its wall are in a state of tone.

The second group of organs composing the digestive system are the **accessory digestive organs**—the teeth, tongue, salivary glands, liver, gallbladder, and pancreas. The accessory digestive organs, except for the tongue and teeth, never come into direct contact with food. They produce or store secretions that aid in the chemical breakdown of food. These secretions flow into the GI tract through ducts.

---

*The relationship of the digestive organs to the nine regions of the abdominopelvic cavity may be reviewed in Figure 1.9b, c.

DIGESTIVE PROCESSES

The digestive system carries out six basic activities:

1. **Ingestion.** Taking food into the mouth (eating) is termed ingestion.

2. **Secretion.** Cells within the walls of the GI tract and accessory digestive organs secrete a total of about 9 liters (about 9.5 quarts) per day of water, acid, buffers, and enzymes into the lumen of the tract.

3. **Propulsion.** Alternating contraction and relaxation of smooth muscle in the walls of the GI tract propel the contents of the gastrointestinal tract toward the anus.

4. **Digestion.** Both mechanical and chemical processes mix secreted fluids with ingested food in order to break down food particles into small molecules. **Mechanical digestion** consists of various movements of the gastrointestinal tract that mix food. The teeth cut and grind food before it is swallowed. Then the smooth muscle of the stomach and small intestine churns the food so it is thoroughly mixed with digestive enzymes. **Chemical digestion** is a series of reactions in which enzymes break down large carbohydrate, lipid, protein, and nucleic acid molecules that are present in food into small molecules and ions.

5. **Absorption.** Most secreted fluids, small molecules, and ions that are products of digestion enter the epithelial cells lining the lumen of the GI tract either by active transport or by passive diffusion. This is absorption. The absorbed nutrients pass into the blood and lymph and circulate to cells throughout the body.

6. **Defecation.** The elimination of variable amounts of indigestible substances and bacteria from the gastrointestinal tract through the anus is termed defecation.

GENERAL HISTOLOGY OF THE GI TRACT

The wall of the GI tract, from the stomach to the anal canal, has the same basic arrangement of tissues. The four layers (tunics) of the tract from the inside out are the mucosa, submucosa, muscularis, and serosa (Figure 24.2).

Mucosa

The **mucosa,** or inner lining of the tract, is a mucous membrane. Three layers compose this membrane in the GI tract: (1) a lining layer of **epithelium** in direct contact with the contents of the GI tract, (2) an underlying layer of areolar connective tissue called the **lamina** (*lamina* = thin, flat plate) **propria,** and (3) a thin layer of smooth muscle called the **muscularis mucosae.**

The epithelial layer in the mouth, esophagus, and anal canal is mainly nonkeratinized stratified squamous epithelium that serves a protective function. The stomach and

Figure 24.1 Organs of the digestive system.

Organs of the gastrointestinal (GI) tract are the mouth, pharynx, esophagus, stomach, small intestine, and large intestine. Accessory digestive organs are the teeth, tongue, salivary glands (parotid, sublingual, and submandibular), liver, gallbladder, and pancreas.

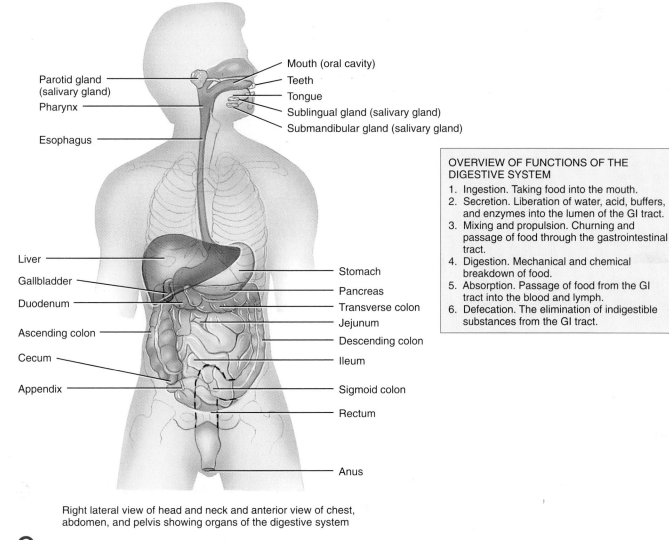

Parotid gland (salivary gland)
Pharynx
Esophagus
Mouth (oral cavity)
Teeth
Tongue
Sublingual gland (salivary gland)
Submandibular gland (salivary gland)

Liver
Gallbladder
Duodenum
Ascending colon
Cecum
Appendix

Stomach
Pancreas
Transverse colon
Jejunum
Descending colon
Ileum
Sigmoid colon
Rectum

Anus

OVERVIEW OF FUNCTIONS OF THE DIGESTIVE SYSTEM

1. Ingestion. Taking food into the mouth.
2. Secretion. Liberation of water, acid, buffers, and enzymes into the lumen of the GI tract.
3. Mixing and propulsion. Churning and passage of food through the gastrointestinal tract.
4. Digestion. Mechanical and chemical breakdown of food.
5. Absorption. Passage of food from the GI tract into the blood and lymph.
6. Defecation. The elimination of indigestible substances from the GI tract.

Right lateral view of head and neck and anterior view of chest, abdomen, and pelvis showing organs of the digestive system

Q *What type of movement propels food along the gastrointestinal tract?*

intestines are lined by simple columnar epithelium, which functions in secretion and absorption. Among the epithelial cells are exocrine cells that secrete mucus and fluid into the lumen of the tract and several types of endocrine cells that secrete hormones into the bloodstream.

The lamina propria is areolar connective tissue containing many blood and lymphatic vessels and scattered lymphatic nodules. This layer supports the epithelium, binds it to the muscularis mucosae, and provides it with a blood and lymph supply. The blood and lymphatic vessels are the avenues by which nutrients in the GI tract reach the other tissues of the body.

The lamina propria also contains most of the cells of the **mucosa-associated lymphoid tissue (MALT).** These patches of lymphatic tissues contain cells of the immune system that protect against disease. They are prevalent all along the GI tract, especially in the palatine and lingual tonsils, small intestine, appendix, and large intestine. It is estimated that there are as many immune cells associated with the GI tract as in all the rest of the body. This makes sense when you realize that the GI tract is in contact with the out-

Figure 24.2 Composite of various sections of the gastrointestinal tract seen in a three-dimensional drawing depicting the various layers and related structures.

The four layers of the GI tract from deep to superficial are the mucosa, submucosa, muscularis, and serosa.

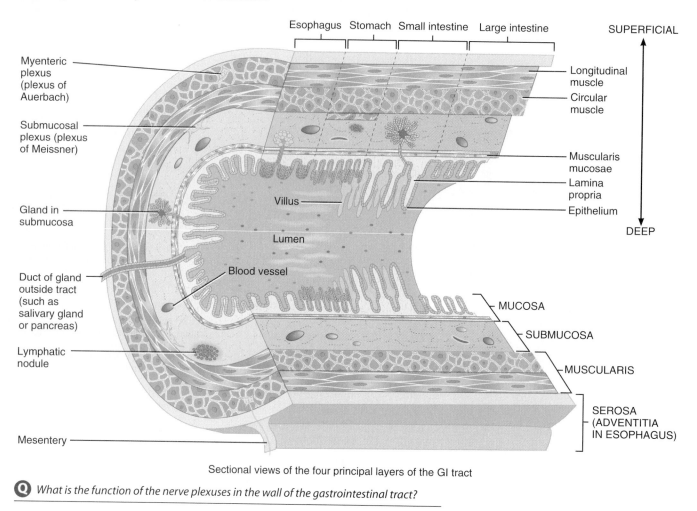

Sectional views of the four principal layers of the GI tract

Q *What is the function of the nerve plexuses in the wall of the gastrointestinal tract?*

side environment and contains food that often carries harmful pathogens. The lymphocytes and macrophages in MALT mount immune responses against pathogens, such as bacteria, that penetrate the mucous membrane.

The muscularis mucosae contains smooth muscle fibers typically arranged into an inner circular layer and an outer longitudinal layer that throw the mucous membrane of the stomach and small intestine into folds. The folds increase the surface area for digestion and absorption, whereas the movements of the muscularis mucosae ensure that all absorptive cells are fully exposed to the contents of the GI tract.

Submucosa

The **submucosa** consists of areolar connective tissue that binds the mucosa to the third layer, the muscularis. It is highly vascular and contains a portion of the **submucosal**

plexus (plexus of Meissner), which is part of the autonomic nerve supply to the smooth muscle cells of the muscularis mucosae and blood vessels. Accordingly, the plexus regulates movements of the mucosa and vasoconstriction of blood vessels. The plexus also innervates secretory cells of mucosal glands and thus is important in controlling secretions by the GI tract. The submucosa may also contain glands and lymphatic tissue.

Muscularis

The **muscularis** of the mouth, pharynx, and superior esophagus consists in part of *skeletal muscle* that produces voluntary swallowing. Skeletal muscle also forms the external anal sphincter, which permits voluntary control of defecation. Throughout the rest of the tract, the muscularis consists

Figure 24.3 Relationship of the peritoneal extensions to each other and to organs of the digestive system. The size of the peritoneal cavity is exaggerated for emphasis.

The peritoneum is the largest serous membrane in the body.

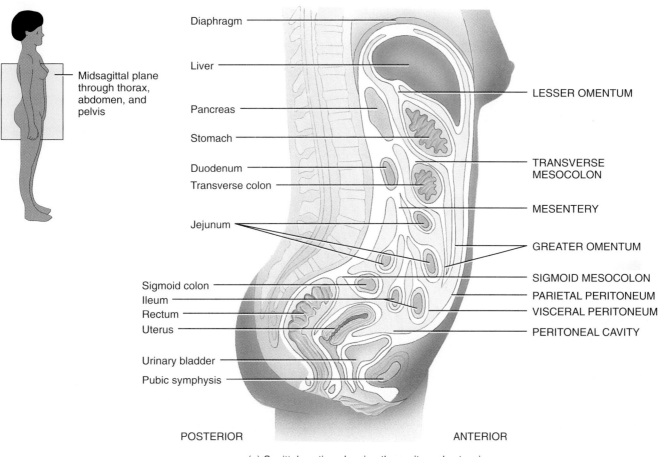

(a) Sagittal section showing the peritoneal extensions

Figure continues

of *smooth muscle* that is generally found in two sheets: an inner sheet of circular fibers and an outer sheet of longitudinal fibers. Involuntary contractions of the smooth muscle help break down food physically, mix it with digestive secretions, and propel it along the tract. The muscularis also contains the major nerve supply to the gastrointestinal tract—the **myenteric** (*myo* = muscle; *enteron* = intestine) **plexus (plexus of Auerbach),** which consists of fibers from both divisions of the autonomic nervous system. This plexus mostly controls GI tract **motility**—that is, the frequency and strength of contraction of the muscularis.

Serosa

The **serosa** is the superficial layer of those portions of the GI tract that are suspended in the abdominopelvic cavity. It is a serous membrane composed of connective tissue and sim-

ple squamous epithelium (mesothelium). As you will see shortly, the esophagus, which passes through the mediastinum, has a superficial layer called the adventitia composed of areolar connective tissue. Inferior to the diaphragm, the serosa is also called the **visceral peritoneum** and forms a portion of the peritoneum, which we will now describe in detail.

PERITONEUM

The **peritoneum** (per′-i-tō-NĒ-um; *peri* = around; *tonos* = tension) is the largest serous membrane of the body. Serous membranes also surround the heart (pericardium) and lungs (pleurae). Serous membranes consist of a layer of simple squamous epithelium and an underlying supporting layer of connective tissue. Whereas the **parietal peritoneum** lines the

Figure 24.3 (continued)

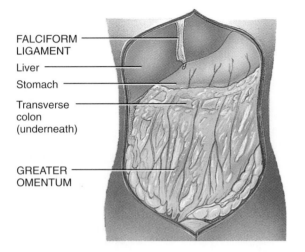

FALCIFORM LIGAMENT

Liver

Stomach

Transverse colon (underneath)

GREATER OMENTUM

(b) Greater omentum, anterior view

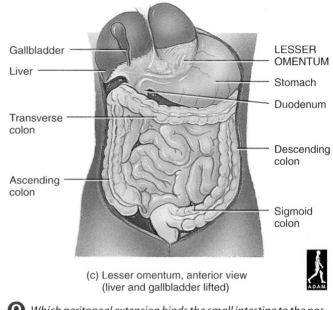

Gallbladder

Liver

Transverse colon

Ascending colon

LESSER OMENTUM

Stomach

Duodenum

Descending colon

Sigmoid colon

(c) Lesser omentum, anterior view (liver and gallbladder lifted)

Q *Which peritoneal extension binds the small intestine to the posterior abdominal wall?*

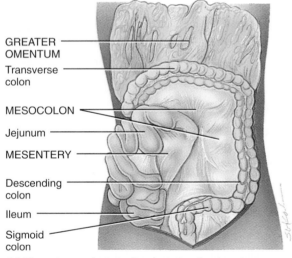

GREATER OMENTUM

Transverse colon

MESOCOLON

Jejunum

MESENTERY

Descending colon

Ileum

Sigmoid colon

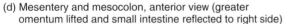

(d) Mesentery and mesocolon, anterior view (greater omentum lifted and small intestine reflected to right side)

wall of the abdominopelvic cavity, the **visceral peritoneum** covers some of the organs in the cavity and is their serosa. The potential space between the parietal and visceral portions of the peritoneum is called the **peritoneal cavity** and contains serous fluid (Figure 24.3a). In certain diseases, the peritoneal cavity may become distended by several liters of fluid so that it forms an actual space. Such an accumulation of serous fluid is called **ascites** (a-SĪ-tēz).

As you will see later, some organs lie on the posterior abdominal wall and are covered by peritoneum on their anterior surfaces only. Such organs, including the kidneys and pancreas, are said to be **retroperitoneal** (*retro* = backward or located behind).

Unlike the pericardium and pleurae, which smoothly cover the heart and lungs, the peritoneum contains large folds that weave between the viscera. The folds bind the organs to each other and to the walls of the abdominal cavity and contain blood and lymphatic vessels and nerves that supply the abdominal organs. One extension of the peritoneum is called the **mesentery** (MEZ-en-ter′ē; *meso* = middle). It is an outward fold of the serous coat of the small intestine (Figure 24.3d). The tip of the fold is attached to the posterior abdominal wall. The mesentery binds the small intestine to the posterior abdominal wall. A fold of peritoneum, called the **mesocolon** (mez′-ō-KŌ-lon), binds the large intestine to the posterior abdominal wall. It also carries blood and lymphatic vessels to the intestines. The mesentery and mesocolon hold the intestines loosely in place and allow for a great amount of movement as muscular contractions mix and move the contents along the GI tract.

Other important peritoneal folds are the falciform ligament, lesser omentum, and greater omentum. The **falciform** (FAL-si-form; *falx* = a sickle; hence, sickle-shaped) **ligament** attaches the liver to the anterior abdominal wall and diaphragm (Figure 24.3b). The liver is the only digestive organ that is attached to the anterior abdominal wall. The **lesser omentum** (ō-MENT-um; *omentum* = fat skin) arises as two folds in the serosa of the stomach and duodenum suspending the stomach and duodenum from the liver (Figure 24.3c). The **greater omentum** is the largest peritoneal fold that drapes over the transverse colon and coils of the small intestine (Figure 24.3b, d).

Because the greater omentum contains large quantities of adipose tissue, it looks like a "fatty apron" draped over the viscera. It is a double sheet that folds upon itself and thus is a four-layered structure. It attaches along the stomach and duodenum, passes inferiorly over the small intestine, and then turns superiorly and attaches to the transverse colon. The greater omentum contains many lymph nodes. If an infection occurs in the intestine, plasma cells formed in the lymph nodes may produce antibodies to combat the infection and help prevent it from spreading to the peritoneum.

Clinical Application

Peritonitis

Peritonitis is an acute inflammation of the peritoneum. As in pleurisy and pericarditis, rubbing together of the inflamed peritoneal surfaces can cause considerable pain. One possible cause is contamination of the peritoneum by pathogenic bacteria from the external environment. This contamination could result from accidental or surgical wounds in the abdominal wall or from perforation or rupture of abdominal organs. For example, the large intestine contains colonies of bacteria that live on undigested nutrients and break them down so they can be eliminated. But if the bacteria enter the peritoneal cavity through an intestinal perforation or rupture of the appendix, they can produce an acute, life-threatening infection.

MOUTH (ORAL CAVITY)

The **mouth,** also referred to as the **oral** or **buccal** (BUK-al; *bucca* = cheeks) **cavity,** is formed by the cheeks, hard and soft palates, and tongue (Figure 24.4). Forming the lateral walls of the oral cavity are the **cheeks**—muscular structures covered externally by skin and lined on the inside by nonkeratinized stratified squamous epithelium. The anterior portions of the cheeks terminate in the superior and inferior lips.

The **lips (labia)** are fleshy folds surrounding the opening of the mouth. They are covered externally by skin and internally by a mucous membrane. The transition zone where the two kinds of covering tissue meet is called the **vermilion** (ver-MIL-yon). This portion of the lips is nonkeratinized, and the color of the blood in the underlying blood vessels is visible through the transparent surface layer. The inner surface of each lip is attached to its corresponding gum by a midline fold of mucous membrane called the **labial frenulum** (LĀ-bē-al FREN-yoo-lum; *labium* = fleshy border; *frenulum* = small bridle).

The orbicularis oris muscle and connective tissue lie between the external integumentary covering and the internal mucosal lining. During chewing, the cheeks and lips help keep food between the upper and lower teeth. They also assist in speech.

The **vestibule** (*vestibule* = entrance to a canal) of the oral cavity is a space bounded externally by the cheeks and lips and internally by the gums and teeth. The **oral cavity proper** is a space that extends from the gums and teeth to the **fauces** (FAW-sēs; *fauces* = passages), the opening between the oral cavity and the pharynx or throat.

The **hard palate,** the anterior portion of the roof of the mouth, is formed by the maxillae and palatine bones. It is covered by mucous membrane and forms a bony partition between the oral and nasal cavities. The **soft palate** forms the posterior portion of the roof of the mouth. It is an arch-shaped muscular partition between the oropharynx and nasopharynx and is lined by mucous membrane.

Hanging from the free border of the soft palate is a conical muscular process called the **uvula** (YOU-vyoo-la; *uvula* = little grape). During swallowing, the soft palate and uvula are drawn superiorly, closing off the nasopharynx. This prevents swallowed foods and liquids from entering the nasal cavity. Lateral to the base of the uvula are two muscular folds that run down the lateral side of the soft palate. Anteriorly, the **palatoglossal arch (anterior pillar)** extends to the side of the base of the tongue. Posteriorly, the **palatopharyngeal** (PAL-a-tō-fa-rin′-jē-al) **arch (posterior pillar)** extends to the side of the pharynx. The palatine tonsils are situated between the arches, and the lingual tonsil is situated at the base of the tongue. At the posterior border of the soft palate, the mouth opens into the oropharynx through the fauces (Figure 24.4).

Tongue

The **tongue,** together with its associated muscles, forms the floor of the oral cavity. It is an accessory digestive organ composed of skeletal muscle covered with mucous membrane (Figure 24.5a). The tongue is divided into symmetrical lateral halves by a median septum that extends throughout its entire length and is attached inferiorly to the hyoid bone. Each half of the tongue consists of an identical complement of extrinsic and intrinsic muscles.

The **extrinsic muscles** of the tongue originate outside the tongue (to bones in the area) and insert into connective tissues in the tongue. They include the hyoglossus, genioglossus, and styloglossus (see Figure 10.7). The extrinsic muscles move the tongue from side to side and in and out. These movements maneuver food for chewing, shape the food into a rounded mass, and force the food to the back of the mouth for swallowing. They also form the floor of the mouth and hold the tongue in position. The **intrinsic muscles** originate and insert into connective tissues within the tongue and alter the shape and size of the tongue for

Figure 24.4 Structures of the mouth (oral cavity).

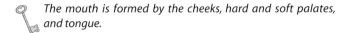 *The mouth is formed by the cheeks, hard and soft palates, and tongue.*

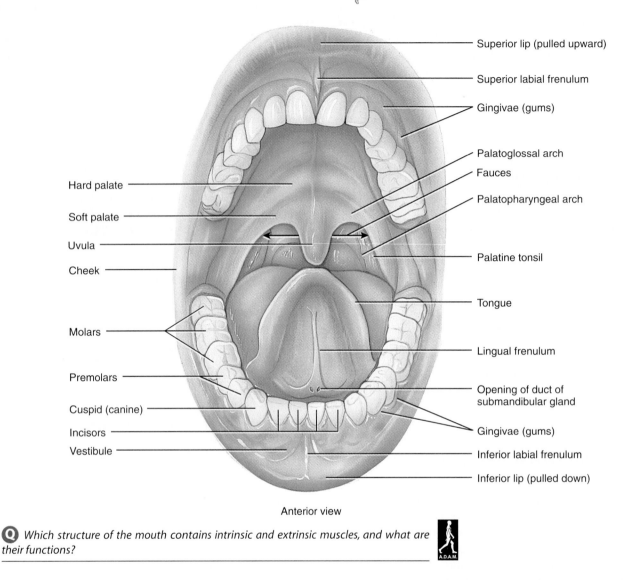

Superior lip (pulled upward)

Superior labial frenulum

Gingivae (gums)

Palatoglossal arch

Fauces

Palatopharyngeal arch

Palatine tonsil

Tongue

Lingual frenulum

Opening of duct of submandibular gland

Gingivae (gums)

Inferior labial frenulum

Inferior lip (pulled down)

Hard palate

Soft palate

Uvula

Cheek

Molars

Premolars

Cuspid (canine)

Incisors

Vestibule

Anterior view

Q *Which structure of the mouth contains intrinsic and extrinsic muscles, and what are their functions?*

speech and swallowing. The intrinsic muscles include the longitudinalis superior, longitudinalis inferior, transversus linguae, and verticalis linguae (see Figure 24.5a). The **lingual** (*lingua* = tongue) **frenulum,** a fold of mucous membrane in the midline of the undersurface of the tongue, is attached to the floor of the mouth and aids in limiting the movement of the tongue posteriorly (see Figure 24.4).

The dorsum (upper surface) and lateral surfaces of the tongue are covered with **papillae** (pa-PIL-ē; *papilla* = nipple-shaped projection), projections of the lamina propria covered with epithelium (Figure 24.5b, c). **Filiform** (*filum* = thread) **papillae** are conical projections distributed in parallel rows over the anterior two-thirds of the tongue. They are whitish and contain no taste buds. **Fungiform** (*fungiform* = shaped like a mushroom) **papillae** are mushroom-

like elevations distributed among the filiform papillae and are more numerous near the tip of the tongue. They appear as red dots on the dorsum of the tongue, and most of them contain taste buds. **Circumvallate** (*circum* = around; *vallare* = to wall) **papillae** are arranged in the form of an inverted V on the posterior aspect of the dorsum of the tongue, and all of them contain taste buds. Ducts of lingual (von Ebner's) glands, which are serous glands, surround the circumvallate papillae. The taste zones of the tongue are illustrated in Figure 20.2a.

Glands on the dorsum of the tongue secrete a digestive enzyme called **lingual lipase,** which initiates digestion of triglycerides (fats) into fatty acids and monoglycerides.

Figure 24.5 Tongue. The frontal section in (a) shows the intrinsic muscles in the left side of the tongue.

The tongue consists of skeletal muscle covered by a mucous membrane and forms the floor of the oral cavity.

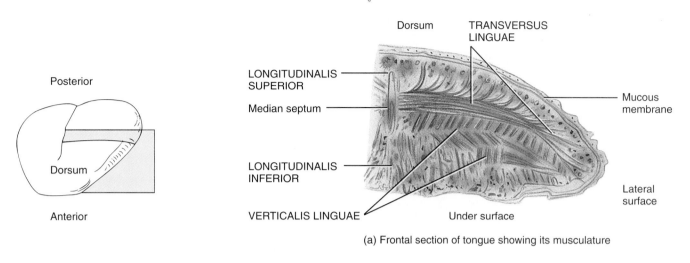

(a) Frontal section of tongue showing its musculature

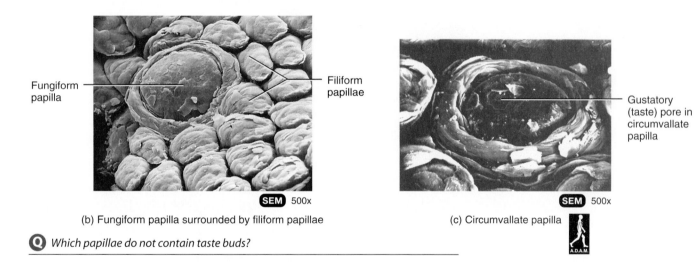

(b) Fungiform papilla surrounded by filiform papillae

(c) Circumvallate papilla

Q *Which papillae do not contain taste buds?*

Salivary Glands

When food enters the mouth, secretion of saliva increases. It lubricates, dissolves, and begins the chemical breakdown of food. The mucous membrane lining the mouth contains many small glands. The **buccal glands** and minor salivary glands secrete small amounts of saliva. However, most saliva is secreted by the **salivary glands (major salivary glands),** accessory digestive organs that lie outside the mouth. Their secretions pour into ducts that empty into the oral cavity. There are three pairs of salivary glands: parotid, submandibular, and sublingual glands (Figure 24.6a; see also Figure 10.4d).

The **parotid** (*para* = near; *otid* = ear) **glands** are located inferior and anterior to the ears between the skin and the masseter muscles. Each secretes into the oral cavity vestibule via a duct, called the **parotid (Stensen's) duct,** that pierces the buccinator muscle to open into the vestibule opposite the second maxillary (upper) molar tooth (see Figure 24.6a). The **submandibular glands** are found inferior to the base of the tongue in the posterior part of the floor of the mouth. Their ducts, the **submandibular (Wharton's) ducts,** run superficially inferior to the mucosa on either side of the midline of the floor of the mouth and enter the oral cavity proper on either side of the lingual frenulum. The **sublingual glands** are superior to the submandibular glands. Their ducts, the **lesser sublingual (Rivinus') ducts,** open into the floor of the mouth in the oral cavity proper.

Figure 24.6 Salivary glands. The submandibular gland shown in the photomicrograph in (b) consists mostly of serous acini (serous fluid–secreting exocrine portion of gland) and a few mucous acini (mucus-secreting exocrine portion of gland). The parotid glands consist of all serous acini, and the sublingual glands consist of mostly mucous acini and a few serous acini.

Saliva lubricates and dissolves foods and begins the chemical breakdown of carbohydrates and lipids.

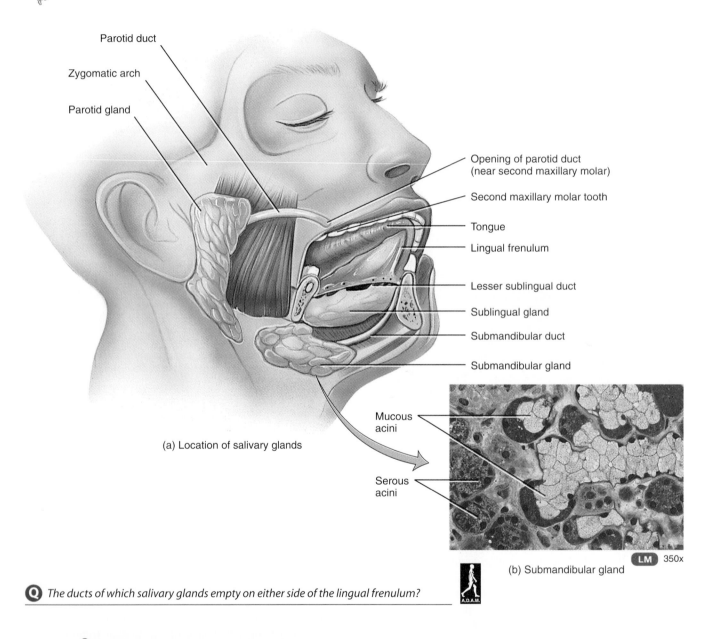

(a) Location of salivary glands

(b) Submandibular gland

LM 350x

Q *The ducts of which salivary glands empty on either side of the lingual frenulum?*

Clinical Application

Mumps

Although any of the salivary glands may be the target of a nasopharyngeal infection, the mumps virus (myxovirus) typically attacks the parotid glands. **Mumps** is an inflammation and enlargement of the parotid glands accompanied by moderate fever, malaise, and extreme pain in the throat, especially when swallowing sour foods or acidic juices. Swelling occurs on one or both sides of the face, just anterior to the ramus of the mandible. In about 30% of males past puberty, the testes may also become inflamed. Sterility rarely occurs because testicular involvement is usually unilateral (one testis only). ▪

The parotid gland receives its blood supply from branches of the external carotid artery and is drained by vessels that are tributaries of the external jugular vein. The submandibular gland is supplied by branches of the facial artery and drained by tributaries of the facial vein. The sublingual gland is supplied by the sublingual branch of the lingual artery and the submental branch of the facial artery and is drained by tributaries of the sublingual and submental veins.

The salivary glands receive both sympathetic and parasympathetic innervation. The sympathetic fibers form plexuses on the blood vessels that supply the glands and initiate vasoconstriction, which decreases the production of saliva. The parotid gland receives sympathetic fibers from the plexus on the external carotid artery, whereas the submandibular and sublingual glands receive sympathetic fibers that contribute to the sympathetic plexus and accompany the facial artery to the glands. The parasympathetic fibers of the glands produce vasodilation and thus increase the production of saliva.

The fluids secreted by the buccal glands, minor salivary glands, and the three pairs of major salivary glands constitute **saliva.** Amounts of saliva secreted daily vary considerably but range from 1000 to 1500 ml (1 to 1.6 qt). Chemically, saliva is 99.5% water and 0.5% solutes and has a slightly acidic pH (6.35 to 6.85). The solute portion includes mucus, the bacteriolytic enzyme lysozyme, the digestive enzymes salivary amylase and lingual lipase, and traces of salts, proteins, and other organic compounds. **Salivary amylase** initiates the breakdown of starch. **Lingual lipase** is secreted by glands on the dorsum of the tongue. This enzyme, which is active in the stomach, can digest as much as 30% of dietary triglycerides (fats) into simpler fatty acids and monoglycerides.

Salivation is entirely under nervous control. Normally, parasympathetic stimulation promotes continuous secretion of a moderate amount of saliva. Sympathetic stimulation dominates during stress, resulting in dryness of the mouth. During dehydration, the salivary and buccal glands stop secreting saliva to conserve water. Chemicals in the food stimulate receptors in taste buds on the tongue. Impulses are conveyed from the receptors to two salivary nuclei in the brain stem (**superior** and **inferior salivatory nuclei**). Returning parasympathetic impulses in fibers of the facial (VIII) and glossopharyngeal (IX) nerves stimulate the secretion of saliva. The smell, sight, sound, or memory of food may also stimulate increased saliva secretion.

Teeth

The **teeth (dentes)** are accessory digestive organs located in sockets of the alveolar processes of the mandible and maxillae. The alveolar processes are covered by the **gingivae** (jin-JI-vē), or gums (Figure 24.7), which extend slightly into each socket forming the gingival sulcus. (**Gingiva** is singu-

lar.) The sockets are lined by the **periodontal** (*peri* = around; *odous* = tooth) **ligament,** which consists of dense fibrous connective tissue and is attached to the socket walls and the cemental surface of the roots. It anchors the teeth in position and is also a shock absorber during chewing.

A typical tooth consists of three principal portions. The **crown** is the exposed portion above the gums. Embedded in the socket are one to three **roots.** The **neck** is the constricted junction line of the crown and the root near the gum line.

Teeth are composed primarily of **dentin,** a calcified connective tissue that gives the tooth its basic shape and rigidity. It is harder than bone because of its higher content of calcium salts (70% of dry weight). The dentin encloses a cavity. The enlarged part of the cavity, the **pulp cavity,** lies in the crown and is filled with **pulp,** a connective tissue containing blood vessels, nerves, and lymphatic vessels. Narrow extensions of the pulp cavity run through the root of the tooth and are called **root canals.** Each root canal has an opening at its base, the **apical foramen.** Blood vessels, lymphatic vessels, and nerves extend through the foramen.

The dentin of the crown is covered by **enamel** that consists primarily of calcium phosphate and calcium carbonate. Enamel is the hardest substance in the body and the richest in calcium salts (about 95% of dry weight). It protects the tooth from the wear of chewing. It is also a barrier against acids that easily dissolve the dentin. The dentin of the root is covered by **cementum,** another bonelike substance, which attaches the root to the periodontal ligament.

The arteries that supply blood to the teeth are distributed to the pulp cavity and surrounding periodontal ligament. These include the anterior and posterior superior alveolar branches of the maxillary artery and the incisive and dental branches of the inferior alveolar artery.

The teeth receive sensory fibers from branches of the maxillary and mandibular divisions of the trigeminal (V) nerve—the maxillary teeth from branches of the maxillary division and the mandibular teeth from branches of the mandibular division.

Humans have two **dentitions,** or sets of teeth, deciduous and permanent. The first of these—the **deciduous** (*deciduus* = falling out) or **primary teeth, milk teeth,** or **baby teeth**—begin to erupt at about 6 months of age, and one pair appears at about each month thereafter until all 20 are present. Figure 24.8a illustrates the deciduous teeth. The incisors, which are closest to the midline, are chisel-shaped and adapted for cutting into food. They are referred to as either **central** or **lateral incisors** on the basis of their position. Next to the incisors, moving posteriorly, are the **cuspids (canines),** which have a pointed surface called a cusp. Cuspids are used to tear and shred food. The incisors and cuspids have only one root apiece. Posterior to them lie the **first** and **second molars,** which have four cusps. Upper molars have three roots; lower molars have two roots. The molars crush and grind food.

Figure 24.7 Parts of a typical tooth.

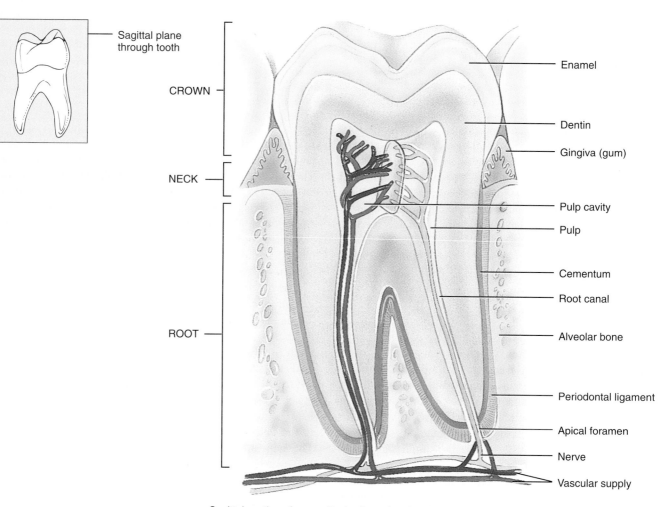

Sagittal plane through tooth

CROWN

NECK

ROOT

Enamel

Dentin

Gingiva (gum)

Pulp cavity

Pulp

Cementum

Root canal

Alveolar bone

Periodontal ligament

Apical foramen

Nerve

Vascular supply

Sagittal section of a mandibular (lower) molar showing the parts of a tooth

Q *What type of tissue is the main component of teeth?*

All the deciduous teeth are lost—generally between 6 and 12 years of age—and are replaced by the **permanent (secondary) teeth** (Figure 24.8b). The permanent dentition contains 32 teeth that erupt between age 6 and adulthood. It resembles the deciduous dentition with the following exceptions. The deciduous molars are replaced with the **first** and **second premolars (bicuspids),** which have two cusps and one root (upper bicuspids have two roots) and are used for crushing and grinding. The permanent molars erupt into the mouth posterior to the bicuspids. They do not replace any deciduous teeth and erupt as the jaw grows to accommodate them—the **first molars** at age 6, the **second molars** at age 12, the **third molars (wisdom teeth)** after age 17.

The human jaw often does not afford enough room posterior to the second molars for the eruption of the third molars. In this case, the third molars remain embedded in the alveolar bone and are said to be "impacted." Often they cause pressure and pain and must be surgically removed. In some people, third molars may be dwarfed in size or may not develop at all.

The branch of dentistry that is concerned with the prevention, diagnosis, and treatment of diseases that affect the pulp, root, periodontal ligament, and alveolar bone is known as **endodontics** (en′dō-DON-tiks; *endo* = within; *odous* = tooth). **Orthodontics** (or′-thō-DON-tiks; *ortho* = straight), by contrast, is that branch of dentistry that is concerned with

Figure 24.8 Dentitions and times of eruptions (indicated in parentheses).

Deciduous teeth begin to erupt at 6 months of age, and one pair appears about each month thereafter.

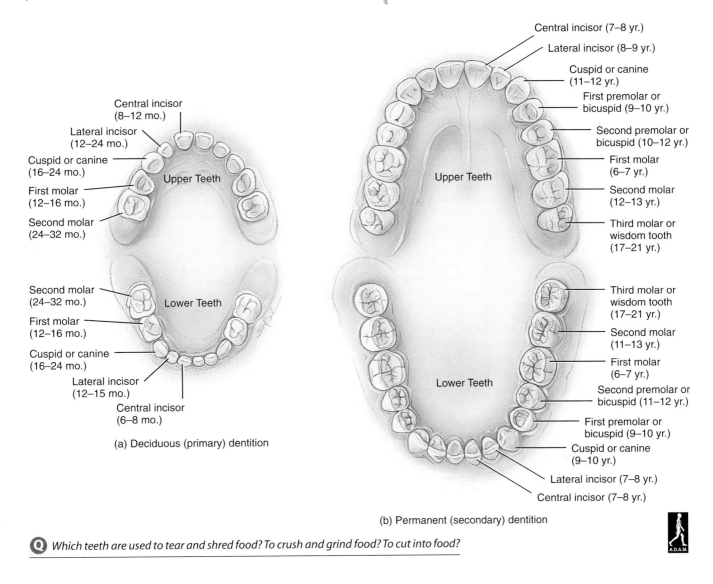

(a) Deciduous (primary) dentition

(b) Permanent (secondary) dentition

Q *Which teeth are used to tear and shred food? To crush and grind food? To cut into food?*

the prevention and correction of abnormally aligned teeth. See Table 24.1 for a list of terms used to describe the surface orientation of teeth.

Clinical Application

Root Canal Therapy

Root canal therapy refers to a procedure, accomplished in several phases, in which all traces of pulp tissue are removed from the pulp cavity and root canals of a badly diseased tooth. After a hole is made in the tooth, the root canals are filed out and irrigated to remove bacteria. Then the canals are treated with medication and sealed tightly. The damaged crown is then repaired. ■

PHARYNX

Through chewing, or **mastication** (mas′ti-KĀ-shun; *masticare* = to chew), the tongue manipulates food, the teeth grind it, and the food is mixed with saliva. As a result, the food is reduced to a soft, flexible mass called a **bolus** (*bolos* = lump) that is easily swallowed. When food is first swallowed, it passes from the mouth into the pharynx.

The **pharynx** (*pharynx* = throat) is a funnel-shaped tube that extends from the internal nares to the esophagus posteriorly and the larynx anteriorly (see Figure 23.3). The pharynx is composed of skeletal muscle and lined by mucous membrane. Whereas the nasopharynx functions only in respiration, both the oropharynx and laryngopharynx have

Table 24.1 *Surface Orientation of Teeth*

TOOTH SURFACE	DESCRIPTION
Labial	Contacting the lips.
Buccal	Contacting or facing the cheeks.
Lingual	Facing the tongue (teeth of the mandible only).
Palatal	Facing the palate (teeth of the maxillae only).
Mesial	Anterior or medial side relative to dental arch.
Distal	Posterior or lateral side relative to dental arch.
Occlusal	The biting surface.

digestive as well as respiratory functions. Food that is swallowed passes from the mouth into the oropharynx and laryngopharynx before passing into the esophagus. Muscular contractions of the oropharynx and laryngopharynx help propel food into the esophagus.

Swallowing, or **deglutition** (dē-gloo-TISH-un), is a mechanism that moves food from the mouth to the stomach. It is helped by saliva and mucus and involves the mouth, pharynx, and esophagus.

ESOPHAGUS

The **esophagus** (e-SOF-a-gus; *oisein* = to carry; *phagema* = food) is a muscular, collapsible tube that lies posterior to the trachea and is about 25 cm (10 in.) long. It begins at the inferior end of the laryngopharynx, passes through the mediastinum anterior to the vertebral column, pierces the diaphragm through an opening called the **esophageal hiatus,** and ends in the superior portion of the stomach. Sometimes, a portion of the stomach protrudes above the diaphragm through the esophageal hiatus. This condition is termed **hiatal hernia** (HER-nē-ah).

The arteries of the esophagus are derived from the arteries along its length: inferior thyroid, thoracic aorta, intercostal arteries, phrenic, and left gastric arteries. It is drained by the adjacent veins. Innervation of the esophagus is by recurrent laryngeal nerves, the cervical sympathetic chain, and vagus (X) nerves.

Histology

The **mucosa** of the esophagus consists of nonkeratinized stratified squamous epithelium, lamina propria (areolar connective tissue), and a muscularis mucosae (smooth muscle).

Refer to Table 3.1 on page 63 to view a photomicrograph of nonkeratinized stratified squamous epithelium. Near the stomach, the mucosa of the esophagus also contains mucous glands. The stratified squamous epithelium associated with the lips, mouth, tongue, oropharynx, laryngopharynx, and esophagus affords considerable protection against abrasion and wear-and-tear of food particles that are chewed, mixed with secretions, and swallowed. The **submucosa** contains areolar connective tissue, blood vessels, and mucous glands. The **muscularis** of the superior third of the esophagus is skeletal, the intermediate third is skeletal and smooth, and the inferior third is smooth muscle. The superficial layer is known as the **adventitia** (ad-ven-TISH-ya) rather than the serosa because the areolar connective tissue of this layer is not covered by epithelium (mesothelium) and because the connective tissue merges with the connective tissue of surrounding structures of the mediastinum through which it passes. The adventitia attaches the esophagus to surrounding structures.

Functions

The esophagus secretes mucus and transports food to the stomach. It does not produce digestive enzymes and does not carry on absorption. The passage of food from the laryngopharynx into the esophagus is regulated by a sphincter (a circular band or rind of muscle that is normally contracted so that there is no opening at the center) at the entrance to the esophagus called the **upper esophageal** (e-sof'a-JĒ-al) **sphincter.** It consists of the cricopharyngeus muscle attached to the cricoid cartilage. Elevation of the larynx causes the sphincter to relax, and the bolus enters the esophagus. The sphincter also relaxes during expiration.

Food is pushed through the esophagus by involuntary muscular movements called **peristalsis** (per'-is-STAL-sis; *peri* = around; *stalsis* = contraction). In the esophagus, peristalsis is controlled by the medulla oblongata. In the section of the esophagus lying just superior to the bolus, the circular muscle fibers contract (Figure 24.9), which constricts the esophageal wall and squeezes the bolus inferiorly. Meanwhile, longitudinal fibers inferior to the bolus also contract. Contracting of the longitudinal fibers shortens this inferior section, pushing its walls outward so it can receive the bolus. The contractions are repeated in a wave that pushes the food toward the stomach. Passage of the bolus is facilitated by glands that secrete mucus.

Just superior to the level of the diaphragm, the esophagus slightly narrows. This narrowing is a physiological sphincter in the inferior part of the esophagus known as the **lower esophageal** (**gastroesophageal** or **cardiac**) **sphincter.** (A *physiological sphincter* is a section of a tubular structure, in this case the esophagus, that functions like a sphincter even though none is actually present.) The lower esophageal sphincter relaxes during swallowing and thus allows the bolus to pass from the esophagus into the stomach.

Figure 24.9 Peristalsis during deglutition (swallowing).

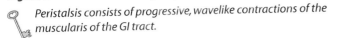

Peristalsis consists of progressive, wavelike contractions of the muscularis of the GI tract.

Esophagus
Relaxed muscularis
Circular muscles contract
Longitudinal muscles contract
Relaxed muscularis
Bolus
Stomach
Lower esophageal sphincter

Anterior view of peristalsis in esophagus

Q *Does peristalsis "push" or "pull" food along the gastrointestinal tract?*

Clinical Application

Achalasia and Heartburn

If the lower esophageal sphincter fails to relax normally as food approaches, the condition is called **achalasia** (ak'a-LĀ-zē-a; *a* = without; *chalasis* = relaxation). As a result, the movement of food from the esophagus into the stomach is greatly impeded. A whole meal may become lodged in the esophagus and enter the stomach very slowly. Distension of the esophagus results in chest pain that is often confused with pain originating from the heart. The condition is caused by malfunction of the myenteric plexus (plexus of Auerbach).

In contrast, if the lower esophageal sphincter fails to close adequately after food has entered the stomach, the stomach contents can enter the inferior portion of the esophagus. Hydrochloric acid (HCl) from the stomach contents can irritate the esophageal wall, resulting in a burning sensation. The sensation is known as **heartburn** because it is experienced in a region very near the heart, although it is not related to any cardiac problem. ■

STOMACH

The **stomach** is a J-shaped enlargement of the GI tract directly inferior to the diaphragm in the epigastric, umbilical, and left hypochondriac regions of the abdomen (see Figure 1.9b, c). The stomach connects the esophagus to the duodenum, the first part of the small intestine (Figure. 24.10). The position and size of your stomach vary continually. For instance, the diaphragm pushes it inferiorly with each inspiration and pulls it superiorly with each expiration. Empty, it is about the size of a large sausage, but it can stretch to accommodate a large quantity of food.

Anatomy

The stomach has four main areas: cardia, fundus, body, and pylorus (Figure 24.10a). The **cardia** (CAR-dē-a) surrounds the superior opening of the stomach. The rounded portion superior and to the left of the cardia is the **fundus** (FUN-dus). Inferior to the fundus is the large central portion of the stomach, called the **body.** The inferior region of the stomach that connects to the duodenum is the **pylorus** (pī-LOR-us; *pyle* = gate; *ouros* = guard). It has two parts, the **pyloric antrum** (AN-trum; *antrum* = cave), which connects to the body of the stomach, and the **pyloric canal,** which leads into the duodenum. When the stomach is empty, the mucosa lies in large folds, called **rugae** (ROO-gē; *ruga* = wrinkle), that can be seen with the naked eye. The pylorus communicates with the duodenum of the small intestine via a sphincter called the **pyloric sphincter (valve).** The concave medial border of the stomach is called the **lesser curvature,** and the convex lateral border is the **greater curvature.**

The arterial supply of the stomach is derived from the celiac artery. The right and left gastric arteries form an anastomosing arch along the lesser curvature, and the right and left gastroepiploic arteries form a similar arch on the greater curvature. Short gastric arteries supply the fundus. The veins of the same name accompany the arteries and drain, directly or indirectly, into the hepatic portal vein.

The vagus (X) nerves convey parasympathetic fibers to the stomach. These fibers form synapses within the submucosal plexus (plexus of Meissner) in the submucosa and the myenteric plexus (plexus of Auerbach) in the muscularis. The sympathetic nerves arise from the celiac ganglia, and the nerves reach the stomach along the branches of the celiac artery.

Clinical Application

Pylorospasm and Pyloric Stenosis

Two abnormalities of the pyloric sphincter can occur in infants. **Pylorospasm** is characterized by failure of the muscle fibers of the sphincter to relax normally. Food does not pass easily from

Figure 24.10 External and internal anatomy of the stomach. The dashed lines in (a) indicate approximate boundaries for the various areas of the stomach.

The four regions of the stomach are the cardia, fundus, body, and pylorus.

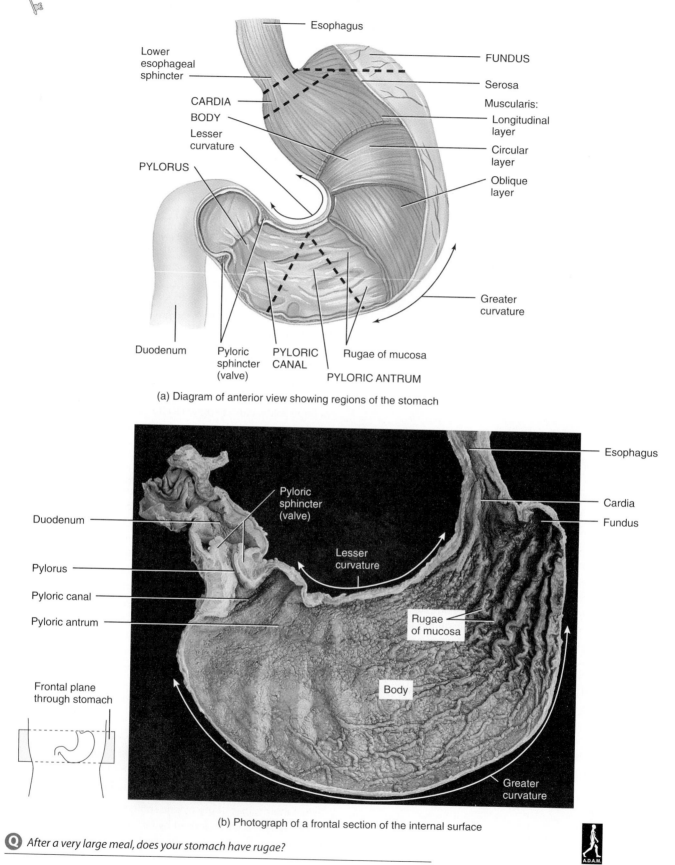

(a) Diagram of anterior view showing regions of the stomach

(b) Photograph of a frontal section of the internal surface

Q *After a very large meal, does your stomach have rugae?*

the stomach to the small intestine, the stomach becomes overly full, and the infant vomits frequently to relieve the pressure. Pylorospasm is treated by drugs that relax the muscle fibers of the sphincter.

Pyloric stenosis is a narrowing of the pyloric sphincter caused by a tumorlike mass that apparently is formed by enlargement of the circular muscle fibers. The hallmark symptom is projectile vomiting—the spraying of liquid vomitus some distance from the infant.

Histology

The stomach wall is composed of the same four basic layers as the rest of the GI tract, with certain modifications (Figure 24.11). The surface of the **mucosa** is a layer of simple columnar epithelial cells called **mucous surface cells** (Figure 24.11b). The mucosa contains a **lamina propria** (areolar connective tissue) and a **muscularis mucosae** (smooth muscle). Epithelial cells extend down into the lamina propria, forming many narrow channels called **gastric pits** and columns of secretory cells called **gastric glands.** Secretions from several gastric glands flow into each gastric pit and then into the lumen of the stomach.

The gastric glands include three types of *exocrine gland* cells that secrete their products into the stomach lumen: mucous neck cells, chief cells, and parietal cells. Both mucous surface cells and **mucous neck cells** secrete mucus (see Figure 24.11b). The **chief (zymogenic) cells** secrete pepsinogen and gastric lipase. **Parietal (oxyntic) cells** produce hydrochloric acid, which helps convert pepsinogen to pepsin and kills microbes in food, and intrinsic factor, involved in the absorption of vitamin B_{12}. The secretions of the mucous, chief, and parietal cells are collectively called **gastric juice,** which totals about 2000–3000 ml (roughly 2–3 qt) per day. In addition, gastric glands include one type of enteroendocrine cell. An **enteroendocrine** (*enteron* = intestine) **cell** is a hormone-producing cell in the gastrointestinal mucosa. One such cell is a **G cell,** located mainly in the pyloric antrum, that secretes the hormone gastrin into the bloodstream. Gastrin stimulates secretion of HCl and pepsinogen, contracts the lower esophageal sphincter, increases motility of the stomach, and relaxes the pyloric sphincter.

Three additional tissue layers lie deep to the mucosa. The **submucosa** of the stomach is composed of areolar connective tissue. The **muscularis** has three rather than two layers of smooth muscle: an outer longitudinal layer, a middle circular layer, and an inner oblique layer. The oblique layer is limited mostly to the body of the stomach. The **serosa** (simple squamous epithelium and areolar connective tissue) covering the stomach is part of the visceral peritoneum. At the lesser curvature, the visceral peritoneum extends superiorly to the liver as the lesser omentum. At the greater curvature, the visceral peritoneum continues inferiorly as the greater omentum and drapes over the intestines.

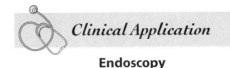

Clinical Application

Endoscopy

The general term **endoscopy** (en-DOS-kō-pē; *endo* = inside; *shopein* = to examine) refers to visual inspection of any cavity of the body using an endoscope, an illuminated tube with lenses. Endoscopes can be used to visualize the entire gastrointestinal tract, as well as other systems of the body. Endoscopes can also be fitted with grasping devices to remove foreign objects from the esophagus and stomach, cutting blades to biopsy lesions and remove polyps from the colon, and cauterizing tips to stop bleeding. Endoscopic examination of the stomach is called **gastroscopy** (gas-TROS-kō-pē). The examiner can view the interior of the stomach directly to evaluate an ulcer, tumor, inflammation, or bleeding source.

Functions

A meal can be eaten much more quickly than the intestines can digest and absorb it. Thus one of the functions of the stomach is to serve as a mixing area and holding reservoir. At appropriate intervals after food is ingested, the stomach forces a small quantity of material into the first portion of the small intestine.

Several minutes after food enters the stomach, gentle, rippling, peristaltic movements called **mixing waves** pass over the stomach every 15 to 25 seconds. These waves macerate food, mix it with gastric juice, and reduce it to a thin liquid called **chyme** (KĪM; *chymos* = juice). As digestion proceeds in the stomach, more vigorous mixing waves begin at the body of the stomach and intensify as they reach the pylorus. As food reaches the pylorus, each mixing wave forces several milliliters of the gastric contents into the duodenum through the pyloric sphincter. Most of the food is forced back into the body of the stomach, where it is subjected to further mixing. The next wave pushes it forward again and forces a little more into the duodenum. The forward and backward movements of the gastric contents are responsible for almost all the mixing in the stomach.

The enzymatic digestion of proteins begins in the stomach. In the adult, this is achieved mainly through the enzyme **pepsin,** secreted by chief cells (secreted in an inactive form called *pepsinogen*). Pepsin breaks certain peptide bonds between the amino acids making up proteins. Thus a protein chain of many amino acids is broken down into smaller fragments called **peptides.** Pepsin also brings about the clumping and digestion of milk proteins. Another enzyme of the stomach is **gastric lipase.** Gastric lipase splits the short-chain triglycerides (fats) in butterfat molecules found in milk. The enzyme has a limited role in the adult stomach. To digest fats, adults rely almost exclusively on lingual lipase and **pancreatic lipase,** an enzyme secreted by the pancreas into the small intestine.

Figure 24.11 Histology of the stomach.

The muscularis of the stomach has three layers of smooth muscle tissue.

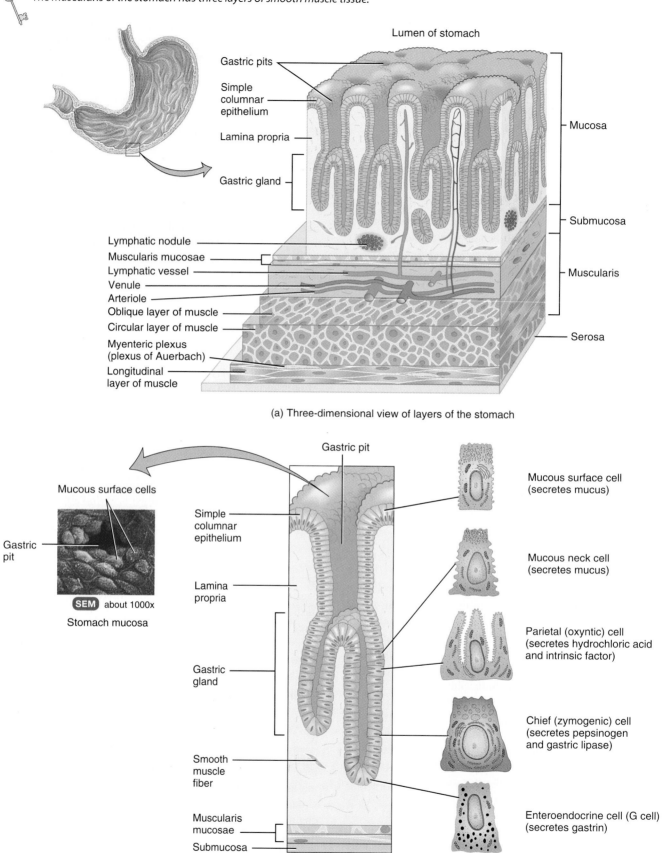

(a) Three-dimensional view of layers of the stomach

(b) Stomach mucosa showing gastric glands and cell types

Figure 24.11 (continued)

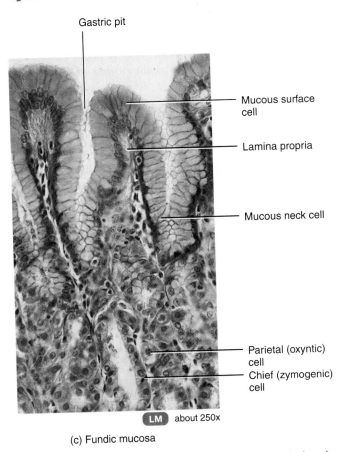

Gastric pit

Mucous surface cell

Lamina propria

Mucous neck cell

Parietal (oxyntic) cell

Chief (zymogenic) cell

LM about 250x

(c) Fundic mucosa

Q *What types of cells are found in gastric glands, and what do they secrete?*

The stomach empties all its contents into the duodenum two to six hours after ingestion. Food rich in carbohydrate leaves the stomach in a few hours. Protein-rich foods pass somewhat slower, and emptying is slowest after a meal containing large amounts of triglycerides.

Clinical Application

Vomiting

Vomiting is the forcible expulsion of the contents of the upper GI tract (stomach and sometimes duodenum) through the mouth. Basically, vomiting involves squeezing the stomach between the diaphragm and abdominal muscles and expelling of the contents through the open esophageal sphincters. The strongest stimuli for vomiting are irritation and distension of the stomach. Other stimuli include unpleasant sights, dizziness, and certain drugs such as morphine and derivatives of digitalis. Nerve impulses are transmitted to the vomiting center in the medulla oblongata, and returning impulses to the upper GI tract organs, diaphragm, and abdominal muscles bring about the vomiting act. Prolonged vomiting, especially in infants and elderly people, can be serious because the loss of gastric juice and fluids can lead to disturbances in fluid and acid–base balance. ▪

The stomach wall is impermeable to the passage of most materials into the blood, so most substances are not absorbed until they reach the small intestine. However, the stomach does participate in the absorption of some water, electrolytes, certain drugs (especially aspirin), and alcohol. The absorption of alcohol by the stomach of females is faster than that in males. The difference is attributed to smaller amounts of the enzyme alcohol dehydrogenase in the stomachs of females. The enzyme breaks down alcohol in the stomach, reducing the amount of alcohol that enters the blood.

PANCREAS

The next organ of the GI tract involved in the breakdown of food is the small intestine. Chemical digestion in the small intestine depends not only on its own secretions but also on secretions of three accessory digestive organs outside the gastrointestinal tract: the pancreas, liver, and gallbladder. We first consider the activities of the accessory digestive organs and then examine their contributions to digestion in the small intestine.

Anatomy

The **pancreas** (*pan* = all; *kreas* = flesh) is a retroperitoneal gland about 12 to 15 cm (5–6 in.) long and 2.5 cm (1 in.) thick. It lies posterior to the greater curvature of the stomach and is connected, usually by two ducts, to the duodenum (Figure 24.12). The pancreas is divided into a head, body, and tail. The **head** is the expanded portion near the C-shaped curve of the duodenum. Located superior and to the left of the head are the central **body** and the tapering **tail.**

Pancreatic secretions pass from the secreting cells in the pancreas into small ducts. They unite to form two larger ducts that convey the secretions into the small intestine. The larger of the two ducts is called the **pancreatic duct (duct of Wirsung).** In most people, the pancreatic duct joins the common bile duct from the liver and gallbladder and enters the duodenum as a common duct called the **hepatopancreatic ampulla (ampulla of Vater;** see inset in Figure 24.12). The ampulla opens on an elevation of the duodenal mucosa known as the **major duodenal papilla,** about 10 cm (4 in.) inferior to the pyloric sphincter of the stomach. The smaller

canals that empty into small bile ducts (Figure 24.14a). These small ducts eventually merge to form the larger **right** and **left hepatic ducts,** which unite and exit the liver as the **common hepatic duct** (see Figures 24.12 and 24.13a). Farther on, the common hepatic duct joins the **cystic duct** from the gallbladder to form the **common bile duct.** Bile enters the cystic duct and is temporarily stored in the gallbladder. After a meal, various stimuli cause contraction of the gallbladder, which releases stored bile into the common bile duct. The common bile duct and pancreatic duct enter the duodenum in a common duct called the **hepatopancreatic ampulla (ampulla of Vater).**

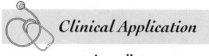

Clinical Application

Jaundice

Jaundice (*jaune* = yellow) is a yellowish coloration of the sclerae (white of the eyes), skin, and mucous membranes due to a buildup in the body of a yellow compound called bilirubin. After bilirubin is formed from the breakdown of the heme pigment in aged red blood cells, it is transported to the liver, where it is processed and eventually excreted into bile. The three main categories of jaundice are (1) *prehepatic jaundice,* due to excess production of bilirubin; (2) *hepatic jaundice,* due to congenital liver disease, cirrhosis of the liver, or hepatitis; and (3) *extrahepatic jaundice,* due to blockage of bile drainage by gallstones or cancer of the bowel or the pancreas.

Because the liver of a newborn functions poorly for the first week or so, many babies experience a mild form of jaundice called *neonatal (physiological) jaundice* that disappears as the liver matures. Usually, it is treated by exposing the infant to blue light, which converts bilirubin into substances the kidneys can excrete.

Functions

The liver performs various vital functions, many of which are related to metabolism. Among the functions of the liver are the following:

1. **Carbohydrate metabolism.** The liver is especially important in maintaining a normal blood glucose level by converting glycogen, amino acids, and other sugars to glucose, and glucose to glycogen and triglycerides.

2. **Lipid metabolism.** The liver stores some triglycerides and synthesizes lipoproteins and cholesterol.

3. **Protein metabolism.** Without the role of the liver in protein metabolism, death would occur in a few days. Hepatocytes synthesize most plasma proteins, such as alpha and beta globulins, albumin, prothrombin, and fibrinogen. It also converts toxic ammonia (NH_3) into the much less toxic urea for excretion in urine.

4. **Removal of drugs and hormones.** The liver can detoxify substances such as alcohol or excrete into bile drugs such as penicillin, erythromycin, and sulfonamides.

5. **Excretion of bilirubin.** Bilirubin, derived from the heme of worn-out red blood cells, is absorbed by the liver from the blood and secreted into bile. Most of the bilirubin in bile is metabolized in the intestine by bacteria and eliminated in feces.

6. **Synthesis of bile salts.** Bile salts are used in the small intestine for the emulsification and absorption of fats, cholesterol, phospholipids, and lipoproteins.

7. **Storage.** In addition to glycogen, the liver stores vitamins (A, B_{12}, D, E, and K) and minerals (iron and copper).

8. **Phagocytosis.** The stellate reticuloendothelial (Kupffer's) cells of the liver phagocytize worn-out red and white blood cells and some bacteria.

9. **Activation of vitamin D.** The skin, liver, and kidneys participate in activating vitamin D.

GALLBLADDER

The **gallbladder** (*galla* = bile) is a pear-shaped sac about 7 to 10 cm (3 to 4 in.) long. It is located in a depression on the posterior surface of the liver and hangs from the anteroinferior margin of the liver (see Figures 24.12 and 24.13a).

The gallbladder is supplied by the cystic artery, which usually arises from the right hepatic artery. The cystic veins drain the gallbladder. The nerves to the gallbladder include branches from the celiac plexus and the vagus (X) nerve.

Histology

The mucosa of the gallbladder consists of simple columnar epithelium arranged in rugae resembling those of the stomach. The gallbladder lacks a submucosa. The middle, muscular coat of the wall consists of smooth muscle fibers. Contraction of these fibers by hormonal stimulation ejects the contents of the gallbladder into the **cystic** (*kystis* = bladder) **duct.** The outer coat is the visceral peritoneum.

Functions

The functions of the gallbladder are to store and concentrate bile (up to tenfold) until it is needed in the small intestine. When bile is needed, the smooth muscle in the wall of the gallbladder contracts and forces bile into the cystic duct, through the common bile duct, and into the small intestine. When the small intestine is empty, a valve around the hepatopancreatic ampulla (ampulla of Vater) called the **sphincter of the hepatopancreatic ampulla (sphincter of Oddi)** closes, and the backed-up bile flows into the cystic duct to the gallbladder for storage (see Figure 24.12).

SMALL INTESTINE

The major events of digestion and absorption occur in a long tube called the **small intestine.** Since almost all the digestion and absorption of nutrients occur in the small intestine, its structure is specially adapted for this function. Its length alone provides a large surface area for digestion and absorption, and that area is further increased by folds, villi, and microvilli. The small intestine begins at the pyloric sphincter of the stomach, coils through the central and inferior part of the abdominal cavity, and eventually opens into the large intestine. It averages 2.65 cm (1 in.) in diameter. The length is about 3 m (10 ft) in a living person and about 6.4 m (21 ft) in a cadaver due to loss of smooth muscle tone after death.

The arterial blood supply of the small intestine is from the superior mesenteric artery and the gastroduodenal artery, coming from the hepatic artery of the celiac trunk. Blood is returned by way of the superior mesenteric vein, which, with the splenic vein, forms the hepatic portal vein.

The nerves to the small intestine are supplied by the superior mesenteric plexus. The branches of the plexus contain postganglionic sympathetic fibers, preganglionic parasympathetic fibers, and sensory fibers. The sensory fibers are both vagal and of spinal nerves. In the wall of the small intestine are two autonomic plexuses: the myenteric plexus between the muscular layers and the submucosal plexus in the submucosa. The nerve fibers are derived chiefly from the sympathetic division of the autonomic nervous system and partly from the vagus (X) nerve.

Anatomy

The small intestine is divided into three segments (Figure 24.15). The **duodenum** (doo′-ō-DĒ-num) is the shortest segment and is retroperitoneal. *Duodenum* means "12"; the structure is 12 fingers' breadth in length. It starts at the pyloric sphincter of the stomach and extends about 25 cm (10 in.) until it merges with the jejunum. The **jejunum** (jē-JOO-num) is about 1 m (3 ft) long and extends to the ileum. *Jejunum* means "empty," because at death it is found empty. The final portion of the small intestine, the **ileum** (IL-ē-um; *eileos* = twisted), measures about 2 m (6 ft) and joins the large intestine at the **ileocecal** (il′-ē-ō-SĒ-kal) **sphincter (valve).**

Projections called **circular folds,** or **plicae circulares** (PLĪ-kē-SER-kyoo-lar-es), are permanent ridges in the mucosa, about 10 mm (0.4 in.) high (see Figure 24.16c). Some extend all the way around the circumference of the intestine, and others extend only part of the way around. The circular folds begin near the proximal portion of the duodenum and end at about the midportion of the ileum. They enhance absorption by increasing surface area and causing the chyme to spiral, rather than to move in a straight line, as it passes through the small intestine.

Figure 24.15 Divisions of the small intestine.

🔑 *Most digestion and absorption occur in the small intestine.*

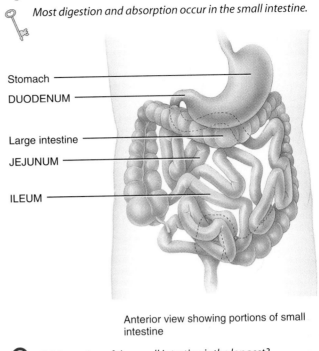

Stomach

DUODENUM

Large intestine

JEJUNUM

ILEUM

Anterior view showing portions of small intestine

Q *Which portion of the small intestine is the longest?*

Histology

The wall of the small intestine is composed of the same four coats that make up most of the GI tract. However, special features of both the mucosa and the submucosa facilitate the processes of digestion and absorption (Figure 24.16). The mucosa forms a series of **villi** (= tuft of hair; **villus** is singular). These projections are 0.5 to 1 mm long and give the intestinal mucosa a velvety appearance. The large number of villi (20 to 40 per square millimeter) vastly increases the surface area of the epithelium available for absorption and digestion. Each villus has a core of lamina propria. Embedded in this connective tissue are an arteriole, a venule, a capillary network, and a **lacteal** (LAK-tē-al), which is a lymphatic capillary. Nutrients being absorbed by the epithelial cells covering the villus pass through the wall of a capillary or a lacteal to enter blood or lymph, respectively.

The epithelium of the mucosa consists of simple columnar epithelium and contains absorptive cells, goblet cells, enteroendocrine cells, and Paneth cells. The apical (free) membrane of the absorptive cells features **microvilli** (mī′-krō-VIL-ī), which are microscopic, fingerlike projections of the plasma membrane that contain actin filaments. In a photomicrograph taken through a light microscope, the microvilli are too small to be seen individually. They form a fuzzy line, called the **brush border,** at the apical surface of the absorptive cells, next to the lumen of the small intestine

Figure 24.16 Small intestine.

Circular folds, villi, and microvilli increase the surface area of the small intestine for digestion and absorption.

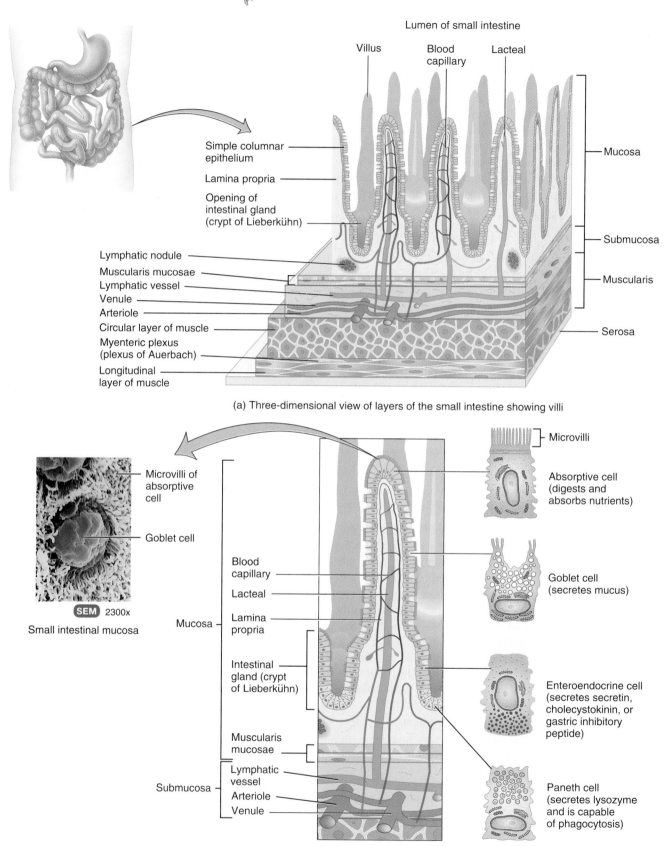

Lumen of small intestine

Villus

Blood capillary

Lacteal

Simple columnar epithelium

Lamina propria

Opening of intestinal gland (crypt of Lieberkühn)

Lymphatic nodule
Muscularis mucosae
Lymphatic vessel
Venule
Arteriole
Circular layer of muscle
Myenteric plexus (plexus of Auerbach)
Longitudinal layer of muscle

Mucosa

Submucosa

Muscularis

Serosa

(a) Three-dimensional view of layers of the small intestine showing villi

Microvilli of absorptive cell

Goblet cell

SEM 2300x
Small intestinal mucosa

Blood capillary

Lacteal

Lamina propria

Mucosa

Intestinal gland (crypt of Lieberkühn)

Muscularis mucosae

Lymphatic vessel
Submucosa
Arteriole
Venule

Microvilli

Absorptive cell (digests and absorbs nutrients)

Goblet cell (secretes mucus)

Enteroendocrine cell (secretes secretin, cholecystokinin, or gastric inhibitory peptide)

Paneth cell (secretes lysozyme and is capable of phagocytosis)

(b) Enlarged villus showing lacteal, capillaries, and intestinal gland and cell types

Figure continues

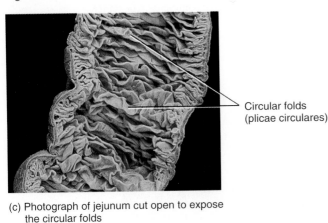

Figure 24.16 (continued)

Circular folds
(plicae circulares)

(c) Photograph of jejunum cut open to expose
the circular folds

Q *What are the major functions of the small intestine?*

(see Figure 24.17b, c). Larger amounts of digested nutrients can diffuse into the absorptive cells of the intestinal wall because the microvilli greatly increase the surface area of the plasma membrane. It is estimated that there are about 200 million microvilli per square millimeter of small intestine.

The mucosa contains many cavities lined with glandular epithelium. Cells lining the cavities form the **intestinal glands (crypts of Lieberkühn)** and secrete intestinal juice. **Paneth cells** are found in the deepest parts of the intestinal glands. They secrete lysozyme, a bactericidal enzyme, and are also capable of phagocytosis. They may have a role in regulating the microbial population in the intestines. The submucosa of the duodenum contains **duodenal (Brunner's) glands** (see Figure 24.17a). They secrete an alkaline mucus that helps neutralize gastric acid in the chyme. Some of the epithelial cells in the mucosa are goblet cells, which secrete additional mucus.

The lamina propria of the small intestine has an abundance of mucosa-associated lymphoid tissue (MALT) in the form of lymphatic nodules, masses of lymphatic tissue not surrounded by a capsule (see Figure 24.16a). **Solitary lymphatic nodules** are most numerous in the lower part of the ileum. Groups of lymphatic nodules, referred to as **aggregated lymphatic follicles (Peyer's patches),** are numerous in the ileum. The muscularis mucosae consist of smooth muscle.

The **muscularis** of the small intestine consists of two layers of smooth muscle. The outer, thinner layer contains longitudinally arranged fibers. The inner, thicker layer contains circularly arranged fibers. Except for a major portion of the duodenum, the serosa (or visceral peritoneum) completely surrounds the small intestine. Additional histological aspects of the small intestine are shown in Figure 24.17.

Functions

Chyme entering the small intestine contains partially digested carbohydrates, proteins, and lipids (mostly triglycerides). The completion of the digestion of carbohydrates, proteins, and lipids is a collective effort of pancreatic juice, bile, and intestinal juice in the small intestine.

Intestinal juice is a clear yellow fluid secreted in amounts of 1 to 2 liters (about 1 to 2 qt) a day. It has a pH of 7.6, which is slightly alkaline, and contains water and mucus. Together, pancreatic and intestinal juice provide a vehicle for the absorption of substances from chyme as they come in contact with the villi.

The absorptive epithelial cells that line the villi synthesize several digestive enzymes, called **brush border enzymes,** and insert them in the plasma membrane of the microvilli. Thus some digestion by enzymes of the small intestine occurs at the surface of the epithelial cells that line the villi, rather than in the lumen exclusively, as in other parts of the GI tract. As small intestinal cells slough off into the lumen of the intestine, they break apart and release enzymes that digest food in the chyme.

The movements of the small intestine are divided into two types: segmentation and peristalsis. **Segmentation** is the major movement of the small intestine. It is strictly a localized contraction in areas containing food. It mixes chyme with the digestive juices and brings the particles of food into contact with the mucosa for absorption. Segmentation occurs 12 to 16 times a minute, sloshing the chyme back and forth, and is similar to alternately squeezing opposite ends of a tube of toothpaste. **Peristalsis** propels the chyme onward through the intestinal tract.

All the chemical and mechanical phases of digestion from the mouth through the small intestine are directed toward changing food into forms that can pass through the epithelial cells lining the mucosa into the underlying blood and lymphatic vessels. These forms are monosaccharides (glucose, fructose, and galactose) from carbohydrates; single amino acids, dipeptides, and tripeptides from proteins; fatty acids, glycerol, and monoglycerides from lipids; and pentoses and nitrogenous bases from nucleic acids. Passage of these digested nutrients from the gastrointestinal tract into the blood or lymph is called **absorption.** Absorption occurs by diffusion, facilitated diffusion, osmosis, and active transport.

About 90% of all absorption of nutrients takes place in the small intestine. The other 10% occurs in the stomach and large intestine. Any undigested or unabsorbed material left in the small intestine passes on to the large intestine.

LARGE INTESTINE

The overall functions of the large intestine are the completion of absorption, the manufacture of certain vitamins, the formation of feces, and the expulsion of feces from the body

Figure 24.17 Histology of the small intestine.

The lamina propria of the small intestine has an abundance of mucosa-associated lymphoid tissue (MALT).

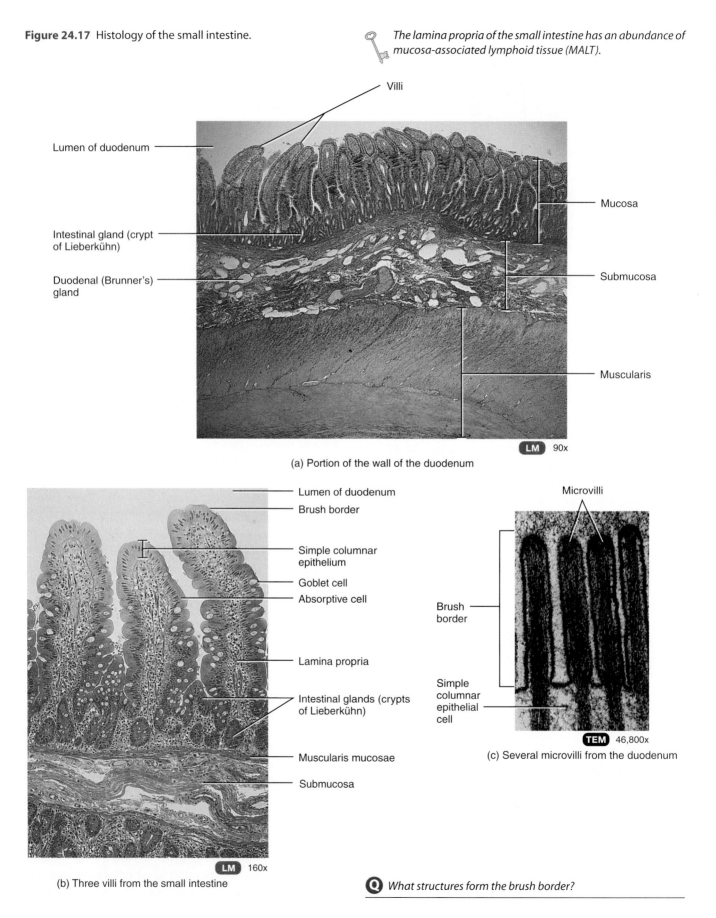

(a) Portion of the wall of the duodenum

(b) Three villi from the small intestine

(c) Several microvilli from the duodenum

Q *What structures form the brush border?*

Figure 24.18 Anatomy of the large intestine.

The subdivisions of the large intestine are the cecum, colon, rectum, and anal canal

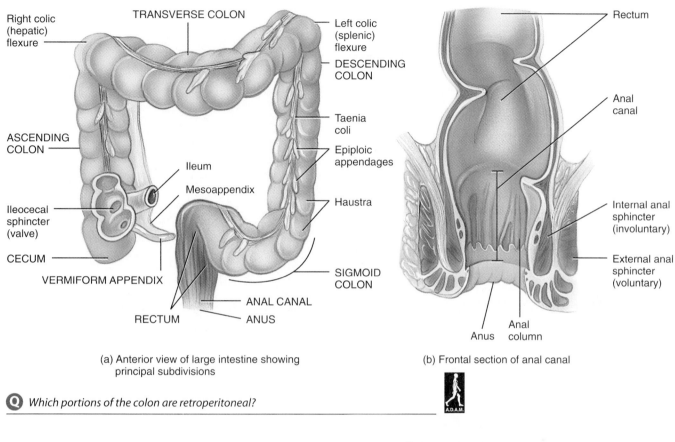

(a) Anterior view of large intestine showing principal subdivisions

(b) Frontal section of anal canal

Q Which portions of the colon are retroperitoneal?

Anatomy

The **large intestine** is about 1.5 m (5 ft) long and 6.5 cm (2.5 in.) in diameter. It extends from the ileum to the anus and is attached to the posterior abdominal wall by its **mesocolon** of visceral peritoneum. Structurally, the large intestine is divided into four principal regions: cecum, colon, rectum, and anal canal (Figure 24.18a).

The opening from the ileum into the large intestine is guarded by a fold of mucous membrane called the **ileocecal sphincter (valve).** This structure allows materials from the small intestine to pass into the large intestine. Hanging below the ileocecal valve is the **cecum,** a blind pouch about 6 cm (2.5 in.) long. Attached to the cecum is a twisted, coiled tube, measuring about 8 cm (3 in.) in length, called the **appendix** or **vermiform appendix** (*vermis* = worm; *appendix* = appendage). Because the appendix has a large amount of lymphatic tissue, it is a component of MALT and assumes a role in immunity. The mesentery of the appendix, called the **mesoappendix,** attaches the appendix to the inferior part of the ileum and adjacent part of the posterior abdominal wall.

Clinical Application

Appendicitis

Inflammation of the appendix is termed **appendicitis.** It is preceded by obstruction of the lumen of the appendix by fecal material, inflammation, a foreign body, carcinoma of the cecum, stenosis, or kinking of the organ. The infection that follows may result in edema, ischemia, gangrene, and perforation. Rupture of the appendix develops into peritonitis. Typically, appendicitis begins with referred pain in the umbilical region of the abdomen, followed by anorexia (lack or loss of appetite for food), nausea, and vomiting. After several hours, the pain localizes in the right lower quadrant and is continuous, dull or severe, and intensified by coughing, sneezing, or body movements. Early appendectomy (removal of the appendix) is recommended in all suspected cases because it is safer to operate than to risk gangrene, rupture, and peritonitis. ▪

The open end of the cecum merges with a long tube called the **colon** (*kolon* = food passage). The colon is divided into ascending, transverse, descending, and sigmoid portions. The **ascending colon,** which is retroperitoneal, ascends on

Figure 24.19 Histology of the large intestine.

🔑 *Intestinal glands formed by simple columnar epithelium and goblet cells extend the full thickness of the mucosa.*

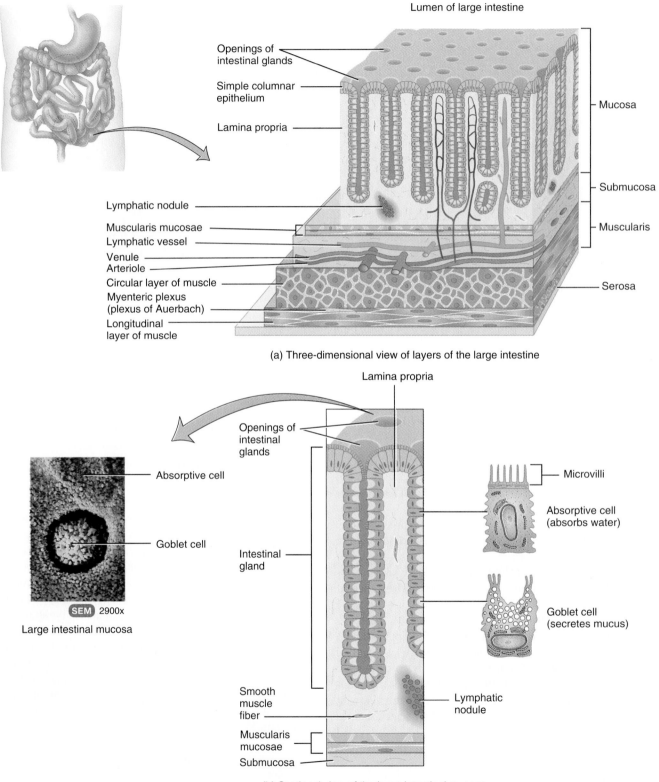

(a) Three-dimensional view of layers of the large intestine

(b) Sectional view of the large intestinal mucosa showing intestinal glands and cell types

SEM 2900x
Large intestinal mucosa

Figure continues

Figure 24.19 (continued)

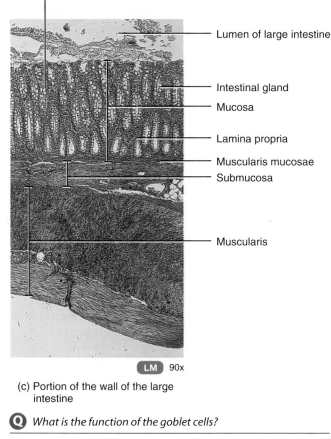

(c) Portion of the wall of the large intestine

Q *What is the function of the goblet cells?*

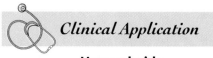

Clinical Application

Hemorrhoids

Varicosities in veins are regions that are enlarged and inflamed. In the rectal veins, varicosities are known as **hemorrhoids (piles).** Hemorrhoids develop when the veins are put under pressure and become engorged with blood. If the pressure continues, the wall of the vein stretches. Such a distended vessel oozes blood, and bleeding or itching is usually the first sign that a hemorrhoid has developed. Stretching of a vein also favors clot formation, further aggravating swelling and pain. Hemorrhoids may be caused by constipation, which may be brought on by low-fiber diets. Also, repeated straining during defecation forces blood down into the rectal veins, increasing pressure in these veins and possibly causing hemorrhoids. ▓

The arterial supply of the cecum and colon is derived from branches of the superior mesenteric and inferior mesenteric arteries. The venous return is by way of the superior and inferior mesenteric veins ultimately to the hepatic portal vein and into the liver. The arterial supply of the rectum and anal canal is derived from the superior, middle, and inferior rectal arteries. The rectal veins correspond to the rectal arteries.

The nerves to the large intestine consist of sympathetic, parasympathetic, and sensory components. The sympathetic innervation is derived from the celiac, superior, and inferior mesenteric ganglia and superior and inferior mesenteric plexuses. The fibers reach the viscera by way of the thoracic and lumbar splanchnic nerves. The parasympathetic innervation is derived from the vagus (X) and pelvic splanchnic nerves.

Histology

The wall of the large intestine differs from that of the small intestine in several respects. No villi or permanent circular folds are found in the mucosa. The **mucosa** consists of simple columnar epithelium, lamina propria (areolar connective tissue), and muscularis mucosae (smooth muscle). The epithelium contains mostly absorptive and goblet cells (Figure 24.19). The absorptive cells function primarily in water absorption. The goblet cells secrete mucus that lubricates the colonic contents as they pass through. Both absorptive and goblet cells are located in long, straight, tubular intestinal glands that extend the full thickness of the mucosa. Solitary lymphatic nodules are also found in the mucosa. The **submucosa** of the large intestine is similar to that found in the rest of the GI tract. The **muscularis** consists of an external layer of longitudinal muscles and an internal layer of circular muscles. Unlike other parts of the GI tract, portions of the longitudinal muscles are thickened, forming three conspicuous longitudinal bands called **taeniae coli** (TĒ-nē-ē KŌ-lī;

the right side of the abdomen, reaches the inferior surface of the liver, and turns abruptly to the left. Here it forms the **right colic (hepatic) flexure.** The colon continues across the abdomen to the left side as the **transverse colon.** It is not retroperitoneal and curves beneath the inferior end of the spleen on the left side as the **left colic (splenic) flexure** and passes inferiorly to the level of the iliac crest as the **descending colon,** also a retroperitoneal structure. The **sigmoid colon** begins near the left iliac crest, projects medially to the midline, and terminates as the rectum at about the level of the third sacral vertebra.

The **rectum,** the last 20 cm (8 in.) of the GI tract, lies anterior to the sacrum and coccyx. The terminal 2 to 3 cm (1 in.) of the rectum is called the **anal canal** (Figure 24.18b). The mucous membrane of the anal canal is arranged in longitudinal folds called **anal columns** that contain a network of arteries and veins. The opening of the anal canal to the exterior is called the **anus.** It is guarded by an internal sphincter of smooth muscle (involuntary) and an external sphincter of skeletal muscle (voluntary). Normally the anus is closed except during the elimination of the wastes of digestion.

taenia = flat band), alternating with a wall section with less or no longitudinal muscle (see Figure 24.18a). Each band runs the length of most of the large intestine. Tonic contractions of the bands gather the colon into a series of pouches called **haustra** (HAWS-tra; singular is **haustrum** = shaped like a pouch), which give the colon a puckered appearance. There is a single layer of circular muscle between taeniae coli. The **serosa** of the large intestine is part of the visceral peritoneum. Small pouches of visceral peritoneum filled with fat are attached to taeniae coli and are called **epiploic appendages.**

Clinical Application

Diverticulitis

Diverticula are saclike outpouchings of the wall of the colon in places where the muscularis has become weak. The development of diverticula is called **diverticulosis.** Many people who develop diverticulosis are asymptomatic and experience no complications. About 15% of people with diverticulosis will eventually develop an inflammation within the diverticula, a condition known as **diverticulitis.** It may be characterized by pain, either constipation or increased frequency of defecation, nausea, vomiting, and low-grade fever. Because diets low in fiber contribute to development of diverticulitis, patients who change to high-fiber diets show marked relief of symptoms. In severe cases, affected portions of the colon may require surgical removal. ▨

Functions

The passage of chyme from the ileum into the cecum is regulated by the action of the ileocecal sphincter. Normally, the valve remains partially closed so that the passage of chyme into the cecum is usually a slow process. Immediately after a meal, ileal peristalsis intensifies, the sphincter relaxes, and chyme is forced from the ileum into the cecum. As food passes through the ileocecal sphincter, it fills the cecum and accumulates in the ascending colon and movements of the colon begin.

One movement characteristic of the large intestine is **haustral churning.** In this process, the haustra remain relaxed and distended while they fill up. When the distension reaches a certain point, the wall contracts and squeezes the contents into the next haustrum. **Peristalsis** also occurs, although at a slower rate (3 to 12 contractions per minute) than in other portions of the tract. A final type of movement is **mass peristalsis,** a strong peristaltic wave that begins at about the middle of the transverse colon and quickly drives the colonic contents into the rectum. Mass peristalsis usually take place three or four times a day, during or immediately after a meal.

The last stage of digestion occurs through bacterial, not enzymatic, action because no enzymes are secreted. Up to 40 percent of fecal mass is bacteria. Bacteria ferment any remaining carbohydrates and release hydrogen, carbon dioxide, and methane gas, which contribute to flatus (gas) in the colon. They also convert remaining proteins to amino acids and break down the amino acids into simpler substances: indole, skatole, hydrogen sulfide, and fatty acids. Some of these are carried off in the feces and contribute to their odor. The rest is absorbed and transported to the liver, converted to less toxic compounds, and excreted in the urine. Bacteria also decompose bilirubin to simpler pigments (such as urobilinogen), which give feces their brown color. Several vitamins needed for normal metabolism, including some B vitamins and vitamin K, are synthesized by bacterial action and absorbed.

By the time the chyme has remained in the large intestine 3 to 10 hours, it has become solid or semisolid as a result of absorption principally of water and is now known as **feces.** Chemically, feces consist of water, inorganic salts, sloughed-off epithelial cells from the mucosa of the gastrointestinal tract, bacteria, products of bacterial decomposition, and undigested parts of food.

Although most water absorption occurs in the small intestine, the large intestine absorbs enough to make it an important organ in maintaining the body's water balance. Of the 0.5 to 1.0 liter of water that enters the large intestine, all but about 100 to 200 ml is absorbed. The absorption is greatest in the cecum and ascending colon. The large intestine also absorbs ions, including sodium and chloride, and some vitamins.

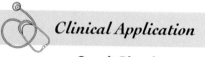

Clinical Application

Occult Blood

The term **occult blood** refers to blood that is hidden; it is not detectable by the human eye. The main diagnostic value of occult blood testing is to screen for colorectal cancer. Two substances frequently examined for occult blood are feces and urine. Several types of products are available for at-home testing for hidden blood in feces. The tests are based on color changes when reagents are added to feces. The presence of occult blood in urine may be detected at home by using dip-and-read reagent strips. ▨

As mass peristaltic movements push fecal material from the sigmoid colon into the rectum, the resulting distension of the rectal wall stimulates stretch receptors, initiating a reflex for **defecation,** the emptying of the rectum. Reflex contraction of the longitudinal rectal muscles shortens the rectum, thereby increasing the pressure inside it. The pressure, along with voluntary contractions of the diaphragm and abdominal muscles and parasympathetic stimulation, open the internal sphincter, and the feces are expelled through the anus.

Table 24.2 *Summary of Organs of the Digestive System and Their Functions*

ORGANS	FUNCTIONS
Tongue	Maneuvers food for mastication, shapes food into a bolus, maneuvers food for deglutition, detects sensations for taste, and initiates digestion of triglycerides.
Salivary glands	Saliva softens, moistens, and dissolves foods; cleanses mouth and teeth; initiates the digestion of starch.
Teeth	Cut, tear, and pulverize food to reduce solids to smaller particles for swallowing.
Pancreas	Pancreatic juice buffers acidic gastric juice in chyme, stops the action of pepsin from the stomach, creates the proper pH for digestion in the small intestine, and participates in the digestion of carbohydrates, proteins, triglycerides, and nucleic acids.
Liver	Produces bile, which is required for the emulsification and absorption of lipids in the small intestine.
Gallbladder	Stores and concentrates bile and releases it into the small intestine.
Mouth	See the functions of the tongue, salivary glands, and teeth, all of which are in the mouth. Additionally, the lips and cheeks keep food between the teeth during mastication, and buccal glands lining the mouth produce saliva.
Pharynx	Receives a bolus from the oral cavity and passes it into the esophagus.
Esophagus	Receives a bolus from the pharynx and moves it into the stomach. This requires relaxation of the upper esophageal sphincter, and secretion of mucus.
Stomach	Mixing waves combine saliva, food, and gastric juice, which initiates protein digestion, activates pepsin, kills microbes in food, helps absorb vitamin B_{12}, contracts the lower esophageal sphincter, increases stomach motility, relaxes the pyloric sphincter, and moves chyme into the small intestine.
Small intestine	Segmentation mixes chyme with digestive juices; peristalsis propels chyme toward the ileocecal sphincter; digestive secretions from the small intestine, pancreas, and liver complete the digestion of carbohydrates, proteins, lipids, and nucleic acids; circular folds, villi, and microvilli help absorb about 90% of digested nutrients.
Large intestine	Haustral churning, peristalsis, and mass peristalsis drive the colonic contents into the rectum; bacteria produce some B vitamins and vitamin K; absorption of some water, ions, and vitamins occurs; defecation.

The external sphincter is voluntarily controlled. If it is voluntarily relaxed, defecation occurs; if it is voluntarily constricted, defecation can be postponed. Voluntary contractions of the diaphragm and abdominal muscles aid defecation by increasing the pressure inside the abdomen, which pushes the walls of the sigmoid colon and rectum inward. If defecation does not occur, the feces back into the sigmoid colon until the next wave of mass peristalsis again stimulates the stretch receptors, creating the desire to defecate. In infants, the defecation reflex causes automatic emptying of the rectum because they have not yet developed voluntary control of the external anal sphincter.

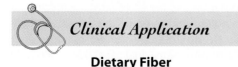

Clinical Application

Dietary Fiber

A dietary deficiency that has received recent attention is lack of adequate **dietary fiber** (bulk or roughage). Dietary fiber consists of indigestible plant substances, such as cellulose, lignin, and pectin, found in fruits, vegetables, grains, and beans. **Insoluble fiber,** which does not dissolve in water, includes the woody or structural parts of plants such as fruit and vegetable skins and the bran coating around wheat and corn kernels. Insoluble fiber passes through the GI tract largely unchanged and speeds up the passage of material through the tract. **Soluble fiber** dissolves in water and forms a gel. It is found in abundance in beans, oats, barley, broccoli, prunes, apples, and citrus fruits. It tends to slow the passage of material through the tract.

People who choose a fiber-rich, unrefined diet may reduce their risk of developing obesity, diabetes, atherosclerosis, gallstones, hemorrhoids, diverticulitis, and appendicitis. Each of these conditions is directly related to the digestion and metabolism of food and the operation of the digestive system. There is also evidence that insoluble fiber may help protect against colorectal cancer and that soluble fiber may help lower blood cholesterol level. ▪

A summary of the digestive organs and their functions is presented in Table 24.2.

Figure 24.20 Development of the digestive system. The inset adjacent to (d) is an external view of a 28-day old embryo.

The gastrointestinal tract is derived from endoderm and mesoderm.

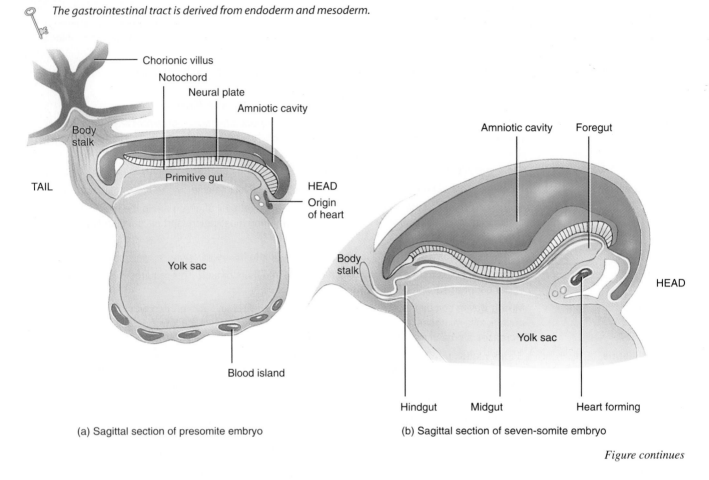

(a) Sagittal section of presomite embryo

(b) Sagittal section of seven-somite embryo

Figure continues

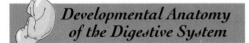

Developmental Anatomy of the Digestive System

About the fourteenth day after fertilization, the cells of the endoderm form a cavity referred to as the **primitive gut** (Figure 24.20). Soon after the mesoderm forms and splits into two layers (somatic and splanchnic), the splanchnic mesoderm associates with the endoderm of the primitive gut. Thus the primitive gut has a double-layered wall. The **endodermal layer** gives rise to the *epithelial lining* and *glands* of most of the gastrointestinal tract. The **mesodermal layer** produces its *smooth muscle* and *connective tissue* of the gastrointestinal tract.

The primitive gut elongates, and about the latter part of the third week, it differentiates into an anterior **foregut,** an intermediate **midgut,** and a posterior **hindgut.** Until the fifth week of development, the midgut opens into the yolk sac. After that time, the yolk sac constricts and detaches from the midgut, and the midgut seals. In the region of the foregut, a depression consisting of **ectoderm,** the **stomodeum,**

appears. This develops into the *oral cavity.* The **oral membrane** that separates the foregut from the stomodeum ruptures during the fourth week of development, so that the foregut is continuous with the outside of the embryo through the oral cavity. Another depression consisting of **ectoderm,** the **proctodeum,** forms in the hindgut and goes on to develop into the *anus.* The **cloacal membrane,** which separates the hindgut from the proctodeum, ruptures, so that the hindgut is continuous with the outside of the embryo through the anus. Thus the GI tract forms a continuous tube from mouth to anus.

The foregut develops into the *pharynx, esophagus, stomach,* and a *portion of the duodenum.* The midgut is transformed into the *remainder of the duodenum,* the *jejunum,* the *ileum,* and *portions of the large intestine* (cecum, appendix, ascending colon, and most of the transverse colon). The hindgut develops into the *remainder of the large intestine,* except for a portion of the anal canal that is derived from the proctodeum.

Figure 24.20 (continued)

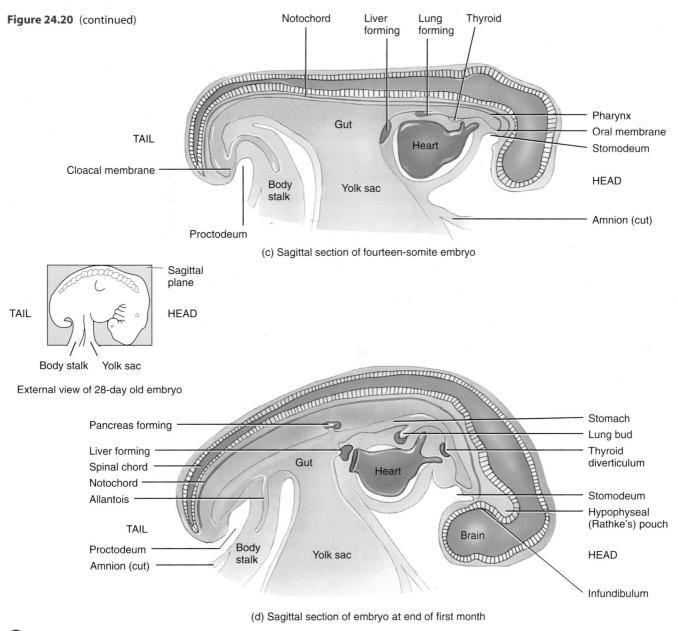

(c) Sagittal section of fourteen-somite embryo

External view of 28-day old embryo

(d) Sagittal section of embryo at end of first month

Q *When does the gastrointestinal tract begin to develop in a fetus?*

As development progresses, the endoderm at various places along the foregut develops into hollow buds that grow into the mesoderm. These buds develop into the *salivary glands, liver, gallbladder,* and *pancreas.* Each gland retains a connection with the gastrointestinal tract through ducts. ■

Aging and the Digestive System

Overall changes associated with aging of the digestive system include decreased secretory mechanisms, decreased motility of the digestive organs, loss of strength and tone of the muscular tissue and its supporting structures, changes in neurosensory feedback regarding enzyme and hormone release, and diminished response to pain and internal sensations. In the upper portion of the GI tract, common changes include reduced sensitivity to mouth irritations and sores, loss of taste, periodontal disease, difficulty in swallowing, hiatal hernia, gastritis, and peptic ulcer disease (see *Applications to Health,* p. 23). Changes that may appear in the small intestine include duodenal ulcers, appendicitis, malabsorption, and maldigestion. Other pathologies that increase in incidence with age are gallbladder problems, jaundice, cirrhosis, and acute pancreatitis. Large intestinal changes such as constipation, hemorrhoids, and diverticular disease may also occur. Cancer of the colon or rectum (colorectal cancer; see *Applications to Health,* p. 23) is quite common.

Key Medical Terms
Associated with the Digestive System

Borborygmus (bor′-bō-RIG-mus) A rumbling noise caused by the propulsion of gas through the intestines.

Canker (KANG-ker) **sore** Painful ulcer on the mucous membrane of the mouth that affects females more often than males and usually occurs between ages 10 and 40; may be an autoimmune reaction or a food allergy.

Cholecystitis (kō′-lē-sis-TĪ-tis; *chole* = bile; *kystis* = bladder; *itis* = inflammation of) Inflammation of the gallbladder. It can be caused by an autoimmune inflammation of the gallbladder or by obstruction of the cystic duct with bile stones.

Colitis (ko-LĪ-tis) Inflammation of the mucosa of the colon and rectum in which absorption of water and salts is reduced, producing watery, bloody feces and, in severe cases, dehydration and salt depletion. Spasms of the irritated muscularis produce cramps. It is thought to be an autoimmune condition.

Colostomy (kō-LOS-tō-mē; *stomoun* = provide an opening) The diversion of the fecal stream through an opening in the colon, creating a surgical "stoma" (artificial opening) that is affixed to the exterior of the abdominal wall. This opening serves as a substitute anus through which feces are eliminated into a bag.

Constipation (kon-sti-PĀ-shun; *con* = together; *stipo* = to press) Infrequent or difficult defecation caused by decreased motility of the intestines; feces become hard and dry due to excessive water reabsorption because the feces remain in the colon for prolonged periods of time.

Dysphagia (dis-FĀ-jē-a; *dys* = abnormal; *phagein* = to eat) Difficulty in swallowing that may be caused by inflammation, paralysis, obstruction, or trauma.

Enteritis (en′-ter-Ī-tis; *enteron* = intestine) An inflammation of the intestine, particularly the small intestine.

Flatus (FLĀ-tus) Air (gas) in the stomach or intestine, usually expelled through the anus. If it is expelled through the mouth, it is called **eructation** or **belching** (burping). Flatus may result from gas released during the breakdown of foods or from swallowing air or gas-containing substances such as carbonated drinks.

Gastrectomy (gas-TREK-tō-mē; *gastro* = stomach; *tome* = excision) Removal of all or a portion of the stomach.

Hernia (HER-nē-a) Protrusion of an organ or part of an organ through a membrane or cavity wall, usually the abdominal cavity. *Diaphragmatic (hiatal) hernia* is the protrusion of the lower esophagus, stomach, or intestine into the thoracic cavity through the opening in the diaphragm (esophageal hiatus) that allows passage of the esophagus. *Umbilical hernia* is usually a mild defect that contains the protrusion of a portion of peritoneum through the navel area of the abdominal wall. *Inguinal hernia* is the protrusion of the hernial sac into the inguinal opening. It may contain a portion of the bowel in an advanced stage and may extend into the scrotal compartment in males, strangling of the herniated part.

Inflammatory bowel (in-FLAM-a tō′-rē BOW-el) **disease** Disorder that exists in two forms: (1) Crohn's disease (inflammation of the gastrointestinal tract, especially the distal ileum and proximal colon, in which the inflammation may extend from the mucosa through the serosa) and (2) ulcerative colitis (inflammation of the mucosa of the gastrointestinal tract, usually limited to the large intestine and usually accompanied by rectal bleeding).

Irritable bowel (IR-i-ta-bul BOW-el) **syndrome (IBS)** Disease of the entire gastrointestinal tract in which a person reacts to stress by developing symptoms such as cramping and abdominal pain associated with alternating patterns of diarrhea and constipation. Excessive amounts of mucus may appear in the stools, and other symptoms include flatulence, nausea, and loss of appetite. The condition is also known as **irritable colon** or **spastic colitis.**

Malocclusion (mal′-ō-KLOO-zhun; *mal* = disease; *occlusio* = to fit together) Condition in which the maxillary (upper) and mandibular (lower) teeth do not close together.

Nausea (NAW-sē-a; *nausia* = seasickness) Discomfort characterized by a loss of appetite and the sensation of impending vomiting. Its causes include local irritation of the gastrointestinal tract, a systemic disease, brain disease or injury, overexertion, or the effects of medication or drug overdosage.

Study Outline

Organization (p. 708)

1. The organs of digestion are usually divided into two main groups: those composing the gastrointestinal (GI) tract, and accessory digestive organs.
2. The GI tract is a continuous tube running through the ventral body cavity from the mouth to the anus.
3. The accessory digestive organs include the teeth, tongue, salivary glands, liver, gallbladder, and pancreas.

Digestive Processes (p. 708)

1. The breaking down of larger food molecules into smaller molecules is called digestion; the passage of these smaller molecules into blood and lymph is termed absorption.

2. The functions of the digestive system are ingestion, secretion, propulsion, digestion, absorption, and defecation.

General Histology of the GI Tract (p. 708)

1. The basic arrangement of layers in most of the gastrointestinal tract from the deep to superficial is the mucosa, submucosa, muscularis, and serosa (visceral peritoneum).

Peritoneum (p. 711)

1. Extensions of the peritoneum include the mesentery, mesocolon, falciform ligament, lesser omentum, and greater omentum.

Mouth (Oral Cavity) (p. 713)

1. The mouth is formed by the cheeks, hard and soft palates, lips, and tongue, which aid mechanical digestion.
2. The vestibule is the space between the cheeks and lips and teeth and gums.
3. The oral cavity proper extends from the vestibule to the fauces.

Tongue (p. 713)

1. The tongue, together with its associated muscles, forms the floor of the oral cavity. It is composed of skeletal muscle covered with mucous membrane.
2. The upper surface and sides of the tongue are covered with papillae. Some papillae contain taste buds.

Salivary Glands (p. 715)

1. The major portion of saliva is secreted by the salivary glands, which lie outside the mouth and pour their contents into ducts that empty into the oral cavity.
2. There are three pairs of salivary glands: parotid, submandibular (submaxillary), and sublingual glands.
3. Saliva lubricates food and starts the chemical digestion of carbohydrates.

Teeth (p. 717)

1. The teeth (dentes) project into the mouth and are adapted for mechanical digestion.
2. A typical tooth consists of three principal portions: crown, root, and neck.
3. Teeth are composed primarily of dentin and are covered by enamel, the hardest substance in the body.
4. There are two dentitions—deciduous and permanent.
5. Through mastication, food is mixed with saliva and shaped into a bolus.
6. Salivary amylase initiates the breakdown of starch.

Pharynx (p. 719)

1. Through mastication, food is reduced to a soft, flexible mass called a bolus.
2. Deglutition, or swallowing, moves a bolus from the mouth to the stomach.

Esophagus (p. 720)

1. The esophagus is a collapsible, muscular tube that connects the pharynx to the stomach.
2. It passes a bolus into the stomach by peristalsis.
3. It contains an upper and lower esophageal sphincter.

Stomach (p. 721)

1. The stomach begins at the inferior portion of the esophagus and ends at the pyloric sphincter.
2. The principal subdivisions of the stomach include the cardia, fundus, body, and pylorus.
3. Adaptations of the stomach for digestion include rugae; gastric glands that produce mucus, hydrochloric acid, a protein-digesting enzyme, intrinsic factor, and stomach gastrin; and a three-layered muscularis for efficient mechanical movement.
4. Mechanical digestion consists of mixing waves.

5. Chemical digestion consists mostly of the breakdown of proteins by pepsin.
6. Among the substances absorbed are some water, certain electrolytes and drugs, and alcohol.

Pancreas (p. 725)

1. The pancreas is divisible into a head, body, and tail and is connected to the duodenum via the pancreatic duct (duct of Wirsung) and accessory duct (duct of Santorini).
2. Pancreatic islets (islets of Langerhans) secrete hormones, and acini secrete pancreatic juice.
3. Pancreatic juice contains enzymes that digest starch (pancreatic amylase), proteins (trypsin, chymotrypsin, and carboxypeptidase), triglycerides (pancreatic lipase), and nucleic acids (ribonuclease and deoxyribonuclease).

Liver (p. 727)

1. The liver is divisible into left and right lobes; associated with the right lobe are the caudate and quadrate lobes.
2. The lobes of the liver are made up of lobules that contain hepatocytes, sinusoids, stellate reticuloendothelial (Kupffer's) cells, and a central vein.
3. Hepatocytes produce bile that is transported by a duct system to the gallbladder for concentration and temporary storage.
4. Bile's contribution to digestion is the emulsification of dietary triglycerides.
5. The liver also functions in carbohydrate, lipid, and protein metabolism; removal of drugs and hormones; excretion of bilirubin; synthesis of bile salts; storage of vitamins and minerals; phagocytosis; and activation of vitamin D.

Gallbladder (p. 730)

1. The gallbladder is a sac located in a depression on the posterior surface of the liver.
2. The gallbladder stores and concentrates bile.

Small Intestine (p. 731)

1. The small intestine extends from the pyloric sphincter to the ileocecal sphincter.
2. It is divided into duodenum, jejunum, and ileum.
3. It is highly adapted for digestion and absorption. Its glands produce enzymes and mucus, and the microvilli, villi, and circular folds of its wall provide a large surface area for digestion and absorption.
4. Intestinal enzymes, pancreatic enzymes, and bile contribute to the complete digestion of carbohydrates into monosaccharides; proteins into amino acids; triglycerides into fatty acids, glycerol, and glycerides; and nucleic acids into pentoses and nitrogenous bases in the small intestine.
5. Mechanical digestion in the small intestine involves segmentation and peristalsis.

Large Intestine (p. 733)

1. The large intestine extends from the ileocecal sphincter to the anus.
2. Its subdivisions include the cecum, colon, and rectum.
3. The mucosa contains numerous goblet cells, and the muscularis consists of taeniae coli and haustra.

4. Mechanical movements of the large intestine include haustral churning, peristalsis, and mass peristalsis.
5. The last stages of chemical digestion occur in the large intestine through bacterial, rather than enzymatic, action. Substances are further broken down, and some vitamins are synthesized.
6. The large intestine absorbs water, electrolytes, and vitamins.
7. Feces consist of water, inorganic salts, epithelial cells, bacteria, and undigested foods.
8. The elimination of feces from the rectum is called defecation.
9. Defecation is a reflex action aided by voluntary contractions of the diaphragm and abdominal muscles.

Developmental Anatomy of the Digestive System (p. 740)

1. The endoderm of the primitive gut forms the epithelium and glands of most of the gastrointestinal tract.
2. The mesoderm of the primitive gut forms the smooth muscle and connective tissue of the gastrointestinal tract.

Aging and the Digestive System (p. 741)

1. General changes include decreased secretory mechanisms, decreased motility, and loss of tone.
2. Specific changes include loss of taste, pyorrhea, hernias, ulcers, constipation, hemorrhoids, and diverticular diseases.

Review Questions

1. Define digestion, and describe the six phases involved. Distinguish between chemical and mechanical digestion. (p. 708)
2. Identify, in sequence, the organs of the gastrointestinal (GI) tract. How does the gastrointestinal tract differ from the accessory digestive organs? (p. 708)
3. Describe the structure of each of the four layers of the gastrointestinal tract and how they vary in different organs. (p. 708)
4. What is the peritoneum? Describe the location and function of the mesentery, mesocolon, falciform ligament, lesser omentum, and greater omentum. (p. 711)
5. What structures form the mouth (oral cavity)? (p. 713)
6. Make a simple diagram of the tongue. Indicate the location of the three types of papillae and the four taste zones. (p. 715)
7. Distinguish buccal from salivary glands. (p. 715)
8. Describe the location of the salivary glands and their ducts. (p. 715)
9. What are the principal parts of a typical tooth? What are the functions of each part? (p. 717)
10. Compare deciduous and permanent dentitions with regard to number of teeth and time of eruption. (p. 717)
11. Contrast the functions of incisors, cuspids, premolars, and molars. (p. 717)
12. What is a bolus? How is it formed? (p. 719)
13. Define deglutition. Describe the passing of a bolus from the mouth to the stomach. (p. 720)
14. Describe the location and histology of the esophagus. What is its role in digestion? (p. 720)
15. Describe the location of the stomach. List and briefly explain the anatomical features of the stomach. (p. 721)
16. What is the importance of rugae, mucous cells, chief cells, parietal cells, and G cells in the stomach? (p. 723)
17. Describe mechanical digestion in the stomach. (p. 723)
18. What is the role of pepsin and hydrochloric acid? (p. 723)
19. Describe the role of the stomach in absorption. (p. 725)

20. Where is the pancreas located? Describe the duct system connecting the pancreas to the duodenum. (p. 725)
21. What are pancreatic acini? Contrast their functions with those of the pancreatic islets (islets of Langerhans). (p. 727)
22. Where is the liver located? What are its principal functions? (p. 727)
23. Describe the anatomy of the liver. Draw a labeled diagram of a liver lobule. (p. 727)
24. How is blood carried to and from the liver? (p. 727)
25. Once bile has been formed by the liver, how is it collected and transported to the gallbladder for storage? (p. 729)
26. Where is the gallbladder located? How is it connected to the duodenum? (p. 730)
27. Describe the function of the gallbladder. (p. 730)
28. What are the subdivisions of the small intestine? How are the mucosa and submucosa of the small intestine adapted for digestion and absorption? (p. 731)
29. Describe the movements in the small intestine. (p. 733)
30. Define absorption. How are the end-products of carbohydrate and protein digestion absorbed? How are the end-products of lipid digestion absorbed? (p. 733)
31. What are the principal subdivisions of the large intestine? How does the muscularis of the large intestine differ from that of the rest of the gastrointestinal tract? What are haustra? (p. 735–738)
32. Describe the mechanical movements that occur in the large intestine. (p. 738)
33. Explain the activities of the large intestine that change its contents into feces. (p. 738)
34. Define defecation. How does it occur? (p. 738)
35. Describe the development of the digestive system. (p. 740)
36. Describe the effects of aging on the digestive system. (p. 741)
37. Refer to the glossary of key medical terms associated with the digestive system. Be sure that you can define each term. (p. 742)

Self Quiz

Complete the following:

1. A fold of mucous membrane called the ___ attaches the tongue to the floor of the oral cavity.
2. Place an E (for endoderm) or an M (for mesoderm) in the blanks to indicate which germ layer gives rise to the following structures.
   (a) epithelial lining and digestive glands of the GI tract: ___;(b) liver, gallbladder, and pancreas: ___; (c) muscularis layer and connective tissue of submucosa: ___.
3. First and second premolars and first, second, and third molars are characteristic of the ___ dentition.
4. Emptying of feces from the rectum is called ___.
5. The duct that conveys bile from the gallbladder is the ___ duct.
6. The ___ is the opening between the oral cavity and the pharynx.
7. Permanent folds in the mucosa and submucosa of the small intestine are called ___.
8. The pouches of the large intestine are called ___.
9. The ___ border of the stomach is the greater curvature.
10. The liver consists of two larger lobes, the right and left lobes, and two smaller lobes, the ___ and ___ lobes.
11. Place numbers in the blanks to arrange the following layers of the gastrointestinal tract in order from deepest to most superficial.
    (a) mucosa: ___; (b) muscularis: ___; (c) serosa: ___; (d) submucosa.
12. Place numbers in the blanks to indicate the correct order for the flow of bile.
    (a) bile canaliculi: ___; (b) common bile duct: ___; (c) common hepatic duct: ___; (d) right and left hepatic ducts: ___; (e) hepatopancreatic ampulla (ampulla of Vater) and duodenum: ___.
13. Match the following terms with their descriptions:
    (1) parotid glands   ___ (a) ducts open into the
    (2) sublingual glands   vestibule
    (3) submandibular   ___ (b) ducts open into the floor of
        glands   the oral cavity
                              ___ (c) ducts open on either side
                                       of the lingual frenulum

Choose the one best answer to the following questions:

14. The cells of the gastric glands that produce secretions directly involved in chemical digestion are
    a. mucous neck cells   d. pancreatic islets
    b. parietal cells   (islets of Langerhans)
    c. chief cells   e. goblet cells
15. Which of the following anatomical regions of the stomach is closest to the esophagus?
    a. body   d. cardia
    b. pylorus   e. pyloric sphincter
    c. fundus
16. An organ of the digestive system that assumes a role in phagocytosis, manufacture of plasma proteins, detoxification, and interconversions of nutrients is the
    a. spleen   d. pancreas
    b. liver   e. gallbladder
    c. small intestine

17. Which of the following four statements is/are characteristic of the large intestine?
    (1) It is divided into cecum, colon, rectum, and anal canal.
    (2) It contains bacteria that synthesize certain nutritional factors such as vitamins.
    (3) It serves as the main absorptive surface for digesting foods.
    (4) It absorbs much of the water remaining in the waste materials.
    a. 1 only   d. 4 only
    b. 2 only   e. 1, 2, and 4
    c. 3 only
18. Which of the following digestive juices contains enzymes that digest carbohydrates, lipids, and proteins?
    a. pancreatic juice   c. saliva
    b. bile   d. gastric juice
    e. none of the above (because no one digestive juice contains enzymes for digesting all three classes of foods)
19. Which of the following three statements about the esophagus is/are true?
    (1) It extends from the pharynx to the stomach.
    (2) It is approximately 25 cm (9 in.) long in adults.
    (3) It is posterior to the trachea, anterior to the vertebral column, and passes through the diaphragm.
    a. 1 only   d. all of the above
    b. 2 only   e. 1 and 2
    c. 3 only
20. Villi
    a. are present in the lining of the stomach
    b. cover the surface of the circular folds of the small intestine
    c. line the appendix
    d. are present in the cecum
    e. are present superior to the esophagus
21. The roots of a tooth are covered by a tissue that is harder and denser than bone and is called
    a. gingivae   d. periodontal ligament
    b. cementum   e. enamel
    c. dentin
22. Which of the following three statements about the parotid glands is/are true?
    (1) They are the largest of the salivary glands.
    (2) They are located just anterior to the ear and deep to and posterior to the mandibular ramus.
    (3) They empty into the mouth by a duct that passes through a cheek muscle.
    a. 1 only   d. all of the above
    b. 2 only   e. 1 and 3
    c. 3 only
23. Which of the following statements is/are true for *every* part of the GI tract from the inferior third of the esophagus to the anus?
    a. Enzyme-secreting cells are present in the mucosa.
    b. The muscular wall consists of smooth muscle.
    c. Lymphoid follicles (Peyer's patches) are present in the submucosa.
    d. Epiploic appendages are present on the outer serosa.
    e. All of the above are true.

24. Match the following terms with their descriptions:
(1) falciform ligament
(2) greater omentum
(3) lesser omentum
(4) mesentery
(5) mesocolon

___ (a) attaches liver to anterior abdominal wall
___ (b) binds small intestine to posterior abdominal wall
___ (c) binds part of large intestine to posterior abdominal wall
___ (d) "fatty apron"; covers and helps prevent infection in small intestine
___ (e) suspends stomach and duodenum from liver

Are the following statements true or false?

25. The submucosal plexus (plexus of Meissner) is important in controlling secretions of the GI tract, whereas the myenteric plexus (plexus of Auerbach) controls GI tract motility.

26. Clusters of cells called acini make up the endocrine portions of the pancreas.

27. The ascending colon is located in the right side of the abdominopelvic cavity.

28. The middle portion of the small intestine is the ileum.

29. The teeth, tongue, and pancreas are all accessory organs of digestion.

30. The deciduous dentition consists of 20 teeth.

Critical Thinking Questions

1. Kyoko's first two permanent teeth came in just in time to fill the space in the top of her smile before school began. Unfortunately, the day before her class picture, she crashed her bike and recreated the gap. Use the proper anatomical terms to describe the teeth she lost and the ones she has remaining.
HINT: *A complete set of deciduous teeth includes a different number than a complete set of permanent teeth.*

2. The surgeon explained to the woman that he was going to remove the smaller lobe of her liver and use it to replace her child's diseased liver. Use proper anatomical terminology to describe the location of this section of the mother's liver.
HINT: *If the hands are positioned incorrectly when performing CPR, the xiphoid process may pierce the liver.*

3. Oksana and her friend were giggling over their milk and fries at her birthday party. In mid-giggle, milk started coming out of Oksana's nose. How did this happen?
HINT: *It has something to do with that little thing that hangs from the roof of your mouth and wiggles when you say "ahh."*

4. Remember Phil from Chapter 11? It turns out he really did have appendicitis. Where was his appendix located? Where would you expect to see his scar after surgery?
HINT: *Look back to Chapter 11 for the description of where his pain is localized.*

5. Umberto had just eaten a super-size steak sandwich with extra cheese and a large order of fried onions when he felt severe pain in his right side, below the ribs. It wasn't the right spot for heartburn, and it felt a lot worse. The emergency room doctor suspects gallstones. How could gallstones cause his problem?
HINT: *Bile salts will be needed to emulsify all the triglycerides in Umberto's huge fast-food meal.*

6. The small intestine is the primary site of digestion and absorption in the gastrointestinal tract. What structural modifications are unique to the small intestine, and what are their functions?
HINT: *Why are pants with an elastic waistband good to wear to a Thanksgiving Day feast?*

Answers to Figure Questions

24.1 Peristalsis.
24.2 They help regulate secretions and motility of the tract.
24.3 Mesentery.
24.4 The tongue. Extrinsic muscles move the tongue during eating and swallowing; intrinsic muscles alter the shape of the tongue for speech production and swallowing.
24.5 Filiform papillae.
24.6 Submandibular glands.
24.7 Connective tissue, specifically dentin.
24.8 Cuspids; molars; incisors.
24.9 Food is pushed along by contraction of smooth muscle superior to the bolus and relaxation of smooth muscle inferior to it.
24.10 Probably not, because as the stomach fills, they stretch out.

24.11 Mucous neck cells secrete mucus; chief cells secrete pepsinogen and gastric lipase; parietal cells secrete HCl and intrinsic factor; G cells secrete gastrin.
24.12 Pancreatic juice (fluid and digestive enzymes); bile; pancreatic juice plus bile.
24.13 Epigastric.
24.14 Hepatic portal vein → branch of hepatic portal vein → sinusoid → central vein → hepatic vein → inferior vena cava.
24.15 Ileum.
24.16 It accounts for about 90% of all digestion and absorption.
24.17 Microvilli.
24.18 Ascending and descending colons.
24.19 They secrete mucus.
24.20 At about 14 days after fertilization.

Chapter 25

The Urinary System

Student Objectives

1. Identify the external and internal gross anatomical features of the kidneys.

2. Describe the blood and nerve supply to the kidneys.

3. Define the structural features of a nephron.

4. Describe the location, structure, histology, and function of the ureters.

5. Describe the location, structure, histology, and function of the urinary bladder.

6. Describe the location, structure, histology, and function of the urethra.

7. Describe the development of the urinary system.

8. Describe the effects of aging on the urinary system.

The **urinary system** consists of two kidneys, two ureters, one urinary bladder, and a single urethra (Figure 25.1).

The kidneys have several functions:

1. **Regulation of blood volume and composition.** They regulate the composition and volume of the blood and remove wastes from the blood. In the process, urine is formed. They also excrete selected amounts of various wastes including excess H$^+$, which helps control blood pH.

2. **Regulation of blood pressure.** They help regulate blood pressure by secreting the enzyme renin, which activates the renin–angiotensin pathway. This results in an increase in blood pressure.

3. **Contribution to metabolism.** The kidneys contribute to metabolism by (1) synthesizing new glucose molecules (gluco-neogenesis) during periods of fasting or starvation, (2) secreting erythropoietin, a hormone that stimulates red blood cell production, and (3) participating in synthesis of vitamin D.

Urine is excreted from each kidney through its ureter and is stored in the urinary bladder until it is expelled from the body through the urethra.

The specialized branch of medicine that deals with the structure, function, and diseases of the kidney is known as **nephrology** (nef-ROL-ō-jē; *nephros* = kidney; *logos* = study of). The branch of medicine related to the male and female urinary systems and the male reproductive system is called **urology** (yoo-ROL-ō-jē; *urina* = urine).

Figure 25.1 Organs of the female urinary system in relation to surrounding structures.

Urine formed by the kidneys passes into the ureters then to the urinary bladder for storage and finally through the urethra for elimination from the body.

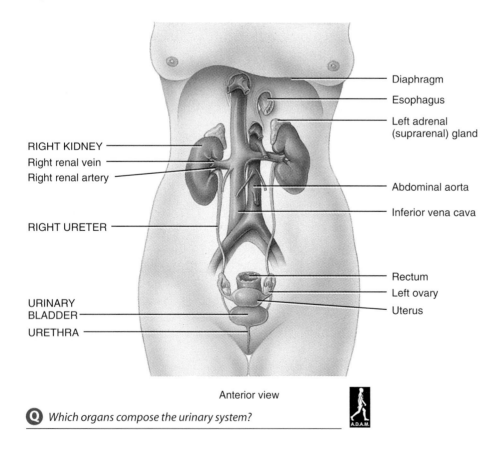

RIGHT KIDNEY
Right renal vein
Right renal artery

RIGHT URETER

URINARY BLADDER

URETHRA

Diaphragm
Esophagus
Left adrenal (suprarenal) gland

Abdominal aorta
Inferior vena cava

Rectum
Left ovary
Uterus

Anterior view

OVERVIEW OF FUNCTIONS OF THE URINARY SYSTEM

1. The kidneys regulate blood volume and composition, help regulate blood pressure, synthesize glucose, release erythropoietin, which stimulates production of red blood cells, and participate in the synthesis of vitamin D.
2. The ureters transport urine from the kidneys to the urinary bladder.
3. The urinary bladder stores urine.
4. The urethra discharges urine from the body.

Q *Which organs compose the urinary system?*

Figure 25.2 Position and coverings of the kidneys. The arrow in the inset indicates the direction from which the abdomen is viewed (inferior).

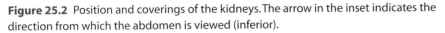 *The kidneys are surrounded by a renal capsule, adipose capsule, and renal fascia.*

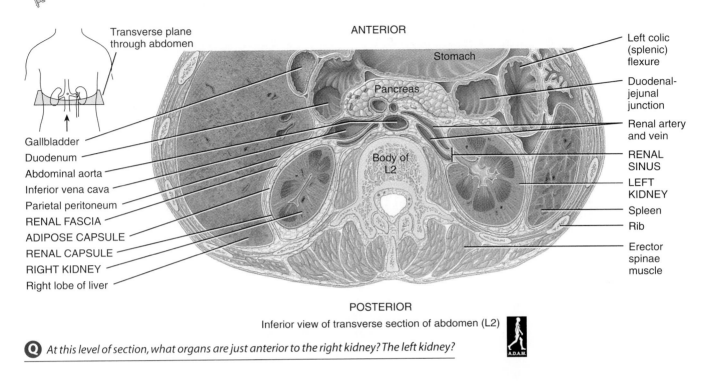

Inferior view of transverse section of abdomen (L2)

Q *At this level of section, what organs are just anterior to the right kidney? The left kidney?*

KIDNEYS

The paired **kidneys** are reddish organs shaped like kidney beans. They are located just above the waist between the parietal peritoneum and the posterior wall of the abdomen (Figure 25.2). Since their position is behind the parietal peritoneum of the abdominal cavity, they are said to be **retroperitoneal** (re′-trō-per-i-tō-NĒ-al; *retro* = behind) organs. Other retroperitoneal structures include the ureters and adrenal (suprarenal) glands. Relative to the vertebral column, the kidneys are located laterally between the levels of the last thoracic and third lumbar vertebrae. They are partially protected by the eleventh and twelfth pairs of ribs. The right kidney is slightly lower than the left because the liver occupies a large area on the right side above the kidney (see Figure 25.1).

External Anatomy

An average adult kidney measures about 10 to 12 cm (4 to 5 in.) long, 5.0 to 7.5 cm (2 to 3 in.) wide, and 2.5 cm (1 in.)

thick. Its concave medial border faces the vertebral column (Figure 25.3). Near the center of the concave border is a notch called the **renal** (*renalis* = kidney) **hilus,** through which the ureter leaves the kidney. Blood and lymphatic vessels and nerves also enter and exit the kidney through the renal hilus. The renal hilus is the entrance to a cavity in the kidney called the **renal sinus** (see Figure 25.4).

Three layers of tissue surround each kidney (see Figure 25.2a). The deep layer, the **renal capsule,** is a smooth, transparent, fibrous membrane that is continuous with the outer coat of the ureter. It serves as a barrier against trauma and helps to maintain the shape of the kidney. The intermediate layer, the **adipose capsule (perirenal fat),** is a mass of fatty tissue surrounding the renal capsule. It also protects the kidney from trauma and holds it firmly in place within the abdominal cavity. The superficial layer, the **renal fascia,** is a thin layer of dense, irregular connective tissue that anchors the kidney to its surrounding structures and to the abdominal wall. On the anterior surface of the kidneys, the renal fascia is deep to the peritoneum.

Figure 25.3 External and internal anatomy of the kidneys.

The kidneys are located between the levels of the last thoracic and third lumbar vertebrae.

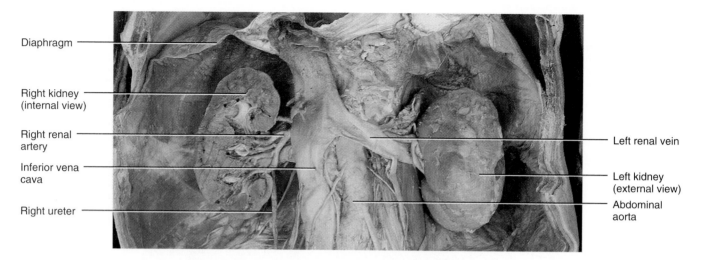

Diaphragm

Right kidney
(internal view)

Right renal
artery

Inferior vena
cava

Right ureter

Left renal vein

Left kidney
(external view)

Abdominal
aorta

(a) Photograph of anterior view of kidneys and associated structures

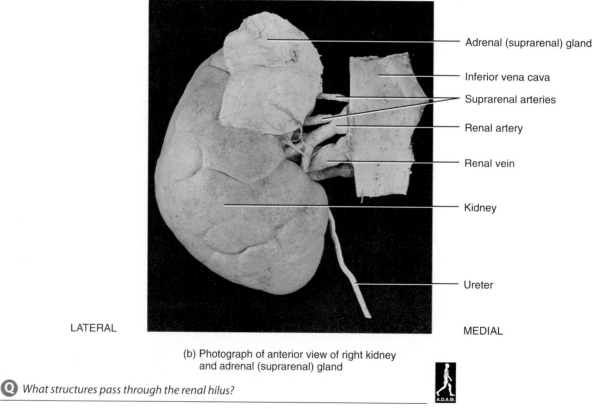

Adrenal (suprarenal) gland

Inferior vena cava

Suprarenal arteries

Renal artery

Renal vein

Kidney

Ureter

LATERAL

MEDIAL

(b) Photograph of anterior view of right kidney
and adrenal (suprarenal) gland

Q *What structures pass through the renal hilus?*

Figure 25.4 Internal anatomy of the kidneys.

The two main regions of the kidney parenchyma are the renal cortex and the renal medulla.

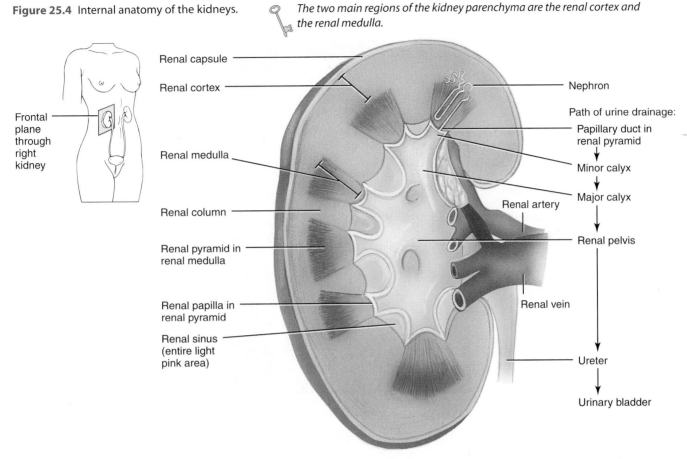

(a) Diagram of frontal section of right kidney

Figure continues

Internal Anatomy

A frontal section through a kidney reveals an outer reddish area called the **renal cortex** (*cortex* = rind or bark) and an inner reddish-brown region called the **renal medulla** (*medulla* = inner portion) (Figure 25.4a, b). Within the renal medulla are 8 to 18 cone-shaped structures termed **renal (medullary) pyramids.** The base of each renal pyramid faces the renal cortex, and its apex, called a **renal papilla** (plural is **papillae**), points toward the center of the kidney. The term **renal lobe** is applied to a renal pyramid and its overlying area of renal cortex. The renal cortex is the smooth-textured area extending from the renal capsule to the bases of the renal pyramids and into the spaces between them. It is divided into an outer *cortical zone* and an inner *juxtamedullary zone*. Those portions of the cortex that extend between renal pyramids are called the **renal columns.**

Together, the renal cortex and renal medulla constitute the functional portion or **parenchyma** of the kidney. Within

Figure 25.4 (continued)

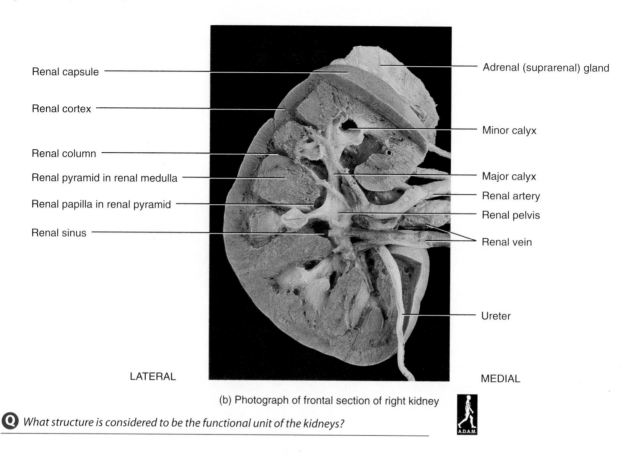

Renal capsule

Renal cortex

Renal column

Renal pyramid in renal medulla

Renal papilla in renal pyramid

Renal sinus

Adrenal (suprarenal) gland

Minor calyx

Major calyx

Renal artery

Renal pelvis

Renal vein

Ureter

LATERAL   MEDIAL

(b) Photograph of frontal section of right kidney

Q *What structure is considered to be the functional unit of the kidneys?*

the parenchyma are about 1 million microscopic structures called **nephrons** (NEF-rons), which are the functional units of the kidney. The number of nephrons is constant from birth; an increase in kidney size is due solely to the growth of individual nephrons. If nephrons are injured or become diseased, new ones cannot form. The remaining functional nephrons, however, may gradually be able to handle a larger than normal load. For example, surgical removal of one kidney results in hypertrophy (enlargement) of the remaining kidney. Eventually, the one remaining kidney is able to filter blood at 80% of the rate of two normal kidneys.

Urine formed by the nephrons ultimately drains into large ducts called **papillary ducts.** They lead to cuplike structures called **minor** and **major calyces** (KĀ-li-sēz; *calyx* = cup). Each kidney has 8–18 minor calyces and 2–3 major calyces. A minor calyx receives urine from the papillary ducts of one renal pyramid and delivers urine to a major calyx. From the major calyces, the urine drains into a large cavity called the **renal pelvis** (*pelvis* = basin) and then out through the ureter to the urinary bladder.

Blood and Nerve Supply

Because the kidneys remove wastes from the blood and regulate its fluid and electrolyte content, it is not surprising that they are abundantly supplied with blood vessels. Although the kidneys make up just 1% of the total body mass, they receive 20–25% of the resting cardiac output via the right and left **renal arteries** (Figure 25.5). This amounts to about 1200 ml of blood per minute.

Within the kidney, the renal artery divides into several **segmental arteries,** each of which supplies one renal segment (region). Each segmental artery gives off several branches that enter the parenchyma and pass as the **interlobar** (*inter* = between; *lobar* = renal lobe) **arteries** in the renal columns between the renal lobes. At the bases of the renal pyramids, the interlobar arteries arch between the renal medulla and cortex; here they are known as the **arcuate** (*arcuatus* = shaped like a bow) **arteries.** Divisions of the

Figure 25.5 Blood supply of the kidneys.

The renal arteries deliver 20–25% of the resting cardiac output to the kidneys.

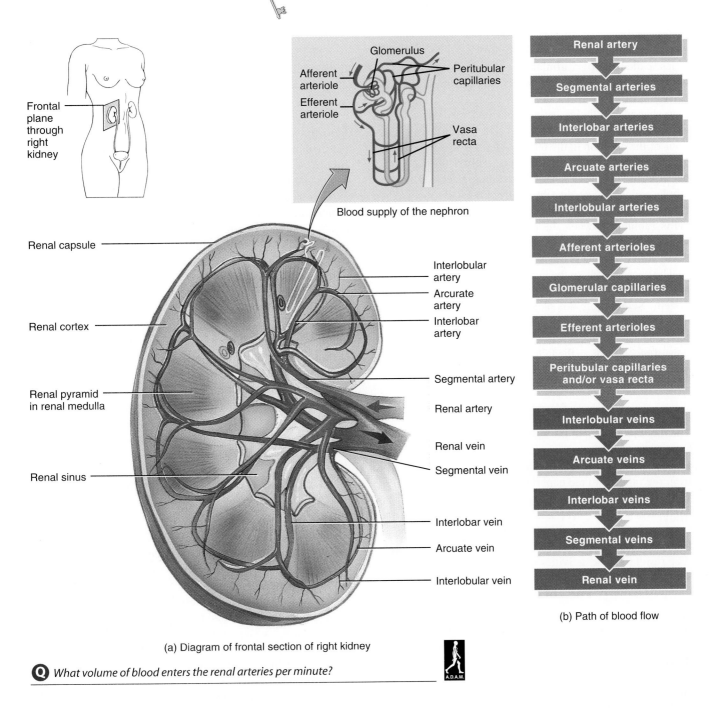

Blood supply of the nephron

(a) Diagram of frontal section of right kidney

(b) Path of blood flow

Q *What volume of blood enters the renal arteries per minute?*

arcuate arteries produce a series of **interlobular** (*lobular* = renal lobule) **arteries,** which enter the renal cortex between lobules and give off branches called **afferent** (*affere* = to carry toward) **arterioles.**

Each nephron receives one afferent arteriole, which divides into a tangled, ball-shaped capillary network called

the **glomerulus** (glō-MER-yoo-lus; *glomus* = ball; *ulus* = small; plural is **glomeruli;** see inset, Figure 25.5a). The glomerular capillaries then reunite to form an **efferent** (*effere* = to carry away) **arteriole** that drains blood out of the glomerulus. The afferent–efferent arteriole situation is unique because blood usually flows out of capillaries into venules and not into other arterioles. Because they are capillary networks, the glomeruli are part of the cardiovascular system as well as the urinary system.

The efferent arterioles divide to form a network of capillaries, called the **peritubular** (*peri* = around) **capillaries,** which surround tubular portions of the nephron in the *renal cortex.* Extending from some efferent arterioles are long loop-shaped capillaries called **vasa recta** (VĀ-sa REK-ta; *vasa* = vessels; *recta* = straight) that supply tubular portions of the nephron in the *renal medulla* (see Figure 25.6b).

The peritubular capillaries eventually reunite to form **peritubular venules** and then **interlobular veins.** The interlobular veins also receive blood from the vasa recta. Then the blood drains through the **arcuate veins** to the **interlobar veins** running between the renal pyramids, and on to the **segmental veins.** Blood leaves the kidney through a single **renal vein** that exits at the renal hilus.

The nerve supply to the kidneys is derived from the **renal plexus** of the sympathetic division of the autonomic nervous system. Nerves from this plexus accompany the renal arteries and their branches and are distributed to the blood vessels. Because these are vasomotor nerves, they regulate the flow of blood through the kidney by regulating the diameters of the arterioles.

Nephron

The functional units of the kidneys are the **nephrons,** which have three basic functions—filtration, secretion, and reabsorption. In *filtration,* some substances are permitted to pass from the blood into the nephrons, while others are kept out. Then, as the filtered liquid (filtrate) moves through the nephrons, it gains some additional materials (wastes and excess substances); this is *secretion.* Other substances (useful materials) are returned to the blood; this is *reabsorption.* As a result of these activities of nephrons, urine is formed.

Parts of a Nephron

A nephron consists of two portions: a **renal corpuscle** (KOR-pus-sul; *corpus* = body; *cle* = tiny) where plasma is filtered, and a **renal tubule** into which the filtered fluid (filtrate) passes (Figure 25.6a).

A renal corpuscle has two components—a tuft of capillary loops, called the **glomerulus,** and a double-walled epithelial cup that surrounds the glomerulus, called the **glomerular (Bowman's) capsule.** Their arrangement is analogous to a fist (glomerulus) punched into a limp balloon (glomerulus capsule) until the fist is covered by two layers of balloon with a space in between. Blood *enters* a glomerulus through an **afferent** (*ad* = toward; *ferre* = to lead) **arteriole** and *exits* through an **efferent** (*efferens* = to bring out) **arteriole.**

The outer wall, or *parietal layer,* of the glomerular capsule is separated from the inner wall, known as the *visceral layer,* by the **capsular (Bowman's) space** (see Figure 25.7a). As blood flows through the glomerular capillaries, water and most kinds of solutes filter from blood plasma into the capsular space. Large plasma proteins and the formed elements in blood do not normally pass through.

From the capsular space, filtered fluid passes into the renal tubule, which has three main sections. In the order that fluid passes through them, the renal tubule consists of a (1) **proximal convoluted tubule (PCT),** (2) **loop of Henle (nephron loop),** and (3) **distal convoluted tubule (DCT).** *Convoluted* means the tubule is coiled rather than straight. *Proximal* signifies the tubule portion attached to the glomerular capsule, and *distal* the portion that is farther away. The renal corpuscle and both convoluted tubules lie in the renal cortex of the kidney, whereas the loop of Henle extends into the renal medulla, makes a hairpin turn, then returns to the renal cortex.

The distal convoluted tubules of several nephrons empty into a single **collecting duct.** Within a renal lobe, a collecting duct and all the nephrons that drain into it constitute a **renal lobule.** Collecting ducts, in turn, unite and converge until eventually there are only several hundred large **papillary ducts,** which drain into minor calyces. The collecting ducts and papillary ducts stretch from the renal cortex through the renal medulla to the renal pelvis. On average, there are about 30 papillary ducts per renal papilla. Each kidney has about 1 million nephrons but a much smaller number of collecting ducts and even fewer papillary ducts.

Cortical and Juxtamedullary Nephrons

In a nephron, the loop of Henle connects the proximal and distal convoluted tubules. The first portion of the loop of Henle dips into the renal medulla, where it is called the **descending limb of the loop of Henle.** It then bends in a U-shape and returns to the renal cortex as the **ascending limb of the loop of Henle.** About 80–85% of the nephrons have short loops of Henle that penetrate only into the superficial region of the renal medulla (Figure 25.6a). These nephrons usually have glomeruli in the superficial region of the renal cortex and are termed **cortical nephrons.** They receive their blood supply from peritubular capillaries that arise from efferent arterioles. The remaining 15–20% are **juxtamedullary nephrons** (*juxta* = beside; Figure 25.6b). They have glomeruli deep in the renal cortex close to the renal medulla and long loops of Henle that stretch through the renal medulla, almost to the renal papilla. They receive their blood supply from peritubular capillaries and vasa recta that arise from efferent arterioles. In addition, the ascending limb of the loop of Henle of juxtamedullary nephrons consists of two portions; the first part is the **thin ascending limb**

Figure 25.6 Nephrons (colored gold).

Nephrons are the functional units of the kidneys.

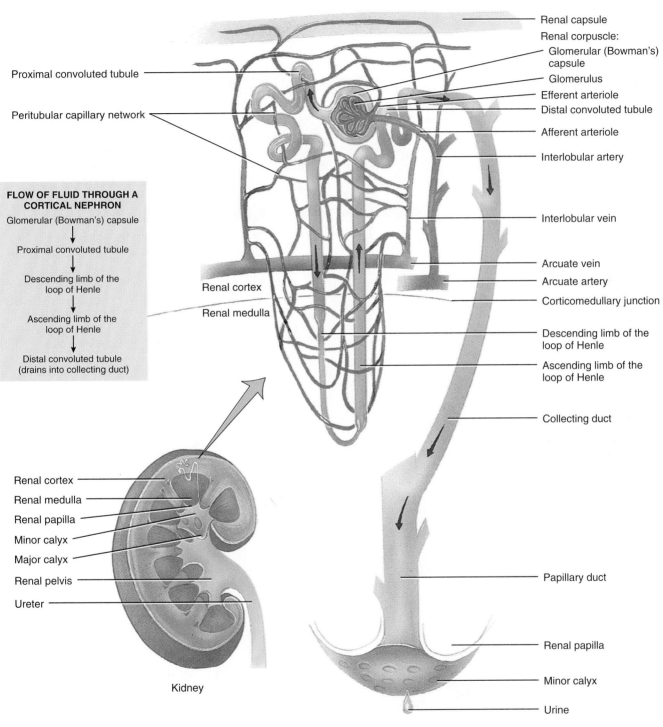

FLOW OF FLUID THROUGH A CORTICAL NEPHRON

Glomerular (Bowman's) capsule
↓
Proximal convoluted tubule
↓
Descending limb of the loop of Henle
↓
Ascending limb of the loop of Henle
↓
Distal convoluted tubule (drains into collecting duct)

Renal capsule

Renal corpuscle:
  Glomerular (Bowman's) capsule
  Glomerulus
Efferent arteriole
Distal convoluted tubule
Afferent arteriole
Interlobular artery

Interlobular vein

Arcuate vein
Arcuate artery
Corticomedullary junction

Descending limb of the loop of Henle

Ascending limb of the loop of Henle

Collecting duct

Papillary duct

Renal papilla

Minor calyx

Urine

Proximal convoluted tubule

Peritubular capillary network

Renal cortex

Renal medulla

Renal cortex
Renal medulla
Renal papilla
Minor calyx
Major calyx
Renal pelvis
Ureter

Kidney

(a) Parts of a cortical nephron and its vascular supply

Figure continues

Figure 25.6 (continued)

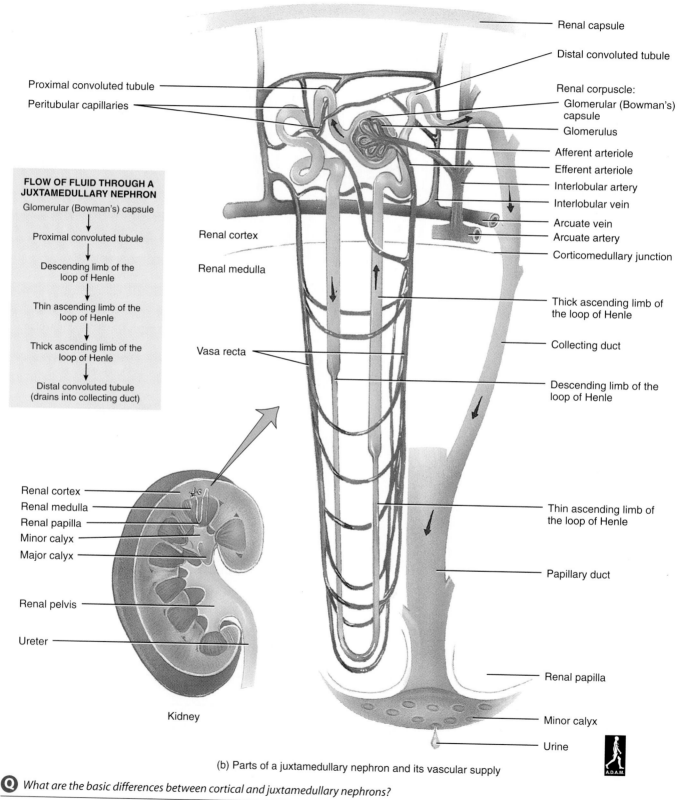

FLOW OF FLUID THROUGH A JUXTAMEDULLARY NEPHRON

Glomerular (Bowman's) capsule

↓

Proximal convoluted tubule

↓

Descending limb of the loop of Henle

↓

Thin ascending limb of the loop of Henle

↓

Thick ascending limb of the loop of Henle

↓

Distal convoluted tubule (drains into collecting duct)

(b) Parts of a juxtamedullary nephron and its vascular supply

Q *What are the basic differences between cortical and juxtamedullary nephrons?*

Figure 25.7 Endothelial–capsular (filtration) membrane. The size of the endothelial fenestrations and filtration slits in (b) have been exaggerated for emphasis.

The endothelial–capsular membrane permits some materials to pass from the blood into the glomerular filtrate but prevents the passage of others.

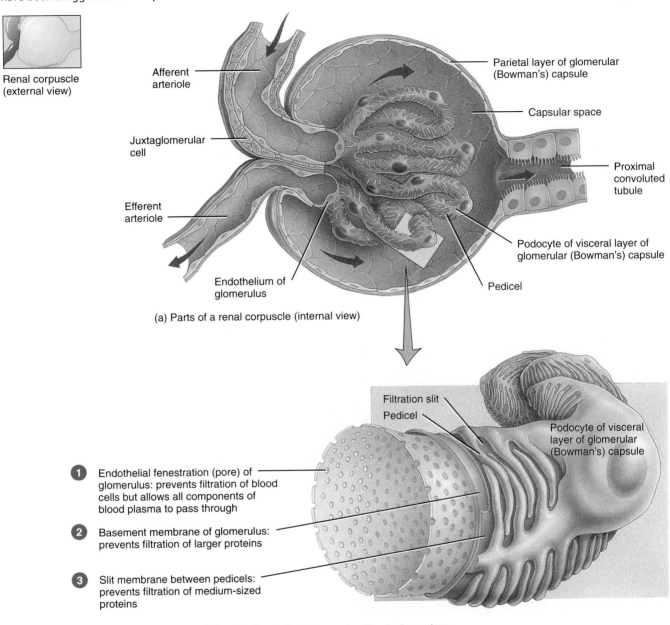

Renal corpuscle (external view)

Afferent arteriole

Juxtaglomerular cell

Efferent arteriole

Endothelium of glomerulus

Parietal layer of glomerular (Bowman's) capsule

Capsular space

Proximal convoluted tubule

Podocyte of visceral layer of glomerular (Bowman's) capsule

Pedicel

(a) Parts of a renal corpuscle (internal view)

Filtration slit
Pedicel

Podocyte of visceral layer of glomerular (Bowman's) capsule

1 Endothelial fenestration (pore) of glomerulus: prevents filtration of blood cells but allows all components of blood plasma to pass through

2 Basement membrane of glomerulus: prevents filtration of larger proteins

3 Slit membrane between pedicels: prevents filtration of medium-sized proteins

(b) Details of endothelial–capsular (filtration) membrane

Figure continues

and the second part is the **thick ascending limb** (Figure 25.6b). These long-loop nephrons enable the kidneys to excrete a very dilute or very concentrated urine.

Histology of a Nephron and Associated Structures

Each portion of a nephron and collecting duct system has distinctive histological features that reflect its particular functions.

Figure 25.7 (continued)

Slit membrane   Filtration slit   Pedicel of podocyte

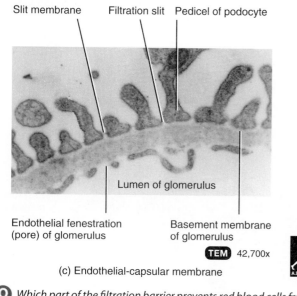

Lumen of glomerulus

Endothelial fenestration
(pore) of glomerulus

Basement membrane
of glomerulus

TEM 42,700x

(c) Endothelial-capsular membrane

Q *Which part of the filtration barrier prevents red blood cells from entering the capsular space?*

Histology of the Filter The visceral layer of the glomerular capsule and the endothelium of glomerular capillaries form an **endothelial–capsular membrane** that acts as a filter (Figure 25.7a). It lets some materials from the blood pass and restricts passage of others. Filtered substances pass through the three layers of this membrane in the following order (Figure 25.7b):

❶ Endothelial fenestrations (pores) of the glomerulus. The single layer of endothelial cells has large fenestrations (pores) that prevent filtration of blood cells but allow all components of blood plasma to pass through.

❷ Basement membrane of the glomerulus. This layer of extracellular material lies between the endothelium and the visceral layer of the glomerular capsule. It consists of fibrils in a glycoprotein matrix and prevents filtration of larger proteins.

❸ Slit membranes between pedicels. The specialized epithelial cells that cover the glomerular capillaries are called **podocytes** (*podos* = foot). Extending from each podocyte are thousands of footlike structures called **pedicels** (PED-i-sels; *pediculus* = little foot). The pedicels cover the basement membrane, except for spaces between them, which are called **filtration slits.** A thin membrane, the **slit membrane,** extends across filtration slits and prevents filtration of medium-sized proteins (Figure 25.7c).

Histology of the Renal Tubule A single layer of epithelial cells, resting on a basement membrane, forms the wall of the entire renal tubule, collecting duct, and papillary duct (Figure 25.8). In the proximal convoluted tubule (PCT), the epithelial cells are cuboidal and have a prominent brush border of microvilli on their apical surface (surface facing the lumen). These microvilli, like those of the small intestine, increase the surface area available for reabsorption and secretion. About 65% of the water and up to 100% of some solutes that pass through the endothelial–capsular membrane return to the bloodstream from the PCT.

The descending limb of the loop of Henle and the first part of the ascending limb of the loop of Henle (the thin ascending limb) are simple squamous epithelium. Cortical (short-loop) nephrons lack the thin portion of the ascending limb. The second part of the ascending limb of the loop of Henle (the thick ascending limb) is cuboidal to low columnar epithelium.

The cells of the distal convoluted tubule (DCT) and collecting ducts are cuboidal epithelium with few microvilli. Up to the DCT, the cells of a given tubule segment are all alike. Beginning in the DCT and continuing into the collecting ducts, however, two different cell types are present. Most are **principal cells,** which are sensitive to antidiuretic hormone (ADH) and aldosterone, two hormones that regulate kidney functions. A few are **intercalated cells,** which can secrete H^+ to rid the body of excess acids.

Cells of the large papillary ducts are simple columnar epithelium.

Histology of the Juxtaglomerular Apparatus (JGA) In each nephron, the final portion of the ascending limb of the loop of Henle makes contact with the afferent arteriole serving its own renal corpuscle. The cells of the renal tubule in this region are tall and crowded together. Collectively, they are known as the **macula densa** (*macula* = spot; *densa* = dense) (Figure 25.8a). These cells monitor the Na^+ and Cl^- concentration of fluid in the tubule lumen. Next to the macula densa, smooth muscle fibers in the wall of the afferent ateriole (and sometimes efferent arteriole) are modified in several ways. Their nuclei are round (instead of long), and their cytoplasm includes renin-containing granules (instead of myofibrils). These modified smooth muscle fibers are called **juxtaglomerular (JG) cells.** Together with the macula densa, they constitute the **juxtaglomerular apparatus,** or **JGA.** The JGA helps regulate blood pressure by secreting renin, an enzyme that starts a sequence of reactions that raises blood pressure and the rate of blood filtration by the kidneys. The DCT begins a short distance past the macula densa.

Figure 25.8 Histology of a nephron, juxtaglomerular apparatus, and collecting duct.

A single layer of epithelial cells forms the two layers of the renal corpuscle and the entire renal tubule.

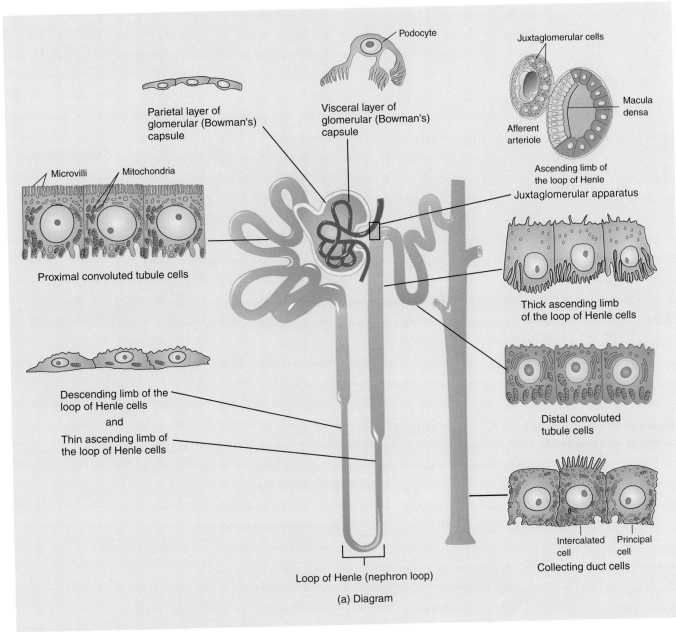

(a) Diagram

Figure continues

URINE FORMATION

Urine formation involves three principal processes: filtration, reabsorption, and secretion. Filtration is the job of the renal corpuscle; reabsorption and secretion occur along selected portions of the renal tubule (Figure 25.9).

❶ In **glomerular filtration,** blood entering the glomerulus is filtered by the endothelial–capsular membrane. The filtered fluid (**filtrate**) consists of all the substances in blood except for the formed elements and most proteins. The blood pressure in the glomerulus is the main force involved in filtering the blood.

❷ As the filtrate flows through the renal tubules, about 99 percent of it moves from the renal tubules back into blood in the peritubular capillaries or vasa recta, a process called **tubular reabsorption.** Among the substances reabsorbed

Figure 25.8 (continued)

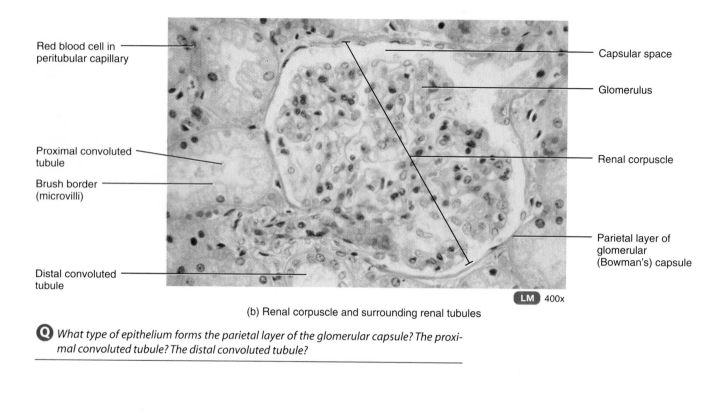

Red blood cell in peritubular capillary

Capsular space

Glomerulus

Proximal convoluted tubule

Renal corpuscle

Brush border (microvilli)

Distal convoluted tubule

Parietal layer of glomerular (Bowman's) capsule

LM 400x

(b) Renal corpuscle and surrounding renal tubules

Q *What type of epithelium forms the parietal layer of the glomerular capsule? The proximal convoluted tubule? The distal convoluted tubule?*

Figure 25.9 Overview of nephron structure and its three basic functions: filtration, reabsorption, and secretion.

Filtration occurs in the renal corpuscle, whereas reabsorption and secretion occur all along the renal tubule. Secreted substances are those that remain in the urine and leave the body.

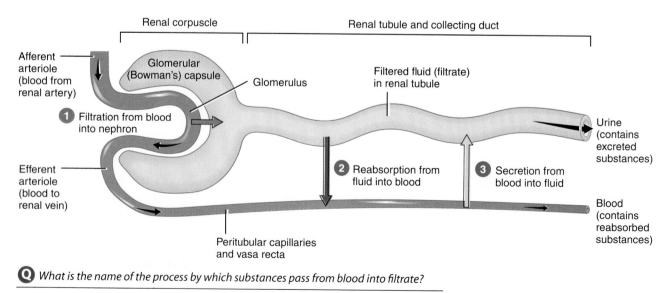

Renal corpuscle

Renal tubule and collecting duct

Afferent arteriole (blood from renal artery)

Glomerular (Bowman's) capsule

Glomerulus

Filtered fluid (filtrate) in renal tubule

1 Filtration from blood into nephron

Urine (contains excreted substances)

Efferent arteriole (blood to renal vein)

2 Reabsorption from fluid into blood

3 Secretion from blood into fluid

Peritubular capillaries and vasa recta

Blood (contains reabsorbed substances)

Q *What is the name of the process by which substances pass from blood into filtrate?*

are water, glucose, amino acids, and ions such as sodium (Na$^+$), potassium (K$^+$), calcium (Ca^{2+}), chloride (Cl$^-$), bicarbonate (HCO$_3^-$), and phosphate (HPO$_4^{2-}$). Tubular reabsorption permits the body to retain most of its nutrients.

❸ In **tubular secretion,** substances pass from the blood into filtrate. Among these are hydrogen ions (H$^+$), ammonium ions (NH$_4^+$), creatine, and certain drugs such as penicillin. Tubular secretion rids the body of certain materials and helps maintain normal pH.

URETERS

Urine drains through papillary ducts into the minor calyces. They join to become major calyces that unite to form the renal pelvis. From the renal pelvis, urine drains into the ureters and then into the urinary bladder (see Figure 25.4a). From the urinary bladder, urine is discharged from the body through the single urethra.

Structure

There are two **ureters** (YOO-re-ters; *ureter* = pertaining to urine)—one for each kidney. Each ureter is an extension of the renal pelvis and stretches 25 to 30 cm (10 to 12 in.) to the urinary bladder (see Figure 25.1). The ureters are thick-walled, narrow tubes that are not uniform in diameter, ranging from 1 mm to 1 cm. Like the kidneys, the ureters are retroperitoneal. At the base of the urinary bladder, the ureters turn medially and enter the posterior aspect of the urinary bladder (see Figure 25.10a).

Although there is no anatomical valve at the opening of each ureter into the urinary bladder, there is a physiological one that is quite effective. The ureters pass obliquely through the wall of the urinary bladder. As the urinary bladder fills with urine, pressure inside the urinary bladder compresses the ureteral openings and prevents backup of urine into the ureters. When this physiological sphincter is not operating, it is possible for microbes to travel up the ureters from the urinary bladder to infect one or both kidneys.

Blood and Nerve Supply

The arterial supply of the ureters is from the renal, testicular or ovarian, common iliac, and inferior vesical arteries (arising from the internal iliac artery, a trunk with the internal pudendal and superior gluteal arteries, or a branch of the internal pudendal artery). The veins terminate in the corresponding trunks.

The ureters are innervated by the renal plexuses, which are supplied by sympathetic and parasympathetic fibers from the lesser and lowest splanchnic nerves.

Histology

Three coats of tissue form the wall of the ureters. The inner coat, or **mucosa,** is a mucous membrane with **transitional epithelium** (see Table 3.1, transitional epithelium on page 67) and an underlying **lamina propria** of areolar connective tissue with considerable collagen and elastic fibers and lymphoid tissue. Transitional epithelium is able to stretch—a marked advantage for any organ that must continually inflate and deflate. The solute concentration and pH of urine differ drastically from the internal environment of cells that form the walls of the ureters. Mucus secreted by the mucosa prevents the cells from coming in contact with urine. Throughout most of the length of the ureters, the middle coat, the **muscularis,** is composed of inner longitudinal and outer circular layers of smooth muscle fibers. The muscularis of the distal third of the ureters also contains an outer layer of longitudinal muscle fibers. Peristalsis is the major function of the muscularis. The external coat of the ureters is the **adventitia,** a layer of areolar connective tissue that contains blood vessels, lymphatics, and nerves that serve the muscularis and mucosa. The adventitia blends in with surrounding connective tissue and anchors ureters in place.

Function

The ureters transport urine from the renal pelvis into the urinary bladder. Peristaltic contractions of the muscular walls of the ureters push urine toward the bladder, but hydrostatic pressure and gravity also contribute. Peristaltic waves pass from the kidney to the urinary bladder, varying in rate from one to five per minute, depending on the rate of urine formation.

Clinical Application

Renal Calculi

Occasionally, the salts present in urine may solidify into stones called **renal calculi** (singular is *calculus* = pebble) or **kidney stones.** Common components of kidney stones are calcium oxalate, uric acid, and calcium phosphate crystals. Calculi may form in any portion of the urinary tract. Conditions leading to calculus formation include the ingestion of excessive calcium, a decrease in water intake, abnormally alkaline or acidic urine, and overactivity of the parathyroid glands. When a stone gets stuck in a ureter, the pain can be excruciating.

A technique called **shock wave lithotripsy** (LITH-ō-trip'-sē; *litho* = stone; *tribein* = to rub) offers an alternative to surgical removal of kidney stones. A device, called a lithotripter, delivers brief, high-intensity sound waves through a water bath or water-filled cushion. Over a period of 30–60 minutes, 1000 or more hydraulic shock waves pulverize the stone until the fragments are small enough to wash out in the urine. ■

URINARY BLADDER

The **urinary bladder** is a hollow muscular organ situated retroperitoneally in the pelvic cavity posterior to the pubic symphysis. In the male, it is directly anterior to the rectum. In the female, it is anterior to the vagina and inferior to the uterus. It is a freely movable organ held in position by folds of the peritoneum. The shape of the urinary bladder depends on how much urine it contains. Empty, it is collapsed. It becomes spherical when slightly distended. As urine volume increases, it becomes pear-shaped and rises into the abdominal cavity. In general, urinary bladder capacity is smaller in females because the uterus occupies the space just superior to the bladder.

Structure

In the floor of the urinary bladder is a small triangular area, the **trigone** (TRĪ-gōn; *trigonium* = triangle; Figure 25.10a). The two posterior corners of the trigone contain the two ureteral openings, whereas the opening into the urethra, the **internal urethral orifice,** lies in the anterior corner. Because its mucosa is firmly bound to the muscularis, the trigone has a smooth appearance.

Blood and Nerve Supply

The arteries of the urinary bladder are the superior vesical (arises from the umbilical artery), the middle vesical (arises from the umbilical artery or a branch of the superior vesical), and the inferior vesical (arises from the internal iliac artery, a trunk with the internal pudendal and superior gluteal arteries, or a branch of the internal pudendal artery). The veins from the urinary bladder pass to the internal iliac trunk.

The nerves are derived partly from the hypogastric sympathetic plexus and partly from the second and third sacral nerves (pelvic splanchnic nerve).

Histology

Three coats make up the wall of the urinary bladder (Figure 25.10c). The inner **mucosa** is a mucous membrane composed of **transitional epithelium** and an underlying **lamina propria** similar to that of the ureters. Rugae (folds in the mucosa) are also present (Figure 25.10a). Surrounding the mucosa is the middle **muscularis,** which is called the **detrusor** (de-TROO-ser; *detrudere* = to push down) **muscle.** It consists of three layers of smooth muscle fiber: inner longitudinal, middle circular, and outer longitudinal. Around the opening to the urethra, the circular fibers form an **internal urethral sphincter.** Inferior to the internal sphincter is the **external urethral sphincter,** which is composed of skeletal muscle and is a modification of the urogenital diaphragm muscle (see Figure 10.13). The outer coat of the urinary blad-

der is the **adventitia,** a layer of areolar connective tissue that is continuous with that of the ureters. Over the superior surface of the urinary bladder is a layer of visceral peritoneum called the **serosa.**

Function

Urine is expelled from the urinary bladder by an act called **micturition** (mik′-too-RISH-un; *micturire* = to urinate), commonly known as urination or voiding. This response is brought about by a combination of involuntary and voluntary nerve impulses. The average capacity of the urinary bladder is 700 to 800 ml. When the amount of urine in the urinary bladder exceeds 200 to 400 ml, stretch receptors in the wall transmit nerve impulses to the inferior portion of the spinal cord. These impulses, by way of sensory tracts to the cortex, initiate a conscious desire to expel urine and, by way of a center in the sacral spinal cord, a reflex called the **micturition (urination) reflex.** In this reflex arc, parasympathetic fibers via the pelvic splanchnic nerves from the micturition reflex center of the spinal cord (spinal segments S2 and S3) conduct motor impulses to the urinary bladder wall and internal urethral sphincter. The nerve impulses cause contraction of the detrusor muscle and relaxation of the internal urethral sphincter. Urination does not yet occur, however, because the external urethral sphincter, which is skeletal muscle, remains closed. When nerve impulses from the cerebral cortex of the brain inhibit activity in motor neurons to the external urethral sphincter (skeletal muscle), there is voluntary relaxation of this muscle and urination takes place. Although emptying the urinary bladder is a reflex, it may be initiated voluntarily and stopped at will because of cerebral cortical control of the external urethral sphincter and certain muscles of the urogenital (pelvic) diaphragm.

Clinical Application

Urinary Incontinence and Retention

A lack of voluntary control over micturition is referred to as **urinary incontinence.** In infants less than 2 years old, urinary incontinence is normal because neurons to the external urethral sphincter are not completely developed. Infants void whenever the urinary bladder is sufficiently distended to stimulate the reflex. Involuntary micturition in adults may occur as a result of unconsciousness, injury to the spinal cord or spinal nerves controlling the urinary bladder, irritation due to abnormal constituents in urine, disease of the urinary bladder, damage to the external urethral sphincter, and inability of the detrusor muscle to relax due to emotional stress.

Urinary retention, a failure to completely or normally void

Figure 25.10 Ureters, urinary bladder, and urethra in the female.

Urine is stored in the urinary bladder until it is expelled by an act called micturition.

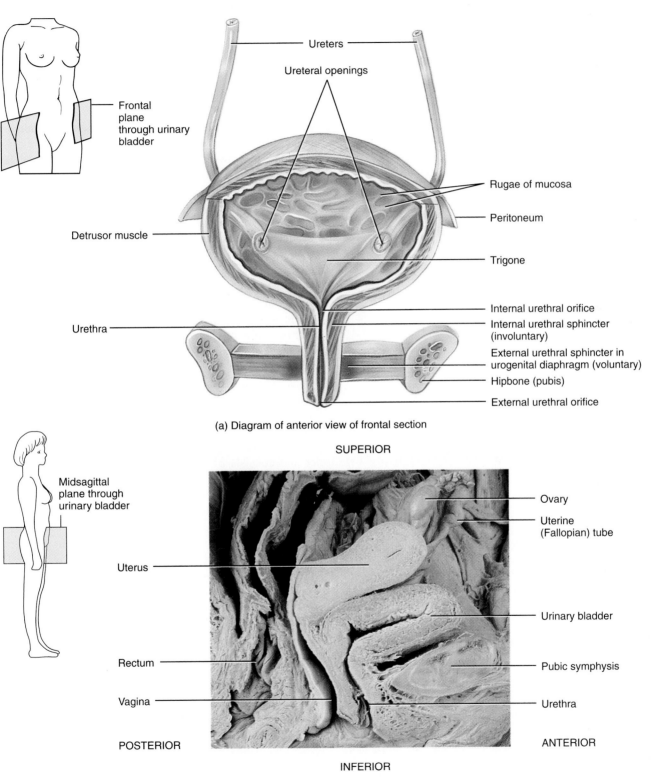

Frontal plane through urinary bladder

Ureters

Ureteral openings

Rugae of mucosa

Peritoneum

Detrusor muscle

Trigone

Urethra

Internal urethral orifice

Internal urethral sphincter (involuntary)

External urethral sphincter in urogenital diaphragm (voluntary)

Hipbone (pubis)

External urethral orifice

(a) Diagram of anterior view of frontal section

SUPERIOR

Midsagittal plane through urinary bladder

Ovary

Uterine (Fallopian) tube

Uterus

Urinary bladder

Rectum

Pubic symphysis

Vagina

Urethra

POSTERIOR

ANTERIOR

INFERIOR

(b) Photograph of midsagittal section

Figure continues

Figure 25.10 (continued)

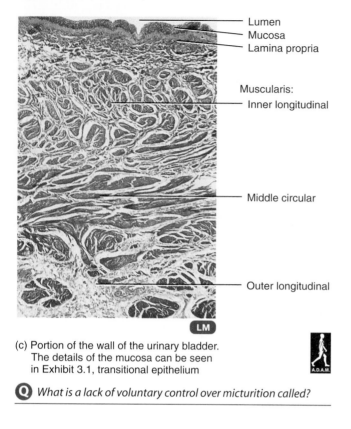

Lumen
Mucosa
Lamina propria

Muscularis:
Inner longitudinal

Middle circular

Outer longitudinal

LM

(c) Portion of the wall of the urinary bladder.
The details of the mucosa can be seen
in Exhibit 3.1, transitional epithelium

Q *What is a lack of voluntary control over micturition called?*

urine, may be due to an obstruction in the urethra or neck of the urinary bladder, nervous contraction of the urethra, or lack of sensation to urinate. ▪

URETHRA

The **urethra** is a small tube leading from the internal urethral orifice in the floor of the urinary bladder to the exterior of the body (see Figure 25.10a, b).

Structure

In females, the urethra lies directly posterior to the pubic symphysis and is embedded in the anterior wall of the vagina (see Figure 25.10b). Its length is approximately 4 cm (1.5 in.). The female urethra is directed obliquely, inferiorly, and anteriorly. The opening of the urethra to the exterior, the **external urethral orifice,** is located between the clitoris and vaginal opening.

In males, the urethra also extends from the internal urethral orifice to the exterior, but its length and passage through the body are considerably different than in females. The male urethra is about 15–20 cm (6–8 in.) long (see Figures 26.1 and 26.10a). From its origin, it passes through the prostate gland, urogenital diaphragm, and finally the penis.

Histology

The wall of the female urethra consists of an inner **mucosa** and outer **muscularis.** The mucosa is a mucous membrane composed of **epithelium** and **lamina propia** (areolar connective tissue with elastic fibers and a plexus of veins). The muscularis consists of circularly arranged smooth muscle fibers and is continuous with that of the urinary bladder. Near the urinary bladder, the mucosa contains transitional epithelium that is continuous with that of the ureter, a potential route for microbes to reach the kidneys. Near the external urethral orifice, the epithelium is nonkeratinized stratified squamous epithelium. Between these areas, the mucosa contains stratified columnar or pseudostratified columnar epithelium.

The male urethra also consists of an inner **mucosa** and an outer **muscularis.** However, their structure is quite variable. The male urethra is subdivided into three anatomical regions: (1) the **prostatic urethra** passes through the prostate gland, (2) the **membranous urethra,** the shortest portion, passes through the urogenital diaphragm, and; (3) the **spongy (penile) urethra,** the longest portion, passes through the penis (see Figure 26.10a). The epithelium of the prostatic urethra is continuous with that of the urinary bladder and consists of transitional epithelium that becomes stratified columnar or psuedostratified columnar epithelium more distally. The mucosa of the membranous urethra contains stratified columnar or pseudostratified columnar epithelium. The epithelium of the spongy urethra is stratified columnar or pseudostratified columnar epithelium, except near the external urethral orifice, which is nonkeratinized stratified squamous epithelium.

The **lamina propria** of the male urethra, like that of the female, is areolar connective tissue with elastic fibers and a plexus of veins.

The muscularis of the prostatic urethra is composed of wisps of mostly circular smooth muscle fibers outside the lamina propria that help to form the internal urethral sphincter of the urinary bladder. The muscularis of the membranous urethra consists of circularly arranged skeletal muscle fibers of the urogenital diaphragm that help form the external urethral sphincter of the urinary bladder.

Several glands and other structures associated with reproduction deliver their contents into the male urethra. The prostatic urethra contains the openings of (1) ducts that transport secretions from the **prostate gland** and (2) the **seminal vesicles** and **ductus (vas) deferens,** which deliver sperm into the urethra and provide secretions that both neutralize the acidity of the female reproductive tract and contribute to sperm motility and viability. The openings of the ducts of the **bulbourethral (Cowper's) glands** empty into the spongy urethra. They deliver an alkaline substance prior to ejaculation that neutralizes the acidity of the urethra. The glands also secrete mucus, which lubricates the end of the penis during sexual arousal. Throughout the urethra, but especially in the spongy urethra, the openings of the ducts of **urethral (Littré) glands** discharge mucus during sexual arousal or ejaculation.

Function

The urethra is the terminal portion of the urinary system. It serves as the passageway for discharging urine from the body. The male urethra also serves as the duct through which reproductive fluid (semen) is discharged from the body.

Developmental Anatomy of the Urinary System

Starting in the third week of development, a portion of the mesoderm along the posterior side of the embryo, the **intermediate mesoderm,** differentiates into the kidneys. Three pairs of kidneys form within the intermediate mesoderm in successive time periods: pronephros, mesonephros, and metanephros (Figure 25.11). Only the last pair remains as the functional kidneys of the newborn.

The first kidney to form, the **pronephros,** is the most superior of the three. Associated with its formation is a tube, the **pronephric duct.** This duct empties into the **cloaca,** which is the dilated caudal end of the gut derived from **endoderm.** The pronephros begins to degenerate during the fourth week and is completely gone by the sixth week. The pronephric ducts, however, remain.

The second kidney, the **mesonephros,** replaces the pronephros. The retained portion of the pronephric duct, which connects to the mesonephros, becomes known as the **mesonephric duct.** The mesonephros begins to degenerate by the sixth week and is almost gone by the eighth week.

At about the fifth week, a mesodermal outgrowth, called a **ureteric bud,** develops from the distal end of the mesonephric duct near the cloaca. The **metanephros,** or ultimate kidney, develops from the ureteric bud and metanephric mesoderm. The ureteric bud forms the *collecting ducts, calyces, renal pelvis,* and *ureter.* The **metanephric mesoderm** forms the nephrons of the kidneys. By the third month, the fetal kidneys begin excreting urine into the surrounding amniotic fluid. In fact, fetal urine makes up most of the amniotic fluid.

During development, the cloaca divides into a **urogenital sinus,** into which urinary and genital ducts empty, and a *rectum* that discharges into the anal canal. The *urinary bladder* develops from the urogenital sinus. In the female, the *urethra* develops from lengthening of the short duct that extends from the urinary bladder to the urogenital sinus. The *vestibule,* into which the urinary and genital ducts empty, is also derived from the urogenital sinus. In the male, the urethra is considerably longer and more complicated but is also derived from the urogenital sinus. ■

Aging and the Urinary System

As one grows older, kidney function diminishes, and by age 70 the filtering mechanism is only about half as effective as it was at age 40. Because water balance is altered and the sensation of thirst diminishes with age, older individuals are susceptible to dehydration. Consequently, urinary tract infections are more common among the elderly. Other problems may include polyuria (excessive urine production), nocturia (excessive urination at night), increased frequency or urination, dysuria (painful urination), urinary retention or incontinence, and hematuria (blood in the urine).

Kidney diseases that become more common as one ages include acute and chronic kidney inflammations and renal calculi (kidney stones). The prostate gland is often implicated in various disorders of the urinary tract, and cancer of the prostate is the most common malignancy in elderly males (see *Applications to Health,* p. 27). Because the prostate gland encircles part of the urethra (prostatic urethra), an enlarged prostate gland may cause difficulty in urination.

Key Medical Terms
Associated with the Urinary System

Azotemia (az-ō-TĒ-mē-a; *azo* = nitrogen-containing; *emia* = condition of blood) Presence of urea or other nitrogenous elements in the blood.

Cystocele (SIS-tō-sēl; *cyst* = bladder; *cele* = cyst) Hernia of the urinary bladder.

Diuresis (dī-yoo-RĒ-sis; *diourein* = to urinate) Increased excretion of urine.

Dysyria (dis-YOO-rē-a; *dys* = painful; *uria* = urine) Painful urination.

Enuresis (en'-yoo-RĒ-sis; *enourein* = to void urine) Bedwetting; may be due to faulty toilet training, to some psychological or emotional disturbance, or rarely to some physical disorder such as diabetes insipidis (lack of ADH). Also referred to as **nocturia.**

Figure 25.11 Development of the urinary system.

🔑 *Three pairs of kidneys form within intermediate mesoderm in successive time periods: pronephros, mesonephros, and metanephros.*

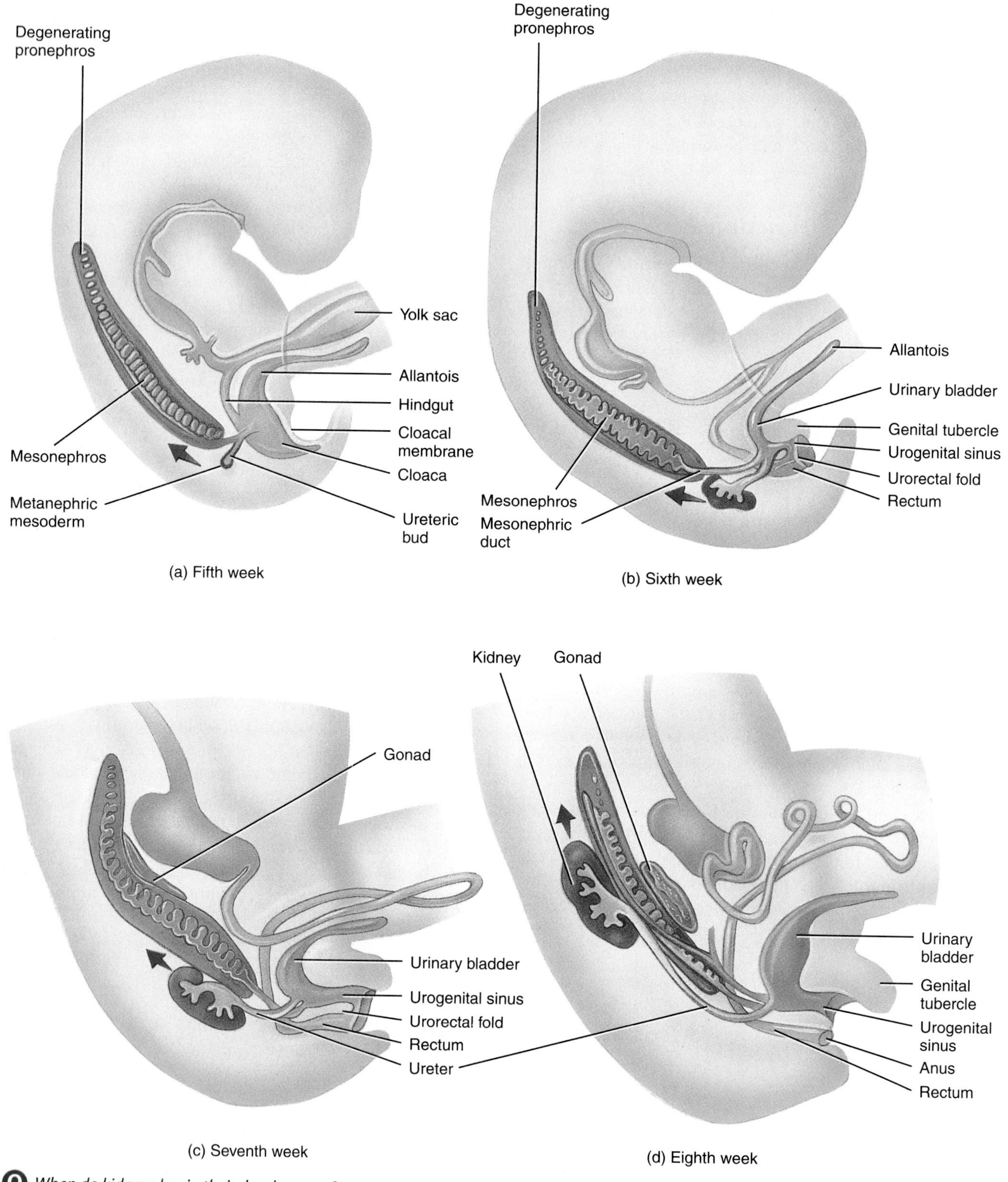

(a) Fifth week

(b) Sixth week

(c) Seventh week

(d) Eighth week

Q *When do kidneys begin their development?*

Intravenous pyelogram (in′tra-VĒ-nus PĪ-e-lō-gram′; *intra* = within; *veno* = vein; *pyelo* = pelvis of kidney; *gram* = written or recorded) or **IVP** X-ray film of the kidneys after venous injection of a dye.

Polyuria (pol′-ē-YOO-rē-a; *poly* = much) Excessive urine formation.

Stricture (STRIK-chur) Narrowing of the lumen of a canal or hollow organ, as may occur in the ureter, urethra, or any other tubular structure in the body.

Uremia (yoo-RĒ-mē-a; *emia* = condition of blood) Toxic levels of urea in the blood resulting from severe malfunction of the kidneys.

Study Outline

Introduction (p. 748)

1. The organs of the urinary system are the kidneys, ureters, urinary bladder, and urethra.
2. The kidneys regulate the volume and composition of blood, blood pressure, and some aspects of metabolism.

Kidneys (p. 749)

1. The kidneys are retroperitoneal organs attached to the posterior abdominal wall.
2. Three layers of tissue surround the kidneys: renal capsule, adipose capsule, and renal fascia.
3. Internally, the kidneys consist of a renal cortex, renal medulla, renal pyramids, renal papillae, renal columns, calyces, and a renal pelvis.
4. Blood flow through the kidney begins in the renal artery and ends in the renal vein.
5. The nerve supply to the kidney is derived from the renal plexus.
6. The nephron is the functional unit of the kidneys. A nephron consists of a renal corpuscle (glomerulus and glomerular or Bowman's capsule) and a renal tubule.
7. A renal tubule consists of a proximal convoluted tubule, loop of Henle, distal convoluted tubule, and collecting duct (shared by several nephrons). The loop of Henle consists of a descending limb of the loop of Henle and ascending limb of the loop of Henle.
8. A cortical nephron has its glomerulus in the outer third of the cortex and a short loop that dips only into the outer region of the renal medulla; a juxtamedullary nephron has its glomerulus deep in the renal cortex near the renal medulla and a long loop of Henle that stretches through the renal medulla almost to its renal papilla.
9. The filtering unit of a nephron is the endothelial–capsular membrane. It consists of the glomerular endothelium, glomerular basement membrane, and slit membranes between pedicels of podocytes.
10. The wall of the entire renal tubule consists of a single layer of epithelial cells and a basement membrane. The epithelium is modified in different portions of the tubule.
11. The juxtaglomerular apparatus (JGA) consists of the juxtaglomerular cells of an afferent ateriole and the macula densa of the renal tubule.

Urine Formation (p. 759)

1. Nephrons are the functional units of the kidneys. They regulate blood composition and volume, regulate blood pH, and remove toxic wastes from blood.
2. The nephrons form urine by glomerular filtration, tubular reabsorption, and tubular secretion.

Ureters (p. 761)

1. The ureters are retroperitoneal and consist of a mucosa, muscularis, and adventitia.
2. The ureters transport urine from the renal pelvis to the urinary bladder, primarily by peristalsis.

Urinary Bladder (p. 762)

1. The urinary bladder is a retroperitoneal organ in the pelvic cavity posterior to the pubic symphysis. Its function is to store urine prior to micturition.
2. Histologically, the urinary bladder consists of a mucosa (with rugae), a muscularis (detrusor muscle), and an adventitia (serosa over posterior surface).
3. A lack of control over micturition is called urinary incontinence; failure to void urine completely or normally is referred to as urinary retention.

Urethra (p. 764)

1. The urethra is a tube leading from the floor of the urinary bladder to the exterior. Its anatomy and histology differ in males and females.
2. Its function is to discharge urine from the body in both sexes; it also discharges semen in males.

Developmental Anatomy of the Urinary System (p. 765)

1. The kidneys develop from intermediate mesoderm.
2. They develop in the following sequence: pronephros, mesonephros, metanephros.

Aging and the Urinary System (p. 765)

1. After age 40, kidney function decreases.
2. Common problems related to aging include incontinence, urinary tract infections, prostate disorders, and renal calculi.

Review Questions

1. What organs compose the urinary system? What are the functions of the organs? (p. 748)
2. Describe the location of the kidneys. Why are they said to be retroperitoneal? (p. 749)
3. Prepare a labeled diagram that illustrates the principal external and internal features of the kidney. What is a nephroptosis? (p. 750)
4. What is a nephron? List and describe the parts of a nephron. (p. 754)
5. Describe the structure of the endothelial–capsular membrane. (p. 758)
6. Describe the histology of the various portions of a renal tubule. (p. 758)
7. Distinguish between cortical and juxtamedullary nephrons. How are nephrons supplied with blood? (p. 754)
8. Describe the structure and importance of the juxtaglomerular apparatus (JGA). (p. 758)
9. Describe the structure, histology, and function of the ureters. (p. 761)
10. How is the urinary bladder adapted to its storage function? (p. 762)
11. What is micturition? Describe the micturition reflex. (p. 762)
12. Contrast the causes of urinary incontinence and retention. (p. 762)
13. Compare the location, length, and histology of the urethra in the male and female. (p. 764)
14. Describe the development of the urinary system. (p. 765)
15. Describe the effects of aging on the urinary system. (p. 765)
16. Refer to the glossary of key medical terms associated with the urinary system. Be sure that you can define each term. (p. 765)

Self Quiz

Complete the following:

1. The cells of the endothelial–capsular membrane that form filtration slits are called ___.
2. The musculature of the urinary bladder is called the ___ muscle.
3. The three subdivisions of the male urethra, from proximal to distal, are ___ urethra, ___ urethra, and ___ urethra.
4. The functional unit of the kidney, known as a ___, consists of two main parts, a renal ___ and a renal ___.
5. The three layers of tissue surrounding each kidney, from superficial to deep, are named ___, ___, and ___.
6. The triangular-shaped structures inside the renal medulla of a kidney are called the ___.
7. The ___ carries urine from the kidney to the urinary bladder.
8. Place numbers in the blanks to arrange the following vessels in order.
   (a) arcuate arteries: ___; (b) interlobular arteries: ___; (c) renal arteries: ___; (d) peritubular capillaries and vasa recta: ___; (e) glomerular capillaries: ___; (f) efferent arteriole: ___; (g) afferent arteriole: ___; (h) interlobar arteries: ___; (i) venules and veins: ___.
9. Place numbers in the blanks to arrange the following structures in the correct sequence for the flow of glomerular filtrate.
   (a) ascending limb of loop of Henle: ___; (b) descending limb of loop of Henle: ___; (c) collecting duct: ___; (d) papillary duct: ___; (e) distal convoluted tubule: ___; (f) proximal convoluted tubule: ___.
10. The process of emptying the urinary bladder is called ___.

Choose the one best answer to the following questions:

11. Which of the following is *not* a function of the kidneys?
    a. participation in the formation of the active form of vitamin D
    b. regulation of volume and composition of the blood
    c. removal of wastes from the blood in the form of urine
    d. production of red blood cells
    e. regulation of blood pressure by secreting renin, which activates the renin–angiotensin pathway
12. Which of the following statements is *not* true?
    a. The kidneys are located posterior to the peritoneum (i.e., retroperitoneally).
    b. The left kidney is usually lower than the right kidney.
    c. The hilus is on the concave medial border of the kidney.
    d. The cavity of the kidney contains the renal pelvis, which represents the superior expanded portion of the ureter.
    e. The kidney exhibits an inner darkened area, the renal medulla, and an outer pale area, the renal cortex.
13. Urine leaving the distal convoluted tubule passes through various structures in which of the following sequences?
    a. collecting duct, renal hilus, calyx, ureter
    b. collecting duct, calyx, renal pelvis, ureter
    c. calyx, collecting duct, renal pelvis, ureter
    d. calyx, renal hilus, renal pelvis, ureter
    e. collecting duct, renal hilus, ureter, calyx
14. The trigone, a landmark in the urinary bladder, is a triangular area bounded by
    a. the orifices of the ejaculatory ducts and the urethra
    b. the internal urethral orifice and the inferior border of the detrusor muscle
    c. the urethral and the internal urethral orifices
    d. the superior portion of the fundus and the urethral orifices
    e. the major and minor calyxes

15. The portion of a kidney that includes a collecting duct and all the nephrons that drain into it is called a renal
    a. pyramid
    b. tubule
    c. lobule
    d. column
    e. lobe
16. The notch on the medial surface of the kidney through which blood vessels enter and exit is called the
    a. renal medulla
    b. renal column
    c. renal hilus
    d. minor calyx
    e. major calyx
17. The epithelium of the urinary bladder that permits distention is
    a. stratified squamous
    b. transitional
    c. simple squamous
    d. pseudostratified columnar
    e. simple cuboidal

Are the following statements true or false?

18. The kidneys are partially protected by the two pairs of floating ribs.
19. The efferent arteriole normally has a larger diameter than the afferent arteriole.
20. Principal cells of the distal convoluted tubule form part of the juxtaglomerular apparatus (JGA).
21. The external urethral sphincter is composed of voluntary skeletal muscle, whereas the internal urethral sphincter is composed of involuntary smooth muscle.
22. Pronephros, mesonephros, and metanephros are the names of the three pairs of kidneys that develop successively in fetal development.

Critical Thinking Questions

1. "It's true!" Calvin said as he poked at his chili. "I saw the can. These little beanie things are really kidneys!" How do kidney beans resemble the real organ?
   HINT: *It's not the taste!*
2. Tran flipped his ATV, smashing his back against a large rock in a dry streambed. When the EMTs got to Tran after the accident, they suspected that a broken rib had pierced a vital organ. The bruise on his back was located at the level of the twelfth rib, on the left side of the body. Which organ is likely to be damaged?
   HINT: *A rib is more likely to pierce a retroperitoneal organ.*
3. While riding down the interstate, little Caitlin told her father that she needed him to stop the car NOW! "I'm so full, I'm gonna burst!" Her dad thinks she can make it to the next rest stop. What structures of the urinary bladder will help her "hold it"?
   HINT: *Children often visit the emergency room with broken bones, but rarely with exploding bladders.*

4. As Emily focused her microscope on her own urine sample during her Human Anatomy lab, she was concerned when she found many cells in the field of view. The instructor told her that these cells were normally found in urine samples. Identify the cells.
   HINT: *The urinary system sheds and renews its lining continuously.*
5. Although a urinary catheter comes in one length only, the distance it is inserted to release urine differs significantly between males and females. Why?
   HINT: *Why do women sit and men usually stand when they urinate?*
6. One technique used to monitor kidney function is an IVP (intravenous pyelography), in which a contrast medium is first injected intravenously and then followed by x-ray as it is cleared from the blood by the kidneys. List the structures through which the contrast medium will pass as it is cleared from the blood.
   HINT: *The contrast medium is cleared by glomerular filtration.*

Answers to Figure Questions

25.1 Kidneys, ureters, urinary bladder, and urethra.
25.2 Right kidney: liver and duodenum; left kidney: colon and pancreas.
25.3 Blood and lymphatic vessels, nerves, ureter.
25.4 The nephron.
25.5 About 1200 ml.
25.6 Cortical nephrons have glomeruli in the superficial renal cortex, and their short loops of Henle penetrate only into the superficial renal medulla; juxtamedullary nephrons have glomeruli deep in the renal cortex, and their long loops of Henle extend through the renal medulla nearly to the renal papilla.
25.7 Endothelium of glomerulus.
25.8 Simple squamous; simple cuboidal with numerous microvilli; simple cuboidal with few microvilli.
25.9 Tubular secretion.
25.10 Urinary incontinence.
25.11 In the third week.

Chapter 26

The Reproductive Systems

Student Objectives

1. Define reproduction, and classify the organs of reproduction by function.

2. Explain the structure, histology, and functions of the testes.

3. Define meiosis, and explain the principal events of spermatogenesis.

4. Describe the seminiferous tubules, straight tubules, and rete testis as components of the duct system of the testes.

5. Describe the location, structure, histology, and functions of the ductus epididymis, ductus (vas) deferens, and ejaculatory duct.

6. Explain the location and functions of the accessory sex glands: seminal vesicles, prostate gland, and bulbourethral (Cowper's) glands.

7. Explain the structure and functions of the penis.

8. Describe the location, histology, and functions of the ovaries.

9. Describe the principal events of oogenesis.

10. Explain the location, structure, histology, and functions of the uterine (Fallopian) tubes.

11. Describe the location, structure, and functions of the uterus and vagina.

12. Describe the components of the vulva, and explain their functions.

13. Explain the structure and histology of the mammary glands.

14. Describe the events and importance of the uterine and ovarian cycles.

15. Describe the development of the reproductive systems.

16. Describe the effects of aging on the reproductive systems.

Sexual reproduction is a process in which organisms produce offspring by means of sex cells called **gametes.** In humans, the male gamete (sperm cell) fuses with the female gamete (secondary oocyte, potential mature ovum), an event called fertilization. Males and females have anatomically distinct reproductive organs that are adapted for producing sex cells, facilitating fertilization, and (in the female) sustaining the growth of the embryo and fetus.

The male and female reproductive organs can be grouped by function. The testes (male) and ovaries (female), collectively called **gonads,** produce gametes and secrete sex hormones. Various **ducts** store and transport the gametes. **Accessory sex glands** produce substances that protect the gametes and facilitate their movement. Finally, **supporting structures** facilitate the delivery and joining of gametes and, in the female, the growth of the fetus during pregnancy.

The specialized branch of medicine concerned with the diagnosis and treatment of diseases of the female reproductive system is **gynecology** (gī-ne-KOL-ō-jē; *gyneco* = woman). As mentioned in the previous chapter, **urology** (u-ROL-ō-jē) is the study of the urinary system. Urologists also treat diseases and disorders of the male reproductive system.

MALE REPRODUCTIVE SYSTEM

The organs of the male reproductive system are the testes, a system of ducts, accessory sex glands, and several supporting structures, including the penis (Figure 26.1). The testes (male gonads) produce sperm and also secrete hormones. A system of ducts stores sperm and conveys them to the exterior. Together with the sperm, secretions provided by accessory sex glands constitute **semen.**

Scrotum

The **scrotum** (SKRŌ-tum; *scrotum* = bag) is the supporting structure for the testes. It is a sac consisting of loose skin and superficial fascia that hangs from the root (attached portion) of the penis (Figure 26.1). Externally, the scrotum looks like a single pouch of skin separated into lateral portions by a median ridge called the **raphe** (RĀ-fē; *rafe* = seam; Figure 26.2). Internally, the **scrotal septum** divides the scrotum into two sacs, each containing a single testis. The septum consists of superficial fascia and muscle tissue called the **dartos** (DAR-tōs; *dartos* = skinned), which contains bundles of smooth muscle fibers. Dartos muscle is also found in the subcutaneous tissue of the scrotum and is directly continuous with the subcutaneous tissue of the abdominal wall. When it contracts, the dartos muscle causes wrinkling of the skin of the scrotum.

The location of the scrotum and contraction of its muscle fibers regulate the temperature of the testes. Because the scrotum is outside the pelvic cavity, it keeps the temperature of the testes about 3°C below body temperature. This cooler temperature is required for reproduction and survival of sperm. The **cremaster** (krē-MAS-ter; *kre-master* = suspender) **muscle** is a small band of skeletal muscle in the spermatic cord that is a continuation of the internal oblique muscle. It elevates the testes during sexual arousal and on exposure to cold. This action moves the testes closer to the pelvic cavity where they can absorb body heat. Exposure to warmth reverses the process. The dartos also contracts in response to cold and relaxes in response to warmth.

The blood supply of the scrotum is derived from the internal pudendal branch of the internal iliac artery, the cremasteric branch of the inferior epigastric artery, and the external pudendal artery from the femoral artery. The scrotal veins follow the arteries.

The scrotal nerves are derived from the pudendal nerve, posterior cutaneous nerve of the thigh, and ilioinguinal nerves.

Testes

The **testes,** or **testicles,** are paired oval glands measuring about 5 cm (2 in.) in length and 2.5 cm (1 in.) in diameter (Figure 26.3 on p. 774). Each weighs between 10 and 15 g. The testes develop high on the embryo's posterior abdominal wall and usually begin their descent into the scrotum through the inguinal canals (passageways in anterior abdominal wall) during the latter half of the seventh month of fetal development (see Figure 26.2).

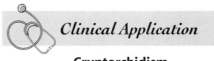

Clinical Application

Cryptorchidism

When the testes do not descend, the condition is called **cryptorchidism** (krip-TOR-ki-dizm; *kryptos* = hidden; *orchis* = testis). The condition occurs in about 3% of fullterm infants and about 30% of premature infants. Untreated cryptorchidism on both sides results in sterility because the cells involved in the initial development of sperm cells are destroyed by the higher temperature of the pelvic cavity. The chance of testicular cancer is 30 to 50 times greater in cryptorchid testes. The testes of about 80% of boys with cryptorchidism will descend spontaneously during the first year of life. When the testes remain undescended, the condition can be corrected surgically (orchiopexy), ideally before 18 months of age. ■

Figure 26.1 Male organs of reproduction and surrounding structures.

Reproductive organs are adapted to produce new individuals and pass on genetic material from one generation to the next.

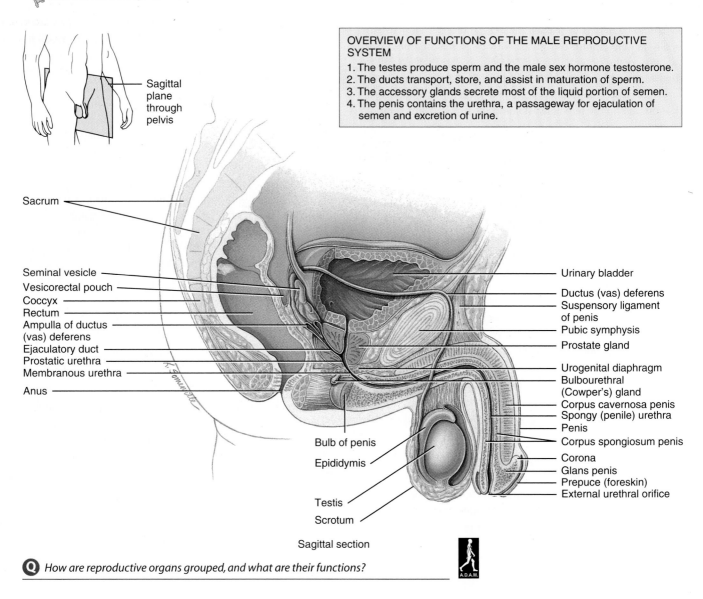

OVERVIEW OF FUNCTIONS OF THE MALE REPRODUCTIVE SYSTEM
1. The testes produce sperm and the male sex hormone testosterone.
2. The ducts transport, store, and assist in maturation of sperm.
3. The accessory glands secrete most of the liquid portion of semen.
4. The penis contains the urethra, a passageway for ejaculation of semen and excretion of urine.

Sagittal plane through pelvis

Sacrum

Seminal vesicle
Vesicorectal pouch
Coccyx
Rectum
Ampulla of ductus (vas) deferens
Ejaculatory duct
Prostatic urethra
Membranous urethra
Anus

Urinary bladder
Ductus (vas) deferens
Suspensory ligament of penis
Pubic symphysis
Prostate gland
Urogenital diaphragm
Bulbourethral (Cowper's) gland
Corpus cavernosa penis
Spongy (penile) urethra
Penis
Corpus spongiosum penis
Corona
Glans penis
Prepuce (foreskin)
External urethral orifice

Bulb of penis
Epididymis

Testis
Scrotum

Sagittal section

Q *How are reproductive organs grouped, and what are their functions?*

The testes are partially covered by a serous membrane called the **tunica** (*tunica* = sheath) **vaginalis,** which is derived from the peritoneum and is formed during the descent of the testes (Figures 26.2 and 26.3a, b). Internal to the tunica vaginalis is a dense white fibrous capsule, the **tunica albuginea** (al'-byoo-JIN-ē-a; *albus* = white). It extends inward, forming septa that divide each testis into a series of internal compartments called **lobules.** Each of the 200 to 300 lobules contains one to three tightly coiled **seminiferous** (*semen* = seed; *ferre* = to carry) **tubules.** Here sperm are produced by a process called **spermatogenesis,** which will be considered shortly.

Spermatogenic cells are sperm-forming cells in various stages that undergo mitosis and differentiation to eventually produce sperm. Together with supporting cells, they line the seminiferous tubules (Figure 26.4 on p. 775). The most immature spermatogenic cells are called **spermatogonia** (sper'-ma-tō-GŌ-nē-a; *sperm* = seed; *gonium* = generation or offspring; singular is **spermatogonium**). They lie next to the basement membrane. Toward the lumen of the tubule are layers of progressively more mature cells. In order of advancing maturity, these are primary spermatocytes, secondary spermatocytes, spermatids, and sperm. By the time a **sperm**

Figure 26.2 Scrotum and testes in relation to the spermatic cord and inguinal canal.

The scrotum consists of loose skin and superficial fascia and supports the testes.

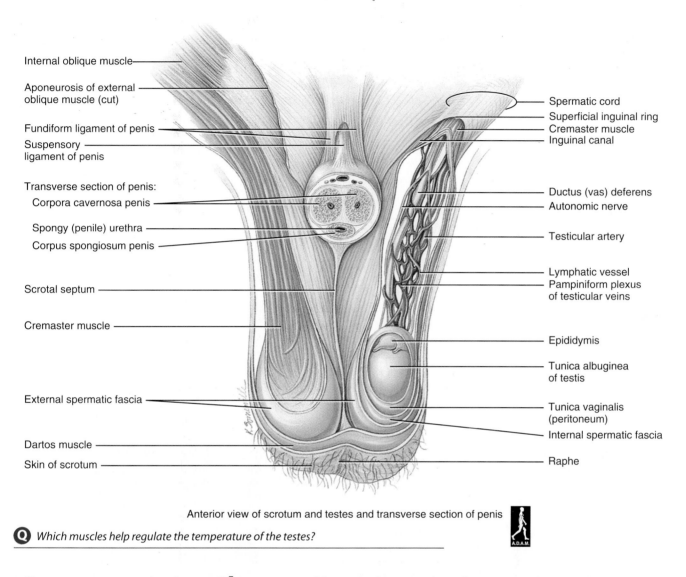

Internal oblique muscle

Aponeurosis of external oblique muscle (cut)

Fundiform ligament of penis

Suspensory ligament of penis

Transverse section of penis:
Corpora cavernosa penis

Spongy (penile) urethra

Corpus spongiosum penis

Scrotal septum

Cremaster muscle

External spermatic fascia

Dartos muscle

Skin of scrotum

Spermatic cord
Superficial inguinal ring
Cremaster muscle
Inguinal canal

Ductus (vas) deferens
Autonomic nerve

Testicular artery

Lymphatic vessel
Pampiniform plexus of testicular veins

Epididymis

Tunica albuginea of testis

Tunica vaginalis (peritoneum)
Internal spermatic fascia

Raphe

Anterior view of scrotum and testes and transverse section of penis

Q *Which muscles help regulate the temperature of the testes?*

cell, or **spermatozoon** (sper′-ma-tō-ZŌ-on; *zoon* = life; plural is **sperm** or **spermatozoa**), has nearly reached maturity, it is released into the lumen of the seminiferous tubule.

Embedded among the spermatogenic cells in the tubules are larger **sustentacular** (sus′-ten-TAK-yoo-lar; *sustentare* = to support) **cells (Sertoli cells)** that extend from the basement membrane to the lumen of the tubule. Just internal to the basement membrane, tight junctions join neighboring sustentacular cells to one another. These tight junctions form the **blood–testis barrier.** To reach the developing gametes, substances must first pass through the sustentacular cells. This barrier is important because spermatogenic cells have surface antigens that are recognized as foreign by the immune system. The barrier prevents an immune response against the surface antigens by isolating the spermatogenic cells from the blood.

Sustentacular cells support and protect developing spermatogenic cells; nourish spermatocytes, spermatids, and sperm; phagocytize excess spermatid cytoplasm as development proceeds; and mediate the effects of testosterone and follicle-stimulating hormone (FSH). Sustentacular cells also control movements of spermatogenic cells and the release of sperm into the lumen of the seminiferous tubule. They produce fluid for sperm transport and secrete the hormone inhibin, which helps regulate sperm production by inhibiting secretion of FSH.

In the spaces between adjacent seminiferous tubules are clusters of cells called **interstitial endocrinocytes** or **Leydig cells.** These cells secrete testosterone, the most important androgen (male sex hormone). (See Table 22.7.)

Figure 26.3 Internal and external anatomy of the testes.

The testes are the male gonads, which produce haploid sperm.

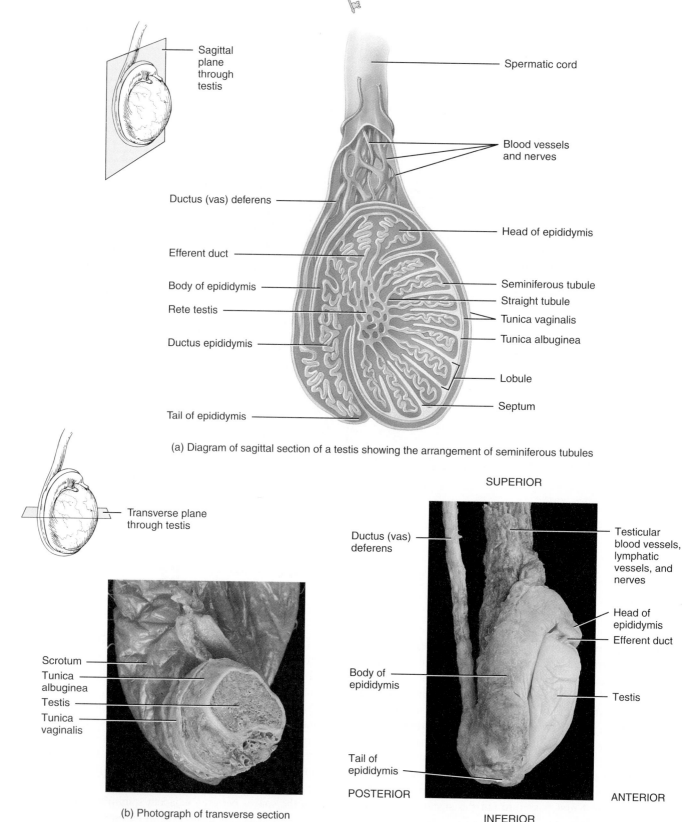

Sagittal plane through testis

Spermatic cord

Blood vessels and nerves

Ductus (vas) deferens

Head of epididymis

Efferent duct

Body of epididymis

Seminiferous tubule

Straight tubule

Rete testis

Tunica vaginalis

Ductus epididymis

Tunica albuginea

Lobule

Septum

Tail of epididymis

(a) Diagram of sagittal section of a testis showing the arrangement of seminiferous tubules

Transverse plane through testis

SUPERIOR

Ductus (vas) deferens

Testicular blood vessels, lymphatic vessels, and nerves

Head of epididymis

Efferent duct

Body of epididymis

Scrotum

Tunica albuginea

Testis

Testis

Tunica vaginalis

Tail of epididymis

POSTERIOR

ANTERIOR

(b) Photograph of transverse section

INFERIOR

(c) Photograph of testis and associated structures (lateral view)

Q *What tissue layers cover and protect the testes?*

Figure 26.4 Histology of the seminiferous tubules, and the stages of spermatogenesis (sperm production). Arrows in (b) indicate the progression from least to most mature spermatogenic cells. The (*n*) and (*2n*) refer to haploid and diploid chromosome number, respectively, to be described shortly.

🔑 *Spermatogenesis occurs in the seminiferous tubules of the testes.*

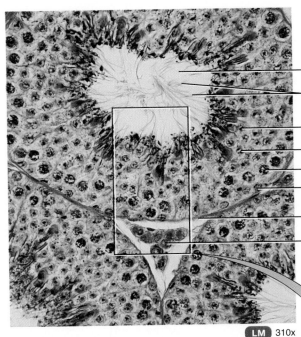

Lumen of seminiferous tubule
Spermatozoa (*n*) (sperm cells)
Spermatid (*n*)
Secondary spermatocyte (*n*)
Primary spermatocyte (*2n*)
Spermatogonium (*2n*) (stem cell)
Basement membrane
Interstitial endocrinocyte (Leydig cell)

Transverse plane through seminiferous tubules

LM 310x

(a) Transverse section of several seminiferous tubules

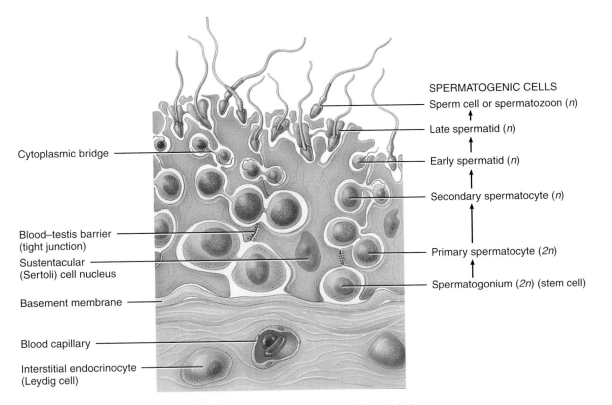

Cytoplasmic bridge

Blood–testis barrier (tight junction)
Sustentacular (Sertoli) cell nucleus
Basement membrane
Blood capillary
Interstitial endocrinocyte (Leydig cell)

SPERMATOGENIC CELLS
Sperm cell or spermatozoon (*n*)
↑
Late spermatid (*n*)
↑
Early spermatid (*n*)
↑
Secondary spermatocyte (*n*)
↑
Primary spermatocyte (*2n*)
↑
Spermatogonium (*2n*) (stem cell)

(b) Transverse section of a portion of a seminiferous tubule

Q *Which cells produce testosterone?*

Spermatogenesis

The process by which the seminiferous tubules of the testes produce haploid (*n*) spermatozoa is called **spermatogenesis** (sper'-ma-tō-JEN-e-sis; *spermatos* = sperm; *genesis* = to produce). Before reading the following discussion of spermatogenesis, keep in mind the following key points.

In sexual reproduction, a new organism is produced by the union and fusion of two different cells, one produced by each parent. These cells, called **gametes** (*gameto* = to marry), are the secondary oocyte (potential mature ovum) produced in the female gonads (ovaries) and the sperm produced in the male gonads (testes). The union and fusion of gametes is called **fertilization,** and the cell thus produced is known as a **zygote** (ZĪ-gōt; *zygosis* = a joining). The zygote contains a mixture of chromosomes (DNA) from the two parents and, through its repeated mitotic divisions, develops into a new organism.

Gametes differ from all other body cells (somatic cells) in that they contain the **haploid** (*haploos* = single; *eidos* = form) **chromosome number,** 23, a single set of chromosomes. It is symbolized by ***n***. Somatic cells, such as brain, stomach, kidney, and so on, contain 46 chromosomes in their nuclei. Of the 46 chromosomes, 23 are a complete set from one parent that contains one copy of all the genes necessary for carrying out the activities of the cell. The other 23 chromosomes are another set from the other parent, which codes for the same traits. Because somatic cells contain two sets of chromosomes, they are referred to as **diploid** (DIP-loyd; *diploos* = double) **cells,** symbolized as ***2n***. In a diploid cell, two chromosomes that belong to a pair are called **homologous** (hō-MOL-ō-gus; *homo* = same) **chromosomes,** or **homologues.**

If gametes had the same number of chromosomes as somatic cells, the zygote formed from their fusion would have double the diploid number, or 92, and with every succeeding generation, the number of chromosomes would continue to double, and normal development could not occur. The chromosome number does not double with each generation because of a special nuclear division called **meiosis** (*meio* = less) that occurs only in the production of gametes. In meiosis, a developing sperm cell or secondary oocyte gives up its duplicate set of chromosomes so that the mature gamete has only 23.

In humans, spermatogenesis takes about 74 days. It begins in the spermatogonia, which contain the diploid (*2n*) chromosome number (Figures 26.4 and 26.5). These cells develop from **primordial** (*primordialis* = primitive or early form) **germ cells** that arise from yolk sac endoderm and enter the testes early in development. In the embryonic testes, the primordial germ cells differentiate into spermatogonia but remain dormant until they begin to undergo mitotic proliferation at puberty.

Figure 26.5 Spermatogenesis. *2n* indicates diploid cells (46 chromosomes); *n* indicates haploid cells (23 chromosomes).

Spermiogenesis involves the maturation of spermatids into sperm.

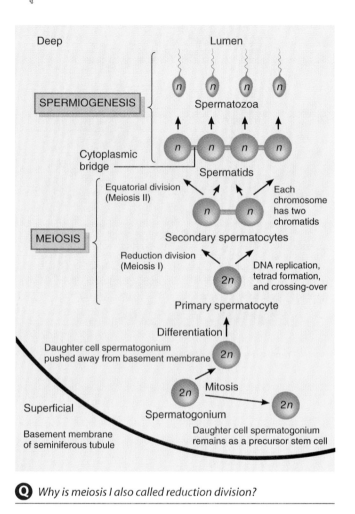

Q *Why is meiosis I also called reduction division?*

Some spermatogonia remain near the basement membrane in an undifferentiated state to serve as a reservoir of cells for future mitosis and subsequent sperm production. The rest lose contact with the basement membrane of the seminiferous tubule, undergo developmental changes, and differentiate into **primary spermatocytes** (SPER-ma-tō-sītz'). Primary spermatocytes, like spermatogonia, are diploid (*2n*); that is, they have 46 chromosomes.

Reduction Division (Meiosis I) Each primary spermatocyte enlarges before dividing. Then two nuclear divisions take place as part of meiosis. In the first, DNA replicates and 46 chromosomes (each made up of two identical chromatids from the replicated DNA) form and move toward the equatorial plane of the cell. There they line up in homologous pairs so that there are 23 pairs of duplicated chromosomes in the center of the cell. At this point, portions of a chromatid

Figure 26.6 Sperm cell (spermatozoon).

About 300 million sperm mature each day.

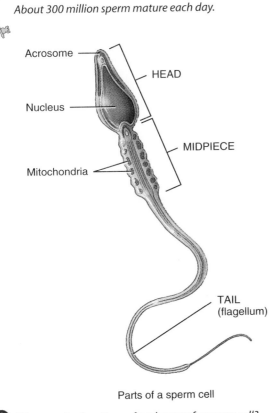

Parts of a sperm cell

Q *What are the functions of each part of a sperm cell?*

from the maternal homologue may be exchanged with the corresponding portion of the paternal homologue. This process, called **crossing-over,** permits an exchange of genes among maternal and paternal chromosomes. Thus the sperm eventually produced are genetically unlike each other and unlike the cell that produced them—one reason for the great genetic variation among individuals.

Next, the meiotic spindle forms and pulls one duplicated chromosome of each pair to an opposite pole of the dividing cell. The random assortment of maternally derived and paternally derived chromosomes toward opposite poles is another reason for genetic variation among sperm. The cells formed by the first nuclear division (reduction division) are called **secondary spermatocytes.** Each cell has 23 chromosomes—the haploid number. Each chromosome within a secondary spermatocyte, however, is made up of two chromatids (two copies of the DNA) still attached by a centromere.

Equatorial Division (Meiosis II) The second nuclear division of meiosis is **equatorial division.** There is no replication of DNA. The chromosomes (each composed of two chromatids) line up in single file along the equatorial plane,

and the chromatids of each chromosome separate from each other. The cells formed from the equatorial division are called **spermatids.** Each has half the original chromosome number, or 23 chromosomes, and is haploid. Each primary spermatocyte therefore produces four spermatids by meiosis through two rounds of cell division (reduction division and equatorial division). Spermatids lie close to the lumen of the seminiferous tubule.

Spermiogenesis The final stage of spermatogenesis, called **spermiogenesis** (sper'-mē-ō-JEN-e-sis), involves the maturation of spermatids into sperm. Each spermatid develops a head with an acrosome (enzyme-containing granule) and a flagellum (tail). Because there is no cell division in spermiogenesis, each spermatid develops into a single **sperm cell (spermatozoon).** The release of a sperm cell from its connection to a sustentacular cell is known as **spermiation.** Sperm then enter the lumen of the seminiferous tubule and flow toward ducts of the testes.

Sperm

Sperm mature at the rate of about 300 million per day and, once ejaculated, most probably do not survive more than 48 hours within the female reproductive tract. A sperm cell is highly adapted for reaching and penetrating a secondary oocyte (a mature ovum). It is composed of a head, a midpiece, and a tail (Figure 26.6). The **head** contains the nuclear material (DNA) and a lysosome-like structure called an **acrosome** (*acro* = atop), which contains enzymes that help the sperm cell penetrate into the ovum. Numerous mitochondria in the **midpiece** carry on the metabolism that provides ATP for locomotion. The **tail,** a typical flagellum, propels the sperm cell along its way.

Ducts

Ducts of the Testis

After their release into the lumen of the twisted seminiferous tubules, sperm and fluid flow to the **straight tubules** (see Figure 26.3a). The continual release of sperm and fluid produces pressure that moves the contents into the straight tubule. The fluid is produced by the sustentacular (Sertoli) cells. The straight tubules lead to a network of ducts in the testis called the **rete** (RĒ-tē; *rete* = network) **testis.** The sperm next move through the rete testis and out of the testis into an adjacent organ, the epididymis. From the rete testis sperm move into a series of coiled **efferent ducts** in the epididymis that empty into a single tube called the **ductus epididymis** (*ducere* = to lead).

Epididymis

The **epididymis** (ep'-i-DID-i-mis; *epi* = above; *didymos* = testis) is a comma-shaped organ about 4 cm (1½ in.) long that lies along the posterior border of each testis (see Figures

Figure 26.7 Histology of the ductus epididymis.

🔑 *Stereocilia increase the surface area for reabsorption of degenerated sperm.*

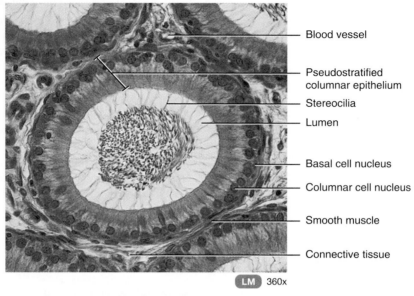

Blood vessel

Pseudostratified columnar epithelium

Stereocilia

Lumen

Basal cell nucleus

Columnar cell nucleus

Smooth muscle

Connective tissue

LM 360x

Transverse section of the ductus epididymis

Q *What are the functions of the ductus epididymis?*

26.2 and 26.3b). The plural is **epididymides** (ep'-i-did-ĪM-i-dēs). Each epididymis consists mostly of the tightly coiled **ductus epididymis.** The larger, superior portion of the epididymis is known as the **head.** In the head, the efferent ducts from the testis join the ductus epididymis. The **body** is the narrow midportion of the epididymis. The **tail** is the smaller, inferior portion. At its distal end, the tail of the epididymis continues as the ductus (vas) deferens (discussed shortly).

The ductus epididymis is a tightly coiled structure that would measure about 6 m (20 ft) in length if it were straightened out. It is lined with pseudostratified columnar epithelium and encircled by layers of smooth muscle. The free surfaces of the columnar cells contain long, branching microvilli called **stereocilia** (Figure 26.7). They increase surface area for reabsorption of degenerated sperm.

Functionally, the ductus epididymis is the site of sperm maturation. Over a 10–14-day period, their motility increases. The ductus epididymis also stores sperm and helps propel them by peristaltic contraction of its smooth muscle into the ductus (vas) deferens. Sperm may remain in storage in the ductus epididymis for a month or more.

Ductus Deferens

Within the tail of the epididymis, the ductus epididymis becomes less convoluted and its diameter increases. After this point, the duct is referred to as the **ductus deferens, vas deferens,** or **seminal duct** (see Figure 26.3b). The ductus deferens, about 45 cm (18 in.) long, ascends along the posterior border of the epididymis, penetrates the inguinal canal (see Figure 26.2), and enters the pelvic cavity. There, it loops over the ureter and over the side and down the posterior surface of the urinary bladder (see Figure 26.1). The dilated terminal portion of the ductus deferens is known as the **ampulla** (am-POOL-la; *ampulla* = little jar). (See Figure 26.9.) The ductus deferens is lined with pseudostratified columnar epithelium (Figure 26.8) and contains a heavy coat of three layers of muscle. The inner and outer layers are longitudinal and the middle layer is circular.

Functionally, the ductus deferens stores sperm. Here, they can remain viable for up to several months. The ductus deferens also conveys sperm from the epididymis toward the urethra by peristaltic contractions of the muscular coat. Sperm that are not ejaculated are ultimately reabsorbed.

Figure 26.8 Histology of the ductus (vas) deferens.

The ductus (vas) deferens enters the pelvic cavity through the inguinal canal.

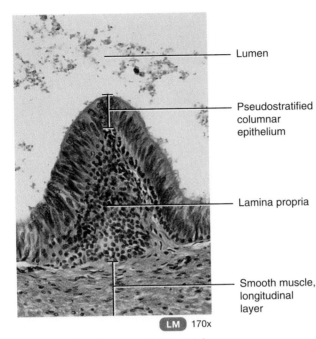

Lumen

Pseudostratified columnar epithelium

Lamina propria

Smooth muscle, longitudinal layer

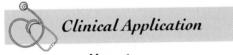

LM 170x

Transverse section of mucosa of ductus deferens

Q *What is the function of the ductus (vas) deferens?*

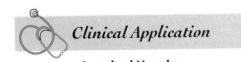

Clinical Application

Vasectomy

The principal method for sterilization of males is a **vasectomy** (vas-EK-tō-mē; *tome* = incision), in which a portion of each ductus deferens is removed. An incision is made in the posterior side of the scrotum, the ducts are located and cut, each is tied in two places, and the portion between the ties is removed. Although sperm production continues in the testes, sperm can no longer reach the exterior. The sperm degenerate and are destroyed by phagocytosis. Because the blood vessels are not cut, testosterone levels in the blood remain normal, so vasectomy has no effect on sexual desire and performance. If done correctly, it is close to 100% effective. The procedure can be reversed, but the chance of regaining fertility is only 30–40%. ▪

The ductus deferens ascends the **spermatic cord** in the scrotum (see Figure 26.2). The spermatic cord encloses the testicular artery, autonomic nerves, veins that drain the testes and carry testosterone into circulation (pampiniform plexus), lymphatic vessels, and the cremaster muscle.

The spermatic cord and ilioinguinal nerve pass through the **inguinal** (IN-gwin-al; *inguina* = groin) **canal.** The canal is an oblique passageway in the anterior abdominal wall just superior and parallel to the medial half of the inguinal ligament. The canal is about 4–5 cm (about 2 in.) in length. It originates at the **deep (abdominal) inguinal ring,** a slitlike opening in the aponeurosis of the transversus abdominis muscle. The canal ends at the **superficial (subcutaneous) inguinal ring,** a somewhat triangular opening in the aponeurosis of the external oblique muscle. In females, the round ligament of the uterus and ilioinguinal nerve pass through the inguinal canal.

Clinical Application

Inguinal Hernias

The inguinal region is a weak area in the abdominal wall. It often is the site of an **inguinal hernia**—a rupture or separation of a portion of the inguinal area of the abdominal wall resulting in the protrusion of a part of the small intestine. In an *indirect inguinal hernia,* a part of the small intestine protrudes through the deep inguinal ring and enters the inguinal canal. It can then pass through the superficial inguinal ring and down into the scrotum. In a *direct inguinal hernia,* a portion of the small intestine pushes into the posterior wall of the inguinal canal and usually causes a localized bulging in the wall of the canal. Inguinal hernias are much more common in males than in females because the inguinal canals in males are large and represent weak points in the abdominal wall. ▪

Ejaculatory Ducts

Each **ejaculatory** (e-JAK-yoo-la-tō'-rē; *ejectus* = to throw out) **duct** (Figure 26.9) is about 2 cm (1 in.) long and is formed by the union of the duct from the seminal vesicle and the ampulla of the ductus deferens. The ejaculatory ducts pass through the prostate gland and eject sperm into the urethra just before **ejaculation,** the powerful propulsion of semen from the urethra to the exterior. They also transport and eject secretions of the seminal vesicles (described shortly).

Urethra

In males the **urethra** is the shared terminal duct of the reproductive and urinary systems. It serves as a passageway for both semen (a mixture of sperm and various fluids) and urine. The urethra passes through the prostate gland, the urogenital diaphragm, and the penis. It measures about 20 cm

Figure 26.9 Male reproductive organs in relation to surrounding structures. The prostate gland, urethra, and penis have been sectioned to show internal details.

The male urethra has three subdivisions: prostatic, membranous, and spongy.

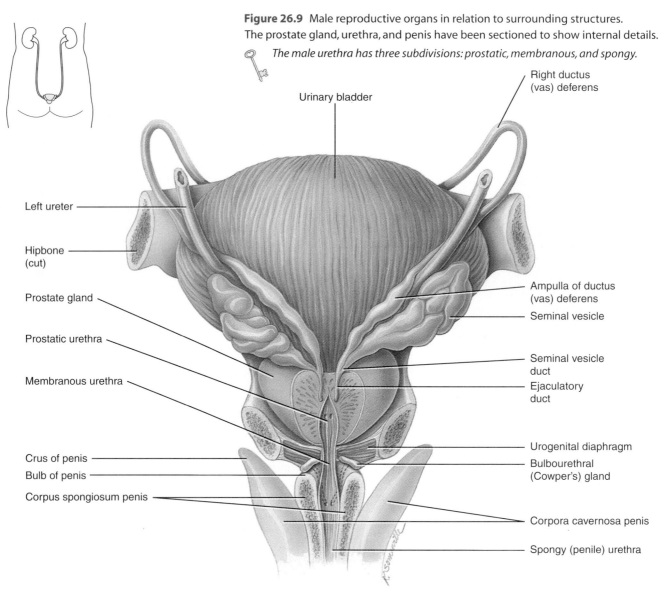

(a) Diagram of posterior view of male accessory organs of reproduction

(8 in.) in length and is subdivided into three parts (see Figures 26.1 and 26.9). The **prostatic urethra** is 2–3 cm (1 in.) long and passes through the prostate gland. It continues inferiorly, and as it passes through the urogenital diaphragm, a muscular partition between the two ischial and pubic rami, it is known as the **membranous urethra.** The membranous urethra is about 1 cm (½ in.) in length. As it passes through the corpus spongiosum of the penis, it is known as the **spongy (penile) urethra.** This portion is about 15–20 cm (6–8 in.) long. The spongy urethra ends at the **external urethral orifice.**

Accessory Sex Glands

Whereas the ducts of the male reproductive system store and transport sperm cells, the **accessory sex glands** secrete most of the liquid portion of semen. Among the accessory sex glands are the seminal vesicles, prostate gland, and bulbourethral glands.

Seminal Vesicles

The paired **seminal** (*seminalis* = pertaining to seed) **vesicles** (VES-i-kuls) are convoluted pouchlike structures, about 5 cm (2 in.) in length, lying posterior to and at the base of the

Figure 26.9 (continued)

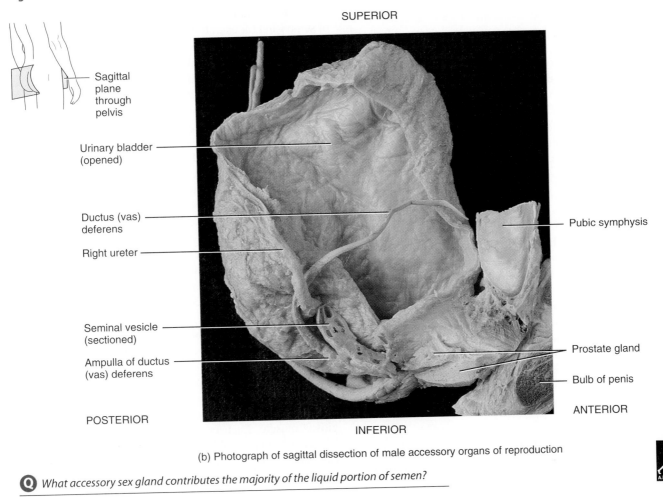

SUPERIOR

Sagittal
plane
through
pelvis

Urinary bladder
(opened)

Ductus (vas)
deferens

Right ureter

Seminal vesicle
(sectioned)

Ampulla of ductus
(vas) deferens

Pubic symphysis

Prostate gland

Bulb of penis

ANTERIOR

POSTERIOR

INFERIOR

(b) Photograph of sagittal dissection of male accessory organs of reproduction

Q *What accessory sex gland contributes the majority of the liquid portion of semen?*

urinary bladder anterior to the rectum (Figure 26.9). They secrete an alkaline, viscous fluid that contains *fructose* (a monosaccharide sugar), *prostaglandins,* and clotting proteins unlike those found in blood. The alkaline nature of the fluid helps to neutralize acid in the female reproductive tract. This acid would inactivate and kill sperm if not neutralized. The fructose is used for ATP production by sperm. Prostaglandins contribute to sperm motility and viability and may also stimulate muscular contraction within the female reproductive tract. Fluid secreted by the seminal vesicles normally constitutes about 60% of the volume of semen.

Prostate Gland

The **prostate** (PROS-tāt) **gland** is a single, doughnut-shaped gland about the size and shape of a chestnut (Figure 26.9). It is inferior to the urinary bladder and surrounds the prostatic urethra. The prostate secretes a milky, slightly acidic fluid (pH about 6.5) that contains a variety of acidic compounds

and enzymes. Prostate gland secretions enter the prostatic urethra through many prostatic ducts. These secretions make up about 25% of the volume of semen and contribute to sperm motility and viability. The prostate gland slowly increases in size from birth to puberty, and then a rapid growth spurt occurs. The size attained by age 30 remains stable until about age 45, when further enlargement of the prostate gland may occur.

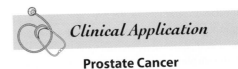

Clinical Application

Prostate Cancer

Prostate cancer is the leading cause of death from cancer in men in the United States (having surpassed lung cancer in 1991). In 1996 about 200,000 U.S. men were diagnosed, and 40,000 died.

A blood test can measure the level of prostate-specific antigen (PSA) in the blood. This substance is an enzyme produced only by prostate epithelial cells. The amount of PSA increases with enlargement of the prostate gland and may indicate infection, benign enlargement, or prostate cancer. For males over 40, the American Cancer Society recommends annual examination of the prostate gland by a **digital rectal exam,** in which the physician palpates the gland through the rectum with the fingers (digits). Many physicians also recommend an annual PSA test for males over 50. Treatment for prostate cancer may involve surgery, radiation, hormonal therapy, and chemotherapy. Because many prostate cancers grow very slowly, some urologists recommend "watchful waiting" before treating small tumors in men over age 70. ■

Bulbourethral Glands

The paired **bulbourethral** (bul′bō-yoo-RĒ-thral), or **Cowper's, glands** are each about the size of a pea. They lie inferior to the prostate gland on either side of the membranous urethra within the urogenital diaphragm and have ducts that open into the spongy urethra (Figure 26.9a). During sexual arousal, the bulbourethral glands pave the way for arrival of sperm by secreting an alkaline substance that protects sperm by neutralizing acids in the urethra. At the same time, they also secrete mucus that lubricates the end of the penis and the lining of the urethra, which decreases the number of sperm injured during ejaculation.

Semen

Semen (*semen* = seed) is a mixture of sperm and **seminal fluid,** which is the liquid portion of semen that consists of the secretions of the seminiferous tubules, seminal vesicles, prostate gland, and bulbourethral glands. The average volume of semen in an ejaculation is 2.5–5 ml, with a sperm count (concentration) of 50–150 million sperm per milliliter. When the number of sperm falls below 20 million/ml, the male is likely to be infertile. The very large number is required because only a tiny fraction ever reach the secondary oocyte. And although only a single sperm cell fertilizes a secondary oocyte, its penetration requires the combined action of a large number of sperm. Their acrosomes release hyaluronidase and proteinases to digest the materials surrounding the secondary oocyte. A single sperm does not produce enough of these enzymes to dissolve the barrier, but the combined action of many permits the entrance of one.

Despite the slight acidity of prostatic fluid, semen has a slightly alkaline pH of 7.2–7.7 due to the higher pH and larger volume of fluid from the seminal vesicles. The prostatic secretion gives semen a milky appearance, whereas fluids from the seminal vesicles and bulbourethral glands give it a sticky consistency. Semen provides sperm with a transportation medium and nutrients. It neutralizes the hostile acidic environment of the male urethra and the female vagina.

Penis

The **penis** contains the urethra, a passageway for ejaculation of semen and for excretion of urine. It is cylindrical in shape and consists of a body, root, and glans penis (Figure 26.10). The **body** of the penis is composed of three cylindrical masses of tissue, each bound by fibrous tissue called the **tunica albuginea.** The paired dorsolateral masses are called the **corpora cavernosa penis** (*corpus* = body; *caverna* = hollow). The smaller midventral mass, the **corpus spongiosum penis,** contains the spongy urethra and functions in keeping the spongy urethra open during ejaculation. All three masses are enclosed by fascia and skin and consist of erectile tissue permeated by blood sinuses. With sexual stimulation, the arteries supplying the penis dilate, and large quantities of blood enter the blood sinuses. Expansion of these spaces compresses the veins draining the penis, so more blood that enters is trapped. These vascular changes result in an **erection,** a parasympathetic reflex. The penis returns to its flaccid state when the arteries constrict and pressure on the veins is relieved.

The **root** of the penis is the attached portion (proximal portion) and consists of the **bulb of the penis,** the expanded portion of the base of the corpus spongiosum penis, and the **crura** (*crus* = leg; singular is **crus**) **of the penis,** the two separated and tapered portions of the corpora cavernosa penis. The bulb of the penis is attached to the inferior surface of the urogenital diaphragm and enclosed by the bulbospongiosus muscle. Each crus of the penis is attached to the ischial and inferior pubic rami and surrounded by the ischiocavernosus muscle (see Figure 10.14). Contraction of these skeletal muscles aids ejaculation.

The distal end of the corpus spongiosum penis is a slightly enlarged, acorn-shaped region called the **glans penis** (*glandes* = acorn). The margin of the glans penis is termed the **corona.** The distal urethra enlarges within the glans penis and forms a terminal slitlike opening, the **external urethral orifice.** Covering the glans in an uncircumsized penis is the loosely fitting **prepuce** (PRĒ-pyoos), or **foreskin.** The weight of the penis is supported by two ligaments that are continuous with the fascia of the penis. These are the **fundiform ligament,** which arises from the inferior part of the linea alba, and the **suspensory ligament of the penis,** which arises from the pubic symphysis (see Figure 26.2).

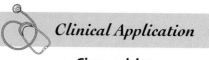

Clinical Application

Circumcision

Circumcision (*circumcido* = to cut around) is a surgical procedure in which part or all of the prepuce is removed. It is usually performed just after delivery, 3–4 days after birth, or on the eighth day as part of a Jewish religious rite. Some health-care

Figure 26.10 Internal structure of the penis. The inset in (b) shows details of the skin and fascia.

The penis contains a pathway for ejaculation of semen and excretion of urine.

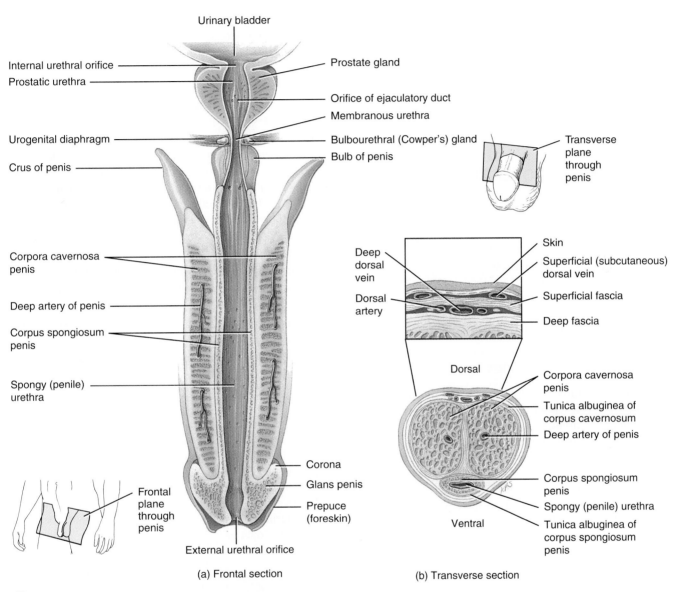

(a) Frontal section

(b) Transverse section

Q *Which tissue masses form the erectile tissue in the penis, and why do they become rigid during sexual arousal?*

professionals think there is no medical justification for circumcision. Others feel that it has benefits such as a lower risk of urinary tract infections, protection against penile cancer, and possibly a lower risk for sexually transmitted diseases. ■

The penis has a very rich blood supply from the internal pudendal artery and the femoral artery. The veins drain into corresponding vessels.

The sensory nerves to the penis are branches from the pudendal and ilioinguinal nerves. The corpora have a sympathetic and parasympathetic supply. As a result of parasympathetic stimulation, the blood vessels dilate, increasing the flow of blood into the erectile tissue. The result is that blood is trapped within the penis and erection is maintained. At ejaculation, sympathetic stimulation causes the smooth muscle located in the walls of the ducts and accessory sex glands of the reproductive tract to contract and propel the semen along its course. The musculature of the penis, which is supplied by the pudendal nerve, also contracts at ejaculation. The muscles include the bulbospongiosus muscle, which overlies the bulb of the penis, the ischiocavernosus muscles on either side of the penis, and the superficial transverse perineus muscles on either side of the bulb of the penis (see Figure 10.14).

FEMALE REPRODUCTIVE SYSTEM

The female organs of reproduction (Figure 26.11) include the ovaries, which produce secondary oocytes (cells that develop into mature ova only after fertilization) and hormones; the uterine (Fallopian) tubes or oviducts, which transport secondary oocytes and fertilized ova to the uterus; the uterus, in which embryonic and fetal development occur; the vagina; and external organs that constitute the vulva, or pudendum. The mammary glands also are considered part of the female reproductive system.

Ovaries

The **ovaries** (*ovarium* = egg receptacle), or female gonads, are paired glands resembling unshelled almonds in size and shape. They are homologous to the testes. (*Homologous* means that two organs have the same embryonic origin.) The ovaries descend to the brim of the pelvis during the third month of development. They lie in the superior portion of the pelvic cavity, one on each side of the uterus. A series of ligaments holds the ovaries in position (Figure 26.12 on page 787). The **broad ligament** of the uterus, which is itself part of the parietal peritoneum, attaches to the ovaries by a double-layered fold of peritoneum called the **mesovarium.** The **ovarian ligament** anchors the ovaries to the uterus, and the **suspensory ligament** attaches them to the pelvic wall. Each ovary contains a **hilus,** the point of entrance for blood vessels and nerves and along which the mesovarium is attached.

Each ovary consists of the following parts (Figure 26.13 on page 789).

1. **Germinal epithelium.** A layer of simple epithelium (low cuboidal or squamous) that covers the surface of the ovary and is continuous with the mesothelium that covers the mesovarium. The term *germinal epithelium* is a misnomer because it does not give rise to ova, although at one time it was believed that it did. It is now known that

the cells that give rise to ova arise from the endoderm of the yolk sac and migrate to the ovaries during embryonic development.

2. **Tunica albuginea.** A whitish capsule of dense, irregular connective tissue immediately deep to the germinal epithelium.

3. **Ovarian cortex.** A region just deep to the tunica albuginea that consists of dense connective tissue and contains ovarian follicles (described shortly).

4. **Ovarian medulla.** A region deep to the ovarian cortex that consists of loose connective tissue and contains blood vessels, lymphatics, and nerves.

5. **Ovarian follicles** (*folliculus* = little bag) lie in the cortex and consist of **oocytes** in various stages of development and their surrounding cells. When the surrounding cells form a single layer, they are called **follicular cells.** Later in development, when they form several layers, they are referred to as **granulosa cells.** The surrounding cells nourish the developing oocyte and begin to secrete estrogens as the follicle grows larger.

6. A **mature (Graafian) follicle** is a large, fluid-filled follicle that soon will rupture and expel a secondary oocyte, a process called **ovulation.**

7. A **corpus luteum** (= yellow body) contains the remnants of an ovulated mature follicle. The corpus luteum produces hormones such as progesterone, estrogens, relaxin, and inhibin until it degenerates and turns into fibrous tissue called a **corpus albicans** (= white body). See Table 22.7.

The ovarian blood supply is furnished by the ovarian arteries, which anastomose with branches of the uterine arteries. The ovaries are drained by the ovarian veins. On the right side they drain into the inferior vena cava, and on the left side they drain into the renal vein.

Sympathetic and parasympathetic nerve fibers to the ovaries terminate on the blood vessels and enter the substance of the ovaries.

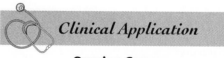

Clinical Application

Ovarian Cancer

Ovarian cancer is the sixth most common form of cancer in females and is the leading cause of death from all gynecological malignancies (excluding breast cancer). This is because it is difficult to detect before it metastasizes (spreads) beyond the ovaries. Risk factors associated with ovarian cancer include age (usually over age 50); race (Caucasians are at highest risk); family history of ovarian cancer; more than 40 years of active ovulation; nulliparity (never having had children) or first pregnancy after age 30; high-fat, low-fiber, vitamin A-deficient diet; and prolonged exposure to asbestos and talc. Early ovarian cancer has no

Figure 26.11 Female organs of reproduction and surrounding structures.

The female organs of reproduction include the ovaries, uterine (Fallopian) tubes, uterus, vagina, vulva, and mammary glands.

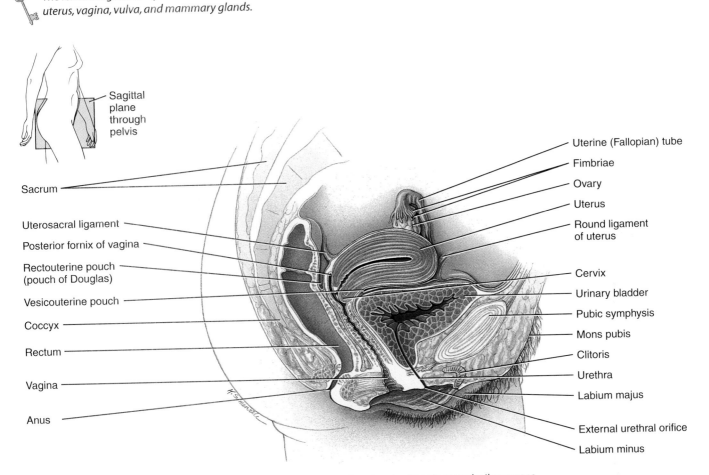

(a) Diagram of sagittal section showing female reproductive organs

OVERVIEW OF FUNCTIONS OF THE FEMALE REPRODUCTIVE SYSTEM
1. The ovaries produce secondary oocytes and hormones, including progesterone and estrogens (female sex hormones), inhibin, and relaxin.
2. The uterine tubes transport a secondary oocyte to the uterus and normally are the sites where fertilization occurs.
3. The uterus is the site of implantation of a fertilized ovum, development of the fetus during pregnancy, and labor.
4. The vagina receives the penis during sexual intercourse and is a passageway for childbirth.
5. The mammary glands synthesize, secrete, and eject milk for nourishment of the newborn.

Figure continues

symptoms or only mild ones associated with other common problems, such as abdominal discomfort, heartburn, nausea, loss of appetite, bloating, and flatulence. Later-stage signs and symptoms include an enlarged abdomen, abdominal and/or pelvic pain, persistent gastrointestinal disturbances, urinary complications, menstrual irregularities, and heavy menstrual bleeding. ▪

Oogenesis

The stages involved in the development of haploid (*n*) secondary oocytes in the ovaries constitute the **ovarian cycle.** It involves **oogenesis** (ō'-ō-JEN-e-sis; *oon* = egg; *genesis* = production), and, like spermatogenesis, oogenesis involves meiosis.

Figure 26.11 (continued)

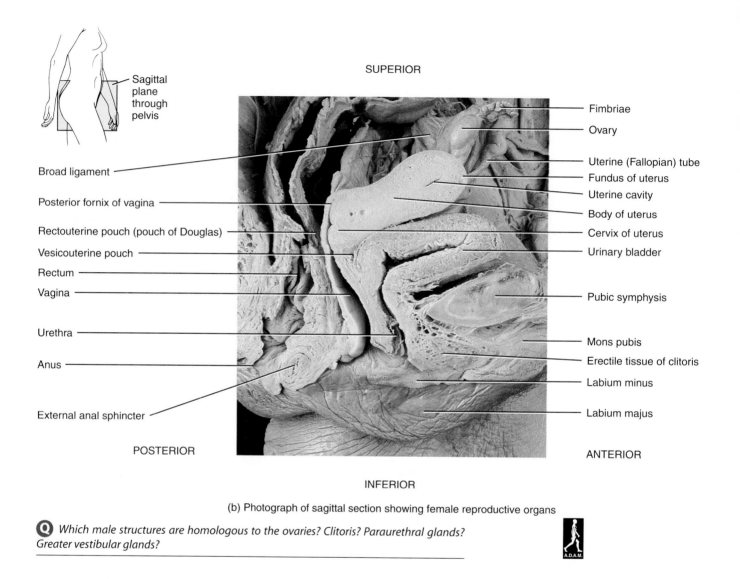

Sagittal plane through pelvis

SUPERIOR

Broad ligament

Posterior fornix of vagina

Rectouterine pouch (pouch of Douglas)

Vesicouterine pouch

Rectum

Vagina

Urethra

Anus

External anal sphincter

POSTERIOR

Fimbriae

Ovary

Uterine (Fallopian) tube

Fundus of uterus

Uterine cavity

Body of uterus

Cervix of uterus

Urinary bladder

Pubic symphysis

Mons pubis

Erectile tissue of clitoris

Labium minus

Labium majus

ANTERIOR

INFERIOR

(b) Photograph of sagittal section showing female reproductive organs

Q *Which male structures are homologous to the ovaries? Clitoris? Paraurethral glands? Greater vestibular glands?*

Reduction Division (Meiosis I) During early fetal development, primordial (primitive) germ cells migrate from the endoderm of the yolk sac to the ovaries. There, germ cells differentiate within the ovaries into **oogonia** (ō′-ō-GŌ-nē-a; singular is **oogonium;** ō′-ō-GŌ-nē-um). Oogonia are diploid (2*n*) cells that divide mitotically to produce millions of germ cells. Even before birth, many of these germ cells degenerate, a process known as **atresia.** A few develop into larger cells called **primary** (*primus* = first) **oocytes** (Ō-ō-sītz) that enter prophase of reduction division (meiosis I)

during fetal development but do not complete it until after puberty. At birth 200,000–2,000,000 oogonia and primary oocytes remain in each ovary. Of these, about 400 will mature and ovulate during a woman's reproductive lifetime; the remainder undergo atresia.

Each primary oocyte is surrounded by a single layer of follicular cells, and the entire structure is called a **primordial follicle** (Figure 26.13b). A few primordial follicles periodically start to grow (even during childhood) and become **primary (preantral) follicles,** which are surrounded first by one layer of cuboidal-shaped follicular cells and then by six

Text continues on page 790

Figure 26.12 Ovaries and uterus with associated structures. In (a), the left side of the uterine (Fallopian) tube and uterus has been sectioned to show internal structures. In (b), part of the posterior wall of the uterus has been removed. The arrow in the inset in (c) indicates the viewing position (superior).

Ligaments holding the ovaries in position are the mesovarium, the ovarian ligament, and the suspensory ligament.

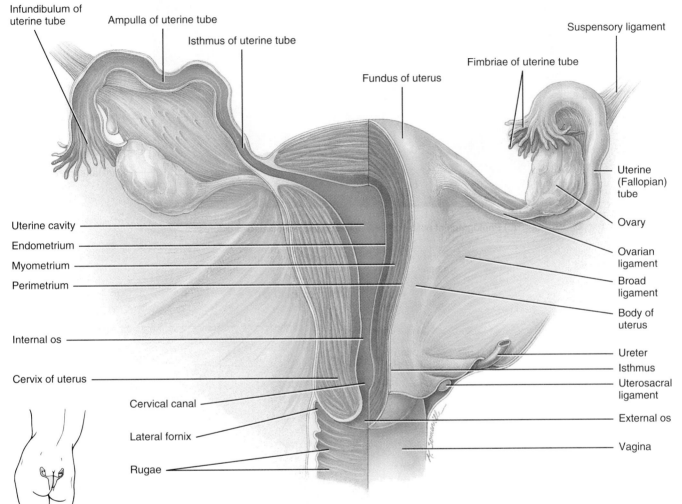

(a) Diagram of posterior view of uterus and associated structures

Figure continues

to seven layers of cuboidal and low-columnar cells called **granulosa cells.** As a follicle grows, it forms a clear glycoprotein layer, called the **zona pellucida** (pe-LOO-si-da) between the primary oocyte and the granulosa cells. The innermost layer of granulosa cells becomes firmly attached to the zona pellucida and is called the **corona radiata** (*corona* = crown; *radiata* = radiation). The outermost granulosa cells rest on a basement membrane that separates them from the surrounding structures. This outer region is called the **theca folliculi.** As the primary follicle continues to grow, the theca differentiates into two layers: (1) the **theca interna,** a vascularized internal layer of secretory cells, and (2) the **theca externa,** an outer layer of connective tissue cells. The granulosa cells begin to secrete follicular fluid, which builds up in a cavity called the **antrum** in the center of the follicle. The follicle is now termed a **secondary (antral) follicle** (Figure 26.13c). During childhood, primordial and developing follicles continue to undergo atresia.

After puberty, under the influence of the gonadotropin hormones secreted by the anterior pituitary gland, each month meiosis resumes in one secondary follicle (Figure 26.14). The diploid primary oocyte completes reduction division (meiosis I), and two haploid cells of unequal size, both with 23 chromosomes (n) of two chromatids each, are produced. The smaller cell produced by meiosis I, called the **first polar body,** is essentially a packet of discarded nuclear material. The larger cell, known as the **secondary oocyte,** receives most of the cytoplasm. Once a secondary oocyte is formed, it proceeds to the metaphase of equatorial division (meiosis II) and then stops at this stage. The follicle in which these events are taking place, termed the **mature (Graafian) follicle** (also called a **vesicular ovarian follicle**) will soon rupture and release its secondary oocyte.

Equatorial Division (Meiosis II) At ovulation, usually one secondary oocyte (with the first polar body and corona radiata) is expelled into the pelvic cavity. Normally, the cells are swept into the uterine (Fallopian) tube. If fertilization does not occur, the secondary oocyte degenerates. If sperm are present in the uterine tube and one penetrates the secondary oocyte (fertilization), however, equatorial division (meiosis II) resumes. The secondary oocyte splits into two haploid (n) cells of unequal size. The larger cell is the **ovum,** or mature egg; the smaller one is the **second polar body.** The ovum is homologous to a sperm cell. The nuclei of the sperm cell and the ovum then unite, forming a diploid ($2n$) **zygote.** The first polar body may also undergo another division to produce two polar bodies. If it does, the primary oocyte ultimately gives rise to a single haploid (n) ovum and three haploid (n) polar bodies, which all degenerate. Thus one oogonium gives rise to a single gamete (ovum), whereas one spermatogonium produces four gametes (sperm).

Figure 26.14 Oogenesis. $2n$ indicates diploid (46 chromosomes); n indicates haploid (23 chromosomes).

🗝 *In an oocyte, equatorial division is completed only if fertilization occurs.*

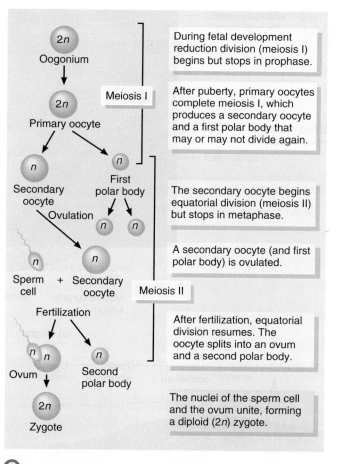

Oogonium	During fetal development reduction division (meiosis I) begins but stops in prophase.
Primary oocyte — Meiosis I	After puberty, primary oocytes complete meiosis I, which produces a secondary oocyte and a first polar body that may or may not divide again.
Secondary oocyte / First polar body	The secondary oocyte begins equatorial division (meiosis II) but stops in metaphase.
Ovulation	A secondary oocyte (and first polar body) is ovulated.
Sperm cell + Secondary oocyte — Meiosis II / Fertilization	After fertilization, equatorial division resumes. The oocyte splits into an ovum and a second polar body.
Ovum / Second polar body / Zygote	The nuclei of the sperm cell and the ovum unite, forming a diploid ($2n$) zygote.

Q *How does the age of a primary oocyte in a female compare with the age of a primary spermatocyte in a male?*

Uterine (Fallopian) Tubes

Females have two **uterine (Fallopian) tubes,** also called **oviducts,** that extend laterally from the uterus. They transport secondary oocytes and fertilized ova from the ovaries to the uterus (see Figure 26.12). Measuring about 10 cm (4 in.) long, the tubes lie between the folds of the broad ligaments of the uterus. The open, funnel-shaped portion of each tube, called the **infundibulum,** is close to the ovary. It ends in a fringe of fingerlike projections called **fimbriae** (FIM-brē-ē; *fimbrae* = fringe, singular is **fimbria**), one of which is attached to the lateral end of the ovary. From the infundibulum, the uterine tube extends medially and inferiorly and attaches to the superior lateral angle of the uterus. The **ampulla** of the uterine tube is the widest, longest portion,

Figure 26.15 Histology of the uterine (Fallopian) tube. Note the ciliated epithelium.

Peristaltic contractions of the muscularis and cilliary action of the mucosa help move the oocyte or fertilized ovum toward the uterus.

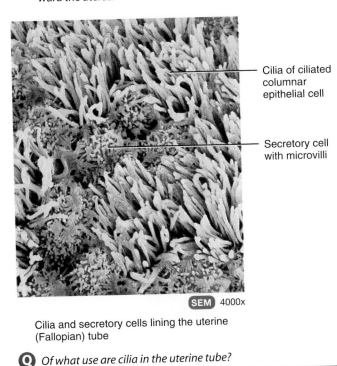

— Cilia of ciliated columnar epithelial cell

— Secretory cell with microvilli

SEM 4000x

Cilia and secretory cells lining the uterine (Fallopian) tube

Q *Of what use are cilia in the uterine tube?*

making up about the lateral two-thirds of its length. The **isthmus** of the uterine tube is the more medial, short, narrow, thick-walled portion that joins the uterus.

Histologically, the uterine tubes are composed of three layers. The internal **mucosa** contains ciliated columnar epithelial cells, which help move the fertilized ovum (or secondary oocyte) along the tube, and secretory cells, which have microvilli and may provide nutrition for the ovum (Figure 26.15). The middle layer, the **muscularis,** is composed of an inner, thick, circular region of smooth muscle and an outer, thin, longitudinal region of smooth muscle, Peristaltic contractions of the muscularis and the ciliary action of the mucosa help move the oocyte or fertilized ovum toward the uterus. The outer layer of the uterine tubes is a serous membrane, the **serosa.**

After ovulation, local currents produced by movements of the fimbriae, which surround the surface of the mature follicle just before ovulation occurs, sweep the secondary oocyte into the uterine tube. A sperm cell usually encounters and fertilizes a secondary oocyte in the ampulla of the uterine tube, any time up to about 24 hours after ovulation. The zygote arrives in the uterus about 7 days after ovulation.

The uterine tubes are supplied by branches of the uterine and ovarian arteries (see Figure 26.17). Venous return is via the uterine veins.

The uterine tubes are supplied with sympathetic and parasympathetic nerve fibers from the hypogastric plexus and the pelvic splanchnic nerves. The fibers are distributed to the muscular coat of the tubes and their blood vessels.

Uterus

The **uterus** (womb) serves as part of the pathway for sperm to reach the uterine tubes. It is also the site of menstruation, implantation of a zygote, development of the fetus during pregnancy, and labor. Situated between the urinary bladder and the rectum, the uterus is the size and shape of an inverted pear (see Figures 26.11 and 26.12). In a female who has never been pregnant, it measures approximately 7.5 cm (3 in.) long, 5 cm (2 in.) wide, and 2.5 cm (1 in.) thick. It is larger in females who have recently been pregnant and smaller (atrophied) when female sex hormones are low, as occurs when women are taking birth control pills or after menopause.

Anatomy

Anatomical subdivisions of the uterus include: (1) the dome-shaped portion superior to the uterine tubes called the **fundus,** (2) the major tapering central portion called the **body,** and (3) the inferior narrow portion opening into the vagina called the **cervix.** Between the body of the uterus and the cervix is the **isthmus** (IS-mus), a constricted region about 1 cm (½ in.) long. The interior of the body of the uterus is called the **uterine cavity,** and the interior of the narrow cervix is called the **cervical canal.** The cervical canal opens into the uterine cavity at the **internal os** (*os* = mouth) and into the vagina at the **external os.**

Normally, the body of the uterus projects anteriorly and superiorly over the urinary bladder in a position called **anteflexion.** The cervix projects inferiorly and posteriorly and enters the anterior wall of the vagina at nearly a right angle (see Figure 26.11). Several ligaments that are either extensions of the parietal peritoneum or fibromuscular cords maintain the position of the uterus (see Figure 26.12b). The paired **broad ligaments** are double folds of peritoneum attaching the uterus to either side of the pelvic cavity. The paired **uterosacral ligaments,** also peritoneal extensions, lie on either side of the rectum and connect the uterus to the sacrum. The **cardinal (lateral cervical) ligaments** extend inferior to the bases of the broad ligaments between the pelvic wall and the cervix and vagina. The **round ligaments** are bands of fibrous connective tissue between the layers of the broad ligament. They extend from a point on the uterus just inferior to the uterine tubes to a portion of the labia majora of the external genitalia. Although the ligaments normally maintain the anteflexed position of the uterus, they also afford the uterine body some movement. As a result, the uterus may become malpositioned. A posterior tilting of the uterus is called **retroflexion** (*retro* = backward).

Clinical Application

Uterine Prolapse

A condition called **uterine prolapse** (*prolapses* = falling down or downward displacement) may result from weakening of supporting ligaments and pelvic musculature associated with age or disease, traumatic vaginal delivery, chronic straining from coughing or difficult bowel movements, or pelvic tumors. The prolapse may be characterized as *first degree (mild)*, in which the cervix remains within the vagina; *second degree (marked)*, in which the cervix protrudes to the exterior through the vagina; and *third degree (complete)*, in which the entire uterus is outside the vagina. Depending on the degree of prolapse, treatment may involve pelvic exercises, dieting if a patient is overweight, stool softeners to minimize straining during defecation, pessary therapy (placement of a rubber device around the uterine cervix that helps prop up the uterus), and surgery. ▩

Histology

Histologically, the uterus consists of three layers of tissue: the perimetrium, myometrium, and endometrium (Figure 26.16). The outer layer, the **perimetrium** (*peri* = around; *metron* = uterus) or **serosa,** is part of the visceral peritoneum. It is composed of simple squamous epithelium and areolar connective tissue. Laterally, it becomes the broad ligament. Anteriorly, it covers the urinary bladder and forms a shallow pouch, the **vesicouterine** (ves′-i-kō-YOO-ter-in; *vesico* = bladder) **pouch** (see Figure 26.11b). Posteriorly, it covers the rectum and forms a deep pouch, the **rectouterine** (rek-tō-YOO-ter-in; *recto* = rectum) **pouch (pouch of Douglas)**—the most inferior point in the pelvic cavity.

The middle layer of the uterus, the **myometrium** (*myo* = muscle), forms the bulk of the uterine wall. This layer consists of three layers of smooth muscle fibers and is thickest in the fundus and thinnest in the cervix. The thicker middle layer is circular, whereas the inner and outer layers are longitudinal or oblique. During childbirth, coordinated contractions of the muscles (labor) in response to oxytocin from the posterior pituitary gland help expel the fetus from the body of the uterus.

The inner layer of the uterus, the **endometrium** (*endo* = within), is highly vascular and is composed of an innermost layer of simple columnar epithelium (ciliated and secretory cells) that lines the lumen; an underlying endometrial stroma that is a very thick region of lamina propria (areolar connective tissue); and endometrial (uterine) glands that develop as invaginations of the luminal epithelium and extend almost to the myometrium. The endometrium is divided into two layers. (1) The **stratum functionalis (functional layer),** the layer closer to the uterine cavity, is shed during menstruation. (2) The deeper layer, the **stratum basalis (basal layer),** is permanent. It gives rise to a new stratum functionalis after each menstruation.

Figure 26.16 Histology of the uterus.

🔑 *The three layers of the uterus, from superficial to deep, are the perimetrium (serosa), myometrium, and endometrium.*

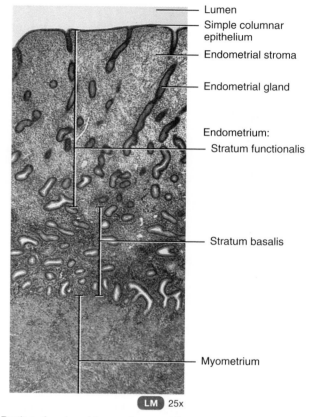

Lumen
Simple columnar epithelium
Endometrial stroma
Endometrial gland
Endometrium:
Stratum functionalis
Stratum basalis
Myometrium

LM 25x

Portion of endometrium and myometrium

Q *What structural features of the endometrium and myometrium contribute to their functions?*

The secretory cells of the mucosa of the cervix produce a secretion called **cervical mucus,** a mixture of water, glycoprotein, serum-type proteins, lipids, enzymes, and inorganic salts. Females of reproductive age secrete 20–60 ml of cervical mucus per day. Cervical mucus is more receptive to sperm at or near the time of ovulation because it is then less viscous and more alkaline (pH 8.5). At other times, viscous mucus forms a plug (cervical plug) that physically impedes sperm penetration. The mucus also supplements the energy needs of sperm. Both the cervix and mucus serve as a sperm reservoir, protect sperm from the hostile environment of the vagina, and protect sperm from phagocytes. They may also play a role in *capacitation*—a functional change that sperm undergo in the female reproductive tract before they can fertilize a secondary oocyte.

Figure 26.17 Blood supply of the uterus. The inset to the right shows the histological details of the blood vessels of the endometrium.

Straight arterioles supply the necessary materials for regeneration of the stratum functionalis.

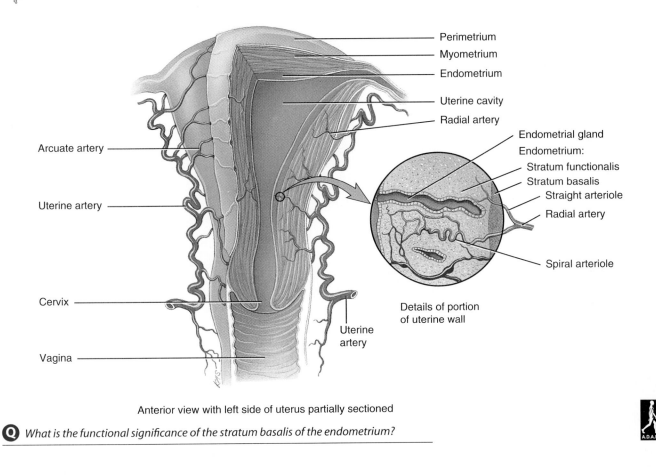

Perimetrium
Myometrium
Endometrium
Uterine cavity
Radial artery

Endometrial gland
Endometrium:
Stratum functionalis
Stratum basalis
Straight arteriole
Radial artery

Spiral arteriole

Details of portion
of uterine wall

Arcuate artery

Uterine artery

Cervix

Vagina

Uterine
artery

Anterior view with left side of uterus partially sectioned

Q *What is the functional significance of the stratum basalis of the endometrium?*

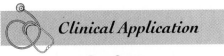

Clinical Application

Pap Smear

Early diagnosis of cancer of the cervix of the uterus is accomplished by a **Pap smear** (Papanicolaou test). A few cells are removed from the part of the vagina surrounding the cervix, and the cervix itself and examined microscopically. Malignant cells have a characteristic appearance and indicate an early stage of cancer, even before symptoms occur. ■

Blood Supply

Blood is supplied to the uterus by branches of the internal iliac artery called **uterine arteries** (Figure 26.17). Branches called **arcuate** (*arcuatus* = shaped like a bow) **arteries** are arranged in a circular fashion in the myometrium and give off **radial arteries** that penetrate deeply into the myometrium. Just before the branches enter the endometrium, they divide into two kinds of arterioles. The **straight arterioles**

supply the basalis with the materials needed to regenerate the functionalis. The **spiral arterioles** supply the functionalis and change markedly during the menstrual cycle. The uterus is drained by the **uterine veins** into the internal iliac veins. The extensive blood supply of the uterus is essential to support regrowth of a new stratum functionalis after menstruation, implantation of a fertilized ovum, and development of the placenta.

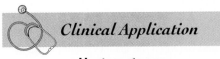

Clinical Application

Hysterectomy

Hysterectomy (hiss-te-RECT-tō-mē; *hyster* = uterus; *ectomy* = surgical removal) refers to surgical removal of the uterus and is the most common gynecological operation. It may be indicated in conditions such as endometriosis, pelvic inflammatory disease, recurrent ovarian cysts, excessive uterine bleeding, and

Figure 26.18 Components of the vulva (pudendum).

The vulva refers to the external genitals of the female.

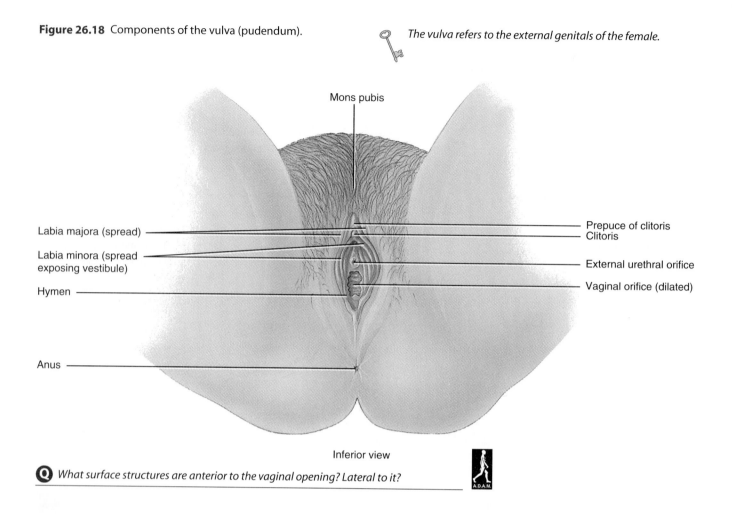

Inferior view

Q *What surface structures are anterior to the vaginal opening? Lateral to it?*

cancer of the cervix, uterus, or ovaries. In a *partial* or *subtotal hysterectomy,* the body of the uterus is removed but the cervix is left in place. A *complete hysterectomy* is the removal of the body and cervix of the uterus. A *radical hysterectomy* includes removal of the body and cervix of the uterus, uterine tubes, possibly the ovaries, superior portion of the vagina, pelvic lymph nodes, and supporting structures, such as ligaments. ■

Vagina

The **vagina** (*vagina* = sheath) serves as a passageway for the menstrual flow and childbirth. It also receives semen from the penis during sexual intercourse. It is a tubular, fibromuscular organ lined with mucous membrane and measures about 10 cm (4 in.) in length (see Figures 26.11 and 26.12). Situated between the urinary bladder and the rectum, the vagina is directed superiorly and posteriorly, where it attaches to the uterus. A recess, called the **fornix** (*fornix* = arch or vault), surrounds the vaginal attachment to the cervix.

The **mucosa** of the vagina is continuous with that of the uterus. Histologically, it consists of nonkeratinized stratified

squamous epithelium and areolar connective tissue (see Table 3.1, Stratified Squamous Epithelium, on page 65). The mucosa is arranged in a series of transverse folds called **rugae** (see also Figure 26.12). The mucosa of the vagina contains large stores of glycogen, which upon decomposition produce organic acids. These acids create a low pH environment that retards microbial growth. However, the acidity is also harmful to sperm. Alkaline components of semen, mainly from the seminal vesicles, neutralize the acidity of the vagina and increase viability of the sperm.

The **muscularis** is composed of an outer circular layer and an inner longitudinal layer of smooth muscle that can stretch considerably to receive the penis during sexual intercourse and allow for birth of a fetus.

The **adventitia** is the superficial layer of the vagina. It consists of areolar connective tissue and anchors the vagina to adjacent organs such as the urethra and urinary bladder anteriorly and the rectum and anal canal posteriorly.

At the inferior end of the vaginal opening to the exterior, the **vaginal orifice,** there may be a thin fold of vascularized mucous membrane called the **hymen** (*hymen* = membrane). It forms a border around the orifice, partially closing it (see Figure 26.18). Sometimes the hymen completely covers the

Figure 26.19 Female perineum. (Figure 10.13 shows the male perineum.)

The perineum is a diamond-shaped area medial to the thighs and buttocks that contains the external genitals and anus.

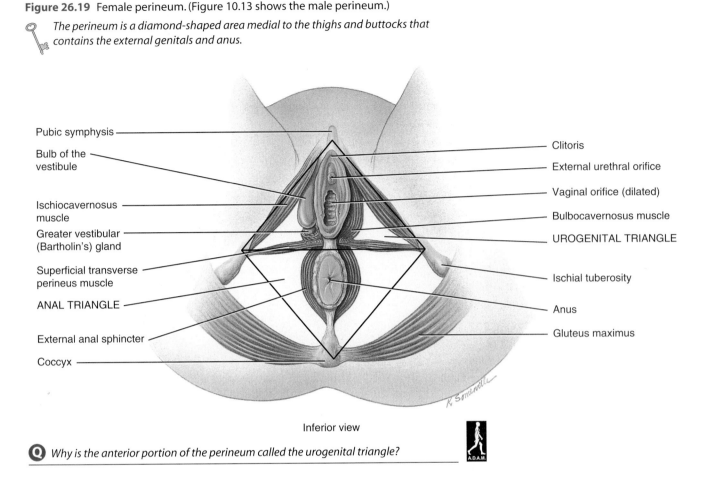

Pubic symphysis

Bulb of the vestibule

Ischiocavernosus muscle

Greater vestibular (Bartholin's) gland

Superficial transverse perineus muscle

ANAL TRIANGLE

External anal sphincter

Coccyx

Clitoris

External urethral orifice

Vaginal orifice (dilated)

Bulbocavernosus muscle

UROGENITAL TRIANGLE

Ischial tuberosity

Anus

Gluteus maximus

Inferior view

Q *Why is the anterior portion of the perineum called the urogenital triangle?*

orifice, a condition called **imperforate** (im-PER-fō-rāt) **hymen,** which may require surgery to open the orifice and permit the discharge of the menstrual flow.

Clinical Application

Vulvovaginal Candidiasis

Candida albicans is a yeastlike fungus that commonly grows on mucous membranes of the gastrointestinal and genitourinary tracts. The organism is responsible for **vulvovaginal candidiasis** (vul'-vō-VAJ-i-nal can'-di-DĪ-a-sis), the most common form of vaginitis. It is characterized by severe itching; a thick, yellow, cheesy discharge; a yeasty odor; and pain. The disorder, experienced at least once by about 75% of females, is usually a result of proliferation of the fungus following antibiotic therapy for another condition. Predisposing conditions include use of oral contraceptives, cortisone-like medications, pregnancy, and diabetes. ▪

Vulva

The term **vulva** (VUL-va; *volvere* = to wrap around), or **pudendum** (pyoo-DEN-dum), refers to the external genitalia of the female (Figure 26.18). Its components are as follows:

1. **Mons pubis.** Anterior to the vaginal and urethral openings is the **mons pubis** (MONZ PŪ-bis; *mons* = mountain). It is an elevation of adipose tissue covered by skin and coarse pubic hair that cushions the pubic symphysis.

2. **Labia majora.** From the mons pubis, two longitudinal folds of skin, the **labia majora** (LĀ-bē-a ma-JŌ-ra; *labium* = lip; singular is **labium majus**), extend inferiorly and posteriorly. The labia majora are homologous to the scrotum and are covered by pubic hair. They contain an abundance of adipose tissue and sebaceous (oil) and sudoriferous (sweat) glands.

3. **Labia minora.** Medial to the labia majora are two smaller folds of skin called the **labia minora** (mī-NŌ-ra; singular is **labium minus**). Unlike the labia majora, the labia minora are devoid of pubic hair and fat and have few sudoriferous (sweat) glands. They do, however, contain many sebaceous (oil) glands. The labia minora are homologous to the spongy (penile) urethra.

4. **Clitoris.** The **clitoris** (KLI-to-ris) is a small, cylindrical mass of erectile tissue and nerves. It is located at the anterior junction of the labia minora. A layer of skin called the **prepuce** (foreskin) is formed at the point where the labia minora unite and covers the body of the clitoris. The exposed portion of the clitoris is the **glans.** The clitoris is homologous to the glans penis of the male. Like the penis, the clitoris is capable of enlargement upon tactile stimulation and has a role in sexual excitement of the female.

5. **Vestibule.** The region between the labia minora is called the **vestibule.** It is homologous to the membranous urethra of the male. Within the vestibule are the hymen (if still present), vaginal orifice, external urethral orifice, and the openings of the ducts of several glands (described shortly). The **vaginal orifice,** the opening of the vagina to the exterior, occupies the greater portion of the vestibule and is bordered by the hymen. Anterior to the vaginal orifice and posterior to the clitoris is the **external urethral orifice,** the opening of the urethra to the exterior. On either side of the external urethral orifice are the openings of the ducts of the **paraurethral (Skene's) glands.** They are embedded in the wall of the urethra and secrete mucus. The paraurethral glands are homologous to the male prostate gland. On either side of the vaginal orifice itself are the **greater vestibular (Bartholin's) glands** (see Figure 26.19). These glands open by ducts into a groove between the hymen and labia minora and produce a small quantity of mucus during sexual arousal and intercourse that adds to cervical mucus and provides lubrication. The greater vestibular glands are homologous to the male bulbourethral (Cowper's) glands. Several **lesser vestibular glands** also open into the vestibule.

6. **Bulb of the vestibule.** The **bulb of the vestibule** (see Figure 26.19) consists of two elongated masses of erectile tissue just deep to the labia on either side of the vaginal orifice. The bulb of the vestibule becomes engorged with blood during sexual arousal, narrowing the vaginal orifice and placing pressure on the penis during intercourse. The bulb of the vestibule is homologous to the corpus spongiosum penis and bulb of the penis in the male.

A summary of homologous structures of the male and female reproductive systems is presented in Table 26.1.

Perineum

The **perineum** (per'-i-NĒ-um) is the diamond-shaped area medial to the thighs and buttocks of both males and females that contains the external genitals and anus (Figure 26.19). It is bounded anteriorly by the pubic symphysis, laterally by the ischial tuberosities, and posteriorly by the coccyx. A transverse line drawn between the ischial tuberosities

Table 26.1 Summary of Homologous Structures of the Male and Female Reproductive Systems

MALE STRUCTURES	FEMALE HOMOLOGUES
Testes	Ovaries
Sperm cell	Ovum
Scrotum	Labia majora
Spongy (penile) urethra	Labia minora
Membranous urethra	Vestibule
Corpus spongiosum penis and bulb of penis	Bulb of vestibule
Prostate gland	Paraurethral glands
Bulbourethral (Cowper's) glands	Greater vestibular glands

divides the perineum into an anterior **urogenital** (yoo'-rō-JEN-i-tal) **triangle** that contains the external genitalia and a posterior **anal triangle** that contains the anus.

Clinical Application

Episiotomy

During childbirth, the emerging fetus stretches the perineal region. To prevent undue stretching and even tearing of this region, a physician sometimes performs an **episiotomy** (e-piz'-ē-OT-ō-mē; *epision* = pubic region; *tome* = incision), a perineal cut made with surgical scissors. This cut enlarges the vaginal opening to make room for the fetus to pass. In effect, a controlled cut is substituted for a jagged, uncontrolled tear. The incision is closed in layers with a continuous suture that is absorbed within a few weeks, so that stitches do not have to be removed. ■

Mammary Glands

The two **mammary** (*mamma* = breast) **glands** are modified sudoriferous (sweat) glands that produce milk. They lie over the pectoralis major and serratus anterior muscles and are attached to them by a layer of deep fascia (dense irregular connective tissue; Figure 26.20).

Anatomy and Histology

Each breast has one pigmented projection, the **nipple,** that has a series of closely spaced openings. These are the external openings of ducts called **lactiferous ducts** where milk emerges. The circular pigmented area of skin surrounding

Figure 26.20 Mammary glands.

The mammary glands function in the synthesis, secretion, and ejection of milk (lactation).

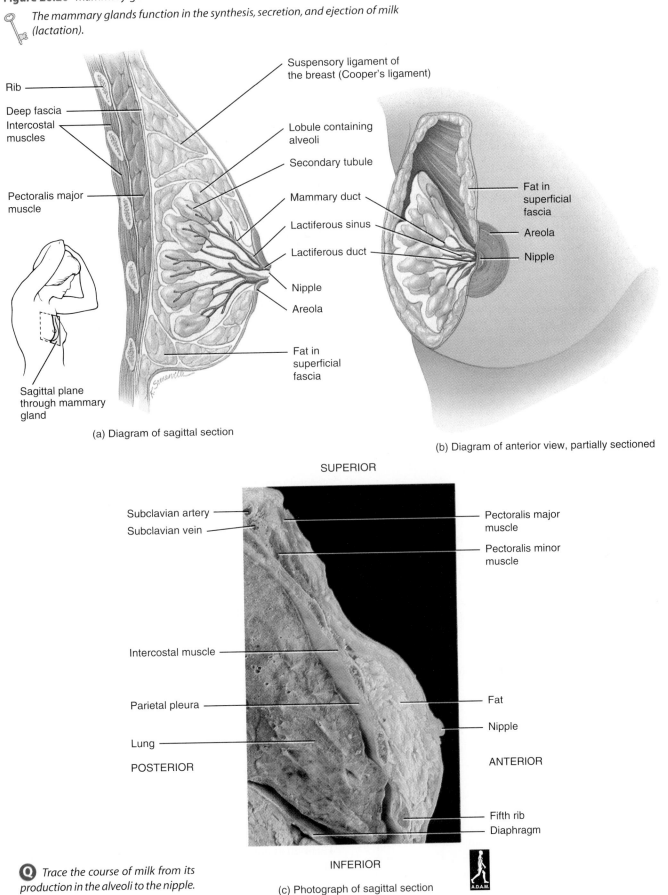

Rib

Deep fascia

Intercostal muscles

Pectoralis major muscle

Sagittal plane through mammary gland

Suspensory ligament of the breast (Cooper's ligament)

Lobule containing alveoli

Secondary tubule

Mammary duct

Lactiferous sinus

Lactiferous duct

Nipple

Areola

Fat in superficial fascia

(a) Diagram of sagittal section

Fat in superficial fascia

Areola

Nipple

(b) Diagram of anterior view, partially sectioned

SUPERIOR

Subclavian artery

Subclavian vein

Intercostal muscle

Parietal pleura

Lung

POSTERIOR

Pectoralis major muscle

Pectoralis minor muscle

Fat

Nipple

ANTERIOR

Fifth rib

Diaphragm

INFERIOR

(c) Photograph of sagittal section

Q *Trace the course of milk from its production in the alveoli to the nipple.*

the nipple is called the **areola** (a-RĒ-ō-la; *areola* = small space). It appears rough because it contains modified sebaceous (oil) glands. Strands of connective tissue called the **suspensory ligaments of the breast (Cooper's ligaments)** run between the skin and deep fascia and support the breast. These ligaments become looser with age or with undue stress, as occurs in long-term jogging or high-impact aerobics. Wearing a supportive bra slows the appearance of "Cooper's droop."

Within each breast, the mammary gland consists of 15–20 **lobes,** or compartments, separated by adipose tissue. The amount of adipose tissue, not the amount of milk produced, determines the size of the breasts. In each lobe are several smaller compartments called **lobules,** composed of grapelike clusters of milk-secreting glands termed **alveoli** (*alveolus* = small cavity) embedded in connective tissue (Figure 26.21). Surrounding the alveoli are spindle-shaped cells called **myoepithelial cells,** whose contraction helps propel milk toward the nipples. When milk is being produced, it passes from the alveoli into a series of **secondary tubules.** From here the milk enters the **mammary ducts.** Close to the nipple the mammary ducts expand to form sinuses called **lactiferous** (*lact* = milk; *ferre* = to carry) **sinuses,** where some milk may be stored before draining into a lactiferous duct. Each lactiferous duct carries milk from one of the lobes to the exterior, although some may join before reaching the surface.

Functions

The functions of the mammary glands are milk synthesis, secretion, and ejection, which are associated with pregnancy and childbirth and together are called **lactation.** Milk production is stimulated largely by the hormone prolactin, with contributions from progesterone and estrogens. The ejection of milk occurs in the presence of oxytocin, which is released from the posterior pituitary gland in response to the sucking action of an infant on the mother's nipple.

UTERINE AND OVARIAN CYCLES

During their reproductive years, nonpregnant females normally experience a cyclical sequence of changes in the ovaries and uterus. Each cycle takes about a month and involves both oogenesis and preparation of the uterus to receive a fertilized ovum. The principal events are all hormonally controlled. The **ovarian cycle** is a series of events associated with the maturation of a secondary oocyte. The **uterine (menstrual) cycle** is a series of changes in the endometrium of the uterus. Each month, the endometrium is prepared for arrival of a fertilized ovum that will develop in the uterus until birth. If fertilization does not occur, the stratum functionalis portion of the endometrium is shed. The

Figure 26.21 Histology of the mammary glands.

Contraction of myoepithelial cells helps propel milk toward the nipples.

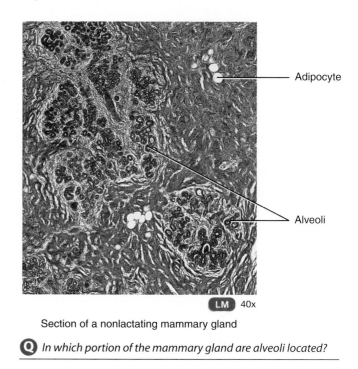

Adipocyte

Alveoli

LM 40x

Section of a nonlactating mammary gland

Q *In which portion of the mammary gland are alveoli located?*

general term **female reproductive cycle** refers to the ovarian and uterine cycles, the hormonal changes that regulate them, and cyclic changes in the breasts and cervix.

The uterine cycle and ovarian cycle are controlled by gonadotropin releasing hormone (GnRH) from the hypothalamus (see Figure 26.22). GnRH stimulates the release of follicle-stimulating hormone (FSH) and luteinizing hormone (LH) from the anterior pituitary gland. FSH stimulates the initial secretion of estrogens by growing follicles. LH stimulates the further development of ovarian follicles and their full secretion of estrogen, brings about ovulation, promotes formation of the corpus luteum, and stimulates the production of estrogens, progesterone, relaxin, and inhibin by the corpus luteum. (See Table 22.7.)

The duration of the uterine and ovarian cycles normally ranges from 24 to 35 days. For this discussion, we will assume a duration of 28 days. Events occurring during the cycles may be divided into four events: the menstrual phase, preovulatory phase, ovulation, and postovulatory phase (Figure 26.22).

Menstrual Phase (Menstruation)

The **menstrual** (MEN-stroo-al) **phase,** also called **menstruation** (men'-stroo-Ā-shun) or **menses** (*mens* = monthly), lasts for roughly the first five days of the cycle. (By convention, the first day of menstruation marks the first day of a new cycle.)

Figure 26.22 Correlation of ovarian and uterine cycles with hypothalamic and anterior pituitary gland hormones. In the cycle shown, fertilization and implantation have not occurred.

The length of the female reproductive cycle typically is 24–36 days; the preovulatory phase is more variable in length than the other phases.

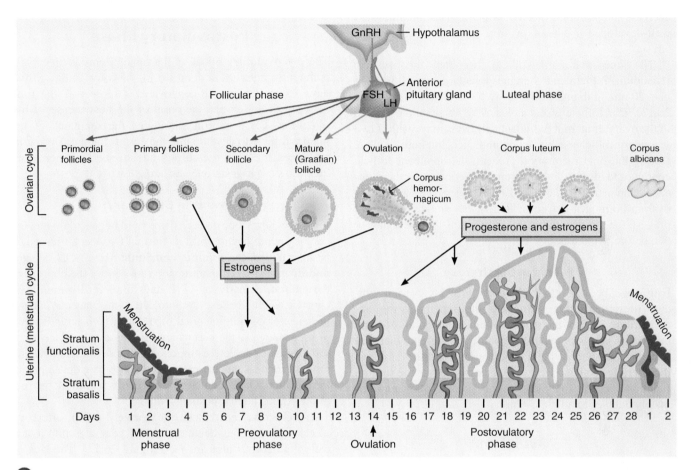

Q *Which hormones stimulate proliferation of the endometrium? Ovulation? Growth of the corpus luteum? The surge of LH at midcycle?*

Events in the Ovaries

During the menstrual phase, 20 or so small secondary (antral) follicles, some in each ovary, begin to enlarge. Follicular fluid, secreted by the granulosa cells and oozing from blood capillaries, accumulates in the enlarging antrum while the oocyte remains near the edge of the follicle (see Figure 26.13c).

Events in the Uterus

Menstrual flow from the uterus consists of 50–150 ml of blood, tissue fluid, mucus, and epithelial cells derived from the endometrium. This discharge occurs because the declining level of estrogens and progesterone causes the uterine spiral arteries to constrict. As a result the cells they supply become ischemic (deficient in blood) and start to die. Eventually, the entire stratum functionalis sloughs off. At this time the endometrium is very thin because only the stratum basalis remains. The menstrual flow passes from the uterine cavity to the cervix and through the vagina to the exterior.

Preovulatory Phase

The **preovulatory phase,** the second phase of the female reproductive cycle, is the time between menstruation and ovulation. The preovulatory phase of the cycle is more variable in length than the other phases and accounts for most of the difference when cycles are shorter or longer than 28 days. It lasts from days 6 to 13 in a 28-day cycle.

Events in the Ovaries

Under the influence of FSH, the group of about 20 secondary follicles continues to grow and begins to secrete estrogen and inhibin. By about day 6, one follicle in one ovary has outgrown all the others and is thus the dominant follicle. Estrogens and inhibin secreted by the dominant follicle decrease the secretion of FSH, which causes the other less well-developed follicles to stop growing and undergo atresia.

The one dominant follicle becomes the **mature (Graafian) follicle** that continues to enlarge until it is more than 20 mm in diameter and ready for ovulation (see Figure 26.13). This follicle forms a blisterlike bulge on the surface of the ovary. Fraternal (nonidentical) twins may result if two secondary follicles achieve codominance and both ovulate. During the final maturation process, the dominant follicle continues to increase its estrogen production under the influence of an increasing level of LH. Estrogens are the primary ovarian hormones before ovulation, but small amounts of progesterone are produced by the mature follicle a day or two before ovulation.

Events in the Uterus

Estrogens liberated into the blood by growing ovarian follicles (described next) stimulate the repair of the endometrium. Cells of the stratum basalis undergo mitosis and produce a new stratum functionalis. As the endometrium thickens, the short, straight endometrial glands develop and the arterioles coil and lengthen as they penetrate the stratum functionalis. The thickness of the endometrium approximately doubles to about 4–6 mm.

Ovulation

Ovulation, the rupture of the mature (Graafian) follicle with release of the secondary oocyte into the pelvic cavity, usually occurs on day 14 in a 28-day cycle. During ovulation, the secondary oocyte remains surrounded by its zona pellucida and corona radiata. It generally takes a total of about 20 days (spanning the last six days of the previous cycle and the first 14 days of the current cycle) for a secondary follicle to develop into a fully mature follicle. During this time the primary oocyte completes reduction division (meiosis I) to become a secondary oocyte, which begins equatorial division (meiosis II) but then halts in metaphase. The high levels of estrogens produced during the latter part of the preovulatory phase stimulate the release of LH by the anterior pituitary gland. This LH surge brings about ovulation. An over-the-counter home test may be used to detect the LH surge and can predict ovulation a day in advance.

After ovulation, the mature follicle collapses, and blood within it forms a clot due to minor bleeding during rupture of the follicle to become the **corpus hemorrhagicum** (*hemo =* blood; *rhegnynai =* to burst forth). See Figure 26.13. The clot is absorbed by the remaining follicular cells, which enlarge, change character, and form the corpus luteum under the influence of LH. Stimulated by LH, the corpus luteum secretes progesterone, estrogen, relaxin, and inhibin.

Postovulatory Phase

The **postovulatory phase** of the female reproductive cycle is the most constant in duration and lasts for 14 days, from days 15 to 28 in a 28-day cycle. It represents the time between ovulation and the onset of the next menses. After ovulation, LH secretion stimulates the remnants of the mature follicle to develop into the corpus luteum. During its two-week lifespan, the corpus luteum secretes increasing quantities of progesterone and some estrogens.

Events in One Ovary

If the secondary oocyte is fertilized and begins to divide, the corpus luteum persists past its normal two-week lifespan. It is maintained by **human chorionic** (kō-rē-ON-ik) **gonadotropin (hCG),** a hormone produced by the chorion of the embryo as early as 8–12 days after fertilization. The chorion eventually develops into part of the placenta, and the presence of hCG in maternal blood or urine is an indication of pregnancy. As the pregnancy progresses, the placenta itself begins to secrete estrogens to support pregnancy, and progesterone to support pregnancy and breast development for lactation. Once the placenta begins its secretion, the role of the corpus luteum becomes minor.

If hCG does not rescue the corpus luteum, after two weeks its secretions decline and it degenerates into a scar called the corpus albicans (see Figure 26.13). The lack of progesterone and estrogens due to degeneration of the corpus luteum then causes menstruation. In addition, the decreased levels of progesterone, estrogens, and inhibin promote the release of GnRH, FSH, and LH, which stimulate follicular growth, and a new ovarian cycle begins.

Events in the Uterus

Progesterone and estrogens produced by the corpus luteum promote growth and coiling of the endometrial glands, which begin to secrete glycogen; vascularization of the superficial endometrium; thickening of the endometrium; and an increase in the amount of tissue fluid. These preparatory changes are maximal about one week after ovulation, corresponding to the time of possible arrival of a fertilized ovum.

Amenorrhea (ā-men′-ō-RĒ-a; *a* = without; *men* = month; *rhein* = to flow) is the absence of menstruation. It may be caused by a hormone imbalance, obesity, extreme weight loss, or very low body fat as may occur during rigorous athletic training.

Dysmenorrhea (dis′-men-ō-RĒ-a; *dys* = difficult) refers to pain associated with menstruation and is usually reserved to describe an individual with menstrual symptoms that are severe enough to prevent her from functioning normally for one or more days each month. Some cases are caused by uterine tumors, ovarian cysts, pelvic inflammatory disease, or intrauterine devices.

Abnormal uterine bleeding includes menstruation of excessive duration or excessive amount, diminished menstrual flow, too frequent menstruation, intermenstrual bleeding, and post-menopausal bleeding. These abnormalities may be caused by disordered hormonal regulation, emotional factors, fibroid tumors of the uterus, and systemic diseases. ■

Developmental Anatomy of the Reproductive System

The *gonads* develop from the **intermediate mesoderm.** By the sixth week, they appear as bulges that protrude into the ventral body cavity (Figure 26.23). The gonads develop near the **mesonephric (Wolffian) ducts.** A second pair of ducts, the **paramesonephric (Müller's) ducts,** develop lateral to the mesonephric ducts. Both sets of ducts empty into the urogenital sinus. An early embryo contains primitive gonads that have the potential to differentiate into either testes or ovaries. The male pattern of differentiation depends on the presence of a master gene on the Y chromosome called *SRY,* which stands for *S*ex-determining *R*egion of the *Y* chromosome, and the release of testosterone. During the seventh week, primitive sustentacular (Sertoli) cells start to appear in the gonadal tissues of male embryos, which have one X and one Y chromosome. Stimulated by hCG, primitive interstitial endocrinocytes (Leydig cells) begin to secrete testosterone. Differentiation of gonadal tissue into ovaries in embryos with two X chromosomes depends on the absence of *SRY* and the absence of testosterone.

In the male embryo, the *testes* connect to the mesonephric duct through a series of tubules. These tubules become the *seminiferous tubules.* Stimulated by testosterone, the mesonephric duct on each side continues to develop and gives rise to the *epididymis, ductus (vas) deferens, ejaculatory duct,* and *seminal vesicle.* The developing sustentacular cells also secrete a second hormone called Müllerian-inhibiting substance (MIS), which causes the death of cells within the paramesonephric (Müllerian) ducts. As a result, those cells do not contribute any functional structures to the male reproductive system. The *prostate* and *bulbourethral (Cowper's) glands* are **endodermal** outgrowths of the urethra.

In the female embryo, the gonads develop into *ovaries.* In the absence of MIS, the paramesonephric ducts flourish. Their distal ends fuse to form the *uterus* and *vagina.* The unfused proximal portions become the *uterine (Fallopian) tubes.* The mesonephric ducts in the female degenerate without contributing any functional structures to the female reproductive system. The *greater (Bartholin's)* and *lesser vestibular glands* develop from **endodermal** outgrowths of the vestibule.

The *external genitals* of both male and female embryos also remain undifferentiated until about the eighth week. Before differentiation, all embryos have an elevated region, the **genital tubercle.** This is a point between the tail (future coccyx) and the umbilical cord where the mesonephric and paramesonephric ducts open to the exterior (Figure 26.24). The tubercle consists of the **urethral groove** (opening into the urogenital sinus), paired **urethral folds,** and paired **labioscrotal swellings.**

In the male embryo, a powerful androgen (hormone) called dihydrotestosterone (DHT) stimulates development of the urethra, prostate gland, and external genitals (scrotum and penis). A portion of the genital tubercle elongates and develops into a *penis.* Fusion of the urethral folds forms the *spongy (penile) urethra* and leaves an opening to the exterior only at the distal end of the penis, the *external urethral orifice.* The labioscrotal swellings develop into the *scrotum.* In the absence of DHT, the genital tubercle gives rise to the *clitoris* in a female embryo. The urethral folds remain open as the *labia minora,* and the labioscrotal swellings become the *labia majora.* The urethral groove becomes the *vestibule.* After birth, androgen levels decline because hCG is no longer present to stimulate secretion of testosterone.

Aging and the Reproductive System

During the first decade of life, the reproductive system is in a juvenile state. At about age 10, hormone-directed changes start to occur in both sexes. **Puberty** (PŪ-ber-tē; *puber* = marriageable age) refers to the period of time when secondary sexual characteristics begin to develop and the potential for sexual reproduction is reached.

In females, the reproductive cycle normally occurs once each month from **menarche** (me-NAR-kē), the first menses, to **menopause** (*mens* = monthly; *pausis* = to cease), the permanent cessation of menses. Between the ages of 40 and 50 the ovaries become less responsive to the stimulation of

Text continues on page 804

Figure 26.23 Development of the internal reproductive systems.

The gonads develop from intermediate mesoderm.

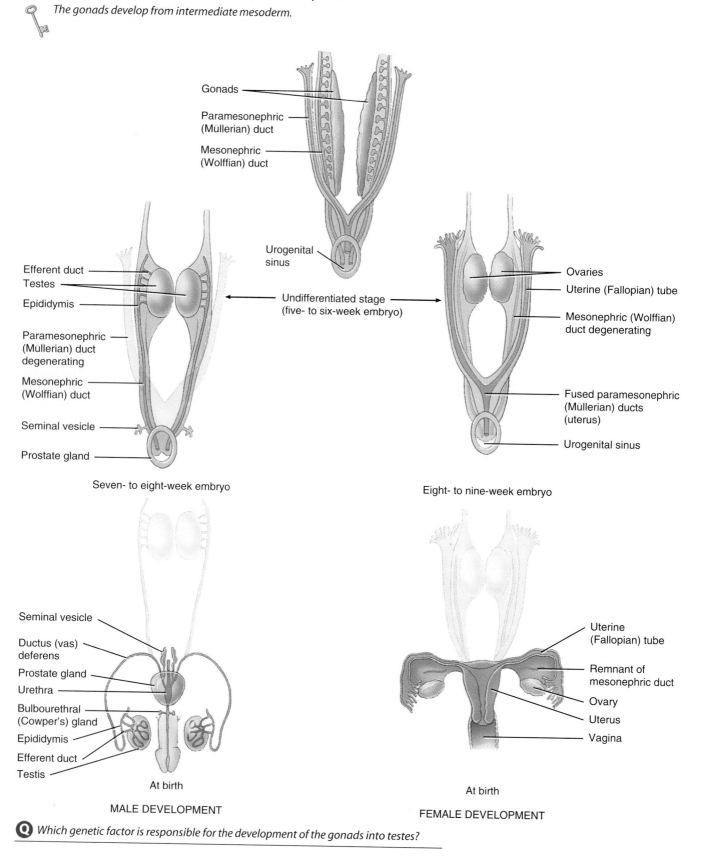

MALE DEVELOPMENT

FEMALE DEVELOPMENT

Q Which genetic factor is responsible for the development of the gonads into testes?

Figure 26.24 Development of the external genitals.

The external genitals of male and female embryos remain undifferentiated until about the eighth week.

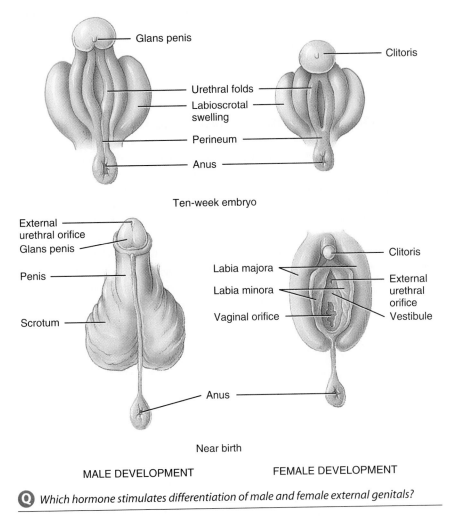

Undifferentiated stage (about five-week embryo)

Ten-week embryo

Near birth

MALE DEVELOPMENT   FEMALE DEVELOPMENT

Q *Which hormone stimulates differentiation of male and female external genitals?*

gonadotropic hormones from the anterior pituitary gland. As a result, the production of estrogens declines, and follicles do not undergo normal development. Changes in GnRH release patterns and decreased responsiveness to it by cells of the anterior pituitary gland that secrete LH also contribute to the onset of menopause. Some women experience hot flashes, copious sweating, headache, hair loss, muscular pains, vaginal dryness, insomnia, depression, weight gain, and mood swings. In the postmenopausal woman there will be some atrophy of the ovaries, uterine tubes, uterus, vagina, external genitalia, and breasts. Osteoporosis is also a possible occurrence due to the diminished level of estrogens (see page 4 in Applications to Health). Sexual desire (libido) does not show a parallel decline, perhaps supported by adrenal sex steroids.

The female reproductive system has a time-limited span of fertility between menarche and menopause. Fertility declines with age, possibly as a result of less frequent ovulation and the declining ability of the uterine (Fallopian) tubes and uterus to support the young embryo. Uterine cancer peaks at about 65 years of age, but cervical cancer is more common in younger women.

In males, declining reproductive function is much more subtle than in the female. Healthy men often retain reproductive capacity into their 80s or 90s. At about age 55 a decline in testosterone synthesis leads to less muscle strength, fewer viable sperm, and decreased sexual desire. However, abundant sperm may be found even in old age. A common problem of older men is enlargement (hypertrophy) of the prostate gland. As the prostate enlarges, it squeezes the urethra. Thus difficulty in urinating often is an early sign of prostatic enlargement. The enlargement may be benign, but cancer of the prostate is quite common.

Key Medical Terms
Associated with the Reproductive Systems

Castration (kas-TRĀ-shun; *castrare* = to prune) Removal, inactivation, or destruction of the gonads.

Culdoscopy (kul-DOS-kō-pē; *skopein* = to examine) A procedure in which a culdoscope (endoscope) is used to view the female pelvic cavity. The approach is through the vagina.

Endocervical curettage (ku-re-TAZH; *curette* = scraper) The cervix is dilated and the endometrium of the uterus is scraped with a spoon-shaped instrument called a curette. This procedure is commonly called a **D and C (dilation and curettage).**

Hermaphroditism (her-MAF-rō-dī-tizm′) Presence of both male and female sex organs in one individual.

Hypospadias (hī-pō-SPĀ-dē-as; *hypo* = below; *span* = to draw) A displaced urethral opening. In the male, the opening may be on the underside of the penis, at the penoscrotal junction, between the scrotal folds, or in the perineum. In the female, the urethra opens into the vagina.

Leukorrhea (loo′-kō-RĒ-a; *leuco* = white; *rrhea* = discharge) A nonbloody vaginal discharge that may occur at any age and affects most women at some time.

Oophorectomy (ō′-of-ō-REK-tō-mē; *oophoro* = bearing eggs; *ektome* = excision) Removal of the ovaries.

Ovarian cyst Fluid-containing sac in the ovary that may arise in follicles or the corpus luteum.

Salpingectomy (sal′-pin-JEK-tō-mē; *salpingo* = tube) Removal of a uterine (Fallopian) tube.

Smegma (SMEG-ma; *smegma* = soap) The secretion, consisting principally of desquamated epithelial cells, found chiefly about the external genitalia and especially under the foreskin of the male.

Vaginitis (vaj′-i′-NĪ-tis) Inflammation of the vagina.

Study Outline

Male Reproductive System (p. 771)

1. Reproduction is the process by which new individuals of a species are produced and the genetic material is passed from generation to generation.
2. The organs of reproduction are grouped as gonads (produce gametes), ducts (transport and store gametes), accessory sex glands (produce materials that support gametes), and supporting structures (have various roles in reproduction).
3. The male structures of reproduction include the testes, ductus epididymis, ductus (vas) deferens, ejaculatory duct, urethra, seminal vesicles, prostate gland, bulbourethral (Cowper's) glands, and penis.

Scrotum (p. 771)

1. The scrotum is a sac that hangs from the root of the penis and consists of loose skin and superficial fascia. It supports the testes.
2. It regulates the temperature of the testes by contraction of the cremaster muscle and dartos, which elevates them and brings them closer to the pelvic cavity or relaxes causing the testes to move farther from the pelvic cavity.

Testes (p. 771)

1. The testes are oval-shaped glands (gonads) in the scrotum containing seminiferous tubules, in which sperm cells are made; sustentacular (Sertoli) cells, which nourish sperm cells and

secrete inhibin; and interstitial endocrinocytes (interstitial cells of Leydig), which produce the male sex hormone testosterone.

2. The testes descend into the scrotum through the inguinal canals during the seventh month of fetal development. Failure of the testes to descend is called cryptorchidism.

3. Secondary oocytes and sperm are collectively called gametes, or sex cells, and are produced in gonads.

4. Somatic cells divide by mitosis, the process in which each daughter cell is identical and receives the full complement of 23 chromosome pairs (46 chromosomes). Somatic cells are said to be diploid (2n).

5. Germ cells divide by meiosis to produce nonidentical gametes, in which the pairs of chromosomes are split so that the gamete has only 23 chromosomes. It is said to be haploid (n).

6. Spermatogenesis occurs in the testes. It results in the formation of four haploid sperm from each primary spermatocyte.

7. Spermatogenesis is a process in which immature spermatogonia develop into mature sperm. The spermatogenesis sequence includes reduction division (meiosis I), equatorial division (meiosis II), and spermiogenesis.

8. Mature sperm consist of a head, midpiece, and tail. Their function is to fertilize a secondary oocyte.

Ducts (p. 777)

1. The duct system of the testes includes the seminiferous tubules, straight tubules, and rete testis.

2. Sperm are transported out of the testes through the efferent ducts.

3. The ductus epididymis is the site of sperm maturation and storage.

4. The ductus (vas) deferens stores sperm and propels them toward the urethra during ejaculation.

5. Cutting and tying-off of the ductus (vas) deferens to achieve sterility by blocking entry of sperm into semen is called vasectomy.

6. Each ejaculatory duct is formed by the union of the duct from the seminal vesicle and ductus (vas) deferens. It is the passageway for ejection of sperm and secretions of the seminal vesicles into the first portion of the urethra, the prostatic urethra.

7. The male urethra is subdivided regionally into three portions: prostatic, membranous, and spongy (penile).

Accessory Sex Glands (p. 780)

1. The seminal vesicles secrete an alkaline, viscous fluid that constitutes about 60% of the volume of semen and contributes to sperm viability.

2. The prostate gland secretes a slightly acidic fluid that constitutes about 25% of the volume of semen and contributes to sperm motility.

3. The bulbourethral (Cowper's) glands secrete mucus for lubrication and an alkaline substance that neutralizes acid.

Semen (p. 782)

1. Semen is a mixture of sperm and accessory sex gland secretions that provides the fluid in which sperm are transported, provides nutrients, and neutralizes the acidity of the male urethra and female vagina.

Penis (p. 782)

1. The penis consists of a root, body, and glans penis.

2. Expansion of its blood sinuses under the influence of sexual excitation is called erection.

Female Reproductive System (p. 784)

1. The female organs of reproduction include the ovaries (gonads), uterine (Fallopian) tubes, uterus, vagina, and vulva.

2. The mammary glands are considered part of the reproductive system.

Ovaries (p. 784)

1. The ovaries are female gonads located in the upper pelvic cavity, on either side of the uterus.

2. They produce secondary oocytes, discharge secondary oocytes (ovulation), and secrete estrogens, progesterone, inhibin, and relaxin.

3. Oogenesis (production of haploid secondary oocytes) begins in the ovaries. The oogenesis sequence includes reduction division (meiosis I) and equatorial division (meiosis II), which goes to completion only after an ovulated secondary oocyte is fertilized by a sperm cell.

Uterine (Fallopian) Tubes (p. 790)

1. The uterine (Fallopian) tubes transport secondary oocytes and fertilized ova from the ovaries to the uterus and are the normal sites of fertilization.

2. Ciliated cells and peristaltic contractions help move a secondary oocyte or fertilized ovum toward the uterus.

Uterus (p. 791)

1. The uterus is an organ the size and shape of an inverted pear that functions in menstruation, implantation of a fertilized ovum, development of a fetus during pregnancy, and labor. It also is part of the pathway for sperm to reach the uterine tubes to fertilize a secondary oocyte.

2. The uterus is normally held in position by a series of ligaments.

3. Histologically, the layers of the uterus are an outer perimetrium (serosa), middle myometrium, and inner endometrium.

Vagina (p. 794)

1. The vagina is a passageway for sperm and the menstrual flow, the receptacle for the penis during sexual intercourse, and the inferior portion of the birth canal.

2. It is capable of considerable distension.

Vulva (p. 795)

1. The vulva is a collective term for the external genitals of the female.

2. It consists of the mons pubis, labia majora, labia minora, clitoris, and vestibule. Within the vestibule are the hymen (if present), vaginal and urethral orifices, and the openings of the ducts of the paraurethral (Skene's), greater vestibular (Bartholin's), and lesser vestibular glands.

Perineum (p. 796)

1. The perineum is a diamond-shaped area at the inferior end of the trunk medial to the thighs and buttocks.

2. An incision in the female perineum before delivery is called an episiotomy.

Mammary Glands (p. 796)

1. The mammary glands are modified sweat glands lying superficial to the pectoralis major muscles. Their function is to synthesize, secrete, and eject milk (lactation).

2. Milk secretion is mainly due to prolactin, and milk ejection is stimulated by oxytocin.

Uterine and Ovarian Cycles (p. 798)

1. The function of the ovarian cycle is to develop a secondary oocyte. The function of the uterine (menstrual) cycle is to prepare the endometrium each month for the reception of a fertilized egg.
2. During the menstrual phase, the stratum functionalis of the endometrium is shed, discharging blood, tissue fluid, mucus, and epithelial cells.
3. During the preovulatory phase of the ovarian cycle, a group of follicles begins to grow and undergo final maturation. One follicle outgrows the others and becomes dominant while the other follicles ultimately degenerate. Also during this phase, a vesicular ovarian (Graafian) follicle develops, and endometrial repair occurs.
4. Ovulation is the rupture of a vesicular ovarian (Graafian) follicle and the release of a secondary oocyte into the pelvic cavity.
5. During the postovulatory phase, the endometrium thickens in readiness for implantation, and both estrogens and progesterone are secreted in large quantity by the corpus luteum.
6. If fertilization and implantation do not occur, the corpus luteum degenerates, and the resulting low levels of estrogens and progesterone allow discharge of the endometrium followed by initiation of another uterine and ovarian cycle.

7. If fertilization and implantation do occur, the corpus luteum is maintained by placental hCG, and the corpus luteum and later the placenta secrete progesterone and estrogens to support pregnancy and breast development for lactation.

Developmental Anatomy of the Reproductive Systems (p. 801)

1. The gonads develop from intermediate mesoderm. In the presence of *SRY* and the hormone testosterone, the gonads begin to differentiate into testes during the seventh week of fetal development. The gonads differentiate into ovaries when testosterone is absent.
2. The external genitals develop from the genital tubercle and are stimulated to develop into typical male structures by the hormone dihydrotestosterone (DHT). The external genitals appear female when testosterone is not produced (the normal situation in female embryos) or when an enzyme that converts testosterone to DHT is deficient due to a genetic defect.

Aging and the Reproductive Systems (p. 801)

1. In the male, decreased levels of testosterone decrease muscle strength, sexual desire, and viable sperm; prostate disorders are common.
2. In the female, levels of progesterone and estrogens decrease, resulting in changes in menstruation; uterine and breast cancer increase in incidence.

Review Questions

1. Define reproduction. Describe how the reproductive organs are classified, and list the male and female organs of reproduction. (p. 771)
2. Describe the function of the scrotum in protecting the testes from temperature fluctuations. What is cryptorchidism? (p. 771)
3. Describe the internal structure of a testis. Where are the sperm cells made? (p. 772)
4. Describe the principal events of spermatogenesis. Why is meiosis important? Distinguish between haploid *(n)* and diploid (2*n*) cells. (p. 776)
5. Identify the principal parts of a sperm cell. List the functions of each. (p. 777)
6. Which ducts are involved in transporting sperm *within* the testes? (p. 777)
7. Describe the location, structure, and functions of the ductus epididymis, ductus (vas) deferens, and ejaculatory duct. (p. 778)
8. What is the spermatic cord? What is an inguinal hernia? (p. 779)
9. Give the location of the three subdivisions of the male urethra. (p. 780)
10. Trace the course of sperm through the male system of ducts from the seminiferous tubules through the urethra. (p. 777)
11. Briefly explain the locations and functions of the seminal vesicles, prostate gland, and bulbourethral (Cowper's) glands. How is cancer of the prostate gland detected? (p. 780)
12. What is semen? What is its function? (p. 782)
13. How is the penis structurally adapted as an organ of copulation? How does an erection occur? What is circumcision? (p. 782)
14. How are the ovaries held in position in the pelvic cavity? (p. 784)
15. Describe the microscopic structure of an ovary. What are the functions of the ovaries? (p. 784)
16. Describe the principal events of oogenesis. (p. 785)
17. Where are the uterine (Fallopian) tubes located? What is their function? (p. 790)
18. Diagram the principal parts of the uterus. What is a Pap smear? (p. 791)
19. Describe the arrangement of ligaments that hold the uterus in its normal position. What is retroflexion? (p. 791)
20. Describe the histology of the uterus. (p. 792)
21. Discuss the blood supply to the uterus. Why is an abundant blood supply important? (p. 793)
22. What is the function of the vagina? Describe its histology. (p. 794)
23. List the parts of the vulva and the functions of each part. What is an episiotomy? (p. 795)
24. Describe the structure of the mammary glands. How are they supported? (p. 796)
25. Briefly outline the major events of each phase of the uterine cycle and correlate them with the events of the ovarian cycle. (p. 798)

26. Describe the development of the reproductive systems. (p. 801)

27. Explain the effects of aging on the reproductive systems. (p. 801)

28. Refer to the glossary of key medical terms at the end of the chapter. Be sure you can define each term. (p. 804)

Self Quiz

Complete the following:

1. Fill in the blanks with the male homologues for each of the following.
(a) labia majora: ___; (b) clitoris: ___; (c) paraurethral (Skene's) glands: ___; (d) greater vestibular (Bartholin's) glands: ___.

2. Gametes have ___ chromosomes, compared to all other body cells, which contain ___ chromosomes. Therefore, we refer to gametes as ___, and to other body cells as ___.

3. The layer of endometrium that is shed is the stratum ___.

4. The ___ is a comma-shaped organ located on the posterior border of the testis.

5. The thickest layer in the uterine wall is the ___, which is composed of three layers of ___.

6. If fertilization and implantation do not occur, the corpus luteum degenerates and becomes the ___.

7. The two dorsolateral masses of tissue within the penis are called the ___ penis.

8. The ovary is attached to the uterus by the ___ ligament.

9. The rupture of the vesicular ovarian (Graafian) follicle with release of a secondary oocyte into the pelvic cavity is called ___.

10. Place numbers in the blanks to arrange the following structures in correct sequence from anterior to posterior.
(a) anus: ___; (b) vaginal orifice: ___; (c) clitoris: ___; (d) urethral orifice: ___.

11. Place numbers in the blanks to arrange the following structures in the correct sequence for the pathway of milk in the breasts.
(a) lactiferous sinuses: ___; (b) alveoli: ___; (c) secondary tubules: ___; (d) mammary ducts: ___; (e) lactiferous ducts: ___.

12. Place numbers in the blanks to arrange the following structures in the correct sequence for the pathway of sperm.
(a) ejaculatory duct: ___;
(b) testis: ___;
(c) urethra: ___;
(d) ductus (vas) deferens: ___;
(e) epididymis: ___.

13. Match the following terms to their descriptions:

(1) interstitial endocrinocytes (cells of Leydig)
(2) ejaculatory duct
(3) sustentacular (Sertoli) cells
(4) seminiferous tubules
(5) ductus (vas) deferens
(6) ductus epididymis
(7) lobule

___ (a) 200–300 of these per testis

___ (b) tightly coiled tubes (1–3 per lobule) composed of cells that develop into sperm

___ (c) cells between developing sperm cells that form the blood–testis barrier and provide nourishment

___ (d) cells located between seminiferous tubules that secrete testosterone

___ (e) highly coiled duct in which sperm are stored for maturation

___ (f) ejects sperm and fluid into the urethra just before ejaculation

___ (g) enters the pelvic cavity through the inguinal canal

Choose the one best answer to the following questions:

14. Which of the following cells is *least* mature?
a. primary spermatocyte
b. spermatid
c. spermatogonium
d. secondary spermatocyte
e. primordial germ cell

15. Which of the following correctly orders the sequence of events in spermatogenesis?
(1) spermiogenesis
(2) mitosis
(3) reduction division
(4) equatorial division
a. 1, 2, 3, 4
b. 2, 3, 4, 1
c. 3, 4, 1, 2
d. 4, 3, 2, 1
e. 2, 4, 3, 1

16. The efferent ducts open into the
a. rete testis
b. epididymus
c. ductus (vas) deferens
d. straight tubules
e. seminiferous tubules

17. Which of the following four statements about ovaries are true?
    (1) They are female gonads.
    (2) They produce secondary oocytes and female hormones.
    (3) They are homologous to the testes.
    (4) They are attached to the pelvic wall by the suspensory ligaments.
    a. 1, 3, 4
    b. 2, 3, 4
    c. 1, 2, 4
    d. 1 and 4
    e. all of the above

18. Which of the following structures is the male gonad?
    a. seminal vesicle
    b. scrotum
    c. testis
    d. prostate gland
    e. penis

19. Which of the following produces a secretion that helps maintain the motility and viability of sperm?
    a. prostate gland
    b. penis
    c. bulbourethral (Cowper's) glands
    d. ejaculatory duct
    e. all of the above

20. Which of the following four statements about the vagina is/are true?
    (1) It is the anterior part of the uterus, sometimes called the cervix.
    (2) It is a thin-walled organ whose external opening lies between two folds of skin, the labia minora.
    (3) It is the site of the growth of the fetus; it is also called the womb.
    (4) Its inner mucosa sloughs off each month as the menstrual flow.
    a. 1 only
    b. 2 only
    c. 3 only
    d. 4 only
    e. 3 and 4

21. The secondary sex gland(s) that contribute(s) the largest volume to the semen is/are the
    a. seminal vesicles
    b. bulbourethral (Cowper's) glands
    c. paraurethral (Skene's) glands
    d. vestibular (Bartholin's) glands
    e. prostate gland

Are the following statements true or false?

22. Contraction of the dartos muscle is responsible for elevation of the testes during sexual arousal or on exposure to cold.

23. The tunica albuginea is a capsule of connective tissue associated with both the ovaries and the testes.

24. The acrosome is the portion of a sperm cell that contains the genetic material.

25. Ovarian follicles are located in the ovarian cortex.

Critical Thinking Questions

1. Trace the route of the ductus (vas) deferens from its origin to its junction with the ducts of the seminal vesicles. Where along this route would a vasectomy be performed?
   HINT: *An ice pack in the jock strap could be used to ease any postsurgical discomfort.*

2. Imagine that you are a sperm cell preparing for your journey toward the ovulated secondary oocyte. How does the sperm's structure and the components of the semen help the sperm cell survive the journey and fertilize the oocyte?
   HINT: *Think about what a fish needs to swim upstream to spawn.*

3. Phil has promised his wife that he will get a vasectomy after the birth of their next child. However, he's concerned about possible effects on his virility. How would you respond to Phil's concerns?
   HINT: *A vasectomy is used for birth control, not sex control.*

4. Consuela, age 39, has been advised to have a hysterectomy due to medical problems. She is worried that the procedure will cause menopause. Is this a valid concern?
   HINT: *Women often take estrogens to ease the symptoms of menopause.*

5. PMS (premenstrual syndrome) has been blamed on many things, including vitamin deficiencies, hormones, stress, and men. What changes normally occur in the body before menstruation?
   HINT: *PMS should not be a problem in postmenopausal women.*

6. When Sean was two days old, his parents were still debating whether or not to have a circumcision performed. What does this procedure entail? Why is it performed on some babies?
   HINT: *Parents don't have to make this decision when they're told, "It's a girl!"*

Answers to Figure Questions

26.1 Gonads: produce gametes and hormones; ducts: transport, store, and receive gametes; accessory sex glands: produce materials that support gametes.

26.2 Cremaster and dartos.

26.3 Tunica vaginalis and tunica albuginea.

26.4 Interstitial endocrinocytes (Leydig cells).

26.5 Because the number of chromosomes in each cell is reduced by half.

26.6 Head contains DNA and enzymes for penetration of secondary oocyte; midpiece contains mitochondria for ATP production; tail is a flagellum that provides propulsion.

26.7 Sperm maturation, sperm storage, and propulsion of sperm into ductus (vas) deferens.

26.8 Stores sperm and conveys sperm toward urethra.

26.9 Seminal vesicles.

26.10 Two corpora cavernosa penis and one corpus spongiosum penis contain blood sinuses that fill with blood that cannot flow out of the penis as quickly as it flows in. The trapped blood stiffens the tissue. The corpus spongiosum penis keeps the spongy urethra open so that ejaculation can occur.

26.11 Testes; penis; prostate gland; bulbourethral glands.

26.12 Mesovarium, to broad ligament of the uterus and the uterine tube; ovarian ligament, to uterus; suspensory ligament, to pelvic wall.

26.13 Ovarian follicles secrete estrogens, and a corpus luteum secretes estrogens, progesterone, relaxin, and inhibin.

26.14 Primary oocytes are present in the ovary at birth, so they are as old as the woman is. In males, primary spermatocytes are continually being formed from stem cells (spermatogonia) and thus are only a few days old.

26.15 They help move the secondary oocyte toward the uterus.

26.16 Endometrium: highly vascular, secretory epithelium that provides oxygen and nutrients to sustain a fertilized egg; myometrium: thick smooth muscle layer that contracts to expel fetus at birth.

26.17 The stratum basalis provides cells to replace those shed (stratum functionalis) during each menstruation.

26.18 Anterior: mons pubis, clitoris, and prepuce. Lateral: labia minora and labia majora.

26.19 Its borders form a triangle that encloses the urethral (uro-) and vaginal (-genital) orifices.

26.20 Alveoli → secondary tubules → mammary ducts → lactiferous sinuses → lactiferous ducts → nipple.

26.21 Lobules.

26.22 Estrogens; LH; LH; estrogens.

26.23 The *SRY* gene on the Y chromosome.

26.24 Dihydrotestosterone (DHT).

Chapter 27

Developmental Anatomy

Student Objectives

1. Explain the processes associated with fertilization, morula formation, blastocyst development, and implantation.

2. Discuss the formation of the primary germ layers, embryonic membranes, placenta, and umbilical cord as the principal events of the embryonic period.

3. List representative body structures produced by the primary germ layers.

4. Discuss the principal body changes associated with fetal growth.

5. Describe some of the structural and functional changes associated with gestation.

6. Describe several prenatal diagnostic tests.

7. Explain the events associated with the three stages of labor.

Developmental anatomy is the study of the sequence of events from the fertilization of a secondary oocyte to the formation of an adult organism. As we look at the sequence from fertilization to birth, we will consider fertilization, implantation, placental development, embryonic development, fetal growth, gestation, labor, and parturition (birth). The **embryonic period** is the first two months of development. During this period the developing human is called an **embryo** (*bryein* = grow). The study of development from the fertilized egg through the eighth week *in utero* is termed **embryology** (em-brē-OL-ō-jē). The months of development after the second month are considered the **fetal period,** and during this time the developing human is called a **fetus** (*feo* = to bring forth). The development of embryo and fetus is a wonderfully complex and precisely coordinated series of events.

DEVELOPMENT DURING PREGNANCY

Once sperm and a secondary oocyte have developed through meiosis and maturation, and the sperm have been deposited in the vagina, pregnancy can occur. **Pregnancy** is a sequence of events that normally includes fertilization, implantation, embryonic growth, and fetal growth that ends with birth about 38 weeks later.

Fertilization

During **fertilization** (fer-til-i-ZĀ-shun; *fertilis* = fruitful) the genetic material from a sperm cell and secondary oocyte merges into a single nucleus. Fertilization normally occurs in the uterine (Fallopian) tube about 12–24 hours after ovulation. Because sperm remain viable in the vagina for about 48 hours and a secondary oocyte is viable for about 24 hours after ovulation, there typically is a 3-day window during which pregnancy can occur—from 2 days before to 1 day after ovulation.

The process leading to fertilization begins as peristaltic contractions and the action of cilia transport the oocyte through the uterine tube. Whiplike movements of their tails (flagella) propel sperm through the female reproductive tract. The acrosome of a sperm cell produces an enzyme called **acrosin** that stimulates sperm motility and migration within the female reproductive tract. Also, muscular contractions of the uterus, stimulated by prostaglandins in semen, probably aid sperm movement toward the uterine tube. Finally, the oocyte is thought to secrete a chemical substance that attracts sperm.

Although sperm mature in the epididymis, they are not able to fertilize an oocyte until they have been in the female reproductive tract for several hours. The female reproductive tract induces a series of functional changes in sperm—a process called **capacitation**—that allows them to fertilize a secondary oocyte. During this process, the acrosome releases several enzymes that penetrate the **corona radiata,** a ring of cells that surrounds the oocyte, and a clear glycoprotein layer internal to the corona radiata called the **zona pellucida** (pe-LOO-si-da; *pellucida* = translucent; Figure 27.1). Normally only one sperm cell penetrates and enters a secondary oocyte, an event called **syngamy** (*syn* = together; *gamos* = marriage). Syngamy causes the cell membrane of the oocyte to depolarize, which triggers the release of calcium ions inside the cell. Calcium ions stimulate the release of granules by the oocyte that, in turn, promote changes in the zona pellucida that block entry of other sperm. This prevents **polyspermy,** fertilization by more than one sperm cell.

Once a sperm cell enters a secondary oocyte, the oocyte completes equatorial division (meiosis II). It divides into a larger ovum (mature egg) and a smaller second polar body that fragments and disintegrates (see Figure 26.14). The sperm cell sheds its tail, and the nucleus in the head develops into the **male pronucleus.** The nucleus of the fertilized ovum develops into a **female pronucleus.** After the pronuclei (Figure 27.1c) are formed, they fuse to produce a **segmentation nucleus.** The segmentation nucleus is diploid because it contains 23 chromosomes (*n*) from the male pronucleus and 23 chromosomes (*n*) from the female pronucleus. Thus the fusion of the haploid (*n*) pronuclei restores the diploid number (2*n*). The fertilized ovum, consisting of a segmentation nucleus, cytoplasm, and zona pellucida, is called a **zygote** (ZĪ-gōt; *zygosis* = a joining).

Dizygotic (fraternal) twins are produced from the independent release of two secondary oocytes and the subsequent fertilization of each by different sperm. They are the same age and are in the uterus at the same time, but they are genetically as dissimilar as any other siblings. Dizygotic twins may or may not be the same sex. **Monozygotic (identical) twins** develop from a single fertilized ovum that splits at an early stage in development, and so they contain exactly the same genetic material and are always the same sex. On rare occasions, monozygotic twins may be joined in varying degrees, from slight skin fusion to sharing of limbs, trunks, and viscera. Such twins are called **conjoined twins.**

Formation of the Morula

After fertilization, rapid mitotic cell divisions of the zygote take place. These early divisions of the zygote are called **cleavage.** The first cleavage begins about 24 hours after fertilization and is completed about 30 hours after fertilization. Each succeeding division takes slightly less time (Figure 27.2a, b). By the second day after fertilization, the second cleavage is completed. By the end of the third day, there are 16 cells. The progressively smaller cells produced by

Figure 27.1 Fertilization. A drawing (a) and a photomicrograph (b) of a sperm cell penetrating the corona radiata and zona pellucida around a secondary oocyte. (c) Male and female pronuclei.

During fertilization, genetic material from a sperm cell merges with that of an oocyte to form a single nucleus.

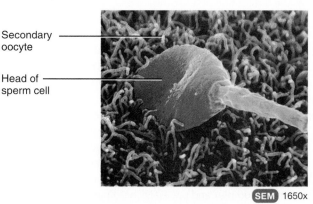

Secondary oocyte

Head of sperm cell

SEM 1650x

(b) Sperm cell in contact with a secondary oocyte

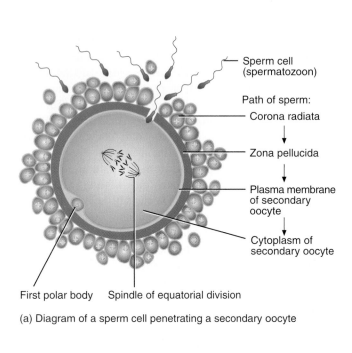

Sperm cell (spermatozoon)

Path of sperm:
Corona radiata

Zona pellucida

Plasma membrane of secondary oocyte

Cytoplasm of secondary oocyte

First polar body   Spindle of equatorial division

(a) Diagram of a sperm cell penetrating a secondary oocyte

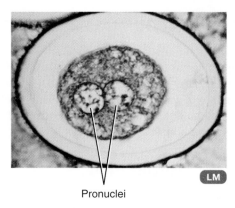

Pronuclei

(c) Male and female pronuclei

Q *What is capacitation?*

cleavage are called **blastomeres** (BLAS-tō-mērz; *blast* = germ, sprout; *meros* = part). Successive cleavages produce a solid sphere of cells, still surrounded by the zona pellucida, called the **morula** (MOR-yoo-la; *morula* = mulberry; Figure 27.2c). The morula is about the same size as the original zygote.

Development of the Blastocyst

By the end of the fourth day, the number of cells in the morula increases, and it continues to move through the uterine (Fallopian) tube. At 4.5–5 days, the dense cluster of cells has developed into a hollow ball of cells and enters the uterine cavity; it is now called a **blastocyst** (*kystis* = bag; Figure 27.2d, e). The blastocyst has an outer covering of cells called the **trophoblast** (TRŌF-ō-blast; *troph* = nourish), an **inner**

cell mass (embryoblast), and an internal fluid-filled cavity called the **blastocele** (BLAS-tō-sēl; *koilos* = hollow). The trophoblast and part of the inner cell mass ultimately form the membranes composing the fetal portion of the placenta; the rest of the inner cell mass develops into the embryo.

Clinical Application

Ectopic Pregnancy

Ectopic (*ektos* = outside; *topos* = place) **pregnancy** is the development of an embryo or fetus outside the uterine cavity. Most occur in the uterine (Fallopian) tube, usually in the ampullar and infundibular portions, although some occur in the abdominal cavity or uterine cervix. An ectopic pregnancy usually occurs when passage of the fertilized ovum through the uterine tube is

Figure 27.2 Cleavage and formation of the morula and blastocyst.

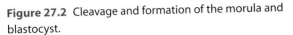

Cleavage refers to the early, rapid mitotic divisions in a zygote.

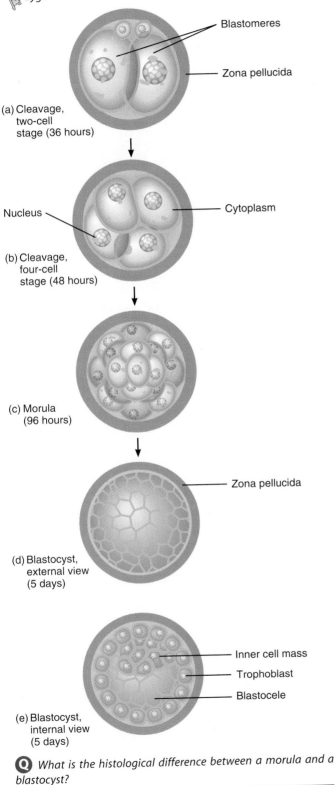

(a) Cleavage, two-cell stage (36 hours)

Blastomeres

Zona pellucida

Nucleus

Cytoplasm

(b) Cleavage, four-cell stage (48 hours)

(c) Morula (96 hours)

Zona pellucida

(d) Blastocyst, external view (5 days)

Inner cell mass

Trophoblast

Blastocele

(e) Blastocyst, internal view (5 days)

Q *What is the histological difference between a morula and a blastocyst?*

impaired. The cause may be decreased motility of the uterine tube smooth muscle or abnormal anatomy. In comparison with nonsmokers, women who smoke and become pregnant are twice as likely to have an ectopic pregnancy. It is thought that the nicotine in cigarette smoke paralyzes the cilia in the lining of the uterine tube (as it does in the respiratory passageways). Scars from pelvic inflammatory disease (PID), previous uterine tube surgery, and previous ectopic pregnancy may hinder movement of the fertilized ovum.

Ectopic pregnancy may be characterized by one or two missed menstrual cycles, followed by bleeding and acute abdominal and pelvic pain. Unless removed, the developing embryo can rupture the tube, often resulting in death of the mother. ■

Implantation

The blastocyst remains free within the cavity of the uterus for a short time before it attaches to the uterine wall. During this time, the zona pellucida disintegrates and the blastocyst enlarges. The blastocyst receives nourishment from glycogen-rich secretions of endometrial (uterine) glands, sometimes called uterine milk. About 6 days after fertilization the blastocyst attaches to the endometrium, a process called **implantation** (Figure 27.3a). At this time, the endometrium is in its postovulatory phase (see Figure 26.22).

As the blastocyst implants, usually on the posterior wall of the fundus or body of the uterus, it is oriented so that the inner cell mass is toward the endometrium (Figure 27.3b). The trophoblast develops two layers in the region of contact between the blastocyst and endometrium. These layers are a **syncytiotrophoblast** (sin-sīt′-ē-ō-TRŌF-ō-blast; *syn* = joined; *cyto* = cell) that contains no cell boundaries and a **cytotrophoblast** (sī-tō-TRŌF-ō-blast) between the inner cell mass and syncytiotrophoblast that is composed of distinct cells (Figure 27.3c). These two layers of trophoblast become part of the chorion (one of the fetal membranes) as they undergo further growth (see Figure 27.5). During implantation, the syncytiotrophoblast secretes enzymes that enable the blastocyst to penetrate the uterine lining. The enzymes digest and liquefy the endometrial cells. The fluid and nutrients further nourish the burrowing blastocyst for about a week after implantation. Eventually, the blastocyst becomes buried in the endometrium (see Figure 27.5a). The trophoblast also secretes human chorionic gonadotropin (hCG), which prevents the corpus luteum in the ovary from degenerating and sustains its secretion of progesterone and estrogens. Thus menstruation does not begin.

Because a developing embryo, and later a fetus, contains genes from the father as well as the mother, it is essentially foreign tissue. Thus it is surprising that it is not rejected by the mother's immune system. The trophoblast is the only tissue of the developing organism that contacts the mother's uterus. Even though trophoblast cells have paternal antigens

Figure 27.3 Implantation. Shown is the blastocyst in relation to the endometrium of the uterus at various time intervals after fertilization.

Implantation refers to the attachment of a blastocyst to the endometrium, which occurs about 6 days after fertilization.

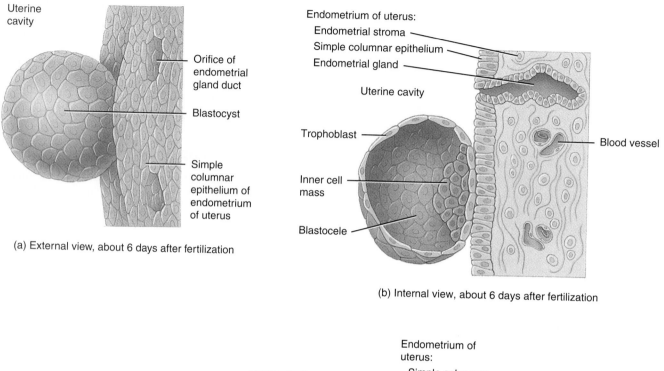

(a) External view, about 6 days after fertilization

(b) Internal view, about 6 days after fertilization

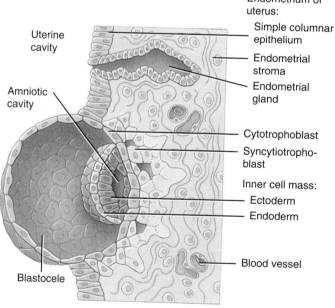

(c) Internal view, about 7 days after fertilization

Q *How does the blastocyst merge with and burrow into the endometrium?*

Figure 27.4 Summary of events associated with fertilization and implantation.

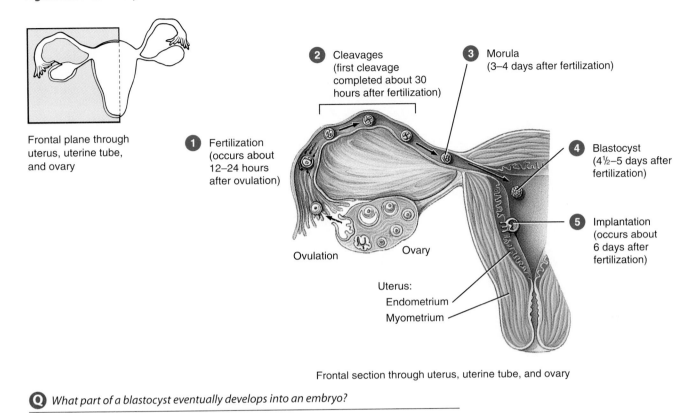

Frontal plane through uterus, uterine tube, and ovary

2 Cleavages (first cleavage completed about 30 hours after fertilization)

3 Morula (3–4 days after fertilization)

1 Fertilization (occurs about 12–24 hours after ovulation)

4 Blastocyst (4½–5 days after fertilization)

5 Implantation (occurs about 6 days after fertilization)

Ovulation

Ovary

Uterus:
Endometrium
Myometrium

Frontal section through uterus, uterine tube, and ovary

Q *What part of a blastocyst eventually develops into an embryo?*

that could provoke a rejection response, some mechanism in the uterus prevents this to permit development of the fetus to term.

A summary of the principal events associated with fertilization and implantation is shown in Figure 27.4.

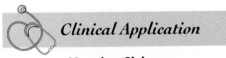

Clinical Application

Morning Sickness

In the early months of pregnancy, **emesis gravidarum (morning sickness)** may occur, characterized by episodes of nausea and possibly vomiting that are most likely to occur in the morning. The cause is unknown, but the high levels of human chorionic gonadotropin (hCG) secreted by the placenta and progesterone secreted by the ovaries have been implicated. In some women the severity of these symptoms requires hospitalization for intravenous feeding, and the condition is then known as **hyperemesis gravidarum.**

EMBRYONIC DEVELOPMENT

As noted earlier, the **embryonic period** is the first two months of development, and the developing human is called an **embryo.** By the end of the embryonic period, the rudiments of all the principal adult organs are present and the embryonic membranes are developed. By the end of the third month, the placenta, which is the site of exchange of nutrients and wastes between the mother and the fetus, is functioning.

Beginnings of Organ Systems

The first major event of the embryonic period is called **gastrulation** (gas'-troo-LĀ-shun; *gastrula* = little belly). The inner cell mass of the blastocyst differentiates into three **primary germ layers:** ectoderm, endoderm, and mesoderm. These are the major embryonic tissues from which all tissues and organs of the body will develop.

Within 8 days after fertilization, the cells of the inner cytotrophoblast proliferate and form the amnion (a fetal membrane) and a space, the **amniotic** (am-nē-OT-ik; *amnion* = lamb) **cavity,** adjacent to the inner cell mass (Figure 27.5a). The layer of cells of the inner cell mass that

Figure 27.5 Formation of the primary germ layers and associated structures.

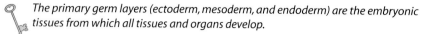

The primary germ layers (ectoderm, mesoderm, and endoderm) are the embryonic tissues from which all tissues and organs develop.

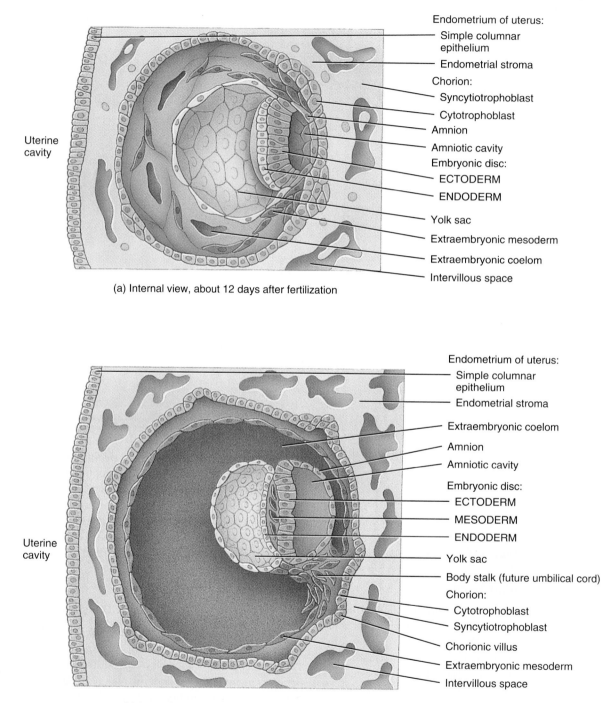

(a) Internal view, about 12 days after fertilization

(b) Internal view, about 14 days after fertilization

Figure 27.5 (continued)

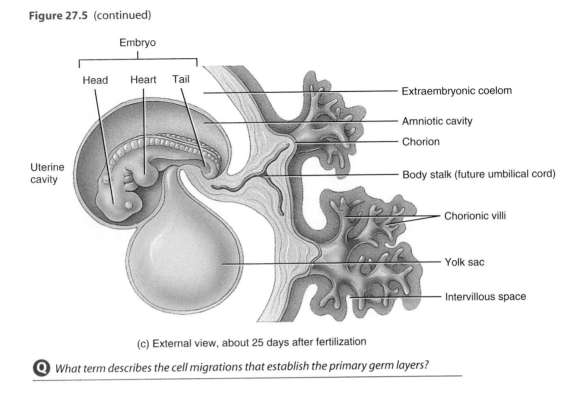

(c) External view, about 25 days after fertilization

Q *What term describes the cell migrations that establish the primary germ layers?*

is closer to the amniotic cavity develops into the **ectoderm** (*ecto* = outside; *derm* = skin). The layer of the inner cell mass that borders the blastocele develops into the **endoderm** (*endo* = inside). As the amniotic cavity forms, the inner cell mass at this stage is called the **embryonic disc.** It will form the embryo. At this stage, the embryonic disc contains ectodermal and endodermal cells; the mesodermal cells are scattered external to the disc.

About the 12th day after fertilization, striking changes appear (Figure 27.5a). The cells of the endodermal layer have been dividing rapidly, so that groups of them now extend around in a circle, forming the yolk sac, another fetal membrane (described shortly). Cells of the cytotrophoblast give rise to a loose connective tissue, the **extraembryonic mesoderm** (*meso* = middle). This completely fills the space between the cytotrophoblast and yolk sac. Soon large spaces develop in the extraembryonic mesoderm and come together to form a single, larger cavity called the **extraembryonic coelom** (SĒ-lōm; *koiloma* = cavity), the future ventral body cavity (Figure 27.5b).

About the 14th day, the cells of the embryonic disc differentiate into three distinct layers; the ectoderm, mesoderm (intraembryonic layer), and endoderm (Figure 27.5b). As the embryo develops (Figure 27.5c), the endoderm becomes the epithelial lining of the gastrointestinal tract, respiratory tract, and several other organs. The mesoderm forms muscle, bone and other connective tissues, and the peritoneum. The

ectoderm develops into the epidermis of the skin and the nervous system. Table 27.1 provides more details about the fates of these primary germ layers.

Embryonic Membranes

A second major event that occurs during the embryonic period is the formation of the **embryonic (extraembryonic) membranes.** These membranes lie outside the embryo and protect and nourish the embryo and, later, the fetus. (Recall that a fetus begins to develop after the third month.) The membranes are the yolk sac, amnion, chorion, and allantois.

In many species such as birds, the **yolk sac** is a membrane that is the origin of blood vessels that transport nutrients to the embryo (Figure 27.5c). However, the human embryo receives nutrients from the endometrium. The yolk sac remains small and functions as an early site of blood formation. The yolk sac also contains cells that migrate into the gonads and differentiate into the primitive germ cells (spermatogonia and oogonia).

The **amnion** is a thin, protective membrane that forms by the eighth day after fertilization and initially overlies the embryonic disc (Figure 27.5a, b). As the embryo grows, the amnion entirely surrounds the embryo, creating a cavity that becomes filled with **amniotic fluid** (Figure 27.6). Most amniotic fluid is initially derived from a filtrate of maternal blood. Later, the fetus makes daily contributions to the fluid by excreting urine into the amniotic cavity. Amniotic fluid serves as a shock absorber for the fetus, helps regulate

Figure 27.6 Embryonic membranes.

Embryonic membranes are outside the embryo; they protect and nourish the embryo and, later, the fetus.

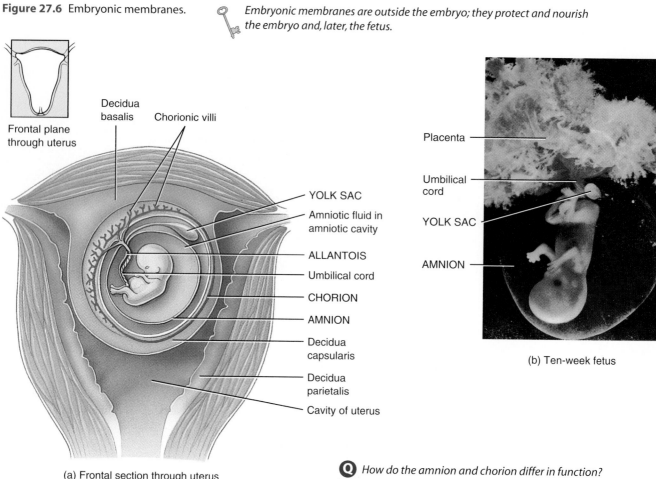

Frontal plane through uterus

Decidua basalis

Chorionic villi

YOLK SAC

Amniotic fluid in amniotic cavity

ALLANTOIS

Umbilical cord

CHORION

AMNION

Decidua capsularis

Decidua parietalis

Cavity of uterus

(a) Frontal section through uterus

Placenta

Umbilical cord

YOLK SAC

AMNION

(b) Ten-week fetus

Q *How do the amnion and chorion differ in function?*

Table 27.1 *Structures Produced by the Three Primary Germ Layers*

ENDODERM	MESODERM	ECTODERM
Epithelial lining of gastrointestinal tract (except the oral cavity and anal canal) and the epithelium of its glands.	All skeletal, most smooth, and all cardiac muscle.	All nervous tissue.
	Cartilage, bone, and other connective tissues.	Epidermis of skin.
Epithelial lining of urinary bladder, gallbladder, and liver.	Blood, red bone marrow, and lymphatic tissue.	Hair follicles, arrector pili muscles, nails, and epithelium of skin glands (sebaceous and sudoriferous).
Epithelial lining of pharynx, auditory (Eustachian) tubes, tonsils, larynx, trachea, bronchi, and lungs.	Endothelium of blood vessels and lymphatic vessels.	Lens, cornea, and internal eye muscles.
Epithelium of thyroid gland, parathyroid glands, pancreas, and thymus gland.	Dermis of skin.	Internal and external ear.
	Fibrous tunic and vascular tunic of eye.	Neuroepithelium of sense organs.
Epithelial lining of prostate and bulbourethral (Cowper's) glands, vagina, vestibule, urethra, and associated glands such as the greater (Bartholin's) vestibular and lesser vestibular glands.	Middle ear.	Epithelium of oral cavity, nasal cavity, paranasal sinuses, salivary glands, and anal canal.
	Mesothelium of ventral body cavity.	Epithelium of pineal gland, pituitary gland (hypophysis), and adrenal medullae.
	Epithelium of kidneys and ureters.	
	Epithelium of adrenal cortex.	
	Epithelium of gonads and genital ducts.	

Figure 27.7 Regions of the decidua.

The decidua is a modified endometrium that develops after implantation.

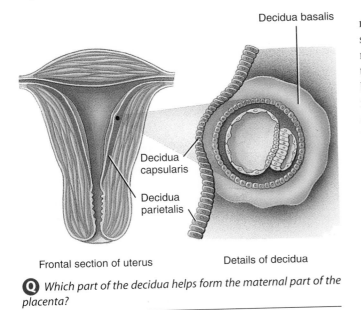

Frontal section of uterus   Details of decidua

Q *Which part of the decidua helps form the maternal part of the placenta?*

fetal body temperature, and prevents adhesions between the skin of the fetus and surrounding tissues. Embryonic cells are sloughed off into amniotic fluid; they can be examined in the procedure called **amniocentesis** (am′-nē-ō-sen-TĒ-sis), which is described on page 826. The amnion usually ruptures just before birth and with its fluid constitutes the "bag of waters."

The **chorion** (KŌR-ē-on) is derived from the trophoblast of the blastocyst and the mesoderm that lines the trophoblast. It surrounds the embryo and, later, the fetus. Eventually, the chorion becomes the principal embryonic part of the placenta, the structure for exchange of materials between mother and fetus. It also produces human chorionic gonadotropin (hCG). The amnion, which also surrounds the fetus, eventually fuses to the inner layer of the chorion.

The **allantois** (a-LAN-tō-is; *allas* = sausage) is a small, vascularized outpouching of the hindgut. It serves as an early site of blood formation. Later its blood vessels serve as the umbilical connection in the placenta between mother and fetus. This connection is the umbilical cord.

Placenta and Umbilical Cord

Development of the **placenta** (pla-SEN-ta; *placenta* = flat cake), the third major event of the embryonic period, is accomplished by the third month of pregnancy. The placenta has the shape of a flat cake when fully developed and is formed by the chorion of the embryo and a portion of the endometrium of the mother (see Figure 27.8b). Functionally,

the placenta allows oxygen and nutrients to diffuse into fetal blood from maternal blood. Simultaneously, carbon dioxide and wastes diffuse from fetal blood into maternal blood at the placenta.

The placenta also is a protective barrier because most microorganisms cannot cross it. However, certain viruses, such as those that cause AIDS, German measles, chickenpox, measles, encephalitis, and poliomyelitis, may pass through the placenta. The placenta also stores nutrients such as carbohydrates, proteins, calcium, and iron, which are released into fetal circulation as required. Finally, the placenta produces several hormones that are necessary to maintain pregnancy. Almost all drugs, including alcohol, and many substances that can cause birth defects pass freely through the placenta.

If implantation occurs, a portion of the endometrium becomes modified and is known as the **decidua** (dē-SID-yoo-a; *deciduus* = falling off; Figure 27.7). The decidua includes all but the stratum basalis layer of the endometrium and separates from the endometrium after the fetus is delivered as it does in normal menstruation. Different regions of the decidua, which are all areas of the stratum functionalis, are named based on their positions relative to the site of the implanted blastocyst (Figure 27.7). The **decidua basalis** is the portion of the endometrium between the chorion and the stratum basalis of the uterus. It becomes the maternal part of the placenta. The **decidua capsularis** is the portion of the endometrium that covers the embryo and is located between the embryo and the uterine cavity. The **decidua parietalis** (par-rī-e-TAL-is) is the remaining modified endometrium that lines the noninvolved areas of the entire pregnant uterus. As the embryo and later the fetus enlarges, the decidua capsularis bulges into the uterine cavity and initially fuses with the decidua parietalis, thus obliterating the uterine cavity. By about 27 weeks, the decidua capsularis degenerates and disappears.

During embryonic life, fingerlike projections of the chorion, called **chorionic villi** (kō′-rē-ON-ik VIL-ī), grow into the decidua basalis of the endometrium (Figure 27.8a). These will contain fetal blood vessels of the allantois. They continue growing until they are bathed in maternal blood sinuses called **intervillous** (in-ter-VIL-us) **spaces.** Thus maternal and fetal blood vessels are brought into proximity. It should be noted, however, that maternal and fetal blood do not normally mix. Rather, oxygen and nutrients in the blood of the mother's intervillous spaces diffuse across the cell membranes into the capillaries of the villi while waste products diffuse in the opposite direction. From the capillaries of the villi, nutrients and oxygen enter the fetus through the umbilical vein. Wastes leave the fetus through the umbilical arteries, pass into the capillaries of the villi, and diffuse into the maternal blood.

Figure 27.8 Placenta and umbilical cord.

The placenta is formed by the chorion of the embryo and part of the endometrium of the mother.

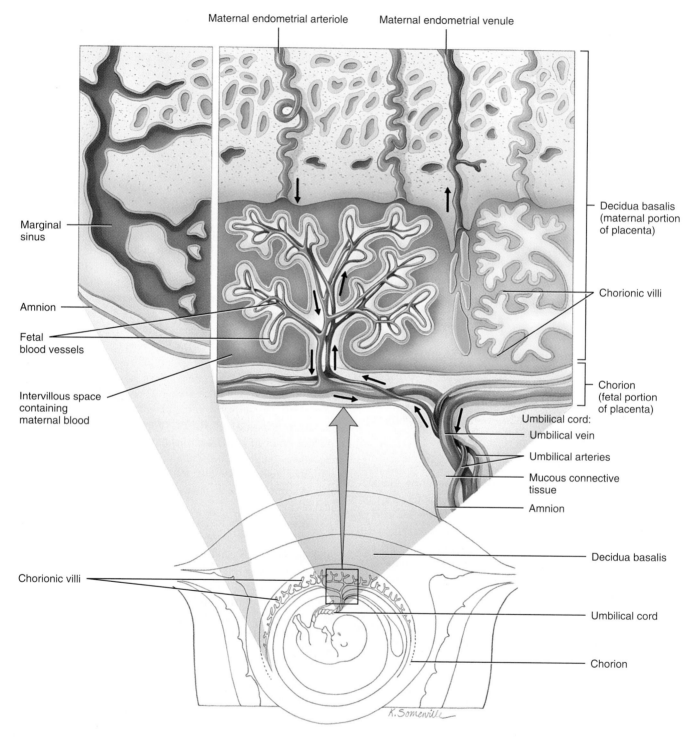

Maternal endometrial arteriole

Maternal endometrial venule

Decidua basalis (maternal portion of placenta)

Marginal sinus

Chorionic villi

Amnion

Fetal blood vessels

Intervillous space containing maternal blood

Chorion (fetal portion of placenta)

Umbilical cord:

Umbilical vein

Umbilical arteries

Mucous connective tissue

Amnion

Decidua basalis

Chorionic villi

Umbilical cord

Chorion

(a) Overall structure

Figure 27.8 (continued)

Umbilical cord   Umbilical veins

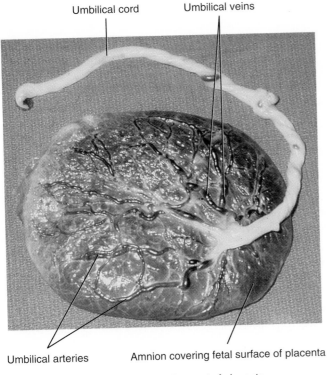

Umbilical arteries   Amnion covering fetal surface of placenta

(b) Photograph of fetal aspect of placenta

Q *What is the function of the placenta?*

The **umbilical** (um-BIL-i-kul) **cord** is a vascular connection between mother and fetus. It consists of two umbilical arteries that carry deoxygenated fetal blood to the placenta, one umbilical vein that carries oxygenated blood into the fetus, and supporting mucous connective tissue called Wharton's jelly from the allantois. The entire umbilical cord is surrounded by a layer of amnion (Figure 27.8a).

After the birth of the baby, the placenta detaches from the uterus and is termed the **afterbirth.** At this time, the umbilical cord is severed, leaving the baby on its own. The small portion (about an inch) of the cord that remains still attached to the infant begins to wither and falls off, usually within 12–15 days after birth. The area where the cord was attached becomes covered by a thin layer of skin, and scar tissue forms. The scar is the **umbilicus (navel).**

Pharmaceutical companies use human placentas to harvest hormones, drugs, and blood. Portions of placentas are also used for burn coverage. The placental and umbilical cord veins can also be used in blood vessel grafts.

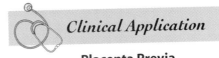
Placenta Previa

In some cases, part or all of the placenta becomes implanted in the inferior portion of the uterus, near or over the internal os of the cervix. This condition is called **placenta previa** (PRĒ-vē-a; *previa* = in front of). It may lead to spontaneous abortion and occurs in approximately 1 in 250 live births. It is dangerous to the fetus because it may cause premature birth and intrauterine hypoxia due to maternal bleeding. Maternal mortality is increased due to hemorrhage and infection. The most important symptom is sudden, painless, bright red vaginal bleeding in the third trimester. Cesarean section is the preferred method of delivery in placenta previa. ■

Throughout the text we have discussed developmental anatomy of the various body systems in their respective chapters. The following list of these sections is presented for your review:

FETAL GROWTH

During the **fetal period,** the period of development after the second month, organs established by the primary germ layers grow rapidly, and the fetus takes on a human appearance. A summary of changes associated with the embryonic and fetal periods is presented in Table 27.2.

Fetal Surgery

Fetal surgery is a new medical field that had its beginnings in 1985. In a pioneering operation, a team of surgeons removed a 23-week-old fetus from its mother's uterus, operated to correct a blocked urinary tract, and then returned the fetus to the uterus.

Table 27.2 Changes Associated with Embryonic and Fetal Growth

END OF MONTH	APPROXIMATE SIZE AND WEIGHT	REPRESENTATIVE CHANGES
1	0.6 cm (3/16 in.)	Eyes, nose, and ears are not yet visible. Vertebral column and vertebral canal form. Small buds that will develop into limbs form. Heart forms and starts beating. Body systems begin to form. The central nervous system appears at the start of the third week.
2	3 cm (1¼ in.) 1 g (1/30 oz)	Eyes are far apart, eyelids fused. Nose is flat. Ossification begins. Limbs become distinct, and digits are well formed. Major blood vessels form. Many internal organs continue to develop.
3	7½ cm (3 in.) 30 g (1 oz)	Eyes are almost fully developed but eyelids are still fused, nose develops a bridge, and external ears are present. Ossification continues. Limbs are fully formed and nails develop. Heartbeat can be detected. Urine starts to form. Fetus begins to move, but it cannot be felt by mother. Body systems continue to develop.
4	18 cm (6½–7 in.) 100 g (4 oz)	Head is large in proportion to rest of body. Face takes on human features and hair appears on head. Many bones are ossified, and joints begin to form. Rapid development of body systems occurs.
5	25–30 cm (10–12 in.) 200–450 g (½–1 lb)	Head is less disproportionate to rest of body. Fine hair (lanugo) covers body. Brown fat forms and is the site of heat production. Fetal movements are commonly felt by mother (quickening). Rapid development of body systems occurs.
6	27–35 cm (11–14 in.) 550–800 g (1¼–1½ lb)	Head becomes even less disproportionate to rest of body. Eyelids separate and eyelashes form. Substantial weight gain occurs. Skin is wrinkled. Type II alveolar cells begin to produce surfactant.
7	32–42 cm (13–17 in.) 1100–1350 g (2½–3 lb)	Head and body are more proportionate. Skin is wrinkled. Seven-month fetus (premature baby) is capable of survival. Fetus assumes an upside-down position. Testes start to descend into scrotum.
8	41–45 cm (16½–18 in.) 2000–2300 g (4½–5 lb)	Subcutaneous fat is deposited. Skin is less wrinkled.
9	50 cm (20 in.) 3200–3400 g (7–7½ lb)	Additional subcutaneous fat accumulates. Lanugo is shed. Nails extend to tips of fingers and maybe even beyond.

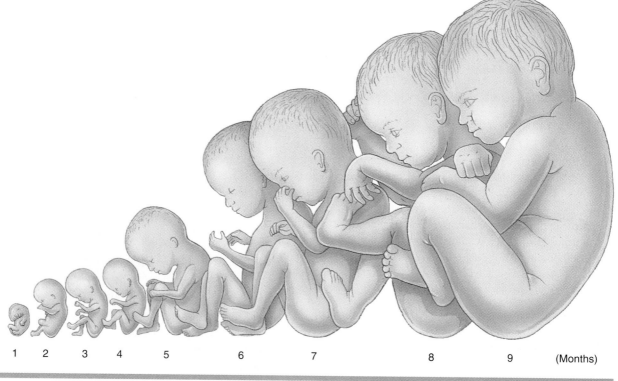

1   2   3   4   5   6   7   8   9   (Months)

Nine weeks later, a healthy baby was born. Since then, surgeons have performed procedures to repair diaphragmatic hernias and are experimenting on animals to try to correct spina bifida (see page 5 of Applications to Health) and hydrocephalus. It has been observed that surgery on fetuses does not leave any scars, although the reason is not known. It is hoped that surgeons can perform craniofacial surgery before birth to correct conditions such as cleft lip without leaving scars. ■

STRUCTURAL AND FUNCTIONAL CHANGES DURING PREGNANCY

The time a zygote, embryo, or fetus is carried in the female reproductive tract is called **gestation** (jes-TĀ-shun; *gestare* = bear). The human gestation period is about 38 weeks, counted from the estimated day of fertilization. The specialized branch of medicine that deals with pregnancy, labor, and the period of time immediately following delivery (about 42 days) is called **obstetrics** (ob-STET-riks; *obstetrix* = midwife).

By about the end of the third month of gestation, the uterus occupies most of the pelvic cavity. As the fetus continues to grow, the uterus extends higher and higher into the abdominal cavity. Toward the end of a full-term pregnancy, the uterus fills nearly all the abdominal cavity, reaching above the costal margin nearly to the xiphoid process of the sternum (Figure 27.9). It pushes the maternal intestines, liver, and stomach superiorly, elevates the diaphragm, and widens the thoracic cavity. Pressure on the stomach may force the stomach contents superiorly into the esophagus, resulting in heartburn. In the pelvic cavity, there is compression of the ureters and urinary bladder.

Besides the anatomical changes associated with pregnancy, there are also pregnancy-induced functional changes. General changes include weight gain due to the fetus, amniotic fluid, placenta, uterine enlargement, and increased total body water. Also, there is increased storage of proteins, triglycerides, and minerals; marked breast enlargement in preparation for lactation; and lower back pain due to lordosis (swayback).

With respect to the maternal cardiovascular system, there is an increase in the volume of blood pumped by the heart due to increased maternal blood flow to the placenta and increased metabolism; an increase in heart rate of about 10–15%; and an increase in blood volume of 30–50%, mostly during the second half of pregnancy. These increases are necessary to meet the additional demands of the fetus for nutrients and oxygen. When a pregnant female is lying on her back, the enlarged uterus may compress the aorta, resulting in diminished blood flow to the uterus. Compression of the inferior vena cava also decreases venous return, which leads to edema in the lower limbs and may produce varicose veins. Compression of the renal artery can lead to renal hypertension.

Pulmonary function is also altered during pregnancy to meet the added demands of the fetus. There is an increase in total body oxygen consumption by about 10–20%. Dyspnea (difficult breathing) also occurs.

With regard to the gastrointestinal tract, there is an increase in appetite and a general decrease in motility that can result in constipation and a delay in gastric emptying time. Nausea, vomiting, and heartburn can also occur.

Pressure on the urinary bladder by the enlarging uterus can produce urinary symptoms, such as increased frequency and urgency and stress incontinence. An increase in renal plasma flow increases renal filtering capacity, which allows faster elimination of the extra wastes produced by the fetus.

Changes in the skin during pregnancy are more apparent in some women than others. Included are increased pigmentation around the eyes and cheekbones in a masklike pattern (chloasma), in the areolae of the breasts, and in the linea alba of the lower abdomen (linea nigra). Striae (stretch marks) over the abdomen can occur as the uterus enlarges, and hair loss also increases.

Changes in the reproductive system include edema and increased vascularity of the vulva and increased pliability and vascularity of the vagina. The uterus increases in weight from its nonpregnant state of 60–80 g to 900–1200 g at term. This increase is due to hyperplasia of muscle fibers (cells) in the myometrium in early pregnancy and hypertrophy of muscle fibers during the second and third trimesters.

Clinical Application

Pregnancy-Induced Hypertension

About 10–15% of all pregnant women in the United States experience **pregnancy-induced hypertension (PIH),** elevated blood pressure associated with pregnancy. The major cause is **preeclampsia** (prē-e-KLAMP-sē-a), an abnormal condition of pregnancy characterized by sudden hypertension, large amounts of protein in the urine, and generalized edema. It typically appears after the 20th week of gestation, and there are large amounts of protein in the urine. Other signs and symptoms are generalized edema, blurred vision, and headaches. It might be related to an autoimmune or allergic reaction due to the presence of a fetus. When the condition is also associated with convulsions and coma, it is termed **eclampsia.** Other forms of PIH are not associated with protein in the urine. ■

LABOR

Labor (*labor* = toil, suffering) is the process by which the fetus is expelled from the uterus through the vagina to the outside. The term **parturition** (par'-too-RISH-un; *parturitio* = childbirth) also means giving birth. The onset of labor is

Figure 27.9 Normal fetal position at the end of a full-term pregnancy.

The gestation period is the time interval (about 38 weeks) from fertilization to birth.

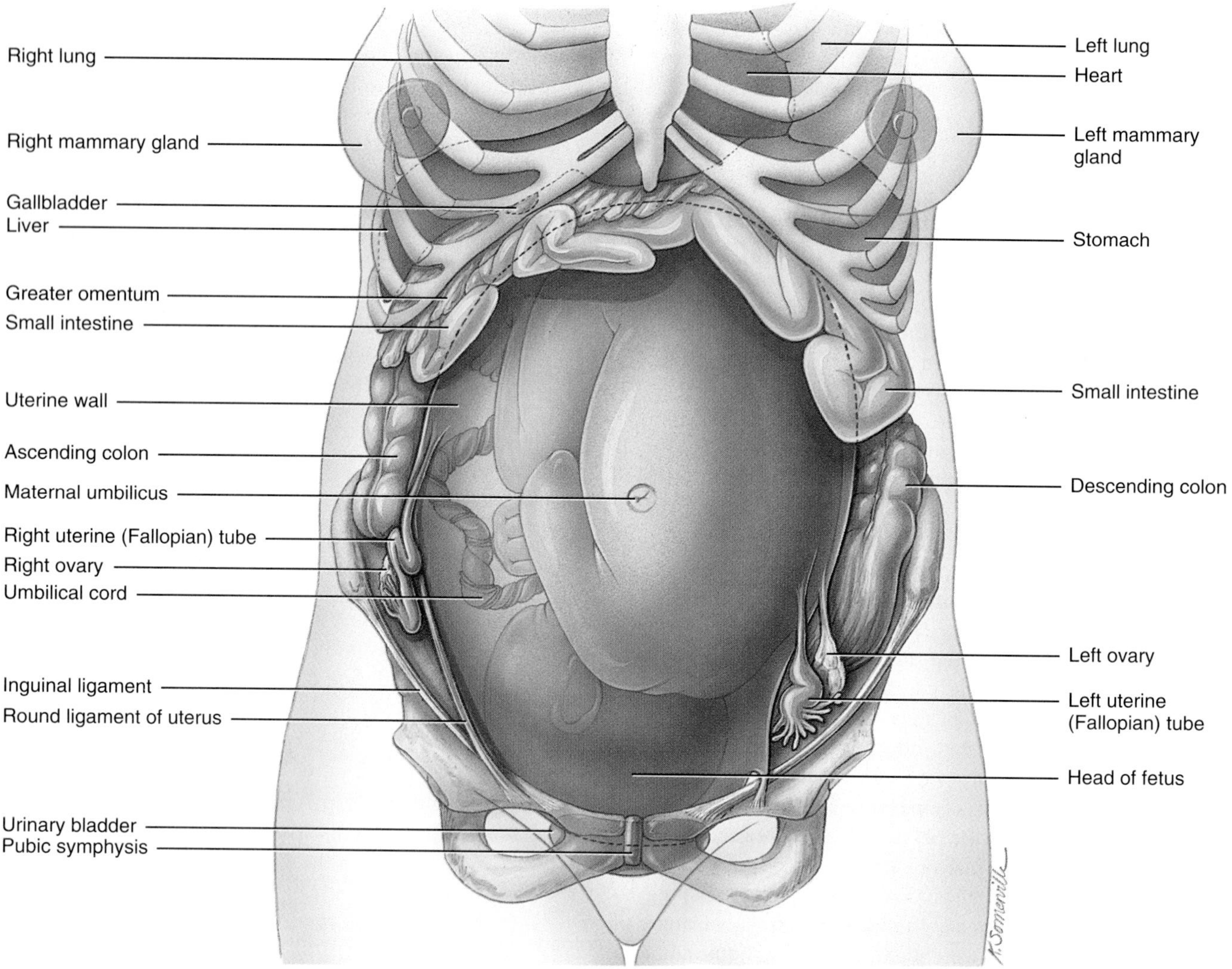

Right lung

Right mammary gland

Gallbladder
Liver

Greater omentum
Small intestine

Uterine wall

Ascending colon

Maternal umbilicus

Right uterine (Fallopian) tube
Right ovary
Umbilical cord

Inguinal ligament
Round ligament of uterus

Urinary bladder
Pubic symphysis

Left lung
Heart

Left mammary
gland

Stomach

Small intestine

Descending colon

Left ovary

Left uterine
(Fallopian) tube

Head of fetus

Anterior view of position of organs at end of full-term pregnancy

Q *What hormone increases the flexibility of the pubic symphysis and helps dilate the cervix of the uterus to ease delivery of the baby?*

related to a complex interaction of many factors. Just before birth, the muscles of the uterus contract rhythmically and forcefully. Both placental and ovarian hormones seem to play a role in these contractions. Because progesterone inhibits uterine contractions, labor cannot take place until its effects are diminished. At the end of gestation, progesterone level falls, the level of estrogens in the mother's blood is sufficient to overcome the inhibiting effects of progesterone, and labor commences. It has been suggested that cortisol released by the fetus overcomes the inhibiting effects

of progesterone so that estrogens can exert their effect. Prostaglandins may also play a role in labor. Oxytocin (OT) from the posterior pituitary gland stimulates uterine contractions, and relaxin assists by relaxing the pubic symphysis and helping to dilate the uterine cervix.

Uterine contractions occur in waves, quite similar to peristaltic waves, that start at the top of the uterus and move downward. These waves expel the fetus. **True labor** begins when uterine contractions occur at regular intervals, usually

Figure 27.10 Stages of true labor.

The term parturition *refers to birth.*

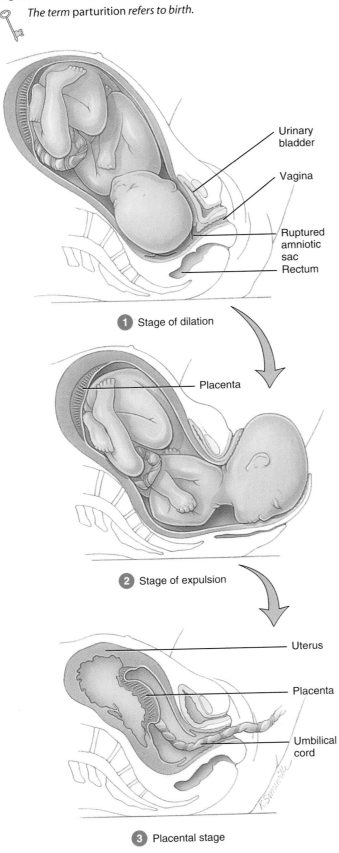

1 Stage of dilation

- Urinary bladder
- Vagina
- Ruptured amniotic sac
- Rectum

2 Stage of expulsion

- Placenta

3 Placental stage

- Uterus
- Placenta
- Umbilical cord

producing pain. As the interval between contractions shortens, the contractions intensify. Another sign of true labor in some females is localization of pain in the back, which is intensified by walking. A reliable indication of true labor is the "show" and dilation of the cervix. The "show" is a discharge of a blood-containing mucus that accumulates in the cervical canal during labor. In **false labor,** pain is felt in the abdomen at irregular intervals. The pain does not intensify and is not altered significantly by walking. There is no "show" and no cervical dilation. True labor can be divided into three stages (Figure 27.10).

1 **Stage of dilation.** The time from the onset of labor to the complete dilation of the cervix is the **stage of dilation.** During this stage, which typically lasts 6–12 hours, there are regular contractions of the uterus, usually a rupturing of the amniotic sac, and complete dilation (10 cm) of the cervix. If the amniotic sac does not rupture spontaneously, it is done deliberately.

2 **Stage of expulsion.** The time (10 minutes to several hours) from complete cervical dilation to delivery of the baby is the **stage of expulsion.**

3 **Placental stage.** The time (5–30 minutes or more) after delivery until the placenta or "afterbirth" is expelled by powerful uterine contractions is the **placental stage.** These contractions also constrict blood vessels that were torn during delivery. In this way, the chance of hemorrhage is reduced.

As a rule, labor lasts longer with first babies (typically 14 hours). In a female who has previously given birth, an average time for labor is 8 hours, although there is considerable variation from person to person. During labor, the fetus may be squeezed through the birth canal (cervix and vagina) for up to several hours. As a result, the fetal head is compressed, and there is some degree of intermittent hypoxia due to compression of the umbilical cord and placenta during uterine contractions. Thus the fetus is stressed during and by the birth process. In response to this stress, the fetal adrenal medullae secrete very high levels of epinephrine and norepinephrine, the "fight-or-flight" hormones. Much of the protection afforded against the stresses of the birth process and preparation of the infant to survive extrauterine life are provided by the adrenal medullary hormones. Among other functions, the hormones clear the lungs and alter their physiology for breathing outside the uterus, mobilize readily usable nutrients for cellular metabolism, and promote an increased blood flow to the brain and heart.

About 7% of pregnant females have not delivered by 2 weeks after their due date. In such cases, there is increased risk of brain damage to the fetus and even fetal death. This is

Q *What event marks the beginning of the stage of expulsion?*

due to inadequate supply of oxygen and nutrients from an aging placenta. Post-term deliveries may be facilitated by induced labor (initiated by artificial means such as the administration of oxytocin) or surgical delivery.

After delivery of the baby and placenta, there is a period of time that lasts about 6 weeks during which the maternal reproductive organs and physiology return to the prepregnancy state. This is called the **puerperium** (pyoo'-er-PE-rē-um). Through a process of tissue catabolism, the uterus undergoes a remarkable reduction in size (especially in lactating females), called **involution** (in'-vō-LOO-shun). It is related primarily to a decrease in cytoplasm and cell size of myometrial cells. The cervix loses its elasticity and regains its prepregnancy firmness. For 2–4 weeks after delivery, there is a uterine discharge called **lochia** (LŌ-kē-a), which consists initially of blood and later serous fluid derived from the former placental site.

Clinical Application

Dystocia and Cesarean Section

Dystocia (dis-TŌ-sē-a; *dys* = difficult; *tokos* = birth), or difficult labor, may result from abnormal position (presentation) of the fetus, or a birth canal of inadequate size to permit vaginal birth. For example, in a **breech presentation** the fetal buttocks or lower limbs, rather than the head, enter the birth canal first; this occurs most often in premature births. If fetal or maternal distress prevents a vaginal birth, the baby may be delivered surgically through an abdominal incision termed a **cesarean section (C-section).** A low, horizontal cut is made through the abdominal wall and lower portion of the uterus, through which the baby and placenta are removed. ■

PRENATAL DIAGNOSTIC TESTS

Several tests are available to detect genetic disorders and assess fetal well-being. Here we will describe fetal ultrasonography, amniocentesis, and chorionic villi sampling (CVS).

Fetal Ultrasonography

If there is a question about the normal progress of a pregnancy, **fetal ultrasonography** (ul'-tra-son-OG-ra-fē) may be performed. By far the most common use of diagnostic ultrasound is to determine true fetal age when the date of conception is uncertain. It is also used to evaluate fetal viability and growth, determine fetal position, ascertain multiple pregnancies, identify fetal–maternal abnormalities, and serve as an adjunct to special procedures such as amniocentesis. Ultrasound is not used routinely to determine the gender of a fetus; it is performed only for a specific medical indication.

An instrument (transducer) that emits high-frequency sound waves is passed back and forth over the abdomen. The reflected sound waves from the developing fetus are picked up by the transducer and converted to an image on a screen. This image is called a **sonogram** (see Table 1.5 on page 23). Because the urinary bladder serves as a landmark during the procedure, the patient needs to drink liquids and not void to maintain a full bladder.

Amniocentesis

Amniocentesis (am'-nē-ō-sen-TĒ-sis; *amnio* = amnion; *kentesis* = puncture) involves withdrawing some of the amniotic fluid that bathes the developing fetus and analyzing the fetal cells and dissolved substances. It is used to test for the presence of certain genetic disorders, such as Down syndrome (DS), spina bifida, hemophilia, Tay–Sachs disease, sickle-cell anemia, and certain muscular dystrophies, or to determine fetal maturity and well-being near the time of delivery. To detect suspected genetic abnormalities, the test is usually done at 14–16 weeks of gestation. To assess fetal maturity, it is usually done after the 35th week of gestation. About 300 chromosomal disorders and over 50 biochemical defects can be detected through amniocentesis. It can also reveal gender. This information is important for diagnosis of sex-linked disorders, in which an abnormal gene is carried by the mother but affects only her male offspring. If the fetus is female, it will not be afflicted unless the father also carries the defective gene.

During amniocentesis, the position of the fetus and placenta is first determined using ultrasound and palpation. After the skin is prepared with an antiseptic, a local anesthetic is given, a hypodermic needle is inserted through the mother's abdominal wall and uterus into the amniotic cavity, and about 10 ml of fluid are aspirated (Figure 27.11a). The fluid and suspended cells are subjected to microscopic examination and biochemical testing. Elevated levels of alphafetoprotein (AFP) and acetylcholinesterase may indicate failure of the nervous system to develop properly, for example, anencephaly (absence of the cerebrum) or spina bifida. Chromosome studies, which require growing the cells for 2–4 weeks in a culture medium, may reveal rearranged, missing, or extra chromosomes. There is about a 0.5% chance of spontaneous abortion after the test.

Chorionic Villi Sampling

Chorionic (ko-rē-ON-ik) **villi** (VIL-ī) **sampling (CVS)** can determine the same defects as amniocentesis because chorion cells and fetal cells contain the same genome. Moreover, CVS offers several advantages over amniocentesis. It can be performed as early as 8 weeks of gestation and test

Figure 27.11 Amniocentesis and chorionic villi sampling.

To detect genetic abnormalities, amniocentesis is performed at 14–16 weeks of gestation, whereas chorionic villi sampling may be performed as early as 8 weeks of gestation.

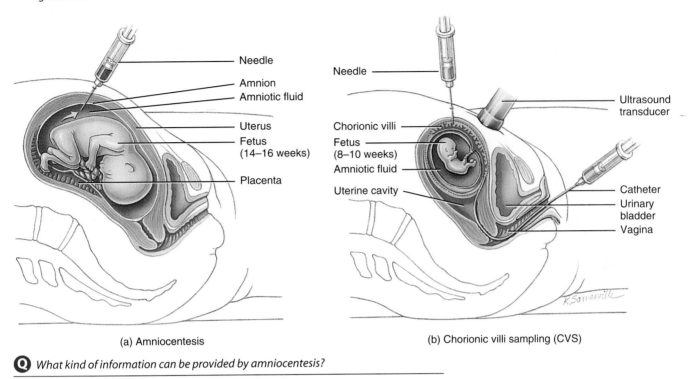

Needle
Amnion
Amniotic fluid
Uterus
Fetus (14–16 weeks)
Placenta

Needle
Chorionic villi
Fetus (8–10 weeks)
Amniotic fluid
Uterine cavity
Ultrasound transducer
Catheter
Urinary bladder
Vagina

(a) Amniocentesis

(b) Chorionic villi sampling (CVS)

Q *What kind of information can be provided by amniocentesis?*

results are available in a few days, which permit an earlier decision on whether or not to continue the pregnancy. In addition, the procedure does not require penetration of the abdomen, uterus, or amniotic cavity by a needle. However, the procedure is slightly riskier than amniocentesis; there is a 1–2% chance of spontaneous abortion after the test.

During CVS, a catheter is placed through the vagina and cervix of the uterus and then advanced to the chorionic villi under ultrasound guidance (Figure 27.11b). About 30 mg of tissue are suctioned out and prepared for chromosomal analysis. Alternatively, the chorionic villi can be sampled by inserting a needle through the abdominal cavity, as for amniocentesis.

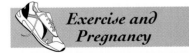
Exercise and Pregnancy

Because pregnancy results in so many major body changes, it has an impact on a female's ability to exercise. In early pregnancy, there are only a few changes that affect exercise. A pregnant woman may tire earlier than usual, or she may become lethargic. Morning sickness may also curtail regular exercise. As the pregnancy progresses, weight is gained and

posture changes. As a result, more energy is needed to perform activities, and certain maneuvers (sudden stopping, changes in direction, rapid movements) are difficult to execute. In addition, certain joints, especially the pubic symphysis, become less stable in response to the increased level of the hormone relaxin. As compensation, many mothers-to-be walk with widely spread legs and a shuffling motion.

Although during exercise blood shifts from viscera (including the uterus) to the muscles and skin, there is no evidence of inadequate blood flow to the placenta. The heat generated during exercise may cause dehydration and further increase body temperature. During early pregnancy, especially, excessive exercise and heat buildup should be avoided because elevated body temperature has been implicated in neural tube defects. Exercise has no known effect on lactation, provided the female remains hydrated and wears a bra with good support. Moderate physical activity does not endanger the fetuses of healthy females who have a normal pregnancy.

Among the benefits of exercise during pregnancy are improvement in oxygen capacity, greater sense of well-being, and fewer minor complaints.

Key Medical Terms
Associated with Developmental Anatomy

Abortion (a-BOR-shun) Premature expulsion from the uterus of the products of conception—embroyo or nonviable fetus. May be caused by abnormal development of the embryo, placental abnormalities, endocrine disturbances, certain diseases, trauma, and stress. **Induced (nonspontaneous) abortions** are brought on intentionally by methods such as vacuum aspiration (suction curettage), up to the 12th week of pregnancy; dilation and evacuation (D & E), commonly performed between the 13th and 15th weeks of pregnancy and sometimes up to 20 weeks; and use of saline or prostaglandin preparations to induce labor and delivery, usually after the 15th week of pregnancy. **Spontaneous abortions (miscarriages)** occur without apparent cause. Recent evidence suggests that as many as one-third of all successful fertilizations end in spontaneous abortions, most of which occur even before a female or her physician is aware that she is pregnant.

Breech presentation A malpresentation in which the fetal buttocks or lower limbs present into the maternal pelvis; most common cause is prematurity.

Karyotype (KAR-ē-ō-tīp; *karyon* = nucleus) The chromosomal elements typical of a cell, drawn in their true proportions, based on the average of measurements determined in a number of cells. Useful in judging whether or not chromosomes are normal in number and structure.

Lethal gene (LĒ-thal jēn; *lethum* = death) A gene that, when expressed, results in death either in the embryonic state or shortly after birth.

Mutation (myoo-TĀ-shun; *mutare* = change) A permanent heritable change in a gene that causes it to have a different effect than it had previously.

Puerperal (pyoo-ER-per-al; *puer* = child, *parere* = to bring forth) **fever** Infectious disease of childbirth, also called puerperal sepsis and childbed fever. The disease results from an infection originating in the birth canal and affects the endometrium. It may spread to other pelvic structures and lead to septicemia.

Teratogen (ter-AT-ō-jen; *teratos* = monster; *gennan* = to produce) Any agent or influence that causes developmental defects in an embryo and results in a severely deformed fetus. Examples of teratogens include alcohol, LSD, marijuana, cocaine, x-rays, antibiotics, pesticides, and diethylstilbestrol (DES).

Study Outline

Development During Pregnancy (p. 811)

1. Pregnancy is a sequence of events that starts with fertilization.
2. Its various events are hormonally controlled.
3. Fertilization refers to the penetration of a secondary oocyte by a sperm cell to form a zygote and the subsequent union of the sperm and oocyte pronuclei.
4. Penetration is facilitated by enzymes produced by sperm acrosomes.
5. Normally, only one sperm fertilizes a secondary oocyte.
6. Early rapid cell division of a zygote is called cleavage, and the cells produced by cleavage are called blastomeres.
7. The solid mass of cells produced by cleavage is a morula.
8. The morula develops into a blastocyst, a hollow ball of cells differentiated into a trophoblast (future embryonic membranes) and inner cell mass (future embryo).
9. The attachment of a blastocyst to the endometrium is called implantation; it occurs by enzymatic degradation of the endometrium.

Embryonic Development (p. 815)

1. During embryonic growth, the primary germ layers and embryonic membranes are formed.
2. The primary germ layers—ectoderm, mesoderm, and endoderm—form all tissues of the developing organism.
3. Embryonic membranes include the yolk sac, amnion, chorion, and allantois.
4. Fetal and maternal materials are exchanged through the placenta.

Fetal Growth (p. 821)

1. During the fetal period, organs established by the primary germ layers grow rapidly.
2. The principal changes associated with fetal growth are summarized in Table 27.2.

Structural and Functional Changes During Pregnancy (p. 823)

1. The time an embryo or fetus is carried in the uterus is called gestation.
2. Human gestation lasts about 38 weeks, counted from the estimated day of fertilization.
3. During gestation, several structural and functional changes occur in the mother.

Labor (p. 823)

1. True labor involves dilation of the cervix, expulsion of the fetus, and delivery of the placenta.
2. Parturition refers to birth.

Prenatal Diagnostic Tests (p. 826)

1. In fetal ultrasonography, an image of a fetus is displayed on a screen.
2. Amniocentesis is the withdrawal of amniotic fluid. It can be used to diagnose inherited biochemical defects and chromosomal disorders, such as hemophilia, Tay–Sachs disease, sickle-cell anemia, and Down syndrome.
3. Chorionic villi sampling involves withdrawal of chorionic villi for chromosomal analysis. It can be done earlier than amniocentesis, and the results are available more quickly, but it is also riskier.

Exercise and Pregnancy (p. 827)

1. Moderate physical activity does not endanger the fetus of a healthy female who has a normal pregnancy.

2. Benefits of exercise during pregnancy include increased oxygen capacity, greater sense of well-being, and fewer minor complaints.

Review Questions

1. Define developmental anatomy. (p. 811)

2. Define fertilization. Where does it normally occur? How is a morula formed? (p. 811)

3. Explain how dizygotic (fraternal) and monozygotic (identical) twins are produced. (p. 811)

4. Describe the components of a blastocyst. (p. 812)

5. What is implantation? How does the fertilized ovum implant itself? Why is an implanted fertilized ovum usually not rejected by the mother? (p. 813)

6. Define the embryonic period and the fetal period. (p. 811)

7. List several body structures formed by the endoderm, mesoderm, and ectoderm. (p. 818)

8. What is an embryonic membrane? Describe the functions of the four embryonic membranes. (p. 817)

9. Explain the importance of the placenta and umbilical cord to fetal growth. (p. 819)

10. Outline some of the major developmental changes during fetal growth. (p. 822)

11. Define gestation and parturition. (p. 823)

12. Describe several structural and functional changes that occur in the mother during gestation. (p. 823)

13. Distinguish between false and true labor. Describe what happens during the state of dilation, the stage of expulsion, and the placental stage of delivery. (p. 823)

14. Explain the procedure and diagnostic value of the following: fetal ultrasonography, amniocentesis, and chorionic villi sampling (p. 826).

15. Explain the effects of pregnancy on exercise and exercise on pregnancy. (p. 827)

16. Refer to the glossary of key medical terms associated with development. Be sure that you can define each term. (p. 828)

Self Quiz

1. Match the following terms with their descriptions:

(1) yolk sac
(2) amnion
(3) chorion
(4) allantois

___ (a) originally formed from ectoderm; encloses fluid that acts as a shock absorber for the developing baby

___ (b) derived from mesoderm and trophoblast; becomes the principal embryonic part of the placenta

___ (c) serves as an early site of blood cell formation (two answers)

___ (d) a small membrane that forms umbilical blood vessels

___ (e) the "bag of waters" that ruptures just before birth

___ (f) produces human chorionic gonadotropin (hCG)

Complete the following:

2. Successive divisions of a zygote produce a solid mass of cells called the ___.

3. The early rapid division of a zygote is called ___. It results in the production of smaller cells called ___.

4. In a developing embryo, the layer of cells of the inner cell mass near the amniotic cavity develops into ___, whereas the layer of the inner cell mass that borders the blastocele develops into ___.

5. The inner cell mass of the implanted blastocyst will form the ___.

6. The embryonic tissues from which all tissues and organs develop are called ___ layers.

7. Place numbers in the blanks to arrange the following in correct sequence.

(a) morula: ___; (b) ovum: ___; (c) blastocyst: ___; (d) zygote: ___; (e) fetus: ___.

8. Place numbers in the blanks to arrange the following events in the correct sequence.

(a) the stage of expulsion, from complete cervical dilation through delivery of the baby: ___; (b) the placental stage, the time after the delivery of the baby until the placenta ("afterbirth") is expelled: ___; (c) the stage of dilation, from the onset of labor to complete dilation of the cervix: ___.

Choose the one best answer to the following questions:

9. Which of the following four statements are true?

(1) Capacitation occurs in the male reproductive tract.
(2) The acrosome produces enzymes that help a sperm cell penetrate a secondary oocyte.
(3) A segmentation nucleus contains the diploid ($2n$) number of chromosomes.
(4) The zona pellucida is a gelatinous covering around the acrosome of a sperm cell.

a. 1, 2, and 3
b. 2, 3, and 4
c. 2 and 3
d. 1 and 3
e. 1, 2, 3, and 4

10. How long after fertilization does implantation of an embryo usually occur?
    a. 3 weeks
    b. 1 week
    c. 1 day
    d. 3 days
    e. 30 minutes
11. Dizygotic (fraternal) twins result from
    a. two secondary oocytes and two sperm
    b. two secondary oocytes and one sperm cell
    c. one secondary oocyte and two sperm
    d. one secondary oocyte and one sperm cell
12. Which of the following four statements are correct concerning maternal changes during pregnancy?
    (1) The uterus extends superiorly into the abdominal cavity.
    (2) Increased heart rate occurs.
    (3) Total oxygen consumption decreases.
    (4) Increased frequency and urgency of urination may occur.
    a. 1, 2, and 4
    b. 1, 2, and 3
    c. 2, 3, and 4
    d. 2 and 3
    e. 1, 2, 3, and 4
13. The stage of development that becomes attached to the lining of the uterus is the
    a. fetus
    b. morula
    c. zygote
    d. blastomere
    e. blastocyst
14. The placenta is formed by the union of the decidua basalis of the endometrium with the
    a. yolk sac
    b. amnion
    c. chorion
    d. umbilicus
    e. allantois
15. Oxygen and nutrients from the maternal blood must pass through which of the following structures before entering the fetal blood?
    a. umbilical vein
    b. umbilical artery
    c. decidua capsularis
    d. intervillous spaces
    e. allantois

16. Match the following:
    ___ (a) epithelial lining of all gastrointestinal, respiratory, and genitourinary tracts except near openings to the exterior of the body
    ___ (b) epidermis of skin, epithelial lining of entrances to the body (such as mouth, nose, and anus), hair, nails
    ___ (c) all of the skeletal system (bone, cartilage, joint cavities)
    ___ (d) muscle (skeletal, most smooth, and cardiac)
    ___ (e) blood and all blood and lymphatic vessels
    ___ (f) entire nervous system
    ___ (g) thyroid, parathyroid, thymus, and pancreas

    (1) develop(s) from ectoderm
    (2) develop(s) from endoderm
    (3) develop(s) from mesoderm

Are the following statements true or false?

17. The enzyme acrosin stimulates the movement and migration of sperm in the female reproductive tract.
18. The fusion of the male and female pronuclei is known as syngamy.
19. The first two months of development are called the embryonic period.
20. The primary germ layers that give rise to all tissues and organs are present at the time of implantation.
21. The umbilical cord contains two umbilical arteries that carry oxygenated blood to the fetus, and one umbilical vein that carries fetal blood to the placenta.
22. Match the following:
    ___ (a) portion of endometrium that initially separates the embryo from the uterine cavity
    ___ (b) portion of the modified endometrium that lines the region of the uterus not involved with the implanted embryo
    ___ (c) layer of the endometrium between the chorion and the stratum basalis
    ___ (d) fuses with the decidua parietalis, thus obliterating the uterine cavity
    ___ (e) becomes the maternal portion of the placenta
    ___ (f) layer of decidua that disappears by about 27 weeks

    (1) decidua basalis
    (2) decidua parietalis
    (3) decidua capsularis

Critical Thinking Questions

1. Your neighbors down the street put up a sign to announce the birth of their twins—a girl and a boy. Your next-door neighbor said, "Oh, how sweet! I wonder if they're identical?" Without even seeing the twins, what can you tell her?
   HINT: *Would you expect the parents to dress them in identical pink or blue outfits?*

2. Some nervous system disorders may also cause particular skin problems. For example, a person with a type of nervous system tumor may show coffee-colored spots on the skin. How does the structure of the nervous system relate to that of the skin?
   HINT: *There are four main tissue types in adults, but only three primary germ layers in the embryo.*

3. Selena, in the last two weeks of her first pregnancy, anxiously called her doctor to ask if she should leave for the hospital. She was experiencing irregular "labor" pains that fortunately were eased by walking. She had no other signs to report. The doctor told Selena to stay home—it wasn't time yet. Why did the doctor tell her to stay home?
   HINT: *There's a saying in showbiz: "Another opening, another show."*

4. It is said that "all life began in the sea." How do fetal conditions reflect this statement?
   HINT: *One sign that birth is imminent occurs when the "waters" break.*

5. Tony couldn't help laughing after the last Human Anatomy examination. "Did you see that true-or-false question about yolk? It must have been a mistake! This isn't Poultry 101!" Actually, it was a mistake for Tony to have spent all night watching cable TV. What does yolk have to do with human development?
   HINT: *The yolk is surrounded by a membrane.*

6. The *Daily Inquiring Minds News* reported, "Baby born without belly button. Mom blames alien abduction." What is the origin of the belly button?
   HINT: *Maybe aliens hatch from an egg, but human fetuses are very attached to their mothers.*

Answers to Figure Questions

27.1 Functional changes experienced by sperm after they have been deposited in the female's vagina that allow them to fertilize a secondary oocyte.

27.2 Morula is a solid ball of cells; blastocyst is the rim of cells surrounding a cavity (blastocele) plus an inner cell mass.

27.3 It secretes digestive enzymes that eat away the endometrial lining at the site of implantation.

27.4 Inner cell mass.

27.5 Gastrulation.

27.6 The amnion functions as a shock absorber, whereas the chorion forms the placenta.

27.7 Decidua basalis.

27.8 Exchange of materials between fetus and mother.

27.9 Relaxin.

27.10 Complete dilation of cervix.

27.11 It is used mainly to detect genetic disorders but also gives evidence of the maturity (and survivability) of the fetus.

Appendix I Measurements

Metric Units of Length and Some U.S. Equivalents

Metric Unit	Meaning of Prefix	Metric Equivalent	U.S. Equivalent
1 kilometer (km)	kilo = 1,000	$1,000m = 10^3m$	3,280.84 ft = 0.62 mi
			1 mi = 1.61 km
1 hectometer (hm)	hecto = 100	$100m = 10^2m$	328 ft
1 dekameter (dam)	deka = 10	$10m = 10^1m$	32.8 ft
1 meter (m)		Standard unit of length	39.37 in. = 3.28 ft = 1.09 yd
1 decimeter (dm)	deci = 1/10	$0.1m = 10^{-1}m$	3.94 in.
1 centimeter (cm)	centi = 1/100	$0.01 = 10^{-2}m$	0.394 in.
			1 in. = 2.54 cm
1 millimeter (mm)	milli = 1/1,000	$0.001 = 1/10$ cm	0.0394 in.
		$= 10^{-3}m$	
1 micrometer (μm)	micro = 1/1,000,000	$0.000,001m = 1/10,000$ cm	3.94×10^{-5} in.
[formerly micron (μ)]		$= 10^{-6}$ m	
1 nanometer (nm)	nano = 1/1,000,000,000	$0.000,000,001m = 1/10,000,000$ cm	3.94×10^{-8} in.
[formerly millimicron (mμ)]		$= 10^{-9}$ m	

Metric Units of Volume and Some U.S. Equivalents

Metric Unit	Metric Equivalent	U.S. Equivalent
1 liter (l)	1,000 ml	33.81 fl oz = 1.057 qt
		946 ml = 1 qt
1 millimeter (ml)	0.001 liter	30 ml = 1 fl oz
		5 ml = 1 teaspoon
1 cubic centimeter (cm^3)	1.0 ml	0.0338 fl oz

Metric Units of Mass and Some U.S. Equivalents

Metric Unit	Metric Equivalent	U.S. Equivalent
1 kilogram (kg)	1,000 g	2,205 lb
		1 ton
		= 907 kg
1 hectogram (hg)	100 g	
1 dekagram (dag)	10 g	0.0353 oz
1 gram (g)	1 g	1 lb
		= 453.6 g
		1 oz
		= 28.35 g
1 decigram (dg)	0.1 g	
1 centigram (cg)	0.01 g	
1 milligram (mg)	0.001 g	0.115 g
1 microgram (μg)	0.000,001 g	
1 nanogram (ng)	0.000,000,001 g	
1 picogram (pg)	0.000,000,000,001 g	

Appendix II Answers

ANSWERS TO SELF QUIZZES

Chapter 1
1. epithelia, connective, muscular, nervous **2.** pleura **3.** hypochondriac, lumbar, iliac **4.** frontal or coronal **5.** organs **6.** ipsi **7.** F **8.** F **9.** T **10.** F **11.** T **12.** T **13.** a3, b6, c5, d1, e2, f8, g7, h4 **14.** a **15.** e **16.** b **17.** b **18.** c **19.** b **20.** e

Chapter 2
1. a9, b11, c3, d6, e5, f12, g8, h7, i10, j1, k4, l2 **2.** phagocytosis **3.** phospholipids **4.** cytology **5.** selectively permeable **6.** c **7.** d **8.** a **9.** a **10.** b **11.** c **12.** a **13.** e **14.** d **15.** b **16.** F **17.** F **18.** T **19.** T **20.** a1, b4, c5, d3, e2

Chapter 3
1. b **2.** d **3.** b **4.** e **5.** b **6.** e **7.** d **8.** a **9.** b **10.** d **11.** T **12.** T **13.** F **14.** a7, b8, c5, d9, e1, f2, g6, h3, i4 **15.** osseous, lamellae **16.** exocrine **17.** vascular, plasma **18.** mesenchyme **19.** lamina propria **20.** a3, b5, c1, d6, e4, f2

Chapter 4
1. ceruminous glands, cerumen **2.** dermis, connective **3.** ecto, fourth, meso **4.** sudoriferous **5.** T **6.** F **7.** T **8.** F **9.** a4, b1, c6, d2, e3, f5 **10.** c **11.** b **12.** e **13.** a **14.** a3, b1, c5, d4, e2

Chapter 5
1. endochondral **2.** collagen **3.** matrix **4.** meso, mesenchyme **5.** greenstick **6.** a1, b4, c3, d5, e2 **7.** e **8.** c **9.** b **10.** e **11.** b **12.** e **13.** F **14.** F **15.** T **16.** F **17.** T **18.** T **19.** T

Chapter 6
1. a12, b1, c7, d14, e6, f8, g13, h5, i4, j2, k9, l10, m11, n3 **2.** T **3.** F **4.** F **5.** T **6.** F **7.** b **8.** b **9.** c **10.** c **11.** d **12.** a **13.** d **14.** a **15.** b **16.** d **17.** a **18.** e **19.** c **20.** b **21.** a1, b1, c3, d5, e3, f5, g2, h4

Chapter 7
1. subscapular fossa **2.** the head of metacarpal II **3.** posterior, femur **4.** false **5.** b **6.** e **7.** d **8.** b **9.** e **10.** c **11.** d **12.** b **13.** c **14.** e **15.** b **16.** F **17.** T **18.** T **19.** T **20.** a7, b5, c10, d8, e4, f1, g9, h2, i6, j3

Chapter 8
1. amphiarthrosis **2.** synarthrosis **3.** diarthrosis **4.** articular or hyaline cartilage **5.** bursae **6.** a3, b5, c7, d2, e4, f8, g9, h1, i6 **7.** T **8.** T **9.** F **10.** F **11.** F **12.** T **13.** a5, b3, c1, d4, e2 **14.** b **15.** d **16.** a **17.** d **18.** e **19.** a **20.** d **21.** e **22.** a **23.** a4, b5, c3, d1, e6, f2

Chapter 9
1. myoglobin **2.** triad **3.** fiber, sarcolemma, sarcoplasm **4.** s,u,u,s,s **5.** T **6.** F **7.** F **8.** F **9.** T **10.** a3, b1, c2, d3, e1, f3, g2 **11.** b **12.** b **13.** e **14.** c **15.** e **16.** c **17.** c **18.** d **19.** a2, b5, c1, d3, e4

Chapter 10
1. a3, b2, c1, d3, e1, f2, g3 **2.** a2 and 3, b3 and 6, c4 and 3, d5, e1 and 6 **3.** lever, fulcrum **4.** insertion **5.** prime mover (agonist), synergists **6.** parallel, fusiform, pennate, circular, triangular **7.** sphincters **8.** a platysma, b orbicularis oculi, c buccinator, d frontalis **9.** inferior oblique **10.** digastric **11.** mastoid process **12.** sternocleidomastoid **13.** quadratus lumborum, iliacus, psoas major **14.** central **15.** urogenital **16.** trapezius, rhomboideus major, rhomboideus minor **17.** triceps brachii **18.** thenar, hypothenar **19.** erector spinae, lumbar **20.** gluteus medius **21.** anterior, four, tibia, extend **22.** ischial tuberosity, extend, flex **23.** gastrocnemius, soleus **24.** plantar **25.** a4, b7, c1, d3, e6, f5, g2 **26.** T **27.** T **28.** F **29.** F **30.** T **31.** d **32.** b **33.** b **34.** e **35.** a **36.** d **37.** e **38.** d **39.** b **40.** a3, b6, c1, d11, e8, f2, g10, h5, i7, j9, k4

Chapter 11
1. facial, cranial **2.** sternocleidomastoid **3.** brachial **4.** a anterior b superior c distal d superior e anterior f lateral **5.** a5, b7, c2, d4, e1, f3, g6 **6.** T **7.** T **8.** T **9.** T **10.** T **11.** F **12.** F **13.** F **14.** a5, b1, c4, d7, e3, f2, g6 **15.** e **16.** a **17.** a4, b3, c7, d8, e2, f1, g6, h5

Chapter 12
1. albumins, globulins, fibrinogen **2.** differential white blood cell **3.** 55, 45 **4.** erythropoiesis **5.** hematology **6.** b **7.** b **8.** a **9.** a **10.** d **11.** e **12.** b **13.** c **14.** e **15.** c **16.** a **17.** e **18.** b **19.** e **20.** T **21.** T **22.** T **23.** F **24.** a5, b3, c3, d1, e2, f4, g4 and 5

Chapter 13
1. meso, first, primitive heart **2.** gap junctions, desmosomes **3.** atrioventricular **4.** SA node **5.** atria, ventricles **6.** right coronary artery, left coronary artery **7.** bicuspid **8.** open, closed **9.** a3, b4, c2, d1 **10.** T **11.** T **12.** F **13.** T **14.** F **15.** F **16.** T **17.** T **18.** F **19.** T **20.** T **21.** F **22.** T **23.** T **24.** a **25.** e **26.** d **27.** d **28.** a **29.** a

Chapter 14
1. aorta, left **2.** fourth lumbar, common iliac **3.** hepatic portal **4.** anastomosis **5.** media **6.** vasoconstriction **7.** common hepatic artery, left gastric artery, splenic artery **8.** internal jugular **9.** a10,

833

b7, c8, d2, e11, f1, g5, h3, i9, j6, k4 **10.** T **11.** F **12.** T **13.** T **14.** F **15.** c **16.** c **17.** c **18.** b **19.** a, c **20.** e **21.** c **22.** e **23.** b **24.** e **25.** d **26.** a

Chapter 15

1. red bone marrow, thymus gland **2.** cysterna chyli **3.** meso **4.** MALT (mucous-associated lymphatic tissue) **5.** red pulp, white pulp **6.** T **7.** T **8.** T **9.** F **10.** T **11.** T **12.** a4, b7, c1, d8, e2, f6, g5, h3 **13.** e **14.** c **15.** c **16.** e **17.** d **18.** d **19.** e **20.** d **21.** e

Chapter 16

1. spinal, cranial **2.** central **3.** dendrite **4.** synapse **5.** synaptic vesicles **6.** diverging **7.** a3, b6, c7, d8, e4, f5, g1, h2 **8.** c **9.** c **10.** c **11.** d **12.** e **13.** a3, b1, c1, d2, e3, f2, g1 **14.** F **15.** T **16.** T **17.** T **18.** F **19.** F

Chapter 17

1. medulla oblongata, second lumbar vertebra **2.** pia mater, arachnoid, dura mater **3.** endo, peri, epi **4.** subdural, subarachnoid **5.** anterior, lateral **6.** 31, 8, 12, 5, 5, 1 **7.** a2, b6, c5, d1, e3, f4, **8.** c **9.** a **10.** d **11.** b **12.** c **13.** b **14.** c **15.** T **16.** F **17.** F **18.** F **19.** T **20.** T **21.** T **22.** a1, b2, c1, d5, e2, f4, g3

Chapter 18

1. a2, b5, c4, d2, e5, f3, g1 **2.** ecto, neural **3.** tentorium cerebelli **4.** cuneatus **5.** chiasm **6.** inferior **7.** basal ganglia, cerebrum **8.** association **9.** cortex, gray, gyri, sulci **10.** wgggwg **11.** a accessory, b vestibulocochlear, c vagus, d trigeminal, e trigeminal, f oculomotor, g facial, h facial and glossopharyngeal, i olfactory, optic, and vestibulocohlear **12.** bac **13.** cab **14.** acfbde **15.** idgecfbhaj **16.** F **17.** F **18.** T **19.** F **20.** T **21.** T **22.** a7, b6, c3, d4, e2, f1, g5, h2, i5, j1, k4, l3 **23.** c **24.** c **25.** a **26.** c **27.** b

Chapter 19

1. selectivity **2.** exteroceptors, interoceptors (visceroceptors), proprioceptors **3.** mechano **4.** brain stem, association neurons **5.** gdbfeahc **6.** bcaed **7.** a3, b1, c2, d1 **8.** a4, b1, c3, d2 **9.** F **10.** T **11.** F **12.** T **13.** a2, b3, c1, d1, e3, f2 **14.** c **15.** a **16.** c **17.** a **18.** b **19.** a4, b3, c1, d2, e6, f5

Chapter 20

1. olfactory receptors, supporting cells, basal cells **2.** papillae, fungiform **3.** ophthalmology **4.** otoliths, vestibular apparatus **5.** ciliary, scleral venous **6.** sclera, choroid, retina **7.** conjunctiva **8.** pigment **9.** bcade **10.** afcdbe **11.** cabgedf **12.** a1, b3, c1, d2 **13.** c **14.** b **15.** a **16.** b **17.** b **18.** e **19.** d **20.** a11, b6, c7, d1, e3, f10, g2, h9, i5, j4, k8, l12 **21.** T **22.** F **23.** T **24.** F

Chapter 21

1. general visceral sensory, general visceral motor **2.** dual **3.** epinephrine, norepinephrine **4.** alpha, beta **5.** nicotinic, muscarinic **6.** parasympathetic **7.** a1, b2, c1, d2, e1, f1, g1, h1, i3, j1, **8.** b **9.** a **10.** e **11.** d **12.** a **13.** e **14.** c **15.** b **16.** c **17.** F **18.** T **19.** T **20.** F **21.** F

Chapter 22

1. cbad **2.** feab **3.** tropic, target **4.** alpha, increase **5.** infundibulum **6.** neurohypophysis **7.** parafollicular or C, principal or chief **8.** a4, b3, c2, d6, e5, f5, g1 **9.** T **10.** F **11.** F **12.** F **13.** T **14.** a **15.** c **16.** a **17.** c **18.** d **19.** e **20.** c **21.** e

Chapter 23

1. base, clavicle, ribs **2.** epiglottis **3.** endo **4.** bronchopulmonary segment **5.** cba **6.** jkdbmioeacnfhgl **7.** a1, b2, c2, d1, e1 **8.** a **9.** b

10. a **11.** d **12.** b **13.** c **14.** d **15.** a1, b5, c2, d3, e4, f4, g2,3, h2 **16.** F **17.** T **18.** T **19.** F **20.** a1, b3, c1, d2, e1, f2, g1, h4

Chapter 24

1. lingual frenulum **2.** aE, bE, cM **3.** permanent **4.** defecation **5.** cystic **6.** fauces **7.** plicae circulares (circular folds) **8.** haustra **9.** lateral **10.** quadrate, caudate **11.** adbc **12.** adcbe **13.** a1, b2, c3, **14.** c **15.** d **16.** b **17.** e **18.** a **19.** d **20.** b **21.** b **22.** e **23.** b **24.** a1, b4, c5, d2, e3, **25.** T **26.** F **27.** T **28.** F **29.** T **30.** T

Chapter 25

1. podocytes **2.** detrusor **3.** prostatic, membranous, spongy (penile) **4.** nephron, corpuscle, tubule **5.** renal fascia, adipose capsule, renal capsule **6.** renal pyramids **7.** ureter **8.** chabgefdi **9.** fbaecd **10.** micturition **11.** d **12.** b **13.** b **14.** c **15.** c **16.** c **17.** b **18.** T **19.** F **20.** F **21.** T **22.** T

Chapter 26

1. a scrotum, b penis, c prostate gland, d bulbourethral (Cowper's glands **2.** 23, 46, haploid, diploid **3.** functionalis **4.** epididymis **5.** myometrium, smooth muscle **6.** corpus albicans **7.** corpora cavernosa **8.** ovarian **9.** ovulation **10.** cdba **11.** bcdae **12.** bedac **13.** a7, b4, c3, d1, e6, f2, g5 **14.** e **15.** b **16.** b **17.** e **18.** c **19.** a **20.** b **21.** a **22.** F **23.** T **24.** F **25.** T

Chapter 27

1. a2, b3, c1 and 4, d4, e2, f3 **2.** morula **3.** cleavage, blastomeres **4.** ectoderm, endoderm **5.** embryo **6.** primary germ **7.** bdace **8.** cab **9.** c **10.** b **11.** a **12.** a **13.** e **14.** c **15.** d **16.** a2, b1, c3, d3, e3, f1, g2 **17.** T **18.** F **19.** T **20.** F **21.** F **22.** a3, b2, c1, d3, e1, f3

ANSWERS TO CRITICAL THINKING QUESTIONS

Chapter 1

1. The hypo (below) chondriac (rib) region lies near the heart, stomach, and gallbladder. Someone suffering from temporary digestive discomfort might imagine that their discomfort indicates problems with other organs in this region.

2. Ming's arm will be extended lateral to the trunk, hand inferior and distal to the elbow, palm facing anteriorly, thumb lateral, and radius lateral to the ulna.

3. The appendix is located in the right iliac (inguinal) region of the abdominopelvic cavity, inferior to the cecum of the large intestine.

4. A living organism would exhibit some or all of the following characteristics: metabolism, responsiveness, movement, growth, differentiation, or reproduction.

5. The navel is located on the anterior side of the body, in line with the midsagittal plane, in the umbilical region of the abdominopelvic cavity or at the intersection of the imaginary lines separating the abdominopelvic cavity into four quadrants.

6. Positron emission tomography (PET), used to detect metabolic changes in the brain, is one way of studying brain function.

Chapter 2

1. Route of mucin: synthesis by the ribosome on rough endoplasmic reticulum, to Golgi complex, to secretory vesicles, to the cell membrane, where the mucin is released by exocytosis.

2. The tail is digested by lysosomes in the process of autolysis and cellular material is reabsorbed.

3. Paralysis of cilia by nicotine will cause the movement of mucus in the respiratory passages to stop. The mucus with its trapped pollutants will collect in the passageways until it irritates the tissue enough to be coughed out.

4. Sixty-four (2^6) daughter cells. Start with 1 parent on Sunday, 2^1 (2 daughters from 1 division) on Monday, 2^2 (4 daughters from 2 divisions) on Tuesday, until there are 2^6 cells on Saturday.

5. Arsenic will poison the mitochondrial enzymes involved in cellular respiration. This will halt production of ATP by the mitochondria. Cells with a high metabolic rate such as muscle will be particularly affected.

6. Disruption of microtubules would halt cell division. Microtubules are also involved in cell motility.

Chapter 3

1. Some tissues found in a chicken leg would include: epithelium in the skin, areolar and adipose tissue in the subcutaneous layer under the skin, skeletal muscle (the meat), blood, hyaline cartilage, and bone.

2. If one student started eating more calories than needed for his activity level, he could have gained 20 pounds of fat stored as adipose tissue. Another possibility is one student grew during his first year at college and gained weight along with height.

3. Functions of the mucous membrane include secretion of mucus to prevent desiccation and to trap foreign particles. Portions of the nasal cavity also contain cilia to move mucus along with the trapped particles.

4. The gelatin portion of the salad represents the ground substance of the connective tissue matrix. The shredded carrots are the fibers of the matrix. The grapes embedded in the gelatin are cells such as fibroblasts.

5. The tissue is keratinized stratified squamous epithelium. Epithelium covers all surfaces and is avascular.

6. Exocrine glands include those that produce mucus (protects against drying and traps foreign particles), sweat (controls body temperature), oil (lubricates skin), ear wax (produces protective lubricant for ear canal), and many others that produce essential secretions.

Chapter 4

1. Chronic exposure to UV light causes damage to elastin and collagen, promoting wrinkles. The suspicious growth may be skin cancer or a precancerous lesion caused by UV photodamage to skin cells.

2. Brad will most likely suffer a sunburn due to overexposure to UV rays. In a typical first-degree burn, the damaged skin will appear red and may peel in a few days. Brad should have used sunscreen and clothing for protection from the sun's rays.

3. Hair growth and replacement is cyclic. The hormonal changes of pregnancy may cause increased rate of hair loss and should return to normal 3 to 4 months after childbirth.

4. The surface layer of the skin is keratinized stratified squamous epithelium. The cells are dead and filled with keratin, a waterproofing protein.

5. Sebum (oil) may collect in enlarged sebaceous glands. The oxidized sebum and melanin give the blackhead its dark color. The junk food diet does not affect sebum production.

6. From most superficial: Epidermis = keratinized stratified squamous epithelium; waterproof barrier, UV protection, immunity, touch receptors (tactile or Merkel disc). Papillary layer of dermis = areolar CT, capillaries and touch receptors (Meissner's corpuscles). Reticular layer of dermis = dense, irregular CT, hair follicles, oil glands. SubQ = areolar and adipose CT, lamellated (Pacinian) corpuscles.

Chapter 5

1. Lynda's diet lacks several necessary nutrients such as essential vitamins (e.g., A, D), minerals (e.g., calcium), and amino acids. Her lack of exercise will weaken bone strength. Her age and smoking habit may result in a lack of estrogens, causing demineralization of bone.

2. The physician will want an X ray of the bone to observe the epiphyseal plate. Laboratory tests may be ordered to check hormone levels such as human growth hormone, thyroid hormones, insulin, and calcitonin. Diet will be analyzed for proper nutrition. Parents' height should be checked for the hereditary component of height.

3. A greenstick fracture occurs only in children due to the greater flexibility of their bones. The bone breaks on one side but bends on the other side, resembling what happens when one tries to break a green (not dry) stick. The child probably broke her fall with her extended arm, breaking the radius or ulna.

4. At Aunt Edith's age, the production of several hormones necessary for bone remodeling, such as estrogens and human growth hormone, is likely to be decreased. She may have osteoporosis and an increased susceptibility to fractures, resulting in damage to the vertebrae and loss of height.

5. Exercise causes mechanical stress on bones, but because there is no gravity in space, the pull of gravity on bones is missing. The lack of stress results in demineralization of bone and weakness.

6. Chantal has stress fractures in her tibia. Aerobic dancing caused repeated, increased stress to the bone, which resulted in fractures. Chantal's footwear had inadequate cushioning and support.

Chapter 6

1. Fontanels, the soft spots between cranial bones, are fibrous connective tissue membranes. They allow the infant's head to be molded during its passage through the birth canal and allow for brain and skull growth in the infant.

2. Barbara probably broke her coccyx or tailbone.

3. The infant was born with a single concave curve. Adults have four curves in their spinal cord in the cervical, thoracic, lumbar, and sacral regions.

4. The sutures are joints between the skull bones.

5. The X rays were taken to view the sinuses—the "fuzzy looking holes." John probably had sinusitis due to infected sinuses.

6. Harry herniated (slipped) a disc in his lower back. Pressure on the roots of the sciatic nerve is causing the pain in his lower limb.

Chapter 7

1. There are 14 phalanges in each hand: 2 bones in the thumb and 3 in each finger. Farmer Dole has lost 5 phalanges on his left hand so he has 9 remaining on his left and 14 remaining on his right for a total of 23.

2. 14 phalanges (toes), tibia (shinbone) and fibula, patella (kneecap), femur (thigh).

3. Pelvic and pectoral girdles are necessary to connect the upper and lower limbs (extremities) to the axial skeleton and are therefore part of the appendicular skeleton.

4. Drew broke the proximal phalanx on the most medial toe. The toes are distal to the metatarsals.

5. Grandmother Amelia has probably fractured the neck of her femur. This is a common fracture in the elderly and is often called a "broken hip."

6. Mary has patellofemoral syndrome (runner's knee). Running daily in the same direction on a banked track has stressed the downhill knee. The patella is tracking laterally to its proper position, resulting in pain after exercise.

Chapter 8

1. Structurally, the hip joint is a simple ball-and-socket synovial joint composed of the head of the femur (ball) and the acetabulum of the coxal bone (socket). The knee is really three joints: patellofemoral, lateral tibiofemoral, and medial tibiofemoral. It is a combined gliding and hinge joint.

2. In TMJ, the mandibular condyle articulates with the mandibular fossa and articular tubercle of the temporal bone. This synovial joint is a combined gliding and hinge joint allowing biaxial movement—side to side, protraction/retraction, depression/elevation, and slight rotation of the mandible.

3. Damage to the knee caused the "water on the knee" (swollen knee). Immediate swelling is caused by blood leaking from damaged vessels in the tissues (ligaments, bone) in the joint area. Excess synovial fluid due to damage to the synovial membrane will contribute to the later swelling.

4. Flexion at the knee, extension at the hip, extension of the vertebral column except for hyperextension at the neck, abduction at the shoulder, flexion of the fingers, adduction of the thumb, flexion/extension at the elbow, and depression of the mandible.

5. No more body surfing for Tish. She has a dislocated shoulder. The head of the humerus was displaced from the acromioclavicular joint, causing tearing of the supporting ligaments and tendons (rotator cuff) of the shoulder joint.

6. Periodontal disease affects the periodontal ligament, which holds the teeth in their sockets in the maxillae and mandible. The joint is a gomphosis—a fibrous synarthrosis.

Chapter 9

1. White meat is fast fibers. Its color is due to low amounts of myoglobin, mitochondria, and capillaries. It is fast contracting and easily fatigued. Dark (red) meat is slow and intermediate fibers. It is fatigue resistant, with abundant mitochondria, myoglobin, and capillaries.

2. Martina's muscles in the casted leg were not being used, so they decreased in size due to loss of myofibrils. Martina's leg shows disuse atrophy.

3. Acetylcholine (ACh) is the neurotransmitter used to "bridge the gap" at the neuromuscular junction. If ACh release is blocked, the neuron cannot send a signal to the muscle and the muscle will not contract.

4. The entire intact rope is comparable to an entire skeletal muscle. As the rope is unwound, the three to four bundles of fibers become separate. These bundles of fibers are comparable to the fascicles. When the fascicles are frayed so that the thin string-like fibers are separated, they would be comparable to the muscle fibers (cells).

5. The rigid body is in rigor mortis. The lack of ATP prevents the myosin cross-bridges from releasing from the actin. The muscle fibers cannot relax and remain fixed in position. After about 24 hours, rigor mortis ceases due to decomposition of the tissues.

6. Bruno volunteered for an electromyogram (EMG). The electrical activity of skeletal muscle is being recorded.

Chapter 10

1. The orbicularis oris protrudes the lips. The genioglossus protracts the tongue. The buccinator aids in sucking.

2. One likely possibility is the rhomboideus major muscle.

3. Masseter, temporalis, medial pterygoid, lateral pterygoid.

4. Gluteus maximus: origin = iliac crest, sacrum, coccyx, aponeurosis of sacrospinalis, and insertion = iliotibial tract and linea aspera; gluteus medius: origin = ilium, and insertion = greater trochanter of femur; gluteus minimus: origin = ilium, and insertion = greater trochanter of femur.

5. Some of the antagonistic pairs for the regions are: arm-biceps brachii vs. triceps brachii; thigh-quadriceps femoris vs. hamstrings; torso-rectus abdominus vs. erector spinae; the leg-gastrocnemius vs. tibialis anterior.

6. The rotator (musculocutaneous) cuff is made up of the tendons of four deep muscles of the shoulder; subscapularis, supraspinatus, infraspinatus, and teres minor. The tendons form almost a complete circle around the joint of the scapula and the humerus.

Chapter 11

1. The spine of C7 forms a prominence along the midline at the base of the neck. The spine of T1 forms a second prominence slightly inferior to C7.

2. Palpation is touching or examining using the hands. Palpitation is a forceful beating of the heart that can often be felt by the patient.

3. The nurse pushed on McBurney's point, which is located two-thirds of the way down along a line connecting the anterior superior iliac spine to the umbilicus. Pressure on McBurney's point produces pain at the site in cases of appendicitis.

4. The xiphoid process is the most inferior portion of the sternum. The superior edge of the costal margin will intersect with the xiphoid process at the level of the seventh costal cartilage.

5. The commonly used site for a blood donation is the median cubital vein, which crosses the cubital fossa at an angle.

6. The skin superior to the buttocks has a dimple or depression on each side of the spine due to the attachment of the skin and fascia to bone.

Chapter 12

1. The average person's blood supply weighs about 8% of their total body weight. Felicia's blood weighs 146 lb. × 0.08 (8%) = 11.7 lb.

2. Blood doping is used by some athletes to improve performance, but the practice strains the heart. A blood test would reveal polycythemia, an increased number of red blood cells.

3. Basophils appear as bluish-black granular cells in a stained blood smear. An elevated number of basophils may suggest an allergic response such as "hay fever."

4. Becky may have nutritional anemia. A CBC would determine red blood cell count, hemoglobin, and hematocrit.

5. Males generally have a higher percentage of red blood cells in blood than females. The average hematocrit is 47 for males and 42 for females. More red cells means more hemoglobin, so Antonio's blood is redder.

6. "A pos." refers to his blood type for the two major blood groups: ABO and Rh. Blood type A has antigen A on the red blood cell surface. "Pos." refers to the presence of the Rh (D) antigen on the surface of the red blood cell.

Chapter 13

1. The fabric of the parachute "catches" the air like the cusps of the valves "catch" the blood. The parachute is anchored to the chutist by the chute's lines. The cusps are anchored to the papillary muscles by the chordae tendineae.

2. The SA node acts as the natural pacemaker to set the heart's rhythm. It is located just inferior to the junction of the superior vena cava with the right atrium.

3. The anterior interventricular branch is located in the anterior interventricular sulcus and supplies both ventricles. The circumflex branch is located in the coronary sulcus and supplies the left ventricle and the left atrium.

4. The pericardium surrounds the heart. The fibrous pericardium anchors the heart to the diaphragm, sternum, and major blood vessels. The intercalated discs contain desmosomes, which hold the cardiac muscle fibers together.

5. Angina pectoris causes severe chest pain due to ischemia (reduced blood flow) to the heart muscle. The tissue may not be killed. In an MI or heart attack, the myocardial tissue dies due to lack of blood caused by a blockage in a vessel.

6. The closure of the aortic semilunar valve is best heard at the second intercostal space near the junction of the costal cartilages and the ribs, to the right side of the sternum. The valve is located at the level of the third intercostal space posterior to the left side of the sternum.

Chapter 14

1. The left ventricle must generate more force to pump blood to the extensive systemic circulation. The right ventricle requires less force to pump to the closer and less extensive pulmonary circulation. Also, pulmonary vessels are larger and less elastic than systemic vessels, which reduces resistance.

2. The foramen ovale and ductus arteriosus close to establish the (pulmonary vs. systemic) circulation. The umbilical artery and veins close because the placenta is no longer functioning. The ductus venosus closes so that the liver is no longer bypassed.

3. An anastomosis provides alternate routes for blood flow to an organ, in this case to the brain. The cerebral arterial circle is formed by the anterior and posterior cerebral arteries, the anterior and posterior communicating arteries, and the internal carotid arteries.

4. Some of the unpaired vessels are the brachiocephalic trunk, celiac, splenic, common hepatic, superior mesenteric, inferior mesenteric, and median sacral.

5. A vascular sinus is a type of vein that lacks smooth muscle in its tunica media. The tunica media and externa are replaced by dense connective tissue as in the intracranial sinuses.

6. Varicose veins are caused by faulty valves. The weak valves allow pooling of blood in the veins. In the peripheral systemic circulation, valves are found only in the veins and not in the arteries.

Chapter 15

1. Lymph capillaries to lymphatics to popliteal nodes to superficial inguinal nodes to the right lumbar trunk to the cisterna chyli to the thoracic duct.

2. The palatine tonsils are in the lateral, posterior oral cavity. The lingual tonsils are at the base of the tongue. The tonsils function in the immune response to inhaled or ingested foreign invaders.

3. Kate has allergies. Her body is hypersensitive to allergens that are common in the spring, such as tree pollen. The allergic reaction is causing her itchy eyes and stuffy nose.

4. Following a splenectomy, the liver and red bone marrow take over two functions of the spleen: phagocytosis of bacteria, platelets, and red blood cells and red blood cell production.

5. The pectoral group of axillary lymph nodes drains most of the breast tissue. Cancer cells may metastasize from the original tumor to nearby lymph nodes. Other axillary nodes may also be involved.

6. AIDS is caused by infection with the human immunodeficiency virus (HIV), which is transmitted in human body fluids. HIV initially attacks the T4 lymphocytes. The virus also infects other cells and uses macrophages and dendritic cells as reservoirs. The virus eventually destroys the T4 cell population, and the patient acquires opportunistic infections and diseases.

Chapter 16

1. The traffic pattern resembles a converging circuit in which input (traffic) is received from several presynaptic neurons (the five roads) synapsing with one postsynaptic neuron (the intersection).

2. This situation reminds Mei-Ling of a reverberating circuit in that a signal will be repeated over and over again. Reverberating circuits function in breathing, memory, waking, and coordinating muscle activity.

3. Gray matter appears gray in color due to the absence of myelin. It is composed of cell bodies, dendrites, and unmyelinated axons. White matter's color is due to the presence of myelin.

4. The somatic, sensory, peripheral division would detect smell. Signals would travel along a cranial nerve to the brain (central division) and then back out along cranial nerves in the autonomic, motor, peripheral division to the salivary gland.

5. In the central nervous system, the myelin sheath is produced by the oligodendrocytes. They do not form a neurolemma over the sheath. In the peripheral nervous system, the neurolemmocyte produces the myelin sheath and forms a neurolemma. A neurolemmocyte myelinates a section of one axon, while the oligodendrocyte myelinates sections of several axons.

6. The pericentriolar area of a centrosome organizes the mitotic spindle during cell division. Without centrosomes, the neuron cannot undergo mitosis.

Chapter 17

1. The senior damaged the cervical region of his spinal cord. Recovery is unlikely.

2. The needle will pierce the epidermis and dermis of the skin, next the subcutaneous layer, then go between the vertebrae into the epidural space, which contains fat and connective tissue.

3. The spinal cord is connected to the periphery by the spinal nerves. The nerves enter through the intervertebral foramina.

The dura mater of the spinal cord fuses with the epineurium. The nerves are connected to the cord by the posterior and anterior roots.

4. The spinal cord is anchored in place by the filum terminale and denticulate ligaments.

5. The spinal cord itself ends at the level between the first and second lumbar vertebrae. A spinal tap is usually performed between the third and fourth or between the fourth and fifth lumbar vertebrae. This is a relatively safe location for a spinal tap.

6. The reflex arc includes a (pain) receptor, a sensory neuron, an integrating (spinal cord) center, a motor neuron, and an effector (skeletal muscle).

Chapter 18

1. Movement of the right arm is controlled by the left hemisphere's primary motor area, located in the postcentral gyrus. Speech is controlled by Broca's area in the left hemisphere's frontal lobe just superior to the lateral cerebral sulcus.

2. The brain is enclosed by the cranial bones and meninges. The temporal bone houses the middle and inner ear, separating these from the brain.

3. The outer ear (auricle) → temporal bone → dura mater → arachnoid → subarachnoid space with cerebrospinal fluid → pia mater → cerebrum → diencephalon → repeat to other ear.

4. The medulla contains nuclei for several vital functions, including the cardiovascular center and the respiratory center, as well as centers for vomiting, swallowing, coughing, and others. The crash victim is not likely to survive the injuries.

5. The gray matter is located on the outer surface of the cerebrum. Because the cerebrum functions in intelligence, thought, mathematical ability, creativity, and so on, Beatriz was being insulted when she was told that her gray matter is thin.

6. The dentist has injected anesthetic into the inferior alveolar nerve, a branch of the mandibular nerve, numbing the lower teeth and the lower lip. The tongue is numbed by blocking the lingual nerve. The upper teeth and lip are numbed by injecting the superior alveolar nerve, a branch of the maxillary branch.

Chapter 19

1. Olfactory receptors are rapidly adapting. The receptors rapidly become less sensitive to the odor in the laboratory, even though the odor doesn't change.

2. Pain provides protection against tissue damage. Pain receptors detect noxious stimuli that may damage the body. Pain also results from excess stimuli of any kind that may cause damage (for example, high heat). Without pain, the body could unknowingly be damaged (for example, burned) and serious illness could occur (for example, appendicitis) without major symptoms as warnings.

3. Tickle receptors are free-nerve endings in the foot. The impulses travel along a first-order neuron to the posterior gray horn to synapse with a second-order neuron, cross to the other side of the cord, and run up the anterior spinothalamic tract to the thalamus. The third-order neuron runs from the thalamus to the "foot" area of the primary somatosensory area of the cerebral cortex.

4. The primary motor area of the cerebral cortex initiates voluntary movements. The basal ganglia help control the automatic, semivoluntary motions. Sensory input related to bike riding

will be received by the brain. The cerebellum and other brain regions will integrate this input to further coordinate the bicycling movements.

5. In an infant or very young child, the myelination of the central nervous system is not complete. The corticospinal tracts, which control fine, voluntary motor movements, are not fully myelinated until the child is about 2 years old. An infant would not be able to manipulate a knife or fork safely due to lack of complete motor control.

6. Ahmad's perception of feeling in his amputated foot is called phantom limb sensation. Impulses from the remaining proximal section of the sensory neuron are perceived by the brain as still coming from the amputated foot.

Chapter 20

1. Mariko is probably responding to olfactory stimulation, because what is perceived as taste is closely related to smell. The chemical odors from the food dissolve in the mucosal surface of the nasal cavity. The olfactory receptors will be stimulated and the signal will be sent along the olfactory nerves to the olfactory bulbs to the olfactory tracts to the brain.

2. Lacrimal fluid (tears) move medially across the anterior surface of the eyeball to the lacrimal puncta to the lacrimal canals to the nasolacrimal duct to the nasal cavity to the nostrils.

3. The stick punctured the transparent corneal portion of the fibrous tunic. The anterior chamber of the anterior cavity lies deep to the cornea. It is filled with watery aqueous humor. The stick then pierced the iris (the colored ring), which is part of the vascular tunic.

4. The route is auricle to external auditory meatus to tympanic membrane to malleus to incus to stapes to oval window to vestibule to cochlea, which contains the spiral organ (organ of Corti).

5. Damage to the left cranial nerve II: the left peripheral portion of the field of vision would be lost. The medial field would still be seen by the right eye. Damage to the left occipital lobe: the right peripheral field of vision would be lost. The right lobe would still "see" the medial field.

6. The eardrum (tympanic membrane) is moved by the vibrations of air (sound waves). The stapes pushes on the oval window and so functions like the mallet used to hit a drum. The secondary tympanic membrane in the round window will be vibrated by the waves in the perilymph of the cochlea.

Chapter 21

1. The exciting and potentially dangerous activities activate the sympathetic nervous system, resulting in a "fight or flight" reaction. The sympathetic division of the ANS stimulates release of epinephrine (adrenaline) and norepinephrine from the adrenal medulla. These hormones prolong the "fight or flight" response.

2. Digestion and relaxation are controlled by the parasympathetic division of the ANS. The salivary glands, pancreas, and liver will show increased secretion; the stomach and intestines will have increased activity; the gallbladder will have increased contractions.

3. The spinal nerve is the E-W highway. The white ramus is the off-ramp leading to the intersection (= the sympathetic chain ganglion). From the ganglion, the fibers can connect to the N-S

highway (= the sympathetic chain) or take the on-ramp (= the gray ramus) back to the E-W highway (= the spinal nerve).

4. The receptor in the colon is stimulated → general visceral sensory neuron → sacral region of spinal cord → preganglionic neuron → terminal ganglion → postganglionic neuron → smooth muscle in colon/rectum.

5. For parasympathetic: feed and breed or rest and digest or R&R (rest and repose) or defecate and procreate. For sympathetic: Dry mouth and damp hands mean a warm (fast) heart.

6. Cranial nerve X is called vagus because it wanders from its origin in the brainstem to a diverse group of effectors located all over the body, including the eyes, salivary glands, lungs, heart, and gastrointestinal tract.

Chapter 22

1. Melatonin is produced by the pineal gland. The gland is located deep to the cerebrum, superior to the third ventricle.

2. Mandy has an enlarged thyroid, also called a goiter. The goiter is probably due to hypothyroidism, which is causing her symptoms such as weight gain and tiredness.

3. Indira's sympathetic division of the autonomic nervous system will stimulate secretion of the hormones epinephrine and norepinephrine by the adrenal medulla. The hormones are released in response to stress and prolong the fight or flight response.

4. The large amount of sugar in the meal will cause an increase in blood glucose levels. The pancreas will increase secretion of insulin from the beta cells of the pancreatic islets (islets of Langerhans). The insulin will lower blood glucose levels. The secretion of other hormones that influence digestion may also be increased.

5. The pituitary is located in the sella turcica in the base of the sphenoid bone. The sphenoid bone is posterior to the ethmoid bone, which makes up a major portion of the walls and roof of the nasal cavity. Due to its proximity to the nasal cavity, the pituitary could have been affected by the radiation.

6. The two adrenal glands are located just superior to the two kidneys. The adrenal gland is really two glands: an outer cortex and an inner medulla. The cortex is composed of three layers: zona glomerulosa, zone fasiculata, and zona reticularis. The medulla is composed of chromaffin cells.

Chapter 23

1. The septum is composed of hyaline cartilage covered with mucous membrane. The lateral wall of the nose is skin and muscle, lined with mucous membrane. (The upper ear is skin covering elastic cartilage.)

2. Due to the influence of androgens, males generally have larger cartilages of the larynx and vocal folds than females. The larger, thicker size of the folds causes them to vibrate more slowly, resulting in a lower pitch of the voice in males.

3. Skin → subcutaneous layer muscle (depending on location may be serratus anterior, external oblique, intercostals) → parietal pleura → pleural cavity → perhaps visceral pleura and lung.

4. The cartilages of the larynx will move while the tenor sings because the vocal cords are attached to the arytenoid cartilages. The "bobbing" structure is probably the thyroid cartilage, which is prominent in males. The thyroid cartilage is a large, shield-shaped cartilage on the anterior aspect of the neck.

5. Three bronchopulmonary segments make up the superior lobe (about the upper third) of the right lung. The apical segment is located at the apex of the lung. The anterior segment is inferior to the apical segment, anterior to the posterior segment, superior to the horizontal fissure. The posterior segment is inferior to the apical segment, posterior to the anterior segment, and bordered by the horizontal and oblique fissures.

6. Locate the larynx on the anterior neck at the level of C4–C6. Locate the large shield-shaped cartilage—the thyroid cartilage. The cartilage inferior to the thyroid cartilage, on the anterior surface of the neck, is the cricoid cartilage. A tracheostomy will require an incision through the wall of the trachea inferior to the cricoid cartilage, between the cricoid cartilage and the trachea's cartilage ring.

Chapter 24

1. Kyoko lost the two upper central incisors of her permanent (secondary) teeth. The rest of her teeth are still deciduous (primary); these include the lower central incisors, an upper and lower pair of lateral incisors, an upper and lower pair of cuspids, an upper and lower pair of first molars, and upper and lower pair of second molars.

2. The smaller left lobe of the liver is separated from the larger right lobe by the falciform ligament. The left lobe is inferior to the diaphragm in the epigastric region of the abdominopelvic cavity.

3. During swallowing, the soft palate and uvula close off the nasopharynx. If the nervous system sends conflicting signals (e.g., breathe, swallow, or giggle), the palate may be in the wrong position to block food from entering the nasal cavity.

4. When present, the appendix is attached to the medial side of the cecum, inferior to the junction of the ileum with the cecum. The surgical incision was probably made on the medial side of the right iliac region.

5. The gallbladder stores and concentrates bile. Gallstones may form from the cholesterol in bile. If the gallstones block any of the ducts leading from the gallbladder to the duodenum, the increased pressure in the gallbladder from the buildup of bile may cause pain.

6. Structural modifications of the small intestine that increase surface area include its overall length, plicae circulares, villi, and microvilli (brush border). The plicae circulares also enhance absorption by imparting a spiraling motion to the chyme.

Chapter 25

1. The kidney is bean-shaped, with the hilus on the medial side. A fibrous renal capsule covers the kidney, similar to the papery covering over the bean. The kidneys' reddish-brown color, due to blood, mitochondria, and other pigments, is similar to the bean's color. The embryonic bean plant's position would be comparable to the location of the pelvis in the kidney.

2. The left kidney is located a bit higher than the right at about the level of the eleventh and twelfth ribs. It is lateral to the position of vertebrae T12 to L3. The kidney is retroperitoneal, held against the body wall by the adipose capsule and the renal fascia.

3. The urinary bladder can stretch considerably due to the presence of transitional epithelium, rugae, and three layers of smooth muscle (detrusor muscle).

4. Urine samples contain an abundance of epithelial cells that are shed from the mucosal lining of the urinary system. (Small numbers of other cell types such as white blood cells may also be present.)

5. In females, the urethra is about 4 cm long. In males, the urethra is about 15–20 cm long, including its passage through the penis, urogenital diaphragm, and prostate gland.

6. The contrast material will be visible as it is cleared from the blood in this order: nephron, minor calyces, major calyces, renal pelvis, ureter, urinary bladder, urethra.

Chapter 26

1. The ductus (vas) deferens originates at the tail of the epididymis, runs posterior to the epididymis, up through the inguinal canal, over the ureter, posterior to the urinary bladder, to the ampulla of the ductus deferens, to join with the seminal vesicle duct. A vasectomy is performed on the posterior side of the scrotum.

2. Sperm cells: acrosomes contain enzymes needed for penetration of secondary oocyte; midpiece contains mitochondria for ATP; flagellum for motion. Semen: alkali to neutralize acidity.

3. A vasectomy cuts only the ductus (vas) deferens and leaves the testis untouched. Male secondary sex characteristics and libido (sex drive) are maintained by androgens, including testosterone. The secretion of these hormones by the testis is not interrupted by the vasectomy.

4. A hysterectomy will not cause menopause. A hysterectomy is the removal of the uterus, which does not produce hormones. The ovaries are left intact and continue to produce estrogens and progesterone.

5. The time before menstruation is the postovulatory phase. The corpus luteum secretes high amounts of estrogens and progesterone, which stimulate growth of the endometrium (secretory phase). The mammary glands and cervix are also affected by the increased hormone levels.

6. A circumcision is the surgical removal of the prepuce (foreskin), usually in an infant male. It may be performed for religious reasons or for social reasons. In later years, the lack of a prepuce may help decrease the chances of urinary tract and sexually transmitted infections, as well as reduce the risk of penile cancer.

Chapter 27

1. Identical (monozygotic) twins develop from the same fertilized ovum or zygote. They are essentially the same person genetically and therefore must be the same gender. These must be fraternal (dizygotic) twins.

2. All tissues and organs form from the embryo's three primary germ layers. The ectoderm, the outermost germ layer, develops into the nervous system and the epithelial layer of the skin. This early embryological connection may sometimes result in disorders showing signs in both tissues.

3. Darla is in false labor. In true labor, the contractions are regular and the pain may localize in the back. The pain may be intensified by walking. True labor is indicated by the "show" of bloody mucus and cervical dilation.

4. The fetus is encased in the amniotic sac. The amniotic cavity that surrounds the fetus is filled with amniotic fluid, a "salty" fluid formed from maternal blood filtrate and fetal urine. The fetus floats in the protective fluid throughout fetal life.

5. The yolk sac provides blood vessels to transport nutrients from the endometrium to the embryo. The germ cells (oogonia and spermatogonia) migrate from the yolk sac.

6. The belly button or navel is a small scar on the surface of the skin in the center of the umbilical region. It is formed when the stump of the umbilical cord withers and falls off after the baby's birth.

Glossary

Pronunciation Key

1. The most strongly accented syllable appears in capital letters, for example, bilateral (bī-LAT-er-al) and diagnosis (dī-ag-NŌ-sis).
2. If there is a secondary accent, it is noted by a prime ('), for example, constitution (kon′-sti-TOO-shun) and physiology (fiz′-ē-OL-ō-jē). Any additional secondary accents are also noted by a prime, for example, decarboxylation (dē′-kar-bok′-si-LĀ-shun).
3. Vowels marked by a line above the letter are pronounced with the long sound as in the following common words:
    - *ā* as in *māke*
    - *ē* as in *bē*
    - *ī* as in *īvy*
    - *ō* as in *pōle*
4. Vowels not so marked are pronounced with the short sound as in the following words:
    - *a* as in *above*
    - *e* as in *bet*
    - *i* as in *sip*
    - *o* as in *not*
    - *u* as in *bud*
5. Other phonetic symbols are used to indicate the following sounds:
    - *oo* as in *sue*
    - *yoo* as in *cute*
    - *oy* as in *oil*

Abdomen (ab-DŌ-men or AB-do-men) The area between the diaphragm and pelvis.

Abdominal (ab-DOM-i-nal) **cavity** Superior portion of the abdominopelvic cavity that contains the stomach, spleen, liver, gallbladder, pancreas, small intestine, and most of the large intestine.

Abdominopelvic (ab-dom′-i-nō-PEL-vic) **cavity** Inferior component of the ventral body cavity that is subdivided into a superior abdominal cavity and an inferior pelvic cavity.

Abduction (ab-DUK-shun) Movement away from the axis or midline of the body or one of its parts usually in the frontal plane.

Abscess (AB-ses) A localized collection of pus and liquefied tissue in a cavity.

Absorption (ab-SORP-shun) The taking up of liquids by solids or of gases by solids or liquids; intake of fluids or other substances by cells of the skin or mucous membranes; the passage of digested foods from the gastrointestinal tract into blood or lymph.

Accessory duct A duct of the pancreas that empties into the duodenum about 2.5 cm (1 in.) superior to the ampulla of Vater (hepatopancreatic ampulla). Also called the **duct of Santorini** (san′-tō-RE-ne).

Acetabulum (as′-e-TAB-yoo-lum) The rounded cavity on the external surface of the hipbone that receives the head of the femur.

Acetylcholine (as′-ē-til-KŌ-lēn) **(ACh)** A neurotransmitter liberated by many peripheral nervous system neurons and some central nervous system neurons. It is excitatory at neuromuscular junctions but inhibitory at some other synapses (slows heart rate).

Achilles tendon *See* **Calcaneal tendon.**

Acini (AS-i-nē) Masses of cells in the pancreas that secrete digestive enzymes.

Acoustic (a-KOOS-tik) Pertaining to sound or the sense of hearing.

Acrosome (AK-rō-sōm) A dense granule in the head of a sperm cell that contains enzymes that facilitate the penetration of a sperm cell into a secondary oocyte.

Actin (AK-tin) The contractile protein that makes up thin myofilaments in muscle fiber (cell).

Action potential A wave of negativity that self-propagates along the outside surface of the membrane of a neuron or muscle fiber (cell); a rapid change in membrane potential that involves a depolarization following a repolarization. Also called a **nerve action potential (nerve impulse)** as it relates to a neuron and a **muscle action potential** as it relates to a muscle fiber (cell).

Active transport The movement of substances, usually ions, across cell membranes, against a concentration gradient, requiring the expenditure of energy (ATP).

Acupuncture (AK-yoo-punk´-chur) The insertion of a needle into a tissue for the purpose of drawing fluid or relieving pain. It is also an ancient Chinese practice employed to cure illnesses by inserting needles into specific locations of the skin.

Acute (a-KYOOT) Having rapid onset, severe symptoms, and a short course; not chronic.

Adam's apple *See* **Thyroid cartilage.**

Adduction (ad-DUK-shun) Movement toward the axis or midline of the body or one of its parts usually in the frontal plane.

Adenoids (AD-e-noyds) Inflamed and enlarged pharyngeal tonsils.

Adenosine triphosphate (a-DEN-ō-sēn trī-FOS-fāt) **(ATP)** The universal energy-carrying molecule manufactured in all living cells as a means of capturing and storing energy. It consists of the purine base *adenine* and the five-carbon sugar *ribose,* to which are added, in linear array, three *phosphate* molecules.

Adhesion (ah-HĒ-zhun) Abnormal joining of parts to each other.

Adipocyte (AD-i-pō-sīt) Fat cell, derived from a fibroblast.

Adrenal cortex (a-DRĒ-nal KOR-teks) The outer portion of an adrenal gland, divided into three zones, each of which has a different cellular arrangement and secretes different hormones.

Adrenal (a-DRĒ-nal) **glands** Two glands located superior to each kidney. Also called the **suprarenal** (soo´-pra-RĒ-nal) **glands.**

Adrenal medulla (me-DULL-a) The inner portion of an adrenal gland, consisting of cells that secrete epinephrine and norepinephrine (NE) in response to the stimulation of preganglionic sympathetic neurons.

Adrenergic (ad´-ren-ER-jik) **fiber** A nerve fiber that when stimulated releases norepinephrine (noradrenaline) at a synapse.

Adrenocorticotropic (ad-rē´-nō-kor-ti-kō-TRŌP-ik) **hormone (ACTH)** A hormone produced by the anterior lobe of the pituitary gland that influences the production and secretion of certain hormones of the adrenal cortex.

Adventitia (ad-ven-TISH-ya) The outermost covering of a structure or organ.

Afferent arteriole (AF-er-ant ar-TĒ-rē-ōl) A blood vessel of a kidney that breaks up into the capillary network called a glomerulus; there is one afferent arteriole for each glomerulus.

Agglutination (a-gloo´-ti-NĀ-shun) Clumping of microorganisms or blood cells; typically an antigen-antibody reaction.

Agglutinin (a-GLOO-ti-nin) A specific antibody in blood serum that reacts with a specific agglutinogen and causes the clumping of bacteria, blood cells, or particles. Also called an **isoantibody.**

Agglutinogen (ag´-loo-TIN-ō-gen) A genetically determined antigen located on the surface of red blood cells; basis for the ABO grouping and Rh system of blood classification. Also called an **isoantigen.**

Aggregated lymphatic follicles Aggregated lymph nodules that are most numerous in the ileum. Also called **Peyer's** (PĪ-erz) **patches.**

Agnosia (ag-NŌ-zē-a) A loss of the ability to recognize the meaning of stimuli from the various senses (visual, auditory, touch).

Albumin (al-BYOO-min) The most abundant (60 percent) and smallest of the plasma proteins, which functions primarily to regulate osmotic pressure of plasma.

Aldosterone (al-do-STĒR-ōn) A mineralocorticoid produced by the adrenal cortex that brings about sodium and water reabsorption and potassium excretion.

Allantois (a-LAN-tō-is) A small, vascularized membrane between the chorion and amnion of the fetus that serves as an early site for blood formation.

Allele (AL-ēl) Genes that control the same inherited trait (such as height or eye color) that are located on the same position (locus) on homologous chromosomes.

Alpha (AL-fa) **cell** A cell in the pancreatic islets (islets of Langerhans) in the pancreas that secretes glucagon.

Alpha receptor Receptor found on visceral effectors innervated by most sympathetic postganglionic axons.

Alveolar-capillary (al-VĒ-ō-lar) **membrane** Structure in the lungs consisting of the alveolar wall and basement membrane and a capillary endothelium and basement membrane through which the diffusion of respiratory gases occurs. Also called the **respiratory membrane.**

Alveolar duct Branch of a respiratory bronchiole around which alveoli and alveolar sacs are arranged.

Alveolar macrophage (MAK-rō-fāj) Cell found in the alveolar walls of the lungs that is highly phagocytic. Also called a **dust cell.**

Alveolar sac A collection or cluster of alveoli that share a common opening.

Alveolus (al-VĒ-ō-lus) A small hollow or cavity; an air sac in the lungs; milk-secreting portion of a mammary gland. *Plural,* **alveoli** (al-VĒ-ō-lī).

Amenorrhea (a-men-ō-RĒ-a) Absence of menstruation.

Amnesia (am-NĒ-zē-a) A lack or loss of memory.

Amniotic (am´-nē-OT-ik) **fluid** Fluid in the amniotic cavity, the space between the developing embryo (or fetus) and amnion; the fluid is initially produced as a filtrate from maternal blood and later from fetal urine.

Amphiarthrosis (am′-fē-ar-THRŌ-sis) A slightly movable articulation midway between a diarthrosis and synarthrosis, in which the articulating bony surfaces are separated by fibrous connective tissue or fibrocartilage to which both are attached.

Ampulla (am-POOL-la) A saclike dilation of a canal.

Ampulla of Vater *See* **Hepatopancreatic ampulla.**

Anabolism (a-NAB-ō-lizm) Synthetic energy-requiring reactions whereby small molecules are built up into larger ones.

Anal (Ā-nal) **canal** The terminal 2 or 3 cm (1 in.) of the rectum; opens to the exterior through the anus.

Anal column A longitudinal fold in the mucous membrane of the anal canal that contains a network of arteries and veins.

Analgesia (an-al-JĒ-zē-a) Pain relief.

Anal triangle The subdivision of the female or male perineum that contains the anus.

Anaphase (AN-a-fāz) Third stage of mitosis in which the chromatids that have separated at the centromeres move to opposite poles of the cell.

Anastomosis (a-nas-tō-MŌ-sis) An end-to-end union or joining together of blood vessels, lymphatic vessels, or nerves.

Anatomic dead space The volume of air that is inhaled but remains in spaces in the upper respiratory system and does not reach the alveoli to participate in gas exchange; about 150 ml.

Anatomical (an′-a-TOM-i-kal) **position** A position of the body universally used in anatomical descriptions in which the body is erect, facing the observer with the head level and the eyes facing forward, the upper limbs are at the sides, the palms are facing forward, and the feet are on the floor and directed forward.

Anatomy (a-NAT-ō-mē) The structure or study of structure of the body and the relation of its parts to each other.

Androgen (AN-drō-jen) Substance producing or stimulating male characteristics, such as the male hormone testosterone.

Anesthesia (an′-es-THĒ-zē-a) A total or partial loss of feeling or sensation, usually defined with respect to loss of pain sensation; may be general or local.

Ankylosis (ang′-ki-LO-sus) Severe or complete loss of movement at a joint.

Anomaly (a-NOM-a-lē) An abnormality that may be a developmental (congenital) defect; a variant from the usual standard.

Anoxia (an-OK-sē-a) Deficiency of oxygen.

Antagonist (an-TAG-ō-nist) A muscle that has an action opposite that of the prime mover (agonist) and yields to the movement of the prime mover.

Anterior (an-TER-ē-or) Nearer to or at the front of the body. Also called **ventral.**

Anterior pituitary (pi-TOO-i-tar′-ē) **gland** Anterior portion of the pituitary gland. Also called the **adenohypophysis** (ad′-e-nō-hī-POF-i-sis).

Anterior root The structure composed of axons of motor fibers that emerges from the anterior aspect of the spinal cord and extends laterally to join a posterior root, forming a spinal nerve. Also called a **ventral root.**

Antibody (AN-ti-bod′-ē) A protein produced by certain cells in the body in the presence of a specific antigen; the antibody combines with that antigen to neutralize, inhibit, or destroy it. Also called an **immunoglobulin** (im-yoo-nō-GLOB-yoo-lin) or **Ig.**

Antidiuretic hormone (ADH) Hormone produced by neurosecretory cells in the paraventricular and supraoptic nuclei of the hypothalamus that stimulates water reabsorption from kidney cells into the blood and vasoconstriction of arterioles. Also called **vasopressin** (vāz-ō-PRESS-in).

Antigen (AN-ti-jen) Any substance that when introduced into the tissues or blood induces the formation of antibodies and reacts only with those specific antibodies.

Antrum (AN-trum) Any nearly closed cavity or chamber, especially one within a bone, such as a sinus.

Anulus fibrosus (AN-yoo-lus fī-BRŌ-sus) A ring of fibrous tissue and fibrocartilage that encircles the pulpy substance (nucleus pulposus) of an intervertebral disc.

Anus (Ā-nus) The distal end and outlet of the rectum.

Aorta (ā-OR-ta) The main systemic trunk of the arterial system of the body; emerges from the left ventricle.

Aortic (ā-OR-tik) **body** Receptor on or near the arch of the aorta that responds to alterations in blood levels of oxygen, carbon dioxide, and hydrogen ions.

Aperture (AP-er-chur) An opening or orifice.

Apex (Ā-peks) The pointed end of a conical structure, such as the apex of the heart.

Apneustic (ap-NOO-stik) **area** Portion of the respiratory center in the pons that sends stimulatory nerve impulses to the inspiratory area that activate and prolong inspiration and inhibit expiration.

Aponeurosis (ap′-ō-noo-RŌ-sis) A sheetlike tendon joining one muscle with another or with bone.

Apoptosis (a-pōp-TŌ-sis) Orderly, genetically programmed cell death.

Appendage (a-PEN-dij) A structure attached to the body.

Aqueous humor (AK-wē-us HYOO-mor) The watery fluid, similar in composition to cerebrospinal fluid, that fills the anterior cavity of the eye.

Arachnoid (a-RAK-noyd) The middle of the three coverings (meninges) of the brain or spinal cord.

Arachnoid villus (VIL-us) Berrylike tuft of arachnoid that protrudes into the superior sagittal sinus and through which cerebrospinal fluid is reabsorbed into the bloodstream.

Arbor vitae (AR-bor VĒ-tē) The treelike appearance of the white matter tracts of the cerebellum when seen in midsagittal section. A series of branching ridges within the cervix of the uterus.

Arch of the aorta (ā-OR-ta) The most superior portion of the aorta, lying between the ascending and descending segments of the aorta.

Areola (a-RĒ-ō-la) Any tiny space in a tissue. The pigmented ring around the nipple of the breast.

Arousal (a-ROW-sal) Awakening from a deep sleep; a response controlled by the thalamic part of the reticular activating system.

Arrector pili (a-REK-tor PI-lē) Smooth muscle attached to hairs; contraction pulls the hairs into a more vertical position, resulting in "goose bumps."

Arrhythmia (a-RITH-mē-a) Irregular heart rhythm. Also called a **dysrhythmia.**

Arteriole (ar-TĒ-rē-ōl) A small, almost microscopic, artery that delivers blood to a capillary.

Artery (AR-ter-ē) A blood vessel that carries blood away from the heart.

Arthritis (ar-THRĪ-tis) Inflammation of a joint.

Arthrology (ar-THROL-ō-jē) The study or description of joints.

Articular (ar-TIK-yoo-lar) **capsule** Sleevelike structure around a synovial joint composed of a fibrous capsule and a synovial membrane.

Articular cartilage (KAR-ti-lij) Hyaline cartilage attached to articular bone surfaces.

Articular disc Fibrocartilage pad between articular surfaces of bones of some synovial joints. Also called a **meniscus** (men-IS-cus).

Articulation (ar-tik′-yoo-LĀ-shun) A joint; a point of contact between bones, cartilage and bones, or teeth and bones.

Arytenoid (ar′-i-TĒ-noyd) **cartilages** A pair of small, pyramidal cartilages of the larynx that attach to the vocal folds and intrinsic pharyngeal muscles and can move the vocal folds.

Ascending colon (KŌ-lon) The portion of the large intestine that passes superiorly from the cecum to the inferior edge of the liver where it bends at the right colic (hepatic) flexure to become the transverse colon.

Ascites (as-SĪ-tēz) Serous fluid in the peritoneal cavity.

Aseptic (ā-SEP-tik) Free from any infectious or septic material.

Association area A portion of the cerebral cortex connected by many motor and sensory fibers to other parts of the cortex. The association areas are concerned with motor patterns, memory, concepts of word-hearing and word-seeing, reasoning, will, judgment, and personality traits.

Association neuron (NOO-ron) A nerve cell lying completely within the central nervous system that carries nerve impulses from sensory neurons to motor neurons. Also called a **connecting neuron.**

Astereognosis (as-ter′-ē-ōg-NŌ-sis) Inability to recognize objects or forms by touch.

Asthenia (as-THĒ-nē-a) Lack or loss of strength.

Astrocyte (AS-trō-sīt) A neuroglial cell having a star shape that supports neurons in the brain and spinal cord and attaches the neurons to blood vessels.

Atresia (a-TRĒ-zē-a) Abnormal closure of a passage, or absence of a normal body opening.

Atrial natriuretic (na′-trē-yoo-RET-ik) **peptide (ANP)** Peptide hormone produced by the atria of the heart in response to their stretching that inhibits aldosterone production and thus lowers blood pressure.

Atrioventricular (AV) (ā′-trē-ō-ven-TRIK-yoo-lar) **bundle** The portion of the conduction system of the heart that begins at the atrioventricular (AV) node, passes through the cardiac skeleton separating the atria and the ventricles, then runs a short distance down the interventricular septum before splitting into right and left bundle branches. Also called the **bundle of His** (HISS).

Atrioventricular (AV) node The portion of the conduction system of the heart made up of a compact mass of conducting cells located near the orifice of the coronary sinus in the right atrial wall.

Atrioventricular (AV) valve A structure made up of membranous flaps or cusps that allows blood to flow in one direction only, from an atrium into a ventricle.

Atrium (Ā-trē-um) A superior chamber of the heart.

Auditory ossicle (AW-di-tō-rē OS-si-kul) One of the three small bones of the middle ear called the malleus, incus, and stapes.

Auditory tube The tube that connects the middle ear with the nose and nasopharynx region of the throat. Also called the **Eustachian** (yoo-STĀ-kē-an) **tube.**

Auscultation (aws-kul-TĀ-shun) Examination by listening to sounds in the body.

Autoimmunity An immunologic response against a person's own tissue antigen.

Autolysis (aw-TOL-i-sis) Spontaneous self-destruction of cells by their own digestive enzymes at death or a pathological process or during normal embryological development.

Autonomic ganglion (aw′-tō-NOM-ik GANG-lē-on) A cluster of sympathetic or parasympathetic cell bodies located outside the central nervous system.

Autonomic nervous system (ANS) Visceral motor neurons, both sympathetic and parasympathetic, that transmit nerve impulses from the central nervous system to smooth muscle, cardiac muscle, and glands; so named because this portion of the nervous system was thought to be self-governing or spontaneous.

Autonomic plexus (PLEK-sus) An extensive network of sympathetic and parasympathetic fibers; the cardiac, celiac, and pelvic plexuses are located in the thorax, abdomen, and pelvis, respectively.

Autophagy (aw-TOF-a-jē) Process by which worn-out organelles are digested within lysosomes.

Autosome (AW-tō-sōm) Any chromosome other than the pair of sex chromosomes.

Axilla (ak-SIL-a) The small hollow beneath the arm where it joins the body at the shoulders. Also called the **armpit.**

Axon (AK-son) The usually single, long process of a nerve cell that carries a nerve impulse away from the cell body.

Axon terminal Terminal branch of an axon and its collateral. Also called a **telodendrium** (tel-ō-DEN-drē-um).

Azygos (AZ-ī-gos) An anatomical structure that is not paired; occurring singly.

Ball-and-socket joint A synovial joint in which the rounded surface of one bone moves within a cup-shaped depression or fossa of another bone, as in the shoulder or hip joint. Also called a **spheroid** (SFĒ-roid) **joint.**

Baroreceptor (bar′-ō-re-SEP-tor) Nerve cell capable of responding to changes in blood, air, or fluid pressure. Also called a **pressoreceptor.**

Bartholin's glands *See* **Greater vestibular glands.**

Basal ganglia (GANG-glē-a) Paired clusters of cell bodies that make up the central gray matter in each cerebral hemisphere, including the caudate nucleus, lentiform nucleus, claustrum, and amygdaloid body. Also called **cerebral nuclei** (SER-e-bral NOO-klē-ī).

Basement membrane Thin, extracellular layer consisting of a basal lamina secreted by epithelial cells and a reticular lamina secreted by connective tissue cells.

Basilar (BAS-i-lar) **membrane** A membrane in the cochlea of the inner ear that separates the cochlear duct from the scala tympani and on which the spiral organ (organ of Corti) rests.

Basophil (BĀ-sō-fil) A type of white blood cell that is characterized by a pale nucleus and large granules that stain readily with basic dyes.

B cell A lymphocyte that develops into an antibody-producing plasma cell or a memory cell.

Belly The abdomen. The gaster or prominent, fleshy part of a skeletal muscle.

Beta (BĀ-ta) **cell** A cell in the pancreatic islets (islets of Langerhans) in the pancreas that secretes insulin.

Beta receptor Receptor found on visceral effectors innervated by most sympathetic postganglionic axons; in general, stimulation of beta receptors leads to inhibition.

Bicuspid (bī-KUS-pid) **valve** Atrioventricular (AV) valve on the left side of the heart. Also called the **mitral valve.**

Bifurcate (bī-FUR-kāt) Having two branches or divisions; forked.

Bilateral (bī-LAT-er-al) Pertaining to two sides of the body.

Bile (BĪL) A secretion of the liver consisting of water, bile salts, bile pigments, cholesterol, lecithin, and several ions; it assumes a role in emulsification of triglycerides prior to their digestion.

Biliary (BIL-ē-er-ē) Relating to bile, the gallbladder, or the bile ducts.

Blastocele (BLAS-tō-sēl) The fluid-filled cavity within the blastocyst.

Blastocyst (BLAS-tō-sist) In the development of an embryo, a hollow ball of cells that consists of blastocele (the internal cavity), trophoblast (outer cells), and inner cell mass.

Blastomere (BLAS-tō-mēr) One of the cells resulting from the cleavage of a fertilized ovum.

Blastula (BLAS-tyoo-la) An early stage in the development of a zygote.

Blind spot Area in the retina at the end of the optic (II) nerve in which there are no light receptor cells.

Blood–brain barrier (BBB) A barrier consisting of specialized blood capillaries and astrocytes that prevents the passage of certain substances from the blood to the cerebrospinal fluid and brain.

Blood island Isolated mass and cord of mesenchyme in the mesoderm from which blood vessels develop.

Blood–testis barrier A barrier formed by sustentacular (Sertoli) cells that prevents an immune response against antigens produced by sperm cells and developing cells by isolating the cells from the blood.

Body cavity A space within the body that contains various internal organs.

Bolus (BŌ-lus) A soft, rounded mass, usually food, that is swallowed.

Bony labyrinth (LAB-i-rinth) A series of cavities within the petrous portion of the temporal bone forming the vestibule, cochlea, and semicircular canals of the inner ear.

Bowman's capsule *See* **Glomerular capsule.**

Brachial plexus (BRĀ-kē-al PLEK-sus) A network of nerve fibers of the anterior rami of spinal nerves C5, C6, C7, C8, and T1. The nerves that emerge from the brachial plexus supply the upper limb.

Brain A mass of nervous tissue located in the cranial cavity.

Brain stem The portion of the brain immediately superior to the spinal cord, which is made up of the medulla oblongata, pons, and midbrain.

Broad ligament A double fold of parietal peritoneum attaching the uterus to the side of the pelvic cavity.

Broca's (BRŌ-kaz) **area** Motor area of the brain in the frontal lobe that translates thoughts into speech. Also called the **motor speech area.**

Bronchi (BRONG-kē) Branches of the respiratory passageway including primary bronchi (the two divisions of the trachea), secondary or lobar bronchi (divisions of the primary that are distributed to the lobes of the lung), and tertiary or segmental bronchi (divisions of the secondary that are distributed to bronchopulmonary segments of the lung).

Bronchial tree The trachea, bronchi, and their branching structures.

Bronchiole (BRONG-kē-ōl) Branch of a tertiary bronchus further dividing into terminal bronchioles (distributed to lobules of the lung), which divide into respiratory bronchioles (distributed to alveolar sacs).

Bronchopulmonary (brong'-kō-PUL-mō-ner-ē) **segment** One of the smaller divisions of a lobe of a lung supplied by its own branches of a bronchus.

Bronchus (BRONG-kus) One of the two large branches of the trachea. *Plural,* **bronchi** (BRONG-kē).

Brunner's gland *See* **Duodenal gland.**

Buccal (BUK-al) Pertaining to the cheek or mouth.

Bulb of penis Expanded portion of the base of the corpus spongiosum penis.

Bulbourethral (bul'-bō-yoo-RĒ-thral) **gland** One of a pair of glands located inferior to the prostate gland on either side of the urethra that secretes an alkaline fluid into the cavernous urethra. Also called a **Cowper's** (KOW-perz) **gland.**

Bundle branch One of the two branches of the atrioventricular (AV) bundle made up of specialized muscle fibers (cells) that transmit electrical impulses to the ventricles.

Bundle of His *See* **Atrioventricular (AV) bundle.**

Bursa (BUR-sa) A sac or pouch of synovial fluid located at friction points, especially about joints.

Bursitis (bur-SĪ-tis) Inflammation of a bursa.

Calcaneal tendon The tendon of the soleus, gastrocnemius, and plantaris muscles at the back of the heel. Also called the **Achilles** (a-KIL-ēz) **tendon.**

Calcification (kal-si-fi-KĀ-shun) Deposition of mineral salts, primarily hydroxyapatite, in a framework formed by collagen fibers in which the tissue hardens. Also called **mineralization.**

Calcitonin (kal-si-TŌ-nin) **(CT)** A hormone produced by the thyroid gland that lowers the calcium and phosphate levels of the blood by inhibiting bone breakdown and accelerating calcium absorption by bones.

Calyx (KĀL-iks) Any cuplike division of the kidney pelvis. *Plural,* **calyces** (KĀ-li-sēz).

Canaliculus (kan'-a-LIK-yoo-lus) A small channel or canal, as in bones, where they connect lacunae. *Plural,* **canaliculi** (kan'-a-LIK-yoo-lī).

Canal of Schlemm *See* **Scleral venous sinus.**

Capillary (KAP-i-lar'-ē) A microscopic blood vessel located between an arteriole and venule through which materials are exchanged between blood and body cells.

Carcinogen (kar-SIN-ō-jen) Any substance that causes cancer.

Cardiac (KAR-dē-ak) **cycle** A complete heartbeat consisting of systole (contraction) and diastole (relaxation) of both atria plus systole and diastole of both ventricles.

Cardiac muscle An organ specialized for contraction, composed of striated muscle fibers (cells), forming the wall of the heart, and stimulated by an intrinsic conduction system and visceral motor neurons.

Cardiac notch An angular notch in the anterior border of the left lung.

Cardinal ligament A ligament of the uterus, extending laterally from the cervix and vagina as a continuation of the broad ligament.

Cardiology (kar-dē-OL-ō-jē) The study of the heart and diseases associated with it.

Cardiovascular (kar-dē-ō-VAS-kyoo-lar) **center** Groups of neurons scattered within the medulla oblongata that regulate heart rate, force of contraction, and blood vessel diameter.

Carina (ka-RĪ-na) A ridge on the inside of the division of the right and left primary bronchi.

Carotene (KAR-o-tēn) Yellow-orange pigment found in the stratum corneum of the epidermis that, together with melanin, accounts for the yellowish coloration of skin.

Carotid (ka-ROT-id) **body** Receptor on or near the carotid sinus that responds to alterations in blood levels of oxygen, carbon dioxide, and hydrogen ions.

Carotid sinus A dilated region of the internal carotid artery immediately superior to the branching of the common carotid artery that contains receptors that monitor blood pressure.

Carpus (KAR-pus) A collective term for the eight bones of the wrist.

Cartilage (KAR-ti-lij) A type of connective tissue consisting of chondrocytes in lacunae embedded in a dense network of collagen and elastic fibers and a matrix of chondroitin sulfate.

Cartilaginous (kar′-ti-LAJ-i-nus) **joint** A joint without a synovial (joint) cavity where the articulating bones are held tightly together by cartilage, allowing little or no movement.

Cast A small mass of hardened material formed within a cavity in the body and then discharged from the body; can originate in different areas and be composed of various materials.

Cauda equina (KAW-da ē-KWĪ-na) A tail-like collection of roots of spinal nerves at the inferior end of the spinal canal.

Cecum (SĒ-kum) A blind pouch at the proximal end of the large intestine to which the ileum is attached.

Celiac plexus (SĒ-lē-ak PLEK-sus) A large mass of ganglia and nerve fibers located at the level of the superior part of the first lumbar vertebra. Also called the **solar plexus.**

Cell The basic structural and functional unit of all organisms; the smallest structure capable of performing all the activities vital to life.

Cell division Process by which a cell reproduces itself that consists of a nuclear division (mitosis) and a cytoplasmic division (cytokinesis); types include somatic and reproductive cell division.

Cell inclusion A lifeless, often temporary constituent in the cytoplasm of a cell as opposed to an organelle.

Cellular respiration *See* **Oxidation.**

Cementum (se-MEN-tum) Calcified tissue covering the root of a tooth.

Central canal A circular channel running longitudinally in the center of an osteon (Haversian system) of mature compact bone, containing blood and lymphatic vessels and nerves. Also called a **Haversian** (ha-VER-shun) **canal.** A microscopic tube running the length of the spinal cord in the gray commissure.

Central fovea (FŌ-vē-a) A cuplike depression in the center of the macula lutea of the retina, containing cones only; the area of clearest vision.

Central nervous system (CNS) That portion of the nervous system that consists of the brain and spinal cord.

Centrioles (SEN-trē-ōlz) Paired, cylindrical structures within a centrosome, each consisting of a ring of microtubules and arranged at right angles to each other. Assume a role in the formation and regeneration of flagella and cilia.

Centromere (SEN-trō-mēr) The clear, constricted portion of a chromosome where the two chromatids are joined; serves as the point of attachment for the chromosomal microtubules.

Centrosome (SEN-trō-sōm) Organelle that consists of a dense area of cytoplasm (pericentriolar area) that forms microtubules in nondividing cells, the mitotic spindle in dividing cells, and centrioles that assume a role in the formation and regeneration of flagella and cilia.

Cerebellar peduncle (ser-e-BEL-ar pe-DUNG-kul) A bundle of nerve fibers connecting the cerebellum with the brain stem.

Cerebellum (ser-e-BEL-um) The portion of the brain lying posterior to the medulla oblongata and pons, concerned with coordination of movements.

Cerebral aqueduct (SER-ē-bral AK-we-dukt) A channel through the midbrain connecting the third and fourth ventricles and containing cerebrospinal fluid.

Cerebral arterial circle A ring of arteries forming an anastomosis at the base of the brain between the internal carotid and basilar arteries and arteries supplying the brain. Also called the **circle of Willis.**

Cerebral cortex The surface of the cerebral hemispheres, 2–4 mm thick, consisting of six layers of nerve cell bodies (gray matter) in most areas.

Cerebral peduncle (pe-DUNG-kul) One of a pair of nerve fiber bundles located on the ventral surface of the midbrain, conducting nerve impulses between the pons and the cerebral hemispheres.

Cerebrospinal (se-rē′-brō-SPĪ-nal) **fluid (CSF)** A fluid produced in the choroid plexuses and ependymal cells of the ventricles of the brain that circulates in the ventricles and the subarachnoid space around the brain and spinal cord.

Cerebrum (SER-ē-brum) The two hemispheres of the forebrain, making up the largest part of the brain.

Cerumen (se-ROO-men) Waxlike secretion produced by ceruminous glands in the external auditory meatus (ear canal).

Ceruminous (se-ROO-mi-nus) **gland** A modified sudoriferous (sweat) gland in the external auditory meatus that secretes cerumen (ear wax).

Cervical ganglion (SER-vi-kul GANG-glē-on) A cluster of nerve cell bodies of postganglionic sympathetic neurons located in the neck, near the vertebral column.

Cervical plexus (PLEX-us) A network of neuron fibers formed by the anterior rami of the first four cervical nerves.

Cervix (SER-viks) Neck; any constricted portion of an organ, such as the inferior cylindrical part of the uterus.

Chemoreceptor (kē′-mō-rē-SEP-tor) Receptor outside the central nervous system on or near the carotid and aortic bodies that detects the presence of chemicals.

Chiasm (kī-AZM) A crossing; especially the crossing of the optic (II) nerve fibers.

Chief cell The secreting cell of a gastric gland that produces pepsinogen, the precursor of the enzyme pepsin, and the enzyme gastric lipase. Also called a **zymogenic** (zī-mō-JEN-ik) **cell.**

Chiropractic (kī-rō-PRAK-tik) A system of treating disease by using one's hands to manipulate body parts, mostly the vertebral column.

Cholesterol (kō-LES-te-rol) Classified as a lipid, the most abundant steroid in animal tissues; located in cell membranes and used for the synthesis of steroid hormones and bile salts.

Cholinergic (kō′-lin-ER-jik) **fiber** A nerve ending that liberates acetylcholine at a synapse.

Chondrocyte (KON-drō-sīt) Cell of mature cartilage.

Chordae tendineae (KOR-dē TEN-din-nē-ē) Tendonlike, fibrous cords that connect the heart valves with the papillary muscles.

Chorion (KŌ-rē-on) The most superficial fetal membrane that becomes the principal embryonic portion of the placenta; serves a protective and nutritive function.

Chorionic villus (kō′-rē-ON-ik VIL-lus) Fingerlike projection of the chorion that grows into the decidua basalis of the endometrium and contains fetal blood vessels.

Chorionic villi sampling (CVS) The removal of a sample of chorionic villus tissue by means of a catheter to analyze the tissue for prenatal genetic defects.

Choroid (KŌ-royd) One of the vascular coats of the eyeball.

Choroid plexus (PLEX-sus) A vascular structure located in the roof of each of the four ventricles of the brain; produces cerebrospinal fluid.

Chromaffin (krō-MAF-in) **cell** Cell that has an affinity for chrome salts, due in part to the presence of the precursors of the neurotransmitter epinephrine; found, among other places, in the adrenal medulla.

Chromatid (KRŌ-ma-tid) One of a pair of identical connected nucleoprotein strands that are joined at the centromere and separate during cell division, each becoming a chromosome of one of the two daughter cells.

Chromatin (KRŌ-ma-tin) The threadlike mass of the genetic material consisting principally of DNA, which is present in the nucleus of a nondividing or interphase cell.

Chromatolysis (krō′-ma-TOL-i-sis) The breakdown of chromatophilic substance (Nissl bodies) into finely granular masses in the cell body of a central or peripheral neuron whose process (axon or dendrite) has been damaged.

Chromatophilic substance Rough endoplasmic reticulum in the cell bodies of neurons that functions in protein synthesis. Also called **Nissl bodies.**

Chromosomal microtubule (mī-krō-TOOB-yool) Microtubule formed during prophase of mitosis that originates from centromeres, extends from a centromere to a pole of the cell, and assists in chromosomal movement; constitutes a part of the mitotic spindle.

Chromosome (KRŌ-mo-sōm) One of the 46 small, dark-staining bodies that appear in the nucleus of a human diploid ($2n$) cell during cell division.

Chronic (KRON-ik) Long-term or frequently recurring; applied to a disease that is not acute.

Chyle (KĪL) The milky fluid found in the lacteals of the small intestine after digestion.

Chyme (KĪM) The semifluid mixture of partly digested food and digestive secretions found in the stomach and small intestine during digestion of a meal.

Ciliary (SIL-ē-ar′-ē) **body** One of the three portions of the vascular tunic of the eyeball, the others being the choroid and the iris; includes the ciliary muscle and ciliary processes.

Ciliary ganglion (GANG-glē-on) A very small parasympathetic ganglion whose preganglionic fibers come from the oculomotor (III) nerve and whose postganglionic fibers carry nerve impulses to the ciliary muscle and the sphincter muscle of the iris.

Cilium (SIL-ē-um) A hair or hairlike process projecting from a cell that may be used to move the entire cell or to move substances along the surface of the cell.

Circle of Willis *See* **Cerebral arterial circle.**

Circular folds Permanent, deep, transverse folds in the mucosa and submucosa of the small intestine that increase the surface area for absorption. Also called **plicae circulares** (PLĪ-kē SER-kyoo-lar-ēs).

Circumduction (ser′-kum-DUK-shun) A movement at a synovial joint in which the distal end of a bone moves in a circle while the proximal end remains relatively stable.

Circumvallate papilla (ser′-kum-VAL-āt pa-PIL-a) One of the circular projections that is arranged in an inverted V-shaped row at the posterior portion of the tongue; the largest of the elevations on the upper surface of the tongue containing taste buds.

Cisterna chyli (sis-TER-na KĪ-lē) The origin of the thoracic duct.

Cleavage The rapid mitotic divisions following the fertilization of a secondary oocyte, resulting in an increased number of progressively smaller cells, called blastomeres, so that the overall size of the zygote remains the same.

Clitoris (KLI-tor-is) An erectile organ of the female located at the anterior junction of the labia minora that is homologous to the male penis.

Clone (KLŌN) A population of cells identical to itself.

Coccyx (KOK-six) The fused bones at the inferior end of the vertebral column.

Cochlea (KŌK-lē-a) The winding, cone-shaped tube forming a portion of the inner ear and containing the spiral organ (organ of Corti).

Cochlear duct The membranous cochlea consisting of a spirally arranged tube enclosed in the bony cochlea and lying along its outer wall. Also called the **scala media** (SCA-la MĒ-dē-a).

Collagen (KOL-a-jen) A protein that is the main organic constituent of connective tissue.

Collateral circulation The alternate route taken by blood through an anastomosis.

Colliculus (ko-LIK-yoo-lus) A small elevation.

Colon (KŌ-lon) The division of the large intestine consisting of ascending, transverse, descending, and sigmoid portions.

Colostrum (kō-LOS-trum) A thin, cloudy fluid secreted by the mammary glands a few days prior to or after delivery before true milk is secreted.

Column (KOL-um) Group of white matter tracts in the spinal cord.

Commissure (KOM-i-shūr) The angular junction of the eyelids at either corner of the eyes.

Common bile duct A tube formed by the union of the common hepatic duct and the cystic duct that empties bile into the duodenum at the hepatopancreatic ampulla (ampulla of Vater).

Compact (dense) bone tissue Bone tissue that contains few spaces between osteons (Haversian systems); forms the external portion of all bones and the bulk of the diaphysis (shaft) of long bones.

Concha (KONG-ka) A scroll-like bone found in the skull. *Plural,* **conchae** (KONG-kē). Also called a **turbinate** (TUR-bi-nāt).

Conduction myofiber Muscle fiber (cell) in the subendocardial tissue of the heart specialized for conducting an action potential to the myocardium; part of the conduction system of the heart. Also called a **Purkinje** (pur-KIN-jē) **fiber.**

Conduction system An intrinsic regulating system composed of specialized muscle tissue that generates and distributes electrical impulses that stimulate cardiac muscle fibers (cells) to contract.

Conductivity (kon′-duk-TIV-i-tē) The ability to carry the effect of a stimulus from one part of a cell to another; highly developed in nerve and muscle fibers (cells).

Condyloid (KON-di-loid) **joint** A synovial joint structured so that an oval-shaped condyle of one bone fits into an elliptical cavity of another bone, permitting side-to-side and back-and-forth movements, as at the joint at the wrist between the radius and carpals. Also called an **ellipsoidal** (e-lip-SOY-dal) **joint.**

Cone The light-sensitive receptor in the retina concerned with color vision.

Congenital (kon-JEN-i-tal) Present at the time of birth.

Conjunctiva (kon′-junk-TĪ-va) The delicate membrane covering the eyeballs and lining the eyes.

Connective tissue The most abundant of the four basic tissue types in the body, performing the functions of binding and supporting; consists of relatively few cells in a great deal of intercellular substance.

Consciousness (KON-shus-nes) A state of wakefulness in which an individual is fully alert, aware, and oriented as a result of feedback between the cerebral cortex and reticular activating system.

Contractility (kon′-trak-TIL-i-tē) The capacity of the contractile elements in muscle tissue to shorten and develop tension.

Contralateral (kon′-tra-LAT-er-al) On the opposite side; affecting the opposite side of the body.

Conus medullaris (KŌ-nus med-yoo-LAR-is) The tapered portion of the spinal cord inferior to the lumbar enlargement.

Convergence (con-VER-jens) Anatomical arrangement in which the synaptic end bulbs of several presynaptic neurons terminate on one postsynaptic neuron. Medial movement of the two eyeballs so that both are directed toward a near object being viewed to produce a single image.

Cornea (KOR-nē-a) The nonvascular, transparent fibrous coat through which the iris can be seen.

Corona (kō-RŌ-na) Margin of the glans penis.

Coronal (kō-RŌ-nal) **plane** A plane that runs vertical to the ground and divides the body into anterior and posterior portions. Also called **frontal plane.**

Corona radiata Deepest layer of granulosa cells around an oocyte attached to the zona pellucida.

Coronary circulation The pathway followed by the blood from the ascending aorta through the blood vessels supplying the heart and returning to the right atrium. Also called **cardiac circulation.**

Coronary sinus (SĪ-nus) A wide venous channel on the posterior surface of the heart that collects the blood from the coronary circulation and returns it to the right atrium.

Corpora quadrigemina (KOR-por-a kwad-ri-JEM-in-a) Four small elevations (superior and inferior colliculi) on the dorsal region of the midbrain concerned with visual and auditory functions.

Corpus (KOR-pus) The principal part of any organ; any mass or body.

Corpus albicans (AL-bi-kanz) A white fibrous patch in the ovary that forms after the corpus luteum regresses.

Corpus callosum (kal-LŌ-sum) The great commissure of the brain between the cerebral hemispheres.

Corpuscle of touch The sensory receptor for the sensation of touch; found in the dermal papillae, especially in palms and soles. Also called a **Meissner's** (MĪS-nerz) **corpuscle.**

Corpus luteum (LOO-tē-um) A yellow endocrine gland in the ovary formed when a follicle has discharged its secondary oocyte; secretes estrogens, progesterone, and relaxin.

Corpus striatum (strī-Ā-tum) An area in the interior of each cerebral hemisphere composed of the caudate and lentiform nuclei of the basal ganglia and white matter of the internal capsule, arranged in a striated manner.

Cortex (KOR-teks) An outer layer of an organ. The convoluted layer of gray matter covering each cerebral hemisphere.

Costal cartilage (KOS-tal KAR-ti-lij) Hyaline cartilage that attaches a rib to the sternum.

Cowper's gland *See* **Bulbourethral gland.**

Cranial (KRĀ-nē-al) **cavity** A subdivision of the dorsal body cavity formed by the cranial bones and containing the brain.

Cranial nerve One of 12 pairs of nerves that leave the brain, pass through foramina in the skull, and supply the head, neck, and part of the trunk; each is designated by a Roman numeral and a name.

Craniosacral (krā-nē-ō SĀ-kral) **outflow** The fibers of parasympathetic preganglionic neurons, which have their cell bodies located in nuclei in the brain stem and in the lateral gray matter of the sacral portion of the spinal cord.

Cranium (KRĀ-nē-um) The skeleton of the skull that protects the brain and the organs of sight, hearing, and balance; includes the frontal, parietal, temporal, occipital, sphenoid, and ethmoid bones.

Crista (KRIS-ta) A crest or ridged structure. A small elevation in the ampulla of each semicircular duct that serves as a receptor for dynamic equilibrium.

Crossing-over The exchange of a portion of one chromatid with another in a tetrad during meiosis. It permits an exchange of genes among chromatids and is one factor that results in genetic variation.

Crus (KRUS) **of penis** Separated, tapered portion of the corpora cavernosa penis. *Plural,* **crura** (KROO-ra).

Crypt of Lieberkühn *See* **Intestinal gland.**

Cupula (KUP-yoo-la) A mass of gelatinous material covering the hair cells of a crista, a receptor in the ampulla of a semicircular canal stimulated when the head moves.

Cutaneous (kyoo-TĀ-nē-us) Pertaining to the skin.

Cystic (SIS-tik) **duct** The duct that transports bile from the gallbladder to the common bile duct.

Cystitis (sis-TĪ-tis) Inflammation of the urinary bladder.

Cytokinesis (sī′-tō-ki-NĒ-sis) Division of the cytoplasm.

Cytology (sī-TOL-ō-jē) The study of cells.

Cytolysis (sī-TOL-i-sis) The destruction of living cells in which the contents leak out.

Cytoplasm (SĪ-tō-plazm) All the cellular contents between the plasma membrane and nucleus.

Cytoskeleton Complex internal structure of cytoplasm consisting of microfilaments, microtubules, and intermediate filaments.

Cytosol (SĪ-tō-sol) The semifluid portion of cytoplasm in which organelles (except the nucleus) and inclusions are suspended and solutes are dissolved. Also called **intracellular fluid.**

Dartos (DAR-tōs) The contractile tissue under the skin of the scrotum.

Debility (dē-BIL-i-tē) Weakness of tonicity in functions or organs of the body.

Decidua (dē-SID-yoo-a) That portion of the endometrium of the uterus (all but the deepest layer) that is modified for pregnancy and shed after childbirth.

Deciduous (dē-SID-yoo-us) Falling off or being shed seasonally or at a particular stage of development. In the body, referring to the first set of teeth.

Decussation (dē′-ku-SĀ-shun) A crossing-over; usually refers to the crossing of most of the fibers in the large motor tracts to opposite sides in the medullary pyramids.

Deep Away from the surface of the body.

Deep fascia (FASH-ē-a) A sheet of connective tissue wrapped around a muscle to hold it in place.

Deep inguinal (IN-gwi-nal) **ring** A slitlike opening in the aponeurosis of the transversus abdominis muscle that represents the origin of the inguinal canal.

Deep-venous thrombosis (DVT) The presence of a thrombus in a vein, usually a deep vein of the lower limbs.

Defecation (def-e-KĀ-shun) The discharge of feces from the rectum.

Degeneration (dē-jen′-er-Ā-shun) A change from a higher to a lower state; a breakdown in structure.

Deglutition (dē-gloo-TISH-un) The act of swallowing.

Delta cell A cell in the pancreatic islets (islets of Langerhans) in the pancreas that secretes somatostatin.

Dendrite (DEN-drīt) A nerve cell process that carries a nerve impulse toward the cell body.

Dendritic (den-DRIT-ik) **cell** One type of antigen-presenting cell with long branchlike projections, for example, Langerhans cells in the epidermis.

Dens (DENZ) Tooth.

Dental caries (KA-rēz) Gradual demineralization of the enamel and dentin of a tooth that may invade the pulp and alveolar bone. Also called **tooth decay.**

Denticulate (den-TIK-yoo-lāt) Finely toothed or serrated; characterized by a series of small, pointed projections.

Dentin (DEN-tin) The osseous tissues of a tooth enclosing the pulp cavity.

Dentition (den-TI-shun) The eruption of teeth. The number, shape, and arrangement of teeth.

Deoxyribonucleic (dē-ok′-sē-rī′-bō-nyoo-KLĒ-ik) **acid (DNA)** A nucleic acid in the shape of a double helix constructed of nucleotides consisting of one of four nitrogenous bases (adenine, cytosine, guanine, or thymine), deoxyribose, and a phosphate group; encoded in the nucleotides is genetic information.

Depression (dē-PRESS-shun) Movement in which a part of the body moves inferiorly.

Dermal papilla (pa-PILL-a) Fingerlike projection of the papillary region of the dermis that may contain blood capillaries or corpuscles of touch (Meissner's corpuscles).

Dermatology (der-ma-TOL-ō-jē) The medical specialty dealing with diseases of the skin.

Dermatome (DER-ma-tōm) The cutaneous area developed from one embryonic spinal cord segment and receiving most of its innervation from one spinal nerve.

Dermis (DER-mis) A layer of dense connective tissue lying deep to the epidermis; the true skin or corium.

Descending colon (KŌ-lon) The part of the large intestine descending from the left colic (splenic) flexure to the level of the left iliac crest.

Detrusor (de-TROO-ser) **muscle** Muscle in the wall of the urinary bladder.

Developmental anatomy The study of development from the fertilized egg to the adult form. The branch of anatomy called embryology is generally restricted to the study of development from the fertilized egg through the eighth week in utero.

Diagnosis (dī-ag-NŌ-sis) Distinguishing one disease from another or determining the nature of a disease from signs and symptoms by inspection, palpation, laboratory tests, and other means.

Diaphragm (DĪ-a-fram) Any partition that separates one area from another, especially the dome-shaped skeletal muscle between the thoracic and abdominal cavities. Also a dome-shaped structure that fits over the cervix, usually with a spermicide, to prevent conception.

Diaphysis (dī-AF-i-sis) The shaft of a long bone.

Diarthrosis (dī′-ar-THRŌ-sis) Articulation in which opposing bones move freely, as in a hinge joint.

Diastole (dī-AS-tō-lē) In the cardiac cycle, the phase of relaxation or dilation of the heart muscle, especially of the ventricles.

Diencephalon (dī′-en-SEF-a-lon) A part of the brain consisting primarily of the thalamus and the hypothalamus.

Differentiation (dif′-e-ren′-shē-Ā-shun) Acquisition of specific functions different from those of the original general type.

Diffusion (dif-YOO-zhun) A passive process in which there is a net or greater movement of molecules or ions from a region of high concentration to a region of low concentration until equilibrium is reached.

Dilate (DĪ-lāt) To expand or swell.

Diploid (DIP-loyd) Having the number of chromosomes characteristically found in the somatic cells of an organism. Symbolized 2*n*.

Diplopia (di-PLŌ-pē-a) Double vision.

Direct pathways Motor pathways that convey information from the brain down the spinal cord; related to precise, voluntary movements of skeletal muscles. Also called **pyramidal** (pi-RAM-i-dal) **pathways.**

Disease Any change from a state of health.

Distal (DIS-tal) Farther from the attachment of a limb to the trunk or a structure; farther from the point of origin.

Divergence (di-VER-jens) An anatomical arrangement in which the synaptic end bulbs of one presynaptic neuron terminate on several postsynaptic neurons.

Diverticulum (dī-ver-TIK-yoo-lum) A sac or pouch in the wall of a canal or organ, especially in the colon.

Dorsal body cavity Cavity near the dorsal surface of the body that consists of a cranial cavity and vertebral canal.

Dorsal ramus (RĀ-mus) A branch of a spinal nerve containing motor and sensory fibers supplying the muscles, skin, and bones of the posterior part of the head, neck, and trunk.

Dorsiflexion (dor′-si-FLEK-shun) Bending the foot in the direction of the dorsum (upper surface).

Duct of Santorini *See* **Accessory duct.**

Duct of Wirsung *See* **Pancreatic duct.**

Ductus arteriosus (DUK-tus ar-tē-rē-Ō-sus) A small vessel connecting the pulmonary trunk with the aorta; found only in the fetus.

Ductus (vas) deferens (DEF-er-ens) The duct that conducts sperm cells from the epididymis to the ejaculatory duct. Also called the **seminal duct.**

Ductus epididymis (ep′-i-DID-i-mis) A tightly coiled tube inside the epididymis, distinguished into a head, body, and tail, in which sperm cells undergo maturation.

Ductus venosus (ve-NŌ-sus) A small vessel in the fetus that helps the circulation bypass the liver.

Duodenal (doo′-ō-DĒ-nal) **gland** Gland in the submucosa of the duodenum that secretes an alkaline mucus to protect the lining of the small intestine from the action of enzymes and to help neutralize the acid in chyme. Also called **Brunner's** (BRUN-erz) **gland.**

Duodenal papilla (pa-PILL-a) An elevation on the duodenal mucosa that receives the hepatopancreatic ampulla (ampulla of Vater).

Duodenum (doo′-ō-DĒ-num) The first 25 cm (10 in.) of the small intestine.

Dura mater (DYOO-ra MA-ter) The outer membrane (meninx) covering the brain and spinal cord.

Dynamic equilibrium (ē-kwi-LIB-rē-um) The maintenance of body position, mainly the head, in response to sudden movements such as rotation.

Dysfunction (dis-FUNK-shun) Absence of complete normal function.

Dystrophia (dis-TRO-fē-a) Progressive weakening of a muscle.

Ectoderm The most superficial of the three primary germ layers that gives rise to the nervous system and the epidermis of skin and its derivatives.

Ectopic (ek-TOP-ik) Out of the normal location, as in ectopic pregnancy.

Edema (e-DĒ-ma) An abnormal accumulation of interstitial fluid.

Effector (e-FEK-tor) The organ of the body, either a muscle or a gland, that responds to a motor neuron impulse.

Efferent arteriole (EF-er-ent ar-TĒ-rē-ōl) A vessel of the renal vascular system that transports blood from the glomerulus to the peritubular capillary.

Efferent ducts A series of coiled tubes that transport sperm cells from the rete testis to the epididymis.

Ejaculation (e-jak-yoo-LĀ-shun) The reflex ejection or expulsion of semen from the penis.

Ejaculatory (e-JAK-yoo-la-tō′-rē) **duct** A tube that transports sperm cells from the ductus (vas) deferens to the prostatic urethra.

Elasticity (e-las-TIS-i-tē) The ability of a tissue to return to its original shape after contraction or extension.

Elevation (el-e-VĀ-shun) Movement in which a part of the body moves superiorly.

Embolism (EM-bō-lizm) Obstruction or closure of a vessel by an embolus.

Embolus (EM-bō-lus) A blood clot, bubble of air, fat from broken bones, mass of bacteria, or other debris or foreign material transported by the blood.

Embryo (EM-brē-ō) The young of any organism in an early stage of development; in humans, the developing organism from fertilization to the end of the eighth week in utero.

Embryology (em′-brē-OL-ō-jē) The study of development from the fertilized egg to the end of the eighth week in utero.

Emigration (em′-i-GRĀ-shun) The passage of white blood cells through intact blood vessel walls.

Emission (ē-MISH-un) Propulsion of sperm cells into the urethra in response to peristaltic contractions of the ducts of the testes, epididymides, and ductus (vas) deferens as a result of sympathetic stimulation.

Emphysema (em′-fi-SĒ-ma) A swelling or inflation of air passages due to loss of elasticity in the alveoli.

Emulsification (ē-mul′-si-fi-KĀ-shun) The dispersion of large triglyceride globules to smaller, uniformly distributed particles in the presence of bile.

Enamel (e-NAM-el) The hard, white substance covering the crown of a tooth.

Endocardium (en-dō-KAR-dē-um) The layer of the heart wall, composed of endothelium and smooth muscle, that lines the inside of the heart and covers the valves and tendons that hold the valves open.

Endochondral ossification (en′-dō-KON-dral os′-i-fi-KĀ-shun) The replacement of cartilage by bone. Also called **intracartilaginous** (in′-tra-kar′-ti-LAJ-i-nus) **ossification.**

Endocrine (EN-dō-krin) **gland** A gland that secretes hormones into the blood; a ductless gland.

Endocrinology (en′-dō-kri-NOL-ō-jē) The science concerned with the structure and functions of endocrine glands and the diagnosis and treatment of disorders of the endocrine system.

Endoderm The deepest of the three primary germ layers of the developing embryo that gives rise to the gastrointestinal tract, urinary bladder and urethra, and respiratory tract.

Endodontics (en′-dō-DON-tiks) The branch of dentistry concerned with the prevention, diagnosis, and treatment of diseases that affect the pulp, root, periodontal ligament, and alveolar bone.

Endogenous (en-DOJ-e-nus) Growing from or beginning within the organism.

Endolymph (EN-dō-lymf′) The fluid within the membranous labyrinth of the inner ear.

Endometriosis (en′-dō-MĒ-trē-ō′-sis) The growth of endometrial tissue outside the uterus.

Endometrium (en′-dō-MĒ-trē-um) The mucous membrane lining the uterus.

Endomysium (en′-dō-MĪZ-ē-um) Invagination of the perimysium separating each individual muscle fiber (cell).

Endoneurium (en′-dō-NYOO-rē-um) Connective tissue wrapping around individual nerve fibers (cells).

Endoplasmic reticulum (en′-dō-PLAZ-mik re-TIK-yoo-lum) **(ER)** A network of channels running through the cytoplasm of a cell that serves in intracellular transportation, support, storage, synthesis, and packaging of molecules. Portions of ER where ribosomes are attached to the outer surface are called **rough (granular) ER;** portions that have no ribosomes are called **smooth (agranular) ER.**

End organ of Ruffini *See* **Type II cutaneous mechanoreceptor.**

Endosteum (en-DOS-tē-um) The membrane that lines the medullary cavity of bones, consisting of osteoprogenitor cells and scattered osteoclasts.

Endothelial-capsular (en-dō-THĒ-lē-ar) **membrane** A filtration membrane in a nephron of a kidney consisting of the endothelium and basement membrane of the glomerulus and the epithelium of the visceral layer of the glomerular (Bowman's) capsule.

Endothelium (en′-dō-THĒ-lē-um) The layer of simple squamous epithelium that lines the cavities of the heart, blood vessels, and lymphatic vessels.

Enteroendocrine (en-ter-ō-EN-dō-krin) **cell** A hormone-producing cell in the gastrointestinal mucosa. An example is a G cell in the gastric mucosa that produces the hormone gastrin.

Enzyme (EN-zīm) A substance that affects the speed of chemical changes; an organic catalyst, usually a protein.

Eosinophil (ē′-ō-SIN-ō-fil) A type of white blood cell characterized by granular cytoplasm readily stained by eosin.

Ependymal (e-PEN-de-mal) **cell** Neuroglial cell that lines ventricles of the brain and probably assists in the circulation of cerebrospinal fluid (CSF).

Epicardium (ep′-i-KAR-dē-um) The thin outer layer of the heart wall, composed of serous tissue and mesothelium. Also called the **visceral pericardium.**

Epidemic (ep′-i-DEM-ik) A disease that occurs above the expected level among individuals in a population.

Epidemiology (ep′-i-dē-mē-OL-ō-jē) Medical science concerned with the occurrence and distribution of disease in human populations.

Epidermis (ep-i-DERM-is) The outermost, thinner layer of skin, composed of keratinized stratified squamous epithelium.

Epididymis (ep′-i-DID-i-mis) A comma-shaped organ that lies along the posterior border of the testis and contains the ductus epididymis, in which sperm undergo maturation. *Plural,* **epididymides** (ep′-i-DID-i-mi-dēz).

Epidural (ep′-i-DOO-ral) **space** A space between the spinal dura mater and the vertebral canal, containing areolar connective tissue and a plexus of veins.

Epiglottis (ep′-i-GLOT-is) A large, leaf-shaped piece of cartilage lying on top of the larynx, with its "stem" attached to the thyroid cartilage and its "leaf" portion unattached and free to move up and down to cover the glottis (vocal folds and rima glottidis).

Epimysium (ep′-i-MĪZ-ē-um) Fibrous connective tissue around muscles.

Epinephrine (ep-ē-NEF-rin) Hormone secreted by the adrenal medulla that produces actions similar to those that result from sympathetic stimulation. Also called **adrenaline** (a-DREN-a-lin).

Epineurium (ep′-i-NYOO-rē-um) The most superficial covering around the entire nerve.

Epiphyseal (ep′-i-FIZ-ē-al) **line** The remnant of the epiphyseal plate in a long bone.

Epiphyseal plate The cartilaginous plate between the epiphysis and diaphysis that is responsible for the lengthwise growth of long bones.

Epiphysis (ē-PIF-i-sis) The end of a long bone, usually larger in diameter than the shaft (diaphysis).

Epiphysis cerebri (se-RĒ-brē) Pineal gland.

Epithalamus (ep′-i-THAL-a-mus) Part of the diencephalon superior and posterior to the thalamus, comprising the pineal gland and associated structures.

Epithelial (ep′-i-THĒ-lē-al) **tissue** The tissue that forms glands or the superficial part of the skin and lines blood vessels, hollow organs, and passages that lead externally from the body.

Eponychium (ep′-ō-NIK-ē-um) Narrow band of stratum corneum at the proximal border of a nail that extends from the margin of the nail wall. Also called the **cuticle.**

Erection (ē-REK-shun) The enlarged and stiff state of the penis (or clitoris) resulting from the engorgement of the spongy erectile tissue with blood.

Erythema (er′-e-THĒ-ma) Skin redness usually caused by engorgement of the capillaries in the deeper layers of the skin.

Erythrocyte (e-RITH-rō-sīt) Red blood cell.

Erythropoiesis (e-rith′-rō-poy-Ē-sis) The process by which erythrocytes (red blood cells) are formed.

Erythropoietin (e-rith′-rō-POY-ē-tin) A hormone formed from a plasma protein that stimulates erythrocyte (red blood cell) production.

Esophagus (e-SOF-a-gus) A hollow muscular tube connecting the pharynx and the stomach.

Estrogens (ES-tro-jens) Female sex hormones produced by the ovaries concerned with the development and maintenance of female reproductive structures and secondary sex characteristics, fluid and electrolyte balance, and protein anabolism. Examples are β-estradiol, estrone, and estriol.

Etiology (ē′-tē-OL-ō-jē) The study of the causes of disease, including theories of origin and the organisms, if any, involved.

Euphoria (yoo-FŌR-ē-a) A subjectively pleasant felling of well-being marked by confidence and assurance.

Eupnea (YOOP-nē-a) Normal quiet breathing.

Eustachian tube *See* **Auditory tube.**

Eversion (ē-VER-zhun) The movement of the sole laterally at the ankle joint.

Exacerbation (eg-zas′-er-BĀ-shun) An increase in the severity of symptoms or of a disease.

Excitability (ek-sīt′-a-BIL-i-tē) The ability of muscle tissue to receive and respond to stimuli; the ability of nerve cells to respond to stimuli and convert them into nerve impulses.

Excretion (eks-KRĒ-shun) The process of eliminating waste products from a cell, tissue, or the entire body; or the products excreted.

Exocrine (EK-sō-krin) **gland** A gland that secretes substances into ducts that empty at covering or lining epithelium or directly onto a free surface.

Exocytosis (ex′-ō-sī-TŌ-sis) A process of discharging cellular products too big to go through the membrane. Particles for export are enclosed by Golgi membranes when they are synthesized. Vesicles pinch off from the Golgi complex and carry the enclosed particles to the interior surface of the cell membrane, where the vesicle membrane and plasma membrane fuse and the contents of the vesicle are discharged.

Exogenous (ex-SOJ-e-nus) Originating outside an organ or part.

Extensibility (ek-sten′-si-BIL-i-tē) The ability of muscle tissue to be stretched when pulled.

Extension (ek-STEN-shun) An increase in the angle between two bones, usually in the sagittal plane; restoring a body part to its anatomical position after flexion.

External Located on or near the surface.

External auditory (AW-di-tōr-ē) **canal** or **meatus** (mē-Ā-tus) A curved tube in the temporal bone that leads to the middle ear.

External ear The outer ear, consisting of the auricle, external auditory canal, and tympanic membrane or eardrum.

External nares (NA-rēz) The external nostrils, or the openings into the nasal cavity on the exterior of the body.

External respiration The exchange of respiratory gases between the lungs and blood. Also called **pulmonary respiration.**

Exteroceptor (eks′-ter-ō-SEP-tor) A receptor adapted for the reception of stimuli from outside the body.

Extracellular fluid (ECF) Fluid outside body cells, such as interstitial fluid and plasma.

Extravasation (eks-trav-a-SĀ-shun) The escape of fluid, especially blood, lymph, or serum, from a vessel into the tissues.

Extrinsic (ek-STRIN-sik) Of external origin.

Exudate (EKS-yoo-dāt) Escaping fluid or semifluid material that oozes from a space that may contain serum, pus, and cellular debris.

Falciform ligament (FAL-si-form LIG-a-ment) A sheet of parietal peritoneum between the two principal lobes of the liver.

The ligamentum teres, or remnant of the umbilical vein, lies within its fold.

Fallopian tube *See* **Uterine tube.**

Falx cerebelli (FALKS ser′-e-BEL-lē) A small triangular process of the dura mater attached to the occipital bone in the posterior cranial fossa and projecting inward between the two cerebellar hemispheres.

Falx cerebri (SER-e-brē) A fold of the dura mater extending deep into the longitudinal fissure between the two cerebral hemispheres.

Fascia (FASH-ē-a) A fibrous membrane covering, supporting, and separating muscles.

Fascicle (FAS-i-kul) A small bundle or cluster, especially of nerve or muscle fibers (cells). Also called a **fasciculus** (fa-SIK-yoo-lus). *Plural,* **fasciculi** (fa-SIK-yoo-lī).

Fauces (FAW-sēz) The opening from the mouth into the pharynx.

F-cell A cell in the pancreatic islets (islets of Langerhans) of the pancreas that secretes pancreatic polypeptide.

Feeding (hunger) center A cluster of neurons in the lateral nuclei of the hypothalamus that, when stimulated, brings about feeding.

Female reproductive cycle General term for the ovarian and uterine cycles, the hormonal changes that accompany them, and cyclic changes in the breasts and cervix.

Fertilization (fer′-ti-li-ZĀ-shun) Penetration of a secondary oocyte by a sperm cell and subsequent union of the nuclei of the cells.

Fetal (FĒ-tal) **circulation** The cardiovascular system of the fetus, including the placenta and special blood vessels involved in the exchange of materials between fetus and mother.

Fetus (FĒ-tus) The latter stages of the developing young of an animal; in humans, the developing organism in utero from the beginning of the ninth week to birth.

Fever An elevation in body temperature above its normal temperature of 37°C (98.6°F).

Fibroblast (FĪ-brō-blast) A large, flat cell that forms collagen and elastic fibers and intercellular substance of areolar connective tissue.

Fibrocyte (FĪ-brō-sīt) A mature fibroblast that no longer produces fibers or intercellular substance in connective tissue.

Fibrosis (fī-BRŌ-sis) Abnormal formation of fibrous tissue.

Fibrous (FĪ-brus) **joint** A joint that allows little or no movement, such as a suture and syndesmosis.

Fibrous tunic (TOO-nik) The outer coat of the eyeball, made up of the posterior sclera and the anterior cornea.

Fight-or-flight response The effects of the stimulation of the sympathetic division of the autonomic nervous system.

Filiform papilla (FIL-i-form pa-PIL-a) One of the conical projections that are distributed in parallel rows over the anterior two-thirds of the tongue and contain no taste buds.

Filtrate (fil-TRĀT) The fluid produced when blood is filtered by the endothelial-capsular membrane.

Filtration (fil-TRĀ-shun) The passage of a liquid through a filter or membrane that acts like a filter.

Filum terminale (FĪ-lum ter-mi-NAL-ē) Nonnervous fibrous tissue of the spinal cord that extends inferiorly from the conus medullaris to the coccyx.

Fimbriae (FIM-brē-ē) Fingerlike structures, especially the lateral ends of the uterine (Fallopian) tubes.

Fissure (FISH-ur) A groove, fold, or slit that may be normal or abnormal.

Fistula (FIS-choo-la) An abnormal passage between two organs or between an organ cavity and the outside.

Fixator A muscle that stabilizes the origin of the prime mover so that the prime mover can act more efficiently.

Fixed macrophage (MAK-rō-fāj) Stationary phagocytic cell found in the liver, lungs, brain, spleen, lymph nodes, subcutaneous tissue, and bone marrow. Also called a **histiocyte** (HISS-tē-ō-sīt).

Flagellum (fla-JEL-um) A hairlike, motile process on the extremity of a sperm cell, bacterium, or protozoan. *Plural,* **flagella** (fla-JEL-a).

Flexion (FLEK-shun) A folding movement in which there is a decrease in the angle between two bones, usually in the sagittal plane.

Follicle (FOL-i-kul) A small secretory sac or cavity.

Follicle-stimulating hormone (FSH) Hormone secreted by the anterior lobe of the pituitary gland that initiates development of ova and stimulates ovaries to secrete estrogens in females and initiates sperm production in males.

Fontanel (fon′-ta-NEL) A membrane-covered spot where bone formation is not yet complete, especially between the cranial bones of an infant's skull.

Foramen (fo-RĀ-men) A passage or opening; a communication between two cavities of an organ or a hole in a bone for passage of vessels or nerves. *Plural,* **foramina** (fo-RĀ-men-a).

Foramen ovale (ō-VAL-ē) An opening in the fetal heart in the septum between the right and left atria. A hole in the greater wing of the sphenoid bone that transmits the mandibular branch of the trigeminal (V) nerve.

Fornix (FOR-niks) An arch or fold; a tract in the brain made up of association fibers, connecting the hippocampus with the mammillary bodies; a recess around the cervix of the uterus where it protrudes into the vagina.

Fossa (FOS-a) A furrow or shallow depression.

Fourth ventricle (VEN-tri-kul) A cavity within the brain lying between the cerebellum and the medulla oblongata and pons.

Frenulum (FREN-yoo-lum) A small fold of mucous membrane that connects two parts and limits movement.

Frontal plane A plane at a right angle to a midsagittal plane that divides the body or organs into anterior and posterior portions. Also called a **coronal** (kō-RŌ-nal) **plane.**

Fundus (FUN-dus) The part of a hollow organ farthest from the opening.

Fungiform papilla (FUN-ji-form pa-PIL-a) A mushroomlike elevation on the dorsum of the tongue appearing as a red dot; most contain taste buds.

Furuncle (FYOOR-ung-kul) A boil; painful nodule caused by bacterial infection and inflammation of a hair follicle or sebaceous (oil) gland.

Gallbladder A small pouch that stores bile, located inferior to the liver, which is filled with bile and emptied via the cystic duct.

Gamete (GAM-ēt) A male or female reproductive cell; the spermatozoon or ovum.

Ganglion (GANG-glē-on) A group of nerve cell bodies that lie outside the central nervous system. *Plural,* **ganglia** (GANG-glē-a).

Gastric (GAS-trik) **glands** Glands in the mucosa of the stomach that empty in narrow channels called gastric pits. The glands contain chief cells (secrete pepsinogen), parietal cells (secrete hydrochloric acid and intrinsic factor), mucous neck cells (secrete mucus), and G cells (secrete gastrin).

Gastroenterology (gas′-trō-en′-ter-OL-ō-jē) The medical specialty that deals with the structure, function, diagnosis, and treatment of diseases of the stomach and intestines.

Gastrointestinal (gas-trō-in-TES-ti-nal) **(GI) tract** A continuous tube running through the ventral body cavity extending from the mouth to the anus. Also called the **alimentary** (al′-i-MEN-tar-ē) **canal.**

Gastrulation (gas′-troo-LĀ-shun) The various movements of groups of cells that lead to the establishment of the primary germ layers.

Gene (JĒN) Biological unit of heredity; an ultramicroscopic, self-reproducing DNA particle located in a definite position on a particular chromosome.

Genetic engineering The manufacture and manipulation of genetic material.

Genetics The study of heredity.

Genitalia (jen′-i-TĀL-ya) Reproductive organs.

Genome (JĒ-nōm) The complete gene complement of an organism.

Genotype (JĒ-nō-tīp) The total hereditary information carried by an individual; the genetic makeup of an organism.

Geriatrics (jer′-ē-AT-riks) The branch of medicine devoted to the medical problems and care of elderly persons.

Gestation (jes-TĀ-shun) The period of intrauterine development.

Gingivae (jin-JI-vē) Gums. They cover the alveolar processes of the mandible and maxilla and extend slightly into each socket.

Gland Single or group of specialized epithelial cells that secrete substances.

Glans penis (glanz PĒ-nis) The slightly enlarged region at the distal end of the penis.

Gliding joint A synovial joint having articulating surfaces that are usually flat, permitting only side-to-side and back-and-forth movements, as between carpal bones, tarsal bones, and the scapula and clavicle. Also called an **arthrodial** (ar-THRŌ-dē-al) **joint.**

Glomerular (glō-MER-yoo-lar) **capsule** A double-walled globe at the proximal end of a nephron that encloses the glomerulus. Also called **Bowman's** (BŌ-manz) **capsule.**

Glomerulus (glō-MER-yoo-lus) A rounded mass of nerves or blood vessels, especially the microscopic tuft of capillaries that is surrounded by the glomerular (Bowman's) capsule of each kidney tubule.

Glottis (GLOT-is) The vocal folds (true vocal cords) in the larynx and the space between them (rima glottidis).

Glucagon (GLOO-ka-gon) A hormone produced by the alpha cells of the pancreas that increases the blood glucose level.

Glucocorticoids (gloo-kō-KOR-ti-koyds) A group of hormones of the adrenal cortex.

Glucose (GLOO-kōs) A six-carbon sugar, $C_6H_{12}O_6$; the major energy source for every cell type in the body. Its metabolism is possible by every known living cell for the production of ATP.

Glycogen (GLĪ-kō-jen) A highly branched polymer of glucose containing thousands of subunits; functions as a compact store of glucose molecules in liver and muscle fibers (cells).

Gnostic (NŌS-tik) **area** Sensory area of the cerebral cortex that receives and integrates sensory input from various parts of the brain so that a common thought can be formed.

Goblet cell A goblet-shaped unicellular gland that secretes mucus. Also called a **mucous cell.**

Golgi (GOL-jē) **complex** An organelle in the cytoplasm of cells consisting of four to eight flattened channels, stacked upon one another, with expanded areas at their ends; functions in packaging secreted proteins, lipid secretion, and carbohydrate synthesis.

Golgi tendon organ *See* **Tendon organ.**

Gomphosis (gom-FŌ-sis) A fibrous joint in which a cone-shaped peg fits into a socket.

Gonad (GŌ-nad) A gland that produces gametes and hormones; the ovary in the female and the testis in the male.

Gonadocorticoids (gō-na-dō-KOR-ti-koydz) Sex hormones (estrogens and androgens) secreted by the adrenal cortex. Also called **adrenal sex hormones.**

Gonadotropic (gō′-nad-ō-TRŌ-pik) **hormone** A hormone that regulates the functions of the gonads.

Gout (GOWT) Hereditary condition associated with excessive uric acid in the blood; the acid crystallizes and deposits in joints, kidneys, and soft tissue.

Graafian follicle *See* **Vesicular ovarian follicle.**

Gray matter Area in the central nervous system and ganglia consisting of nonmyelinated nerve tissue.

Gray ramus communicans (RĀ-mus kō-MYOO-ni-kans) A short nerve containing postganglionic sympathetic fibers; the cell bodies of the fibers are in a sympathetic chain ganglion, and the nonmyelinated axons run by way of the gray ramus to a spinal nerve and then to the periphery to supply smooth muscle in blood vessels, arrector pili muscles, and sweat glands. *Plural,* **rami communicantes** (RĀ-mē kō-myoo-ni-KAN-tēz).

Greater omentum (ō-MEN-tum) A large fold in the serosa of the stomach that hangs down like an apron anterior to the intestines.

Greater vestibular (ves-TIB-yoo-lar) **glands** A pair of glands on either side of the vaginal orifice that open by a duct into the space between the hymen and the labia minora. Also called **Bartholin's** (BAR-to-linz) **glands.**

Groin (GROYN) The depression between the thigh and the trunk; the inguinal region.

Gross anatomy The branch of anatomy that deals with structures that can be studied without using a microscope. Also called **macroscopic anatomy.**

Gynecology (gī′-ne-KOL-ō-jē) The branch of medicine dealing with the study and treatment of disorders of the female reproductive system.

Gyrus (JĪ-rus) One of the folds of the cerebral cortex of the brain. *Plural,* **gyri** (JĪ-rī). Also called a **convolution.**

Hair follicle (FOL-li-kul) Structure composed of epithelium surrounding the root of a hair from which hair develops.

Hair root plexus (PLEK-sus) A network of dendrites arranged around the root of a hair as free or naked nerve endings that are stimulated when a hair shaft is moved.

Haploid (HAP-loyd) Having half the number of chromosomes characteristically found in the somatic cells of an organism; characteristic of mature gametes. Symbolized *n.*

Hard palate (PAL-at) The anterior portion of the roof of the mouth, formed by the maxillae and palatine bones and lined by mucous membrane.

Haustra (HAWS-tra) The sacculated elevations of the colon.

Haversian canal *See* **Central canal.**

Haversian system *See* **Osteon.**

Heart murmur (MER-mer) An abnormal sound that consists of a flow noise that is heard before the normal lubb-dupp or that may mask normal heart sounds.

Hematology (hē′-ma-TOL-ō-jē) The study of blood.

Hematoma (hē-ma-TŌ-ma) A tumor or swelling filled with blood.

Hemoglobin (hē′-mō-GLŌ-bin) **(Hb)** A substance in erythrocytes (red blood cells) consisting of the protein globin and the iron-containing red pigment heme and constituting about 33 percent of the cell volume; involved in the transport of oxygen and carbon dioxide.

Hemopoiesis (hē-mō-poy-Ē-sis) Blood cell production occurring in the red bone marrow of bones. Also called **hematopoiesis** (hem′-a-tō-poy-Ē-sis).

Hemopoietic (hē-mō-poy-E-tic) **stem cell** Red bone marrow cell derived from mesenchyme that gives rise to all formed elements in blood.

Hepatic (he-PAT-ik) **duct** A duct that receives bile from the bile capillaries. Small hepatic ducts merge to form the larger right and left hepatic ducts that unite to leave the liver as the common hepatic duct.

Hepatic portal circulation The flow of blood from the gastrointestinal organs to the liver before returning to the heart.

Hepatopancreatic (hep′-a-tō-pan′-krē-A-tic) **ampulla** A small, raised area in the duodenum where the combined common bile duct and main pancreatic duct empty into the duodenum. Also called the **ampulla of Vater** (VA-ter).

Heterocrine (HET-er-ō-krin) **gland** A gland, such as the pancreas, that is both an exocrine and an endocrine gland.

Hiatus (hī-A-tus) An opening; a foramen.

Hilus (HĪ-lus) An area, depression, or pit where blood vessels and nerves enter or leave an organ. Also called a **hilum.**

Hinge joint A synovial joint in which a convex surface of one bone fits into a concave surface of another bone, such as the elbow, knee, ankle, and interphalangeal joints. Also called a **ginglymus** (JIN-gli-mus) **joint.**

Histamine (HISS-ta-mēn) Substance found in many cells, especially mast cells, basophils, and platelets, released when the cells are injured; results in vasodilation, increased permeability of blood vessels, and bronchiole constriction.

Histology (hiss-TOL-ō-jē) Microscopic study of the structure of tissues.

Holocrine (HŌL-ō-krin) **gland** A type of gland in which the entire secreting cell, along with its accumulated secretions, makes up the secretory product of the gland, as in the sebaceous (oil) glands.

Homeostasis (hō′-mē-ō-STĀ-sis) The condition in which the body's internal environment remains relatively constant, within physiological limits.

Homologous (hō-MOL-ō-gus) Correspondence of two organs in structure, position, and origin.

Homologous chromosomes Two chromosomes that belong to a pair. Also called **homologues.**

Hormone (HOR-mōn) Secretion of endocrine tissue that alters the physiological activity of target cells of the body.

Horn Principal area of gray matter in the spinal cord.

Human chorionic gonadotropin (kō-rē-ON-ik gō-nad-ō-TRŌ-pin) **(hCG)** A hormone produced by the developing placenta that maintains the corpus luteum.

Human chorionic somatomammotropin (sō-mat-ō-mam-ō-TRŌ-pin) **(hCS)** A hormone produced by the chorion of the placenta that may stimulate breast tissue for lactation, enhance body growth, and regulate metabolism.

Human growth hormone (hGH) Hormone secreted by the anterior lobe of the pituitary that brings about growth of body tissues, especially skeletal and muscular. Also known as **somatotropin** and **somatotropic hormone (STH).**

Human leukocyte associated (HLA) antigens Surface proteins on white blood cells and other nucleated cells that are unique for each person (except for identical twins) and are used to type tissues and help prevent rejection.

Hyaluronic (hī-a-loo-RON-ik) **acid** A viscous, amorphous extracellular material that binds cells together, lubricates joints, and maintains the shape of the eyeballs.

Hymen (HĪ-men) A thin fold of vascularized mucous membrane at the vaginal orifice.

Hypersecretion (hī-per-se-KRĒ-shun) Overactivity of glands resulting in excessive secretion.

Hypersensitivity (hī-per-sen-si-TI-vi-tē) Overreaction to an allergen that results in pathological changes in tissues. Also called **allergy.**

Hypertonia (hī-per-TŌ-nē-a) Increased muscle tone that is expressed as spasticity or rigidity.

Hypophyseal (hī-pō-FIZ-ē-al) **pouch** An outgrowth of ectoderm from the roof of the stomodeum (mouth) from which the anterior lobe of the pituitary gland develops.

Hypophysis (hī-POF-i-sis) Pituitary gland.

Hypoplasia (hī-pō-PLĀ-zē-a) Defective development of tissue.

Hyposecretion (hī′-pō-se-KRĒ-shun) Underactivity of glands resulting in diminished secretion.

Hypothalamus (hī′-pō-THAL-a-mus) A portion of the diencephalon, lying beneath the thalamus and forming the floor and part of the wall of the third ventricle.

Hypotonic (hī′-pō-TON-ik) Having an osmotic pressure lower than that of a solution with which it is compared.

Ileocecal (il′-ē-ō-SĒ-kal) **sphincter** A fold of mucous membrane that guards the opening from the ileum into the large intestine. Also called the **ileocecal valve.**

Ileum (IL-ē-um) The terminal portion of the small intestine.

Immunoglobulin (im-yoo-nō-GLOB-yoo-lin) **(Ig)** An antibody synthesized by plasma cells derived from B lymphocytes in response to the introduction of antigen. Immunoglobulins are divided into five kinds (IgG, IgM, IgA, IgD, IgE) based primarily on the larger protein component present in the immunoglobulin.

Immunology (im′-yoo-NOL-ō-jē) The branch of science that deals with the responses of the body when challenged by antigens.

Imperforate (im-PER-fō-rāt) Abnormally closed.

Implantation (im-plan-TĀ-shun) The insertion of a tissue or a part into the body. The attachment of the blastocyst to the lining of the uterus 7–8 days after fertilization.

Inclusion (in-KLOO-zhun) Temporary structure that contains secretions and storage products of cells.

Indirect pathway Motor pathways that convey information from the brain down the spinal cord; related to automatic movements, coordination of body movements with visual stimuli, skeletal muscle tone and posture, and equilibrium.

Infarction (in-FARK-shun) The presence of a localized area of necrotic tissue, produced by inadequate oxygenation of the tissue.

Infection (in-FEK-shun) Invasion and multiplication of microorganisms in body tissues, which may be inapparent or characterized by cellular injury.

Inferior (in-FĒR-ē-or) Away from the head or toward the lower part of a structure. Also called **caudal** (KAW-dal).

Inferior vena cava (VĒ-na CĀ-va) **(IVC)** Large vein that collects blood from parts of the body inferior to the heart and returns it to the right atrium.

Inflammation (in′-fla-MĀ-shun) Localized, protective response to tissue injury designed to destroy, dilute, or wall off the infecting agent or injured tissue; characterized by redness, pain, heat, swelling, and sometimes loss of function.

Infundibulum (in′-fun-DIB-yoo-lum) The stalklike structure that attaches the pituitary gland to the hypothalamus of the brain. The funnel-shaped, open, distal end of the uterine (Fallopian) tube.

Inguinal (IN-gwi-nal) Pertaining to the groin.

Inguinal canal An oblique passageway to the anterior abdominal wall just superior and parallel to the medial half of the inguinal ligament that transmits the spermatic cord and ilioinguinal nerve in the male and the round ligament of the uterus and ilioinguinal nerve in the female.

Inheritance The acquisition of body characteristics and qualities by transmission of genetic information from parents to offspring.

Inhibin A male sex hormone secreted by sustentacular (Sertoli) cells that inhibits FSH release by the anterior pituitary gland and thus spermatogenesis.

Inner cell mass A region of cells of a blastocyst that differentiates into the three primary germ layers—ectoderm, mesoderm, and endoderm—from which all tissues and organs develop; also called an **embryoblast.**

Insertion (in-SER-shun) The manner or place of attachment of a muscle to the bone that it moves.

Inspiration (in-spi-RĀ-shun) The act of drawing air into the lungs.

Insula (IN-su-la) A triangular area of cerebral cortex that lies deep within the lateral cerebral fissure, under the parietal, frontal, and temporal lobes, and cannot be seen in an external view of the brain. Also called the **island** or **isle of Reil** (RĪL).

Insulin (IN-su-lin) A hormone produced by the beta cells of the pancreas that decreases the blood glucose level.

Intercalated (in-TER-ka-lāt-ed) **disc** An irregular transverse thickening of sarcolemma that holds cardiac muscle fibers (cells) together and aids in conduction of muscle action potentials.

Intercostal (in′-ter-KOS-tal) **nerve** A nerve supplying a muscle located between the ribs.

Intermediate (in′-ter-MĒ-dē-it) **filament** Protein filament ranging from 8 to 12 nm in diameter that may provide structural reinforcement, hold organelles in place, and give shape to a cell.

Internal capsule A tract of projection fibers connecting various parts of the cerebral cortex and lying between the thalamus and the caudate and lentiform nuclei of the basal ganglia.

Internal ear The inner ear or labyrinth, lying inside the temporal bone, containing the organs of hearing and balance.

Internal nares (NA-rēz) The two openings posterior to the nasal cavities opening into the nasopharynx. Also called the **choanae** (kō-A-nē).

Internal respiration The exchange of respiratory gases between blood and body cells. Also called **tissue respiration.**

Interphase (IN-ter-fāz) The period during its life cycle when a cell is carrying on every life process except division; the stage between two mitotic divisions. Also called **metabolic phase.**

Interstitial cell of Leydig *See* **Interstitial endocrinocyte.**

Interstitial (in′-ter-STISH-al) **endocrinocyte** A cell that is located in the connective tissue between seminiferous tubules in a mature testis that secretes testosterone. Also called an **interstitial cell of Leydig** (LĪ-dig).

Interstitial fluid The portion of extracellular fluid that fills the microscopic spaces between the cells of tissues; the internal environment of the body. Also called **intercellular** or **tissue fluid.**

Interstitial growth Growth from within, as in the growth of cartilage. Also called **endogenous** (en-DOJ-e-nus) **growth.**

Interventricular (in′-ter-ven-TRIK-yoo-lar) **foramen** A narrow, oval opening through which the lateral ventricles of the brain communicate with the third ventricle. Also called the **foramen of Monro.**

Intervertebral (in′-ter-VER-te-bral) **disc** A pad of fibrocartilage located between the bodies of two vertebrae.

Intestinal gland Gland that opens onto the surface of the intestinal mucosa and secretes digestive enzymes. Also called a **crypt of Lieberkühn** (LĒ-ber-kyoon).

Intracellular (in′-tra-SEL-yoo-lar) **fluid (ICF)** Fluid located within cells.

Intrafusal (in′-tra-FYOO-zal) **fibers** Three to ten specialized muscle fibers (cells), partially enclosed in a connective tissue capsule that is filled with lymph; the fibers compose muscle spindles.

Intramembranous ossification (in′-tra-MEM-bra-nus os′-i′-fi-KĀ-shun) The method of bone formation in which the bone is formed directly in membranous tissue.

Intraocular (in-tra-OC-yoo-lar) **pressure (IOP)** Pressure in the eyeball, produced mainly by aqueous humor.

Intrinsic (in-TRIN-sik) Of internal origin; for example, the intrinsic factor, a mucoprotein formed by the gastric mucosa that is necessary for the absorption of vitamin B_{12}.

Intrinsic factor (IF) A glycoprotein synthesized and secreted by the parietal cells of the gastric mucosa that facilitates vitamin B_{12} absorption in the small intestine.

Invagination (in-vaj′-i-NĀ-shun) The pushing of the wall of a cavity into the cavity itself.

Inversion (in-VER-zhun) The movement of the sole medially at the ankle joint.

Ipsilateral (ip′-si-LAT-er-al) On the same side of the midline, affecting the same side of the body.

Iris The colored portion of the eyeball seen through the cornea that consists of circular and radial smooth muscle; the black hole in the center of the iris is the pupil.

Ischemia (is-KĒ-mē-a) A lack of sufficient blood to a part due to obstruction of circulation.

Island of Reil *See* **Insula.**

Islet of Langerhans *See* **Pancreatic islet.**

Isthmus (IS-mus) A narrow strip of tissue or narrow passage connecting two larger parts.

Jejunum (jē-JOO-num) The middle portion of the small intestine.

Joint kinesthetic (kin′-es-THET-ik) **receptor** A proprioceptive receptor located in a joint, stimulated by joint movement.

Juxtaglomerular (juks-ta-glō-MER-yoo-lar) **apparatus (JGA)** Consists of the macula densa (cells of the distal convoluted tubule adjacent to the afferent and efferent arteriole) and juxtaglomerular cells (modified cells of the afferent and sometimes efferent arteriole); secretes renin when blood pressure starts to fall.

Keratinocyte (ker-A-tin′-ō-sīt) The most numerous of the epidermal cells that function in the production of keratin.

Kidney (KID-nē) One of the paired reddish organs located in the lumbar region that regulates the composition and volume of blood and produces urine.

Kinesiology (ki-nē′-sē-OL-ō-jē) The study of the movement of body parts.

Kinesthesia (kin-is-THĒ-szē-a) Ability to perceive the extent, direction, or weight of movement; muscle sense.

Kinetochore (ki-NET-ō-kor) Protein complex attached to the outside of a centromere to which kinetochore microtubules attach.

Kupffer's cell *See* **Stellate reticuloendothelial cell.**

Kyphosis (kī-FŌ-sis) An exaggeration of the thoracic curve of the vertebral column, resulting in a "round-shouldered" or hunchback appearance.

Labial frenulum (LĀ-bē-al FREN-yoo-lum) A medial fold of mucous membrane between the inner surface of the lip and the gums.

Labia majora (LĀ-bē-a ma-JO-ra) Two longitudinal folds of skin extending downward and backward from the mons pubis of the female.

Labia minora (min-OR-a) Two small folds of mucous membrane lying medial to the labia majora of the female.

Labyrinth (LAB-i-rinth) Intricate communicating passageway, especially in the internal ear.

Lacrimal (LAK-ri-mal) **canal** A duct, one on each eyelid, commencing at the punctum at the medial margin of an eyelid and conveying tears medially into the nasolacrimal sac.

Lacrimal gland Secretory cells located at the superior anterolateral portion of each orbit that secrete tears into excretory ducts that open onto the surface of the conjunctiva.

Lacrimal sac The superior expanded portion of the nasolacrimal duct that receives the tears from a lacrimal canal.

Lactation (lak-TĀ-shun) The secretion and ejection of milk by the mammary glands.

Lacteal (LAK-tē-al) One of many intestinal lymphatic vessels in villi that absorb triglycerides from digested food.

Lacuna (la-KOO-na) A small, hollow space, such as that found in bones in which the osteoblasts lie. *Plural,* **lacunae** (la-KOO-nē).

Lambdoid (LAM-doyd) **suture** The line of union in the skull between the parietal bones and the occipital bone; sometimes contains sutural bones.

Lamellae (la-MEL-ē) Concentric rings found in compact bone.

Lamellated corpuscle Oval pressure receptor located in subcutaneous tissue and consisting of concentric layers of connective tissue wrapped around a sensory nerve fiber. Also called a **Pacinian** (pa-SIN-ē-an) **corpuscle.**

Lamina (LAM-i-na) A thin, flat layer or membrane, as the flattened part of either side of the arch of a vertebra. *Plural,* **laminae** (LAM-i-nē).

Lamina propria (PRŌ-prē-a) The connective tissue layer of a mucous membrane.

Langerhans (LANG-er-hans) **cell** Epidermal cell that arises from bone marrow that presents antigens to helper T cells, thus activating T cells as part of the immune response.

Lanugo (lan-YOO-gō) Fine downy hairs that cover the fetus.

Large intestine The portion of the gastrointestinal tract extending from the ileum of the small intestine to the anus, divided structurally into the cecum, colon, rectum, and anal canal.

Laryngopharynx (la-rin′-gō-FAR-inks) The inferior portion of the pharynx, extending downward from the level of the hyoid bone to divide posteriorly into the esophagus and anteriorly into the larynx. Also called the **hypopharynx.**

Laryngotracheal (la-rin′-gō-TRĀ-kē-al) **bud** An outgrowth of endoderm of the foregut from which the respiratory system develops.

Larynx (LAR-inks) The voice box, a short passageway that connects the pharynx with the trachea.

Lateral (LAT-er-al) Farther from the midline of the body or a structure.

Lateral ventricle (VEN-tri-kul) A cavity within a cerebral hemisphere that communicates with the lateral ventricle in the other cerebral hemisphere and with the third ventricle by way of the interventricular foramen.

Lens A transparent organ constructed of proteins (crystallins) lying posterior to the pupil and iris of the eyeball and anterior to the vitreous body.

Lesion (LĒ-zhun) Any localized, abnormal change in tissue formation.

Lesser omentum (ō-MEN-tum) A fold of the peritoneum that extends from the liver to the lesser curvature of the stomach and the commencement of the duodenum.

Lesser vestibular (ves-TIB-yoo-lar) **gland** One of the paired mucus-secreting glands that have ducts that open on either side of the urethral orifice in the vestibule of the female.

Leukocyte (LOO-kō-sīt) A white blood cell.

Libido (li-BĒ-dō) The sexual drive, conscious or subconscious.

Ligament (LIG-a-ment) Dense, regularly arranged connective tissue that attaches bone to bone.

Limbic system A portion of the forebrain, sometimes termed the visceral brain, concerned with various aspects of emotion and behavior, that includes the limbic lobe, denate gyrus, amygdaloid body, septal nuclei, mammillary bodies, anterior thalamic nucleus, olfactory bulbs, and bundles of myelinated axons.

Lingual frenulum (LIN-gwal FREN-yoo-lum) A fold of mucous membrane that connects the tongue to the floor of the mouth.

Lingual lipase (LĪ-pās) Digestive enzyme secreted by glands on the dorsum of the tongue that digests triglycerides.

Lipase (LĪ-pās) A fat-splitting enzyme.

Lipid An organic compound composed of carbon, hydrogen, and oxygen that is usually insoluble in water, but soluble in alcohol, ether, and chloroform; examples include triglycerides, phospholipids, steroids, and prostaglandins.

Liver Large gland under the diaphragm that occupies most of the right hypochondriac region and part of the epigastric region; functionally, it produces bile salts, heparin, and plasma proteins; converts one nutrient into another; detoxifies substances; stores glycogen, minerals, and vitamins; carries on phagocytosis of blood cells and bacteria; and helps activate vitamin D.

Lobe (lōb) A curved or rounded projection.

Locus coeruleus (LŌ-kus sē-ROO-lē-us) A group of neurons in the brain stem where norepinephrine (NE) is concentrated.

Lordosis (lor-DŌ-sis) An exaggeration of the lumbar curve of the vertebral column.

Lower limb (extremity) The appendage attached at the pelvic (hip) girdle, consisting of the thigh, knee, leg, ankle, foot, and toes.

Lumbar (LUM-bar) Region of the back and side between the ribs and pelvis; loin.

Lumbar plexus (PLEK-sus) A network formed by the anterior branches of spinal nerves L1 through L4.

Lumen (LOO-men) The space within an artery, vein, duct, or hollow organ.

Lung One of the two main organs of respiration, lying on either side of the heart in the thoracic cavity.

Lunula (LOO-nyoo-la) The moon-shaped white area at the base of a nail.

Luteinizing (LOO-tē-in′-īz-ing) **hormone (LH)** A hormone secreted by the anterior lobe of the pituitary gland that stimulates ovulation, progesterone secretion by the corpus luteum, and readies the mammary glands for milk secretion in females and stimulates testosterone secretion by the testes in males.

Lymph (LIMF) Fluid confined in lymphatic vessels and flowing through the lymphatic system to be returned to the blood.

Lymphatic (lim-FAT-ik) **capillary** Blind-ended microscopic lymph vessel that begins in spaces between cells and converges with other lymph capillaries to form lymphatic vessels.

Lymphatic tissue A specialized form of reticular tissue that contains large numbers of lymphocytes.

Lymphatic vessel A large vessel that collects lymph from lymph capillaries and converges with other lymphatic vessels to form the thoracic and right lymphatic ducts.

Lymph node An oval or bean-shaped structure located along lymphatic vessels.

Lymphocyte (LIM-fō-sīt) A type of white blood cell, found in lymph nodes, associated with the immune system.

Lysosome (LĪ-sō-sōm) An organelle in the cytoplasm of a cell, enclosed by a single membrane and containing powerful digestive enzymes.

Lysozyme (LĪ-sō-zīm) A bactericidal enzyme found in tears, saliva, and perspiration.

Macrophage (MAK-rō-fāj) Phagocytic cell derived from a monocyte. May be fixed or wandering.

Macula (MAK-yoo-la) A discolored spot or a colored area. A small, thickened region on the wall of the utricle and saccule that serves as a receptor for static equilibrium.

Macula lutea (LOO-tē-a) The yellow spot in the center of the retina.

Malaise (ma-LĀYZ) Discomfort, uneasiness, and indisposition, often indicative of infection.

Malignant (ma-LIG-nant) Referring to diseases that tend to become worse and cause death; especially the invasion and spreading of cancer.

Mammary (MAM-ar-ē) **gland** Modified sudoriferous (sweat) gland of the female that secretes milk for the nourishment of the young.

Mammillary (MAM-i-ler-ē) **bodies** Two small rounded bodies posterior to the tuber cinereum that are involved in reflexes related to the sense of smell.

Marrow (MAR-ō) Soft, spongelike material in the cavities of bone. Red bone marrow produces blood cells; yellow bone marrow, formed mainly of fatty tissue, has no blood-producing function.

Mast cell A cell found in areolar connective tissue along blood vessels that produces heparin, a dilator of small blood vessels during inflammation. The name given to a basophil after it has left the bloodstream and entered the tissues.

Matrix (MĀ-trix) Ground substance and fibers external to cells of a connective tissue.

Meatus (mē-Ā-tus) A passage or opening, especially the external portion of a canal.

Mechanoreceptor (me-KAN-ō-rē′-sep-tor) Receptor that detects mechanical deformation of the receptor itself or adjacent cells; stimuli so detected include those related to touch, pressure, vibration, proprioception, hearing, equilibrium, and blood pressure.

Medial (MĒ-dē-al) Nearer the midline of the body or a structure.

Medial lemniscus (lem-NIS-kus) A flat band of myelinated nerve fibers extending through the medulla oblongata, pons, and midbrain and terminating in the thalamus on the same side. Sensory neurons in this tract transmit impulses for proprioception, fine touch, pressure, and vibration sensations.

Median aperture (AP-er-choor) One of the three openings in the roof of the fourth ventricle through which cerebrospinal fluid enters the subarachnoid space of the brain and cord. Also called the **foramen of Magendie.**

Median plane A vertical plane dividing the body into equal right and left sides. Situated in the middle.

Mediastinum (mē′-dē-as-TĪ-num) A broad, median partition, actually a mass of tissue found between the pleurae of the lungs that extends from the sternum to the vertebral column.

Medulla (me-DULL-la) An inner layer of an organ, such as the medulla of the kidneys.

Medulla oblongata (ob′-long-GA-ta) The most inferior part of the brain stem.

Medullary (MED-yoo-lar′-ē) **cavity** The space within the diaphysis of a bone that contains yellow bone marrow. Also called the **marrow cavity.**

Medullary rhythmicity (rith-MIS-i-tē) **area** Portion of the respiratory center in the medulla oblongata that controls the basic rhythm of respiration.

Meibomian gland *See* **Tarsal gland.**

Meiosis (mī-Ō-sis) A type of cell division restricted to sex-cell production involving two successive nuclear divisions that result in daughter cells with the haploid (*n*) number of chromosomes.

Meissner's corpuscle *See* **Corpuscle of touch.**

Melanin (MEL-a-nin) A dark black, brown, or yellow pigment found in some parts of the body such as the skin.

Melanocyte (MEL-a-nō-sīt′) A pigmented cell located between or beneath cells of the deepest layer of the epidermis that synthesizes melanin.

Melanocyte-stimulating hormone (MSH) A hormone secreted by the anterior lobe of the pituitary gland that stimulates the dispersion of melanin granules in melanocytes in amphibians; continued administration produces darkening of skin in humans.

Melatonin (mel-a-TŌN-in) A hormone secreted by the pineal gland that may inhibit reproductive activities.

Membrane A thin, flexible sheet of tissue composed of an epithelial layer and an underlying connective tissue layer, as in an epithelial membrane, or of areolar connective tissue only, as in a synovial membrane.

Membranous labyrinth (mem-BRA-nus LAB-i-rinth) The portion of the labyrinth of the inner ear that is located inside the bony labyrinth and separated from it by the perilymph; made up of the membranous semicircular canals, the saccule and utricle, and the cochlear duct.

Memory The ability to recall thoughts; commonly classified as short-term (activated) and long-term.

Menarche (me-NAR-kē) Beginning of the menstrual function.

Meninges (me-NIN-jēz) Three membranes covering the brain and spinal cord, called the dura mater, arachnoid, and pia mater. *Singular,* **meninx** (MEN-inks).

Menopause (MEN-ō-pawz) The termination of the menstrual cycles.

Menstrual (MEN-stroo-al) **cycle** A series of changes in the endometrium of a nonpregnant female that prepares the lining of the uterus to receive a fertilized ovum.

Menstruation (men′-stroo-Ā-shun) Periodic discharge of blood, tissue fluid, mucus, and epithelial cells that usually lasts for 5 days; caused by a sudden reduction in estrogens and progesterone. Also called the **menstrual phase** or **menses.**

Merkel (MER-kel) **cell** Cell found in the epidermis of hairless skin that makes contact with a tactile (Merkel) disc that functions in touch.

Merocrine (MER-ō-krin) **gland** A secretory cell that remains intact throughout the process of formation and discharge of the secretory product, as in the salivary and pancreatic glands.

Mesenchyme (MEZ-en-kīm) An embryonic connective tissue from which all other connective tissues arise.

Mesentery (MEZ-en-ter′-ē) A fold of peritoneum attaching the small intestine to the posterior abdominal wall.

Mesocolon (mez′-ō-KŌ-lon) A fold of peritoneum attaching the colon to the posterior abdominal wall.

Mesoderm The middle of the three primary germ layers that gives rise to connective tissues, blood and blood vessels, and muscles.

Mesothelium (mez′-ō-THĒ-lē-um) The layer of simple squamous epithelium that lines serous cavities.

Mesovarium (mez′-ō-VAR-ē-um) A short fold of peritoneum that attaches an ovary to the broad ligament of the uterus.

Metabolism (me-TAB-ō-lizm) The sum of all the biochemical reactions that occur within an organism, including the synthetic (anabolic) reactions and decomposition (catabolic) reactions.

Metacarpus (met′-a-KAR-pus) A collective term for the five bones that make up the palm of the hand.

Metaphase (MET-a-fāz) The second stage of mitosis in which chromatid pairs line up on the equatorial plane of the cell.

Metaphysis (me-TAF-i-sis) Growing portion of a bone.

Metarteriole (met′-ar-TĒ-rē-ōl) A blood vessel that emerges from an arteriole, traverses a capillary network, and empties into a venule.

Metatarsus (met′-a-TAR-sus) A collective term for the five bones located in the foot between the tarsals and the phalanges.

Microfilament (mī-krō-FIL-a-ment) Rodlike cytoplasmic structure about 6 nm in diameter; comprises contractile units in muscle fibers (cells) and provides support, shape, and movement in nonmuscle cells.

Microglia (mī-krō-GLĒ-a) Neuroglial cells that carry on phagocytosis. Also called **brain macrophages** (MAK-rō-fāj-ez).

Microphage (MĪK-rō-fāj) Granular leukocyte that carries on phagocytosis, especially neutrophils and eosinophils.

Microtubule (mī-krō-TOOB-yool′) Cylindrical cytoplasmic structure, ranging in diameter from 18 to 30 nm, consisting of the protein tubulin; provides support, structure, and transportation.

Microvilli (mī-krō-VIL-ē) Microscopic, fingerlike projections of the cell membranes of small intestinal cells that increase surface area for absorption.

Micturition (mik′-too-RISH-un) The act of expelling urine from the urinary bladder. Also called **urination** (yoo-ri-NĀ-shun).

Midbrain The part of the brain between the pons and the diencephalon. Also called the **mesencephalon** (mes′-en-SEF-a-lon).

Middle ear A small, epithelial-lined cavity hollowed out of the temporal bone, separated from the external ear by the eardrum and from the internal ear by a thin bony partition containing the oval and round windows; extending across the middle ear are the three auditory ossicles. Also called the **tympanic** (tim-PAN-ik) **cavity.**

Midline An imaginary vertical line that divides the body into equal left and right sides.

Midsagittal plane A vertical plane through the midline of the body that divides the body or organs into equal right and left sides. Also called a **median plane.**

Mineralocorticoids (min′-er-al-ō-KOR-ti-koyds) A group of hormones of the adrenal cortex.

Mitochondrion (mī′-tō-KON-drē-on) A double-membraned organelle that plays a central role in the production of ATP; known as the "powerhouse" of the cell.

Mitosis (mī-TŌ-sis) The orderly division of the nucleus of a cell that ensures that each new daughter nucleus has the same number and kind of chromosomes as the original parent nucleus. The process includes the replication of chromosomes and the distribution of the two sets of chromosomes into two separate and equal nuclei.

Mitotic spindle Collective term for a football-shaped assembly of microtubules (nonkinetochore, kinetochore, and aster) produced by the pericentriolar area of a centrosome and that is responsible for the movement of chromosomes during cell division.

Modality (mō-DAL-i-tē) Any of the specific sensory entities, such as vision, smell, or taste.

Modiolus (mō-DĪ-ō′-lus) The central pillar or column of the cochlea.

Monoclonal antibody (MAb) Antibody produced by in vitro clones of B cells hybridized with cancerous cells.

Monocyte (MON-ō-sīt′) A type of white blood cell characterized by agranular cytoplasm; the largest of the leukocytes.

Mons pubis (MONZ PYOO-bis) The rounded, fatty prominence over the pubic symphysis, covered by coarse pubic hair.

Morphology (mor-FOL-o-jē) The study of the form and structure of things.

Morula (MOR-yoo-la) A solid mass of cells produced by successive cleavages of a fertilized ovum a few days after fertilization.

Motor area The region of the cerebral cortex that governs muscular movement, particularly the precentral gyrus of the frontal lobe.

Motor end plate Portion of the sarcolemma of a muscle fiber (cell) in close approximation with an axon terminal.

Motor neuron (NOO-ron) A neuron that conveys nerve impulses from the brain and spinal cord to effectors that may be either muscles or glands. Also called an **efferent neuron.**

Motor unit A motor neuron together with the muscle fibers (cells) it stimulates.

Mucous (MYOO-kus) **cell** A unicellular gland that secretes mucus. Also called a **goblet cell.**

Mucous membrane A membrane that lines a body cavity that opens to the exterior. Also called the **mucosa** (myoo-KŌ-sa).

Mucus The thick fluid secretion of mucous glands and mucous membranes.

Muscarinic (mus′-ka-RIN-ik) **receptor** Receptor found on all effectors innervated by parasympathetic postganglionic axons and some effectors innervated by sympathetic postganglionic axons; so named because the actions of acetylcholine (ACh) on such receptors are similar to those produced by muscarine.

Muscle An organ composed of one of three types of muscle tissue (skeletal, cardiac, or visceral), specialized for contraction to produce voluntary or involuntary movement of parts of the body.

Muscle spindle An encapsulated receptor in a skeletal muscle, consisting of specialized muscle fiber (cell) and nerve endings, stimulated by changes in length or tension of muscle fibers; a proprioceptor. Also called a **neuromuscular** (noo-rō-MUS-kyoo-lar) **spindle.**

Muscle tissue A tissue specialized to produce motion in response to muscle action potentials by its qualities of contractility, extensibility, elasticity, and excitability. Types include skeletal, cardiac, and smooth.

Muscle tone A sustained, partial contraction of portions of a skeletal muscle in response to activation of stretch receptors.

Muscularis (MUS-kyoo-la′-ris) A muscular layer (coat or tunic) of an organ.

Muscularis mucosae (myoo-KŌ-sē) A thin layer of smooth muscle fibers (cells) located in the most superficial layer of the mucosa of the gastrointestinal tract, underlying the lamina propria of the mucosa.

Myelin (MĪ-e-lin) **sheath** A white, phospholipid, segmented covering, formed by neurolemmocytes (Schwann cells), around the axons and dendrites of many peripheral neurons.

Myenteric plexus A network of nerve fibers from both autonomic divisions located in the muscularis coat of the small intestine. Also called the **plexus of Auerbach** (OW-er-bak).

Myocardium (mī′-ō-KAR-dē-um) The middle layer of the heart wall, made up of cardiac muscle, comprising the bulk of the heart, and lying between the epicardium and the endocardium.

Myofibril (mī′-ō-FĪ-bril) A threadlike structure, running longitudinally through a muscle fiber (cell) consisting mainly of thick myofilaments (myosin) and thin myofilaments (actin).

Myoglobin (mī-ō-GLŌ-bin) The oxygen-binding, iron-containing conjugated protein complex present in the sarcoplasm of muscle fibers (cells); contributes the red color to muscle.

Myology (mī-OL-ō-jē) The study of muscles.

Myometrium (mī′-ō-MĒ-trē-um) The smooth muscle layer of the uterus.

Myopia (mī-Ō-pē-a) Defect in vision so that objects can be seen distinctly only when very close to the eyes; nearsightedness.

Myosin (MĪ-ō-sin) The contractile protein that makes up the thick myofilaments of muscle fibers (cells).

Myxedema (mix-e-DĒ-ma) Condition caused by hypothyroidism during the adult years characterized by swelling of facial tissues.

Nail A hard plate, composed largely of keratin, that develops from the epidermis of the skin to form a protective covering on the dorsal surface of the distal phalanges of the fingers and toes.

Nail matrix (MĀ-triks) The part of the nail beneath the body and root from which the nail is produced.

Nasal (NĀ-zal) **cavity** A mucosa-lined cavity on either side of the nasal septum that opens onto the face at an external naris and into the nasopharynx at an internal naris.

Nasal septum (SEP-tum) A vertical partition composed of bone (perpendicular plate of ethmoid and vomer) and cartilage, covered with a mucous membrane, separating the nasal cavity into left and right sides.

Nasolacrimal (nā′-zō-LAK-ri-mal) **duct** A canal that transports the lacrimal secretion (tears) from the nasolacrimal sac into the nose.

Nasopharynx (nā′-zō-FAR-inks) The superior portion of the pharynx, lying posterior to the nose and extending inferiorly to the soft palate.

Nephron (NEF-ron) The functional unit of the kidney.

Nerve A cordlike bundle of nerve fibers (axons and/or dendrites) and their associated connective tissue coursing together outside the central nervous system.

Nerve fiber General term for any process (axon or dendrite) projecting from the cell body of a neuron.

Nerve impulse A wave of negativity (depolarization) that self-propagates along the outside surface of the plasma membrane of a neuron; also called a **nerve action potential.**

Nervous tissue Tissue that initiates and transmits nerve impulses to coordinate body activities.

Neural plate A thickening of ectoderm that forms early in the third week of development and represents the beginning of the development of the nervous system.

Neuroeffector (noo-rō-e-FEK-tor) **junction** Collective term for neuromuscular and neuroglandular junctions.

Neurofibral (noo-rō-FĪ-bral) **node** A space, along a myelinated nerve fiber, between the individual neurolemmocytes (Schwann cells) that form the myelin sheath and the neurolemma. Also called **node of Ranvier** (ron-VĒ-ā).

Neurofibril (noo-rō-FĪ-bril) One of the delicate threads that forms a complicated network in the cytoplasm of the cell body and processes of a neuron.

Neuroglandular (noo-rō-GLAND-yoo-lar) **junction** Area of contact between a motor neuron and a gland.

Neuroglia (noo-RŌG-lē-a) Cells of the nervous system that are specialized to perform the functions of connective tissue. The neuroglia of the central nervous system are the astrocytes, oligodendrocytes, microglia, and ependyma; neuroglia of the peripheral nervous system include the neurolemmocytes (Schwann cells) and the ganglion satellite cells. Also called **glial** (GLĒ-al) **cells.**

Neurohypophyseal (noo′-rō-hī-po-FIZ-ē-al) **bud** An outgrowth of ectoderm located on the floor of the hypothalamus that gives rise to the posterior lobe of the pituitary gland.

Neurolemma (noo-rō-LEM-ma) The peripheral, nucleated cytoplasmic layer of the neurolemmocyte (Schwann cell). Also called **sheath of Schwann** (SCHVON).

Neurolemmocyte A neuroglial cell of the peripheral nervous system that forms the myelin sheath and neurolemma of a nerve fiber by wrapping around a nerve fiber in a jelly-roll fashion. Also called a **Schwann** (SCHVON) **cell.**

Neurology (noo-ROL-ō-jē) The branch of science that deals with the normal functioning and disorders of the nervous system.

Neuromuscular (noo-rō-MUS-kyoo-lar) **junction** The area of contact between the axon terminal of a motor neuron and a portion of the sarcolemma of a muscle fiber (cell). Also called a **myoneural** (mī-ō-NOO-ral) **junction.**

Neuron (NOO-ron) A nerve cell, consisting of a cell body, dendrites, and an axon.

Neurosecretory (noo-rō-SĒC-re-tō-rē) **cell** A cell in a nucleus (paraventricular and supraoptic) in the hypothalamus that produces oxytocin (OT) or antidiuretic hormone (ADH), hormones stored in the posterior lobe of the pituitary gland.

Neurotransmitter One of a variety of molecules synthesized within the nerve axon terminals, released into the synaptic cleft in response to a nerve impulse, and affecting the membrane potential of the postsynaptic neuron. Also called a **transmitter substance.**

Neutrophil (NOO-trō-fil) A type of white blood cell characterized by granular cytoplasm that stains as readily with acid or basic dyes.

Nicotinic (nik′-ō-TIN-ik) **receptor** Receptor found on both sympathetic and parasympathetic postganglionic neurons so named because the actions of acetylcholine (ACh) in such receptors are similar to those produced by nicotine.

Nipple A pigmented, wrinkled projection on the surface of the mammary gland that is the location of the openings of the lactiferous ducts for milk release.

Nissl bodies *See* **Chromatophilic substance.**

Nociceptor (nō′-sē-SEP-tor) A free (naked) nerve ending that detects pain.

Node of Ranvier *See* **Neurofibral node.**

Norepinephrine (nor′-ep-ē-NEF-rin) **(NE)** A hormone secreted by the adrenal medulla that produces actions similar to those that result from sympathetic stimulation. Also called **noradrenaline** (nor-a-DREN-a-lin).

Notochord (NŌ-tō-cord) A flexible rod of embryonic tissue that lies where the future vertebral column will develop.

Nucleic (noo-KLĒ-ik) **acid** An organic compound that is a long polymer of nucleotides, with each nucleotide containing a pentose sugar, a phosphate group, and one of four possible nitrogenous bases (adenine, cytosine, guanine, and thymine or uracil).

Nucleolus (noo-KLĒ-ō-lus) Nonmembranous spherical body within the nucleus composed of protein, DNA, and RNA that functions in the synthesis and storage of ribosomal RNA.

Nucleosome (NOO-klē-ō-sōm) Elementary structural subunit of a chromosome consisting of histones and DNA.

Nucleus (NOO-klē-us) A spherical or oval organelle of a cell that contains the hereditary factors of the cell, called genes. A cluster of unmyelinated nerve cell bodies in the central nervous system. The central portion of an atom made up of protons and neutrons.

Nucleus cuneatus (kyoo-nē-Ā-tus) A group of nerve cells in the inferior portion of the medulla oblongata in which fibers of the fasciculus cuneatus terminate.

Nucleus gracilis (gras-I-lis) A group of nerve cells in the inferior portion of the medulla oblongata in which fibers of the fasciculus gracilis terminate.

Nucleus pulposus (pul-PŌ-sus) A soft, pulpy, highly elastic substance in the center of an intervertebral disc, a remnant of the notochord.

Oblique (ō-BLĒK) **plane** A plane that passes through the body or an organ at an angle between the transverse plane and either the midsagittal, parasagittal, or frontal plane.

Obstetrics (ob-STET-riks) The specialized branch of medicine that deals with pregnancy, labor, and the period of time immediately following delivery (about 42 days).

Obturator (OB-tyoo-rā′-ter) Anything that obstructs or closes a cavity or opening.

Olfactory bulb A mass of gray matter at the termination of an olfactory (I) nerve, lying inferior to the frontal lobe of the cerebrum on either side of the crista galli of the ethmoid bone.

Olfactory receptor cell A bipolar neuron with its cell body lying between supporting cells located in the mucous membrane lining the superior portion of each nasal cavity; converts odor into neural signals.

Olfactory tract A bundle of axons that extends from the olfactory bulb posteriorly to the olfactory portion of the cortex.

Oligodendrocyte (o-lig-ō-DEN-drō-sīt) A neuroglial cell that supports neurons and produces a phospholipid myelin sheath around axons of neurons of the central nervous system.

Oliguria (ol′-i-GYOO-rē-a) Daily urinary output usually less than 250 ml.

Olive A prominent oval mass on each lateral surface of the superior part of the medulla oblongata.

Oncogene (ONG-kō-jēn) Gene that has the ability to transform a normal cell into a cancerous cell when it is inappropriately activated.

Oncology (ong-KOL-ō-jē) The study of tumors.

Oogenesis (ō′-ō-JEN-e-sis) Formation and development of the ovum.

Ophthalmologist (of′-thal-MOL-ō-jist) A physician who specializes in the diagnosis and treatment of eye disorders with drugs, surgery, and corrective lenses.

Ophthalmology (of′-thal-MOL-ō-jē) The study of the structure, function, and diseases of the eye.

Ophthalmoscopy (of′-thal-MOS-kō-pē) Examination of the interior fundus of the eyeball to detect retinal changes associated with hypertension, diabetes mellitus, atherosclerosis, and increased intracranial pressure.

Optic chiasm (kī-AZM) A crossing point of the optic (II) nerves, anterior to the pituitary gland.

Optic disc A small area of the retina containing openings through which the fibers of the ganglion neurons emerge as the optic (II) nerve. Also called the **blind spot.**

Optician (op-TISH-an) A technician who fits, adjusts, and dispenses corrective lenses on prescription of an ophthalmologist or optometrist.

Optic tract A bundle of axons that transmits nerve impulses from the retina of the eye between the optic chiasm and the thalamus.

Optometrist (op-TOM-e-trist) Specialist with a doctorate degree in optometry who is licensed to examine and test the eyes and treat visual defects by prescribing corrective lenses.

Ora serrata (Ō-ra ser-RĀ-ta) The irregular margin of the retina lying internal and slightly posterior to the junction of the choroid and ciliary body.

Orbit (OR-bit) The bony, pyramid-shaped cavity of the skull that holds the eyeball.

Organ A structure composed of two or more different kinds of tissues with a specific function and usually a recognizable shape.

Organelle (or-gan-EL) A permanent structure within a cell with characteristic morphology that is specialized to serve a specific function in cellular activities.

Orgasm (OR-gazm) Sensory and motor events involved in ejaculation for the male and involuntary contraction of the perineal muscles in the female at the climax of sexual arousal.

Orifice (OR-i-fis) Any aperture or opening.

Origin (OR-i-jin) The place of attachment of a muscle to the more stationary bone, or the end opposite the insertion.

Oropharynx (or′-ō-FAR-inks) The intermediate portion of the pharynx, lying posterior to the mouth and extending from the soft palate inferiorly to the hyoid bone.

Orthopedics (or′-thō-PĒ-diks) The branch of medicine that deals with the preservation and restoration of the skeletal system, articulations, and associated structures.

Osmosis (os-MŌ-sis) The net movement of water molecules through a selectively permeable membrane from an area of high water concentration to an area of lower water concentration until an equilibrium is reached.

Osseous (OS-ē-us) Bony.

Ossicle (OS-si-kul) Small bone, as in the middle ear (malleus, incus, stapes).

Ossification (os′-i-fi-KĀ-shun) Formation of bone. Also called **osteogenesis.**

Osteoblast (OS-tē-ō-blast′) Cell formed from an osteoprogenitor cell that participates in bone formation by secreting some organic components and inorganic salts.

Osteoclast (OS-tē-ō-clast′) A large multinuclear cell that destroys or resorbs bone tissue.

Osteocyte (OS-tē-ō-sīt′) A mature bone cell that maintains the daily activities of bone tissue.

Osteology (os′-tē-OL-ō-jē) The study of bones.

Osteon (OS-tē-on) The basic unit of structure in adult compact bone, consisting of a central (Haversian) canal with its concentrically arranged lamellae, lacunae, osteocytes, and canaliculi. Also called a **Haversian** (ha-VER-shun) **system.**

Osteoprogenitor (os′-tē-ō-prō-JEN-i-tor) **cell** Stem cell derived from mesenchyme that has mitotic potential and the ability to differentiate into an osteoblast.

Otolith (Ō-tō-lith) A particle of calcium carbonate embedded in the otolithic membrane that functions in maintaining static equilibrium.

Otolithic (ō-tō-LITH-ik) **membrane** Thick, gelatinous, glycoprotein layer located directly over hair cells of the macula in the saccule and utricle of the inner ear.

Otorhinolaryngology (ō′-tō-rī-nō-lar′-in-GOL-ō-jē) The branch of medicine that deals with the diagnosis and treatment of diseases of the ears, nose, and throat.

Oval window A small opening between the middle ear and inner ear into which the footplate of the stapes fits. Also called the **fenestra vestibuli** (fe-NES-tra ves-TIB-yoo-lē).

Ovarian (ō-VAR-ē-an) **cycle** A monthly series of events in the ovary associated with the maturation of an ovum.

Ovarian follicle (FOL-i-kul) A general name for oocytes (immature ova) in any stage of development, along with their surrounding epithelial cells.

Ovarian ligament (LIG-a-ment) A rounded cord of connective tissue that attaches the ovary to the uterus.

Ovary (Ō-var-ē) Female gonad that produces ova and the hormones estrogen, progesterone, and relaxin.

Ovulation (ō-vyoo-LĀ-shun) The rupture of a vesicular ovarian (Graafian) follicle with discharge of a secondary oocyte into the pelvic cavity.

Ovum (Ō-vum) The female reproductive or germ cell; an egg cell.

Oxidation (ok-si-DĀ-shun) The removal of electrons and hydrogen ions (hydrogen atoms) from a molecule or, less commonly, the addition of oxygen to a molecule that results in a decrease in the energy content of the molecule. The oxidation of glucose in the body is also called **cellular respiration.**

Oxyhemoglobin (ok′-sē-HĒ-mō-glō-bin) **(HbO$_2$)** Hemoglobin combined with oxygen.

Oxyphil cell A cell found in the parathyroid gland that secretes parathyroid hormone (PTH).

Oxytocin (ok′-sē-TŌ-sin) **(OT)** A hormone secreted by neurosecretory cells in the paraventricular and supraoptic nuclei of the hypothalamus that stimulates contraction of the smooth muscle fibers (cells) in the pregnant uterus and contractile cells around the ducts of mammary glands.

Pacinian corpuscle *See* **Lamellated corpuscle.**

Palate (PAL-at) The horizontal structure separating the oral and the nasal cavities; the roof of the mouth.

Pancreas (PAN-krē-as) A soft, oblong organ lying along the greater curvature of the stomach and connected by a duct to the duodenum. It is both exocrine (secreting pancreatic juice) and endocrine (secreting insulin, glucagon, somatostatin, and pancreatic polypeptide).

Pancreatic (pan′-krē-AT-ik) **duct** A single, large tube that unites with the common bile duct from the liver and gallbladder and drains pancreatic juice into the duodenum at the hepatopancreatic ampulla (ampulla of Vater). Also called the **duct of Wirsung.**

Pancreatic islet A cluster of endocrine gland cells in the pancreas that secretes insulin, glucagon, somatostatin, and pancreatic polypeptide. Also called an **islet of Langerhans** (LANG-er-hanz).

Paranasal sinus (par′-a-NĀ-zal SĪ-nus) A mucus-lined air cavity in a skull bone that communicates with the nasal cavity. Paranasal sinuses are located in the frontal, maxillary, ethmoid, and sphenoid bones.

Paraplegia (par-a-PLĒ-jē-a) Paralysis of both lower limbs.

Parasagittal plane A vertical plane that does not pass through the midline and that divides the body or organs into *unequal* left and right portions.

Parasympathetic (par′-a-sim-pa-THET-ik) **division** One of the two subdivisions of the autonomic nervous system, having cell bodies of preganglionic neurons in nuclei in the brain stem and in the lateral gray matter of the sacral portion of the spinal cord; primarily concerned with activities that conserve and restore body energy. Also called the **craniosacral** (krā-nē-ō-SĀ-kral) **division.**

Parathyroid (par′-a-THĪ-royd) **gland** One of usually four small endocrine glands embedded on the posterior surfaces of the lateral lobes of the thyroid gland.

Parathyroid hormone (PTH) A hormone secreted by the parathyroid glands that decreases blood phosphate level and increases blood calcium level.

Paraurethral (par′-a-yoo-RĒ-thral) **gland** Gland embedded in the wall of the urethra whose duct opens on either side of the urethral orifice and secretes mucus. Also called **Skene's** (SKĒNZ) **gland.**

Parenchyma (par-EN-ki-ma) The functional parts of any organ, as opposed to tissue that forms its stroma or framework.

Parietal (pa-RĪ-e-tal) **cell** The secreting cell of a gastric gland that produces hydrochloric acid and intrinsic factor. Also called an **oxyntic cell.**

Parietal pleura (PLOO-ra) The outer layer of the serous pleural membrane that encloses and protects the lungs; the layer that is attached to the wall of the pleural cavity.

Parotid (pa-ROT-id) **gland** One of the paired salivary glands located inferior and anterior to the ears connected to the oral cavity via a duct (Stensen's) that opens into the inside of the cheek opposite the maxillary second molar tooth.

Pars intermedia A small avascular zone between the anterior and posterior pituitary glands.

Pathological anatomy The study of structural changes caused by disease.

Pectinate (PEK-ti-nāt) **muscles** Projecting muscle bundles of the anterior atrial walls and the lining of the auricles.

Pediatrician (pē′-dē-a-TRISH-un) A physician who specializes in the care and treatment of children and their illnesses.

Pedicel (PED-i-sel) Footlike structure, as on podocytes of a glomerulus.

Pelvic (PEL-vik) **cavity** Inferior portion of the abdominopelvic cavity that contains the urinary bladder, sigmoid colon, rectum, and internal female and male reproductive structures.

Pelvic splanchnic (SPLANGK-nik) **nerves** Preganglionic paraysmpathetic fibers from the levels of S2, S3, and S4 that supply the urinary bladder, reproductive organs, and the descending and sigmoid colon and rectum.

Pelvimetry (pel-VIM-e-trē) Measurement of the size of the inlet and outlet of the birth canal.

Pelvis The basinlike structure that is formed by the two hipbones, the sacrum, and the coccyx. The expanded, proximal portion of the ureter, lying within the kidney and into which the major calyces open.

Penis (PĒ-nis) The male copulatory organ, used to introduce sperm cells into the female vagina.

Pepsin Protein-digesting enzyme secreted by chief (zymogenic) cells of the stomach as the inactive form pepsinogen, which is converted to active pepsin by hydrochloric acid.

Perforating canal A minute passageway by means of which blood vessels and nerves from the periosteum penetrate into compact bone. Also called **Volkmann's** (FŌLK-manz) **canal.**

Pericardial (per′-i-KAR-dē-al) **cavity** Small potential space between the visceral and parietal layers of the serous pericardium that contains pericardial fluid.

Pericardium (per′-i-KAR-dē-um) A loose-fitting membrane that encloses the heart, consisting of a superficial fibrous layer and a deep serous layer.

Perichondrium (per′-i-KON-drē-um) The membrane that covers cartilage.

Perikaryon (per′-i-KAR-ē-on) The nerve cell body that contains the nucleus and other organelles. Also called a **soma.**

Perilymph (PER-i-lymf) The fluid contained between the bony and membranous labyrinths of the inner ear.

Perimetrium (per-i-MĒ-trē-um) The serosa of the uterus.

Perimysium (per′-i-MĪZ-ē-um) Invagination of the epimysium that divides muscles into bundles.

Perineum (per′-i-NĒ-um) The pelvic floor; the space between the anus and the scrotum in the male and between the anus and the vulva in the female.

Perineurium (per′-i-NYOO-rē-um) Connective tissue wrapping around fascicles in a nerve.

Periodontal (per-ē-ō-DON-tal) **ligament** The periosteum lining the alveoli (sockets) for the teeth in the alveolar processes of the mandible and maxillae.

Periosteum (per′-ē-OS-tē-um) The membrane that covers bone and consists of connective tissue, osteoprogenitor cells, and osteoblasts and is essential for bone growth, repair, and nutrition.

Peripheral (pe-RIF-er-al) Located on the outer part or a surface of the body.

Peripheral nervous system (PNS) The part of the nervous system that lies outside the central nervous system—nerves and ganglia.

Peristalsis (per′-i-STAL-sis) Successive muscular contractions along the wall of a hollow muscular structure.

Peritoneum (per′-i-tō-NĒ-um) The largest serous membrane of the body that lines the abdominal cavity and covers the viscera within the cavity.

Peroxisome (pe-ROKS-ī-sōm) Organelle similar in structure to a lysosome that contains enzymes that use molecular oxygen to oxidize various organic compounds. Such reactions produce hydrogen peroxide; abundant in liver cells.

Perspiration Substance produced by sudoriferous (sweat) glands containing water, salts, urea, uric acid, amino acids, ammonia, sugar, lactic acid, and ascorbic acid; helps maintain body temperature and eliminate wastes.

Peyer's patches *See* **Aggregated lymphatic follicles.**

pH A symbol of the measure of the concentration of hydrogen ions (H^+) in a solution. The pH scale extends from 0 to 14, with a value of 7 expressing neutrality, values lower than 7 expressing increasing acidity, and values higher than 7 expressing increasing alkalinity.

Phagocytosis (fag′-ō-sī-TŌ-sis) The process by which cells (phagocytes) ingest particulate matter; especially the ingestion and destruction of microbes, cell debris, and other foreign matter.

Phalanx (FĀ-lanks) The bone of a finger or toe. *Plural,* **phalanges** (fa-LAN-jēz).

Pharmacology (far′-ma-KOL-ō-jē) The science that deals with the effects and uses of drugs in the treatment of disease.

Pharynx (FAR-inks) The throat; a tube that starts at the internal nares and runs partway down the neck where it opens into the esophagus posteriorly and the larynx anteriorly.

Phenotype (FĒ-nō-tīp) The observable expression of genotype; physical characteristics of an organism determined by genetic makeup and influenced by interaction between genes and internal and external environmental factors.

Phospholipid (fos'-fō-LIP-id) **bilayer** Arrangement of phospholipid molecules in two parallel rows in which the hydrophilic "heads" face outward and the hydrophobic "tails" face inward.

Photopigment A substance that can absorb light and undergo structural changes that can lead to the development of a receptor potential. An example is rhodopsin. Also called **visual pigment.**

Photoreceptor Receptor that detects light on the retina of the eye.

Physiology (fiz'-ē-OL-ō-jē) Science that deals with the functions of an organism or its parts.

Pia mater (PĪ-a MA-ter) The inner membrane (meninx) covering the brain and spinal cord.

Pineal (PĪN-ē-al) **gland** The cone-shaped gland located in the roof of the third ventricle. Also called the **epiphysis cerebri** (ē-PIF-i-sis se-RĒ-brē).

Pinealocyte (pin-ē-AL-ō-sīt) Secretory cell of the pineal gland that produces hormones.

Pinna (PIN-na) The projecting part of the external ear composed of elastic cartilage and covered by skin and shaped like the flared end of a trumpet. Also called the **auricle** (OR-i-kul).

Pinocytosis (pi'-nō-sī-TŌ-sis) the process by which cells ingest liquid.

Pituicyte (pi-TOO-i-sīt) Supporting cell of the posterior lobe of the pituitary gland.

Pituitary (pi-TOO-i-tar'-ē) **gland** A small endocrine gland lying in the sella turcica of the sphenoid bone and attached to the hypothalamus by the infundibulum; nicknamed the "master gland." Also called the **hypophysis** (hī-POF-i-sis).

Pivot joint A synovial joint in which a rounded, pointed, or conical surface of one bone articulates with a ring formed partly by another bone and partly by a ligament, as in the joint between the atlas and axis and between the proximal ends of the radius and ulna. Also called a **trochoid** (TRŌ-koid) **joint.**

Placenta (pla-SEN-ta) The special structure through which the exchange of materials between fetal and maternal circulations occurs. Also called the **afterbirth.**

Plantar flexion (PLAN-tar FLEX-shun) Bending the foot in the direction of the plantar surface (sole).

Plaque (plak) A cholesterol-containing mass in the tunica media of arteries. A mass of bacterial cells, dextran (polysaccharide), and other debris that adheres to teeth.

Plasma (PLAZ-ma) The extracellular fluid found in blood vessels; blood minus the formed elements.

Plasma cell Cell that produces antibodies and develops from a B cell (lymphocyte).

Plasma (cell) membrane Outer, limiting membrane that separates the cell's internal parts from extracellular fluid and the external environment.

Platelet (PLĀT-let) A fragment of cytoplasm enclosed in a cell membrane and lacking a nucleus; found in the circulating blood; plays a role in blood clotting.

Pleura (PLOOR-a) The serous membrane that covers the lungs and lines the walls of the chest.

Pleural cavity Small potential space between the visceral and parietal pleurae.

Plexus (PLEX-sus) A network of nerves, veins, or lymphatic vessels.

Plexus of Auerbach *See* **Myenteric plexus.**

Plexus of Meissner *See* **Submucosal plexus.**

Pneumotaxic (noo-mō-TAK-sik) **area** Portion of the respiratory center in the pons that sends inhibitory nerve impulses to the inspiratory area that limit inspiration and facilitate expiration.

Podiatry (pō-DĪ-a-trē) The diagnosis and treatment of foot disorders.

Polar body The smaller cell resulting from the unequal division of cytoplasm during the meiotic divisions of an oocyte. The polar body has no function and is resorbed.

Polyuria (pol'-ē-YOO-rē-a) An excessive production of urine.

Pons (ponz) The portion of the brain stem that forms a "bridge" between the medulla oblongata and the midbrain, anterior to the cerebellum.

Postcentral gyrus *See* **Primary somatosensory area.**

Posterior (pos-TĒR-ē-or) Nearer to or at the back of the body. Also called **dorsal.**

Posterior column tracts Sensory tracts (fasciculus gracilis and fasciculus cuneatus) that convey information up the spinal cord to the brain related to proprioception, discriminative touch, two-point discrimination, pressure, and vibrations.

Posterior pituitary gland Posterior portion of the pituitary gland. Also called the **neurohypophysis** (noo-rō-hī-POF-i-sis).

Posterior root The structure composed of sensory fibers lying between a spinal nerve and the dorsolateral aspect of the spinal cord. Also called the **dorsal (sensory) root.**

Posterior root ganglion (GANG-glē-on) A group of cell bodies of sensory (afferent) neurons and their supporting cells located along the posterior root of a spinal nerve. Also called a **dorsal (sensory) root ganglion.**

Postganglionic neuron (pōst'-gang-lē-ON-ik NOO-ron) The second visceral motor neuron in an autonomic pathway, having its cell body and dendrites located in an autonomic ganglion and its unmyelinated axon ending at cardiac muscle, smooth muscle, or a gland.

Postsynaptic (pōst-sin-AP-tik) **neuron** The nerve cell that is activated by the release of a neurotransmitter substance from another neuron and carries nerve impulses away from the synapse.

Pouch of Douglas *See* **Rectouterine pouch.**

Precapillary sphincter (SFINGK-ter) A ring of smooth muscle fibers (cells) at the site of origin of true capillaries that regulate blood flow into true capillaries.

Precentral gyrus *See* **Primary motor area.**

Preganglionic (prē-gang-lē-ON-ik) **neuron** The first visceral motor neuron in an autonomic pathway, with its cell body and dendrites in the brain or spinal cord and its myelinated axon ending at an autonomic ganglion, where it synapses with a postganglionic neuron.

Prepuce (PRĒ-pyoos) The loose-fitting skin covering the glans of the penis and clitoris. Also called the **foreskin.**

Presynaptic (prē-sin-AP-tik) **neuron** A nerve cell that carries nerve impulses toward a synapse.

Prevertebral ganglion (prē-VERT-e-bral GANG-lē-on) A cluster of cell bodies of postganglionic sympathetic neurons anterior to the spinal column and close to large abdominal arteries. Also called a **collateral ganglion.**

Primary germ layer One of three layers of embryonic tissue, called ectoderm, mesoderm, and endoderm, that give rise to all tissues and organs of the organism.

Primary motor area A region of the cerebral cortex in the precentral gyrus of the frontal lobe of the cerebrum that controls specific muscles or groups of muscles.

Primary somatosensory area A region of the cerebral cortex posterior to the central sulcus in the postcentral gyrus of the parietal lobe of the cerebrum that localizes exactly the points of the body where somatic sensations originate.

Prime mover The muscle directly responsible for producing the desired motion. Also called an **agonist** (AG-ō-nist).

Primitive gut Embryonic structure composed of endoderm and mesoderm that gives rise to most of the gastrointestinal tract.

Primordial (prī-MOR-dē-al) Existing first; especially primordial egg cells in the ovary.

Principal cell Cell found in the parathyroid glands that secretes parathyroid hormone (PTH). Also called a **chief cell.**

Proctology (prok-TOL-ō-jē) The branch of medicine that treats the rectum and its disorders.

Progesterone (prō-JES-te-rōn) A female sex hormone produced by the ovaries that helps prepare the endometrium for implantation of a fertilized ovum and the mammary glands for milk secretion.

Prolactin (prō-LAK-tin) **(PRL)** A hormone secreted by the anterior lobe of the pituitary gland that initiates and maintains milk secretion by the mammary glands.

Prolapse (PRŌ-laps) A dropping or falling down of an organ, especially the uterus or rectum.

Proliferation (pro-lif'-er-Ā-shun) Rapid and repeated reproduction of new parts, especially cells.

Pronation (prō-NĀ-shun) A movement of the forearm in which the palm of the hand is turned posteriorly or inferiorly.

Prophase (PRŌ-fāz) The first stage of mitosis during which chromatid pairs are formed and aggregate around the equatorial plane region of the cell.

Proprioceptor (prō'-prē-ō-SEP-tor) A receptor located in muscles, tendons, or joints that provides information about body position and movements.

Prostaglandin (pros'-ta-GLAN-din) **(PG)** A membrane-associated lipid composed of 20-carbon fatty acids with 5 carbon atoms joined to form a cyclopentane ring; synthesized in small quantities and basically mimics the activities of hormones.

Prostate (PROS-tāt) **gland** A doughnut-shaped gland inferior to the urinary bladder that surrounds the superior portion of the male urethra and secretes a slightly acid solution that contributes to sperm motility and viability.

Protein An organic compound consisting of carbon, hydrogen, oxygen, nitrogen, and sometimes sulfur and phosphorus, and made up of amino acids linked by peptide bonds.

Proto-oncogene (prō'-tō-ONG-kō-jēn) Gene responsible for some aspect of normal growth and development; it may transform into an oncogene, a gene capable of causing cancer.

Protraction (prō-TRAK-shun) The movement of the mandible or shoulder girdle forward on a plane parallel with the ground.

Proximal (PROK-si-mal) Nearer the attachment of a limb to the trunk or a structure; nearer to the point of origin.

Pseudopods (SOO-dō-pods) Temporary, protruding projections of cytoplasm.

Pterygopalatine ganglion (ter'-i-gō-PAL-a-tīn GANG-glē-on) A cluster of cell bodies of parasympathetic postganglionic neurons ending at the lacrimal and nasal glands.

Pubic (PYOO-bik) **symphysis** A slightly movable cartilaginous joint between the anterior surfaces of the hipbones.

Pudendum (pyoo-DEN-dum) A collective designation for the external genitalia of the female.

Pulmonary (PUL-mo-ner'-ē) Concerning or affected by the lungs.

Pulmonary circulation The flow of deoxygenated blood from the right ventricle to the lungs and the return of oxygenated blood from the lungs to the left atrium.

Pulmonary ventilation The inflow (inspiration) and outflow (expiration) of air occurring between the atmosphere and the lungs. Also called **breathing.**

Pulp cavity A cavity within the crown and neck of a tooth, filled with pulp, a connective tissue containing blood vessels, nerves, and lymphatic vessels.

Pulse (PULS) The rhythmic expansion and elastic recoil of an artery with each contraction of the left ventricle.

Pupil The hole in the center of the iris, the area through which light enters the posterior cavity of the eyeball.

Purkinje fiber *See* **Conduction myofiber.**

Pus The liquid product of inflammation containing leukocytes or their remains and debris of dead cells.

Pyloric (pī-LOR-ik) **sphincter** A thickened ring of smooth muscle through which the pylorus of the stomach communicates with the duodenum. Also called the **pyloric valve.**

Pyogenesis (pī'-ō-JEN-e-sis) Formation of pus.

Pyorrhea (pī-ō-RĒ-a) A discharge or flow of pus, especially in the alveoli (sockets) and the tissues of the gums.

Pyramid (PIR-a-mid) A pointed or cone-shaped structure; one of two roughly triangular structures on the ventral side of the medulla oblongata composed of the largest motor tracts that run from the cerebral cortex to the spinal cord; a triangular-shaped structure in the renal medulla composed of the straight segments of renal tubules.

Pyuria (pī-YOO-rē-a) The presence of leukocytes and other components of pus in urine.

Quadrant (KWOD-rant) One of four parts.

Quadriplegia (kwod'-ri-PLĒ-jē-a) Paralysis of the two upper and two lower extremities.

Radiographic (rā'-dē-ō-GRAF-ik) **anatomy** Diagnostic branch of anatomy that includes the use of X rays.

Rami communicantes (RĀ-mē ko-myoo-ni-KAN-tēz) Branches of a spinal nerve. *Singular,* **ramus communicans** (RĀ-mus ko-MYOO-ni-kans).

Rathke's pouch *See* **Hypophyseal pouch.**

Receptor A specialized cell or a nerve cell terminal modified to respond to some specific sensory modality, such as touch, pressure, cold, light, or sound, and convert it to a nerve impulse by way of a generator or receptor potential. A specific molecule or arrangement of molecules organized to accept only molecules with a complementary shape.

Receptor-mediated endocytosis A highly selective process in which cells take up large molecules or particles. In the process, successive compartments called endocytic vesicles and endosomes form. The molecules or particles are eventually broken down by enzymes in lysosomes.

Recombinant DNA Synthetic DNA, formed by joining a fragment of DNA from one source to a portion of DNA from another.

Rectouterine pouch A pocket formed by the parietal peritoneum as it moves posteriorly from the surface of the uterus and is reflected onto the rectum; the most inferior point in the pelvic cavity. Also called the **pouch** or **cul de sac of Douglas.**

Rectum (REK-tum) The last 20 cm (7 in.) of the gastrointestinal tract, from the sigmoid colon to the anus.

Recumbent (re-KUM-bent) Lying down.

Red nucleus A cluster of cell bodies in the midbrain, occupying a large portion of the tectum and sending fibers into the rubroreticular and rubrospinal tracts.

Red pulp That portion of the spleen that consists of venous sinuses filled with blood and cords of splenic tissue called splenic (Billroth's) cords.

Reflex Fast response to a change (stimulus) in the internal or external environment that attempts to restore body activities within normal limits; passes over a reflex arc.

Reflex arc The most basic conduction pathway through the nervous system, connecting a receptor and an effector and consisting of a receptor, a sensory neuron, a center in the central nervous system for a synapse, a motor neuron, and an effector.

Regeneration (rē-jen′-er-Ā-shun) The natural renewal of a structure.

Regimen (REJ-i-men) A strictly regulated scheme of diet, exercise, or activity designed to achieve certain ends.

Regional anatomy The division of anatomy dealing with a specific region of the body, such as the head, neck, chest, or abdomen.

Relapse (RĒ-laps) The return of a disease weeks or months after its apparent cessation.

Relaxin A female hormone produced by the ovaries that relaxes the pubic symphysis and helps dilate the uterine cervix to facilitate delivery.

Remodeling Replacement of old bone by new bone tissue.

Renal corpuscle (KOR-pus′-l) A glomerular (Bowman's) capsule and its enclosed glomerulus.

Renal pelvis A cavity in the center of the kidney formed by the expanded, proximal portion of the ureter, lying within the kidney, and into which the major calyces open.

Renal pyramid A triangular structure in the renal medulla composed of the straight segments of renal tubules.

Reproduction (rē′-prō-DUK-shun) Either the formation of new cells for growth, repair, or replacement, or the production of a new individual.

Reproductive cell division Type of cell division in which sperm and egg cells are produced; consists of meiosis and cytokinesis.

Respiration (res-pi-RĀ-shun) Overall exchange of gases between the atmosphere, blood, and body cells consisting of pulmonary ventilation, external respiration, and internal respiration.

Respiratory center Neurons in the reticular formation of the brain stem that regulate the rate of respiration.

Rete (RĒ-tē) **testis** The network of ducts in the testes.

Reticular (re-TIK-yoo-lar) **activating system (RAS)** An extensive network of branched nerve cells running through the core of the brain stem. When these cells are activated, a generalized alert or arousal behavior results.

Reticular formation A network of small groups of nerve cells scattered among bundles of fibers beginning in the medulla oblongata as a continuation of the spinal cord and extending upward through the central part of the brain stem.

Reticulocyte (re-TIK-yoo-lō-sīt) An immature red blood cell.

Reticulum (re-TIK-yoo-lum) A network.

Retina (RET-i-na) The deep coat of the eyeball, lying only in the posterior portion of the eye and consisting of nervous tissue and a pigmented layer composed of epithelial cells lying in contact with the choroid. Also called the **nervous tunic** (TOO-nik).

Retraction (rē-TRAK-shun) The movement of a protracted part of the body posteriorly on a plane parallel to the ground, as in pulling the lower jaw back in line with the upper jaw.

Retroflexion (re-trō-FLEK-shun) A malposition of the uterus in which it is tilted posteriorly.

Retrograde degeneration (RE-trō-grād dē-jen-er-Ā-shun) Changes that occur in the proximal portion of a damaged axon only as far as the first neurofibral node (node of Ranvier); similar to changes that occur during Wallerian degeneration.

Retroperitoneal (re′-trō-per-i-tō-NĒ-al) External to the peritoneal lining of the abdominal cavity.

Rh factor An inherited agglutinogen (antigen) on the surface of red blood cells.

Rhinology (rī-NOL-ō-jē) The study of the nose and its disorders.

Ribonucleic (rī′-bō-nyoo-KLĒ-ik) **acid (RNA)** A single-stranded nucleic acid constructed of nucleotides consisting of one of four possible nitrogenous bases (adenine, cytosine, guanine, or uracil), ribose, and a phosphate group; three types are messenger RNA (mRNA), transfer RNA (tRNA), and ribosomal RNA (rRNA), each of which cooperates with DNA for protein synthesis.

Ribosome (RĪ-bō-sōm) An organelle in the cytoplasm of cells, composed of ribosomal RNA and ribosomal proteins, that synthesizes proteins; nicknamed the "protein factory."

Right lymphatic (lim-FAT-ik) **duct** A vessel of the lymphatic system that drains lymph from the upper right side of the body and empties it into the right subclavian vein.

Rigidity (ri-JID-i-tē) Hypertonia characterized by increased muscle tone, but reflexes are not affected.

Rima glottidis (RĪ-ma GLOT-ti-dis) Space between the vocal folds (true vocal cords).

Rod A visual receptor in the retina of the eye that is specialized for vision in dim light.

Root canal A narrow extension of the pulp cavity lying within the root of a tooth.

Root of penis Attached portion of penis that consists of the bulb and crura.

Rotation (rō-TĀ-shun) Moving a bone around its own axis, with no other movement.

Round ligament (LIG-a-ment) A band of fibrous connective tissue enclosed between the folds of the broad ligament of the uterus, emerging from a point on the uterus just inferior to the uterine (Fallopian) tube, extending laterally along the pelvic wall, and penetrating the abdominal wall through the deep inguinal ring to end in the labia majora.

Round window A small opening between the middle and inner ear, directly inferior to the oval window, covered by the secondary tympanic membrane. Also called the **fenestra cochlea** (fe-NES-tra KŌK-lē-a).

Rugae (ROO-jē) Large folds in the mucosa of an empty hollow organ, such as the stomach and vagina.

Saccule (SAK-yool) The inferior and smaller of the two chambers in the membranous labyrinth inside the vestibule of the inner ear containing a receptor organ for static equilibrium.

Sacral hiatus (hī-Ā-tus) Inferior entrance to the vertebral canal formed when the laminae of the fifth sacral vertebra (and sometimes fourth) fail to meet.

Sacral plexus (PLEK-sus) A network formed by the anterior branches of spinal nerves L4 through S3.

Sacral promontory (PROM-on-tor′-ē) The superior surface of the body of the first sacral vertebra that projects anteriorly into the pelvic cavity; a line from the sacral promontory to the superior border of the pubic symphysis divides the abdominal and pelvic cavities.

Saddle joint A synovial joint in which the articular surface of one bone is saddle-shaped and the articular surface of the other bone fits into the saddle as a sitting rider would, as in the joint between the trapezium and the metacarpal of the thumb. Also called a **sellaris** (sel-LA-ris) **joint.**

Sagittal (SAJ-i-tal) **plane** A vertical plane that divides the body or organs into left and right portions. Such a plane may be **mid-sagittal (median),** in which the divisions are equal, or **parasagittal,** in which the divisions are unequal.

Saliva (sa-LĪ-va) A clear, alkaline, somewhat viscous secretion produced by the three pairs of salivary glands; contains various salts, mucin, lysozyme, and salivary amylase.

Salivary amylase (SAL-i-ver-ē AM-i-lās) An enzyme in saliva that initiates the chemical breakdown of starch, mostly in the mouth.

Salivary gland One of three pairs of glands that lie external to the mouth and pour their secretory product (called saliva) into ducts that empty into the oral cavity; the parotid, submandibular, and sublingual glands.

Sarcolemma (sar′-kō-LEM-ma) The cell membrane of a muscle fiber (cell), especially of a skeletal muscle fiber.

Sarcoma (sar-KŌ-ma) A connective tissue tumor, often highly malignant.

Sarcomere (SAR-kō-mēr) A contractile unit in a striated muscle fiber (cell) extending from one Z disc to the next Z disc.

Sarcoplasm (SAR-kō-plazm) The cytoplasm of a muscle fiber (cell).

Sarcoplasmic reticulum (sar′-kō-PLAZ-mik re-TIK-yoo-lum) A network of saccules and tubes surrounding myofibrils of a muscle fiber (cell), comparable to endoplasmic reticulum; functions to reabsorb calcium ions during relaxation and to release them to cause contraction.

Satiety center A collection of nerve cells located in the ventromedial nuclei of the hypothalamus that, when stimulated, brings about the cessation of eating.

Scala tympani (SKA-la TIM-pan-ē) The inferior spiral-shaped channel of the bony cochlea, filled with perilymph.

Scala vestibuli (ves-TIB-yoo-lē) The superior spiral-shaped channel of the bony cochlea, filled with perilymph.

Schwann cell *See* **Neurolemmocyte.**

Sclera (SKLE-ra) The white coat of fibrous tissue that forms the superficial protective covering over the eyeball except in the most anterior portion; the posterior portion of the fibrous tunic.

Scleral venous sinus A circular venous sinus located at the junction of the sclera and the cornea through which aqueous humor drains from the anterior chamber of the eyeball into the blood. Also called the **canal of Schlemm** (SHLEM).

Sclerosis (skle-RŌ-sis) A hardening with loss of elasticity of tissues.

Scoliosis (skō′-lē-Ō-sis) An abnormal curvature from the normal vertical line of the backbone.

Scrotum (SKRŌ-tum) A skin-covered pouch that contains the testes and their accessory structures.

Sebaceous (se-BĀ-shus) **gland** An exocrine gland in the dermis of the skin, almost always associated with a hair follicle, that secretes sebum. Also called an **oil gland.**

Sebum (SĒ-bum) Secretion of sebaceous (oil) glands.

Secondary sex characteristic A feature characteristic of the male or female body that develops at puberty under the stimulation of sex hormones but is not directly involved in sexual reproduction, such as distribution of body hair, voice pitch, body shape, and muscle development.

Secretion (se-KRĒ-shun) Production and release from a gland cell of a fluid, especially a functionally useful product as opposed to a waste product.

Selective permeability (per′-mē-a-BIL-i-tē) The property of a membrane by which it permits the passage of certain substances but restricts the passage of others.

Sella turcica (SEL-a TUR-si-ka) A depression on the superior surface of the sphenoid bone that houses the pituitary gland.

Semen (SĒ-men) A fluid discharged at ejaculation by a male that consists of a mixture of sperm cells and the secretions of the seminal vesicles, prostate gland, and bulbourethral (Cowper's) glands.

Semicircular canals Three bony channels (anterior, posterior, lateral), filled with perilymph, in which lie the membranous semicircular canals filled with endolymph. They contain receptors for equilibrium.

Semicircular ducts The membranous semicircular canals filled with endolymph and floating in the perilymph of the bony semicircular canals. They contain cristae that are concerned with dynamic equilibrium.

Semilunar (sem′-ē-LOO-nar) **valve** A valve guarding the entrance into the aorta or the pulmonary trunk from a ventricle of the heart.

Seminal vesicle (SEM-i-nal VES-i-kul) One of a pair of convoluted, pouchlike structures, lying posterior and inferior to the urinary bladder and anterior to the rectum, that secrete a component of semen into the ejaculatory ducts.

Seminiferous tubule (sem′-i-NI-fer-us TOO-byool) A tightly coiled duct, located in a lobule of the testis, where spermatozoa are produced.

Sensory area A region of the cerebral cortex concerned with the interpretation of sensory impulses.

Sensory neuron (NOO-ron) A neuron that carries nerve impulses toward the central nervous system. Also called an **afferent neuron.**

Septal defect An opening in the septum (interatrial or interventricular) between the left and right sides of the heart.

Septum (SEP-tum) A wall dividing two cavities.

Serosa (ser-Ō-sa) Any serous membrane. The external layer of an organ formed by a serous membrane. The membrane that lines the pleural, pericardial, and peritoneal cavities.

Serous (SIR-us) **membrane** A membrane that lines a body cavity that does not open to the exterior. Also called the **serosa** (se-RŌ-sa).

Serum Plasma minus its clotting proteins.

Sesamoid (SES-a-moyd) **bones** Small bones usually found in tendons.

Sex chromosomes The twenty-third pair of chromosomes, designated X and Y, which determines the genetic sex of an individual; in males, the pair is XY; in females, XX.

Sheath of Schwann *See* **Neurolemma.**

Sigmoid colon (SIG-moyd KŌ-lon) The S-shaped portion of the large intestine that begins at the level of the left iliac crest, projects inward to the midline, and terminates at the rectum at about the level of the third sacral vertebra.

Sign Any objective evidence of disease that can be observed or measured such as a lesion, swelling, or fever.

Sinoatrial (si-nō-Ā-trē-al) **(SA) node** A compact mass of cardiac muscle fibers (cells) specialized for conduction, located in the right atrium inferior to the opening of the superior vena cava. Also called the **sinuatrial node** or **pacemaker.**

Sinus (SĪ-nus) A hollow in a bone (paranasal sinus) or other tissue; a channel for blood (vascular sinus); any cavity having a narrow opening.

Sinusoid (SĪN-yoo-soyd) A microscopic space or passage for blood in certain organs such as the liver or spleen.

Skeletal muscle An organ specialized for contraction, composed of striated muscle fibers (cells), supported by connective tissue, attached to a bone by a tendon or an aponeurosis, and stimulated by somatic motor neurons.

Skene's gland *See* **Paraurethral gland.**

Skull The skeleton of the head consisting of the cranial and facial bones.

Sliding-filament mechanism The most commonly accepted explanation for muscle contraction in which actin and myosin myofilaments move into interdigitation with each other, decreasing the length of the sarcomeres.

Small intestine A long tube of the gastrointestinal tract that begins at the pyloric sphincter of the stomach, coils through the central and inferior part of the abdominal cavity, and ends at the large intestine; divided into three segments: duodenum, jejunum, and ileum.

Smooth muscle An organ specialized for contraction, composed of smooth muscle fibers (cells), located in the walls of hollow internal structures, and innervated by a visceral motor neuron.

Soft palate (PAL-at) The posterior portion of the roof of the mouth, extending posteriorly from the palatine bones and ending at the uvula. It is a muscular partition lined with mucous membrane.

Somatic cell division Type of cell division in which a single starting cell (parent cell) duplicates itself to produce two identical cells (daughter cells); consists of mitosis and cytokinesis.

Somatic (sō-MAT-ik) **nervous system (SNS)** The portion of the peripheral nervous system made up of the somatic motor fibers that run between the central nervous system and the skeletal muscles and skin.

Somite (SŌ-mīt) Block of mesodermal cells in a developing embryo that is distinguished into a myotome (which forms most of the skeletal muscles), dermatome (which forms connective tissues), and sclerotome (which forms the vertebrae).

Spermatic (sper-MAT-ik) **cord** A supporting structure of the male reproductive system, extending from a testis to the deep inguinal ring, that includes the ductus (vas) deferens, arteries, veins, lymphatic vessels, nerves, cremaster muscle, and connective tissue.

Spermatogenesis (sper′-ma-tō-JEN-e-sis) The formation and development of sperm cells in the seminiferous tubules of the testes.

Spermatozoon (sper′-ma-tō-ZŌ-on) A mature sperm cell.

Sperm cell *See* **Spermatozoon.**

Spermiogenesis (sper′-mē-ō-JEN-e-sis) The maturation of spermatids into sperm cells.

Spincter (SFINGK-ter) A circular muscle constricting an orifice.

Sphincter of Oddi *See* **Sphincter of the hepatopancreatic ampulla.**

Sphincter of the hepatopancreatic ampulla A circular muscle at the opening of the common bile and main pancreatic ducts in the duodenum. Also called the **sphincter of Oddi** (OD-ē).

Spinal (SPĪ-nal) **cord** A mass of nerve tissue located in the vertebral canal from which 31 pairs of spinal nerves originate.

Spinal nerve One of the 31 pairs of nerves that originate on the spinal cord from posterior and anterior roots.

Spinothalamic (spi-nō-THAL-a-mik) **tracts** Sensory tracts that convey information up the spinal cord to the brain related to pain, temperature, crude touch, and deep pressure.

Spinous (SPĪ-nus) **process** A sharp or thornlike process or projection. Also called a **spine.** A sharp ridge running diagonally across the posterior surface of the scapula.

Spiral organ The organ of hearing, consisting of supporting cells and hair cells that rest on the basilar membrane and extend into the endolymph of the cochlear duct. Also called the **organ of Corti** (KOR-tē).

Splanchnic (SPLANK-nik) Pertaining to the viscera.

Spleen (SPLĒN) Large mass of lymphatic tissue between the fundus of the stomach and the diaphragm that functions in phagocytosis, production of lymphocytes, and blood storage.

Spongy (cancellous) bone tissue Bone tissue that consists of an irregular latticework of thin columns of bone called trabeculae; spaces between trabeculae of some bones are filled with red bone marrow; found inside short, flat, and irregular bones and in the epiphyses (ends) of long bones.

Sputum (SPYOO-tum) Substance ejected from the mouth containing saliva and mucus.

Squamous (SKWĀ-mus) Scalelike.

Stasis (STĀ-sis) Stagnation or halt of normal flow of fluids, as blood, urine, or of the intestinal fluid.

Static equilibrium (ē-kwi-LIB-rē-um) The maintenance of posture in response to changes in the orientation of the body, mainly the head, relative to the ground.

Stellate reticuloendothelial (STEL-āt re-tik′-yoo-lō-en′-dō-THĒ-lē-al) **cell** Phagocytic cell that lines a sinusoid of the liver. Also called a **Kupffer's** (KOOP-ferz) **cell.**

Stereocilia (ste′-rē-ō-SIL-ē-a) Groups of extremely long, slender, nonmotile microvilli projecting from epithelial cells lining the epididymis.

Stomach The J-shaped enlargement of the gastrointestinal tract directly under the diaphragm in the epigastric, umbilical, and left hypochondriac regions of the abdomen, between the esophagus and small intestine.

Straight tubule (TOO-byool) A duct in a testis leading from a convoluted seminiferous tubule to the rete testis.

Stratum (STRĀ-tum) A layer.

Stratum basalis (STRĀ-tum ba-SAL-is) The superficial layer of the endometrium, next to the myometrium, that is maintained during menstruation and gestation and produces a new functionalis following menstruation or parturition.

Stratum functionalis (funk′-shun-AL-is) The deep layer of the endometrium, the layer next to the uterine cavity, that is shed during menstruation and that forms the maternal portion of the placenta during gestation.

Stretch receptor Receptor in the walls of bronchi, bronchioles, and lungs that sends impulses to the respiratory center that prevents overinflation of the lungs.

Stroma (STRŌ-ma) The tissue that forms the ground substance, foundation, or framework of an organ, as opposed to its functional parts.

Subarachnoid (sub′-a-RAK-noyd) **space** A space between the arachnoid and the pia mater that surrounds the brain and spinal cord and through which cerebrospinal fluid circulates.

Subcutaneous (sub′-kyoo-TĀ-nē-us) Beneath the skin. Also called **hypodermic** (hī-pō-DER-mik).

Subcutaneous layer A continuous sheet of areolar connective tissue and adipose tissue between the dermis of the skin and the deep fascia of the muscles. Also called the **superficial fascia** (FASH-ē-a) and hypodermis.

Subdural (sub-DOO-ral) **space** A space between the dura mater and the arachnoid of the brain and spinal cord that contains a small amount of fluid.

Sublingual (sub-LING-gwal) **gland** One of a pair of salivary glands situated in the floor of the mouth deep to the mucous membrane and to the side of the lingual frenulum, with a duct (Rivinus's) that opens into the floor of the mouth.

Submandibular (sub′-man-DIB-yoo-lar) **gland** One of a pair of salivary glands found inferior to the base of the tongue under the mucous membrane in the posterior part of the floor of the mouth, posterior to the sublingual glands, with a duct (Wharton's) situated to the side of the lingual frenulum. Also called the **submaxillary** (sub′-MAK-si-ler-ē) **gland.**

Submucosa (sub-myoo-KŌ-sa) A layer of connective tissue located deep to a mucous membrane, as in the gastrointestinal tract or the urinary bladder; the submucosa connects the mucosa to the muscularis layer.

Submucosal plexus A network of autonomic nerve fibers located in the superficial portion of the submucous layer of the small intestine. Also called the **plexus of Meissner** (MĪS-ner).

Subserous fascia (sub-SE-rus FASH-ē-a) A layer of connective tissue internal to the deep fascia, lying between the deep fascia and the serous membrane that lines the body cavities.

Subthalamus (sub-THAL-a-mus) Part of the diencephalon inferior to the thalamus; the substantia nigra and red nucleus extend from the midbrain into the subthalamus.

Sudoriferous (soo′-dor-IF-er-us) **gland** An eccrine or merocrine exocrine gland in the dermis or subcutaneous layer that produces perspiration. Also called a **sweat gland.**

Sulcus (SUL-kus) A groove or depression between parts, especially between the convolutions of the brain. *Plural,* **sulci** (SUL-sē).

Superficial (soo′-per-FISH-al) Located on or near the surface of the body.

Superficial fascia (FASH-ē-a) A continuous sheet of fibrous connective tissue between the dermis of the skin and the deep fascia of the muscles. Also called **subcutaneous** (sub′-kyoo-TĀ-nē-us) **layer.**

Superficial inguinal (IN-gwi-nal) **ring** A triangular opening in the aponeurosis of the external oblique muscle that represents the termination of the inguinal canal.

Superior (soo-PĒR-ē-or) Toward the head or upper part of a structure. Also called **cephalic** (SEF-a-lic) or **cranial.**

Superior vena cava (VĒ-na CĀ-va) **(SVC)** Large vein that collects blood from parts of the body superior to the heart and returns it to the right atrium.

Supination (soo-pī-NĀ-shun) A movement of the forearm in which the palm is turned anteriorly or superiorly.

Suppuration (sup′-yoo-RĀ-shun) Pus formation and discharge.

Supraopticohypophyseal (soo′-pra-op′-tik-ō-hī-po-FIZ-ē-al) **tract** A bundle of nerve processes made up of fibers that have their cell bodies in the hypothalamus but release their neurosecretions in the posterior pituitary gland.

Surface anatomy The study of the structures that can be identified from the outside of the body by visualization and palpation.

Surfactant (sur-FAK-tant) A phospholipid substance produced by the lungs that decreases surface tension.

Susceptibility (sus-sep′-ti-BIL-i-tē) Lack of resistance of a body to the deleterious or other effects of an agent such as pathogenic microorganisms.

Suspensory ligament (sus-PEN-so-rē LIG-a-ment) A fold of peritoneum extending laterally from the surface of the ovary to the pelvic wall.

Sustentacular (sus′-ten-TAK-yoo-lar) **cell** A supporting cell of seminiferous tubules that produces secretions for supplying nutrients to sperm cells and the hormone inhibin. Also called a **Sertoli** (ser-TŌ-lē) **cell.**

Sutural (SOO-cher-al) **bone** A small bone located within a suture between certain cranial bones.

Suture (SOO-cher) An immovable fibrous joint in the skull where bone surfaces are closely united.

Sympathetic (sim′-pa-THET-ik) **division** One of the two subdivisions of the autonomic nervous system, having cell bodies of preganglionic neurons in the lateral gray columns of the thoracic segment and first two or three lumbar segments of the spinal cord; it is primarily concerned with processes that involve the expenditure of energy. Also called the **thoracolumbar** (thō′-ra-kō-LUM-bar) **division.**

Sympathetic trunk ganglion (GANG-lē-on) A cluster of cell bodies of postganglionic sympathetic neurons lateral to the vertebral column, close to the body of a vertebra. These ganglia extend inferiorly through the neck, thorax, and abdomen to the coccyx on both sides of the vertebral column and are connected to one another to form a chain on each side of the vertebral column. Also called **lateral, or sympathetic, chain** or **vertebral chain ganglia.**

Symphysis (SIM-fi-sis) A line of union. A slightly movable cartilaginous joint such as the pubic symphysis between the anterior surfaces of the hipbones.

Symptom (SIMP-tum) A subjective change in body function not apparent to an observer, such as fever or nausea, that indicates the presence of a disease or disorder of the body.

Synapse (SIN-aps) The functional junction between two neurons or between a neuron and an effector, such as a muscle or gland; may be electrical or chemical.

Synapsis (sin-AP-sis) The pairing of homologous chromosomes during prophase I of meiosis.

Synaptic (sin-AP-tik) **cleft** The narrow gap that separates the axon terminal of one nerve cell from another nerve cell or muscle fiber (cell) and across which a neurotransmitter diffuses to affect the postsynaptic cell.

Synaptic end bulb Expanded distal end of an axon terminal that contains synaptic vesicles. Also called a **synaptic knob** or **end foot.**

Synaptic gutter Invaginated portion of a sarcolemma under an axon terminal. Also called a **synaptic trough** (TROF).

Synaptic vesicle Membrane-enclosed sac in a synaptic end bulb that stores neurotransmitters.

Synarthrosis (sin′-ar-THRŌ-sis) An immovable joint.

Synchondrosis (sin′-kon-DRŌ-sis) A cartilaginous joint in which the connecting material is hyaline cartilage.

Syndesmosis (sin′-dez-MŌ-sis) A fibrous joint in which articulating bones are united by dense fibrous tissue.

Syndrome (SIN-drōm) A group of signs and symptoms that occur together in a pattern that is characteristic of a particular disease or abnormal condition.

Synergist (SIN-er-jist) A muscle that assists the prime mover by preventing unwanted movements in intermediate joints or otherwise aiding the movement of the prime mover.

Synostosis (sin′-os-TŌ-sis) A joint in which the dense fibrous connective tissue that unites bones at a suture has been replaced by bone, resulting in a complete fusion across the suture line.

Synovial (si-NŌ-vē-al) **cavity** The space between the articulating bones of a synovial (diarthrotic) joint, filled with synovial fluid. Also called a **joint cavity.**

Synovial fluid Secretion of synovial membranes that lubricates joints and nourishes articular cartilage.

Synovial joint A fully movable or diarthrotic joint in which a synovial (joint) cavity is present between the two articulating bones.

Synovial membrane The deeper of the two layers of the articular capsule of a synovial joint, composed of areolar connective tissue that secretes synovial fluid into the synovial (joint) cavity.

System An association or organs that have a common function.

Systemic (sis-TEM-ik) Affecting the whole body; generalized.

Systemic anatomy The study of particular systems of the body, such as the skeletal, muscular, nervous, cardiovascular, or urinary systems.

Systemic circulation The routes through which oxygenated blood flows from the left ventricle through the aorta to all the organs of the body and deoxygenated blood returns to the right atrium.

Systole (SIS-tō-lē) In the cardiac cycle, the phase of contraction of the heart muscle, especially of the ventricles.

Systolic (sis-TO-lik) **blood pressure** The force exerted by blood on arterial walls during ventricular contraction; the highest pressure measured in the large arteries, about 120 mm Hg under normal conditions for a young, adult male.

Tachycardia (tak′-i-KAR-dē-a) A rapid heartbeat or pulse rate.

Tactile disc Modified epidermal cell in the stratum basale of hairless skin that functions as a cutaneous receptor for discriminative touch. Also called **Merkel's** (MER-kelz) **disc.**

Taenia coli (TĒ-nē-a KŌ-lī) One of three flat bands of thickened, longitudinal muscles running the length of the large intestine.

Target cell A cell whose activity is affected by a particular hormone.

Tarsal gland Sebaceous (oil) gland that opens on the edge of each eyelid. Also called a **Meibomian** (mī-BŌ-mē-an) **gland.**

Tarsal plate A thin, elongated sheet of connective tissue, one in each eyelid, giving the eyelid form and support. The aponeurosis of the levator palpebrae superioris is attached to the tarsal plate of the superior eyelid.

Tarsus (TAR-sus) A collective term for the seven bones of the ankle.

T cell A lymphocyte that becomes mature in the thymus gland and differentiates into one of several kinds of cells that function in cell-mediated immunity.

Tectorial (tek-TŌ-rē-al) **membrane** A gelatinous membrane projecting over and in contact with the hair cells of the spiral organ (organ of Corti) in the cochlear duct.

Telophase (TEL-ō-fāz) The final stage of mitosis in which the daughter nuclei become established.

Temporomandibular joint (TMJ) syndrome A disorder of the temporomandibular joint (TMJ) characterized by dull pain around the ear, tenderness of jaw muscles, a clicking or popping noise when opening or closing the mouth, limited or abnormal opening of the mouth, headache, tooth sensitivity, and abnormal wearing of the teeth.

Tendon (TEN-don) A white fibrous cord of dense, regularly arranged connective tissue that attaches muscle to bone.

Tendon organ A proprioceptive receptor sensitive to changes in muscle tension and force of contraction, found chiefly near the junction of tendons and muscles. Also called a **Golgi** (GOL-jē) **tendon organ.**

Tentorium cerebelli (ten-TŌ-rē-um ser′-e-BEL-ē) A transverse shelf of dura mater that forms a partition between the occipital lobe of the cerebral hemispheres and the cerebellum and that covers the cerebellum.

Terminal ganglion (TER-mi-nal GANG-glē-on) A cluster of cell bodies of postganglionic parasympathetic neurons either lying very close to the visceral effectors or located within the walls of the visceral effectors supplied by the postganglionic fibers.

Testis (TES-tis) Male gonad that produces sperm and the hormones testosterone and inhibin. Also called a **testicle.**

Testosterone (tes-TOS-te-rōn) A male sex hormone (androgen) secreted by interstitial endocrinocytes (cells of Leydig) of a mature testis; controls the growth and development of male sex organs, secondary sex characteristics, sperm cells, and body growth.

Tetralogy of Fallot (tet-RAL-ō-jē of fal-Ō) A combination of four congenital heart defects: (1) constricted pulmonary semilunar valve, (2) interventricular septal opening, (3) emergence of the aorta from both ventricles instead of from the left only, and (4) enlarged right ventricle.

Thalamus (THAL-a-mus) A large, oval structure located superior to the midbrain, consisting of two masses of gray matter covered by a thin layer of white matter.

Thermoreceptor (THER-mō-rē-sep-tor) Receptor that detects changes in temperature.

Third ventricle (VEN-tri-kul) A slitlike cavity between the right and left halves of the thalamus and between the lateral ventricles.

Thirst center A cluster of neurons in the hypothalamus that is sensitive to the osmotic pressure of extracellular fluid and brings about the sensation of thirst.

Thoracic (thō-RAS-ik) **cavity** Superior component of the ventral body cavity that contains two pleural cavities, the mediastinum, and the pericardial cavity.

Thoracic duct A lymphatic vessel that begins as a dilation called the cisterna chyli, receives lymph from the left side of the head, neck, and chest, the left upper limb, and the entire body below the ribs, and empties into the left subclavian vein. Also called the **left lymphatic** (lim-FAT-ik) **duct.**

Thoracolumbar (thō′-ra-kō-LUM-bar) **outflow** The fibers of the sympathetic preganglionic neurons, which have their cell bodies in the lateral gray columns of the thoracic segment and first two or three lumbar segments of the spinal cord.

Thorax (THŌ-raks) The chest.

Thrombus A clot formed in an unbroken blood vessel, usually a vein.

Thymus (THĪ-mus) **gland** A bilobed organ, located in the superior mediastinum posterior to the sternum and between the lungs, that plays a role in the immune mechanism of the body.

Thyroid cartilage (THĪ-royd KAR-ti-lij) The largest single cartilage of the larynx, consisting of two fused plates that form the anterior wall of the larynx. Also called the **Adam's apple.**

Thyroid colloid (KOL-loyd) A complex in thyroid follicles consisting of thyroglobulin and stored thyroid hormones.

Thyroid follicle (FOL-i-kul) Spherical sac that forms the parenchyma of the thyroid gland and consists of follicular cells that produce thyroxine (T_4) and triiodothyronine (T_3) and parafollicular cells that produce calcitonin (CT).

Thyroid gland An endocrine gland with right and left lateral lobes on either side of the trachea connected by an isthmus located anterior to the trachea just inferior to the cricoid cartilage.

Thyroid-stimulating hormone (TSH) A hormone secreted by the anterior lobe of the pituitary gland that stimulates the synthesis and secretion of hormones produced by the thyroid gland.

Thyroxine (thī-ROK-sēn) (T_4) A hormone secreted by the thyroid gland that regulates organic metabolism, growth and development, and the activity of the nervous system.

Tissue A group of similar cells and their inercellular substance joined together to perform a specific function.

Tissue macrophage system General term that refers to wandering and fixed macrophages.

Tissue rejection Phenomenon by which the body recognizes the protein (HLA antigens) in transplanted tissues or organs as foreign and produces antibodies against them.

Tonsil (TON-sil) A multiple aggregation of large lymphatic nodules embedded in mucous membrane.

Topical (TOP-i-kal) Applied to the surface rather than ingested or injected.

Trabecula (tra-BEK-yoo-la) Irregular latticework of thin columns of spongy bone. Fibrous cord of connective tissue serving as supporting fiber by forming a septum extending into an organ from its wall or capsule. *Plural,* **trabeculae** (tra-BEK-yoo-lē).

Trabeculae carneae (tra-BEK-yoo-lē KAR-nē-ē) Ridges and folds of the myocardium in the ventricles.

Trachea (TRĀ-kē-a) Tubular air passageway extending from the larynx to the fifth thoracic vertebra. Also called the **windpipe.**

Tract A bundle of nerve fibers in the central nervous system.

Transverse colon (trans-VERS KŌ-lon) The portion of the large intestine extending across the abdomen from the right colic (hepatic) flexure to the left colic (splenic) flexure.

Transverse fissure (FISH-er) The deep cleft that separates the cerebrum from the cerebellum.

Transverse plane A plane that runs parallel to the ground and divides the body or organs into superior and inferior portions. Also called a **horizontal plane.**

Transverse tubules (TOO-byools) **(T tubules)** Minute, cylindrical invaginations of the muscle fiber (cell) membrane that carry the muscle action potentials deep into the muscle fiber.

Trauma (TRAW-ma) An injury, either a physical wound or psychic disorder, caused by an external agent or force, such as a physical blow or emotional shock; the agent or force that causes the injury.

Triad (TRĪ-ad) A complex of three units in a muscle fiber (cell) composed of a transverse tubule and the segments of sarcoplasmic reticulum on both sides of it.

Tricuspid (trī-KUS-pid) **valve** Atrioventricular (AV) valve on the right side of the heart.

Trigone (TRĪ-gon) A triangular area at the base of the urinary bladder.

Triiodothyronine (trī-ī-ōd-ō-THĪ-rō-nēn) (T_3) A hormone produced by the thyroid gland that regulates organic metabolism, growth and development, and the activity of the nervous system.

Trophoblast (TRŌF-ō-blast) The superficial covering of cells of the blastocyst.

Tropic (TRŌ-pik) **hormone** A hormone whose target is another endocrine gland.

Trunk The part of the body to which the upper and lower limbs are attached.

Tunica albuginea (TOO-ni-ka al´-byoo-JIN-ē-a) A dense layer of white fibrous tissue covering a testis or deep to the surface of an ovary.

Tunica externa (eks-TER-na) The superficial coat of an artery or vein, composed mostly of elastic and collagen fibers. Also called the **adventitia.**

Tunica interna (in-TER-na) The deep coat of an artery or vein, consisting of a lining of endothelium, basement membrane, and internal elastic lamina. Also called the **tunica intima** (IN-ti-ma).

Tunica media (MĒ-dē-a) The middle coat of an artery or vein, composed of smooth muscle and elastic fibers.

Tympanic antrum (tim-PAN-ik AN-trum) An air space in the posterior wall of the middle ear that leads into the mastoid air cells or sinus.

Tympanic (tim-PAN-ik) **membrane** A thin, semitransparent partition of fibrous connective tissue between the external auditory meatus and the middle ear. Also called the **eardrum.**

Type II cutaneous mechanoreceptor A receptor embedded deeply in the dermis and deeper tissues that detects heavy and continuous touch sensations. Also called an **end organ of Ruffini.**

Umbilical (um-BIL-i-kal) **cord** The long, ropelike structure, containing the umbilical arteries and vein that connect the fetus to the placenta.

Umbilicus (um-BIL-i-kus or um-bil-Ī-kus) A small scar on the abdomen that marks the former attachment of the umbilical cord to the fetus. Also called the **navel.**

Upper limb (extremity) The appendage attached at the shoulder girdle, consisting of the arm, forearm, wrist, hand, and fingers.

Uremia (yoo-RĒ-mē-a) Accumulation of toxic levels of urea and other nitrogenous waste products in the blood, usually resulting from severe kidney malfunction.

Ureter (YOO-re-ter) One of two tubes that connect the kidney with the urinary bladder.

Urethra (yoo-RĒ-thra) The duct from the urinary bladder to the exterior of the body that conveys urine in females and urine and semen in males.

Urinary (YOO-ri-ner-ē) **bladder** A hollow, muscular organ situated in the pelvic cavity posterior to the pubic symphysis.

Urine The fluid produced by the kidneys that contains wastes or excess materials and is excreted from the body through the urethra.

Urogenital (yoo´-rō-JEN-i-tal) **triangle** The region of the pelvic floor inferior to the pubic symphysis, bounded by the pubic symphysis and the ischial tuberosities and containing the external genitalia.

Urology (yoo-ROL-ō-jē) The specialized branch of medicine that deals with the structure, function, and diseases of the male and female urinary systems and the male reproductive system.

Uterine (YOO-ter-in) **tube** Duct that transports ova from the ovary to the uterus. Also called the **Fallopian** (fal-LŌ-pē-an) **tube** or **oviduct.**

Uterosacral ligament (yoo´-ter-ō-SĀ-kral LIG-a-ment) A fibrous band of tissue extending from the cervix of the uterus laterally to attach to the sacrum.

Uterovesical (yoo´-ter-ō-VES-i-kal) **pouch** A shallow pouch formed by the reflection of the peritoneum from the anterior surface of the uterus, at the juntion of the cervix and the body, to the posterior surface of the urinary bladder.

Uterus (YOO-te-rus) The hollow, muscular organ in females that is the site of menstruation, implantation, development of the fetus, and labor. Also called the **womb.**

Utricle (YOO-tri-kul) The larger of the two divisions of the membranous labyrinth located inside the vestibule of the inner ear, containing a receptor organ for static equilibrium.

Uvea (YOO-vē-a) The three structures that together make up the vascular tunic of the eye.

Uvula (YOO-vyoo-la) A soft, fleshy mass, especially the V-shaped pendant part, descending from the soft palate.

Vacuole (VAK-yoo-ōl) Membrane-bound organelle that, in animal cells, frequently functions in temporary storage or transportation.

Vagina (va-JĪ-na) A muscular, tubular organ that leads from the uterus to the vestibule, situated between the urinary bladder and the rectum of the female.

Valvular stenosis (VAL-vyoo-lar ste-NŌ-sis) A narrowing of a heart valve, usually the bicuspid (mitral) valve.

Variocele (VAR-i-kō-sēl) A twisted vein; especially, the accumulation of blood in the veins of the spermatic cord.

Varicose (VAR-i-kōs) Pertaining to an unnatural swelling, as in the case of a varicose vein.

Vas A vessel or duct.

Vasa recta (VĀ-sa REK-ta) Extensions of the efferent arteriole of a juxtaglomerular nephron that run alongside the loop of the nephron (Henle) in the medullary region.

Vascular (VAS-kyoo-lar) Pertaining to or containing many blood vessels.

Vascular spasm Contraction of the smooth muscle in the wall of a damaged blood vessel to prevent blood loss.

Vascular tunic (TOO-nik) The middle layer of the eyeball, composed of the choroid, ciliary body, and iris. Also called the **uvea** (YOO-vē-a).

Vascular (venous) sinus A vein with a thin endothelial wall that lacks a tunica media and externa and is supported by surrounding tissue.

Vasectomy (va-SEK-tō-mē) A means of sterilization of males in which a portion of each ductus (vas) deferens is removed.

Vasoconstriction (vāz-ō-kon-STRIK-shun) A decrease in the size of the lumen of a blood vessel caused by contraction of the smooth muscle in the wall of the vessel.

Vasodilation (vās´-ō-DĪ-lā-shun) An increase in the size of the lumen of a blood vessel caused by relaxation of the smooth muscle in the wall of the vessel.

Vasomotor (vā-sō-MŌ-tor) **center** A cluster of neurons in the medulla oblongata that controls the diameter of blood vessels, especially arteries.

Vein A blood vessel that conveys blood from tissues back to the heart.

Vena cava (VĒ-na KĀ-va) One of two large veins that open into the right atrium, returning to the heart all of the deoxygenated blood from the systemic circulation except from the coronary circulation.

Venereal (ve-NĒ-rē-al) **disease (VD)** *See* **Sexually transmitted disease.**

Ventral (VEN-tral) Pertaining to the anterior or front side of the body; opposite of dorsal.

Ventral body cavity Cavity near the ventral aspect of the body that contains viscera and consists of a superior thoracic cavity and an inferior abdominopelvic cavity.

Ventral ramus (RĀ-mus) The anterior branch of a spinal nerve, containing sensory and motor fibers to the muscles and skin of the anterior surface of the head, neck, trunk, and the limbs.

Ventricle (VEN-tri-kul) A cavity in the brain or an inferior chamber of the heart.

Ventricular fibrillation (ven-TRIK-yoo-lar fib-ri-LĀ-shun) Asynchronous ventricular contractions that result in cardiovascular failure.

Venule (VEN-yool) A small vein that collects blood from capillaries and delivers it to a vein.

Vermiform appendix (VER-mi-form a-PEN-diks) A twisted, coiled tube attached to the cecum.

Vermilion (ver-MIL-yon) The area of the mouth where the skin on the outside meets the mucous membrane on the inside.

Vermis (VER-mis) The central constricted area of the cerebellum that separates the two cerebellar hemispheres.

Vertebral (VER-te-bral) **canal** A cavity within the vertebral column formed by the vertebral foramina of all the vertebrae and containing the spinal cord. Also called the **spinal canal.**

Vertebral column The 26 vertebrae; encloses and protects the spinal cord and serves as a point of attachment for the ribs and back muscles.

Vesicle (VES-i-kul) A small bladder or sac containing liquid.

Vesicular ovarian follicle A relatively large, fluid-filled follicle that will rupture and release a secondary oocyte, a process called ovulation. Also called a **Graafian** (GRAF-ē-an) **follicle.**

Vestibular (ves-TIB-yoo-lar) **apparatus** Collective term for the organs for equilibrium, which includes the saccule, utricle, and semicircular ducts.

Vestibular (ves-TIB-yoo-lar) **membrane** The membrane that separates the cochlear duct from the scala vestibuli.

Vestibule (VES-ti-byool) A small space or cavity at the beginning of a canal, especially the inner ear, larynx, mouth, nose, and vagina.

Villus (VIL-lus) A projection of the intestinal mucosal cells containing connective tissue, blood vessels, and a lymphatic vessel; functions in the absorption of the end products of digestion. *Plural,* **villi** (VIL-ī).

Viscera (VIS-er-a) The organs inside the ventral body cavity. *Singular,* **viscus** (VIS-kus).

Visceral effector (e-FEK-tor) Cardiac muscle, smooth muscle, and glandular epithelium.

Visceral muscle An organ specialized for contraction, composed of smooth muscle fibers (cells), located in the walls of hollow internal structures, and stimulated by visceral motor neurons.

Visceral pleura (PLOO-ra) The deep layer of the serous membrane that covers the lungs.

Visceroceptor (vis′-er-ō-SEP-tor) Receptor that provides information about the body's internal environment.

Vitreous (VIT-rē-us) **body** A soft, jellylike substance that fills the vitreous chambers of the eyeball, lying between the lens and the retina.

Vocal folds Pair of mucous membrane folds below the ventricular folds that function in voice production. Also called **true vocal cords.**

Volkmann's canal *See* **Perforating canal.**

Vulva (VUL-va) Collective designation for the external genitalia of the female. Also called the **pudendum** (poo-DEN-dum).

Wallerian (wal-LE-rē-an) **degeneration** Degeneration of the portion of the axon and myelin sheath of a neuron distal to the site of injury.

Wandering macrophage (MAK-rō-fāj) Phagocytic cell that develops from a monocyte, leaves the blood, and migrates to infected tissues.

White matter Aggregations or bundles of myelinated axons located in the brain and spinal cord.

White pulp The portion of the spleen composed of lymphatic tissue, mostly lymphocytes, arranged around central arteries; in some areas of white pulp, lymphocytes are thickened into lymphatic nodules called splenic nodules (Malpighian corpuscles).

White ramus communicans (RĀ-mus ko-MYOO-ni-kans) The portion of a preganglionic sympathetic nerve fiber that branches away from the anterior ramus of a spinal nerve to enter the nearest sympathetic trunk ganglion.

Xiphoid (ZĪ-foyd) Sword-shaped. The most inferior portion of the sternum.

Yolk sac An extraembryonic membrane that connects with the midgut during early embryonic development, but is nonfunctional in humans.

Zona fasciculata (ZŌ-na fa-sik′-yoo-LA-ta) The middle zone of the adrenal cortex that consists of cells arranged in long, straight cords and that secretes glucocorticoid hormones.

Zona glomerulosa (glo-mer′-yoo-LŌ-sa) The external zone of the adrenal cortex, directly under the connective tissue covering, that consists of cells arranged in arched loops or round balls and that secretes mineralocorticoid hormones.

Zona pellucida (pe-LOO-si-da) Clear glycoprotein layer between a secondary oocyte and granulosa cells to the corona radiata that surrounds a secondary oocyte.

Zona reticularis (ret-ik′-yoo-LAR-is) The internal zone of the adrenal cortex, consisting of cords of branching cells that secrete sex hormones, chiefly androgens.

Zygote (ZĪ-gōt) The single cell resulting from the union of a male and female gamete; the fertilized ovum.

Credits

Illustration Credits

Chapter 1: Table 1.2: Kevin Somerville. Table 1.5: Kevin Somerville. 1.1: Kevin Somerville. 1.2: Kevin Somerville. 1.3: Kevin Somerville. 1.4: Lynn O'Kelley. 1.6: Kevin Somerville. 1.7: Kevin Somerville. 1.8: Kevin Somerville. 1.9: Kevin Somerville. 1.10: Kevin Somerville.

Chapter 2: Table 2.2: Lauren Keswick. 2.1: Hilda Muinos. 2.2: Lauren Keswick. 2.3: Lauren Keswick. 2.4: Nadine Sokol. 2.5: Jared Schneidman Design. 2.6: Jared Schneidman Design. 2.7: Nadine Sokol. 2.8: Jared Schneidman Design. 2.10: Lauren Keswick. 2.11a: Lauren Keswick. 2.11d: Hilda Muinos. 2.12: Lauren Keswick. 2.13: Hilda Muinos. 2.14a: Lauren Keswick. 2.14d: Hilda Muinos. 2.15: Hilda Muinos. 2.16: Jared Schneidman Design. 2.17: Lauren Keswick.

Chapter 3: Table 3.1: Nadine Sokol, Kevin Somerville. Table 3.2: Nadine Sokol, Kevin Somerville. Table 3.3: Nadine Sokol, Kevin Somerville, Leonard Dank, Lauren Keswick. Table 3.4: Nadine Sokol, Lauren Keswick. Table 3.5: Nadine Sokol. 3.1: Hilda Muinos. 3.2: Nadine Sokol. 3.3: Nadine Sokol.

Chapter 4: 4.1: Lauren Keswick. 4.2: Kevin Somerville. 4.3: Nadine Sokol. 4.4: Lauren Keswick. 4.5: Kevin Somerville.

Chapter 5: 5.1: Leonard Dank. 5.2: Lauren Keswick. 5.3: Leonard Dank. 5.4: Lauren Keswick. 5.5: Leonard Dank. 5.7: Leonard Dank. 5.8: Leonard Dank. 5.9: Leonard Dank.

Chapter 6: Table 6.1: Nadine Sokol. Table 6.3: Leonard Dank. Table 6.5: Leonard Dank. Table 6.6: Leonard Dank. 6.1: Leonard Dank. 6.2: Leonard Dank. 6.3: Leonard Dank. 6.4: Leonard Dank. 6.5: Leonard Dank. 6.6: Leonard Dank. 6.7: Leonard Dank. 6.8: Leonard Dank. 6.9a: Leonard Dank. 6.10: Leonard Dank. 6.11: Leonard Dank. 6.12: Leonard Dank. 6.13: Leonard Dank. 6.14: Leonard Dank. 6.15: Leonard Dank. 6.16a, b, d: Leonard Dank. 6.16c: Nadine Sokol. 6.17: Leonard Dank. 6.18: Leonard Dank. 6.19: Leonard Dank. 6.20: Leonard Dank. 6.21: Leonard Dank. 6.22: Nadine Sokol. 6.23: Leonard Dank, Nadine Sokol. 6.24: Leonard Dank. 6.25: Leonard Dank.

Chapter 7: Table 7.1: Leonard Dank. 7.1: Leonard Dank. 7.2: Leonard Dank, Nadine Sokol. 7.3: Leonard Dank, Nadine Sokol. 7.4: Leonard Dank. 7.5: Leonard Dank, Nadine Sokol. 7.6: Leonard Dank, Nadine Sokol. 7.7: Leonard Dank, Nadine Sokol. 7.8: Leonard Dank, Nadine Sokol. 7.9: Leonard Dank. 7.10: Leonard Dank. 7.11: Leonard Dank. 7.12: Leonard Dank, Nadine Sokol. 7.13: Leonard Dank, Nadine Sokol. 7.14: Leonard Dank, Nadine Sokol. 7.15: Leonard Dank, Nadine Sokol. 7.16: Leonard Dank.

Chapter 8: 8.1: Leonard Dank. 8.7: Leonard Dank. 8.8: Leonard Dank. 8.9: Leonard Dank, Nadine Sokol. 8.10: Leonard Dank. 8.11: Leonard Dank. 8.12: Leonard Dank.

Chapter 9: Table 9.3: Nadine Sokol, Lauren Keswick. 9.1: Nadine Sokol. 9.2: Leonard Dank. 9.3: Jared Schneidman Design. 9.4: Beth Willert. 9.5: Nadine Sokol/Precision Graphics. 9.7: Hilda Muinos. 9.8: Hilda Muinos. 9.9: Beth Willert. 9.10: Beth Willert. 9.11: Beth Willert.

Chapter 10: Table 10.1: Kevin Somerville. 10.1: Leonard Dank. 10.2: Leonard Dank. 10.3: Leonard Dank. 10.4: Leonard Dank. 10.5: Leonard Dank. 10.6: Leonard Dank. 10.7: Leonard Dank. 10.8: Leonard Dank. 10.9: Leonard Dank. 10.10: Leonard Dank. 10.11: Leonard Dank. 10.12: Leonard Dank. 10.13: Leonard Dank. 10.14: Leonard Dank. 10.15: Leonard Dank. 10.16: Leonard

Dank. 10.17: Leonard Dank. 10.18: Leonard Dank. 10.19a-e: Leonard Dank. 10.19f: Kevin Somerville. 10.20: Leonard Dank. 10.21: Leonard Dank. 10.22: Leonard Dank. 10.23: Leonard Dank. 10.24: Leonard Dank.

Chapter 11:   11.6: Leonard Dank.

Chapter 12:   Table 12.2: Nadine Sokol. 12.1: Hilda Muinos. 12.2: Nadine Sokol. 12.3: Nadine Sokol.

Chapter 13:   13.1: Nadine Sokol. 13.2a: Kevin Somerville. 13.2b: Hilda Muinos. 13.2c: Keith Ciociola. 13.3a: Hilda Muinos. 13.3b: Beth Willert. 13.3c: Kevin Somerville. 13.4: Hilda Muinos. 13.5a: Hilda Muinos. 13.5c: Kevin Somerville. 13.6: Hilda Muinos. 13.7: Hilda Muinos. 13.8: Nadine Sokol. 13.9: Hilda Muinos. 13.10: Hilda Muinos. 13.11a: Hilda Muinos. 13.11b: Com Com. 13.12: Hilda Muinos. 13.13: Hilda Muinos. 13.15: Hilda Muinos.

Chapter 14:   Exhibit 14.2: Keith Ciociola. Exhibit 14.3: Keith Ciociola. Exhibit 14.4: Keith Ciociola. Exhibit 14.5: Keith Ciociola. Exhibit 14.6: Keith Ciociola. Exhibit 14.8: Keith Ciociola. Exhibit 14.9: Keith Ciociola. Exhibit 14.10: Keith Ciociola. Exhibit 14.11: Keith Ciociola. Exhibit 14.12: Keith Ciociola. 14.1: Hilda Muinos. 14.3: Nadine Sokol. 14.4: Hilda Muinos. 14.5: Leonard Dank. 14.6: Jared Schneidman Design. 14.7: Hilda Muinos. 14.8: Hilda Muinos. 14.9: Hilda Muinos. 14.10: Hilda Muinos. 14.11: Hilda Muinos. 14.12: Hilda Muinos. 14.13: Hilda Muinos. 14.14: Hilda Muinos. 14.15: Hilda Muinos. 14.16: Hilda Muinos. 14.17: Hilda Muinos. 14.19: Hilda Muinos. 14.20a: Kevin Somerville. 14.20b: Nadine Sokol. 14.21a: Hilda Muinos. 14.21b: Nadine Sokol. 14.22a, b: Hilda Muinos. 14.22c: Keith Ciociola. 14.23: Hilda Muinos.

Chapter 15:   15.1: Sharon Ellis. 15.2: Sharon Ellis. 15.3: Nadine Sokol. 15.4: Sharon Ellis. 15.5: Sharon Ellis. 15.6: Sharon Ellis. 15.8: Sharon Ellis. 15.9: Sharon Ellis. 15.10: Sharon Ellis. 15.11: Sharon Ellis. 15.12: Sharon Ellis. 15.13: Sharon Ellis. 15.14: Nadine Sokol.

Chapter 16:   Table 16.1: Kevin Somerville. 16.1: Hilda Muinos. 16.2: Sharon Ellis. 16.3: Nadine Sokol. 16.4: Sharon Ellis. 16.5: Nadine Sokol. 16.6: Sharon Ellis. 16.7a Nadine Sokol. 16.8: Sharon Ellis. 16.9: Jared Schneidman Design. 16.10: Nadine Sokol.

Chapter 17:   Table 17.1: Jared Schneidman Design. 17.1: Sharon Ellis. 17.2: Sharon Ellis. 17.3: Sharon Ellis. 17.4: Sharon Ellis. 17.5: Sharon Ellis. 17.7: Sharon Ellis/Nadine Sokol. 17.8: Sharon Ellis. 17.9: Sharon Ellis. 17.10a: Sharon Ellis. 17.10b: Sharon Ellis/Precision Graphics.

17.11: Precision Graphics. 17.12: Sharon Ellis, Precision Graphics. 17.13: Sharon Ellis.

Chapter 18:   Table 18.1: Nadine Sokol. Table 18.2: Hilda Muinos. 18.1: Sharon Ellis. 18.2: Sharon Ellis. 18.3: Sharon Ellis. 18.4: Sharon Ellis. 18.5: Sharon Ellis. 18.6: Sharon Ellis. 18.7: Sharon Ellis. 18.8: Sharon Ellis. 18.9: Sharon Ellis. 18.10: Sharon Ellis. 18.11: Sharon Ellis. 18.13: Sharon Ellis. 18.14: Sharon Ellis. 18.15: Sharon Ellis. 18.16: Hilda Muinos. 18.17: Sharon Ellis. 18.18: Sharon Ellis. 18.19: Sharon Ellis. 18.20: Sharon Ellis. 18.21: Sharon Ellis. 18.22: Sharon Ellis. 18.23: Sharon Ellis. 18.24: Sharon Ellis. 18.25: Kevin Somerville. 18.26: Kevin Somerville.

Chapter 19:   Table 19.1: Kevin Somerville. Table 19.2: Kevin Somerville. 19.1: Lynn O'Kelley. 19.2: Kevin Somerville. 19.3: Beth Willert. 19.4: Sharon Ellis. 19.5: Sharon Ellis. 19.6: Lynn O'Kelley. 19.7: Sharon Ellis.

Chapter 20:   Table 20.1: Kevin Somerville. Table 20.2: Kevin Somerville/Keith Ciociola. 20.1: Lynn O'Kelley. 20.2: Lynn O'Kelley. 20.3: Sharon Ellis. 20.4: © Biagio John Melloni. 20.6: Lynn O'Kelley. 20.7: Lynn O'Kelley. 20.9: © Biagio John Melloni. 20.10a: © Biagio John Melloni/Kevin Somerville. 20.10b, c: Lynn O'Kelley. 20.11: © Biagio John Melloni/Kevin Somerville. 20.12: © Biagio John Melloni/Kevin Somerville. 20.13: © Biagio John Melloni. 20.14: Sharon Ellis. 20.15: Sharon Ellis.

Chapter 21:   21.1: Jared Schneidman Design. 21.2: Hilda Muinos. 21.3: Kevin Somerville. 21.4: Sharon Ellis. 21.05: Jared Schneidman Design.

Chapter 22:   22.1: Lynn O'Kelley. 22.2: Lynn O'Kelley. 22.3: Nadine Sokol. 22.4: Lynn O'Kelley. 22.5: Lynn O'Kelley. 22.6: Lynn O'Kelley. 22.7: Lynn O'Kelley. 22.8: Lynn O'Kelley. 22.9: Lynn O'Kelley.

Chapter 23:   23.1: Lynn O'Kelley. 23.2a: Kevin Somerville. 23.2b: Lynn O'Kelley, Nadine Sokol. 23.3: Lynn O'Kelley, Nadine Sokol. 23.4: Lynn O'Kelley. 23.6: Lynn O'Kelley. 23.7: Lynn O'Kelley. 23.8: Nadine Sokol. 23.9a-d: Lynn O'Kelley. 23.9e: Nadine Sokol. 23.10: Leonard Dank. 23.11: Lynn O'Kelley. 23.12: Kevin Somerville. 23.13: Kevin Somerville. 23.14: Jared Schneidman Design. 23.15: Jared Schneidman Design.

Chapter 24:   24.1: Nadine Sokol. 24.2: Hilda Muinos. 24.3: Nadine Sokol. 24.4: Nadine Sokol. 24.5: Nadine Sokol. 24.6: Nadine Sokol. 24.7: Nadine Sokol. 24.8: Nadine Sokol. 24.9: Nadine Sokol. 24.10: Nadine Sokol. 24.11: Kevin Somerville. 24.12: Nadine Sokol. 24.13: Nadine Sokol. 24.14: Nadine Sokol. 24.15: Kevin

Somerville. 24.16: Kevin Somerville. 24.18: Nadine Sokol. 24.19: Kevin Somerville. 24.20: Nadine Sokol.

Chapter 25: 25.1: Kevin Somerville. 25.2: Kevin Somerville. 25.4: Nadine Sokol. 25.5: Nadine Sokol. 25.6: Nadine Sokol. 25.7: Kevin Somerville. 25.8: Nadine Sokol. 25.9: Nadine Sokol. 25.10: Nadine Sokol. 25.11: Nadine Sokol.

Chapter 26: 26.1: Kevin Somerville. 26.2: Kevin Somerville. 26.3: Kevin Somerville. 26.4: Kevin Somerville. 26.5: Jared Schneidman Design. 26.6: Kevin Somerville. 26.9: Kevin Somerville. 26.10: Kevin Somerville. 26.11: Kevin Somerville. 26.12: Kevin Somerville. 26.13: Kevin Somerville. 26.14: Jared Schneidman Design. 26.17: Kevin Somerville. 26.18: Kevin Somerville. 26.19: Kevin Somerville. 26.20: Kevin Somerville. 26.22: Jared Schneidman Design. 26.23: Kevin Somerville. 26.24: Kevin Somerville.

Chapter 27: Table 27.2: Nadine Sokol. 27.1: Kevin Somerville. 27.2: Jared Schneidman Design. 27.3: Kevin Somerville. 27.4: Kevin Somerville. 27.5: Kevin Somerville. 27.6: Kevin Somerville. 27.7: Kevin Somerville. 27.8: Kevin Somerville. 27.9: Kevin Somerville. 27.10: Kevin Somerville. 27.11: Kevin Somerville.

Photo Credits

Page 14 top: Stephen A. Kieffer and E. Robert Heitzman, *An Atlas of Cross-Sectional Anatomy.* Harper & Row, Publishers, Inc. New York, 1979; page 14 center: Lester Bergman; page 14 bottom: © 1988, Martin Rotker; page 17: Mark Nielsen/Addison Wesley Longman, Inc.; page 20: © Biophoto/Photo Researchers; page 21 top: Simon Fraser/SPL/Photo Researchers; page 21 bottom: Robb/Mayo Foundation; page 22 top: The Bayer Company; page 22 bottom: © Howard Sochurek/Medical Images, Inc.; page 23 top: Technicare Corp.; page 23 bottom: courtesy of Anthony Gerard Tortora 12/4/94; page 30: © Dr. Donald Fawcett/SPL/Photo Researchers; page 34 both: Abbott Laboratories; page 36 left: Lennart Nilsson, *The Body Victorious*, Delacorte Press, © Boehringer Ingleheim International GmBH.; page 36 right: © CNRI/ SPL/Photo Researchers; page 38: Lennart Nilsson, *The Body Victorious*, Delacorte Press, © Boehringer Ingleheim International GmBH.; page 39: © M. Weinstock/ SPL/Photo Researchers; page 40 all: Lennart Nilsson, *The Body Victorious*, Delacorte Press, © Boehringer Ingleheim International GmBH.; page 42: Daniel S. Friend, Harvard Medical School; page 43 top: Lennart Nilsson, *The Body Victorious*, Delacorte Press, © Boehringer Ingleheim International GmBH.; page 43 bottom: Lester Bergman and Associates; page 45: Kent

McDonald; page 48 all: CABISCO/Phototake NY; page 63 top left: Biophoto/Photo Researchers; page 63 top right: Douglas Merrill; page 63 bottom: © Ed Reschke; page 64 top: © Ed Reschke; page 64 bottom: Douglas Merrill; page 65 top: Biophoto/Photo Researchers; page 65 bottom: Biophoto/Photo Researchers; page 66 top: Biophoto/Photo Researchers; page 66 bottom: Andrew J. Kuntzman; page 67: © Ed Reschke; page 68 top: Bruce Iverson; page 68 bottom: Lester Bergman; page 73 top: © R. Kessell/VU; page 73 bottom: Lester Bergman; page 74: Biophoto/Photo Researchers; page 75 top: © Ed Reschke; page 75 bottom: Biophoto/Photo Reserachers; page 76 top: Andrew J. Kuntzman; page 76 bottom: © Ed Reschke; page 77 all: Biophoto/Photo Researchers; page 78: Biophoto/Photo Researchers; page 79 top: Biophoto/Photo Researchers; page 79 bottom: © Ed Reschke; pages 80–84: © Ed Reschke; page 93: Lester Bergman; page 98: Brain/SPL/Photo Researchers; page 116 top: Courtesy of Arthur Provost, R.T.; page 116 bottom: Biophoto/Photo Researchers; pages 133–139: Mark Nielsen/Addison Wesley Longman, Inc.; page 145: © Omikron/SPL/Photo Researchers; page 146 left: © H. Hubtaka/Mauritius GmBH/Phototake NY; page 146 right: D. Gentry Steele; pages 153–192 all: Mark Nielsen/Addison Wesley Longman, Inc.; page 196 a: © 1994 Evan J. Collela; page 196 b: © Addison Wesley Longman, Inc.; page 196 c: © 1994 Evan J. Collela; page 196 d: © 1994 Evan J. Collela; page 196 e: © Addison Wesley Longman, Inc.; page 196 f: © 1983 by Gerard J. Tortora. Courtesy of Lynn and James Borghesi; page 197 top left: © 1994 Evan J. Collela; page 197 top right: © Addison Wesley Longman, Inc.; page 197 bottom left: © 1994 Evan J. Collela; page 197 bottom right: © 1983 by Gerard J. Tortora. Courtesy of Lynn and James Borghesi; page 198 left: 1994 Evan J. Collela; page 198 right: © Addison Wesley Longman, Inc.; page 198 a bottom : Evan J. Collela; page 198 b bottom: © Addison Wesley Longman, Inc.; page 198 c bottom: © Addison Wesley Longman, Inc.; page 199 a: Lynn Marie Borghesi; page 199 b: Lynn Marie Borghesi; page 199 c and d: © 1992 Scott, Foresman photo by John Moore; page 199 e and f: © 1983 by Gerard J. Tortora. Courtesy of Lynn and James Borghesi; page 199 g: Evan J. Collela; page 199 h: © 1983 Gerard J. Tortora. Courtesy of Lynn and James Borghesi; pages 213–216: Mark Nielsen/Addison Wesley Longman, Inc.; page 231: Fujita; page 232: © Ed Reschke; page 233: H. E. Huxley; page 237: Biophoto/Photo Researchers; page 239: © Ed Reschke; page 241: Andrew J. Kuntzman; pages 263–339: Mark Nielsen/Addison Wesley Longman, Inc.; page 351 all: © Addison Wesley Longman, Inc.; page 352 all: © Addison Wesley Longman, Inc.; page 353: © Addison Wesley Longman, Inc.; page 354: © Scott, Foresman; page 355: © 1982 Gerard J. Tortora. Courtesy of Lynn and James Borghesi; page 356 all: © Addison Wesley Longman, Inc.;

Index

NOTE: Page numbers in **boldface** type indicate a major discussion. A *t* following a page number indicates tabular material, an *f* following a page number indicates a figure, and an *e* following a page number indicates an exhibit.